Algebra

FACTORING AND SOLVING EQUATIONS

(Sections 0.2.2 and 0.2.3)

$a^2 - b^2 = (a - b)(a + b)$

$a^3 - b^3 = (a - b)(a^2 + ab + b^2)$

$a^n - b^n = (a - b)(a^{n-1} + a^{n-2}b + a^{n-3}b^2 + \cdots$
$\qquad\qquad + a^2 b^{n-3} + ab^{n-2} + b^{n-1})$

Quadratic Formula:

The solutions to $ax^2 + bx + c = 0$ are $x = \frac{-b \pm \sqrt{b^2 - 4ac}}{2a}$.

$AB = 0 \iff A = 0 \text{ or } B = 0$

$\frac{A}{B} = 0 \iff A = 0 \text{ and } B \neq 0$

RULES OF FRACTIONS (Section 0.2.3)

$\frac{a}{b} + \frac{c}{d} = \frac{ad + bc}{bd} \qquad\qquad \left(\frac{a}{b}\right)\left(\frac{c}{d}\right) = \frac{ac}{bd}$

$\frac{a}{b} - \frac{c}{d} = \frac{ad - bc}{bd} \qquad\qquad c\left(\frac{a}{b}\right) = \frac{ac}{b}$

$\frac{\left(\frac{a}{b}\right)}{\left(\frac{c}{d}\right)} = \left(\frac{a}{b}\right)\left(\frac{d}{c}\right) = \frac{ad}{bc} \qquad \frac{a}{\left(\frac{b}{c}\right)} = \frac{ac}{b} \qquad \frac{\left(\frac{a}{b}\right)}{c} = \frac{a}{bc}$

INEQUALITIES (Sections 0.3.1–0.3.4)

If $a < b$ and $c > 0$, then $ac < bc$.

If $a < b$ and $c < 0$, then $ac > bc$.

If $a < b$, then for any c, $a + c < b + c$.

If $0 < a < b$, then $\frac{1}{a} > \frac{1}{b}$.

If A and B have the same sign, then $AB > 0$.

If A and B have opposite signs, then $AB < 0$.

$|x - c| < \delta \iff x \in (c - \delta, c + \delta)$

$|x - c| > \delta \iff x \in (-\infty, c - \delta) \cup (c + \delta, \infty)$

$x^2 < c^2 \iff x \in (-c, c)$.

$x^2 > c^2 \iff x \in (-\infty, -c) \cup (c, \infty)$.

ABSOLUTE VALUES (Sections 0.1.5 and 0.3.5)

$|x| = \begin{cases} x, & \text{if } x \geq 0 \\ -x, & \text{if } x < 0 \end{cases} \qquad |ab| = |a||b| \qquad \left|\frac{a}{b}\right| = \frac{|a|}{|b|}$

Triangle Inequality: $|a + b| \leq |a| + |b|$

RULES OF EXPONENTS (Sections 4.4.1–4.1.3 and 8.1.2)

$a^{-b} := \frac{1}{a^b} \qquad a^{\frac{1}{b}} := \sqrt[b]{a} \qquad a^0 := 1 \text{ (for } a \neq 0)$

$c^a c^b = c^{a+b} \qquad (ab)^c = a^c b^c \qquad c^{ab} = (c^a)^b = (c^b)^a$

$\frac{c^a}{c^b} = c^{a-b} \qquad \left(\frac{a}{b}\right)^c = \frac{a^c}{b^c} \qquad c^{\frac{a}{b}} = \sqrt[b]{c^a} = (\sqrt[b]{c})^a$

RULES OF LOGARITHMS (Sections 9.1.1–9.1.4)

$\ln x := \log_e x \qquad\qquad \log x := \log_{10} x$

$\log_b(x^a) = a \log_b x \qquad \log_b(xy) = \log_b x + \log_b y$

$\log_b x = \frac{\log_a x}{\log_a b} \qquad \log_b(b^x) = x$

$\log_b 1 = 0 \qquad\qquad \log_b(b) = 1$

$\log_b(\frac{1}{x}) = -\log_b x \qquad \log_b(\frac{x}{y}) = \log_b x - \log_b y$

$b^{\log_b x} = x \qquad\qquad \log_b x = y \iff b^y = x$

TRIGONOMETRIC IDENTITIES

(Sections 10.2.2 and 10.3.2–10.3.4)

$\csc x = \frac{1}{\sin x} \qquad\qquad \sec x = \frac{1}{\cos x}$

$\tan x = \frac{\sin x}{\cos x} \qquad\qquad \cot x = \frac{1}{\tan x} = \frac{\cos x}{\sin x}$

Pythagorean: *Even/Odd:*

$\sin^2 x + \cos^2 x = 1 \qquad \sin(-x) = -\sin x$

$\tan^2 x + 1 = \sec^2 x \qquad \cos(-x) = \cos x$

$1 + \cot^2 x = \csc^2 x \qquad \tan(-x) = -\tan x$

Sum:

$\sin(\alpha + \beta) = \sin\alpha \cos\beta + \sin\beta \cos\alpha$

$\cos(\alpha + \beta) = \cos\alpha \cos\beta - \sin\alpha \sin\beta$

Difference:

$\sin(\alpha - \beta) = \sin\alpha \cos\beta - \sin\beta \cos\alpha$

$\cos(\alpha - \beta$

Double-

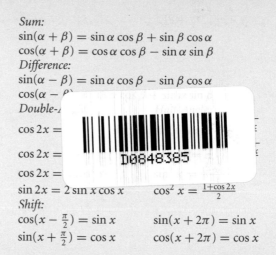

$\cos 2x =$

$\cos 2x =$

$\cos 2x =$

$\sin 2x = 2 \sin x \cos x \qquad \cos^2 x = \frac{1 + \cos 2x}{2}$

Shift:

$\cos(x - \frac{\pi}{2}) = \sin x \qquad \sin(x + 2\pi) = \sin x$

$\sin(x + \frac{\pi}{2}) = \cos x \qquad \cos(x + 2\pi) = \cos x$

Geometry

DISTANCES (Sections 0.1.5 – 0.1.6)

- The *distance* between two real numbers a and b is $|b - a|$.
- *Distance Formula:* The distance between two points (x_1, y_1) and (x_2, y_2) in the plane is $\sqrt{(x_2 - x_1)^2 + (y_2 - y_1)^2}$.
- *Midpoint Formula:* The midpoint between two points (x_1, y_1) and (x_2, y_2) in the plane is the point $\left(\frac{x_1 + x_2}{2}, \frac{y_1 + y_2}{2}\right)$.

TRIANGLES (Section 7.4.2)

- *Pythagorean Theorem:* If a right triangle has legs of length a and b, and a hypotenuse of length c, then $a^2 + b^2 = c^2$.
- *Law of Similar Triangles:* Suppose a right triangle has legs of length A and B, and hypotenuse of length C. For any right triangle similar to the first, with corresponding legs and hypotenuse of lengths a, b, and c, we have $\frac{a}{b} = \frac{A}{B}$, $\frac{b}{c} = \frac{B}{C}$, and $\frac{a}{c} = \frac{A}{C}$.

LENGTHS, AREAS, AND VOLUMES (Section 7.4.2)

- A *rectangle* with length l and width w has area $A = lw$ and perimeter $P = 2l + 2w$.
- A *triangle* with base b and height h has area $A = \frac{1}{2}bh$.
- A *circle* with radius r has area $A = \pi r^2$ and circumference $C = 2\pi r$. (See also Section 12.1.1)
- A *trapezoid* with base b and heights h_1 and h_2 has area $A = b\left(\frac{h_1 + h_2}{2}\right)$. (See also Section 12.2.3)
- A *rectangular prism* with length l, width w, and height h has volume $V = lwh$ and surface area $SA = 2lw + 2wh + 2lh$.
- A *sphere* with radius r has volume $V = \frac{4}{3}\pi r^3$ and surface area $SA = 4\pi r^2$.
- A *right circular cylinder* with radius r and height h has volume $V = \pi r^2 h$, surface area $SA = 2\pi rh = 2\pi r^2$, and lateral surface area $LSA = 2\pi rh$.
- A *right circular cone* with radius r and height h has volume $V = \frac{1}{3}\pi r^2 h$, surface area $SA = \pi r\sqrt{r^2 + h^2} + \pi r^2$, and lateral surface area $LSA = \pi r\sqrt{r^2 + h^2}$.
- A *shell* (*i.e.* a right circular cylinder with an interior cylinder of the same height removed) with small radii r_1, large radius r_2, and height h has volume $V = \pi(r_2^2 - r_1^2)h$. A shell with average radius $r = \frac{r_1 + r_2}{2}$, thickness $\Delta x = r_2 - r_1$, and height h has volume $V = 2\pi rh \Delta x$. (See also Section 15.3.2)

Limits

DEFINITIONS (Sections 2.2.1 and 2.2.3)

- *Definition of limit:* $\lim\limits_{x \to c} f(x) = L$ means that for all $\epsilon > 0$, there exists a $\delta > 0$ such that $0 < |x - c| < \delta \Rightarrow |f(x) - L| < \epsilon$.
- *Right limit:* $\lim\limits_{x \to c^+} f(x) = L$ means that for all $\epsilon > 0$, there exists a $\delta > 0$ such that $x \in (c, c + \delta) \Rightarrow |f(x) - L| < \epsilon$.
- *Left limit:* $\lim\limits_{x \to c^-} f(x) = L$ means that for all $\epsilon > 0$, there exists a $\delta > 0$ such that $x \in (c - \delta, c) \Rightarrow |f(x) - L| < \epsilon$.
- *Infinite limit:* $\lim\limits_{x \to c^+} f(x) = \infty$ means that for all $M > 0$, there exists a $\delta > 0$ such that $x \in (c, c + \infty) \Rightarrow f(x) > M$.
- *Limit at infinity:* $\lim\limits_{x \to \infty} f(x) = L$ means that for all $\epsilon > 0$, there exists an $N > 0$ such that $x > N \Rightarrow |f(x) - L| < \epsilon$.
- *Infinite limit at infinity:* $\lim\limits_{x \to \infty} f(x) = \infty$ means that for all $M > 0$, there exists an $N > 0$ such that $x > N \Rightarrow f(x) > M$.

INDETERMINATE FORMS

- $\frac{0}{0}$ or $\frac{\infty}{\infty}$: Apply algebra (Sections 2.5.2, 2.5.4, 6.2.1, 6.2.3, and 8.3.4) or L'Hôpital's Rule (Section 8.7.3)
- $0 \cdot \infty$: Use algebra to convert to $\frac{0}{0}$ or $\frac{\infty}{\infty}$ (Section 8.7.3)
- $\infty - \infty$: Use algebra to rewrite as another form (Sections 4.2.4 and 8.7.3)
- 0^0, 1^∞, or ∞^0: Investigate the logarithm of the limit (Section 9.3.2)

NON-INDETERMINATE FORMS

- $\frac{1}{0^+} \to \infty$ $\frac{1}{0^-} \to -\infty$ $\frac{1}{\infty} \to 0$ $\frac{1}{-\infty} \to 0$ (Sections 2.5.3 and 8.7.1)
- $\frac{0}{1} \to 0$ $\frac{\infty}{1} \to \infty$ $\frac{0}{\infty} \to 0$ $\frac{\infty}{0} \to \infty$
  $\infty + \infty \to \infty$ $\infty \cdot \infty \to \infty$ (Section 8.7.1)

- $\infty^\infty \to \infty$ $0^\infty \to 0$ $\infty^1 \to \infty$
  $1^0 \to 1$ $0^1 \to 0$ (Section 9.3.2)

SOME IMPORTANT LIMITS

- If $k > 0$, then $\lim\limits_{x \to \infty} x^k = \infty$ and $\lim\limits_{x \to \infty} x^{-k} = 0$. (Section 4.2.3)
- If $f(x)$ is a polynomial of even degree and positive leading coefficient, then $\lim\limits_{x \to \infty} f(x) = \infty$ and $\lim\limits_{x \to -\infty} f(x) = \infty$. (Section 5.2.2)
- If $f(x)$ is a polynomial of odd degree and positive leading coefficient, then $\lim\limits_{x \to \infty} f(x) = \infty$ and $\lim\limits_{x \to -\infty} f(x) = -\infty$. (Section 5.2.2)
- If $f(x) = \frac{a_n x^n + a_{n-1} x^{n-1} \cdots + a_1 x + a_0}{b^m x^m b_{m-1} x^{m-1} + \cdots + b_1 x + b_0}$ is a rational function, then $\lim\limits_{x \to \infty} f(x)$ is equal to: 0, if $n > m$; $\frac{a_n}{b_m}$, if $n = m$; or ∞, if $n > m$. (Section 6.2.3)
- $\lim\limits_{h \to 0} (1 + h)^{\frac{1}{h}} = e$, $\lim\limits_{n \to \infty} (1 + \frac{1}{n})^n = e$, and $\lim\limits_{h \to 0} \frac{e^h - 1}{h} = 1$. (Section 8.2.1)
- $\lim\limits_{n \to \infty} (1 + \frac{1}{n})^{nr} = e^r$, and $\lim\limits_{n \to \infty} (1 + \frac{r}{n})^n = e^r$. (Section 8.6.3)
- If $k > 0$, then $\lim\limits_{x \to \infty} e^{kx} = \infty$, $\lim\limits_{x \to -\infty} e^{kx} = 0$, $\lim\limits_{x \to \infty} e^{-kx} = 0$, and $\lim\limits_{x \to -\infty} e^{-kx} = \infty$. (Section 8.3.3)
- If $b > 1$, then $\lim\limits_{x \to \infty} b^x = \infty$ and $\lim\limits_{x \to -\infty} b^x = 0$. If $0 < b < 1$, then $\lim\limits_{x \to \infty} b^x = 0$ and $\lim\limits_{x \to -\infty} b^x = \infty$. (Section 8.3.3)
- If $r > 0$ and $k > 0$, then $\lim\limits_{x \to \infty} \frac{e^{rx}}{x^k} = \infty$ and $\lim\limits_{x \to \infty} \frac{x^k}{e^{rx}} = 0$. (Section 8.7.4)
- If $b > 1$, then $\lim\limits_{x \to 0^+} \log_b x = -\infty$ and $\lim\limits_{x \to \infty} \log_b x = \infty$. If $0 < b < 1$, then $\lim\limits_{x \to 0^+} \log_b x = \infty$ and $\lim\limits_{x \to \infty} \log_b x = -\infty$. (Section 9.2.2)
- If $b > 1$ and $k > 0$, then $\lim\limits_{x \to \infty} \frac{\log_b x}{x^k} = 0$ and $\lim\limits_{x \to \infty} \frac{x^k}{\log_b x} = \infty$. (Section 9.2.4)
- $\lim\limits_{x \to 0} \frac{\sin x}{x} = 1$ and $\lim\limits_{x \to 0} \frac{1 - \cos x}{x} = 0$. (Section 10.4.1)

LIMIT RULES

Limit of a constant	$\lim\limits_{x \to c} k = k$	(Section 2.4.1)
Limit of the identity function	$\lim\limits_{x \to c} x = c$	(Section 2.4.1)
Limit of a linear function	$\lim\limits_{x \to c} (mx + b) = mc + b$	(Section 2.4.2)
Constant multiple rule for limits	$\lim\limits_{x \to c} k f(x) = k \lim\limits_{x \to c} f(x)$	(Section 2.4.3)
Sum rule for limits	$\lim\limits_{x \to c} (f(x) + g(x)) = \lim\limits_{x \to c} f(x) + \lim\limits_{x \to c} g(x)$	(Section 2.4.4)
Difference rule for limits	$\lim\limits_{x \to c} (f(x) - g(x)) = \lim\limits_{x \to c} f(x) - \lim\limits_{x \to c} g(x)$	(Section 2.4.4)
Product rule for limits	$\lim\limits_{x \to c} f(x) g(x) = \lim\limits_{x \to c} f(x) \lim\limits_{x \to c} g(x)$	(Section 2.4.5)
Quotient rule for limits	$\lim\limits_{x \to c} \frac{f(x)}{g(x)} = \frac{\lim\limits_{x \to c} f(x)}{\lim\limits_{x \to c} g(x)}$, if $\lim\limits_{x \to c} g(x) \neq 0$	(Section 2.4.5)
Limit of a composition	$\lim\limits_{x \to c} f(g(x)) = f\left(\lim\limits_{x \to c} g(x)\right)$, if $\lim\limits_{x \to c} g(x) = L$ and f is continuous at L	(Section 2.6.5)

(*Reference Pages continue on page R-4*)

INTEGRATED CALCULUS

INTEGRATED CALCULUS

Calculus with Precalculus and Algebra

Laura Taalman

James Madison University

Houghton Mifflin Company

Boston   New York

Publisher: Jack Shira
Sponsoring Editor: Lauren Schultz
Senior Development Editor: Claire Y. Boivin
Editorial Associate: Kasey McGarrigle
Project Editor: Kathleen Deselle
Senior Production/Design Coordinator: Carol Merrigan
Senior Manufacturing Manager: Priscilla J. Bailey
Senior Marketing Manager: Danielle Potvin
Marketing Associate: Nicole Mollica

Cover illustration © Eileen Hoff. The cover illustration is a picture of the **Borromean rings.** Although all three rings are linked, if any one ring is removed, then the remaining two rings are not linked. For this book, the three rings represent the study of calculus, precalculus, and algebra. Together the three make a coherent and integrated whole, but if any one is lacking, then the rest fall apart. In particular, a firm grounding in both algebra and precalculus is necessary for success in calculus!

Printed in the U.S.A.

Library of Congress Control Number: 2003109848

ISBN: 0-618-21950-1

123456789-DOW-08 07 06 05 04

CONTENTS

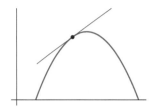

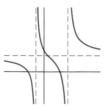

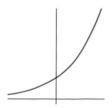

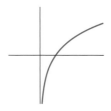
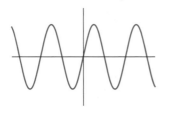
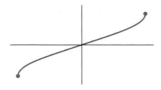

PART IV   *INTEGRATION*   713

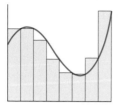

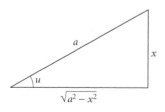

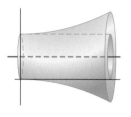

15 *Applications of Integration* 847

PREFACE

To the Student

This textbook is written for you. That means it is written so that you can read it and understand it. Of course, reading a mathematics textbook is not the same as reading a novel; when reading mathematics you will often have to stop, think about the material, do some examples, and reread the material again. In this book we will spend a lot of time on algebra and precalculus topics; this will give you the knowledge and confidence you need to succeed in calculus, regardless of your previous background.

This textbook encourages you to understand calculus, rather than just memorize facts or equations. You will be responsible for understanding *why* the mathematics works, not just *how* to apply it. In the homework exercises, "Concepts" problems will help you test your understanding of the material, "Skills" and "Applications" exercises will help you apply that material to computational and real-world problems, and the exercises in the "Proofs" section will help you review the important proofs from the reading and ask you to extend the theory beyond what is given in the section.

Suggestions for Students

- Read each section carefully, and test your understanding as you go by answering the "Questions" in the reading. Read the material before attempting the exercises.

- Pay careful attention to the "Cautions" given in the reading, as they will help you avoid common mistakes.

- Each section of the homework exercises begins with a "Problem 0" that asks you to write your own summary of the material; this summary will help you organize your thoughts and prepare for quizzes and tests.

- Answers for about half of the homework exercises are given in the back of the book. It is good practice to check your own answers; the "Checking" notes in the reading will help you do this.

To the Instructor

This textbook is different from other calculus books on the market for three reasons: First, it integrates calculus with precalculus and algebra material in a manner suitable even for math and science majors. Second, it is written in a way that students can understand (without watering down the material). Third, the homework exercises hold students accountable for the theory and the reading and test conceptual understanding as well as computational skill.

The exercises in this book address two problems common to many calculus courses: first, that students believe that everything they need to know is covered in the homework exercises, and second, that students often complain that quizzes and tests do not reflect the homework problems they did. *Integrated Calculus* includes many homework problems that ask students to explain definitions, concepts, theorems, or proofs from the reading, as well as problems that test their computational skills and problems that apply what they have learned to real-world situations. There is also a "Problem 0" in each homework section that asks students to summarize the material. If you like, you can use this homework problem to test whether students read the section before class or, alternatively, to test whether the students have understood the material by the following day.

What Is Integrated Calculus?

Integrated Calculus is designed for a two-semester course that integrates differential and basic integral calculus with precalculus and algebra material. The material includes all of what is commonly referred to as "Calculus I," some of the material known as "Calculus II," and a good deal of algebra and precalculus material. All material, even the algebra and precalculus, is handled in a fashion that is appropriate for math and science majors.

This textbook has four major parts. Part I covers elementary algebra, logic, and proofs and the concepts of functions, limits, and derivatives. The focus is on concepts rather than calculations; most examples are computationally simple or based on graphs. Parts II and III of the text use the concepts and techniques learned in Part I to investigate the properties of algebraic and transcendental functions. Calculations get progressively more difficult as students progress through the types of functions in Parts II and III. Part IV covers basic integration skills and applications.

Goals and Features of the Text

- **A readable text for students** This book is meant for students to read. The text is readable while still being rigorous and is full of examples, questions, and cautions. Proofs in the text are written with the student (rather than just the instructor) in mind. The book is also organized so as to be a good reference text when a student needs to review a topic.

- **Integrates calculus, precalculus, and algebra in a meaningful way** Calculus, precalculus, and algebra are integrated throughout this textbook. Precalculus and algebra are used to investigate calculus, and calculus is used to investigate precalculus (specifically graphical features of functions).

- **Inverted structure** From *Integrated Calculus*, students will learn the basic theories and techniques of calculus and then use this knowledge to explore different types of functions (power, polynomial, exponential, and so on). This approach differs from the usual structure of a calculus book where students learn one technique or concept after another, each time applying their knowledge to *all* types of functions.

- **Systematic development of calculation skills** The first part of the book covers functions, limits, and derivatives through graphic and simple calculation examples, concentrating more on concepts than on symbolic manipulation. In the second and third parts of the book, students apply their knowledge to more and more complicated types of algebraic and transcendental functions, building calculation skills along the way.

- **Beneficial repetition** Because of the inverted structure of this book, students will revisit topics many times as they study different types of functions. For example, "definition of derivative" problems don't appear just in one section, they appear in each chapter that investigates a certain class of functions.

- **Focus on algebra** Many calculus students struggle with even simple algebraic manipulations. *Integrated Calculus* covers algebraic material from simple factoring and manipulation of exponents to complex techniques such as polynomial long division and manipulation of sums in sigma notation.

- **Rigorous treatment and encouragement of mathematical thought** Although *Integrated Calculus* covers basic topics in precalculus and algebra, it is not a "dumbed-down" or "slow" treatment of calculus. Every topic is treated rigorously; students learn about logic and proofs early in the text and use these skills throughout. Proof by induction, delta-epsilon proofs, and general Riemann sums are included in the reading and exercises as well as many short, easy proofs.

● **Exercises that test more than computational skills**   The exercises in *Integrated Calculus* hold students accountable for the reading by asking conceptual as well as computational problems. Each exercise set begins with "Concepts" questions that test the students' knowledge of definitions, theorems, and concepts from the reading. The remaining exercises are divided into "Skills," "Applications," and "Proofs" sections.

Comments and Suggestions for Instructors

● This textbook is written with an "inverted structure." *Integrated Calculus* applies each skill to all types of functions in contrast to most calculus textbooks that are organized by skills; that is, they cover one skill after another (for example, differentiation) and apply each skill to all types of functions. In Part I students learn the basic skills they need for the course. Then in each section of Parts II and III, they use these skills to investigate particular types of functions (for example, polynomial functions). This "inverted structure" provides for a lot of beneficial repetition; for example, students see the definition of derivative over and over again as they study different types of functions. This approach also allows for a systematic development of calculational skills. Students first see the definition of derivative in a simple calculational context. Then, as the textbook progresses through more complicated types of functions, they apply the definition of derivative in more challenging contexts. This is particularly helpful to students with a weak background in algebra.

● Limits and derivatives are introduced using only simple polynomial functions, piecewise functions, and graphical examples. The power rule is introduced in Chapter 3 using only positive integer powers; in Chapter 4 it is extended to negative integer powers and rational powers; then the general power rule is proved in Chapter 7. The quotient rule is presented in Chapter 6, and the product and chain rules presented in Chapter 7. This allows for a careful development of differentiation rules and is reflected in exercises that ask students to identify whether a given function can be differentiated given the rules developed so far.

● Other topics also get covered with increasing complexity as the book progresses. Optimization is presented first in Chapter 5 using polynomial functions, and then in Chapter 7 with general algebraic functions. Curve sketching gets progressively more sophisticated as more complicated types of functions are introduced. Similarly, algebraic techniques, domain computations, limits, and derivative calculations progress from simple to challenging throughout the book.

● Because this calculus course is geared toward students with weak algebra and precalculus backgrounds, logarithmic functions are defined as inverses of exponential functions rather than the more rigorous definition in terms of integration.

● Part IV of *Integrated Calculus* includes more integration topics than are normally included in a "Calculus I" course. This provides the background and experience with integration that is needed by students in science and math-related majors who might not need all of "Calculus II," but who do need more integration techniques than are typically covered in "Calculus I."

● It is possible to cover all sections of the book in a two-semester sequence of courses. However, it is also possible to customize the course for various needs. Certain sections can be omitted if you prefer. For example, the integration techniques in Sections 14.2, 14.3, and 14.4 can be skipped without much trouble; even Chapter 15 is covered later (homework problems will have to be selected carefully). You might also consider Sections 4.5, 7.4, 11.1, and 11.2 to be optional (assuming you will skip Sections 14.4, 15.1, 15.2, 15.3, or 15.4). For a less rigorous course, you could omit much of the section on proofs (Section 0.5) and the section on delta-epsilon proofs (Section 2.2). You could also combine two or more sections into one lesson by omitting some proofs. As one example, the sections on limits and derivatives of exponential functions (Sections 8.3 and 8.4) could be combined into one lesson if you were to omit the proofs concerning the continuity of and differentiation rules for exponential functions.

● Throughout the textbook, students are encouraged to use graphing calculators to check their answers and investigate graphs of functions. However, it is also possible to use this book in a course that does not require graphing calculators. In addition, all references to calculators in this textbook are nonspecific, that is, they do not assume any particular type of graphing calculator. Any calculator that can graph functions and approximate extrema, points of intersection, slopes, and areas should be sufficient.

Acknowledgments

I would like to thank all of my students over the years. Every semester they teach me something new. I would especially like to thank my students from the 2001–2002, 2002–2003, and 2003–2004 school years at James Madison University. These students were the guinea pigs for the preliminary versions of this textbook. Many of them gave valuable comments and suggestions that led to improvements in this book. In particular, the following students deserve extra thanks for finding many typographical errors:

> Krista Adamovich, Andrew Boryan, Carlee Brueser, Melissa Coggins, Ashley Crawford, Tamara East, Katie Eppig, Alex Haueter, Cricket Jenkins, Emily Kitamura, Brittany Klass, Cheryl Lock, Kay Magnuson, Emily Milam, Steven Mondziel, Shannon Mooney, Brian Neal, Melissa Pantalo, Nathan Powell, Jen Purcell, Alex Ross, Ashanti Samuel, Sarah Seal, Allison Watts, and Kimberly Witkowski.

A large number of people from James Madison University have been extremely helpful during the development of this textbook and the integrated course we teach at JMU, especially

> Chuck Cunningham, Brett Enge, Reid Harris, Robert Hanson, Irma Housden, John Klippert, Peter Kohn, Terry LePera, Ed Parker, Gary Peterson, Jason Rosenhouse, Elaine Smith, James Sochacki, Len VanWyk, Paul Warne, and Brenda Wilkinson.

I am particularly grateful to David Carothers, Chair of Department of Mathematics and Statistics, and David Brakke, Dean of the College of Science and Mathematics, for allowing me to create an integrated calculus course at JMU. I would also like to thank the Biology and Chemistry departments at JMU for incorporating our integrated course into their major program requirements. I am also indebted to Jack Bookman and Lewis Blake from Duke University, who helped me get started as a teacher.

I also want to thank Houghton Mifflin for believing in this project and giving me the chance to publish this textbook. Many people at Houghton Mifflin have been extremely helpful and supportive throughout this project, in particular

> Nancy Blodget, Claire Boivin, Kathleen Deselle, Dr. Donna Flint, Kyle Henderson, Marika Hoe, Teresa Hoens, Kasey McGarrigle, Paul Murphy, Lauren Schultz, Lauri Semarne, Jack Shira, Danielle Steiner, Danielle Wilkinson, Laura Wheel, and many other people behind the scenes that I have not had the pleasure of meeting but that I know have worked hard to make this book the best it can be.

I am indebted to the reviewers and class-testers of this textbook; their comments were extremely helpful and led to significant revisions.

Reviewers

Jack Bookman, Duke University; Christopher Brazfield, Lebanon Valley College, Pennsylvania; Priscilla Bremser, Middlebury College, Vermont; David A. Cox, Amherst College, Massachusetts; David Cruz-Uribe, Trinity College, Connecticut; David Erwin, Trinity College, Connecticut; Frances Gulick, University of Maryland College, Maryland; Major Darryl Lanford, United States Military Academy, New York; Carmen M. Latterell, University of Minnesota, Duluth, Minnesota; Melissa Shepard Loe, University of St. Thomas, Minnesota; Dr. Carl V. Lutzer, Rochester Institute of Technology, New York; Jeffery T. McLean, University of St. Thomas, Minnesota; Elaine Brooks Rinaldo, Hudson Valley Community College, New York; Melanie Stein, Trinity College, Connecticut; Elaine Terry, St. Joseph's University, Pennsylvania; and Barbara Wahl, Hanover College, Indiana

Class Testers

David Cruz-Uribe, Trinity College, Connecticut; Brett Enge, James Madison University, Virginia; David Erwin, Trinity College, Connecticut; Robert Hanson, James Madison University, Virginia; Peter Kohn, James Madison University, Virginia; Jason Rosenhouse, James Madison University, Virginia; and Len VanWyk, James Madison University, Virginia

Finally, thanks go out to all the friends and family who have supported me, and in particular to my sisters Linda and Alina, who were kind enough to lend me their names for many homework exercises. Last but certainly not least, I want to thank my husband, Phil Riley, who has been helpful, supportive, and understanding during my long hours in front of the computer screen writing this book. Without him this book would not have been possible. Phil, thanks for listening to me talk endlessly about this project and for putting up with me in my lesser moments.

Comments and Suggestions

I welcome comments and suggestions from instructors and students using this textbook. In particular, I am interested in hearing about any typographical, mathematical, or formatting errors you find in this book; any topics or sections you would like to see added to the book; any topics or sections you think could be improved; and any pedagogical or stylistic comments. I would also be happy to answer any questions you might have about this book or the type of course it is for. I can be reached by email at taal@math.jmu.edu or by post at

Dr. Laura Taalman
James Madison University
Department of Mathematics and Statistics
Burruss Hall, MSC 7803
Harrisonburg, VA 22807

Look for information and updates at the official website for this book, *math.college.hmco.com,* and at my personal website, *www.math.jmu.edu/~taal.* Thank you in advance for your comments!

TRUE INTEGRATION OF CALCULUS, ALGEBRA, AND PRECALCULUS

This text is **NOT** simply a cut-and-paste "just-in-time" combination of calculus and precalculus. Precalculus and algebra topics are woven **THROUGHOUT** the book and connected with the ideas of calculus. This provides a fresh perspective for students who are repeating these topics.

Inverted Structure with Functions as the Unifying Theme

Integrated Calculus is structured around the study of functions. Students learn basic theories and techniques of calculus and then use the knowledge to explore different types of functions. For example, after learning the basics of differentiation in Chapter 3, students apply what they know to increasingly complex functions in Chapters 4–12 as they investigate the algebra, precalculus, and calculus of each type of function. This reinforces skills through application to each new class of functions and is especially helpful to students with weak algebra skills.

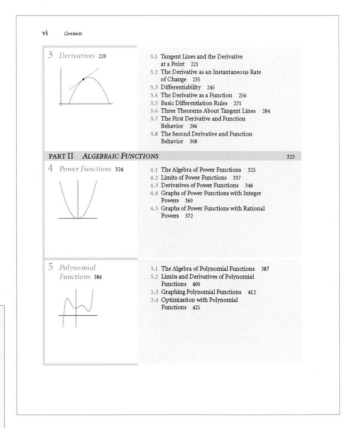

Handy Reference Pages

The Reference Pages provide a comprehensive summary of **ALL** symbols, formulas, theorems, and major topics that are used or developed throughout the text. Cross-references to appropriate sections of the book are cited to maximize effectiveness as a reference tool.

Chapter Openers

Each chapter opens with an example of mathematics in art or nature and a short introduction to the chapter. A detailed table of contents describes the sections and subsections of the chapter for easy reference.

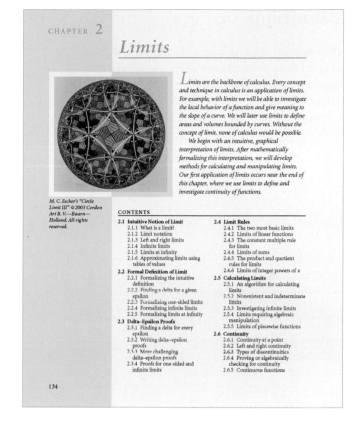

CHAPTER **2**

Limits

Limits are the backbone of calculus. Every concept and technique in calculus is an application of limits. For example, with limits we will be able to investigate the local behavior of a function and give meaning to the slope of a curve. We will later use limits to define areas and volumes bounded by curves. Without the concept of limit, none of calculus would be possible.

We begin with an intuitive, graphical interpretation of limits. After mathematically formalizing this interpretation, we will develop methods for calculating and manipulating limits. Our first application of limits occurs near the end of this chapter, where we use limits to define and investigate continuity of functions.

134

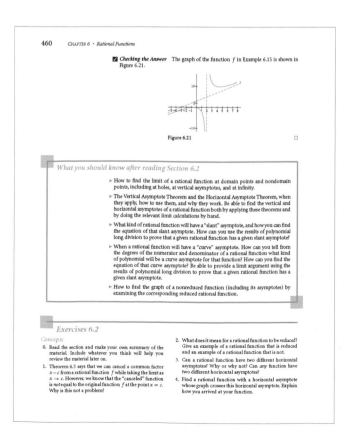

460   CHAPTER 6 · *Rational Functions*

☑ *Checking the Answer*   The graph of the function f in Example 6.15 is shown in Figure 6.21.

Figure 6.21

What you should know after reading Section 6.2

▸ How to find the limit of a rational function at domain points and nondomain points, including at holes, at vertical asymptotes, and at infinity.

▸ The Vertical Asymptote Theorem and the Horizontal Asymptote Theorem, when they apply, how to use them, and why they work. Be able to find the vertical and horizontal asymptotes of a rational function both by applying these theorems and by doing the relevant limit calculations by hand.

▸ What kind of rational function will have a "slant" asymptote, and how you can find the equation of that slant asymptote. How can you use the results of polynomial long division to prove that a given rational function has a given slant asymptote?

▸ When a rational function will have a "curve" asymptote. How can you tell from the degrees of the numerator and denominator of a rational function what kind of polynomial will be a curve asymptote for that function? How can you find the equation of that curve asymptote? Be able to provide a limit argument using the results of polynomial long division to prove that a given rational function has a given slant asymptote.

▸ How to find the graph of a nonreduced function (including its asymptotes) by examining the corresponding reduced rational function.

Exercises 6.2

Concepts

0. Read the section and make your own summary of the material. Include whatever you think will help you review the material later on.

1. Theorem 6.3 says that we can cancel a common factor $x - c$ from a rational function f while taking the limit as $x \to c$. However, we know that the "canceled" function is *not* equal to the original function f at the point $x = c$. Why is this not a problem?

2. What does it mean for a rational function to be *reduced*? Give an example of a rational function that is reduced and an example of a rational function that is not.

3. Can a rational function have two different horizontal asymptotes? Why or why not? Can *any* function have two different horizontal asymptotes?

4. Find a rational function with a horizontal asymptote whose graph crosses this horizontal asymptote. Explain how you arrived at your function.

Section Summaries

Each section ends with a *What you should know* summary that describes the main points of the section. This summary helps students test their understanding of the reading before going on to the homework exercises and helps them review for quizzes and tests.

Definitions

All important definitions are given titles and clearly set apart from the text for ease of reference. These definitions are worked into the exposition in the text and are often preceded by motivation and followed by clarification, examples, and illustrations.

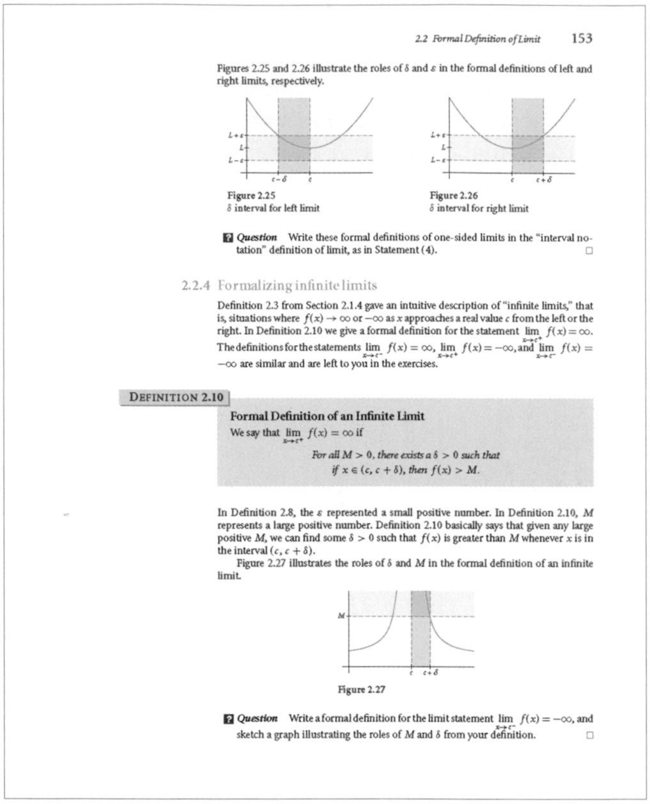

Figures 2.25 and 2.26 illustrate the roles of δ and ε in the formal definitions of left and right limits, respectively.

Figure 2.25
δ interval for left limit

Figure 2.26
δ interval for right limit

? **Question**   Write these formal definitions of one-sided limits in the "interval notation" definition of limit, as in Statement (4).   □

2.2.4 Formalizing infinite limits

Definition 2.3 from Section 2.1.4 gave an intuitive description of "infinite limits," that is, situations where $f(x) \to \infty$ or $-\infty$ as x approaches a real value c from the left or the right. In Definition 2.10 we give a formal definition for the statement $\lim_{x \to c^+} f(x) = \infty$. The definitions for the statements $\lim_{x \to c^+} f(x) = -\infty$, $\lim_{x \to c^-} f(x) = -\infty$, and $\lim_{x \to c^-} f(x) = \infty$ are similar and are left to you in the exercises.

DEFINITION 2.10

Formal Definition of an Infinite Limit
We say that $\lim_{x \to c^+} f(x) = \infty$ if

For all $M > 0$, there exists a $\delta > 0$ such that
if $x \in (c, c + \delta)$, then $f(x) > M$.

In Definition 2.8, the ε represented a small positive number. In Definition 2.10, M represents a large positive number. Definition 2.10 basically says that given any large positive M, we can find some $\delta > 0$ such that $f(x)$ is greater than M whenever x is in the interval $(c, c + \delta)$.

Figure 2.27 illustrates the roles of δ and M in the formal definition of an infinite limit.

Figure 2.27

? **Question**   Write a formal definition for the limit statement $\lim_{x \to c^-} f(x) = -\infty$, and sketch a graph illustrating the roles of M and δ from your definition.   □

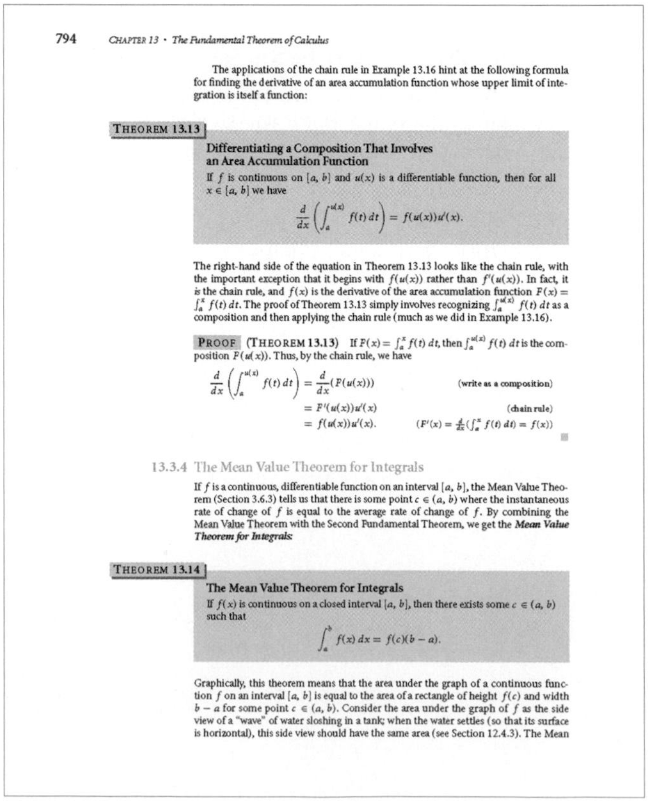

The applications of the chain rule in Example 13.16 hint at the following formula for finding the derivative of an area accumulation function whose upper limit of integration is itself a function:

THEOREM 13.13

Differentiating a Composition That Involves an Area Accumulation Function
If f is continuous on $[a, b]$ and $u(x)$ is a differentiable function, then for all $x \in [a, b]$ we have

$$\frac{d}{dx}\left(\int_a^{u(x)} f(t)\, dt\right) = f(u(x))u'(x).$$

The right-hand side of the equation in Theorem 13.13 looks like the chain rule, with the important exception that it begins with $f(u(x))$ rather than $f'(u(x))$. In fact, it *is* the chain rule, and $f(x)$ is the derivative of the area accumulation function $F(x) = \int_a^x f(t)\, dt$. The proof of Theorem 13.13 simply involves recognizing $\int_a^{u(x)} f(t)\, dt$ as a composition and then applying the chain rule (much as we did in Example 13.16).

PROOF   (**THEOREM 13.13**)   If $F(x) = \int_a^x f(t)\, dt$, then $\int_a^{u(x)} f(t)\, dt$ is the composition $F(u(x))$. Thus, by the chain rule, we have

$$\frac{d}{dx}\left(\int_a^{u(x)} f(t)\, dt\right) = \frac{d}{dx}(F(u(x)))$$ (write as a composition)
$$= F'(u(x))u'(x)$$ (chain rule)
$$= f(u(x))u'(x).$$ $(F'(x) = \frac{d}{dx}(\int_a^x f(t)\, dt) = f(x))$

13.3.4 The Mean Value Theorem for Integrals

If f is a continuous, differentiable function on an interval $[a, b]$, the Mean Value Theorem (Section 3.6.3) tells us that there is some point $c \in (a, b)$ where the instantaneous rate of change of f is equal to the average rate of change of f. By combining the Mean Value Theorem with the Second Fundamental Theorem, we get the *Mean Value Theorem for Integrals*:

THEOREM 13.14

The Mean Value Theorem for Integrals
If $f(x)$ is continuous on a closed interval $[a, b]$, then there exists some $c \in (a, b)$ such that

$$\int_a^b f(x)\, dx = f(c)(b - a).$$

Graphically, this theorem means that the area under the graph of a continuous function f on an interval $[a, b]$ is equal to the area of a rectangle of height $f(c)$ and width $b - a$ for some point $c \in (a, b)$. Consider the area under the graph of f as the side view of a "wave" of water sloshing in a tank; when the water settles (so that its surface is horizontal), this side view should have the same area (see Section 12.4.3). The Mean

Theorems

Theorems are also named and set apart from the text, with motivation and explanation in the text before and after each theorem. Proofs of theorems are written with the student in mind, with clear steps and discussion of the method of proof.

Examples

Numerous examples in the text illustrate important concepts and techniques and are named for easy reference. Students are encouraged to reflect on examples in the following ways:

- *Question* encourages students to be critical and active readers.

- *Caution* points out common pitfalls to avoid.

- *Checking the Answer* stresses the importance of verifying the correctness of answers independently and provides students with the skills to do so.

3.1 Tangent Lines and the Derivative at a Point 225

EXAMPLE 3.2 **Using the definition of derivative to find the exact slope of a tangent line**

Find the exact slope of the tangent line to the graph of $f(x) = x^2$ at $x = 3$.

Solution We use Definition 3.1 and find that the derivative of f at $x = 3$ is equal to

$$f'(3) = \lim_{h \to 0} \frac{f(3+h) - f(3)}{h} = \lim_{h \to 0} \frac{(3+h)^2 - (3)^2}{h}.$$

It remains only to calculate this limit. Note that if we "plug in" $h = 0$ at this point, we get the indeterminate form $\frac{0}{0}$, so we must do some algebra before evaluating this limit. We calculate

$$f'(3) = \lim_{h \to 0} \frac{f(3+h) - f(3)}{h} = \lim_{h \to 0} \frac{(3+h)^2 - (3)^2}{h}$$

$$= \lim_{h \to 0} \frac{(9 + 6h + h^2) - 9}{h} = \lim_{h \to 0} \frac{6h + h^2}{h}$$

$$= \lim_{h \to 0} \frac{h(6 + h)}{h} = \lim_{h \to 0}(6 + h) = 6.$$

Because the derivative of $f(x) = x^2$ at $x = 3$ is equal to 6, the line tangent to the graph of $y = x^2$ at the point $(3, 9)$ has slope 6. □

? Question Which limit rules did we use implicitly in Example 3.2? □

◆ Caution Be careful that you do not "drop the limits" when doing a calculation like the one in Example 3.2. Since you are calculating a *limit* when you use the definition of derivative, each step of your calculation (until you actually "take the limit," of course) should include the expression "$\lim_{h \to 0}$." □

☑ Checking the Answer To check whether the solution to Example 3.2 is reasonable, sketch the graph of f and see whether its tangent line at $x = 3$ appears to have a slope of 6. Figure 3.10 shows this graph. Note that the slope of the tangent line at $x = 3$ is positive, and considering the scale of the graph, a slope of 6 seems reasonable.

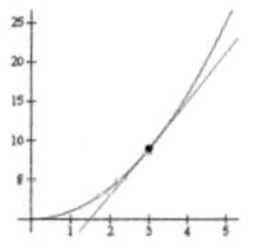

Figure 3.10

Alternatively, you could check the answer by using a graphing calculator to approximate the slope of the tangent line to $f(x) = x^2$ at $x = 3$. □

◆ Caution You may have seen "rules" for differentiating in a previous course. Do *not* use these differentiation shortcuts at this point. Later, we will develop rules for quickly calculating derivatives, but at this point we must use the definition of derivative, as in Example 3.2. □

---

804 *CHAPTER 14 · Basic Integration Techniques*

ALGORITHM 14.1

Integration by Substitution
1. Choose some part of the integrand to be $u(x)$; usually a good choice for $u(x)$ is an "inside" function of a composition, whose derivative $u'(x)$ appears as a multiplicative factor of the integrand.
2. Differentiate the equation for $u(x)$ to find du; remember that $\frac{du}{dx} = u'(x)$, so $du = u'(x)\,dx$.
3. Change variables; that is, use the equations for $u(x)$ and du to rewrite the integrand, *including* the dx, *entirely* in terms of u and du.
4. Integrate the new, hopefully simpler integral.
5. Change variables back to x by substituting the formula for $u(x)$ for each u in the answer.

Algorithm 14.1 works for any integral in the form of Equation (1). The following example uses the algorithm of integration by substitution to find the integral from Example 14.1 and Example 14.2. (We keep repeating the same example so you can see how the substitution algorithm is derived from the chain rule.)

EXAMPLE 14.3 **Using integration by substitution**

Use integration by substitution to find $\int \cos x^2(2x)\,dx$.

Solution Since x^2 is the "inside" of a composition, and its derivative $2x$ appears elsewhere in the integrand, we'll try setting $u = x^2$. Now we must differentiate $u(x)$ to find du:

$$u = x^2 \implies \frac{du}{dx} = 2x \implies du = 2x\,dx.$$

By replacing x^2 by u (so $\cos(x^2) = \cos u$) and $2x\,dx$ by du (using the boxed equations), we have

$$\int \cos x^2(2x)\,dx = \int \cos u\,du \quad \text{(since } u = x^2 \text{ and } du = 2x\,dx\text{)}$$

$$= \sin u + C \quad \text{(integrate the new integral)}$$

$$= \sin x^2 + C. \quad \text{(change variables back to } x\text{)} \quad \square$$

◆ Caution In Example 14.3, the "new" integral $\int \cos u\,du$ was written *entirely* in terms of u and du. If there had been any x's left in the integral after we changed variables, it would have meant that we had made a bad choice for $u(x)$. We *cannot* integrate if the integrand involves both x and u. □

The method of integration by substitution works even if the integrand is not *exactly* in the form $f'(u(x))u'(x)$. The following example shows that we can use integration by substitution even when a constant multiple is "missing" from the integrand.

EXAMPLE 14.4 **Using integration by substitution when a constant multiple is "missing"**

Use integration by substitution to find $\int x^2 e^{x^3+1}\,dx$.

Algorithms

While much of the book focuses on a conceptual and intuitive understanding of mathematics, important techniques are clearly outlined step-by-step and highlighted with detailed algorithms to enhance student understanding.

Readability

Integrated Calculus uses a linear, uncluttered presentation style that increases readability. Figures are included within the text itself, never banished to cluttered margins (where students may not look for them). The tone of the book is readable and conversational without the loss of an appropriate level of detail or mathematical precision.

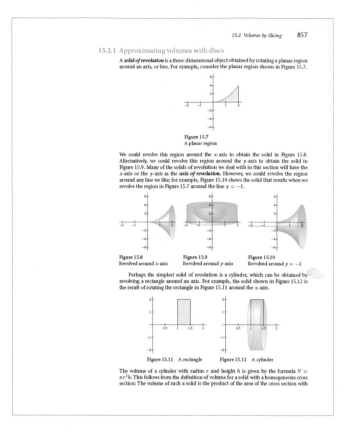

15.2 Volumes by Slicing   857

15.2.1 Approximating volumes with discs

A *solid of revolution* is a three-dimensional object obtained by rotating a planar region around an axis, or line. For example, consider the planar region shown in Figure 15.7.

Figure 15.7
A planar region

We could revolve this region around the x-axis to obtain the solid in Figure 15.8. Alternatively, we could revolve this region around the y-axis to obtain the solid in Figure 15.9. Many of the solids of revolution we deal with in this section will have the x-axis or the y-axis as the *axis of revolution*. However, we could revolve the region around any line we like; for example, Figure 15.10 shows the solid that results when we revolve the region in Figure 15.7 around the line $y = -1$.

Figure 15.8
Revolved around x-axis

Figure 15.9
Revolved around y-axis

Figure 15.10
Revolved around $y = -1$

Perhaps the simplest solid of revolution is a cylinder, which can be obtained by revolving a rectangle around an axis. For example, the solid shown in Figure 15.12 is the result of rotating the rectangle in Figure 15.11 around the x-axis.

Figure 15.11 A rectangle

Figure 15.12 A cylinder

The volume of a cylinder with radius r and height h is given by the formula $V = \pi r^2 h$. This follows from the definition of volume for a solid with a homogeneous cross section: The volume of such a solid is the product of the area of the cross section with

Side-by-Side Intuitive and Formal Definitions

Intuitive and graphical approaches are presented side-by-side with the formal and rigorous mathematical definitions to complete students' understanding of terminology and concepts.

1.2 Graphs of Functions   77

Question   Approximate the largest value of δ for which $f(1) \geq f(x)$ for all x in $(1 - \delta, 1 + \delta)$ in Figure 1.20. Then find δ-intervals for the local minima at $x = -1$ and $x = 2$.

The mathematical and verbal descriptions of extrema are summarized in Table 1.5.

Terminology	Verbal Description	Math Description
f has a global maximum at $x = c$	The point $(c, f(c))$ is the highest point on the graph of f	$f(c) \geq f(x)$ for all x in the domain of f
f has a global minimum at $x = c$	The point $(c, f(c))$ is the lowest point on the graph of f	$f(c) \leq f(x)$ for all x in the domain of f
f has a local maximum at $x = c$	The point $(c, f(c))$ is *locally* the highest point on the graph of f	for some $\delta > 0$, $f(c) \geq f(x)$ for all $x \in (c - \delta, c + \delta)$
f has a local minimum at $x = c$	The point $(c, f(c))$ is *locally* the lowest point on the graph of f	for some $\delta > 0$, $f(c) \leq f(x)$ for all $x \in (c - \delta, c + \delta)$

Table 1.5

◆ *Caution*   Remember, we always identify the location of an extremum by its x-value. The y-value of the extremum is called the "value" of the extremum. For example, the graph in Figure 1.20 has a local maximum at $x = 1$, and this local maximum has a value of $f(1) = 3.25$.

 EXAMPLE 1.33   On a constant function, every point is a global maximum and a global minimum

For the constant function $f(x) = 2$, *every* point is a global maximum, and *every* point is a global minimum (see Figure 1.21). For example, $x = 1$ is a global maximum of $f(x) = 2$ because $f(1) = 2$ is at least as high as every other point on the graph.

Figure 1.21

We say that a function f has an *inflection point* at $x = c$ if c is in the domain of f and the graph of f changes concavity at $x = c$; in other words, $x = c$ is an inflection point of f if the graph of f is concave down to the immediate left of $x = c$ and concave up to the immediate right of c, or vice versa. More precisely, a function f has an inflection point at $x = c$ if there is some positive number δ such that f is concave up (or down) on the interval $(c - \delta, c)$ and has the opposite concavity on the interval $(c, c + \delta)$.

Question   There are four types of inflection points. Can you sketch a graph illustrating each of these four types of inflection points? (*Hint:* One type of inflection point can be seen in a graph that changes from concave down to concave up while decreasing; see Figure 1.22.)

Since we don't yet have a formal mathematical description of concavity, we can't algebraically find the exact values of the inflection points of a function f. For the moment we will estimate these values by inspecting the graph of f.

Exercises: Concepts and Skills

Each exercise set begins with "Problem 0" that stresses the importance of reading the text before attempting the exercises and helps students gather their thoughts before attempting problems. A great many "Concepts" problems in each section hold students accountable to the definitions and concepts from the reading and often include True/False problems and problems involving writing. "Skills" problems then provide practice and development for any calculational skills presented in the section.

---

668   CHAPTER 10 · Trigonometric Functions

Exercises 10.4

Concepts

0. Read the section and make your own summary of the material. Include whatever you think will help you review the material later on.

1. Use a diagram of the unit circle to explain why $\sin\theta \approx \theta$ when θ is a small positive or negative angle.

2. Use a table of values to estimate $\lim_{\theta\to 0}\frac{\sin 2\theta}{\theta}$. Then check your approximation by examining a graph of $f(\theta)=\frac{\sin 2\theta}{\theta}$ near $\theta = 0$.

3. Use a table of values to estimate $\lim_{\theta\to 0}\frac{1-\cos 2\theta}{\theta}$. Then check your approximation by examining a graph of $f(\theta)=\frac{1-\cos 2\theta}{\theta}$ near $\theta = 0$.

4. What type of discontinuity does the function $f(x)=\frac{\sin x}{x}$ have at $x = 0$, and why?

5. Sketch a graph of $y=\sin x$ from memory. (*Hint*: Thinking about the unit circle may help you remember the features of the graph.) Using what you know about transformations of functions, use this graph to sketch the graphs of (a) $y=\sin x + 2$, (b) $y=\sin(x+2)$, (c) $y=2\sin x$, and (d) $y=\sin 2x$.

6. Sketch a graph of $y=\tan x$ from memory. (*Hint*: Thinking about the unit circle may help you remember the features of the graph.) Using what you know about transformations of functions, use this graph to sketch the graphs of (a) $y=\tan x + 2$, (b) $y=\tan(x+2)$, (c) $y=2\tan x$, and (d) $y=\tan 2x$.

7. Sketch a graph of $y=\sec x$ from memory. (*Hint*: Thinking about the unit circle may help you remember the features of the graph.) Using what you know about transformations of functions, use this graph to sketch the graphs of (a) $y=\sec x + 2$, (b) $y=\sec(x+2)$, (c) $y=2\sec x$, and (d) $y=\sec 2x$.

8. List the periods of all of the six trigonometric functions.

9. Suppose you know that $\lim_{x\to 0}\sin x=0$; explain why this means that $f(x)=\sin x$ is continuous at $x=0$.

10. Suppose you know that $\lim_{x\to 0}(\cos x - 1)=0$; explain why this means that $f(x)=\cos x$ is continuous at $x=0$.

11. Explain why the limit $\lim_{x\to c}\cos x$ is equivalent to the limit $\lim_{h\to 0}\cos(c+h)$.

12. Suppose $x=c$ is a real number in the domain of $f(x)=\sec x$. How can we calculate $\lim_{x\to c}\sec x$, and why?

13. Guess the value of the limit $\lim_{x\to 0}e^x\sin x$. Then check your answer by sketching a *good* graph. You should use both your calculator and your knowledge of the sine and exponential functions to sketch your graph.

14. Guess the value of the limit $\lim_{x\to\infty}\frac{\sin x}{e^x}$. Then check your answer by sketching a *good* graph. You should use both your calculator and your knowledge of the sine and exponential functions to sketch your graph.

15. Guess the value of the limit $\lim_{x\to\infty}\frac{e^x}{\sin x}$. Then check your answer by sketching a *good* graph. You should use both your calculator and your knowledge of the sine and exponential functions to sketch your graph.

■ Determine whether each of the following statements is true or false. If a statement is true, explain why. If a statement is false, explain why or provide a counterexample.

16. True or False: When θ is a small angle, the quantities $\sin\theta$ and θ are equal.

17. True or False: $y=\tan x$ is continuous at $x=0$.

18. True or False: $y=\tan x$ is continuous at $x=\frac{\pi}{2}$.

19. True or False: The graph of $y=\csc x$ has vertical asymptotes at $x=k\pi$, for any integer k.

20. True or False: The graph of $y=\tan x$ intersects the x-axis at every integer multiple of π.

21. True or False: For any real number x, $\sin(x+2\pi)=\sin x$.

Skills

■ Calculate the following limits (note that you will not be able to apply L'Hôpital's Rule, since we do not yet know how to differentiate trigonometric functions). Show all work, and then check your answers with graphs.

22. $\lim_{x\to\infty}\frac{\cos x}{\sin x + 2}$

23. $\lim_{x\to\infty}\frac{1}{\frac{1}{3}\sin x + \cos x}$

24. $\lim_{x\to 0}\frac{1+\sin x}{1-\cos x}$

25. $\lim_{x\to 0}\frac{x}{\sin x}$

26. $\lim_{x\to 0}\frac{x}{1-\cos x}$

27. $\lim_{x\to 0}\frac{\sin 2x}{2x}$

28. $\lim_{x\to 0}\frac{\sin 3x}{5x}$

29. $\lim_{x\to 0}\frac{1-\cos 2x}{7x}$

30. $\lim_{x\to 0}\frac{3\sin x \mid x}{x}$

31. $\lim_{x\to 0}\frac{\sin^2 3x}{x^3-x}$

32. $\lim_{x\to 0}\frac{\sin(3x^2)}{x^3-x}$

33. $\lim_{x\to 0}\left(\frac{1}{2x}-\frac{2\cos 5x}{4x}\right)$

34. $\lim_{x\to 1^+}\frac{1-\cos\sqrt{x-1}}{\sqrt{x-1}}$

35. $\lim_{x\to 0^+}\frac{x^2\csc 3x}{\cos 2x}$

36. $\lim_{x\to\frac{\pi}{2}}\frac{\sin(x-\frac{\pi}{2})}{\frac{\pi}{2}-x}$

37. $\lim_{x\to 0}\frac{x^2\cot x}{\sin x}$

38. $\lim_{x\to 0}\frac{\sec x\tan x}{x}$

39. $\lim_{x\to 0}3x^2\cot^2 x$

10.4 Limits of Trigonometric Functions   **669**

40. $\lim_{x\to 2}\frac{x^2-4}{\sin(x-2)}$

41. $\lim_{x\to 1}\frac{\sin^2(x-1)}{x^2-1}$

42. $\lim_{x\to 2}\frac{1-\cos(x-2)}{x^2+x-6}$

43. $\lim_{x\to\pi}\tan x$

44. $\lim_{x\to\frac{\pi}{2}}\tan x$

45. $\lim_{x\to\pi}\csc^2 x$

46. $\lim_{x\to 0}\cos x$

47. $\lim_{x\to 0}x\cos x$

48. $\lim_{x\to\infty}\frac{\cos x}{x}$

49. $\lim_{x\to 0}\sin\frac{1}{x}$

50. $\lim_{x\to 0}x\sin\frac{1}{x}$

51. $\lim_{x\to 0}x^2\sin\frac{1}{x}$

■ Use limits and the definition of continuity to determine whether each of the following functions is continuous at $x=0$. Then use a graph to support your answer.

52. $f(x)=\begin{cases}\sin(\frac{1}{x}), & \text{if } x\neq 0 \\ 0, & \text{if } x=0\end{cases}$

53. $f(x)=\begin{cases}x\sin(\frac{1}{x}), & \text{if } x\neq 0 \\ 0, & \text{if } x=0\end{cases}$

54. $f(x)=\begin{cases}x^2\sin(\frac{1}{x}), & \text{if } x\neq 0 \\ 0, & \text{if } x=0\end{cases}$

Applications

55. Consider a mass hanging from the ceiling at the end of a spring. If you pull down on the mass and let go, it will oscillate up and down according to the equation

$$s(t)=A\sin\left(\sqrt{\tfrac{k}{m}}\,t\right)+B\cos\left(\sqrt{\tfrac{k}{m}}\,t\right).$$

where $s(t)$ is the distance of the mass from its equilibrium position, m is the mass of the bob on the end of the spring, and k is a "spring coefficient" that measures how tight or stiff the spring is. The constants A and B depend on initial conditions—specifically, how far you pull down the mass (s_0) and the velocity at which you release the mass (v_0). This equation does *not* take into account any friction due to air resistance.

(a) Determine whether or not the limit of $s(t)$ as $t\to\infty$ exists. What does this say about the long-term behavior of the mass on the end of the spring?

(b) Explain how this limit is related to the fact that the equation above does *not* take friction due to air resistance into account.

(c) Suppose the bob at the end of the spring has a mass of 2 grams and that the coefficient for the spring is $k=9$. Suppose also that the spring is released in such a way that $A=\sqrt{2}$ and $B=2$. Use a calculator to graph the function $s(t)$ that describes

the distance of the mass from its equilibrium position. Use your graph to support your answer to part (a).

56. In Problem 55 we gave an equation describing spring motion without air resistance. If we take into account friction due to air resistance, the mass will oscillate up and down according to the equation

$$s(t)=e^{-\frac{f}{2m}t}\left(A\sin\left(\frac{\sqrt{4km-f^2}}{2m}t\right)+B\cos\left(\frac{\sqrt{4km-f^2}}{2m}t\right)\right).$$

where m, k, A, and B are the constants described in Problem 55, and f is a positive "friction coefficient" that measures the amount of friction due to air resistance.

(a) Find the limit of $s(t)$ as $t\to\infty$. What does this say about the long-term behavior of the mass on the end of the spring?

(b) Explain how this limit is related to the fact that the equation above *does* take friction due to air resistance into account.

(c) Suppose the bob at the end of the spring has a mass of 2 grams, the coefficient for the spring is $k=9$, and the friction coefficient is $f=6$. Suppose also that the spring is released in such a way that $A=4$ and $B=2$. Use a calculator to graph the function $s(t)$ that describes the distance of the mass from its equilibrium position. Use your graph to support your answer to part (a).

Proofs

57. Use the fact that $\lim_{x\to 0}\frac{\sin x}{x}=1$ to prove that $\lim_{x\to 0}\sin x=\sin 0$. Explain why this means that $f(x)=\sin x$ is continuous at $x=0$.

58. Use the fact that $\lim_{x\to 0}\frac{1-\cos x}{x}=0$ to prove that $\lim_{x\to 0}(\cos x-1)=0$. Explain why this means that $f(x)=\cos x$ is continuous at $x=0$.

59. Use Theorem 10.12 and the sum identity for the sine function to prove that $\lim_{h\to 0}\sin(c+h)=\sin c$ for all real numbers c. Explain why this means that $f(x)=\sin x$ is continuous everywhere.

60. Use Theorem 10.12 and the sum identity for the cosine function to prove that $\lim_{h\to 0}\cos(c+h)=\cos c$ for all real numbers c. Explain why this means that $f(x)=\cos x$ is continuous everywhere.

61. Use the fact that the cosine function is continuous to prove that $f(x)=\sec x$ is continuous everywhere on its domain.

62. Use the fact that the sine and cosine functions are continuous to prove that $f(x)=\tan x$ is continuous everywhere on its domain.

Exercises: Applications

"Applications" problems link the material in the section to the real world. Some sections of the text are almost entirely devoted to applications. "Proofs" problems include many simple, easy proofs that students will be able to handle with confidence, as well as problems that ask students to repeat proofs from the reading.

PART I

FUNCTIONS, LIMITS, AND DERIVATIVES

In this part of the book we introduce the main theorems and definitions of calculus. We present these theorems and definitions with graphical investigations and simple algebraic examples. We also give a short review of basic algebra and discuss proofs and logic. In Parts II and III we will use these theorems and definitions to investigate progressively more complicated types of functions.

The Basics

"Roads Untaken," a sculpture by George Hart, www.georgehart.com

*T*his chapter reviews and introduces basic concepts concerning numbers, sets, equations, and inequalities. We also present a brief introduction to elementary logic, and discuss methods for constructing and writing proofs. These basic skills will be used and built upon throughout the entire text.

CONTENTS

0.1 | Numbers and Sets

In this section we present basic definitions, notations, and graphical representations for numbers and sets of numbers.

0.1.1 Number systems

We begin with some basic definitions of different types of numbers.

DEFINITION 0.1

Types of Numbers

The *integers* consist of the positive and negative whole numbers and the number zero. For example, $0, 1, 2, 3, \ldots$ and $-1, -2, -3, \ldots$ are integers.

A *rational number* is a number that can be written in the form $\frac{p}{q}$, where p and q are integers and $q \neq 0$. In decimal form, a rational number will always terminate or repeat. For example, $\frac{2}{5}, -\frac{3}{6}, 3.0167, 0.3823232\overline{3}$, and 4 are rational numbers.

An *irrational number* is a number that can be represented in decimal notation but cannot be written in the form $\frac{p}{q}$ for any integers p and q. Therefore, irrational numbers are nonterminating, nonrepeating decimals. For example, $-0.2319834783239\ldots$ (never terminating or repeating), π, and $\sqrt{2}$ are irrational numbers.

The *real numbers* consist of all rational and irrational numbers. For example, $0, -3, 1.5, \frac{3}{17}, 3.\overline{21}$, and $\sqrt{2}$ are real numbers.

A *complex number* is a number that can be written in the form $a + bi$, where a and b are real numbers and $i = \sqrt{-1}$. For example, $\sqrt{-1}, 3 + 4i, 32, \sqrt{-63}$, and $-i$ are complex numbers.

In this text we will work primarily with real numbers. Unless we say otherwise, when we refer to "a number," we always mean "a *real* number." In Section 0.2.1, you will see that complex numbers arise naturally as solutions to equations such as $x^2 + 1 = 0$.

? *Question* Is every irrational number a real number? Is every rational number a real number? Is every integer a rational number? Is every real number a complex number? Explain why or why not. □

$\otimes$ *Caution* Some decimal numbers have a recognizable pattern but are not repeating. For example, the number $0.101001000100001\ldots$ has a pattern but is not a repeating decimal. Therefore (assuming that the pattern continues and does not terminate), this number is *irrational*. □

EXAMPLE 0.1

Identifying rational numbers

Which of the following are rational numbers? Express each rational number in the form $\frac{p}{q}$ for some integers p and q.

  (a) 0.001 **(b)** 5 **(c)** 3.258 **(d)** $-\frac{11}{7}$ **(e)** $0.33\overline{3}$

Solution   All of the numbers above are rational. The first number is "one one-thousandth" and is thus represented by the fraction $\frac{1}{1000}$. The number 5 can be written as $\frac{5}{1}$ and is thus a rational number. The third number is the sum of 3 and 0.258, or $\frac{3000}{1000} + \frac{258}{1000} = \frac{3258}{1000}$. The fourth number could be represented in the form $\frac{p}{q}$ either by $\frac{-11}{7}$ or by $\frac{11}{-7}$. Finally, the last number is represented by the fraction $\frac{1}{3}$ (try dividing 3 into 1 to see why).   □

Question   It is an interesting but unusual fact that $0.99\overline{9}$ and 1 represent the same number! Can you think of a reason why $0.99\overline{9} = 1$? Given this fact, can you think of any other real numbers that have two different decimal representations?   □

We say that a real number c is **positive** if c is greater than zero ($c > 0$) and **negative** if $c < 0$. The number zero is neither positive nor negative. We say that a real number c is **nonnegative** if it is positive or zero, that is, if $c \geq 0$. Similarly, a number c is **nonpositive** if $c \leq 0$. The number zero is the only real number that is both nonnegative and nonpositive.

0.1.2 Sets and set notation

A **set** is a collection of objects (these objects could be real numbers, or integers, or books, or just about anything). The objects in a set are called **elements,** or **members,** of the set. For example, the collection of all the science fiction books in your house is a "set," and the collection of all real numbers greater than 5 is a set. The collection of all real numbers is a set, and we will use this set so often that we have a special notation for it:

(1)   The symbol "$\mathbb{R}$" represents the set of all real numbers.

The following notation gives us a compact way to say whether or not a given object is a member of a given set:

NOTATION 0.1

Set Membership

If an object x is a member of a set A, then we write

$$x \in A.$$

If x is not a member of the set A, then we write

$$x \notin A.$$

For example, because $\mathbb{R}$ is the set of all real numbers, we have $1.5 \in \mathbb{R}$, $-\frac{1}{3} \in \mathbb{R}$, and $\pi \in \mathbb{R}$, but $\sqrt{-1} \notin \mathbb{R}$.

A set can be described by specifying two things: first, a *category* (or larger set) of objects to consider; second, a *test* that describes when an object in that category is in the set.

EXAMPLE 0.2   **Two examples of sets**

The set A of all the science fiction books in your house has the category "books in your house" and the test "science fiction book." If a book in your house is a science fiction book, then it passes the test and is in the set A. If a book in your house is not a science fiction book, it is not a member of the set A. As another example, the set B of all real numbers greater than 5 consists of objects from the category "real numbers"; the test "greater than 5" tests whether or not a given real number is in the set B; if a number is greater than 5, then it is in the set B, and if it is not greater than 5, then it is not in the set B.   □

We can describe sets in compact notation by using ***set notation.*** This compact notation enables us to describe sets quickly and precisely, using mathematical notation.

NOTATION 0.2

Set Notation

A set S can often be described in the following notation:

$$S = \{x \in category \mid test\ determining\ whether\ x\ is\ in\ the\ set\}.$$

For example, the set B consisting of all real numbers greater than 5 is written in set notation as

$$B = \{x \in \mathbb{R} \mid x > 5\}.$$

This notation is pronounced "real numbers x such that $x > 5$," or sometimes "x an element of the real numbers such that $x > 5$." In particular, note that the vertical bar means "such that." When the category is all real numbers, we often simply write

$$B = \{x \mid x > 5\}.$$

In other words, if a category is not given, we assume that all of the objects are real numbers.

EXAMPLE 0.3   **Using set notation**

Suppose L is the set of all even positive real numbers. Express the set L in set notation.

Solution   The set L consists of real numbers that are positive and even; thus in set notation we can write

$$L = \{x \in \mathbb{R} \mid x > 0 \text{ and } x \text{ is even}\}.$$

Because the category is the set of all real numbers, we could also write simply

$$L = \{x \mid x > 0 \text{ and } x \text{ is even}\}.$$ □

Sometimes we can describe a set by listing all of its elements. If a set S has a small finite number of elements, then we simply list them all inside curly brackets; for example, the set $\{0, 3, 6, 7\}$ is a set with exactly four elements. If a set S has an infinite number of elements, or a large number of elements, we can list them "all" only if they have a recognizable pattern; for example, $\{2, 4, 6, 8, 10, 12, \ldots\}$ represents the set of all positive even numbers.

We say that one set is a ***subset*** of a second set if it is contained within the second set. In other words, a set A is a subset of a set B if every member x of A is also a member of B. For example, the set B of all real numbers greater than 5 is a subset of the set of real numbers, since each element in B is also an element of the real numbers. The set A

of all the science fiction books in your house is a subset of the set of all books in your house, which is in turn a subset of the set of all books in the world.

Perhaps the strangest set is the set containing no elements. This set is called the **empty set,** or the **null set,** and is denoted by the symbol ∅. For example, the set S of real numbers that are both negative and positive is empty, since no real number is both strictly less than and strictly greater than zero:

$$S = \{x \mid x < 0 \text{ and } x > 0\} = \emptyset.$$

0.1.3 The real number line and interval notation

The real numbers can be represented graphically on a **number line** like the one in Figure 0.1. Every real number can be represented as a point on a number line. For example, the real number 1.3 is marked on the number line in Figure 0.1. The arrows at the "ends" of the number line indicate that the number line continues indefinitely in either direction.

Figure 0.1

In this course we will often be interested in subsets of the real numbers that are **intervals** on the real number line. An interval is a connected subset of the real number line. For example, the set $S = \{x \mid 0 \le x \le 2\}$ is an interval of real numbers; this interval is depicted by the thick line in Figure 0.2.

Figure 0.2

This interval can be represented in **interval notation** as $[0, 2]$. The square brackets indicate that the interval is **closed** (that is, the interval includes the endpoints $x = 0$ and $x = 2$). In Figure 0.2, the fact that the dots at $x = 0$ and at $x = 2$ are "solid" or "filled in" indicates that the interval is closed.

An **open** interval is an interval that does not contain its endpoints. For example, the open interval $\{x \mid 0 < x < 2\}$ consists of all real numbers between, but not including, zero and 2. In interval notation we represent an open interval with round brackets. The interval $\{x \mid 0 < x < 2\}$ is written as $(0, 2)$ in interval notation. On a number line we indicate an open interval by a thick line that ends in *open* dots, as in Figure 0.3.

Figure 0.3

An interval can also be **half-closed** (or, equivalently, **half-open**). Such an interval contains one of its endpoints but not the other. For example, the interval $\{x \mid 0 < x \le 2\}$ is a half-closed interval, since it contains its right endpoint, $x = 2$, but not its left endpoint, $x = 0$. In interval notation we would write this interval as $(0, 2]$. On the real number line we would represent this interval as shown in Figure 0.4.

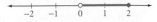

Figure 0.4

We also want to consider ***infinite intervals***—that is, intervals of real numbers that extend to ∞ or to $-\infty$. For example, the set $\{x \mid x \geq 0\}$ of real numbers that are greater than or equal to zero is an infinite interval. In interval notation we would write $[0, \infty)$. Note that we use an open, or "round," bracket at the infinite end of the interval; because "infinity" is not a real number, it is never included in the interval. Figure 0.5 shows the interval $[0, \infty)$ on a number line.

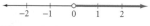

Figure 0.5

The entire real number line is represented in interval notation as $(-\infty, \infty)$. As one might expect, this interval can be represented on the real number line by a thick line covering the entire number line. The interval $(-\infty, \infty)$ is considered an open interval, because neither $-\infty$ nor ∞ is actually contained in the interval.

Sometimes an interval contains exactly one element. For example, consider the set of all numbers both greater than or equal to 2 and less than or equal to 2. This set consists of exactly one element, namely the number 2. This one number is indeed a connected subset of the real numbers, and thus technically it is an interval (although we usually won't think of it that way!). We do *not* write this set in interval notation as $[2]$. Instead we write $\{2\}$, which is shorthand for the set $\{x \in \mathbb{R} \mid x = 2\}$, containing only the number 2.

In general, there are 11 possible types of intervals of real numbers. If a and b are any real numbers with $a < b$, these intervals are $[a, b]$, $[a, b)$, $(a, b]$, (a, b), $[a, \infty)$, (a, ∞), $(-\infty, b]$, $(-\infty, b)$, $(-\infty, \infty)$, $\{a\}$, and $\emptyset$.

> ❓ ***Question***   Express each of the 11 types of intervals in set notation, in words, and on a real number line.   □

0.1.4   Unions and intersections

Given two sets A and B, we can consider the set of all objects that are in *both* A and B, and the set of all objects that are in *either* A or B. These new sets are called the ***intersection*** and the ***union*** of the sets A and B, respectively.

DEFINITION 0.2

Intersection and Union

Given sets A and B,

(a) The ***intersection*** of A and B is the set $A \cap B := \{x \mid x \in A \text{ and } x \in B\}$.

(b) The ***union*** of A and B is the set $A \cup B := \{x \mid x \in A \text{ or } x \in B\}$.

The ":=" sign means that the expression on the left is *defined* to be the expression on the right. It may help you distinguish $\cup$ and $\cap$ if you notice that the symbol "$\cup$" for union looks like a "u" and that the symbol "$\cap$" for intersection looks like a lowercase "n" (for the "n" in "intersection").

> 🚫 ***Caution***   When we say "or" in mathematics, we mean it "inclusively." In other words, we consider x to be an element of "A or B" (that is, $x \in A \cup B$) if x is in A, or if x is in B, or if x is in *both* A and B. For example, if A is the set of all green Volvos, and B is the set of all Volvo station wagons, then $A \cap B$ is the set of all green Volvo station wagons, and $A \cup B$ is the set of all Volvos that are either green, or station wagons, or both.   □

If we represent the sets *A* and *B* by overlapping circles as in Figures 0.6 and 0.7, then $A \cap B$ is the intersection of the two circles (Figure 0.8), and $A \cup B$ is the union of the two circles (Figure 0.9). Diagrams such as those in Figures 0.6–0.9 are called **Venn diagrams.**

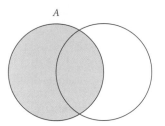

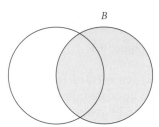

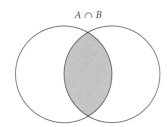

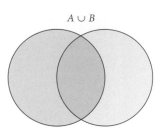

Figure 0.6   Figure 0.7   Figure 0.8   Figure 0.9

We will often consider unions and intersections of intervals and how they appear on a real number line.

<table>
<tr><td>EXAMPLE 0.4</td></tr>
</table>

Finding the union of two intervals

Consider the intervals $[-3, 1]$ and $(-1, 3)$. Express the union of these two intervals three ways: (**a**) on a real number line, (**b**) in interval notation, and (**c**) in set notation.

Solution   The union of $[-3, 1]$ and $(-1, 3)$ is the set of all real numbers that are either in the interval $[-3, 1]$ or in the interval $(-1, 3)$. For example, the number -2 is in the first interval (but not the second) and is thus in the union of the two intervals. Figure 0.10 shows the interval $[-3, 1]$ above the number line and the interval $(-1, 3)$ below the number line. Figure 0.11 shows the union of these two intervals.

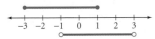

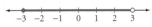

Figure 0.10   Figure 0.11

In interval notation, we would write

$$[-3, 1] \cup (-1, 3) = [-3, 3).$$

In set notation, $[-3, 1] = \{x \mid -3 \leq x \leq 1\}$ and $(-1, 3) = \{x \mid -1 < x < 3\}$. Thus the union $\{x \mid -3 \leq x \leq 1\} \cup \{x \mid -1 < x < 3\}$ of these two sets is

$$\{x \mid -3 \leq x \leq 1 \text{ or } -1 < x < 3\} = \{x \mid -3 \leq x < 3\}. \qquad \square$$

? Question   Repeat Example 0.4 for the *intersection* of the intervals. $\square$

0.1.5 Absolute value and distance

Intuitively, the ***absolute value*** of a real number *a* is the magnitude, or size, of *a* and is denoted $|a|$. For example, $|3| = 3$ and $|-3| = 3$. The absolute value of *a* "makes the number *a* positive." A much more precise way to say this is given in Definition 0.3.

DEFINITION 0.3

Absolute Value

The ***absolute value*** of a real number a is defined as follows:

$$|a| := \begin{cases} a, & \text{if } a \geq 0 \\ -a, & \text{if } a < 0. \end{cases}$$

Definition 0.3 defines the absolute value of a number in two cases. The first case means that if a is positive or zero, then $|a| = a$ (for example, $|3| = 3$). The second case means that if a is negative, then $|a| = -a$ (for example, $|-3| = -(-3) = 3$). The absolute value $|a|$ of any real number a is the distance between a and 0 on the real number line. Note that the absolute value of any real number is always positive, with the exception that $|0| = 0$.

※ *Caution* The quantity $-a$ is not necessarily negative! For example, if $a = -3$, then $-a = -(-3) = 3$ is positive. The sign of $-a$ depends on whether a itself is positive or negative. In Definition 0.3, $|a| = -a$ only if a is negative, in which case $-a$ is a *positive* number. □

We can use absolute values to define the distance between any two real numbers a and b. Intuitively, the distance between a and b is the length of the line segment between a and b on the real number line. For example, the distance between $a = 2$ and $b = -1$ is shown on the number line in Figure 0.12 and is equal to 3 units.

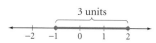

Figure 0.12

DEFINITION 0.4

Distance Between Two Real Numbers

The ***distance*** between two real numbers a and b is given by the absolute value of the difference of a and b. In other words,

$$\text{dist}(a, b) := |b - a|.$$

We could have equivalently defined the distance between a and b to be $|a - b|$, because for any real numbers a and b, the quantities $|b - a|$ and $|a - b|$ are the same. We'll "prove" this fact in Section 0.5. For now, pick three different values of a and three different values of b, and check that $|b - a| = |a - b|$ for those values.

EXAMPLE 0.5 | **Finding the distance between two real numbers**

The distance between the numbers $a = -1$ and $b = 2$ is

$$|b - a| = |-1 - 2| = |-3| = -(-3) = 3.$$

This corresponds with our intuitive notion of the distance between -1 and 2 on the number line in Figure 0.12. □

0.1.6 The Cartesian plane

An ***ordered pair*** of real numbers a and b is denoted (a, b). For example, $(2, 5)$ is an ordered pair. The "order" matters; in other words, $(2, 5)$ is a different ordered pair from $(5, 2)$.

> ⚛ ***Caution***   Unfortunately, the notation (a, b) has two possible meanings. Earlier we used this notation to represent the open interval from a to b; now we are using it to represent an ordered list of two numbers a and b. It should be clear from the context which of these two meanings we intend when we write (a, b).   □

Ordered pairs (a, b) will be very important when we work with equations and functions in subsequent sections. In particular, we need a way to graphically represent ordered pairs (a, b). We do this by plotting the point (a, b) on the ***Cartesian plane,*** as follows. The Cartesian plane (named for the famous mathematician and philosopher René Descartes) is a two-dimensional plane with two axes, one horizontal and one vertical, that are perpendicular and intersect at a point we call the ***origin*** of the plane. We assign ***coordinates*** (a, b) to each point in the plane according to its (positive or negative) horizontal distance a and vertical distance b from the origin. The positive direction on the horizontal axis is to the right of the origin, and the positive direction on the vertical axis is above the origin. The Cartesian plane is naturally divided into four quadrants. See Figures 0.13–0.16.

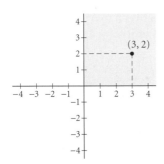

Figure 0.13
First quadrant

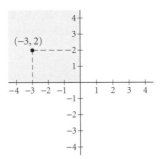

Figure 0.14
Second quadrant

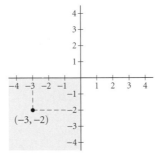

Figure 0.15
Third quadrant

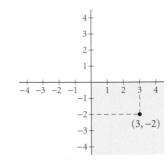

Figure 0.16
Fourth quadrant

The origin has coordinates $(0, 0)$ because both its horizontal and its vertical distance from itself are zero. Often we will use the letter x to represent the horizontal coordinate and call the horizontal axis in the plane the ***x-axis.*** Similarly, we will use y to represent the vertical coordinate and call the vertical axis the ***y-axis.*** The coordinates of a point in the plane will then be given by (x, y). We will often call the Cartesian plane simply the plane or the xy-plane.

We know that the distance between two real numbers a and b is given by $|b - a|$. We now define the "distance" between two ordered pairs of numbers (x_1, y_1) and (x_2, y_2), that is, the length of the line segment connecting these two points in the plane.

DEFINITION 0.5

Distance Between Two Points in the Plane

The ***distance*** between two points $P = (x_1, y_1)$ and $Q = (x_2, y_2)$ in the plane is given by the ***distance formula:***

$$\text{dist}(P, Q) := \sqrt{(x_2 - x_1)^2 + (y_2 - y_1)^2}.$$

In Section 0.5 we will use the **Pythagorean Theorem** to "prove" that the distance formula represents our intuitive notion of distance. For now we will just use the formula (see Example 0.6).

Finally, we present a formula that finds the **midpoint** between any two points in the plane. Given two points P and Q, this formula finds the coordinates of the point M that is at the midpoint of the line segment connecting P and Q.

DEFINITION 0.6

Midpoint Between Two Points in the Plane

The coordinates of the **midpoint** M between two points $P = (x_1, y_1)$ and $Q = (x_2, y_2)$ in the plane are given by the **midpoint formula:**

$$\text{midpoint}(P, Q) = M := \left(\frac{x_1 + x_2}{2}, \ \frac{y_1 + y_2}{2} \right).$$

It makes intuitive sense that these coordinates are midway between the points P and Q, because the x-coordinate is the average of the x-coordinates of P and Q, and the y-coordinate is the average of the y-coordinates of P and Q. We'll "prove" that the point M defined in Definition 0.6 is the midpoint of P and Q in Section 0.5.

EXAMPLE 0.6

Using the distance and midpoint formulas

Plot the points $P = (-1, 3)$ and $Q = (4, -2)$ on a set of axes, and sketch the line segment connecting these two points. Use the distance formula to find the length of this line segment, and use the midpoint formula to find the midpoint of this line segment.

Solution The points $P = (-1, 3)$ and $Q = (4, -2)$, and the line segment between them, are shown in Figure 0.17.

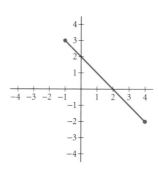

Figure 0.17

The distance between P and Q is

$$\text{dist}(P, Q) = \sqrt{(4 - (-1))^2 + (-2 - 3)^2} = \sqrt{5^2 + (-5)^2} = \sqrt{50} \approx 7.071 \text{ units.}$$

The midpoint of the line segment from P to Q has coordinates

$$\text{midpoint}(P, Q) = \left(\frac{-1 + 4}{2}, \ \frac{3 + (-2)}{2} \right) = \left(\frac{3}{2}, \frac{1}{2} \right).$$

☑ *Checking the Answer*   We can check whether our answer seems correct by plotting the point $(\frac{3}{2}, \frac{1}{2})$ in the plane; see Figure 0.18.

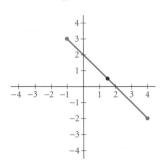

Figure 0.18

From the graph it seems right that the midpoint of the line segment is at the point $(\frac{3}{2}, \frac{1}{2})$. Moreover, it does not seem unreasonable that this line segment is just over 7 units long.   □

What you should know after reading Section 0.1

▶ Definitions and examples of integers, rational numbers, irrational numbers, and real numbers. How to write rational numbers in the form $\frac{p}{q}$, where p and q are integers.

▶ What it means for a real number to be positive, negative, nonpositive, or nonnegative. Which of these words describes the number zero?

▶ The definition of a set, set notation, and how to tell if a given object is a member of a given set. The definitions of a subset and of the empty set.

▶ Interval notation, including closed, open, half-closed, and infinite intervals. How to represent such intervals on a real number line. How can we represent the set or interval consisting of only one point?

▶ The definitions of and notations for intersections and unions of sets. How to express intersections and unions of sets in set notation, interval notation, and on a real number line.

▶ The mathematical definition of the absolute value of a real number, and how to use this definition to define and determine the distance between any two real numbers.

▶ The definition of and vocabulary concerning the Cartesian plane, and how to plot ordered pairs as points in the plane. In which quadrants are the coordinates of a point positive or negative? How to find the distance between any two points in the plane, and the coordinates of the midpoint of the line segment connecting those two points.

Exercises 0.1

Concepts

0. Read the section and make your own summary of the material. Include whatever you think will help you review the material later on.

1. Give an example of a number that is:

   (**a**) an integer   (**b**) rational

   (**c**) irrational   (**d**) real

2. Give an example of a number that is *not:*
   - (a) an integer
   - (b) rational
   - (c) irrational
   - (d) real

3. Define what it means for a real number to be a *rational number,* without referring to decimal notation. What does it mean for a number to be *irrational?*

4. Explain why every integer is a rational number.

5. Explain why your calculator cannot possibly tell you that the number $\sqrt{2}$ is irrational.

6. Suppose a is a real number whose first 24 decimal places are $a = 0.123123123123123123123123\ldots$. Explain why this information does not necessarily show that a is a rational number.

7. Assuming that the pattern continues, is the number $0.919911999111999991111\ldots$ a rational number?

8. Is the number $\frac{1.7}{2.3}$ a rational number? If not, why not? If so, write this number in the form $\frac{p}{q}$ for some integers p and q.

9. Use long division to argue that $0.33\overline{3} = \frac{1}{3}$.

10. Use the fact that $0.33\overline{3} = \frac{1}{3}$ to convince yourself that $0.99\overline{9} = 1$.

11. What do the following symbols mean or represent?
    - (a) $\mathbb{R}$
    - (b) $\in$
    - (c) $\emptyset$
    - (d) $:=$
    - (e) $\cap$
    - (f) $\cup$

12. Is 1.5 in the set $\{x \in \mathbb{R} \mid x - 1 > 0\}$?

13. Name an element of the set $\{x \in \mathbb{R} \mid x^2 = x\}$.

14. Give an example of a set that is empty. Use a different example from the one given in the reading.

15. What does it mean to say that a subset of the real numbers is an *interval?*

16. Give an example of each of the following types of intervals. Use different examples from those given in the reading.
    - (a) an open interval
    - (b) a closed interval
    - (c) a half-closed interval
    - (d) an infinite interval

■ Suppose a and b are real numbers with $a < b$. Express each of the following intervals (a) in set notation, (b) in words, and (c) on a real number line.

17. $[a, b]$
18. $[a, b)$
19. $(a, b]$
20. (a, b)
21. $[a, \infty)$
22. (a, ∞)
23. $(-\infty, b]$
24. $(-\infty, b)$
25. $(-\infty, \infty)$
26. $\{a\}$

27. Explain the meaning of the Venn diagrams in Figures 0.6–0.9 in the reading.

28. Use set notation to define (a) the intersection $A \cap B$ and (b) the union $A \cup B$ of two sets A and B.

29. If A and B are any two sets, is $A \cap B$ necessarily a subset of A? Is $A \cap B$ necessarily a subset of $A \cup B$?

■ For each set A and B given below, describe (a) the intersection $A \cap B$ and (b) the union $A \cup B$.

30. $A =$ the set of all red hats, $B =$ the set of all baseball hats.

31. $A =$ the set of all cats, $B =$ the set of all dogs.

32. $A =$ the set of all record albums, $B =$ the set of all Elvis record albums.

33. Give a formal mathematical definition for the *absolute value* $|a|$ of a real number a. Then explain in words what this definition means, and give some examples.

34. Give a formal mathematical definition for the *distance* between any two real numbers a and b. Then explain in words what this definition means, and give some examples.

35. In the text discussion we claim that $|b - a| = |a - b|$. Show that this is true for four examples, one with a and b both positive, one with a positive and b negative, one with a negative and b positive, and one with a and b both negative.

36. Draw a set of axes for the plane, and label the first, second, third, and fourth quadrants of the plane with the roman numerals I, II, III, and IV.

37. Out of context, the notation $(3, 5)$ could mean two different things. What are those two things?

38. State the distance formula and the midpoint formula.

■ Determine whether each of the following statements is true or false. Explain your answers.

39. True or False: Every real number is either rational or irrational, but not both.

40. True or False: Every nonpositive number is greater than or equal to zero.

41. True or False: No real number is both nonpositive and nonnegative.

42. True or False: If a is a real number, then $-a$ is negative.

43. True or False: The absolute value of zero is equal to zero.

44. True or False: The distance between any point in the coordinate plane and itself is zero.

45. True or False: If (a, b) is in the second quadrant of the Cartesian plane, then $a > 0$ and $b < 0$.

Skills

■ Each of the following is a rational number. Express each number in the form $\frac{p}{q}$ for some integers p and q.

46. 0.25
47. 0.13
48. 0.00001
49. 5.29
50. -11.7
51. $-\frac{2}{15}$
52. -6
53. 0
54. $-0.\overline{666}$
55. $0.\overline{999}$

■ In each problem below, a subset of the real numbers is described in words. Describe each set using **(a)** set notation, **(b)** interval notation (if possible), and **(c)** the real number line.

56. The set of all real numbers that are less than or equal to -1.5.

57. The set of all real numbers strictly between -2 and 3.

58. The set of all real numbers greater than 0.75 and less than or equal to 4.

59. The set of all real numbers whose product with zero is zero.

60. The set of all real numbers that are either greater than 3 or less than 2.

61. The set of all real numbers that are both greater than 3 and less than 2.

62. The set of all odd negative integers.

63. The set of all real numbers that are both less than or equal to -1 and greater than or equal to -1.

64. The set containing only the numbers -3, 4, 7, and 102.

65. The set of all real numbers whose squares are less than 4.

■ Each of the sets below is some combination of the sets $A = [-2, 3)$, $B = (-\infty, 0]$, $C = (-1, 3]$, $D = \{x \mid x \geq -1\}$, $E = \{x \mid 0 \leq x < 2\}$, and $F = \{x \mid x < -2\}$. Write each set three ways: **(a)** in set notation; **(b)** in interval notation; and **(c)** on a real number line.

66. $A \cup B$

67. $C \cap D$

68. $C \cup D$

69. $E \cup F$

70. $C \cap F$

71. $B \cap E$

72. $A \cap B \cap C$

73. $C \cup E \cup F$

74. $(B \cap C) \cup F$

75. $(E \cup F) \cap B$

76. $(B \cup E) \cap (D \cup F)$

77. $(A \cap C) \cap (D \cap E)$

■ Assuming that a is a positive number and b is a negative number, write each of the following expressions without absolute values.

78. $|-1 + 3|$

79. $|3 - 1|$

80. $|3a|$

81. $|3b|$

82. $|-2b|$

83. $|ab|$

84. $|b^2|$

85. $|ab - 1|$

■ Calculate the distance between each of the following pairs of real numbers.

86. 3 and 17

87. -8 and -21

88. -2 and 6

89. 1.72 and -3.81

■ For each pair of points P and Q, sketch P and Q in the Cartesian plane, and sketch the line segment connecting P and Q. Then calculate the length of that line segment and the midpoint of that line segment. Use your picture to check whether your answers are reasonable.

90. $P = (2, 3)$, $Q = (1, 4)$

91. $P = (0, 0)$, $Q = (3, 2)$

92. $P = (-2, -1)$, $Q = (1, -3)$

93. $P = (-5, 2)$, $Q = (2, 5)$

94. $P = (-1, 0)$, $Q = (3, -2)$

95. $P = (0, 8)$, $Q = (-2, 0)$

96. $P = (-3, 4)$, $Q = (-3, -6)$

97. $P = (1.7, -3)$, $Q = (-1, 1.8)$

98. $P = (3, 3^2)$, $Q = (-2, (-2)^2)$

0.2   Equations

Although algebra is covered throughout this text, there are some basic algebra skills that you must have at the start. In this section we review methods for solving simple equations, including those that involve complicated fractions and those that involve simple factoring. We also discuss methods for solving systems of two or more simple equations.

0.2.1   Equations and solutions

An **equation** consists of two mathematical expressions connected by an equals sign. For example,

$$x^2 - 1 = 0, \qquad x^2 - x = y + 2, \qquad \text{and} \qquad 3u + 2v - 8x = 5$$

are equations. We will be primarily concerned with equations involving one or two **variables.** Usually we will use letters near the end of the alphabet, such as x, y, z, u, and v, to represent variables and will reserve letters near the beginning of the alphabet, such as *a*, *b*, and *c*, for **constants.** A variable represents a changing or unknown quantity, whereas a constant represents a *fixed* number.

⬧ ***Caution*** Although we typically use letters like *x* and *y* for variables, and letters like *a* and *b* for constants, that will not always be the case. *Any* letter can be used to represent a variable, and *any* letter can be used to represent a constant. You'll have to tell from the context whether a given letter is meant to represent a variable or a constant. □

An equation may be true for some values of the variables involved and false for others. For example, the equation $x^2 - 1 = 0$ is clearly not true for all values of *x*. In fact, this equation is only true when $x = 1$ or when $x = -1$. The set of values for which an equation is true is called the **solution set** (or simply the **solution**) for that equation. Thus the solution for the equation $x^2 - 1 = 0$ is the set $\{1, -1\}$.

If an equation involves two variables, then its solution set consists of *ordered pairs* of numbers. For example, if an equation involves two variables *x* and *y*, then each ordered pair (x, y) that makes the equation true is a solution.

EXAMPLE 0.7

Solutions to two-variable equations

For example, the equation $x^2 - 1 = y$ involves two variables, *x* and *y*. Each value (x, y) for which the equation is true is a solution to the equation. For example, is $(x, y) = (1, 2)$ a solution? No, because this ordered pair does not make the equation true: $x^2 - 1 = 1^2 - 1$ is not equal to 2. What about $(x, y) = (-2, 3)$? Because $(-2)^2 - 1 = 3$, this ordered pair is a solution to the equation $x^2 - 1 = y$. The set of solutions to this equation is the set

$$\{(x, y) \mid x^2 - 1 = y\}.$$ □

To represent the solutions to a two-variable equation graphically, we use the Cartesian plane. If the variables in the equation are *x* and *y*, we can plot each solution (x, y) on the *xy* plane. For example, the solution set $\{(x, y) \mid x^2 - 1 = y\}$ to the two-variable equation $x^2 - 1 = y$ is represented graphically in Figure 0.19.

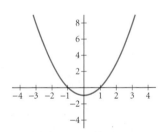

Figure 0.19

Each point (x, y) on the *graph* of the solution set in Figure 0.19 is a solution to the equation $x^2 - 1 = y$. Each point (x, y) that is *not* on the graph is *not* a solution to the equation. Therefore, this graph is a complete picture of all the solutions to the equation $x^2 - 1 = y$. For example, the ordered pair $(x, y) = (1, 2)$ is not a solution to the equation, and thus it is not included in the graph. On the other hand, the ordered pair $(x, y) = (-2, 3)$ is a solution to the equation, so it is a point on the graph in Figure 0.19. Every point on the graph is of the form $(x, x^2 - 1)$ for some real value of *x*; for example, the point $(1.25, (1.25)^2 - 1) = (1.25, 0.5625)$ is a solution to the equation and a point on the graph.

We'll learn more about how to find graphical representations of solution sets when we talk about functions in Chapter 1. For now, you can use your graphing calculator. For example, if you graph the equation $y = x^2 - 1$ with your calculator, you will see the graph of solutions shown in Figure 0.19.

An equation that is *always* true, for all values of the variables involved, is called an *identity.* For example, the equation $(x + 1)^2 = x^2 + 2x + 1$ is true for all real numbers x, and so it is an identity. Identities will be particularly important when we learn about trigonometric functions in Part III. On the other hand, some equations have *no* real solutions. For example, the equation $x^2 = -1$ has no real solution, because there is no real number whose square is -1.

0.2.2 Factoring

Many equations can be solved easily by the method of *factoring.* To factor an algebraic expression, you write that expression as a product of simpler expressions. For example, the expression $3x^2 - 6x$ can be factored as $3x(x - 2)$.

The reason why factoring can help solve an equation is that a product can be equal to zero only if one of the factors in the product is zero. This is a basic property of the set of real numbers:

THEOREM 0.1

A Product Is Zero If and Only If One of the Factors Is Zero
If A and B are any real numbers, then $AB = 0$ if and only if $A = 0$ or $B = 0$.

"If and only if" is shorthand for saying that the implication goes "both ways." That is, if AB is zero, then one of A or B (or both) must be zero, and if A or B is zero, then AB will also be zero. This theorem can be generalized to products involving more than two factors; for example, a product $ABCD = 0$ if and only if one (or more) of the factors A, B, C, or D is zero.

Theorem 0.1 helps us solve complicated equations by turning them into a series of simpler equations, as shown in the following example.

EXAMPLE 0.8   **Solving an equation by examining factors**

Suppose we want to solve the equation $3x^2 - 6x = 0$. This equation factors as $3x(x - 2) = 0$. By Theorem 0.1, this equation is true if and only if one of the factors is zero. Consider $3x = 0$ and $x - 2 = 0$. Solving both, we get $x = 0$ and $x = 2$. The solution set to the original equation $3x^2 - 6x = 0$ is $\{0, 2\}$.   □

§ *Caution*   Theorem 0.1 applies only to equations that are "set to zero"—that is, only to equations that are in the form $AB = 0$. The theorem does *not* say, for example, that $3x(x - 2) = 2$ if and only if $3x = 2$ or $x - 2 = 2$. Why not?   □

Many of the equations we will factor at this point are *quadratic* equations—equations of the form $ax^2 + bx + c = 0$ for some constants a, b, and c. These equations can often be factored by doing the "FOIL" ("First-Outside-Inside-Last") method backwards. For example, the equation $x^2 - 3x - 4 = 0$ can be factored as $(x - 4)(x + 1) = 0$. (We will assume that you know how to do this already; if not, go find your old algebra textbook.)

Other quadratic expressions, for example, $3x^2 - 7x + 1 = 0$, are not so easy to factor. For these equations we can apply the *quadratic formula* to determine the solutions.

THEOREM 0.2

The Quadratic Formula

The solutions to any quadratic equation $ax^2 + bx + c = 0$ are of the form

$$x = \frac{-b \pm \sqrt{b^2 - 4ac}}{2a}.$$

The "$\pm$" symbol means "plus or minus" and is a shorthand way of saying that there are *two* solutions to the equation $ax^2 + bx + c = 0$, namely

$$x = \frac{-b + \sqrt{b^2 - 4ac}}{2a} \quad \text{and} \quad x = \frac{-b - \sqrt{b^2 - 4ac}}{2a}.$$

Sometimes both of these solutions are distinct real numbers (if $b^2 - 4ac$ is positive), sometimes both solutions are actually the same real number (if $b^2 - 4ac = 0$), and sometimes the solutions are not real numbers at all, but are complex numbers (if $b^2 - 4ac < 0$). In this book we will consider only *real* solutions; for example, the quadratic equation $x^2 - 1 = 0$ will be said to have "no solutions," because neither of its solutions ($\sqrt{-1}$ and $-\sqrt{-1}$) is a real number.

The following formulas may also help with factoring:

THEOREM 0.3

Factoring Formulas

For all real numbers a and b,

$$a^2 - b^2 = (a - b)(a + b) \qquad\qquad a^2 + 2ab + b^2 = (a + b)^2$$
$$a^3 - b^3 = (a - b)(a^2 + ab + b^2) \quad a^3 + 3a^2 b + 3ab^2 + b^3 = (a + b)^3$$

EXAMPLE 0.9

Solving equations by factoring

Solve the following equations using Theorems 0.1–0.3.

  (a) $2x^3 - 5x^2 - 3x = 0$ **(b)** $3x^2 = 7x - 1$ **(c)** $2x^5 - 32x = 0$

Solution

  (a) This equation can easily be factored:

$$2x^3 - 5x^2 - 3x = 0$$
$$x(2x^2 - 5x - 3) = 0$$
$$x(2x + 1)(x - 3) = 0$$

By Theorem 0.1, the expression $2x^3 - 5x^2 - 3x$ is zero if and only if one of x, $2x + 1$, or $x - 3$ is zero. In other words, the solutions to the equation $2x^3 - 5x^2 - 3x = 0$ are $x = 0$, $x = -\frac{1}{2}$, and $x = 3$. The solution set is thus $\{-\frac{1}{2}, 0, 3\}$.

  (b) We first need to write the equation $3x^2 = 7x - 1$ in the general form of a quadratic equation: $3x^2 - 7x + 1 = 0$. This equation cannot be factored using the reverse "FOIL" method, so we'll apply the quadratic formula. In this example we have $a = 3$, $b = -7$, and $c = 1$, so the solutions of $3x^2 - 7x + 1 = 0$ are

$$x = \frac{-(-7) \pm \sqrt{(-7)^2 - 4(3)(1)}}{2(3)} = \frac{7 \pm \sqrt{49 - 12}}{6} = \frac{7 \pm \sqrt{37}}{6}.$$

The solutions of $3x^2 - 7x + 1 = 0$ are $x = \frac{1}{6}(7 + \sqrt{37})$ and $x = \frac{1}{6}(7 - \sqrt{37})$. Clearly we couldn't have figured that out by doing the "FOIL" method backwards!

(c)   This time the factoring will involve one of the formulas from Theorem 0.3:

$$2x^5 - 32x = 0$$

$$2x(x^4 - 16) = 0$$

$$2x(x^2 - 4)(x^2 + 4) = 0 \qquad \text{(formula for } a^2 - b^2 \text{ with } a = x^2 \text{ and } b = 16\text{)}$$

$$2x(x - 2)(x + 2)(x^2 + 4) = 0 \qquad \text{(formula for } a^2 - b^2 \text{ with } a = x \text{ and } b = 4\text{)}$$

Thus the solutions to the original equation $2x^5 - 32x = 0$ are the solutions to the following four equations:

$$
\begin{array}{cccc}
2x = 0 & x - 2 = 0 & x + 2 = 0 & x^2 + 4 = 0 \\
x = 0 & x = 2 & x = -2 & \varnothing
\end{array}
$$

Therefore, the solution set of the equation $2x^5 - 32x = 0$ is $\{-2, 0, 2\}$. ☐

✔ *Checking the Answer*   To check the answers in Example 0.9, simply substitute each proposed solution into the original equation (it should make the equation true). Of course, this won't tell you whether you've *missed* any solutions, but it will tell you whether the solutions you found are correct. ☐

The following factoring formula will be particularly handy in Section 3.5.2 and Section 4.3. It generalizes the first two factoring formulas in Theorem 0.3.

THEOREM 0.4

Factoring $a^n - b^n$

For any real numbers a and b, and any positive integer n,

$$a^n - b^n = (a - b)(a^{n-1} + a^{n-2}b + a^{n-3}b^2 + a^{n-4}b^3 + \cdots + a^2 b^{n-3} + ab^{n-2} + b^{n-1}).$$

The ellipses (the "dot-dot-dot") in the formula indicate that we are to continue the pattern for as long as necessary. Note that the powers of each of the terms in the second factor add up to $n - 1$; for example, the term $a^{n-3}b^2$ has a power of $n - 3$ and a power of 2, which add up to $(n - 3) + 2 = n - 1$. It is important to note that the formula in Theorem 0.4 does *not* necessarily completely factor the expression $a^n - b^n$, but it does pull out one simple factor $a - b$. (We will learn more factoring techniques in Section 5.1.) Theorem 0.4 can easily be verified simply by multiplying out the left-hand side of the equation; you will do so in the exercises.

EXAMPLE 0.10

Pulling a linear factor out of $a^n - b^n$

Using the formula in Theorem 0.4 with $a = x$, $b = 2$, and $n = 5$, we have

$$x^5 - 2^5 = (x - 2)(x^4 + x^3(2) + x^2(2^2) + x(2^3) + 2^4)$$

$$= (x - 2)(x^4 + 2x^3 + 4x^2 + 8x + 16).$$ ☐

0.2.3 Fractions

In this section we present a quick review of fraction operations. Some of the most common fraction rules are given in Theorem 0.5.

THEOREM 0.5

Algebraic Rules for Fractions

Let a, b, c, and d be any real numbers. Then (assuming that no denominator in any expression is zero),

$$\frac{a}{b} + \frac{c}{d} = \frac{ad + bc}{bd} \qquad \left(\frac{a}{b}\right)\left(\frac{c}{d}\right) = \frac{ac}{bd} \qquad \frac{\left(\frac{a}{b}\right)}{\left(\frac{c}{d}\right)} = \left(\frac{a}{b}\right)\left(\frac{d}{c}\right) = \frac{ad}{bc}$$

You can also add fractions by forming the lowest common denominator. As a simple example, note that it is easier to calculate $\frac{1}{6} + \frac{2}{3} = \frac{1}{6} + \frac{4}{6} = \frac{5}{6}$ than it is to calculate $\frac{1}{6} + \frac{2}{3} = \frac{(1)(3)+(6)(2)}{(6)(3)} = \frac{3+12}{18} = \frac{15}{18} = \frac{5}{6}$.

As immediate consequences of the rules in Theorem 0.5, we also have the following rules:

THEOREM 0.6

More Algebraic Rules for Fractions

Let a, b, c, and d be any real numbers. Then (assuming that no denominator in any expression is zero),

$$\frac{a}{b} - \frac{c}{d} = \frac{ad - bc}{bd} \qquad c\left(\frac{a}{b}\right) = \frac{ac}{b} \qquad \frac{\left(\frac{a}{b}\right)}{c} = \frac{a}{bc} \qquad \frac{a}{\left(\frac{b}{c}\right)} = \frac{ac}{b}$$

We will accept the rules in Theorem 0.5 as basic properties of real numbers. The rules in Theorem 0.6 can be easily derived from the rules in Theorem 0.5; we will do so when we cover proofs in Section 0.5.

The following theorem tells us when we can cancel a ***common factor*** in the numerator and the denominator of a fraction. The key point is that if the factor you are canceling involves a variable, then the cancellation is not valid when that factor is equal to zero.

THEOREM 0.7

Cancellation When Variables Are Involved

Given any x and c,

$$\frac{x - c}{x - c} = \begin{cases} 1, & \text{if } x \neq c \\ \text{undefined}, & \text{if } x = c. \end{cases}$$

This theorem means that you can cancel a common factor $x - c$ from the numerator and denominator of a fraction, and the canceled expression will equal the original expression for all values of x except $x = c$.

EXAMPLE 0.11   **Cancellation when variables are involved**

For example, consider the expression

$$\frac{x^2 - 1}{x + 1} = \frac{(x - 1)(x + 1)}{x + 1}.$$

Can we cancel the factor of $x + 1$ in the numerator with the factor of $x + 1$ in the denominator? Theorem 0.7 says that we can, as long as $x \neq -1$. For $x = -1$ the expression is undefined. For $x \neq -1$ the expression is equal to $x - 1$.   □

EXAMPLE 0.12   **Writing an expression as a simple fraction**

Write the following expression in the form of a simple fraction (that is, in the form $\frac{A}{B}$ for some expressions A and B that do not themselves involve fractions and do not have any common factors).

$$\frac{3\left(\frac{1}{x+2}\right) - \frac{1}{x}}{x - 1}$$

Solution   Using the rules for fractions,

$$\frac{3\left(\frac{1}{x+2}\right) - \frac{1}{x}}{x - 1} = \frac{\frac{3}{x+2} - \frac{1}{x}}{x - 1} = \frac{\frac{3x - 1(x+2)}{(x+2)(x)}}{x - 1} = \frac{\frac{3x - x - 2}{x(x+2)}}{x - 1} = \frac{\frac{2x - 2}{x(x+2)}}{x - 1}$$

$$= \frac{2x - 2}{x(x + 2)(x - 1)} = \frac{2(x - 1)}{x(x + 2)(x - 1)} = \frac{2}{x(x + 2)}.$$

The last equality is true only for $x \neq 1$. Can you list each of the algebraic rules used in the calculation?   □

Theorem 0.1 told us when a product AB can be equal to zero. The following theorem tells us when a quotient $\frac{A}{B}$ can be equal to zero.

THEOREM 0.8

When Is a Quotient Equal to Zero?
If A and B are any real numbers, then $\dfrac{A}{B} = 0$ if and only if $A = 0$ and $B \neq 0$.

✿ *Caution*   Note in particular that $B = 0$ does *not* imply that $\frac{A}{B} = 0$. In fact, if $B = 0$, then $\frac{A}{B}$ is undefined. That is why we require not only that $A = 0$ but also that $B \neq 0$.   □

An equation involving fractions can often be simplified using Theorems 0.5, 0.6, and 0.7. Sometimes it is convenient to simplify an equation involving fractions by *cross-multiplying* using the following theorem.

THEOREM 0.9

Cross-Multiplication
If a, b, c, and d are any real numbers, and if $b \neq 0$ and $d \neq 0$, then
$$\frac{a}{b} = \frac{c}{d} \quad \text{if and only if} \quad ad = bc.$$

This theorem can be justified using Theorem 0.8 and the rules of fractions we have already established. (Set the equation equal to zero, rewrite as one quotient, and then apply Theorem 0.8.)

EXAMPLE 0.13

Solving equations involving quotients

Find the solution sets of the following equations.

(a) $\dfrac{x+1}{x-1} = 4$ (b) $\dfrac{x^3 + x^2 - 6x}{x^2 - 3x + 2} = 0$

Solution

(a) By cross-multiplying (and thinking of $4 = \frac{4}{1}$ as a fraction), we find that the equation $\frac{x+1}{x-1} = \frac{4}{1}$ is equivalent to the equation $(x+1)(1) = 4(x-1)$, which can be solved easily:

$$x + 1 = 4(x - 1)$$
$$x + 1 = 4x - 4$$
$$5 = 3x$$
$$x = \tfrac{5}{3}.$$

Thus the solution set of the equation $\frac{x+1}{x-1} = 4$ is $\{\frac{5}{3}\}$.

(b) We will use Theorem 0.8 to solve this equation, but first we need to factor the numerator and the denominator:

$$\frac{x^3 + x^2 - 6x}{x^2 - 3x + 2} = \frac{x(x^2 + x - 6)}{x^2 - 3x + 2} = \frac{x(x-2)(x+3)}{(x-2)(x-1)}.$$

This expression is equal to zero only when the numerator is equal to zero (and the denominator is *not* zero). The numerator of the expression is zero at $x = 2$, but the denominator is also zero at $x = 2$, so the expression is not zero at $x = 2$ (it is undefined). The solution to the original equation consists only of those values of x for which the numerator is zero while the denominator is nonzero, namely $x = 0$ and $x = -3$.   □

0.2.4 Systems of equations

A *system of equations* is a list of two or more equations. A *solution* to such a system of equations is a solution to each of the equations in the system simultaneously.

EXAMPLE 0.14

Solutions for systems of equations

Consider the following system of equations.

$$\begin{cases} x^2 - 1 = 0 \\ x^2 - x - 2 = 0 \end{cases}$$

The first equation can be written as $x^2 = 1$, so its solutions are $x = \pm\sqrt{1}$, that is, $x = 1$ and $x = -1$. The second equation can be factored as $(x - 2)(x + 1) = 0$, and thus its solutions are $x = 2$ and $x = -1$. The only number that is a solution to *both* equations is $x = -1$, so the solution set for the system of equations is $\{-1\}$.   □

Systems of equations often involve more than one variable. For example, the solution set for a system of two or more equations in the variables x and y will consist of the ordered pairs (x, y) that are solutions to *all* of the equations in the system.

To solve a system of equations, sometimes it is a good strategy to use one of the equations to solve for one of the variables in terms of the other variables, and then use this information to simplify the other equations. Occasionally, it is convenient to add or subtract a multiple of one equation to or from another equation. We give examples of both techniques in the following example.

EXAMPLE 0.15

Solving systems of equations

Solve the following systems of equations.

(a) $\begin{cases} x^2 + y = 3 \\ 2x + y = 0 \end{cases}$

(b) $\begin{cases} a + 2b - c = 0 \\ a + b = c - 1 \\ a - 2b + c = 2 \end{cases}$

Solution

(a) We'll start by using the second equation to solve for y in terms of x. Since $2x + y = 0$, we know that $y = -2x$. Now we can substitute this into the first equation (note that this new equation has only one variable), and solve:

$$x^2 + y = 3$$
$$x^2 - 2x = 3$$
$$x^2 - 2x - 3 = 0$$
$$(x - 3)(x + 1) = 0$$
$$x = 3 \text{ or } x = -1.$$

We now know that any solution (x, y) of this system has the property that $x = 3$ or $x = -1$. Since we know that $y = -2x$, we can use these values of x to solve for the corresponding values of y. If $x = 3$, then $y = -2(3) = -6$. If $x = -1$, then $y = -2(-1) = 2$. Thus the solutions to the system of equations are $(x, y) = (3, -6)$ and $(x, y) = (-1, 2)$.

(b) This system of equations has three equations and three variables, a, b, and c. The solutions to this system will be the triples (a, b, c) that are solutions to all three equations.

Since the first and last equations look somewhat similar, we'll try adding them together; in other words, we will add the left-hand side of the last equation to the left-hand side of the first equation and, similarly, add the right-hand side of the last equation to the right-hand side of the first equation. (Why is this legal?) If we do this, we get the new equation

$$(a + 2b - c) + (a - 2b + c) = 0 + 2$$
$$2a + 0b + 0c = 2$$
$$2a = 2$$
$$a = 1.$$

This immediately tells us that any solution (a, b, c) to this system must have $a = 1$. Replacing a with 1 in the second equation gives us the equation $1 + b = c - 1$, which implies that $b = c - 2$. Using $a = 1$ and $b = c - 2$, we can rewrite the first equation so that it involves only one variable:

$$a + 2b - c = 0$$
$$1 + 2(c - 2) - c = 0$$
$$1 + 2c - 4 - c = 0$$
$$c - 3 = 0$$
$$c = 3.$$

Now, because $b = c - 2$, we know that $b = 3 - 2 = 1$. There is only one solution to the system of equations, namely $(a, b, c) = (1, 1, 3)$. □

☑ *Checking the Answer*   To check our solutions, we need only substitute them into each equation in the system of equations to see whether they work. For example, both $(x, y) = (3, -6)$ and $(x, y) = (-1, 2)$ satisfy both of the equations in the first system (part (a) in Example 0.15). In the second system (part (b) in Example 0.15), substituting $a = 1$, $b = 1$, and $c = 3$ makes each of the three equations true. Of course, checking your answer this way won't tell you whether you have missed any solutions, but it will tell you whether your proposed solutions are correct.   □

◈ *Caution*   Don't forget that a solution to an equation in two variables will be an ordered *pair* of numbers, the solution to an equation in three variables will be an ordered *triple* of numbers, and so on. When solving the system in Example 0.15, one might be tempted to stop after discovering $x = 3$ and $x = -1$. However, that is only half of the solution; we also need to know the values of y that correspond to those values of x.   □

What you should know after reading Section 0.2

▶ What is an equation? What is a solution to a one-variable equation? To a two-variable equation? To a three-variable equation? What is the "solution set" of an equation?

▶ How to represent solutions to one-variable equations on a number line, and how to represent solutions to two-variable equations on a Cartesian plane.

▶ How to solve equations of the form $AB = 0$ and $\frac{A}{B} = 0$. Basic factoring skills such as the "reverse FOIL method," the quadratic formula, and other basic factoring formulas.

▶ Basic rules for manipulating and simplifying fractions, when common factors can be canceled in a fraction, cross-multiplying, and recognizing when a quotient is equal to zero.

▶ How to solve simple systems of equations. What is a "solution" to a system of equations in one variable? In two variables? In three variables?

Exercises 0.2

Concepts

0. Read the section and make your own summary of the material. Include whatever you think will help you review the material later on.

■ Explain in your own words the meaning of each of the following terms.

1. equation

2. solution

3. solution set

4. identity

5. factoring

6. quadratic expression

7. common factor

8. system of equations

9. solution to a system of equations

10. Give an example (different from those in the chapter or exercises) of an equation in:
   (a) one variable   (b) two variables
   (c) three variables

11. Consider the equation $y = 2 - x + 3x^5$.
   (a) Is $(x, y) = (1, 4)$ a solution to this equation? Why or why not?
   (b) Is $(x, y) = (2, 1)$ a solution to this equation? Why or why not?
   (c) Find three more solutions (x, y) to this equation.
   (d) Use a graphing calculator to make a graph of the solutions to this equation, and check that the solutions you found appear on this graph.

12. Give an example (different from the example given in the chapter) of an equation that is an *identity*.

13. Give an example (different from those in the chapter or exercises) of a system of equations with:
  (a) 2 equations, 2 variables
  (b) 2 equations, 3 variables
  (c) 3 equations, 2 variables
  (d) 3 equations, 3 variables
  (e) 3 equations, 4 variables
  (f) 4 equations, 4 variables

14. Consider the following system of equations:
$$\begin{cases} y^3 + 3 = x \\ x^2 - 7y = 11. \end{cases}$$
Is $(x, y) = (4, 1)$ a solution to this system of equations? Why or why not? Is $(x, y) = (2, -1)$ a solution to this system of equations? Why or why not?

15. Show that the factoring formula in Theorem 0.4 is accurate by multiplying out the right-hand side of the formula and showing that it is equal to the left-hand side of the formula. (*Hint:* It may help to do this in a particular example, such as $n = 5$, before doing the general case.)

■ Determine whether each of the following statements is true or false. Explain your answers.

16. True or False: If $AB = 0$, then A and B must be zero.

17. True or False: $\dfrac{A}{B} = 0$ if and only if $A = 0$.

18. True or False: Every quadratic equation $ax^2 + bx + c = 0$ has exactly two real solutions.

19. True or False: If $ax^2 + bx + c = 0$ is a quadratic equation with $b^2 - 4ac = 0$, then the only real solution to $ax^2 + bx + c = 0$ is $x = \frac{-b}{2a}$.

20. True or False: For all values of x, $\dfrac{(x+1)(x+2)}{x+1} = x+2$.

21. True or False: If $b \neq 0$ and $d \neq 0$, then $k\left(\dfrac{a}{b} + \dfrac{c}{d}\right) = \dfrac{adk + bck}{bd}$.

22. True or False: If $a \neq 0$ and $b \neq 0$, then $\dfrac{12}{\left(\frac{a}{b}\right)} = \dfrac{12b}{a}$.

Skills

■ Factor each of the following expressions as much as possible.

23. $x^2 + 2x - 15$ **24.** $x^2 - 9x + 18$
25. $2x - 48 + x^2$ **26.** $3x^2 - 11x - 4$
27. $6x^2 - x - 2$ **28.** $6x^3 + 2x^2 - 20x$
29. $x^2 - 9$ **30.** $8x^2 - 50$
31. $x^2 + 16$ **32.** $16 - x^4$
33. $81x^4 - 1$ **34.** $x^4 + 81x$
35. $x^3 + 3x^2 + 3x + 1$ **36.** $x^3 - 8$
37. $27 - 64x^3$

■ Use a factoring formula to take out one linear factor from each expression.

38. $x^5 - 243$ **39.** $a^6 - x^6$ **40.** $t^7 - x^7$

■ Use the quadratic formula to solve each of the following quadratic equations.

41. $2x^2 - x - 3 = 0$ **42.** $2x^2 + 4x + 2 = 0$
43. $x^2 + 2x + 3 = 0$ **44.** $4x^2 - 12x + 9 = 0$
45. $5 - 2x - x^2 = 0$ **46.** $x - 2 - 3x^2 = 0$

■ Solve each of the following equations by factoring as much as possible and using Theorem 0.1.

47. $x^2 + 2x - 8 = 0$ **48.** $6 - 7x + x^2 = 0$
49. $3x^2 + 5x = -2$ **50.** $6x^2 + 3 = -11x$
51. $4x(x^2 - 1) = 6x^2$ **52.** $10x - 12x^2 = 20x^2 - 6x^3$
53. $x^2 - 2x - 4 = 0$ **54.** $2x(x - 1) = 3$
55. $4x^2 - 12x = -9$ **56.** $(x-1)(x+2) = x(2x-3)$
57. $x^3 - 9x = 0$ **58.** $8 - x^3 = 0$
59. $16x^4 - 81 = 0$ **60.** $x^3 - 3x^2 + 3x = 1$
61. $4x^3 - 9x = 0$

■ Write each of the following expressions in the form of a simple fraction (in the form $\frac{A}{B}$ for some expressions A and B that do not themselves involve fractions and do not have any common factors). Is your simplified expression equal to the original expression for all values of x? If not, for which values of x does your simplified expression differ from the original expression?

62. $\dfrac{\frac{x+1}{x} - \frac{1}{x-1}}{x+1}$ **63.** $\dfrac{x^2 - 1}{\frac{1}{x+1} - \frac{1}{x}}$ **64.** $\dfrac{\left(\frac{x^2-4}{x+1}\right)}{\left(\frac{x+2}{x-1}\right)}$

65. $\dfrac{x^2}{x-2}\left(\dfrac{x+1}{x-2} + \dfrac{3}{x+1}\right)$

66. $3x\left(\dfrac{2-x}{x^2-1}\right) - \dfrac{x}{x-1}$ **67.** $\dfrac{\frac{1}{x^2-4} - \frac{3}{x}}{\frac{x-2}{x^2-x}}$

■ Find the solution sets of each of the following equations.

68. $\dfrac{x-1}{x+1} = 3$ **69.** $\dfrac{x^2-1}{x^2-4} = 0$
70. $\dfrac{x^3 + x^2 - 2x}{x^2 - 4x + 3} = 0$ **71.** $\dfrac{2}{x^2 - x} = \dfrac{1}{1-x}$
72. $\dfrac{2x-1}{x-3} = \dfrac{x-2}{1+x}$ **73.** $\dfrac{4x^3 - 1}{2x^2 + x - 1} = 0$
74. $\dfrac{(x-2)^2}{3x-2} = x - 5$ **75.** $\dfrac{1}{3x^2 - 2x - 1} = 2x - 1$
76. $\dfrac{1}{x} = \dfrac{1}{x+1}$

■ Find the solution sets of each of the following systems of equations.

77. $\begin{cases} 2x + y = 4 \\ 3x + 2y = 0 \end{cases}$ **78.** $\begin{cases} 3a + b = -1 \\ a + 5b = 3 \end{cases}$

79. $\begin{cases} x^2 + y = 0 \\ y - x = 2 \end{cases}$ **80.** $\begin{cases} r^2 - s = 0 \\ r^2 + s = 1 \end{cases}$

81. $\begin{cases} 3t - u^2 = -3 \\ u - 1 = t \end{cases}$ **82.** $\begin{cases} 3x + y - z = 2 \\ x - y + 2z = 0 \end{cases}$

83. $\begin{cases} 2a + 3b - c = 4 \\ a - b - c = 3 \\ 2a + b - c = 2 \end{cases}$ **84.** $\begin{cases} x - 2y + z = 0 \\ x + y - 3z = 1 \\ 3x - 2y + z = 2 \end{cases}$

85. $\begin{cases} x + 2y = 1 \\ 2y = 1 - 3x \\ 3x - y = 2 \end{cases}$

Applications

86. Len works for a company that produces cans that are 6 inches tall and vary in radius. The cost of producing a can with radius r is $C = 10\pi r^2 + 24\pi r$ cents. If Len wants to spend 300 cents on each can, how large should he make the radius? (*Hint:* Set up and solve an equation; you will need the quadratic formula.)

87. Suppose x and y are positive real numbers whose sum is 25 and whose product is 136. Construct and solve a system of equations to find x and y.

88. A sheet of paper with height h and width w is to have 2-inch margins on all four sides. Suppose the perimeter of the paper is 38 inches, and the area of the printable part of the paper (inside the margins) is 45 square inches. Construct and solve a system of equations to find h and w.

0.3   Inequalities

An *inequality* consists of two mathematical expressions related by a $>$, $<$, $\geq$, or $\leq$ sign. The *solution set* of an inequality is the set of all values that make the inequality true. The techniques used to solve inequalities are similar to those used for solving equations, except that certain operations (for example, multiplying by a negative constant) may change the direction of the inequality. Whereas the solution set of an equation often consists of a finite number of points, the solution set of an inequality is often an *interval* of real numbers.

0.3.1 Basic properties of inequalities

The following theorem describes what happens to inequalities when they are inverted or multiplied by a nonzero constant. This information will be useful when solving inequalities.

THEOREM 0.10

Algebraic Rules for Inequalities
For any nonzero real numbers a, b, and c,
(a) If $a < b$ and $c > 0$, then $ac < bc$. **(b)** If $a < b$ and $c < 0$, then $ac > bc$.
(c) If $0 < a < b$, then $\dfrac{1}{a} > \dfrac{1}{b}$. **(d)** If $a < b$, then $a + c < b + c$.

In particular, notice that part (b) implies that multiplying an inequality by a negative number "flips" the inequality; for example, if $a < b$, then $-a > -b$. Part (d) illustrates that we may add any number to, or subtract any number from, both sides of an inequality without changing the direction of the inequality. Theorem 0.10 is written in terms of *strict inequalities* $>$ and $<$, but similar results apply for the inequalities $\geq$ and $\leq$.

❖ *Caution*   Since multiplying both sides of an inequality by a negative number "flips" the inequality, you *cannot* multiply both sides of an inequality by a quantity whose sign you do not know. For example, you can't multiply both sides of the inequality $\frac{3}{x} > 2$ by x, because x may be positive or negative. Therefore, you don't know whether or not multiplying both sides of an inequality by x will change the direction of the inequality.   □

EXAMPLE 0.16

Solving a simple inequality

Find the solution set of the inequality $3 - 2x \geq 7$.

Solution   Using parts (b) and (d) of Theorem 0.10,

$$3 - 2x \geq 7$$

$$-2x \geq 7 - 3$$

$$-2x \geq 4$$

$$x \leq \frac{4}{-2} \qquad \text{(change the direction of the inequality)}$$

$$x \leq -2.$$

The solution set of this inequality is $\{x \mid x \leq -2\}$. This set can also be described in interval notation as $(-\infty, -2]$, or simply as $x \leq -2$ (shorthand for the set notation). We could also indicate the solution as a graph on the real number line.   □

0.3.2 Inequalities involving products and quotients

Theorem 0.1 in Section 0.2 tells us what happens when a product of numbers is equal to zero; the following theorem tells us how to determine whether a product of numbers is less than or greater than zero.

THEOREM 0.11

The Sign of a Product Depends on the Signs of Its Factors

If A and B are any nonzero real numbers, then

(a) If A and B have the same sign, then $AB > 0$.

(b) If A and B have opposite signs, then $AB < 0$.

An easy corollary to this theorem is that a product AB is greater than *or equal to* zero either if A and B have the same sign or if one of A or B is zero. Theorem 0.11 also holds for quotients; that is, a quotient $\frac{A}{B}$ of nonzero real numbers is greater than zero if A and B have the same sign and is less than zero if A and B have opposite signs.

Theorem 0.11 is just a fancy way of saying that the product of two positive numbers is positive, the product of two negative numbers is positive, and the product of a negative number and a positive number is negative. We can state Theorem 0.11 more generally as follows:

THEOREM 0.12

The Sign of a Product Depends on the Number of Negative Factors

A product or quotient of nonzero real numbers is positive if it has an even number of negative factors and is negative if it has an odd number of negative factors.

For example, the product $(-2)(-1)(2)(-6)$ is negative because it has an odd number of negative factors. On the other hand, $\frac{(-3)(2)(8)}{(7)(-4)}$ is positive because it has an even number of negative factors.

EXAMPLE 0.17 | **Solving an inequality by examining signs of factors**

Solve the inequality $\dfrac{3}{x+1} > 2$.

Solution Although it is tempting, we can't multiply both sides of the inequality by $x + 1$, because $x + 1$ may be positive or may be negative (depending on the value of x). Instead, we will turn this inequality into one that involves a product or quotient of factors on one side, and zero on the other.

$$\frac{3}{x+1} > 2$$

$$\frac{3}{x+1} - 2 > 0$$

$$\frac{3}{x+1} - \frac{2(x+1)}{x+1} > 0$$

$$\frac{3 - 2(x+1)}{x+1} > 0$$

$$\frac{3 - 2x - 2}{x+1} > 0$$

$$\frac{1 - 2x}{x+1} > 0$$

The expression $\frac{1-2x}{x+1}$ is positive if $1 - 2x$ and $x + 1$ both have the same sign; in other words, the solution set of this inequality consists of all values x for which

$$(1 - 2x > 0 \text{ and } x + 1 > 0) \quad \text{or} \quad (1 - 2x < 0 \text{ and } x + 1 < 0).$$

Solving these four simple inequalities, we get

$$\left(x < \tfrac{1}{2} \text{ and } x > -1\right) \quad \text{or} \quad \left(x > \tfrac{1}{2} \text{ and } x < -1\right).$$

The values of x for which $x < \frac{1}{2}$ and $x > -1$ are $-1 < x < \frac{1}{2}$. However, there are *no* values of x that satisfy the second pair of inequalities (a number cannot be simultaneously greater than $\frac{1}{2}$ and less than -1). Thus the solution set of the original inequality is $\{x \mid -1 < x < \frac{1}{2}\}$, or, in interval notation, $(-1, \frac{1}{2})$. □

Ⓢ *Caution* It is best not to write the solution set in the form $\frac{1}{2} > x > -1$. Instead, when writing these compound inequalities, start with the least number on the left, and end with the greatest number on the right. □

Ⓢ *Caution* Suppose an inequality has solution set $\{x \mid x < 3 \text{ or } x > 5\}$. In interval notation, this set is written as $(-\infty, 3) \cup (5, \infty)$. Be sure that you *never* write $5 < x < 3$ to represent this set. In general, compound inequality notations cannot be used for unions (i.e., with "or"). □

EXAMPLE 0.18 | **Solving an inequality that eventually involves three factors**

Find the solution set of the inequality $\dfrac{1}{x-2} \le \dfrac{1}{6-x}$.

Solution   We *cannot* cross-multiply here, because we do not know whether $x-2$ and $6-x$ are positive or negative. Instead we rewrite:

$$\frac{1}{x-2} \le \frac{1}{6-x}$$

$$\frac{1}{x-2} - \frac{1}{6-x} \le 0$$

$$\frac{(6-x)-(x-2)}{(x-2)(6-x)} \le 0$$

$$\frac{8-2x}{(x-2)(6-x)} \le 0$$

$$\frac{2(4-x)}{(x-2)(6-x)} \le 0.$$

There are three factors in the expression above (in the numerator or denominator), namely $4-x$, $x-2$, and $6-x$. The inequality will be true when an odd number of these factors are negative or when the numerator $2(4-x)$ is equal to zero. (This happens only when $x=4$.) This presents the following possibilities:

$$4-x<0, \quad x-2>0, \quad \text{and } 6-x>0$$

$$\text{OR} \quad 4-x>0, \quad x-2<0, \quad \text{and } 6-x>0$$

$$\text{OR} \quad 4-x>0, \quad x-2>0, \quad \text{and } 6-x<0$$

$$\text{OR} \quad 4-x<0, \quad x-2<0, \quad \text{and } 6-x<0$$

Solving each of these simple inequalities, we get

$$x>4, \quad x>2, \quad \text{and } x<6$$

$$\text{OR} \quad x<4, \quad x<2, \quad \text{and } x<6$$

$$\text{OR} \quad x<4, \quad x>2, \quad \text{and } x>6$$

$$\text{OR} \quad x>4, \quad x<2, \quad \text{and } x>6$$

The solution sets of these four sets of inequalities are, respectively, $(4, 6)$, $(-\infty, 2)$, $\emptyset$, and $\emptyset$ (it may help to think of the number line). Taking the union of these four solution sets, we get $(-\infty, 2) \cup (4, 6)$. Don't forget that $x=4$ is also a solution to the original inequality (we saw already that $x=4$ is the solution where the inequality is an *equality*). Therefore, the solution to the original inequality is $(-\infty, 2) \cup [4, 6)$. $\square$

✓ *Checking the Answer*   After all that work it is a very good idea to check your answer! In Example 0.18, we found the solution set for the inequality $\frac{1}{x-2} \le \frac{1}{6-x}$ to be the set $(-\infty, 2) \cup [4, 6)$. To test whether this is reasonable, try evaluating the inequality at a few values in the solution set and at a few values outside the solution set. For example, the inequality should be true for $x=0$ and $x=5$ (these are in the solution set) but false for $x=3$ and for $x=8$ (check this).

You can also check the answer to Example 0.18 by using a graphing calculator to graph the equation $y = \frac{1}{x-2} - \frac{1}{6-x}$. The graph of this equation should have y-values that are less than or equal to zero for $x \in (-\infty, 2) \cup [4, 6)$. For all other values of x, the y-values should be positive. Figure 0.20 shows the graph of this equation. Check whether it verifies the solution set we found in Example 0.18.

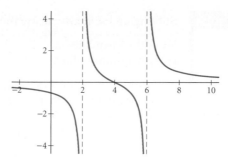

Figure 0.20   □

❓ *Question*   Consider the inequality $(x - 2)^2(x - 1) > 0$. The left-hand side of this inequality has three factors, namely $(x - 2)$, $(x - 2)$, and $(x - 1)$. Do you have to look at the signs of all of these factors to solve this particular inequality? Why or why not? (*Hint:* Think about the sign of $(x - 2)^2$.)   □

0.3.3 Inequalities and absolute values

In Section 0.1.6 we saw that the distance between two real numbers x and c is by definition $|x - c|$. Therefore, the solutions to inequalities of the form $|x - c| < \delta$ and $|x - c| > \delta$ describe how close or far away x is from c. (The symbol δ is the Greek letter delta and is one of the traditional letters used to represent a distance.)

On the real number line, we can graphically represent the solution sets to $|x - c| < \delta$ and $|x - c| > \delta$ as follows:

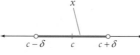

Figure 0.21 $|x - c| < \delta$

Figure 0.22 $|x - c| > \delta$

The number line in Figure 0.21 shows that if the distance between x and c is less than δ, then x must lie between $c - \delta$ and $c + \delta$. Figure 0.22 shows the values of x whose distance from the point c is greater than δ (note that x can be in the right or left segment of the solution set). These number lines suggest the following theorem:

THEOREM 0.13

Inequalities That Express Distances

For any real numbers x and c, and any positive real number δ,

(a) If $|x - c| < \delta$, then the distance between x and c is less than δ:

$$|x - c| < \delta \quad \text{if and only if} \quad c - \delta < x < c + \delta.$$

(b) If $|x - c| > \delta$, then the distance between x and c is greater than δ:

$$|x - c| > \delta \quad \text{if and only if} \quad x < c - \delta \text{ or } x > c + \delta.$$

In interval notation, this means that

$$|x - c| < \delta \quad \text{if and only if} \quad x \in (c - \delta, c + \delta),$$

and

$$|x - c| > \delta \quad \text{if and only if} \quad x \in (-\infty, c - \delta) \cup (c + \delta, \infty).$$

EXAMPLE 0.19

Solving an inequality using the distance interpretation of an absolute value

Solve the inequality $|x - 2| > 5$. Express the solution set in words, in interval notation, and on a number line.

Solution   The solution set of $|x - 2| > 5$ is the set of all values x whose distance from 2 is greater than 5; in other words, it is the set of values x that are less than $2 - 5 = -3$ or greater than $2 + 5 = 7$. (In this example, $c = 2$ and $\delta = 5$.) In interval notation this is the set $(-\infty, -3) \cup (7, \infty)$. Figure 0.23 shows this solution set on the real number line.

Figure 0.23

For more complicated inequalities like $|2x - 5| > 3$ or $|x^2 - 4| < 12$, the following method of solution is helpful.

THEOREM 0.14

Replacing an Absolute Value Inequality with Two Simpler Inequalities

Suppose δ is any positive number and A is any mathematical expression.

(a) $|A| < \delta$ if and only if $A < \delta$ and $A > -\delta$ (that is, $-\delta < A < \delta$).

(b) $|A| > \delta$ if and only if $A > \delta$ or $A < -\delta$.

The words "and" and "or" in Theorem 0.14 are essential. The "and" represents an *intersection*, whereas the "or" represents a *union*. In other words,

$$|A| < \delta \quad \text{if and only if} \quad A \text{ is in } (-\delta, \infty) \cap (-\infty, \delta) = (-\delta, \delta),$$

and

$$|A| > \delta \quad \text{if and only if} \quad A \text{ is in } (-\infty, -\delta) \cup (\delta, \infty).$$

We will "prove" this theorem in Section 0.5; for now we will simply make use of it.

? *Question*   A similar theorem is true for $|A| \leq \delta$ and $|A| \geq \delta$. Can you write these inequalities in interval notation?

EXAMPLE 0.20

Using cases to solve inequalities that involve absolute values

Solve the following inequalities.

(a) $|2x - 5| > 3$ (b) $|x^2 - 4| \leq 12$

Solution

(a) By Theorem 0.14, $|2x - 5| > 3$ if and only if

$$\begin{array}{ccc} 2x - 5 > 3 & & 2x - 5 < -3 \\ 2x > 8 & \text{or} & 2x < 2 \\ x > 4 & & x < 1 \end{array}$$

The solution set is $\{x \mid x < 1 \text{ or } x > 4\}$, or, in interval notation, $(-\infty, 1) \cup (4, \infty)$.

(b) $|x^2 - 4| \leq 12$ if and only if

$$x^2 - 4 \leq 12 \qquad\qquad x^2 - 4 \geq -12$$
$$x^2 - 16 \leq 0 \qquad \text{and} \qquad x^2 + 8 \geq 0$$
$$(x - 4)(x + 4) \leq 0 \qquad\qquad x \in (-\infty, \infty)$$

The second inequality has solution set $(-\infty, \infty)$ because $x^2 + 8$ is *always* positive. It remains to solve the first inequality. We have $(x - 4)(x + 4) \leq 0$ if the factors $x - 4$ and $x + 4$ have opposite signs or if $x = \pm 4$ (in which case the inequality is an equality). Thus we must solve for the values of x with

$$(x - 4 > 0 \text{ and } x + 4 < 0) \quad \text{or} \quad (x - 4 < 0 \text{ and } x + 4 > 0).$$

This is equivalent to the set of values x where

$$(x > 4 \text{ and } x < -4) \quad \text{or} \quad (x < 4 \text{ and } x > -4).$$

This is the set $\emptyset \cup (-4, 4) = (-4, 4)$. Therefore, the solution set for $(x - 4)(x + 4) \leq 0$ is the set $[-4, 4]$.

Finally, we can say that the solution set to the original inequality $|x^2 - 4| \leq 12$ is $[-4, 4] \cap (-\infty, \infty) = [-4, 4]$. $\quad\square$

✅ *Checking the Answer* To check the solutions for Example 0.20(a) and 0.20(b), test points inside and outside of the proposed solution set to see whether they make the inequality true. Alternatively, you could graph the equations $y = |2x - 5| - 3$ and $y = |x^2 - 4| - 12$ and check the solution sets by looking at the graphs. $\quad\square$

◈ *Caution* Do not confuse the cases used for inequalities with absolute values (as in Example 0.20) with the cases used in Section 0.3.2 for inequalities involving products or quotients. Although both methods of solving inequalities involve cases, they are very different and apply in different situations. $\quad\square$

0.3.4 Solving simple inequalities quickly

Solving inequalities with cases by examining the number of positive or negative factors can be time-consuming. In Section 2.7.4 we will learn a faster way to solve inequalities using number lines and the Intermediate Value Theorem. However, two simple inequalities appear so often that it is worthwhile for us immediately to learn a fast method for solving them:

THEOREM 0.15

Solutions for Two Simple Inequalities

For any positive real number c,

(a) The solution set for the inequality $x^2 < c^2$ is $(-c, c)$.

(b) The solution set for the inequality $x^2 > c^2$ is $(-\infty, -c) \cup (c, \infty)$.

Theorem 0.15 also applies to inequalities involving $\leq$ or $\geq$, but the solution sets will be closed sets. For example, the solution to $x^2 \leq c$ is $[-c, c]$. If you think about this theorem for a moment, it should make some sense. For example, consider the inequality $x^2 < 9$. Clearly $x = 2$ is a solution (since $2^2 < 9$), but $x = 4$ is not (since 4^2 is not less than 9). On the other hand, $x = -2$ is a solution, but $x = -4$ is not. Any positive or negative number with a magnitude strictly less than 3 will be a solution; thus the solution should be $(-3, 3)$. The solution sets described in Theorem 0.15 are pictured in Figures 0.24 and 0.25.

Figure 0.24
Solution set for $x^2 < c^2$

Figure 0.25
Solution set for $x^2 > c^2$

If we were to use the methods of Section 0.3.2 to solve the inequalities in Theorem 0.15, we would first use algebra to get an inequality involving zero, then factor and examine cases depending on the number of positive or negative factors. For example, to solve $x^2 < 9$ we would rewrite the inequality as $x^2 - 9 < 0$ and then factor the left-hand side to get $(x - 3)(x + 3) < 0$. The solution to this inequality is equivalent to the solutions to

$$(x - 3 < 0 \text{ and } x + 3 > 0) \quad \text{or} \quad (x - 3 > 0 \text{ and } x + 3 < 0).$$

If you solve this combination of inequalities, you will eventually arrive at the solution set $(-3, 3)$. In the exercises, you will justify the solution sets given in Theorem 0.15 by using the techniques of Section 0.3.2 as we just did for the example $x^2 < 9$.

⑨ *Caution*   You *cannot* solve the inequality $x^2 < c^2$ by taking the square root of both sides of the inequality. In general, taking the square root of both sides of an *equation* is okay, but we cannot take the square root of both sides of an *inequality*.   □

EXAMPLE 0.21

Using Theorem 0.15 to solve simple inequalities

Use Theorem 0.15 to solve the inequality $5 - x^2 \leq 0$.

Solution   If we were to use the techniques of Section 0.3.2 to solve this inequality, we would first factor the left-hand side to get $(\sqrt{5} - x)(\sqrt{5} + x) \leq 0$ and then consider cases. Using Theorem 0.15, we can solve this inequality much faster. We start by putting our inequality into the form of Theorem 0.15:

$$5 - x^2 \leq 0$$
$$-x^2 \leq -5$$
$$x^2 \geq 5 \qquad \text{(multiply by } -1 \text{ and flip the inequality)}$$

This inequality is now in the form $x^2 \geq c^2$, with $c = \sqrt{5}$. Thus, by the second part of Theorem 0.15 (with non-strict inequality), the solution to our inequality is $(-\infty, -\sqrt{5}] \cup [\sqrt{5}, \infty)$.   □

0.3.5 The triangle inequality

In this section we discuss some rules that help simplify or rewrite expressions involving absolute values. The most important of the inequalities we present here is called the ***triangle inequality,*** so called because it is related to the fact that a given side of a triangle can never be longer than the sum of the lengths of its other two sides. Algebraically, the triangle inequality says that the absolute value of a sum of numbers is always less than or equal to the sum of the absolute values of those numbers.

THEOREM 0.16

The Triangle Inequality

Given any real numbers a and b,

$$|a + b| \leq |a| + |b|.$$

We will often use the triangle inequality to help us prove other theorems. We will prove the triangle inequality in Section 0.5.

❓ Question　Test the triangle inequality for values a and b in the following four cases: where a and b are both positive; where a is positive and b is negative; where a is negative and b is positive; and where a and b are both negative.　　□

THEOREM 0.17

Three Inequalities That Follow from the Triangle Inequality

Given any real numbers a and b,

(a) $|a - b| \leq |a| + |b|$,　(b) $|a - b| \geq |a| - |b|$,　(c) $|a - b| \geq ||a| - |b||$.

The first inequality follows easily from the triangle inequality and can be interpreted as saying that the distance $|a - b|$ between any two real numbers a and b is less than or equal to the sum $|a| + |b|$ of the distances of the numbers a and b from zero. You will prove the inequalities in Theorem 0.17 in the exercises of Section 0.5.

❓ Question　Test each of the inequalities in Theorem 0.17 for various values of a and b. (Consider both positive and negative values.)　　□

Theorems 0.16 and 0.17 show you how to relate the absolute value of a sum or difference to the sum or difference of absolute values. Things are much simpler when dealing with absolute values of products and quotients; the absolute value of a product of two numbers is *equal* to the product of the absolute values of those two numbers.

THEOREM 0.18

Absolute Values Commute with Products and Quotients

Given any real numbers a and b,

(a) $|ab| = |a||b|$,　　　　　　　(b) $\left|\dfrac{a}{b}\right| = \dfrac{|a|}{|b|}$.

EXAMPLE 0.22

Using the triangle inequality

Use the inequalities in Theorems 0.16–0.18 to show that $|2x + 1| < 2|x| + 1$ (for all values of x).

Solution   We will use the triangle inequality and the first part of Theorem 0.18:

$$|2x + 1| < |2x| + |1| \qquad \text{(triangle inequality)}$$
$$= |2||x| + 1 \qquad \text{(Theorem 0.18)}$$
$$= 2|x| + 1.$$

Therefore, $|2x + 1| < 2|x| + 1$.   □

◈ *Caution*   It is very important to conclude the solution to Example 0.22 by saying that $|2x + 1| < 2|x| + 1$. Otherwise, one might mistakenly read the calculation to say that $|2x + 1|$ and $2|x| + 1$ are *equal*, since an equality is in the last line of the calculation. Remember that each inequality or equality connects to the previous line of the calculation. In other words, the multi-line calculation in the example says the same thing as the one-line calculation:

$$|2x + 1| < |2x| + |1| = |2||x| + 1 = 2|x| + 1.$$

Since the inequality "$<$" is the weakest link in the string of equalities and inequalities, we have $|2x + 1| < 2|x| + 1$.   □

What you should know after reading Section 0.3

▶ The definition of an inequality and its solution set, and basic facts about manipulating and simplifying inequalities.

▶ How to solve inequalities by turning them into inequalities involving products and/or quotients that are compared to zero. How to tell when a product or quotient of factors is positive, negative, or zero (this often involves examining many cases and understanding when to use "and" and when to use "or").

▶ How to check that a solution set to a given inequality is reasonable both by testing values inside and outside the solution set and by inspecting the graph of an appropriate equation.

▶ How to express inequalities of the form $|x - c| < \delta$ and $|x - c| > \delta$ many ways, including in words, in terms of distances, in set notation, and in interval notation. How to solve these simple inequalities using the distance interpretation.

▶ How to solve inequalities of the form $|A| < \delta$ and $|A| > \delta$ (where A is some mathematical expression) by splitting the problem into cases. Again, this involves understanding of "and" (intersections), "or" (unions), and when to use them.

▶ Basic inequalities such as the triangle inequality and how to verify and use these inequalities. How is the absolute value of a sum, difference, product, or quotient of numbers related to the absolute values of those numbers?

Exercises 0.3

Concepts

0. Read the section and make your own summary of the material. Include whatever you think will help you review the material later on.

1. Explain why you can't solve the inequality $\frac{2}{x+1} < 7$ by multiplying both sides of the inequality by $x + 1$.

2. Can you solve the inequality $x^2 > 4$ by taking the square root of both sides of the inequality? What is the solution to this inequality?

3. Explain why a product $abcd$ of four numbers is positive if it has an even number of negative factors and is negative if it has an odd number of negative factors.

4. Suppose a, b, c, d, m, and n are nonzero real numbers. List all of the possible ways that $\frac{abcd}{mn}$ could have an odd number of negative factors. (For example, one way is to have $a < 0$ while the rest of the numbers are positive.)

5. Explain the difference between the sets $\{x \mid x < 2$ or $x > 0\}$ and $\{x \mid x < 2$ and $x > 0\}$. Use a number line and the words "union" and "intersection" in your explanation.

6. Answer the following two questions without solving any inequalities or using your graphing calculator: Is $x = 2$ a solution to the inequality $x^4 \leq 11x - 1$? What about $x = 3$? Explain your answers.

7. State the triangle inequality.

■ Express the following sentences as inequalities that involve absolute values.

8. The distance between x and 5 is less than 6.

9. The distance between a and -3 is greater than or equal to 2.

10. The distance between the square of a number c and the number -1.5 is less than 3.8.

11. The number r is between 3.3 and 4.7.

12. The number x is either less than 1.9 or greater than 2.1.

■ Sketch the solution set of each of the following inequalities on an appropriately labeled real number line. The letters b, c, r, and δ represent positive real numbers.

13. $|x - c| \leq \delta$ 14. $|x - c| \geq \delta$

15. $|4 - x| < c$ 16. $|b - x| > \delta$

17. $|x + c| \leq r$ 18. $|x + c| \geq r$

■ Use interval notation to fill in the blanks.

19. $|x - c| \leq \delta$ if and only if x is in _____.

20. $|x - c| \geq \delta$ if and only if x is in _____.

21. $|A| \leq \delta$ if and only if A is in _____.

22. $|A| \geq \delta$ if and only if A is in _____.

23. $|x - 2| < 0.1$ if and only if x is in _____.

24. $|A| > 4$ if and only if A is in _____.

25. Suppose c is a positive real number. Use the techniques of Section 0.3.2 (factoring and cases) to verify the first part of Theorem 0.15. In other words, use factoring and cases to show that the solution set of the inequality $x^2 < c^2$ is the interval $(-c, c)$.

26. Suppose c is a positive real number. Use the techniques of Section 0.3.2 (factoring and cases) to verify the second part of Theorem 0.15. In other words, use factoring and cases to show that the solution set of the inequality $x^2 > c^2$ is the set $(-\infty, -c) \cup (c, \infty)$.

■ Suppose A, B, C, and D are real numbers. For each string of inequalities and equalities given, what is the most that we can say about the relationship between A and D? (For example, can we say that $A < D$, $A \leq D$, or $A = D$? Or can we guarantee anything about which of A or D is larger?)

27. $A \leq B < C = D$

28. $A = B < C = D$

29. $A < B = C \geq D$

■ For each pair of values a and b, verify that the following inequalities are true: (a) the triangle inequality $|a + b| \leq |a| + |b|$; (b) the inequality $|a - b| \leq |a| + |b|$; (c) the inequality $|a - b| \geq |a| - |b|$; (d) the inequality $|a - b| \geq ||a| - |b||$.

30. $a = 3$, $b = 1$ 31. $a = -3$, $b = -1$

32. $a = 3$, $b = -1$ 33. $a = -3$, $b = -1$

34. $a = 1$, $b = 3$ 35. $a = -1$, $b = 3$

36. $a = 1$, $b = -3$ 37. $a = -1$, $b = -3$

■ Determine whether each of the following statements is true or false. Explain your answers.

38. True or False: If $a \leq b$, then $-3a \leq -3b$.

39. True or False: If $a < b$, then $-4a < b$.

40. True or False: If $x < -2$, then $\frac{1}{x} < -\frac{1}{2}$.

41. True or False: If a, b, and c are negative real numbers, then $-2abc$ is negative.

42. True or False: $|x - c| > \delta$ if and only if $x < c - \delta$ and $x > c + \delta$.

43. True or False: The set $\{x \mid x > 4$ and $x < 6\}$ is equal to the set $(4, \infty) \cup (-\infty, 6)$.

44. True or False: For any real numbers a and b, $|a - b| \leq |a| - |b|$.

Skills

■ Solve each of the following inequalities two ways: (a) by using factoring and cases as in Section 0.3.2, and (b) by using Theorem 0.15.

45. $x^2 - 1 \leq 3$

46. $x^2 - 7 \geq 0$

47. $3 - x^2 < 0$

■ Find the solution set of each of the following inequalities. Express the solution sets (a) in "inequality" notation, (b) in interval notation, and (c) on a real number line. Check your answers by graphing an appropriate equation.

48. $2x + 1 \geq 4$ 49. $1 - 5x > 11$

50. $0 \leq \frac{1}{x} \leq 3$ 51. $(x + 1)(x - 2) > 0$

52. $2x^2 - 7x + 3 \leq 0$ 53. $x^3 - x \geq 0$

54. $x^3 \leq 4x$ 55. $6x^2 > x^3 + 9x$

56. $x^4 > 16$ 57. $\dfrac{2}{x + 1} \geq 3$

58. $x + 2 \le \dfrac{1}{x + 2}$

59. $\dfrac{1}{x - 1} < \dfrac{2}{x + 2}$

60. $\dfrac{1}{1 + x} < \dfrac{1}{x + x^2}$

61. $1 > \dfrac{3}{x^2 - 1}$

62. $\dfrac{1}{x - 1} \ge \dfrac{x - 1}{x + 5}$

63. $|x - 3| < 8$

64. $|x + 2| > 4$

65. $|x + 8| \le 1$

66. $|3x + 1| \le 4$

67. $|4 - 2x| > 6$

68. $|-8x + 3| \ge 2$

69. $|x^2 - 1| \ge 3$

70. $|x^2 + 4x - 1| > 4$

71. $|x^3 + x^2 - 3x| \le 3x$

■ Use a graphing calculator to approximate the solution sets of the following inequalities.

72. $x^5 \le 2x^3 + 5$

73. $x^3 - 4x^2 > 1$

74. $3x^4 - 1 \ge x^5$

75. $\dfrac{x - 1}{x^2 + 5} < \dfrac{x}{x + 1}$

76. $\dfrac{2 - x^2}{x + 1} \ge \dfrac{x - 1}{x^2 + 1}$

77. $\sqrt{x + 1} > x^2 - 3x$

■ Use the triangle inequality and Theorems 0.17 and 0.18 to show that the following inequalities are true for all *x*. *Hint:* Start from the left side of the inequality, and use Theorems 0.16–0.18 to work toward the right-hand side.

78. $|3x + 5| \le 3|x| + 5$

79. $|3x + 5| \ge 3|x| - 5$

80. $|3x + 5| \ge |3|x| - 5|$

81. $|3x - 5| \le 3|x| + 5$

82. $|3x - 5| \ge 3|x| - 5$

83. $|3x - 5| \ge |3|x| - 5|$

84. $|x^2 + x - 6| = |x - 2||x + 3|$

85. $|x + 2x^2| \le |x|(1 + 2|x|)$

86. $\left| \dfrac{3x}{x + 1} \right| \ge 3 \dfrac{|x|}{|x| + 1}$ $(x \ne -1)$

Applications

87. The formula for converting degrees Fahrenheit (F) to degrees Celsius (C) is $C = \frac{5}{9}(F - 32)$.

   (a) For water to be between the freezing point and the boiling point, its temperature in degrees Fahrenheit has to be between 32° and 212° (that is, $32 < F < 212$). What is the corresponding range in degrees Celsius?

   (b) Ella says that the temperature will be between 19° Celsius and 23° Celsius (that is, $19 < C < 23$). What is the corresponding range of temperatures in degrees Fahrenheit?

88. The instructions that came with Byron's stereo claim that the optimal volume range v for his speakers satisfies the inequality $|v - 85| < 35$, where v is measured in decibels. What is the lowest optimal volume for Byron's speakers? The greatest?

89. For her summer job, Alina fills large buckets with pre-sliced pickles. The label on the buckets indicates that each bucket should contain 20 pounds of pickles. Her boss says it is okay to be over or under by as much as half a pound. Express the acceptable weight W of a full bucket of pickles using an absolute value inequality.

90. The cylindrical cans that Len's company produces are 6 inches tall and vary in radius. The cost of producing a can with radius r is $C = 10\pi r^2 + 24\pi r$ cents.

   (a) Suppose Len's boss wants him to construct the cans so that the cost is between 250 and 450 cents. Write this condition as an absolute value inequality involving C.

   (b) Given the cost restrictions described in part (a), what is the acceptable range of values for r? Write your answer two ways: using interval notation and as an absolute value inequality involving r.

0.4   Logic

In this section we give a short introduction to elementary logic. Throughout this text we will encounter many theorems and arguments based on this elementary logic. Logic will be particularly important in Section 0.5, where we learn how to construct logical arguments that argue, or "prove," mathematical facts.

0.4.1 Quantifiers

We will often deal with statements that describe *all* the objects with a particular property or ensure the existence of *at least one* object with a particular property. Two examples are:

(1)   *"Every integer is a rational number."*

(2)   *"There is at least one real number that is greater than 12."*

Both of these statements involve **quantifiers**. There are two types of quantifiers: the quantifier **for all** and the quantifier **there exists**.

EXAMPLE 0.23   **Writing statements with quantifiers**

To write Statements (1) and (2) as precisely as possible using the quantifiers "for all" and "there exists," we could write

(3)   "*For all integers x, x is a rational number.*"

(4)   "*There exists a real number x such that x > 12.*"   □

Suppose S is a set of objects (for example, the set of all real numbers). A statement of the form "For all $x \in S$, x has property P" is true only if *every* member x of the set S has property P. A statement of the form "There exists $x \in S$ such that x has property P" is true if there is *at least one* member x of the set S that has property P. Statements (3) and (4) both happen to be true. However, if we change the quantifier in Statement (4), we get the statement "For all real numbers x, $x > 12$," which is false! The choice of quantifier makes a big difference in the meanings of these statements.

Some statements involve more than one quantifier. As Examples 0.24 and 0.25 illustrate, the order of quantifiers is important if more than one quantifier is present.

EXAMPLE 0.24   **A statement with multiple quantifiers**

Consider the statement

(5)   "*For all x > 0, there exists y > 0 such that y > x.*"

This statement means that for any positive number x, you can find some positive number y that is greater than x. Statement (5) is true. For example, given $x = 25$, the number $y = 26$ is greater. In general, whatever number you choose for x, there is some greater number you can choose for y.   □

EXAMPLE 0.25   **Reversing the order of quantifiers**

If we reverse the order of the quantifiers in Statement (5), we get a much different statement:

(6)   "*There exists y > 0 such that for all x > 0, we have y > x.*"

This statement says that there is some positive number y with the property that y is greater than *every* positive number x. This is clearly false, since no matter what is chosen for y, the number $x = y + 1$ will be greater than y (and thus y cannot be larger than *all* numbers x).   □

0.4.2 Negations and counterexamples

Every logical statement can be "negated." In other words, for every statement there is another statement, called the **negation,** that means the exact logical opposite of the original statement.

DEFINITION 0.7

The Negation of a Statement

Given a statement A, the **negation** of A, denoted "Not A," is a statement that is false whenever A is true, and true whenever A is false.

The negation of a statement involving the quantifier "for all" will involve the quantifier "there exists." On the other hand, the negation of a statement involving the quantifier "there exists" will involve the quantifier "for all." This is illustrated in Table 0.1.

Statement	Negation
"For all x, we have P"	"There exists x such that (Not P)"
"There exists x such that P"	"For all x, we have (Not P)"

Table 0.1

EXAMPLE 0.26

Negating statements with quantifiers

Write down **(a)** the negation of Statement (3): "For all integers x, x is a rational number," and **(b)** the negation of Statement (4): "There exists a real number x such that $x > 12$."

Solution   The negation of Statement (3) is the statement that means the "opposite." It may help to start with the simpler form of the statement: "Every integer is a rational number." The negation of this statement is "There is some integer that is not a rational number." In more mathematical notation, we would write

"*There exists an integer x such that x is not a rational number.*"

Note that Statement (3) is true, and thus its negation is false.

Statement (4) says that *some* real number x is greater than 12. The negation of this statement would say that *no* real numbers are greater than 12. In other words, the negation would say that *all* real numbers are *not* greater than 12 (that is, are less than or equal to 12). The negation of Statement (4) is thus

"*For all real numbers x, we have $x \leq 12$.*"

Of course, this statement is false (not all numbers are less than or equal to 12); this makes sense because the original statement was true.   □

⊛ *Caution*   You can't negate a statement just by adding "not" here and there in the statement. Remember that the negation of a statement is supposed to be the logical opposite of the original statement. Think about what a statement *means* before you attempt to write down its negation.   □

EXAMPLE 0.27

Negating statements with multiple quantifiers

Write down the negation of Statement (5): "For all $x > 0$, there exists $y > 0$ such that $y > x$."

Solution   Let P stand for the end of Statement (5); that is, let P represent "There exists $y > 0$ such that $y > x$." Then Statement (5) says "For all $x > 0$, we have P." By Table 0.1, the negation of this statement is "There exists $x > 0$ such that (Not P)." The negation of P is "For all $y > 0$, we have $y \leq x$." Therefore, the negation of Statement (5) is

"*There exists $x > 0$ such that, for all $y > 0$, we have $y \leq x$.*"   □

❓ *Question*   Follow the method used in Example 0.27 to write down the negation of Statement (6).   □

A statement of the form "For all x, we have P" is true only if P is true for *all* values of x. On the other hand, a statement of this form is false if it is false for even *one* value of x, which we call a ***counterexample.***

DEFINITION 0.8

Counterexamples

Suppose the statement "For all x, we have P" is false. A ***counterexample*** to this statement is a value of x for which P is false.

EXAMPLE 0.28

Finding a counterexample

The statement "For all real numbers x, we have $x \leq 12$" is false. Some values of x make the conclusion of this statement true (for example, $x = 5$ is less than or equal to 12). On the other hand, some values of x make the conclusion false, and each of these values is a counterexample. For example, $x = 13$ is a counterexample because $x = 13$ is not less than or equal to 12. The existence of just one counterexample shows that the original statement is false. ☐

EXAMPLE 0.29

A statement that is true has no counterexamples

Consider the statement "For all real numbers x, we have $x^2 \geq 0$." This statement is true because it has *no* counterexamples. There is no number x for which x^2 fails to be greater than or equal to zero. ☐

☢ ***Caution*** The statement "There exists an integer x such that x is not a rational number" is false (since every integer is rational). However, we cannot produce a counterexample for this statement. The reason is that this statement involves the quantifier "there exists." It is true that, for example, $x = 3$ is an integer that is a rational number. However, this does not contradict our original statement that there is *some* integer that is not rational, so it is not a counterexample. ☐

0.4.3 "And" and "or"

Consider the statement

(7)   *"x is positive and x is rational."*

This statement is true for $x = \frac{1}{2}$, since $\frac{1}{2}$ is both positive and rational. It is false for $x = -\frac{1}{2}$, since $-\frac{1}{2}$ fails to be positive. It is also false for $x = \sqrt{2}$, since $\sqrt{2}$ is not rational. Statement (7) is false for a value x if x is negative, or if x is irrational, or both. Therefore, the negation of Statement (7) is

(8)   *"x is negative or x is irrational."*

Note that Statement (7) involves an "and," whereas its negation in Statement (8) involves an "or."

Now consider the statement obtained by replacing the "and" in Statement 7 with an "or":

(9)   *"x is positive or x is rational."*

This statement is true for values of x that are positive, or rational, or both. For example, it is true for $x = \sqrt{2}$, $x = -\frac{1}{2}$, and $x = \frac{1}{2}$. Statement (9) is false only when x fails to be positive *and* fails to be rational (for example, when $x = -\sqrt{2}$). In other words, the negation of Statement (9) is

(10)   *"x is negative and x is irrational."*

Table 0.2 illustrates the general procedure for negating statements involving "and" and statements involving "or."

Statement	Negation
"*A* and *B*"	"(Not *A*) or (Not *B*)"
"*A* or *B*"	"(Not *A*) and (Not *B*)"

Table 0.2

⊛ *Caution*   The statement "Not(*A* and *B*)" is *not* the same as the statement "(Not *A*) and (Not *B*)" (see Table 0.2). We cannot distribute the "Not" through the parentheses. (You may have thought to do this by analogy, since $3(a + b) = 3a + 3b$.) □

⊛ *Caution*   In daily language we often use the **exclusive** "or," in which "*A* or *B*" is true if either *A* or *B*, but *not both*, are true. For example, if Phil says, "I'll meet you in the lobby or at the coffee bar," clearly he means that he will be in one or the other of these places, but not both. In mathematics, however, we always use the **inclusive** "or," in which the statement "*A* or *B*" is true if *A* is true, or if *B* is true, or if *both* *A* and *B* are true. □

EXAMPLE 0.30   **Negating a statement involving "or"**

Write down the negation of the statement "For all real numbers *a* and *b*, either $a < b$ or $a > b$." Simplify the negation as much as possible. Which is true: the original statement, or its negation? Provide a counterexample for whichever statement is false.

Solution   The negation of "$a < b$ or $a > b$" is

"*(Not $a < b$) and (Not $a > b$)*,"   or in other words,   "$a \geq b$ and $a \leq b$."

If *a* is both less than or equal to *and* greater than or equal to *b*, then we must have $a = b$. Therefore, the negation of the original statement is

There exist real numbers a and b such that $a = b$.

Clearly the negation is true (for example, take $a = 2$ and $b = 2$; then $a = b$). Therefore, the original statement must have been false. The case where $a = 2$ and $b = 2$ is a counterexample to the original statement "$a < b$ or $a > b$", since 2 is not less than 2 or greater than 2. □

❓ *Question*   Would the original statement in Example 0.30 be true if we replaced the strict inequalities with $\leq$ and $\geq$? What would the negation be in this case?   □

0.4.4 Implications

Statements of the form "If *A*, then *B*" are called **implications.** Most of the theorems in this book have this form. In a statement of the form "If *A*, then *B*," the statement *A* is called the **hypothesis,** and the statement *B* is called the **conclusion.** Implications occur so often in this course that we have a shorthand notation for them, namely "$A \Rightarrow B$" (pronounced "*A* implies *B*").

EXAMPLE 0.31

A true statement involving an implication

Consider the statement

(11)   "*If x > 2, then x > 0.*"

(In the "arrow" notation, we would write this as "$x > 2 \Rightarrow x > 0$.") The hypothesis of the statement is "$x > 2$," and the conclusion is "$x > 0$." The implication asserts that if we know that x is greater than 2, we can conclude that x must be greater than 0; this implication is true. We can also write Statement (11) in terms of a quantifier, rather than an implication, as follows: "For all $x > 2$, we have $x > 0$." □

A statement of the form "$A \Rightarrow B$" is false only when A is true but B is false. Therefore, a counterexample to "For all x, $A \Rightarrow B$" is a value of x for which A is true and B is false. In addition, the negation of "$A \Rightarrow B$" is the statement "A and (Not B)," as illustrated in Table 0.3.

Statement	Negation
"$A \Rightarrow B$"	"A and (Not B)"
"$B \Rightarrow A$"	"B and (Not A)"

Table 0.3

🔊 *Caution*　A statement of the form "$A \Rightarrow B$" implies nothing about the truth or falsehood of A; it merely asserts that *if* the statement A happens to be true, *then* the statement B must also be true. Moreover, if A is false, the statement "$A \Rightarrow B$" does not imply anything about statement B; if A is false, then B could be either true or false. However, if A is true, then the implication $A \Rightarrow B$ asserts that B is also true. □

EXAMPLE 0.32

Investigating an implication and its negation

Write down the negation of the statement "For all x, $x > 2 \Rightarrow x > 0$. Simplify the negation as much as possible. Which is true: the original statement, or the negation?

Solution　Before we take the negation, let's investigate the original statement. This statement asserts that every real number x that is greater than 2 must also be greater than 0, which is true. Therefore, there are no counterexamples to this statement. Note that although we can think of values of x for which the "$x > 2$" part of the statement is false, these values of x are not counterexamples to the original statement. We can also think of values of x for which "$x > 0$" is false, but none of these values would be greater than 2, so they would not contradict the statement either.

The negation of the implication "$x > 2 \Rightarrow x > 0$" is "$x > 2$ and Not($x > 0$)." Because "Not($x > 0$)" means that $x \leq 0$, the negation of the original statement is

"*There exists x such that x > 2 and x ≤ 0.*"

Since the original statement is true, the negation must be false. There is no number x that is both greater than 2 and less than or equal to 0. (*Note:* We cannot provide a counterexample since this is an *existence* statement, not a "for all" statement.) □

EXAMPLE 0.33

Finding a counterexample for a false implication

The statement "For all x, if x is positive, then x is an integer" is false. Exhibit a counterexample.

Solution   This statement is false because not every positive number is an integer. One counterexample is $x = \frac{1}{2}$, since $\frac{1}{2}$ is positive but not an integer. (Note that $x = -2$ is *not* a counterexample, even though -2 is not positive. Why?)   □

The statement "$A \Longleftrightarrow B$," or "A ***iff*** B," means "A is true ***if and only if*** B is true." The phrase "if and only if" means that the implication goes both ways: A if B, and A only if B. Said another way, this means "If A then B, and if B then A."

DEFINITION 0.9

If and Only If

The statement "$A \Longleftrightarrow B$" means "$(A \Rightarrow B)$ and $(B \Rightarrow A)$."

EXAMPLE 0.34   **An "if and only if" statement**

Consider the statement "For all x, x is even if and only if x is divisible by 2." This statement is true and means two things. First, if x is even, then x must be divisible by 2. Second, if x is divisible by 2, then x is even. The "if and only if" means that the statements "x is even" and "x is divisible by 2" are equivalent.   □

0.4.5 Converses and contrapositives

We will often be interested in two variants of the implication $A \Rightarrow B$ called the ***converse*** and the ***contrapositive.***

In the converse of a statement, the roles of the hypothesis and the conclusion are reversed. This results in a much different statement; if an implication is true, its converse may *or may not* be true.

DEFINITION 0.10

The Converse of an Implication

The ***converse*** of the implication $A \Rightarrow B$ is the statement $B \Rightarrow A$.

Notice that the "if and only if" statement $A \Longleftrightarrow B$ asserts both the implication $A \Rightarrow B$ and its converse $B \Rightarrow A$.

EXAMPLE 0.35   **Finding the converse of an implication**

The converse of the statement "If x is an integer, then x is a rational number" (which is true) is the statement "If x is a rational number, then x is an integer" (which is clearly false).   □

❓ *Question*   Can you think of a statement $A \Rightarrow B$ whose converse is true?   □

◈ *Caution*   The statement "If A, then B" is equivalent to the statement "B if A." For example, it means the same thing to say "If $x > 2$, then $x > 0$" as to say "$x > 0$ if $x > 2$." Don't make the common mistake of thinking that the second statement is the converse of the first; in this example, the converse of the original statement is "If $x > 0$, then $x > 2$."   □

In the contrapositive of a statement, the hypothesis and conclusion switch places *and* are negated.

DEFINITION 0.11

The Contrapositive of an Implication

The *contrapositive* of the implication $A \Rightarrow B$ is the statement $(\text{Not } B) \Rightarrow (\text{Not } A)$.

If A is the hypothesis and B is the conclusion of the original statement, then the contrapositive asserts that if the conclusion fails to be true, then the hypothesis must also fail to be true. The contrapositive of a statement is *always* logically equivalent to the original statement. One way to see this is to notice that $A \Rightarrow B$ is false if and only if A is true and B is false; on the other hand, the statement $(\text{Not } B) \Rightarrow (\text{Not } A)$ is false if and only if "Not B" is true (so B is false) and "Not A" is false (so A is true).

EXAMPLE 0.36   Finding the contrapositive of an implication

The contrapositive of the statement "If x is an integer, then x is a rational number" is the statement "If x is not a rational number, then x is not an integer." These two statements are logically equivalent (and happen to be true). □

❓ *Question* What is the contrapositive of the statement "If x is greater than 5, then x is greater than 2"? Is the contrapositive logically equivalent to the original statement? What is the converse of this statement? Is it logically equivalent to the original statement? □

What you should know after reading Section 0.4

▶ How to write statements with the quantifiers "for all" and "there exists." How to interpret the meanings of such statements, including those that involve multiple quantifiers.

▶ How to negate logical statements (including those that involve quantifiers, "and," "or," or implications) and how to exhibit counterexamples to statements that are false.

▶ The meanings of $A \Rightarrow B$, $B \Rightarrow A$, and $A \Longleftrightarrow B$.

▶ How to write the converse and contrapositive of an implication.

Exercises 0.4

Concepts

0. Read the section and make your own summary of the material. Include whatever you think will help you review the material later on.

1. Explain the difference in meaning between the "inclusive or" and the "exclusive or." Give a real-world example of a sentence where the "or" is meant to be inclusive, and give an example where the "or" is meant to be exclusive. Which type of "or" do we always use in mathematics?

2. If A is true and B is false, what can you say about the truth or falsehood of the statement "A or B"? What if A is true and B is true?

3. If A is true and B is false, what can you say about the truth or falsehood of the statement "A and B"? What if A is false and B is false?

4. Suppose the implication "$C \Rightarrow D$" is true. If C is true, what can you say about D? If C is false, what can you say about D?

5. Consider the implication "$P \Rightarrow Q$." Is this implication necessarily false if P is false?

6. Suppose the implication "$R \Rightarrow S$" is false. What does this mean about statements R and S?

7. Explain the difference between the "implies" symbol "$\Rightarrow$" and the "if and only if" symbol "$\Longleftrightarrow$."

8. Explain what a counterexample is.

9. Which of the following statements are logically equivalent to the statement "Not(A and B)" and why?

   (a) (Not A) and (Not B)
   (b) (Not A) or (Not B)
   (c) $A \Rightarrow$ (Not B)

10. Which of the following statements are logically equivalent to the statement "Not(A or B)" and why?

    (a) (Not A) and (Not B)
    (b) (Not A) or (Not B)
    (c) Not((Not A) $\Rightarrow B$)

11. At some restaurants, you are given a choice of "soup or salad and potato or vegetable." This statement is mathematically imprecise, and in addition it uses the "exclusive" rather than the "inclusive" or. Write down what the restaurant really means, using a precise, logical statement.

12. Consider the statement "Every positive real number is greater than -2." Write this statement using the quantifier "for all." Then write a statement that is logically equivalent but uses "if ..., then ..." instead of quantifiers.

13. Consider the statement "The square of any real number is nonnegative." Write this statement using the quantifier "for all." Then write a statement that is logically equivalent but uses "if ..., then ..." instead of quantifiers.

14. What is the converse of the statement $C \Rightarrow D$? Is the converse logically equivalent to the original statement? Why or why not?

15. What is the contrapositive of the statement $P \Rightarrow Q$? Is the contrapositive logically equivalent to the original statement? Why or why not?

16. Consider the statement "Every square is a rectangle." Is this statement true? Write down the converse and the contrapositive of this statement, and determine whether they are true or false.

17. Consider the statement "For all x, there is some y with $y = x + 1$." Use complete sentences to explain what this statement means, and determine whether it is true or false.

18. Consider the statement "There exists an x such that, for all y, we have $y = x + 1$." Use complete sentences to explain what this statement means, and determine whether it is true or false.

Skills

■ Write down the negation of each statement, and simplify if possible. (The letter S represents a set, and the letters A, B, C, D, P, and Q represent statements about real numbers x and/or y.)

19. "There exists $x \in S$ such that P is true."

20. "For all $y \in S$, P is true."

21. "P is true and Q is false."

22. "P is false or Q is true."

23. "$C \Rightarrow$ (Not D)."

24. "$A \Rightarrow (B$ or $C)$."

25. "For all x, $B \Rightarrow C$."

26. "There is some x such that A and B are true."

27. "For all x, there is some y such that C is true."

28. "For all x, there is some y such that $A \Rightarrow B$."

■ Determine whether each of the following statements is true or false, and explain your answers. Then write down the negation of each statement, and simplify the negation as much as possible.

29. "There is some real number between 2 and 3."

30. "Every real number is a rational number."

31. "All real numbers are either greater than zero or less than zero."

32. "No real number is both rational and irrational."

33. "Every real number is either rational or irrational."

34. "No rational number is both less than $\frac{1}{2}$ and greater than $\frac{1}{3}$."

35. "For all x, there is some y with $y = x^2$."

36. "For all x, there is some y with $x = y^2$."

37. "There exists an x such that, for all y, we have $y = x + 1$."

38. "If x is an integer greater than 1, then $x \geq 2$."

39. "For all real numbers a and b, if $a < b$ then $3a + 1 < 3b + 1$."

■ Write down the negation of each of the following statements. (These types of statements will be an important part of Chapter 2. The symbols ε and δ are Greek letters that represent real numbers.)

40. "For all $M > 0$, there exists an $N > 0$ such that for all x, if $x > N$, then $x^2 > M$."

41. "For all $\varepsilon > 0$, there exists a $\delta > 0$ such that for all x, if $0 < |x - 2| < \delta$, then $|x^2 - 4| < \varepsilon$."

42. "For all $\varepsilon > 0$, there exists a $\delta > 0$ such that for all x, if $0 < |x - 4| < \delta$, then $|\sqrt{x} - 2| < \varepsilon$."

■ Determine whether each statement below is true or false. If the statement is false, provide a counterexample. If the statement is true, write down the negation of the statement.

43. "For all x, if x is an integer then x is rational."

44. "For all x, if x is negative then x is irrational."

45. "For all x, if $x < -2$ then $x^2 > 4$."

46. "For all x, either $x \geq 2$ or $x \leq 1$."

47. "For all x, either x is even or x is odd."

48. "For all real numbers x, $x^2 \geq 0$ and $|x| \geq 0$."

49. "For all x and y, if $x < y$ then $2x - 1 < 2y - 1$."

50. "For all x and y, if $x < y$ then $x^2 < y^2$."

51. "For all x, there is some y such that $x < y$."

52. "For all x, there is some y such that $x = y^2$."

53. "For all x and y, $xy = 0$ if and only if $x = 0$ or $y = 0$."

54. "For all x and y, $\frac{x}{y} = 0$ if and only if $x = 0$."

55. "For all x and y, $\frac{x}{y} = 0$ if and only if $x = 0$ and $y \neq 0$."

■ Determine whether each statement below is true or false. If the statement is true, give an example that shows it is true. If the statement is false, write down the negation of the statement.

56. "There exists x such that $x \leq 1$ or $x \geq 2$."

57. "There exists x such that $x \leq 1$ and $x \geq 2$."

58. "There exists x such that $x > 0$ and $x^2 > 10$."

59. "There exist x and y such that $x + y = 4$."

60. "There exists $x < 0$ and $y < 0$ such that $xy < 0$."

61. "There exists $x > 0$ and $y > 0$ such that $xy = 0$."

62. "There exists x such that, for all y, $y > x$."

63. "There exists x such that, for all y, $|y| > x$."

■ Suppose A and B represent logical statements. Write (a) the converse and (b) the contrapositive of each statement below. Simplify each of your statements, if possible.

64. $B \Rightarrow A$

65. (Not A) $\Rightarrow B$

66. $A \Rightarrow$ (Not B)

67. (Not B) $\Rightarrow$ (Not A)

68. (A and B) $\Rightarrow C$

69. (A or B) $\Rightarrow C$

70. $A \Rightarrow$ (B and C)

71. $A \Rightarrow$ (B or C)

72. (A and (Not B)) $\Rightarrow C$

■ Write (a) the converse and (b) the contrapositive of each statement. Simplify your statements as much as possible. If the converse is false for any x, provide a counterexample. If the contrapositive is false for any x, provide a counterexample.

73. "If x is a real number, then x is rational."

74. "If $x \geq 2$, then $x \geq 1$."

75. "If $x \geq 2$, then $x \geq 3$."

76. "If $x > 2$, then $x \geq 3$."

77. "If x is negative, then $\sqrt{x}$ is not a real number."

78. "If x is rational, then x is not irrational."

79. "If x is not zero, then $x^2 > x$."

80. "If $x \leq 0$, then $|x| = -x$."

81. "If $x < -2$, then $|x| = -x$."

82. "If x is positive and rational, then $x - 1$ is positive and rational."

83. "If x is a real number, then x is even or x is odd."

84. "If x is even, then $\frac{x}{2}$ is an integer."

85. "If x is odd, then there is some integer n such that $x = 2n + 1$."

Applications

■ Use logic to solve each of the following three puzzles. (*Hints:* Don't worry if you come to a few dead ends and have to start over; these puzzles aren't easy! It may help to write the statements in more mathematical language or to make tables or notes to keep track of the information in each statement.)

86. Xena, Yolanda, and Zeke each have different favorite fruits. One of them likes apples the best, one of them likes bananas the best, and one of them has cantaloupes as his or her favorite fruit (though not necessarily in that order). Use the statements below to determine which person prefers which fruit.

▶ Xena likes bananas better than apples.

▶ Zeke is allergic to cantaloupes.

▶ Bananas are the favorite fruit of one of the girls.

▶ Yolanda likes bananas better than cantaloupes.

87. Linda, Alina, Phil, and Stuart are each wearing a hat of a different color (red, yellow, green, or blue, not necessarily in that order). From the statements below, determine which person is wearing which hat.

▶ Neither boy wears a red hat.

▶ The oldest person wears a green hat.

▶ Linda is older than Alina.

▶ Alina never wears yellow or red.

▶ Stuart is the youngest and hates blue.

88. There are three clubs in the Springfield Insectology Society: the Ant Association, the Beetle Band, and the Caterpillar Club. Dave, Ed, and Fran are members of at least one club each. Given the following statements, determine which club (or clubs) each of the three insectologists belongs to.

▶ Everyone in the Ant Association is also a member of the Beetle Band.

▶ Dave and Ed are not in any clubs together.

▶ No member of the Beetle Band is also a member of the Caterpillar Club.

▶ Dave is in exactly two clubs.

▶ Fran is in only one club, which Ed is not a member of.

■ **On the island of Michi there are exactly two types of people: those who always tell the truth ("truth-tellers") and those who always lie ("liars"). You have been sent to the island to determine which Michians are liars and which are truth-tellers.**

89. Any inhabitant of Michi (whether truth-teller or liar) can say the statement "I am a truth-teller." Explain why.

90. Can any Michian utter the statement "I am a liar"? If so, who? If not, why not?

91. On your first day on the island of Michi, you encounter Amanda, Brittany, and Chris sitting under a shade tree. They tell you the following:

Amanda: "At least one of Brittany or Chris tells the truth."

Brittany: "Amanda is a liar."

Chris: "Brittany is a liar."

Determine whether each of the three people is a liar or a truth-teller. (*Hint for problems of this type:* Suppose that one of the three people is a liar (or a truth-teller). Given that fact, what can you say about the other two people? Do you arrive at a contradiction? If you arrive at a contradiction, then your original assumption must have been false. There is only *one* possible answer.)

92. On the second day you meet Liz, Rein, and Zubin on the beach. Given their statements below, determine whether each person is a liar or a truth-teller.

Liz: "We all tell the truth."

Rein: "Exactly two of us tell the truth."

Zubin: "Liz and Rein always lie."

93. On the third day, you find Hyun, Jaan, and Kathleen in the coffee shop. Given the two statements below, determine whether each of the three people is a liar or a truth-teller.

Kathleen: "Hyun and Jaan are liars."

Hyun: "Kathleen always tells the truth."

94. Make up a "truth-teller" problem that has only *one* possible solution. *Hints:* Make up three names, and decide whether each person is a liar or a truth-teller. Then make up things for each person to say that is consistent with the type of person he or she is and with the types of the remaining two people. Check to see whether it is possible to solve the puzzle (with only *one* possible solution) given the statements you wrote; if not, revise the statements and check again.

0.5 │ Proofs

Almost every theorem in this book can be "proved" using previous theorems or definitions. Since a major goal of this course is that you *understand* rather than just *memorize* the theorems of calculus, you will be expected not only to know and use, but also to *prove*, these theorems. In this section you will learn what proofs are, how to write them, and how to understand them.

0.5.1 What is a proof? Why do we care?

A mathematical **proof** is a logical argument. A proof of the statement or theorem $A \Rightarrow B$ is a logical argument that starts with the hypothesis A and leads to the conclusion B. There are many methods of proof; some are no more than computations, some are "direct," and some are more exotic, such as proof by contradiction and proof by induction.

Why would we want to prove anything? Isn't calculus just about calculating things? Certainly not; calculus, like mathematics in general, is about building up a logical system of definitions, facts, and theorems that can be used to investigate functions and describe real-world phenomena. In mathematics it is very important that each new theorem be proved using previous theorems and definitions. It is not enough simply to rely on our intuition of what "ought" to be true; we must make sure that every fact or theorem that we state is mathematically and logically sound. Moreover, being able to prove mathematical theorems and facts requires a solid understanding of the definitions and properties of functions and other mathematical objects, and isn't that the whole point?

There are fancy calculators and computer programs that can handle even the most difficult calculations, but these calculators have no understanding of what they are doing or why it works. If your knowledge of calculus is only at this "calculator level," then your knowledge is worth only as much as a good calculator, say $170.00. Think of how much money you could save if you just bought a calculator and didn't have to pay your college tuition! Moreover, how could these calculators ever get programmed to do what they do if there weren't people who understood the theory of calculus and how to implement it?

If you're wondering "when you'll ever use this stuff," perhaps the best answer is that learning calculus and the theory behind it will teach you how to *think*. Depending on your choice of career, you may or may not use the computational skills of calculus later in life. For example, if you pursue a career in science, then you might use calculus to model or analyze real-world situations. On the other hand, you may *never* need the practical skills of calculus in your chosen career. However, no matter what you choose to do, the ability to think logically and solve problems will be an invaluable asset. That is one reason why it is so important not only to learn the calculational mechanics of calculus, or how to apply calculus to real-world problems, but also to *understand* the theory (and thus the proofs) behind calculus.

0.5.2 The structure of a proof

Whatever the method, proofs should be written with great attention to precision and clarity. A good way to start is to state the hypotheses (what is "given"), in mathematical notation, if possible. Then state what it is that you are trying to prove. Now that you're ready to start the actual proof, write "Proof" so the reader knows that your argument is about to start. If appropriate, you can state what method of proof you intend to use (for example, "by contradiction" or "by induction").

The body of your proof should consist of a clear, concise, logical argument. Support any statements using definitions, facts, and theorems. Use words and phrases such as "therefore," "then," "thus," "if... then ...," "since," and "because" so that the logic of your argument is clear. Above all, make sure that your argument is clear and logically valid!

It is often a good idea to finish a proof by clearly stating what it is that you have shown. You can indicate that the proof is over by making a box "■" or by writing "QED" (which represents the Latin phrase *quod erat demonstrandum*, meaning "which was to be demonstrated"). Keep in mind that a well-written proof has a beginning, a middle, and an end.

The following simple proof is really nothing more than a computation written in the form of a logical argument. It illustrates a good example of the process we have just discussed.

EXAMPLE 0.37

Writing a simple computational proof

Prove that for all real numbers a and b,

$$a^3 - b^3 = (a - b)(a^2 + ab + b^2).$$

(Note that this is one of our "factoring formulas.")

Solution First, we clearly write what we are given and what it is that we are trying to show.

Given: a and b are any real numbers.

Show: $a^3 - b^3 = (a - b)(a^2 + ab + b^2)$.

Now we can write the proof itself, which is just a computation (from the right-hand side to the left-hand side of what we are trying to show) with supporting reasons:

> **PROOF**   For any real numbers a and b,
>
> $$(a-b)(a^2+ab+b^2) = a^3 + a^2b + ab^2 - a^2b - ab^2 - b^3 \qquad \text{(multiply out)}$$
>
> $$= a^3 - b^3. \qquad \text{(simplify)}$$
>
> Therefore, $a^3 - b^3 = (a-b)(a^2+ab+b^2)$. ∎

Computational proofs often have the form shown above; to prove $A = B$, we simply start from one side and work toward the other, justifying our calculations along the way.

⚜ **Caution**   The computational proof in Example 0.37 represents the *one-directional* calculation

$$(a-b)(a^2+ab+b^2) = a^3 + a^2b + ab^2 - a^2b - ab^2 - b^3 = a^3 - b^3.$$

In general, this is *not* the same as "working on both sides" of the equation $a^3 - b^3 = (a-b)(a^2+ab+b^2)$. For example, a *bad* proof of this fact would look like

$$(a-b)(a^2+ab+b^2) = a^3 - b^3$$

$$a^3 + a^2b + ab^2 - a^2b - ab^2 - b^3 = a^3 - b^3$$

$$a^3 - b^3 = a^3 - b^3.$$

What is wrong with this "bad" proof is that we *started* with what we were trying to show, namely $(a-b)(a^2+ab+b^2) = a^3 - b^3$, and then manipulated that equation until we arrived at something that we know to be true, namely $a^3 - b^3 = a^3 - b^3$. This makes no logical sense; what the "bad" proof shows is that *if* the equality $A = B$ is true, then something obvious is also true. This is not what we wanted to prove! *Never* start a proof by assuming what it is that you are trying to show! When doing a calculational proof of an identity $A = B$, always start with one side (A or B) and work toward the other side. Do *not* start with $A = B$ and reduce that equation to something obvious. ☐

Remember that the point of a proof is not to convince *yourself* that something is true; it's to convince *another reader* that it is true. Structure your proof clearly and effectively so that the logic of your argument is clear. The "final answer" of a proof is not the conclusion. It is the proof itself, and the structure and validity of the argument!

0.5.3 Direct proof

The calculational proof in Example 0.37 is extremely simple. Many of the proofs we see in this course are much more complicated and will require a good deal of thought. The proof in Example 0.37 is simply a calculation with supporting reasons for each step. A ***direct*** proof requires a simple but logical argument. To prove the statement "$A \Rightarrow B$" directly, we simply start by assuming statement A, and then build a string of implications (with supporting reasons) until we arrive at the conclusion B. In other words, a direct proof has the general structure

$$A \Rightarrow \cdots \Rightarrow \cdots \Rightarrow B.$$

This is much like the structure of a computational proof, but with implications ("$\Rightarrow$") instead of equality signs ("$=$").

When constructing a proof, it often helps to do some draft work and example computations first. Convince yourself that you *believe* what it is that you are trying to prove. Then ask yourself *why* you believe it! Often the intuitive reasons you come up with can become the backbone of a proof. It is also often helpful to look at previous theorems or definitions to see whether you can use them in your proof.

As a simple example, we can prove that every real number that is divisible by 10 (that is, any real number x where $\frac{x}{10}$ is an integer) must be an even number.

EXAMPLE 0.38

Writing a direct proof

Prove that any real number that is divisible by 10 must be even.
Solution   Again we begin by clearly writing out what we are given, and what it is that we are trying to show, in mathematical notation.

   Given: x is a real number that is divisible by 10.

   Show: x is even.

You might start by trying a few examples: 100 is divisible by 10, and indeed 100 is an even number; 20 is divisible by 10 and indeed it is even, and so on. After a while you might be reasonably convinced that the statement is true. However, you've been asked to prove it is true for *all* real numbers that are divisible by 10, and no matter how many examples you write down, you won't have shown that it is true for *all* real numbers.

Why is this happening? Why is it that being divisible by 10 means that a number must be even? One possible answer is that if a number is divisible by 10, then it must also be divisible by 2 (and thus must be even). How can we argue this using facts that we already know? One way is as follows:

PROOF   Suppose that x is a real number that is divisible by 10. This means that $\frac{x}{10}$ is an integer. Since $\frac{x}{10}$ is an integer, the number $5(\frac{x}{10}) = \frac{x}{2}$ must also be an integer (since the product of two integers is always an integer). Therefore, x must be divisible by 2, which is exactly what it means for x to be even. ∎

The proof in Example 0.38 made use of the definition of what it means for a number to be divisible by 10 or 2 and the definition of what it means for a number to be even. The only tricky step in the proof is getting from the fact that $\frac{x}{10}$ is an integer to the fact that $\frac{x}{2}$ is an integer. Notice that words and phrases like "suppose," "this means," "since," and "therefore" appear in this proof and that they help give structure to the argument.

We could also write the proof in a more abstract way, without using complete sentences. A proof should not be like an essay; it should be as brief as possible and as clear and logical as possible. Often a proof is more clear and easier to read when written more abstractly, as in the following proof.

PROOF   Suppose x is any real number.

$$x \text{ is divisible by } 10 \implies \frac{x}{10} \text{ is an integer} \qquad \text{(definition of "divisible")}$$

$$\implies 5\left(\frac{x}{10}\right) \text{ is an integer} \qquad \text{(product of integers is an integer)}$$

$$\implies \frac{x}{2} \text{ is an integer} \qquad (5(\tfrac{x}{10}) = \tfrac{x}{2})$$

$$\implies x \text{ is divisible by } 2 \qquad \text{(definition of "divisible")}$$

$$\implies x \text{ is even.} \qquad \text{(definition of "even")}$$

Thus, if x is divisible by 10, then x must be even. ∎

This proof contains the same argument as the "paragraph style" proof but is written in a more direct, concise way. In this proof the symbol "$\Rightarrow$" is used in place of the word "then" or "therefore" or "thus."

0.5.4 Proof by contradiction

Proof *by contradiction* is perhaps the strangest method of proof, since you start by assuming that what you want to prove is *false* and then show that something ridiculous happens because of it. This shows that your original assumption must have been false, and thus that what you wanted to prove must be true. A **contradiction** is a statement or collection of statements that are logically contradictory. For example, taken together, the statements $x > 2$ and $x < 2$ are contradictory, since they cannot both be true at the same time.

Suppose you wish to use "proof by contradiction" to show that a statement S is true. You start the proof by supposing that the statement S is *false*. Then you make logical arguments and conclusions that follow from this supposition until you arrive at a contradiction (for example, two logically contradictory statements or a statement that contradicts one of the assumptions of the problem). Since the assumption that S is false led to a contradiction, the statement "S is false" must itself be false. In other words, the statement S must be true!

Proof by contradiction is a good tool for proving theorems for which direct proof is difficult. Usually we use proof by contradiction only as a last resort, if we are unable to come up with a direct proof.

For example, suppose you know that the sum of two rational numbers is a rational number (you'll prove this in the exercises). The next proof uses this fact, and proof by contradiction, to prove that the sum of a rational number and an irrational number must be irrational.

EXAMPLE 0.39

Writing a proof by contradiction

Prove that the sum of a rational number and an irrational number is irrational. (You may use the fact that the sum of two rational numbers is rational.)

Solution   We start by using mathematical notation to write down what we are given and what we are trying to show. We will have to make up some notation, namely a name r for an arbitrary rational number and a name x for an arbitrary irrational number.

Given: r is a rational number and x is an irrational number.

Show: $r + x$ is irrational.

After a few minutes of trying to figure out how to prove this statement, you'll probably realize that it's pretty difficult to *prove* that a given number is irrational. That's because, by definition, a number is irrational if it is a real number that is not rational. We can easily show that a number is rational by writing it in the form $\frac{p}{q}$ for some integers p and q, but to show that a number is irrational, we would have to show that it *cannot* be written in that form, which is more difficult.

Because it is difficult to prove this statement directly, we'll try a proof by contradiction. This means that we will assume the *opposite* of what we wish to show and then argue that something impossible happens given that assumption. The argument will involve showing that if $r + x$ is rational, then so is the sum of $r + x$ and $-r$. This will mean that $(r + x) + (-r) = x$ is a rational number, which contradicts our hypothesis that x is an *irrational* number. We write this proof out carefully as follows:

PROOF   Suppose r is a rational number and x is an irrational number. Seeking a contradiction, suppose that the sum $r + x$ is a *rational* number.

Since r is rational, so is $-r$. We know that the sum of two rational numbers is always rational, and thus, since $r + x$ is assumed to be rational, $(r + x) + (-r)$ must also be rational (it is the sum of two rational numbers). However, since $(r + x) + (-r) = r + x - r = x$, this means that x is a rational number. This contradicts our hypothesis that x is an *irrational* number. Thus our assumption that $r + x$ is rational must have been false. Therefore, $r + x$ must be an irrational number, which is what we were trying to show.

0.5.5 Proof by induction

Inductive proofs are a clever way of proving that a statement holds for all positive integers. For example, consider the following statement P concerning positive integers n:

(1) $$1 + 2 + 3 + \cdots + n = \frac{n(n+1)}{2}.$$

For which positive integers n is this statement true? By trying a few examples (like $n = 2$ or $n = 8$), you might start to believe that it is true for *all* positive integers. Of course, we can't show this statement is true for *all* positive integers just by testing it out for a few examples. Also, since the left-hand side of Equation (1) involves the sum of an unspecified number of integers (it involves a "$\cdots$"), we can't prove this statement by using only algebra. In other words, we can't just simplify or rewrite the left-hand side and work toward the right-hand side (as we would in a strictly "computational" proof). Instead we will prove this statement by **induction,** as follows.

Induction is like climbing a ladder, where each rung of the ladder represents a positive integer (the first rung is $n = 1$, the second rung of the ladder is $n = 2$, and so on). How can you climb the whole ladder? You need two skills to climb the ladder: First, you need to be able to get on the ladder (by stepping on the first rung); second, you need to know how to get from any given step to the next higher step. If you can do these two things, then you can climb the ladder.

Now let's go back to the statement P in Equation (1). The first step in proving P inductively is to show that P is true for some small value of n, like $n = 1$ (this is "getting on the ladder"). It is easy to show that statement P is true for $n = 1$ by evaluating the left and right sides of Equation (1) at $n = 1$ (check this). Now suppose we knew that whenever P is true for some positive integer $n - 1$, then it is also true for the next higher integer n (this is "stepping up one rung"). Since P is true for $n = 1$, it must also be true for $n = 2$; then since P is true for $n = 2$, it must also be true for $n = 3$; and so on for *all* positive integers n. In general, this is called the **Induction Axiom:**

THEOREM 0.19

The Induction Axiom

Suppose P is a statement concerning positive integers n such that

(a) P is true for $n = 1$, and

(b) for all positive integers $n > 1$, if P is true for $n - 1$, then P is also true for n.

Then P is true for all positive integers n.

? *Question* Actually, we don't have to start with $n = 1$ in Theorem 0.19; we could start with $n = 2$, or $n = 5$, or $n = 8$. However, if we started with $n = 5$, then knowing that P is true for n whenever it is true for $n - 1$ would only show that P is true for all positive integers that are greater than or equal to $n = 5$. What would happen if we started with $n = 8$? □

To prove property P by induction, we first show that the statement is true for $n = 1$. The heart of the proof is then showing that:

(2)   *If P is true for $n - 1$, then P is true for n.*

To do this, you must first assume that the statement P is true for $n - 1$. This assumption is called the ***inductive hypothesis.*** Given this assumption, you must then show that P is true for the next integer n. This last step usually involves some algebra, and it *always* involves using the inductive hypothesis. Then, by the Induction Axiom, you will have proved that P is true for *all* positive integers n.

EXAMPLE 0.40

Writing an inductive proof

Use induction to prove that for all positive integers n,

$$1 + 2 + 3 + \cdots + n = \frac{n(n+1)}{2}.$$

Solution   Our inductive proof will contain the following steps: First we will show that the statement is true for $n = 1$; then we will assume that it is true for $n - 1$ and show that, given this fact, it is also true for n; finally, we will refer to the Induction Axiom to state that what we have done proves the statement for all positive integers.

PROOF   First, note that the statement is true for $n = 1$, since in that case the left-hand side of the equation is 1, and the right-hand side of the equation is $\frac{1(2)}{2} = 1$.

Now let n be any positive integer greater than 1, and assume that the statement is true for $n - 1$ (this is the inductive hypothesis):

$$1 + 2 + 3 + \cdots + (n - 1) = \frac{(n-1)((n-1)+1)}{2}.$$

(Note that we replaced each n in the original equation with $n - 1$ to get this statement.) Given this assumption, we wish to show that the statement is true for n, in other words, that

$$1 + 2 + 3 + \cdots + n = \frac{n(n+1)}{2}.$$

Using the inductive hypothesis, we have

$$\begin{aligned}
1 + 2 + 3 + \cdots + n &= 1 + 2 + 3 + \cdots + (n-1) + n \\
&= (1 + 2 + 3 + \cdots + (n-1)) + n \\
&= \frac{(n-1)((n-1)+1)}{2} + n \qquad \text{(inductive hypothesis)} \\
&= \frac{n^2 - n}{2} + n = \frac{n^2 - n + 2n}{2} \\
&= \frac{n^2 + n}{2} = \frac{n(n+1)}{2}.
\end{aligned}$$

Thus, by induction, $1 + 2 + 3 + \cdots + n = \frac{n(n+1)}{2}$ for all positive integers n. ∎

In a sense, an inductive proof is just a calculational proof where you are allowed to assume that the statement you are trying to prove is true for $n - 1$. Then, if you also show that the statement is true for $n = 1$, the Induction Axiom implies that the statement is true for all positive integers n.

Inductive proofs are more complicated than calculational or direct proofs, or even proofs by contradiction, but they *always* have the format used in Example 0.40. In general, an inductive proof will always have the following structure (with the blanks filled in appropriately).

PROOF   The statement is true for $n = 1$, since _____.

Given that n is a positive integer greater than 1, assume the statement is true for $n - 1$; that is, assume that _____.

Given the inductive hypothesis, we now show that the statement is true for n; that is, we show that _____:

(Now prove the statement for n using algebra and the inductive hypothesis)

Therefore, by the Induction Axiom, _____ for all positive integers n.

What you should know after reading Section 0.5

▶ What a "proof" is, and why proofs are important in mathematics (and for you). What is the basic structure of a proof (that is, what should appear at the beginning, the middle, and the end of a proof)?

▶ How to write "calculational" proofs. Remember, you can *never* begin a proof by assuming what it is that you wish to show; you can't prove an algebraic identity by "working on both sides." Why?

▶ How to write direct proofs and proofs by contradiction. What is the logical structure of each of these types of proofs? Be able to write proofs clearly and succinctly, with clear logical arguments.

▶ The Induction Axiom and what it can be used for. How to structure a "proof by induction," and why this structure makes logical sense.

Exercises 0.5

Concepts

0. Read the section and make your own summary of the material. Include whatever you think will help you review the material later on.

1. What is a "proof"? Why are proofs important in mathematics?

2. Consider the statement that $(a + b)^3 = a^3 + 3a^2b + 3ab^2 + b^3$ for all real numbers a and b. This statement is true, and in this problem we examine how to prove (and how not to prove) this statement.

(a) Why is the following a *bad* proof of the statement? (*Note:* None of the algebra is wrong; it's the *style* and *logic* of the proof that are wrong.)

PROOF   For all real numbers a and b,

$$(a + b)^3 = a^3 + 3a^2b + 3ab^2 + b^3$$
$$(a + b)(a + b)(a + b) = a^3 + 3a^2b + 3ab^2 + b^3$$
$$(a + b)(a^2 + 2ab + b^2) = a^3 + 3a^2b + 3ab^2 + b^3$$
$$a^3 + 2a^2b + ab^2 + a^2b$$
$$+ 2ab^2 + b^3 = a^3 + 3a^2b + 3ab^2 + b^3$$
$$a^3 + 3a^2b + 3ab^2 + b^3 = a^3 + 3a^2b + 3ab^2 + b^3.$$

(b) The calculations in the "bad" proof are the same calculations that you must perform to write a "good" proof. Write a *good* proof that $(a + b)^3 = a^3 + 3a^2b + 3ab^2 + b^3$ for all real numbers a and b.

3. Explain in your own words how a proof "by contradiction" works.

4. By giving examples, prove that the sum of two irrational numbers can be either rational or irrational. Why is it okay to prove "by example" here if it is not okay to prove "by example" in general?

5. State the Induction Axiom.

6. Explain in your own words what the Induction Axiom means. You may use an analogy concerning climbing a ladder if it clarifies your description.

7. Suppose a statement P that concerns positive integers n is true for $n = 4$. Suppose also that if P is true for a positive integer $n - 1$, then it is also true for the positive integer n. For which positive integers must P be true?

8. Suppose a statement P that concerns positive integers n is true for $n = 1$. Suppose also that if P is true for a positive integer n, then it is also true for the positive integer $n + 1$. For which positive integers must P be true? (*Note:* Some people take this to be the Induction Axiom; it is equivalent to the Induction Axiom we are using in this book.)

9. Suppose a statement P that concerns positive integers n is true for $n = 3$. Suppose also that if P is true for a positive integer n, it is also true for the positive integer $n + 2$. For which positive integers must P be true?

10. Suppose a statement P that concerns positive integers n is true for $n = 2$. Suppose also that if P is true for a positive integer $n - 1$, then P is also true for the positive integer $2(n - 1)$. For which positive integers must P be true?

■ **Verify that each of the following equations is true for $n = 1$, $n = 4$, and $n = 9$.**

11. $1 + 2 + 3 + \cdots + n = \frac{n(n+1)}{2}$

12. $2 + 4 + 6 + \cdots + 2n = n(n + 1)$

13. $1 + 3 + 5 + \cdots + (2n - 1) = n^2$

14. $1^2 + 2^2 + 3^2 + \cdots + n^2 = \frac{n(n+1)(2n+1)}{6}$

■ **Each of the following statements is true for every positive integer n (you will show this later on in the exercises). Write down what each of these statements says about a number $n - 1$.**

15. $1 + 2 + 3 + \cdots + n = \frac{n(n+1)}{2}$

16. $2 + 4 + 6 + \cdots + 2n = n(n + 1)$

17. $1 + 3 + 5 + \cdots + (2n - 1) = n^2$

18. $1^2 + 2^2 + 3^2 + \cdots + n^2 = \frac{n(n+1)(2n+1)}{6}$

Proofs

19. Suppose that A, B, and C are statements such that A is true, (Not C) $\Rightarrow$ B, and $A \Rightarrow$ (Not B). Prove that, under these conditions, statement C must be true.

20. Suppose that A, B, C, and D are statements such that (A or B) $\Rightarrow$ (Not C), (Not D) $\Rightarrow$ B, and C is true. Prove that D must be true.

21. Suppose that A, B, C, and D are statements such that $A \Rightarrow$ (B and C), $D \Rightarrow$ (Not B), A is true, and $C \Rightarrow D$. Show that this combination of statements produces a contradiction.

22. Suppose that A, B, and C are statements such that C is true, $C \Rightarrow$ (Not B), and (Not A) $\Rightarrow$ B.
    (a) Prove *directly* that statement A must be true.
    (b) Prove *by contradiction* that statement A must be true.

23. Prove that the sum of any two rational numbers is rational.

24. Prove that if x is irrational and r is rational, then the difference $x - r$ must be an irrational number.

25. Prove that $a^2 - b^2 = (a - b)(a + b)$ for all real numbers a and b.

26. Prove that $|b - a| = |a - b|$ for any real numbers a and b.

27. The Pythagorean Theorem states that if a right triangle has legs of length a and b and a hypotenuse of length c, then $a^2 + b^2 = c^2$. Use the Pythagorean Theorem and the definition of the distance between two real numbers to prove that the distance between any two points $P = (x_1, y_1)$ and $Q = (x_2, y_2)$ in the plane is given by the "distance formula" $\sqrt{(x_2 - x_1)^2 + (y_2 - y_1)^2}$. (*Hint:* Draw an example of two points P and Q in the plane, label their coordinates, and use an appropriate right triangle.)

28. Prove that the midpoint $(\frac{x_1+x_2}{2}, \frac{y_1+y_2}{2})$ between the points $P = (x_1, y_1)$ and $Q = (x_2, y_2)$ is equidistant from P and Q.

29. Prove that the numbers $x = \frac{-b+\sqrt{b^2-4ac}}{2a}$ and $x = \frac{-b-\sqrt{b^2-4ac}}{2a}$ found by the quadratic formula are solutions to the quadratic equation $ax^2 + bx + c = 0$.

30. Use the fact that $(\frac{a}{b})(\frac{c}{d}) = \frac{ac}{bd}$ if b and d are nonzero to prove that $c(\frac{a}{b}) = \frac{ac}{b}$ if b is nonzero.

31. Use the fact that $(\frac{a}{b})/(\frac{c}{d}) = \frac{ad}{bc}$ if b, c, and d are nonzero to prove that:
    (a) $\dfrac{\left(\frac{a}{b}\right)}{c} = \dfrac{a}{bc}$, if b and c are nonzero.
    (b) $\dfrac{a}{\left(\frac{b}{c}\right)} = \dfrac{ac}{b}$, if b and c are nonzero.

32. Prove that if b and d are nonzero, then $\frac{a}{b} = \frac{c}{d}$ if and only if $ad = bc$. (*Hint:* Use algebra to set the equation equal to zero, combine fractions, and then use what you know about when a quotient is equal to zero.)

33. Prove the triangle inequality ($|a + b| \le |a| + |b|$ for any real numbers a and b) by following the steps below.
    (a) Argue that for any real number x, $|x| = \sqrt{x^2}$.
    (b) Show that $(a + b)^2 \le (|a| + |b|)^2$. (*Hint:* Start on the left-hand side, multiply out the expression, and use the fact that $a \le |a|$ and $b \le |b|$.)
    (c) Take the square root of both sides of the inequality from part (b) (this is legal because both sides are positive), and use part (a), to show that $|a + b| \le |a| + |b|$.

34. Use the triangle inequality to prove that $\left| \dfrac{2x}{x + 3} \right| \ge 2\dfrac{|x|}{|x| + 3}$.

■ **Use the triangle inequality to prove the following three inequalities, for any real numbers a and b.**

35. $|a - b| \le |a| + |b|$

36. $|a - b| \ge |a| - |b|$

37. $|a - b| \ge ||a| - |b||$

■ Use the definition of absolute value, and systems of inequalities, to prove that for any real number A and any positive real number δ:

38. $|A| < \delta$ if and only if $A < \delta$ and $A > -\delta$.

39. $|A| > \delta$ if and only if $A > \delta$ or $A < -\delta$.

40. $|A| \leq \delta$ if and only if $A \leq \delta$ and $A \geq -\delta$.

41. $|A| \geq \delta$ if and only if $A \geq \delta$ or $A \leq -\delta$.

■ Use the definition of absolute value, and systems of inequalities, to prove that for any real numbers x and c and any positive real number δ:

42. $|x - c| < \delta$ if and only if $x \in (c - \delta, c + \delta)$.

43. $|x - c| > \delta$ if and only if $x \in (-\infty, c - \delta) \cup (c + \delta, \infty)$.

44. $|x - c| \leq \delta$ if and only if $x \in [c - \delta, c + \delta]$.

45. $|x - c| \geq \delta$ if and only if $x \in (-\infty, c - \delta] \cup [c + \delta, \infty)$.

■ Use induction to prove that each of the following statements is true for all positive integers n.

46. $1 + 2 + 3 + \cdots + n = \frac{n(n+1)}{2}$

47. $2 + 4 + 6 + \cdots + 2n = n(n + 1)$

48. $1 + 3 + 5 + \cdots + (2n - 1) = n^2$

49. $1^2 + 2^2 + 3^2 + \cdots + n^2 = \frac{n(n+1)(2n+1)}{6}$

50. $1^3 + 2^3 + 3^3 + \cdots + n^3 = \frac{n^2(n+1)^2}{4}$

51. $1^3 + 2^3 + 3^3 + \cdots + n^3 = (1 + 2 + 3 + \cdots + n)^2$

CHAPTER **1**

Functions

Islamic tiles. © Gerard Degeorge/Corbis

A *function is a relation between two or more variables or sets of data. Functions can be represented by formulas, tables, diagrams, graphs, or words.*

In this chapter we will learn what functions are and investigate their properties and graphs. We will explore linear, proportional, and piecewise functions and develop a basic library of functions that will serve as examples throughout the chapter and the text. We will see how to combine functions to make new ones and then investigate the properties of these combination functions. Finally, we will discuss symmetry and inverses of functions.

CONTENTS

1.1   What Is a Function?

Calculus is the study of ***functions*** and how they behave. A function is a special kind of rule that describes a relationship between two variables. Such relationships are often described with equations but can also be defined using tables, diagrams, graphs, and verbal descriptions. In this section we will present the formal definition of a function, the mathematical notation of functions, and the concepts of domain and range. We will then discuss the two simplest types of functions: constant functions and identity functions.

1.1.1 Definition of function

What is a function? You may have seen functions in a previous math class, and if so, perhaps you can even give some examples of functions. In this section we will present a precise, mathematical definition of functions.

? *Question*   If you have worked with functions before, try to write your own definition of a function before reading Definition 1.1.   □

DEFINITION 1.1

Function, Domain, and Target

Suppose A and B are sets. A ***function*** f from A to B is a rule that assigns to *each* element of A a *unique* element of B. The set A is called the ***domain*** of f. The set B is called the ***target.***

In Definition 1.1, "unique" means *exactly one;* to each element of the domain A we assign *one and only one* element of the target B. We will often use the notation Domain(f) and Target(f) to represent the domain set and target set, respectively, of a function f.

EXAMPLE 1.1

A simple function from $\mathbb{R}$ to $\mathbb{R}$

Let A and B both be the set of all real numbers $\mathbb{R}$. Let f be the rule assigning to each real number its square. Since *each* real number has *exactly one* square, f is a function from the set of real numbers to the set of real numbers.   □

EXAMPLE 1.2

A real-world example of a function

Let P be the set of all people, and let W be the set of all women who have ever lived. Then the rule m assigning to each person in the world his or her biological mother is a function from the domain P to the target W, since *each* person has *exactly one* biological mother. More than one person may have the same mother, and some women are not mothers; however, these facts do not keep the rule m from being a function from P to W.   □

EXAMPLE 1.3   **Two rules that are not functions**

If P and W are the sets described in Example 1.2, then the rule g assigning to each person his or her biological grandmothers is *not* a function, since each person has *two* biological grandmothers. This rule violates the "unique" part of Definition 1.1. Additionally, the rule s assigning to each person his or her oldest sister is *not* a function from P to W, since some people do not have any sisters. Since there are some elements of P to which no element of W can be assigned, this rule violates the "each" part of Definition 1.1.   □

Caution   In this text we will assume that the domain and target sets of a function are subsets of the real numbers unless we specify otherwise (as we did in Examples 1.2 and 1.3).   □

1.1.2 Function notation

The following notation expresses the concept of a function assigning elements of a domain set A to elements of a target set B.

NOTATION 1.1

Function Notation with Sets

A function f from A to B can be mathematically described using the compact notation

$$f: A \to B.$$

EXAMPLE 1.4   **Using function notation with sets**

The squaring function described in Example 1.1 is a function $f: \mathbb{R} \to \mathbb{R}$. The "mother" function m from Example 1.2 is a function $m: P \to W$. We could also write the function as

m: {people in the world} $\to$ {all women who have ever lived}.   □

Caution   The notation "$f: A \to B$" does not tell you what the "rule" of the function is; it tells you only what the domain and target of the function are. To describe the "rule" of a function, we need another kind of notation.   □

NOTATION 1.2

Function Notation with Variables

If x is a variable representing elements of a set A, and f is a function from A to B, then we write $f(x)$ for the element of B that is assigned to x under the rule f.

By Definition 1.1, given *any* $x \in \text{Domain}(f)$, $f(x)$ always represents a *unique* element of Target(f).

Caution   The notation $f(x)$ does *not* represent "f multiplied by x." When we write $f(x)$, we mean that the function f is *applied* to x, that is, x is the input of the function f. We say "f of x" when we write $f(x)$.   □

Caution   The notation $f(x)$ tells you the rule for the function but does not explicitly tell you the domain and target of the function.   □

EXAMPLE 1.5

Using function notation with variables

For example, the squaring function $f: \mathbb{R} \to \mathbb{R}$ can be described in the notation $f(x) = x^2$. Each element x of the domain set $\mathbb{R}$ is assigned to the unique real number $f(x) = x^2$. □

Although it is common for a function to be described in terms of an equation, not all functions can be described this way.

EXAMPLE 1.6

A real-world example using function notation

The "mother" function m from Example 1.2 must be described verbally rather than with an equation:

$$m(x) = \text{biological mother of person } x.$$ □

We can think of a function as a "black box," where inputs x are fed into the function "box," and outputs $f(x)$ come out the other side (see Figure 1.1).

Figure 1.1   **Figure 1.2**

By the definition of function, we can put any element of the domain into the function box, and everything that comes out of the function box will be an element of the target, determined by the rule that defines the function f. In the squaring example where $f(x) = x^2$, if $x = 3$ is put into the function box, then $f(3) = 3^2 = 9$ will come out. In this example, the value $x = 3$ is the ***input*** of the function and $f(3) = 9$ is the ***output*** (see Figure 1.2).

EXAMPLE 1.7

Evaluating a function

If $f(x) = x^2 + x + 1$, find $f(2)$, $f(-1)$, $f(a)$, $f(3x + 1)$, and $f(1 - x^3)$.

Solution   Whatever appears inside the parentheses is the "input" for f and takes the place of the x in the equation $f(x) = x^2 + x + 1$. We can write this visually as

$$f(\square) = (\square)^2 + \square + 1;$$

whatever goes into the "input box" gets put into each of the boxes in the expression. Thus we can calculate

$$f(2) = (2)^2 + 2 + 1 = 4 + 2 + 1 = 7$$
$$f(-1) = (-1)^2 + (-1) + 1 = 1 - 1 + 1 = 1$$
$$f(a) = (a)^2 + a + 1$$
$$f(3x + 1) = (3x + 1)^2 + (3x + 1) + 1$$
$$= 9x^2 + 6x + 1 + 3x + 1 + 1 = 9x^2 + 9x + 3$$
$$f(1 - x^3) = (1 - x^3)^2 + (1 - x^3) + 1$$
$$= 1 - 2x^3 + x^6 + 1 - x^3 + 1 = x^6 - 3x^3 + 3.$$ □

Sometimes it is convenient to have variable names for both the domain set and the target set of a function, as in Notation 1.3.

NOTATION 1.3

Function Notation with Input and Output Variables

Given a function $f: A \to B$, let x be a variable representing elements of the domain A, and let y be a variable representing elements of the target B. We can denote the function f with the notation

$$y = f(x).$$

We can also say simply that **y is a function of x** or use the notation $y(x)$ to represent the function $y = f(x)$.

DEFINITION 1.2

Independent and Dependent Variables

Given a function $y = f(x)$, the variable x is called the ***independent variable*** and the variable y is called the ***dependent variable.***

The variable y is called the *dependent* variable because its value *depends* on the value of the variable x.

❖ *Caution*   Although x and y are the letters most commonly used to represent the independent and dependent variables of a function, they are not the only letters that can be used. Don't be afraid to use other letters to represent the variables, especially in real-world problems (see Example 1.8).   □

EXAMPLE 1.8   **A real-world example with independent and dependent variables**

Alina's height can be thought of as a function of her age. The independent variable is Alina's age (in years), and the dependent variable is her height (in feet). Given an age t as "input," the function f would produce the height h that Alina was at age t. We can express the fact that height h is a function of age t by writing $h = f(t)$, or simply $h(t)$. Note that Alina's height h (the dependent variable) *depends* on the value of her age t.   □

1.1.3 Functions defined by tables and diagrams

So far we have seen functions defined by verbal descriptions and equations. Functions can also be described by tables, diagrams, and graphs. We will put off a discussion of graphs until Section 1.2 and discuss tables and diagrams here.

EXAMPLE 1.9   **Describing a function with a list**

Consider a function f from the domain $\{1, 2, 3, 4\}$ to the target $\{2, 5, 6, 9\}$. To describe such a function, we need to determine *how* the function f assigns to each element of the domain a unique element of the target. Suppose we let $f: \{1, 2, 3, 4\} \to \{2, 5, 6, 9\}$ be the function defined by

(1) $$f(1) = 5, \quad f(2) = 2, \quad f(3) = 9, \quad f(4) = 9.$$

Since we have described a *unique* output in the target $\{2, 5, 6, 9\}$ for *each* input from the domain $\{1, 2, 3, 4\}$, we have defined a function.   □

⊗ *Caution* Although the output 9 in the target set is assigned to *two* different elements of the domain (since $f(3) = 9$ and $f(4) = 9$), the rule described in this example *is* a function. Although two "inputs" have the same "output," each individual "input" produces a *unique* "output." If you put the input $x = 3$ into the function f, you get one and only one output (namely 9), and if you put the input $x = 4$ into the function f, you get one and only one output (which happens also to be 9). □

EXAMPLE 1.10

Describing a function with a table

We can describe the function $f(x)$ defined in Example 1.9 a bit more efficiently by arranging the values in a table (Table 1.1).

x	1	2	3	4
$f(x)$	5	2	9	9

Table 1.1 □

EXAMPLE 1.11

Describing a function with a diagram

Alternatively, we could describe the function from Example 1.9 schematically using the diagram in Figure 1.3.

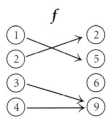

Figure 1.3

Figure 1.3 indicates that $x = 1$ is sent to $y = 5$ (i.e., that the input $x = 1$ is assigned to the output $y = 5$), $x = 2$ is sent to $y = 2$, and $x = 3$ and $x = 4$ are both sent to $y = 9$ by the rule f. This diagram describes the same function we described with the list and the table previously; in general, we will use whatever method of description is most convenient. □

Function diagrams provide a good way to examine the definition of function more closely. For example, consider the three rules assigning elements of {1, 2, 3, 4} to elements of {2, 5, 6, 9} shown in Figures 1.4, 1.5, and 1.6.

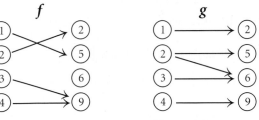 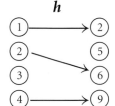

Figure 1.4   **Figure 1.5**   **Figure 1.6**

EXAMPLE 1.12

Determining when a diagram describes a function

Determine which of the rules described in Figures 1.4–1.6 are functions and which fail to be functions.

Solution   The rule f defined by the diagram in Figure 1.4 is a function; *every* element of the domain $\{1, 2, 3, 4\}$ is assigned to *exactly one* element of the target $\{2, 5, 6, 9\}$.

The rule g shown in Figure 1.5 is *not* a function; one of the elements in the domain ($x = 2$) is sent to *two* elements of the target ($y = 5$ and $y = 6$).

The rule shown in Figure 1.6 is *not* a function; one of the elements of the domain ($x = 3$) is not assigned to anything in the target, so h does not define a rule for *every* element in its domain. (However, if we instead considered h as a rule from the restricted domain $\{1, 2, 4\}$ to the target $\{2, 5, 6, 9\}$, it would be a function.)   □

1.1.4 Domain and range

When the rule of a function is written as an equation, the domain of the function is not explicitly stated. We then assume that the domain is the largest subset of the real numbers for which the function is defined.

EXAMPLE 1.13   **Division by zero and negative numbers under square roots**

The function $f(x) = \frac{1}{x}$ is defined only for *nonzero* real numbers x, since division by zero is undefined. Therefore, the domain of the function $f(x) = \frac{1}{x}$ is the set $(-\infty, 0) \cup (0, \infty)$. As another example, consider the function $g(x) = \sqrt{x}$. Since we cannot take the square root of a negative number, $g(x)$ is defined only for nonnegative values of x; in other words, $\text{Domain}(g) = [0, \infty)$.   □

Example 1.13 illustrates the *only* two types of domain problems we will run into at this point: division by zero, and negative numbers under square roots. Later on, as we discuss more complicated types of functions, we will have to consider more complicated domain restrictions (such as negative numbers under even roots or negative numbers inside logarithms).

EXAMPLE 1.14   **Finding domains**

Find the domains of the following functions.

(a) $f(x) = \dfrac{3x}{x + 1}$ (b) $f(x) = 3\sqrt{x - 5}$

Solution

(a) Since division by zero is undefined, $f(x)$ is defined only when $x + 1 \neq 0$, that is, for real numbers $x \neq -1$. In interval notation, $\text{Domain}(f) = (-\infty, -1) \cup (-1, \infty)$.

(b) Since we cannot take the square root of a negative number, $f(x) = 3\sqrt{x - 5}$ is defined only when $x - 5 \geq 0$, that is, when $x \geq 5$. In interval notation, $\text{Domain}(f) = [5, \infty)$.   □

EXAMPLE 1.15   **Some elements of the target of a function may not be "hit"**

Consider the function $f : \{1, 2, 3, 4\} \to \{2, 5, 6, 9\}$ defined by the diagram in Figure 1.7. This function f is defined for each element in its domain, but not every element of the target set is an output of the function. The number $y = 6$ is an element of the target but is not an output of the function, because $y = 6$ is not assigned to any element of the domain.

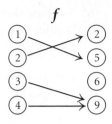

Figure 1.7

The subset of the target that is actually "hit" by the function f is called the ***range*** of the function.

DEFINITION 1.3

The Range of a Function

Given a function $y = f(x)$, the ***range*** of f is the set of elements of Target(f) that are outputs of f. In set notation,

$$\text{Range}(f) := \{y \in \text{Target}(f) \mid y = f(x) \text{ for some } x \in \text{Domain}(f)\}.$$

❓ *Question* Make sure that you examine and think about the definition of range (in set notation) until you truly understand what it means. Can you "read" the set notation out loud? Describe the range of a function in your own words (while being as mathematically precise as you can).

EXAMPLE 1.16 **Examining the range of a function**

In Figure 1.7, $y = 6$ is not in the range of f because there is no number x in the domain that is sent to $y = 6$ by f. In other words, $f(x) \neq 6$ for any x in the domain $\{1, 2, 3, 4\}$. On the other hand, 2, 5, and 9 are all in the range of f.

For the squaring function $f(x) = x^2$, we gave the target as $\mathbb{R}$. However, not everything in this target is in the range of $f(x) = x^2$. In particular, because the square of any real number is nonnegative, no negative number can be in the range of the squaring function.

EXAMPLE 1.17 **Testing whether elements are in the range, and finding the range**

Consider the function $f(x) = x^2$.

(a) Show that $y = -4$ is not in the range of $f(x)$.

(b) Show that $y = 5$ is in the range of $f(x)$.

(c) Show that the range of the function $f(x) = x^2$ is $[0, \infty)$.

Solution

(a) To see whether $y = -4$ is the range of f we must see whether any number x in the domain of f "hits" $y = -4$. However, there is no real number x for which $x^2 = -4$ (since the square of any real number is nonnegative). The number $y = -4$ is not in the range of $f(x) = x^2$ because there is no number x in $\mathbb{R}$ (the domain of f) for which $f(x) = -4$.

(b) To show that $y = 5$ is in the range of $f(x) = x^2$ we must show that there is some real number x for which $x^2 = 5$. In other words, we must find an element of the domain $\mathbb{R}$ that is sent to $y = 5$ by f. Choose $x = \sqrt{5}$; then $f(\sqrt{5}) = (\sqrt{5})^2 = 5$, and thus $y = 5$ is in the range of $f(x) = x^2$. (We could also have observed that $f(-\sqrt{5}) = 5$.)

(c) By Definition 1.3, the range of the function $f(x) = x^2$ is the set

$$\{y \in \mathbb{R} \mid y = x^2 \text{ for some } x \in \mathbb{R}\}.$$

A real number y is in the range of $f(x)$ if and only if there is some real number x such that $x^2 = y$. In general, given any nonnegative number y, we can choose $x = \sqrt{y}$, and with this choice we will have $f(x) = x^2 = (\sqrt{y})^2 = y$. Thus any nonnegative number y is in the range of f. On the other hand, no negative number can be in the range, since the square of a real number can never be negative. Therefore, the domain of $f(x) = x^2$ is the set $[0, \infty)$. □

❖ Caution   Now that we know that the range of $f(x) = x^2$ is $[0, \infty)$, we can think of f as a function $f : \mathbb{R} \to [0, \infty)$ rather than a function $f : \mathbb{R} \to \mathbb{R}$. Note that, in general, we cannot assume that the target set is the range of the function. □

1.1.5 The two most basic functions

Perhaps the simplest kind of function is a ***constant function***, whose output is always the same, regardless of the input.

DEFINITION 1.4

Constant Function

A ***constant function*** is a function of the form $f(x) = c$. The range of a constant function is the set $\{c\}$ that consists of one element (the value c).

EXAMPLE 1.18   **A constant function from $\mathbb{R}$ to $\mathbb{R}$**

The function $f : \mathbb{R} \to \mathbb{R}$ defined by the equation $f(x) = 3$ is a constant function. No matter what we put in for x, the output $f(x)$ will always be equal to 3: $f(10) = 3$, $f(-1.67) = 3$, $f(0) = 3$, and so on. The domain of f is all real numbers. The range is the set $\{3\}$, consisting only of the number 3. □

EXAMPLE 1.19   **Real-world examples of constant functions**

Let P be the set of people in your mathematics class. Construct two constant functions $g : P \to P$ and $h : P \to P$.

Solution   A constant function g must assign to *each* element of the domain P (i.e., to each person in your class) *exactly one* element of the target P (i.e., exactly one person in your class). One way to do this is as follows: For each input (person in the class), define the output to be yourself (since you are a person in your class).

To construct a second constant function h, suppose that Linda is a person in your class. You could define a constant function $h : P \to P$ whose output is always Linda, regardless of the input. □

You can visualize the constant functions from Example 1.19 by picturing everyone in your class pointing at his or her "output." For the constant function g, everyone in your class (including yourself) would point to you. The domain of this function is the set of people in your class; the range is the set containing only one student, namely you. For the constant function h, everyone in your class (including Linda) would point to Linda. The range of h is the set containing only the element "Linda."

> **? Question**   Describe a constant function from the set of all cars to the set of all colors. Your constant function does *not* have to make "real-life" sense; it just has to be a function. □

Another simple, but extremely important, type of function is the *identity function*.

DEFINITION 1.5

Identity Function

The function $f\colon A \to A$ that sends each element of the domain to itself is called the ***identity function*** for the set A. If f is the identity function, then $f(x) = x$ for every element x in the domain.

> **⊗ Caution**   A function f can be an identity function only if the domain and range of f are the same. □

If $f\colon A \to A$ is the identity function, then f sends each element x of A to itself. In the "function box" model, if you put an element of A into the function box, it just comes right out the other side as itself. We call such a function an "identity" function because the output is "identical" to the input.

EXAMPLE 1.20

The identity function from $\mathbb{R}$ to $\mathbb{R}$

The function $f\colon \mathbb{R} \to \mathbb{R}$ defined by $f(x) = x$ is the identity function for $\mathbb{R}$. Each real number is sent to itself under this function; for example, $f(3) = 3$, $f(1.599) = 1.599$, $f(-750) = -750$, and so on. The domain and range of the identity function $f(x) = x$ are both $\mathbb{R}$. □

EXAMPLE 1.21

A real-world example of an identity function

The identity function for the set $P = \{\text{people in your class}\}$ is the function that assigns to each person in your class himself or herself. Given any person in the domain P as an "input," the identity function returns that same person as the "output." To visualize this function, picture all the people in your class pointing at themselves. The domain of this function is P and the range of this function is P. □

> **? Question**   Suppose $P = \{\text{people in your class}\}$. Can you think of another identity function from the set P to the set P? Or is there only one? □

Constant and identity functions are so simple that they are what mathematicians call "trivial." This does not mean that they are not important, just that they are extremely *simple* examples of functions. Despite (or, more accurately, because of) the fact that constant functions and identity functions are so basic, they will be important examples and building blocks in future sections. In Section 1.7, the identity function will be used to define inverse functions. In Chapter 3, constant functions will be used to introduce important facts about differentiation.

> ## What you should know after reading Section 1.1
>
> ▶ The definition of a function and function notation. Be able to construct and identify examples of relationships that are functions and relationships that are not.
>
> ▶ How to work with functions defined by words, formulas, tables, or diagrams.
>
> ▶ How to find the domain and range of a function. In particular, be able to describe the range of a function in set notation.
>
> ▶ How to recognize and construct constant and identity functions.

Exercises 1.1

Concepts

0. Read the section and make your own summary of the material. Include whatever you think will help you review the material later on.

1. State the mathematical definition of a function. Then describe (in your own words) what this definition means. To support your description, give an example of a relationship that is a function and an example of a relationship that is not a function.

2. If $u = w(v)$ describes a function, which letter represents the independent variable? Which letter represents the dependent variable? Which letter represents the "name" of the function itself?

3. Suppose P is the set of people alive today, and C is the set of possible eye colors. Let $f: P \to C$ be the rule that assigns to each person his or her eye color. Is f a function? Why or why not? (*Note:* There may be more than one way to answer this question!)

4. Give an example of a function $f: \mathbb{R} \to \mathbb{R}$ that you can define with an equation $y = f(x)$. Then give an example of an equation that does *not* define a function from the real numbers to the real numbers.

5. Use set notation to write down the mathematical definition of the "range" of a function. Then use the same notation to express the range of the function $f(x) = x^2$.

6. Is the range of a function always equal to the target of that function? Why or why not? Support your answer with an example.

■ Let P be the set of all people living in the United States, and let S be the set of all U.S. states. Each of Problems 7–10 specifies either P or S as a domain set and either P or S as a target set. Complete parts (a)–(d) for each of these problems.

    (a) Define a function f with the indicated domain and target. Then explain why the rule you describe is a function, and find the range of your function.

    (b) Describe a rule between the given domain and target that is *not* a function, and explain why your rule fails to be a function.

    (c) Describe a constant function from the given domain to the given target.

    (d) Describe the identity function from the given domain to the given target (if possible).

7. $f: \{\text{people in U.S.}\} \to \{\text{U.S. states}\}$

8. $f: \{\text{U.S. states}\} \to \{\text{people in U.S.}\}$

9. $f: \{\text{people in U.S.}\} \to \{\text{people in U.S.}\}$

10. $f: \{\text{U.S. states}\} \to \{\text{U.S. states}\}$

■ Let $A = \{2, 4, 6, 8, 10\}$ and $B = \{1, 2, 3, 4\}$. Each of Problems 11–14 specifies either A or B as a domain set and either A or B as a target set. Complete parts (a)–(c) for each of these problems.

    (a) Use a table to define a function f with the indicated domain and target. Then represent this same function in a diagram (like the one in Figure 1.3). What is the range of your function?

    (b) Use a table to define a rule from the given domain to the given target that is *not* a function because it violates the "each" part of Definition 1.1. Also represent your function using a diagram.

    (c) Use a table to define a rule from the given domain to the given target that is *not* a function because it violates the "unique" part of Definition 1.1. Also represent your function using a diagram.

11. $f: A \to B$ **12.** $f: B \to A$

13. $f: A \to A$ **14.** $f: B \to B$

15. The following questions involve the range of the function $f(x) = x^2 + 1$.

    (a) Explain why $y = 5$ is in the range of $f(x)$.

    (b) Explain why $y = 0$ is *not* in the range of $f(x)$.

    (c) Argue that the range of $f(x) = x^2 + 1$ is the set $[1, \infty)$.

16. The following questions involve the range of the function $f(x) = \sqrt{x-2}$.
   (a) Explain why $y = 3$ is in the range of $f(x)$.
   (b) Explain why $y = -1$ is *not* in the range of $f(x)$.
   (c) Argue that the range of $f(x) = \sqrt{x-2}$ is the set $[0, \infty)$.

17. Let P be the set of all people living in the United States. Describe (a) a constant function $f: P \to P$, and (b) the identity function $g: P \to P$. What is the range of each of these functions?

18. If P is the set of all people living in the United States, and $\mathbb{R}$ is the set of all real numbers, describe a constant function $f: P \to \mathbb{R}$. What is the range of your constant function? Is there an identity function with domain P and target $\mathbb{R}$? Why or why not?

19. If $f: A \to B$ is a constant function, is the domain set A necessarily the same as the target set B? If $f: A \to B$ is an identity function, is the domain set A necessarily the same as the target set B?

Skills

■ Find the domains of the following functions.

20. $f(x) = \dfrac{1}{x-8}$

21. $f(x) = \sqrt{3x+1}$

22. $f(x) = \dfrac{1}{\sqrt{3x+1}}$

23. $f(x) = \sqrt{(x-1)(2x+3)}$

24. $f(x) = \dfrac{1}{(x-1)(2x+3)}$

25. $f(x) = \dfrac{\sqrt{x}}{3x-5}$

26. $f(x) = \dfrac{1}{\sqrt{(x-1)(2x+3)}}$

27. $f(x) = \dfrac{\sqrt{x^2-1}}{x^2-9}$

28. $f(x) = \dfrac{\sqrt{x^2-1}}{\sqrt{x^2-9}}$

29. $f(x) = \sqrt{\dfrac{1}{2x^2-x-3}}$

30. $f(x) = \sqrt{3x^2-5x-2} + \sqrt{x}$

31. $f(x) = \dfrac{1}{\sqrt{x-1}} - \dfrac{\sqrt{x}}{x-2}$

32. If $f(x) = (x+1)^2$, find:
   (a) $f(27)$ (b) $f(1+h)$
   (c) $f(x+h)$ (d) $f((x+1)^2)$

33. If $f(x) = \dfrac{x}{1-x}$, find:
   (a) $f(-2)$ (b) $f(q)$
   (c) $f(a+1)$ (d) $f(x^2+x)$

34. If $f(x) = \dfrac{1}{1-x}$, find:
   (a) $f(2)$ (b) $f(2+h)$
   (c) $\dfrac{f(2+h)-f(2)}{h}$ (d) $f(f(2))$

35. If $f(x) = x^2+1$, find:
   (a) $\dfrac{f(1+h)-f(1)}{h}$ (b) $\dfrac{f(t)-f(1)}{t-1}$
   (c) $f(f(-2)+1)$ (d) $f(f(x))$

Applications

■ Each problem below describes a real-world situation. In each situation, identify and name (with variables) a dependent variable and an independent variable. Find reasonable domain and range sets for each example. Then use function notation to write the dependent variable in terms of the independent variable. (*Note:* There may be more than one correct way to assign the dependent and independent variables in these problems.)

36. The area of a circular tarp depends on the length of its radius.

37. Linda's commute to work in the morning takes 20 minutes if she does not encounter any red lights. Each red light she has to stop at costs her 2 minutes.

38. A federal research study has determined that every time you yell at your pets, you lose 10 hairs from your head.

39. In 1990 there were 25 groundhogs living at the Highland Campgrounds. Each year there are 3 more groundhogs living at the campgrounds.

40. Nathan's starting salary is $12,000 a year, and he will get a raise of $500 each year.

41. The lima-bean-green paint that Alina just has to have costs $1.50 per square foot. How much it costs her to paint the walls of her perfectly cubical dining room this color depends on how high her ceilings are. (*Hint:* You will need to know how to find the surface area of a cubical room.)

1.2 **Graphs of Functions**

By studying the graphs of functions, we can develop an intuition for how they behave. In this section we will develop a vocabulary for describing the graphical behavior of functions. In later chapters we will develop techniques that will help us determine this behavior algebraically.

1.2.1 Graphing functions

A function $y = f(x)$ can be represented as a graph in the xy-plane, as the collection of ordered pairs of the form $(x, f(x))$. For example, the graph of the function $f(x) = x^2$ consists of the points (x, x^2) for all real numbers x. In general,

---

DEFINITION 1.6

The Graph of a Function

The *graph* of a function $y = f(x)$ is a collection of ordered pairs:

$$\text{Graph}(f) := \{(x, f(x)) \mid x \in \text{Domain}(f)\}.$$

We usually represent the graph of f as a picture in the xy-plane by plotting the ordered pairs of the form $(x, f(x))$.

EXAMPLE 1.22   **The graph of $f(x) = x^2$ is the collection of ordered pairs (x, x^2)**

The graph of $f(x) = x^2$ contains the points $(-1, 1)$ and $(2, 4)$, since $f(-1) = 1$ and $f(2) = 4$. On the other hand, the point $(1, 2)$ is *not* on the graph of $f(x) = x^2$ since although $x = 1$ is in the domain of $f(x) = x^2$, $f(1) \neq 2$. Figure 1.8 shows the complete graph of $f(x) = x^2$ and the three ordered pairs discussed above (including one that is *not* part of the graph of $f(x)$).

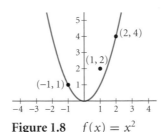

Figure 1.8 $f(x) = x^2$

Since the domain of $f(x) = x^2$ is $\mathbb{R}$, we can describe this graph as the set of ordered pairs $\{(x, x^2) \mid x \in \mathbb{R}\}$.   □

Definition 1.6 basically says that the graph of a function is a picture of its solutions. A *solution* to an equation $y = f(x)$ is an ordered pair (a, b) that makes the equation true; (a, b) is a solution to the equation $y = f(x)$ if and only if $f(a) = b$. Each of these ordered pairs (a, b) is represented as a point on the graph of $f(x)$.

In Chapters 4–7 you will learn how to graph various types of functions. For now you will graph functions with the help of a graphing calculator. Be sure that your calculator has the capacity to graph functions in different graphing windows and has a "zoom" feature.

❖ *Caution*   When you are graphing functions using a calculator, the "window" you choose is very important! You will have to use a bit of trial and error to get an effective window.   □

EXAMPLE 1.23

Choosing a "good" window for a graph

Figures 1.9–1.11 are examples of misleading windows for $f(x) = 0.3(x^3 - 6x^2 - x + 6)$.

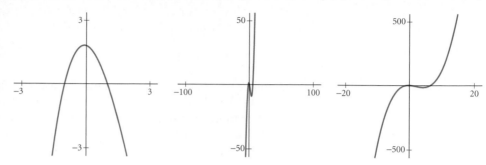

Figure 1.9   **Figure 1.10**   **Figure 1.11**

Compare the graphs in Figures 1.9–1.11 with the graph of f in Figure 1.12. The window used in Figure 1.9 is too small; this graph does not display the right-hand behavior of the function. On the other hand, the windows in Figures 1.10 and 1.11 are large enough to capture the total behavior of f, but they are too large to represent the behavior of f accurately near the origin or the x-intercepts.

Figure 1.12 captures all the features of the graph of f: The local behavior of the graph of f is clear, and the global behavior is accurately represented (the "ends" of the graph keep going in the directions indicated).

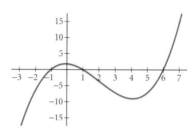

Figure 1.12

For now we will use trial and error to find an effective graphing window. After we have learned more about the behavior of various types of functions, we will be able to be more systematic about choosing a graphing window—and more confident that our graph has captured the true behavior of a function. □

1.2.2 The Vertical Line Test

If $f(x)$ is a function, then by definition *each* value x in the domain of f is sent to *exactly one* value y in the range of f. For each value x in the domain, there can be one and only one value $f(x)$. This means that there is *exactly one* point $(x, f(x))$ on the graph of f corresponding to each x-value in the domain.

EXAMPLE 1.24

A graph that is not a function

The graph shown in Figure 1.13 could *not* be the graph of a function. There are *two* points on the graph that correspond to $x = 2$, namely $(2, 1)$, and $(2, 3)$. In other words, there are *two* outputs corresponding to the input $x = 2$. □

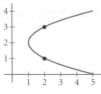

Figure 1.13

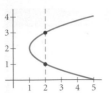

Figure 1.14

Another way to say that the graph in Figure 1.13 has more than one point corresponding to the input value $x = 2$ is to say that a vertical line at $x = 2$ passes through the graph at more than one point (see Figure 1.14). We can determine whether a given graph represents a function by applying this "Vertical Line Test."

THEOREM 1.1

The Vertical Line Test

A graph represents a function if and only if every vertical line passes through the graph at *at most one* point.

If every vertical line passes through at most one point on the graph, we say that the graph "passes" the Vertical Line Test. A graph represents a function if and only if it passes the Vertical Line Test.

⊗ *Caution* The Vertical Line Test states that each vertical line has to hit the graph in *at most* one point. It is okay for a vertical line to pass through the graph at *no* points (i.e., for the line not to intersect the graph). In fact, given a function f, a vertical line at any nondomain point will fail to intersect the graph of f. □

PROOF (THEOREM 1.1) We will first show that the graph of a function f always passes the Vertical Line Test. We will do this by showing that every vertical line $x = c$ intersects the graph of f either at one point or not at all.

Let c be a number in the domain of f. By the definition of function, there is a *unique* value $f(c)$. By the definition of the graph of a function, this means that $(c, f(c))$ is the only point on the graph of f whose x-coordinate is c. Thus the vertical line $x = c$ intersects the graph of f only at *one* point, namely $(c, f(c))$.

On the other hand, let c be a number that is *not* in the domain of f. Then by Definition 1.6 there is *no* point on the graph of $f(x)$ whose x-coordinate is c, so the vertical line $x = c$ will not intersect the graph of $f(x)$ at all.

Now we must show that if a graph passes the Vertical Line Test, then it is the graph of some function f (remember, Theorem 1.1 is an "if-and-only-if" theorem, so we need to prove both directions of the implication). Given that a graph passes the Vertical Line Test, we will construct a function f that will have that graph.

By hypothesis, each vertical line $x = c$ passes through the given graph at one point or not at all. Let the domain of f be the set of all values c for which the vertical line $x = c$ intersects the given graph. At each of these points, define $f(c)$ to be the y-coordinate of the point of intersection between the vertical line $x = c$ and the given graph. Since the graph passes the Vertical Line Test, the relationship $f(x)$ that we have constructed will be a function. ■

❓ *Question* Did you read that proof? It's not hard; in fact, it follows directly from the definition of function and the definition of the graph of a function. Try sketching some graphs while you follow the argument. □

EXAMPLE 1.25   **The graph of a function passes the Vertical Line Test**

The graph of the function $f(x) = \sqrt{x}$ passes the Vertical Line Test; *every* vertical line passes through the graph at one point or not at all. Figure 1.15 shows the graph of $f(x) = \sqrt{x}$ and a number of these vertical lines. Of course, it would be impossible to show *all* the vertical lines $x = c$ because there are infinitely many!

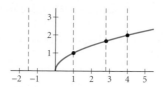

Figure 1.15

Note that the vertical line $x = -1.5$ does not pass through the graph of f at all, since $x = -1.5$ is not in the domain of $f(x) = \sqrt{x}$. On the other hand, any vertical line $x = c$ with $c \geq 0$ will pass through the graph of f at exactly one point, namely $(c, \sqrt{c})$. □

❓ *Question*   Although a graph that passes the Vertical Line Test must represent some function f, it is not always possible to find an equation for that function f. To convince yourself of this, sketch a relatively complicated graph that passes the Vertical Line Test. What function did you just sketch? Although there may be an equation that has the graph you just sketched, it would be difficult to find. On the other hand, it may not be possible to write down an equation for your graph at all! This doesn't mean that your graph isn't a function, just that we don't have a convenient way of writing down that function algebraically. □

1.2.3 Intercepts

The values at which a graph crosses the x-axis or the y-axis are called **intercepts.** The **y-intercept** of the graph of a function f is the height at which the graph crosses the y-axis (if at all). The graph of a function can cross the y-axis at most once (or else it would fail the Vertical Line Test), so every function graph has at most one y-intercept. Since the y-axis is the collection of points in the plane where $x = 0$, the graph of a function always has its y-intercept at $x = 0$, with height $f(0)$ (assuming that the function is defined at $x = 0$).

The **x-intercepts** of a graph are the places where the graph crosses the x-axis. The graph of a function may have many x-intercepts, since it can cross the x-axis many times without violating the Vertical Line Test. On the other hand, the graph might have *no* x-intercepts. Since the x-axis is the collection of points in the plane where $y = 0$, the x-intercepts of a function are the x-values for which $f(x) = 0$. We will also refer to x-intercepts as **roots** and **zeros.**

Table 1.2 summarizes this information. This table gives the formal terminology, a verbal description, and a mathematical description for each type of intercept; be sure you know all three.

Terminology	Verbal Description	Math Description
y-intercept	Height where the graph of f crosses the y-axis	$f(0)$
x-intercept	Places where the graph of f crosses the x-axis	Solutions of $f(x) = 0$

Table 1.2

§ *Caution*   It is very important that you be able to go back and forth among mathematical terminology, verbal descriptions, and math descriptions. The terminology is really a *language* that you must learn in order to read and discuss mathematics. The verbal descriptions tell you what is intuitively "meant" by the terminology. The math descriptions tell you what the terminology means *mathematically*—that is, precisely and rigorously. The math descriptions are what you will use when you need to find, solve for, or prove things about mathematical objects.   □

EXAMPLE 1.26

Finding the *x*-intercepts and *y*-intercepts of a function

Find the *y*-intercept and *x*-intercepts of the function $f(x) = x^2 - 1$.

Solution   The *y*-intercept of $f(x) = x^2 - 1$ is $f(0) = 0^2 - 1 = -1$. The *x*-intercepts of $f(x) = x^2 - 1$ are the values of *x* for which $f(x) = x^2 - 1 = 0$. Solving this equation, we get $x^2 = 1$, or $x = \pm 1$, so the *x*-intercepts of $f(x)$ are at $x = 1$ and $x = -1$. The graph of $f(x) = x^2 - 1$ and its intercepts are shown in Figure 1.16.

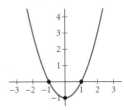

Figure 1.16   □

1.2.4 Positive and negative graphs

Given the graph of a function f, there are three basic characteristics that describe the location and behavior of the graph. First, the graph could be above or below the *x*-axis. Second, the graph could be "moving upward" or "moving downward" as we look from left to right. Third, the graph could have an upward or a downward curvature. We will now give proper terminology and mathematical descriptions of the first of these graphical characteristics.

For simplicity we restrict our attention to a small piece of the graph, say the piece of the graph over some interval *I*. We will say that the graph of a function f is **positive** on *I* if it is above the *x*-axis (i.e., if $f(x) > 0$) for all values *x* in *I*. Similarly, the graph of f is **negative** on *I* if it is below the *x*-axis (i.e., if $f(x) < 0$) for all $x \in I$. A given function might alternate between being positive and negative (or zero) as we look at the graph from left to right.

§ *Caution*   We will use interchangeably such statements as "f is positive" and "the graph of f is positive."   □

EXAMPLE 1.27

Graphically determining where a function is positive or negative

Find the intervals on which the function in Figure 1.17 is positive and the intervals on which it is negative.

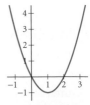

Figure 1.17

Solution By inspecting the graph, we can see that the function is positive on the intervals $(-\infty, 0)$ and $(2, \infty)$ and negative on the interval $(0, 2)$. ☐

◈ *Caution* The solution to Example 1.27 assumes that the given graph is representative of the "global behavior" of the function, and that the "ends" of the graph keep going in the directions shown. Because we will often make assumptions like these, it is especially important that we graph each function in a "window" that is representative of its total behavior. ☐

In Example 1.27, we estimated the intervals on which a function was positive or negative by looking at the graph. We can also use techniques for solving inequalities to find these intervals algebraically.

EXAMPLE 1.28

Algebraically determining where a function is positive or negative

Find the intervals on which the function $f(x) = x^2 - 2x$ is negative and the intervals on which it is positive.

Solution To find the intervals on which $f(x)$ is negative, we must solve the inequality $f(x) < 0$. We follow the method from Section 0.3.2.

$$f(x) < 0$$
$$x^2 - 2x < 0$$
$$x(x - 2) < 0$$

A real number x is a solution to this inequality if it makes the first factor of $x(x-2)$ positive and the second factor negative, or vice versa. In other words, the solutions are

$$\begin{pmatrix} x < 0 & \text{and} & x - 2 > 0 \\ x < 0 & \text{and} & x > 2 \\ & \emptyset & \end{pmatrix} \quad \text{OR} \quad \begin{pmatrix} x > 0 & \text{and} & x - 2 < 0 \\ x > 0 & \text{and} & x < 2 \\ & (0, 2) & \end{pmatrix}$$

Thus $f(x)$ is negative on the interval $(0, 2)$. A similar calculation shows that $f(x)$ is positive on the intervals $(-\infty, 0)$ and $(2, \infty)$. In fact, $f(x) = x^2 - 2x$ is the function graphed previously in Figure 1.17. ☐

◈ *Caution* Whenever possible, find your answers algebraically (as in Example 1.28) and then check them graphically (as in Example 1.27). ☐

For reference, we now summarize our discussion in Table 1.3. Suppose I is an interval and f is a function defined on that interval.

Terminology	Verbal Description	Math Description
f is positive on I	The graph of f is above the x-axis on I	$f(x) > 0$ for all $x \in I$
f is negative on I	The graph of f is below the x-axis on I	$f(x) < 0$ for all $x \in I$

Table 1.3

Recall from Section 0.1 that a number is nonnegative if it is positive or zero. We will use this same terminology here and say that f is ***nonnegative*** on I if $f(x) \geq 0$ for all $x \in I$. Similarly, f is ***nonpositive*** on I if $f(x) \leq 0$ for all $x \in I$.

❓ *Question* Can you make a table like Table 1.3 that describes the terminology *nonnegative* and *nonpositive*?   □

1.2.5 Increasing and decreasing graphs

A function f is said to be **increasing** on an interval I if the graph "moves upward" as we follow the graph from left to right. Mathematically we describe this behavior by saying that $f(b) > f(a)$ whenever $b > a$ (input values farther to the right have higher output values) for all a and b in I. A function f is **decreasing** on I if the graph "moves downward," that is, if $f(b) < f(a)$ whenever $b > a$ for all a and b in I.

EXAMPLE 1.29   **Graphically determining where a function is increasing or decreasing**

The function graphed in Figure 1.17 is decreasing on the interval $(-\infty, 1)$ (the graph "moves down" as we follow it from left to right on this interval of x-values) and increasing on the interval $(1, \infty)$.   □

Showing algebraically that a function is increasing or decreasing on an interval I is a bit more complicated than showing that it is positive or negative. The presentation of derivatives in Chapter 3 will make this process easier.

EXAMPLE 1.30   **Algebraically proving that a function is increasing**

Show that the function $f(x) = 2x + 1$ is increasing on the interval $(-\infty, \infty)$.

Solution   We must show that for all $b > a$ in $(-\infty, \infty)$, we have $f(b) > f(a)$. Starting with $b > a$, we have

$$b > a \quad \Rightarrow \quad 2b > 2a \quad \Rightarrow \quad 2b + 1 > 2a + 1 \quad \Rightarrow \quad f(b) > f(a).\qquad \square$$

Figure 1.18 shows the graph of $f(x) = 2x + 1$. Consider $a = 0$ and $b = 2$ (therefore $b > a$). Then $f(a) = f(0) = 1$ and $f(b) = f(2) = 5$, so indeed we have $f(b) > f(a)$. What we showed in Example 1.30 is that this is true for *all* a and b in $(-\infty, \infty)$.

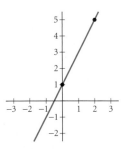

Figure 1.18

The descriptions of what it means for a function f to be increasing or decreasing on an interval I are summarized in Table 1.4.

Terminology	Verbal Description	Math Description
f is increasing on I	The graph of f is "moving up" as we look from left to right on I	$f(b) > f(a)$ for all $b > a$ in I
f is decreasing on I	The graph of f is "moving down" as we look from left to right on I	$f(b) < f(a)$ for all $b > a$ in I

Table 1.4

◈ *Caution*   When we say that a graph is ***increasing at a point*** (or decreasing at a point), we mean that the graph is increasing (or decreasing) in a small interval *around* that point. When we say simply that *f* is an ***increasing function*** (or a decreasing function) we mean that the function *f* is increasing (or decreasing) over its *entire* domain.   ☐

◈ *Caution*   When we want to describe "where" a function is positive, negative, increasing, or decreasing, we describe the set of *x*-values for which the graph of the function has the given property. For example, we would say that the graph in Figure 1.17 is increasing on $(1, \infty)$, since that is the set of *x*-values for which the graph is moving in the upward direction.   ☐

1.2.6 Concavity

The graph of a function *f* is said to be ***concave up*** on an interval *I* if it has an upward curvature (i.e., the "inside" of the curve is on the top) on that interval. A function *f* is ***concave down*** on *I* if it has a downward curvature on *I*. We must postpone the mathematical descriptions of concavity until after we have seen the derivative in Chapter 3. For now we will just recognize concavity graphically. For example, the function graphed in Figure 1.17 (on page 72) is concave up on its entire domain.

EXAMPLE 1.31

Graphically determining where a function is concave up or down

The function graphed in Figure 1.19 is concave down on the interval $(0, 2)$ and concave up on the interval $(2, 4)$.

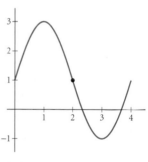

Figure 1.19   ☐

❓ *Question*   In Example 1.31, we used the intervals $(0, 2)$ and $(2, 4)$ instead of the intervals $(-\infty, 2)$ and $(2, \infty)$. Why do you think we did this? Do you think the graph suggests that the "ends" will continue in the directions indicated?   ☐

◈ *Caution*   As Example 1.31 illustrates, it can be difficult to tell visually when a graph changes from upward to downward concavity, or vice versa. In Section 3.8 we will learn how to find exact values for where these changes in concavity occur, but for now we will have to estimate.   ☐

One way to think of curvature is to imagine that a piece of a graph that is concave up will "hold water," whereas a piece that is concave down will not. Another way is to cup your hand to the graph: If you hold your hand palm side up while following the graph, it is concave up; if your hand has to be palm side down to match the curvature of the graph, then the graph is concave down.

1.2.7 Extrema and inflection points

The *extrema* of a graph are the high and low points of the graph. There are four kinds of extrema: global maxima, global minima, local maxima, and local minima.

A *global maximum* of a function f is a point on the graph of f that is at least as high as every other point on the graph. Mathematically, we say that f has a global maximum at $x = c$ if $f(c) \geq f(x)$ for all x in the domain of f. When f has a global maximum at $x = c$, we say that f has a *maximum value* of $f(c)$. Similarly, f has a *global minimum* at the point $x = c$ if $f(c) \leq f(x)$ for all x in the domain of f, and $f(c)$ is called the *minimum value.*

✿ Caution   A function can have more than one global maximum. A function can also have *no* global maximum. Can you draw a function that has two global maxima? A function that has *no* global maximum?   ☐

A function f has a *local maximum* at $x = c$ if the point $(c, f(c))$ is at least as high as every *sufficiently nearby* point on the graph. By "sufficiently nearby" we mean that there is a neighborhood $(c - \delta, c + \delta)$ for some positive number δ such that $f(c) \geq f(x)$ for all x in the interval $(c - \delta, c + \delta)$. Similarly, $x = c$ is a *local minimum* of f if $f(c) \leq f(x)$ for all x in some interval $(c - \delta, c + \delta)$ around c. The local maxima and minima of a function are often called the *local extrema* of the function, or, alternatively, the *turning points* of the function (since the function "turns around" at these points).

❓ Question   Every global maximum or minimum is *also* a local maximum or minimum. Can you show this using the mathematical descriptions of local and global extrema?   ☐

In Chapter 3 we will see how to find the locations of extrema algebraically. For now we will find them graphically.

EXAMPLE 1.32

Graphically determining the locations of local and global extrema

The function f graphed in Figure 1.20 has a local maximum at $x = 1$ with value $f(1) = 3.25$ and local minima at $x = -1$ and $x = 2$ with values $f(-1) = -4.75$ and $f(2) = 2$, respectively. There is a global minimum at $x = -1$ and no global maximum (the function gets infinitely large at the "ends," so there is no point at which the function is highest).

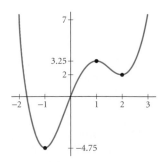

Figure 1.20

What about the δ intervals described in the definition of local extrema? Let's consider the local maximum at $x = 1$. According to the mathematical definition of a local maximum, there is some positive number δ such that $f(1) \geq f(x)$ for all x in the interval $(1 - \delta, 1 + \delta)$. For example, $\delta = 0.5$ works, since $f(1) \geq f(x)$ for all x in the interval $(0.5, 1.5)$.   ☐

? *Question*   Approximate the largest value of δ for which $f(1) \geq f(x)$ for all x in $(1 - \delta, 1 + \delta)$ in Figure 1.20. Then find δ-intervals for the local minima at $x = -1$ and $x = 2$. $\square$

The mathematical and verbal descriptions of extrema are summarized in Table 1.5.

Terminology	Verbal Description	Math Description
f has a global maximum at $x = c$	The point $(c, f(c))$ is the highest point on the graph of f	$f(c) \geq f(x)$ for all x in the domain of f
f has a global minimum at $x = c$	The point $(c, f(c))$ is the lowest point on the graph of f	$f(c) \leq f(x)$ for all x in the domain of f
f has a local maximum at $x = c$	The point $(c, f(c))$ is *locally* the highest point on the graph of f	for some $\delta > 0$, $f(c) \geq f(x)$ for all $x \in (c - \delta, c + \delta)$
f has a local minimum at $x = c$	The point $(c, f(c))$ is *locally* the lowest point on the graph of f	for some $\delta > 0$, $f(c) \leq f(x)$ for all $x \in (c - \delta, c + \delta)$

Table 1.5

✸ *Caution* Remember, we always identify the location of an extremum by its x-value. The y-value of the extremum is called the "value" of the extremum. For example, the graph in Figure 1.20 has a local maximum at $x = 1$, and this local maximum has a value of $f(1) = 3.25$. $\square$

EXAMPLE 1.33   **On a constant function, every point is a global maximum and a global minimum**

For the constant function $f(x) = 2$, *every* point is a global maximum, and *every* point is a global minimum (see Figure 1.21). For example, $x = 1$ is a global maximum of $f(x) = 2$ because $f(1) = 2$ is at least as high as every other point on the graph.

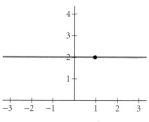

Figure 1.21 $\square$

We say that a function f has an ***inflection point*** at $x = c$ if c is in the domain of f and the graph of f changes concavity at $x = c$; in other words, $x = c$ is an inflection point of f if the graph of f is concave down to the immediate left of $x = c$ and concave up to the immediate right of c, or vice versa. More precisely, a function f has an inflection point at $x = c$ if there is some positive number δ such that f is concave up (or down) on the interval $(c - \delta, c)$ and has the opposite concavity on the interval $(c, c + \delta)$.

? *Question*   There are four types of inflection points. Can you sketch a graph illustrating each of these four types of inflection points? (*Hint:* One type of inflection point can be seen in a graph that changes from concave down to concave up while decreasing; see Figure 1.22.) $\square$

Since we don't yet have a formal mathematical description of concavity, we can't algebraically find the exact values of the inflection points of a function f. For the moment we will estimate these values by inspecting the graph of f.

EXAMPLE 1.34   **Graphically determining the locations of inflection points**

The function graphed in Figure 1.22 is concave down on $(0, 2)$ and concave up on $(2, 4)$. We approximate that the point $x = 2$ is an inflection point of f, since the graph of f changes concavity (from down to up) at approximately $x = 2$.

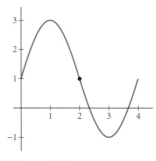

Figure 1.22

1.2.8 Vertical and horizontal asymptotes

The rough definition of an **asymptote** of a graph is a line that the graph gets infinitely close to. An asymptote is called a **vertical asymptote** if it is a vertical line; it is called a **horizontal asymptote** if it is a horizontal line. The formal mathematical definition of an asymptote, which involves limits, will be given in Chapter 2.

EXAMPLE 1.35   **Graphically determining the locations of vertical and horizontal asymptotes**

The function f graphed in Figure 1.23 has vertical asymptotes at $x = -2$ and $x = 2$ and a horizontal asymptote at $y = 1$.

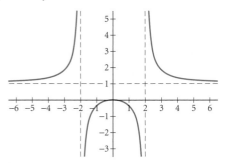

Figure 1.23

Consider the vertical asymptote at $x = -2$. As we follow the graph of f approaching $x = -2$ from the left-hand side, the graph gets closer and closer to (but never quite reaches) the vertical line $x = -2$. Similarly, if we follow the graph of f approaching $x = -2$ from the right-hand side, the graph gets closer and closer to the vertical line $x = -2$.

◈ *Caution*   A function *can* cross one of its horizontal asymptotes, as long as the graph of the function *eventually* gets infinitely close to the asymptote. Can you draw a picture of such a function? Can a function ever cross a vertical asymptote? Why or why not?

Later we will see other types of asymptotes besides vertical and horizontal asymptotes; in Chapter 6 we will see "slant" asymptotes (asymptotes that are lines but are neither horizontal nor vertical) and even asymptotes that are curves!

> *What you should know after reading Section 1.2*
>
> ▶ The definition of the graph of a function, in set notation and in terms of solutions to the equation $y = f(x)$.
>
> ▶ How to use your graphing calculator to graph functions. This includes being able to choose sensible and informative viewing windows when you are graphing!
>
> ▶ How to use the Vertical Line Test (and explain why it works).
>
> ▶ Verbal and mathematical descriptions of x- and y-intercepts and of functions that are positive, negative, nonpositive, nonnegative, increasing, decreasing, nonincreasing, or nondecreasing on an interval I. Verbal and graphical descriptions of concavity.
>
> ▶ How to determine graphically and algebraically the intervals on which a function is positive or negative and any y-intercepts or roots of a function. How to show algebraically that a function is increasing or decreasing on an interval I. How to estimate graphically the intervals on which a function is concave up or down.
>
> ▶ Verbal and mathematical descriptions of local and global maxima and minima of a function. Verbal descriptions of inflection points and vertical and horizontal asymptotes. How to estimate these extrema, inflection points, and asymptotes graphically.

Exercises 1.2

Concepts

0. Read the section and make your own summary of the material. Include whatever you think will help you review the material later on.

■ **Use set notation to describe the graphs of the following functions as sets of ordered pairs.**

1. $f(x) = 3x + 2$

2. $f(x) = \sqrt{x}$

3. $f(x) = \dfrac{1}{x - 1}$

4. Determine whether the given points lie on the graph of $f(x) = \sqrt{x + 1}$. Justify your answers *without* referring to a picture of the graph of f.

  (a) $(3, 2)$ **(b)** $(1, 1)$ **(c)** $(-5, 2)$

5. Explain in your own words why every vertical line must pass through the graph of a function at most once.

6. For each part below, sketch a graph that has the given characteristics. (*Note:* Your graph does not necessarily have to be one connected curve.)

  (a) The graph fails the vertical line test *everywhere* in its domain.

  (b) The graph fails the vertical line test at exactly *one* point in its domain.

  (c) The graph fails the vertical line test because a certain vertical line intersects the graph at *infinitely many* points.

■ **For each graph in Problems 7–9, locate, estimate, and/or identify the characteristics listed in (a)–(g) below.**

  (a) List any y-intercepts and any roots of the graph.

  (b) Find the intervals on which the graph is positive and the intervals on which it is negative. Where is the graph nonnegative? Where is it nonpositive?

  (c) Find the intervals on which the graph is increasing and the intervals on which it is decreasing.

  (d) Find the intervals on which the graph is concave up and the intervals on which it is concave down.

  (e) Locate all local and global extrema of the graph.

  (f) Approximate the locations of any inflection points of the graph.

  (g) List any vertical or horizontal asymptotes of the graph.

7.

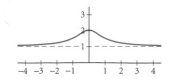

8.

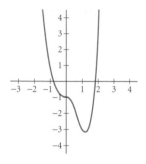

9.

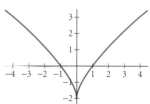

10. For each part, sketch, if possible, the graph of a function with domain $\mathbb{R}$ that has each of the listed characteristics on its entire domain.

   (a) Concave up and increasing
   (b) Concave down and increasing
   (c) Concave up and decreasing
   (d) Concave down and decreasing
   (e) Concave up, increasing, and negative
   (f) Concave down, increasing, and negative

11. Each part below describes a graph $y = f(x)$ that has an inflection point at $x = 2$. Sketch the graph described in each part.

   (a) f is concave up on $(-\infty, 2)$, is concave down on $(2, \infty)$, and is always increasing.
   (b) f is concave down on $(-\infty, 2)$, is concave up on $(2, \infty)$, and is always increasing.
   (c) f is concave up on $(-\infty, 2)$, is concave down on $(2, \infty)$, and is always decreasing.
   (d) f is concave down on $(-\infty, 2)$, is concave up on $(2, \infty)$, and is always decreasing.

■ **For each problem below, sketch the graph of a function f that has the given characteristics.**

12. f is always increasing and has two horizontal asymptotes, one at $y = -2$ and one at $y = 2$.

13. f has domain $(0, \infty)$ and is always negative and always increasing.

14. f has four roots but no y-intercept.

15. f is concave down on $(-\infty, 0)$ and concave up on $(0, \infty)$ but does *not* have an inflection point at $x = 0$.

■ **Make a table (like Table 1.2) with verbal and mathematical descriptions of the following terminologies.**

16. f is nonnegative on an interval I.

17. f is nonpositive on an interval I.

18. f is increasing on an interval I.

19. f is decreasing on an interval I.

20. f is increasing at the point $x = c$.

21. f is decreasing at the point $x = c$.

22. Sketch three different graphs of functions that go through the points shown. Explain what this means about the limitations of graphing functions by "plotting points."

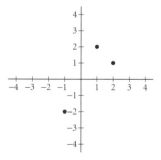

23. Use the graph to find values of δ for each local extremum (as described in Table 1.5). For example, if f has a local maximum at $x = c$, find a $\delta > 0$ such that $f(c) \geq f(x)$ for all x in $(c - \delta, c + \delta)$. Then try to approximate the *largest* such value of δ for each local extremum.

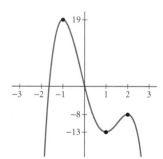

Skills

■ **Use a graphing calculator to sketch graphs of the following functions. Use trial and error to find a graphing window such that your graph represents the local and global behavior of the function. Be sure to include the x and y ranges of your window in your answer.**

24. $f(x) = (x^2 - 5)^7$

25. $f(x) = x^2 - 0.1$

26. $f(x) = x^3 - 11x^2 + 10x$

■ **Find any roots and y-intercepts of the following functions algebraically. Then check your answers by graphing the functions with your calculator.**

27. $f(x) = 3x - 5$

28. $f(x) = x^2 - 2x - 3$

29. $f(x) = 3$

■ **Find (algebraically, that is, by solving inequalities) the intervals on which each of the following functions is** positive and the intervals on which each is negative. Then check your answers by graphing the functions with your calculator.

30. $f(x) = 3x - 5$

31. $f(x) = x^2 - 2x - 3$

32. $f(x) = x^2 + 1$

■ Show that the given functions are increasing or decreasing on the intervals given. Use the mathematical descriptions of what it means for a function to be increasing or decreasing.

33. $f(x) = 1 - 3x$, decreasing on $(-\infty, \infty)$

34. $f(x) = \dfrac{1}{3 - x}$, increasing on $(-\infty, 3)$

Applications

■ For each problem below, sketch and label the graph of a function that describes the given situation.

35. Susie is late for calculus class and leaves her dorm in a panic. She hurries toward the math building, but about halfway there, she realizes she has left her notebook in her room. She sprints back to her dorm and gets her notebook. Coming out of the dorm, she sprains her ankle, and so the best she can do is limp as fast as she can to her classroom. Graph Susie's distance from the math building as a function of time.

36. In Lake Erie there is an island with a growing population of warthogs. At first the warthog population grows quickly, but as space and food become sparse on the island, the warthogs choose to have fewer baby warthogs. Eventually the population of the island levels off at 512 warthogs. Sketch a graph of the warthog population over time.

37. After you drink a cup of coffee, the amount of caffeine in your body decreases by half every hour. Assuming that you do not have any more coffee, sketch a graph of the amount of caffeine in your body as a function of time.

38. On a dare, you go skydiving. Because of gravity, you fall faster and faster as you plummet toward the ground. When you open your parachute, your speed is drastically reduced. After opening your parachute, you approach the ground at a constant speed. Graph your distance from the ground as a function of time.

39. Graph your velocity as a function of time as you skydive as described in Problem 38.

1.3 Linear Functions

Besides constant functions and the identity function, the simplest functions are *linear* functions. In fact, the identity function and constant functions are special cases of linear functions.

1.3.1 What is a linear function?

Functions with constant rates of change are called **linear** functions. As we will soon see, the graphs of linear functions are lines (as their name would suggest). We begin with an algebraic description of these functions.

DEFINITION 1.7

Linear Functions

A function f is a **linear function** if it can be written in the form

$$f(x) = mx + b$$

for some constants m and b.

Note that all constant functions are linear functions (with $m = 0$) and that the identity function is a linear function (with $m = 1$ and $b = 0$).

🕭 *Caution* Definition 1.7 says that a function is linear if it *can* be written in the form $f(x) = mx + b$. Sometimes, however, you will have to do some algebra to determine whether a function can be written in this form. ☐

EXAMPLE 1.36   **A linear function in disguise**

The function $f(x) = \frac{3x+1}{2}$ is a linear function, since $f(x) = \frac{3}{2}x + \frac{1}{2}$ ($m = \frac{3}{2}$ and $b = \frac{1}{2}$).   □

EXAMPLE 1.37   **Another disguised linear function**

The equation $2y - 1 = 6x$ defines a linear function. Solve for y to get $y = 3x + \frac{1}{2}$ ($m = 3$ and $b = \frac{1}{2}$).   □

The domain of every linear function is $\mathbb{R}$, since *any* real number x can be input into the expression $mx + b$, no matter what m and b are (why?). The range of a linear function is also $\mathbb{R}$, *unless $m = 0$*. (If $m = 0$, then the function $f(x) = (0)x + b = b$ is a constant function, and its range is the set $\{b\}$.)

❓ *Question*   Explain why, if $m \neq 0$, the range of the linear function $f(x) = mx + b$ is $\mathbb{R}$. Use the definition of range: $\text{Range}(f) = \{y \in \mathbb{R} \mid y = f(x) \text{ for some } x \in \text{Domain}(f)\}$.   □

1.3.2 Slope and average rates of change

What we mean when we say that a linear function $f(x) = mx + b$ has a constant rate of change is that for every unit of change in the independent variable x, the dependent variable $f(x)$ changes by a constant amount.

THEOREM 1.2

Interpreting the Rate of Change of a Linear Function

If $f(x) = mx + b$, then a unit change in the independent variable always produces a change of m units in the dependent variable.

PROOF (**THEOREM 1.2**)   Suppose $f(x) = mx + b$ is a linear function. We have to show that for every unit change in x, the dependent variable $f(x)$ changes by m units. In other words, we have to show that $f(x + 1) = f(x) + m$. By evaluating $f(x + 1)$, we have

$$f(x + 1) = m(x + 1) + b = mx + m + b = (mx + b) + m = f(x) + m. ■$$

To state a more general fact about the rate of change of a linear function, we first need the following definition.

DEFINITION 1.8

The Average Rate of Change of a Function on an Interval

The *average rate of change* from $x = a$ to $x = b$ of a function f is the quotient

$$\frac{f(b) - f(a)}{b - a}.$$

We also call this the *average rate of change of f on the interval* $I = [a, b]$.

⑨ *Caution*   Definition 1.8 applies to *any* function f, not just to linear functions. We will use this definition again in Section 3.2.2.   □

EXAMPLE 1.38 **Calculating the average rate of change of a function on an interval**

Calculate the average rate of change of the linear function $f(x) = 4x - 7$ **(a)** from $x = 1$ to $x = 3$, and **(b)** from $x = -2$ to $x = 1$.

Solution Using Definition 1.8, we find

$$\frac{f(3) - f(1)}{3 - 1} = \frac{(4(3) - 7) - (4(1) - 7)}{2} = \frac{5 - (-3)}{2} = \frac{8}{2} = 4.$$

A similar calculation shows that the average rate of change of $f(x)$ from $x = -2$ to $x = 1$ is $\frac{f(1) - f(-2)}{1 - (-2)} = 4$. ☐

THEOREM 1.3

Linear Functions Have a Constant Average Rate of Change

The average rate of change of a linear function $f(x) = mx + b$ is constant. More precisely, the average rate of change of $f(x) = mx + b$ on *any* interval I is always equal to m.

PROOF **(THEOREM 1.3)** To avoid confusion, we will use the notation $I = [c, d]$ instead of $I = [a, b]$, since we are already using the letter b in our notation for linear functions. Let $f(x) = mx + b$ be any linear function and let $I = [c, d]$ be any closed interval. Then the average rate of change of f on I is

$$\frac{f(d) - f(c)}{d - c} = \frac{(m(d) + b) - (m(c) + b)}{d - c} = \frac{md + b - mc - b}{d - c} = \frac{m(d - c)}{d - c} = m.$$

Since the average rate of change of a linear function is constant (that is, the same on *any* interval I), we can give it a name.

DEFINITION 1.9

The Slope of a Linear Function

The *slope* of a linear function $f(x) = mx + b$ is its constant average rate of change.

Moreover, since if $f(x) = mx + b$ is a linear function, then $f(0) = m(0) + b = b$, and thus the constant b is the y-intercept of the linear function $f(x) = mx + b$. This fact, Definition 1.9, and Theorem 1.3 prove the following theorem.

THEOREM 1.4

The Slope and y-intercept of a Linear Function

If $f(x) = mx + b$ is a linear function, then the graph of $y = f(x)$ is a line with slope m and y-intercept b.

It is precisely the fact that the average rate of change of a linear function is constant that ensures that the graph of a linear function is a line. It is easy to sketch the graph of any linear function $f(x) = mx + b$ since we know that the graph will be a line with slope m and y-intercept b.

We often think of the slope of a linear function $f(x) = mx + b$ as the "rise over run" of the graph of f. The "rise" is a distance traveled in the y direction corresponding to a "run," or change, in the x direction. On the graph of a line with slope m, a run of 1 unit will correspond to a rise of m units. We will denote the run by the symbol Δx (this is the traditional notation for a change in x) and the rise by the symbol Δy (the corresponding change in y).

NOTATION 1.4

The Δx and Δy Notation

$$\frac{\Delta y}{\Delta x} = \frac{\text{rise}}{\text{run}} = \frac{f(b) - f(a)}{b - a} = \text{average rate of change of } f \text{ on } [a, b],$$

where $\Delta x = b - a$ and $\Delta y = f(b) - f(a)$.

EXAMPLE 1.39

The graphical meanings of Δy and Δx

The graph of the linear function $f(x) = 3x - 5$ is shown in Figure 1.24. On the interval $I = [1, 3]$, we have a run of $3 - 1 = 2$ units. From the point $(1, f(1)) = (1, -2)$ to the point $(3, f(3)) = (3, 4)$, we have a corresponding rise of $4 - (-2) = 6$ units.

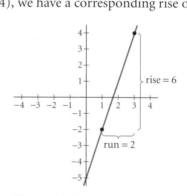

Figure 1.24

Thus the average rate of change of f on the interval $[1, 3]$ is

$$\frac{\Delta y}{\Delta x} = \frac{f(3) - f(1)}{3 - 1} = \frac{4 - (-2)}{3 - 1} = \frac{6}{2} = 3.$$

The average rate of change of this linear function on $[1, 3]$, and in fact on *any* interval, is given by the slope $m = 3$.   □

You should be able to estimate the slope of a line just by looking at its graph. Lines with positive slope are increasing, lines with negative slope are decreasing, and lines with zero slope are horizontal. Vertical lines have "undefined" slope (and fail the Vertical Line Test, so are not functions).

❓ Question   Can you justify the statements about slope in the paragraph above? Think about the definition of the slope of a linear function as an average rate of change, and remember the mathematical descriptions of increasing and decreasing functions.   □

❓ Question   Label each of the lines graphed in Figure 1.25 with one of the following slopes (think about the "rise over run" that would produce each of these slopes):

$$m = 1, \quad m = -1, \quad m = 0, \quad m = 2, \quad m = -2, \quad m = \tfrac{1}{2}, \quad m = -\tfrac{1}{2}.$$

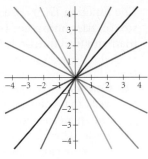

Figure 1.25   □

1.3.3 Proportional functions

A linear function that passes through the origin is called a ***proportional function.***

> ### Proportional Function
>
> A ***proportional function*** is a function that can be written in the following form, for some real number k:
>
> $$f(x) = kx.$$

In Definition 1.10, we used the constant k (rather than m) to represent the slope. We could use any letter we want, but for a proportional function the letter is traditionally k. By definition, a proportional function is just a linear function whose y-intercept is $b = 0$. Thus every proportional function is linear (but not every linear function is proportional).

Proportional functions are used often in describing how two real-world quantities compare to each other. In word problems we will often say that one variable or quantity y is "proportional to" another variable or quantity x. When we say this, we mean that y is a proportional function of x, that is, that y is some constant multiple of x.

> ### Proportionality and the Proportionality Constant
>
> If y is ***proportional to*** x, then $y = kx$ for some constant k. The constant k is called the ***proportionality constant.***

A real-world example of proportionality

Some people would argue that your grade in a calculus course is proportional to the time you spend reading the textbook. In symbols, if we let G represent your grade point average in this calculus course and r represent the number of hours each week that you spend reading this book, then G is a proportional function of r. In other words,

$$G = kr \quad \text{for some constant } k.$$

Presumably the constant k would be positive, so that spending more time reading this textbook would result in a higher grade. For example, suppose that $k = \frac{2}{3}$. If $G(r) = \frac{2}{3}r$, then spending $r = 2$ hours a week reading this textbook would result in a grade point average of $G(2) = \frac{2}{3}(2) \approx 1.33$. According to the function $G = \frac{2}{3}r$, reading this book for $r = 6$ hours a week would give you a grade point average of $G(6) = \frac{2}{3}(6) = 4.0$.

The function $G(r)$ described here makes "real-world" sense only for $0 \leq r \leq 6$ since we consider only grade point averages between 0.0 and 4.0 (even though, as an abstract mathematical function, $G(r) = kr$ makes sense for all real numbers r). ☐

⑤ Caution   Your grade will not be computed by multiplying the number of hours per week that you read this book by the constant $\frac{2}{3}$, although it is almost certainly true that spending more time reading will lead to greater understanding and thus a better grade. The function whose output is your grade point average probably has *many* input variables: the time you spend reading the book, the amount of homework you do, the amount of time you spend talking and thinking about calculus, how much you come to class, and so on. ☐

❓ Question   If y is proportional to x with proportionality constant k, is x necessarily proportional to y? If so, what is the proportionality constant? What if $k = 0$? ☐

1.3.4 Equations for lines

By definition, if y is a linear function of x, then $y(x)$ can be written in the form $y(x) = mx + b$ for some constants m and b. This form of the equation for a line is called the **slope-intercept** form, since it specifies the y-intercept b of the line and the slope m. Sometimes it will be more convenient to represent linear functions with other equations.

Clearly, a line is uniquely determined by its slope and its y-intercept. That is, if we are given values for the slope m and y-intercept b, then *one and only one* line has that slope and y-intercept. What other information could we use to uniquely determine a line?

❓ Question   Think about this for a minute. What information would you need to know to be able to uniquely specify a line? There are a few possible answers. ☐

One possible answer is that a line is determined by its slope and *any* point (not necessarily the y-intercept). In other words, given a slope and any point in the plane, there is one and only one line that passes through that point and has that slope. In general, if a linear function has slope m and goes through the point (x_0, y_0), then for *any* point (x,y) on the line (since the average rate of change is constant),

$$m = \frac{y - y_0}{x - x_0}.$$

Rearranging this equation, we get

(1) $$y - y_0 = m(x - x_0).$$

This is the **point-slope** form of a line. We could also write this form as $y = m(x - x_0) + y_0$.

⑤ Caution   In Equation (1), the letters x_0 and y_0 represent *fixed* points (constants), and the x and y represent variables. (We pronounce x_0 as "x-naught" and y_0 as "y-naught.") ☐

On the other hand, a line in the plane can be determined by any two points. Given any two points in the plane, there is one and only one line that passes through those two points. If a linear function $y = f(x)$ passes through the points (x_0, y_0) and (x_1, y_1), then its slope is given by the quotient

$$m = \frac{\Delta y}{\Delta x} = \frac{y_1 - y_0}{x_1 - x_0}.$$

We can substitute this expression for m into Equation (1) to get

(2)
$$y - y_0 = \left(\frac{y_1 - y_0}{x_1 - x_0} \right) (x - x_0).$$

We will call this the **two-point** form of a line. Note that x_0, y_0, x_1, and y_1 all represent *constants*.

Table 1.6 describes the three forms of linear functions we have discussed. Each form represents a set of information that determines a line in the plane. Keep in mind that the two-point form for a line is nothing more than the point-slope form of a line where we first have to compute the slope.

Name of Form	Information Given	Equation of Line
Slope-intercept form	Slope m, y-intercept b	$y = mx + b$
Point-slope form	Slope m and any point (x_0, y_0) on the line	$y - y_0 = m(x - x_0)$
Two-point form	Any two points (x_0, y_0) and (x_1, y_1) on the line	$y - y_0 = \left(\frac{y_1 - y_0}{x_1 - x_0} \right) (x - x_0)$

Table 1.6

EXAMPLE 1.41

Finding a line given its slope and one point on the line

Find an equation for the line that has slope -2 and passes through the point $(3, 1)$. Then put the equation into slope-intercept form.

Solution Using the point-slope form with $m = -2$ and $(x_0, y_0) = (3, 1)$, we get the equation $y - 1 = -2(x - 3)$. To put this into slope-intercept form, we do a little algebra:

$$y - 1 = -2(x - 3) \quad \Rightarrow \quad y - 1 = -2x + 6 \quad \Rightarrow \quad y = -2x + 7.$$

This line has slope -2 and y-intercept 7; it is shown in Figure 1.26. □

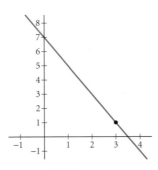

Figure 1.26

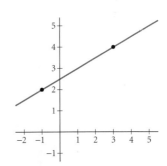

Figure 1.27

EXAMPLE 1.42

Finding a line given two points on the line

Find an equation for the line that passes through the points $(-1, 2)$ and $(3, 4)$. Then put the equation into slope-intercept form.

Solution We will use the two-point form with $(x_0, y_0) = (-1, 2)$ and $(x_1, y_1) = (3, 4)$. The slope of the line is $\frac{4-2}{3-(-1)} = \frac{2}{4} = \frac{1}{2}$. Using this slope and the point $(-1, 2)$, we get

$$y - 2 = \tfrac{1}{2}(x - (-1)) \quad \Rightarrow \quad y - 2 = \tfrac{1}{2}(x + 1) \quad \Rightarrow \quad y = \tfrac{1}{2}x + \tfrac{5}{2}.$$

This means that the line has slope $\frac{1}{2}$ and y-intercept $\frac{5}{2}$. The graph of this line is shown in Figure 1.27. □

? *Question*   After computing the slope, we could have used the point $(3, 4)$ instead of the point $(-1, 2)$ to get the point-slope form of the line in Example 1.42. Would the resulting point-slope form of the equation have been the same? Would the equation represent the same line?   □

✓ *Checking the Answer*   To verify the answer in Example 1.42, we could check to see whether the graph of the function $y = \frac{1}{2}x + \frac{5}{2}$ actually passes through the points $(-1, 2)$ and $(3, 4)$. We could do this by using a graphing calculator or checking algebraically that $y(-1) = 2$ and $y(3) = 4$.   □

1.3.5 Parallel and perpendicular lines

Two distinct lines in a plane are said to be *parallel* if they never intersect and are said to be *perpendicular* if they intersect in a right angle. Mathematically, two lines are parallel if their slopes are equal, and are perpendicular if one slope is the negative reciprocal of the other.

DEFINITION 1.12

Parallel and Perpendicular Lines

Suppose l_1 and l_2 are lines with slopes m_1 and m_2, respectively. The lines l_1 and l_2 are *parallel* if $m_1 = m_2$. The lines l_1 and l_2 are *perpendicular* if $m_2 = -\frac{1}{m_1}$.

EXAMPLE 1.43

The graphs of parallel and perpendicular lines

The graphs of the linear functions $f(x) = 2x - 1$ and $g(x) = 2x + 2$ are parallel (see Figure 1.28), since they have the same slope. On the other hand, the graphs of the linear functions $f(x) = 2x - 1$ and $g(x) = -\frac{1}{2}x + 1$ are perpendicular (see Figure 1.29). Note that the value of the y-intercept b does not have any effect on whether or not two lines of the form $y = mx + b$ are parallel or perpendicular.

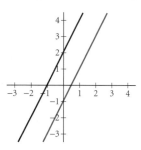   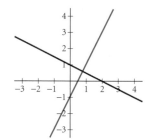

Figure 1.28   **Figure 1.29**   □

EXAMPLE 1.44

Finding a line through a given point that is perpendicular to another line

Find an equation for the line that passes through the point $(2, 0)$ and is perpendicular to the line $y = \frac{1}{3}x + 1$.

Solution   If the line we are looking for is perpendicular to $y = \frac{1}{3}x + 1$, then it has slope $m = -\frac{1}{1/3} = -3$. The line we are looking for has slope $m = -3$ and $(x_0, y_0) = (2, 0)$, so we can use the point-slope form to find its equation: $y - 0 = -3(x - 2)$. In slope-intercept form this is written $y = -3x + 6$.   □

☑ *Checking the Answer*   Sketch graphs of $y = \frac{1}{3}x + 1$ and $y = -3x + 6$ to check that they are in fact perpendicular and that the second line passes through $(2, 0)$. (*Note:* Depending on your graphing window, perpendicular lines may not appear perpendicular when you graph them on your graphing calculator. Why not?)   □

What you should know after reading Section 1.3

► The definition, domain, and range of linear functions. In particular, the graphical meanings of m and b in the slope-intercept form $y = mx + b$.

► The definition of average rate of change, and how to find it algebraically. What can you say about the average rate of change of a linear function? What is the definition of the *slope* of a linear function? How to visually approximate positive, negative, zero, and undefined slopes.

► The definition of a proportional function and its constant of proportionality. How to recognize proportional relationships in verbal descriptions.

► How to find the equation of a line given its slope and y-intercept, or given its slope and any point on the line, or given any two points on the line. How to tell whether or not two lines are parallel or perpendicular.

Exercises 1.3

Concepts

0. Read the section and make your own summary of the material. Include whatever you think will help you review the material later on.

1. Sometimes it is convenient to use an equation of the form $Ay + Bx + C = 0$ to describe a linear function. Show that any equation of this form is a linear function, and find formulas for the slope and y-intercept in terms of the constants A, B, and C.

2. If $y = 3x - 2$, then a unit change in x results in a change of _____ units in y. A change of 6 units in x will result in a change of _____ units in y.

3. Suppose $f(x) = -3x + 1$ and that over a certain interval you have $\Delta x = 4$. What is the corresponding value of Δy? (*Hint:* Think about the relationship among Δy, Δx, and the slope of a linear function.)

4. Illustrate on a graph of $f(x) = -5x + 1$ that the average rate of change of f on $[-1, 3]$ is -5. (*Hint:* Think about "rise over run.")

5. Without a calculator, and on the same set of coordinate axes, sketch graphs of the functions $f(x) = 3x - 1$, $g(x) = \frac{1}{10}x + 8$, and $h(x) = 10x$.

6. Is every proportional function a linear function? Why or why not? Is every linear function a proportional function? Why or why not?

7. If y is proportional to x with proportionality constant k, is x necessarily proportional to y? If so, what is the proportionality constant? What if $k = 0$?

8. If $f(0) = 1$, $f(1) = 2$, and $f(4) = 5$, explain why f could not be a proportional function. Could f be a linear function? If not, explain why not. If so, find a linear equation for $f(x)$.

9. If $f(1) = 4$, $f(2) = 5$, and $f(4) = 6$, could f be a linear function? If not, explain why not. If so, find a linear equation for $f(x)$.

10. Assuming that f is a linear function, use the values given in the table to fill in the missing values.

x	1	3		7		10
$f(x)$	1	-5	-8		-23	

Skills

■ Determine whether the following functions are linear. For each linear function, find the slope and y-intercept.

11. $f(x) = \dfrac{3 - x}{8}$

12. $f(x) = 3x(2x + 1)$

13. $f(x) = -2(1 + x) + 7$

14. $x - 2y = \frac{1}{2}$

15. $y - 3 = 2(x + 5)$

16. $\dfrac{4 + x}{2y} = 3$

17. Is $y = 5$ in the range of $f(x) = 3 - 4x$? Justify your answer using the set notation definition of range from Section 1.1.4. Then sketch a graph illustrating your answer.

■ Compute the average rates of change of the following functions on the indicated intervals.

18. $f(x) = 2x + 542$, $[-1, 1]$

19. $f(x) = -0.5 + 4.2x$, $[1, 3.5]$

20. $f(x) = 3$, $[-100, 100]$

21. $f(x) = x^2 - 10$, $[0, 2]$

22. $f(x) = x^2 - 10$, $[2, 4]$

23. $f(x) = \sqrt{x + 1}$, $[1, 9]$

■ Write the following linear functions in point-slope form (see Table 1.6). Is there a *unique* way to do this? If possible, find a second way to write each linear function in point-slope form.

24. $y = 3x + 1$

25. $2y + x = 7$

26. $y = 3$

■ In Problems 27–30, find linear functions that have the given properties.

27. Slope 6 and y-intercept -3

28. Slope -1, passes through $(3, -2)$

29. Passes through $(-2, 1)$ and $(3, -4)$

30. y-intercept 5, passes through $(1, 2)$

31. Perpendicular to $y = -0.5x + 2$, passes through the point $(1, 0)$

32. Parallel to $y = 15x - 7$, passes through the point $(-1, 8)$

33. Parallel to $y = 2x + 1$, passes through the point $(1, 3)$

Applications

■ For each scenario described below, identify and name (with variables and units) the dependent and independent variables and use function notation to write down a function describing the situation.

34. The number of hours per week that Tim spends watching television is proportional to the amount of credit card debt that he has.

35. Alina drops a grapefruit from the top of the Empire State Building. The velocity of the grapefruit (in meters per second) is proportional to the number of seconds it has been falling, with proportionality constant 9.8.

36. Linda captured 15 fireflies last Tuesday and put them in a jar. Since then there have been three new fireflies in the jar every day.

37. The population on the island of Whimsy can be expressed as a linear function of time. In 1985 there were 35 Whimsians on the island. In the year 2000 the population of Whimsy had increased to 91.

■ Solve the following word problems. Identify and name all variables that you use.

38. A recent study shows that the number of Twinkies produced in the United States each year is proportional to the amount (in dollars) that the government spends on national security. If last year 500 million Twinkies were produced in the United States, and the United States government spent 25 billion dollars on national security, then what is the proportionality constant relating Twinkies to U.S. national security?

39. The Lake Crystal police have determined that a disgruntled pet store owner abandoned an unknown number of wombats on a small island in the center of the lake in 1996. Since then it has been determined that the average rate of change of the wombat population was four wombats per year and that the wombat population was a linear function of time. In the year 2001 when the abandoned wombats were discovered, there were 376 wombats on the island. How many wombats did the disgruntled pet store owner originally leave on the island?

40. The table below describes the monthly rent for Linda's New York City apartment in the years 1995–2000. Represent these data as a linear function, if possible. If not possible, explain why. If it is possible to construct a linear function to match these data, use it to predict the rent for Linda's apartment this year.

Year	1995	1996	1997	1998	1999	2000
Rent	$1475	$1537	$1599	$1661	$1723	$1785

41. The table below shows the numbers of compact discs that Alina owned at the ends of certain months last year. Represent these data as a linear function, if possible. If it is not possible, explain why. If it is possible to construct a linear function to match these data, use it to determine how many more months after December it will take before Alina has a collection of 200 CDs.

Month	Jan.	Feb.	April	June	July	Sept.	Dec.
CDs	51	54	60	66	69	75	84

42. A department store is having trouble selling a particular jacket. In desperation, store managers reduce the price of the jacket each Friday in hopes that someone will eventually find it a bargain and buy it. The table below shows the price of the jacket t weeks after they start to mark it down. Represent these data as a linear function, if possible. If it is not possible, explain why. If it is possible to construct a linear function to match these data, use it to determine when the jacket will cost only $50.

Week	1	2	4	5	7	8
Price	$150	$135	$120	$105	$90	$75

43. The table below describes Paul's weight in pounds t days after starting his diet. Represent these data as a linear function, if possible. If not possible, explain why. If it is possible to construct a linear function to match these data, use it to predict the number of days Paul has to stay on the diet to reach his target weight of 155 pounds.

Days	4	8	16	20	32	44
Weight	194	193	191	190	187	184

Proofs

44. Prove that the range of the function $f(x) = 3x - 1$ is $\mathbb{R}$. Use the definition of range given in Section 1.1.4.

45. Prove that the range of *any* linear function $f(x) = mx + b$ with $m \neq 0$ is $\mathbb{R}$. Use the definition of range given in Section 1.1.4. At which point do you use the fact that m is nonzero?

46. Prove that the average rate of change of the linear function $f(x) = -2x + 4$ on *any* interval I is always equal to -2.

47. Prove that a function whose graph is a line with positive slope is an increasing function. Use the definition of average rate of change and the mathematical description of increasing functions given in Table 1.4 in Section 1.2.5.

48. Show that every linear equation in point-slope form $y - y_0 = m(x - x_0)$ can be written in slope-intercept form. What are the slope and y-intercept in terms of x_0 and y_0?

49. Show that if line l_2 is perpendicular to line l_1, and line l_3 is perpendicular to line l_2, then lines l_1 and l_3 are parallel.

1.4 A Basic Library of Functions

In this section we present a basic stock of functions that will serve as examples throughout the text. In Part II we will examine algebraic functions in more detail, and in Part III we will investigate transcendental functions. For now we will memorize a few basic graphs and use a graphing calculator to sketch graphs of more complicated functions.

1.4.1 Algebraic functions

A function is **algebraic** if it can be expressed in terms of a variable x using only arithmetic operations ($+$, $-$, $\times$, and $\div$) and rational powers. For example,

$$f(x) = \frac{3x^2 - 5}{1 - x} \quad \text{and} \quad f(x) = \left(\sqrt{x} + \sqrt[3]{x} \right)^7$$

are algebraic, but $f(x) = 2^x$ and $f(x) = \sin x$ are not.

◈ *Caution* The function 2^x is *not* algebraic because the independent variable appears in the exponent. An algebraic function involves powers and roots of the independent variable x; that is, the independent variable has to be in the *base* and not in the exponent. □

There are four basic types of algebraic functions: linear functions, power functions, polynomial functions, and rational functions. More complicated algebraic functions can be constructed using combinations of these four basic types. Table 1.7 gives a general form and examples for each of these types of algebraic functions.

❓ *Question* There is some overlap among the four basic types of algebraic functions in Table 1.7. Is every linear function a polynomial function? Is every polynomial function a rational function? Is every power function a polynomial function? Which types of functions include constant functions? □

We will examine power functions, polynomial functions, rational functions, and more complicated algebraic functions in detail in subsequent chapters. For now we will use these functions as examples but refer to a graphing calculator to obtain most of

Type	General Form	Examples
Linear	$f(x) = mx + b$, where m and b are any real numbers	$f(x) = 2x - 1$, $f(x) = -3x + 5$, $f(x) = 0.12x$, $f(x) = 4$.
Power	$f(x) = Ax^k$, where $A \neq 0$ is any real number and k is any rational number	$f(x) = 3x^3$, $f(x) = -2x^{-\frac{1}{5}}$, $f(x) = 1.7x^{3.22}$, $f(x) = 2$.
Polynomial	$f(x) = a_n x^n + a_{n-1} x^{n-1} + \cdots + a_1 x + a_0$, where n is a positive integer and each a_i is a real number	$f(x) = 3x^5 - 2x^3 + x - 6$, $f(x) = -1.2x^2 + 0.7x + 3.9$, $f(x) = 3x + 1$, $f(x) = 7.2$.
Rational	$f(x) = \frac{p(x)}{q(x)}$, where $p(x)$ and $q(x)$ are polynomial functions	$f(x) = \frac{3x^2 - 1}{x^5 + x^3 + 1}$, $f(x) = \frac{x^3 - 1}{x - 1}$, $f(x) = \frac{1}{x}$, $f(x) = 3x^2 + 1$.
Other	Any other algebraic function that is not one of the four types above	$f(x) = \frac{1 + \sqrt{x}}{1 + x}$, $f(x) = \sqrt[3]{x^{\frac{1}{3}} + 1}$, $f(x) = x^{-2} + x^{-3}$.

Table 1.7

their graphs. However, the four graphs in Figures 1.30–1.33 are so basic, and we will refer to them so often, that you should know them from memory *without* referring to a graphing calculator.

❓ *Question*   To familiarize yourself with the graphs in Figures 1.30–1.33, sketch them by hand (without looking back at the graphs in Figures 1.30–1.33 or using your calculator) and label five points (x, y) on each graph.   ☐

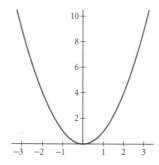

Figure 1.30 $f(x) = x^2$

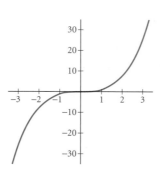

Figure 1.31 $f(x) = x^3$

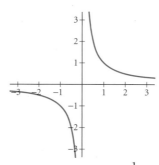

Figure 1.32 $f(x) = \dfrac{1}{x}$

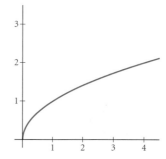

Figure 1.33 $f(x) = \sqrt{x}$

1.4.2 Transcendental functions

Functions that involve more than just sums, differences, products, and quotients of integer powers and roots of x are not algebraic and are called **transcendental** functions. We will investigate the algebra, precalculus, and calculus of various types of transcendental functions in Part III.

In this book we will investigate four basic types of transcendental functions: exponential functions, logarithmic functions, trigonometric functions, and inverse trigonometric functions. We will not go into the definitions or general forms of such functions right now, but Table 1.8 gives examples of each of these four types of functions. Not all transcendental functions are of these four types; some are combinations of the four types, and some are combinations of basic transcendental and algebraic functions. There are even functions that cannot be expressed as combinations of these elementary algebraic and transcendental functions (that is, functions without a "formula" we can write down)!

✧ *Caution* Table 1.8 is provided only to give you an idea of the wealth of transcendental functions that we will be studying in later chapters. For the moment you do *not* have to be able to define or give examples of these types of functions, although occasionally we will use one of these types of functions as an example.   □

Type	Examples
Exponential	$f(x) = 2^x$, $f(x) = 3e^{4x}$, $f(x) = 1.2(3.4)^x$
Logarithmic	$f(x) = \log_{10} x$, $f(x) = \ln x$, $f(x) = \log_2 x$
Trigonometric	$f(x) = \sin x$, $f(x) = \cos x$, $f(x) = \cot x$
Inverse trigonometric	$f(x) = \arcsin x$, $f(x) = \cos^{-1} x$, $f(x) = \arctan x$
Other	$f(x) = x + \sin x$, $f(x) = \ln(\sqrt{x} + 12)$, $f(x) = 2^x \arctan x$

Table 1.8

Although we will not study these transcendental functions until Part III, there will be a few instances before that where it will be necessary to use transcendental functions as examples; when we do so, we will rely on a graphing calculator to provide us with their graphs. In other words, right now we will think of functions like $\ln x$ and $\sin x$ simply as functions that our calculator knows how to calculate and graph. Your calculator should have built-in function keys for most of these functions.

1.4.3 Piecewise functions and absolute values

A *piecewise function* is a function that is defined in pieces. In other words, a piecewise function is given by different formulas on different parts of its domain.

Recall from Section 0.1.5 that the absolute value $|x|$ of a real number is equal to x if x is positive (or zero) and is equal to $-x$ if x is negative. For example, $|2| = 2$ and $|-2| = -(-2) = 2$. In symbols, we can write the function $f(x) = |x|$ as a piecewise function by splitting the definition of $|x|$ into two cases: $x \geq 0$ and $x < 0$.

DEFINITION 1.13

The Absolute Value Function

The *absolute value function* $f(x) = |x|$ is defined to be

$$|x| := \begin{cases} x, & \text{if } x \geq 0 \\ -x, & \text{if } x < 0. \end{cases}$$

Since we know the graphs of the linear functions $y = x$ and $y = -x$, we can graph $|x|$ by graphing $y = x$ on $[0, \infty)$ and $y = -x$ on $(-\infty, 0)$. Thus the graph of $f(x) = |x|$ is as shown in Figure 1.34.

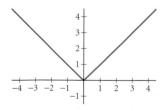

Figure 1.34

In general, the absolute value $|f(x)|$ of a *function* is equal to $f(x)$ for all x where $f(x)$ is positive (or zero) and is equal to $-f(x)$ for all x where $f(x)$ is negative.

DEFINITION 1.14

The Absolute Value of a Function

The *absolute value of a function* $f(x)$ is defined to be

$$|f(x)| := \begin{cases} f(x), & \text{for all } x \text{ with } f(x) \geq 0 \\ -f(x), & \text{for all } x \text{ with } f(x) < 0. \end{cases}$$

EXAMPLE 1.45

Writing the absolute value of a function as a piecewise function

Write the function $f(x) = |x^2 - 1|$ as a piecewise function. Then use this piecewise function to calculate $f(-2)$, $f(0)$, and $f(1)$.

Solution   By definition, $f(x)$ is the function

(1) $$|x^2 - 1| = \begin{cases} x^2 - 1, & \text{for all } x \text{ with } x^2 - 1 \geq 0 \\ -(x^2 - 1), & \text{for all } x \text{ with } x^2 - 1 < 0. \end{cases}$$

Although this is one way to write $|x^2 - 1|$ as a piecewise function, it is a little difficult to work with. For example, to evaluate $f(3)$, we need to know whether $x^2 - 1 \geq 0$ or $x^2 - 1 < 0$ when $x = 3$. To simplify this piecewise function, we need to rewrite the conditions as *intervals* of x.

If we solve the inequalities $x^2 - 1 \geq 0$ and $x^2 - 1 < 0$, we have

$$x^2 - 1 \geq 0 \quad \Rightarrow \quad x^2 \geq 1 \quad \Rightarrow \quad x \geq 1 \text{ or } x \leq -1.$$

Similarly, $x^2 - 1 < 0$ if $-1 < x < 1$. We now have

(2) $$|x^2 - 1| = \begin{cases} x^2 - 1, & \text{if } x \leq -1 \\ -(x^2 - 1), & \text{if } -1 < x < 1 \\ x^2 - 1, & \text{if } x \geq 1. \end{cases}$$

Now we can easily calculate f when $x = -2$, $x = 0$, and $x = 1$.

$$f(-2) = (-2)^2 - 1 = 3 \qquad \text{(use first piece since } -2 \leq -1)$$

$$f(0) = -((0)^2 - 1) = 1 \qquad \text{(use middle piece since } -1 < 0 < 1)$$

$$f(1) = (1)^2 - 1 = 0 \qquad \text{(use last piece since } 1 \geq 1)$$

We could have just substituted these values of x into the expression $|x^2 - 1|$, but the purpose of this example is to illustrate how absolute value functions can be written as piecewise functions.   □

The graph of $y = |f(x)|$ will match the graph of $y = f(x)$ for all x where $f(x)$ is positive (above the x-axis). For the values of x where $f(x)$ is negative (below the x-axis), the graph of $y = |f(x)|$ will have height $-f(x)$, or *negative* the height of the original function $f(x)$. This has the effect of reflecting the negative parts of the graph of $f(x)$ over the x-axis but leaving the positive parts of the graph of $f(x)$ alone.

EXAMPLE 1.46

Graphing the absolute value of a function

Sketch the graph of $y = |2x - 1|$.

Solution   The function $y = |2x - 1|$ is equal to

$$|2x - 1| = \begin{cases} 2x - 1, & \text{for all } x \text{ with } 2x - 1 \geq 0 \\ -(2x - 1), & \text{for all } x \text{ with } 2x - 1 < 0. \end{cases}$$

Since $2x - 1 \geq 0$ when $x \geq \frac{1}{2}$, and $2x - 1 < 0$ for $x < \frac{1}{2}$, the function $y = |2x - 1|$ can be written

$$|2x - 1| = \begin{cases} 2x - 1, & \text{if } x \geq \frac{1}{2} \\ -(2x - 1), & \text{if } x < \frac{1}{2}. \end{cases}$$

The graph of the linear function $y = 2x - 1$ has slope 2 and y-intercept -1. The graph of $y = -(2x - 1) = -2x + 1$ has slope -2 and y-intercept 1 (see Figures 1.35 and 1.36). If we graph $y = 2x - 1$ on the interval $[0.5, \infty)$ and $y = -(2x - 1)$ on the interval $(-\infty, 0.5)$, we get the graph of $y = |2x - 1|$ shown in Figure 1.37.

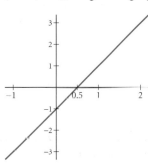 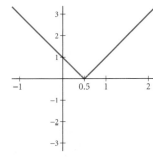

Figure 1.35
$y = 2x - 1$

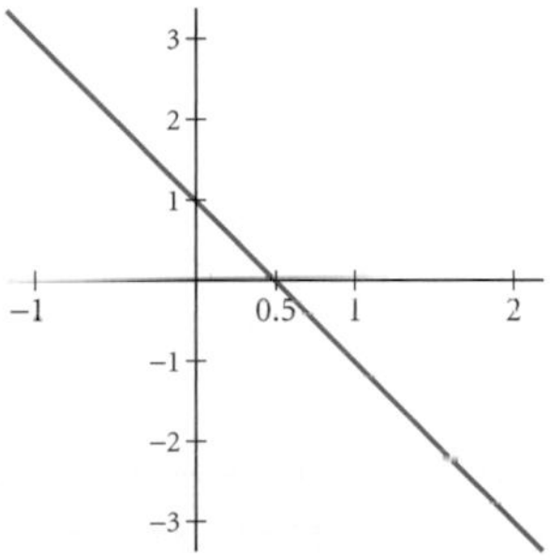

Figure 1.36
$y = -(2x - 1) = -2x + 1$

Figure 1.37
$y = |2x - 1|$

We could also obtain the graph of $y = |2x - 1|$ from the graph of $y = 2x - 1$ by reflecting all the negative parts of the graph in Figure 1.35 over the x-axis.   ☐

> **❓ Question**   Use the techniques in Example 1.46, and the work that we did in Example 1.45, to graph $f(x) = |x^2 - 1|$. You should get the graph shown in Figure 1.38.   ☐

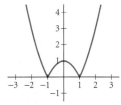

Figure 1.38

Piecewise notation can also be used to describe functions that do not involve absolute values. (Note that all absolute value functions can be written in piecewise notation, but not all piecewise functions can be written using absolute values.)

EXAMPLE 1.47

Graphing a piecewise function

Sketch a graph of the function $f(x) = \begin{cases} x^2, & \text{if } x \leq -1 \\ 2x, & \text{if } x > -1. \end{cases}$

Solution   We start by graphing the functions $y = x^2$ and $y = 2x$ that are used in the definition of $f(x)$; see Figures 1.39 and 1.40.

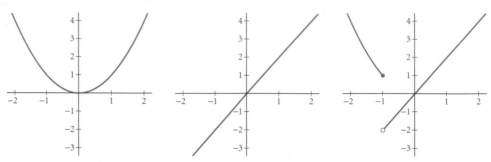

Figure 1.39 $y = x^2$ **Figure 1.40** $y = 2x$ **Figure 1.41** $y = f(x)$

To graph $f(x)$, we must restrict the graph of $y = x^2$ to the interval $(-\infty, -1]$ and restrict $y = 2x$ to $(-1, \infty)$. Whether these intervals are open or closed is important. To find $f(-1)$, we use the *first* equation $y = x^2$: $f(-1) = (-1)^2 = 1$. When we sketch the graph of f, we use open and closed dots to represent the ends of open and closed intervals. See Figure 1.41.   ☐

💲 *Caution*   Notice that the graph in Figure 1.41 has two "pieces." Be aware that a piecewise-defined function may have a graph with pieces that don't "connect." We'll learn more about this when we discuss limits and continuity in Section 2.6.   ☐

What you should know after reading Section 1.4

▶ The different types of algebraic functions (linear, power, polynomial, and rational) and examples and definitions of each type. Also know the difference between an algebraic function and a transcendental function, and be able to give examples of each.

▶ The shapes of the graphs of $y = x^2$, $y = x^3$, $y = \frac{1}{x}$, and $y = \sqrt{x}$ (without using your graphing calculator!). Be able to use a calculator to graph $y = \sin x$, $y = \cos x$, $y = 2^x$, $y = (\frac{1}{2})^x$, and $y = \ln x$.

▶ How to write $|x|$ and $|f(x)|$ as piecewise functions defined on intervals, and how to find values of and graph these functions. How to find values of and graph piecewise functions in general.

Exercises 1.4

Concepts

0. Read the section and make your own summary of the material. Include whatever you think will help you review the material later on.

1. Give an example of an algebraic function that does *not* appear in the reading for this section. Then give an example of a transcendental function.

2. For each part below, give two examples of algebraic functions: one that *is* of the type given and one that is *not* of the type given. Use examples other than the ones in the reading.
   (a) Linear   **(b)** Power
   (c) Polynomial   **(d)** Rational

3. Give an example of an algebraic function that is *not* any of the four basic types listed in Table 1.7.

4. For each part below, give an example of a transcendental function of the type listed.

   (a) Exponential   (b) Logarithmic
   (c) Trigonometric   (d) Inverse trigonometric

5. Give an example of a transcendental function that is *not* any of the four basic types listed in Problem 4.

6. Explain in your own words why the function $f(x) = 3^x$ is *not* an algebraic function.

7. Is every power function also a polynomial function? Is every polynomial function also a rational function? Explain your answers.

■ Decide whether each of the functions given is algebraic or transcendental. Then figure out what *type* of algebraic or transcendental function it is (see Tables 1.7 and 1.8 in the chapter).

8. $f(x) = \dfrac{1}{x+1}$ 9. $f(x) = 1 - x - x^2 - x^3$

10. $f(x) = \dfrac{\ln x}{\sin x}$ 11. $f(x) = 1 + x + 2^x$

12. $f(x) = \dfrac{x^2 - 1}{\sqrt{x} + 1}$ 13. $f(x) = 3$

14. An alternative definition of the absolute value function is $|x| := \sqrt{x^2}$. Explain in your own words why this is equivalent to the definition of the absolute value function given in Definition 1.13. (Note that this problem shows that the absolute value function is algebraic.)

■ For each graph of $f(x)$ given below, sketch the graph of $y = |f(x)|$.

15.

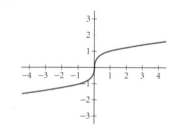

16.

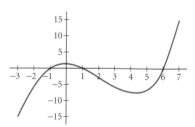

17.

18. Suppose that $f(x)$ is the function defined in pieces using some functions $g(x)$ and $h(x)$ as shown in Equation (3) below. Will the piece of the graph on $(-\infty, 2]$ always "match up" with the piece of the graph on $(2, \infty)$? Why or why not? Use graphs to illustrate your answer.

$$(3) \qquad f(x) = \begin{cases} g(x), & \text{if } x \le 2 \\ h(x), & \text{if } x > 2. \end{cases}$$

Skills

■ Sketch labeled graphs of the functions below *without* using your calculator. Label five points (x, y) on each graph.

19. $f(x) = x^2$ 20. $f(x) = x^3$

21. $f(x) = \sqrt{x}$ 22. $f(x) = \dfrac{1}{x}$

■ Use your calculator to graph each of the functions below. Find the best possible window for each graph, and make a careful, labeled sketch based on your calculator graph.

23. $f(x) = x^2 - 17x - 18$

24. $f(x) = x^5 - 3x^4 - 7x + 5$

25. $f(x) = \dfrac{2x^2 - 2}{x^2 - 3x - 5}$

■ For each piecewise function $f(x)$, calculate $f(-2)$, $f(-1)$, $f(0)$, $f(1)$, and $f(2)$.

26. $f(x) = \begin{cases} 3x + 2, & \text{if } x < 0.5 \\ -4x, & \text{if } x \ge 0.5 \end{cases}$

27. $f(x) = \begin{cases} x^2, & \text{if } x < -1 \\ 2x + 1, & \text{if } x \ge -1 \end{cases}$

28. $f(x) = \begin{cases} 3x + 1, & \text{if } x \le 0 \\ 4, & \text{if } 0 < x \le 1 \\ x^3, & \text{if } x > 1 \end{cases}$

29. $f(x) = \begin{cases} 4x - 1, & \text{if } x < 0 \\ 2, & \text{if } x = 0 \\ -3x + 5, & \text{if } x > 0 \end{cases}$

■ Write each function as a piecewise function where each piece is defined on an *interval* of x-values. In other words, use Definition 1.14 to write $f(x)$ as a piecewise function, and then solve the appropriate inequalities to write $f(x)$ in a form similar to Equation (2) in Example 1.45. Calculate $f(-2)$, $f(1)$, and $f(5)$ for each function.

30. $f(x) = |5 - 3x|$ 31. $f(x) = |1.5x + 2.3|$

32. $f(x) = |x^2 - 4|$ 33. $f(x) = |x^2 + 1|$

34. $f(x) = |x^2 - 3x - 4|$ 35. $f(x) = |3 - 4x + x^2|$

■ For each function $f(x) = |g(x)|$ in the previous block of problems, sketch a graph of $g(x)$ (with the aid of your calculator for the nonlinear functions). Then use what you know about absolute values to sketch (without your calculator) the graph of $f(x) = |g(x)|$.

36. $f(x)$ from Problem 30   **37.** $f(x)$ from Problem 31
38. $f(x)$ from Problem 32   **39.** $f(x)$ from Problem 33
40. $f(x)$ from Problem 34   **41.** $f(x)$ from Problem 35

■ Sketch graphs of the functions below *by hand;* do not use a calculator.

42. $f(x) = |x^3|$ **43.** $f(x) = \left| \dfrac{1}{x} \right|$

44. $f(x) = |3x + 1|$

■ Sketch the graphs of the indicated functions *by hand;* do not use a calculator.

45. $f(x)$ as defined in Problem 26
46. $f(x)$ as defined in Problem 27
47. $f(x)$ as defined in Problem 28
48. $f(x)$ as defined in Problem 29

Applications

49. In your first job after graduating from college you make $36,000 a year before taxes. After 4 years you get a raise of $2500. After 2 more years you change jobs and go to work for a company that pays you $49,000 a year. Write a piecewise function that describes your yearly pre-tax income t years after you graduate from college.

50. Consider again the situation described in Problem 49. Write down a piecewise function that describes the total amount of money you have earned t years after graduating from college. How many years after graduating from college will you have earned a total of one million pre-tax dollars?

51. The accompanying table shows the Federal Tax Rate Schedule for single filers for the year 2000.

(a) How much tax would you owe if you made $18,000 of taxable income in the year 2000? If you made $180,000?

(b) What percentage of your taxable income did you owe in taxes if your taxable income in the year 2000 was $18,000? What if your taxable income was $180,000?

(c) Write a piecewise function describing the dollar amount of tax T owed by a single person with m dollars of taxable yearly income. Each piece of your function will be linear. Do the pieces "match up"? Does this make financial sense?

For Taxable Income That Is:		The Federal Tax Owed Is:		Of the
Over	But Not Over	This Amount	Plus This %	Amount Over
$0	$26,250	$0	15%	$0
$26,250	$63,550	$3,937.50	28%	$26,250
$63,550	$132,600	$14,381.50	31%	$63,550
$132,600	$288,350	$35,787.00	36%	$132,600
$288,350	—	$91,857.00	39.6%	$288,350

Table for Problem 51

1.5   Combinations of Functions

In this section we investigate sums, differences, products, quotients, and compositions of functions. We also define and examine *inverses* of functions, when they exist.

1.5.1 Arithmetic combinations of functions

One way to think of functions is as *operators;* that is, they take an input and "operate" on it to produce an output. In this section we think of functions another way: as *objects* that can be added to, subtracted from, and multiplied or divided by one another.

Real numbers are a simple example of such objects; given any two real numbers, we can add them together and get a new real number. Similarly, given any two functions, we can add them together and get a new function. How this new function "operates" on inputs will depend on how the original two functions operated.

? *Question*   Given $f(x) = x^2$ and $g(x) = \sqrt{x + 1}$, what do you think the function $f + g$ should do to an input x? What would be the formula for the function $(f + g)(x)$?   □

We have to state precisely what we want to mean by the combination functions kf (where k is a real number), $f + g$, $f - g$, $f \cdot g$, and $\frac{f}{g}$. For example, to define the function $f + g$, we must say what $f + g$ does to each input x. The "obvious" way to define $f + g$ is as the function whose output at x is the sum of the outputs of f and g at x.

DEFINITION 1.15

Arithmetic Combinations of Functions

For all x in the domain of f, and any real number k,

$$(kf)(x) := kf(x).$$

For all x in the domain of f *and* in the domain of g,

$$(f + g)(x) := f(x) + g(x),$$

$$(f - g)(x) := f(x) - g(x),$$

$$(f \cdot g)(x) := f(x) \cdot g(x).$$

For all x in the domain of f *and* in the domain of g *and* for which $g(x) \neq 0$,

$$\left(\frac{f}{g}\right)(x) := \frac{f(x)}{g(x)}.$$

❓ *Question* What are the domains of the functions kf, $f + g$, $f - g$, $f \cdot g$, and $\frac{f}{g}$ (in terms of the domains of f and g)? Can you write these domains in set notation? □

EXAMPLE 1.48

Calculating values of algebraic combinations of functions

If $f(1) = 5$, $f(2) = 3$, and $f(3) = 1$ while $g(1) = 6$, $g(2) = 0$, and $g(3) = 2$, calculate $(5f)(1)$, $(f + g)(2)$, $(f - g)(1)$, $(f \cdot g)(3)$, and $\left(\frac{f}{g}\right)(3)$.

Solution By Definition 1.15,

$$(5f)(1) = 5(f(1)) = 5(5) = 25$$

$$(f + g)(2) = f(2) + g(2) = 3 + 0 = 3$$

$$(f - g)(1) = f(1) - g(1) = 5 - 6 = -1$$

$$(f \cdot g)(3) = f(3) \cdot g(3) = (1)(2) = 2$$

$$\left(\frac{f}{g}\right)(3) = \frac{f(3)}{g(3)} \text{ is undefined.}$$

Note that since $g(2) = 0$, the function $\frac{f}{g}$ is not defined at $x = 2$. □

EXAMPLE 1.49

Describing algebraic combinations of functions and their domains

Given $f(x) = x^2$ and $g(x) = \sqrt{x + 1}$, describe the functions $f + g$, $f - g$, $f \cdot g$, and $\frac{f}{g}$. What are the domains of these combination functions?

Solution   The domain of f is $\mathbb{R}$ and the domain of g is $[-1, \infty)$. A value x is in *both* the domain of f and the domain of g if $x \in (-\infty, \infty) \cap [-1, \infty) = [-1, \infty)$. By Definition 1.15, the first three algebraic combinations are defined for all $x \in [-1, \infty)$ and can be described as

$$(f + g)(x) = f(x) + g(x) = x^2 + \sqrt{x + 1}$$
$$(f - g)(x) = f(x) - g(x) = x^2 - \sqrt{x + 1}$$
$$(f \cdot g)(x) = f(x) \cdot g(x) = x^2\sqrt{x + 1}.$$

We have $g(x) = 0$ only if $x = -1$, so for all $x \in (-1, \infty)$,

$$\left(\frac{f}{g}\right)(x) = \frac{f(x)}{g(x)} = \frac{x^2}{\sqrt{x + 1}}. \qquad \square$$

1.5.2 Graphs and tables of combinations of functions

Remember that the graph of a function f is the collection of ordered pairs $(x, f(x))$ for all x in the domain of f. Given a point $x = c$ in the domain of f, the value $f(c)$ is the height of the graph of f at the point $x = c$. This means that the graph of $(f + g)(x) = f(x) + g(x)$ consists of the ordered pairs $(x, f(x) + g(x))$ for all x in the domain of $f + g$. In a similar fashion we can describe the graphs of αf, $f - g$, $f \cdot g$, and $\frac{f}{g}$ in terms of the graphs of f and g.

EXAMPLE 1.50   **Graphing the sum of two functions**

Figures 1.42 and 1.43 show the graphs of two functions f and g with various marked points. Use these graphs (and the marked points) to sketch the graph of the function $f + g$.

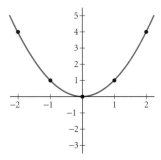

Figure 1.42 $f(x)$

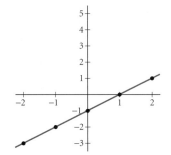

Figure 1.43 $g(x)$

Solution   A good place to start is to calculate the values of $(f + g)(x) = f(x) + g(x)$ for the values of x where we know $f(x)$ and $g(x)$. Doing this, we get

$$f(-2) + g(-2) = 4 + (-3) = 1$$
$$f(-1) + g(-1) = 1 + (-2) = -1$$
$$f(0) + g(0) = 0 + (-1) = -1$$
$$f(1) + g(1) = 1 + 0 = 1$$
$$f(2) + g(2) = 4 + 1 = 5.$$

Now we just have to plot these points and "connect the dots." The height of the graph of $f + g$ at $x = c$ is the sum of the height of the graph of f and the height of the graph of g at $x = c$ (see Figure 1.44). Figure 1.45 shows the resulting graph of $(f + g)(x)$.

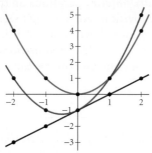

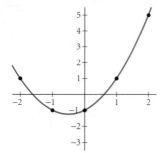

Figure 1.44

$f(x), g(x),$ and $(f + g)(x)$

Figure 1.45

$(f + g)(x)$ ☐

? *Question* Can you graph the functions $2f$, $f - g$, and $f \cdot g$, where f and g are the graphs from Example 1.50? Start by plotting some points on the graph of each function. ☐

✓ *Checking the Answer* To use a graphing calculator to check these graphs, you need to know the equations for the functions f and g. So that you can check, we will tell you that the function in Figure 1.42 is $f(x) = x^2$, and the function in Figure 1.43 is $g(x) = x - 1$. ☐

EXAMPLE 1.51

Completing a table of values for algebraic combinations of functions

Consider the functions f and g defined in Table 1.9. Add columns to the table for the functions $3f$, $f + g$, $f - g$, $f \cdot g$, and $\frac{f}{g}$.

x	$f(x)$	$g(x)$
0	5	3
1	3	2
2	1	0
3	2	5
4	4	4
5	8	3

Table 1.9

Solution Table 1.10 shows the values of $(3f)(x)$, $(f + g)(x)$, $(f - g)(x)$, $(fg)(x)$, and $\left(\frac{f}{g}\right)(x)$ for each value of x in the table. (*Note:* Since $g(2) = 0$, $\frac{f}{g}(x)$ is not defined for $x = 2$.)

x	$f(x)$	$g(x)$	$(3f)(x)$	$(f + g)(x)$	$(f - g)(x)$	$(fg)(x)$	$\left(\frac{f}{g}\right)(x)$
0	5	3	15	8	2	15	$\frac{5}{3}$
1	3	2	9	5	1	6	$\frac{3}{2}$
2	1	0	3	1	1	0	*Undef.*
3	2	5	6	7	-3	10	$\frac{2}{5}$
4	4	4	12	8	0	16	1
5	8	3	24	11	5	24	$\frac{8}{3}$

Table 1.10 ☐

1.5.3 Compositions of functions

There are four *arithmetic* operations on the real numbers: addition, subtraction, multiplication, and division. We have just seen that we can combine functions with these arithmetic operations. There is an additional operation on functions that we do not have for numbers; it is called *composition.* We *compose* two functions f and g by taking the output from one function as the input for the other, as the following definition indicates.

DEFINITION 1.16

The Composition of Two Functions

The *composition* $f \circ g$ of two functions f and g is defined to be the function

$$(f \circ g)(x) := f(g(x)),$$

for all x such that x is in the domain of g and $g(x)$ is in the domain of f.

The notation $(f \circ g)$ is pronounced " f composed with g" or sometimes " f circle g." The notation $f(g(x))$ is pronounced " f of g of x."

Caution   You may have noticed that the notation for composition looks a bit like multiplication. However, composition and multiplication of functions are very different. When we want to denote multiplication, we use a small "closed" dot or no dot at all. To denote composition, we always use an "open circle."   □

The composition $f \circ g$ of two functions f and g is a function that applies first the function g, and *then* the function f, to an input x. This can be represented schematically by

$$x \xrightarrow{\;g\;} g(x) \xrightarrow{\;f\;} f(g(x)).$$

For example, if $f(x) = x + 1$ and $g(x) = x^2$, then the composition $(f \circ g)(x)$ squares x and then adds 1:

$$x \xrightarrow{\;g\;} x^2 \xrightarrow{\;f\;} x^2 + 1.$$

On the other hand, with the same functions f and g, the composition $(g \circ f)(x)$ adds 1 to x and then squares the result:

$$x \xrightarrow{\;f\;} x + 1 \xrightarrow{\;g\;} (x + 1)^2.$$

EXAMPLE 1.52

Calculating values of compositions of functions

Given $f(x) = 3x + 1$ and $g(x) = x^2 - 1$, we can calculate values of the compositions $f \circ g$ and $g \circ f$ by using values of f and g:

$$(f \circ g)(2) = f(g(2)) = f(2^2 - 1) = f(3) = 3(3) + 1 = 10$$

$$(g \circ f)(2) = g(f(2)) = g(3(2) + 1) = g(7) = 7^2 - 1 = 48. \qquad □$$

Caution   The compositions $f \circ g$ and $g \circ f$ are *not* in general the same function! In Example 1.52, $f \circ g$ and $g \circ f$ have different values at $x = 2$ and thus are different functions. The operation of composition is *not* commutative.   □

⑤ *Caution* Consider a composition $(f \circ g)(x)$ of functions. When we read this composition from left to right, we see the function f first. However, when we evaluate $(f \circ g)(2)$, we apply the functions in the reverse order: We first apply g to $x = 2$, and then apply f to the result $g(2)$.   □

EXAMPLE 1.53

Finding a formula for a composition of functions

Given the functions $f(x) = 3x + 1$ and $g(x) = x^2 - 1$, we can easily find formulas for the compositions $f \circ g$ and $g \circ f$:

$$(f \circ g)(x) = f(g(x)) = f(x^2 - 1) = 3(x^2 - 1) + 1$$

$$= 3x^2 - 3 + 1 = 3x^2 - 2$$

$$(g \circ f)(x) = g(f(x)) = g(3x + 1) = (3x + 1)^2 - 1$$

$$= 9x^2 + 6x + 1 - 1 = 9x^2 + 6x.$$

Again note that $(f \circ g)(x) \neq (g \circ f)(x)$!   □

Finding the domain of a composition of functions can be harder than finding the domain of a sum or product of functions. In order for a value x to be in the domain of the composition $(f \circ g)(x) = f(g(x))$, it must first be in the domain of g. Then the value $g(x)$ must be in the domain of the function f. In set notation, the domain of the composition $(f \circ g)(x)$ is

(1) $\text{Domain}(f \circ g) = \{x \mid x \in \text{Domain}(g) \text{ and } g(x) \in \text{Domain}(f)\}.$

❓ *Question* Write the domain of the composition $g \circ f$ in set notation (in terms of the domains of f and g).   □

EXAMPLE 1.54

Finding the domain of a composition

If $f(x) = \frac{3}{x^2 - 4}$ and $g(x) = \sqrt{x - 1}$, find the domain of $f \circ g$.

Solution The domain of $f(x)$ is $x \neq \pm 2$ and the domain of $g(x)$ is $x \geq 1$. Therefore, the domain of $f \circ g$ is

$$\text{Domain}(f \circ g) = \{x \mid x \geq 1 \text{ and } \sqrt{x - 1} \neq \pm 2\}.$$

Note that $\sqrt{x - 1} \neq \pm 2 \Rightarrow x - 1 \neq 4 \Rightarrow x \neq 5$. Thus we have

$$\text{Domain}(f \circ g) = \{x \mid x \geq 1 \text{ and } x \neq 5\} = [1, 5) \cup (5, \infty).$$

Alternatively, we could find an equation for the composition $f \circ g$ and then find its domain (provided that we do not simplify the equation before doing so). We have

$$(f \circ g)(x) = f(g(x)) = f(\sqrt{x - 1}) = \frac{3}{(\sqrt{x - 1})^2 - 4}.$$

Since we cannot take the square root of a negative number, we must have $x \geq 1$. Since we cannot divide by zero, we must also have

$$(\sqrt{x - 1})^2 - 4 \neq 0 \quad \Rightarrow \quad x - 1 \neq 4 \quad \Rightarrow \quad x \neq 5.$$

Combining these conditions, we see that the domain of $f \circ g$ is $[1, 5) \cup (5, \infty)$.   □

⟡ *Caution*   Be careful that you do *not* simplify $f \circ g$ before attempting to find the domain. In Example 1.54 we have $(f \circ g)(x) = \frac{3}{(\sqrt{x-1})^2-4}$, which is equal to the function $\frac{3}{(x-1)-4} = \frac{3}{x-5}$ only at the points where it is defined. The domain of the simplified function is $x \neq 5$, which is not equal to the domain of $f \circ g$. □

1.5.4 Tables and graphs of compositions of functions

In this section we give examples of finding the table or graph for a composition $f \circ g$ or $g \circ f$, given tables or graphs for functions f and g.

EXAMPLE 1.55

Completing a table of values for a composition of functions

Table 1.11 defines two functions f and g. Add columns for the functions $f \circ g$ and $g \circ f$.

x	$f(x)$	$g(x)$
0	5	3
1	3	2
2	1	0
3	2	5
4	4	4
5	8	3

Table 1.11

From Table 1.11 we can calculate certain values of the functions $f \circ g$ and $g \circ f$. For example, $(f \circ g)(2) = f(g(2)) = f(0) = 5$. Table 1.12 shows each value of $f \circ g$ and $g \circ f$ that we can calculate from Table 1.11.

x	$f(x)$	$g(x)$	$(f \circ g)(x)$	$(g \circ f)(x)$
0	5	3	2	3
1	3	2	1	5
2	1	0	5	2
3	2	5	8	0
4	4	4	4	4
5	8	3	2	??

Table 1.12

Note that we cannot calculate $(g \circ f)(5) = g(f(5)) = g(8)$ because the table does not give a value for $g(8)$. In this example, Domain$(g \circ f) = \{0, 1, 2, 3, 4\}$. □

EXAMPLE 1.56

Graphing a composition of functions

To graph a composition $f \circ g$ given graphs of f and g, we use the given graphs to approximate values of f and g. To find the value of $f \circ g$ at some $x = c$, first find the value of $g(c)$ using the graph of g. Then use the graph of f to find the value of f at this new value. After plotting points, try to "connect the dots" in a reasonable way. Figures 1.46 and 1.47 are graphs of functions f and g. Figure 1.48 is the graph of the composition $f \circ g$.

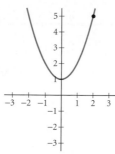

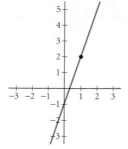

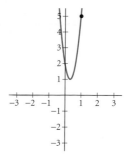

Figure 1.46   **Figure 1.47**   **Figure 1.48**
$f(x)$ $g(x)$ $(f \circ g)(x)$

The marked points illustrate how to find the point $(f \circ g)(1)$. Since $g(1) = 2$ and $f(g(1)) = f(2) = 5$, we know that $(f \circ g)(1) = 5$. You will have to plot quite a few points on the graph of $f \circ g$ to get an accurate sketch. □

❓ Question   Can you graph the composition $g \circ f$ given the graphs of f and g in Example 1.56? □

1.5.5 Decomposing functions

Given a function f, can we find functions g and h such that $f = g \circ h$? Being able to **decompose** functions in this manner will be especially important in Section 7.2.2. As we will see in the next example, there can be more than one way to decompose a given function f.

EXAMPLE 1.57   **Decomposing a function**

Let $f(x) = (x^2 + x + 1)^2$. Find functions g and h such that $f = g \circ h$.

Solution   There are many possible answers. Perhaps the most obvious is to let $g = x^2$ (the "outside part" of the function $f(x)$) and to let $h(x) = x^2 + x + 1$ (the "inside part" of $f(x)$). Then

$$(g \circ h)(x) = g(h(x)) = g(x^2 + x + 1) = (x^2 + x + 1)^2 = f(x).$$

Another solution is to let $h(x) = x^2 + x$ and $g(x) = (x + 1)^2$. Then we have

$$(g \circ h)(x) = g(h(x)) = g(x^2 + x) = ((x^2 + x) + 1)^2 = (x^2 + x + 1)^2 = f(x).$$

Can you think of any other ways to decompose $f(x) = (x^2 + x + 1)^2$? □

In general, given *any* function $f(x)$, it is possible to write f as a trivial composition $f = g \circ h$. If $g(x) = f(x)$ and $h(x) = x$, then

$$(g \circ h)(x) = g(h(x)) = g(x) = f(x).$$

On the other hand, if $g(x) = x$ and $h(x) = f(x)$, then

$$(g \circ h)(x) = g(h(x)) = g(f(x)) = f(x).$$

Although these are not very useful decompositions, they can always be done.

What you should know after reading Section 1.5

▶ The definitions and domains of kf, $f + g$, $f - g$, fg, $\frac{f}{g}$, $f \circ g$, and $g \circ f$. How to calculate values of, or formulas for, these combination functions, given values or formulas for f and g.

▶ How to sketch the graph of (or make a table for) an arithmetic combination or composition of f and g, given the graphs of (or tables for) f and g.

▶ How to write a complicated function as a composition of two or more simpler functions (also known as "decomposing" a function).

Exercises 1.5

Concepts

0. Read the section and make your own summary of the material. Include whatever you think will help you review the material later on.

■ Suppose f has domain $[1, \infty)$, and g has domain $[-4, 4]$. Suppose also that $f(x)$ is zero only at $x = 2$ and $x = 5$ and that $g(x)$ is zero only at $x = -1$ and $x = 1$. Find the domains of the following functions (if possible). For one of these problems there is not enough information to find the domain; explain why.

1. $3f$ **2.** $f - g$ **3.** fg **4.** $3f + 4g$

5. $\dfrac{f}{g}$ **6.** $\dfrac{g}{f}$ **7.** $\dfrac{1}{fg}$ **8.** $\dfrac{1}{f + g}$

■ Describe the domains of the following functions in terms of the domains of f, g, and h.

9. $6g$ **10.** $f + g$ **11.** $\dfrac{f}{g}$

12. $\dfrac{1}{fg}$ **13.** $f \circ g$ **14.** $g \circ f$

15. $f \circ f$ **16.** $f \circ g \circ h$ **17.** $h \circ f \circ h$

18. $(f + g) \circ h$ **19.** $(f \circ h) - (h \circ g)$

20. $\dfrac{f \circ g}{gh}$

■ Given that f is a function with domain $[0, \infty)$ and g is a function with domain $(-5, 5)$, find the domain of each of the following functions. (*Hint:* These functions are compositions.)

21. $f(x^2 - 1)$ **22.** $f\left(\dfrac{1}{x - 1}\right)$

23. $g\left(\dfrac{1}{x}\right)$ **24.** $g(\sqrt{x} - 6)$

■ Define the following combinations of functions in terms of the functions f, g, and h. That is, describe how each of these functions operates on an input x, as we did in Definition 1.15. If $f(2) = 0$, $g(2) = 8$, and $h(2) = -1$, calculate the value of each function at $x = 2$.

25. $3f + 4g$ **26.** $f + gh$

27. $\dfrac{f + g}{gh}$ **28.** $2(h + 3)$

■ If $f(0) = 2$, $f(1) = 2$, and $f(2) = 0$ while $g(0) = 1$, $g(1) = 0$, and $g(2) = 1$, calculate the value of each function below at $x = 1$.

29. $f \circ g$ **30.** $g \circ f$

31. $f \circ f$ **32.** $g \circ f \circ g$

33. Suppose f is a function with domain $[-2, \infty)$ and range $[-3, 3]$, and let g be a function with domain $[-10, \infty)$ and range $[0, \infty)$.

(a) What is the domain of the composition $g \circ f$ (in interval notation)? Justify your answer.

(b) It is not possible to determine the domain of $f \circ g$ in this example; explain why not. What extra information would you have to know to be able to determine the domain of $f \circ g$?

(c) Is there enough information here to determine the domain of the composition $f \circ f$? What about the domain of the function $g \circ g$?

34. If f and g are functions, describe the ordered pairs on the graph of the function $\frac{f}{g}$ (using set notation). How is the height of the graph of $\frac{f}{g}$ at $x = 2$ related to the heights of the graphs of f and g at $x = 2$?

35. Explain how to find the height of the graph of a composition $g \circ f$ at $x = 2$ by using the graphs of f and g.

36. Let f be the function defined by the accompanying graph on the left. Suppose the graph on the right is the graph of the product fg of the function f with another function g. Use the graphs to estimate the values of $g(-1)$, $g(0)$, and $g(1)$.

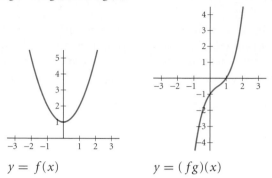

$$y = f(x) \qquad\qquad y = (fg)(x)$$

37. Find functions f and g that illustrate that $f \circ g \neq g \circ f$. Can you think of any examples where $f \circ g$ is equal to $g \circ f$?

38. Use the values given in the table at the bottom of the page to fill in the missing values. (*Note:* There is only one correct way to fill in the table. Don't make up values; use logic to deduce the missing values from the given values.)

39. Use the values given in the table at the bottom of the page to fill in the missing values. (*Note:* There is only one correct way to fill in the table. *Also note:* The functions f, g, and h are *not* necessarily one-to-one, so values in the columns for f, g, and h *can* be repeated. Don't give up; this is a hard problem, but it can be done! You won't necessarily be able to fill in the table "in order.")

40. Explain why Problem 39 is so much harder than Problem 38.

41. Use compositions to answer each of the following questions.

(a) If $g(x) = x^2$ and $f(g(x)) = \dfrac{1}{x^2 + 1}$, what is $f(x)$?

(b) If $h(x) = x^2 - 1$ and $h(l(x)) = \dfrac{1}{x^4} - 1$, what is $l(x)$?

(c) If $r(t) = 1 + t + t^3$ and $r(s(t)) = 1 + t + t^3$, what is $s(t)$?

(d) If $u(x) = \dfrac{1}{1-x}$ and $y(u(x)) = \dfrac{1}{1-x}$, what is $y(x)$?

Skills

■ Suppose that $f(x) = x^2 + 1$, $g(x) = \frac{1}{x}$, and $h(x) = 1 - 2x + x^2$. For each problem below, describe the function given, find the domain, and calculate the value of the function at $x = 1$.

42. $f - 2g$ **43.** fgh **44.** $\dfrac{f}{g}$

45. $\dfrac{3g - h}{f}$ **46.** $\dfrac{f-1}{h}$ **47.** $2g^2 - f$

48. $(h - x^3)^2$ **49.** $xg - 1$ **50.** $f \circ g$

51. $g \circ f$ **52.** $f \circ h$ **53.** $h \circ g$

54. $f \circ f$ **55.** $g \circ g$

56. $g \circ h \circ f$ **57.** $h \circ g \circ g$

58. $(f + h) \circ g$ **59.** $(f \circ g) + (h \circ g)$

60. $\dfrac{f \circ h}{g}$ **61.** $\dfrac{f \circ g}{h \circ g}$

■ The table below defines three functions f, g, and h with domain $\{0, 1, 2, 3, 4, 5, 6\}$. In Problems 62–69, create for this table another row that describes the given function.

x	0	1	2	3	4	5	6
$f(x)$	0	-1	-2	-1	1	5	6
$g(x)$	2	4	6	3	-1	-4	-3
$h(x)$	4	3	2	1	0	-1	-2

62. $g - f$ **63.** $3f + g - 2h$ **64.** $1 - g$

65. fh **66.** $\dfrac{f}{g}$ **67.** $\dfrac{fh}{g}$

68. $\dfrac{g - f}{h}$ **69.** $\dfrac{1}{fh}$

x	$f(x)$	$g(x)$	$h(x)$	$(-2f)(x)$	$(f + 2g - h)(x)$	$(fg)(x)$	$\left(\frac{g}{h}\right)(x)$
0	1	0			4		
1			1	4			3
2		-2				2	-1

Table for Problem 38

x	$f(x)$	$g(x)$	$h(x)$	$(f \circ g)(x)$	$(h \circ f)(x)$	$(g \circ g)(x)$	$(g \circ f \circ h)(x)$
1	4	3			3		2
2	3		2	4			
3	1	1	4				
4	2	2			2		

Table for Problem 39

■ The table below defines three functions f, g, and h with domain $\{0, 1, 2, 3, 4, 5, 6\}$. In Problems 70–77, create for this table another row that describes the given function.

x	0	1	2	3	4	5	6
$f(x)$	2	4	6	5	3	1	0
$g(x)$	5	3	2	2	0	1	4
$h(x)$	1	0	3	4	6	4	5

70. $f \circ g$ **71.** $g \circ f$ **72.** $h \circ f$

73. $g \circ h$ **74.** $h \circ g \circ f$ **75.** $f \circ h \circ h$

76. $g \circ g$ **77.** $f \circ f \circ f$

■ The table below defines three functions f, g, and h with domain $\{0, 1, 2, 3, 4, 5, 6\}$. In Problems 78–81, create for this table another row that describes the given function.

x	0	1	2	3	4	5	6
$f(x)$	0	1	3	2	3	0	2
$g(x)$	1	0	1	1	0	1	0
$h(x)$	3	2	0	3	2	3	1

78. $f \circ (g + h)$ **79.** $(f \circ g) + (f \circ h)$

80. $(h \circ f) + g$ **81.** $(fg) \circ h$

■ Using the graphs of f, g, and h given below, sketch the graphs of the functions in Problems 82–97. Label at least four points on each graph you sketch.

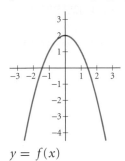

$y = f(x)$

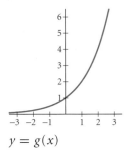

$y = g(x)$

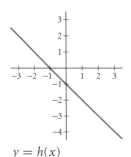

$y = h(x)$

82. $f + g$ **83.** $h - g$ **84.** $2f - h$ **85.** $-0.5f$

86. fg **87.** $\dfrac{1}{f}$ **88.** $\dfrac{h}{g}$ **89.** $\dfrac{g}{h}$

90. $f \circ h$ **91.** $h \circ g$ **92.** $f \circ f$ **93.** $f \circ g \circ h$

94. $(f + g) \circ h$ **95.** $h - (g \circ f)$ **96.** $f \circ \left(\dfrac{1}{g}\right)$

97. $\dfrac{g \circ g}{g}$

■ Find *two* nontrivial ways to write each of the following functions f as a composition $f = g \circ h$. "Nontrivial" means that you should *not* choose $g(x) = x$ or $h(x) = x$.

98. $f(x) = (3x + 5)^2$ **99.** $f(x) = 3x^2 + 5$

100. $f(x) = 3(x + 5)^2$ **101.** $f(x) = \dfrac{6}{x + 1}$

102. $f(x) = \dfrac{6x^2}{6x^2 + 1}$ **103.** $f(x) = \dfrac{(x - 4)^2}{1 + (x - 4)^2}$

■ Write each function as a composition $g \circ h$ for some functions g and h.

104. $f(x) + 1$ **105.** $f(x + 1)$ **106.** $2f(x)$ **107.** $f(2x)$

■ Suppose $f(x)$ and $g(x)$ are the piecewise functions defined below. For each problem, find the value of the indicated combination or composition at $x = -1$, $x = 0$, $x = 1$, $x = 2$, and $x = 3$. Then write the given combination or composition as a piecewise function.

$$f(x) := \begin{cases} 2x + 1, & \text{if } x \leq 0 \\ x^2, & \text{if } x > 0 \end{cases} \qquad g(x) := \begin{cases} -x, & \text{if } x < 2 \\ 5, & \text{if } x \geq 2 \end{cases}$$

108. $3f$ **109.** $f + g$ **110.** $f \circ g$ **111.** $g \circ f$

Applications

112. CarpetKing charges \$4.25 per square foot for its Deluxe ThriftySoft carpet, plus a flat fee of \$200.00 for delivery and installation.

   (a) If a square room measures x feet on a side, and S is the number of square feet of floor in the room, write down S as a function of x.

   (b) Write down a function that describes the cost C of carpeting a room enclosing S square feet.

   (c) Write down a function that describes the cost C of carpeting a square room that measures x feet on a side. Explain how this function is a composition.

113. The first table below shows the numbers of deer in the Happyland Forest Park from 1990 to 1995. The number of deer seems to affect the number of ticks in the park; the second table shows the numbers of ticks that can be expected for various numbers of deer in the park.

   (a) Use the tables to estimate the number of ticks that were in the park in 1995.

   (b) Make a table that predicts the numbers of ticks in the park from 1990 to 1995.

   (c) Explain how your table represents a composition of the deer and tick tables.

Year	1990	1991	1992	1993	1994	1995
Deer	183	180	177	179	184	181

Deer	177	178	179	180	181	182	183	184
Ticks	850	855	860	865	870	875	880	885

1.6 Transformations and Symmetry

We will now examine the graphical effects of adding constants to, and multiplying constants by, the independent and dependent variables of a function $y = f(x)$. These **transformations** of $y = f(x)$ are special cases of the combinations and compositions covered in the previous section.

1.6.1 Translations

If we add a constant to the independent or the dependent variable of a function $y = f(x)$, how does the graph change? As we will soon see, adding a constant to the dependent variable results in a vertical shift of the graph, and adding a constant to the independent variable causes a horizontal shift of the graph. In either case, we say that the new graph is a **translation** of the original graph.

THEOREM 1.5

Vertical and Horizontal Translations

Let f be a function and C be any nonzero real number.

(a) The graph of the function $f(x) + C$ is shifted vertically C units from the graph of $f(x)$. The graph shifts up if $C > 0$ and down if $C < 0$.

(b) The graph of the function $f(x + C)$ is shifted horizontally C units from the graph of $f(x)$. The graph shifts to the right if $C < 0$ and to the left if $C > 0$.

❧ Caution   Notice that the graph of $f(x + C)$ is shifted to the *right* if C is negative and to the *left* if C is positive. This may be the opposite of what you expected! You'll see why this happens in the proof of Theorem 1.6 and in Examples 1.60 and 1.61.   ☐

❧ Caution   Just because we write $f(x) + C$ doesn't mean that C is necessarily positive. We could have, for example, $C = -2$.   ☐

EXAMPLE 1.58

The graphs of vertical and horizontal translations

Consider the function $f(x)$ graphed in Figure 1.49. Figure 1.50 shows the graph of $y = f(x) + 1$ (shifted one unit up from the graph of $y = f(x)$). Figure 1.51 shows the graph of $y = f(x + 1)$ (shifted one unit to the left of $y = f(x)$).

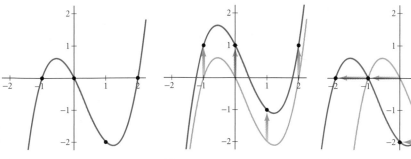

Figure 1.49
$y = f(x)$

Figure 1.50
$y = f(x) + 1$

Figure 1.51
$y = f(x + 1)$ ☐

Before we can prove Theorem 1.5, we have to specify exactly what we mean when we say that a graph is "shifted" up, down, or to the right or left, C units. A **vertical shift** of the graph of $f(x)$ is a graph consisting of ordered pairs with the same x-coordinates as the graph of $f(x)$ but whose corresponding y-coordinates are each C units higher or lower. A **horizontal shift** of the graph of $f(x)$ is a graph consisting of ordered pairs with the same y-coordinates as the graph of $f(x)$ but whose corresponding x-coordinates are each C units greater (for a right shift) or less (for a left shift).

With these definitions, Theorem 1.5 is equivalent to the following theorem:

THEOREM 1.6

Algebraic Descriptions of Vertical and Horizontal Shifts

Let f be a function and C be any nonzero real number.

(a) If (x, y) is a point on the graph of $f(x)$, then $(x, y + C)$ is a point on the graph of $f(x) + C$.

(b) If (x, y) is a point on the graph of $f(x)$, then $(x - C, y)$ is a point on the graph of $f(x + C)$.

PROOF (**THEOREM 1.6**)   Part (a) of the theorem is obvious, since the domains of $f(x)$ and $f(x) + C$ are the same, and by definition each point on the graph of $f(x) + C$ is of the form $(x, f(x) + C)$.

To prove part (b), let $g(x) = f(x + C)$. Suppose (x, y) is a point on the graph of $f(x)$ (i.e., suppose $y = f(x)$). Then $g(x - C) = f((x - C) + C) = f(x) = y$, so $(x - C, y)$ is on the graph of $g(x) = f(x + C)$. ∎

EXAMPLE 1.59

Graphs of vertical translations

Consider the function $f(x) = x^2$. According to Theorem 1.5, the graph of $f(x) + 2 = x^2 + 2$ should be exactly the same as the graph of $f(x) = x^2$ but shifted up two units. Similarly, the graph of $f(x) - 2 = x^2 - 2$ will be shifted down two units from the graph of $f(x) = x^2$. See Figures 1.52, 1.53, and 1.54.

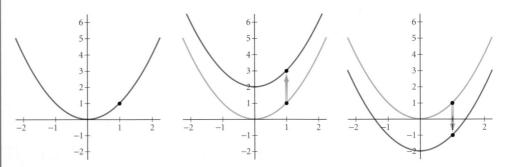

Figure 1.52
$f(x) = x^2$

Figure 1.53
$f(x) + 2 = x^2 + 2$

Figure 1.54
$f(x) - 2 = x^2 - 2$

The point $(1, 1)$ on the original graph becomes the point $(1, 3)$ on the graph of $y = x^2 + 2$. This is because we are adding two units to the dependent variable y in this transformation, so if $f(1) = 1$, then $f(1) + 2 = 1 + 2 = 3$. Each point on the graph in Figure 1.52 is translated up two units in Figure 1.53. In Figure 1.54, the point $(1, 1)$ on the graph of $f(x) = x^2$ gets translated to the point $(1, -1)$, two units down on the graph of $f(x) - 2 = x^2 - 2$. □

EXAMPLE 1.60

Graphs of horizontal translations

On the other hand, if we add or subtract 2 from the *independent* variable of the function $f(x) = x^2$, we get a *horizontal* shift of the graph. The graph of $f(x - 2) = (x - 2)^2$ will be shifted to the right two units from the graph of $f(x) = x^2$. The graph of $f(x + 2) = (x + 2)^2$ will be shifted two units to the left. Figures 1.55, 1.56, and 1.57 show the original graph of $f(x) = x^2$ and these horizontal translations.

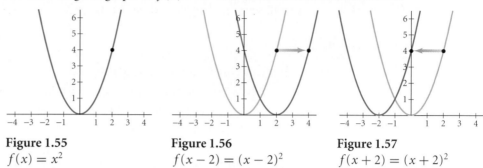

Figure 1.55
$f(x) = x^2$

Figure 1.56
$f(x - 2) = (x - 2)^2$

Figure 1.57
$f(x + 2) = (x + 2)^2$

Note: These graphs have a different scale than those in Example 1.59 so that we can better illustrate the horizontal translations.

Why, then, did the graph move to the *right* when we *subtracted* 2 from the independent variable? Notice that the graph of $y = x^2$ contains the point $(2, 4)$. Now consider the function $y = (x - 2)^2$; what would we have to substitute for x to get a height of 4 for y? Since two units will be subtracted from any x that we put into $y = (x - 2)^2$, we have to choose a number that is two units *greater* than $x = 2$ to get a y-value of 4. Since $(2, 4)$ is on the graph of $f(x)$, the point $(4, 4)$ is on the graph of $f(x - 2)$; each point gets shifted to the *right* two units. Similarly, each point on the graph of $f(x + 2)$ is shifted to the *left* two units from the corresponding point on the graph of $f(x)$. □

Perhaps the reason why the graph of the translation $f(x - C)$ shifts C units to the *right* of the graph of $f(x)$ is better illustrated with a table.

EXAMPLE 1.61

Completing a table of values for a horizontal translation

Consider the function $f(x)$ defined by Table 1.13. Make a table that describes the translation $f(x - 2)$ of $f(x)$.

x	0	1	2	3	4
$f(x)$	4	8	10	12	14

Table 1.13

Solution   Your first thought might be to make a table with x-values 0, 1, 2, 3, and 4. However, a quick calculation shows that this won't work. What is $f(x - 2)$ when $x = 0$? In order to know this we would have to know the value of $f(0 - 2) = f(-2)$, which is not given in Table 1.13.

Since we will have to subtract two units from every value of x we choose before applying the function f, our table for $f(x - 2)$ will have to involve the x-values 2, 3, 4, 5, and 6. Now when $x = 2$, we have $f(x - 2) = f(2 - 2) = f(0) = 4$ (using Table 1.13). Calculate $f(x - 2)$ for each of our x-values to produce Table 1.14.

x	2	3	4	5	6
$f(x - 2)$	4	8	10	12	14

Table 1.14

The x-values are shifted to the right two units, whereas the y-values for $f(x - 2)$ are the same as the values for $f(x)$ given in Table 1.13. Graph the data for $f(x)$ in Table 1.13 and the data for $f(x - 2)$ in Table 1.14 to see the horizontal translation to the right. □

1.6.2 Stretches

We now know what happens when we *add* a constant to the independent or dependent variable of a function. In this section we examine what happens when we *multiply* the independent or dependent variable by a constant. The graphs of $Af(x)$ and $f(Ax)$ will be vertical and horizontal stretches, respectively, of the graph of $f(x)$. For now we consider only the cases where the constant A is a *positive* real number; in Section 1.6.4 we will see what happens when A is negative.

THEOREM 1.7

Vertical and Horizontal Stretches

Let f be a function and A be any *positive* real number.

(a) The graph of the function $Af(x)$ is a vertical stretch (or shrink) of the graph of $f(x)$. The graph of f stretches vertically if $A > 1$ and shrinks vertically if $0 < A < 1$.

(b) The graph of the function $f(Ax)$ is a horizontal stretch (or shrink) of the graph of $f(x)$. The graph of f shrinks horizontally if $A > 1$ and stretches horizontally if $0 < A < 1$.

◈ **Caution**   Once again, modifying the independent variable leads to a result that may be the opposite of what you expected. Multiplying the independent variable by a number A that is *larger* than 1 causes the graph of f to *shrink* horizontally! For example, if $A = 3$, then the graph of $f(3x)$ is *shrunk* horizontally by a factor of 3 when compared to the graph of $f(x)$ (see Example 1.62).   □

EXAMPLE 1.62

Graphs of vertical and horizontal stretches and shrinks

Consider the graph of the function $f(x)$ shown in Figure 1.58. The graph of $y = 3f(x)$ is obtained from this graph by stretching the graph vertically by a factor of 3. The point $(1, -1.2)$ on the graph of $f(x)$ in Figure 1.58 becomes the point $(1, -3.6)$ on the graph of $3f(x)$ in Figure 1.59. On the other hand, the graph of $y = f(3x)$ is obtained from the graph of $f(x)$ by shrinking the graph horizontally by a factor of 3 (Figure 1.60). In this graph, the y-coordinates are the same as those in Figure 1.58, but each corresponding x-coordinate is three times closer to the y-axis.

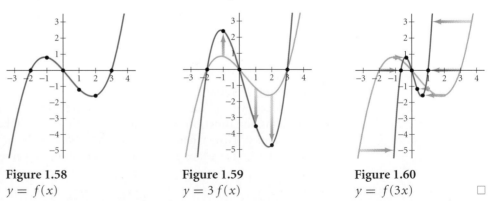

Figure 1.58
$y = f(x)$

Figure 1.59
$y = 3f(x)$

Figure 1.60
$y = f(3x)$ □

Again we must be mathematically precise about what we mean by a "stretch" or a "shrink" of the graph of a function. A ***vertical stretch by a factor of A*** of the graph of $f(x)$ is a graph consisting of ordered pairs with the same x-coordinates as the graph of $f(x)$ but whose corresponding y-coordinates are A times larger. A ***vertical shrink*** *by a factor of A* of the graph of $f(x)$ is a graph consisting of ordered pairs with the same x-coordinates as the graph of $f(x)$ but whose corresponding y-coordinates are A times smaller. Despite the fact that we use different terminology for a "stretch" and a "shrink," in both cases we really have the same situation; the x-coordinates stay the same, and the y-coordinates are all multiplied by the constant A. Horizontal stretches and shrinks can be defined in a similar manner.

THEOREM 1.8

Algebraic Descriptions of Vertical and Horizontal Stretches and Shrinks

Let f be a function and A be any real number.

(a) If (x, y) is a point on the graph of $f(x)$, then (x, Ay) is a point on the graph of $Af(x)$.

(b) If (x, y) is a point on the graph of $f(x)$, then $(\frac{1}{A}x, y)$ is a point on the graph of $f(Ax)$.

Theorem 1.8 applies regardless of whether A is positive or negative (or greater than or less than 1). The proof of Theorem 1.8 is left to you in the exercises.

EXAMPLE 1.63

Graphs of vertical stretches and shrinks

Consider the function $f(x) = 2 - x^2$. If we multiply the dependent variable of this function by $A = 2$, we get the function $2f(x) = 2(2 - x^2)$. According to Theorem 1.7, the graph of $2f(x)$ will be a vertical stretch by a factor of 2 of the graph $f(x)$. This means that each height on the graph of $2f(x)$ is 2 times the height of the corresponding point on the graph of $f(x)$. In general, the point $(x, f(x))$ on the graph of f becomes the point $(x, 2f(x))$ on the graph of $2f(x)$.

Similarly, if we multiply the independent variable by the constant $A = \frac{1}{2}$, each point on the graph of $\frac{1}{2}f(x)$ will be half as high as the corresponding point on the graph of $f(x)$. In other words, the point $(x, f(x))$ on the graph of f will become the point $(x, \frac{1}{2}f(x))$ on the graph of the transformation $\frac{1}{2}f(x)$. We call this a vertical shrink of the graph by a factor of $\frac{1}{1/2} = 2$.

Figures 1.61–1.63 show the graphs of $f(x)$ and the two transformations we have described. Note what happens to the point $(1, 1)$ on the graph of $f(x)$ under each transformation.

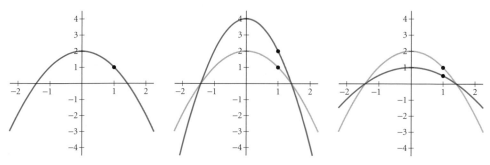

Figure 1.61
$f(x) = 2 - x^2$

Figure 1.62
$2f(x) = 2(2 - x^2)$

Figure 1.63
$\frac{1}{2}f(x) = \frac{1}{2}(2 - x^2)$ □

❓ *Question*   What happens to points on the graph of $f(x)$ that lie on the x-axis when we stretch the graph of f vertically by a factor of 2? (*Hint:* What is the "height" of each of these points?) □

◈ *Caution*   The transformations shown in Figures 1.62 and 1.63 are *not* translations. For example, in Figure 1.62, points above the *x*-axis move farther up and points below the *x*-axis move farther down (by different amounts, but each by a factor of 2). If this transformation were a translation, all the points would move up by the same amount (or down by the same amount).   □

EXAMPLE 1.64   **Graphs of horizontal stretches and shrinks**

If we multiply the *independent* variable of the function $f(x) = 2 - x^2$ by a constant, we get a *horizontal* stretch or shrink. For example, by Theorem 1.7, if $A = 2$ then the graph of $f(2x) = 2 - (2x)^2$ should be a horizontal *shrink* of the graph of $f(x)$ by a factor of 2.

Why is the graph of $f(2x)$ a horizontal *shrink* of the graph of $f(x)$? Again, modifying the independent variable does the opposite of what you might expect. Note that the point $(1, 1)$ is on the graph of $f(x)$. What value of x will make $f(2x)$ equal to the same height of one unit? If we choose $x = \frac{1}{2}$, then we will have $2x = 2(\frac{1}{2}) = 1$ and thus $f(2(\frac{1}{2})) = f(1) = 1$. We have to use a value for x that is *half* of the value we want to input into $f(x)$.

Similarly, if we choose $A = \frac{1}{2}$, then we get the transformation $f(\frac{1}{2}x) = 2 - (\frac{1}{2}x)^2$. By Theorem 1.7, the graph of $f(\frac{1}{2}x)$ should be a horizontal *stretch* of the graph of $f(x)$ by a factor of $\frac{1}{1/2} = 2$. In other words, the point $(x, f(x))$ on the graph of $f(x)$ will become the point $(2x, f(x))$ under this transformation.

Figures 1.64–1.66 illustrate the functions $f(x)$, $f(2x)$, and $f(\frac{1}{2}x)$. Note in particular what happens to the point $(1, 1)$ under each transformation.

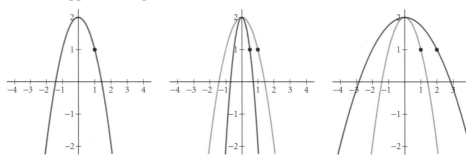

Figure 1.64   **Figure 1.65**   **Figure 1.66**
$f(x) = 2 - x^2$ $f(2x) = 2 - (2x)^2$ $f(\frac{1}{2}x) = 2 - (\frac{1}{2}x)^2$

Note: We are using the same function that we used in Example 1.63, but with a different viewing window so we can better see these horizontal transformations.   □

❓ *Question*   What happens to points on the graph of $f(x)$ that lie on the *y*-axis under these horizontal stretches and shrinks? Why?   □

❓ *Question*   Can you explain why these horizontal stretches and shrinks are *not* translations? (*Hint:* See what happens to two different points on the graph of $f(x)$ under such a transformation and use your answer to show that the transformation is not a translation.)   □

1.6.3 Multiple transformations

You have to be careful when combining horizontal stretches and translations and when combining vertical stretches and translations. For example, given a function $f(x)$, the transformations $2f(x) + 1$ and $2(f(x) + 1)$ are in general different functions and therefore will have different graphs.

EXAMPLE 1.65

The order of vertical transformations makes a difference

Suppose $f(x) = x^2$ (see Figure 1.67). Figures 1.68 and 1.69 illustrate the transformations $2f(x) + 1$ and $2(f(x) + 1)$, respectively. In this example, $2f(x) + 1 = 2x^2 + 1$, but $2(f(x) + 1) = 2(x^2 + 1) = 2x^2 + 2$.

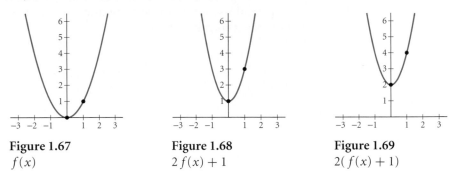

Figure 1.67
$f(x)$

Figure 1.68
$2f(x) + 1$

Figure 1.69
$2(f(x) + 1)$

The graph of the function $2f(x) + 1$ is obtained from the graph of $f(x) = x^2$ by *first* stretching vertically by a factor of 2 and *then* shifting the graph up by one unit. On the other hand, the graph of $2(f(x) + 1)$ is obtained from the graph of $f(x) = x^2$ by shifting the graph of $f(x)$ up one unit and *then* stretching the graph vertically by a factor of 2. Note in particular what happens to the points $(0, 0)$ and $(1, 1)$ on the graph of $f(x)$ under these two different transformations. □

In general, the transformation $Af(x) + C$ is a stretch followed by a shift, whereas $A(f(x) + C)$ is a shift followed by a stretch. To see why, consider that if $g(x) = Ax$ and $h(x) = x + C$, then $Af(x) + C = h(g(f(x)))$. Under this multiple transformation, we *first* apply the horizontal stretch g and *then* apply the horizontal shift h. On the other hand, $A(f(x) + C) = g(h(f(x)))$, so in this case the horizontal shift h is applied first.

Similarly, the functions $f(2x + 1)$ and $f(2(x + 1))$ are different functions, with different graphs. As the following example illustrates, $f(2x + 1)$ is a horizontal shift followed by a horizontal shrink, while $f(2(x + 1))$ is a horizontal shrink followed by a horizontal shift.

◈ *Caution* In the example described in the paragraph above (and in Example 1.66), the order of transformations may be the opposite of what you expected. Remember that this has often been the case with transformations of the independent variable! □

EXAMPLE 1.66

The order of horizontal transformations makes a difference

Consider the function $f(x) = x^2$ shown in Figure 1.70. The graphs of $f(2x + 1)$ and $f(2(x + 1))$ are shown in Figures 1.71 and 1.72. Here $f(2x + 1) = (2x + 1)^2$ but $f(2(x + 1)) = (2(x + 1))^2 = (2x + 2)^2$.

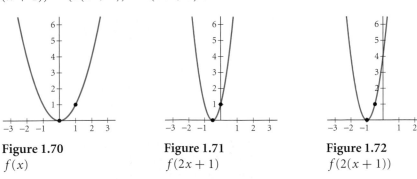

Figure 1.70
$f(x)$

Figure 1.71
$f(2x + 1)$

Figure 1.72
$f(2(x + 1))$

The graph of $f(2x+1)$ is obtained from the graph of $f(x)$ by *first* shifting the graph of $f(x)$ to the left one unit and *then* shrinking the graph horizontally by a factor of 2. Follow the points $(0, 0)$ and $(1, 1)$ on the graph of $f(x) = x^2$ under these transformations in Figure 1.71. On the other hand, the graph of the function $f(2(x + 1))$ is obtained from the graph of $f(x)$ by shrinking the graph horizontally by a factor of 2 and *then* shifting the graph to the left one unit. Note in particular what happens to the marked points in this transformation.   □

In general, the transformation $f(Ax + C)$ is a shift followed by a stretch, while $f(A(x + C))$ is a stretch followed by a shift. We justify this statement by writing these transformations in terms of compositions of f with $g(x) = Ax$ and $h(x) = x + C$. On one hand, we have

$$f(h(g(x))) = f(h(Ax)) = f(Ax + C).$$

This means that to modify $f(x)$ to get $f(Ax + C)$, we *first* apply the horizontal shift h and *then* apply the stretch g. On the other hand,

$$f(g(h(x))) = f(g(x + C)) = f(A(x + C)),$$

so the transformation $f(A(x + C))$ is *first* a horizontal stretch or shrink and *then* a horizontal shift.

Although the order in which you apply two horizontal or two vertical transformations makes a significant difference, a horizontal transformation and a vertical transformation can be applied in either order. For example, first stretching a graph horizontally by a factor of 2 and then shifting up one unit is equivalent to first shifting the graph up one unit and then stretching the graph horizontally by a factor of 2. In either case, the transformation described is $f(\frac{1}{2}x) + 1$.

1.6.4 Reflections

We have seen what happens when we multiply the independent or dependent variable of a function by a *positive* constant. In this section we will investigate what happens when the constant is negative. We begin by examining the effect of multiplying the independent or dependent variable by -1.

THEOREM 1.9

Vertical and Horizontal Reflections

(a) The graph of the function $-f(x)$ is the graph of $f(x)$ reflected over the x-axis.

(b) The graph of the function $f(-x)$ is the graph of $f(x)$ reflected over the y-axis.

In mathematical notation, Theorem 1.9 translates into the following:

THEOREM 1.10

Algebraic Descriptions of Vertical and Horizontal Reflections

(a) If (x, y) is a point on the graph of $y = f(x)$, then $(x, -y)$ is a point on the graph of $y = -f(x)$.

(b) If (x, y) is a point on the graph of $y = f(x)$, then $(-x, y)$ is a point on the graph of $y = f(-x)$.

The proof of Theorem 1.10 is identical to the proof of Theorem 1.8 (but with $A = -1$) and is left to you in the exercises.

EXAMPLE 1.67

Graphs of vertical and horizontal reflections

Consider the function $f(x) = x^2 - 2x + 2$ (see Figure 1.73). Each point $(x, f(x))$ on the graph of $f(x)$ corresponds to the point $(x, -f(x))$ on the graph of $-f(x) = -(x^2 - 2x + 2)$. In other words, we multiply each height on the graph of $f(x)$ by the constant -1. This has the effect of reflecting the graph of $f(x)$ over the x-axis. For example, the point $(2, 2)$ on the graph of $f(x)$ becomes the point $(2, -2)$ on the graph of $-f(x)$ (see Figure 1.74).

On the other hand, consider the graph of $f(-x) = (-x)^2 - 2(-x) + 2$. Every point on the graph of $f(-x)$ is of the form $(x, f(-x))$. This is equivalent to saying that each point has the form $(-x, f(x))$ (why?). For example, if $x = -2$ then $(-2, f(2))$ is on the graph of $f(-x)$. In general, the point $(x, f(x))$ on the graph of f gets sent to the point $(-x, f(x))$ under the transformation $f(-x)$. This means that the graph of $f(-x)$ is the graph of $f(x)$ reflected over the y-axis. See Figure 1.75.

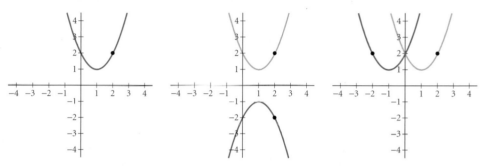

Figure 1.73
$f(x) = x^2 - 2x + 2$

Figure 1.74
$-f(x) = -(x^2 - 2x + 2)$

Figure 1.75
$f(-x) = (-x)^2 - 2(-x) + 2$

☐

A transformation that multiplies the dependent or independent variable of $y = f(x)$ by a negative number is a combination of a reflection and a stretch or shrink. For example, the graph of $y = -2f(x)$ is obtained from the graph of $y = f(x)$ by reflecting the graph over the x-axis and then stretching vertically by a factor of 2. In general, we can perform such transformations in the opposite order and obtain the same result; reflections commute with stretches and shrinks.

1.6.5 Even and odd functions

In this section we examine functions that have y-axis symmetry or rotational symmetry. We call such functions ***even functions*** and ***odd functions,*** respectively.

DEFINITION 1.17

Even and Odd Functions

$f(x)$ is an ***even function*** if $f(-x) = f(x)$ for all $x \in \text{Domain}(f)$.

$f(x)$ is an ***odd function*** if $f(-x) = -f(x)$ for all $x \in \text{Domain}(f)$.

Even functions are symmetric about the y-axis; if we reflect the graph of an even function $f(x)$ over the y-axis, its graph remains unchanged. Odd functions have rotational symmetry; if we rotate the graph of an odd function by 180 degrees around the origin, we end up with the original graph. Theorem 1.11 describes these symmetries in mathematical notation.

THEOREM 1.11

Even and Odd Functions Have Graphical Symmetries

(a) If (x, y) is a point on the graph of an *even* function $f(x)$, then $(-x, y)$ is also on the graph of $f(x)$.

(b) If (x, y) is a point on the graph of an *odd* function $f(x)$, then $(-x, -y)$ is also on the graph of $f(x)$.

The proof of Theorem 1.11 follows directly from the definitions of odd and even functions and is left to you in the exercises. Notice how we have described "rotational" symmetry in part (b) of Theorem 1.11. Figure 1.76 illustrates why that description is equivalent to rotational symmetry about the origin. If we rotate a point (a, b) in the plane 180 degrees around the origin, we arrive at the point $(-a, -b)$.

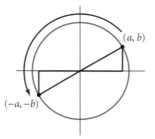

Figure 1.76

EXAMPLE 1.68

Graphically determining whether a function is even, odd, or neither

Determine whether the functions in Figures 1.77–1.79 are even, odd, or neither by examining the symmetry (or lack of symmetry) in their graphs.

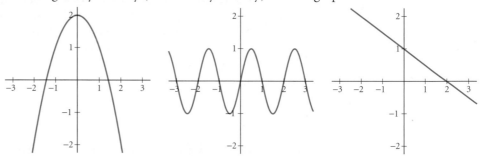

Figure 1.77   **Figure 1.78**   **Figure 1.79**

Solution   The graph in Figure 1.77 is symmetric about the y-axis and is therefore the graph of an even function. The graph in Figure 1.78 has rotational symmetry and is the graph of an odd function. If we rotate the graph 180 degrees around the origin, we end up with the same graph. Finally, the graph in Figure 1.79 has neither even nor odd symmetry and thus is neither an even function nor an odd function.   □

To determine algebraically whether a function is even or odd (or neither), we must use Definition 1.17. In the following example, we use this definition to show that $f(x) = x^2$ is an even function and $g(x) = x^3$ is an odd function. We also exhibit a function that is neither even nor odd.

EXAMPLE 1.69

An even function, an odd function, and a function that is neither even nor odd

The function $f(x) = x^2$ is an even function. For all real numbers x we have $f(-x) = (-x)^2 = x^2 = f(x)$. On the other hand, the function $g(x) = x^3$ is an odd function because $g(-x) = (-x)^3 = -x^3 = -g(x)$ for all real numbers x.

Not every function is either even or odd. For example, the function $h(x) = x^3 + 1$ is neither even nor odd because $h(-x) = (-x)^3 + 1 = -x^3 + 1$, which is neither equal to $h(x) = x^3 + 1$ nor equal to $-h(x) = -x^3 - 1$.

Figures 1.80–1.82 show the graphs of $f(x)$, $g(x)$, and $h(x)$. Note the y-axis symmetry in Figure 1.80 and the rotational symmetry in Figure 1.81.

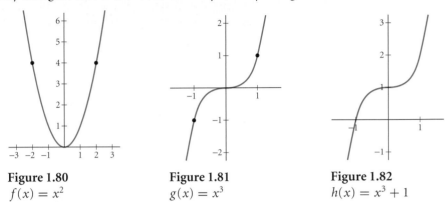

Figure 1.80
$f(x) = x^2$

Figure 1.81
$g(x) = x^3$

Figure 1.82
$h(x) = x^3 + 1$

For example, in Figure 1.80 the points $(2, 4)$ and $(-2, 4)$ are both on the graph of $y = f(x)$. In fact, for each point (x, x^2) on the graph of f there is a corresponding point $(-x, x^2)$. In Figure 1.81, the graph contains both the point $(1, 1)$ and the point $(-1, -1)$. In general, every point (x, x^3) on the graph of g has a corresponding point $(-x, -x^3)$. The graph in Figure 1.82 has neither even nor odd symmetry. For example, the point $(1, 2)$ is on the graph of h, but neither $(-1, 2)$ nor $(-1, -2)$ is on the graph. □

◆ *Caution* In Example 1.69, we saw that $f(x) = x^2$ is an even function and $f(x) = x^3$ is an odd function. In Chapter 4 we will see that if a function is of the form $f(x) = x^k$ for some positive integer k, then f is even if k is even, and f is odd if k is odd. However, not *all* even functions are of the form x^k, nor are *all* odd functions of the form x^k. Moreover, just because a function includes an even power does not mean that the function is even. Similarly, the presence of an odd power somewhere in a function does not necessarily imply that the function is odd. □

EXAMPLE 1.70

Algebraically determining whether a function is even, odd, or neither

Determine algebraically whether the following functions are even, odd, or neither.

(a) $f(x) = \dfrac{1}{x}$ (b) $g(x) = x^4 - x^2$ (c) $h(x) = \dfrac{2 + x}{1 + x^2}$

Solution

(a) To determine whether $f(x)$ is even or odd (or neither), we must calculate $f(-x)$ and determine whether it is equal to $f(x)$, $-f(x)$, or neither. We calculate

$$f(-x) = \frac{1}{-x} = -\frac{1}{x} = -f(x),$$

so $f(x) = \frac{1}{x}$ is an odd function.

(b) The function $g(x)$ is even because
$$g(-x) = (-x)^4 - (-x)^2 = x^4 - x^2 = g(x).$$

(c) Since we have
$$h(-x) = \frac{2 + (-x)}{1 + (-x)^2} = \frac{2 - x}{1 + x^2},$$
which is not equal to $h(x)$, nor equal to $-h(x)$, this function is neither even nor odd. ☐

✓ *Checking the Answer*   You can use your calculator to check these answers. Figures 1.83–1.85 show the functions considered in Example 1.70.

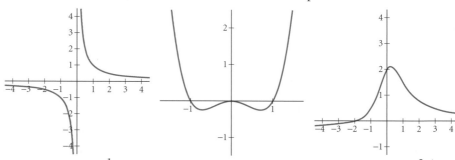

Figure 1.83 $y = \dfrac{1}{x}$ **Figure 1.84** $y = x^4 - x^2$ **Figure 1.85** $y = \dfrac{2 + x}{1 + x^2}$

Check the answers that we found algebraically in Example 1.70 by examining the symmetry (or lack of symmetry) in these graphs. ☐

What you should know after reading Section 1.6

▶ What happens to the graph of a function $y = f(x)$ if we add or subtract a constant to or from the dependent variable? The independent variable? What is the mathematical description of a horizontal or vertical shift?

▶ What happens to the graph of a function $y = f(x)$ if we multiply the dependent or independent variable by a positive constant? What is the mathematical description of a horizontal or vertical stretch or shrink by a factor of A?

▶ Given the graph of $y = f(x)$, be able to sketch graphs of $y = f(x) + C$, $y = f(x + C)$, $y = Af(x)$, and $y = f(Ax)$ for various constants A and C. Also be able to construct tables of values for these translations given a table for f.

▶ How does multiplying the dependent or independent variable of a function by a negative number (in particular, -1) change the graph of $y = f(x)$? What is the mathematical description of a reflection over the x- or y-axis? Be able to sketch the graphs of $y = -f(x)$ and $y = f(-x)$ given the graph of $y = f(x)$.

▶ When we are combining transformations, which transformation takes effect "first"? Does order matter for all types of multiple transformations or just for some? If just for some, which ones?

▶ The definitions of even functions and odd functions, and mathematical descriptions of even and odd functions. What kind of symmetry does an even function have? An odd function? Be able to determine graphically *and* algebraically whether a given function is even, odd, or neither.

Exercises 1.6

Concepts

0. Read the section and make your own summary of the material. Include whatever you think will help you review the material later on.

■ **Fill in the blanks with ordered pairs of numbers.**

1. If $(3, 5)$ is on the graph of $y = f(x)$, then _____ is on the graph of $y = f(x) - 7$.

2. If $(4, -2)$ is on the graph of $y = f(x)$, then _____ is on the graph of $y = -3f(x)$.

3. If $(-2, 3)$ is on the graph of $y = f(x)$, then _____ is on the graph of $y = f(4x)$.

4. If _____ is on the graph of $y = f(x)$, then $(4, 2)$ is on the graph of $y = f(x - 3)$.

5. If $(3, -2)$ is on the graph of $y = f(x)$, then _____ is on the graph of $f(2x - 1)$.

6. If $(-2, 1)$ is on the graph of an *even* function $y = f(x)$, then _____ is also on the graph of $y = f(x)$.

7. If $(2, 5)$ is on the graph of an *odd* function $y = f(x)$, then _____ is also on the graph of $y = f(x)$.

■ **Describe what the following sentences mean (a) in your own words and (b) in mathematical notation (in the spirit of the theorems in this section). Then sketch the graph of a function $f(x)$ and an example of each type of transformation (for some C or A).**

8. The graph of g is a *vertical translation* of the graph of f by C units.

9. The graph of g is a *horizontal translation* of the graph of f by C units.

10. The graph of g is a *vertical stretch* of the graph of f by a factor of A.

11. The graph of g is a *horizontal stretch* of the graph of f by a factor of A.

12. The graph of g is a *vertical shrink* of the graph of f by a factor of A.

13. The graph of g is a *horizontal shrink* of the graph of f by a factor of A.

■ **Let f be a function and let C and A be real numbers. Write each transformation as a composition $g \circ h$ of two functions g and h. Write the domain of $g \circ h$ in set notation.**

14. $f(x) + C$ **15.** $f(x + C)$

16. $Af(x)$ **17.** $f(Ax)$

■ **Suppose $g(x) = 2x$, $h(x) = x - 3$, and $k(x) = -x$. Each composition below describes a transformation of the function $f(x) = x^2$. Find a formula for each**

composition, and describe in terms of transformations how the graph of the given composition could be obtained from the graph of $f(x) = x^2$.

18. $f(k(g(x)))$ **19.** $k(f(g(x)))$

20. $g(f(k(x)))$ **21.** $f(h(k(x)))$

22. $h(g(f(k(x))))$ **23.** $k(f(g(h(x))))$

■ **Let f be a function and let C and A be real numbers. Write each transformation as a composition $g \circ h \circ k$ of three functions g, h, and k. Then describe the transformation graphically.**

24. $2f(x) - 3$ **25.** $2(f(x) - 3)$

26. $f(9x) + 2$ **27.** $3f(x + 5)$

28. $f(4x + 2)$ **29.** $f(4(x + 2))$

30. If $y = f(x)$ has the graph shown on the left, describe the graph on the right as a transformation of $f(x)$.

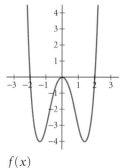
$f(x)$

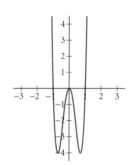
A transformation of $f(x)$

31. If $y = f(x)$ has the graph shown on the left, describe the graph on the right as a transformation of $f(x)$.

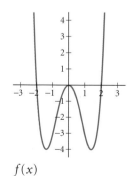
$f(x)$

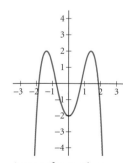
A transformation of $f(x)$

32. If $f(0) = 2$, can f be an odd function? If so, sketch the graph of an odd function with $f(0) = 2$. If not, explain why not.

33. Suppose that $f(x)$ is a function with domain $\mathbb{R}$ and that the graph below shows the right-hand side of the graph of $f(x)$.

(a) Sketch the left-hand side of the graph so that $f(x)$ is an even function.

(b) Sketch the left-hand side of the graph so that $f(x)$ is an odd function.

(c) Sketch the left-hand side of the graph so that $f(x)$ is neither an even nor an odd function.

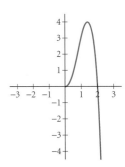

34. Complete the entries in the table below so that the function described by the table is (a) even and (b) odd.

x	-3	-2	-1	0	1	2	3
$f(x)$	4				-2	1	

35. Can you think of a function that is *both* even *and* odd? Is there only one such function?

Skills

■ Let f be the function graphed below. In Problems 36–50, sketch the graph of each transformation described. Label where the three marked points go under each transformation.

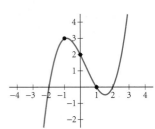

36. $f(x) - 2$ **37.** $2f(x)$

38. $f(2x)$ **39.** $-f(x)$

40. $f(-x)$ **41.** $-f(-x)$

42. $-2f(x)$ **43.** $f(-2x)$

44. $f\left(-\frac{1}{2}x\right)$ **45.** $f(2x) + 1$

46. $2f(x) + 1$ **47.** $2(f(x) + 1)$

48. $2f(x + 1)$ **49.** $f(2x + 1)$

50. $f(2(x + 1))$

■ Graph the following functions without using your graphing calculator. Use your knowledge of transformations

and the graphs of basic functions. Label three points on each graph.

51. $f(x) = x^2 - 3$ **52.** $f(x) = (x + 3)^2$

53. $f(x) = -\frac{1}{3}x^2$ **54.** $f(x) = (-2x)^3$

55. $f(x) = 2\sqrt{x} + 1$ **56.** $f(x) = 2(\sqrt{x} + 1)$

57. $f(x) = \dfrac{2}{x + 3}$ **58.** $f(x) = \dfrac{1}{x - 1} + 1$

59. $f(x) = \dfrac{-1}{3x + 1}$

■ Let f be the function defined by the table below. Construct tables that describe each of the transformations in Problems 60–65.

x	-3	-2	-1	0	1	2	3
$f(x)$	8	5	4	-2	0	3	4

60. $\frac{1}{2}f(x)$ **61.** $f(-2x)$ **62.** $f(x - 3)$

63. $2f(x) + 1$ **64.** $f(2x + 1)$ **65.** $2f(x + 1)$

■ Determine graphically whether the following functions are even, odd, or neither.

66.

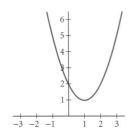

67.

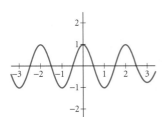

68.

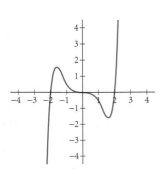

■ Determine algebraically whether the following functions are even, odd, or neither.

69. $f(x) = 1 - 2x$ **70.** $f(x) = -15x$

71. $f(x) = x^4 + 1$ **72.** $f(x) = x^3 + x^2$

73. $f(x) = \dfrac{1}{x+1}$ **74.** $f(x) = \dfrac{x^3}{x^2+1}$

■ Suppose $f(x)$ is the piecewise function defined below. For each problem, sketch the graph of f and the graph of the given transformation (on the same set of axes). Then express each transformation as a piecewise function.

$$f(x) = \begin{cases} 2x+1, & \text{if } x \le 1 \\ x^2, & \text{if } x > 1 \end{cases}$$

75. $-2f(x)$ **76.** $f(x) - 1$

77. $f(x+3)$ **78.** $f\left(\tfrac{1}{2}x\right)$

Proofs

79. Prove that every odd function that is defined at $x = 0$ must pass through the origin.

■ Suppose f is a function and A is any positive real number. The next two problems prove Theorem 1.8.

80. Prove that if (x, y) is a point on the graph of $f(x)$, then (x, Ay) is a point on the graph of $Af(x)$.

81. Prove that if (x, y) is a point on the graph of $f(x)$, then $(\tfrac{1}{A}x, y)$ is a point on the graph of $f(Ax)$.

■ Suppose f is a function and A is any positive real number. The next two problems prove Theorem 1.10.

82. Prove that if (x, y) is a point on the graph of $f(x)$, then $(x, -y)$ is on the graph of $-f(x)$.

83. Prove that if (x, y) is a point on the graph of $f(x)$, then $(-x, y)$ is on the graph of $f(-x)$.

■ Suppose f is a function and A is any positive real number. The next two problems prove Theorem 1.11.

84. Prove that if (x, y) is a point on the graph of an *even* function $f(x)$, then $(-x, y)$ is a point on the graph of $f(x)$.

85. Prove that if (x, y) is a point on the graph of an *odd* function $f(x)$, then $(-x, -y)$ is a point on the graph of $f(x)$.

1.7   Inverse Functions

The *inverse* of a function f is a function that "undoes" the action of f. For example, the function that adds 1 to each real number can be "undone" by subtracting 1 from each real number. As we will soon see, not all functions have inverses.

1.7.1 One-to-one functions

Suppose f is a function with domain A and range B. By the definition of function, f assigns *each* element of the domain A to *exactly one* element of the range B. If in addition *each* element of the range B is hit by *exactly one* element of the domain, the function f is said to be **one-to-one.**

DEFINITION 1.18

One-to-One Function

A function f is **one-to-one** if for each a and b in the domain of f,

$$a \ne b \implies f(a) \ne f(b).$$

In other words, a function f is one-to-one if no two distinct elements of the domain get sent to the same element in the range. An alternative, equivalent definition of one-to-one is the contrapositive of the statement in Definition 1.18: A function f is **one-to-one** if for any elements a and b in the domain of f, we have

(1) $$f(a) = f(b) \implies a = b.$$

Statement (1) says that if f is a one-to-one function and a and b both get sent to the same element in the range of f, then a and b must have been the same.

EXAMPLE 1.71

A function that is not one-to-one, and a function that is one-to-one

Consider the functions f and g defined by the diagrams in Figures 1.86 and 1.87.

Figure 1.86   Figure 1.87

The domain of f is $\{1, 2, 3, 4\}$ and the range of f is $\{2, 5, 9\}$. Note that f is indeed a function; *each* element of the domain is sent to *exactly one* element of the range. However, the function f is *not* one-to-one since two distinct elements of the domain (namely 3 and 4) are sent to the *same* element of the range (9).

On the other hand, the function g in Figure 1.87 *is* one-to-one, since each element in the range $\{2, 5, 6, 9\}$ of g is hit by *exactly one* element from the domain. This example illustrates why we call such a function one-to-one; there is a one-to-one correspondence between elements in the domain and elements in the range.   □

Graphically, we can tell whether a function f is one-to-one by checking to see whether every horizontal line meets the graph of f at most once.

THEOREM 1.12

The Horizontal Line Test

A function f is one-to-one if and only if every horizontal line passes through the graph of f in *at most one* point.

❓ *Question*   Why does the Horizontal Line Test measure "one-to-one-ness"? Can you use Definition 1.18 to argue that a function is one-to-one if and only if it passes the Horizontal Line Test? For inspiration, look at Figures 1.88 and 1.89.

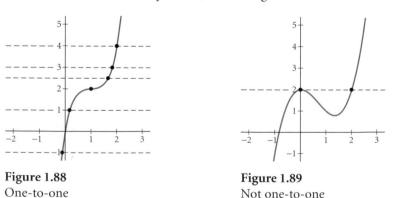

Figure 1.88
One-to-one

Figure 1.89
Not one-to-one   □

To tell algebraically whether a function is one-to-one we must use Definition 1.18 or its equivalent form, Statement (1).

EXAMPLE 1.72

Algebraically proving that a function is one-to-one

Show that the function $f(x) = 3x - 1$ is one-to-one.

Solution It is usually easier to use Statement (1) than Definition 1.18. We will show that if $f(a) = f(b)$, then a and b must be equal:

$$f(a) = f(b) \quad \Rightarrow \quad 3a - 1 = 3b - 1 \quad \Rightarrow \quad 3a = 3b \quad \Rightarrow \quad a = b.$$

In other words, the only way that a and b can have the same outputs $f(a) = f(b)$ is if a and b are actually the same number. No two *different* elements a and b get sent to the same value by the function f, so f is one-to-one. $\quad\square$

There is another way that we can sometimes (not always!) determine if a function $f(x)$ is one-to-one. If a function f is increasing on its entire domain, or decreasing on its entire domain, we say that f is **monotonic.** Such functions always pass the Horizontal Line Test, and therefore they are one-to-one.

THEOREM 1.13

Monotonic Functions Are One-to-One

If f is monotonic, then f is one-to-one.

? Question You will prove Theorem 1.13 in the exercises. Can you convince yourself that it is true? Try drawing some graphs of monotonic functions. $\quad\square$

⊛ Caution If a function is *not* always increasing (or always decreasing), it *may or may not* be one-to-one. In other words, not every one-to-one function is necessarily monotonic. Can you sketch an example of a one-to-one function that is not always increasing? (*Hint:* Consider a piecewise function.) $\quad\square$

Sometimes a function is not one-to-one but can be defined on a restricted domain so that it *is* one-to-one.

EXAMPLE 1.73

Restricting the domain of a function so that it becomes one-to-one

The function $f(x) = x^2$ is *not* one-to-one because it is possible for two values of the domain to be sent to the same element of the range. For example, $x = 2$ and $x = -2$ are both sent to 4, since $f(2) = 2^2 = 4$ and $f(-2) = (-2)^2 = 4$. Figure 1.90 shows the graph of $f(x) = x^2$. Note that it clearly does not pass the Horizontal Line Test.

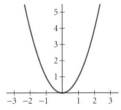

Figure 1.90
$f(x) = x^2$ is not one-to-one

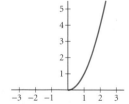

Figure 1.91
$f(x) = x^2$ on $[0, \infty)$ is one-to-one

If we restrict the domain of $f(x) = x^2$ to the interval $[0, \infty)$, then the restricted function *is* one-to-one, since no two positive real numbers have the same square. Figure 1.91 shows the graph of $f(x) = x^2$ restricted to the domain $[0, \infty)$. (Note that we also could have used the restricted domain $(-\infty, 0]$.) $\quad\square$

1.7.2 Definition of inverses

Two functions are said to be *inverses* of each other if each "undoes" the other. As a real-world example, consider the function that converts temperature in degrees Celsius to temperature in degrees Fahrenheit. The inverse of this function is the function that converts Fahrenheit to Celsius. Note that if we convert $0°$ Celsius to $32°$ Fahrenheit, and then convert this temperature in degrees Fahrenheit back into Celsius, we get what we started with, namely $0°$ Celsius. Similarly, if we convert $212°$ Fahrenheit (the boiling point of water) to $100°$ Celsius and then back into Fahrenheit, we still have $212°$ Fahrenheit.

DEFINITION 1.19

The Inverse of a Function

A function g is the *inverse* of a function f if

$$g(f(x)) = x \quad \text{for all } x \text{ in the domain of } f, \text{ and}$$

$$f(g(x)) = x \quad \text{for all } x \text{ in the domain of } g.$$

When g is the inverse of f, we use the notation $g = f^{-1}$.

We can also state the conditions in Definition 1.19 using the inverse function notation $f^{-1}(x)$:

$$f^{-1}(f(x)) = x \quad \text{for all } x \text{ in the domain of } f, \text{ and}$$

$$f(f^{-1}(x)) = x \quad \text{for all } x \text{ in the domain of } f^{-1}.$$

⊗ Caution   Although we use the notation x^{-1} to denote the reciprocal $\frac{1}{x}$, the notation f^{-1} does *not* stand for the reciprocal $\frac{1}{f}$ of f.   □

? Question   In Definition 1.19 we talk about *the* inverse of f. If a function has an inverse, it has *only one*. Can you explain why?   □

EXAMPLE 1.74   **Algebraically verifying that two functions are inverses of each other**

Show that $g(x) = x - 1$ is the inverse of the function $f(x) = x + 1$.

Solution   Intuitively, subtracting 1 from a number is the inverse of adding 1 to a number. However, we want to use the definition of inverses to verify our intuition. By Definition 1.19, we must calculate $g(f(x))$ and $f(g(x))$ and show that they are both equal to x:

$$g(f(x)) = g(x + 1) = (x + 1) - 1 = x$$

$$f(g(x)) = f(x - 1) = (x - 1) + 1 = x.$$ □

EXAMPLE 1.75   **A real-world example of two functions that are inverses**

Show that the function $C(f) = \frac{5}{9}(f - 32)$ that converts degrees Fahrenheit to degrees Celsius is the inverse of the function $F(c) = \frac{9}{5}c + 32$ that converts Celsius to Fahrenheit.

Solution   The notation here is a little different, but the concept is the same. In particular, we are using the capital letters F and C to denote the *names* of the functions and using the lowercase letters f and c to denote the variables (in the notation of Definition 1.19, both variables f and c would be represented by x). We will show that $C(F(c)) = c$ and $F(C(f)) = f$:

$$C(F(c)) = C\left(\tfrac{9}{5}c + 32\right) = \tfrac{5}{9}\left(\left(\tfrac{9}{5}c + 32\right) - 32\right) = \tfrac{5}{9}\left(\tfrac{9}{5}c\right) = c$$

$$F(C(f)) = F\left(\tfrac{5}{9}(f - 32)\right) = \tfrac{9}{5}\left(\tfrac{5}{9}(f - 32)\right) + 32 = (f - 32) + 32 = f. \qquad \square$$

1.7.3 Properties of inverses

If $f: A \to B$ is a function with an inverse, then its inverse f^{-1} takes elements of the range B and sends them to the elements in the domain A that they "came from." In other words, the domain of f^{-1} is the range of f, and the range of f^{-1} is the domain of f. Another way to say this is that the dependent variable for f^{-1} is the independent variable for f, and vice versa. Moreover, if $f(a) = b$ then $f^{-1}(b) = a$ (since f^{-1} must "undo" the function f).

THEOREM 1.14

Properties of Inverses

If f^{-1} is the inverse of f, then

(a) Domain(f^{-1}) = Range(f), and Range(f^{-1}) = Domain(f).

(b) $f^{-1}(b) = a$ if and only if $f(a) = b$.

PROOF (THEOREM 1.14)   We will leave the proof of part (a) to the exercises and prove part (b). We first show that if $f(a) = b$ then $f^{-1}(b) = a$, using the definition of inverse. Since $f^{-1}(f(x)) = x$ for all x in the domain of f, we have

$$f^{-1}(b) = f^{-1}(f(a)) \qquad \text{(since } f(a) = b \text{ by hypothesis)}$$
$$= a. \qquad \text{(by the definition of inverse)}$$

Similarly, if $f^{-1}(b) = a$ then $f(a) = f(f^{-1}(b)) = b$. (Can you fill in the reasons?)   ▨

EXAMPLE 1.76

Diagrams of a function and its inverse

The function g defined by the diagram in Figure 1.92 has inverse g^{-1} shown in Figure 1.93. For example, since $g(1) = 5$, we must have $g^{-1}(5) = 1$.

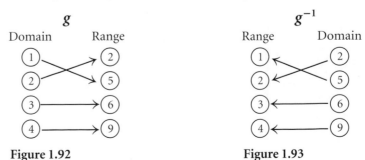

Figure 1.92   **Figure 1.93** $\square$

Example 1.76 suggests that in order for us to be able to "invert" a function, we need to have a one-to-one correspondence between the elements of the domain and the elements of the range. Otherwise, we would not be able to construct an inverse function g^{-1} that sends each element of the range "back where it came from."

For example, consider the function f defined previously in Figure 1.86 on page 124. In this example, the elements 3 and 4 in the domain *both* get sent to the element 9 in the range by f. If we were to have an inverse g for this f, we would have to have *both* $g(9) = 3$ *and* $g(9) = 4$ (since $f(3) = 9$ *and* $f(4) = 9$). Clearly this relationship g would not be a function! Thus the function f from Figure 1.86 does not have an inverse.

If a function f has an inverse f^{-1}, then we say that f is an ***invertible function.*** The following theorem states that every one-to-one function is invertible and every invertible function is one-to-one.

THEOREM 1.15

One-to-One Functions Are Always Invertible, and Vice Versa

A function is invertible if and only if it is one-to-one.

PROOF **(THEOREM 1.15)**   We first prove that if a function is invertible, it must be one-to-one. If f is an invertible function, then it has an inverse f^{-1}. If $f(a) = f(b)$, we must show that $a = b$. Using properties of inverses, this is easy:

$$f(a) = f(b) \quad \Rightarrow \quad f^{-1}(f(a)) = f^{-1}(f(b)) \qquad \text{(apply } f^{-1} \text{ to both sides)}$$
$$\Rightarrow \quad a = b \qquad\qquad\qquad\qquad \text{(definition of inverse)}$$

On the other hand, if f is one-to-one, then it must have an inverse. If f is one-to-one, then each element b in the range of f is hit by *exactly one* element a in the domain. Simply define $f^{-1}(b)$ to be this element a. Since f is one-to-one, this new relationship f^{-1} will be a function (can you explain why?). Moreover, we will have $f^{-1}(b) = a$ if and only if $f(a) = b$.   ∎

A function f is invertible if and only if it is one-to-one, and f is one-to-one if and only if it passes the Horizontal Line Test. Therefore, f is invertible if and only if it passes the Horizontal Line Test.

Sometimes a function is not invertible on its largest domain but *is* invertible on some smaller "restricted" domain. To find a restricted domain on which a function is one-to-one, we must find an interval or intervals on which the graph passes the Horizontal Line Test.

EXAMPLE 1.77   **Graphically finding a restricted domain on which a function is invertible**

Find a restricted domain on which the function graphed in Figure 1.94 is invertible.

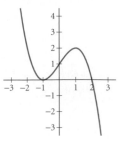

Figure 1.94

Solution   Clearly the function in Figure 1.94 does not pass the Horizontal Line Test and thus is not one-to-one. If we restrict the domain of the function to $[-1, 1]$, the restricted function will be invertible (it will pass the Horizontal Line Test). Alternatively, we could restrict the domain to the interval $(-\infty, -1]$ or to the interval $[1, \infty)$. Figures 1.95, 1.96, and 1.97 show the graphs of $f(x)$ on these restricted domains. All three of these restricted graphs pass the Horizontal Line Test and thus are invertible.

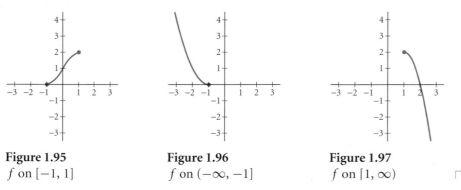

Figure 1.95
f on $[-1, 1]$

Figure 1.96
f on $(-\infty, -1]$

Figure 1.97
f on $[1, \infty)$

As the next example illustrates, since $f^{-1}(b) = a$ if and only if $f(a) = b$, the point (b, a) is on the graph of f^{-1} if and only if the point (a, b) is on the graph of f.

EXAMPLE 1.78

Graphing the inverse of a function

Consider the function $f(x) = 2x - 2$ graphed in Figure 1.98. Since this graph passes the Horizontal Line Test, f is one-to-one and thus invertible. Since $f(1) = 0$, the point $(1, 0)$ is on the graph of f; therefore, $f^{-1}(0) = 1$ and the point $(0, 1)$ is on the graph of f^{-1}. Whenever (a, b) is a point on the graph of f, the point (b, a) is a point on the graph of f^{-1}. Doing this for a number of points and "connecting the dots," we get the graph of f^{-1} in Figure 1.99.

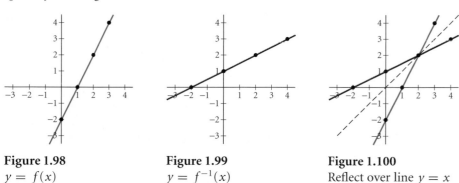

Figure 1.98
$y = f(x)$

Figure 1.99
$y = f^{-1}(x)$

Figure 1.100
Reflect over line $y = x$

If we compare the graphs of f and f^{-1}, we see that the graph of $y = f^{-1}(x)$ is the graph of $y = f(x)$ reflected over the line $y = x$ (see Figure 1.100).

THEOREM 1.16

The Relationship Between the Graph of a Function and Its Inverse
The graph of $y = f^{-1}(x)$ is the graph of $y = f(x)$ reflected over the line $y = x$.

You will prove this theorem in the exercises by showing that the line connecting the point (a, b) on the graph of $y = f(x)$ and the point (b, a) on the graph of $y = f^{-1}(x)$ is perpendicular to and bisected by the line $y = x$.

1.7.4 Calculating inverses

Given a one-to-one function $f(x)$, we can use Theorem 1.16 to sketch the graph of $f^{-1}(x)$. If we want to find an *algebraic* expression for the function $f^{-1}(x)$, we have to solve the equation $y = f(x)$ for x. Then by Theorem 1.14 we will have a formula for $x = f^{-1}(y)$. Since we normally use x to represent the independent variable of a function, sometimes we switch the variables x and y at this point to get an equation $y = f^{-1}(x)$.

EXAMPLE 1.79   **Algebraically finding the inverse of a function**

The function $f(x) = 3x + 1$ is one-to-one because it is always increasing (see Theorem 1.13). Therefore, it has an inverse. Find an equation for $f^{-1}(x)$.

Solution   Let $y = f(x) = 3x + 1$ so that y is a function of x. To find the inverse f^{-1}, we must solve for x as a function of y:

$$y = 3x + 1 \quad \Rightarrow \quad y - 1 = 3x \quad \Rightarrow \quad \tfrac{1}{3}(y - 1) = x.$$

Since $y = f(x)$, we have $x = f^{-1}(y)$ so $f^{-1}(y) = \tfrac{1}{3}(y - 1)$. Since we would rather represent the independent variable of f^{-1} by the traditional letter x, replace the y's in the equation with x's to get $f^{-1}(x) = \tfrac{1}{3}(x - 1)$.

Alternatively, we can *first* switch the variables x and y to get $x = 3y + 1$ and then solve for y as a function of x:

$$x = 3y + 1 \quad \Rightarrow \quad x - 1 = 3y \quad \Rightarrow \quad \tfrac{1}{3}(x - 1) = y.$$

Now, since we started with $f(y) = x$, we have an equation for $f^{-1}(x) = y$. Therefore, we have $f^{-1}(x) = \tfrac{1}{3}(x - 1)$.   □

✓ *Checking the Answer*   You can check the answer to Example 1.79 algebraically by verifying that $f(f^{-1}(x)) = x$ and $f^{-1}(f(x)) = x$ for all x in the appropriate domains. A less rigorous but still effective method is to graph $y = f(x)$ and $y = f^{-1}(x)$ and check that their graphs are symmetric about the line $y = x$.   □

◈ *Caution*   In Example 1.79 we saw two ways to find the inverse of a function (solving for x in terms of y, or switching the variables and then solving for y in terms of x). Either way is acceptable unless the names of the variables have specific meanings in the problem (see Example 1.80), in which case you should *not* switch the variables.   □

EXAMPLE 1.80   **Algebraically finding an inverse without switching variables**

The function that converts degrees Celsius to degrees Fahrenheit is $F(C) = \tfrac{9}{5}C + 32$. Find the inverse of this function—that is, the function $C(F)$ that converts Fahrenheit to Celsius.

Solution   We are given $F(C) = \tfrac{9}{5}C + 32$ and we wish to find the function $C(F)$. (In the "inverse notation" this is the function F^{-1}.) To do this we start with our expression of F as a function of C and solve for C as a function of F:

$$F = \tfrac{9}{5}C + 32 \quad \Rightarrow \quad F - 32 = \tfrac{9}{5}C \quad \Rightarrow \quad \tfrac{5}{9}(F - 32) = C.$$

Therefore, the inverse of $F(C)$, that is, the function that converts degrees Fahrenheit to degrees Celsius, is the function $C(F) = \tfrac{5}{9}(F - 32)$.   □

> ## What you should know after reading Section 1.7
>
> ▶ What it means for a function to be one-to-one, and the contrapositive of this definition. How to restrict the domain of a function so that it is one-to-one.
>
> ▶ How to show whether a function is one-to-one algebraically (using the definition), graphically, or by showing that the function is always increasing or always decreasing.
>
> ▶ The definition of the *inverse* of a function, and how to use it to show that two functions g and f are inverses. When is a function invertible? What are the domain, range, and graph of the inverse?
>
> ▶ Given an invertible function f, be able to calculate algebraically the inverse f^{-1} of f. When is it appropriate to "switch variables" during this calculation? When is it not appropriate?

Exercises 1.7

Concepts

0. Read the section and make your own summary of the material. Include whatever you think will help you review the material later on.

1. Describe in your own words what it means for a function f to be *one-to-one*. Then give a mathematical description.

2. Construct a table defining a function f that is one-to-one. Then change one value in the table so that f is *not* one-to-one.

3. Is every linear function one-to-one? If so, explain why. If not, find a linear function that is *not* one-to-one.

4. Use values of $f(x) = x^2 + 1$ (*not* the graph) to show that the function $f(x) = x^2 + 1$ is *not* one-to-one.

5. Explain in your own words why an invertible function f can have only *one* inverse.

6. If an invertible function f has domain $[-1, 1)$ and range $(-\infty, 3]$, what are the domain and range of f^{-1}? Sketch the graph of a function f with the domain and range given, and on the same set of axes sketch the graph of its inverse f^{-1}.

■ Given that f is an invertible function, fill in the blanks.

7. If $f(-1) = 0$, then $f^{-1}(0) = $ _____.

8. If $(2, 3)$ is on the graph of f, then _____ is on the graph of f^{-1}.

9. If _____ is on the graph of f, then $(-2, 4)$ is on the graph of f^{-1}.

10. Use a diagram (like the one in Figure 1.92) to define a function from $\{1, 2, 3, 4\}$ to $\{2, 5, 6, 9\}$ that is *not* invertible. Explain why your function fails to have an inverse.

11. Sketch the graph of a function that is *not* one-to-one. Show that if you reflect this graph across the $y - x$ line, the resulting graph fails to pass the Vertical Line Test (and thus is not a function). What does this have to do with inverse functions?

12. Suppose f is an invertible function with inverse f^{-1}. What is $(f^{-1})^{-1}$? Explain your answer.

13. If g is the inverse of f, is f necessarily the inverse of g? Explain.

14. Can a function be its own inverse? What would have to be true about the graph of such a function? Sketch the graphs of two functions that are their own inverses.

Skills

■ Use Definition 1.18 or Statement (1) to show that the following functions are one-to-one. Then check each answer with a graph.

15. $f(x) = 12x$

16. $f(x) = 2 - 4x$

17. $f(x) = x^3 + 1$

18. $f(x) = x^2 + 1$, for $x \geq 0$

19. $f(x) = \dfrac{1}{x}$, for $x > 0$

20. $f(x) = \dfrac{1}{x}$, for $x < 0$

■ Use Theorem 1.13 to argue that the following functions are one-to-one.

21. $f(x) = 2x - 5$

22. $f(x) = 1 + 15x$

23. $f(x) = x^2$, for $x > 0$

■ In each problem, use Definition 1.19 to show that g is the inverse of f.

24. $f(x) = 2 - 3x$, $g(x) = -\frac{1}{3}x + \frac{2}{3}$

25. $f(x) = x^2$ restricted to $[0, \infty)$, $g(x) = \sqrt{x}$

26. $f(x) = \dfrac{1}{2x}$, $g(x) = \dfrac{1}{2x}$

27. $f(x) = \dfrac{x}{1-x}$, $g(x) = \dfrac{x}{1+x}$

■ Each of the functions below is invertible. Find the inverse of each function. Check your answers by graphing f and f^{-1}.

28. $f(x) = 1 - 5x$

29. $f(x) = \dfrac{1 - 5x}{2}$

30. $f(x) = \dfrac{2}{x+1}$

31. $f(x) = 1 + \dfrac{1}{x}$

32. $f(x) = \dfrac{x - 2}{x}$

33. $f(x) = \dfrac{x - 1}{x + 1}$

■ For each of the functions below, decide whether the function f is invertible. If not, find a horizontal line that intersects the graph of f in more than one point. If the function is invertible, sketch the graph of f^{-1} and label three points on its graph.

34.

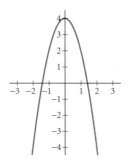

35.

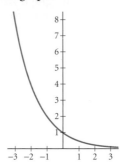

36.

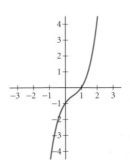

■ For each function, find a restricted domain on which the function is invertible. Then sketch the inverse of the restricted function and label two points on its graph.

37.

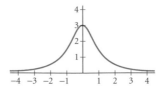

38.

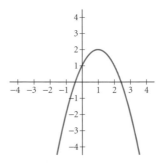

39.

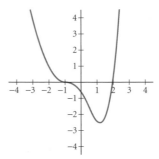

Applications

40. Give a real-world example of a one-to-one function. Describe the inverse of your function in real-world terms.

41. Give a real-world example of a function that is *not* one-to-one. Explain in real-world terms why this function is not invertible.

42. Juri's custom printing T-shirt shop charges $12.00 per T-shirt plus a $20.00 setup fee.

   (a) Write a function that describes the cost C of printing n T-shirts at Juri's store.

   (b) Show that $C(n)$ is an invertible function.

   (c) Find the inverse of $C(n)$. What does this inverse function $n(C)$ represent in terms of C and n?

   (d) Use the function you found in part (c) to calculate the number of shirts you can produce for $150.

43. The math teachers at Pinnacle High School have discovered a relationship between the number of hours their students watch television and their students' scores on math tests. The accompanying table shows this relationship.

   (a) Find a linear function that describes the grade g on a student's math test as a function of the number of hours t of television that he or she watches every day.

(b) Show that your function $g(t)$ is invertible.

(c) Find the inverse of $g(t)$. What does this function $t(g)$ represent in terms of t and g?

(d) Use the function $t(g)$ to predict the number of hours that Alina watched television each day if she scored a 40 on her math test. How much television can she be allowed to watch each day if her father wants her to get an 85 on her next test?

Hours of Daily TV Watching	0	1	2	3	4	5	6
Grade on Math Test	92	86	80	74	68	62	56

Proofs

44. Use the definition of *one-to-one* to argue that a function is one-to-one if and only if it passes the Horizontal Line Test.

45. Prove Theorem 1.13: If a function f is increasing on its entire domain (or decreasing on its entire domain), then f is one-to-one. Use the definitions of *one-to-one* and *increasing*.

46. Let f be an invertible function with inverse f^{-1}. Prove that if $f^{-1}(b) = a$, then $f(a) = b$.

47. Let f be an invertible function with inverse f^{-1}. Use Definition 1.19 to argue that the domain of f^{-1} is the range of f and that the range of f^{-1} is the domain of f.

48. If f is an invertible function with inverse f^{-1}, show that f^{-1} is invertible. (*Hint:* Suppose it wasn't; what would that tell you about f? In other words, try using proof by contradiction.)

49. In this problem you will show that if f is an invertible function, then the graph of $f^{-1}(x)$ is the graph of $f(x)$ reflected over the line $y = x$. You may find it helpful to use the graph given below.

(a) Suppose (a, b) is a point on the graph of $f(x)$. What point do you know is on the graph of $f^{-1}(x)$?

(b) Sketch the line connecting (a, b) and (b, a), and prove that it is perpendicular to the line $y = x$.

(c) Find the coordinates of the point where the line from (a, b) to (b, a) intersects the line $y = x$.

(d) Show that (a, b) and (b, a) are equidistant from the point of intersection found in part (c).

(e) Explain why parts (a)–(d) above show that the graph of $f^{-1}(x)$ is the graph of $f(x)$ reflected across the line $y = x$.

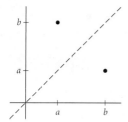

Limits

*L*imits are the backbone of calculus. Every concept and technique in calculus is an application of limits. For example, with limits we will be able to investigate the local behavior of a function and give meaning to the slope of a curve. We will later use limits to define areas and volumes bounded by curves. Without the concept of limit, none of calculus would be possible.

We begin with an intuitive, graphical interpretation of limits. After mathematically formalizing this interpretation, we will develop methods for calculating and manipulating limits. Our first application of limits occurs near the end of this chapter, where we use limits to define and investigate continuity of functions.

CONTENTS

2.1 Intuitive Notion of Limit

Limits are the key to understanding the local behavior of functions. Continuity, the derivative, and the definite integral will all be defined in terms of limits of functions. In this section we will develop an intuitive understanding of the concept of a limit. In Section 2.2 we will develop a rigorous mathematical definition of limit.

2.1.1 What is a limit?

Consider the functions $f(x) = x + 1$ and $g(x) = \frac{x^2 - 1}{x - 1}$, graphed in Figures 2.1 and 2.2.

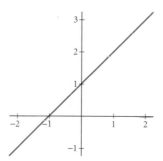

Figure 2.1 $f(x) = x + 1$

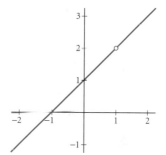

Figure 2.2 $g(x) = \dfrac{x^2 - 1}{x - 1}$

These two functions are identical except at one point: At $x = 1$ the value of f is $f(1) = 1 + 1 = 2$, whereas the value of g is $g(1) = \frac{1^2 - 1}{1 - 1}$, which is undefined (since $\frac{0}{0}$ is undefined). When $x \neq 1$ we have $g(x) = \frac{x^2 - 1}{x - 1} = \frac{(x-1)(x+1)}{x - 1} = x + 1 = f(x)$.

Although f and g do not agree *at* the point $x = 1$, both functions have values that *approach* $y = 2$ as x approaches 1. In other words, as x gets closer to the value $x = 1$, both $f(x)$ and $g(x)$ get closer to the value $y = 2$. In this case, we say that both f and g have a **limit** of 2 as x approaches 1. Figures 2.3 and 2.4 illustrate these limits.

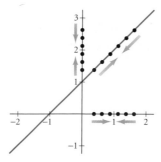

Figure 2.3

$f(x) = x + 1$ approaches 2 as x approaches 1

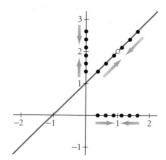

Figure 2.4

$g(x) = \dfrac{x^2 - 1}{x - 1}$ approaches 2 as x approaches 1

✧ *Caution*   These two functions have the same *limit* at $x = 1$ even though they do not have the same *value* at $x = 1$. The limits of f and g at $x = 1$ concern the values of f and g only at points *near*, but not *at*, $x = 1$.   □

DEFINITION 2.1

Intuitive Description of Limit

If $f(x)$ approaches a real number L as x approaches a real number c, we say that L is the **limit** of f as x approaches c.

More precisely, L is the **limit** of f as x approaches c if we can make $f(x)$ as close as we like to a number L by making x sufficiently close (but not equal) to c.

We will give a rigorous mathematical description of limit in Section 2.2. This rigorous description will come from the second part of Definition 2.1. For now, we focus on the intuitive meaning of limit, primarily making use of the first part of Definition 2.1.

2.1.2 Limit notation

We will use limits often so it is convenient to have a compact notation for them.

NOTATION 2.1

Limit Notation

If L is the limit of f as x approaches c, we will write

$$\lim_{x \to c} f(x) = L.$$

We will also sometimes write this limit as

$$f(x) \to L \text{ as } x \to c.$$

The first notation is read "the limit of $f(x)$ as x approaches c is L." The second notation is read "$f(x)$ approaches L as x approaches c." We could also turn this second notation around to write the equivalent statement "as x approaches c, $f(x)$ approaches L" (i.e., as $x \to c$, $f(x) \to L$).

EXAMPLE 2.1

Using limit notation

If $f(x) = x + 1$ (see Figure 2.1 on page 135), then the limit of $f(x)$ as x approaches $c = 1$ is $L = 2$. In limit notation, we would write

$$\lim_{x \to 1}(x + 1) = 2.$$

In other words, $x + 1 \to 2$ as $x \to 1$.

As we saw in Figure 2.2, even though $g(x) = \frac{x^2 - 1}{x - 1}$ is not defined at $x = 1$, it has a limit of $L = 2$ as x approaches $c = 1$. In limit notation, we write

$$\lim_{x \to 1} \frac{x^2 - 1}{x - 1} = 2.$$

We could also write $\frac{x^2-1}{x-1} \to 2$ as $x \to 1$.   □

EXAMPLE 2.2 **Graphically approximating limits**

Given the graph of $f(x)$ in Figure 2.5, approximate the following limits:

(a) $\lim\limits_{x \to 2} f(x)$ **(b)** $\lim\limits_{x \to 1} f(x)$ **(c)** $\lim\limits_{x \to -1} f(x)$ **(d)** $\lim\limits_{x \to -2} f(x)$

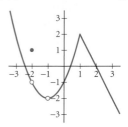

Figure 2.5

Solution   Note from the graph that the value of $f(x)$ at $x = -2$ is $f(-2) = 1$ and that $f(x)$ is undefined at the point $x = -1$.

   (a) As x approaches 2, the height of the graph of $y = f(x)$ approaches zero. Thus $f(x) \to 0$ as $x \to 2$, so $\lim\limits_{x \to 2} f(x) = 0$. In this case the *value* of $f(x)$ at $x = 2$ is also equal to zero (although this does not affect the *limit* of $f(x)$ as $x \to 2$).

   (b) As $x \to 1$, $f(x)$ approaches the height $y = 2$. Thus $\lim\limits_{x \to 1} f(x) = 2$. The fact that the graph of $f(x)$ has a sharp corner at $x = 1$ does not affect the limit of $f(x)$ as $x \to 1$.

   (c) As x approaches -1, the value of $f(x)$ approaches -2. It does not matter that $f(x)$ is not defined *at $x = -1$*, since we are concerned only with values of x that are *close* to $x = -1$. Thus $f(x) \to -2$ as $x \to -1$, so $\lim\limits_{x \to -1} f(x) = -2$.

   (d) As x gets closer to -2, $f(x)$ gets closer to -1. Remember that the value $f(-2)$ is not relevant when we are calculating the limit as x approaches $x = -2$. We have $f(x) \to -1$ as $x \to -2$, so $\lim\limits_{x \to -2} f(x) = -1$.   □

2.1.3 Left and right limits

When we say that "x approaches c" in Definition 2.1, we mean to consider values of x that are close to c from either the right or the left. In other words, when trying to find $\lim\limits_{x \to c} f(x)$, we consider values of x that are slightly to the left of c as well as values of x that are slightly to the right of c. Sometimes it is convenient to consider these two cases separately.

EXAMPLE 2.3 **Graphically approximating left and right limits**

Consider the function $f(x)$ shown in Figure 2.6. As x approaches $c = 1$ from the "left" (consider values such as $x = 0.9$, $x = 0.99$, etc.), the height of $f(x)$ gets closer and closer to $y = 2$. On the other hand, as x approaches 1 from the "right" (consider values such as $x = 1.1$, $x = 1.01$, etc.), the height of $f(x)$ approaches $y = 1$. In this case we

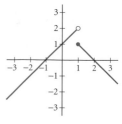

Figure 2.6

say that the limit of $f(x)$ as x approaches 1 *from the left* is equal to 2 and that the limit of $f(x)$ as x approaches 1 *from the right* is equal to 1.

Since we get different limits for $f(x)$ depending on whether we approach $c = 1$ from the left or the right, there is no *one* real number that represents the limit of $f(x)$ as x approaches 1. In this case we say that $\lim_{x \to 1} f(x)$ does not exist. Limits for $f(x)$ *do* exist as x approaches 1 from the left side or as x approaches 1 from the right side. □

We use the following notation to denote the **one-sided limits** of a function at $x = c$, that is, the limits from the left and from the right of the point $x = c$.

DEFINITION 2.2

> ### Intuitive Definitions of Left and Right Limits
>
> If $f(x)$ approaches a value L as x approaches c from the left, we say that L is the **left-hand limit** of $f(x)$ as x approaches c. In this case we write
>
> $$\lim_{x \to c^-} f(x) = L.$$
>
> If $f(x)$ approaches a value R as x approaches c from the right, we say that R is the **right-hand limit** of $f(x)$ as x approaches c. In this case we write
>
> $$\lim_{x \to c^+} f(x) = R.$$

We may also use the notations $f(x) \to L$ as $x \to c^-$ and $f(x) \to R$ as $x \to c^+$ to denote left-hand and right-hand limits. In Example 2.3, we have $c = 1$, $L = 2$, and $R = 1$.

§ *Caution*   The notation $x \to c^-$ means that x approaches c from the left. This notation does not imply anything about whether c is a positive or a negative number. □

EXAMPLE 2.4   **Limit notation for left and right limits**

In Example 2.3 we would say that the left-hand limit of $f(x)$ as x approaches 1 is 2 and that the right-hand limit of $f(x)$ as x approaches 1 is 1. In limit notation we would write this as

$$\lim_{x \to 1^-} f(x) = 2 \quad \text{and} \quad \lim_{x \to 1^+} f(x) = 1. \qquad \square$$

A function $f(x)$ has a well-defined limit (i.e., a limit that "exists") as x approaches c exactly when the corresponding left-hand and right-hand limits exist and are equal.

THEOREM 2.1

> ### For a Limit to Exist, the Left and Right Limits Must Exist and Be Equal
>
> L is the **limit** of $f(x)$ as x approaches c if and only if the left-hand and right-hand limits of $f(x)$ as x approaches c both exist and are equal to L. In limit notation,
>
> $$\lim_{x \to c} f(x) = L \iff \left(\lim_{x \to c^-} f(x) = L \quad \text{and} \quad \lim_{x \to c^+} f(x) = L \right).$$

As we will see in Section 2.2, this theorem will follow from the rigorous mathematical descriptions of one-sided and two-sided limits.

EXAMPLE 2.5 **An example where the left and right limits exist but are not equal**

Consider again the function $f(x)$ in Examples 2.3 and 2.4 (see Figure 2.6 on page 137). For this function, $\lim\limits_{x\to 1^-} f(x) = 2$ and $\lim\limits_{x\to 1^+} f(x) = 1$, so the two-sided limit $\lim\limits_{x\to 1} f(x)$ does not exist. □

2.1.4 Infinite limits

Suppose we are given a function f and a real number $x = c$. As x approaches c, the values of $f(x)$ may or may not approach any specific number. Sometimes, as x approaches c, the values $f(x)$ increase without bound; this means that the values of $f(x)$ eventually "approach infinity" as x approaches c.

DEFINITION 2.3

Intuitive Description of Infinite Limits

If $f(x)$ increases without bound as x approaches c from the left, we say that $f(x) \to \infty$ as $x \to c^-$. Similarly, if $f(x)$ increases without bound as x approaches c from the right, we say that $f(x) \to \infty$ as $x \to c^+$. In limit notation we express these limits as

$$\lim\limits_{x\to c^-} f(x) = \infty \quad \text{and} \quad \lim\limits_{x\to c^+} f(x) = \infty.$$

If $f(x)$ decreases without bound as x approaches c from the left, we say that $f(x) \to -\infty$ as $x \to c^-$. Similarly, if $f(x)$ decreases without bound as x approaches c from the right, we say that $f(x) \to -\infty$ as $x \to c^+$. In limit notation we express these limits as

$$\lim\limits_{x\to c^-} f(x) = -\infty \quad \text{and} \quad \lim\limits_{x\to c^+} f(x) = -\infty.$$

EXAMPLE 2.6 **Graphically identifying infinite limits**

Consider the function $f(x)$ graphed in Figure 2.7.

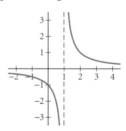

Figure 2.7

As x approaches 1 from the right, the values of $f(x)$ get larger and larger, without bound. In this case we say that the limit of $f(x)$ as $x \to 1^+$ is infinite, and we write $f(x) \to \infty$ as $x \to 1^+$. On the other hand, as x approaches 1 from the left, $f(x)$ decreases without bound. In other words, $f(x) \to -\infty$ as $x \to 1^-$. Therefore,

$$\lim\limits_{x\to 1^+} f(x) = \infty \quad \text{and} \quad \lim\limits_{x\to 1^-} f(x) = -\infty.$$

Since the right-hand and left-hand limits of $f(x)$ as x approaches 1 are not equal, the two-sided limit of $f(x)$ as $x \to 1$ does not exist. In limit notation,

$$\lim\limits_{x\to 1} f(x) \quad \text{does not exist.} \qquad \square$$

Caution Technically, we could say that limits of ∞ or $-\infty$ "do not exist," because ∞ and $-\infty$ are not numbers. However, whenever possible we will indicate the *way* that the limit does not exist by writing ∞ or $-\infty$, as appropriate. ☐

We will say that a function $f(x)$ has a limit of "infinity" at a point $x = c$ if $f(x)$ grows without bound as x approaches c from the right *and* the left.

DEFINITION 2.4

Two-Sided Infinite Limits

Given a function $f(x)$ and a real number c,

$$\lim_{x \to c} f(x) = \infty \iff \left(\lim_{x \to c^-} f(x) = \infty \quad \text{and} \quad \lim_{x \to c^+} f(x) = \infty \right),$$

and

$$\lim_{x \to c} f(x) = -\infty \iff \left(\lim_{x \to c^-} f(x) = -\infty \quad \text{and} \quad \lim_{x \to c^+} f(x) = -\infty \right).$$

EXAMPLE 2.7

An example where the left and right limits are infinite with opposite signs

The function $f(x)$ in Figure 2.7 on the previous page does *not* have a limit of ∞ or $-\infty$ at $c = 1$. This function $f(x)$ approaches $-\infty$ as $x \to 1^-$ but approaches ∞ as $x \to 1^+$. ☐

EXAMPLE 2.8

An example where the left and right limits are infinite with the same sign

On the other hand, consider the function $g(x)$ graphed in Figure 2.8.

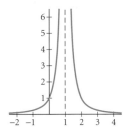

Figure 2.8

In this example, $f(x) \to \infty$ as $x \to 1^-$ and as $x \to 1^+$, so $\lim_{x \to 1} f(x) = \infty$. ☐

We can use limit notation to give a mathematical description of a vertical asymptote.

DEFINITION 2.5

Vertical Asymptote

A function $f(x)$ has a ***vertical asymptote*** at $x = c$ if one or more of the following are true:

$$\lim_{x \to c^+} f(x) = \infty, \quad \lim_{x \to c^-} f(x) = \infty, \quad \lim_{x \to c^+} f(x) = -\infty, \quad \text{or} \quad \lim_{x \to c^-} f(x) = -\infty.$$

EXAMPLE 2.9   **Graphically identifying vertical asymptotes**

The function graphed in Figure 2.8 has a vertical asymptote at $x = 1$ because $\lim\limits_{x \to 1^-} f(x) = \infty$ (we could also use the fact that $\lim\limits_{x \to 1^+} f(x) = \infty$). Similarly, the function shown in Figure 2.7 has a vertical asymptote at $x = 1$ because $\lim\limits_{x \to 1^-} f(x) = -\infty$ (or because $\lim\limits_{x \to 1^+} f(x) = \infty$). $\square$

2.1.5 Limits at infinity

So far we have discussed limits of functions as the independent variable x approaches some real number c. We are also interested in examining limits where x gets larger and larger without bound (that is, where $x \to \infty$) and limits where $x \to -\infty$.

DEFINITION 2.6

Intuitive Description of Limits at Infinity

If $f(x)$ approaches a value L as x gets larger and larger, we say that $f(x) \to L$ as $x \to \infty$. In limit notation we write

$$\lim_{x \to \infty} f(x) = L.$$

Similarly, if $f(x)$ approaches a value L as x becomes a negative number of larger and larger magnitude, we say that $f(x) \to L$ as $x \to -\infty$. In limit notation we write

$$\lim_{x \to -\infty} f(x) = L.$$

❓ *Question*   Explain the difference between what we called "infinite limits" in Definition 2.3 and what we called a "limit at infinity" in Definition 2.6. $\square$

EXAMPLE 2.10   **Graphically identifying a limit as $x \to \infty$**

Consider the function $f(x)$ shown in Figure 2.9.

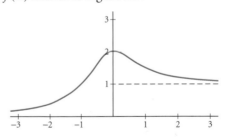

Figure 2.9

As $x \to \infty$, the value of $f(x)$ gets closer and closer to 1 (assuming that Figure 2.9 represents the global behavior of $y = f(x)$, that is, that the height of the graph continues to approach $y = 1$ as x gets larger and larger). On the other hand, as $x \to -\infty$, the value of $f(x)$ gets closer and closer to zero. Therefore,

$$\lim_{x \to \infty} f(x) = 1 \quad \text{and} \quad \lim_{x \to -\infty} f(x) = 0. \qquad \square$$

❓ *Question*   Combine Definitions 2.3 and 2.6 to describe the meaning of the statement

$$\lim_{x \to \infty} f(x) = -\infty.$$

Then sketch the graph of a function that has this limit. $\square$

We can use limits at infinity to give a mathematical description of horizontal asymptotes.

DEFINITION 2.7

Horizontal Asymptotes

A function $f(x)$ has a ***horizontal asymptote*** at $y = L$ if one or both of the following are true:

$$\lim_{x \to \infty} f(x) = L \quad \text{or} \quad \lim_{x \to -\infty} f(x) = L.$$

By convention, if $f(x)$ is *equal* to L as $x \to \infty$ or as $x \to -\infty$, we do not consider $f(x)$ to have a horizontal asymptote at $y = L$.

EXAMPLE 2.11

Graphically identifying a horizontal asymptote

The function $f(x)$ shown in Figure 2.9 (on the previous page) has a horizontal asymptote at $y = 1$, since $\lim_{x \to \infty} f(x) = 1$. This function also has a horizontal asymptote at $y = 0$, since $\lim_{x \to -\infty} f(x) = 0$. Note that this function has two horizontal asymptotes: one on the left, and one on the right. □

EXAMPLE 2.12

A constant function does not have a horizontal asymptote

The constant function $f(x) = 2$ does *not* have a horizontal asymptote at $y = 2$, since as $x \to \infty$ and as $x \to -\infty$ (in fact, for all x), $f(x)$ is constantly equal to 2. □

❓ Question   The function $f(x)$ in Figure 2.9 has two horizontal asymptotes. Can a function have *more* than two horizontal asymptotes? Why or why not? □

2.1.6 Approximating limits using tables of values

So far we have examined limits graphically. In Sections 2.4 and 2.5 we will see how to evaluate limits algebraically. In this section we will approximate limits using tables of values.

Given a function $f(x)$ and a value $x = c$, we can approximate $\lim_{x \to c} f(x)$ by looking at values of $f(x)$ when x is close to c. By constructing a table of values for $f(x)$, we can often make an educated guess for the limit of $f(x)$ as x approaches c. We will get the values in the table by evaluating $f(x)$ at x-values to the right and left of $x = c$.

EXAMPLE 2.13

Using a table to approximate a limit

Consider the function $f(x) = x + 1$ and the point $c = 1$. What is $\lim_{x \to 1}(x + 1)$? Table 2.1 shows values of $f(x) = x + 1$ for some values of x approaching $c = 1$ from the left and the right.

x	0.9	0.99	0.999	1	1.001	1.01	1.1
$f(x)$	1.9	1.99	1.999	*	2.001	2.01	2.1

Table 2.1

As x approaches 1 from the left, the values of $f(x)$ approach 2. Similarly, as x approaches 1 from the right, the values of $f(x)$ approach 2. Note that although we know the value of $f(x)$ at $x=1$ is $f(1)=1+1=2$, we do not include that value in the table (instead we wrote "$*$," to indicate that we are not interested in the value *at* $x=1$), since $\lim_{x\to 1} f(x)$ does not depend on the value of $f(x)=x+1$ at $x=1$. (For example, we could be dealing with a function like the one in Figure 2.2 on page 135.) From Table 2.1 we would approximate that $\lim_{x\to 1}(x+1)=2$. $\qquad\square$

⑤ *Caution*   Table 2.1 does *not* definitively show that $\lim_{x\to 1} f(x)$ is equal to 2, since it does not show the values of $f(x)$ for *every* value of x near $c=1$. For example, considering only the values in the table, it is possible that $\lim_{x\to 1} f(x)$ is 2.00001 or 1.99999 (we chose 2 in Example 2.13 simply because it is a nice round number). It is also possible that the limit is nowhere near 2; perhaps $f(x)$ approaches some other value entirely when we get even closer to $x=1$. $\qquad\square$

❷ *Question*   Find a function $f(x)$ whose limit as $x\to 1$ is 2.00001. What does this have to do with Example 2.13? $\qquad\square$

⑤ *Caution*   For some functions you can make a reasonable guess for $\lim_{x\to c} f(x)$ by evaluating $f(x)$ at just one value of x that is very close to c. However, in general this is not a good strategy; the limit of a function depends on the values of $f(x)$ as x gets closer and closer to c, and testing only one value might be misleading. When using a calculator to approximate the limit of a function, it is always preferable to construct a table of values (rather than just testing one value) for $f(x)$. $\qquad\square$

We can approximate the limit of a function $f(x)$ as $x\to\infty$ by looking at a table of values of $f(x)$ for larger and larger values of x.

EXAMPLE 2.14   **Using a table to approximate a limit at infinity**

Use a table of values to approximate $\lim_{x\to\infty} \frac{x}{x-1}$.

Solution   Table 2.2 shows values of $f(x)=\frac{x}{x-1}$ for various large values of x.

x	25	50	100	1000	10,000
$f(x)$	1.04167	1.02041	1.0101	1.001	1.0001

Table 2.2

There is no reason why we chose these particular values of x; any sequence of progressively larger values of x will do. From the table it appears that as $x\to\infty$ (i.e., as x gets larger and larger without bound), the value of $f(x)$ gets closer and closer to 1. Thus we would approximate $\lim_{x\to\infty} \frac{x}{x-1}=1$. $\qquad\square$

⑤ *Caution*   What we did in Example 2.14 was only an approximation. We will learn how to calculate limits exactly in Sections 2.4 and 2.5. Also note that we didn't just look at the value of $f(x)$ at *one* large value of x; instead we examined the behavior of $f(x)$ as x got larger and larger without bound. $\qquad\square$

The following example shows what can go wrong when we try to approximate a limit from one value or from a table of values. This example involves the sine function, which you can find on your calculator (perhaps with a button labeled SIN). Be sure to have your calculator set to "radian" mode (rather than "degree" mode).

EXAMPLE 2.15

A function whose limit as $x \to 0$ does not exist (and is not infinite)

Suppose we want to approximate $\lim\limits_{x \to 0} \sin(\frac{1}{x})$. Let's first try to approximate this limit by evaluating $f(x)$ at just one value of x that is very close to $c = 0$. To do this we could use a calculator (in "radian" mode) to find, for example, $f(0.01) = \sin(\frac{1}{0.01}) \approx -0.50637$. Since 0.01 is so close to zero, we might guess that the limit of $\sin(\frac{1}{x})$ as x approaches zero is equal to something very close to -0.50637; perhaps it is approaching -0.5. Although this method gives a reasonable approximation in some examples, in this example it does not.

Consider Table 2.3, which lists approximate values for $f(x) = \sin(\frac{1}{x})$.

x	-0.001	-0.0001	-0.00001	0	0.00001	0.0001	0.001
$\sin\left(\frac{1}{x}\right)$	-0.82688	0.30561	-0.03575	$*$	0.03575	-0.30561	0.82688

Table 2.3

In this table we didn't have a choice about leaving the space for $f(0)$ blank, since $\sin(\frac{1}{x})$ is undefined at $x = 0$ (compare with Example 2.13). From the table we see that as x approaches zero from the left and from the right, the values of $f(x)$ seem to jump around! It is not clear whether or not the function $f(x) = \sin(\frac{1}{x})$ approaches any particular value as $x \to 0$. (In case you're thinking that the values are approaching zero as $x \to 0$, note that $f(0.000001) \approx -0.34999$.) It is impossible to make an educated guess for $\lim\limits_{x \to 0} \sin(\frac{1}{x})$ from Table 2.3.

Figures 2.10 and 2.11 show the graph of $f(x) = \sin(\frac{1}{x})$ in two different viewing windows. In fact, this function does not have a well-defined limit as $x \to 0$. This function oscillates between -1 and 1 faster and faster as we approach $x = 0$ from either the left or the right.

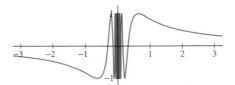

Figure 2.10

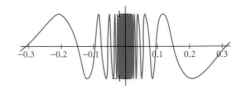

Figure 2.11

❓ Question   Try looking at the graph of $f(x) = \sin(\frac{1}{x})$ in different windows on your graphing calculator to get a sense for what it looks like. In particular, "zoom in" near $x = 0$ (while keeping y between -1 and 1).

❓ Question   In Example 2.15, we saw that the values of $f(x) = \sin(\frac{1}{x})$ do not seem to approach any particular limit as $x \to 0$. It is possible that we simply did not look at values that are close *enough* to zero. Try making a table of values for $\lim\limits_{x \to 0} \sin(\frac{1}{x})$ using numbers x that are even closer to zero. Does the function appear to have a limit as $x \to 0$, or does it continue to "jump around"?

What you should know after reading Section 2.1

▶ The graphical and intuitive concept of a limit, and notation for expressing limits. How to approximate limits graphically and with tables of values. What are the limitations of approximating limits in these ways?

▶ Graphical and intuitive descriptions of left-hand and right-hand limits, notation for these one-sided limits, and how to approximate one-sided limits graphically and with tables of values. What can you say about the limit of a function at a point if both of its one-sided limits exist and are equal?

▶ Graphical and intuitive descriptions of infinite limits (both one-sided and two-sided), and notation for such limits. How to recognize an infinite limit from a graph or table of values. What does it mean to say that a limit does not exist? What is the connection between infinite limits and vertical asymptotes?

▶ Limits as $x \to \infty$ graphically and intuitively, and notation for these limits at infinity. How could you approximate a limit as $x \to \infty$ by using a graph or table of values? What is the connection between limits at infinity and horizontal asymptotes?

Exercises 2.1

Concepts

0. Read the section and make your own summary of the material. Include whatever you think will help you review the material later on.

1. Use your calculator to graph the function $f(x) = \frac{x-2}{x^2-x-2}$. Use this graph to argue that $f(x) \to \frac{1}{3}$ as $x \to 2$, even though $f(x)$ is not defined at $x = 2$.

2. Explain graphically what it means to say that L is the limit of f as x approaches c.

3. If $f(x) = 1 - x^2$, then the limit of $f(x)$ as x approaches 2 is equal to -3. Express this fact in two different ways using limit notation.

■ **Explain the meanings of the following mathematical statements.**

4. $\lim\limits_{x \to 3^-} f(x) = 4$

5. $\lim\limits_{x \to -\infty} f(x) = \infty$

6. $\lim\limits_{x \to 1^+} f(x) = -\infty$

7. Explain the differences between the statements $\lim\limits_{x \to \infty} f(x) = 3$ and $\lim\limits_{x \to 3} f(x) = \infty$.

■ **For each problem below, sketch the graph of a function that has the indicated limits and values.**

8. $\lim\limits_{x \to 2} f(x) = -4$, $\lim\limits_{x \to -\infty} f(x) = -\infty$

9. $\lim\limits_{x \to -\infty} f(x) = 3$, $\lim\limits_{x \to \infty} f(x) = -\infty$

10. $\lim\limits_{x \to 0^+} f(x) = \infty$, $\lim\limits_{x \to 0^-} f(x) = \infty$

11. $\lim\limits_{x \to -\infty} f(x) = -2$, $\lim\limits_{x \to 3} f(x) = \infty$, $f(0) = -5$

12. $\lim\limits_{x \to 2^-} f(x) = 2$, $\lim\limits_{x \to 2^+} f(x) = -1$, $f(2) = 2$

13. $\lim\limits_{x \to 2^-} f(x) = 3$, $\lim\limits_{x \to 2^+} f(x) = 3$, $f(2) = 0$

14. If $\lim\limits_{x \to 1^-} f(x) = 5$ and $\lim\limits_{x \to 1^+} f(x) = 5$, what can you say about $\lim\limits_{x \to 1} f(x)$? What can you say about $f(1)$?

15. If $\lim\limits_{x \to 0^+} f(x) = -2$, $\lim\limits_{x \to 0^-} f(x) = 3$, and $f(0) = -2$, what can you say about $\lim\limits_{x \to 0} f(x)$?

16. If $\lim\limits_{x \to 2^+} f(x) = 8$ but $\lim\limits_{x \to 2} f(x)$ does not exist, what can you say about $\lim\limits_{x \to 2^-} f(x)$?

17. If $\lim\limits_{x \to -1^+} f(x) = -\infty$ and $\lim\limits_{x \to -1^-} f(x) = -\infty$, what can you say about $\lim\limits_{x \to -1} f(x)$?

18. If $\lim\limits_{x \to -\infty} f(x) = \infty$, $\lim\limits_{x \to \infty} f(x) = 3$, and $\lim\limits_{x \to 1^+} f(x) = \infty$, what can you say about any horizontal and vertical asymptotes of $f(x)$?

19. Can a function have more than two horizontal asymptotes? More than two vertical asymptotes? Why or why not?

20. Sketch a function that has the following table of values but whose limit as $x \to \infty$ is equal to $-\infty$.

x	100	200	500	1,000	10,000
$f(x)$	50	55	56.2	56.89	56.99

21. Sketch a function that has the following table of values but whose limit as $x \to 2$ does not exist.

x	1.9	1.99	1.999	2	2.001	2.01	2.1
$f(x)$	3.12	3.09	3.01	*	2.99	2.92	2.87

22. Investigate the graphs of $f(x) = \sin(\frac{1}{x})$ and $g(x) = x\sin(\frac{1}{x})$ near $x = 0$. Use the graphs to make educated guesses for $\lim_{x \to 0} f(x)$ and $\lim_{x \to 0} g(x)$. (*Hint:* Be sure to have your calculator set to "radian" mode; remember that your calculator has a SIN button that you can use to input this function.)

Skills

23. Given the graph of $y = f(x)$ below, approximate the following limits.

(a) $\lim_{x \to -2} f(x)$ (b) $\lim_{x \to -1} f(x)$

(c) $\lim_{x \to -\infty} f(x)$ (d) $\lim_{x \to \infty} f(x)$

(e) $\lim_{x \to 0} f(x)$ (f) $\lim_{x \to 2^-} f(x)$

(g) $\lim_{x \to 2^+} f(x)$ (h) $\lim_{x \to 2} f(x)$

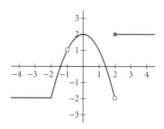

24. Given the graph of $y = g(x)$ below, approximate the following limits.

(a) $\lim_{x \to -\infty} g(x)$ (b) $\lim_{x \to -1^-} g(x)$

(c) $\lim_{x \to -1^+} g(x)$ (d) $\lim_{x \to -1} g(x)$

(e) $\lim_{x \to 2^-} g(x)$ (f) $\lim_{x \to 2^+} g(x)$

(g) $\lim_{x \to 2} g(x)$ (h) $\lim_{x \to \infty} g(x)$

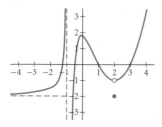

■ Use graphs to approximate each of the following limits.

25. $\lim_{x \to 4}(3 - 4x - 5x^2)$ 26. $\lim_{x \to -1}(x^3 - 2)$

27. $\lim_{x \to 0} \dfrac{1}{x}$ 28. $\lim_{x \to 2} \dfrac{x+1}{x-2}$

29. $\lim_{x \to 1} \dfrac{x-1}{x^2-1}$ 30. $\lim_{x \to -2} \dfrac{x^2+x-2}{x+2}$

■ Use tables of values to approximate each of the following limits.

31. $\lim_{x \to 2^-}(x^2 + x + 1)$ 32. $\lim_{x \to 3} \dfrac{1}{x^2 - 2x - 3}$

33. $\lim_{x \to 1} \dfrac{1}{1-x}$ 34. $\lim_{x \to -\infty}(3x + 1)$

35. $\lim_{x \to \infty}\left(1 + \dfrac{1}{x}\right)$ 36. $\lim_{x \to \infty} \dfrac{1+2x}{x-1}$

Applications

37. There are four squirrels currently living in Linda's attic. If she does nothing to evict these squirrels, the number of squirrels in her attic after t days will be given by the formula

$$S(t) = \frac{12 + 5.5t}{3 + 0.25t}.$$

(a) Verify that the formula for $S(t)$ indicates that there are four squirrels in Linda's attic at time $t = 0$.

(b) Use $S(t)$ to determine the number of squirrels in Linda's attic after 30 days, 60 days, and 1 year.

(c) Approximate $\lim_{t \to \infty} S(t)$ with a table of values. Explain what this limit means in real-world terms.

(d) Graph $S(t)$ with a graphing calculator, and use the graph to verify your answer to part (c).

38. The graph below describes the temperature $T(t)$ of a potato in an oven. The temperature T is measured in degrees Fahrenheit, and time t is measured in minutes.

(a) Use the graph of $T(t)$ to approximate the temperature of the potato when it is first put in the oven.

(b) Use the graph to approximate $\lim_{t \to \infty} T(t)$.

(c) What is the temperature of the oven, and why?

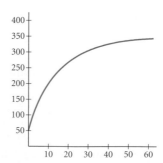

39. In 1960, H. von Forester suggested that the human population could be measured by the function

$$P(t) = \frac{179 \times 10^9}{(2027 - t)^{0.99}}.$$

Here P is the size of the human population. The time t is measured in years, where $t = 1$ corresponds to A.D. 1, time $t = 1973$ corresponds to A.D. 1973, and so on.

(a) Use a graphing calculator to graph this function. You will have to be very careful when choosing a graphing window!

(b) Use the graph you found in part (a) to approximate $\lim\limits_{t \to 2027^-} P(t)$.

(c) This population model is sometimes called the Doomsday Model. Why do you think this is? What year is "doomsday," and why?

(d) In part (b), we considered only the *left* limit of $P(t)$ as $x \to 2027$. Why? What is the real-world meaning of the part of the graph that is to the right of $t = 2027$?

2.2   Formal Definition of Limit

In the previous section we gave intuitive descriptions of limits. In this section we give rigorous mathematical definitions for various types of limits. We begin by formally defining what we mean by the statement $\lim\limits_{x \to c} f(x) = L$. After doing some examples that illustrate this formal definition, we will give rigorous descriptions of left and right limits, infinite limits, and limits at infinity.

2.2.1   Formalizing the intuitive definition

Throughout this section we will assume that $f(x)$ is a function defined on a **punctured interval** $(c - \delta, c) \cup (c, c + \delta)$ about some point c, where $\delta > 0$ represents a small distance. A punctured interval about c is some open interval that is centered on c but *does not include* the point c itself (see Figure 2.12).

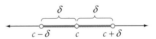

Figure 2.12
A punctured interval around c
with a "radius" of δ

Remember that we want the limit of $f(x)$ as x approaches c to depend only on values *near* c, not on the value of $f(x)$ at $x = c$. Whenever we discuss limits of functions, we will be working on such a punctured interval.

Recall that the intuitive definition of the statement $\lim\limits_{x \to c} f(x) = L$ is

(1) $f(x)$ *approaches L as x approaches c.*

More precisely, we say that $\lim\limits_{x \to c} f(x) = L$ if:

(2)   *We can make $f(x)$ as close as we like to L by making x sufficiently close (but not equal) to c.*

The "but not equal" part of Statement (1) refers to the fact that the value of $f(x)$ *at* $x = c$ is not relevant to the limit of $f(x)$ as x *approaches c*.

What does Statement (2) mean? Saying that we can make $f(x)$ "as close as we like" to the number L means that we can choose x so that the distance between $f(x)$ and L is as small as we like. Let the Greek letter ε (epsilon) represent the distance between $f(x)$ and L. On the other hand, requiring that x be "sufficiently close to" c is equivalent to requiring that the distance between x and c be sufficiently small. We will let the Greek letter δ (delta) represent the distance between x and c. With this new notation, Statement (2) says that we can make $f(x)$ within ε of L by making x within δ of c.

More precisely,

(3)   *For any small distance ε around L, there is a small distance δ around c such that if x is within δ of c (but not equal to c), then $f(x)$ is within ε of L.*

The following example illustrates the graphical roles of c, L, δ, and ε.

EXAMPLE 2.16   **Graphical interpretations of ε and δ intervals**

Consider the graph of $y = f(x)$ shown in Figure 2.13. Let ε be a small positive number. All the y-values within ε of L are in the bar between $L - \varepsilon$ and $L + \varepsilon$ in Figure 2.14.

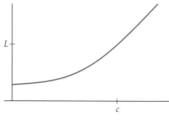

Figure 2.13   **Figure 2.14**

How close to c do we have to choose x so that $f(x)$ is within ε of L? In other words, which values of x will have $f(x)$ values between $L - \varepsilon$ and $L + \varepsilon$? Figure 2.15 shows a value of x for which $f(x)$ is within ε of L. Figure 2.16 shows a value of x for which $f(x)$ is *not* within ε of L.

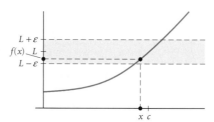
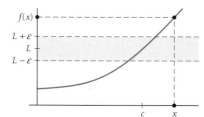

Figure 2.15   **Figure 2.16**

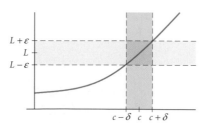

Figure 2.17

Consider the punctured interval $(c - \delta, c) \cup (c, c + \delta)$ shown in Figure 2.17 and the vertical "bar" in the plane defined by this interval. Each real number x in this punctured interval has a value $f(x)$ that is within ε of L (i.e., within the horizontal "bar" defined by $(L - \varepsilon, L + \varepsilon)$).

To summarize, given the small distance ε around L, we found a small distance δ around c such that every x within δ of c has an $f(x)$ value within ε of L. Saying that $\lim\limits_{x \to c} f(x) = L$ is equivalent to saying that we can *always* find such a δ for *any* given ε, no matter how small ε is.   □

❖ *Caution*   In Example 2.16, it happens to be true that the real number c has a value $f(c)$ that is in the horizontal "bar" (since $f(c) = L$). However, in general, the value exactly at $x = c$ may lie outside this horizontal bar. When discussing a limit, we do not consider what happens *at $x = c$*, only what happens *near $x = c$*.   □

❓ *Question*   If $\lim_{x \to c} f(x) = L$, then given any ε, we can always find some δ such that every x within δ of c corresponds to a value $f(x)$ within ε of L. Consider the ε shown in Figure 2.17. What will the δ interval have to be like if we choose a smaller value of ε? What if we choose a larger value of ε? Do you think we can find a δ that works for *every ε* with the graph in Figure 2.17? What does this have to do with the intuitive definition of a limit?   □

We are now ready to write Statement (3) in a more mathematical way, but first we need to explain the phrase "x is within δ of c." This phrase means that the distance between x and c is less than δ, that is, that $|x - c| < \delta$ (see Section 0.1.5). In addition, we want to omit the case when $x = c$ (i.e., the case when the distance between x and c is zero). Thus we will also require that $0 < |x - c|$. Putting the two inequalities together, we get $0 < |x - c| < \delta$.

Similarly, $f(x)$ is within ε of L when the distance between $f(x)$ and L is less than ε. We can write this as $|f(x) - L| < \varepsilon$. The formal definition of a limit is Statement (3) written in this mathematical notation.

DEFINITION 2.8

Formal Definition of Limit

We say that a real number L is the *limit* of a function $f(x)$ as x approaches a real value c, and write $\lim_{x \to c} f(x) = L$, if

For all $\varepsilon > 0$, there exists a $\delta > 0$ such that
if $0 < |x - c| < \delta$, then $|f(x) - L| < \varepsilon$.

If we write the inequalities as intervals, then Definition 2.8 says that $\lim_{x \to c} f(x) = L$ if

(4)
For all $\varepsilon > 0$, there exists a $\delta > 0$ such that
if $x \in (c - \delta, c) \cup (c, c + \delta)$, then $f(x) \in (L - \varepsilon, L + \varepsilon)$.

❖ *Caution*   It is very important that you can accurately state *and understand* Definition 2.8. Try phrasing this definition in your own words, being as mathematically precise as possible. Can you explain the meanings of all the variables and the notation in this definition?   □

❓ *Question*   Why do we have $0 < |x - c| < \delta$ in Definition 2.8 rather than just $|x - c| < \delta$? Think about what it means about x to require that $0 < |x - c|$.   □

We mentioned in Section 2.1 that limits are *unique*. This means that $\lim_{x \to c} f(x)$, if it exists, can be equal to one and only one real number.

THEOREM 2.2

Uniqueness of Limits

If $\lim_{x \to c} f(x) = L$ and $\lim_{x \to c} f(x) = M$, then $L = M$.

The proof of this theorem involves the formal definition of limit given in Definition 2.8 and can be found in the appendix at the end of this chapter (Section 2.8).

2.2.2 Finding a delta for a given epsilon

By Definition 2.8, if $\lim_{x \to c} f(x) = L$, then for *every* $\varepsilon > 0$ we can find *some* $\delta > 0$ such that $f(x)$ is within ε of L whenever x is within δ of c (but not equal to c). In this section we focus on finding a value of δ satisfying Definition 2.8 given a *particular* value of ε. In Section 2.3 we will see how to find a δ (as a function of ε) that works for *any* given ε.

The following example shows (in a real-world context) how to find a value of δ that "works" given a particular value of ε.

EXAMPLE 2.17

A real-world example of finding δ given ε

The efficiency at which a car uses fuel depends on the speed at which the car is driven. A typical car runs at 100% fuel efficiency when driven at 55 miles per hour. Suppose that fuel efficiency E is the function of speed s (in mph) shown in Figure 2.18, with equation $E(s) = -0.033058(s^2 - 110s)$.

Most of the time it is not possible to drive at exactly 55 miles per hour. However, the limit of $E(s)$ as $s \to 55$ is equal to 100% efficiency, so by Definition 2.8, you can ensure a fuel efficiency as close to 100% as you please by making sure that you drive at a speed sufficiently close to 55 miles per hour.

If you want your car to run with at least 95% fuel efficiency, how close to 55 miles per hour do you have to drive? In the language of limits we are asking: if $\varepsilon = 5$, what is δ? In other words, given that we want to be within 5% of 100% fuel efficiency, how much can our speed differ from 55 miles per hour? Figure 2.19 shows the "ε bar" that we want to be in. We need to figure out which s-values have E-values that lie within that ε bar.

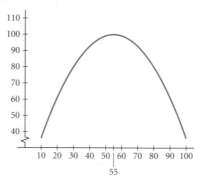

Figure 2.18   Figure 2.19

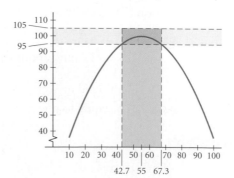

Figure 2.20

From Figure 2.19, we see that we need to know the values of s for which $E(s) = 95$; any values of s between those two values will have E-values in the ε bar (look at the figure until you are convinced of this). We can solve the equation $-0.033058(s^2 - 110s) = 95$ by hand (using the quadratic formula), or we can use a calculator to graph E and approximate where $E(s) = 95$. In either case, we find $s \approx 42.7$ and $s \approx 67.3$. Figure 2.20 illustrates that every value of s between 42.7 and 67.3 has an E-value within 5% of 100%.

Finally, from all this work we see that you can drive anywhere from 42.7 to 67.3 miles per hour and still get at least 95% fuel efficiency. In the language of limits, to be within $\varepsilon = 5$ of 100% fuel efficiency, you must drive within $\delta = 12.3$ miles per hour of 55 miles per hour (since $42.7 = 55 - 12.3$ and $67.3 = 55 + 12.3$). ☐

The technique used in Example 2.17 for finding δ given a particular ε works in general. Suppose that $\lim_{x \to c} f(x) = L$ and that we wish to find the largest possible δ interval for a given ε. We can often find such a δ by solving the equations $f(x) = L - \varepsilon$ and $f(x) = L + \varepsilon$ or by tracing along the graph of $f(x)$ to approximate these x-values. (Of course you should always sketch a graph to make sure that this method is appropriate for the function in question.) As the next example shows, however, the δ interval of x-values for which $f(x)$ is within ε of L will not always be evenly centered on the point c.

| EXAMPLE 2.18 | **Graphically approximating the smallest δ that "works" for a given ε** |

Consider the function $f(x) = x^2$. From the graph of f in Figure 2.21 it certainly appears that $x^2 \to 1$ as $x \to 1$. How does one *prove* that $\lim_{x \to 1} x^2 = 1$? To do so, we would have to show that for *every* $\varepsilon > 0$ we can find some $\delta > 0$ satisfying Definition 2.8. Such proofs are challenging, and we will see how to do them in the next section. For now, we will consider a *particular* ε, namely $\varepsilon = 0.5$, and show that there is a δ such that $|x^2 - 1| < 0.5$ whenever $|x - 1| < \delta$. Figure 2.22 shows the $\varepsilon = 0.5$ interval around $y = 1$.

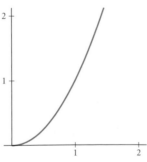

Figure 2.21 $y = x^2$

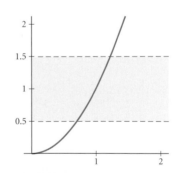

Figure 2.22 ε interval

From the graph in Figure 2.22, it is clear that we need to know the numbers x for which $f(x) = 0.5$ and those for which $f(x) = 1.5$. Solving the equations $x^2 = 0.5$ and $x^2 = 1.5$, we get $x = \sqrt{0.5} \approx 0.707$ and $x = \sqrt{1.5} \approx 1.22$, respectively.

Figure 2.23 on the next page shows this interval $(0.707, 1.22)$ of x-values. (*Note:* Since we do not consider what happens when x is *equal* to 1, we are actually considering the punctured interval $(0.707, 1) \cup (1, 1.22)$ here.) Each value of x in this interval has a value of $f(x)$ in the ε interval $(0.5, 1.5)$.

What is the δ that "works" for the given $\varepsilon = 0.5$ in this example? The interval $(0.707, 1.22)$ is *not* centered on the point $c = 1$. In fact, $|1 - 0.707| = 0.293$ and $|1 - 1.22| = 0.22$, so in order to pick an x such that $f(x)$ is in $(0.5, 1.5)$ we must pick x within 0.293 unit to the left of $c = 1$ or within 0.22 unit to the right of $c = 1$.

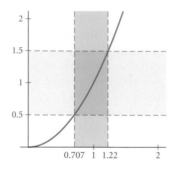

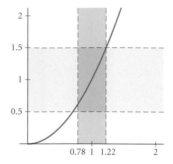

Figure 2.23
ε interval with *all* "working" x

Figure 2.24
ε interval and δ interval

The smaller of these two numbers is our δ. If $\delta = 0.22$, then any x within 0.22 unit on either side of $c = 1$ will have an $f(x)$-value within 0.5 of $L = 1$. Therefore, our punctured δ interval is $(1 - \delta, 1) \cup (1, 1 + \delta) = (1 - 0.22, 1) \cup (1, 1 + 0.22) = (0.78, 1) \cup (1, 1.22)$. This interval is shown in Figure 2.24 (although we cannot see the "puncture" at $x = 1$).

Of course, we were using approximations above; the actual punctured interval of x-values for which $f(x)$ is between 0.5 and 1.5 is $(\sqrt{0.5}, 1) \cup (1, \sqrt{1.5})$. Therefore, the δ corresponding to $\varepsilon = 0.5$ is $\delta = \sqrt{1.5} - 1$ (the exact answer), which is approximately equal to $\delta \approx 0.224745$ (a better approximation than the 0.22 we had found earlier). □

⊛ Caution   Example 2.18 does not *prove* that $\lim\limits_{x \to 1} x^2 = 1$. To do that, we would have to show that for *every* $\varepsilon > 0$ we can find a $\delta > 0$ satisfying Definition 2.8. In Example 2.18, we showed this for only one ε, namely $\varepsilon = 0.5$. We will see how to find a $\delta > 0$ for every ε in Section 2.3. □

⊛ Caution   At this time, we do not know how to calculate the values of limits algebraically. We can recognize limits graphically using the intuitive definition, and we can discuss limits using the language of the formal definition in Definition 2.8. We won't calculate limits algebraically until Section 2.4. The focus right now is on understanding the concept and definition of limit, not on calculating limits. □

2.2.3 Formalizing one-sided limits

In this section we will put the intuitive definitions from Sections 2.1.3–2.1.5 into formal mathematical notation.

In Definition 2.2 from Section 2.1.3, we gave an intuitive definition of left-hand and right-hand limits. The formal definitions for these one-sided limits are exactly the same as the formal definition in Definition 2.8 except that the δ intervals are one-sided.

DEFINITION 2.9

Formal Definitions of Left and Right Limits

We say that a real number L is the **left limit** of a function $f(x)$ as x approaches a real value c (from the left), and write $\lim\limits_{x \to c^-} f(x) = L$, if

> *For all $\varepsilon > 0$, there exists a $\delta > 0$ such that*
> *if $x \in (c - \delta, c)$, then $|f(x) - L| < \varepsilon$.*

We say that a real number L is the **right limit** of a function $f(x)$ as x approaches a real value c (from the right), and write $\lim\limits_{x \to c^+} f(x) = L$, if

> *For all $\varepsilon > 0$, there exists a $\delta > 0$ such that*
> *if $x \in (c, c + \delta)$, then $|f(x) - L| < \varepsilon$.*

Figures 2.25 and 2.26 illustrate the roles of δ and ε in the formal definitions of left and right limits, respectively.

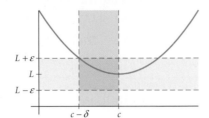

Figure 2.25
δ interval for left limit

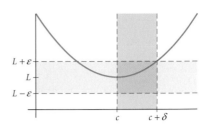

Figure 2.26
δ interval for right limit

❓ *Question*   Write these formal definitions of one-sided limits in the "interval notation" definition of limit, as in Statement (4).   □

2.2.4 Formalizing infinite limits

Definition 2.3 from Section 2.1.4 gave an intuitive description of "infinite limits," that is, situations where $f(x) \to \infty$ or $-\infty$ as x approaches a real value c from the left or the right. In Definition 2.10 we give a formal definition for the statement $\lim\limits_{x \to c^+} f(x) = \infty$. The definitions for the statements $\lim\limits_{x \to c^-} f(x) = \infty$, $\lim\limits_{x \to c^+} f(x) = -\infty$, and $\lim\limits_{x \to c} f(x) = -\infty$ are similar and are left to you in the exercises.

DEFINITION 2.10

Formal Definition of an Infinite Limit
We say that $\lim\limits_{x \to c^+} f(x) = \infty$ if

For all $M > 0$, there exists a $\delta > 0$ such that
if $x \in (c, c + \delta)$, then $f(x) > M$.

In Definition 2.8, the ε represented a small positive number. In Definition 2.10, M represents a large positive number. Definition 2.10 basically says that given any large positive M, we can find some $\delta > 0$ such that $f(x)$ is greater than M whenever x is in the interval $(c, c + \delta)$.

Figure 2.27 illustrates the roles of δ and M in the formal definition of an infinite limit.

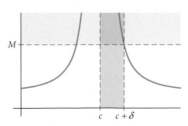

Figure 2.27

❓ *Question*   Write a formal definition for the limit statement $\lim\limits_{x \to c^-} f(x) = -\infty$, and sketch a graph illustrating the roles of M and δ from your definition.   □

2.2.5 Formalizing limits at infinity

We now formalize the intuitive description of the limit statement $\lim\limits_{x \to \infty} f(x) = L$ discussed in Definition 2.6 in Section 2.1.5. The formal definition of $\lim\limits_{x \to -\infty} f(x) = L$ is similar, and its construction is left to the exercises.

DEFINITION 2.11

Formal Definition of a Limit at Infinity

We say that $\lim\limits_{x \to \infty} f(x) = L$ if

For all $\varepsilon > 0$, there exists an $N > 0$ such that
if $x > N$, then $|f(x) - L| < \varepsilon$.

This definition basically says that we can make $f(x)$ as close as we like to L by choosing sufficiently large values of x. Figure 2.28 illustrates the roles of N and ε in this definition.

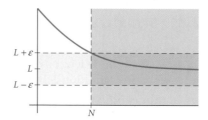

Figure 2.28

Finally, we give a formal definition of an infinite limit at infinity, that is, a limit of the form $\lim\limits_{x \to \infty} f(x) = \infty$. Formal definitions for the limit statements $\lim\limits_{x \to \infty} f(x) = -\infty$, $\lim\limits_{x \to -\infty} f(x) = \infty$, and $\lim\limits_{x \to -\infty} f(x) = -\infty$ are similar and are left to you in the exercises.

DEFINITION 2.12

Formal Definition of an Infinite Limit at Infinity

We say that $\lim\limits_{x \to \infty} f(x) = \infty$ if

For all $M > 0$, there exists an $N > 0$ such that
if $x > N$, then $f(x) > M$.

In this definition both M and N represent large numbers. Definition 2.12 basically says that we can make $f(x)$ as large as we want (i.e., larger than M) by choosing x sufficiently large (i.e., larger than N). Figure 2.29 illustrates the roles of M and N in this definition.

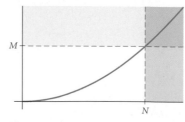

Figure 2.29

❓ *Question*   Give a formal mathematical definition of the statement $\lim\limits_{x \to -\infty} f(x) = -\infty$. Sketch a graph illustrating the roles of M and N from your definition.   □

What you should know after reading Section 2.2

▶ The formal (delta–epsilon) definition of a limit, and how it is developed from the intuitive definition of limit. Be able to state and explain this formal definition, including sketching a labeled graph.

▶ How to use your graphing calculator to approximate a δ that "works" for a given ε in a limit statement. Consider both abstract and real-world examples.

▶ The formal definitions for one-sided limits, infinite limits, and limits at infinity. Be able to state and explain these formal definitions, including sketching labeled graphs.

Exercises 2.2

Concepts

0. Read the section and make your own summary of the material. Include whatever you think will help you review the material later on.

1. What is a "punctured interval," and why do we need to use punctured intervals when discussing limits? Describe the punctured interval around $x = 2$ that has a radius of 3 and the punctured interval around $x = 4$ that has a radius of 0.25.

2. Find punctured intervals around **(a)** $x = 1.5$, **(b)** $x = 0.5$, **(c)** $x = 1$, and **(d)** $x = 0.25$ on which the function $f(x) = \frac{1}{x^2 - x}$ is defined.

■ Write each of the three limit statements below in the following five ways: **(a)** the form of Statement (1); **(b)** the form of Statement (2); **(c)** the form of Statement (3); **(d)** the form of Definition 2.8; and **(e)** the form of Statement (4).

3. $\lim\limits_{x \to -3} \sqrt{x + 7} = 2$ **4.** $\lim\limits_{x \to 2} \dfrac{x^2 - 4}{x + 2} = 0$

5. $\lim\limits_{h \to 0} \dfrac{(2 + h)^2 - 4}{h} = 4$

6. Describe in your own words, using complete sentences, the mathematical meaning of the limit statement $\lim\limits_{x \to c} f(x) = L$. Use the descriptions of limit given in the chapter for inspiration, but try to find a phrasing that makes the greatest sense to *you* (while still being mathematically accurate and complete, of course!). Sketch and label a graph that illustrates your description.

7. Write down a mathematical expression that expresses the sentence "x is not equal to 5, and the distance between x and 5 is less than 0.01." Then write an expression that means "the distance between $f(x)$ and -2 is less than 0.5."

8. Why do we have $0 < |x - c| < \delta$ instead of just $|x - c| < \delta$ in Definition 2.8?

■ Write each of the following inequalities using interval notation.

9. $0 < |x - 2| < 0.1$ **10.** $0 < |x + 3| < 0.05$

11. $|(3x + 1) - 2| < 0.1$ **12.** $|(x^2 - 1) + 3| < 0.5$

13. $0 < |x - c| < \delta$ **14.** $|f(x) - L| < \varepsilon$

■ Use interval notation to fill in the blanks below. Your answers may involve δ, ε, N, and/or M.

15. If $\lim\limits_{x \to 2} f(x) = 5$, then for all $\varepsilon > 0$, there is some $\delta > 0$ such that if $x \in$ _____, then $f(x) \in$ _____.

16. If $\lim\limits_{x \to 3^-} f(x) = 1$, then for all $\varepsilon > 0$, there is some $\delta > 0$ such that if $x \in$ _____, then $f(x) \in$ _____.

17. If $\lim\limits_{x \to \infty} f(x) = 2$, then for all $\varepsilon > 0$, there is some $N > 0$ such that if $x \in$ _____, then $f(x) \in$ _____.

18. If $\lim\limits_{x \to 1^+} f(x) = \infty$, then for all $M > 0$, there is some $\delta > 0$ such that if $x \in$ _____, then $f(x) \in$ _____.

19. If $\lim\limits_{x \to \infty} f(x) = -\infty$, then for all $M > 0$, there is some $N > 0$ such that if $x \in$ _____, then $f(x) \in$ _____.

■ Problems 20 and 21 are related to the fact that $\lim\limits_{x \to -2} x^2 = 4$.

20. Sketch and label a graph that illustrates the following true statement: If $0 < |x + 2| < 0.075$, then $|x^2 - 4| < 0.4$. Make your graph as accurate as possible.

21. Is the following statement true or false? Justify your answer with a graph. The statement is: If $0 < |x + 2| < 0.1$, then $|x^2 - 4| < 0.4$.

22. Sketch the graph of a function $f(x)$ with the following two properties. First, for all $\varepsilon > 0$, there is some $\delta > 0$ such that if $0 < |x - 2| < \delta$, then $|f(x) - 3| < \varepsilon$. Second, $f(2) = 5$. What is $\lim\limits_{x \to 2} f(x)$?

23. Use your graphing calculator to answer the following: If $0 < |x + 2| < 0.5$, what is the smallest interval

$I = (4 - \varepsilon, 4 + \varepsilon)$ for which we can guarantee that $x^2 \in I$?

■ Write each limit below as a formal statement involving δ or N, and ε or M, as in Definitions 2.9, 2.10, and 2.11.

24. $\lim\limits_{x \to 1}(x^2 - 3) = -2$

25. $\lim\limits_{x \to 3} f(x) = 10$

26. $\lim\limits_{x \to a} g(x) = k$

27. $\lim\limits_{x \to 3^-}(4 - x^2) = -5$

28. $\lim\limits_{x \to 0^+} \sqrt{x} = 0$

29. $\lim\limits_{x \to 2^+} \dfrac{1}{x - 2} = \infty$

30. $\lim\limits_{x \to 1^+} \dfrac{1}{1 - x} = -\infty$

31. $\lim\limits_{x \to \infty} \dfrac{x}{1 - 2x} = -0.5$

32. $\lim\limits_{x \to -\infty}(1 - 3x) = \infty$

■ Write down the logical negation of each of the following statements. Then explain what the statement means in terms of the intuitive definition of limit.

33. For all $\varepsilon > 0$, there exists $\delta > 0$ such that if $0 < |x - 2| < \delta$, then $|f(x) - 1| < \varepsilon$.

34. For all $\varepsilon > 0$, there exists $\delta > 0$ such that if $x \in (3, 3 + \delta)$, then $f(x) \in (7 - \varepsilon, 7 + \varepsilon)$.

35. For all $M > 0$, there exists $\delta > 0$ such that if $0 < |x - c| < \delta$, then $f(x) > M$.

36. For all $\varepsilon > 0$, there exists $N > 0$ such that if $x > N$, then $|f(x) - 3| < \varepsilon$.

37. It is *false* that $\lim\limits_{x \to 1}(x + 1.01) = 2$. Express this fact in a mathematical sentence involving δ and ε. (*Hint:* Negate the statement of the formal definition of limit.)

38. It is *false* that $\lim\limits_{x \to \infty} \dfrac{1000}{x} = \infty$. Express this fact in a mathematical sentence involving M and N. (*Hint:* Negate the statement of the formal definition of limit.)

39. Use the formal definition of limit to show that the limit as $x \to 2$ of $f(x) = \sqrt{x - 1.1}$ is *not* equal to 1. (*Hint:* Find an $\varepsilon > 0$ for which there is no corresponding $\delta > 0$ satisfying the formal definition of limit.)

■ For each limit statement listed below, sketch a graph illustrating the roles of δ, ε, M, and/or N, as appropriate. (In other words, sketch the graph of the appropriate function f, a representative ε or M interval, and a δ or N interval that "works" for this choice of ε or M.)

40. $\lim\limits_{x \to -1}(x^3 - 2) = -3$

41. $\lim\limits_{x \to 1} \sqrt{1 - x} = 0$

42. $\lim\limits_{x \to 0^-} \dfrac{1}{x} = -\infty$

43. $\lim\limits_{x \to -2^+} \dfrac{1}{x + 2} = \infty$

44. $\lim\limits_{x \to -\infty} \dfrac{1}{x^2 + 1} = 0$

45. $\lim\limits_{x \to \infty} x^3 + x^2 + x + 1 = \infty$

■ Modify Definitions 2.10 and 2.11 to construct formal definitions of each of the following limit statements. Sketch a graph illustrating the roles of the letters ε, δ, M, and/or N that may appear in your formal definitions.

46. $\lim\limits_{x \to c^-} f(x) = \infty$

47. $\lim\limits_{x \to c^+} f(x) = -\infty$

48. $\lim\limits_{x \to c^-} f(x) = -\infty$

49. $\lim\limits_{x \to -\infty} f(x) = L$

50. $\lim\limits_{x \to \infty} f(x) = -\infty$

51. $\lim\limits_{x \to -\infty} f(x) = \infty$

Skills

■ For each $f(x)$, L, c, and ε given below, find the largest value of δ such that if $0 < |x - c| < \delta$, then $|f(x) - L| < \varepsilon$. What does this have to do with limits?

52. $f(x) = x^3$, $L = 8$, $c = 2$, $\varepsilon = 0.5$

53. $f(x) = \sqrt{x - 1}$, $L = 2$, $c = 5$, $\varepsilon = 0.2$

54. $f(x) = 2 - x^2$, $L = -7$, $c = 3$, $\varepsilon = 0.01$

55. $f(x) = \dfrac{x^2 - 2x - 3}{x + 1}$, $L = -4$, $c = -1$, $\varepsilon = 1$

56. Approximate the largest value of δ such that if $x \in (1, 1 + \delta)$, then $\dfrac{1}{x^2 - 1} > 1000$. Show your work by sketching and labeling a graph. What does this have to do with limits?

57. Approximate the smallest value of N such that if $x > N$, then $|\frac{3x}{x+1} - 3| < 0.5$. Show your work by sketching and labeling a graph. What does this have to do with limits?

58. Approximate the smallest value of N such that if $x > N$, then $\sqrt{x - 2} > 1,000,000$. Show your work by sketching and labeling a graph. What does this have to do with limits?

Applications

59. Every month, Jack hides $50 under a broken floorboard to save up for a new boat. After t months of saving, he will have $F(t) = 50t$ dollars.

   (a) The boat Jack wants costs at least $7465. How many months does Jack have to save money before he will have enough to pay for the boat? Illustrate this information on a graph of $F(t)$.

   (b) Suppose a different boat costs M dollars. Will there always be a value $t = N$ for which $F(N) > M$? What does this mean in real-world terms?

   (c) Interpret this problem in terms of an infinite limit at infinity, and illustrate the roles of M and N on a graph of $F(t)$.

60. The cylindrical cans that Len's company produces are 6 inches tall, with varying radius. The cost C of producing a can with radius r is $C(r) = 10\pi r^2 + 24\pi r$ cents.

   (a) Use a graphing calculator to sketch a labeled graph of the function $C(r)$. Be sure to choose a sensible graphing window.

   (b) Len's boss wants him to construct the cans so that the cost of each can is within 25 cents of $4.00. Write this requirement as an absolute value inequality involving C.

   (c) Given the cost requirements described in part (b), what is the acceptable range of values for r? Write your answer with an absolute value inequality involving r.

(d) Interpret this problem in terms of δ and ε ranges. Specifically, what is c? What is L? What is ε? Use the range of r-values you found in part (c) to determine the largest δ that will work for the given value of ε. Then illustrate c, L, ε, and δ on the graph from part (a).

(e) Len's boss now says that he wants the cans to cost within 10 cents of $4.00. Repeat parts (c) and (d) with this new requirement.

(f) Assuming that a δ can be found for every ε as you did above, what is the limit statement this problem concerns? Your answer should involve c, L, and $C(r)$.

61. You work for a company that sells velvet Elvis paintings. The function $N(p) = 9.2\,p^2 - 725\,p + 16{,}333$ predicts the number N of velvet Elvis paintings that your company will sell if they are priced at p dollars each (see the graph below).

(a) Use your graphing calculator to estimate the price that company planners should charge per painting if they wish to sell 6000 velvet Elvis paintings.

(b) Find the range of prices that would enable your company to sell between 5000 and 7000 velvet Elvis paintings.

(c) Interpret this problem in terms of δ and ε ranges. Specifically, what is c? What is L? What is ε? Use the range of p-values you found in part (b) to determine the largest δ that will work for the given value of ε. Then illustrate c, L, ε, and δ on the graph of $N(p)$.

(d) Assuming that a δ can be found for every ε as you did above, what is the limit statement this problem concerns? Your answer should involve c, L, and $N(p)$.

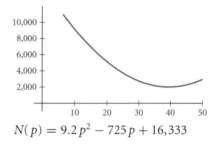

$$N(p) = 9.2\,p^2 - 725\,p + 16{,}333$$

2.3   Delta–Epsilon Proofs

In this section we will see how to prove limit statements using the formal definition of limit. We begin by showing how to find δ in terms of ε, and then we show how to write a logical argument called a delta–epsilon proof using this information.

2.3.1   Finding a delta for every epsilon

Suppose $f(x)$ is a function defined on a punctured interval $(c-\delta, c) \cup (c, c+\delta)$ around some point c. For example, the function $f(x) = 3x - 1$ is defined on a punctured interval around the point $c = 2$. (In fact, $f(x)$ is defined for all of $\mathbb{R}$, but in particular it is defined on a punctured interval, say $(1.5, 2) \cup (2, 2.5)$, around $c = 2$.) Consider the limit statement

(1) $$\lim_{x \to 2}(3x - 1) = 5.$$

We don't yet know how to calculate the limit in Equation (1). However, from a graph or table of values, it seems fairly clear that as we choose values of x closer and closer to 2, the values of $f(x)$ get closer and closer to 5. See Table 2.4 and Figure 2.30.

x	1.9	1.99	1.999	2	2.001	2.01	2.1
$3x - 1$	4.7	4.97	4.997	*	5.003	5.03	5.3

Table 2.4

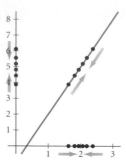

Figure 2.30
$f(x) = 3x - 1$

Although the graph and table both provide convincing arguments, neither the graph nor the table *proves* that the limit of $f(x) = 3x - 1$ as $x \to 2$ is equal to 5. For all we know, the function f could have some behavior we can't see from the graph or the table as x gets very close to 2.

So how do we prove that $\lim\limits_{x \to 2}(3x - 1) = 5$? According to the formal definition of limit, we would have to show that

(2)   *For all $\varepsilon > 0$, there exists a $\delta > 0$ such that
if $0 < |x - 2| < \delta$, then $|(3x - 1) - 5| < \varepsilon$.*

Using the techniques of the previous section, we could find a δ given a *particular* value of ε. For example, if we let ε be 0.5, then $\delta = 0.16$ will work: If $0 < |x - 2| < 0.16$, then $|(3x - 1) - 5| < 0.5$. Check this using your graphing calculator by verifying that if $1.84 < x < 2.16$ (but $x \ne 2$), then $4.5 < (3x - 1) < 5.5$. Note that there are other values of δ that would work for $\varepsilon = 0.5$ (can you find one?). Also, if we chose a different ε, say $\varepsilon = 0.1$, then we'd need another, smaller value of δ. (Can you find one?) The value of δ depends on what we are given for ε; a smaller epsilon might require a smaller value of δ.

However, finding a "working" delta given one particular epsilon still does not prove Statement (2). We must show that given *any ε*, we can *always* find a value of δ such that $|(3x - 1) - 5| < \varepsilon$ for any x with $0 < |x - 2| < \delta$. In other words, for *each* value of ε we have to find *some* value of δ that will work. To do this we will find a way to write δ as a function of ε (i.e., a formula that will give us a working δ given any choice of ε).

EXAMPLE 2.19

A simple example of finding δ in terms of ε

Given $\varepsilon > 0$, find $\delta > 0$ such that if $0 < |x - 2| < \delta$, then $|(3x - 1) - 5| < \varepsilon$.

Solution   Unlike the examples in the previous section, we are not told a particular value for ε. We must find an expression for δ that will work given *any ε*. It is not clear at the outset what this δ should be, but if we start manipulating $|(3x - 1) - 5|$, we will soon find an expression for δ that will work. Consider the calculation

$$|(3x - 1) - 5| = |3x - 1 - 5| = |3x - 6| = |3(x - 2)| = 3|x - 2|.$$

What do we know about $|x - 2|$? We know it will be less than δ (whatever we choose δ to be), since we have $0 < |x - 2| < \delta$. Therefore, $3|x - 2| < 3\delta$. Combining this with the calculation above, we see that if $0 < |x - 2| < \delta$, we must have

$$|(3x - 1) - 5| < 3\delta.$$

What we wanted was to make $|(3x - 1) - 5|$ less than the given ε. What we managed to show is that $|(3x - 1) - 5|$ is less than 3δ. But remember, we get to choose δ! From our earlier work it makes sense to choose $\delta = \varepsilon/3$ (this depends on ε, as we expected). If we do this, then the calculation above becomes

$$|(3x - 1) - 5| < 3\delta = 3(\varepsilon/3) = \varepsilon.$$

Thus given *any* ε, if we choose $\delta = \varepsilon/3$, then we will have $|(3x - 1) - 5| < \varepsilon$ for every x with $0 < |x - 2| < \delta = \varepsilon/3$. Although we have not yet written our argument as a "delta–epsilon proof" (see Section 2.3.2), our work proves that $\lim_{x \to 2}(3x - 1) = 5$.   □

◈ *Caution*   Remember that in any string of equalities and inequalities, the weakest link in the chain prevails. For example, if $A = B < C \leq D = E$, then $A < E$ (since the weakest connector between A and E is a strict inequality).   □

✓ *Checking the Answer*   How can you check whether $\delta = \varepsilon/3$ is a reasonable answer for Example 2.19? Try it out for some values of epsilon! For example, if $\varepsilon = 0.6$, then the solution to Example 2.19 says that a value of $\delta = \frac{0.6}{3} = 0.2$ should "work." That is, it should be true that $|(3x - 1) - 5| < 0.6$ whenever $0 < |x - 2| < 0.2$. Use a graph of $f(x) = 3x - 1$ to check that this is true. What value of δ corresponds to $\varepsilon = 0.01$? Test that on your calculator as well.   □

The process of finding δ in terms of ε will be identical to that used in Example 2.19 whenever f is a linear function. Even if f is not a linear function, the calculation is often similar, as the next example shows.

EXAMPLE 2.20

Finding δ as a function of ε

To show that $\lim_{x \to 2}(x^2 - 4x + 5) = 1$, we must show that for any $\varepsilon > 0$, we can find $\delta > 0$ such that if $0 < |x - 2| < \delta$, $|(x^2 - 4x + 5) - 1| < \varepsilon$. Find δ as a function of ε.

Solution   We begin as we did in the last example, by algebraically manipulating the expression $|(x^2 - 4x + 5) - 1|$ in the hopes of relating it to $|x - 2|$:

$$|(x^2 - 4x + 5) - 1| = |x^2 - 4x + 4| = |(x - 2)^2| = |x - 2|^2.$$

If we assume that $0 < |x - 2| < \delta$, we have

$$|x - 2|^2 < \delta^2.$$

(We *are* allowed to "square both sides" of the inequality $|x - 2| < \delta$, but *only* because both $|x - 2|$ and δ are positive.) To ensure that $|(x^2 - 4x + 5) - 1| < \varepsilon$, we need only set $\delta^2 = \varepsilon$. Since δ must be positive, this means we should choose $\delta = \sqrt{\varepsilon}$. Now with this choice of δ, if $0 < |x - 2| < \delta$, then

$$|(x^2 - 4x + 5) - 1| = |x^2 - 4x + 4| = |x - 2|^2 < \delta^2 = \left(\sqrt{\varepsilon}\right)^2 = \varepsilon.$$

This means that $|(x^2 - 4x + 5) - 1| < \varepsilon$, as desired. We have shown that δ can be chosen in terms of ε as $\delta = \sqrt{\varepsilon}$.   □

❓ *Question*   Using the result in Example 2.20, find a punctured interval of x-values for which $x^5 - 4x + 5$ is within 0.05 of 1.   □

2.3.2 Writing delta–epsilon proofs

In Example 2.19, we did all of the calculational legwork involved in proving the limit statement $\lim_{x \to 2}(3x-1) = 5$. Writing a delta–epsilon proof of this limit statement involves simply writing down our arguments in a logical order. Recall that we must show that

(3)

For all $\varepsilon > 0$, there exists a $\delta > 0$ such that
if $0 < |x - 2| < \delta$, then $|(3x - 1) - 5| < \varepsilon$.

In Example 2.19 we showed that for any $\varepsilon > 0$, a choice of $\delta = \varepsilon/3$ would make the second line of Statement (3) true. Since we have shown that for every ε there is a "working" choice of δ, we have in effect proved that $\lim_{x \to 2}(3x - 1) = 5$. Now we just have to write this information in a clear and concise way called a ***delta–epsilon proof.***

The general argument used in every delta–epsilon proof will be the same, since to prove $\lim_{x \to c} f(x) = L$, we will always have to show that

(4)

For all $\varepsilon > 0$, there exists a $\delta > 0$ such that
if $0 < |x - c| < \delta$, then $|f(x) - L| < \varepsilon$.

To write a delta–epsilon proof, we must first let ε be an arbitrary positive number; then we will choose a value for δ (in terms of ε). We then assume that $0 < |x - c| < \delta$ and use this to prove that $|f(x) - L| < \varepsilon$.

EXAMPLE 2.21

A simple delta–epsilon proof

Give a delta–epsilon proof that $\lim_{x \to 2}(3x - 1) = 5$.

Solution   Recall from Example 2.19 that given any ε, a choice of $\delta = \varepsilon/3$ will "work" in Statement (3). We now use this information to write a formal delta–epsilon proof.

PROOF   Given $\varepsilon > 0$, choose $\delta = \underline{\ \varepsilon/3\ }$.
With this choice of δ, if $0 < |x - 2| < \delta$, then

$$|(3x - 1) - 5| = |3x - 6|$$
$$= |3(x - 2)|$$
$$= 3|x - 2|$$
$$< 3\delta \qquad (|x - 2| < \delta \text{ by hypothesis})$$
$$= 3(\varepsilon/3) \qquad (\text{since we chose } \delta = \varepsilon/3)$$
$$= \varepsilon,$$

and therefore $|(3x - 1) - 5| < \varepsilon$.

In the proof in Example 2.21, we needed to know what to choose for δ on the very first line, before having done any calculations. This time we had done the calculations previously and already knew to choose $\delta = \varepsilon/3$. In general, you will write "choose $\delta = \underline{\ \ \ }$" and leave the choice of δ blank until you figure out what it should be.

The proof that $0 < |x - 2| < \delta$ implies $|(3x - 1) - 5| < \varepsilon$ involves exactly the calculations we did in Example 2.19 when we discovered that we should choose $\delta = \varepsilon/3$. In fact, the fourth line down in the calculation suggests what to choose for δ. In general, you will find out about halfway through your proof what δ should be. When you find that out, write it in the blank you left in the first line and then go on with your argument.

The last line in the proof in Example 2.21 is important because it summarizes the calculation and concludes the argument that if $0 < |x - 2| < \delta$, then $|(3x - 1) - 5| < \varepsilon$.

We now present the work we did in Example 2.20 in the form of a delta–epsilon proof.

EXAMPLE 2.22

Another delta–epsilon proof

Give a delta–epsilon proof that $\lim_{x\to 2}(x^2 - 4x + 5) = 1$.

Solution   Note that in the first line of the proof below we choose δ to be $\sqrt{\varepsilon}$, but we don't actually arrive at this choice until the second to last line of the calculation.

> **PROOF**   Given $\varepsilon > 0$, choose $\delta = \sqrt{\varepsilon}$.
> With this choice of δ, if $0 < |x - 2| < \delta$, then
> $$\begin{aligned}|(x^2 - 4x + 5) - 1| &= |x^2 - 4x + 4| \\ &= |(x-2)^2| \\ &= |x-2|^2 \\ &< \delta^2 && (|x-2| < \delta \text{ by hypothesis}) \\ &= (\sqrt{\varepsilon})^2 && (\text{since we chose } \delta = \sqrt{\varepsilon}) \\ &= \varepsilon, \end{aligned}$$

and therefore $|(x^2 - 4x + 5) - 1| < \varepsilon$.

2.3.3 More challenging delta–epsilon proofs

In each of the delta–epsilon proofs we have done so far, it was clear after a simple calculation what we should choose for δ. This is not always the case. In this section we examine delta–epsilon proofs where it is necessary to choose δ in such a way that we can ensure that $\delta \leq 1$. This added assumption will allow us to bound "left-over" expressions involving x.

EXAMPLE 2.23

A delta–epsilon proof where we have to bound δ

Write a delta–epsilon proof for the limit statement $\lim_{x\to 2} 5x^4 = 80$.

Solution   Recall that, by the formal definition of limit, we need to prove that

(5)   *For all $\varepsilon > 0$, there exists a $\delta > 0$ such that
if $0 < |x - 2| < \delta$, then $|5x^4 - 80| < \varepsilon$.*

The proof will follow the same general structure as the delta–epsilon proofs in Section 2.3.2, but we will have to make an additional assumption about δ.

> **PROOF**   Given any $\varepsilon > 0$, choose $\delta = $ _____. (We will determine what to choose for δ later in the proof.)
> Assume that $0 < |x - 2| < \delta$. We must show that $|5x^4 - 80|$ is less than ε. We begin by calculating
> $$\begin{aligned}|5x^4 - 80| &= |5(x^4 - 16)| = 5|x^4 - 16| \\ &= 5|(x^2 - 4)(x^2 + 4)| && (\text{factoring}) \\ &= 5|(x-2)(x+2)(x^2+4)| && (\text{factoring}) \\ &= 5|x-2|\,|(x+2)(x^2+4)| \\ &< 5\delta\,|(x+2)(x^2+4)|. && (\text{by hypothesis}) \end{aligned}$$

Let's step out of this proof for a moment, since we have now encountered a problem we haven't seen before. Usually at this stage of a delta–epsilon proof it is clear what we should choose for δ. Here, we still have to take care of the factor $|(x+2)(x^2+4)|$, which depends on x. (Look back at the delta–epsilon proofs we did in Examples 2.21 and 2.22. What would we usually do at this point in the proof? Why can't we do it here?) We solve this problem by putting a preliminary bound on δ:

(6) Assume that δ is less than or equal to 1.

We can do this because, once we are given ε, we can choose δ to be whatever we please. The key is that if we assume $\delta \le 1$, then we can bound the troublesome expression $|(x+2)(x^2+4)|$. We'll choose $\delta = \min(1, \underline{\quad})$. (When we figure out what to choose for δ in terms of ε, we'll put it in the blank.) The notation $\min(a, b)$ represents the **minimum** of the numbers a and b. For example, $\min(1, 1.25) = 1$ and $\min(1, 0.75) = 0.75$.

If $\delta \le 1$, then since $0 < |x-2| < \delta$, we have $|x-2| < 1$. Therefore, the distance between x and 2 is less than 1, so $1 < x < 3$ (but $x \ne 2$). On this interval, the expression $|(x+2)(x^2+4)|$ is largest when x is the largest (i.e., when $x = 3$). Thus we can assume that $|(x+2)(x^2+4)| \le |(3+2)(3^2+4)| = |(5)(13)| = 65$. Now we can continue our proof that $|5x^4 - 80|$ can be made less than ε, as follows:

CONTINUATION OF PROOF

$$
\begin{aligned}
|5x^4 - 80| &< 5\delta|(x+2)(x^2+4)| && \text{(from above)} \\
&< 5\delta(65) && (|(x+2)(x^2+4)| < 65 \text{ since } 1 < x < 3) \\
&= 325\delta.
\end{aligned}
$$

Now choose $\delta = \min(1, \frac{\varepsilon}{325})$. (This means that for any given ε, δ is equal to $\frac{\varepsilon}{325}$ or 1, whichever is smaller. This is the δ that goes in the blank at the beginning of this proof. Continuing the computation, we now have

$$
|5x^4 - 80| < 325\delta \le 325\left(\frac{\varepsilon}{325}\right) = \varepsilon.
$$

Therefore, $|5x^4 - 80| < \varepsilon$, and we are done. ∎

EXAMPLE 2.24

Another delta–epsilon proof where we must bound δ

Prove that $\lim\limits_{x \to 3} \frac{1}{x} = \frac{1}{3}$.

Solution   In this proof we will once again have to assume that δ is less than or equal to 1, so that we can bound an expression involving x. We must show that

(7)
> For all $\varepsilon > 0$, *there exists a $\delta > 0$ such that*
> *if $0 < |x-3| < \delta$, then $\left|\frac{1}{x} - \frac{1}{3}\right| < \varepsilon$.*

PROOF   Given any $\varepsilon > 0$, choose $\delta = \underline{\min(1, 6\varepsilon)}$.
Assume that $0 < |x-3| < \delta$. We must show that our δ "works" (i.e., that our choice of δ is small enough so that $\left|\frac{1}{x} - \frac{1}{3}\right| < \varepsilon$). We begin by calculating

$$
\left|\frac{1}{x} - \frac{1}{3}\right| = \left|\frac{3-x}{3x}\right| = \frac{|3-x|}{|3x|} = \frac{|x-3|}{|3x|} < \frac{\delta}{|3x|}.
$$

Assuming $\delta \le 1$, the fact that $|x-3| < \delta$ implies that $2 < x < 4$. This enables us to get a bound on the $\frac{1}{|3x|}$ term. The expression $\frac{1}{|3x|}$ is largest when its denominator $|3x|$ is smallest (which in our δ-interval is when $x = 2$). In other words, for all $2 < x < 4$, we have $\frac{1}{|3x|} < \frac{1}{|3(2)|} = \frac{1}{6}$. Choosing $\delta = \min(1, 6\varepsilon)$, we have

$$\left|\frac{1}{x} - \frac{1}{3}\right| = \frac{|x - 3|}{|3x|} < \frac{\delta}{|3x|} < \frac{\delta}{6} \leq \frac{6\varepsilon}{6} = \varepsilon.$$

We have shown that, given any ε, our choice of $\delta = \min(1, 6\varepsilon)$ will ensure that $\left|\frac{1}{x} - \frac{1}{3}\right| < \varepsilon$ whenever $0 < |x - 3| < \delta$.   ∎

EXAMPLE 2.25

Finding a punctured interval for a given ε

The proof in Example 2.24 tells us exactly how close we need to get to $c = 3$ to ensure that $f(x) = \frac{1}{x}$ is within any ε interval of $\frac{1}{3}$. Use the work in the proof to find a punctured interval about $x = 3$ such that the value of $\frac{1}{x}$ is (a) within 0.01 of $\frac{1}{3}$, and (b) within 0.5 of $\frac{1}{3}$.

Solution

(a) The proof in Example 2.24 shows that given any $\varepsilon > 0$, a choice of $\delta = \min(1, 6\varepsilon)$ will be sufficient. Here we have $\varepsilon = 0.01$, so $\delta = \min(1, 6(0.01)) = \min(1, 0.06) = 0.06$ will work. This means that any x in $(2.94, 3) \cup (3, 3.06)$ will make $\frac{1}{x}$ within 0.01 of $\frac{1}{3}$. (Verify this with your calculator.)

(b) Now ε is given as 0.5, so we take $\delta = \min(1, 6(0.5)) = \min(1, 3) = 1$. Thus any x in $(2, 3) \cup (3, 4)$ will make $\frac{1}{x}$ within 0.5 of $\frac{1}{3}$. (Test this out on your calculator. Is it true? Is it the *largest* interval around x you can take? If not, what is the largest such interval?)   ☐

2.3.4 Proofs for one-sided and infinite limits

We now give examples of proofs of limit statements involving left and right limits, limits at infinity, and infinite limits. We will use the formal definitions from Sections 2.2.3–2.2.5.

EXAMPLE 2.26

A delta–epsilon proof for a one-sided limit

Prove that $\lim\limits_{x \to 1^+} \sqrt{x - 1} = 0$.

Solution   Recall from Definition 2.9 that we must show that

(8)
> *For all $\varepsilon > 0$, there exists a $\delta > 0$ such that*
> *if $x \in (1, 1 + \delta)$, then $\left|\sqrt{x - 1} - 0\right| < \varepsilon$.*

The proof will be much like a regular delta–epsilon proof with the exception that we have $x \in (1, 1 + \delta)$ instead of $0 < |x - 1| < \delta$.

PROOF   Given $\varepsilon > 0$, choose $\delta = \varepsilon^2$.
We begin by manipulating $|\sqrt{x - 1} - 0|$. Since $\sqrt{x - 1}$ is positive, we have

$$\left|\sqrt{x - 1} - 0\right| = \left|\sqrt{x - 1}\right| = \sqrt{x - 1}.$$

We want to relate the expression $\sqrt{x - 1}$ to the fact that x is in the interval $(1, 1 + \delta)$. For all $x \in (1, 1 + \delta)$ we have $1 < x < 1 + \delta$, and thus $0 < x - 1 < \delta$. Therefore, $\sqrt{x - 1} < \sqrt{\delta}$ and we have

$$\left|\sqrt{x - 1} - 0\right| = \left|\sqrt{x - 1}\right| = \sqrt{x - 1} < \sqrt{\delta}.$$

It should now be clear that we want to choose δ such that $\sqrt{\delta} = \varepsilon$, i.e., choose $\delta = \varepsilon^2$. For all $x \in (1, 1 + \delta)$ we have

$$\left|\sqrt{x - 1} - 0\right| = \left|\sqrt{x - 1}\right| = \sqrt{x - 1} < \sqrt{\delta} = \sqrt{\varepsilon^2} = \varepsilon.$$

Therefore, we have $|\sqrt{x - 1} - 0| < \varepsilon$.   ∎

EXAMPLE 2.27

An "$N–\varepsilon$" proof of a limit at infinity

Prove that $\lim\limits_{x\to\infty} \dfrac{1}{x} = 0$.

Solution By Definition 2.11, we must show that

(9)
> *For all $\varepsilon > 0$, there exists an $N > 0$ such that*
> *if $x > N$, then $\left|\frac{1}{x} - 0\right| < \varepsilon$.*

In this example, the role of the small number δ is now played by a *large* number N. For every $\varepsilon > 0$ we must find a large enough value of N (depending on ε) so that $\frac{1}{x}$ is within ε of 0 whenever x is greater than N.

PROOF Given $\varepsilon > 0$, choose $N = \dfrac{1}{\varepsilon}$.

Whenever $x > N$ we have $\frac{1}{x} < \frac{1}{N}$, and thus

$$\left|\frac{1}{x} - 0\right| = \left|\frac{1}{x}\right| = \frac{1}{x} < \frac{1}{N}.$$

(*Note:* We were able to drop the absolute value because $x > N > 0$, and thus $\frac{1}{x} > 0$.) Since we want to show that $\left|\frac{1}{x} - 0\right|$ is less than ε, we should choose N such that $\frac{1}{N} = \varepsilon$ (i.e., choose $N = \frac{1}{\varepsilon}$). With this choice of N, for all $x > N$ we have

$$\left|\frac{1}{x} - 0\right| = \cdots = \frac{1}{x} < \frac{1}{N} = \frac{1}{1/\varepsilon} = \varepsilon,$$

and so $\left|\frac{1}{x} - 0\right| < \varepsilon$, as desired. ■

EXAMPLE 2.28

A "$\delta–M$" proof of an infinite limit

Prove that $\lim\limits_{x\to 3^-} \dfrac{1}{3 - x} = \infty$.

Solution By the left-sided version of Definition 2.10, we must show that

(10)
> *For all $M > 0$, there exists a $\delta > 0$ such that*
> *if $x \in (c - \delta, c)$, then $\frac{1}{3-x} > M$.*

This time it is the small value of ε that is replaced with a *large* value M, since we wish to show that $f(x) = \frac{1}{3-x}$ approaches ∞ as $x \to 3^-$.

PROOF Given any $M > 0$, choose $\delta = \dfrac{1}{M}$.

If $x \in (3 - \delta, 3)$, we have the following inequalities:

$$\begin{aligned}
3 - \delta < x < 3 \quad &\implies \quad -\delta < x - 3 < 0 \\
&\implies \quad \delta > 3 - x > 0 \\
&\implies \quad 0 < 3 - x < \delta \\
&\implies \quad \frac{1}{3 - x} > \frac{1}{\delta}.
\end{aligned}$$

Choosing $\delta = \frac{1}{M}$, we get

$$\frac{1}{3 - x} > \frac{1}{\delta} = \frac{1}{1/M} = M,$$

so $\frac{1}{3-x} > M$ for all $x \in (3 - \delta, 3)$. ■

EXAMPLE 2.29 **An "N–M" proof of an infinite limit at negative infinity**

Prove that $\lim\limits_{x \to -\infty} (1 - 2x) = \infty$.

Solution To prove this limit statement, we must show that

(11)
$$\text{For all } M > 0, \text{ there exists an } N < 0 \text{ such that}$$
$$\text{if } x < N, \text{ then } 1 - 2x > M.$$

In Statement (11), M represents a large positive number, and N represents a negative number of large magnitude. We want to show that we can make $1 - 2x$ arbitrarily large by choosing x to be less than some negative number with sufficiently large magnitude.

PROOF Given any $M > 0$, choose $N = \frac{-M+1}{2}$. (We will see why this is a good choice for N in a moment.)

What can we say about $1 - 2x$ when $x < N$? Whenever $x < N$, we have $-x > -N$ and thus $-2x > -2N$. Therefore, $1 - 2x > 1 - 2N$. Choosing $N = \frac{-M+1}{2}$, we have, for all $x > N$,

$$1 - 2x > 1 - 2N = 1 - 2\left(\frac{-M+1}{2}\right) = 1 + M - 1 = M.$$

Therefore, for all $x > N$, we have $1 - 2x > M$.

What you should know after reading Section 2.3

▶ How to find an expression for δ in terms of ε for a given limit statement. Once you have such an expression for δ, how can you use it to ensure that $f(x)$ is within ε of L for a given value of ε?

▶ How to write a basic delta–epsilon proof. In particular, what is the logic behind a delta–epsilon proof? Be sure you can write these proofs clearly and concisely. You will have to be comfortable working with inequalities to construct these proofs.

▶ How to write more challenging delta–epsilon proofs, namely proofs that involve choosing δ such that it is guaranteed to be less than 1. Be able to use such an expression for δ to find values of δ that work for particular given values of ε.

▶ How to write formal proofs of one-sided and infinite limit statements and proofs for limits at infinity. These proofs involve the formal definitions for these limit statements given in the previous section.

Exercises 2.3

Concepts

0. Read the section and make your own summary of the material. Include whatever you think will help you review the material later on.

1. Suppose you show that $|(1 - 2x) - (-5)| < 0.05$ for all x with $0 < |x - 3| < 0.025$. Explain why this does *not* prove that $\lim\limits_{x \to 3}(1 - 2x) = -5$.

2. Write down the formal delta–epsilon statement you would have to prove in order to prove the limit statement $\lim\limits_{x \to -2} \dfrac{3}{x + 1} = -3$.

3. Use the proof in Example 2.21 to find a punctured interval around 2 such that $3x - 1$ is within 0.25 of 5. Sketch a graph that illustrates your answer.

4. Use the proof in Example 2.24 to find a punctured interval around 3 such that $\frac{1}{x}$ is within 0.2 of $\frac{1}{3}$. Sketch a graph that illustrates your answer.

5. Use the proof in Example 2.26 to find an interval of x-values for which $\sqrt{x-1}$ is within 0.5 of 0. Sketch a graph that illustrates your answer.

6. Use the proof in Example 2.27 to find an interval of x-values for which $\frac{1}{x}$ is within 0.15 of 0. Sketch a graph that illustrates your answer.

7. Use the proof in Example 2.28 to find an interval of x-values for which $\frac{1}{3-x}$ is greater than 1000. Sketch a graph that illustrates your answer.

8. Use the proof in Example 2.29 to find an interval of x-values for which $1 - 2x$ is greater than 500. Sketch a graph that illustrates your answer.

9. In Example 2.22 we proved that $\lim_{x\to 2}(x^2 - 4x + 5) = 1$ by showing that for any $\varepsilon > 0$, a choice of $\delta = \sqrt{\varepsilon}$ would ensure that $|(x^2 - 4x + 5) - 1| < \varepsilon$ for all x with $0 < |x - 2| < \delta$. Use this information to find a "working" δ for (a) $\varepsilon = 1$, (b) $\varepsilon = 0.1$, and (c) $\varepsilon = 0.01$. Check that your choice of δ works by examining a graph of $f(x) = x^2 - 4x + 5$ and sketching appropriate ε and δ intervals.

10. In Example 2.23 we proved that $\lim_{x\to 2} 5x^4 = 80$ by showing that for any $\varepsilon > 0$, a choice of $\delta = \min(1, \frac{\varepsilon}{325})$ would ensure that $|5x^4 - 80| < \varepsilon$ for all x with $0 < |x - 2| < \delta$. Use this information to find a "working" δ for (a) $\varepsilon = 5$, (b) $\varepsilon = 0.01$, and (c) $\varepsilon = 350$. Check that your choice of δ works by examining a graph of $f(x) = 5x^4$ and sketching appropriate ε and δ intervals.

Skills

■ For each limit statement $\lim_{x\to c} f(x) = L$ below, find $\delta > 0$ in terms of $\varepsilon > 0$ such that if $0 < |x - c| < \delta$, then $|f(x) - L| < \varepsilon$. Then, on a graph of $f(x)$, illustrate the roles of c, L, δ, and ε.

11. $\lim_{x\to 1}(2x + 4) = 6$

12. $\lim_{x\to -6}(x + 2) = -4$

13. $\lim_{x\to 2}(3 - 4x) = -5$

14. $\lim_{x\to -3}(1 - x) = 4$

15. $\lim_{x\to 0}(3x^2 + 1) = 1$

16. $\lim_{x\to 3}(x^2 - 6x + 11) = 2$

■ For each limit statement $\lim_{x\to c} f(x) = L$ below, find $\delta > 0$ in terms of $\varepsilon > 0$ such that if $0 < |x - c| < \delta$, then $|f(x) - L| < \varepsilon$. (You may have to assume $\delta \leq 1$.) Then, on a graph of $f(x)$, illustrate the roles of c, L, δ, and ε.

17. $\lim_{x\to 1} \frac{1}{x} = 1$

18. $\lim_{x\to 3}(x^2 - 2x - 3) = 0$

19. $\lim_{x\to 1} 2x^4 = 2$

20. Find $\delta > 0$ in terms of $\varepsilon > 0$ such that if $x \in (-2, -2+\delta)$, then $|(1 + \sqrt{x + 2}) - 1| < \varepsilon$. Write a limit statement

corresponding to this fact, and draw a picture illustrating the roles of δ and ε.

21. Find $\delta > 0$ in terms of $M > 0$ such that if $x \in (1 - \delta, 1)$, then $\frac{1}{1-x} > M$. Write a limit statement corresponding to this fact, and draw a picture illustrating the roles of δ and M.

22. Find $N > 0$ in terms of $\varepsilon > 0$ such that if $x > N$, then $|\frac{x-1}{x} - 1| < \varepsilon$. Write a limit statement corresponding to this fact, and draw a picture illustrating the roles of N and ε.

23. Find $N > 0$ in terms of $M > 0$ such that if $x > N$, then $x^2 + 2 > M$. Write a limit statement corresponding to this fact, and draw a picture illustrating the roles of N and M.

Proofs

■ Give delta–epsilon proofs for the following limit statements.

24. $\lim_{x\to 1}(2x + 4) = 6$

25. $\lim_{x\to -6}(x + 2) = -4$

26. $\lim_{x\to 2}(3 - 4x) = -5$

27. $\lim_{x\to -3}(1 - x) = 4$

28. $\lim_{x\to 0}(3x^2 + 1) = 1$

29. $\lim_{x\to 3}(x^2 - 6x + 11) = 2$

30. $\lim_{x\to 1}(x^2 - 2x + 3) = 2$

31. $\lim_{x\to 1}\frac{x^2 - 1}{x - 1} = 2$

32. $\lim_{x\to 2}\frac{x^2 - 3x + 2}{x - 2} = 1$

33. Prove that $\lim_{x\to c}(mx + b) = mc + b$ for any constants c, m, and b. Note that this means that the limit of any linear function as $x \to c$ is the *value* of that linear function at $x = c$. *Hint:* Choose $\delta = \varepsilon/|m|$.

■ Give delta–epsilon proofs for the following limit statements. You will have to bound δ such that $\delta \leq 1$.

34. $\lim_{x\to 1}\frac{1}{x} = 1$

35. $\lim_{x\to 3}(x^2 - 2x - 3) = 0$

36. $\lim_{x\to 1} 2x^4 = 2$

■ Give delta–epsilon proofs for the following limit statements.

37. $\lim_{x\to 1^-}(2x + 4) = 6$

38. $\lim_{x\to -2^+}(1 + \sqrt{x + 2}) = 1$

39. $\lim_{x\to 3^-}(x^2 - 2x - 3) = 0$

■ Prove the following limit statements.

40. $\lim_{x\to -2^+}\frac{1}{x + 2} = \infty$

41. $\lim_{x\to\infty}\frac{2x - 1}{x} = 2$

42. $\lim_{x\to -\infty} 3x - 5 = -\infty$

43. Use the formal definition of limit and the formal definitions of left and right limits to show that
$\lim_{x\to c} f(x) = L \iff \lim_{x\to c^-} f(x) = L$ and $\lim_{x\to c^+} f(x) = L$.

2.4   Limit Rules

In the past few sections, we have seen how to estimate limits graphically and with tables of values, and we have used the formal definition of limit to write delta–epsilon proofs for certain simple limits. However, until now we have not developed any quick calculational methods for evaluating limits.

In this section we will develop rules for evaluating simple limits and finding limits of sums, constant multiples, products, and quotients of functions. These basic *limit rules* will enable us to calculate simple limits quickly without using approximations or delta–epsilon proofs. We will then develop rules that enable us to calculate limits of products, quotients, and powers.

⬧ *Caution*   At this point you may be tempted to evaluate a limit like $\lim_{x \to 3}(2x + 1)$ by simply "plugging in" $x = 3$ to the expression $2x + 1$. In this particular example it happens that you will get the correct answer if you do this. However, in general, the limit of a function as $x \to c$ is *not* equal to the value of $f(x)$ at $x = c$. The value of $f(x)$ at $x = c$ does not necessarily tell you what $f(x)$ is *approaching* as x approaches c. Think about it: If the limit of $f(x)$ as x approaches c were always equal to the value of $f(x)$ at $x = c$, then why would we need limits at all? Limits are an entirely new computational object to us right now, and we have to *develop* methods to calculate them that are consistent with the formal definition of limit. ☐

2.4.1   The two most basic limits

We start by proving two very basic limit statements. The first limit statement says that the limit of a constant function $f(x) = k$ as $x \to c$ is equal to k. We will prove this limit statement using a general delta–epsilon proof that works for each limit of the form $\lim_{x \to c} k$. Then, when we encounter such limits in the future, we can apply this rule to evaluate each limit rather than going through an entire delta–epsilon proof.

THEOREM 2.3

The Limit of a Constant
If c and k are any real numbers, then $\lim_{x \to c} k = k$.

EXAMPLE 2.30   **Limits of constants**

Theorem 2.3 tells us that

$$\lim_{x \to 2} 3 = 3, \quad \lim_{x \to 100} 0 = 0, \quad \text{and} \quad \lim_{x \to -1.4} \sqrt{7} = \sqrt{7}.$$

This seems sensible; for example, if $f(x) = 3$ is the function that is constantly equal to 3, then as x approaches 2, the values of $f(x)$ approach (in fact are all exactly equal to) 3. Another intuitive way to think of this is that $\lim_{x \to 2} 3 = 3$ because making x approach 2 has no effect on the number 3 (since the number 3 does not involve x). ☐

PROOF (**THEOREM 2.3**)   By the formal definition of limit, we must show that

(1)   *For all ε > 0, there exists a δ > 0 such that*
if 0 < |x − c| < δ, then |k − k| < ε.

But $|k - k| = 0$ is *always* less than ε, no matter what positive value is given for ε or what we choose for δ. Statement (1) is trivially true for all $\varepsilon > 0$. (*Note:* We have just shown that in the case of $f(x) = k$, the value of δ would not depend on ε. Figure 2.31 shows why this makes sense graphically.) ▪

❓ Question   Sketch the graph of the constant function $f(x) = 3$, and use it to illustrate that for all $\varepsilon > 0$, Statement (1) is true for *any* choice of δ.   □

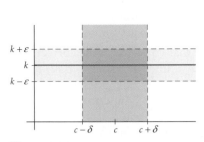

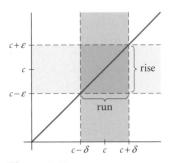

Figure 2.31
$f(x) = k$ is a constant function; *any* δ will "work" for this ε

Figure 2.32
$f(x) = x$ is the identity function; $\frac{\text{rise}}{\text{run}} = 1$, so $\delta = \varepsilon$

Our second basic limit statement says that the limit of the identity function $f(x) = x$ as $x \to c$ is always equal to c. Once we prove this limit (with a delta–epsilon proof), we can use this rule to evaluate limits of $f(x) = x$.

THEOREM 2.4

The Limit of the Identity Function

If c is any constant, then $\lim\limits_{x \to c} x = c$.

If you think about it in terms of the intuitive definition of limit, Theorem 2.4 is almost silly; it simply says that "x approaches c as x approaches c."

EXAMPLE 2.31

Limits of the identity function

By Theorem 2.4, we have

$$\lim_{x \to 2} x = 2, \quad \lim_{x \to -35} x = -35, \quad \text{and} \quad \lim_{x \to 1.7} x = 1.7.$$ □

PROOF (**THEOREM 2.4**)   By the formal definition of limit, we need to show that

(2)   *For all ε > 0, there exists a δ > 0 such that*
if 0 < |x − c| < δ, then |x − c| < ε.

In other words, given any $\varepsilon > 0$, we have to find some $\delta > 0$ such that $|x - c|$ is less than ε whenever $|x - c|$ is less than δ. So we choose δ to be ε. Then, for all x with $|x - c| < \delta$, we have $|x - c| < \delta = \varepsilon$, as desired.   ▪

In this proof we saw that if $f(x) = x$ is the identity function, then the largest possible δ is equal to the given choice of ε. Figure 2.32 on the previous page shows this graphically. Since the function $f(x) = x$ is a linear function with slope 1, its "rise" and "run" are equal over any interval. In particular, this means that the length of the interval $(c - \delta, c + \delta)$ is equal to the length of the interval $(c - \varepsilon, c + \varepsilon)$ (i.e., that $\delta = \varepsilon$).

2.4.2 Limits of linear functions

You may have noticed that each time we take the limit of a linear function as $x \to c$, the limit is equal to the *value* of the linear function at $x = c$. This is in fact true for all linear functions, as the next theorem shows.

THEOREM 2.5

The Limit of a Linear Function

If c, m, and b are any constants, then $\lim_{x \to c} (mx + b) = mc + b$.

Another way to state Theorem 2.5 is to say that if $f(x)$ is a linear function, then $\lim_{x \to c} f(x) = f(c)$. Remember that this is *not* true for all functions! We will investigate functions whose limits are equal to their function values ("continuous" functions) in Section 2.6. Before proving Theorem 2.5, we present a simple example.

EXAMPLE 2.32

A limit of a linear function

Theorem 2.5 tells us that

$$\lim_{x \to 5} (-2x + 3) = -2(5) + 3 = -10 + 3 = 7.$$

In the notation of Theorem 2.5, we have $c = 5$, $m = -2$, and $b = 3$. □

The proof of Theorem 2.5 will be nothing more than a delta–epsilon proof that works for all linear functions at the same time (in terms of c, m, and b).

PROOF (**THEOREM 2.5**) We first handle the case where $m = 0$. If $m = 0$, then $0(x) + b = b$ is a constant, and by the rule for taking the limit of a constant (Theorem 2.3), we have

$$\lim_{x \to c} (mx + b) = \lim_{x \to c} (0(x) + b) = \lim_{x \to c} b = b.$$

On the other hand, if $m = 0$, then $mc + b = 0(c) + b = b$, which equals the expression above. Therefore, if $m = 0$, we have $\lim_{x \to c} (mx + b) = mc + b$.

Now suppose $m \neq 0$. Given any constants c, $m \neq 0$, and b, we have to show that

(3)
$$\textit{For all } \varepsilon > 0, \textit{ there exists a } \delta > 0 \textit{ such that}$$
$$\textit{if } 0 < |x - c| < \delta, \textit{ then } |(mx + b) - (mc + b)| < \varepsilon.$$

Given $\varepsilon > 0$, choose $\delta = \underline{ \varepsilon/|m| }$. For all x with $0 < |x - c| < \delta$, we have

$$|(mx + b) - (mc + b)| = |mx + b - mc - b|$$
$$= |mx - mc|$$
$$= |m(x - c)| = |m||x - c|$$
$$< |m|\delta \qquad \text{(since } |x - c| < \delta)$$
$$= |m|(\varepsilon/|m|) \qquad \text{(since we chose } \delta = \varepsilon/|m|)$$
$$= \varepsilon.$$

Therefore, $|(mx + b) - (mc + b)| < \varepsilon$, as desired.   ■

You will prove Theorem 2.5 in a different way in the exercises, *without* using a delta–epsilon proof; instead you will use the other limit rules proved in this section.

❓ Question   In the proof of Theorem 2.5, why does it make graphical sense that δ is ε divided by the absolute value of the slope? Think graphically about slope, rise over run, and the δ and ε intervals for a linear function. Why will $\delta = \varepsilon/|m|$ always work?   □

❓ Question   Why did we have to consider the case $m = 0$ separately in the proof of Theorem 2.5? What goes wrong in the main argument if we allow m to be zero?   □

2.4.3 The constant multiple rule for limits

We now prove that the limit of a constant multiple of a function is that constant times the limit of the function. More precisely, we will show that if $f(x) \to L$ as $x \to c$ and k is any constant, then $kf(x) \to kL$ as $x \to c$. This means that taking a limit and multiplying by a constant are *commutative* operations; the order in which we apply these operations does not matter. If we multiply a function by a constant k and then take its limit as $x \to c$, we get the same answer we get if we first take the limit of the function as $x \to c$ and then multiply by the constant k.

THEOREM 2.6

The Constant Multiple Rule for Limits

If $\lim\limits_{x \to c} f(x)$ exists, and k is any constant, then

$$\lim_{x \to c} kf(x) = k \lim_{x \to c} f(x).$$

⑂ Caution   Do not confuse this theorem with Theorem 2.3. Theorem 2.6 tells us the limit of a function that is multiplied by a constant, whereas Theorem 2.3 concerns the limit of a constant function.   □

EXAMPLE 2.33

Using the constant multiple rule for limits

Using Theorem 2.6, we can calculate

$$\lim_{x \to -1} 7(3x + 1) = 7 \lim_{x \to -1} (3x + 1) \qquad \text{(constant multiple rule for limits)}$$
$$= 7(3(-1) + 1) = -14. \qquad \text{(limit of a linear function)}$$

On the other hand, we could do a little algebra to write $7(3x + 1)$ as a linear function and avoid having to use the constant multiple rule for limits:

$$\lim_{x \to -1} 7(3x + 1) = \lim_{x \to -1} (21x + 7) \qquad \text{(algebra)}$$
$$= (21(-1) + 7) = -14. \qquad \text{(limit of a linear function)}$$

Although we can calculate constant multiples of linear functions without the constant multiple rule for limits, the constant multiple rule will be very useful for calculating constant multiples of more complicated limits. □

PROOF **(THEOREM 2.6)**   The main argument will require that $k \neq 0$, so we'll assume that k is nonzero and do the case where $k = 0$ afterwards. By hypothesis we know that $\lim_{x \to c} f(x) = L$ for some real number L (since the limit "exists"). By the formal definition of limit, this means we know that

(4)   *For all $\varepsilon_1 > 0$, there exists a $\delta_1 > 0$ such that*
            if $0 < |x - c| < \delta_1$, then $|f(x) - L| < \varepsilon_1$.

We wish to show that $\lim_{x \to c} kf(x) = kL$; in other words, we wish to show that

(5)   *For all $\varepsilon > 0$, there exists a $\delta > 0$ such that*
            if $0 < |x - c| < \delta$, then $|kf(x) - kL| < \varepsilon$.

Given $\varepsilon > 0$, let $\varepsilon_1 = \underline{\;\varepsilon/|k|\;}$ (we'll see why in a moment). Next use Statement (4) to choose δ_1 such that $|f(x) - L| < \varepsilon_1 = \varepsilon/|k|$ for all $0 < |x - c| < \delta_1$. Then let $\delta = \underline{\;\delta_1\;}$. With this choice of δ, for all $0 < |x - c| < \delta$ we have

$$|kf(x) - kL| = |k(f(x) - L)| = |k||f(x) - L|$$
$$< |k|\,\varepsilon_1 \qquad\qquad (\delta = \delta_1 \text{ and Statement (4)})$$
$$= |k|\,(\varepsilon/|k|) \qquad\qquad \textbf{(choice of } \boldsymbol{\varepsilon_1}\textbf{)}$$
$$= \varepsilon,$$

and therefore $|kf(x) - kL| < \varepsilon$. Note that we don't know until the second line in the computation that we should choose δ_1 such that $|f(x) - L| < \varepsilon/|k|$ and then let $\delta = \delta_1$ (although the proof is written as though we knew that in advance).

The argument above required k to be nonzero, since we chose $\varepsilon_1 = \frac{\varepsilon}{|k|}$. We now do the case where $k = 0$ with a different argument. This case follows from Theorem 2.6:

$$\lim_{x \to c} 0(f(x)) = \lim_{x \to c} 0 = 0, \qquad \text{(limit of a constant)}$$

which is equal to $0 \left(\lim_{x \to c} f(x) \right) = 0$. ◾

The constant multiple rule for limits, and in fact every limit rule and algorithm for computing limits, also holds for left and right limits. In general, if a rule holds for two-sided limits as $x \to c$, then in particular it holds as $x \to c^-$ and as $x \to c^+$.

2.4.4 Limits of sums

With Theorems 2.3, 2.4, and 2.5 we can quickly calculate the limit of any constant, identity, or linear function as x approaches any value c. However, these are the *only* limits we can quickly calculate at this point! In this section we will learn how to calculate the limit of a sum of functions whose limits we know.

Intuitively, if $f(x)$ approaches L as $x \to c$, and $g(x)$ approaches M as $x \to c$, then as x approaches c, the function $f(x) + g(x)$ approaches $L + M$. The limit of a sum is the sum of limits. Another way to say this is that sums *commute* with limits; you will get the same answer whether you add two functions together and then take the limit of their sum, or take each of their limits and then add the limits together.

THEOREM 2.7

The Sum Rule for Limits

If $\lim_{x \to c} f(x)$ and $\lim_{x \to c} g(x)$ both exist, then

$$\lim_{x \to c}(f(x) + g(x)) = \lim_{x \to c} f(x) + \lim_{x \to c} g(x).$$

If either the limit of f or the limit of g fails to exist, then Theorem 2.7 does not apply. If both limits exist, say $\lim_{x \to c} f(x) = L$ and $\lim_{x \to c} g(x) = M$ for some numbers L and M, then Theorem 2.7 says that $\lim_{x \to c}(f(x) + g(x)) = L + M$.

? *Question*   It is vital that $\lim_{x \to c} f(x)$ and $\lim_{x \to c} g(x)$ both exist when we apply the sum rule for limits. Illustrate the importance of this hypothesis by finding examples of functions $f(x)$ and $g(x)$ and a real number c where $\lim_{x \to c} f(x) + \lim_{x \to c} g(x)$ is *not* equal to $\lim_{x \to c}(f(x) + g(x))$.   □

EXAMPLE 2.34   **Using the sum rule for limits**

By the sum rule for limits, and Theorem 2.5, we can calculate

$$\lim_{x \to 3}((2x + 1) + (5 - 4x))$$

$$= \lim_{x \to 3}(2x + 1) + \lim_{x \to 3}(5 - 4x) \qquad \text{(sum rule for limits)}$$

$$= (2(3) + 1) + (5 - 4(3)) \qquad \text{(limits of linear functions)}$$

$$= 7 + (-7) = 0.$$

Alternatively, we could have used a little algebra to write $(2x + 1) + (5 - 4x)$ as a linear function and then used Theorem 2.5:

$$\lim_{x \to 3}((2x + 1) + (5 - 4x)) = \lim_{x \to 3}(2x + 1 + 5 - 4x) \qquad \text{(algebra)}$$

$$= \lim_{x \to 3} -2x + 6$$

$$= -2(3) + 6 = 0. \qquad \text{(limit of a linear function)}$$

□

We will now prove the sum rule for limits. The proof will follow directly from the formal definition of a limit.

PROOF   (**THEOREM 2.7**)   Suppose f and g are functions whose limits as $x \to c$ exist. Then, for some real numbers L and M, we have

$$\lim_{x \to c} f(x) = L \quad \text{and} \quad \lim_{x \to c} g(x) = M.$$

In terms of the formal definition of limit, this means that

(6)   *For all $\varepsilon_1 > 0$, there exists a $\delta_1 > 0$ such that if $0 < |x - c| < \delta_1$, then $|f(x) - L| < \varepsilon_1$,*

and

(7)
$$\textit{For all } \varepsilon_2 > 0, \textit{ there exists a } \delta_2 > 0 \textit{ such that}$$
$$\textit{if } 0 < |x - c| < \delta_2, \textit{ then } |g(x) - M| < \varepsilon_2.$$

We wish to show that, given the above, the limit of $f(x) + g(x)$ as $x \to c$ is equal to $L + M$. In other words, we must show that $\lim_{x \to c}(f(x) + g(x)) = L + M$. In terms of the formal definition of limit, this means that we have to show that

(8)
$$\textit{For all } \varepsilon > 0, \textit{ there exists a } \delta > 0 \textit{ such that}$$
$$\textit{if } 0 < |x - c| < \delta, \textit{ then } |(f(x) + g(x)) - (L + M)| < \varepsilon.$$

The key is to prove Statement (8) using Statements (6) and (7).

Now suppose ε is any positive number. By Statement (6), we can choose $\delta_1 > 0$ small enough so that $0 < |x - c| < \delta_1$ implies that $|f(x) - L| < \varepsilon/2$. (We are letting $\varepsilon_1 = \varepsilon/2$ here; it will become apparent why this is a good idea later in the proof.) Similarly, by Statement (7), there is some $\delta_2 > 0$ small enough so that $|g(x) - M| < \varepsilon/2$ whenever $0 < |x - c| < \delta_2$.

Given $\varepsilon > 0$, choose δ_1 and δ_2 as described above, and let $\delta = \underline{\min(\delta_1, \delta_2)}$.

With this choice of δ, for all $0 < |x - c| < \delta$, we have

$$|(f(x) + g(x)) - (L + M)|$$
$$= |f(x) + g(x) - L - M|$$
$$= |(f(x) - L) + (g(x) - M)|$$
$$\leq |f(x) - L| + |g(x) - M| \qquad \text{(triangle inequality)}$$
$$< \varepsilon/2 + \varepsilon/2 \qquad \text{(since } \delta \leq \delta_1 \text{ and } \delta \leq \delta_2\text{)}$$
$$= \varepsilon.$$

Therefore, for all $0 < |x - c| < \delta$, we have $|(f(x) + g(x)) - (L + M)| < \varepsilon$.

Notice that we used the triangle inequality (see Section 0.3.5) to relate the quantity we cared about, $|(f(x) + g(x)) - (L + M)|$, to the quantities we knew about, namely $|f(x) - L|$ and $|g(x) - M|$. What we knew about these quantities was that we could make them as small as we liked by choosing a small enough value for δ. We chose to make each of them smaller than $\varepsilon/2$ since we would be adding them together, and our choice of $\delta = \min(\delta_1, \delta_2)$ ensured that they would both be smaller than $\varepsilon/2$. ∎

⊛ *Caution* The proof of Theorem 2.7 is different from most of the delta–epsilon proofs we have done so far. You might want to read it over a couple of times until you understand the logical arguments involved. □

Using Theorem 2.7, we can find the limit of any function that is the sum of two functions whose limits we know. If we combine the sum rule for limits and the constant multiple rule, we can derive the following theorem, which tells us how to find the limit of a difference of functions.

THEOREM 2.8

The Difference Rule for Limits

If $\lim_{x \to c} f(x)$ and $\lim_{x \to c} g(x)$ both exist, then

$$\lim_{x \to c}(f(x) - g(x)) = \lim_{x \to c} f(x) - \lim_{x \to c} g(x).$$

The proof of this theorem involves writing the difference $f(x) - g(x)$ as a sum $f(x) + (-g(x)) = f(x) + (-1)g(x)$ and applying the sum rule and the constant multiple rule for limits (with $k = -1$). We could also prove Theorem 2.8 directly, using a delta–epsilon argument (much like the one we used to prove the sum rule for limits). You will prove Theorem 2.8 both ways in the exercises.

2.4.5 The product and quotient rules for limits

How do you think the limit of a product of functions f and g is related to the limits of f and g? Once again, the nicest possible thing happens: The limit of a product is the product of limits (if both limits exist). In other words, limits and products *commute*. The same can be said about quotients, provided that the denominator of the quotient is nonzero.

> **THEOREM 2.9**
>
> **The Product and Quotient Rules for Limits**
>
> If $\lim_{x \to c} f(x)$ and $\lim_{x \to c} g(x)$ both exist, then
>
> **(a)** $\lim_{x \to c} f(x)g(x) = \lim_{x \to c} f(x) \lim_{x \to c} g(x)$;
>
> **(b)** $\lim_{x \to c} \dfrac{f(x)}{g(x)} = \dfrac{\lim_{x \to c} f(x)}{\lim_{x \to c} g(x)}$, if $\lim_{x \to c} g(x) \neq 0$.

Proofs for both parts of Theorem 2.9 can be found in the appendix at the end of this chapter, Section 2.8.

The product, quotient, and reciprocal rules for limits enable us to calculate some limits that would otherwise be impossible to compute without delta–epsilon proofs.

> **EXAMPLE 2.35**
>
> **Using limit rules to calculate limits**
>
> Calculate the following limits, citing any limit rules that you use.
>
> **(a)** $\lim_{x \to -2} (3x - 4)(8 - 5x)$ **(b)** $\lim_{x \to 1} \dfrac{7x + 5}{3x - 1}$
>
> *Solution*
>
> **(a)** Since $(3x - 4)(8 - 5x)$ is a product of linear functions, we can calculate its limit using the product rule for limits.
>
> $$\lim_{x \to -2} (3x - 4)(8 - 5x)$$
>
> $$= \lim_{x \to -2} (3x - 4) \lim_{x \to -2} (8 - 5x) \qquad \text{(product rule for limits)}$$
>
> $$= (3(-2) - 4)(8 - 5(-2)) \qquad \text{(limits of linear functions)}$$
>
> $$= (-10)(18) = -180$$
>
> Note that (with the rules we know to this point) we could not have calculated this limit without the product rule! Why not? Why can't we multiply out $(3x - 4)(8 - 5x)$ and then just "plug in" $x = -2$ to find the limit?

(b) Since this limit concerns the quotient of two linear functions, we can calculate the limit using the quotient rule for limits and what we know about the limits of linear functions.

$$\lim_{x \to 1} \frac{7x + 5}{3x - 1} = \frac{\lim\limits_{x \to 1}(7x + 5)}{\lim\limits_{x \to 1}(3x - 1)} \qquad \text{(quotient rule for limits)}$$

$$= \frac{7(1) + 5}{3(1) - 1} \qquad \text{(limits of linear functions)}$$

$$= \frac{12}{2} = 6$$

We couldn't have calculated this limit without the quotient rule for limits. With Theorem 2.9 we can calculate the limit of any function that is a product or quotient of linear functions. $\square$

✓ *Checking the Answer* Sketch a graph or create a table of values to verify that the function $(3x - 4)(8 - 5x)$ approaches -180 as $x \to -2$. Then sketch a graph or create a table to verify that $\frac{7x+5}{3x-1}$ approaches 6 as $x \to 1$. $\square$

2.4.6 Limits of integer powers of x

Our final limit rule concerns the limits of functions of the form $f(x) = x^k$, where k is a positive integer. We will call such functions "power functions with positive integer powers," or ***positive integer power functions.***

THEOREM 2.10

The Power Rule for Limits (Positive Integer Powers)

If k is a positive integer, then $\lim\limits_{x \to c} x^k = c^k$.

Given a particular integer k, say $k = 3$, this theorem is really an iterated form of the product rule for limits, as the next example shows.

EXAMPLE 2.36

The power rule for limits is really a repeated-product rule

By Theorem 2.10, $\lim\limits_{x \to 2} x^3 = 2^3 = 8$. We can also see this by repeatedly applying the product rule for limits:

$$\lim_{x \to 2} x^3 = \lim_{x \to 2} x(x^2) \qquad \text{(algebra)}$$

$$= \lim_{x \to 2} x \, \lim_{x \to 2} x^2 \qquad \text{(product rule for limits)}$$

$$= \lim_{x \to 2} x \, \lim_{x \to 2} x(x) \qquad \text{(algebra)}$$

$$= \lim_{x \to 2} x \, \lim_{x \to 2} x \, \lim_{x \to 2} x \qquad \text{(product rule for limits)}$$

$$= (2)(2)(2) = 8. \qquad \text{(limit of the identity function)} \quad \square$$

The proof of Theorem 2.10 follows by induction (see Section 0.5.5) and the product rule for limits, and it involves a calculation similar to the one in Example 2.36.

PROOF (**THEOREM 2.10**) If $k = 1$, then $\lim\limits_{x \to c} x^1 = \lim\limits_{x \to c} x = c = c^1$ by Theorem 2.4. Now, assume the theorem is true for $k - 1$; that is, assume that $\lim\limits_{x \to c} x^{k-1} = c^{k-1}$

(this is our inductive hypothesis). Given the inductive hypothesis, we have

$$\lim_{x\to c} x^k = \lim_{x\to c} x(x^{k-1}) \qquad \text{(algebra)}$$
$$= \lim_{x\to c} x \, \lim_{x\to c} x^{k-1} \qquad \text{(product rule for limits)}$$
$$= c \lim_{x\to c} x^{k-1} \qquad \text{(limit of the identity function)}$$
$$= c(c^{k-1}) = c^k. \qquad \text{(inductive hypothesis)}$$

By the Induction Axiom, we have shown that $\lim_{x\to c} x^k = c^k$ for all positive integers k. ∎

With the limit rules that we have proved, we can calculate limits of sums, differences, products, and quotients of functions that are linear or of the form x^k (where k is a positive integer).

EXAMPLE 2.37

Calculating limits using the limit rules

Calculate the following limits, citing any limit rules that you use.

(a) $\lim_{x\to 3} x^2(x-5)$ (b) $\lim_{x\to -2}\left(x^5 + \dfrac{1+x}{3x+5}\right)$

Solution

(a) The function $x^2(x-5)$ is the product of a function of the form x^k (where k is the positive integer 2) and a linear function. We can calculate this limit using Theorems 2.5, 2.9, and 2.10:

$$\lim_{x\to 3} x^2(x-5)$$
$$= \lim_{x\to 3} x^2 \, \lim_{x\to 3}(x-5) \qquad \text{(product rule for limits)}$$
$$= 3^2(3-5) \qquad \text{(limit of } x^k \text{ and limit of a linear function)}$$
$$= 9(-2) = -18.$$

(b) This function is the sum of a function of the form x^k (k is the positive integer 5) and a quotient of two linear functions. We can calculate this limit using Theorems 2.5, 2.7, 2.9, and 2.10:

$$\lim_{x\to -2}\left(x^5 + \frac{1+x}{3x+5}\right)$$
$$= \lim_{x\to -2} x^5 + \lim_{x\to -2}\frac{1+x}{3x+5} \qquad \text{(sum rule for limits)}$$
$$= \lim_{x\to -2} x^5 + \frac{\lim_{x\to -2}(1+x)}{\lim_{x\to -2}(3x+5)} \qquad \text{(quotient rule for limits)}$$
$$= (-2)^5 + \frac{1+(-2)}{3(-2)+5} \qquad \text{(limit of } x^k, \text{ limits of linear functions)}$$
$$= -32 + \frac{-1}{-1} = -31. \qquad \square$$

Although the limit rules in this section do not tell us how to calculate *every* limit, they do enable us to do some simple limit calculations, namely limits of simple combinations of linear and positive integer power functions. We will see how to find limits of compositions in Section 2.6 and learn how to calculate limits of more complicated functions in subsequent chapters.

What you should know after reading Section 2.4

▶ How to state (mathematically and verbally) the limit rules. Be able to give examples of the use of each rule. Why do we need these "limit rules" to calculate limits? Be able to give some examples of limits that we can't calculate using only the limit rules from this section.

▶ How to prove the limit rules, with the possible exception of the product and reciprocal rules. Some of these proofs are delta–epsilon proofs, and some follow from previously proved limit rules.

▶ How to use the limit rules to calculate the limits of simple combinations of linear and positive integer power functions. Be able to write out each step of your calculation individually, justifying each step with an appropriate limit rule. Some limits can be calculated in more than one way; be able to calculate such limits in different ways.

Exercises 2.4

Concepts

0. Read the section and make your own summary of the material. Include whatever you think will help you review the material later on.

1. Write down three limit statements that we do *not* know how to calculate using the limit rules in this section. Then use your calculator or a table of values to approximate these limits.

2. Explain in your own words the types of functions whose limits we can calculate with the limit rules in this section.

■ **State each of the following limit rules in mathematical notation, and give an example of a calculation for each rule. For Problem 11, see Theorem 2.27 in Section 2.8.3.**

3. The limit of a constant

4. The limit of the identity function

5. The limit of a linear function

6. The constant multiple rule for limits

7. The sum rule for limits

8. The difference rule for limits

9. The product rule for limits

10. The quotient rule for limits

11. The reciprocal rule for limits

12. The limit of a positive integer power function

13. By the rule for the limit of a constant, we know that $\lim_{x \to c} k = k$ for any constants k and c. In the proof of this limit rule, we used the fact that given any $\varepsilon > 0$, a choice of *any* $\delta > 0$ will "work" in terms of the formal

definition of limit. Sketch a labeled graph involving the constant function $f(x) = k$ that illustrates this fact.

14. By the rule for the limit of the identity function, we know that $\lim_{x \to c} x = c$ for any constant c. In the proof of this limit rule, we used the fact that given any $\varepsilon > 0$, the choice of $\delta = \varepsilon$ will "work" in terms of the formal definition of limit. Sketch a labeled graph involving the identity function $f(x) = x$ that illustrates this fact.

15. Explain the difference between the rule for calculating the limit of a constant (Theorem 2.3) and the constant multiple rule for limits (Theorem 2.6).

16. Explain why we can't calculate every limit $\lim_{x \to c} f(x)$ just by evaluating $f(x)$ at $x = c$. Support your argument with the graph of a function $f(x)$ for which $\lim_{x \to c} f(x) \neq f(c)$. (*Hint:* You can find examples of such functions in the body of Section 2.1.)

17. Although we can't always calculate a limit $\lim_{x \to c} f(x)$ by evaluating $f(x)$ at $x = c$, we can do so for some types of functions $f(x)$. In this section we saw some simple types of functions $f(x)$ for which $\lim_{x \to c} f(x) = f(c)$. Describe two such types of functions.

18. Find functions $f(x)$ and $g(x)$ and a real number c such that $\lim_{x \to c} f(x) + \lim_{x \to c} g(x) \neq \lim_{x \to c} (f(x) + g(x))$. Does this example contradict Theorem 2.7? Why or why not?

■ **Write each of the following limit statements using the formal delta–epsilon definition of limit (don't try to prove anything; just write down the statements). Assume each limit written below exists; then you can use *L* or *M* to denote a given limit if it is notationally** helpful. In Problems 26 and 27 you may assume that $\lim_{x \to c} g(x) \neq 0$.

19. $\lim_{x \to c} k = k$

20. $\lim_{x \to c} x = c$

21. $\lim_{x \to c} (mx + b) = mc + b$

22. $\lim_{x \to c} kf(x) = k \lim_{x \to c} f(x)$

23. $\lim_{x \to c} (f(x) + g(x)) = \lim_{x \to c} f(x) + \lim_{x \to c} g(x)$

24. $\lim_{x \to c} (f(x) - g(x)) = \lim_{x \to c} f(x) - \lim_{x \to c} g(x)$

25. $\lim_{x \to c} (f(x)g(x)) = \lim_{x \to c} f(x) \lim_{x \to c} g(x)$

26. $\lim_{x \to c} (f(x)/g(x)) = \lim_{x \to c} f(x) / \lim_{x \to c} g(x)$

27. $\lim_{x \to c} (1/g(x)) = 1/ \lim_{x \to c} g(x)$

28. $\lim_{x \to c} x^k = c^k$

29. Sketch the graph of the constant function $f(x) = 3$, and use it to illustrate that for all $\varepsilon > 0$, *any* choice of δ will ensure that $|f(x) - 3| < \varepsilon$ for all x with $0 < |x - 2| < \delta$. What limit does this describe?

30. Sketch the graph of the identity function $f(x) = x$, and use it to illustrate that for all $\varepsilon > 0$, if $\delta = \varepsilon$ then $0 < |x + 1| < \delta \Longrightarrow |f(x) + 1| < \varepsilon$. What limit does this describe?

31. Sketch the graph of the linear function $f(x) = 1 - 3x$, and use it to illustrate that for all $\varepsilon > 0$, if $\delta = \varepsilon/3$ then $0 < |x - 4| < \delta \Longrightarrow |f(x) + 11| < \varepsilon$. (*Hint:* Think about rise over run.) What limit does this describe?

■ Suppose f and g are functions such that $\lim_{x \to 3} f(x) = 5$, $\lim_{x \to 4} f(x) = 2$, and $\lim_{x \to 3} g(x) = 4$. Given this information, calculate the limits below, if possible. If it is not possible with the given information, explain why.

32. $\lim_{x \to 3} -2f(x)$

33. $\lim_{x \to 3} (2f(x) - 3g(x))$

34. $\lim_{x \to 3} f(g(x))$

35. $\lim_{x \to 7} f(x)$

36. $\lim_{x \to 4} f(x)g(x)$

37. $\lim_{x \to 3} \dfrac{f(x) - 3}{g(x)}$

Skills

■ Evaluate each of the following limits (a) using the constant multiple and sum rules and (b) without using the constant multiple or sum rule. Do your calculations one step at a time, justifying each step by referring to any limit rules you use.

38. $\lim_{x \to 1} 15(3 - 2x)$

39. $\lim_{x \to 0} ((2x + 1) - 3(x - 5))$

40. $\lim_{x \to -2} \dfrac{9x + 6}{3}$

■ Evaluate each of the following limits. Justify each step of your computations with a limit rule, as appropriate. Check your answer with a graph.

41. $\lim_{x \to -5} 17$

42. $\lim_{x \to 1.5} x$

43. $\lim_{x \to 0} (x^2 - 1)$

44. $\lim_{x \to -3} \dfrac{2}{3x + 1}$

45. $\lim_{x \to 1} (3x^2 + 5x - 2)$

46. $\lim_{x \to 2} (x - 1)(x + 1)(x + 5)$

47. $\lim_{x \to 4} \dfrac{2 - 3x}{x^3}$

48. $\lim_{x \to -1} \left(x^2 - \dfrac{3x}{x + 2} \right)$

49. $\lim_{x \to 2} \dfrac{x^2(x - 5)}{3(1 - 2x)(2 - 3x)}$

Proofs

50. In this section we proved Theorem 2.5 (the rule for limits of linear functions) directly, using a delta–epsilon argument. We could also have proved this theorem given the sum and constant multiple rules and the rule for the limit of the identity function. Do this; in other words, prove Theorem 2.5 using Theorems 2.3, 2.4, 2.6, and 2.7. Your proof will *not* be a delta–epsilon argument.

51. Prove the difference rule for limits (Theorem 2.8) by using Theorems 2.6 and 2.7.

52. Prove the difference rule for limits (Theorem 2.8) by using a direct delta–epsilon argument. (*Hint:* The proof will look much like the proof of Theorem 2.7.)

■ Write *in your own words* a proof for each of the theorems below. Explain each step of each proof so that your argument would be clear to one of your fellow students.

53. Theorem 2.3 (limit of a constant)

54. Theorem 2.4 (limit of the identity function)

55. Theorem 2.5 (limit of a linear function)

56. Theorem 2.6 (constant multiple rule for limits)

57. Theorem 2.7 (sum rule for limits)

58. Theorem 2.8 (difference rule for limits)

59. Theorem 2.9(b) (quotient rule for limits)

60. Theorem 2.10 (power rule for limits)

2.5 Calculating Limits

In the previous section, we developed a library of rules and used them to calculate limits of simple combinations of linear functions and positive integer power functions. In this section we will develop an even faster way of calculating limits. We will also use the limit rules to calculate more complicated limits, including limits that require algebraic manipulation, limits that "do not exist," and limits of piecewise functions.

2.5.1 An algorithm for calculating limits

As we saw in the previous section, if $f(x)$ is a linear function or a power function with a positive integer power, we can find $\lim_{x \to c} f(x)$ simply by evaluating $f(x)$ at $x = c$. If $f(x) = mx + b$ is a linear function, then (by Theorem 2.5 in Section 2.4)

$$\lim_{x \to c} f(x) = \lim_{x \to c}(mx + b) = m(c) + b = f(c),$$

and if $f(x) = x^k$ is a positive integer power function, then (by Theorem 2.10 in Section 2.4)

$$\lim_{x \to c} f(x) = \lim_{x \to c} x^k = c^k = f(c).$$

This means that we can find the limit of a linear function or a positive integer power function simply by evaluating $f(x)$ at $x = c$. The constant multiple, sum, product, and quotient rules for limits tell us how to calculate limits of combinations of these types of functions.

> ⊛ *Caution*   Remember that you cannot calculate *all* limits this way! At this point the only types of functions whose limits we can calculate this way are limits of linear functions and limits of positive integer power functions (and some simple combinations of these functions, such as polynomial functions).   ☐

EXAMPLE 2.38

Calculating limits using the limit rules

Use limit rules to calculate $\lim_{x \to 1}(x^2 - 2(x + 1))$.

Solution   We have

$$\lim_{x \to 1}(x^2 - 2(x+1)) = \lim_{x \to 1} x^2 + \lim_{x \to 1}(-2(x+1)) \qquad \text{(sum rule for limits)}$$

$$= \lim_{x \to 1} x^2 - 2 \lim_{x \to 1}(x+1) \qquad \text{(constant multiple rule for limits)}$$

$$= (1)^2 - 2(1+1) \qquad \text{(limits of power and linear functions)}$$

$$= 1 - 2(2) = -3. \qquad ☐$$

Although the calculation in Example 2.38 is mathematically correct and clearly justifies every step with a limit rule, it is a little long. We don't want to have to write out each step individually every time we calculate a simple limit. You may have noticed that -3 is exactly the answer we would have arrived at had we evaluated the function $f(x) = x^2 - 2(x + 1)$ at $x = 1$ at the very start (since $f(1) = (1)^2 - 2(1 + 1) = -3$). Can we do the "plugging in" sooner? The answer is yes, if we are careful, as explained in the following algorithm.

ALGORITHM 2.1

An Algorithm for Calculating Limits

If $f(x)$ is a function such that

(a) $f(x)$ is a sum, product, or quotient of constant multiples of linear functions and/or positive integer power functions, and

(b) evaluating $f(x)$ at $x = c$ produces a real number $f(c)$,

then $\lim_{x \to c} f(x) = f(c)$, that is, we can calculate $\lim_{x \to c} f(x)$ by evaluating $f(x)$ at $x = c$.

This algorithm is a consequence of the limit rules we developed in the previous section. What it means is that we can often evaluate a limit of a function that is a combination of simple functions simply by evaluating at $x = c$.

> ⓢ *Caution*   If $f(x)$ is a more complicated function than described in Algorithm 2.1, or if $f(c)$ is not a real number, then at this point we *cannot* calculate the limit of $f(x)$ simply by evaluating $f(c)$. For example, we still can't calculate $\lim_{x \to 3} \sqrt{x+1}$ or $\lim_{x \to 1} \frac{1}{x-1}$. (Limits of the second type will be examined in Section 2.5.2.)   □

EXAMPLE 2.39

Calculating limits quickly with Algorithm 2.1

Consider the limit that we calculated in Example 2.38. With Algorithm 2.1 the calculation is much shorter:

$$\lim_{x \to 1}(x^2 - 2(x + 1)) = (1)^2 - 2(1 + 1) = 1 - 2(2) = -3.$$

The function $f(x) = x^2 - 2(x + 1)$ is the sum of the integer power function x^2 and a constant multiple (-2) of the linear function $x + 1$. Moreover, $f(1) = -3$ is a real number. Thus Algorithm 2.1 applies here.   □

What we are really doing when we apply this "shortcut" algorithm is applying limit rules in our head but not writing them down. We are implicitly using the sum and constant multiple rules for limits, as well as the rules for calculating limits of linear functions and limits of positive integer power functions.

> ⓢ *Caution*   Most of the time we will calculate limits without justifying each step with a limit rule. However, it is important that you also be able to do limit calculations carefully, step by step, with appropriate references to limit rules when asked to do so.   □

EXAMPLE 2.40

Various limit calculations

Calculate the following limits using Algorithm 2.1, if possible.

(a) $\lim\limits_{x \to 0} \dfrac{x + 6}{(x - 1)(x + 2)}$ (b) $\lim\limits_{x \to -1} \dfrac{2}{x + 1}$ (c) $\lim\limits_{x \to 2} \sqrt{2x + 5}$

Solution

(a) The function in this example is a quotient of a linear function $(x + 6)$ with a product of linear functions $(x - 1$ and $x + 2)$, so Algorithm 2.1 could apply here. In this particular example, we get a real number if we evaluate the function at $x = 0$, so

$$\lim_{x \to 0} \frac{x + 6}{(x - 1)(x + 2)} = \frac{0 + 6}{(0 - 1)(0 + 2)} = \frac{6}{-2} = -3.$$

(b) The function $f(x) = \frac{2}{x+1}$ is a constant multiple of the reciprocal of a linear function, so Algorithm 2.1 might apply here. However, $f(-1)$ is undefined (division by zero), so we can't find the limit of $f(x)$ as $x \to -1$ using Algorithm 2.1. We'll discuss how to calculate this limit in Section 2.5.2.

(c) The function $f(x) = \sqrt{2x + 5}$ is not one of the combinations of functions listed in Algorithm 2.1 (since it involves a square root). At present, we don't have a quick way to calculate this limit.   □

❓ Question Which limit rules did we implicitly use in the calculation in part (a) of Example 2.40? Can you write out the limit calculations one step at a time, citing any appropriate limit rules when you use them? □

✓ Checking the Answer You can use a graph or a table of values to check the answer to part (a) of Example 2.40. Figure 2.33 shows the graph of the function $f(x) = \frac{x+6}{(x-1)(x+2)}$ near $x = 0$. As you can see, the function appears to approach -3 as $x \to 0$.

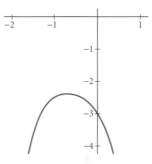

Figure 2.33 □

2.5.2 Nonexistent and indeterminate limits

As we have seen, not all limits exist. In other words, as $x \to c$, a function $f(x)$ might or might not approach a real number; for example, it might approach ∞ or be undefined in some other way. On the other hand, a limit might be "indeterminate," which means that we cannot tell whether or not it exists (at least not without doing some algebra or other work first).

DEFINITION 2.13

We say that a limit $\lim\limits_{x \to c} f(x)$ is **undefined** if there is no real number L that is equal to that limit. We say that $\lim\limits_{x \to c} f(x)$ is in an **indeterminate form** if it is written in such a way that it is not clear whether or not the limit exists.

As we will see in a moment, a limit of the form $\frac{1}{0}$ "does not exist," whereas a limit of the form $\frac{0}{0}$ is said to be "indeterminate."

We begin our investigation of nonexistent and indeterminate limits by stating a more general version of the quotient rule for limits. The following theorem tells us what happens if the limit in the denominator is zero.

THEOREM 2.11

Limits of Quotients of Functions

Suppose $\lim\limits_{x \to c} f(x) = L$ and $\lim\limits_{x \to c} g(x) = M$ for some real numbers L and M.

(a) If $L \neq 0$ and $M \neq 0$, then $\lim\limits_{x \to c} \frac{f(x)}{g(x)} = \frac{L}{M}$.

(b) If $L = 0$ but $M \neq 0$, then $\lim\limits_{x \to c} \frac{f(x)}{g(x)} = 0$.

(c) If $L \neq 0$ but $M = 0$, then $\lim\limits_{x \to c} \frac{f(x)}{g(x)}$ does not exist.

(d) If $L = 0$ and $M = 0$, then $\lim\limits_{x \to c} \frac{f(x)}{g(x)}$ may or may not exist.

◈ *Caution*   Remember that when we say that a limit "exists," we mean that it is equal to some real number, and when we say that a limit "does not exist," we mean that it is *not* equal to any real number. A limit that does not exist is sometimes, but not always, infinite.   □

PROOF (**THEOREM 2.11**)   Parts (a) and (b) are just restatements of the quotient rule for limits (Theorem 2.9 from Section 2.4); in fact, part (b) is really a special case of part (a).

We prove part (c) by contradiction. Suppose $f(x)$ and $g(x)$ are functions such that for some real number c, $\lim_{x \to c} f(x) \neq 0$ and $\lim_{x \to c} g(x) = 0$. Then (seeking a contradiction) suppose that $\lim_{x \to c} \frac{f(x)}{g(x)}$ exists, that is, it is equal to Q for some real number Q. By the product rule for limits, since the limits of $g(x)$ and of $\frac{f(x)}{g(x)}$ both exist, we have

$$\lim_{x \to c} f(x) = \lim_{x \to c} \left(g(x) \frac{f(x)}{g(x)} \right) \qquad \text{(algebra)}$$

$$= \lim_{x \to c} g(x) \lim_{x \to c} \frac{f(x)}{g(x)} \qquad \text{(product rule for limits)}$$

$$= (0)(Q) = 0. \qquad \text{(by hypothesis)}$$

On the other hand, by assumption we have $\lim_{x \to c} f(x) \neq 0$. This is a contradiction. Therefore, our original assumption that $\lim_{x \to c} \frac{f(x)}{g(x)}$ exists must be incorrect.

Part (d) of Theorem 2.11 concerns a limit of the form $\frac{0}{0}$ and is an example of a limit of an "indeterminate form" (meaning that we don't know whether or not the limit exists). Sometimes, after some algebra, we can find the limit. Other times, the limit may be infinite or not exist. To prove part (d), it is sufficient to exhibit an example where the given limit exists and an example where the given limit does not exist. For example, if $f(x) = x^2$ and $g(x) = x$, we have

$$\lim_{x \to 0} \frac{f(x)}{g(x)} = \lim_{x \to 0} \frac{x^2}{x} = \lim_{x \to 0} x = 0,$$

so although $\lim_{x \to 0} x^2 = 0$ and $\lim_{x \to 0} x = 0$, the limit of the quotient $\frac{x^2}{x} = x$ as $x \to 0$ also equals zero (and thus exists). Note that we canceled an x from the numerator and the denominator when we changed $\frac{x^2}{x}$ into x. This cancellation is valid whenever $x \neq 0$, and since the limit as $x \to 0$ does not concern what happens *at* $x = 0$, we can perform the cancellation. We will talk about this more in Section 2.5.4.

On the other hand, if $f(x) = x$ and $g(x) = x^2$, then

$$\lim_{x \to 0} \frac{f(x)}{g(x)} = \lim_{x \to 0} \frac{x}{x^2} = \lim_{x \to 0} \frac{1}{x} \text{ does not exist,}$$

by part (c). In this case, the limits of the numerator and the denominator are both zero, but the limit of the quotient does not exist.

We have shown that in some cases, a limit of the indeterminate form $\frac{0}{0}$ exists, whereas in other cases, it does not. This proves part (d) of Theorem 2.11.   ▪

EXAMPLE 2.41   **Calculating limits of quotients**

Calculate the following limits.

(a) $\displaystyle\lim_{x \to 2} \frac{x - 2}{x + 1}$ (b) $\displaystyle\lim_{x \to -1} \frac{2}{x + 1}$ (c) $\displaystyle\lim_{x \to 1} \frac{x^2 - 1}{x - 1}$

Solution

(a) The limits of the numerator and denominator are

$$\lim_{x \to 2} f(x) = \lim_{x \to 2} (x - 2) = 0 \quad \text{and} \quad \lim_{x \to 2} g(x) = \lim_{x \to 2} (x + 1) = 3 \neq 0.$$

Therefore, by Theorem 2.11, we have $L = 0$ and $M \neq 0$ and thus

$$\lim_{x \to 2} \frac{x - 2}{x + 1} = \frac{0}{3} = 0.$$

(b) This is one of the limits we could not evaluate in Example 2.40. The limits of the numerator and denominator are

$$\lim_{x \to -1} f(x) = \lim_{x \to -1} 2 = 2 \neq 0 \quad \text{and} \quad \lim_{x \to -1} g(x) = \lim_{x \to -1} (x + 1) = 1 + 1 = 0.$$

Since the limit of the numerator is nonzero, and the limit of the denominator is zero, we know that, by part (c) of Theorem 2.11,

$$\lim_{x \to -1} \frac{2}{x + 1} \text{ does not exist.}$$

Although it is not exactly mathematically proper to write this, you should be thinking the following in your head:

$$\lim_{x \to -1} \frac{2}{x + 1} \to \frac{2}{0}, \quad \text{which does not exist.}$$

This is an example where we evaluate at $x = -1$ and "something bad happens," namely that there is a division by zero (so Algorithm 2.1 does not apply). The limit approaches a *nonzero* constant divided by zero, which by part (c) does not exist. In Section 2.5.3 we will more closely investigate what happens in this situation.

(c) This time both L and M are zero (can you say why?), so by Theorem 2.11,

$$\lim_{x \to 1} \frac{x^2 - 1}{x - 1} \text{ is indeterminate (of the form } \tfrac{0}{0}).$$

In Section 2.5.4 we will find out whether or not this particular limit exists. □

❓ *Question*   The last limit in Example 2.41 was indeterminate; we don't know at this point whether it exists or not. Use a graph of $\frac{x^2 - 1}{x - 1}$ to make an educated guess as to whether or not this limit exists. □

2.5.3 Investigating infinite limits

In Section 2.5.2 we encountered limits where the numerator approaches a nonzero constant and the denominator approaches zero. These limits "do not exist," but as we will see in a moment, we can be more specific than that.

EXAMPLE 2.42   **A limit of a quotient where the denominator approaches zero**

In part (b) of Example 2.41, we saw that $\lim_{x \to -1} \dfrac{2}{x + 1}$ approaches the quotient $\frac{2}{0}$ and so does not exist (by Theorem 2.11). □

There are various ways that a limit can fail to exist. One way is illustrated in Example 2.15 in Section 2.1.6; the function could jump around or oscillate as $x \to c$. Another way a limit can fail to exist is by being infinite. Which type of limit do we have in Example 2.42?

The following theorem tells us what happens when a limit approaches a quotient of the form $\frac{L}{0}$ (with $L \neq 0$) as $x \to c^+$. What happens when $x \to c^-$ is similar and will be investigated in the exercises. We will use the notation $g(x) \to 0^+$ as $x \to c^+$ to denote that the function $g(x)$ has small *positive* values as x approaches c from the right. More precisely, $g(x) \to 0^+$ as $x \to c^+$ if there is some interval $(c, c + \delta)$ for which $g(x)$ is always positive. Similarly, we say that $g(x) \to 0^-$ as $x \to c^+$ if $g(x)$ has small *negative* values as x approaches c from the right.

THEOREM 2.12

Limits of the Form $\frac{L}{0}$, Where $L \neq 0$

Suppose $\lim\limits_{x \to c^+} f(x) = L$ and $\lim\limits_{x \to c^+} g(x) = 0$. If $L > 0$, then

(a) If $g(x) \to 0^+$ as $x \to c^+$, then $\lim\limits_{x \to c^+} \dfrac{f(x)}{g(x)} = \infty$.

(b) If $g(x) \to 0^-$ as $x \to c^+$, then $\lim\limits_{x \to c^+} \dfrac{f(x)}{g(x)} = -\infty$.

If $L < 0$, a similar theorem applies. If $L = 0$, then Theorem 2.12 is invalid, and we must instead try to use algebra until we no longer have the indeterminate form $\frac{0}{0}$. Note that this theorem does not tell you what happens if $g(x)$ keeps changing from positive to negative as $x \to c^+$. (The function $g(x) = x^2 \sin(\frac{1}{x})$ is an example of such a function. Use your calculator to graph this function near $c = 0$.) The proof of Theorem 2.12 can be found in the appendix at the end of this chapter (Section 2.8).

The following notation will assist us in calculating limits like the ones in Theorem 2.12.

NOTATION 2.2

Notation for Dealing with Limits of the Form $\frac{1}{0^+}$ and $\frac{1}{0^-}$

We will use the following notational conventions:

$$\frac{1}{0^+} \to \infty \quad \text{and} \quad \frac{1}{0^-} \to -\infty.$$

It is important to stress that the expressions in Notation 2.2 are not mathematically precise (since we cannot divide a number by zero). However, we will use them because they express the behavior of the limits from Theorem 2.12 in a very compact way.

Recall from Definition 2.4 in Section 2.1 that $\lim\limits_{x \to c} f(x) = \infty$ if and only if its left and right limits approach ∞ as $x \to c$. To examine a limit that does not exist because it is of the "constant over zero" form, look at its left and right limits and, if applicable, use Theorem 2.12.

EXAMPLE 2.43   **Investigating both sides of a limit of the form $\frac{2}{0}$**

Evaluate the limit $\lim\limits_{x \to -1} \dfrac{2}{x + 1}$.

Solution   As we saw in Example 2.41(b), $\lim\limits_{x \to -1} \frac{2}{x+1}$ approaches $\frac{2}{0}$ and thus does not exist. Note that $\frac{2}{x+1}$ is the quotient of $f(x) = 2$ and $g(x) = x + 1$, and that $\lim\limits_{x \to -1} 2 = 2$ and $\lim\limits_{x \to -1} (x + 1) = 0$, so Theorem 2.12 applies. We will look at the left and right limits as $x \to -1$ separately, starting with the limit from the left:

$$\lim\limits_{x \to -1^-} \frac{2}{x + 1} = \frac{2}{(-1^-) + 1} \to \frac{2}{0^-} \to -\infty,$$

since if x is slightly less than -1 (maybe -1.1 or -1.01, for example), then $x + 1 = (-1^-) + 1$ is slightly less than zero. In the last step (where we say that $\frac{2}{0^-} \to -\infty$), we are applying Theorem 2.12.

On the other hand, the limit from the right is

$$\lim_{x \to -1^+} \frac{2}{x+1} = \frac{2}{(-1^+) + 1} \to \frac{2}{0^+} \to \infty.$$

Since the limit from the left is $-\infty$ and the limit from the right is ∞, we still can say only that $\lim\limits_{x \to -1} \frac{2}{x+1}$ does not exist.

Investigating the left and right limits did give us a better idea of just what happens to the function $\frac{2}{x+1}$ as x approaches -1. Figure 2.34 shows the graph of this function near $c = -1$. As $x \to -1^-$ the height of the graph decreases without bound, and as $x \to -1^+$ the height of the graph increases without bound.

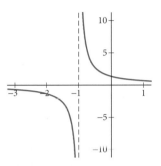

Figure 2.34

⑤ *Caution*   Whenever we encounter a limit of the form $\frac{L}{0}$ ($L/0$), we investigate the left and right limits to see whether they are ∞ or $-\infty$. This gives a great deal more information than just saying that the limit fails to exist.

2.5.4 Limits requiring algebraic manipulation

In Section 2.5.2 we saw that some limits have an *indeterminate* form, that is, they may or may not exist. Here we learn how to handle such limits. We begin by discussing when we can "cancel" terms in a limit.

THEOREM 2.13

Cancellation in a Limit

For any functions $f(x)$ and $g(x)$ and any real number c,

$$\lim_{x \to c} \frac{(x - c)f(x)}{(x - c)g(x)} = \lim_{x \to c} \frac{f(x)}{g(x)}.$$

PROOF (**THEOREM 2.13**)   Algebraically, for all $x \neq c$, we have

(1)
$$\frac{(x - c)f(x)}{(x - c)g(x)} = \frac{f(x)}{g(x)}.$$

If $x = c$, then these two expressions are *not* equal; the one on the left is undefined, whereas the one on the right is not necessarily undefined. However, the limit of a function as $x \to c$ does not depend on what happens *at $x = c$*. Since Equation (1) is true for all $x \neq c$, the limits in Theorem 2.13 are equal. ∎

Theorem 2.13 tells us that we can cancel common factors of $(x-c)$ in the numerator and denominator of a function without affecting the limit of that function as $x \to c$. The indeterminate form $\frac{0}{0}$ frequently arises because of such common factors. Such limits can often be evaluated after we cancel the common factors.

EXAMPLE 2.44

Using cancellation to calculate a limit of the form $\frac{0}{0}$

We can now calculate the limit $\lim\limits_{x \to 1} \dfrac{x^2 - 1}{x - 1}$ from part (c) of Example 2.41. (Recall that this limit was of the form $\frac{0}{0}$ and thus indeterminate.)

$$\lim_{x \to 1} \frac{x^2 - 1}{x - 1} = \lim_{x \to 1} \frac{(x - 1)(x + 1)}{x - 1} \qquad \text{(algebra)}$$

$$= \lim_{x \to 1} \frac{x + 1}{1} \qquad \text{(Theorem 2.13)}$$

$$= \frac{1 + 1}{1} = 2 \qquad \text{(Algorithm 2.1)}$$

Without Theorem 2.13, all we would know about this limit would be that it has the indeterminate form $\frac{0}{0}$. Whenever a limit is in an indeterminate form, more work (like the algebra and cancellation above) must be done to determine whether the limit exists (and if so, what it is equal to). □

Sometimes we need to manipulate a function algebraically so it is written as a sum, product, or quotient of constant multiples of linear or positive integer power functions (recall that these are the types of functions whose limits we have an algorithm for calculating).

EXAMPLE 2.45

Using algebra so that we can apply the limit rules

Find $\lim\limits_{x \to 1} (2x + 1)^2$.

Solution   The function $f(x) = (2x + 1)^2$ is not explicitly written as the sum, product, or quotient of constant multiples of linear or integer power functions. As currently written, it is a *composition* of a linear function and a positive integer power function. (We will see how to take limits of some compositions in Section 2.6.5.) However, with a little algebra we can write $f(x)$ as a sum of constant multiples of positive integer power functions:

(2) $$f(x) = (2x + 1)^2 = 4x^2 + 4x + 1.$$

We can now calculate the limit of this function using Algorithm 2.1 on page 179. With $f(x)$ written as in Equation (2), we have

$$\lim_{x \to 1} (2x + 1)^2 = \lim_{x \to 1} (4x^2 + 4x + 1) \qquad \text{(algebra)}$$

$$= 4(1)^2 + 4(1) + 1 \qquad \text{(Algorithm 2.1)}$$

$$= 4 + 4 + 1 = 9. \qquad □$$

2.5.5 Limits of piecewise functions

In this section we investigate the limits of piecewise-defined functions. At the "break" points of a piecewise function we will need to examine the limits from the left and from the right (since the function is defined differently to the left and to the right of these "break" points). We will then use the fact that the limit of a function as $x \to c$ exists if, and only if, its left and right limits exist and are equal.

EXAMPLE 2.46

Calculating the limit of a piecewise function away from, and at, a break point

Calculate the limit as $x \to 1$ and as $x \to -1$ of the function $f(x) = \begin{cases} x^2, & \text{if } x \le -1 \\ 2x, & \text{if } x > -1 \end{cases}$.

Solution Near $x = 1$, we have $f(x) = 2x$ (since $1 > -1$). Therefore,

$$\lim_{x \to 1} f(x) = \lim_{x \to 1} 2x = 2(1) = 2.$$

The limit of $f(x)$ is more interesting at $x = -1$, since to the left of $x = -1$ we have $f(x) = x^2$, whereas to the right of $x = -1$ we have $f(x) = 2x$. This means that we will have to calculate the left and right limits of $f(x)$ separately:

$$\lim_{x \to -1^-} f(x) = \lim_{x \to -1^-} x^2 = (-1)^2 = 1, \quad \text{and}$$

$$\lim_{x \to -1^+} f(x) = \lim_{x \to -1^+} 2x = 2(-1) = -2.$$

Recall that $\lim_{x \to -1} f(x)$ exists if, and only if, its left and right limits exist and are equal to each other. In this example, both the left and right limits exist, but they are not equal. Therefore,

$$\lim_{x \to -1} f(x) \text{ does not exist.} \qquad \square$$

✓ *Checking the Answer* Figure 2.35 shows the graph of $f(x)$ from Example 2.46. Note that when we look at the graph, it seems true that $f(x) \to 1$ as x approaches -1 from the left and that $f(x) \to -2$ as x approaches -1 from the right. $\qquad \square$

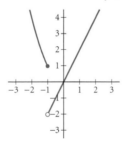

Figure 2.35

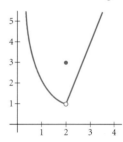

Figure 2.36

EXAMPLE 2.47

The limit of a piecewise function with three cases

Find $\lim_{x \to 2} f(x)$, given $f(x) = \begin{cases} \frac{2}{x}, & \text{if } x < 2 \\ 3, & \text{if } x = 2. \\ 3x - 5, & \text{if } x > 2 \end{cases}$

Solution Since $x = 2$ is a "break" point of the piecewise function $f(x)$, we must calculate the left and right limits separately. As x approaches 2 from the left, $f(x) = \frac{2}{x}$; as x approaches 2 from the right, we have $f(x) = 3x - 5$. Therefore,

$$\lim_{x \to 2^-} f(x) = \lim_{x \to 2^-} \frac{2}{x} = \frac{2}{2} = 1, \quad \text{and}$$

$$\lim_{x \to 2^+} f(x) = \lim_{x \to 2^+} (3x - 5) = 3(2) - 5 = 6 - 5 = 1.$$

Since both limits exist and are equal, we have

$$\lim_{x \to 2} f(x) = 1.$$

Note that the *limit* $\lim_{x \to 2} f(x)$ is not equal to the *value* $f(2) = 3$. In this case, we could not have solved the limit simply by evaluating at $x = 2$. $\qquad \square$

☑ *Checking the Answer*   Figure 2.36 shows the graph of the function $f(x)$ from Example 2.47. Note that $f(x) \to 1$ as x approaches 2 from the left and from the right. Note also that the value $f(2) = 3$ does not affect the limit of $f(x)$ as $x \to 2$.   □

What you should know after reading Section 2.5

▶ How to calculate quickly the limits of sums, products, or quotients of constant multiples of linear functions and positive integer power functions. When does the algorithm for calculating such limits fail to work?

▶ How to deal with limits that approach $\frac{1}{0}$ or $\frac{0}{0}$. This includes investigating infinite limits from the left and right and using cancellation techniques to simplify an indeterminate form.

▶ How to calculate limits of piecewise functions, in particular limits that occur at the "break" points of such functions.

Exercises 2.5

Concepts

0. Read the section and make your own summary of the material. Include whatever you think will help you review the material later on.

1. True or False: The limit of a function $f(x)$ as $x \to c$ is equal to $f(c)$. Explain your answer or provide a counterexample.

2. Multiple Choice: If $\lim_{x \to c} f(x) = 0$ and $\lim_{x \to c} g(x) = 0$, then $\lim_{x \to c}(f(x)/g(x))$ is equal to: (A) 0; (B) 1; (C) infinity; (D) not enough information to tell.

3. Give an example of a limit that *cannot* be calculated using the method of Algorithm 2.1 because it fails condition (a). Then give an example of a function that passes condition (a) but fails condition (b).

4. Explain in your own words what it means for a function $f(x)$ to be a sum, product, or quotient of constant multiples of linear functions and positive integer power functions.

5. Explain how the function $f(x) = \frac{(x^2+1)(4-3x)}{3x^2}$ is a sum, product, or quotient of constant multiples of linear functions and positive integer power functions.

6. Write the function $f(x) = (3x+1)^3$ as a sum, product, or quotient of constant multiples of linear functions and positive integer power functions.

7. In this section it was mentioned that we have not yet developed a way of easily calculating the limit $\lim_{x \to 2} \sqrt{2x+5}$. Why can't we calculate this limit using what we have developed in this section? How could we approximate this limit? What would we have to do to show that this limit is exactly equal to 3?

8. For each part, find functions $f(x)$ and $g(x)$ and a real number c such that $\lim_{x \to c} f(x) = 0$, $\lim_{x \to c} g(x) = 0$, and:

   (a) $\lim_{x \to c} \frac{f(x)}{g(x)} = 0$.

   (b) $\lim_{x \to c} \frac{f(x)}{g(x)}$ does not exist.

   (c) $\lim_{x \to c} \frac{f(x)}{g(x)} = 1$.

   (d) $\lim_{x \to c} \frac{f(x)}{g(x)} = 3$.

9. State a version of Theorem 2.12 where $L < 0$.

10. State a version of Theorem 2.12 where $L > 0$ and $x \to c^-$.

11. Suppose $h(x)$ is a quotient $h(x) = \frac{f(x)}{g(x)}$ of two functions. Suppose also that $\lim_{x \to 3} f(x) = -2$ and that $g(x) \to 0^+$ as $x \to 3^-$ and $g(x) \to 0^-$ as $x \to 3^+$. Sketch a possible graph of $y = h(x)$.

12. For which values of x is the equation $\frac{x^2-x-2}{x^2-5x+6} = \frac{x+1}{x-3}$ true?

13. Explain why the function $f(x) = \frac{(x-1)(x+5)}{x-1}$ is *not* equal to the function $g(x) = x + 5$.

Skills

■ Calculate each of the following limits (a) one step at a time, citing any limit rules you use, and (b) using Algorithm 2.1.

14. $\lim_{x \to 3}(3x + x^2(2x+1))$

15. $\lim\limits_{x \to -1} \dfrac{x - 1}{(x + 4)(x + 2)}$ 16. $\lim\limits_{x \to 0} \dfrac{3}{2x^2 - 4x + 1}$

■ Calculate each of the following limits using the methods of this section. Indicate where you use each of the following: Algorithm 2.1, Theorem 2.11, Theorem 2.12, and Theorem 2.13. If a given limit does not exist, investigate the left and right limits to see whether they are infinite. When you have finished your calculations, check your answers with graphs.

17. $\lim\limits_{x \to 1.7} (3.1x^2 - 4x + 0.8)$ 18. $\lim\limits_{x \to 0} \dfrac{x^2 - 1}{x - 1}$

19. $\lim\limits_{x \to -3} (3x - 1)(x + 2)(x + 45)$

20. $\lim\limits_{x \to 2} \dfrac{4 - 2x}{x + 2}$ 21. $\lim\limits_{x \to -2} \dfrac{4 - 2x}{x + 2}$

22. $\lim\limits_{x \to 2} \dfrac{4 - 2x}{x - 2}$ 23. $\lim\limits_{x \to 1} \dfrac{1}{x - 1}$

24. $\lim\limits_{x \to 1} \dfrac{1}{1 - x}$ 25. $\lim\limits_{x \to 1} \dfrac{1}{x^2 - 1}$

26. $\lim\limits_{x \to 1} \left(x^2 + \dfrac{1}{x}\right)$ 27. $\lim\limits_{x \to -3} (x^2(3x + 1) + x - 4)$

28. $\lim\limits_{x \to -4} \dfrac{x^2 + 8x + 16}{(x + 4)^2(x + 1)}$ 29. $\lim\limits_{x \to 0} \dfrac{x^2 + 1}{x(x - 1)}$

30. $\lim\limits_{x \to 2} \dfrac{x + 1}{(x - 2)^2}$ 31. $\lim\limits_{x \to -4} \dfrac{x + 4}{x^2 + 8x + 16}$

32. $\lim\limits_{x \to 0} \dfrac{x}{x^2 - x}$ 33. $\lim\limits_{x \to 1} \dfrac{x - 1}{x^2 - 2x + 1}$

34. $\lim\limits_{x \to -5} ((x - 1)^2 - x^2)$

■ For each piecewise function $f(x)$ and value c given below, calculate $\lim\limits_{x \to c^-} f(x)$, $\lim\limits_{x \to c^+} f(x)$, and $\lim\limits_{x \to c} f(x)$. Then check your answers with graphs.

35. $f(x) = \begin{cases} x^2 + 1, & \text{if } x \le 0 \\ 1 - x, & \text{if } x > 0 \end{cases}$, $c = 0$

36. $f(x) = \begin{cases} 3x + 2, & \text{if } x < -1 \\ 5 + 4x^3, & \text{if } x \ge -1 \end{cases}$, $c = -1$

37. $f(x) = \begin{cases} x^2 - 3x - 1, & \text{if } x \ne -2 \\ 3, & \text{if } x = -2 \end{cases}$, $c = -2$

38. $f(x) = \begin{cases} \frac{3x^2}{4 + x}, & \text{if } x < 2 \\ x^2 - 2, & \text{if } x \ge 2 \end{cases}$, $c = 2$

39. $f(x) = \begin{cases} \frac{x^2 - 1}{x - 1}, & \text{if } x < 1 \\ 0, & \text{if } x = 1 \\ 3x - 1, & \text{if } x > 1 \end{cases}$, $c = 1$

40. $f(x) = \begin{cases} x + 1, & \text{if } x < 3 \\ 2, & \text{if } x = 3 \\ x^2 - 9, & \text{if } x > 3 \end{cases}$, $c = 3$

■ For each function $f(x)$ and real number c given below, find a real number a such that $\lim\limits_{x \to c} f(x)$ exists (if possible).

41. $f(x) = \begin{cases} 3x + 1, & \text{if } x < 0 \\ 2x + a, & \text{if } x \ge 0 \end{cases}$, $c = 0$

42. $f(x) = \begin{cases} x^2 - a, & \text{if } x \ne 1 \\ 2, & \text{if } x = 1 \end{cases}$, $c = 1$

43. $f(x) = \begin{cases} x + a, & \text{if } x < 2 \\ 0, & \text{if } x = 2 \\ x^2 - a, & \text{if } x > 2 \end{cases}$, $c = 2$

44. $f(x) = \begin{cases} \frac{a - 2}{x^2 + 1}, & \text{if } x < 0 \\ 3, & \text{if } x = 0 \\ x + a, & \text{if } x > 0 \end{cases}$, $c = 0$

Applications

45. In 1960, H. von Forester suggested that the human population could be measured by the function:

$$P(t) = \dfrac{179 \times 10^9}{(2027 - t)^{0.99}}.$$

The time t is measured in years, where $t = 1$ corresponds to A.D. 1, time $t = 1973$ corresponds to A.D. 1973, and so on. (We saw this "Doomsday Model" for population in Problem 39 of Section 2.1.) Use limit techniques to calculate $\lim\limits_{t \to 2027^-} P(t)$. What does this limit mean in real-world terms?

46. Suppose Alina accidentally drops a bowling ball from her window ledge, 350 feet above ground level. After t seconds, the height $B(t)$ of the bowling ball is given by the function

$$B(t) = 350 - 16t^2.$$

In this problem, we are interested in the average rate of change of the height of the bowling ball over short time intervals starting at time $t = 3$.

(a) Calculate the average rate of change of the height of Alina's bowling ball from $t = 3$ seconds to $t = 3.5$ seconds. (See Definition 1.8 in Section 1.3.2 if you need to review the definition of an average rate of change.)

(b) Let h represent a small positive number. Write down a formula for computing the average rate of change of the height of the bowling ball from time $t = 3$ to time $t = 3 + h$. (The only letter in your formula should be h.)

(c) Use the formula you found in part (b) to calculate the average rate of change of the height of Alina's bowling ball from $t = 3$ to $t = 3.25$, from $t = 3$ to $t = 3.1$, and from $t = 3$ to $t = 3.01$.

(d) You may have noticed in part (c) that as the time interval gets shorter (i.e., as h gets smaller), the average rate of change seems to be approaching a limit. Take the limit as $h \to 0^+$ of the average-rate-of-change formula you found in part (b).

(e) What do you think the limit you found in part (d) represents in real-world terms? (*Note:* We will learn more about limits of average rates of change in Chapter 3, when we study "derivatives.")

Proofs

47. Use a limit rule to show that if f is a linear function, then we can calculate $\lim_{x \to c} f(x)$ by evaluating f at $x = c$; in other words, prove that if f is a linear function, then $\lim_{x \to c} f(x) = f(c)$.

48. Use limit rules to prove that Algorithm 2.1 works in the following simple case: If f is a product of linear functions, then for any c, we can calculate $\lim_{x \to c} f(x)$ by evaluating f at $x = c$; in other words, $\lim_{x \to c} f(x) = f(c)$ if f is a product of linear functions.

49. Explain in your own words the proof of part (c) of Theorem 2.11.

2.6 Continuity

Intuitively, a function is **continuous** if its graph has no breaks, jumps, or holes anywhere on its domain; you can sketch the graph of the function "without picking up your pencil." We often want to restrict our attention to continuous functions because they are well behaved (they move "continuously" without breaking or jumping). In this section we will use limits to develop a mathematically precise definition of continuity.

2.6.1 Continuity at a point

Let's start by thinking about what we intuitively want a "continuous" function to look like, and then use limits to define continuity mathematically. We want a continuous function to be one whose graph has no breaks, jumps, or holes. The very simple function f graphed in Figure 2.37 seems to have this property.

Figure 2.37
$y = f(x)$

To understand better what we want to mean by continuous, we will consider some examples of functions that are *not* continuous. The following three graphs are such examples.

Figure 2.38
$y = g(x)$

Figure 2.39
$y = h(x)$

Figure 2.40
$y = k(x)$

Each of these graphs has a **discontinuity** at $x = 1$. The graph of g has two pieces that do not match up at $x = 1$. The graphs of h and k have a hole taken out of them at $x = 1$. None of these three graphs could be drawn without "picking up your pencil."

How can we mathematically quantify what is happening in each graph? Let's investigate the limits of each of these functions as $x \to 1$.

EXAMPLE 2.48

Examining limits at breaks or holes in three graphs

For each function in Figures 2.37–2.40, determine the left, right, and two-sided limits as $x \to 1$, and the value of the function at $x = 1$.

Solution   The following table summarizes the desired information. In the table, DNE stands for "does not exist."

Function	Left Limit	Right Limit	Limit	Value
$f(x)$	$\lim\limits_{x \to 1^-} f(x) = 1$	$\lim\limits_{x \to 1^+} f(x) = 1$	$\lim\limits_{x \to 1} f(x) = 1$	$f(1) = 1$
$g(x)$	$\lim\limits_{x \to 1^-} g(x) = 1$	$\lim\limits_{x \to 1^+} g(x) = 2$	$\lim\limits_{x \to 1} g(x)$ DNE	$g(1) = 2$
$h(x)$	$\lim\limits_{x \to 1^-} h(x) = 1$	$\lim\limits_{x \to 1^+} h(x) = 1$	$\lim\limits_{x \to 1} h(x) = 1$	$h(1) = 2$
$k(x)$	$\lim\limits_{x \to 1^-} k(x) = 1$	$\lim\limits_{x \to 1^+} k(x) = 1$	$\lim\limits_{x \to 1} k(x) = 1$	$k(1)$ DNE

Table 2.5

What can we learn from the calculations in Example 2.48? Take a minute to look at the last two columns of Table 2.5. Remember that the first function in the table is one we wish to consider "continuous" at $x = 1$, whereas the last three functions are examples of functions we think should be called "discontinuous" at $x = 1$.

? Question   Think about this for a minute. What happens in the last two columns for f that doesn't happen in the last two columns for g, h, and k?

Note that $\lim\limits_{x \to 1} f(x)$ and $f(1)$ are both equal to 1. This means that the *limit* that $f(x)$ approaches as $x \to 1$ is equal to the *value* of $f(x)$ at $x = 1$. In other words, $f(x) \to f(1)$ as $x \to 1$. This is not true for the functions g, h, and k. For example, $g(x) \not\to g(1)$ as $x \to 1$, since $\lim\limits_{x \to 1} g(x)$ does not exist. Our definition of **continuity at a point** will measure exactly this property: If a function approaches the value $f(c)$ as $x \to c$, we will say that f is continuous at $x = c$.

DEFINITION 2.14

Continuity of a Function at a Point

A function $f(x)$ is **continuous at $x = c$** if

$$\lim_{x \to c} f(x) = f(c).$$

⚡ Caution   To see whether a function f is continuous at $x = c$, we must check three things: First, the function must be defined at $x = c$ (i.e., $f(c)$ must be a real number); second, the limit of $f(x)$ as $x \to c$ must exist (i.e., $\lim\limits_{x \to c} f(x)$ must be a real number); and third, $f(c)$ and $\lim\limits_{x \to c} f(x)$ must be equal. Note in particular that if the limit and the function both fail to exist at $x = c$, then they are *not* considered to be equal.

EXAMPLE 2.49

Using limits to determine whether a function is continuous at a point

Use Definition 2.14 to show that the function f in Figure 2.37 is continuous at $x = 1$, but the functions g, h, and k in Figures 2.38–2.40 are not continuous at $x = 1$.

Solution   The function $f(x)$ from Figure 2.37 is continuous at $x=1$ because $\lim_{x\to1} f(x)=1$ and $f(1)=1$ are equal. On the other hand, the function $g(x)$ from Figure 2.38 is *not* continuous at $x=1$ because $\lim_{x\to1} g(x)$ does not exist. The function $h(x)$ has $\lim_{x\to1} h(x) \neq h(1)$, and the function $k(x)$ does not exist at $x=1$, so by Definition 2.14, neither $h(x)$ nor $k(x)$ is continuous at $x=1$. $\square$

2.6.2 Left and right continuity

As we saw in the last section, a function is continuous at $x=c$ if it approaches its value $f(c)$ as $x \to c$. If a function approaches its value $f(c)$ as $x \to c$ from the left, we will say that $f(x)$ is **left continuous** at $x=c$. Similarly, a function is **right continuous** at $x=c$ if it approaches its value from the right.

DEFINITION 2.15

Left and Right Continuity at a Point

A function $f(x)$ is **left continuous at $x=c$** if

$$\lim_{x\to c^-} f(x) = f(c).$$

A function $f(x)$ is **right continuous at $x=c$** if

$$\lim_{x\to c^+} f(x) = f(c).$$

A function is continuous at a point $x=c$ if and only if it is both left and right continuous at $x=c$. You will prove this fact in the exercises, and we state it as the following theorem.

THEOREM 2.14

Continuity Is Equivalent to Having Both Left and Right Continuity

A function f is continuous at $x=c$ if and only if it is left continuous at $x=c$ and right continuous at $x=c$.

EXAMPLE 2.50   **Using limits to determine left and right continuity at a point**

Determine the left and right continuity of the functions f, g, h, and k from Example 2.48 at the point $x=1$.

Solution   The function $f(x)$ from Figure 2.37 is both left and right continuous at $x=1$ because it is continuous at $x=1$. From Table 2.5 (or from the graph), we see that $\lim_{x\to1^-} f(x)=f(1)$ and $\lim_{x\to1^+} f(x)=f(1)$.

The function $g(x)$ in Figure 2.38 is right continuous but not left continuous at $x=1$, since $\lim_{x\to1^+} g(x)=2$ is equal to $g(1)=2$, but $\lim_{x\to1^-} g(x)=1$ is not equal to $g(1)=2$.

The functions $h(x)$ and $k(x)$ from Figures 2.39 and 2.40 are neither left nor right continuous at $x=1$. The left and right limits of $h(x)$ as $x \to 1$ are both equal to 1, whereas $h(1)=2$. The left and right limits of $k(x)$ as $x \to 1$ are both equal to 1, but the function $k(x)$ is not defined at $x=1$. $\square$

Sometimes it is convenient to talk about the continuity of a function on an interval. We say that a function is continuous on an open interval if it is continuous at each point in the interval. We say that a function is continuous on a closed or half-closed interval if it is continuous on the interior of the interval and, in addition, is right or left continuous at the left or right closed end of the interval, as described in the following definition.

DEFINITION 2.16

Continuity on Open, Half-Closed, and Closed Intervals

A function $f(x)$ is **continuous on (a, b)** if it is continuous for every $c \in (a, b)$.

A function $f(x)$ is **continuous on $[a, b)$** if it is continuous for all $c \in (a, b)$ and right continuous at a.

A function $f(x)$ is **continuous on $(a, b]$** if it is continuous for all $c \in (a, b)$ and left continuous at b.

A function $f(x)$ is **continuous on $[a, b]$** if it is continuous for all $c \in (a, b)$, right continuous at a, and left continuous at b.

⊗ *Caution*   Notice that a function $f(x)$ is continuous on $[a, b)$ if it is continuous on the open interval (a, b) and *right* continuous at the *left* endpoint $x = a$. This is because as we approach the left-hand side of the interval $[a, b)$ from its interior, we are approaching $x = a$ from the right.   □

We will use the notion of continuity on an interval when we are interested in looking at functions defined on or restricted to an interval. This will be especially important in Section 2.7. The four types of interval continuity are illustrated in Figures 2.41–2.44.

Figure 2.41
Continuous
on $(1, 3)$

Figure 2.42
Continuous
on $[1, 3)$

Figure 2.43
Continuous
on $(1, 3]$

Figure 2.44
Continuous
on $[1, 3]$

2.6.3 Types of discontinuities

When a function is not continuous at a point $x = c$, we say that it is **discontinuous at $x = c$**. In this section we identify the three basic types of discontinuities that a function can have.

By definition, a function $f(x)$ is continuous at $x = c$ if $\lim_{x \to c} f(x) = f(c)$. The contrapositive of this says: If $f(x)$ is discontinuous at $x = c$, then we must have $\lim_{x \to c} f(x) \neq f(c)$ (which may mean that one or both of $f(c)$ and $\lim_{x \to c} f(x)$ fail to exist). The simplest type of discontinuity happens when the limit of $f(x)$ as $x \to c$ exists but is not equal to $f(c)$. We call this type of discontinuity a **removable discontinuity,** since if we could change just one value of f—namely $f(c)$—we could "remove" the discontinuity and make f continuous. (Of course, we aren't allowed to change any of the values of f. However, we can measure how "bad" a discontinuity is by determining how much work it would take to "fix" it. Removable discontinuities don't take much work to "fix.")

? *Question*   If f has a removable discontinuity at $x = c$, then changing one value of $f(x)$ will make the function continuous. The value we would have to change is $f(c)$. What would we have to change it to?   □

On the other hand, $f(x)$ could be discontinuous at $x = c$ because $\lim_{x \to c} f(x)$ does not exist. There are two basic ways that this limit could fail to exist: The first is if the left and right limits exist but are not equal to each other; the second is if either the left or the right limit does not exist. In the first case we say that $f(x)$ has a ***jump discontinuity*** at $x = c$. If either the left or the right limit is infinite at $x = c$, we say that $f(x)$ has an ***infinite discontinuity.***

DEFINITION 2.17

Removable, Jump, and Infinite Discontinuities

Suppose $f(x)$ is discontinuous at $x = c$.

(a) If $\lim\limits_{x \to c} f(x)$ exists, then $x = c$ is a ***removable discontinuity*** of f.

(b) If $\lim\limits_{x \to c^-} f(x)$ and $\lim\limits_{x \to c^+} f(x)$ both exist but are not equal, then $x = c$ is a ***jump discontinuity*** of f.

(c) If one or both of $\lim\limits_{x \to c^-} f(x)$ and $\lim\limits_{x \to c^+} f(x)$ are infinite, then $x = c$ is an ***infinite discontinuity*** of f.

EXAMPLE 2.51

Graphs and limits for removable, jump, and infinite discontinuities

Figures 2.45–2.47 illustrate these three types of discontinuities.

Figure 2.45
Removable
discontinuity

Figure 2.46
Jump discon-
tinuity

Figure 2.47
Infinite discon-
tinuity

The graph $y = f(x)$ in Figure 2.45 has a removable discontinuity since $\lim\limits_{x \to 2} f(x)$ exists (but is not equal to $f(2) = 2$). Changing one value of f would remove this discontinuity; specifically, we would have to change the value of $f(2)$ so that it was equal to the limit of the function as $x \to 2$. In other words, the function

$$g(x) = \begin{cases} f(x), & \text{if } x \neq 2 \\ 1, & \text{if } x = 2 \end{cases}$$

(which is equal to $f(x)$ except at one point) would be continuous.

The graph $y = f(x)$ in Figure 2.46 has a jump discontinuity at $x = 2$ because $\lim\limits_{x \to 2^-} f(x)$ and $\lim\limits_{x \to 2^+} f(x)$ both exist but are not equal. This type of discontinuity could not be fixed by moving just one value of $f(x)$; however, we *could* "fix" it by shifting the right-hand piece of the graph down by one unit.

Finally, the graph $y = f(x)$ in Figure 2.47 has an infinite discontinuity at $x = 2$ because the limit of $f(x)$ as $x \to 2$ is infinite. This kind of discontinuity is the "worst" of the three types, since the function grows without bound as x gets closer to 2, and thus the discontinuity could not be "fixed" even by shifting half of the graph up or down. ☐

◈ *Caution* Definition 2.17 doesn't list *all* the possible types of discontinuities a function could have. For example, the function $f(x) = \sin(\frac{1}{x})$ has a discontinuity at $x = 0$ but does not have a removable, jump, or infinite discontinuity (see Figure 2.10 in Section 2.1.6). Instead, the function $f(x) = \sin(\frac{1}{x})$ oscillates faster and faster between $y = 1$ and $y = -1$ as $x \to 0$. ☐

Another type of discontinuity can occur when a function is defined separately on the rational and the irrational numbers. The traditional way to graph such functions is to use a dotted line to represent the values of the function at rational values and a somewhat thicker dotted line to represent the values of the function at the irrational numbers. (This reflects the fact that in some sense the irrational numbers are more "dense" in the real number line than the rational numbers.) To examine the limit of such a function, we must look at $\lim_{x \to c} f(x)$ for rational values of x separately from $\lim_{x \to c} f(x)$ for irrational values of x. If the limit of $f(x)$ as $x \to c$ is the same whether we choose rational or irrational values of x, then the limit exists. Otherwise, the limit does not exist.

EXAMPLE 2.52　**Continuity of functions defined separately for rational and irrational numbers**

Determine graphically whether the functions f and g in Figures 2.48 and 2.49 are continuous at $x = 0$.

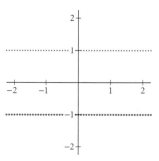

Figure 2.48
$$f(x) = \begin{cases} 1, & \text{if } x \text{ rational} \\ -1, & \text{if } x \text{ irrational} \end{cases}$$

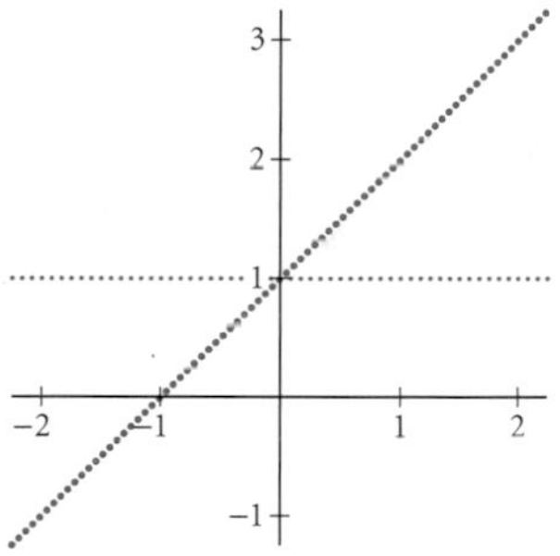

Figure 2.49
$$g(x) = \begin{cases} 1, & \text{if } x \text{ rational} \\ x + 1, & \text{if } x \text{ irrational} \end{cases}$$

Solution　The function $f(x)$ is equal to 1 for all rational values of x and equal to -1 for all irrational values of x. We will consider the limit of $f(x)$ as $x \to 0$ separately for rational and irrational values of x (the "thin" dotted line and the "thick" dotted line, respectively):

$$\text{for rational values of } x: \quad \lim_{x \to 0} f(x) = 1,$$
$$\text{for irrational values of } x: \quad \lim_{x \to 0} f(x) = -1.$$

Since the limit of $f(x)$ as $x \to 0$ is different depending on whether we choose rational or irrational values of x, the limit does not exist. Therefore, $\lim_{x \to 0} f(x)$ does not exist, whereas $f(0) = 1$ (since $x = 0$ is rational). The function f is not continuous at $x = 0$.

On the other hand, looking at the graph of g we see that

$$\text{for rational values of } x: \quad \lim_{x \to 0} g(x) = 1,$$
$$\text{for irrational values of } x: \quad \lim_{x \to 0} g(x) = 1.$$

Therefore, $\lim_{x \to 0} g(x)$ exists and is equal to 1. Since $g(0)$ is also equal to 1, the function $g(x)$ is continuous at $x = 0$. (Note that this is the *only* value at which $g(x)$ is continuous!)　□

？ *Question* Explain in terms of the formal definition of limit (i.e., with δ and epsilon) why in Example 2.52 the function g is continuous whereas the function f is not. (*Hint:* What can you say about $\lim_{x \to 1} f(x)$ and $\lim_{x \to 1} g(x)$ in terms of the formal delta–epsilon definition of limit?)　□

2.6.4 Proving or algebraically checking for continuity

So far our treatment of continuity has been mostly graphical in nature. In this section we will use algebraic limit calculations to determine whether a function is continuous at a given point. We also discuss the delta–epsilon definition of continuity.

EXAMPLE 2.53

Using limits to detect continuity

Use limits to determine whether the piecewise function $f(x)$ is continuous at $x = 1$. If $f(x)$ is discontinuous at $x = 1$, what type of discontinuity does it have?

$$f(x) = \begin{cases} x+1, & \text{if } x < 1 \\ 3-x^2, & \text{if } x \geq 1 \end{cases}$$

Solution   By Definition 2.14, we must see whether $\lim_{x \to 1} f(x)$ is equal to $f(1)$. Clearly we have $f(1) = 3 - (1)^2 = 2$. Since $x = 1$ is the "break" point of the piecewise function $f(x)$, we must calculate the limits from the left and right separately:

$$\lim_{x \to 1^-} f(x) = \lim_{x \to 1^-} (x+1) = 1 + 1 = 2, \quad \text{and}$$

$$\lim_{x \to 1^+} f(x) = \lim_{x \to 1^+} (3 - x^2) = 3 - (1)^2 = 2.$$

Since the left and right limits both exist and are equal to 2, we have $\lim_{x \to 1} f(x) = 2$, which is equal to the value $f(1) = 2$. Therefore, f is continuous at $x = 2$.   □

✓ *Checking the Answer*   We can verify the answer to Example 2.53 by sketching the graph of $f(x)$ (see Figure 2.50). The graph of this function approaches the value $f(1) = 2$ from both the right and the left as $x \to 1$.

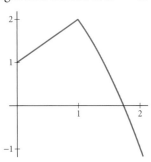

Figure 2.50   □

EXAMPLE 2.54

Using limits to detect a jump discontinuity

Determine whether the piecewise function $f(x)$ is continuous at $x = 2$. If $f(x)$ is discontinuous at $x = 2$, what type of discontinuity does it have? Is $f(x)$ right continuous at $x = 2$? Left continuous?

$$f(x) = \begin{cases} x^3 - 4x^2 + 3x + 1, & \text{if } x < 2 \\ 3, & \text{if } x \geq 2 \end{cases}$$

Solution   Once again we must calculate the left and right limits of $f(x)$ separately:

$$\lim_{x \to 2^-} f(x) = \lim_{x \to 2^-} (x^3 - 4x^2 + 3x + 1)$$

$$= (2)^3 - 4(2)^2 + 3(2) + 1$$

$$= 8 - 16 + 6 + 1 = -1, \quad \text{and}$$

$$\lim_{x \to 2^+} f(x) = \lim_{x \to 2^+} 3 = 3.$$

Since the left and right limits both exist but are not equal, the limit of $f(x)$ as $x \to 2$ does not exist (and thus can't possibly be equal to $f(2) = 3$). Therefore, $f(x)$ is discontinuous at $x = 2$. By Definition 2.17, the function $f(x)$ has a jump discontinuity at $x = 2$. Since $\lim\limits_{x \to 2^+} f(x)$ and $f(2)$ are equal (to 3), the function f is right continuous at $x = 2$. On the other hand, $\lim\limits_{x \to 2^-} f(x) = 2^3 - 4(2)^2 + 3(2) + 1 = -1$ is not equal to $f(2) = 3$, so f is not left continuous at $x = 2$. □

EXAMPLE 2.55

Using limits to detect an infinite discontinuity

Determine whether the function $f(x) = \frac{x}{x-1}$ is continuous at $x = 1$. If the function fails to be continuous, what type of discontinuity does it have?

Solution We must calculate the limit of $f(x)$ as $x \to 1$. We have

$$\lim_{x \to 1} \frac{x}{x-1} \to \frac{1}{0}, \quad \text{which does not exist.}$$

We will calculate the right and left limits separately:

$$\lim_{x \to 1^-} \frac{x}{x-1} = \frac{1^-}{(1^-)-1} = \frac{1^-}{0^-} \to -\infty,$$

$$\lim_{x \to 1^+} \frac{x}{x-1} = \frac{1^+}{(1^+)-1} = \frac{1^+}{0^+} \to \infty.$$

Therefore, $f(x)$ has an infinite discontinuity at $x = 1$. Its graph is shown in Figure 2.51.

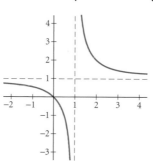

Figure 2.51 □

EXAMPLE 2.56

Calculating limits separately for rational and irrational values

Determine algebraically whether or not the function $g(x) = \begin{cases} 1, & \text{if } x \text{ rational} \\ x+1, & \text{if } x \text{ irrational} \end{cases}$ is continuous at $x = 0$.

Solution As in Example 2.52, we calculate limits of g separately for rational and irrational values of x.

for rational values of x: $\quad \lim\limits_{x \to 0} g(x) = \lim\limits_{x \to 0} 1 = 1,$

for irrational values of x: $\quad \lim\limits_{x \to 0} g(x) = \lim\limits_{x \to 0} (x+1) = 0 + 1 = 1.$

Since $g(x) \to 1$ as $x \to 0$, regardless of whether the x-values are rational or irrational, we have $\lim\limits_{x \to 0} g(x) = 1$. Since this is equal to $g(0) = 1$, the function g is continuous at $x = 0$. (This is the function whose limit we found graphically in Figure 2.49 in Example 2.52; the algebraic process above just repeats our graphical reasoning from Example 2.52.) □

Since the definition of continuity concerns a limit statement, we can write the definition of continuity in terms of the formal definition of limit (i.e., in terms of delta and epsilon). Since by definition a function $f(x)$ is continuous at $x = c$ if $\lim_{x \to c} f(x) = f(c)$, the delta–epsilon definition of continuity is as follows:

THEOREM 2.15

The Delta–Epsilon Interpretation of Continuity

Suppose $f(x)$ is a function and c is a real number. Then $f(x)$ is continuous at $x = c$ if and only if

For all $\varepsilon > 0$, there exists a $\delta > 0$ such that

if $0 < |x - c| < \delta$, then $|f(x) - f(c)| < \varepsilon$.

The proof of this theorem consists of writing down the limit statement in Definition 2.14 in terms of the formal definition of limit. It is left to you in the exercises.

Theorem 2.15 can be loosely interpreted as saying that if $f(x)$ is continuous at a point $x = c$, then small changes in x (near c) result in small changes in $f(x)$. This means that near $x = c$, the function $f(x)$ varies in a continuous way as x changes; it doesn't suddenly jump or break.

EXAMPLE 2.57 **Expressing continuity at a point as a delta–epsilon statement**

If you wanted to prove that the function $f(x) = \frac{1}{x}$ is continuous at $x = 3$, what delta–epsilon statement would you have to prove?

Solution   To prove that $f(x)$ is continuous at $x = 3$, we have to show that $\lim_{x \to 3} \frac{1}{x} = \frac{1}{3}$ (since $f(3) = \frac{1}{3}$). This means that we would have to prove the delta–epsilon statement

(1)   *For all $\varepsilon > 0$, there exists a $\delta > 0$ such that*

if $0 < |x - 3| < \delta$, then $\left|\frac{1}{x} - \frac{1}{3}\right| < \varepsilon$.

We did this delta–epsilon proof in Example 2.24 in Section 2.3.3 (page 162).   □

2.6.5 Continuous functions

We have seen what it means for a function to be continuous at a point or on an interval. In this section we examine functions that are continuous everywhere they are defined.

DEFINITION 2.18

Continuous Functions

A function $f(x)$ is called a ***continuous function*** if it is continuous on its domain.

❦ ***Caution***   We will see in Chapter 4 that the function $f(x) = \sqrt{x}$ is continuous on its domain $[0, \infty)$. This means that $f(x)$ is only *right* continuous at $x = 0$ (since this is what it means to be continuous at the closed end of an interval).   □

EXAMPLE 2.58

An example of a noncontinuous function

The function $h(x)$ in Figure 2.39 (page 190) is *not* a continuous function, since $x = 1$ is in the domain of $h(x)$ but $h(x)$ is not continuous at $x = 1$. □

⊗ Caution Definition 2.18 is a little tricky, because the graph of a continuous function $f(x)$ *can* actually have a jump or break at some point, as long as that point is not in the domain of $f(x)$. For example, the function $f(x) = \frac{1}{x}$ fails to be continuous at $x = 0$ (it has a vertical asymptote there), but the point $x = 0$ is not in the domain of $f(x)$. The function $f(x) = \frac{1}{x}$ is continuous at every point in its domain, and therefore we consider $f(x) = \frac{1}{x}$ to be a continuous function (even though it is not continuous at $x = 0$). □

EXAMPLE 2.59

A function that is continuous everywhere it is defined

The function $k(x)$ in Figure 2.40 (page 190) is a continuous function. Even though k is not continuous at the point $x = 1$, the point $x = 1$ is not in the domain of k. For every point c in the domain $\{x \mid x \neq 1\}$ of k, we do have $\lim_{x \to c} k(x) = k(c)$. □

The point of Example 2.59 is that if c is not in the domain of a function $f(x)$, then we don't care whether or not $f(x)$ is continuous at $x = c$; we're content to consider a function continuous if it is continuous on its domain.

If a function f is continuous at each point in its domain, then by definition $\lim_{x \to c} f(x) = f(c)$ for all $c \in \text{Domain}(f)$. In other words, if f is a continuous function, then for every point in the domain of f, the limit of f as $x \to c$ is equal to the value $f(c)$ of f at $x = c$. This means that we can calculate limits of continuous functions (at points in their domains) simply by "plugging in"! The discussion above proves the following theorem.

THEOREM 2.16

Limit Interpretation of a Continuous Function

If f is a continuous function, then $\lim_{x \to c} f(x) = f(c)$ for all $c \in \text{Domain}(f)$.

During our exploration of limits, we have already seen two types of functions whose limits we can calculate by evaluating at $x = c$ (i.e., two types of functions that are continuous). The following two theorems are direct consequences of Theorems 2.5 and 2.10 in Section 2.4, respectively.

THEOREM 2.17

Every Linear Function Is Continuous

If m and b are any constants, then the function $f(x) = mx + b$ is continuous.

THEOREM 2.18

Every Positive Integer Power Function Is Continuous

If k is a positive integer, then the function $f(x) = x^k$ is continuous.

We can use the limit rules from Section 2.4 to prove that constant multiples, sums, products, and quotients of continuous functions are also continuous.

THEOREM 2.19

Algebraic Combinations of Continuous Functions Are Continuous Functions

If f and g are continuous on their domains, and k is any constant, then the functions kf, $f + g$, $f - g$, fg, and $\frac{f}{g}$ are continuous on their domains.

The proof of this theorem follows directly from Theorems 2.6, 2.7, 2.8, and 2.9 in Section 2.4 and is left to you in the exercises. For example, if f and g are continuous functions, then they are continuous at each point in their domains; to prove that $f + g$ is continuous, you have to use Theorem 2.7 to show that $f + g$ is continuous at each point of its domain, Domain(f) ∩ Domain(g).

❓ *Question*   What do Theorems 2.17, 2.18, and 2.19 have to do with calculating limits of sums, constant multiples, products, and quotients of linear and positive integer power functions? (Think about Algorithm 2.1 in Section 2.5.)   □

EXAMPLE 2.60

Using continuity to calculate limits

Use continuity (Theorems 2.17, 2.18, and 2.19) to calculate $\lim\limits_{x \to -3} \dfrac{x^2}{3(x + 2)}$.

Solution   The linear function $3(x + 2) = 3x + 6$ and the power function x^2 are continuous by Theorems 2.17 and 2.18. The function $f(x) = \frac{x^2}{3(x+2)}$ is a quotient of two continuous functions. Since a quotient of continuous functions is itself continuous, the function $f(x)$ is continuous. Therefore, we know that $\lim\limits_{x \to -3} f(x)$ must be equal to $f(-3)$, provided that $x = -3$ is in the domain of $f(x)$:

$$\lim_{x \to -3} \frac{x^2}{3(x + 2)} = \frac{(-3)^2}{3(-3 + 2)} = \frac{9}{3(-1)} = -3.$$

Of course, we could have calculated this limit using Algorithm 2.1 in Section 2.5. Actually, this is exactly what we did, but we did so using the language of continuity.   □

You may have noticed that we have limit rules for constant multiples, sums, products, and quotients of functions, but not for compositions of functions. Now that we have the notion of continuity, we can state the limit rule for calculating limits of continuous compositions.

THEOREM 2.20

The Limit of a Composition

If $\lim\limits_{x \to c} g(x) = L$ for some real number L, and if f is a function that is continuous at L, then

$$\lim_{x \to c} f(g(x)) = f\left(\lim_{x \to c} g(x) \right) = f(L).$$

The proof of Theorem 2.20 uses the delta–epsilon definition of continuity from Theorem 2.15 and can be found in the appendix at the end of this chapter (Section 2.8). This theorem will enable us to calculate a number of limits that we were not able to calculate previously and will be very important when we investigate more complicated functions in subsequent chapters.

EXAMPLE 2.61

Calculating a limit of a composition of functions

Use Theorem 2.20 to calculate $\lim_{x \to 2}(3x^2 - 4x + 2)^{12}$.

Solution   Without Theorem 2.20 we would have to multiply out the function $f(x) = (3x^2 - 4x + 2)^{12}$ so that it would be written in a form where we could apply Algorithm 2.1 from Section 2.5. Clearly this would be very tedious! However, the function $f(x)$ is a composition $f(x) = h(g(x))$ of the functions $g(x) = 3x^2 - 4x + 2$ and $h(x) = x^{12}$. Since h is a positive integer power function and thus is continuous (at every point), by Theorem 2.20 we have

$$\lim_{x \to 2}(3x^2 - 4x + 2)^{12} = \left(\lim_{x \to 2}(3x^2 - 4x + 2)\right)^{12}$$
$$= (3(2)^2 - 4(2) + 2)^{12}$$
$$= 6^{12} = 2,176,782,336. \qquad \square$$

EXAMPLE 2.62

Proving that limits and absolute values commute

In the exercises (Problem 69) you will prove that the function $f(x) = |x|$ is continuous. Use this fact to prove that for any function $g(x)$ that is continuous at the point $x = c$,

(2)
$$\lim_{x \to c}|g(x)| = \left|\lim_{x \to c}g(x)\right|.$$

Solution   Given that $f(x) = |x|$ is continuous, the proof is just a simple application of Theorem 2.20.

PROOF   By Theorem 2.20, we have

$$\lim_{x \to c}|g(x)| = \lim_{x \to c}f(g(x)) \qquad (f(x) = |x|)$$
$$= f\left(\lim_{x \to c}g(x)\right) \qquad \text{(Theorem 2.20)}$$
$$= \left|\lim_{x \to c}g(x)\right|. \qquad (f(x) = |x|) \qquad \blacksquare$$

We can also phrase Theorem 2.20 in terms of continuity to determine when a composition of two functions is continuous.

THEOREM 2.21

Continuity of a Composition of Functions at a Point

If g is continuous at c, and f is continuous at $g(c)$, then $f \circ g$ is continuous at c.

The proof of this theorem is just a rewording of Theorem 2.20; it is left to you in the exercises.

❓ Question   Note that unlike Theorem 2.19, in Theorem 2.21 we have to talk about the continuity of $f \circ g$ one point at a time (instead of over its entire domain). Why? $\qquad \square$

What you should know after reading Section 2.6

▶ What does it mean for a function to be continuous at a point? Left or right continuous at a point? Continuous on an open, closed, or half-closed interval? Continuous as a function? Know the graphical interpretations as well as the mathematical definitions of these types of continuity.

▶ What are the three basic types of discontinuities? Be able to recognize them graphically as well as algebraically.

▶ How to determine algebraically whether a function is continuous at a point. The delta–epsilon interpretation of continuity, and how to prove that a function is continuous using this interpretation.

▶ What types of functions do we know are continuous? What can you say about calculating the limits of such functions? Be able to prove that combinations (sums, products, etc.) of continuous functions are continuous (using limit rules). Know how to calculate the limit of a composition and how to tell where a composite function is continuous.

Exercises 2.6

Concepts

0. Read the section and make your own summary of the material. Include whatever you think will help you review the material later on.

1. Explain in a sentence what it means for a function $f(x)$ to be continuous at a point $x = c$. Your sentence should include the words *approaches* and *value*. Then write similar sentences describing left and right continuity at $x = c$.

2. Sketch three functions that are discontinuous at $x = -2$: one with a removable discontinuity, one with a jump discontinuity, and one with an infinite discontinuity.

3. Write down formulas for three functions that are discontinuous at $x = 3$: one with a removable discontinuity, one with a jump discontinuity, and one with an infinite discontinuity.

■ Given the information below, determine whether f has a removable discontinuity, a jump discontinuity, or an infinite discontinuity at $x = 2$. Also determine if f is left or right continuous at $x = 2$.

4. $\lim\limits_{x \to 2^-} f(x) = 2$, $\lim\limits_{x \to 2^+} f(x) = 1$, $f(2) = 1$

5. $\lim\limits_{x \to 2^-} f(x) = 2$, $\lim\limits_{x \to 2^+} f(x) = 1$, $f(2) = 3$

6. $\lim\limits_{x \to 2^-} f(x) = 2$, $\lim\limits_{x \to 2^+} f(x) = 2$, $f(2) = 3$

7. $\lim\limits_{x \to 2^-} f(x) = -\infty$, $\lim\limits_{x \to 2^+} f(x) = \infty$, $f(2)$ DNE

■ For each problem below, sketch the graph of a function f that has the properties listed.

8. $\lim\limits_{x \to 2^-} f(x) = 5$, $\lim\limits_{x \to 2^+} f(x) = 3$, $f(2) = 5$

9. $\lim\limits_{x \to -1^-} f(x) = 2$, $\lim\limits_{x \to -1^+} f(x) = 2$, $f(-1) = 1$

10. $\lim\limits_{x \to 0^-} f(x) = -1$, $\lim\limits_{x \to 0^+} f(x) = 1$, $f(0) = 0$

11. $\lim\limits_{x \to 1^-} f(x) = 3$, $\lim\limits_{x \to 1^+} f(x) = 3$, $f(1) = 3$

■ For each problem below, sketch a function $f(x)$ with the properties listed, if possible. If it is not possible for such a function to exist, explain why.

12. f is left continuous at $x = 2$ but not continuous at $x = 2$, and $f(2) = 3$.

13. f is left continuous at $x = 1$ and right continuous at $x = 1$ but is not continuous at $x = 1$, and $f(1) = -2$.

14. f has an infinite discontinuity at 0 but is right continuous at 0, and $f(0) = 1$.

15. f has a jump discontinuity at $x = -1$ and is left continuous at $x = -1$, and $f(-1) = 2$.

16. f has a removable discontinuity at $x = -2$ and is right continuous at $x = -2$, and $f(-2) = 0$.

17. f is continuous on $[0, 2)$ but not on $[0, 2]$.

18. f is continuous at each point in its domain but fails to be continuous at $x = 3$.

■ Each function $f(x)$ given below is undefined at $x = 1$. Define $f(1)$ such that $f(x)$ is continuous at $x = 1$.

19. $f(x) = \dfrac{x^2 - 1}{x - 1}$

20. $f(x) = \dfrac{x^2 - 2x + 1}{x^2 - 6x + 5}$

21. $f(x) = \begin{cases} 3x - 1, & \text{if } x < 1 \\ x^2 + 1, & \text{if } x > 1 \end{cases}$

■ Use the formal definition of limit to express each of the following statements.

22. $f(x)$ is left continuous at $x = c$.

23. $f(x)$ is right continuous at $x = c$.

■ Fill in the blanks with the appropriate values.

24. If $f(x)$ is continuous at $x = 2$, then $f(x^2)$ is continuous at _____ .

25. If $f(x)$ is continuous at _____ , then $f(3x - 1)$ is continuous at $x = 1$.

26. If $g(x)$ is continuous at $x = 3$, then $(g(x))^3$ is continuous at _____ .

27. Define in terms of limits what it means for a function f to have a jump discontinuity at $x = 1$.

28. Define what it means for a function f to be continuous on the half-closed interval $[-2, 5)$.

29. If a function is continuous on $(0, 2]$ and also continuous on $(2, 3)$, is that function necessarily continuous on $(0, 3)$? If so, explain why. If not, explain why not, and sketch an example of a function that is continuous on $(0, 2]$ and $(2, 3)$ but not on $(0, 3)$.

30. True or False: If f is a continuous function, then it is continuous at every $c \in \mathbb{R}$. Justify your answer or provide a counterexample.

31. True or False: If f and g are continuous functions, then $\frac{f}{g}$ is continuous at each point in the set Domain(f) $\cap$ Domain(g). Justify your answer or provide a counterexample.

32. If $f(x)$ is a continuous function and $x = 1$ is in the domain of f, what can you say about $\lim\limits_{x \to 1} f(x)$?

33. What do Theorems 2.17–2.19 have to do with Algorithm 2.1 in Section 2.5? In other words, what do Theorems 2.17–2.19 have to do with calculating limits of sums, constant multiples, products, and quotients of linear and positive integer power functions?

34. In Example 2.56 we discussed a method for determining the limit of a function defined separately on the rational and irrational numbers. Use the definition of *limit* (with δ and ε) to explain why this method is valid.

Skills

■ For each function f given below, determine any values at which f is discontinuous. At each of these points, determine the type of discontinuity (removable, jump, or infinite) and whether f is left or right continuous at that point. On what intervals is f continuous?

35.

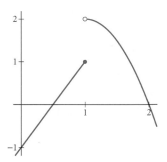

36.

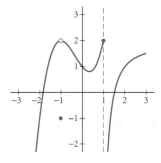

37.

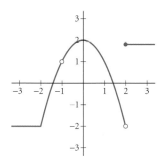

■ The three functions below are defined in terms of the sine function, which you can enter into your calculator using the button SIN (make sure your calculator is set to "radian" mode first). For each of these three functions, use a graphing calculator to examine the graph of f near $x = 0$. Does it appear that f is continuous at $x = 0$? You may need to "zoom in" near $x = 0$ to determine whether or not the limit of f as $x \to 0$ exists.

38. $f(x) = \begin{cases} \sin\left(\frac{1}{x}\right), & \text{if } x \neq 0 \\ 0, & \text{if } x = 0 \end{cases}$

39. $f(x) = \begin{cases} x \sin\left(\frac{1}{x}\right), & \text{if } x \neq 0 \\ 0, & \text{if } x = 0 \end{cases}$

40. $f(x) = \begin{cases} x^2 \sin\left(\frac{1}{x}\right), & \text{if } x \neq 0 \\ 0, & \text{if } x = 0 \end{cases}$

■ For each function f and point c given below, algebraically determine the values of $\lim\limits_{x \to c^-} f(x)$, $\lim\limits_{x \to c^+} f(x)$, $\lim\limits_{x \to c} f(x)$, and $f(c)$. Then use these values to determine whether f

is continuous at c and, if not, what type of discontinuity occurs.

41. $f(x) = x^2 - 3x + 5$, $c = 4$

42. $f(x) = \dfrac{x^2 - 2x - 3}{x - 3}$, $c = 3$

43. $f(x) = \dfrac{x - 1}{x^2 - 2x + 1}$, $c = 1$

44. $f(x) = \dfrac{-3x}{(x + 3)^2}$, $c = -3$

45. $f(x) = \dfrac{x + 2}{x^2 + 4x + 4}$, $c = -2$

46. $f(x) = \begin{cases} \frac{x^2 - 1}{x - 1}, & \text{if } x < 1 \\ 2, & \text{if } x = 1, \quad c = 1 \\ 5x - 1, & \text{if } x > 1 \end{cases}$

■ For each function $f(x)$ and point c given below, algebraically determine whether f is left continuous at c and whether f is right continuous at c.

47. $f(x) = \begin{cases} 3x + 2, & \text{if } x < -1 \\ 5 + 4x^3, & \text{if } x \geq -1 \end{cases}$, $c = -1$

48. $f(x) = \begin{cases} x^2 - 3x - 1, & \text{if } x \neq -2 \\ 3, & \text{if } x = -2 \end{cases}$, $c = -2$

49. $f(x) = \begin{cases} \frac{3x^2}{4 + x}, & \text{if } x < 2 \\ x^2 - 2, & \text{if } x \geq 2 \end{cases}$, $c = 2$

50. $f(x) = \begin{cases} x + 1, & \text{if } x < 3 \\ 2, & \text{if } x = 3, \quad c = 3 \\ x^2 - 9, & \text{if } x > 3 \end{cases}$

■ Use Theorem 2.19 to find the set of values on which the following functions are continuous. (In other words, recognize each function f as a combination of continuous functions, and use Theorem 2.19 and domain considerations to determine where f is continuous.)

51. $f(x) = 3x^2 + 2x - 10$

52. $f(x) = \dfrac{x^3}{4(x - 2)}$

53. $f(x) = x + \dfrac{x}{x + 1}$

■ Evaluate the following limits. Indicate where you use Theorem 2.20 (the limit of a composition).

54. $\lim_{x \to 1} (x^2 + x + 1)^3$

55. $\lim_{x \to 0} |3x^2 - 5|$

56. $\lim_{x \to -2} \dfrac{(x + 1)^7}{(x - 1)^3}$

■ Each of the functions below is defined separately on the rational and irrational numbers. For each function f and value c, determine algebraically whether or not f is

continuous at $x = c$. Then check your answers by sketching the graph of f.

57. $f(x) = \begin{cases} 2 - x, & \text{if } x \text{ rational} \\ x^2, & \text{if } x \text{ irrational} \end{cases}$, $c = 2$

58. $f(x) = \begin{cases} 2 - x, & \text{if } x \text{ rational} \\ x^2, & \text{if } x \text{ irrational} \end{cases}$, $c = 1$

59. $f(x) = \begin{cases} x^2 - 3, & \text{if } x \text{ rational} \\ 3x + 1, & \text{if } x \text{ irrational} \end{cases}$, $c = 0$

60. $f(x) = \begin{cases} x^2 - 3, & \text{if } x \text{ rational} \\ 3x + 1, & \text{if } x \text{ irrational} \end{cases}$, $c = 4$

Applications

61. As a vacuum cleaner salesman, Alex earns a salary of \$8500 a year whether he sells any vacuum cleaners or not. In addition, for every 30 vacuum cleaners he sells, he earns a \$500 commission.

    (a) Construct a piecewise function $M(v)$ that describes the amount of money M that Alex will make in a year if he sells v vacuum cleaners over the course of the year. (Assume that Alex sells between 0 and 100 vacuum cleaners in any given year.)

    (b) Check that your piecewise function makes sense by using it to calculate $M(0)$, $M(30)$, $M(59)$, $M(61)$, and $M(100)$. Then sketch a graph of $M(v)$ on the interval $0 \leq v \leq 100$.

    (c) The piecewise function $M(v)$ is *not* continuous. List all the values at which $M(v)$ fails to be continuous, and support your answers using the definition of continuity.

62. In Problem 51 in Section 1.4 (page 98), you constructed a piecewise function from the 2000 Federal Tax Rate Schedule. Specifically, you found that a person who makes m dollars a year will pay $T(m)$ dollars in tax, given by the formula

$$T(m) =$$
$$\begin{cases} 0.15m, & \text{if } 0 \leq m \leq 26{,}250 \\ 3{,}937.5 + 0.28(m - 26{,}250), & \text{if } 26{,}250 < m \leq 63{,}550 \\ 14{,}381.5 + 0.31(m - 63{,}550), & \text{if } 63{,}550 < m \leq 132{,}600 \\ 35{,}787 + 0.36(m - 132{,}600), & \text{if } 132{,}600 < m \leq 288{,}350 \\ 91{,}857 + 0.396(m - 288{,}350), & \text{if } m \geq 288{,}350. \end{cases}$$

    (a) Calculate each of the following:
    $T(63{,}550)$, $\displaystyle\lim_{m \to 63{,}550^-} T(m)$, and $\displaystyle\lim_{m \to 63{,}550^+} T(m)$.

    (b) Use the information in part (a) to argue that the function $T(m)$ is continuous at $m = 63{,}550$. What does this mean in real-world terms?

    (c) Repeat parts (a) and (b) for $m = 288{,}350$.

Proofs

63. Use the formal definition of limit and the definition of continuity at a point to prove Theorem 2.15.

64. Use what you know about left and right limits to prove that a function $f(x)$ is continuous at a point $x = c$ if and only if it is both left and right continuous at $x = c$.

65. Use a limit rule and the definition of continuity to prove that every linear function is continuous.

66. Use a limit rule and the definition of continuity to prove that every function of the form $f(x) = x^k$, where k is a positive integer, is continuous.

67. Use limit rules and the definition of continuity to prove that every quadratic function (that is, every function of the form $f(x) = ax^2 + bx + c$ for some constants a, b, and c) is continuous.

68. Use Theorem 2.15 to prove (with a delta–epsilon argument) that the function $f(x) = 3x - 5$ is continuous at $x = 2$.

69. Prove that the function $f(x) = |x|$ is continuous. You may find the following inequality useful: For any real numbers a and b, $||a| - |b|| < |a - b|$. (*Hint:* Your proof will involve deltas and epsilons. Why?)

70. Prove that if f and g are continuous functions, then $f + g$ is a continuous function. (*Hint:* Show that $f + g$ is continuous at each point of its domain. What is the domain of $f + g$?)

71. Prove that if f and g are continuous functions, then $\frac{f}{g}$ is a continuous function. (*Hint:* Show that $\frac{f}{g}$ is continuous at each point of its domain. What is the domain of $\frac{f}{g}$?)

72. Prove that if g is continuous at $x = c$, then $\lim_{x \to c} (g(x))^k = (g(c))^k$.

73. Use the formal "delta–epsilon" definition of continuity to prove that the function in Problem 57 is not continuous at $x = 2$.

74. Use the formal "delta–epsilon" definition of continuity to prove that the function in Problem 58 is continuous at $x = 1$.

2.7   Two Theorems About Continuous Functions

In this section we examine two important properties of continuous functions. First, a continuous function on a closed interval has maximum and minimum values on that interval. Second, a continuous function attains every intermediate value between any two of its values $f(a)$ and $f(b)$. These properties of continuous functions are called the *Extreme Value Theorem* and the *Intermediate Value Theorem*. Both theorems will be important in later chapters when we use them to prove theorems relating a function and its derivative, or rate of change. The Intermediate Value Theorem also provides us with a quick way of solving inequalities, which will be particularly useful in Sections 3.7 and 3.8.

2.7.1   The Extreme Value Theorem

The **Extreme Value Theorem** illustrates one of the most basic properties of continuous functions, namely that a continuous function on a closed interval must have a maximum and a minimum value. We will use the Extreme Value Theorem to prove an important theorem called Rolle's Theorem in Section 3.6.2.

THEOREM 2.22

The Extreme Value Theorem

If f is continuous on a closed interval $[a, b]$, then f has both a maximum and a minimum value on $[a, b]$.

Recall that a function f is said to be "continuous on a closed interval $[a, b]$" if it is continuous on the interior (a, b) of the interval, right continuous at $x = a$, and left continuous at $x = b$. The Extreme Value Theorem says that for such a function, there is always some $c \in [a, b]$ such that $f(c) \geq f(x)$ for all $x \in [a, b]$, and some $d \in [a, b]$ such that $f(d) \leq f(x)$ for all $x \in [a, b]$. In this notation, $f(c)$ is the maximum value of f on $[a, b]$, and $f(d)$ is the minimum value of f on $[a, b]$. Figure 2.52 shows the graph of a function f that is continuous on $[1, 4]$ and attains a maximum value of $f(2) = 2.5$ and a minimum value of $f(4) = 0.5$.

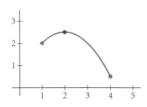

Figure 2.52

The Extreme Value Theorem says something that is intuitively obvious: If f is a continuous function on a closed interval $[a, b]$, then the graph of f continuously connects the points $(a, f(a))$ and $(b, f(b))$ and must have a "highest" and a "lowest" value somewhere along the way. Although the statement of the Extreme Value Theorem seems obvious, the proof of the Extreme Value Theorem is fairly difficult. To prove the Extreme Value Theorem, we would have to introduce a number of new definitions and concepts that we do not have the time to treat properly here (for example, the Least Upper Bound Axiom), and thus we omit the proof.

EXAMPLE 2.63   **A real-world example of the Extreme Value Theorem**

Consider the function $w(t)$ that describes a particular person's weight at t years of age. This should be a continuous function, since a person's weight changes continuously over time and cannot "jump" from one value to another. Consider this person's weight between the ages of 21 and 45. At some point in this time range, the person must have had a maximum weight. He or she must also have had a minimum weight. In other words, since $w(t)$ is continuous, the Extreme Value Theorem tells us that there is some time $c \in [21, 45]$ at which the person's weight was greatest over that time interval, and some point $d \in [21, 45]$ at which that person weighed the least.   ☐

⑤ Caution   The maximum and minimum values referred to in the Extreme Value Theorem are *global* maxima and minima on the interval $[a, b]$. They are not merely local maxima and minima on the interval $[a, b]$; they are the values at which the graph of f is highest or lowest on the *entire* interval $[a, b]$. Note also that these maximum and minimum values on $[a, b]$ may not be the global maximum or minimum values of f on its entire domain.   ☐

⑤ Caution   The Extreme Value Theorem is true only for functions that are continuous on a *closed* interval $[a, b]$. For example, if f is continuous on the half-closed interval $[a, b)$, we cannot say for certain that f has a maximum and a minimum value on the interval $[a, b)$. Sometimes such a function will have such maximum and minimum values, and sometimes it won't. The function f in Figure 2.53 has a maximum value on the interval $[1, 4)$, but it does *not* have a minimum value on that interval. (Why not?)

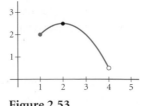

Figure 2.53   ☐

⑤ Caution   The Extreme Value Theorem is a conceptual, "existence" theorem rather than a computational theorem. It tells you that a continuous function on a closed interval attains a maximum and a minimum value, but it doesn't tell you where the maximum or minimum value is located.   ☐

2.7.2 The Intermediate Value Theorem

The Intermediate Value Theorem describes another very basic property of continuous functions, namely that a continuous function must take on every intermediate value between two of its values $f(a)$ and $f(b)$. This important theorem will be used to prove some very important theorems in Chapter 3, and a special case of the Intermediate Value Theorem will give us a simpler method of solving inequalities. We'll start with an intuitive example from "real life."

EXAMPLE 2.64

A real-world example of the Intermediate Value Theorem

Consider a watermelon dropped from a height of 100 feet. If the height of the watermelon is given by the function $s(t)$, and the watermelon takes 2.5 seconds to hit the ground, then $s(0) = 100$ and $s(2.5) = 0$. At time $t = 0$ seconds, the watermelon is 100 feet above the ground. At time $t = 2.5$ seconds, the watermelon is 0 feet above the ground. As the watermelon falls, it must take on every intermediate height between 100 feet and 0 feet during the time interval $[0, 2.5]$ (it can't "skip" any height along the way). For example, there must be some time $c \in (0, 2.5)$ where the height of the watermelon is $s(c) = 50$ feet above the ground. $\square$

The **Intermediate Value Theorem** tells us that a continuous function $y = f(x)$ on a closed interval $[a, b]$ attains all values between $f(a)$ and $f(b)$. In Example 2.64, the continuous function was the height function $s(t)$, with $a = 0$ and $b = 2.5$, and $s(t)$ took on all values between $s(a) = 100$ and $s(b) = 0$ on the interval $[0, 2.5]$. The values between $f(a)$ and $f(b)$ (such as 50, since $0 < 50 < 100$) are the "intermediate" y-values from which the Intermediate Value Theorem gets its name.

THEOREM 2.23

The Intermediate Value Theorem
If f is continuous on a closed interval $[a, b]$, then for any K between $f(a)$ and $f(b)$, there is at least one $c \in (a, b)$ such that $f(c) = K$.

When we say that K is "between" $f(a)$ and $f(b)$, we mean that either $f(a) < K < f(b)$ or $f(b) < K < f(a)$, depending on whether $f(a)$ or $f(b)$ is larger.

Like the Extreme Value Theorem, the Intermediate Value Theorem describes an intuitive and fundamental property of continuous functions but nonetheless is fairly difficult to prove. The proof of the Intermediate Value Theorem also relies on the Least Upper Bound Axiom of the real numbers. We omit the proof in this text and refer the reader to either an advanced calculus text or a set theory text.

Intuitively, however, the Intermediate Value Theorem seems to be pretty obvious. In order to connect the points $(a, f(a))$ and $(b, f(b))$ with a continuous graph, one must pass through all heights K between $f(a)$ and $f(b)$. See Figure 2.54 (which shows the height of the falling watermelon from Example 2.64).

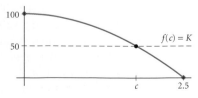

Figure 2.54

◆ *Caution* The Intermediate Value Theorem is another example of an "existence" theorem: It tells you there is some value $c \in (a, b)$ where $f(c) = k$, but it doesn't tell you *what* the value of c is. Also note that there may be more than one such value c in the interval (a, b). □

The Intermediate Value Theorem applies only to *continuous* functions; if a function is discontinuous, then its graph can "skip over" some heights. For example, consider the functions graphed in Figures 2.55 and 2.56.

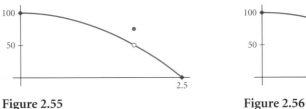

Figure 2.55　　　　　　　　　　　**Figure 2.56**

Neither of these functions ever has a y-value of 50. In other words, for either function f, there is no value $c \in (0, 2.5)$ for which $f(c) = 50$.

❷ *Question* Suppose the graph in Figure 2.55 shows the height of a watermelon t seconds after being dropped from a 100-foot building. Describe how the watermelon falls. Do the same thing for the graph in Figure 2.56. (*Note:* It is not possible for a watermelon to fall in either of these two ways in reality, so you will have to be creative!) □

EXAMPLE 2.65

Applying the Intermediate Value Theorem to a Continuous Function

Use the Intermediate Value Theorem to show that there is some point c at which the function $f(x) = x^3 - 3x + 1$ has the value $f(c) = 2$.

Solution The function $f(x) = x^3 - 3x + 1$ is continuous everywhere because it is a sum of continuous functions. To show that there is some c with $f(c) = 2$, we need to find some numbers a and b such that $K = 2$ is between $f(a)$ and $f(b)$ and apply the Intermediate Value Theorem. By trial and error, we can find such numbers a and b by looking at different values of f until we find a value that is less than 2 and a value that is greater than 2. For example,

$$f(0) = 0^3 - 3(0) + 1 = 1,$$
$$f(2) = 2^3 - 3(2) + 1 = 3.$$

Since f is continuous on $[0, 2]$ and $f(0) < 2 < f(2)$, by the Intermediate Value Theorem there is some value $c \in (0, 2)$ for which $f(c) = 2$. Note that the Intermediate Value Theorem doesn't tell us where c is, only that such a c exists somewhere in the interval $(0, 2)$. □

✔ *Checking the Answer* Figure 2.57 shows the graph of $f(x) = x^3 - 3x + 1$. Note that there is indeed a value $c \in (0, 2)$ for which $f(c) = 2$. In fact, we can use the graph to estimate this value as approximately $c \approx 1.9$.

Although it is easy to graphically approximate values of c with $f(c) = 2$, it would be difficult in this particular example to find such values of c exactly (since it is difficult to solve the equation $x^3 - 3x + 1 = 2$). However, the Intermediate Value Theorem allows us to say with certainty that at least one such c exists.

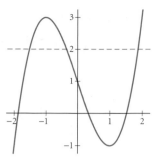

Figure 2.57 □

2.7.3 A special case of the Intermediate Value Theorem

A simple but extremely useful special case of the Intermediate Value Theorem says that if a continuous function is positive at $x = a$ and negative at $x = b$ (or vice versa), then there must be some point c between a and b at which f is zero.

THEOREM 2.24

Special Case of the Intermediate Value Theorem

If f is continuous on a closed interval $[a, b]$, then if $f(a)$ and $f(b)$ have opposite signs (one positive and one negative), there exists a point $c \in (a, b)$ such that $f(c) = 0$.

This is just the Intermediate Value Theorem in the special case where $K = 0$. Since $f(a)$ and $f(b)$ have opposite signs by hypothesis, the value $K = 0$ must be between $f(a)$ and $f(b)$.

Figure 2.58 illustrates Theorem 2.24. To get from $f(a)$, which is positive, to $f(b)$, which is negative, the graph must at some point c cross the x-axis. As you can see in Figure 2.59, there may be more than one value in (a, b) where f crosses the x-axis (in this case, there are three: c_1, c_2, and c_3). Theorem 2.24 applies only to continuous functions; in Figure 2.60 we see that a discontinuous function can travel from the positive value $f(a)$ to the negative value $f(b)$ without crossing the x-axis.

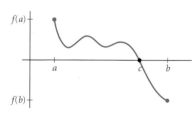

Figure 2.58

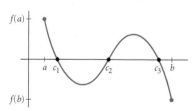

Figure 2.59

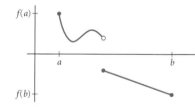

Figure 2.60

EXAMPLE 2.66 **A real-world example of the special case of the Intermediate Value Theorem**

As an intuitive real-world example, consider the temperature on a cold day in Wisconsin. If the temperature was $-5°$ Fahrenheit at 4 a.m. and $15°$ at noon, then there must have been some time between 4 a.m. and noon at which the temperature was zero degrees. The continuous function is the function $F(t)$ that describes the temperature t hours after 4 a.m. At $t = 0$ the temperature is negative, and at $t = 8$ (noon) the temperature is positive, so by Theorem 2.24, there is some time $c \in (0, 8)$ at which $F(c) = 0$ degrees Fahrenheit. The temperature changes continuously and therefore can't just "skip over" zero degrees. □

2.7.4 Using the Intermediate Value Theorem to solve inequalities

Theorem 2.24 provides us with an easy way to solve inequalities, since it guarantees that the only place that a continuous function can change sign is at a root.

THEOREM 2.25

A Function That Is Continuous and Nonzero Cannot Change Sign

If f is continuous and nonzero on (a, b), then either $f(x)$ is positive for all $x \in (a, b)$ or $f(x)$ is negative for all $x \in (a, b)$.

PROOF (THEOREM 2.25)   Let f be a function that is continuous and nonzero on (a, b). Seeking a contradiction, suppose that f changes sign somewhere in the interval (a, b). Specifically, suppose there is a point $c \in (a, b)$ with $f(c) > 0$ and a point $d \in (a, b)$ with $f(d) < 0$. By Theorem 2.24, f must have a root somewhere between c and d. This contradicts our assumption that f is nonzero on (a, b), and therefore our supposition that f changes sign on (a, b) must be false. Thus f must have the same sign on all of (a, b). ∎

The following theorem is a restatement of Theorem 2.25. The proof of Theorem 2.26 is just an application of the contrapositive of Theorem 2.25 and is left to you in the exercises.

THEOREM 2.26

A Function Can Change Sign Only at Roots and Discontinuities

If a function $f(x)$ changes sign at $x = c$ (from positive to negative, or vice versa), then $f(x)$ is either zero or discontinuous at $x = c$.

EXAMPLE 2.67   **Determining when a function changes sign**

Consider the function f graphed in Figure 2.61.

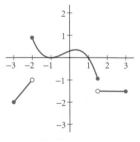

Figure 2.61

The function f is discontinuous at $x = -2$ and at $x = 1.5$ and has roots at $x = -1$ and $x = 1$. By Theorem 2.26, these are the only values of x at which the function f can change sign. However, notice that f does not *necessarily* change sign at these values. According to the graph, f changes sign at $x = -2$ and at $x = 1$ but not at $x = -1$ or at $x = 1.5$. The function f is positive on $[-2, -1) \cup (-1, 1)$ and negative on $[-3, -2) \cup (1, 3]$. □

We can represent the sign information for the function in Example 2.67 on a number line, as shown in Figure 2.62.

Figure 2.62

The values marked on the number line are the points at which the function f is zero or discontinuous. We mark the discontinuous points with the symbol "dc" to distinguish them from the zeros. Each interval between the marked points is labeled with a "+" or a "−" to indicate whether the sign of f is positive or negative on that interval. By Theorem 2.26, we are assured that f will be *always* positive, or *always* negative, on each of these intervals.

Recording signs on a number line can be very useful for solving inequalities, that is, for finding the intervals on which a function is positive or negative. The key point is that f can change sign only at x-values that are roots or discontinuities. On each interval between these values, f is either positive on the whole interval or negative on the whole interval, and we can determine what that sign is by testing the sign of f at just *one point* in the interval! If f is positive at *one* point in such an interval, then f must be positive on the entire interval, since by construction that interval contains no roots or discontinuities of f.

ALGORITHM 2.2

Finding the Intervals Where a Function Is Positive or Negative

To determine the intervals on which a function f is positive or negative,

(a) Find all the x-values where f is zero or discontinuous, and record these values on a number line.

(b) Test the sign of f at some point in each interval, and record this sign as a + or a − on the number line.

(c) Use the number line to write in interval notation the intervals on which f is positive or negative.

As you will see in the next example, Algorithm 2.2 is *much* easier than the method of solving systems of inequalities we used in Section 0.3.

EXAMPLE 2.68

Using Algorithm 2.2 to determine where a function is negative

Find the intervals on which the function $f(x) = 3x^3 + 3x^2 - 6x$ is negative.

Solution Finding the intervals on which f is negative is equivalent to solving the inequality $3x^3 + 3x^2 - 6x < 0$. We begin by finding the values for which f is zero or discontinuous. Since f is a continuous function (it is a sum of continuous functions), we need only find its roots. We do this by factoring:

$$f(x) = 3x^3 + 3x^2 - 6x = 3x(x^2 + x - 2) = 3x(x - 1)(x + 2).$$

We have $f(x) = 0$ at $x = 0$, $x = 1$, and $x = -2$. We record these three values on a number line (see Figure 2.63).

Figure 2.63

The roots of f divide the number line into four intervals: $(-\infty, -2), (-2, 0), (0, 1)$, and $(1, \infty)$. By Theorem 2.26, the function f is always positive, or always negative, on each of these intervals. We need to test the sign of f at only *one* point in each interval. For example,

$$f(-3) = 3(-3)^3 + 3(-3)^2 - 6(-3) = -81 + 27 + 18 = -36 < 0,$$

$$f(-1) = 3(-1)^3 + 3(-1)^2 - 6(-1) = -3 + 3 + 6 = 6 > 0,$$

$$f(0.5) = 3(0.5)^3 + 3(0.5)^2 - 6(0.5) = 0.375 + 0.75 - 3 = -1.875 < 0,$$

$$f(2) = 3(2)^3 + 3(2)^2 - 6(2) = 24 + 12 - 12 = 24 > 0.$$

We record the sign of f on each interval on the number line shown in Figure 2.64.

Figure 2.64

Reading from the number line, we can see that $f(x) = 3x^3 + 3x^2 - 6x$ is negative on the set $(-\infty, -2) \cup (0, 1)$. ☐

⑨ *Caution*   Be sure to show your work (as in Example 2.68) so that it is clear how you determined the sign of f on each interval. ☐

❓ *Question*   In Section 0.3, we would have solved the inequality $3x^3 + 3x^2 - 6x < 0$ by factoring it as $3x(x - 1)(x + 2) < 0$ and then examining the four cases where an odd number of the factors $3x$, $x - 1$, and $x + 2$ are negative. These cases are

$$\begin{pmatrix} 3x < 0 \\ x - 1 > 0 \\ x + 2 > 0 \end{pmatrix} \quad \text{or} \quad \begin{pmatrix} 3x > 0 \\ x - 1 < 0 \\ x + 2 > 0 \end{pmatrix} \quad \text{or} \quad \begin{pmatrix} 3x > 0 \\ x - 1 > 0 \\ x + 2 < 0 \end{pmatrix} \quad \text{or} \quad \begin{pmatrix} 3x < 0 \\ x - 1 < 0 \\ x + 2 < 0 \end{pmatrix}.$$

The solution to $3x^3 + 3x^2 - 6x < 0$ is the union of the solution sets for each of these four systems of inequalities. (The details are left to the reader.) Compare this lengthy calculation to the calculation that we did in Example 2.68. ☐

❓ *Question*   Explain in your own words why testing the sign of a function f at *one point* in an interval with no roots or discontinuities determines the sign of the function f along the *entire* interval. ☐

What you should know after reading Section 2.7

▶ The Extreme Value Theorem and the Intermediate Value Theorem, including mathematical and real-world illustrations of these theorems. Examples that illustrate these theorems when they apply, and examples of discontinuous functions where the conclusions fail.

▶ The statement of the special case of the Intermediate Value Theorem (where $K = 0$). Why is this special case important?

▶ The algorithm for determining the intervals on which a function f is positive or negative, including recording sign information on a real number line. What does this method have to do with the Intermediate Value Theorem?

Exercises 2.7

Concepts

0. Read the section and make your own summary of the material. Include whatever you think will help you review the material later on.

1. In this problem you will examine the statement of the Extreme Value Theorem.

 (a) State the Extreme Value Theorem. Which part of the theorem is the hypothesis? Which part of the theorem is the conclusion?

 (b) What does the Extreme Value Theorem have to do with "extreme values"?

 (c) State the converse of the Extreme Value Theorem. Is it true? Why or why not?

 (d) State the contrapositive of the Extreme Value Theorem. Is it true? Why or why not?

2. In this problem you will construct examples that illustrate the Extreme Value Theorem. When asked to provide examples or graphs, give examples or graphs different from those given in this section.

 (a) Give a real-world example of a continuous function on a closed interval, and describe in real-world terms what the Extreme Value Theorem says about this function.

 (b) Sketch a labeled graph of a function that illustrates the Extreme Value Theorem. Explain how your graph satisfies both the hypothesis and the conclusion of the Extreme Value Theorem.

 (c) Sketch a labeled graph of a function that fails to satisfy the hypothesis *and* the conclusion of the Extreme Value Theorem.

 (d) Sketch a labeled graph of a function that fails to satisfy the hypothesis of the Extreme Value Theorem but for which the conclusion of the Extreme Value Theorem happens to be true. What does the existence of this example tell you about the converse of the Extreme Value Theorem?

3. In this problem you will examine the statement of the Intermediate Value Theorem.

 (a) State the Intermediate Value Theorem. Which part of the theorem is the hypothesis? Which part of the theorem is the conclusion?

 (b) What does the Intermediate Value Theorem have to do with "intermediate values"?

 (c) State the converse of the Intermediate Value Theorem. Is it true? Why or why not?

 (d) State the contrapositive of the Intermediate Value Theorem. Is it true? Why or why not?

4. In this problem you will construct examples that illustrate the Intermediate Value Theorem. When asked to

provide examples or graphs, give examples or graphs different from those given in this section.

 (a) Give a real-world example of a continuous function, and describe in real-world terms what the Intermediate Value Theorem says about this function.

 (b) Sketch a labeled graph of a function that illustrates the Intermediate Value Theorem. Explain how your graph satisfies both the hypothesis and the conclusion of the Intermediate Value Theorem.

 (c) Sketch a labeled graph of a function that fails to satisfy the hypothesis *and* the conclusion of the Intermediate Value Theorem.

 (d) Sketch a labeled graph of a function that fails to satisfy the hypothesis of the Intermediate Value Theorem but for which the conclusion of the Intermediate Value Theorem happens to be true. What does the existence of this example tell you about the converse of the Intermediate Value Theorem?

5. In this problem you will examine the statement of the special case of the Intermediate Value Theorem where the only intermediate value we are interested in is $K = 0$ (Theorem 2.24).

 (a) State this special case. Which part of the theorem is the hypothesis? Which part of the theorem is the conclusion?

 (b) State the converse of this special case. Is it true? Why or why not?

 (c) State the contrapositive of this special case of the Intermediate Value Theorem. Is it true? Why or why not?

6. In this problem you will construct examples that illustrate the special case of the Intermediate Value Theorem given in Theorem 2.24. When asked to provide examples or graphs, give examples or graphs different from those given in this section.

 (a) Give a real-world example of a continuous function, and describe in real-world terms what the special case of the Intermediate Value Theorem says about this function.

 (b) Sketch a labeled graph of a function that illustrates the special case of the Intermediate Value Theorem. Explain how your graph satisfies both the hypothesis and the conclusion of this special case of the Intermediate Value Theorem.

 (c) Sketch a labeled graph of a function that fails to satisfy the hypothesis *and* the conclusion of this special case of the Intermediate Value Theorem.

 (d) Sketch a labeled graph of a function that fails to satisfy the hypothesis of the special case of the Intermediate Value Theorem but for which the conclusion happens to be true. What does the existence of this example tell you about the converse of this special case of the Intermediate Value Theorem?

7. In this problem you will examine Theorem 2.26.

   (a) State Theorem 2.26 in the form "If _____ , then _____ ."

   (b) Use your answer above to state the converse of Theorem 2.26. Is it true?

   (c) The contrapositive of Theorem 2.26 is true (since the contrapositive of a statement is always logically equivalent to the original statement). State the contrapositive, and explain how it can be used to find the intervals on which a function f is positive or negative in Algorithm 2.2.

8. Suppose that f is a function that is continuous and nonzero on the interval $(1, 5)$. If $f(4)$ is negative, then f must be negative on the entire interval $(1, 5)$. Explain in your own words why this is so.

■ Determine whether each of the following statements is true or false. If a statement is true, explain why. If a statement is false, provide a counterexample.

9. True or False: If f is continuous on the interval $(2, 4)$, then f must have a maximum value and a minimum value on $(2, 4)$.

10. True or False: If $f(3) = -5$ and $f(9) = -2$, then there must be a value c at which $f(c) = -3$.

11. True or False: If f is continuous everywhere, and if $f(-2) = 3$ and $f(1) = -2$, then there is some $c \in (-2, 1)$ with $f(c) = 0$.

12. True or False: If f is continuous everywhere, and if f has a zero at $x = 2$, then there must be some value at which f is positive and some value at which f is negative.

13. True or False: If f is continuous everywhere, and if $f(-2) = 3$ and $f(1) = 2$, then f must have a root somewhere in $(-2, 1)$.

14. True or False: If f is continuous everywhere, and if $f(0) = -2$ and $f(4) = 3$, then f must have exactly one root in $(0, 4)$.

15. True or False: If f changes sign at $x = 2$ and $f(2) \neq 0$, then f is discontinuous at $x = 2$.

16. True or False: If f is continuous everywhere, $f(0) = 0$, $f(6) = 0$, and $f(2) > 0$, then f is positive on the entire interval $(0, 6)$.

■ For each number line below, sketch the graph of a function f that has the signs, zeros, and discontinuities indicated on the number line.

17.

18.

Skills

■ For each function f and interval $[a, b]$, (a) use the Extreme Value Theorem to show that f has both a maximum and a minimum value on $[a, b]$. Then (b) use a graphing calculator to approximate values c and d in $[a, b]$ at which f has a maximum and a minimum, respectively.

19. $f(x) = x^4 - 3x^2 - 2$, $[-2, 2]$

20. $f(x) = x^4 - 3x^2 - 2$, $[0, 2]$

21. $f(x) = x^4 - 3x^2 - 2$, $[-1, 1]$

22. $f(x) = 3 - 2x^2 + x^3$, $[-1, 2]$

23. $f(x) = 3 - 2x^2 + x^3$, $[0, 2]$

24. $f(x) = 3 - 2x^2 + x^3$, $[-1, 1]$

■ For each function f, interval $[a, b]$, and value K, (a) use the Intermediate Value Theorem to show that there is some $c \in (a, b)$ for which $f(c) = K$. Then (b) use a graphing calculator to approximate all such values of c.

25. $f(x) = x^2$, $[a, b] = [0, 3]$, $K = 5$

26. $f(x) = 2 + x + x^3$, $[a, b] = [-1, 2]$, $K = 3$

27. $f(x) = x^3 - 3x^2 - 2$, $[a, b] = [-2, 4]$, $K = -4$

28. $f(x) = x^3 - 3x^2 - 2$, $[a, b] = [2, 4]$, $K = -4$

29. $f(x) = 5 - x^4$, $[a, b] = [0, 3]$, $K = 0$

30. $f(x) = 5 - x^4$, $[a, b] = [-2, -1]$, $K = 0$

■ For each function f and value K, (a) use the Intermediate Value Theorem to show that there is some value c at which $f(c) = K$. Then (b) use a graphing calculator to approximate all such values of c.

31. $f(x) = 3x + 1$, $K = -4$

32. $f(x) = -2x^2 + 4$, $K = 0$

33. $f(x) = x^3 + 2$, $K = -15$

34. $f(x) = \dfrac{4}{x^2 + 1}$, $K = 2$

35. $f(x) = (x - 3)^4 - 2$, $K = -1$

36. $f(x) = (x - 3)^4 - 2$, $K = 100$

■ For each f, a, and b, the special case of the Intermediate Value Theorem given in Theorem 2.24 may or may not apply. (a) Determine whether the theorem applies. If so, (b) use the theorem to show that f has a root between $x = a$ and $x = b$, and (c) approximate any such roots with a calculator.

37. $f(x) = x^3 + x^2 - 4x$, $a = 1$, $b = 2$

38. $f(x) = x^3 + x^2 - 4x$, $a = -3$, $b = 1$

39. $f(x) = x^3 + x^2 - 4x$, $a = -1$, $b = 1$

40. $f(x) = x^4 - 3x + 1$, $a = 0$, $b = 1$

41. $f(x) = x^4 - 3x + 1$, $a = 0$, $b = 2$

42. $f(x) = x^4 - 3x + 1$, $a = -1$, $b = 0$

■ Solve each of the following inequalities two ways: first, by using the methods of Section 0.3 (solving systems of inequalities to determine when an even or odd number of factors are negative); second, by using Algorithm 2.2. Show all computations and work clearly.

43. $2x^2 + 9x \le 5$

44. $(x^2 - 1)(3x - 2) \ge 0$

45. $x^3 - 4x < 2x^2 - x$

■ For each function f, use Algorithm 2.2 to find the intervals on which f is positive and the intervals on which f is negative. Express your final answer in interval notation, and check your answer with a graph of f.

46. $f(x) = 3x + 1$

47. $f(x) = x^2 + 2x - 3$

48. $f(x) = 2 + 5x + 2x^2$

49. $f(x) = x^3 - x$

50. $f(x) = \dfrac{(x + 4)(x - 1)}{2x + 3}$

51. $f(x) = \dfrac{(x - 1)^2}{5 - x}$

52. $f(x) = x^3 - 2x^2 - 3x$

53. $f(x) = x^2(x^2 - 3x - 10)$

54. $f(x) = \dfrac{x^2 - 4}{x^2 - 1}$

■ For each piecewise function f below, use Algorithm 2.2 to find the intervals on which f is positive and the intervals on which f is negative. Note that you will have to examine the continuity of each function f at its "break" points; do so algebraically. Express your final answer in interval notation, and check your answer with a graph of f.

55. $f(x) = \begin{cases} 3x + 1, & \text{if } x < 0 \\ x, & \text{if } x \ge 0 \end{cases}$

56. $f(x) = \begin{cases} x - 4, & \text{if } x \le 1 \\ x^2 - 4, & \text{if } x > 1 \end{cases}$

57. $f(x) = \begin{cases} x^2 - 9, & \text{if } x \le -2 \\ x^2 + x - 2, & \text{if } x > -2 \end{cases}$

58. $f(x) = \begin{cases} x^3, & \text{if } x < 2 \\ 8, & \text{if } x = 2 \\ 4x - x^3, & \text{if } x > 2 \end{cases}$

Applications

■ Explain in practical terms what the Extreme Value Theorem says about each continuous function defined in Problems 59, 60, and 61. Then explain in practical terms what the Intermediate Value Theorem says in each situation.

59. Alina hasn't cut her hair for 6 years. Six years ago her hair was just 2 inches long. Now her hair is 42 inches long. Let $H(t)$ be the function that describes the length, in inches, of Alina's hair t years after she stopped cutting it.

60. Linda collects rain in a bucket outside her back door. Since January 1 she has been keeping track of how the amount of water in the bucket changes as rainwater fills the bucket and then evaporates. On January 1 the bucket was empty, and today it contains 4 inches of water. Let $w(t)$ be the height, in inches, of rainwater in the bucket t days after January 1.

61. The number of gallons of gas in Phil's new station wagon t days after he bought it is given by the function $g(t)$. When he purchased the station wagon, the tank had 19 gallons of gas in it. Today he ran out of gas.

62. Peter was 20 inches tall when he was born and was 5 feet 11 inches tall at age 83. Use the Intermediate Value Theorem to show that there must have been some point in Peter's life at which his height (in inches) was equal to his age (in years). (*Hint:* Let $h(t)$ represent Peter's height, and let $a(t)$ represent his age t years after he was born. Apply the special case of the Intermediate Value Theorem (Theorem 2.24) to the function $f(t) = h(t) - a(t)$.)

Proofs

63. Write, in your own words, a proof of Theorem 2.25. Justify or explain each step in your proof. Illustrate your proof with graphs if doing so makes it clearer.

64. Use Theorem 2.25 to prove Theorem 2.26. Be sure that the logic of your argument is clear.

2.8 Appendix: Proofs of Selected Limit Rules

In this appendix we present proofs that were too long or detailed to include in the body of the chapter. In general, these proofs are not much harder than those in the chapter, but they are a little more complicated and do require careful attention to detail.

2.8.1 Uniqueness of limits

We begin by using the formal definition of limit to prove that the limit of a function as $x \to c$, if it exists, is unique. This fact was stated in Theorem 2.2 in Section 2.2.1.

PROOF (**THEOREM 2.2**)   Suppose $\lim_{x \to c} f(x) = L$ and $\lim_{x \to c} f(x) = M$. We will show that $L = M$ by showing that $|L - M|$ is zero.

By the formal definition of limit, we know that

(1)   *For all $\varepsilon_1 > 0$, there exists a $\delta_1 > 0$ such that if $0 < |x - c| < \delta_1$, then $|f(x) - L| < \varepsilon_1$.*

and that

(2)   *For all $\varepsilon_2 > 0$, there exists a $\delta_2 > 0$ such that if $0 < |x - c| < \delta_2$, then $|f(x) - M| < \varepsilon_2$.*

Given $\varepsilon > 0$, choose δ_1 and δ_2 using the statements above so that $|f(x) - L| < \varepsilon/2$ for all $0 < |x - c| < \delta_1$ and $|f(x) - M| < \varepsilon/2$ for all $0 < |x - c| < \delta_2$. Define $\delta = \min(\delta_1, \delta_2)$ and let x be any number with $0 < |x - c| < \delta$. Then

$$|L - M| = |L - f(x) + f(x) - M| \qquad \text{(add and subtract } f(x))$$
$$= |(L - f(x)) + (f(x) - M)|$$
$$\leq |L - f(x)| + |f(x) - M| \qquad \text{(triangle inequality)}$$
$$= |f(x) - L| + |f(x) - M| \qquad (|L - f(x)| = |f(x) - L|)$$
$$< \varepsilon/2 + \varepsilon/2 = \varepsilon. \qquad \text{(by choice of } x \text{ and } \delta \text{ above)}$$

Therefore, for all positive numbers ε, no matter how small, we have $|L - M| < \varepsilon$. Since $|L - M|$ is smaller than any positive number (and must be nonnegative because of the absolute value), we must have $|L - M| = 0$. Thus $L = M$, which is what we wanted to show. ∎

2.8.2 The product rule for limits

In this section we prove the product rule for limits from Section 2.4.5.

PROOF (**PART (A) OF THEOREM 2.9**)   Suppose $\lim_{x \to c} f(x) = L$ and $\lim_{x \to c} g(x) = M$. By the formal definition of limit, this means that

(3)   *For all $\varepsilon_1 > 0$, there exists a $\delta_1 > 0$ such that if $0 < |x - c| < \delta_1$, then $|f(x) - L| < \varepsilon_1$.*

and that

(4)   *For all $\varepsilon_2 > 0$, there exists a $\delta_2 > 0$ such that if $0 < |x - c| < \delta_2$, then $|g(x) - M| < \varepsilon_2$.*

Given the statements above, we want to show that $\lim_{x \to c} f(x)g(x) = LM$, that is, that

(5)   *For all $\varepsilon > 0$, there exists a $\delta > 0$ such that if $0 < |x - c| < \delta$, then $|f(x)g(x) - LM| < \varepsilon$.*

Given $\varepsilon > 0$, we will choose $\delta = \min(\delta_1, \delta_2, \delta_3)$ for some δ_1, δ_2, and δ_3 that we will define below. Given this ε and choice of δ, suppose that $0 < |x - c| < \delta$.

We begin by manipulating $|f(x)g(x) - LM|$ to relate it to Statements (3) and (4):

$$|f(x)g(x) - LM|$$
$$= |f(x)g(x) - f(x)M + f(x)M - LM| \qquad \text{(add and subtract } f(x)M)$$
$$= |f(x)(g(x) - M) + M(f(x) - L)| \qquad \text{(factoring)}$$
$$\leq |f(x)(g(x) - M)| + |M(f(x) - L)| \qquad \text{(triangle inequality)}$$
$$= |f(x)||g(x) - M| + |M||f(x) - L|.$$

We are trying to make the statement above less than ε. We start by bounding $|f(x)|$. Use Statement (3) to choose δ_1 such that $|f(x) - L| < 1$ for all $0 < |x - c| < \delta_1$. Since $|f(x)| - |L| \leq |f(x) - L|$ (this is a consequence of the triangle inequality; Section 0.3), this means that $|f(x)| - |L| < 1$ and thus $|f(x)| < |L| + 1$. Using this in the calculation above, for all $0 < |x - c| < \delta$ we have

$$
\begin{aligned}
&|f(x)g(x) - LM| \\
&\quad \leq |f(x)||g(x) - M| + |M||f(x) - L| &&\text{(from above)} \\
&\quad < (|L| + 1)|g(x) - M| + |M||f(x) - L| &&\text{(since } \delta < \delta_1) \\
&\quad < (|L| + 1)|g(x) - M| + (|M| + 1)|f(x) - L| &&\text{(since } |M| < |M| + 1)
\end{aligned}
$$

It is now clear how we should choose δ_2 and δ_3. Use Statement (4) to choose δ_2 such that $|g(x) - M| < \frac{\varepsilon/2}{|L|+1}$ for all $0 < |x - c| < \delta_2$. Similarly, use Statement (3) to choose δ_3 such that $|f(x) - L| < \frac{\varepsilon/2}{|M|+1}$ for all $0 < |x - c| < \delta_3$. Note that $|L| + 1$ and $|M| + 1$ are never zero, so division by these expressions is always defined. (This is why we changed $|M|$ to $|M| + 1$ in the last line of the calculation above.)

Since $\delta = \min(\delta_1, \delta_2, \delta_3)$, we know that $\delta \leq \delta_2$ and $\delta \leq \delta_3$. Therefore, for all x with $0 < |x - c| < \delta$, we have

$$
\begin{aligned}
|f(x)g(x) - LM| &< (|L| + 1)|g(x) - M| + (|M| + 1)|f(x) - L| &&\text{(above)} \\
&< (|L| + 1)\frac{\varepsilon/2}{|L| + 1} + (|M| + 1)\frac{\varepsilon/2}{|M| + 1} &&\text{(choice of } \delta_2, \delta_3) \\
&= \varepsilon/2 + \varepsilon/2 = \varepsilon.
\end{aligned}
$$

Therefore, for all $0 < |x - c| < \delta$ we have $|f(x)g(x) - LM| < \varepsilon$, which is what we wanted to show. ∎

2.8.3 The quotient rule for limits

The quotient rule for limits (part (b) of Theorem 2.9 from Section 2.4.5) can be easily derived from the product rule for limits, if we first prove the following "reciprocal" rule:

THEOREM 2.27

The Reciprocal Rule for Limits

If $\lim_{x \to c} g(x)$ exists and is nonzero, then

$$
\lim_{x \to c} \frac{1}{g(x)} = \frac{1}{\lim_{x \to c} g(x)}.
$$

PROOF (**THEOREM 2.27**) Suppose $\lim_{x \to c} g(x) = M$ for some nonzero real number M. Then, by the formal definition of limit,

(6)
> *For all $\varepsilon_1 > 0$, there exists a $\delta_1 > 0$ such that*
> *if $0 < |x - c| < \delta_1$, then $|g(x) - M| < \varepsilon_1$.*

We wish to show that $\lim_{x \to c} \frac{1}{g(x)} = \frac{1}{M}$, that is, that

(7)
> *For all $\varepsilon > 0$, there exists a $\delta > 0$ such that*
> *if $0 < |x - c| < \delta$, then $\left| \dfrac{1}{g(x)} - \dfrac{1}{M} \right| < \varepsilon$.*

Given $\varepsilon > 0$, choose $\delta = \min(\delta_1, \delta_2)$, where δ_1 and δ_2 are chosen as shown next.

We start by manipulating $|\frac{1}{g(x)} - \frac{1}{M}|$, with the goal of relating it to $|g(x) - M|$:

$$\left| \frac{1}{g(x)} - \frac{1}{M} \right| = \left| \frac{M - g(x)}{Mg(x)} \right| = \frac{|g(x) - M|}{|Mg(x)|} = \frac{|g(x) - M|}{|M||g(x)|}.$$

We want to replace the $|g(x)|$ in the denominator with something larger but simpler. Use Statement (6) to choose δ_1 such that $|g(x) - M| < \frac{|M|}{2}$. (Here $\varepsilon_1 = \frac{|M|}{2}$.) Since $||g(x)| - |M|| < |g(x) - M| < \frac{|M|}{2}$ (a consequence of the triangle inequality), we have $|M| - \frac{|M|}{2} < |g(x)| < |M| + \frac{|M|}{2}$. Using the leftmost inequality, we see that $|g(x)| > \frac{|M|}{2}$. This means that for all x with $0 < |x - c| < \delta_1$, we have

$$\frac{|g(x) - M|}{|M||g(x)|} < \frac{|g(x) - M|}{|M|\frac{|M|}{2}} = \frac{|g(x) - M|}{\frac{|M|^2}{2}} = \frac{2|g(x) - M|}{|M|^2}.$$

Our goal is to have this string of inequalities end with ε. Use Statement (6) again, this time to choose δ_2 such that $|g(x) - M| < \frac{|M|^2}{2}\varepsilon$. If $\delta = \min(\delta_1, \delta_2)$, then for all x with $0 < |x - c| < \delta$, we have

$$\frac{2|g(x) - M|}{|M|^2} < \frac{2\left(\frac{|M|^2}{2}\varepsilon\right)}{|M|^2} = \varepsilon.$$

Therefore, we have shown $|\frac{1}{g(x)} - \frac{1}{M}| < \varepsilon$, as desired. ∎

Any quotient can be written in terms of products and reciprocals. Since we have already established the product and reciprocal rules for limits, we can use them to prove the quotient rule for limits.

PROOF (**PART (B) OF THEOREM 2.9**)   If $\lim\limits_{x \to c} f(x)$ exists and $\lim\limits_{x \to c} g(x)$ exists and is nonzero, then

$$\lim_{x \to c} \frac{f(x)}{g(x)} = \lim_{x \to c} f(x) \frac{1}{g(x)} \qquad \text{(algebra)}$$

$$= \lim_{x \to c} f(x) \lim_{x \to c} \frac{1}{g(x)} \qquad \text{(product rule for limits)}$$

$$= \lim_{x \to c} f(x) \frac{1}{\lim\limits_{x \to c} g(x)} \qquad \text{(reciprocal rule for limits)}$$

$$= \frac{\lim\limits_{x \to c} f(x)}{\lim\limits_{x \to c} g(x)}. \qquad\qquad ∎$$

2.8.4 Infinite limits

In this section we prove that some limits that "do not exist" are infinite. We will prove the first part of Theorem 2.12 from Section 2.5.3. The proof of the second part is similar.

PROOF (**THEOREM 2.12**)   By hypothesis we know that $\lim\limits_{x \to c^+} f(x) = L > 0$ and that $\lim\limits_{x \to c^+} g(x) = 0$, with $g(x) \to 0^+$ as $x \to c^+$. These three facts translate into the following three mathematical statements:

(8)   *For all $\varepsilon_1 > 0$, there exists a $\delta_1 > 0$ such that if $x \in (c, c + \delta_1)$, then $|f(x) - L| < \varepsilon_1$.*

(9)   *For all $\varepsilon_2 > 0$, there exists a $\delta_2 > 0$ such that if $x \in (c, c + \delta_2)$, then $|g(x)| < \varepsilon_2$.*

(10)   *There is an interval $(c, c + \alpha)$ on which $g(x) > 0$.*

To show that $\lim_{x \to c^+} \frac{f(x)}{g(x)} = \infty$, we must show that

(11)

$$\textit{For all } M > 0, \textit{ there exists a } \delta > 0 \textit{ such that}$$
$$\textit{if } x \in (c, c + \delta), \textit{ then } \frac{f(x)}{g(x)} > M.$$

Given $M > 0$, use Statement (8) to choose δ_1 such that $|f(x) - L| < 1$ for all $x \in (c, c + \delta_1)$, and use Statement (9) to choose δ_2 such that $|g(x)| < \frac{L-1}{M}$ for all $x \in (c, c + \delta_2)$. Let α be some positive number so that (by Statement 10) $g(x) > 0$ for all $x \in (c, c + \alpha)$. Then choose $\delta = \min(\delta_1, \delta_2, \alpha)$.

With this choice of δ, for all $x \in (c, c + \delta)$ we have $|f(x) - L| < 1$ and thus $L - 1 < f(x) < L + 1$; in particular, this means that $f(x) > L - 1$. We also have $|g(x)| < \frac{L-1}{M}$, and since $g(x) > 0$ for all $x \in (c, c+\delta)$, this means that $0 < g(x) < \frac{L-1}{M}$ and thus that $\frac{1}{g(x)} > \frac{1}{(L-1)/M}$.

Therefore, for all $x \in (c, c + \delta)$, we have

$$\frac{f(x)}{g(x)} > \frac{L - 1}{g(x)} \qquad \text{(choice of } \delta_1\text{)}$$
$$> \frac{L - 1}{(L - 1)/M} \qquad \text{(choice of } \delta_2 \text{ and } g(x) > 0\text{)}$$
$$= M.$$

Thus $\frac{f(x)}{g(x)} > M$ for all $x \in (c, c + \delta)$, which is what we were trying to prove. ■

2.8.5 The limit of a composition

Our last proof is the limit rule for compositions of functions in Section 2.6.5.

PROOF (**THEOREM 2.20**) We are given that g is continuous at c, so by Theorem 2.16, $\lim_{x \to c} g(x) = g(c)$. Thus by Theorem 2.15, we know that

(12)

$$\textit{For all } \varepsilon_1 > 0, \textit{ there exists a } \delta_1 > 0 \textit{ such that}$$
$$\textit{if } 0 < |x - c| < \delta_1, \textit{ then } |g(x) - g(c)| < \varepsilon_1.$$

We are also given that f is continuous at L, and thus $\lim_{x \to L} f(x) = f(L)$. By Theorem 2.15, this means that

(13)

$$\textit{For all } \varepsilon_2 > 0, \textit{ there exists a } \delta_2 > 0 \textit{ such that}$$
$$\textit{if } 0 < |x - L| < \delta_2, \textit{ then } |f(x) - f(L)| < \varepsilon_2.$$

We wish to show that $\lim_{x \to c} f(g(x)) = f(L)$. By the formal definition of limit, this means we need to prove that

(14)

$$\textit{For all } \varepsilon > 0, \textit{ there exists a } \delta > 0 \textit{ such that}$$
$$\textit{if } 0 < |x - c| < \delta, \textit{ then } |f(g(x)) - f(L)| < \varepsilon.$$

Given any $\varepsilon > 0$, start by using Statement (13) to choose $\delta_2 > 0$ such that $|f(x) - f(L)| < \varepsilon$ for all $0 < |x - L| < \delta_2$. Then use Statement (12) to choose $\delta_1 > 0$ such that $|g(x) - L| < \delta_2$ for all $0 < |x - c| < \delta_1$. Finally, choose $\delta = \delta_1$.

Now for all $0 < |x - c| < \delta$, we have the following string of implications:

$$|x - c| < \delta \implies |x - c| < \delta_1 \qquad (\delta = \delta_1)$$
$$\implies |g(x) - L| < \delta_2 \qquad (\text{choice of } \delta_1 \text{ and } \varepsilon_1 = \delta_2)$$
$$\implies |f(g(x)) - f(L)| < \varepsilon \qquad (\text{choice of } \delta_2 \text{ and } \varepsilon_2 = \varepsilon)$$

Note that in the last line above we are using $g(x)$ in the place of x in Statement (13). We have now shown that given any $\varepsilon > 0$, we can choose $\delta > 0$ such that $|f(g(x)) - f(L)| < \varepsilon$ for all $0 < |x - c| < \delta$, and we are done. ■

CHAPTER 3

Derivatives

"Costa II," a mathematical sculpture by Helaman Ferguson, www.helasculpt.com; *photo by Claire Ferguson*

*M*easuring the way that a function or quantity changes involves finding the slope, or direction, of the graph at each point. The slope of the line tangent to the graph of a function represents the instantaneous rate of change of the function and is called a derivative.

The definition of derivative will be our first substantial application of limits. Derivatives can be thought of in terms of tangent lines or in terms of instantaneous rates of change. The slope of the tangent line to a graph at a given point will be defined as the limit of slopes of nearby secant lines to that graph. Similarly, the instantaneous rate of change will be defined as a limit of average rates of change.

CONTENTS

3.1 Tangent Lines and the Derivative at a Point

In this section we will learn how to find the slope of the **tangent line** to the graph of a function at a point. Intuitively, given a function f and a value c, the tangent line to f at $x = c$ is the line passing through the point $(c, f(c))$ and pointing in the "direction" of the graph of the function at that point. The slope of this tangent line represents the instantaneous rate of change of the function at the point $x = c$ and is called the **derivative** of f at $x = c$.

We will see many real-world applications of tangent lines and derivatives in later sections, particularly applications to distance, velocity, and acceleration. For now we concentrate on how to define the tangent line to the graph of a function at a given point and on how to find the slope of that tangent line.

3.1.1 Approximating the slope of a tangent line

Suppose f is a function defined on an interval around a point $x = c$. We are interested in calculating the slope of the line through $(c, f(c))$ that just skims the graph of f (if such a line exists). We will call this line the **tangent line** to f at $x = c$. At this point we don't have a formal mathematical definition for what we mean by "skimming" the graph of f, but we do have an intuitive idea of what we are looking for. Figure 3.1 shows the graph of a function f and its tangent line at the point $x = 1$.

How do we calculate the slope of such a tangent line? At first glance this seems like a very simple problem; we just need to calculate the slope of a line, and that's usually easy. However, we have a problem. *We know only one point on this line!* To calculate a slope, we need *two* points, and the only point that we know is $(c, f(c))$.

We do not yet know how to calculate the slope of this tangent line directly. However, we can approximate the slope of the tangent line using the slope of a "nearby" line. Given a function f and a point $x = c$, a line passing through $(c, f(c))$ and any other point $(d, f(d))$ on the graph of f is called a **secant line** for f from c to d. If d is close to c, then the secant line from c to d will be "near" the tangent line at $x = c$. The slope of the secant line will be a close approximation for the slope of the tangent line.

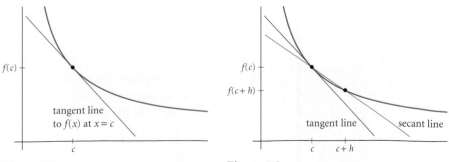

Figure 3.1   **Figure 3.2**

Given a point $x = c$, if we let h be a small number, then the point $d = c + h$ will be close to the point c. The secant line for f from c to $c + h$ will have a slope that is close to the slope of the tangent line to f at c (see Figure 3.2). If we make h smaller, then $c + h$ will be even closer to c, and the secant line from c to $c + h$ will be even closer to the tangent line to f at c.

❓ *Question*   In Figure 3.2, as h gets smaller and $c + h$ approaches c, the secant line from c to $c + h$ rotates, while fixed at the point $(c, f(c))$, and approaches the tangent line. Sketch some secant lines for smaller values of h to help you visualize what is happening as $h \to 0$.   □

This idea forms the basis for our definition of tangent line. In fact, the tangent line to f at c will be defined as a limit of secant lines. Before we give this definition formally, we will review the discussion above with an example.

EXAMPLE 3.1

Approximating the slope of a tangent line with slopes of secant lines

Suppose $f(x) = \frac{2}{x}$ and $c = 1$. How can we approximate the slope of the line tangent to the graph of $f(x)$ at $(1, f(1))$? (See Figure 3.5.) We know only one point on this tangent line (namely $(1, f(1)) = (1, 2)$), so we can't calculate the slope directly.

We can, however, calculate the slopes of nearby secant lines. Figure 3.3 shows the secant line on the graph of $f(x) = \frac{2}{x}$ from $(1, f(1))$ to $(2, f(2))$. The slope of this line is

$$\frac{f(2) - f(1)}{2 - 1} = \frac{\frac{2}{2} - \frac{2}{1}}{1} = 1 - 2 = -1.$$

The point $x = 2$ is one unit away from the point $c = 1$, so here our value for h is 1. Since $x = 2$ is somewhat close to $x = 1$, the slope of this secant line is somewhat close to the slope of the tangent line we are looking for.

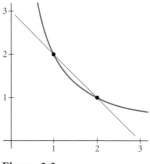 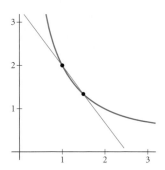 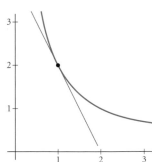

Figure 3.3
Secant line from $(1, f(1))$ to $(2, f(2))$

Figure 3.4
Secant line from $(1, f(1))$ to $(1.5, f(1.5))$

Figure 3.5
Tangent line at $(1, f(1))$

To get a better approximation for the slope of the tangent line, we could choose a smaller h. For example, see Figure 3.4, where $h = 0.5$ and $c + h = 1 + 0.5 = 1.5$. The slope of the secant line from $(1, f(1))$ to $(1.5, f(1.5))$ is

$$\frac{f(1.5) - f(1)}{1.5 - 1} = \frac{\frac{2}{1.5} - \frac{2}{1}}{0.5} \approx -1.33.$$

To get an even better approximation for the slope of the tangent line to $f(x) = \frac{2}{x}$ at $x = 1$, we could choose a very small value of h, say $h = 0.01$. The slope of the secant line from $(1, f(1))$ to $(1.01, f(1.01))$ is

$$\frac{f(1.01) - f(1)}{1.01 - 1} = \frac{\frac{2}{1.01} - \frac{2}{1}}{0.01} \approx -1.98.$$

Try to picture this secant line on the graph in Figure 3.5. Since $x = 1.01$ is so close to $x = 1$, this secant line will look very much like the tangent line we are looking for, so the slope of this secant line will be very close to the slope of the tangent line. Presently, our best approximation for the slope of the tangent line to $f(x) = \frac{2}{x}$ at $x = 1$ is -1.98. Given these calculations, a reasonable guess for the slope of the tangent line in Figure 3.5 would be -2. $\qquad\qquad\qquad\qquad\qquad\qquad\qquad\qquad\qquad\qquad\qquad\qquad\qquad\qquad\square$

3.1.2 The definition of derivative

In Section 3.1.1 we saw that, intuitively, the slope of the tangent line to f at a point $x = c$ can be approximated by the slope of a "nearby" secant line from c to $c + h$ for some small value of h. As $c + h$ approaches c (i.e., as h approaches zero), the slopes of the secant lines will approach the slope of the tangent line. In other words, the slope of the tangent line is a *limit* of slopes of secant lines as h approaches zero.

Given a function f, a point $x = c$, and some small number h, the slope of the secant line from $(c, f(c))$ to $(c + h, f(c + h))$ is

(1)
$$\frac{f(c + h) - f(c)}{(c + h) - c} = \frac{f(c + h) - f(c)}{h}.$$

As h approaches zero, the quantity in Equation (1) approaches the slope of the tangent line to f at c. This suggests the following limit:

(2)
$$\begin{array}{c}\text{The slope of the}\\\text{tangent line to } f \text{ at } c\end{array} = \lim_{h \to 0} \left(\begin{array}{c}\text{The slope of the secant line}\\\text{from } (c, f(c)) \text{ to } (c + h, f(c + h))\end{array} \right).$$

We call the limiting slope (if it exists) the **derivative** of f at c. So that we don't have to write out "the derivative of f at c" every time we wish to refer to it, we will use the compact notation $f'(c)$ to represent the derivative of the function f at the point $x = c$. This notation is pronounced "f prime of c." Putting this notation together with Equations (1) and (2), we arrive at the formal definition of derivative:

DEFINITION 3.1

The Derivative at a Point

Given a function f and a point $x = c$, the **derivative of f at c** is defined to be

$$f'(c) := \lim_{h \to 0} \frac{f(c + h) - f(c)}{h}.$$

We can use the definition of derivative to give a precise mathematical definition of the **tangent line** to f at c.

DEFINITION 3.2

Tangent Lines

Given a function f and a point $x = c$, the **tangent line to f at c** is defined to be the line through the point $(c, f(c))$ whose slope is $f'(c)$.

Figure 3.6 shows the graph of a function f with its tangent line at a given point $x = c$ (the purple line) and the secant line from $(c, f(c))$ to $(c + h, f(c + h))$ when h is a positive number. If h is negative, the picture looks like the one in Figure 3.7. As $h \to 0$ (from the left or the right), the slope of the secant line approaches the slope of the tangent line.

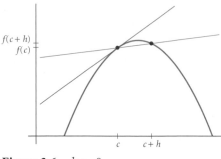

Figure 3.6 $h > 0$ **Figure 3.7** $h < 0$

> ✎ *Caution*   Remember that a limit is a two-sided process. Although we will often use examples where h is positive, in general h can be positive or negative, and if h is negative, then the point $x = c + h$ is to the *left* of $x = c$ as in Figure 3.7.   ☐

> ✎ *Caution*   Keep in mind that the derivative of f at c represents the slope of the line tangent to the graph of f at $x = c$, and that this line is obtained by taking a limit of secant lines. It is important not only that you know and can use the definition of derivative, but also that you understand how we arrived at that definition and what it *means*. This is one of the fundamental ideas of calculus.   ☐

Sometimes it will be convenient for us to use an alternative definition of derivative:

DEFINITION 3.3

Alternative Definition of Derivative at a Point

Given a function f and a real number c, the **derivative of f at c** is defined to be

$$f'(c) := \lim_{x \to c} \frac{f(x) - f(c)}{x - c}.$$

The difference quotient in Definition 3.3 represents the slope of the secant line from $(c,\ f(c))$ to $(x,\ f(x))$. Figure 3.8 shows a secant line when x is to the right of c. Figure 3.9 shows a secant line when x is to the left of c. As x approaches c from either side, the slopes of these secant lines approach the slope of the tangent line at $x = c$.

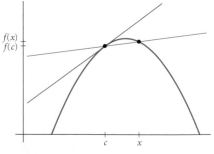

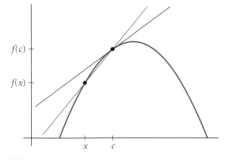

Figure 3.8 $x > c$ **Figure 3.9** $x < c$

Definition 3.3 is equivalent to Definition 3.1, with the change of variables $x := c + h$. With this substitution, the denominator from Definition 3.1 becomes $h = x - c$. As $h \to 0$, the quantity $x = c + h$ approaches $c + 0 = c$.

3.1.3 Calculating the derivative at a point

We can now calculate the *exact* slope of the tangent line to a function at a point by using the definition of derivative.

EXAMPLE 3.2   **Using the definition of derivative to find the exact slope of a tangent line**

Find the exact slope of the tangent line to the graph of $f(x) = x^2$ at $x = 3$.

Solution   We use Definition 3.1 and find that the derivative of f at $x = 3$ is equal to

$$f'(3) = \lim_{h \to 0} \frac{f(3+h) - f(3)}{h} = \lim_{h \to 0} \frac{(3+h)^2 - (3)^2}{h}.$$

It remains only to calculate this limit. Note that if we "plug in" $h = 0$ at this point, we get the indeterminate form $\frac{0}{0}$, so we must do some algebra before evaluating this limit. We calculate

$$\begin{aligned} f'(3) &= \lim_{h \to 0} \frac{f(3+h) - f(3)}{h} = \lim_{h \to 0} \frac{(3+h)^2 - (3)^2}{h} \\ &= \lim_{h \to 0} \frac{(9 + 6h + h^2) - 9}{h} = \lim_{h \to 0} \frac{6h + h^2}{h} \\ &= \lim_{h \to 0} \frac{h(6+h)}{h} = \lim_{h \to 0}(6 + h) = 6. \end{aligned}$$

Because the derivative of $f(x) = x^2$ at $x = 3$ is equal to 6, the line tangent to the graph of $y = x^2$ at the point $(3, 9)$ has slope 6.   ☐

❓ Question   Which limit rules did we use implicitly in Example 3.2?   ☐

⑤ Caution   Be careful that you do not "drop the limits" when doing a calculation like the one in Example 3.2. Since you are calculating a *limit* when you use the definition of derivative, each step of your calculation (until you actually "take the limit," of course) should include the expression "$\lim_{h \to 0}$."   ☐

✓ Checking the Answer   To check whether the solution to Example 3.2 is reasonable, sketch the graph of f and see whether its tangent line at $x = 3$ appears to have a slope of 6. Figure 3.10 shows this graph. Note that the slope of the tangent line at $x = 3$ is positive, and considering the scale of the graph, a slope of 6 seems reasonable.

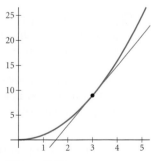

Figure 3.10

Alternatively, you could check the answer by using a graphing calculator to approximate the slope of the tangent line to $f(x) = x^2$ at $x = 3$.   ☐

⑤ Caution   You may have seen "rules" for differentiating in a previous course. Do *not* use these differentiation shortcuts at this point. Later, we will develop rules for quickly calculating derivatives, but at this point we must use the definition of derivative, as in Example 3.2.   ☐

⊛ *Caution*   Be careful when evaluating $f(c + h)$. Avoid the common mistake of accidentally writing $f(c) + h$ instead of $f(c + h)$. For example, if $f(x) = x^2$, then $f(c + h) = (c + h)^2$ (not $c^2 + h$). If $f(x) = x^2 - x$, then $f(c + h) = (c + h)^2 - (c + h)$. (Of course, we arrive at the same answer!)   □

In the following example we repeat the calculation in Example 3.2, using the alternative definition of derivative.

EXAMPLE 3.3

Using the alternative definition of the derivative

Use Definition 3.3 to find the derivative of the function $f(x) = x^2$ at $x = 3$.

Solution   By Definition 3.3 we have

$$f'(c) = \lim_{x \to 3} \frac{f(x) - f(3)}{x - 3}$$

$$= \lim_{x \to 3} \frac{x^2 - 3^2}{x - 3}$$

$$= \lim_{x \to 3} \frac{(x - 3)(x + 3)}{x - 3}$$

$$= \lim_{x \to 3} (x + 3) = 3 + 3 = 6.$$ □

The derivative of a function at a point is the slope of its tangent line at that point. We can easily use this information to find an equation for the tangent line, since we know the slope and one point on the line.

EXAMPLE 3.4

Finding the equation of a tangent line

Find the equation of the tangent line to $f(x) = x^2$ at $x = 3$.

Solution   From Definition 3.2 we know that the tangent line to $f(x) = x^2$ at $x = 3$ passes through the point $(3, f(3)) = (3, 3^2) = (3, 9)$ and has slope $f'(3) = 6$ (see Examples 3.2 and 3.3). Recall that the point-slope form of the line through (x_0, y_0) with slope m is $y - y_0 = m(x - x_0)$. Using this formula, we find that the equation of the line through $(x_0, y_0) = (3, 9)$ with slope $m = 6$ is

$$y - 9 = 6(x - 3) \implies y = 6(x - 3) + 9 \implies y = 6x - 9.$$ □

✓ *Checking the Answer*   If we graph the function $y = x^2$ and the line $y = 6x - 9$, we can see that the line is tangent to the graph of the function at $x = 3$ (we already did this in Figure 3.10).   □

In Example 3.1 we approximated the slope of the tangent line to $f(x) = \frac{2}{x}$ at $x = 1$. In the following example we find the *exact* value of this slope. Much algebraic manipulation will be necessary before we can take the limit in this calculation.

EXAMPLE 3.5

A more challenging example of using the definition of derivative

Find the slope of the line tangent to the graph of $f(x) = \frac{2}{x}$ at $x = 1$.

Solution   The slope of the tangent line to $f(x) = \frac{2}{x}$ at $x = 1$ is

$$f'(1) = \lim_{h \to 0} \frac{f(1 + h) - f(1)}{h} \qquad \text{(Definition 3.1 with } c = 1)$$

$$= \lim_{h \to 0} \frac{\frac{2}{1+h} - \frac{2}{1}}{h} \qquad \text{(since } f(x) = \frac{2}{x}\text{)}$$

Note that we cannot evaluate this limit as it is written right now, since substituting 0 for h at this point would result in the indeterminate form $\frac{0}{0}$. We must first do some algebra; continuing the calculation above, we have

$$f'(1) = \lim_{h \to 0} \frac{\left(\frac{2-2(1+h)}{1+h}\right)}{h} \qquad \text{(combining the fractions in the numerator)}$$

$$= \lim_{h \to 0} \frac{2-2-2h}{h(1+h)} \qquad \text{(since } \frac{a/b}{c} = \frac{a}{bc}\text{)}$$

$$= \lim_{h \to 0} \frac{-2h}{h(1+h)}$$

$$= \lim_{h \to 0} \frac{-2}{1+h} \qquad \text{(canceling the } h\text{'s)}$$

$$= \frac{-2}{1+0} = -2 \qquad \text{(limit rules)}$$

The line tangent to the graph of $f(x) = \frac{2}{x}$ at $x = 1$ has slope -2. In Example 3.1, we used slopes of secant lines to determine that the slope of this tangent line was *approximately* -2; now we know that the slope is in fact *equal to* -2. ☐

◈ **Caution** Every time you use the definition of derivative to calculate some $f'(c)$, you will have a limit statement that initially is of the indeterminate form $\frac{0}{0}$ (can you explain why?). Each time, you will have to do some algebra to rewrite the expression whose limit you are trying to evaluate; usually the goal in rewriting the expression is to get some h's to cancel. ☐

3.1.4 Left and right derivatives

Since the derivative of a function at a point is defined using a limit, we can use left and right limits to define the **left derivative** and the **right derivative** of a function at a point.

DEFINITION 3.4

Left and Right Derivatives

Given a function f and a real number c, the **left derivative of f at c** is

$$f'_-(c) := \lim_{h \to 0^-} \frac{f(c+h) - f(c)}{h},$$

and the **right derivative of f at c** is

$$f'_+(c) := \lim_{h \to 0^+} \frac{f(c+h) - f(c)}{h}.$$

The left derivative of a function f at a point $x = c$ is the limit of secant lines from $(c, f(c))$ to $(c + h, f(c + h))$ as $h \to 0^-$. Each value of h is *negative* in this case (see Figure 3.7 on page 224), and thus $x = c+h$ is to the *left* of $x = c$. On the other hand, the right derivative of f at c is the limit of secant lines from $(c, f(c))$ to $(c + h, f(c + h))$ as $h \to 0^+$. Each value of h is *positive* in this case (see Figure 3.6 on page 224). We can also define left and right derivatives in terms of the alternative definition of derivative, by considering $x \to c^-$ and $x \to c^+$.

Since the limit of the secant lines as $h \to 0$ exists if and only if the left and right limits exist and are equal, the derivative of a function f at a point c will be well defined only if its left and right derivatives exist and are equal.

THEOREM 3.1

The Relationship Between Two-Sided and One-Sided Derivatives

Given a function f and a value $x = c$, the derivative $f'(c)$ exists if and only if $f'_-(c)$ and $f'_+(c)$ exist and are equal.

We will make use of this theorem in Section 3.3 when we discuss differentiability. The proof of Theorem 3.1 follows directly from the definitions of left, right, and "two-sided" derivatives and properties of limits. It is left to you in the exercises.

Left and right derivatives can be useful when examining piecewise functions at their "break" points, as in the next example.

EXAMPLE 3.6

The derivative of a piecewise function at its "break" point

If $f(x) = \begin{cases} 3x + 2, & \text{if } x < 1 \\ 6 - x, & \text{if } x \geq 1 \end{cases}$, calculate $f'_+(1)$ and $f'_-(1)$.

Solution   Using Definition 3.4, the right derivative of f at $x = 1$ is

$$f'_+(1) = \lim_{h \to 0^+} \frac{f(1 + h) - f(1)}{h} \qquad \text{(Definition 3.4, } c = 1\text{)}$$

$$= \lim_{h \to 0^+} \frac{(6 - (1 + h)) - (6 - 1)}{h} \qquad (h \to 0^+ \text{ means } h > 0, \text{ so } 1 + h > 1)$$

$$= \lim_{h \to 0^+} \frac{6 - 1 - h - 6 + 1}{h}$$

$$= \lim_{h \to 0^+} \frac{-h}{h} = \lim_{h \to 0^+} (-1) = -1.$$

On the other hand, the left derivative of f at $x = 1$ is

$$f'_-(1) = \lim_{h \to 0^-} \frac{f(1 + h) - f(1)}{h} \qquad \text{(Definition 3.4, } c = 1\text{)}$$

$$= \lim_{h \to 0^-} \frac{(3(1 + h) + 2) - (6 - 1)}{h} \qquad (h \to 0^- \text{ means } h < 0, \text{ so } 1 + h < 1)$$

$$= \lim_{h \to 0^-} \frac{3 + 3h + 2 - 6 + 1}{h}$$

$$= \lim_{h \to 0^-} \frac{3h}{h} = \lim_{h \to 0^-} 3 = 3.$$

Note that in the second line of the calculation of the left derivative, we used $3x + 2$ to calculate $f(1 + h)$, since as $h \to 0^-$ we have $h < 0$ and thus $1 + h < 1$. However, to calculate $f(1)$ we needed to use $6 - x$, since $1 \geq 1$.

Since the left and right derivatives of f at $x = 1$ are not equal to each other, the derivative of f at $x = 1$ is undefined.   □

⊗ *Caution*   Be sure that you understand when to use each piece of a piecewise-defined function when calculating left and right derivatives. (See the calculation of $f'_-(1)$ in Example 3.6.)   □

☑ *Checking the Answer*   Figure 3.11 shows the graph of the function $f(x)$ from Example 3.6. We can see from the graph that the right derivative of the function at $x = 1$ is negative (think about the secant lines when $h > 0$), while the left derivative at $x = 1$ is positive (and steeper). In fact, since the pieces of this graph are linear, we can use the graph to show that the slopes of the secant lines from the right and the left are -1 and 3, respectively.

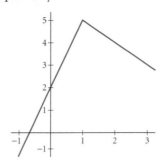

Figure 3.11

Since the left and right derivatives at $x = 1$ do not agree, there is no well-defined tangent line at $x = 1$; the (two-sided) derivative of f at $x = 1$ does not exist. Graphically, there is not a *unique* line that passes through the point $(1, f(1))$ and just "skims" the graph of f.   ☐

3.1.5 Approximating derivatives

Recall that we can approximate $\lim_{x \to c} f(x)$ by using a table of values. The derivative of a function f at a point $x = c$ is the limit (as $h \to 0$) of the slopes of the secant lines from $(c, f(c))$ to $(c + h, f(c + h))$. We can approximate $f'(c)$ by using a table of values of the difference quotient, since

(3) $$f'(c) \approx \frac{f(c + h) - f(c)}{h},$$

for values of h that are close to zero.

§ *Caution*   In the definition of derivative, the function whose limit we are taking is the "slope of the secant line" function (the difference quotient in Equation (3)). The variable that is "moving" in the limit is h. Don't confuse this with limits of the form $\lim_{x \to c} f(x)$, where we arc taking the limit of the function $f(x)$ and where the variable that is "moving" is x. Can you illustrate the difference graphically?   ☐

EXAMPLE 3.7

Approximating a derivative with a table

Use a table of values to approximate $f'(-2)$, where $f(x) = x^3 + 4$.

Solution   Each difference quotient in our table will be of the form

(4) $$\frac{f(c + h) - f(c)}{h} = \frac{((-2 + h)^3 + 4) - ((-2)^3 + 4)}{h}.$$

For example, if $h = 0.01$, then

$$\frac{f(-2 + 0.01) - f(-2)}{0.01} = \frac{((-2 + 0.01)^3 + 4) - ((-2)^3 + 4)}{0.01} \approx 11.94.$$

(*Calculator Hint:* Type in the entire expression at once in order to reduce rounding error; small errors can make an enormous difference in these kinds of calculations.)

From this estimate we might guess that $f'(-2) = 12$, or at least that $f'(-2)$ is very close to 11.94. To get a better estimate we could calculate the difference quotients with even smaller values of h, say $h = 0.001$ or $h = -0.0001$. We collect a number of these estimates in Table 10.3.

h	-0.1	-0.01	-0.001	0	0.001	0.01	0.1
$\frac{f(-2+h)-f(-2)}{h}$	12.61	12.0601	12.0060	*	11.9940	11.9401	11.41

Table 3.1

(*Calculator Hint:* If you type the difference quotient into your calculator as a function of h (see Equation (4)), then you can quickly evaluate the difference quotients for different values of h without retyping each time.)

From Table 10.3 it is reasonable to approximate that as $h \to 0$, the difference quotient approaches 12. Since the derivative $f'(-2)$ is by definition equal to the limit of this difference quotient as $h \to 0$, we approximate that $f'(-2) = 12$.   □

? Question   We can actually use a limit calculation to find $f'(-2)$ exactly with the definition of derivative. Do this, and verify that $f'(-2) = 12$.   □

In Example 3.7, we approximated the derivative of $f(x) = x^3 + 4$ at $x = -2$ by constructing a table of values for $\frac{f(-2+h)-f(-2)}{h}$, using various small values of h. There is an alternative but equivalent way to go about approximating $f'(-2)$. We could also construct a table of values for the function $f(x) = x^3 + 4$, using various values of x close to $x = -2$. Then we could take difference quotients using the information in the table. Table 3.2 is the corresponding table for $f(x) = x^3 + 4$.

x	-2.1	-2.01	-2.001	-2	-1.999	-1.99	-1.9
$f(x)$	-5.621	-4.121	-4.012	-4	-3.988	-3.881	-2.859

Table 3.2

We can now use this table to calculate difference quotients; for example,

$$\frac{f(-1.99) - f(-2)}{0.01} \approx \frac{-3.881 - (-4)}{0.01} = 11.9.$$

Note that this difference quotient involves the same calculation as the second-to-last entry of Table 10.3 (although here we had some extra rounding error).

EXAMPLE 3.8

Left, right, and two-sided approximations of a derivative

Use Table 3.2 to calculate left, right, and two-sided approximations for $f'(-2)$.

Solution   For the left estimate for the derivative of f at $x = -2$, we will use the values of f at $x = -2$ and at the closest left point, $x = -2.001$. According to this estimate,

$$f'(-2) \approx \frac{f(-2.001) - f(-2)}{-0.001} \approx \frac{-4.012 - (-4)}{-0.001} = 12.$$

This is in fact the estimate from the third entry in Table 10.3 (where $h = -0.001$), but with greater rounding error. For the right estimate, we use $x = -2$ and the x-value in the table closest to $x = -2$ on the right:

$$f'(-2) \approx \frac{f(-1.999) - f(-2)}{0.001} \approx \frac{-3.988 - (-4)}{0.001} = 12.$$

This is the estimate from the fifth entry in Table 10.3 (where $h = 0.001$), but again with greater rounding error. Finally, for the two-sided estimate (which is not represented in

Table 10.3), we use the x-values closest to $x = -2$ on either side:

$$f'(-2) \approx \frac{f(-1.999) - f(-2.001)}{0.002} \approx \frac{-3.988 - (-4.012)}{0.002} = 12.$$

Each of our estimates came out to be approximately 12, so we can conclude that 12 is probably a decent estimate for $f'(-2)$. □

The two-sided difference quotient is sometimes called a **symmetric difference.** In general, this kind of difference quotient has the form

$$\frac{f(c + h) - f(c - h)}{2h}$$

for some small positive value of h and represents the slope of the line from the point $(c - h, \ f(c - h))$ to the point $(c + h, \ f(c + h))$. Figure 3.12 illustrates a line whose slope is the symmetric difference about $x = c$ for some small value of h and function f. This line is close in slope to the tangent line at $x = c$ (when such a tangent line exists).

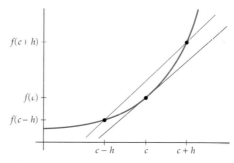

Figure 3.12

⑤ *Caution*   The two-sided "symmetric difference" approximation of the tangent line can be dangerous when there is not a well-defined tangent line to f at c. For example, consider the function $f(x) = |x|$ at $c = 0$. The symmetric difference $\frac{f(0+h) - f(0-h)}{2h} = \frac{|0+h|}{2h} = \frac{|h| - |h|}{2h} = \frac{0}{2h}$ is zero for all nonzero values of h, which would suggest that $f'(0) = 0$; however, the absolute value function does not have a well-defined tangent line at $c = 0$ (since it is "pointy" there). □

What you should know after reading Section 3.1

▶ The definition of derivative (and the alternative definition) as a limit of slopes of secant lines. Be able to explain this definition graphically as well as write it down algebraically. What is the definition of the tangent line to a function at a point?

▶ How to approximate the slope of a tangent line by calculating slopes of "nearby" secant lines, and how to calculate derivatives exactly using the definition (or the alternative definition). How to find the equation of a tangent line.

▶ The algebraic definitions and geometrical interpretations of the left and right derivatives of a function at a point. Be able to calculate left and right derivatives, especially for piecewise functions.

▶ Be able to approximate the derivative of a function at a point using a table of values for the function, including left, right, and two-sided estimates. Also be able to construct a table of difference quotient values and use the table to approximate a derivative.

Exercises 3.1

Concepts

0. Read the section and make your own summary of the material. Include whatever you think will help you review the material later on.

1. Explain why it is not a simple task to calculate the slope of the tangent line to a function f at a point $x = c$. Shouldn't calculating the slope of a line be really easy? What goes wrong here?

2. What is the formal mathematical description of the *tangent line* to a function f at a point $x = c$?

3. What is the formal mathematical description of the *slope* of the tangent line to a function f at a point $x = c$?

4. Explain in your own words (and using mathematical notation) how one can approximate the slope of the tangent line to a function f at a point $x = c$ using a secant line. Use pictures to support your explanation. Then explain how to use this kind of approximation, and a limit, to find the *exact* value of the slope of the tangent line.

■ For each function f and value $x = c$ given, sketch a graph of f near $x = c$. Sketch the tangent line to f at c, and the secant lines from $(c, f(c))$ to $(c + h, f(c + h))$ for $h = 1$, $h = 0.5$, and $h = -0.5$. Which secant line seems to be the best approximation for the tangent line to f at c?

5. $f(x) = x^2$, $c = 0$

6. $f(x) = 3 - x^2$, $c = 1$

7. $f(x) = 3x + 5$, $c = -2$

8. The graph below shows a function $f(x)$ together with a tangent line and a secant line for $f(x)$. The slope of the tangent line is approximated by the slope of the secant line (although not very accurately, in this example). In other words, we have

$$f'(c) \approx \frac{f(c + h) - f(c)}{h}.$$

Sketch a copy of the graph below, and label each of the following quantities on the graph. *Note:* Some of these values are locations, some are slopes, and some are distances; be sure to make clear which are which.

(a) c

(b) h

(c) $c + h$

(d) $f(c)$

(e) $f(c + h)$

(f) $f(c + h) - f(c)$

(g) $\dfrac{f(c + h) - f(c)}{h}$

(h) $f'(c)$

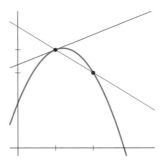

9. Repeat Problem 8 using the graph below.

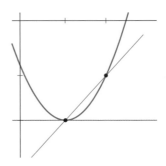

10. Repeat Problem 8 using the graph below.

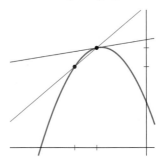

11. Repeat Problem 8 using the alternative definition of derivative from Definition 3.3, as follows. For values of x close to c, the slope of the secant line from $(x, f(x))$ to $(c, f(c))$ approximates the slope of the tangent line to f at c:

$$f'(c) \approx \frac{f(x) - f(c)}{x - c}.$$

Sketch a copy of the graph in Problem 8, and label each of the following quantities on the graph. *Note:* Some of these values are locations, some are slopes, and some are distances; be sure to make clear which are which.

(a) c **(b)** $x - c$

(c) x **(d)** $f(c)$

(e) $f(x)$ **(f)** $f(x) - f(c)$

(g) $\dfrac{f(x) - f(c)}{x - c}$ **(h)** $f'(c)$

12. Make a copy of your labeled graph from Problem 8. Sketch additional secant lines on this graph for three more values of h to illustrate that the secant lines approach the tangent line as $h \to 0$.

13. Make a copy of your labeled graph from Problem 9. Sketch additional secant lines on this graph for three more values of h to illustrate that the secant lines approach the tangent line as $h \to 0$.

14. Make a copy of your labeled graph from Problem 10. Sketch additional secant lines on this graph for three more values of h to illustrate that the secant lines approach the tangent line as $h \to 0$.

15. Use Definition 3.1 to explain why every time you use the definition of derivative to calculate some $f'(c)$, you will have a limit statement that initially is of the indeterminate form $\frac{0}{0}$. What do you have to do in such a situation to evaluate the limit?

16. Explain why the alternative definition of derivative (Definition 3.3) is equivalent to our original definition of derivative (Definition 3.1).

17. Express the left and right derivatives $f'_-(c)$ and $f'_+(c)$ in terms of the alternative definition of derivative in Definition 3.3.

18. Fill in the blanks: The derivative of a function f at a point c will be well defined only if its left and right derivatives _____ and are _____.

19. Write the definition of derivative given in Definition 3.1 as a delta–epsilon statement.

20. Write the alternative definition of derivative given in Definition 3.3 as a delta–epsilon statement.

Skills

■ For each f, c, and x, find the slope of the secant line from $(c, f(c))$ to $(x, f(x))$.

21. $f(x) = 3x^2 + 1$, $\quad c = -1$, $\quad x = -0.5$

22. $f(x) = 1 - x^3$, $\quad c = 2$, $\quad x = 1.9$

23. $f(x) = x^2 - 4$, $\quad c = 1$, $\quad x = 1.01$

24. $f(x) = x^2 - 4$, $\quad c = 1$, $\quad x = 1.001$

25. $f(x) = x^2 - 4$, $\quad c = 1$, $\quad x = 0.99$

26. $f(x) = x^2 - 4$, $\quad c = 1$, $\quad x = 0.999$

■ For each f, c, and h, find the slope of the secant line from $(c, f(c))$ to $(c + h, f(c + h))$.

27. $f(x) = 3x^2 + 1$, $\quad c = -1$, $\quad h = 0.5$

28. $f(x) = 1 - x^3$, $\quad c = 2$, $\quad h = -0.1$

29. $f(x) = x^2 - 4$, $\quad c = 1$, $\quad h = 0.01$

30. $f(x) = x^2 - 4$, $\quad c = 1$, $\quad h = 0.001$

31. $f(x) = x^2 - 4$, $\quad c = 1$, $\quad h = -0.01$

32. $f(x) = x^2 - 4$, $\quad c = 1$, $\quad h = -0.001$

33. $f(x) = 2x - 3$, $\quad c = -2$, $\quad h = -1$

34. $f(x) = \dfrac{x}{x + 1}$, $\quad c = 1$, $\quad h = 0.01$

■ For each function f and value $x = c$, approximate the slope of the tangent line to f at c by calculating the slopes of *four* nearby secant lines, as we did in Example 3.1. You can choose your own values for h, but choose two positive and two negative values. Bonus Question: Why use four secant lines instead of just one?

35. $f(x) = x^2$, $\quad c = 1$

36. $f(x) = 1 - x$, $\quad c = 3$

37. $f(x) = \dfrac{1}{2x}$, $\quad c = -2$

38. $f(x) = 4$, $\quad c = -2$

39. $f(x) = x^4 - 3x^2 - 5$, $\quad c = 0$

40. $f(x) = |x - 1|$, $\quad c = 1$

■ For each function f and value $x = c$, use the definition of derivative to calculate the *exact* value of the slope of the tangent line to f at c, if it exists. Justify each step in your calculation. When you have finished your calculation, use a graph to check that your answer is reasonable.

41. $f(x) = x^2$, $\quad c = 1$

42. $f(x) = 1 - x$, $\quad c = 3$

43. $f(x) = \dfrac{1}{2x}$, $\quad c = -2$

44. $f(x) = 4$, $\quad c = -2$

45. $f(x) = x^4 - 3x^2 - 5$, $\quad c = 0$

46. $f(x) = |x - 1|$, $\quad c = 1$

47. $f(x) = 3$, $\quad c = 2$

48. $f(x) = 3 + x^2$, $\quad c = 1$

49. $f(x) = 1 + x + x^2$, $\quad c = 0$

■ For each f and c, use the definition of derivative to find $f'(c)$.

50. $f(x) = x^2$, $\quad c = 0$

51. $f(x) = x^2$, $\quad c = -3$

52. $f(x) = x^2$, $\quad c = 1$

53. $f(x) = 4x + 3$, $\quad c = -2$

54. $f(x) = -2$, $\quad c = 8$

55. $f(x) = 1 - x^3$, $\quad c = -1$

56. $f(x) = 1 - x - x^2$, $\quad c = 1$

57. $f(x) = \dfrac{1}{x}$, $\quad c = -1$

58. $f(x) = x^4 + 1$, $\quad c = 2$

■ Repeat Problems 50–58 using the alternative definition of derivative (Definition 3.3) to calculate $f'(c)$.

59. f and c from Problem 50

60. f and c from Problem 51

61. f and c from Problem 52

62. f and c from Problem 53

63. f and c from Problem 54

64. f and c from Problem 55

65. f and c from Problem 56

66. f and c from Problem 57

67. f and c from Problem 58

■ Find the equations of the lines described below. You may find your answers from Problems 50–58 useful. After finding the linear equations algebraically, use graphs to support your answers.

68. The tangent line to $f(x) = x^2$ at $x = 0$

69. The tangent line to $f(x) = x^2$ at $x = -3$

70. The line tangent to the graph of $y = 4x + 3$ at the point $(-2, -5)$

71. The line tangent to the graph of $y = 1 - x - x^2$ at the point $(1, -1)$

72. The line that passes through the point $(3, 2)$ and is parallel to the tangent line to $f(x) = \frac{1}{x}$ at $x = -1$

73. The line that is perpendicular to the tangent line to $f(x) = x^4 + 1$ at $x = 2$ and passes through $(-1, 8)$

■ Each of the following limits represents the derivative of some function f at some point $x = c$. Find f and c, and then evaluate the limit to find $f'(c)$. (There may be more than one possible answer.)

74. $\lim\limits_{h \to 0} \dfrac{(3(-1+h)^2 + 1) - 4}{h}$

75. $\lim\limits_{h \to 0} \dfrac{\frac{1}{2+h} - 0.5}{h}$

76. $\lim\limits_{h \to 0} \dfrac{(1-h) - 1}{h}$

77. $\lim\limits_{x \to -1} \dfrac{(3x^2 + 1) - 4}{x + 1}$

78. $\lim\limits_{x \to 3} \dfrac{x^4 - 81}{x - 3}$

79. $\lim\limits_{x \to 0} \dfrac{\frac{1}{x+1} - 1}{x}$

■ For each piecewise function f and value c below, use Definition 3.4 to calculate $f'_-(c)$ and $f'_+(c)$. Use the results to determine the value of $f'(c)$, if it exists. After completing your calculations algebraically, check them with graphs.

80. $f(x) = \begin{cases} x + 4, & \text{if } x < 2 \\ 3x, & \text{if } x \geq 2 \end{cases}$, $c = 2$

81. $f(x) = \begin{cases} x + 4, & \text{if } x \leq 2 \\ 3x, & \text{if } x > 2 \end{cases}$, $c = 2$

82. $f(x) = \begin{cases} -1, & \text{if } x < 0 \\ x - 1, & \text{if } x \geq 0 \end{cases}$, $c = 0$

83. $f(x) = \begin{cases} x^2, & \text{if } x \leq -1 \\ -1 - 2x, & \text{if } x > -1 \end{cases}$, $c = -1$

■ For each function f and value c given below, construct a table of difference quotients (similar to Table 10.3) to approximate $f'(c)$. Then graph $f(x)$ and examine its slope at $x = c$ to check your approximation. (Note that in each of these examples it would be quite tedious, not to mention complicated, to calculate the exact value of the derivative using Definition 3.1.)

84. $f(x) = x^5 - 26$, $c = 3$

85. $f(x) = 1 + x + x^2 + x^3$, $c = 0$

86. $f(x) = \dfrac{1}{x^2 - x + 1}$, $c = -1$

87. $f(x) = (1 + \sqrt{x})^2$, $c = 9$

88. $f(x) = \dfrac{1 - x}{1 + x}$, $c = 1$

89. $f(x) = \dfrac{1}{1 + \sqrt{x}}$, $c = 4$

■ For problems 90–93, calculate the best possible left, right, and two-sided difference quotient estimates for $f'(c)$. Then plot the data for f, and sketch lines whose slopes are these three estimates. Which approximation seems the most accurate?

90. The table below, with $c = 0$

x	−0.5	−0.2	0	0.15	0.4
$f(x)$	1.25	1.04	1	1.023	1.16

91. The table in Problem 90, with $c = 0.15$

92. The table below, with $c = 3$

x	1	2	3	4	5
$f(x)$	2	2	0	−4	−10

93. The table in Problem 92, with $c = 2$

Proofs

94. Use the mathematical definition of a tangent line and the point-slope form of a line to show that, for any function f and point $x = c$ with a well-defined tangent line, that tangent line is given by the equation

$$(5) \qquad y = f'(c)(x - c) + f(c).$$

95. Use Definitions 3.1 and 3.4 and properties of limits to prove Theorem 3.1.

96. Use the definition of derivative to show that if $f(x) = 3x + 1$, then $f'(c) = 3$ for *any* value of c. Why does this make graphical sense?

97. Use the definition of derivative to show that if f is any linear function $f(x) = mx + b$, then $f'(0) = m$.

98. Use the definition of derivative to show that if f is any linear function $f(x) = mx + b$, then $f'(c) = m$ for *all* values of c. Why does this make graphical sense?

3.2   The Derivative as an Instantaneous Rate of Change

In the previous section we learned how to define and calculate the slope of the line tangent to a function at a point. Tangent lines, and thus derivatives, have many important real-world applications. In this section we examine some of these applications and interpret the derivative as an instantaneous rate of change.

3.2.1 Average velocity

Let's start with a real-world example. If you walked 2 miles in 30 minutes, how fast did you travel? In a sense that is an unanswerable question, since you may have traveled slowly for some parts of your journey and quickly during other parts, and just knowing how long it took you to walk 2 miles won't tell *how* you chose to walk along the way.

However, we can answer this question in a limited way. For simplicity, let's assume that you walked along a straight and level road. Given that it took you half an hour to walk 2 miles, it took you 15 minutes, on average, to walk a mile. Said another way, this means that you could walk 4 miles in 1 hour, which means that your *average* velocity was

$$(1) \qquad \text{average velocity} = \frac{2 \text{ miles}}{0.5 \text{ hour}} = 4 \text{ mph}.$$

◈ *Caution*   Velocity is not the same as speed. Speed tells you how fast something is going, while velocity tells you the speed of an object *and its direction*. When talking about velocity we will always have a "positive direction" in mind. When an object travels in that direction, it will have a positive velocity; when it travels in the opposite direction, it will have a negative velocity. Note that although velocity can be negative, speed can never be negative.   □

The calculation in Equation (1) seemed like common sense, but how did we actually get that answer? We used the **distance formula,** which says that the distance d, time t, and rate r of your journey are related by the equation $d = rt$, or "distance equals rate times time." We wanted to find the rate, so we used the equation $r = \frac{d}{t}$, with $d = 2$ miles and $t = 0.5$ hour. The rate in the distance formula is an *average* rate, or the constant rate that you would have to walk if you wanted to travel 2 miles in half an hour.

To be more precise, the distance formula says that your average rate is equal to the quotient of the *difference* of your starting and ending distances and the *difference* of your starting and ending times. If you started at distance d_0 from some reference point at time t_0 and ended at distance d_1 at time t_1, your average velocity would be

$$(2) \qquad r = \frac{d}{t} = \frac{d_1 - d_0}{t_1 - t_0}.$$

Equation (2) is a generalized distance formula. Note that the starting distance and time need not be zero.

EXAMPLE 3.9   **Finding average velocity**

While walking home from a concert, you were 5 miles from home 2 hours after the concert and 3 miles from home 2.5 hours after the concert. Your average velocity on this trip was (using $d_0 = 5$, $d_1 = 3$, $t_0 = 2$, and $t_1 = 2.5$):

$$r = \frac{3 \text{ miles} - 5 \text{ miles}}{2.5 \text{ hours} - 2 \text{ hours}} = \frac{-2 \text{ miles}}{0.5 \text{ hour}} = -4 \text{ mph}. \qquad \square$$

♦ *Caution*   Wait a minute! What does it mean for our answer to be negative? All it means is that you were traveling in a "negative direction," since this example was phrased in terms of "miles from home" and your distance from home was *decreasing* (since you were walking toward home).   □

3.2.2 A review of average rates of change

Recall from Section 1.3 that in general, the average rate of change of a function f on an interval $I = [a, b]$ is given by the following definition:

DEFINITION 3.5

Average Rate of Change

The *average rate of change from x = a to x = b* of a function f is the quotient

$$\frac{f(b) - f(a)}{b - a}.$$

We also call this the *average rate of change of f on the interval I = [a, b]*.

EXAMPLE 3.10

Finding the average rate of change of a function on an interval

Calculate the average rate of change of the function $f(x) = x^2 - x + 1$ on the interval $[-2, 1]$.

Solution   Using Definition 3.5, the average rate of change of f on $[-2, 1]$ is

$$\frac{f(1) - f(-2)}{1 - (-2)} = \frac{(1^2 - 1 + 1) - ((-2)^2 - (-2) + 1)}{3} = \frac{1 - 7}{3} = \frac{-6}{3} = -2.$$

□

We can write the expression for average velocity in Equation (2) as an average rate of change of position. Let $s(t)$ be the function describing your distance (from some reference point) as a function of time t. Then your *average velocity* from time a to time b is the average rate of change of your position from time a to time b.

DEFINITION 3.6

Average Velocity

Suppose an object travels along a straight path with position function $s(t)$. The *average velocity* of the object from $t = a$ to $t = b$ is defined to be the average rate of change of position $s(t)$ from $t = a$ to $t = b$:

$$\begin{array}{ccc} \text{average velocity} \\ \text{from } t = a \text{ to } t = b \end{array} = \frac{s(b) - s(a)}{b - a} = \begin{array}{c} \text{average rate of change of} \\ s(t) \text{ from } t = a \text{ to } t = b. \end{array}$$

The leftmost equality in Definition 3.6 comes from the distance formula in Equation (2). The rightmost equality comes from the definition of the average rate of change of a function on an interval.

? *Question*   How is the definition of average velocity in Definition 3.6 the same as that in Equation (2)?   □

THEOREM 3.2

> ### Graphical Interpretation of Average Rate of Change
> The average rate of change of a function f on an interval $I = [a, b]$ is equal to the slope of the line connecting the points $(a, f(a))$ and $(b, f(b))$.

The proof of Theorem 3.2 follows directly from Definition 3.5 and is left to the exercises.

EXAMPLE 3.11

Seeing an average rate of change as a slope

Figure 3.13 shows the graph of a function $f(x)$ with its average rate of change on $[a, b]$ represented as the slope of the line from $(a, f(a))$ to $(b, f(b))$.

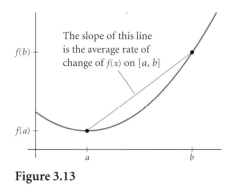

The slope of this line is the average rate of change of $f(x)$ on $[a, b]$

Figure 3.13

Combining Theorem 3.2 with the definition of average velocity, we can say that the average velocity of a moving object over a time interval $t = a$ to $t = b$ is equal to the slope of the line between the two points $(a, s(a))$ and $(b, s(b))$ on the graph of the position function $s(t)$.

3.2.3 Instantaneous rates of change

The average velocity of an object over a particular time interval describes the average speed and direction of the object over that entire time interval. How can we determine how fast an object is going at a particular instant in time? In other words, how can we find the ***instantaneous velocity*** of a moving object?

It can be shown that a watermelon dropped from a height of 100 feet will have position $s(t) = -16t^2 + 100$ feet t seconds after it is dropped (we will see where this equation comes from in Section 3.4.6). This equation sets ground level at $s = 0$ feet and the top of the building at $s = 100$ feet.

How fast does the watermelon fall? This question could mean many things, for example: What is the average velocity of the watermelon over its entire fall? What is the average velocity of the watermelon from $t = 1$ second to $t = 2$ seconds? How fast is the watermelon falling exactly $t = 1$ second after being dropped? The first two questions can be answered using average rates of change; the last question is more difficult and will involve a *limit* of average rates of change.

> ❓ *Question*   Find the average velocity of the watermelon over its entire fall. Note that you will first have to determine how long the watermelon is in the air! You should find that the watermelon takes 2.5 seconds to hit the ground and that its average velocity over that time is -40 feet per second. (The average velocity is negative because the watermelon is falling down, against the positive direction.)

Of course, although the average velocity of the watermelon is -40 feet per second, the watermelon does not fall at a *constant* velocity of -40 feet per second. When the watermelon is first released, its initial velocity is zero. After the watermelon is released, gravity causes it to fall toward the ground and to speed up as it falls.

? Question   Does the equation $s(t) = -16t^2 + 100$ really reflect the fact that the watermelon is speeding up as it falls? Calculate the average velocity of the watermelon over the following half-second time intervals: $[0, 0.5]$, $[0.5, 1]$, $[1, 1.5]$, $[1.5, 2]$, and $[2, 2.5]$. What do you notice about these average velocities?   □

How can we determine the velocity of the watermelon at a *particular* time, for example, 1 second after it is dropped? Intuitively, we know that the watermelon's speed won't change much over a very small time interval. Therefore, it is reasonable to use an average velocity over a small time interval to approximate the *instantaneous* velocity of the watermelon at $t = 1$.

EXAMPLE 3.12

Finding average velocity over shrinking time intervals

Calculate the average velocity of the falling watermelon described above over the time intervals $[1, 2]$, $[1, 1.5]$, and $[1, 1.25]$.

Solution   Over the interval $[1, 2]$, the average velocity of the watermelon is

$$\frac{s(2) - s(1)}{2 - 1} = \frac{(-16(2)^2 + 100) - (-16(1)^2 + 100)}{1}$$
$$= 36 - 84 = -48 \text{ feet per second.}$$

From $t = 1$ to $t = 1.5$, the average velocity of the watermelon is

$$\frac{s(1.5) - s(1)}{1.5 - 1} = \frac{(-16(1.5)^2 + 100) - (-16(1)^2 + 100)}{0.5}$$
$$= \frac{64 - 84}{0.5} = -40 \text{ feet per second.}$$

Finally, the average velocity of the watermelon over the time interval $[1, 1.25]$ is

$$\frac{s(1.25) - s(1)}{1.25 - 1} = \frac{(-16(1.25)^2 + 100) - (-16(1)^2 + 100)}{0.25}$$
$$= \frac{75 - 84}{0.25} = -36 \text{ feet per second.}$$ □

Since $[1, 1.25]$ is the smallest of the three intervals, the average velocity of the watermelon over this interval is likely to be closest to the instantaneous velocity of the watermelon at time $t = 1$ second. To get an even better estimate of the instantaneous velocity, we could calculate the average velocity over an even smaller time interval, say $[1, 1.001]$.

? Question   Show that the average velocity of the watermelon from $t = 1$ to $t = 1.001$ is -32.016 feet per second. Use this average velocity to make an educated guess for the instantaneous velocity of the watermelon at time $t = 1$.   □

Each of the average rates of change in Example 3.12 involved an interval of the form $[1, 1 + h]$ for some value of h. As h approaches zero, these average rates of change should get closer and closer to the instantaneous rate of change. In limit notation, we would write this statement as

$$\begin{pmatrix} \text{instantaneous velocity} \\ \text{at } t = 1 \end{pmatrix} = \lim_{h \to 0} \begin{pmatrix} \text{average velocity} \\ \text{from } t = 1 \text{ to } t = 1 + h \end{pmatrix} = \lim_{h \to 0} \frac{s(1 + h) - s(1)}{(1 + h) - 1}.$$

We now take a graphical look at the calculations in Example 3.12. Recall that average velocity is the average rate of change of position and that average rates of change can be seen as slopes. Figures 3.14–3.16 show the three average rates of change calculated in Example 3.12 as slopes.

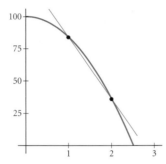

Figure 3.14
Average rate of change of $s(t)$
on $[1, 2]$

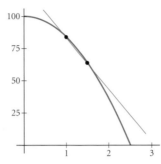

Figure 3.15
Average rate of change of $s(t)$
on $[1, 1.5]$

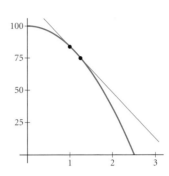

Figure 3.16
Average rate of change of $s(t)$
on $[1, 1.25]$

As $h \to 0$, the lines whose slopes are these average velocities from 1 to $1 + h$ approach the tangent line to the position function $s(t)$ of the watermelon at time $t = 1$ (see Figure 3.17). Therefore, if the instantaneous velocity is the limit as $h \to 0$ of average velocities, then the instantaneous velocity at $t = 1$ should be represented by the tangent line to $s(t)$ at $t = 1$.

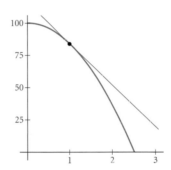

Figure 3.17
Instantaneous rate of change
of $s(t)$ at $t = 1$

This should sound very familiar! Both algebraically and graphically, the instantaneous velocity of the watermelon at $t = 1$ is the *derivative* of the position function at $t = 1$. This is in fact the definition of instantaneous velocity.

DEFINITION 3.7

Instantaneous Velocity

Suppose an object travels along a straight path with position function $s(t)$. The ***instantaneous velocity*** of the object at $t = c$ is defined to be the derivative of the position function at $t = c$:

$$\text{instantaneous velocity at } t = c = \lim_{h \to 0} \frac{s(c + h) - s(c)}{h}.$$

In other words, if $v(t)$ represents the velocity of the object at time t, then given any time $t = c$, we have $v(c) = s'(c)$.

EXAMPLE 3.13 **Calculating instantaneous velocity**

Suppose that a watermelon dropped off a 100-foot building has position $s(t) = -16t^2 + 100$ feet t seconds after being dropped. Find the instantaneous velocity of the watermelon 1 second after it is dropped.

Solution By Definition 3.7, the instantaneous velocity of the watermelon is $s'(1)$. By the definition of derivative,

$$s'(1) = \lim_{h \to 0} \frac{s(1+h) - s(1)}{h} \quad \text{(definition of derivative)}$$

$$= \lim_{h \to 0} \frac{(-16(1+h)^2 + 100) - (-16(1)^2 + 100)}{h} \quad \text{(formula for } s(t))$$

$$= \lim_{h \to 0} \frac{-16 - 32h - 16h^2 + 100 + 16 - 100}{h} \quad \text{(multiply out)}$$

$$= \lim_{h \to 0} \frac{-32h - 16h^2}{h}$$

$$= \lim_{h \to 0} (-32 - 16h) = -32. \quad \text{(limit rules)}$$

The instantaneous velocity of the watermelon at time $t = 1$ is -32 feet per second. Compare this exact answer with our earlier approximations. □

? Question We could also have used the "alternative" definition of derivative, and found $s'(1)$ by calculating the limit

$$\lim_{t \to 1} \frac{s(t) - s(1)}{t - 1} = \frac{(-16t^2 + 100) - (-16(1)^2 + 100)}{t - 1}.$$

Do this alternative calculation; of course you should arrive at the same answer: -32 feet per second. □

In general, we define the ***instantaneous rate of change*** of *any* function f at a point $x = c$ to be the slope of the tangent line to f at $x = c$ (if it exists). This makes some intuitive sense even if f is not a position function, since the tangent line to f at $x = c$ is the "direction" of f at $x = c$ and describes how the graph of f changes at $x = c$.

DEFINITION 3.8

Instantaneous Rate of Change

The ***instantaneous rate of change*** of a function f at a point $x = c$ is the derivative $f'(c)$ of f at $x = c$.

⟁ Caution When f is a position function for a moving object, the instantaneous velocity of the object at time $t = c$ is by definition the instantaneous rate of change of the position of the object. This means that the expressions "instantaneous velocity" and "instantaneous rate of change" can be used interchangeably when f happens to be a position function. □

As the next example shows, we can compare the instantaneous and average rates of change of a function graphically by using the interpretations of these rates of change as tangent and secant lines.

EXAMPLE 3.14

Graphically comparing average and instantaneous rates of change

Given that f is the function shown in Figure 3.18, put the following four quantities in order from least to greatest:

  (a) The average rate of change of f on $[0, 2]$.

  (b) The average rate of change of f from $x = -1$ to $x = 0$.

  (c) The instantaneous rate of change of f at $x = 0$.

  (d) The instantaneous rate of change of f at $x = -1$.

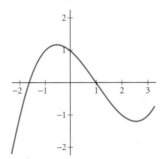

Figure 3.18

Solution   The average rates of change in parts (a) and (b) are the slopes of the green secant lines in Figure 3.19. The instantaneous rates of change in parts (c) and (d) are the slopes of the purple tangent lines in Figure 3.20.

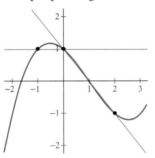

Figure 3.19

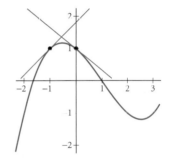

Figure 3.20

Of the four lines, the only one with a positive slope is the tangent line to f at $x = -1$ in Figure 3.20, so (d) is the largest of the four quantities. In Figure 3.19 we see that the average rate of change from $x = -1$ to $x = 0$ is zero (since the slope of the line is horizontal).

The slopes representing (a) and (c) are both negative, but upon inspection it seems that the tangent line at $x = 0$ is less steep than the line from $(0, f(0))$ to $(2, f(2))$. Since these slopes are negative, a steeper slope is a *smaller* slope. Therefore, (a) is smaller than (c). We have now determined that the four quantities can be listed in order from least to greatest as follows: (a) < (c) < (b) < (d).   □

? *Question*   Let f be the function shown in Figure 3.18. At which values of x is the instantaneous rate of change of f the greatest? Where is the instantaneous rate of change of f equal to zero?   □

3.2.4 Applications of instantaneous rates of change

Many real-life applications involve investigating the instantaneous rate of change, or derivative, of a particular quantity or function. For example, we have seen that the (instantaneous) velocity of a moving object at a given time $t = c$ is the derivative of the position function for that object at that point in time. We can also use derivatives to examine the instantaneous rates of change of other quantities, such as the amount of money in a bank account, the population of birds on an island, the rainfall during a hurricane, or the height of a growing oak tree. In such examples the independent variable is often, but not always, time.

If $y = f(x)$ is a function representing how a quantity y varies with respect to a quantity x, then the derivative $f'(c)$ at a particular value $x = c$ represents the instantaneous rate of change of y with respect to x. Since the derivative of a function at a point $x = c$ is by definition a limit of difference quotients $\frac{\Delta y}{\Delta x}$, the units of the derivative $f'(c)$ will be the quotient of the units of y and the units of x.

EXAMPLE 3.15

The units of an instantaneous rate of change

Suppose $y = f(t)$ represents the height (in feet) of a watermelon t seconds after being dropped. Then $f'(2)$ represents the instantaneous rate of change of the height of the watermelon (i.e., the velocity of the watermelon) at time $t = 2$. Since y is measured in feet and t is measured in seconds, the derivative $f'(2)$ is measured in feet per second.   □

EXAMPLE 3.16

Real-world interpretation of an instantaneous rate of change

Suppose $r(h)$ represents the average trunk radius, in centimeters, of an oak tree that is h meters high. The quantity $r(30)$ represents the average radius of a 30-meter oak tree. The derivative $r'(30)$ represents the instantaneous rate of change of the radius of an oak tree with respect to its height, at the instant that $h = 30$. In other words, $r'(30)$ describes how the radius of an oak tree changes as its height changes, at the instant that the tree is 30 meters high.

Since r is measured in centimeters and h in meters, the units of the derivative $r'(30)$ are centimeters per meter. For example, if $r(30) = 24$ centimeters and $r'(30) = 0.8$ centimeter per meter, then the average 30-meter oak tree has a trunk radius of 24 centimeters, and the radius of its trunk is (at that instant) growing at a rate of 0.8 centimeter for every meter of growth in height.   □

We have seen that velocity is by definition the (instantaneous) rate of change of position. The rate of change of velocity is known as **acceleration.**

DEFINITION 3.9

Acceleration

Suppose an object travels along a straight path with velocity function $v(t)$. The **acceleration** of the object at time $t = c$ is defined to be the derivative of the velocity function at $t = c$:

$$\text{acceleration at } t = c = \lim_{h \to 0} \frac{v(c + h) - v(c)}{h}.$$

In other words, if $a(t)$ represents the acceleration of the object at time t, then given any $x = c$, we have $a(c) = v'(c)$.

Thus velocity is the derivative of position, and acceleration is the derivative of velocity.

❓ *Question*   If an object's position in miles t hours after it starts to move is given by the function $s(t)$, what are the units of $s(2)$? If $v(t)$ is the velocity of the object at time t, what are the units of $v(2)$, and how is $v(2)$ related to $s(t)$? If $a(t)$ describes the acceleration of the object after t hours, what are the units of $a(2)$, and how is $a(2)$ related to $v(t)$?   □

> ## What you should know after reading Section 3.2
>
> ▶ How is the distance formula really an expression for an average rate of change? What is the difference between velocity and speed?
>
> ▶ The definition of an average rate of change, how to calculate average rates of change, and the graphical interpretation of an average rate of change as the slope of a secant line. The definition of average velocity as an average rate of change of position.
>
> ▶ The definitions of instantaneous velocity and of instantaneous rates of change as limits of average rates of change. How to approximate such instantaneous rates of change, calculate them exactly, and interpret them graphically.
>
> ▶ How to identify and work with instantaneous rates of change in real-world problems, in particular the units and physical interpretation of an instantaneous rate of change.

Exercises 3.2

Concepts

0. Read the section and make your own summary of the material. Include whatever you think will help you review the material later on.

1. Explain what the expression $\frac{d_1 - d_0}{t_1 - t_0}$ in Equation (2) has to do with the distance formula $d = rt$.

2. Describe the difference between velocity and speed. Make up a real-world example where speed is positive but velocity is negative.

3. Let l be the line connecting two points $(a, f(a))$ and $(b, f(b))$ on the graph of a function f. What does this line l have to do with the average rate of change of f on the interval $[a, b]$, and why?

4. Explain the difference between an average rate of change and an instantaneous rate of change. How can an instantaneous rate of change be approximated using average rates of change? How can we find the *exact* value of the instantaneous rate of change of a function at a given point?

5. For the graph of $f(x)$ shown below, list the following quantities in order from least to greatest: **(a)** the average rate of change of f on $[.9, 1]$; **(b)** $f'(-1)$; **(c)** $\frac{f(2) - f(-1)}{2 - (-1)}$; **(d)** the instantaneous rate of change of f at $x = 1$.

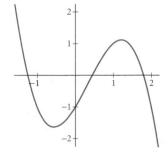

6. For the graph of $f(x)$ shown below, list the following quantities in order from least to greatest: **(a)** the average rate of change of f on $[0, 1]$; **(b)** the instantaneous rate of change of f at $x = 1$; **(c)** $\lim_{h \to 0} \frac{f(-1+h) - f(-1)}{h}$; **(d)** $\frac{f(1) - f(-1)}{2}$.

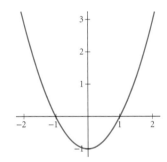

7. Consider the function f from Problem 5. At which values of x on $[-1, 2]$ does f have the greatest instantaneous rate of change? The least? At which values of x is the instantaneous rate of change of f equal to zero?

8. Consider the function f from Problem 6. For which values of x does f have a positive instantaneous rate of change? Negative? Zero?

9. Sketch the graph of a function f with the following three properties: The average rate of change of f on $[0, 2]$ is equal to 3; the average rate of change of f on $[0, 1]$ is equal to -1; and the average rate of change of f on $[-2, 2]$ is equal to 0.

10. Sketch the graph of a function f with the following three properties: The instantaneous rate of change of f at $x = 2$ is zero; the average rate of change of f on $[1, 2]$ is 2; the average rate of change of f on $[2, 4]$ is 1.

Skills

■ For each function f and interval $I = [a, b]$, calculate the average rate of change of f from $x = a$ to $x = b$. Then sketch a graph of f, and draw lines whose slopes are these average rates of change.

11. $f(x) = 1 - x^2$, $[2, 5]$

12. $f(x) = 1 - x^2$, $[-3, 2]$

13. $f(x) = \dfrac{1}{x}$, $[0.9, 1.1]$

14. $f(x) = -3x + 2$, $[-1, 1]$

15. $f(x) = \dfrac{1 - x}{1 + x^3}$, $[0, 0.5]$

16. $f(x) = (x - 2)^2 + \dfrac{3}{x}$, $[-2, 2]$

■ For each function f and value c, use small average rates of change to estimate the instantaneous rate of change of f at c. Sketch lines on a graph of f that represent these average and instantaneous rates of change.

17. $f(x) = 3x + 1$, $c = 2$

18. $f(x) = 4 - x^2$, $c = 1$

19. $f(x) = \dfrac{x}{x - 1}$, $c = 0$

■ For each function f and value c, find the instantaneous rate of change of f at $x = c$. Support your answer with a graph.

20. $f(x) = 3x + 1$, $c = -2$

21. $f(x) = x^2$, $c = 0$

22. $f(x) = \dfrac{1}{x}$, $c = 1$

23. $f(x) = 1 + x + x^2$, $c = 0$

24. $f(x) = 3 - x^2$, $c = -1$

25. $f(x) = \dfrac{1}{x + 1}$, $c = -2$

Applications

26. If your rain-catching bucket starts empty and collects 3 inches of rain during a 6-hour rainstorm, what is the average rate of change of the level of rainwater in the bucket over the 6 hours that it rained? Did the rain necessarily collect in the bucket at a constant rate?

27. Tania is driving home after a conference. If she drives at an average velocity of 58 miles per hour, and the conference is 104 miles away from her home, can she make it home in 2 hours?

28. If Katie walked at 3 miles per hour for 20 minutes and then sprinted at 10 miles an hour for 8 minutes, how fast would Dave have to walk or run to go the same distance as Katie did in the same time, while walking or running at a constant speed? Sketch a graph of Katie's position over time and a graph of Dave's position over time on the same set of axes.

29. If a racecar's position in miles t hours after the start of a race is given by the function $s(t)$, what are the units of $s(1.2)$? $v(1.2)$? $a(1.2)$?

30. Stuart left his house at noon and walked north on Pine Street for 20 minutes. At this point he realized he was late for an appointment at the dentist, whose office was located *south* of Stuart's house on Pine Street. Fearing he would be late, Stuart sprinted south on Pine Street, past his house, and to the dentist's office. When he got there, he found the office closed for lunch; he was 10 minutes early for his 12:40 appointment. Stuart waited at the office for 10 minutes and then found out that his appointment was actually for the next day, so he walked back to his house. Sketch a graph that describes Stuart's position over time. Then sketch a graph that describes Stuart's instantaneous rate of change over time.

31. Last night Phil went jogging along Main Street. His distance (in miles) from the Post Office t minutes after 6:00 p.m. is shown in the graph that follows.

(a) Give a narrative (that matches the graph) of what Phil did on his jog.

(b) Sketch a graph that represents Phil's instantaneous velocity t minutes after 6:00 p.m. Make sure you label the tick-marks on the vertical axis as accurately as you can.

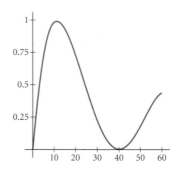

32. The numbers of operating drive-in movie theatres in the state of Virginia in various years are given in the table below. Use the table to answer each of the questions that follow.

(a) Find the average rate of change in the number of drive-in movie theatres in Virginia over each time period between table entries.

(b) Describe the units and real-world significance of the average rates of change you calculated above.

(c) Over which time period was the average rate of change the most drastic? On average, assuming no new drive-in theatres were built, how many drive-ins closed per year during that time period?

Year	1958	1963	1967	1972	1977	1982	1987	1999
Drive-ins	143	115	90	102	87	56	16	9

33. Every morning Linda jogs back and forth on a path in Central Park for 30 minutes. Suppose her distance s in feet from the oak tree on the north side of the park t minutes after she begins her jog is given by the function $s(t)$ shown below, and that she jogs on a straight path leading into the park from the oak tree.

(a) What is the average rate of change of Linda's distance from the oak tree over the entire 30-minute jog? What does this mean in real-world terms?

(b) On which 10-minute interval is the average rate of change of Linda's distance from the oak tree the greatest: the first 10 minutes, the second 10 minutes, or the last 10 minutes?

(c) Use the graph of $s(t)$ to estimate Linda's average velocity during the 5-minute interval from $t = 5$ to $t = 10$. What does the sign of this average velocity tell you in real-world terms?

(d) Approximate the times at which Linda's (instantaneous) velocity is equal to zero. What is the physical significance of these times?

(e) Approximate the time intervals during Linda's jog when her (instantaneous) velocity is negative. What does a negative velocity mean in terms of this physical example?

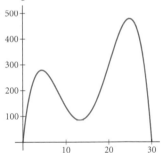

34. If you drop a rotten orange from your fourth-floor dorm window (42 feet above the ground), its height in feet above the ground t seconds after being dropped is given by the position function $s(t) = -16t^2 + 42$.

(a) Find the average velocity of the orange during the time that it is in the air.

(b) Use small average rates of change to approximate the instantaneous velocity of the orange at the moment it hits the ground.

(c) Find the *exact* instantaneous velocity of the orange at the instant it hits the ground.

(d) If the position of the orange is given by $s(t) = -16t^2 + 42$, then the velocity of the orange is $v(t) = -32t$ feet per second t seconds after it is dropped. Use the velocity function for the orange to determine the acceleration of the orange at the time it is dropped. What physical force causes this acceleration? Why is the acceleration negative?

35. Suppose $h(t)$ represents the average height, in feet, of a person who is t years old.

(a) In real-world terms, what does $h(12)$ represent, and what are its units? What does $h'(12)$ represent, and what are its units?

(b) Is $h(12)$ positive or negative, and why? Is $h'(12)$ positive or negative, and why?

(c) At approximately what value of t would $h(t)$ have a maximum, and why? At approximately what value of t would $h'(t)$ have a maximum, and why?

36. Suppose $A(t)$ represents the amount of money in your savings account (in dollars) t years after you graduated from high school. In real-world terms, what does $A(3)$ represent, and what are its units? What does $A'(3)$ represent, and what are its units?

37. Suppose a tomato plant needs $F(t)$ pounds of fertilizer to bear t tomatoes in a growing season. In real-world terms, what does $F(30)$ represent, and what are its units? What does $F'(30)$ represent, and what are its units?

Proofs

38. Prove Theorem 3.2: The average rate of change of a function f on an interval $I = [a, b]$ is equal to the slope of the line connecting the points $(a, f(a))$ and $(b, f(b))$.

3.3 Differentiability

Given a function f and a value $x = c$, there may or may not be a well-defined tangent line to the graph of f at $x = c$. If there is a well-defined tangent line with a finite slope, we say that f is **differentiable at** $x = c$. In this section we will graphically and algebraically investigate points of differentiability and nondifferentiability.

3.3.1 Definition of differentiability

Consider the graph of the absolute value function $f(x) = |x|$ shown in Figure 3.21.

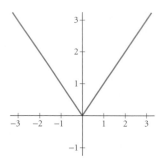

Figure 3.21

The absolute value function has a well-defined tangent line at each value of x except at $x = 0$. For example, near $x = 2$ the function $f(x) = |x|$ is linear, so $f(x)$ has a well-defined tangent line at $x = 2$. On the other hand, at the point $x = 0$ the graph of the absolute value function has a "corner" and thus does not have a well-defined tangent line.

Intuitively, the tangent line to this function at $x = 0$ should be the *unique* line that just skims the function at the point $(0, 0)$, and this line should indicate the "direction" of the graph at that point. However, in this example, there is more than one such line at $x = 0$. In some sense, the graph of $f(x) = |x|$ has *two* directions at $x = 0$: one with slope -1 and one with slope 1. Since there is not a unique tangent line to this graph at $x = 0$, we will say that the function $f(x) = |x|$ is not **differentiable** at $x = 0$. Since there is no well-defined tangent line at $x = 0$, the derivative $f'(0)$ does not exist.

In general, we say that a function f is **differentiable** at a point $x = c$ if its derivative $f'(c)$ exists. Since $f'(c)$ is by definition a limit (of difference quotients), this means that a function f is differentiable at $x = c$ if, and only if, the limit that would define its derivative exists.

DEFINITION 3.10

Differentiability at a Point

A function f is **differentiable at a point** $x = c$ if

$$\lim_{h \to 0} \frac{f(c + h) - f(c)}{h} \text{ exists.}$$

If the limit in Definition 3.10 exists, then by the definition of derivative it is equal to $f'(c)$. Remember that a limit exists if its left and right limits exist (that is, its left and right limits are finite numbers) and are equal to each other. Recall also that the left and right derivatives of f at $x = c$ are defined to be the limits as $h \to 0^-$ and as $h \to 0^+$ of the difference quotient in Definition 3.10.

Given a function f and a value $x = c$, we will say that f is **left differentiable** at $x = c$ if its left derivative $f'_-(c)$ exists. Similarly, f is **right differentiable** if its right derivative $f'_+(c)$ exists. In this language, Theorem 3.3 says that a function f is differentiable at a point $x = c$ if and only if it is both left and right differentiable at $x = c$ and its left and right derivatives at $x = c$ are equal. This theorem will be particularly useful when we investigate the differentiability of piecewise functions.

THEOREM 3.3

The Relationship Between One-Sided and Two-Sided Differentiability

A function f is differentiable at a point $x = c$ if, and only if, its left and right derivatives $f'_-(c)$ and $f'_+(c)$ exist and are equal.

We will define differentiability on an open or closed interval in the same way that we defined continuity on an open or closed interval. A function f is said to be ***differentiable on an open interval*** (a, b) if f is differentiable at every point $x \in (a, b)$. A function f is ***differentiable on a closed interval*** $[a, b]$ if three conditions hold: f is differentiable on the open interval (a, b), f is right differentiable at the left endpoint $x = a$, and f is left differentiable at the right endpoint $x = b$.

As with continuity, we also have a term for functions that are differentiable everywhere they are defined.

DEFINITION 3.11

Differentiable Functions

A function f is a ***differentiable function*** if it is differentiable on its domain.

For example, the absolute value function is not differentiable at $x = 0$ (we will show this using the definition of differentiability in a moment). Since $x = 0$ is in the domain of $f(x) = |x|$, the function $f(x) = |x|$ is not a differentiable function.

Sometimes a function has a well-defined tangent line at a point, but that line happens to be vertical. One such function is shown in Figure 3.22. Since a vertical line has an undefined slope, the derivative of the function does not exist at that point. The derivative is the slope of the tangent line, and in this case the slope is undefined. We sometimes say (somewhat imprecisely) that a vertical line has "infinite" slope. We say that the derivative of the function in Figure 3.22 is infinite (and since there is no real number that could be equal to the derivative, we could say that the derivative "does not exist").

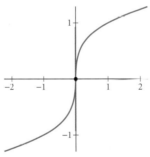

Figure 3.22
Infinite slope at $x = 0$

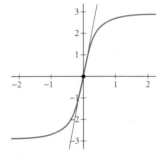

Figure 3.23
Finite slope at $x = 0$

The function in Figure 3.22 has a tangent line with infinite slope at $x = 0$ (and thus is not differentiable at zero). Compare this to the function shown in Figure 3.23, which has a tangent line with a well-defined slope at $x = 0$.

⊗ *Caution*   Be careful when using a graph to determine whether the tangent line to a function at a point has infinite or finite slope. As you can see in Figures 3.22 and 3.23, a line with infinite slope can look very similar to a line with a very steep slope.   □

⊗ *Caution*   The graph in Figure 3.22 passes through the line $x = 0$ only once. This graph passes the Vertical Line Test; it *is* the graph of a function.   □

3.3.2 Graphically determining differentiability

Before we get into the somewhat complicated algebraic and limit computations necessary to determine whether a function is differentiable at a given point, let's do some graphical, intuitive examples.

By inspecting the graph of a function f near a point $x = c$, we can determine intuitively whether or not f is differentiable at $x = c$. There are many ways that a function f can fail to be differentiable at a point $x = c$. For example, the function could have a "corner" or a "cusp" at $x = c$ (Figures 3.24 and 3.25). Or the function could have a vertical tangent line at $x = c$ (Figure 3.26). Another way a function could fail to be differentiable at $x = c$ is if it is discontinuous at that point (Figures 3.27–3.29). (We will prove that discontinuous functions are nondifferentiable in Section 3.3.5.)

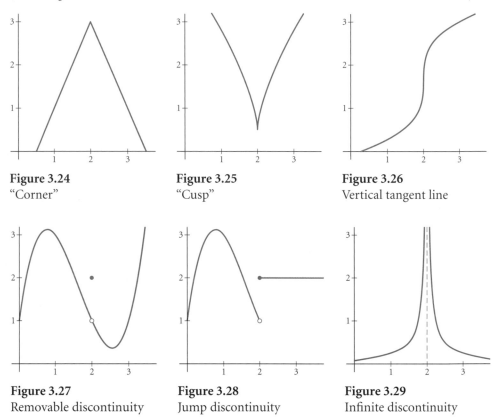

Figure 3.24
"Corner"

Figure 3.25
"Cusp"

Figure 3.26
Vertical tangent line

Figure 3.27
Removable discontinuity

Figure 3.28
Jump discontinuity

Figure 3.29
Infinite discontinuity

Clearly the first three graphs fail to have a well-defined derivative at $x = 2$. Why is there no derivative at $x = 2$ for the last three graphs? To answer this question, we must consider the definition of derivative at $x = 2$:

$$f'(2) = \lim_{h \to 0} \frac{f(2 + h) - f(2)}{h}.$$

Remember that the difference quotient represents the slope of the secant line from $(2, f(2))$ to $(2 + h, f(2 + h))$.

EXAMPLE 3.17

A function is nondifferentiable at a removable discontinuity

Consider the function $f(x)$ shown in Figure 3.27. What happens to the difference quotient $\frac{f(2+h)-f(2)}{h}$ as $h \to 0$?

Solution   In Figure 3.27, $f(2) = 2$ is a point of removable discontinuity. If h is a small positive number, what is the slope of the secant line from c to $c + h$ like? According to the graph in Figure 3.27, the slope of such a secant line would be a large negative number; as $h \to 0^+$ this secant line approaches a vertical line. Similarly, if $h < 0$, then the secant line from c to $c + h$ has a large positive slope, and this slope increases without bound as $h \to 0^-$. Figures 3.30 and 3.31 illustrate some of these secant lines.

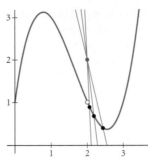

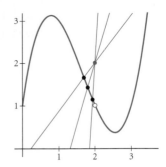

Figure 3.30   **Figure 3.31**

In mathematical notation, what we can see from these graphs is that

$$\lim_{h \to 0^+} \frac{f(2+h) - f(2)}{h} = -\infty,$$

$$\lim_{h \to 0^-} \frac{f(2+h) - f(2)}{h} = \infty.$$

Therefore, the derivative of this function f at $x = 2$ does not exist. The reason we got such strange limits here is that the point $f(2) = 2$ was not connected to the rest of the graph of $y = f(x)$, so the values of $f(2+h)$ were very different from the value $f(2)$, even for small values of h. □

❓ Question   Consider the function $f(x)$ shown in Figure 3.28. Sketch secant lines from $(c, f(c))$ to $(c+h, f(c+h))$ for various small (positive and negative) values of h. You should find that

$$\lim_{h \to 0^+} \frac{f(2+h) - f(2)}{h} = 0,$$

$$\lim_{h \to 0^-} \frac{f(2+h) - f(2)}{h} = \infty,$$

and thus that the derivative at $x = 2$ of this function does not exist. □

EXAMPLE 3.18   **A function is nondifferentiable at an infinite discontinuity**

The function in Figure 3.29 fails to be differentiable at $x = 2$ because $f(2)$ is not defined. We cannot calculate *any* difference quotient $\frac{f(2+h)-f(2)}{h}$, because we can't calculate $f(2)$. □

3.3.3 Algebraically determining differentiability

To see whether a given function is differentiable at some point $x = c$, we need only calculate the limit of difference quotients that defines $f'(c)$. If the limit exists, then it is equal to $f'(c)$. If the limit does not exist, then f is not differentiable at $x = c$.

EXAMPLE 3.19   **Determining differentiability at a point**

Determine whether the function $f(x) = 3x^2$ is differentiable at the point $x = -1$.

Solution   By Definition 3.10, we are interested in whether the limit

$$\lim_{h \to 0} \frac{f(-1+h) - f(-1)}{h} = \lim_{h \to 0} \frac{3(-1+h)^2 - 3(-1)^2}{h}$$

exists. This is a straightforward limit calculation (and is in fact exactly the calculation you would do if you were looking for $f'(-1)$):

$$\lim_{h \to 0} \frac{f(-1+h) - f(-1)}{h} = \lim_{h \to 0} \frac{3(-1+h)^2 - 3(-1)^2}{h}$$

$$= \lim_{h \to 0} \frac{3(1 - 2h + h^2) - 3}{h}$$

$$= \lim_{h \to 0} \frac{3 - 6h + 3h^2 - 3}{h}$$

$$= \lim_{h \to 0} \frac{-6h + 3h^2}{h}$$

$$= \lim_{h \to 0} (-6 + 3h) = -6.$$

The derivative of $f(x) = 3x^2$ exists at $x = -1$ and is equal to $f'(-1) = -6$ (which "exists"). Therefore, the function $f(x) = 3x^2$ is differentiable at $x = -1$. ☐

✓ **Checking the Answer** Figure 3.32 shows the graph of $f(x) = 3x^2$. Note that it does appear to have a well-defined tangent line at $x = -1$ and that the slope of this tangent line could reasonably be -6.

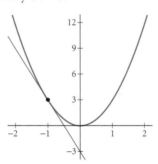

Figure 3.32 ☐

We now show that our intuition was correct about the absolute value function: It does not have a well-defined tangent line at $x = 0$. In other words, $f(x) = |x|$ is not differentiable at $x = 0$.

EXAMPLE 3.20

The absolute value function is not differentiable at $x = 0$

Show that the absolute value function $f(x) = |x|$ is not differentiable at $x = 0$.

Solution By Definition 3.10, we must show that the limit

$$(1) \qquad \lim_{h \to 0} \frac{f(0+h) - f(0)}{h} = \lim_{h \to 0} \frac{|0+h| - |0|}{h} = \lim_{h \to 0} \frac{|h|}{h}$$

does not exist. Since the quantity $\frac{|h|}{h}$ is different depending on whether h is positive or negative (i.e., to the right or left of zero), we will consider the right and left limits separately. The right limit is equal to

$$\lim_{h \to 0^+} \frac{f(0+h) - f(0)}{h} = \lim_{h \to 0^+} \frac{|h|}{h} \qquad \text{(Equation (1))}$$

$$= \lim_{h \to 0^+} \frac{h}{h} \qquad \text{(as } h \to 0^+, h > 0, \text{ so } |h| = h\text{)}$$

$$= \lim_{h \to 0^+} 1 = 1. \qquad \text{(limit of a constant)}$$

On the other hand, the left limit is

$$\lim_{h \to 0^-} \frac{f(0+h) - f(0)}{h} = \lim_{h \to 0^-} \frac{|h|}{h} \qquad \text{(Equation (1))}$$

$$= \lim_{h \to 0^-} \frac{-h}{h} \qquad (\text{as } h \to 0^-, h < 0, \text{ so } |h| = -h)$$

$$= \lim_{h \to 0^-} -1 = -1. \qquad \text{(limit of a constant)}$$

Although the left and right limits both exist, they are not equal to each other. Therefore, the limit in Equation (1) does not exist. This means that the absolute value function $f(x) = |x|$ is not differentiable at $x = 0$. ☐

? *Question* Calculate the slope of the secant line from $(0, f(0))$ to $(0+h, f(0+h))$ for the absolute value function $f(x) = |x|$ first with $h = 0.5$ and then with $h = -0.5$. Sketch these secant lines on a graph of $f(x) = |x|$. What does this have to do with the limit calculation in Example 3.20? ☐

3.3.4 Differentiability of piecewise functions

Piecewise functions often behave strangely at their "break" points. In this section we examine the differentiability of such piecewise functions. We will prove that if a function f is discontinuous at a point $x = c$, then it also fails to be differentiable at $x = c$. Since taking limits of functions is often easier than taking limits of difference quotients, it can be easier to test for continuity than for differentiability. Therefore, when presented with a piecewise function, we will first test for continuity at $x = c$; if the function is discontinuous at $x = c$, we know it is also nondifferentiable at $x = c$. If, on the other hand, the function is continuous at $x = c$, we will take limits of difference quotients to determine whether it is also differentiable at $x = c$.

EXAMPLE 3.21

Investigating the differentiability of a piecewise function

Determine whether or not $f(x) = \begin{cases} x^2, & \text{if } x < 1 \\ 2 - x, & \text{if } x \geq 1 \end{cases}$ is differentiable at $x = 1$.

Solution We first check to see whether this function is continuous at $x = 1$. Recall that f is continuous at $x = 1$ if, and only if, $\lim_{x \to 1} f(x) = f(1)$. We examine the left and right limits separately:

$$\lim_{x \to 1^-} f(x) = \lim_{x \to 1^-} x^2 = 1^2 = 1,$$

$$\lim_{x \to 1^+} f(x) = \lim_{x \to 1^+} 2 - x = 2 - 1 = 1.$$

Thus $\lim_{x \to 1} f(x) = 1$. Since this is equal to $f(1) = 2 - 1 = 1$, f is indeed continuous at $x = 1$. This means that f has at least a *chance* of being differentiable at $x = 1$. However, we must try to calculate the limit of the difference quotients at $x = 1$ to find out. (If we had found out that f was *not* continuous at $x = 1$, we would be finished at this point, for in that case it would not be differentiable at $x = 1$ either.) We tackle the left and right limits separately:

$$\lim_{h \to 0^-} \frac{f(1+h) - f(1)}{h} = \lim_{h \to 0^-} \frac{(1+h)^2 - (2-1)}{h} = \lim_{h \to 0^-} \frac{1 + 2h + h^2 - 1}{h}$$

$$= \lim_{h \to 0^-} \frac{2h + h^2}{h} = \lim_{h \to 0^-} (2 + h) = 2,$$

$$\lim_{h \to 0^+} \frac{f(1+h) - f(1)}{h} = \lim_{h \to 0^+} \frac{(2 - (1+h)) - (2-1)}{h} = \lim_{h \to 0^+} \frac{2 - 1 - h - 1}{h}$$

$$= \lim_{h \to 0^+} \frac{-h}{h} = \lim_{h \to 0^+} (-1) = -1.$$

According to these calculations, we have $f'_-(1) = 2$, and $f'_+(1) = -1$. Since the left and right derivatives are not equal at $x = 1$, the function f is not differentiable at $x = 1$. □

✔ *Checking the Answer*   Figure 3.33 shows the graph of the function f from Example 3.21. The secant lines from 1 to $1 + h$ have positive slope when $h < 0$ but negative slope when $h > 0$. This particular graph has a sort of "half-cusp, half-corner" at $x = 1$ and does not have a well-defined tangent line or derivative at that point.

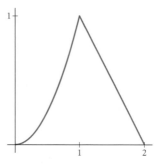

Figure 3.33   □

EXAMPLE 3.22   **Differentiability of a function defined separately for rationals and irrationals**

Determine whether or not $f(x) = \begin{cases} 1, & \text{if } x \text{ rational} \\ x^2 + 1, & \text{if } x \text{ irrational} \end{cases}$ is differentiable at $x = 0$.

Solution   The graph of f is represented in Figure 3.34 (see Section 2.6.3 to review how to graph functions defined separately on the rational and irrational numbers).

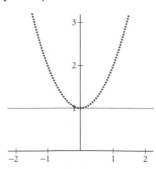

Figure 3.34

Although the graph of this function is strange, it does appear that we might have a well-defined tangent line (with slope zero) at $x = 0$. Both the "thin" dotted line and the "thick" dotted curve seem to have a horizontal tangent line at $x = 0$. We need to show this algebraically. We will use the "alternative" definition of derivative and examine cases where x is rational and x is irrational separately.

For rational values of x we have

$$\lim_{x \to 0} \frac{f(x) - f(0)}{x - 0} = \lim_{x \to 0} \frac{1 - 1}{x} \qquad (f(x) = 1 \text{ for rational } x)$$

$$= \lim_{x \to 0} \frac{0}{x} = \lim_{x \to 0} 0 = 0.$$

On the other hand, for irrational values of x we have

$$\lim_{x \to 0} \frac{f(x) - f(0)}{x - 0} = \lim_{x \to 0} \frac{(x^2 + 1) - 1}{x} \qquad (f(x) = x^2 + 1 \text{ for irrational } x)$$

$$= \lim_{x \to 0} \frac{x^2}{x} = \lim_{x \to 0} x = 0.$$

Since the limit approaches zero whether we consider rational or irrational values of x, the derivative $f'(0)$ exists and is equal to zero. □

❓ *Question*   Sketch secant lines on the graph of f in Figure 3.34 to illustrate that the derivative $f'(0)$ exists and is equal to zero. □

3.3.5 Differentiability implies continuity

Intuitively, if a function is not continuous, then it has absolutely no chance of being differentiable; think of the discontinuous examples from Section 3.3.2 and what happened to the limits of secant lines at the points of discontinuity. In other words, we suspect that

(2)   *not continuous* $\implies$ *not differentiable.*

Recall that the contrapositive of a statement $A \Rightarrow B$ is the statement (not B) $\Rightarrow$ (not A) and that a statement and its contrapositive are always logically equivalent. The contrapositive of Statement (2) is

(3)   *differentiable* $\implies$ *continuous.*

This statement is in fact true in general, and it is important enough to include as a theorem.

THEOREM 3.4

Differentiability Implies Continuity

If f is differentiable at $x = c$, then f is continuous at $x = c$.

✎ *Caution*   The converse of Theorem 3.4 is false; not all continuous functions are differentiable. For example, the absolute value function is continuous at $x = 0$ but not differentiable at $x = 0$. □

PROOF (**THEOREM 3.4**)   If f is differentiable at $x = c$, then by definition,

(4) $$\lim_{x \to c} \frac{f(x) - f(c)}{x - c} \text{ exists}$$

and is in fact equal to the derivative $f'(c)$ of f at c. (Note that we are using the "alternative" definition of derivative.) We need to use Equation (4) to show that f is continuous at $x = c$. By the definition of continuity at a point, this means we need to show that

$$\lim_{x \to c} f(x) = f(c).$$

By the sum rule for limits, this is equivalent to showing that

(5)
$$\lim_{x \to c} (f(x) - f(c)) = 0.$$

We can now calculate

$$\lim_{x \to c}(f(x) - f(c)) = \lim_{x \to c}\left(\frac{f(x) - f(c)}{x - c}(x - c)\right) \qquad (\tfrac{x-c}{x-c} = 1 \text{ if } x \neq c)$$

$$= \left(\lim_{x \to c}\frac{f(x) - f(c)}{x - c}\right)\left(\lim_{x \to c}(x - c)\right) \qquad \text{(product rule for limits)}$$

$$= f'(c)\lim_{x \to c}(x - c) \qquad \text{(definition of derivative)}$$

$$= f'(c)(0) = 0. \qquad \text{(limit rules)}$$

The step where we applied the product rule for limits depended on the fact that the limit in Equation (4) exists. ∎

What you should know after reading Section 3.3

▶ The definition of differentiability at a point, including left and right differentiability. The definition of a differentiable function. What happens if the tangent line to a function is vertical?

▶ Types of nondifferentiability (corners, cusps, vertical tangent lines, and discontinuities), and how to recognize them graphically. How to examine secant lines to approximate left and right derivatives and determine whether a function is differentiable at a point.

▶ Determining differentiability algebraically, by performing a limit calculation to see whether the derivative exists at a given point, and how to determine whether a piecewise function is differentiable at a point.

▶ Differentiability implies continuity, although the converse of this statement is not true. Be able to prove this and to provide examples.

Exercises 3.3

Concepts

0. Read the section and make your own summary of the material. Include whatever you think will help you review the material later on.

1. If $\lim_{x \to c}\frac{f(x)-f(c)}{x-c}$ exists, what can you say about the differentiability of f at $x = c$? What can you say about the continuity of f at $x = c$?

2. List four different ways that a function f could fail to be differentiable at a point $x = c$. Sketch examples of each type of discontinuity (use different examples from those in the reading).

3. Suppose a function f has the following properties:

$$\lim_{x \to 0^+}\frac{f(x)-f(0)}{x} = 3, \quad \lim_{x \to 0^-}\frac{f(x)-f(0)}{x} = -2,$$

$$\lim_{x \to 0^+} f(x) = -1, \quad \lim_{x \to 0^-} f(x) = 1.$$

(a) Is f differentiable at $x = 0$? Why or why not? (b) Is f continuous at $x = 0$? Why or why not? (c) Sketch a possible graph of f.

4. Suppose a function f has the following properties:

$$\lim_{h \to 0^-}\frac{f(1+h)-f(1)}{h} = 2, \quad \lim_{h \to 0^+}\frac{f(1+h)-f(1)}{h} = 0,$$

$$\lim_{x \to 1^-} f(x) = 3, \quad \lim_{x \to 1^+} f(x) = 3.$$

(a) Is f differentiable at $x = 1$? Why or why not? (b) Is f continuous at $x = 1$? Why or why not? (c) Sketch a possible graph of f.

5. Consider the function f graphed below.

(a) Sketch secant lines from $(2, f(2))$ to $(2 + h, f(2 + h))$ on the graph of f given below, for the following values of h: $h = 0.5$, $h = 0.25$, $h = 0.1$, $h = -0.5$, $h = -0.25$, and $h = -0.1$.

(b) Use these secant lines to evaluate graphically $\lim\limits_{h \to 0^+} \frac{f(2+h) - f(2)}{h}$ and $\lim\limits_{h \to 0^-} \frac{f(2+h) - f(2)}{h}$.

(c) Use your answer to part (b) to show that f is not differentiable at $x = 2$.

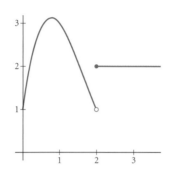

6. Sketch secant lines on a graph of $f(x) = |x|$ and use them to argue that the absolute value function is not differentiable at $x = 0$.

7. The function $f(x) = 4x^3 - 5x + 1$ is differentiable at $x = 2$. Write this fact as a limit statement (see Definition 3.10).

■ Determine whether each of the following statements is true or false. If a statement is true, explain why. If a statement is false, provide a counterexample.

8. True or False: If $f'_-(c)$ and $f'_+(c)$ both exist, then f is differentiable at $x = c$.

9. True or False: If f is both left and right differentiable at $x = c$, then f is differentiable at $x = c$.

10. True or False: If f is continuous at $x = c$, then f is differentiable at $x = c$.

11. True or False: If f is not differentiable at $x = c$, then f is not continuous at $x = c$.

12. True or False: If f is not continuous at $x = c$, then f is not differentiable at $x = c$.

Skills

■ For each function f graphed below, determine the values of x at which f fails to be differentiable. For each of these values, is f left differentiable? Right differentiable? Sketch secant lines supporting your answers.

13.

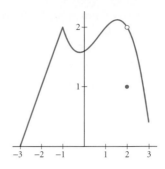

14.

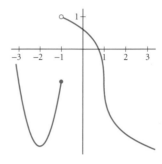

15.

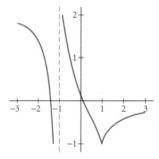

■ Given each function f and value $x = c$ below, determine algebraically whether or not f is differentiable at $x = c$. For those functions that are not differentiable at $x = c$, are they left differentiable? Right differentiable? After doing your calculations, support your answers with graphs.

16. $f(x) = 2x - 5$, $c = -2$

17. $f(x) = -3x^2$, $c = 1$

18. $f(x) = \dfrac{1}{x}$, $c = 0$

19. $f(x) = |3x|$, $c = 0$

20. $f(x) = |x^2 - x - 2|$, $c = 0$

21. $f(x) = |x^2 - x - 2|$, $c = -1$

■ The three functions below are defined in terms of the sine function, which you can enter into your calculator using the button SIN (make sure your calculator is set to "radian" mode first). For each of these three functions, use a graphing calculator the examine the graph of f near $x = 0$. Does it appear that f is differentiable at $x = 0$? You may need to "zoom in" repeatedly near $x = 0$ to determine whether or not a well-defined tangent line seems to exist at $x = 0$.

22. $f(x) = \begin{cases} \sin\left(\frac{1}{x}\right), & \text{if } x \neq 0 \\ 0, & \text{if } x = 0 \end{cases}$

23. $f(x) = \begin{cases} x\sin\left(\frac{1}{x}\right), & \text{if } x \neq 0 \\ 0, & \text{if } x = 0 \end{cases}$

24. $f(x) = \begin{cases} x^2\sin\left(\frac{1}{x}\right), & \text{if } x \neq 0 \\ 0, & \text{if } x = 0 \end{cases}$

■ Given each piecewise function f and value $x = c$ below, determine algebraically (a) whether f is continuous at $x = c$, and (b) whether f is differentiable at $x = c$. After doing your calculations, support your answers with graphs.

25. $f(x) = \begin{cases} -x - 1, & \text{if } x \leq -2 \\ 1 - x^2, & \text{if } x > -2 \end{cases}$, $c = -2$

26. $f(x) = \begin{cases} x^2 - 3, & \text{if } x < 3 \\ x + 2, & \text{if } x \geq 3 \end{cases}$, $c = 3$

27. $f(x) = \begin{cases} x^2, & \text{if } x \leq 1 \\ 2x - 1, & \text{if } x > 1 \end{cases}$, $c = 1$

28. $f(x) = \begin{cases} x^2, & \text{if } x \leq 1 \\ 2x + 4, & \text{if } x > 1 \end{cases}$, $c = 1$

29. $f(x) = \begin{cases} 3x + 1, & \text{if } x < 0 \\ 3, & \text{if } x = 0 \\ 3x + 1, & \text{if } x > 0 \end{cases}$, $c = 0$

30. $f(x) = \begin{cases} x + 1, & \text{if } x < 2 \\ 1, & \text{if } x = 2 \\ x^2 - 1, & \text{if } x > 2 \end{cases}$, $c = 2$

■ For each function f and value $x = c$ given below, sketch a graph representing the function f near $x = c$. Use the definition of continuity to determine whether f is continuous at c. If f is continuous at c, then use the definition of differentiability to determine whether f is differentiable at c. Sketch a graph of f to check your answer.

31. $f(x) = \begin{cases} 1, & \text{if } x \text{ rational} \\ x + 1, & \text{if } x \text{ irrational} \end{cases}$, $c = 1$

32. $f(x) = \begin{cases} x^2, & \text{if } x \text{ rational} \\ 0, & \text{if } x \text{ irrational} \end{cases}$, $c = 0$

33. $f(x) = \begin{cases} 2 - x^2, & \text{if } x \text{ rational} \\ -x, & \text{if } x \text{ irrational} \end{cases}$, $c = 2$

34. $f(x) = \begin{cases} x^2, & \text{if } x \text{ rational} \\ 2x - 1, & \text{if } x \text{ irrational} \end{cases}$, $c = 1$

Applications

35. To save up for a car, you take a job working 10 hours a week at the school library. For the first 6 weeks the library pays you $8.00 an hour. After that you earn $11.50 an hour. You put all of the money you earn each week into a savings account. On the day you start work, your savings account already holds $200.00. Let $S(t)$ be the function that describes the amount in your savings account t weeks after your library job begins.

(a) Find the values of $S(3)$, $S(6)$, and $S(8)$, and describe their meanings in practical terms.
(b) Find the values of $S'(3)$, $S'(6)$, and $S'(8)$, if possible, and describe their meanings in practical terms. (*Hint:* What are the units of $S'(3)$?) If it is not possible to find one of these values, explain why.
(c) Write an equation for the function $S(t)$. (*Hint:* $S(t)$ will be a piecewise function.) Be sure that your equation correctly produces the values you calculated in part (a).
(d) Sketch a labeled graph of $S(t)$.
(e) By looking at the graph of $S(t)$, determine whether $S(t)$ is continuous and whether $S(t)$ is differentiable. Explain the practical significance of your answers.
(f) Show algebraically that $S(t)$ is a continuous function.
(g) Show algebraically that $S(t)$ is not a differentiable function.

Proofs

36. Prove Theorem 3.3: A function f is differentiable at a point $x = c$ if, and only if, its left and right derivatives $f'_-(c)$ and $f'_+(c)$ exist and are equal.

37. Theorem 3.4 was proved in the reading. Use the proof to write *in your own words* a proof of this theorem in the following special case: Suppose f is any function defined at $x = 2$. Prove that if f is differentiable at $x = 2$, then f is continuous at $x = 2$.

3.4   The Derivative as a Function

Given a function f and a point c, we have seen how to calculate the slope of the line tangent to the graph of f at $x = c$ (if such a line exists). In this section we construct a function whose inputs are the values $x = c$ for which f has a well-defined tangent line and whose outputs are the slopes of these tangent lines.

3.4.1 The definition of the derivative function

Consider the graph $y = f(x)$ in Figure 3.35. At each point $x = c$ the graph seems to have a well-defined tangent line; some of these tangent lines are shown in Figure 3.36.

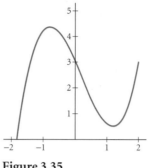

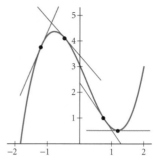

Figure 3.35   **Figure 3.36**

Given an equation for the function f, we could calculate the slopes of all of these tangent lines using the definition of derivative. Each of these calculations would involve a limit, and in each case the calculation would basically be the same, with some numbers changed. As we will see in a moment, we can do all of these calculations at once by not specifying the value of c while we do the calculation.

For each value x, the function f tells you the *height* of the graph of $y = f(x)$ at x. Since there is a well-defined tangent line at each value x, we can also construct a function that, given a value x, returns the *slope* of the line tangent to the graph of $y = f(x)$ at x. If f is differentiable, then this rule will be a function, because for each value x there will be a unique tangent line to the graph of $f(x)$ at that point. We will call this "slope function" the **derivative** of $f(x)$, and denote it by $f'(x)$.

DEFINITION 3.12

The Derivative Function

Given a function f, the **derivative of f** is the function f' that at each point x returns the slope of the tangent line to the graph of f at x. In other words,

$$f'(x) := \lim_{h \to 0} \frac{f(x+h) - f(x)}{h}.$$

The domain of f' is the set of all x at which f is differentiable.

Note that Definition 3.12 is identical to the definition of derivative at a point (Definition 3.1 from Section 3.1), except that the constant c is replaced by the variable x.

Recall from Section 3.1.2 that we have an "alternative" way to define the derivative $f'(c)$ of a function f at a point c. By replacing the constant c with the independent variable x, we obtain the following alternative definition of the derivative function.

DEFINITION 3.13

Alternative Definition of the Derivative Function

Given a function f, the **derivative of f** is the function f' defined as

$$f'(x) := \lim_{t \to x} \frac{f(t) - f(x)}{t - x}.$$

This definition can be obtained from Definition 3.12 by changing variables. If we set $t = x + h$, then as $h \to 0$, the variable t will approach $x + 0 = x$. Moreover, we have $f(t) = f(x + h)$ and $t - x = h$. Note that the interpretation of the derivative as a limit of average rates of change is especially clear in the "alternative" definition of derivative.

⑤ *Caution*   In the alternative definition of the derivative at a point (Definition 3.3), we defined $f'(c)$ as the limit $\lim_{x \to c} \frac{f(x) - f(c)}{x - c}$. In that definition, we could use the variable x to represent the quantity that would approach the real number c. In Definition 3.13 we cannot use the variable x for this purpose, since we wish to use the variable x in place of the constant c. This is why we need another variable (namely t) for the limit in Definition 3.13.   ☐

3.4.2 Calculating derivatives

Before, when we used the definition of derivative to calculate the derivative $f'(c)$ of a function f at some point c, we obtained a *number*. When we use Definition 3.12 to calculate the derivative $f'(x)$, we will obtain a *function*. That is, there will still be an "x" in the answer. After calculating the derivative function, we can quickly calculate the derivatives of f at various points c simply by evaluating $f'(x)$ at $x = c$.

EXAMPLE 3.23   **Finding a derivative function using the definition of derivative**

If $f(x) = x^2$, use Definition 3.12 to find the derivative $f'(x)$.

Solution   Using the definition of the derivative function in Definition 3.12, if $f(x) = x^2$ then $f'(x)$ is given by

$$f'(x) = \lim_{h \to 0} \frac{f(x + h) - f(x)}{h} \qquad \text{(definition of derivative)}$$

$$= \lim_{h \to 0} \frac{(x + h)^2 - x^2}{h} \qquad (f(x) = x^2, \ f(x + h) = (x + h)^2)$$

We now have a limit to calculate; note that the limit involves the variable h, which does not concern the variable x. We can treat the variable x as a constant throughout the limit computation. As usual, if we attempted to evaluate this limit by setting $h = 0$ at this point, we would have the indeterminate form $\frac{0}{0}$. Some algebra will be necessary before we can attempt to evaluate the limit. Continuing our calculation, we have

$$f'(x) = \lim_{h \to 0} \frac{(x + h)^2 - x^2}{h} \qquad \text{(from above)}$$

$$= \lim_{h \to 0} \frac{(x^2 + 2xh + h^2) - x^2}{h} \qquad \text{(multiply out } (x + h)^2)$$

$$= \lim_{h \to 0} \frac{2xh + h^2}{h} = \lim_{h \to 0} \frac{h(2x + h)}{h} \qquad \text{(algebra)}$$

$$= \lim_{h \to 0} (2x + h) = 2x + 0 = 2x. \qquad \text{(cancellation, limit rules)} \qquad ☐$$

⑤ *Caution*   Be sure that you understand the difference between the roles of x and h in Example 3.23. Although x is the independent variable for the derivative function $f'(x)$, it is treated as a constant when we evaluate the limit of the difference quotient as $h \to 0$.   ☐

Example 3.23 shows that the derivative of $f(x) = x^2$ is $f'(x) = 2x$. This means that at any point $x = c$, the derivative of $f(x) = x^2$ at $x = c$ is given by $f'(c) = 2c$. For example, $f'(3) = 2(3) = 6$ is the slope of the tangent line to f at $x = 3$, and $f'(4) = 2(4) = 8$ is the slope of the tangent line to f at $x = 4$.

? *Question*   Compare the calculation from Example 3.2 in Section 3.1.3 with the calculation in Example 3.23. How are they different? How are they the same?   □

In many cases it can be easier to use the "alternative" definition of the derivative in Definition 3.13 than to use Definition 3.12. The following example repeats the calculation of the derivative of $f(x) = x^2$ using the "alternative" definition of derivative.

EXAMPLE 3.24

Finding a derivative function using the alternative definition of derivative

If $f(x) = x^2$, use Definition 3.13 to find the derivative $f'(x)$.

Solution   The derivative function $f'(x)$ is

$$f'(x) = \lim_{t \to x} \frac{f(t) - f(x)}{t - x} \qquad \text{(alternative definition of derivative)}$$

$$= \lim_{t \to x} \frac{t^2 - x^2}{t - x} \qquad \text{(since } f(x) = x^2\text{)}$$

$$= \lim_{t \to x} \frac{(t - x)(t + x)}{t - x} \qquad \text{(factor numerator)}$$

$$= \lim_{t \to x}(t + x) = x + x = 2x. \qquad \text{(cancellation, limit rules)}$$

Note that we *had* to cancel the common factor $t - x$ in the numerator and denominator before calculating the limit. Prior to this cancellation, the limit had the indeterminate form $\frac{0}{0}$.   □

? *Question*   Compare the calculations in Examples 3.23 and 3.24, and decide for yourself which you think is easier.   □

If f is a linear function, then it has the same slope at every point of its graph. Given a linear function $f(x) = mx + b$, the graph of $y = f(x)$ will be its own tangent line at every point $(x, f(x))$. The following example illustrates this fact for a particular linear function. In Section 3.5.1 we will prove that this fact is true in general.

EXAMPLE 3.25

A linear function is its own tangent line at every point

Find the derivative of the function $f(x) = -2x + 3$. Then use this derivative to find the equation for the line tangent to $f(x)$ at $x = 1$.

Solution   By Definition 3.12, we have

$$f'(x) = \lim_{h \to 0} \frac{f(x + h) - f(x)}{h} \qquad \text{(definition of derivative)}$$

$$= \lim_{h \to 0} \frac{(-2(x + h) + 3) - (-2x + 3)}{h} \qquad \text{(definition of } f(x)\text{)}$$

$$= \lim_{h \to 0} \frac{-2x - 2h + 3 + 2x - 3}{h}$$

$$= \lim_{h \to 0} \frac{-2h}{h} = \lim_{h \to 0} -2 = -2. \qquad \text{(limit of a constant)}$$

Therefore, the derivative of the linear function $f(x) = -2x + 3$ is equal to $f'(x) = -2$. The derivative of f is the same at every value of x, since $f'(x) = -2$ is a constant function and does not depend on x.

By definition, the line tangent to $f(x)$ at $x = 1$ has slope $f'(1)$ and passes through the point $(1, f(1))$. By our calculations above, we know that $f'(1) = -2$. Since $(1, f(1)) = (1, -2(1) + 3) = (1, 1)$, we can use the point-slope form of a line to write an equation for the tangent line to f at $x = 1$:

$$y - 1 = -2(x - 1) \implies y = -2x + 2 + 1 \implies y = -2x + 3.$$

Notice that this is what we expected; the linear function $f(x) = -2x + 3$ is its own tangent line at $x = 1$.   □

3.4.3 The graph of the derivative

In general, once we have calculated the derivative $f'(x)$ of a function f, we can use it to find the derivative of f at each point $x = c$ where there is a well-defined tangent line. The next theorem relates the graph of a function f to the graph of its derivative function f'.

THEOREM 3.5

The Graphical Relationship Between a Function and Its Derivative

Suppose f is a function and $x = c$ is a point where the graph of f has a well-defined tangent line. Then the height of the graph of $y = f'(x)$ at $x = c$ is equal to the slope of the line tangent to the graph of $y = f(x)$ at $x = c$.

In particular, f has a horizontal tangent line at $x = c$ if, and only if, f' has a root at $x = c$.

The proof of Theorem 3.5 follows directly from the definitions of tangent lines and derivatives. It is left to you in the exercises.

EXAMPLE 3.26

Graphical interpretation of the derivative

If $f(x) = x^2$, then by Example 3.23, the derivative of f is $f'(x) = 2x$. Use this fact to calculate $f'(3)$, $f'(0)$, and $f'(-1)$, and interpret the results graphically.

Solution   Since $f'(x) = 2x$, we have

$$f'(3) = 2(3) = 6,$$

$$f'(0) = 2(0) = 0,$$

$$f'(-1) = 2(-1) = -2.$$

Since $f'(3) = 6$, the slope of the line tangent to the graph of $f(x) = x^2$ at $x = 3$ is 6. Likewise, the slope of the tangent line to $f(x) = x^2$ at $x = 0$ is 0, and the line tangent to $f(x) = x^2$ at $x = -1$ has a slope of -2. Figure 3.37 shows the graph of $y = f(x)$ with these three tangent lines. Note that the slopes of these lines are equal to the corresponding heights on the graph of $f'(x) = 2x$ in Figure 3.38.

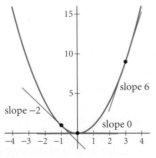

Figure 3.37 $y = f(x)$

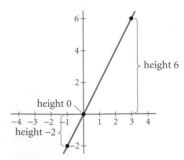

Figure 3.38 $y = f'(x)$ □

Given the graph of a function f, we can sketch an approximate graph of its derivative, using the fact that the height of the derivative function is the slope of the original function. Start by finding the places where the slope of the graph is zero, that is, the x-values where the tangent lines to the graph are horizontal. The graph of the derivative $f'(x)$ will have roots at these x-values. When the graph of f has positive slope, the graph of f' will be above the x-axis; the steeper the slope of f, the higher the value of f'. Similarly, when the graph of f has negative slope, the graph of f' will be below the x-axis.

EXAMPLE 3.27   **Sketching the derivative of a function**

Let f be the function graphed in Figure 3.39. Use the graph of f to sketch the graph of its derivative f'.

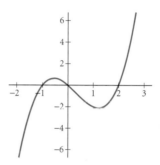

Figure 3.39 $y = f(x)$

Solution   As suggested, we begin by approximating the values of x at which the tangent line to the graph of $y = f(x)$ is horizontal. From the graph of f we approximate that the tangent line is horizontal at approximately $x = -0.5$ and $x = 1.2$. Therefore, the graph of the derivative $y = f'(x)$ should have zeros at approximately $x = -0.5$ and $x = 1.2$; see Figures 3.40 and 3.41.

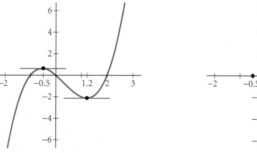
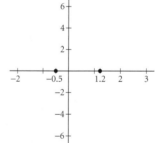

Figure 3.40
Horizontal tangent lines of f

Figure 3.41
Zeros of f'

Now we know two points that should be on the graph of $y = f'(x)$, namely $(-0.5, 0)$ and $(1.2, 0)$. What about the rest of the graph? The graph of the derivative $y = f'(x)$ will be positive wherever the graph of the function $y = f(x)$ has positive slope. Similarly, f' will be negative when f has negative slope. Figures 3.42 and 3.43 illustrate this relationship between the slope of the graph of $y = f(x)$ and the height of the graph of $y = f'(x)$.

With this information we can make a rough sketch of the graph of the derivative function $y = f'(x)$. Note that the graph of the derivative should be farther away from the x-axis when the slope of f is steeper and should be closer to the x-axis when the slope of f is shallower (closer to zero). This is reflected in the sketch of the graph of

$y = f'(x)$ given in Figure 3.45. Figure 3.44 shows the graph of the original function $y = f(x)$ for comparison.

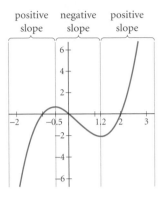

Figure 3.42
Sign of slope of f

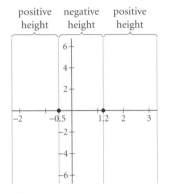

Figure 3.43
Sign of height of f'

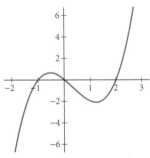

Figure 3.44 $y = f(x)$

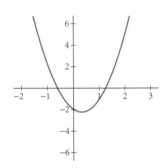

Figure 3.45 $y = f'(x)$ □

3.4.4 Derivative notation

So far our only notation for the derivative of a function f is the "prime notation" f'. In many situations, especially in dealing with real-world applications, it will be more convenient to use other notations to represent derivatives.

NOTATION 3.1

Notations for the Derivative Function

The following are all notations that represent the derivative of a function $y = f(x)$:

$$f'(x) \qquad y'(x) \qquad \frac{df}{dx} \qquad \frac{dy}{dx} \qquad \frac{d}{dx}(f(x)) \qquad \frac{d}{dx}(y(x))$$

There are really only three types of notations here, but each of these notations can be written both in terms of the function f and in terms of the independent variable y.

The third and fourth notations for the derivative of $y = f(x)$ are in what is called **Leibnitz notation.** This type of derivative notation is supposed to remind us of the connection between derivatives and average rates of change. Just as Δx often represents a small change in x, the expression dx in some sense represents an *infinitely small* change in x. Consider the definition of derivative written using Leibnitz notation and the Δx

notation:

(1)
$$\frac{dy}{dx} := \lim_{\Delta x \to 0} \left(\frac{\Delta y}{\Delta x} \right).$$

In Equation (1), Δx plays the role usually played by h. Note that in Definition 3.12, the letter h represents the change in the independent variable from x to $x + h$.

> ⑤ *Caution* The expressions dx and dy are not usually used on their own except for notational purposes (see Part IV of this book). The expression $\frac{dy}{dx}$ does *not* mean "dy divided by dx"; you should consider $\frac{dy}{dx}$ to be *one* symbol that by definition represents the limit of the difference quotient given in Equation (1). □

The last two notations for the derivative in Notation 3.1 use the expression $\frac{d}{dx}$ as an *operator.* Another way to say this is that $\frac{d}{dx}$ is actually a function that takes functions as inputs and returns the derivatives of those functions as outputs. The real advantage of this notation is that we write, for example, $\frac{d}{dx}(x^2 + 1)$ to denote the derivative of the function $f(x) = x^2 + 1$. This is more descriptive than writing $f'(x)$ or $\frac{df}{dx}$ since we can actually see the expression for the function that we are differentiating.

> ⑤ *Caution* Never use the $\frac{d}{dx}$ notation on its own. The $\frac{d}{dx}$ symbol must always be applied to some function. Also, never use the $\frac{dy}{dx}$ notation as an operator; that is, do not write $\frac{dy}{dx}(x^2 + 1)$ when you really mean $\frac{d}{dx}(x^2 + 1)$. □

There are still more notations that are commonly used to represent the derivative of a function $y = f(x)$, for example, y_x and $\dot{y}$. In this book we will focus primarily on the expressions for the derivative given in Notation 3.1.

We will use whatever notation for the derivative suits us at the time. Although the Leibnitz notation and the "operator" notation (which is really just another type of Leibnitz notation) are often more useful and informative than the "prime" notation, they have one drawback. In Leibnitz and operator notation, it is awkward to write down notation for the derivative of a function *at a point.*

NOTATION 3.2

> **Notations for the Derivative at a Point**
>
> The following are all notations that represent the derivative of a function $y = f(x)$ at a point $x = c$:
>
> $$f'(c) \qquad y'(c) \qquad \left.\frac{df}{dx}\right|_c \qquad \left.\frac{dy}{dx}\right|_c \qquad \left.\frac{d}{dx}\right|_c (f(x)) \qquad \left.\frac{d}{dx}\right|_c (y(x))$$

> ⑤ *Caution* We do not write $\frac{d}{dx}(f(c))$ to represent the derivative of f at c. This notation represents the derivative of the *constant* $f(c)$, not the derivative of the function $f(x)$ at $x = c$. Can you explain the difference? □

EXAMPLE 3.28   **Using different notations for the derivative**

Let $y = f(x) = x^2 + 1$. Express the derivative of this function in different ways, as in Notation 3.1. (Don't calculate the derivative; we are only practicing notation here.) Then use the forms in Notation 3.2 to denote the derivative of this function at the point $x = 3$.

Solution   Using the expressions in Notation 3.1, we could express the derivative of the given $y = f(x)$ in the following ways:

$$f'(x) \qquad y'(x) \qquad \frac{df}{dx} \qquad \frac{dy}{dx} \qquad \frac{d}{dx}(f(x)) \qquad \frac{d}{dx}(y(x)).$$

All of the above use either the function name f or the independent variable name y, but none of them refers to the equation that defines $y = f(x)$. To write the derivative of the function while referring to its equation, we could write

$$(x^2 + 1)' \qquad \text{or} \qquad \frac{d}{dx}(x^2 + 1).$$

Of these two notations, the second is usually preferable, for two reasons. First, the second notation refers to the variable with which we are differentiating (namely x). In problems where there is more than one possible variable, this information will be useful. The second reason is that there is no way to write "the derivative of $x^2 + 1$ at $x = 3$" in the first notation, and in the second there is.

Given $y = f(x) = x^2 + 1$, the derivative of $y = f(x)$ at $x = 3$ can be written in the following notations:

$$f'(3) \qquad y'(3) \qquad \frac{df}{dx}\bigg|_3 \qquad \frac{dy}{dx}\bigg|_3 \qquad \frac{d}{dx}\bigg|_3(f(x)) \qquad \frac{d}{dx}\bigg|_3(y(x)).$$

If we want our notation to involve the equation of the function, we can represent the derivative of $f(x) = x^2 + 1$ at the point $x = 3$ as

$$\frac{d}{dx}\bigg|_3(x^2 + 1).$$

$\square$

3.4.5 Higher-order derivatives

Suppose f is a function. Then the derivative f' of f is itself a function. That means that we can try to find the derivative of f', that is, the *derivative of the derivative!* We will call the derivative of the derivative of a function f the **second derivative** of f.

DEFINITION 3.14

The Second Derivative

The **second derivative** of a function f is the derivative of the derivative of f:

$$f''(x) := (f'(x))'.$$

In Leibnitz notation, the second derivative of f is

$$\frac{d^2 f}{dx^2} := \frac{d}{dx}\left(\frac{d}{dx}(f(x))\right).$$

Of course, the second derivative $f''(x)$ is also a function, so it has a derivative as well (assuming it is well defined). The derivative of $f''(x)$ is called the **third derivative** of f and can be written as $f'''(x)$. In general, we can take as many derivatives as we like:

DEFINITION 3.15

Higher-Order Derivatives

The **nth derivative** of a function f is defined to be the function f differentiated n times. In Leibnitz notation,

$$\frac{d^n f}{dx^n} := \underbrace{\frac{d}{dx}\left(\frac{d}{dx}\left(\frac{d}{dx}\cdots\left(\frac{d}{dx}(f(x))\right)\cdots\right)\right)}_{n \text{ times}}.$$

We can represent the nth derivative of a function f in the "prime" notation as $f^{[n]}(x)$ (or, if n is small, by writing n "primes" after the function; for example, $f^{[4]}(x) = f''''(x)$).

In operator notation, we would use $\frac{d^2}{dx^2}(f(x))$ to represent the second derivative of f and $\frac{d^n}{dx^n}(f(x))$ to represent the nth derivative.

⊗ *Caution*   The Leibnitz notation for the nth derivative of a function looks somewhat like it involves a product of n copies of $\frac{d}{dx}$ (or like the nth power of $\frac{d}{dx}$). However, the notation $\frac{d^n}{dx^n}$ is not really a product of n copies of $\frac{d}{dx}$. The notation just works that way so that it is convenient to remember and work with.   □

When we have quick ways of calculating derivatives, we will often calculate and use second, third, and higher-order derivatives. Since our only method of calculating derivatives is currently the definition of derivative, which involves a limit calculation, calculating even the second derivative of a function can be very tedious. We give a simple example anyway.

EXAMPLE 3.29   **Finding higher-order derivatives using the definition of derivative**

Find the first, second, and third derivatives of $f(x) = x^2$.

Solution   We have already found that the derivative of $f(x) = x^2$ is $f'(x) = 2x$ (see Example 3.23). The second derivative of $f(x) = x^2$ is the derivative of $f'(x) = 2x$, which we can easily compute with the definition of derivative:

$$\frac{d}{dx}(2x) = \lim_{h\to 0}\frac{2(x+h)-2x}{h} \qquad \text{(definition of derivative)}$$

$$= \lim_{h\to 0}\frac{2x+2h-2x}{h}$$

$$= \lim_{h\to 0}\frac{2h}{h} = \lim_{h\to 0} 2 = 2. \qquad \text{(limit of a constant)}$$

Thus the second derivative of $f(x) = x^2$ is $f''(x) = 2$. The third derivative of $f(x) = x^2$ is the derivative of this second derivative. Again we can calculate this with the definition of derivative:

$$\frac{d}{dx}(2) = \lim_{h\to 0}\frac{2-2}{h} \qquad \text{(definition of derivative)}$$

$$= \lim_{h\to 0}\frac{0}{h} = \lim_{h\to 0} 0 = 0. \qquad \text{(limit of a constant)}$$

We can summarize our results in "operator" notation as follows:

$$\frac{d}{dx}(x^2) = 2x, \quad \frac{d^2}{dx^2}(x^2) = \frac{d}{dx}\left(\frac{d}{dx}(x^2)\right) = \frac{d}{dx}(2x) = 2,$$

$$\frac{d^3}{dx^3}(x^2) = \frac{d}{dx}\left(\frac{d}{dx}\left(\frac{d}{dx}(x^2)\right)\right) = \frac{d}{dx}\left(\frac{d}{dx}(2x)\right) = \frac{d}{dx}(2) = 0. \qquad \square$$

❓ Question   In Example 3.29, we saw that the third derivative of $f(x) = x^2$ is $f'''(x) = 0$. In fact, if $f(x) = x^2$, then for any integer $n \geq 3$ we will have $f^{[n]}(x) = 0$. For example, show that $f^{[4]}(x) = 0$ when $f(x) = x^2$. $\qquad \square$

3.4.6 Position, velocity, and acceleration functions

Suppose an object is moving on a straight path with position function $s(t)$. In Section 3.2 we saw that the velocity $v(c)$ of such an object at a given time $t = c$ is equal to the derivative $s'(c)$ of the position function at that point. Similarly, the acceleration $a(c)$ of the object is the derivative $v'(c)$ of the object's velocity function at $t = c$.

Now that we have the concept of a derivative *function*, we can state the relationships among position, velocity, and acceleration in terms of functions. If $s(t)$, $v(t)$, and $a(t)$ are the position, velocity, and acceleration functions of a moving object, then

$$v(t) = s'(t) \qquad \textit{Velocity is the derivative of position, and}$$
$$a(t) = v'(t) \qquad \textit{acceleration is the derivative of velocity.}$$

Since this means that acceleration is the derivative of the derivative of position, we can also say that

$$a(t) = s''(t) \qquad \textit{Acceleration is the second derivative of position.}$$

EXAMPLE 3.30   **Using a graph of position to interpret velocity and acceleration**

As you drive along a straight highway, your velocity (in miles per hour) t hours after you pass Dinosaur Land is given by the function $v(t)$ shown in Figure 3.46.

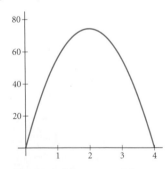

Figure 3.46 $y = v(t)$

Use the graph of $v(t)$ to answer the following questions: At what point in your trip are you driving the fastest? Are you ever decelerating, and if so, when? What is your average acceleration over the 4-hour trip?

Solution   According to the graph of $v(t)$ in Figure 3.46, your velocity has the greatest magnitude (about 75 miles per hour) at approximately $t = 2$ hours, so that is the point at which you are driving the fastest.

You are decelerating when your acceleration is negative, that is, when the derivative of your velocity is negative. The slopes of the tangent lines to the velocity graph in Figure 3.46 are negative on the interval (2, 4). Therefore, you are decelerating for the last 2 hours of your trip.

Your average acceleration is the average rate of change of your velocity over the entire trip. According to the graph, both your initial and your final velocity are zero (since $v(0) = 0$ and $v(4) = 0$). Your average acceleration is

$$\frac{v(4) - v(0)}{4 - 0} = \frac{0 - 0}{4} = 0.$$

□

What you should know after reading Section 3.4

► The definition of the derivative function. In particular: Why is the derivative a function? How does the derivative function differ from the derivative at a point? Be able to calculate derivatives using the definition.

► How to sketch the graph of the derivative of a function by recognizing that the height of the graph of the derivative is the slope of the tangent line to the graph of the function at every point. In particular, what can you say about f' when f has a horizontal tangent line?

► Derivative notation, including Leibnitz and operator notation. How to represent notationally the derivative at a point in these notations.

► Higher-order derivatives (in various notations) and how to compute them. Be especially sure that you understand the second derivative of a function and how it is defined.

► The derivative relationships among position, velocity, and acceleration. How to interpret graphs of position, velocity, or acceleration to describe real-world phenomena.

Exercises 3.4

Concepts

0. Read the section and make your own summary of the material. Include whatever you think will help you review the material later on.

1. Explain what the height of the graph of $y = f'(x)$ at a point $x = c$ tells you about the graph of $f(x)$ at $x = c$.

2. State the formal definition of the derivative function f', and explain what the notation represents.

3. Compare and contrast the definition of derivative at a point (Definition 3.1 from Section 3.1.2) and the definition of the derivative function (Definition 3.12).

4. We now have two ways of calculating the derivative of a function f at a point $x = c$ (using the definition of derivative at a point and the definition of the derivative

as a function). Explain how each method works. Is one way any better than the other?

■ For each function f graphed below, sketch five tangent lines. Use the approximate slopes of these lines to plot five points on the graph of f'. Then use these five points (and the graph of f) to sketch a graph of f'.

5.

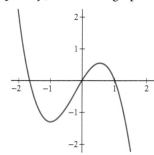

6.

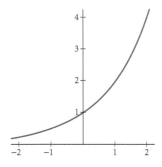

7.

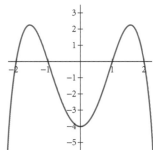

■ For each problem below, sketch the graph of a function f with the given characteristics.

8. $f(1) = 2$, $f'(1) = 0$, $f'(3) = 2$

9. $f'(-3) = 0$, $f'(-1) = 0$, $f'(2) = 0$

10. $f'(-2) = 2$, $f'(0) = 1$, $f(1) = -5$

11. $f(-1) = 2$, $f'(-1) = 3$, $f(1) = -2$, $f'(1) = 3$

12. Consider graphs I, II, and III. They are the graphs of $y = f(x)$, $y = f'(x)$, and $y = f''(x)$ for some function f (although not necessarily in that order). Determine which graph must be which, and support your answers using Theorem 3.5.

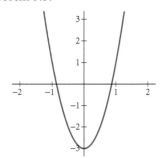

Graph I

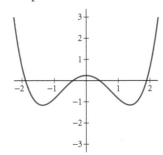

Graph II

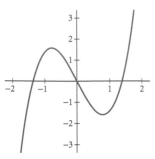

Graph III

13. Sketch the graph of a function $y = f(x)$ whose derivative is given by the graph in the figure below. (*Note:* There is more than one possible correct graph.)

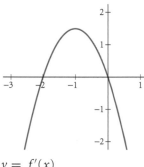

$$y = f'(x)$$

14. Suppose $f(x) = 3x^2 - 1$. Express the derivative of this function at $x = -4$ in each of the following types of notations. (Don't compute the derivative; just express it in the appropriate notation.)

   (a) "Prime" notation

   (b) Leibnitz notation

   (c) Operator notation

15. Suppose $f(x) = 3x^2 - 1$. Express the fourth derivative of this function at $x = -4$ in each of the following types of notations. (Don't compute the fourth derivative; just express it in the appropriate notation.)

   (a) "Prime" notation

   (b) Leibnitz notation

   (c) Operator notation

16. Explain the connection between the Leibnitz notation $\frac{dy}{dx}$ for the derivative and the difference quotient $\frac{\Delta y}{\Delta x}$.

17. Suppose $f(x) = x^3 - 2x + 1$, and let $y = f(x)$. It can be shown that $f'(x) = 3x^2 - 2$. Use this information to determine the following. (*Note:* No differentiation is necessary in this problem, since the derivative has been given. This problem is just a test of your ability to interpret derivative notation.)

   (a) $\frac{dy}{dx}$ **(b)** $\frac{df}{dx}$

   (c) $\frac{d}{dx}(f(x))$ **(d)** $\frac{dy}{dx}(f(x))$

   (e) $\frac{dy}{dx}\big|_{x=2}$ **(f)** $\frac{df}{dx}\big|_{x=2}$

   (g) $\frac{d}{dx}\big|_{x=2}(f(x))$ **(h)** $\frac{dy}{dx}(f(2))$

Skills

- Simplify each of the following until you can cancel a common factor of h from the numerator and the denominator.

18. $\dfrac{(3(x+h)+1)-(3x+1)}{h}$

19. $\dfrac{(x+h)^2-x^2}{h}$

20. $\dfrac{((x+h)^2+1)-(x^2+1)}{h}$

21. $\dfrac{(x+h)^3-x^3}{h}$

22. $\dfrac{(x+h)^4-x^4}{h}$

23. $\dfrac{\frac{1}{x+h}-\frac{1}{x}}{h}$

- Simplify each of the following until you can cancel a common factor of $t-x$ from the numerator and the denominator.

24. $\dfrac{(3t+1)-(3x+1)}{t-x}$ 25. $\dfrac{t^2-x^2}{t-x}$

26. $\dfrac{(t^2+1)-(x^2+1)}{t-x}$ 27. $\dfrac{t^3-x^3}{t-x}$

28. $\dfrac{t^4-x^4}{t-x}$ 29. $\dfrac{\frac{1}{t}-\frac{1}{x}}{t-x}$

- Calculate the following limits. Write out each step individually, citing all limit rules that you use.

30. $\lim\limits_{h\to 0} 2(h+x)$

31. $\lim\limits_{h\to 0}(x^2+hx+3h)$

32. $\lim\limits_{h\to 0}\dfrac{-1}{x(x+h)}$

33. $\lim\limits_{t\to x}(t+x)$

34. $\lim\limits_{t\to x}(t^2+xt+x^2)$

35. $\lim\limits_{t\to x}\dfrac{(3t+1)-(3x+1)}{t-x}$

- Find the derivative of each of the following functions in two ways: **(a)** using the definition of derivative, and **(b)** using the alternative definition of derivative. Justify each step of your calculation (including the steps where you evaluate the limit) in as much detail as in Example 3.23.

36. $f(x)=5x-2$ 37. $f(x)=3x^2$

38. $f(x)=x^2-4$

- Find the derivative of each of the following functions in two ways: **(a)** with Definition 3.12, and **(b)** with Definition 3.13. Then **(c)** use your answer to calculate $f'(-2)$, $f'(0)$, and $f'(3)$.

39. $f(x)=3x+1$ 40. $f(x)=-2x^2$

41. $f(x)=1+x+x^2$ 42. $f(x)=5$

43. $f(x)=3x^2+1$ 44. $f(x)=x^3$

45. $f(x)=x^3+2$ 46. $f(x)=x^4$

47. $f(x)=x^4-1$ 48. $f(x)=x^3+x$

49. $f(x)=\dfrac{1}{x}$ 50. $f(x)=\dfrac{2}{x+1}$

- Use your answers to Problems 39–41 to find the tangent lines to each function $f(x)$ at $x=-2$, $x=0$, and $x=3$. Check that your answers are correct by graphing $y=f(x)$ with these three lines.

51. $f(x)=3x+1$ 52. $f(x)=-2x^2$

53. $f(x)=1+x+x^2$

- For each function f graphed below, sketch a graph of the derivative function f'.

54.

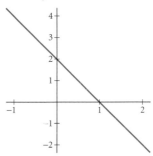

55.

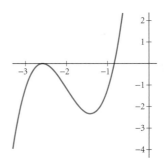

56.

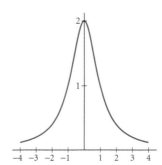

- Each limit below represents the derivative of some function $f(x)$. Find $f(x)$, and calculate the limit to determine $f'(x)$.

57. $\lim\limits_{h\to 0}\dfrac{(1-x-h)-(1-x)}{h}$

58. $\lim\limits_{h\to 0}\dfrac{2(x+h)^3-2x^3}{h}$

59. $\lim\limits_{h \to 0} \dfrac{\frac{1}{3x+3h} - \frac{1}{3x}}{h}$

60. $\lim\limits_{t \to x} \dfrac{(t^3 + 1) - (x^3 + 1)}{t - x}$

61. $\lim\limits_{t \to x} \dfrac{2t^4 - 2x^4}{t - x}$

62. $\lim\limits_{t \to x} \dfrac{\frac{1}{t+1} - \frac{1}{x+1}}{t - x}$

63. Use the definition of derivative (or the alternative definition) to calculate the following derivatives.

   (a) $\dfrac{d}{dx}(2x^3)$ **(b)** $\dfrac{d^2}{dx^2}(2x^3)$

   (c) $\dfrac{d^3}{dx^3}(2x^3)$

64. Use your answers to Problem 63 to evaluate the following derivatives of the function $f(x) = 2x^3$.

   (a) $f'(3)$ **(b)** $\dfrac{d^2 f}{dx^2}\Big|_1$

   (c) $\dfrac{d^3}{dx^3}\Big|_{-2}(2x^3)$

65. If $f^{[3]}(x) = 3x^2 + 1$, find each of the following:

   (a) $f'''(2)$ **(b)** $\dfrac{d^4 f}{dx^4}$

   (c) $\dfrac{d^4 f}{dx^4}\Big|_2$

66. If $\frac{d^2 f}{dx^2} = 2 - x^2$, find each of the following:

   (a) $f'''(x)$ **(b)** $f^{[3]}(4)$

   (c) $\dfrac{d^4}{dx^4}(f(x))$

Applications

67. On a long road trip, you are driving along a straight portion of Route 188. Suppose that t hours after you enter Nevada, your distance from the Donut Hole is $s(t) = -10t^2 - 40t + 120$ miles.

   (a) How long will it take you to reach the Donut Hole after entering Nevada?

   (b) Find your velocity $v(t)$ as you drive toward the Donut Hole.

   (c) Are you accelerating or decelerating as you approach the Donut Hole? At what rate?

68. Suppose you drive north along a straight highway with position $s(t)$ as shown in the graph for this problem. Suppose also that t is measured in hours, s is measured in miles, and $s = 0$ corresponds to the point on the highway where you pass the Donut Shack.

   (a) Sketch a graph of your velocity $v(t)$, and use the graph to describe your velocity over the course of the 2-hour drive.

   (b) Sketch a graph of your acceleration $a(t)$, and use the graph to describe your acceleration over the course of the 2-hour drive.

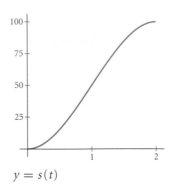

$y = s(t)$

69. While she is on a walk along a straight north-south forest path, Carol's velocity (in feet per minute) after t minutes is given by the graph below. (Suppose the "positive" direction is north.)

   (a) Describe the sort of walk Carol must have taken to have this velocity graph. Be sure your description explains the physical significance of the fact that her velocity is zero at $t = 15$ minutes and the fact that her velocity is negative for the second half of her walk.

   (b) Find Carol's average acceleration over the 30-minute walk. Was her acceleration constant over the duration of her walk? Why or why not?

   (c) What was Carol's average velocity over her entire walk? Why does this make sense?

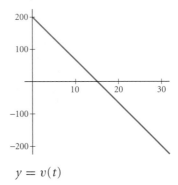

$y = v(t)$

Proofs

70. Use the definition of derivative to prove that our concept of slope for linear functions matches the slope that is defined by the derivative. In other words, show that if $f(x) = mx + b$ is any linear function, then $f'(x) = m$.

71. Use Problem 70 to prove that a linear function is its own tangent line at every point. In other words, show that if $f(x) = mx + b$ is any linear function, the tangent line to f at any point $x = c$ is given by $y = mx + b$.

72. Prove the first part of Theorem 3.5: The slope of the line tangent to the graph of a function $y = f(x)$ at $x = c$ is equal to the height of the graph of its derivative $y = f'(x)$ at $x = c$.

73. The second part of Theorem 3.5 is really a special case of the main statement of the theorem. Prove this special case: f has a horizontal tangent line at $x = c$ if, and only if, f' has a root at $x = c$.

| 3.5 | **Basic Differentiation Rules** |

You may have noticed by now that calculating derivatives can be rather tedious. Moreover, many of the derivative calculations we have seen involve the same types of calculations. Since derivatives will be used often throughout this course, we need to develop a faster method of calculating them. In this section we learn some "rules" or "shortcuts" that will enable us to calculate the derivatives of some simple kinds of functions quickly. In later chapters we will expand our set of differentiation rules to include more complicated functions.

3.5.1 Three basic derivatives

We begin by developing rules for differentiating constant functions, the identity function, and linear functions. To arrive at each rule, we will simply make a general computation involving the definition of derivative and make note of the answer. Then when we encounter constant, identity, or linear functions, we will know their derivatives from memory.

Our first rule tells us that the derivative of any constant is zero.

THEOREM 3.6

The Derivative of a Constant

If $f(x) = c$ is a constant function, then $f'(x) = 0$.

We could also phrase this theorem in Leibnitz notation, as follows: If c is any constant, then

(1) $$\frac{d}{dx}(c) = 0.$$

We can now use Theorem 3.6 to differentiate any constant function from memory, without having to go through the limit calculation for the definition of derivative. For example, Theorem 3.6 tells us that

$$\frac{d}{dx}(3) = 0, \quad \frac{d}{dx}(-1.2) = 0, \quad \text{and} \quad \frac{d}{dx}(\pi^2) = 0.$$

If we think about it for a moment, this is a reasonable theorem. The graph of a constant function is nothing more than a horizontal line, which has a slope of zero at every point. For example, Figure 3.47 shows the graph of the constant function $f(x) = 2$.

PROOF **(THEOREM 3.6)** Suppose $f(x) = c$, where c is any real number. By the definition of derivative,

$$
\begin{aligned}
f'(x) &= \lim_{h \to 0} \frac{f(x+h) - f(x)}{h} && \text{(definition of derivative)} \\
&= \lim_{h \to 0} \frac{c - c}{h} && (f(x) = c \text{ for all inputs } x) \\
&= \lim_{h \to 0} \frac{0}{h} = \lim_{h \to 0}(0) = 0. && \text{(limit of a constant)}
\end{aligned}
$$

Caution In the last step of the proof, we had to calculate $\lim_{h \to 0} \frac{0}{h}$. This is *not* a limit of the form $\frac{0}{0}$, even though h is approaching zero. The reason is that the numerator is equal to zero, so for all nonzero h, the expression $\frac{0}{h}$ is equal to zero *before* the limit is taken. In this case we are really taking the limit $\lim_{h \to 0}(0)$ of the constant 0. □

Our next rule tells us how to differentiate the identity function $f(x) = x$. Intuitively, the graph of the identity function is a line with slope 1, so the derivative of $f(x) = x$ should always be equal to 1; that is, we should have $f'(x) = 1$ (see Figure 3.48). This is in fact the case, as we see in Theorem 3.7.

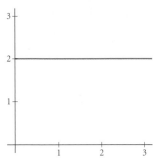

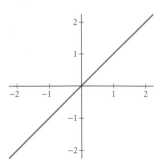

Figure 3.47
$f(x) = 2$ has zero slope
everywhere

Figure 3.48
$f(x) = x$ has slope 1
everywhere

THEOREM 3.7

The Derivative of the Identity Function
If $f(x) = x$ is the identity function, then $f'(x) = 1$.

This theorem can be written in Leibnitz notation as

$$(2) \qquad \frac{d}{dx}(x) = 1.$$

The proof of Theorem 3.7 is just a calculation using the definition of derivative.

PROOF (THEOREM 3.7)   If $f(x) = x$, then

$$
\begin{aligned}
f'(x) &= \lim_{h \to 0} \frac{f(x+h) - f(x)}{h} && \text{(definition of derivative)} \\
&= \lim_{h \to 0} \frac{(x+h) - x}{h} && (f \text{ is the identity function}) \\
&= \lim_{h \to 0} \frac{h}{h} = \lim_{h \to 0} 1 = 1 && \text{(limit of a constant)}
\end{aligned}
$$

We now show that the concept of slope we had for linear functions and the concept of slope defined by the tangent line are one and the same. In other words, if $f(x) = mx + b$ is a linear function, then its derivative $f'(x)$ is always equal to its slope m.

THEOREM 3.8

The Derivative of a Linear Function
If $f(x) = mx + b$ is a linear function, then $f'(x) = m$.

In Leibnitz notation, this derivative rule says that for any constants m and b,

$$(3) \qquad \frac{d}{dx}(mx + b) = m.$$

The proof of this derivative rule is a general calculation using the definition of derivative (without specifying values for m and b).

PROOF (**THEOREM 3.8**) If m and b are any constants, and $f(x) = mx + b$, then

$$f'(x) = \lim_{h \to 0} \frac{f(x+h) - f(x)}{h} \qquad \text{(definition of derivative)}$$

$$= \lim_{h \to 0} \frac{(m(x+h) + b) - (mx + b)}{h} \qquad \text{(use formula for } f)$$

$$= \lim_{h \to 0} \frac{mx + mh + b - mx - b}{h}$$

$$= \lim_{h \to 0} \frac{mh}{h} = \lim_{h \to 0} m = m. \qquad \text{(limit of a constant)} \quad \blacksquare$$

The derivatives of constant functions and the identity function are really just special cases of this more general rule, since constant functions are just linear functions with $m = 0$, whereas the identity function is a linear function with $m = 1$ and $b = 0$.

EXAMPLE 3.31 **Calculating derivatives using Theorems 3.6, 3.7, and 3.8**

Calculate the following derivatives.

(a) $\frac{d}{dx}(3x + 1)$ (b) $\frac{d}{dx}(-4)$ (c) $\frac{d}{dx}(1 - 2x)$

Solution

(a) $f(x) = 3x + 1$ is a linear function with slope $m = 3$, so by Theorem 3.8, $\frac{d}{dx}(3x + 1) = 3$.

(b) The function $f(x) = -4$ is a linear function $f(x) = 0x - 4$ with $m = 0$ and $b = -4$. Thus by Theorem 3.8, we have $f'(x) = 0$. (We could also have used Theorem 3.6 here.)

(c) $f(x) = 1 - 2x$ is a linear function with $m = -2$, so $f'(x) = -2$. $\square$

3.5.2 The power rule (for positive integer powers)

In previous examples or homework exercises, we have seen that the derivative of $f(x) = x^2$ is $f'(x) = 2x$, the derivative of $f(x) = x^3$ is $f'(x) = 3x^2$, and the derivative of $f(x) = x^4$ is $f'(x) = 4x^3$. There is a general pattern to these derivatives: The derivative of $f(x) = x^n$ is $f'(x) = nx^{n-1}$.

THEOREM 3.9

The Power Rule (for Positive Integer Powers)

If n is any positive integer, then the derivative of x^n is

$$\frac{d}{dx}(x^n) = nx^{n-1}.$$

✎ *Caution* In fact, the power rule is much more general and can be used to differentiate x^n for *any* real number n. However, at this point we will use the power rule only for positive integer powers. We will discuss a more general power rule in Section 4.3.2. $\square$

PROOF (**THEOREM 3.9**) Suppose n is a positive integer. We need to apply the definition of derivative to find $\frac{d}{dx}(x^n)$. The "alternative" definition of derivative is easier to use in this case. In the second line of our calculation we will use a factoring formula from Theorem 0.4 in Section 0.2.2.

$$\frac{d}{dx}(x^n)$$

$$= \lim_{t \to x} \frac{t^n - x^n}{t - x} \quad \text{(definition of derivative)}$$

$$= \lim_{t \to x} \frac{(t - x)(t^{n-1} + t^{n-2}x + t^{n-3}x^2 + \cdots + tx^{n-2} + x^{n-1})}{t - x} \quad \text{(factoring formula)}$$

$$= \lim_{t \to x}(t^{n-1} + t^{n-2}x + t^{n-3}x^2 + \cdots + tx^{n-2} + x^{n-1}) \quad \text{(cancellation)}$$

$$= x^{n-1} + x^{n-2}x + x^{n-3}x^2 + \cdots + xx^{n-2} + x^{n-1}. \quad \text{(evaluate limit)}$$

$$= x^{n-1} + x^{n-1} + x^{n-1} + \cdots + x^{n-1} + x^{n-1} \quad \text{(there are } n \text{ of these)}$$

$$= nx^{n-1}.$$

❓ Question   Prove that $\frac{d}{dx}(x^6) = 6x^5$ using the method used in the proof of Theorem 3.9. ☐

❓ Question   Why does the proof of Theorem 3.9 hold only if n is a positive integer? What would go wrong in the proof if we had, for example, $n = -3$ or $n = \frac{1}{2}$? ☐

3.5.3 The constant multiple rule

Using the rules we have developed, we can quickly calculate the derivative of any constant function, identity function, linear function, or power function with a positive integer power. However, at this point there are still quite a few functions we cannot differentiate. For example, we know the derivative of x^4, but what is the derivative of $f(x) = 2x^4$? What about $g(x) = 2x - x^3 + 1$? Both of these functions are combinations of functions that we know how to differentiate, so their derivatives should be related somehow to the derivatives we know. In the following subsections, we will develop rules for differentiating simple combinations of functions whose derivatives we know.

Consider, for example, the function $f(x) = 2x^4$. This is just a constant multiple of the function x^4, so intuitively, its derivative should be related in some simple way to the derivative of x^4. In fact, since $f(x) = 2x^4$ is two times the function x^4, one might suspect that the derivative of $f(x) = 2x^4$ is two times the derivative of the function x^4 (i.e., we might suspect that $\frac{d}{dx}(2x^4) = 2\frac{d}{dx}(x^4) = 2(4x^3) = 8x^3$). This is in fact the case in general: The derivative of a constant multiple k of a function f is always equal to k times the derivative of f.

THEOREM 3.10

The Constant Multiple Rule

If $f(x)$ is any function and k is any constant, then for all x where f is differentiable,

$$\frac{d}{dx}(kf(x)) = k\frac{df}{dx}.$$

In "prime" notation this derivative rule is written

(4) $(kf)' = kf'.$

Theorem 3.10 means that we can "factor out" a constant from a derivative. Differentiation and multiplication by a constant are *commutative* operations; it doesn't matter whether we differentiate first and then multiply by a constant, or vice versa.

◈ *Caution* We already have a "constant multiple rule" for limits; now we have a constant multiple rule for derivatives. When we refer to "the constant multiple rule," we mean the one for derivatives. When we want to refer to the rule for limits, we will explicitly write out "the constant multiple rule for limits." □

With the constant multiple rule, we can now differentiate constant multiples of any function whose derivative we already know. At this point that means we can differentiate constant multiples of linear functions and constant multiples of x^n, where n is a positive integer. Later we will have rules for differentiating many other types of functions, and the constant multiple rule will then tell us how to differentiate constant multiples of those functions. The constant multiple rule is our first "abstract" differentiation rule; rather than telling us how to differentiate a certain *type* of function (such as linear or power functions), it tells us how to *relate* the derivative of a simple transformation of a function f to the derivative of f.

| EXAMPLE 3.32 | **Calculating derivatives using the constant multiple and power rules** |

Find the derivatives of the following functions.

(a) $f(x) = 2x^4$ (b) $f(x) = 72(2x - 384)$ (c) $f(x) = 2$

Solution

(a) By the constant multiple rule and the power rule,

$$\frac{d}{dx}(2x^4) = 2\frac{d}{dx}(x^4) \qquad \text{(constant multiple rule)}$$

$$= 2(4x^3) \qquad \text{(power rule)}$$

$$= 8x^3.$$

(b) By the constant multiple rule and the rule for differentiating linear functions,

$$\frac{d}{dx}(72(2x - 384)) = 72\frac{d}{dx}(2x - 384) \qquad \text{(constant multiple rule)}$$

$$= 72(2) \qquad \text{(derivative of a linear function)}$$

$$= 144.$$

We can find this derivative another way, since a constant multiple of a linear function is itself a linear function. If we multiply out the expression for $f(x)$, write it in slope-intercept form, and then use the rule for differentiating linear functions, we will not need to invoke the constant multiple rule:

$$\frac{d}{dx}(72(2x - 384)) = \frac{d}{dx}(144x - 27{,}648) \qquad \text{(algebra)}$$

$$= 144. \qquad \text{(derivative of a linear function)}$$

We often have a choice of how to differentiate a given function. In part (a), however, we did not have a choice; we had to use the constant multiple rule.

(c) Since $f(x) = 2$ is a constant function, we don't need to apply the constant multiple rule; Theorem 3.6 tells us that the derivative of any constant function is zero. Thus $f'(x) = 0$. See the "Caution" below. □

◈ *Caution* Do not confuse the constant multiple rule with the rule for differentiating constant functions. The constant multiple rule tells us how to relate the derivative of $kf(x)$ to the derivative of $f(x)$, whereas the rule for differentiating constant functions tells us how to differentiate a constant function $f(x) = k$. □

The proof of the constant multiple rule will be, as usual, a general calculation using the definition of derivative. We'll start by using the definition to write out a limit for the derivative of $kf(x)$. Then we'll use algebra to manipulate this expression until we can factor out the k, and use the definition of derivative for $f(x)$ to finish the proof.

PROOF (**THEOREM 3.10**)   Given a function f and a constant k, we wish to show that $\frac{d}{dx}(kf(x)) = k\frac{df}{dx}$. We'll start from the left and work to the right.

$$\frac{d}{dx}(kf(x)) = \lim_{h \to 0} \frac{kf(x+h) - kf(x)}{h} \qquad \text{(definition of derivative for } kf(x))$$

$$= \lim_{h \to 0} \frac{k(f(x+h) - f(x))}{h}$$

$$= \lim_{h \to 0} \left(k \left(\frac{f(x+h) - f(x)}{h} \right) \right)$$

$$= k \left(\lim_{h \to 0} \frac{f(x+h) - f(x)}{h} \right) \qquad \text{(constant multiple rule for limits)}$$

$$= k \frac{df}{dx}. \qquad \text{(definition of derivative for } f(x))$$

Notice that the backbone of the proof of the constant multiple rule for differentiation was the constant multiple rule for limits!

3.5.4 The sum and difference rules

With the differentiation rules we have developed up to this point, we can differentiate the functions $f(x) = 2x^4$ (a constant multiple of x^4) and $g(x) = 2x - 1$ (a linear function). What is the derivative of their sum $f(x) + g(x) = 2x^4 + 2x - 1$? One might hope that the derivative of a sum of functions would be related in the simplest possible way to the derivatives of the component functions: namely, that the derivative of a sum is the sum of derivatives. This is in fact the case! In other words, differentiation and addition of functions are *commutative* operations: We can add two functions and then take the derivative of the sum, or we can differentiate the two functions and then add their derivatives; either way we get the same answer.

THEOREM 3.11

The Sum Rule

If $f(x)$ and $g(x)$ are any functions, then for all x where both f and g are differentiable,

$$\frac{d}{dx}(f(x) + g(x)) = \frac{df}{dx} + \frac{dg}{dx}.$$

In "prime" notation, Theorem 3.11 says that

(5) $$(f + g)' = f' + g'.$$

With the sum rule, we can differentiate a sum of functions whose derivatives we know by differentiating each piece and adding the derivatives together. This is our second "abstract" differentiation rule; it relates the derivative of a sum of functions to the sum of their derivatives.

❖ *Caution*   Again, when we want to refer to the sum rule for derivatives, we will refer simply to "the sum rule," and when we wish to refer to the sum rule for limits, we will write out explicitly "the sum rule for limits." Be sure that you understand the difference between these two types of sum rules. □

EXAMPLE 3.33

Calculating derivatives using the sum rule

Find the derivatives of the following functions.

(a) $f(x) = x^2 + 5$ (b) $f(x) = 2x^4 + 2x - 1$

Solution

(a) The function $f(x) = x^2 + 5$ is the sum of the power function x^2 and the constant function 5. Since we know the derivatives of both of these component functions, we can use the sum rule to determine the derivative of their sum.

$$\frac{d}{dx}(x^2 + 5) = \frac{d}{dx}(x^2) + \frac{d}{dx}(5) \qquad \text{(sum rule)}$$
$$= 2x + 0 = 2x. \qquad \text{(power rule)}$$

Therefore, the derivative of $f(x) = x^2 + 5$ is $f'(x) = 2x$.

(b) We could think of $f(x) = 2x^4 + 2x - 1$ as the sum of $2x^4$ and the linear function $2x - 1$, or we could write f as the sum of the three functions $2x^4$, $2x$, and -1. With the latter decomposition of f, we have

$$\frac{d}{dx}(2x^4 + 2x - 1) = \frac{d}{dx}(2x^4) + \frac{d}{dx}(2x) + \frac{d}{dx}(-1) \qquad \begin{array}{c}\text{(sum rule,}\\\text{twice)}\end{array}$$
$$= 2\frac{d}{dx}(x^4) + 2\frac{d}{dx}(x) + \frac{d}{dx}(-1) \qquad \begin{array}{c}\text{(constant}\\\text{multiple rule)}\end{array}$$
$$= 2(4x^3) + 2(1) + 0 \qquad \begin{array}{c}\text{(derivatives}\\\text{we know)}\end{array}$$
$$= 8x^3 + 2. \qquad \qquad \qquad □$$

The proof of the sum rule uses the definition of derivative and the sum rule for limits.

PROOF (**THEOREM 3.11**)   We wish to show that $\frac{d}{dx}(f(x) + g(x)) = \frac{df}{dx} + \frac{dg}{dx}$ for any functions f and g. We will work from the left to the right; our goal is to use the definition of derivative to write the left-hand statement as a limit and then use algebra and the sum rule for limits to split this limit into two limits, one of which will be the definition of derivative for f, and one the derivative of g.

$$\frac{d}{dx}(f(x) + g(x))$$
$$= \lim_{h \to 0}\frac{(f(x+h) + g(x+h)) - (f(x) + g(x))}{h} \qquad \begin{array}{c}\text{(definition of}\\\text{derivative)}\end{array}$$
$$= \lim_{h \to 0}\frac{f(x+h) + g(x+h) - f(x) - g(x)}{h}$$
$$= \lim_{h \to 0}\frac{(f(x+h) - f(x)) + (g(x+h) - g(x))}{h} \qquad \text{(reordering terms)}$$
$$= \lim_{h \to 0}\left(\frac{f(x+h) - f(x)}{h} + \frac{g(x+h) - g(x)}{h}\right) \qquad \text{(algebra)}$$
$$= \left(\lim_{h \to 0}\frac{f(x+h) - f(x)}{h}\right) + \left(\lim_{h \to 0}\frac{g(x+h) - g(x)}{h}\right) \qquad \text{(sum rule for limits)}$$
$$= \frac{df}{dx} + \frac{dg}{dx}. \qquad \begin{array}{c}\text{(derivative of}\\f \text{ and } g)\end{array}$$

⊛ *Caution*   Be sure that you understand the proofs of the differentiation rules we have covered so far. Each proof is just an application of the definition of derivative. Remember that in this book we are interested not only in *how* to calculate things but also in *why* the methods we use are mathematically valid.   □

Since a difference $f - g$ of functions can be written as a sum $f + (-g)$, we can use the sum rule and the constant multiple rule to prove a "difference rule" for differentiation. This rule says that the derivative of a difference is the difference of derivatives:

THEOREM 3.12

The Difference Rule

If f and g are any functions, then for all x where both f and g are differentiable,

$$\frac{d}{dx}(f(x) - g(x)) = \frac{df}{dx} - \frac{dg}{dx}.$$

In "prime" notation, the difference rule says that

(6) $$(f - g)' = f' - g'.$$

The difference rule follows directly from the sum and constant multiple rules; this means that we will not have to use the definition of derivative in our proof! This is typical in mathematics: As we build up a system of rules, new rules can be proved from old rules without returning all the way to the original definitions.

PROOF   (THEOREM 3.12)   For any functions f and g,

$$\frac{d}{dx}(f(x) - g(x)) = \frac{d}{dx}(f(x) + (-g(x))) \qquad \text{(algebra)}$$
$$= \frac{d}{dx}(f(x)) + \frac{d}{dx}(-g(x)) \qquad \text{(sum rule)}$$
$$= \frac{d}{dx}(f(x)) + (-1)\frac{d}{dx}(g(x)) \qquad \text{(constant multiple rule)}$$
$$= \frac{df}{dx} - \frac{dg}{dx}. \qquad \text{(derivative notation)}$$

❓ *Question*   We can find the derivative of $f(x) = x^2 - x$ in two ways: We can use the difference rule, or we can use the sum and constant multiple rules. (In either case we will also need the power rule.) Write this derivative computation both ways.   □

❓ *Question*   We could have proved the rule for differentiating linear functions from the sum rule and the constant multiple rule (and the rules for differentiating constant and identity functions). Then we wouldn't have had to use the definition of derivative in Theorem 3.8. Write down this alternative proof of the rule for differentiating linear functions.   □

3.5.5 Calculating derivatives using the "rules"

Up until now, we have always written out every step individually when calculating a derivative using differentiation rules. In the future we will not usually write out every step. With practice, you will eventually differentiate functions in a very fast, pattern-recognition kind of way. The next example illustrates what a "quick" differentiation calculation might look like.

EXAMPLE 3.34

Using the differentiation rules

Find the derivative of the function $f(x) = 2x^4 + 2x - 1$. (This is the derivative we calculated in part (b) of Example 3.33.)

Solution   One can quickly calculate that $f'(x) = 2(4x^3) + 2 = 8x^3 + 2$. In fact, most of the time, you won't even write that middle step. Compare this with the detailed calculation we did in part (b) of Example 3.33. What differentiation rules did you use (perhaps unconsciously) while differentiating this function?   □

⊜ *Caution*   Although it is important that you be able to differentiate functions quickly, be sure that you can also write out each step of your calculation when asked to do so (as in Example 3.33). If you fall into the trap of differentiating by pattern recognition alone, you'll get stuck when trying to differentiate even moderately complicated functions in the future.   □

⊜ *Caution*   Since the derivative of a sum is the sum of derivatives, you may be tempted to think that the derivative of a product is the product of derivatives. Unfortunately, this is not the case! For example, consider the function $f(x) = (3x + 1)(2x - 4)$. After expanding the product we see that $f(x) = 6x^2 - 10x - 4$, which has derivative $f'(x) = 12x - 10$. This is clearly not equal to the product of $\frac{d}{dx}(3x + 1) = 3$ and $\frac{d}{dx}(2x - 4) = 2$.   □

⊜ *Caution*   For some reason, many students mistakenly say "derive" when they mean "differentiate." These words are *not* interchangeable. The word "derive" has a meaning very similar to that of the word "prove," and it has nothing to do with differentiation, despite the fact that it contains the first five letters of "derivative."   □

With the differentiation rules we have developed so far, we can now easily differentiate any function that can be written as a sum or constant multiple of power functions with positive integer powers. (Note that the derivatives of constant, identity, and linear functions are also covered by the previous sentence.) In later chapters we will develop rules for differentiating products, quotients, and compositions of functions whose derivatives we know. We will also learn rules for differentiating more complicated types of functions, for example, general power functions, exponential functions, and trigonometric functions.

⊜ *Caution*   There are many functions that we do not yet know how to differentiate. For such functions we must use the definition of derivative.   □

EXAMPLE 3.35

When do the differentiation rules apply?

Which of the following functions can we differentiate using the differentiation rules we have covered so far? (*Note:* Some functions may need to be manipulated algebraically before the rules are applied.)

(a) $f(x) = 3\sqrt{x} + 4$

(b) $f(x) = (x + 1)(x + 2)$

(c) $f(x) = \dfrac{x^2}{x + 1}$

(d) $f(x) = (2x + 1)^3$

(e) $f(x) = (3x^2 + 1)^{100}$

(f) $f(x) = 2^x$

Solution   The functions in parts (b) and (d) need to be rewritten before we can apply the differentiation rules that we know:

$$\frac{d}{dx}((x+1)(x+2)) = \frac{d}{dx}(x^2 + 3x + 2) = 2x + 3,$$

$$\frac{d}{dx}((2x+1)^3) = \frac{d}{dx}(8x^3 + 12x^2 + 6x + 1) = 8(3x^2) + 12(2x) + 6.$$

The function in part (e) *can* be differentiated using the rules we know, but we would first have to multiply out the expression $(3x^2 + 1)^{100}$ until it is a sum of power functions (which would be a very tedious process!).

The remaining functions cannot be differentiated using the differentiation rules we have so far developed (although we could always use the definition of derivative if we needed to find their derivatives). We do not yet know how to differentiate the function $f(x) = 3\sqrt{x} + 4$ from part (a), because it involves a square root, and we don't have a rule for differentiating that function (we would need a more general power rule). Part (c) involves a quotient, which we don't yet know how to handle with differentiation rules. Part (f) involves a type of function we won't know much about until Part III, and we certainly don't have a rule for its derivative at this point (in particular, the power is not a positive integer, it is a variable). ☐

The following theorem makes it easy to tell whether a piecewise function is differentiable at its "break" points:

THEOREM 3.13

Differentiability at the "Break" Point of a Piecewise Function

If g and h are continuous and differentiable at $x = c$, then the piecewise function

$$f(x) = \begin{cases} g(x), & \text{if } x \leq c \\ h(x), & \text{if } x > c \end{cases}$$

is differentiable at $x = c$ if, and only if, $g(c) = h(c)$ and $g'(c) = h'(c)$.

The first condition ($g(c) = h(c)$) basically says that the two pieces of the function $f(x)$ "match up" at $x = c$ (i.e., the function is continuous at $x = c$); the second condition ($g'(c) = h'(c)$) says that the two pieces of the function $f(x)$ have the same slope at $x = c$. Theorem 3.13 also holds if the first inequality is $<$ and the second is $\geq$. You will prove this theorem in the exercises.

EXAMPLE 3.36

When is a piecewise function differentiable?

Consider the function

$$f(x) = \begin{cases} bx + 1, & \text{if } x < 2 \\ ax^2, & \text{if } x \geq 2. \end{cases}$$

Find real numbers a and b such that $f(x)$ is differentiable at $x = 2$ (if possible).

Solution   Since $g(x) = bx + 1$ and $h(x) = ax^2$ are continuous and differentiable at $x = 2$ (for any numbers a and b), the function $f(x)$ will be differentiable at $x = 2$ if, and only if, $g(2) = h(2)$ and $g'(2) = h'(2)$. The first condition means that $b(2) + 1 = a(2)^2$, that is, that $2b + 1 = 4a$. Since $g'(x) = b$ and $h'(x) = 2ax$, the second condition means that $b = 2a(2)$, that is, that $b = 4a$. Solving this system of equations, we find that $a = -\frac{1}{4}$ and $b = -1$. ☐

? *Question* Graph the piecewise function from Example 3.36, with $a = -\frac{1}{4}$ and $b = -1$. Verify that the graph appears continuous and differentiable at $x = 2$. □

⑤ *Caution* The first condition in Theorem 3.13 (that $g(c) = h(c)$) is vital! If the two pieces of the graph of a piecewise function $f(x)$ do not have the same y-value at $x = c$, then $f(x)$ will be discontinuous and thus cannot possibly be differentiable. □

What you should know after reading Section 3.5

▶ The statements and proofs of the rules for differentiating constant, identity, linear, and power functions. The statements and proofs of the constant multiple, sum, and difference rules. Be able to use these rules to calculate derivatives.

▶ What types of functions can we differentiate using the rules in this section? What types of functions *can't* we differentiate?

▶ Be able to calculate derivatives both by writing each step out individually (applying one differentiation rule at a time) and by quick pattern recognition.

Exercises 3.5

Concepts

0. Read the section and make your own summary of the material. Include whatever you think will help you review the material later on.

■ Give formal statements of each of the following differentiation rules, in both Leibnitz and "prime" notation. Then give an example calculation for each rule (different from any examples in the reading).

1. The derivative of a constant

2. The derivative of the identity function

3. The derivative of a linear function

4. The derivative of x^n (n a positive integer)

5. The constant multiple rule

6. The sum rule

7. The difference rule

8. The derivative of a quadratic function

9. The constant multiple rule, sum rule, and difference rule are in some way different from the rules that tell you how to differentiate constant, identity, linear, or power functions. How are they different?

10. List four functions that we *cannot* differentiate by using the rules of this section. Use different examples from the ones given in the text.

11. Why does it make graphical sense that the derivative of a constant is zero? That the derivative of the identity function is constantly equal to 1? That the derivative of a linear function $f(x) = mx + b$ is equal to m?

■ In Problems 12–15 you will differentiate the function $f(x) = 3 - 2x + 5x^2$ in many different ways (of course, the final answer should always be the same!). Differentiate f using *only the rules listed* in each problem below. Write out each step individually, citing where you use each differentiation rule.

12. The derivatives of constant, identity, and power functions, and the sum, difference, and constant multiple rules.

13. The derivatives of constant, identity, and power functions, and the sum and constant multiple rules (no difference rule this time).

14. The derivatives of linear and power functions, and the sum and constant multiple rules.

15. No differentiation "rules" at all; use the definition of derivative.

■ Suppose f, g, and h are functions with the following derivative values: $f'(-2) = 3$, $f'(0) = -2$, $f'(3) = 1$, $g'(-1) = -1$, $g'(0) = 5$, $g'(3) = 0$, $h'(-2) = -1$, $h'(-1) = 8$, and $h'(3) = -2$. Calculate each of the quantities below, if possible. When it is not possible to calculate a quantity, explain why.

16. $(f + g)'(0)$

17. $(3f + 4h)'(-2)$

18. $(fh)'(-2)$

19. $(2f + 3g - h)'(3)$

20. $(2f + 3g - h)'(-2)$

21. $\dfrac{d}{dx}\bigg|_{-1} (x^2 + g(x) - h(x))$

22. Which of the following is a notationally correct way to express the fact that the derivative of the function $y = 4 - x^3$ is $-3x^2$? (Choose *all* that apply.)

(a) $\frac{d}{dx}(4 - x^3) = -3x^2$

(b) $\frac{dy}{dx}(4 - x^3) = -3x^2$

(c) $\frac{dy}{dx} = -3x^2$

23. Find an example that shows that the derivative of a product fg is *not* in general the product of derivatives f' and g'. In other words, find functions $f(x)$ and $g(x)$ such that if we define $h(x) = f(x)g(x)$, then $h'(x) \neq f'(x)g'(x)$.

■ In each problem below, the *derivative* $f'(x)$ of some function $f(x)$ is given. Find a function $f(x)$ that has the derivative shown. Then find *another* function $f(x)$ that could have that derivative.

24. $f'(x) = 2$ **25.** $f'(x) = 1 - 2x$

26. $f'(x) = 3x + 5$ **27.** $f'(x) = 5x^2 + 3x - 1$

28. $f'(x) = x^{17}$ **29.** $f'(x) = (x^3 - 1)(x + 2)$

30. Explain why Theorem 3.13 provides a sufficient condition for determining when some piecewise functions are differentiable. In particular, why don't we need to look at left and right limits or left and right derivatives if the "pieces" g and h are continuous and differentiable?

(a) $f(x) + g(x)$ (b) $3f(x) - 2g(x)$

(c) $f(x)g(x)$ (d) $(f(x))^2$

(e) $f(x) + x^2$ (f) $xg(x)$

Skills

■ Differentiate the following functions using the differentiation rules we have learned so far. Write out every step individually, citing each derivative rule that you use.

31. $f(x) = x^2 - 3$ **32.** $f(x) = 2(1 + 3x^2)$

33. $f(x) = 2x - 3 + 4x^2$ **34.** $f(x) = 4 - 3x^7$

35. $f(x) = 2(3x + 1) - 4x^5$

36. $f(x) = 2x^{100} - 42(3x + 1)$

■ Differentiate each of the following functions using the rules in this section. Each of these functions requires some algebraic manipulation before the differentiation rules can be applied.

37. $f(x) = (x + 2)(x - 1)$ **38.** $f(x) = x^2 + x(2 - 3x^2)$

39. $f(x) = (3x + 2)^3$ **40.** $f(x) = \dfrac{1 - 6x^3}{3}$

41. $f(x) = (3 - x)^2 + 5$ **42.** $f(x) = \dfrac{x^2 - 1}{x + 1}$

■ Some of the following functions can be differentiated using the rules in this section (perhaps after some algebra), and some cannot. Find the derivative of each function either by using the rules for differentiation (when they apply) or by using the definition of derivative (when the rules don't apply).

43. $f(x) = x(x - 3)$ **44.** $f(x) = (x^2 - 3)^2$

45. $f(x) = \dfrac{1}{x}$ **46.** $f(x) = \dfrac{x}{x + 1}$

47. $f(x) = (2x)^3 - 7x + 1$ **48.** $f(x) = \dfrac{x^4 - 7x^3}{2x}$

■ Given that $f(x) = -4x^3 - x(2x + 1)$, find the following derivatives.

49. $f'(x)$ **50.** $f''(x)$

51. $f'''(x)$ **52.** $f'(2)$

53. $f''(-1)$ **54.** $f'''(7)$

■ Calculate the following derivatives using the rules from this section.

55. $\dfrac{d}{dx}(-5x^2 + 3x - 2)$ **56.** $\dfrac{d}{dx}(x(1.7x - 0.4))$

57. $\dfrac{d}{dx}(3x^2 - x(4 + 2x^2))$ **58.** $\dfrac{d}{dx}\bigg|_2 (1 - 3x^5)$

59. $\dfrac{d}{dx}\bigg|_0 (3x^5 + 4 - 7x^2)$ **60.** $\dfrac{d^2}{dx^2}(-4x^5 + 3x^3 - 5)$

61. $\dfrac{d^2}{dx^2}((2 - 3x)^2)$

62. $\dfrac{d^2}{dx^2}\bigg|_{-1} (x(x - 1)(x - 2))$

63. $\dfrac{d^7}{dx^7}(x^7)$

■ For each piecewise function $f(x)$ below, determine whether $f(x)$ is (a) continuous and (b) differentiable at each of its break points. If $f(x)$ fails to be continuous at a point, determine whether it is left or right continuous (or neither) at that point. If $f(x)$ fails to be differentiable at a point, determine whether it is left or right differentiable (or neither) at that point.

64. $f(x) = \begin{cases} x^2, & \text{if } x < 0 \\ 3, & \text{if } x \geq 0 \end{cases}$

65. $f(x) = \begin{cases} 3x + 2, & \text{if } x \leq 2 \\ x^3, & \text{if } x > 2 \end{cases}$

66. $f(x) = \begin{cases} x^2 + 1, & \text{if } x < 4 \\ x^2 - 1, & \text{if } x \geq 4 \end{cases}$

■ For each piecewise function $f(x)$ described below, write $f'(x)$ as a piecewise function. (Pay special attention to whether or not $f'(x)$ exists at each of the break points of

$f(x)$.) If $f(x)$ fails to be differentiable at a point $x = c$, find values for $f'_-(c)$ and $f'_+(c)$, if they exist.

67. $f(x)$ from Problem 64

68. $f(x)$ from Problem 65

69. $f(x)$ from Problem 66

■ For each piecewise function $f(x)$ below, find constants a and b such that $f(x)$ is continuous and differentiable everywhere (if possible).

70. $f(x) = \begin{cases} ax + b, & \text{if } x \leq 3 \\ 2, & \text{if } x > 3 \end{cases}$

71. $f(x) = \begin{cases} 3x + a, & \text{if } x < 1 \\ bx^2 + 1, & \text{if } x \geq 1 \end{cases}$

72. $f(x) = \begin{cases} ax^3 + b, & \text{if } x < -2 \\ bx - a, & \text{if } x \geq -2 \end{cases}$

Applications

73. As a vacuum cleaner salesman, Alex earns a salary of $8,500 a year, whether he sells any vacuum cleaners or not. In addition, for every 30 vacuum cleaners he sells, he earns a $500 commission. Suppose that Alex always sells between 0 and 100 vacuum cleaners every year. The function $M(v)$ describing the amount of money M that Alex will make in a year if he sells v vacuum cleaners over the course of the year is

$$M(v) = \begin{cases} 8,500, & \text{if } 0 \leq v < 30 \\ 9,000, & \text{if } 30 \leq v < 60 \\ 9,500, & \text{if } 60 \leq v < 90 \\ 10,000, & \text{if } 90 \leq v \leq 100. \end{cases}$$

You found the equation for this function in Problem 61 in Section 2.6 (page 204).

(a) Use Theorem 3.13 to show that $M(v)$ is *not* differentiable at $v = 60$. Explain in real-world terms why this makes sense.

(b) Write the derivative $M'(v)$ as a piecewise function. (Be especially careful near the "break" points of $M(v)$.) Interpret $M'(35)$ and $M'(72)$ in real-world terms.

74. In Problem 51 in Section 1.4 (page 98), you constructed a piecewise function from the 2000 Federal Tax Rate Schedule. Specifically, you found that a person who makes m dollars a year will pay $T(m)$ dollars in tax, given by the formula

$T(m) =$

$$\begin{cases} 0.15m, & \text{if } 0 \leq m \leq 26,250 \\ 3,937.5 + 0.28(m - 26,250), & \text{if } 26,250 < m \leq 63,550 \\ 14,381.5 + 0.31(m - 63,550), & \text{if } 63,550 < m \leq 132,600 \\ 35,787 + 0.36(m - 132,600), & \text{if } 132,600 < m \leq 288,350 \\ 91,857 + 0.396(m - 288,350), & \text{if } m \geq 288,350. \end{cases}$$

(a) Use Theorem 3.13 to show that $T(m)$ is continuous, but *not* differentiable, at $m = 63,550$. What does this mean in real-world terms?

(b) Write the derivative $T'(m)$ as a piecewise function. (Be especially careful near the "break" points of $T(m)$.) Interpret $T'(68,000)$ and $T'(150,000)$ in real-world terms.

Proofs

75. Use the (alternative) definition of derivative to prove that $\frac{d}{dx}(x^6) = 6x^5$, following the method used in the proof of Theorem 3.9.

■ Prove each of the derivative rules below in your own words, using the definition of derivative.

76. The derivative of a constant

77. The derivative of the identity function

78. The derivative of a linear function

79. The power rule (positive integer powers)

80. The constant multiple rule

81. The sum rule

82. Prove the rule for differentiating linear functions (Theorem 3.8) without using the definition of derivative. Instead, use the sum and constant multiple rules and the rules for differentiating constant and identity functions.

83. Prove the difference rule (Theorem 3.12) directly, that is, using the definition of derivative.

84. Prove that $\frac{d}{dx}(ax^2 + bx + c) = 2ax + b$ for any real numbers a, b, and c, in two different ways: (a) using the definition of derivative, and (b) using other differentiation rules.

85. Prove that for any constant a we have $\frac{d}{dx}((a - x)^2) = 2x - 2a$, in two different ways: (a) using the definition of derivative, and (b) using other differentiation rules.

86. Prove that for any constant a we have $\frac{d}{dx}((x - a)(x - b)) = 2x - a - b$, in two different ways: (a) using the definition of derivative, and (b) using other differentiation rules.

87. Prove that every quadratic function $f(x) = ax^2 + bx + c$ is differentiable.

88. Prove that if f is differentiable at $x = c$, and k is any real number, then the function kf is differentiable at $x = c$.

89. Prove that if f and g are differentiable functions, then so is their sum $f + g$.

90. Prove Theorem 3.13: If $g(x)$ and $h(x)$ are continuous and differentiable at a point $x = c$, and if $g(c) = h(c)$ and $g'(c) = h'(c)$, then the piecewise function $f(x) = \begin{cases} g(x), & \text{if } x \leq c \\ h(x), & \text{if } x > c \end{cases}$ is differentiable at $x = c$.

3.6 Three Theorems About Tangent Lines

In this section we explore three theorems about tangent lines. The first theorem tells us that the tangent line to a differentiable function at a local minimum or maximum is horizontal. This basic theorem will be used to prove **Rolle's Theorem,** which states that if a continuous, differentiable function has two roots, then that function must have a horizontal tangent line somewhere between those roots. This important theorem is actually a special case of the **Mean Value Theorem.** The Mean Value Theorem is one of the most important theorems in differential calculus, and we will use it to prove many important relationships between functions and their derivatives.

3.6.1 The derivative at a maximum or minimum

Recall that a function $f(x)$ has a local maximum at $x = c$ if $f(c) \geq f(x)$ for all x in some neighborhood $(c - \delta, c + \delta)$ of the point c (and a local minimum if the inequality is reversed). Intuitively, at a local maximum or minimum, the graph of a function f must "turn around." If the function is differentiable, then as it "turns around" it will have a horizontal tangent line. If the function is not differentiable at the "turnaround" point, then it won't have a tangent line there at all. Figures 3.49–3.51 illustrate this phenomenon.

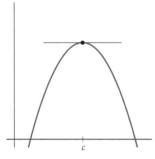

Figure 3.49
Local maximum with horizontal tangent line

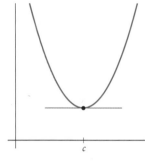

Figure 3.50
Local minimum with horizontal tangent line

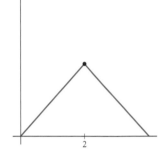

Figure 3.51
Local maximum with no tangent line

In general, if a function has a local extremum at $x = c$, then its derivative at $x = c$ either does not exist or is equal to zero. Although this seems "obvious" graphically, we must show that it follows from our definition of derivative and the definition of a local extremum. This basic fact will be used in many proofs, including the proof of Rolle's Theorem.

THEOREM 3.14

The Derivative at a Local Extremum
If f has a local minimum or maximum at $x = c$, then either $f'(c)$ does not exist or $f'(c) = 0$.

The x-values where the derivative of a function is zero or does not exist are so important that they are given a name:

DEFINITION 3.16

Critical Points
A number c is a **critical point** of a function f if $f'(c) = 0$ or if $f'(c)$ does not exist.

Theorem 3.14 says that if f has a local extremum at $x = c$, then $x = c$ is a critical point of f. (However, as we will soon see, the converse is not true; not every critical point is a local extremum.)

PROOF (**THEOREM 3.14**)   Suppose f has a local maximum at $x = c$. (The proof for a local minimum is entirely similar and is left to you in the exercises.) If f is not differentiable at $x = c$, then we are finished. If f is differentiable at $x = c$, then we must show that $f'(c) = 0$. This is what we will do for the remainder of the proof.

Recall from Section 1.2.7 that by definition, if f has a local maximum at $x = c$, then there is some interval around c on which $f(c)$ is greater than or equal to any other value of $f(x)$. In other words, there is some $\delta > 0$ such that

(1)   *For all $x \in (c - \delta, c + \delta)$ we have $f(c) \geq f(x)$.*

Fix this δ for the remainder of the proof. Statement (1) will be a key part of our argument.

By hypothesis, f is differentiable at $x = c$. Therefore, $f'(c)$ exists, which means that

$$\lim_{x \to c^+} \frac{f(x) - f(c)}{x - c} \quad \text{and} \quad \lim_{x \to c^-} \frac{f(x) - f(c)}{x - c}$$

both exist and are equal to $f'(c)$. Let's examine these two limits one at a time.

On the one hand, for all $x \in (c, c + \delta)$ we have $x > c$ and $f(c) \geq f(x)$ by Statement (1). Thus $x - c > 0$ and $f(x) - f(c) \leq 0$, so

$$\frac{f(x) - f(c)}{x - c} \leq 0 \quad \text{for all } x \in (0, c + \delta).$$

Since this difference quotient is negative for *all* values of x sufficiently close to but greater than c, we have

(2) $$\lim_{x \to c^+} \frac{f(x) - f(c)}{x - c} \leq 0.$$

On the other hand, we can use a similar argument to show that

(3) $$\lim_{x \to c^-} \frac{f(x) - f(c)}{x - c} \geq 0,$$

since for all $x \in (c - \delta, c)$ we have $x < c$ and $f(c) \geq f(x)$.

Since the quantities in Equations (2) and (3) are both equal to $f'(c)$, we have shown both that $f'(c) \leq 0$ and that $f'(c) \geq 0$. Therefore, we must have $f'(c) = 0$, which is what we set out to prove.   ∎

◈ *Caution*   Theorem 3.14 says that every local extremum is a critical point. The converse of this statement is *not* true: It is not true that every critical point is a local extremum. For example, consider the function f shown in Figure 3.52. The point $x = 1$ is a critical point of f, since $f'(1) = 0$ (the graph of f has a horizontal tangent line at $x = 1$). However, $x = 1$ is clearly neither a local minimum nor a local maximum of f.

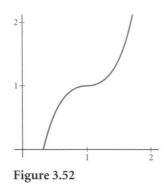

Figure 3.52   □

3.6.2 Rolle's Theorem

Suppose a continuous, differentiable function f has two roots $x = a$ and $x = b$. What can you say about the graph of f between a and b? Figures 3.53–3.55 show the graphs of three continuous, differentiable functions with roots at $x = a$ and $x = b$. What do these three graphs have in common? In each graph the function has at least one turning point somewhere between a and b. More precisely, each of these functions has at least one local extremum between the two roots $x = a$ and $x = b$. By Theorem 3.14, each of these functions has a derivative of zero for at least one value of x between a and b. This is exactly the statement of Rolle's Theorem.

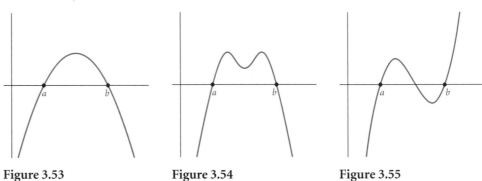

Figure 3.53   **Figure 3.54**   **Figure 3.55**

THEOREM 3.15

Rolle's Theorem

Let f be a function that is continuous on $[a, b]$ and differentiable on (a, b). If $f(a) = f(b) = 0$, then there is at least one value $c \in (a, b)$ for which $f'(c) = 0$.

Actually, in the hypotheses of Rolle's Theorem, we do not have to have $f(a)$ and $f(b)$ equal to *zero*, as long as they are equal to each other. Rolle's Theorem is also true if $f(a) = f(b) = 5$, or if $f(a) = f(b) = 3$, and so on. However, the classic way to state Rolle's Theorem is to assume that $f(a)$ and $f(b)$ are both zero. The case where $f(a)$ and $f(b)$ are nonzero will be included in Theorem 3.16.

EXAMPLE 3.37   **Graphical interpretation of Rolle's Theorem**

Since the functions graphed in Figures 3.53–3.55 are continuous and differentiable with roots at $x = a$ and $x = b$, Rolle's Theorem applies to each of these functions. Therefore, for each of these functions there is at least one value $x = c$ in the interval (a, b) for which $f'(c) = 0$, that is, for which the tangent line to f at $x = c$ is horizontal. Figures 3.56–3.58 show one such value $x = c$ for each of the graphs in Figures 3.53–3.55.

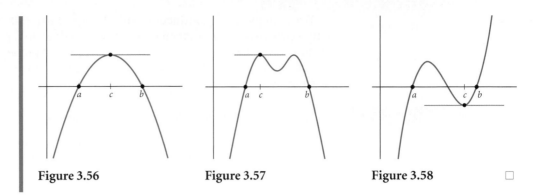

Figure 3.56   **Figure 3.57**   **Figure 3.58**   ☐

⊛ *Caution* If f is continuous and differentiable with roots at $x = a$ and $x = b$, then Rolle's Theorem guarantees the existence of *at least one* value $c \in (a, b)$ for which $f'(c) = 0$. In general, there may be more than one such value of c (see Figures 3.57 and 3.58).   ☐

⊛ *Caution* If f is continuous and differentiable with roots at $x = a$ and $x = b$, then Rolle's Theorem tells us that there is some value $c \in (a, b)$ where the derivative is zero, but it does not tell us *what that value is*! This is a conceptual rather than a computational theorem; it tells us that the function f has a horizontal tangent line somewhere between a and b, but it does not tell us exactly where.   ☐

Rolle's Theorem is an immediate consequence of the Extreme Value Theorem and Theorem 3.14:

PROOF (**THEOREM 3.15**) Suppose f is a function that is continuous on the closed interval $[a, b]$ and differentiable on the open interval (a, b), with $f(a) = f(b) = 0$. By the Extreme Value Theorem (Section 2.7.1), f has a maximum value and a minimum value on $[a, b]$. Suppose one of these maximum or minimum values occurs at a point c in the *interior* (a, b) of the interval $[a, b]$. Since $c \in (a, b)$ and f is differentiable on (a, b), by Theorem 3.14 we must have $f'(c) = 0$.

This proves Theorem 3.15 except in the special case where all the maximum and minimum values of f on $[a, b]$ occur at the *endpoints* of the interval, at $x = a$ or at $x = b$. In this case, since $f(a) = f(b) = 0$, the maximum and minimum values of f are both equal to 0. For all x in $[a, b]$ we would have $0 \leq f(x) \leq 0$; in this case f is the constant function $f(x) = 0$ on $[a, b]$. Since the derivative of a constant function is always zero, in this special case we have $f'(x) = 0$ for *all* values of c in (a, b), and we are finished.   ■

❓ *Question* Could you explain that proof to a classmate? You may have to read it a few times and sketch some pictures to help you understand the argument. In particular, given the graph of a function f that satisfies the hypotheses of Rolle's Theorem and has multiple values of $c \in (a, b)$ with $f'(c) = 0$, which of these values of c are referred to in the proof of Theorem 3.15?   ☐

Just as the Intermediate Value Theorem and the Extreme Value Theorem illustrate basic properties of continuous functions, Rolle's Theorem illustrates a basic property of functions that are both continuous and differentiable. The continuity and the differentiability hypotheses of Rolle's Theorem are essential. If a function f fails to be continuous on $[a, b]$ or fails to be differentiable on (a, b), then the conclusion of Rolle's Theorem does not follow.

For example, if f is a function with $f(a) = f(b) = 0$, but f is not differentiable on the open interval (a, b), then Rolle's Theorem does not apply to that function: There may or may not be a point $c \in (a, b)$ with $f'(c) = 0$. Similarly, if $f(a) = f(b) = 0$ but f is not continuous on the closed interval $[a, b]$, then there is not necessarily a point $c \in (a, b)$ with $f'(c) = 0$.

EXAMPLE 3.38

The hypotheses of Rolle's Theorem are necessary

Each of the functions graphed in Figures 3.59–3.61 has roots at $x = 1$ and $x = 3$ but fails to satisfy one of the hypotheses of Rolle's Theorem. (Note that the graph in Figure 3.61 has a point at $(3, 0)$.)

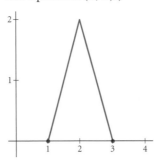 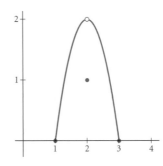 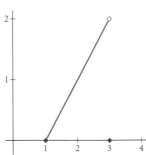

Figure 3.59
Not differentiable on $(1, 3)$

Figure 3.60
Not continuous on $[1, 3]$

Figure 3.61
Not continuous on $[1, 3]$

Since the function f graphed in Figure 3.59 fails to be differentiable on $(1, 3)$, the function can "turn around" without having a horizontal tangent line. In this example, there is no value of $c \in (a, b)$ with $f'(c) = 0$.

The functions graphed in Figures 3.60 and 3.61 fail to be continuous on the closed interval $[a, b]$. In Figure 3.60 the function fails to be continuous at the very place where we would have expected its derivative to be zero. Since this function f is not continuous at $x = 2$, it is also not differentiable at $x = 2$, and there is no value $c \in (1, 3)$ with $f'(c) = 0$.

The function in Figure 3.61 illustrates why it is important for the function f to be continuous on the entire *closed* interval $[1, 3]$. This function f fails to be continuous at the right endpoint $x = 3$ of the interval. There is no value $c \in (1, 3)$ with $f'(c) = 0$; the function never has to "turn around" since it just "jumps" down to the root at $x = 3$.   □

The functions graphed in Figures 3.59–3.61 are examples of functions that fail the hypotheses *and* the conclusion of Rolle's Theorem. It is also possible for a function to fail the hypotheses of Rolle's Theorem and still have a value $c \in (a, b)$ with $f'(c) = 0$ (can you sketch the graph of such a function?). However, there is *no* function that satisfies the hypotheses of the theorem and then fails the conclusion.

◈ *Caution*   Example 3.38 does *not* show that Rolle's Theorem is false! It shows that each hypothesis of Rolle's Theorem is necessary to the conclusion. If a function does not satisfy all of the hypotheses of Rolle's Theorem, then the theorem simply does not apply to that function. In general, if a function does not satisfy the hypotheses of a theorem, then the conclusion of that theorem may or may not be true for that function.   □

EXAMPLE 3.39

Applying Rolle's Theorem

Use Rolle's Theorem to prove that there is some value of c in $(-2, 5)$ at which the function $f(x) = x^2 - 3x - 10$ has a horizontal tangent line. Then use $f'(x)$ to find such a value of c.

Solution   First notice that $f(x) = x^2 - 3x - 10 = (x+2)(x-5)$ has roots at $x = -2$ and at $x = 5$. Since f is a quadratic function (and thus a sum of constant multiples of power functions with positive integer powers), it is continuous and differentiable. In particular, f is continuous on $[-2, 5]$ and differentiable on $(-2, 5)$. Therefore, Rolle's Theorem applies to this function f, and we can conclude that there is some value of $c \in (-2, 5)$ for which $f'(c) = 0$. At this value of c, the graph of f will have a horizontal tangent line.

Rolle's Theorem tells us that there exists some $c \in (-2, 5)$ where $f'(c) = 0$, but it doesn't tell us *what* values of c have this property. However, we can find such a c by solving the equation $f'(x) = 0$. Using differentiation rules, we know that the derivative of $f(x) = x^2 - 3x - 10$ is $f'(x) = 2x - 3$. Since $f'(x) = 2x - 3 = 0$ when $x = \frac{3}{2}$, the function f has a horizontal tangent line at $c = \frac{3}{2}$. Note that this value of c is in the interval $(-2, 5)$.   □

☑ *Checking the Answer*   Figure 3.62 shows the graph of $f(x) = x^2 - 3x - 10$ on $[-2, 5]$. Note that there is indeed a value $c \in (-2, 5)$ at which f has a horizontal tangent line, and our calculation that $c = \frac{3}{2}$ appears to be accurate.

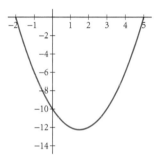

Figure 3.62   □

3.6.3 The Mean Value Theorem

The Mean Value Theorem is a generalization of Rolle's Theorem to continuous, differentiable functions where $f(a)$ and $f(b)$ are not necessarily equal. This important theorem says that if f is continuous and differentiable, then there is some $c \in (a, b)$ where the tangent line to f at c has the same slope as the line representing the average rate of change from $x = a$ to $x = b$.

For example, consider the function f shown in Figure 3.63 and the slope of the secant line from $(a, f(a))$ to $(b, f(b))$. Is there some value $c \in (a, b)$ where the slope of the tangent line to f at $x = c$ is equal to the slope of this secant line? In this example, there is; Figure 3.64 shows such a value c. In this graph, the tangent line at $(c, f(c))$ is parallel to the line connecting $(a, f(a))$ and $(b, f(b))$.

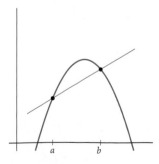

Figure 3.63

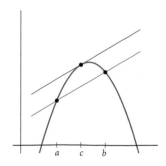

Figure 3.64

The Mean Value Theorem says that if a function is continuous and differentiable, then we can always find such a c. The "mean" in the Mean Value Theorem refers to the average rate of change of f from $x = a$ to $x = b$; this theorem says that we can always find a point where the derivative of f is equal to this "mean" change.

THEOREM 3.16

The Mean Value Theorem

If f is continuous on $[a, b]$ and differentiable on (a, b), then there exists at least one value $c \in (a, b)$ such that

$$f'(c) = \frac{f(b) - f(a)}{b - a}.$$

How is this a generalization of Rolle's Theorem? Turn your head to the side while looking at the graph in Figure 3.63 until the line from $(a, f(a))$ to $(b, f(b))$ appears horizontal. Now think of this line as the x-axis. Rolle's Theorem tells us we can find a $c \in (a, b)$ where the tangent line to f at c is also horizontal (with our heads still turned to the side). This is the line shown in Figure 3.64.

The proof of the Mean Value Theorem will follow exactly the same strategy as the "turning your head" argument. The equivalent of "turning our heads" will be to consider the function $g(x) = f(x) - l(x)$ that is the difference of $f(x)$ and the secant line $l(x)$ from $(a, f(a))$ to $(b, f(b))$. (To visualize this, it may help to look back at the example in Figure 3.63.) The graph of this new function $g(x)$ will have roots at $x = a$ and $x = b$ (so the secant line on $g(x)$ from a to b will be the x-axis, as we wanted). If we apply Rolle's Theorem to this new function, we can find a value of c at which the tangent line to $g(x)$ is horizontal. It will turn out that the graph of the original function f will have a tangent line with the desired slope at exactly this value of c.

PROOF (**THEOREM 3.16**)   Suppose $f(x)$ is a function that is continuous on $[a, b]$ and differentiable on (a, b). Consider the secant line $l(x)$ from $(a, f(a))$ to $(b, f(b))$. Since this line has slope $\frac{f(b)-f(a)}{b-a}$ and passes through the point $(a, f(a))$, the equation of this secant line is

$$l(x) = \left(\frac{f(b) - f(a)}{b - a} \right) (x - a) + f(a).$$

Define a new function $g(x)$ to be the difference between $f(x)$ and the secant line $l(x)$. We then have

$$g(x) := f(x) - l(x)$$
$$= f(x) - \left(\frac{f(b) - f(a)}{b - a} \right) (x - a) - f(a).$$

We wish to apply Rolle's Theorem to this function $g(x)$. To do this we must first make sure that $g(x)$ satisfies all the hypotheses of Rolle's Theorem. The function $g(x)$ is continuous on $[a, b]$ because $f(x)$ is continuous on $[a, b]$. (The function $g(x)$ is a sum of constant multiples of the continuous functions $f(x)$, $(x - a)$, and $f(a)$.) Similarly, $g(x)$ is differentiable on (a, b) because $f(x)$ is differentiable on (a, b). Finally, we do have $g(a) = 0$ and $g(b) = 0$:

$$g(a) = f(a) - \left(\frac{f(b) - f(a)}{b - a} \right) (a - a) - f(a)$$
$$= f(a) - 0 - f(a) = 0,$$

$$g(b) = f(b) - \left(\frac{f(b) - f(a)}{b - a} \right) (b - a) - f(a)$$
$$= f(b) - (f(b) - f(a)) - f(a) = 0.$$

Now that we know that Rolle's Theorem applies to the function $g(x)$, we can use it to state that there exists some $c \in (a, b)$ for which $g'(c) = 0$. How is this related to our original problem? All we need to do is calculate $g'(x)$:

$$g'(x)$$

$$= \frac{d}{dx}\left(f(x) - \left(\frac{f(b) - f(a)}{b - a}\right)(x - a) - f(a)\right)$$

$$= \frac{d}{dx}(f(x)) - \left(\frac{f(b) - f(a)}{b - a}\right)\frac{d}{dx}(x - a) - \frac{d}{dx}(f(a)) \qquad \text{(differentiation rules)}$$

$$= f'(x) - \left(\frac{f(b) - f(a)}{b - a}\right)(1) - 0$$

$$= f'(x) - \frac{f(b) - f(a)}{b - a}.$$

We know by Rolle's Theorem that there exists a $c \in (a, b)$ for which $g'(c) = 0$. By our calculation above, for this value of c we have

$$0 = g'(c) = f'(c) - \frac{f(b) - f(a)}{b - a}.$$

In other words, at this value of c, the derivative of f is equal to the slope of the line from $(a, f(a))$ to $(b, f(b))$:

$$f'(c) = \frac{f(b) - f(a)}{b - a}.$$

◈ *Caution*   Like Rolle's Theorem, the Mean Value Theorem tells us that some $c \in (a, b)$ exists at which f has an interesting property, but it does not tell us what this value of c actually is. The Mean Value Theorem is a conceptual and not a computational theorem. It is possible to find these values of c computationally (see Example 3.40), but the Mean Value Theorem won't help us do that.   ☐

EXAMPLE 3.40   **Applying the Mean Value Theorem**

Use the Mean Value Theorem to show that there is some value $c \in (0, 2)$ where the tangent line to the function $f(x) = x^2 - 2$ has slope 2. Then find such a value c algebraically.

Solution   The function $f(x) = x^2 - 2$ is a quadratic and thus is continuous and differentiable; in particular, it is continuous on $[0, 2]$ and differentiable on $(0, 2)$. Therefore, the Mean Value Theorem applies to this function on the interval $[0, 2]$. The slope of the line from $(0, f(0))$ to $(2, f(2))$ is

$$\frac{f(2) - f(0)}{2 - 0} = \frac{(2^2 - 2) - (0^2 - 2)}{2} = \frac{2 - (-2)}{2} = \frac{4}{2} = 2.$$

By the Mean Value Theorem, we know that there exists a point $c \in (0, 2)$ with $f'(c) = 2$.

Now that we know there exists some $c \in (0, 2)$ with $f'(c) = 2$, let's see if we can find such a c algebraically. The derivative of $f(x) = x^2 - 2$ is $f'(x) = 2x + 0 = 2x$. We want to find $c \in (0, 2)$ with $f'(c) = 2$, so we solve

$$f'(c) = 2 \implies 2c = 2 \implies c = 1.$$

The point $c = 1$ is indeed in the interval $(0, 2)$, and $f'(1) = 2$, so we are done.   ☐

✓ *Checking the Answer*   The graph of $f(x) = x^2 - 2$ is shown in Figure 3.65. The slope of the tangent line at $x = 1$ appears to have the same slope as the secant line from $(0, f(0))$ to $(2, f(2))$.

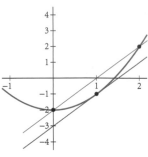

Figure 3.65   □

? *Question*   The continuity and differentiability hypotheses of the Mean Value Theorem are necessary. Can you sketch examples of functions that fail either the continuity or the differentiability hypothesis for which the conclusion of the Mean Value Theorem does not hold? (*Hint:* we did this in Example 3.38 for Rolle's Theorem; try to do something similar here.)   □

? *Question*   There may be more than one value $c \in (a, b)$ that satisfies the conclusion of the Mean Value Theorem for a given function. Sketch a function on an interval $[a, b]$ for which there are two values of $c \in (a, b)$ with $f'(c) = \frac{f(b) - f(a)}{b - a}$.   □

What you should know after reading Section 3.6

► If f has a local extremum at $x = c$, then $f'(c)$ is zero or does not exist. Be able to prove this theorem. What does this theorem say about critical points? The converse of this theorem is not true; be able to provide a counterexample.

► The statement, proof, and graphical examples of Rolle's Theorem. Be able to produce examples illustrating that the continuity and differentiability hypotheses of Rolle's Theorem are necessary.

► The statement, proof, and graphical examples of the Mean Value Theorem. Be able to produce examples illustrating that the continuity and differentiability hypotheses of the Mean Value Theorem are necessary. Why is it called the *Mean* Value Theorem?

► Be able to phrase the three main theorems from this section in terms of tangent lines, and to explain what they mean graphically.

Exercises 3.6

Concepts

0. Read the section and make your own summary of the material. Include whatever you think will help you review the material later on.

1. If f has a local extremum at $x = 1$, then $f'(1)$ is either _____ or _____.

2. If f is differentiable and has a local extremum at $x = 1$, then $f'(1)$ is _____.

3. If f is differentiable everywhere except $x = -2$ and $x = 4$, and the derivative of f is zero only at $x = 0$ and $x = 5$, list all the critical points of f.

4. State Rolle's Theorem and sketch a graph illustrating the theorem.

5. State the Mean Value Theorem and sketch a graph illustrating the theorem.

6. Sketch the graph of a function $f(x)$ on $[-2, 2]$ with $f(-2) = f(2) = 0$ for which there are *three* values $c \in (-2, 2)$ satisfying the conclusion of Rolle's Theorem.

7. Sketch the graph of a function $f(x)$ on $[0, 4]$ for which there are three values $c \in (0, 4)$ satisfying the conclusion of the Mean Value Theorem.

8. What does the Mean Value Theorem have to do with a "mean"?

■ The title of this section is "Three Theorems About Tangent Lines," and each of the three theorems we proved in this section has to do with tangent lines. State each theorem in terms of tangent lines.

9. Theorem 3.14

10. Rolle's Theorem

11. The Mean Value Theorem

■ Determine whether each of the following statements is true or false. If a statement is true, explain why. If a statement is false, provide a counterexample.

12. True or False: If f is a differentiable function with an extremum at $x = -2$, then $f'(-2) = 0$.

13. True or False: If f has a critical point at $x = 1$, then f has a local minimum or maximum at $x = 1$.

14. True or False: If f is any function with $f(2) = 0$ and $f(8) = 0$, then there is some c in the interval $(2, 8)$ such that $f'(c) = 0$.

15. True or False: If f is continuous and differentiable on $[-2, 2]$ with $f(-2) = 4$ and $f(2) = 0$, then there is some $c \in (-2, 2)$ with $f'(c) = -1$.

16. True or False: If f is continuous and differentiable on $[0, 10]$ with $f'(5) = 0$, then f has a local maximum or minimum at $x = 5$.

17. True or False: If f is continuous and differentiable on $[0, 10]$ with $f'(5) = 0$, then there are some values a and b in $(0, 10)$ for which $f(a) = 0$ and $f(b) = 0$.

■ According to Theorem 3.14, if f has a local extremum at $x = c$, then either $f'(c)$ does not exist or $f'(c) = 0$. In the following problems, you will sketch examples of functions that are nondifferentiable at their local extrema.

18. Sketch a function with a local minimum at $x = 3$ that is continuous but not differentiable at $x = 3$.

19. Sketch a function with a local maximum at $x = -2$ that is nondifferentiable at $x = -2$ because of a removable discontinuity.

20. Sketch a function with a local minimum at $x = 1$ that is nondifferentiable at $x = 1$ because of a jump discontinuity.

■ As we saw in this section, the continuity and differentiability hypotheses are essential in Rolle's Theorem and the Mean Value Theorem. In the following problems, you will sketch examples of functions that fail to satisfy these hypotheses (and thus may fail the conclusions) of these theorems.

21. Sketch a function on $[-2, 2]$ with $f(-2) = f(2) = 0$ that is continuous everywhere, is differentiable everywhere except at $x = -1$, and fails the conclusion of Rolle's Theorem.

22. Sketch a function on $[1, 5]$ with $f(1) = f(5) = 0$ that is continuous everywhere except at $x = 2$, is differentiable everywhere except at $x = 2$, and fails the conclusion of Rolle's Theorem.

23. Sketch a function on $[-3, -1]$ with $f(-3) = f(-1) = 0$ that is continuous everywhere except at $x = -1$, is differentiable everywhere except at $x = -1$, and fails the conclusion of Rolle's Theorem.

24. Sketch a function on $[0, 4]$ that is continuous everywhere, is differentiable everywhere except at $x = 2$, and fails the conclusion of the Mean Value Theorem with $a = 0$ and $b = 4$.

25. Sketch a function on $[-3, 3]$ that is continuous everywhere except at $x = 1$, is differentiable everywhere except at $x = 1$, and fails the conclusion of the Mean Value Theorem with $a = -3$ and $b = 3$.

26. Sketch a function on $[-2, 0]$ that is continuous everywhere except at $x = -2$, is differentiable everywhere except at $x = -2$, and fails the conclusion of the Mean Value Theorem with $a = -2$ and $b = 0$.

27. If a continuous, differentiable function f has zeros at $x = -4$, $x = 1$, and $x = 2$, what can you say about f' on $[-4, 2]$?

28. If a continuous, differentiable function f has values $f(-2) = 3$ and $f(4) = 1$, what can you say about f' on $[-2, 4]$?

29. The proof of Rolle's Theorem depends on an important continuity theorem. What is that theorem and how is it used in the proof of Rolle's Theorem?

30. Where in the proof of Rolle's Theorem did we use the hypothesis that f is continuous on $[a, b]$? Where in this proof did we use the hypothesis that f is differentiable on (a, b)?

31. Where in the proof of the Mean Value Theorem did we use the hypotheses that f is continuous on $[a, b]$ and differentiable on (a, b)?

32. The function f graphed on the next page has roots at $x = -2$ and $x = 2$. Since f is continuous and differentiable, Rolle's Theorem says that there is at least one value $c \in (-2, 2)$ with $f'(c) = 0$. Approximate all such

values of *c*. Which of these values of *c* are referred to in the argument used in the proof of Rolle's Theorem, and why?

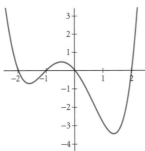

Graph for Problem 32

Skills

■ For each graph of *f* below, approximate all the values $x \in (0, 4)$ at which the derivative of *f* is zero or does not exist. Indicate whether any of these values are local maxima or minima of *f*.

33.

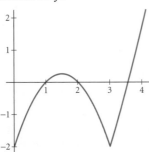

34.

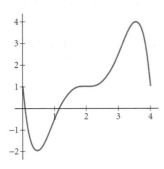

35.

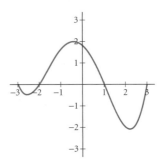

■ For each function *f* below, use derivatives and algebra to find the values at which the derivative of *f* is zero.

36. $f(x) = 3x + 1$

37. $f(x) = x^2 + 5$

38. $f(x) = x^2 + x - 2$

39. $f(x) = (x - 1.7)(x + 3)$

40. $f(x) = 2x^3 + x^2 - 4x + 5$

41. $f(x) = 3x^4 + 8x^3 - 18x^2 - 7$

■ For each function *f* and interval [*a, b*], explain why *f* satisfies the hypotheses of Rolle's Theorem on [*a, b*]. Then approximate the values $c \in (a, b)$ that satisfy the conclusion of Rolle's Theorem.

42.

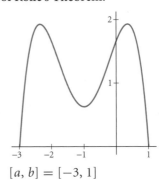

$[a, b] = [-3, 1]$

43.

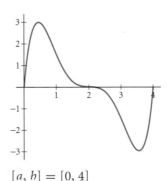

$[a, b] = [-3, 3]$

44.

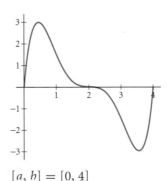

$[a, b] = [0, 4]$

■ For each function *f* and interval [*a, b*], use a graphing calculator to determine whether the hypotheses of Rolle's Theorem are satisfied. If they are, approximate the values $c \in (a, b)$ that satisfy the conclusion of Rolle's Theorem.

45. $f(x) = x^3 - 4x^2 + 3x, \quad [0, 3]$

46. $f(x) = x^4 - 3.24x^2 - 3.04, \quad [-2, 2]$

47. $f(x) = \dfrac{x^2 - 4x}{x^2 - 4x + 3}, \quad [0, 4]$

48. $f(x) = x^5 - 9x^4 + 26x^3 - 20x^2 - 24x + 32, \quad [-1, 4]$

■ For each function f and interval $[a, b]$, show that f satisfies the hypotheses of Rolle's Theorem on $[a, b]$. Then use derivatives and algebra to find the exact values of all $c \in (a, b)$ that satisfy the conclusion of Rolle's Theorem.

49. $f(x) = 4 - x^2$, $[-2, 2]$

50. $f(x) = x^2 - 3x - 4$, $[-1, 4]$

51. $f(x) = x(x - 1.7)$, $[0, 1.7]$

■ For each function f and interval $[a, b]$, explain why f satisfies the hypotheses of the Mean Value Theorem on $[a, b]$. Then approximate the values $c \in (a, b)$ that satisfy the conclusion of the Mean Value Theorem.

52.

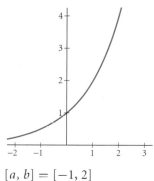

$[a, b] = [-1, 2]$

53.

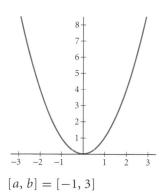

$[a, b] = [-1, 3]$

54.

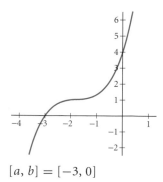

$[a, b] = [-3, 0]$

■ For each function f and interval $[a, b]$, use a graphing calculator to determine whether f satisfies the hypotheses of the Mean Value Theorem on $[a, b]$, and if so,

approximate the values $c \in (a, b)$ that satisfy the conclusion of the Mean Value Theorem.

55. $f(x) = 4 - x^2$, $[-1, 3]$

56. $f(x) = x^2 + \dfrac{1}{x}$, $[-3, 2]$

57. $f(x) = -x^3 + 3x^2 - 7$, $[-2, 3]$

■ For each function f and interval $[a, b]$, show that f satisfies the hypotheses of the Mean Value Theorem on $[a, b]$. Then use derivatives and algebra to find the exact values of all $c \in (a, b)$ that satisfy the conclusion of the Mean Value Theorem.

58. $f(x) = x^2 - 5$, $[-2, 0]$

59. $f(x) = 1 + 2x + 3x^2$, $[-2, 2]$

60. $f(x) = (x - 1)(x + 3)$, $[-3, 2]$

Applications

61. The cost (in cents) of manufacturing a frozen orange juice container is given by $C(h) = h^2 - 7.4h + 13.7$, where h is the height of the orange juice container (in inches). Your boss claims that the orange juice containers will be cheapest to make if they are 4 inches tall. Use Theorem 3.14 to show quickly that he is wrong.

62. Last night at 6 p.m., Linda got up from her blue easy chair. At 8 p.m. she sat back down in her blue easy chair. Let $s(t)$ be the distance between Linda and her easy chair t minutes after 6 p.m. last night.

   (a) Sketch a possible graph of $s(t)$, and describe what Linda did between 6 p.m. and 8 p.m. according to your graph. Please assume that she moves in a continuous and differentiable way!

   (b) Use Rolle's Theorem to show that at some point between 6 and 8 p.m. (not including 6 p.m. or 8 p.m.), Linda's velocity $v(t)$ with respect to the easy chair was zero. Find such a place on the graph of $s(t)$.

63. It took Alina half an hour to drive to the grocery store that is 20 miles from her house.

   (a) Use the Mean Value Theorem to show that at some point during her trip, Alina must have been traveling exactly 40 miles per hour.

   (b) Why does this make sense in real-world terms? (Questions to think about: Could Alina possibly have been traveling under 40 miles per hour for her whole trip? Or over 40 miles an hour for her whole trip? If her velocity was less than 40 miles per hour at some point and more than 40 miles per hour at some other point, can she have avoided going *exactly* 40 miles per hour at some time during her trip? Why or why not?)

Proofs

64. Prove Theorem 3.14: If a function f has a local extremum at $x = c$, then either $f'(c)$ does not exist or $f'(c) = 0$.

65. Prove Rolle's Theorem: If f is continuous on $[a, b]$ and differentiable on (a, b), and $f(a) = f(b) = 0$, then there is some value $c \in (a, b)$ with $f'(c) = 0$.

66. Prove the Mean Value Theorem: If f is continuous on $[a, b]$ and differentiable on (a, b), then there is some value $c \in (a, b)$ with $f'(c) = \frac{f(b) - f(a)}{b - a}$.

67. Use Rolle's Theorem to prove that if f is continuous and differentiable everywhere and has three roots, then its derivative f' has at least two roots.

68. Follow the method of proof used for Rolle's Theorem to prove the following slightly more general theorem: If f is continuous on $[a, b]$ and differentiable on (a, b), and if $f(a) = f(b)$, then there is some value $c \in (a, b)$ with $f'(c) = 0$.

69. Use Rolle's Theorem to prove the slightly more general theorem from Problem 68: If f is continuous on $[a, b]$ and differentiable on (a, b), and if $f(a) = f(b)$, then there is some value $c \in (a, b)$ with $f'(c) = 0$. (*Hint:* Apply Rolle's Theorem to the function $g(x) = f(x) - f(a)$.)

3.7   The First Derivative and Function Behavior

What does the first derivative f' of a function say about the function f? In this section we will discover that the derivative f' can give us quite a lot of information about f. In particular, the derivative of a function tells you where the function is increasing and decreasing and where the local extrema of the function are.

3.7.1   The first derivative and increasing/decreasing functions

Intuitively, the graph of a function f is "moving up" at a point $x = c$ if the slope of its tangent line, that is, its "direction," is positive. In other words, if $f'(c)$ is positive, then f should be increasing at $x = c$. Similarly, if a function has a negative derivative, then the function should be decreasing.

THEOREM 3.17

The Derivative Determines Where a Function Is Increasing or Decreasing

Let f be a function that is differentiable on an interval I.

(a) If f' is positive in the interior of I, then f is increasing on I.

(b) If f' is negative in the interior of I, then f is decreasing on I.

(c) If f' is zero in the interior of I, then f is constant on I.

EXAMPLE 3.41   **The sign of f' determines where f is increasing or decreasing**

If a function f is differentiable (and thus also continuous) on the interval $[0, 4]$, and if $f'(x) > 0$ for all x in the *open* interval $(0, 4)$, then Theorem 3.17 says that f is increasing on the *entire* interval $[0, 4]$.   □

⧆ *Caution*   The converse of Theorem 3.17 is *almost* true. The converse of part (a) of the theorem would say that if f is increasing on an interval I, then its derivative f' is positive on that interval. This is almost the case; what happens in general is that if f is increasing on an interval I, then its derivative f' is greater than *or equal to* zero on that interval. For example, consider the function $f(x) = x^3$ shown in Figure 3.66.

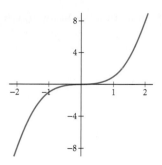

Figure 3.66

Graphically, we can see that f is increasing everywhere on $(-\infty, \infty)$ but that its derivative at $x = 0$ is $f'(0) = 0$ (the graph has a horizontal tangent line at $x = 0$). Thus f is increasing on $(-\infty, \infty)$ while $f'(x)$ is greater than *or equal to* zero on $(-\infty, \infty)$. $\qquad\square$

PROOF **(THEOREM 3.17)**   We'll prove part (a); parts (b) and (c) are similar and are left to you in the exercises. Let f be a function that is differentiable on an interval I. By hypothesis, f' is positive on the interior of I. By definition this means that

(1) $$f'(x) > 0 \text{ for all } x \text{ in the interior of } I.$$

We will use Statement (1) to prove that f is increasing on all of I. By the definition of what it means for a function to be increasing, we must show that

(2) $$\text{Given } a \text{ and } b \text{ in } I, \text{ if } b > a \text{ then } f(b) > f(a).$$

Suppose a and b are in the interval I. The key to our argument will be the Mean Value Theorem, so first we will make sure that the Mean Value Theorem applies to f on the interval $[a, b]$. Since f is differentiable on I, and (a, b) is contained in I, f must be differentiable on (a, b). Moreover, since f is differentiable on I, it must also be continuous on I. Since a and b are in I, the closed interval $[a, b]$ is contained in I; therefore f is continuous on the closed interval $[a, b]$, and the Mean Value Theorem applies.

Now by the Mean Value Theorem, there exists some $c \in (a, b)$ such that

(3) $$f'(c) = \frac{f(b) - f(a)}{b - a}.$$

Since $c \in (a, b)$, c must be in the interior of I. Therefore, by Statement (1) we know that $f'(c) > 0$. Furthermore, since $b > a$ we have $b - a > 0$, and thus

$$f(b) - f(a) = f'(c)(b - a) \qquad\qquad \textbf{(Equation (3))}$$
$$> 0. \qquad\qquad (f'(c) > 0 \text{ and } b - a > 0)$$

Since $f(b) - f(a) > 0$, we have $f(b) > f(a)$ and we are done. $\qquad\blacksquare$

Up to this point we could only approximate graphically the intervals on which a function is increasing or decreasing. With Theorem 3.17 we can find these intervals algebraically, that is, by determining the intervals on which f' is positive or negative.

EXAMPLE 3.42

Determining where a function is increasing or decreasing

Find the intervals on which $f(x) = x^2 - 2x + 1$ is increasing and the intervals on which it is decreasing.

Solution   We can find the intervals on which the derivative f' is positive or negative by solving the inequalities $f'(x) > 0$ and $f'(x) < 0$. The derivative of $f(x) = x^2 - 2x + 1$ is $f'(x) = 2x - 2$, so the inequalities are

$$
\begin{array}{ll}
f'(x) > 0 & \qquad f'(x) < 0 \\
2x - 2 > 0 & \qquad 2x - 2 < 0 \\
2(x - 1) > 0 & \qquad 2(x - 1) < 0 \\
x - 1 > 0 & \qquad x - 1 < 0 \\
x > 1, & \qquad x < 1.
\end{array}
$$

Therefore, f' is positive on $(1, \infty)$ and negative on $(-\infty, 1)$. By Theorem 3.17, this means that f is increasing on $(1, \infty)$ and decreasing on $(-\infty, 1)$.   □

☑ *Checking the Answer*   We can check the solution to Example 3.42 by inspecting the graph of $f(x) = x^2 - 2x + 1$ in Figure 3.67. Note that it does indeed appear that f is decreasing from $-\infty$ to 1 and increasing from 1 to ∞. We can also look at the graph of the derivative $f'(x) = 2x - 2$ (Figure 3.68); this graph is positive to the right of $x = 1$ and negative to the left of $x = 1$.

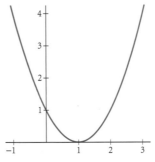

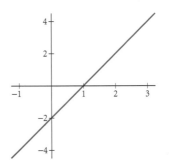

Figure 3.67
$f(x) = x^2 - 2x + 1$

Figure 3.68
$f'(x) = 2x - 2$ □

An interesting consequence of part (c) of Theorem 3.17 is that any two functions with the same derivative for all x must differ by a constant for all x. Graphically, this means that if two functions have the same derivative, then one function is a vertical shift of the other.

THEOREM 3.18

Functions with the Same Derivative Differ by a Constant

If f and g are functions such that $f'(x) = g'(x)$ for all $x \in [a, b]$, then for some constant C we have $f(x) = g(x) + C$ for all $x \in [a, b]$.

The proof of this theorem follows directly from Theorem 3.17 and is left to you in the exercises. The main idea is to show that the derivative of the function $f - g$ is constant and apply part (c) of Theorem 3.17.

3.7.2 Using the Intermediate Value Theorem to find sign intervals

In Example 3.42 we found the intervals on which a function was increasing or decreasing by solving inequalities involving the derivative. In general there is a more efficient way to find these intervals, using the Intermediate Value Theorem.

We start with a simple fact: The derivative function f' can change sign only at values c where $f'(c) = 0$ or $f'(c)$ does not exist. This fact is worded more precisely in the following theorem.

THEOREM 3.19

Determining Where f' Is Positive or Negative

If f' is continuous and nonzero on an interval (a, b), then either $f'(x) > 0$ for all $x \in (a, b)$ or $f'(x) < 0$ for all $x \in (a, b)$.

Note that the hypothesis that f' is continuous on (a, b) implies that f' exists on all of (a, b). Also, Theorem 3.19 is actually true of all functions, not just derivative functions (see Section 2.7.3). We prove it again here (in the context of derivatives) using the Intermediate Value Theorem.

PROOF (**THEOREM 3.19**) Suppose f' is continuous and nonzero on an interval (a, b). We wish to show that f' must have the same sign on all of (a, b). Seeking a contradiction, suppose that f' is positive for one value in the interval (a, b) and negative for some other value in (a, b). Specifically, suppose there are values c and d in (a, b) with $f'(c) > 0$ and $f'(d) < 0$. By assumption, f' is continuous on (a, b) and thus on (c, d). Therefore, the Intermediate Value Theorem applies and tells us that there is a number $x \in (c, d)$ with $f'(x) = 0$. But this contradicts our assumption that f' is nonzero on (a, b), so we are done. ∎

What good is this new theorem? Suppose we are given a function f and wish to find the intervals on which f is increasing or decreasing. That means we have to find the intervals on which f' is positive or negative. If we make a list of all the values of x for which the derivative f' is zero or does not exist (or is discontinuous), then we will have a complete list of all the places where the derivative can change sign. In other words, we can use this list of values to divide the domain of f into intervals, on each of which the derivative is nonzero and continuous. Then we can test the sign of f' at *just one point* on each interval, and by Theorem 3.19 we will know the sign of f' on that *entire* interval! (This will be much more efficient than determining where $f'(x) > 0$ and where $f'(x) < 0$ by solving inequalities, as we did in Example 3.42.)

The following example uses this strategy to find the intervals on which a function is increasing or decreasing. We did this in Example 3.42 by solving inequalities, but now we will use our new algorithm inspired by Theorem 3.19.

EXAMPLE 3.43

Determining where a function is increasing or decreasing

Find the intervals on which $f(x) = x^2 - 2x + 1$ is increasing and the intervals on which it is decreasing.

Solution We must find the intervals on which the derivative f' is positive or negative. Since $f(x) = x^2 - 2x + 1$, we have $f'(x) = 2x - 2$. We first find the values at which $f'(x)$ is zero (it is continuous and thus is always defined):

$$f'(x) = 0 \implies 2x - 2 = 0 \implies 2(x - 1) = 0 \implies x = 1.$$

Now by Theorem 3.19, we know that f' is either always positive or always negative on the interval $(-\infty, 1)$ and, similarly, is always positive or always negative on the interval $(1, \infty)$. We need to test just *one value* of f' in each interval to see whether f' is positive or negative on that interval.

For example, $f'(0) = 2(0) - 2 = -2 < 0$ is negative, and $0 \in (-\infty, 1)$, so f' must be negative on all of $(-\infty, 1)$. On the other hand, $f'(2) = 2(2) - 2 = 2 > 0$ is positive, and $2 \in (1, \infty)$, so f' is positive on all of $(1, \infty)$. Note that these are the same answers we got in Example 3.42 by solving inequalities. □

It is often useful to record the sign of the derivative of a function using a number line (see Figure 3.69). In Example 3.43 we saw that $f'(x) = 0$ only at $x = 1$, so this point is marked on the number line. The real number line is now divided into two intervals, $(-\infty, 1)$ and $(1, \infty)$. On each of these intervals, the derivative has constant sign. After testing each interval (using $x = 0$ and $x = 2$, for example), we record the result as a "$+$" or a "$-$" above that interval.

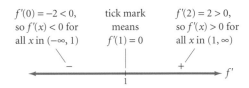

Figure 3.69

(*Note:* At any points where the derivative does not exist, we will write "DNE" above the marked point on the number line; see Example 3.44.)

As the next example shows, number lines can record a great deal of information about f' and thus about the original function f.

EXAMPLE 3.44   **Interpreting the number line for the derivative of a function**

Let f be a function whose derivative f' has the signs indicated on the number line in Figure 3.70. Sketch possible graphs for f and f'.

Figure 3.70

Solution   This number line says that f' is positive on the intervals $(-\infty, -1)$ and $(2, \infty)$ and negative on the interval $(-1, 2)$. It also says that $f'(-1) = 0$ and that $f'(2)$ does not exist. One possible sketch of f' (with the information we just read off the number line) is given in Figure 3.71.

By Theorem 3.17, the sign of f' tells us whether f is increasing or decreasing. Specifically, we know that f is increasing on the intervals $(-\infty, -1)$ and $(2, \infty)$ and that f is decreasing on the interval $(-1, 2)$. Moreover, the graph of f should have a horizontal tangent line at $x = -1$ and should be nondifferentiable at $x = 2$. One possible sketch of f that has these characteristics is given in Figure 3.72.

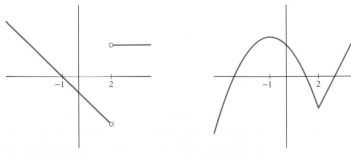

Figure 3.71
Possible graph of f'

Figure 3.72
Possible graph of f □

3.7.3 The first derivative test

As we saw in Section 3.6.1, if a function f has a local extremum at $x = c$, then either $f'(c) = 0$ or $f'(c)$ does not exist. However, if $f'(c) = 0$, then f may or may not have a local extremum at c. In general, if we list all the values $x = c$ for which $f'(c) = 0$ or $f'(c)$ does not exist, then we will have a complete list of all the *possible* local maxima and minima of f. To see which of these locations are local maxima, which are local minima, and which are neither, we will examine the first derivative on either side of the point $x = c$.

Suppose $f'(c) = 0$ and that f is differentiable near c. Assuming that f is not constant, there are four different ways that f can behave near $x = c$ (see Figures 3.73–3.76).

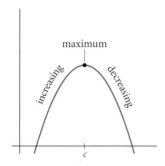

Figure 3.73
f changes from increasing to decreasing at $x = c$

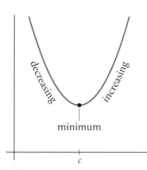

Figure 3.74
f changes from decreasing to increasing at $x = c$

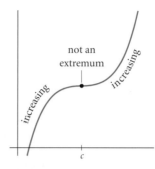

Figure 3.75
f is increasing on both sides of $x = c$

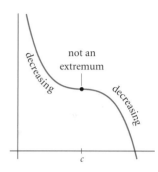

Figure 3.76
f is decreasing on both sides of $x = c$

Intuitively, if f changes from increasing to decreasing at $x = c$, then the point $x = c$ is a local maximum. If f changes from decreasing to increasing at the point $x = c$, then $x = c$ is a local minimum. In either of these cases, the point $x = c$ is called a ***turning point*** of f. On the other hand, if f does not change "direction," that is, if f is increasing (or decreasing) on *both* sides of $x = c$, then $x = c$ is *not* a local extremum of f (and thus is not a "turning point"). The same four situations apply when f is continuous, but not differentiable, at $x = c$ (see Figures 3.77–3.80).

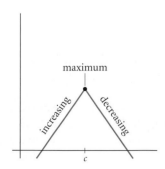

Figure 3.77
f changes from increasing to decreasing at $x = c$

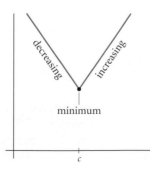

Figure 3.78
f changes from decreasing to increasing at $x = c$

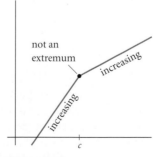

Figure 3.79
f is increasing on both sides of $x = c$

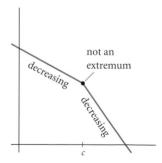

Figure 3.80
f is decreasing on both sides of $x = c$

Recall that $x = c$ is a ***critical point*** of f if $f'(c) = 0$ or if $f'(c)$ does not exist. The list of critical points of f is a complete list of the *possible* extrema of f. We will now develop a method for testing whether a critical point is an extremum. This method will involve determining whether f is increasing or decreasing on each side of $x = c$. By Theorem 3.17, we can determine this information by looking at the sign of the derivative on each side of $x = c$.

THEOREM 3.20

The First Derivative Test

Suppose $x = c$ is a critical point of a function f, and let I be an open interval around c that does not contain any other critical points of f. If f is continuous on I and differentiable at every point of I except possibly at $x = c$, then

(a) If $f'(x) > 0$ to the left of c, and $f'(x) < 0$ to the right of c, then $x = c$ is a local maximum of f.

(b) If $f'(x) < 0$ to the left of c, and $f'(x) > 0$ to the right of c, then $x = c$ is a local minimum of f.

(c) If $f'(x) > 0$ to the left and the right of c, then $x = c$ is not a local extremum of f.

(d) If $f'(x) < 0$ to the left and the right of c, then $x = c$ is not a local extremum of f.

In the statements above, "to the left of c" is understood to mean "for numbers $x \in I$ with $x < c$," and "to the right of c" is understood to mean "for numbers $x \in I$ with $x > c$."

❖ *Caution*   Please don't blindly memorize Theorem 3.20! Think about the pictures in Figures 3.73–3.76 and Figures 3.77–3.80, and use these pictures to help you remember how the first derivative test works.   □

PROOF   (THEOREM 3.20)   We will prove parts (a) and (c). The proofs of parts (b) and (d) are similar and are left to the exercises. Suppose $x = c$ is a critical point of f, and let $I = (a, b)$ be an interval around $x = c$ that satisfies the hypotheses of the theorem.

By Theorem 3.17 and the hypotheses of part (a), we know that f is increasing on $(a, c]$ and decreasing on $[c, b)$. Given this, we will show that $f(c) \geq f(x)$ for all $x \in I$; then we will have shown that $x = c$ is a local maximum of f. Given any $x \in I$, there are three cases to consider. First, if $x = c$, then clearly $f(c) = f(x)$. Second, if $a < x < c$, then since f is increasing on $(a, c]$ we have $f(x) < f(c)$. Third, if $c < x < b$, then since f is decreasing on $[c, b)$ we have $f(c) > f(x)$. In all of these three cases (and thus for all $x \in I$), we have $f(c) \geq f(x)$, and therefore c is a local maximum of f.

(*Technical Note:* In order to make the argument above, it would not have been enough to know that f was increasing on the open interval (a, c). (We would not have been able to state that $f(x) < f(c)$ in the second case.) Similarly, we need to know that f is decreasing on $[c, b)$, not just on (c, b).)

We now prove part (c). If $f'(x) > 0$ at every point of I except for $x = c$, then $f'(x) > 0$ for all $x \in (a, c) \cup (c, b)$. By Theorem 3.17, f is increasing on all of I. The point $x = c$ cannot be a local maximum of f, because for all $x > c$ in I we have $f(x) > f(c)$, since f is increasing on I. On the other hand, $x = c$ cannot be a local minimum of f, because for all $x < c$ in I we have $f(x) < f(c)$. Therefore, $x = c$ is neither a local minimum nor a local maximum of f.   ■

❓ *Question*   Here are two questions to test whether you understood some finer points of the proof: Why does Theorem 3.17 imply that if $f'(x) > 0$ for all $x \in (a, c) \cup (c, b)$, then f is increasing on *all* of the interval I? Why do we need to know that f is increasing on all of I, rather than just on $(a, c) \cup (c, b)$?   □

We can represent the information from the first derivative test on a number line. In Figures 3.81–3.84, the number line is used to record the sign of f' to the left and right of the critical point $x = c$. Above the number line we record the corresponding information about f; the diagonal lines record whether f is increasing or decreasing on each interval. These diagrams are valid whether $f'(c) = 0$ or $f'(c)$ does not exist, as long as f is continuous near $x = c$ and differentiable near (but not necessarily at) $x = c$.

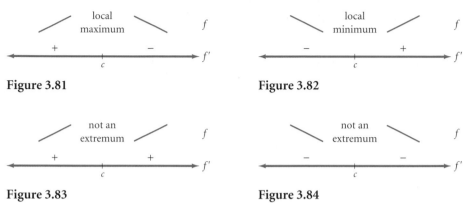

Figure 3.81   **Figure 3.82**

Figure 3.83   **Figure 3.84**

By combining Theorem 3.19 with the first derivative test, we have a quick way of finding the intervals on which a function is increasing or decreasing, as well as the local extrema of the function. In general the algorithm for this process is as follows:

ALGORITHM 3.1

Using the Derivative to Find the Local Extrema of a Function

Suppose f is a continuous function. To determine the local extrema of f, follow these three steps:

1. Calculate $f'(x)$, and determine where $f'(x) = 0$ and where $f'(x)$ does not exist. These critical points are a list of all the *possible* local extrema of f.

2. Determine whether f is increasing or decreasing on each interval between the critical points of f. If f' is continuous, then you can do this by testing the sign of f' at just one point in each interval.

3. Apply the first derivative test to determine whether each critical point of f is a local minimum, a local maximum, or neither.

Algorithm 3.1 assumes that f is defined on an open interval (for example, all real numbers). If f is defined on a closed or half-closed interval, then sometimes local extrema will occur at the "ends" of that interval. We will learn how to handle "endpoint extrema" later on when we study the extrema of polynomial functions in Chapter 5.

EXAMPLE 3.45

Using the derivative to find local extrema

Find all the local extrema of the function $f(x) = x^3 - 3x + 2$. Then use first derivative information to sketch a graph of f.

Solution   We first calculate the derivative: $f'(x) = 3x^2 - 3$ (what derivative rules did we use?). This derivative is always defined and continuous, so the critical points of f are just the places where $f'(x) = 0$:

$$3x^2 - 3 = 0 \implies 3(x^2 - 1) = 0 \implies x^2 = 1 \implies x = \pm 1.$$

Therefore, $x = -1$ and $x = 1$ are the only critical points of f.

The critical points divide the real number line into three intervals, namely $(-\infty, -1)$, $(-1, 1)$, and $(1, \infty)$. We'll test the sign of f' on each interval by testing the sign of f' at one point in each interval, say at $x = -2$, $x = 0$, and $x = 2$:

$$f'(-2) = 3(-2)^2 - 3 = 12 - 3 = 9 > 0,$$
$$f'(0) = 3(0)^2 - 3 = 0 - 3 = -3 < 0,$$
$$f'(2) = 3(2)^2 - 3 = 12 - 3 = 9 > 0.$$

The first derivative information we have collected so far can now be recorded on a number line (see Figure 3.85).

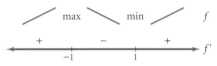

Figure 3.85

By the first derivative test, since f' is positive to the left of the critical point $x = -1$ and negative to the right of $x = -1$, the point $x = -1$ is a local maximum of f. Also, since f' is negative to the left of the critical point $x = 1$ and positive to the right, the point $x = 1$ is a local minimum of f.

With all the first derivative information we have collected, we can sketch a reasonable graph of f by plotting just a few points. As a general rule, it is a good idea to plot the points $(c, f(c))$ for each critical point $x = c$ and then use the derivative information to connect the dots accordingly. In this example, the critical points are $x = -1$ and $x = 1$, so we calculate

$$f(-1) = (-1)^3 - 3(-1) + 2 = -1 + 3 + 2 = 4,$$
$$f(1) = (1)^3 - 3(1) + 2 = 1 - 3 + 2 = 0.$$

Using this information, we can make a sketch of the graph of $f(x) = x^3 - 3x + 2$ (see Figure 3.86). Compare the features of this graph with the information in the number line. (*Note:* The fact that $f(x)$ has a root at $x = -2$ is not obvious from the information we collected, although you can easily check that $f(-2) = (-2)^3 - 3(-2) + 2 = 0$.)

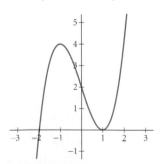

Figure 3.86   □

? Question   Repeat Example 3.45 for the function $f(x) = x^3$. You should find that $x = 0$ is a critical point, but not a turning point, of $f(x)$.   □

⊗ Caution   After a while you will be able to apply the algorithm for finding local extrema very quickly. Be sure that as this happens you do not forget why the algorithm works. For example, at the beginning of the algorithm you find the derivative and set it equal to zero; why do you do this?   □

🖙 *Caution* One more word of warning: Don't forget to look for any places where the derivative does not exist! The critical points of f are by definition the numbers $x = c$ where $f'(c) = 0$ or where $f'(c)$ does not exist. So far the only functions we can differentiate by "rules" are quite simple, so we won't run into many calculational examples of functions that have nondifferentiable points. In later sections when we work with more complicated functions, it will be important to search for places where $f'(x)$ does not exist. □

What you should know after reading Section 3.7

▶ The sign of the derivative f' determines whether a function f is increasing, decreasing, or constant on I. What can you say about the converse? Be able to prove Theorem 3.17 using the Mean Value Theorem.

▶ How to find the intervals on which a function is increasing or decreasing by solving inequalities, and by testing intervals between critical points. Be able to prove that the Intermediate Value Theorem enables us to determine the signs of f' on certain intervals by checking its sign at just one point in each interval. How to record first derivative information on a number line and use this information to sketch a graph of f.

▶ How to prove that two functions with the same derivative must differ by a constant, and how to interpret this fact graphically.

▶ Be able to state and sketch supporting examples for the first derivative test. What is the first derivative test used for and how does it work? How do you prove it? Be able to follow the algorithm for using f' to find the local extrema of a function f, and explain how the algorithm works.

▶ Be able to sketch f' given a graph of f, and f given a graph of f'. Also be able to sketch graphs of f and/or f' given number line information.

Exercises 3.7

Concepts

0. Read the section and make your own summary of the material. Include whatever you think will help you review the material later on.

■ Decide whether each of the following statements is true or false. If a statement is true, explain why. If a statement is false, provide a counterexample.

1. True or False: If $f'(x) < 0$ for all $x \in (0, 3)$, then f is decreasing on $[0, 3]$.

2. True or False: If f is increasing on $(-2, 2)$, then $f'(x) \geq 0$ for all $x \in (-2, 2)$.

3. True or False: If $f'(x) = 2x$, then $f(x) = x^2 + C$ for some constant C.

4. True or False: If $f'(x)$ is continuous on $(1, 8)$ and $f'(3)$ is negative, then f' is negative on all of $(1, 8)$.

5. True or False: If f' changes sign at $x = 3$, then $f'(3) = 0$.

6. True or False: If $f'(-2) = 0$, then $x = -2$ is either a local maximum or a local minimum of f.

7. True or False: If $x = 1$ is the only critical point of f, and $f'(0)$ is positive, then $f'(2)$ must be negative.

8. True or False: If $f'(1)$ is negative and $f'(3)$ is positive, then $x = 2$ is a local minimum of f.

9. Why is it that you can determine the intervals on which a function f is increasing or decreasing by testing the sign of f' at *one point* of each interval between the critical points of f? Specifically, why is testing the sign of f' at just one point sufficient to determine the sign of f' on the whole interval between critical points?

■ Each number line below describes the sign of the derivative f' of some function f. (All zeros of f' are marked on the number line, and x-values where f' does not exist are labeled DNE.) For each number line, sketch possible graphs for f and f'.

10.

11.

12.

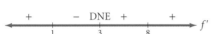

13.

14.

15.

16. Describe what a "critical point" is, intuitively and in mathematical language. Then describe what a "turning point" is. How are these two types of points related?

17. Can a point $x = c$ be both a turning point and a critical point of a differentiable function f? Both an inflection point and a critical point? Both an inflection point and a turning point? Sketch examples, or explain why such a point cannot exist.

18. Repeat Problem 17 given that f may or may not be differentiable.

19. Describe what the first derivative test is for and how to use it. Sketch graphs and number lines to illustrate your description.

20. Sketch the graph of a function f with the following properties: f is continuous and defined for all real numbers, $f'(-2) = 0$, $f(-2) = -3$, $f'(1)$ does not exist, $f(0) = 5$, f is increasing on $(-2, 1)$, and f' is negative on $(-\infty, -2)$ and $(1, \infty)$.

21. Sketch the graph of a function f with the following properties: f has critical points at $x = -3$, $x = 0$, and $x = 5$, and f has inflection points at $x = -3$, $x = -1$, and $x = 2$.

■ For each function f graphed below, sketch a possible graph of its derivative f'.

22.

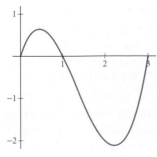

23.

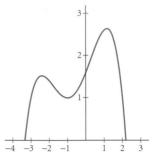

24.

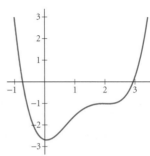

■ Each graph given below is the graph of the *derivative* of some function f. Use the sign and zeros of the derivative to sketch a possible graph of f.

25.

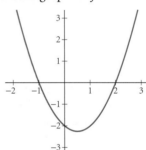

26.

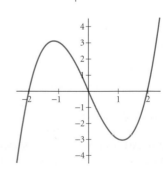

27.

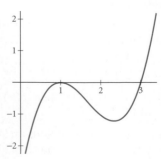

28. Suppose $f(x)$ is a function that is continuous and differentiable everywhere, and that the *derivative* of $f(x)$ is $f'(x) = \sqrt{1 + x^2} - 4$. What are the critical points of $f(x)$? Find the intervals on which $f(x)$ is increasing or decreasing, and use the first derivative test to determine the local extrema of $f(x)$.

Skills

■ For each function f below, solve inequalities involving the first derivative to determine the intervals on which f is increasing and the intervals on which it is decreasing. Represent your answers in interval notation.

29. $f(x) = x^2 - x + 1$ 30. $f(x) = 2x^3 - 9x^2 + 1$

31. $f(x) = x^3 + 4x^2 + 4x - 5$

■ For each function f below, find the intervals on which f is increasing, and the intervals on which it is decreasing, by finding the critical points of f and testing the sign of f' on each interval. Represent your answers in interval notation *and* on a number line.

32. $f(x) = x^2 - x + 1$

33. $f(x) = 2x^3 - 9x^2 + 1$

34. $f(x) = x^3 + 4x^2 + 4x - 5$

35. $f(x) = x^3 - x^2 - x - 1$

36. $f(x) = x(x + 3)(x - 1)$

37. $f(x) = (x - 1)^3 - 2x^2 + 8$

■ For each function f below, use the first derivative test to find any local extrema of f. Show your work clearly! Then sketch (by hand) a possible graph of f.

38. $f(x) = x^2 - 5x - 6$

39. $f(x) = 3 - 4x$

40. $f(x) = -x^2 - 10x - 33$

41. $f(x) = 9 - x^3$

42. $f(x) = x^3 - 6x^2 + 12x - 5$

43. $f(x) = (x - 2)^2 - 4x^3$

Applications

44. Linda throws an orange straight up in the air to see what will happen. The distance in feet between the orange and the ground after t seconds is given by the equation $s(t) = -16t^2 + 90t + 5$. Use this equation to answer the questions below.

   (a) What is the initial height of the orange? What is the initial velocity of the orange? What is the initial acceleration of the orange?

   (b) What is the maximum height of the orange?

   (c) When will the orange hit the ground?

45. Dr. Alina is interested in the behavior of rats trapped in a long tunnel. Her rat Bubbles is released from the left-hand side of the tunnel and runs back and forth in the tunnel for 4 minutes. The rat's *velocity* $v(t)$, in feet per minute, is given by the graph below. Use the graph to answer the questions that follow.

   (a) On which time intervals is Bubbles moving toward the right-hand side of the tunnel?

   (b) At which point in time is Bubbles farthest away from the left-hand side of the tunnel, and why? Do you think Bubbles ever comes back to the left-hand side of the tunnel?

   (c) On which time intervals does Bubbles have a positive acceleration?

   (d) Find an interval on which Bubbles has a negative velocity but a positive acceleration. Describe what Bubbles is doing during this time period.

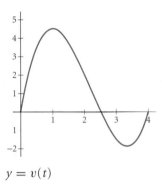

$$y = v(t)$$

Proofs

46. Prove part (b) of Theorem 3.17: Suppose f is differentiable on an interval I; if f' is negative in the interior of I, then f is decreasing on I.

47. Prove part (c) of Theorem 3.17: Suppose f is differentiable on an interval I; if f' is zero in the interior of I, then f is constant on I. (*Hint:* Use the Mean Value Theorem to show that any two numbers a and b in I must be equal.)

48. Prove Theorem 3.18: If f and g are functions with $f'(x) = g'(x)$ for all $x \in [a, b]$, then for some constant C we have $f(x) = g(x) + C$ for all $x \in [a, b]$. (*Hint:* Use Theorem 3.17 to show that the function $f - g$ is constant.)

49. Describe in your own words how Theorem 3.19 follows from the Intermediate Value Theorem.

50. Prove part (b) of Theorem 3.20: With hypotheses as stated in the theorem, if $x = c$ is a critical point of f, where $f'(x) < 0$ to the left of c, and $f'(x) > 0$ to the right of c, then $x = c$ is a local minimum of f.

51. Prove part (d) of Theorem 3.20: With hypotheses as stated in the theorem, if $x = c$ is a critical point of f, where $f'(x) < 0$ to the left and to the right of c, then $x = c$ is not a local extremum of f.

52. Prove that every nonconstant linear function is either always increasing or always decreasing.

53. Prove that every quadratic function has exactly one turning point.

3.8   The Second Derivative and Function Behavior

In the previous section we saw that the first derivative of a function f gives us a great deal of information about the graph of f. By looking at the sign and zeros of the derivative, we could determine the intervals on which f is increasing or decreasing and the local extrema of f.

However, the first derivative cannot tell us *how* a function is increasing or decreasing; for example, f' cannot tell us whether a function is curving up or down as it increases. This information about the concavity of f is given by the second derivative. Higher-order derivatives provide even more detailed information about f, but the first and second derivatives provide so much information about the graph of f that we usually do not need to consider any higher-order derivatives.

3.8.1   Formal definition of concavity

In Section 1.2.6 we gave an informal definition of concavity: the graph of a function is **concave up** if it "curves upward" and is **concave down** if it "curves downward." We are now ready to give a more formal definition of concavity.

DEFINITION 3.17

Concavity

Suppose f is a function that is twice-differentiable on an interval I.

(a) f is **concave up** on I if f' is increasing on I.

(b) f is **concave down** on I if f' is decreasing on I.

The statement in Definition 3.17 that f is "twice-differentiable" on I simply means that f is differentiable on I and that f' is differentiable on I. Such a twice-differentiable function f has a well-defined second derivative f'' on all of I.

How does the formal definition of concavity correspond with our intuitive notion of concavity? Consider the functions graphed in Figures 3.87 and 3.88.

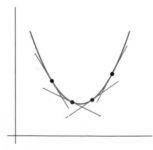

Figure 3.87
f is concave up and f' is increasing

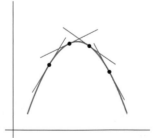

Figure 3.88
f is concave down and f' is decreasing

Consider the four tangent lines drawn in Figure 3.87. As we move from left to right, the slopes of these lines increase. The leftmost tangent line has a negative slope. The second tangent line also has a negative slope but is less steep than the first line; therefore, the second tangent line has a *greater* slope than the first tangent line. (For example, the first tangent line might have a slope of -2, and the second tangent line might have a slope of something like $-\frac{1}{2}$, which is *larger* than -2.) The third tangent line has a small positive slope (perhaps slope 1), and the fourth has a larger positive slope (perhaps slope 3). As we move from left to right, the slopes increase (perhaps from -2 to $-\frac{1}{2}$ to 1 to 3), and thus the derivative of f is increasing.

> **? Question** Argue that the slopes of the tangent lines in Figure 3.88 decrease as we move from left to right. □

How does the fact that f' is increasing cause the graph to curve upward? Consider the right-hand side of the graph in Figure 3.87. In this graph, the function f is increasing *and* its derivative f' is increasing. This means that f is getting larger at a progressively higher rate, which causes the graph of f to curve upward as it increases. On the other hand, if f is decreasing while its derivative f' is increasing (see the left-hand side of the graph in Figure 3.87), then f is decreasing at a progressively lower rate; this also causes the graph of f to curve upward.

3.8.2 The second derivative and concavity

As we saw in Section 3.7, a function is increasing if its derivative is positive. Similarly, the derivative function f' is increasing if *its* derivative function (which is the second derivative f'') is positive. Thus we can check whether a function f is concave up or concave down by looking at the sign of its second derivative.

THEOREM 3.21

The Second Derivative Determines Concavity
Let f be a function that is twice-differentiable on an interval I.
(a) If f'' is positive on I, then f is concave up on I.
(b) If f'' is negative on I, then f is concave down on I.

PROOF (**THEOREM 3.21**)   We prove part (a) and leave part (b) to the exercises. Suppose that f is twice-differentiable on I and that $f''(x) > 0$ for all x in I. By Theorem 3.17 in Section 3.7, this means that $f'(x)$ is increasing on I (since the derivative of f' is f''; here we are using f' as the function and f'' as its derivative). By Definition 3.17, this means that f is concave up on the interval I. ▪

> **⊗ Caution** The converse of Theorem 3.21 is *almost* true. The converse of part (a) of the theorem would say that if f is concave up on I, then f'' is positive on I. This is *nearly* the case, with one small change: What happens in general is that if f is concave up on an interval I, then its second derivative f'' is greater than *or equal to* zero on that interval. For example, the function $f(x) = x^4$ is sufficiently flat at the origin that its second derivative is zero there (see Figure 3.91). □

> **? Question** We could use part (c) of Theorem 3.17 in Section 3.7 to add a third part to Theorem 3.21 that would tell us what it means when f'' is zero in the interior of an interval I. Fill in the blank accordingly: If f'' is zero on I, then f is _____ on I. Can you prove it? □

EXAMPLE 3.46   **Using f'' to determine where a function is concave up or down**

Find the intervals on which $f(x) = x^3 - 3x + 2$ is concave up and the intervals on which it is concave down.

Solution   First we find the second derivative of $f(x) = x^3 - 3x + 2$. Since $f'(x) = 3x^2 - 3$, we know that $f''(x) = 3(2x) = 6x$. (Note that f is the same function we examined in Example 3.45 in Section 3.7.)

   To find the intervals on which f'' is positive and the intervals on which it is negative, we will first find the places where f'' is zero (or does not exist). The second derivative $f''(x) = 6x$ is zero only at $x = 0$ and always exists. This splits the number line into two intervals, $(-\infty, 0)$ and $(0, \infty)$. By the Intermediate Value Theorem (and a second derivative version of Theorem 3.19 from Section 3.7), we can test whether f'' is positive or negative on each of these intervals by testing the sign of f'' at *one point* in each interval. We'll use $x = -1$ and $x = 1$:

$$f''(-1) = 6(-1) = -6 < 0,$$

$$f''(1) = 6(1) = 6 > 0.$$

The number line in Figure 3.89 is used to record the sign of the second derivative (and is correspondingly labeled f''). Above the number line we record the consequences for f: The function f is concave down on the interval $(-\infty, 0)$ and concave up on the interval $(0, \infty)$.

Figure 3.89

☑ *Checking the Answer*   Figure 3.90 shows the graph of $f(x) = x^3 - 3x + 2$. The graph is concave down to the left of $x = 0$ and concave up to the right of $x = 0$. Compare this graph with the information recorded on the number line.   □

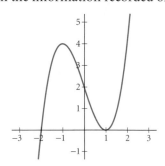

Figure 3.90

3.8.3 Inflection points

Recall from Section 1.2.7 that the *inflection points* of a function f are the points in the domain of f at which the concavity changes. Thus $x = c$ is an inflection point of f if f'' changes from positive to negative, or from negative to positive, at $x = c$.

EXAMPLE 3.47   **Inflection points occur where f'' changes sign**

The work in Example 3.46 shows that $x = 0$ is an inflection point of the function $f(x) = x^3 - 3x + 2$. Since f'' is negative on $(-\infty, 0)$ and positive on $(0, \infty)$, the function f changes from concave down to concave up at $x = 0$ (see Figure 3.90).   □

If a function f changes concavity at a point $x = c$, then the sign of its second derivative changes. If the second derivative is continuous, then the only places it can change sign are where it is zero or where it does not exist. Therefore, to find the inflection points of a function, we must find all the places where f'' is zero or does not exist and must test each of these points to see whether the concavity of f changes there.

? Question The numbers $x = c$ where $f''(c) = 0$ or where $f''(c)$ does not exist are the critical points of the function f'. Why? ☐

⊘ Caution Just as not every critical point is a local extremum, not every place where the second derivative is zero is an inflection point. One example is the function $f(x) = x^4$ (see Figure 3.91). This function is so flat at $x = 0$ that its second derivative (which in some sense measures curvature) is actually zero at that point (rather than some positive number, as you might have expected). Thus $f''(0) = 0$, but $x = 0$ is not an inflection point of $f(x)$. ☐

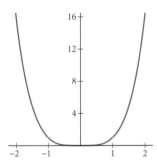

Figure 3.91

If $x = c$ is an inflection point of f where $f''(c) = 0$, then near $x = c$, the graph of f could appear in one of the four ways shown in Figures 3.92–3.95 depending on how f changes concavity and on whether f is increasing or decreasing.

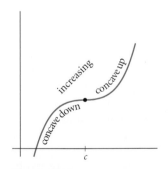
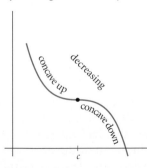
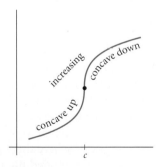
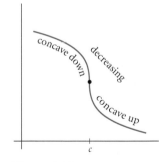

Figure 3.92   **Figure 3.93**   **Figure 3.94**   **Figure 3.95**

It is interesting to note that in each of these graphs, the function f shown crosses its tangent line at the inflection point $x = c$. (Sketch the tangent line in each case at $x = c$.)

Another interesting fact about the inflection points of a function f is that they are extrema of the first derivative f'. For example, you can see this by inspecting the tangent lines to the graph of f in Figure 3.92. The tangent line at $x = c$ has slope 0, whereas all the other tangent lines for f have positive slope; therefore, f' has a minimum at $x = c$. We can also argue this more precisely using the first derivative test (again considering Figure 3.92). Since $f''(c) = 0$, we know that $x = c$ is a critical point of f'. Moreover, f'' is negative to the left of $x = c$ and positive to the right of $x = c$. This means that f' (whose derivative is f'') is decreasing to the left of $x = c$ and increasing to the right of c. By the first derivative test, the function f' has a local minimum at $x = c$.

? *Question*   Discuss the behavior of f' and f'' for Figures 3.93–3.95. In particular, what kind of local extrema does the function f' have at $x = c$ in each case?   □

? *Question*   Consider the function f in Figure 3.92. At $x = c$, the function f has an inflection point, the function f' has a local minimum, and the function f'' has a zero. Use this information, and the graph of f, to sketch rough graphs of f' and f''.   □

◈ *Caution*   Because we will often discuss three functions at once (f, f', and f''), it is important that we always specify which function we are referring to. For example, in the "Question" above it would not be clear simply to say that "it" has a local minimum at $x = c$, for then it would not be clear whether we meant to refer to f, f', or f''.   □

3.8.4 The second derivative test

In Section 3.7.3 we saw how to use the first derivative to test critical points to see whether they are local maxima, minima, or neither. We can also use the second derivative to test critical points. Which test is easier to use depends on the function.

Suppose f is a differentiable function, and $x = c$ is a critical point of f. As we saw in Section 3.7.3, there are four possible ways that f can behave near $x = c$. In Figures 3.96–3.99 we examine the second derivative at the critical point $x = c$ in each case.

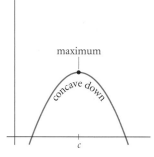

Figure 3.96
$f''(c) < 0$ or $f''(c) = 0$

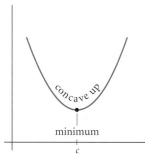

Figure 3.97
$f''(c) > 0$ or $f''(c) = 0$

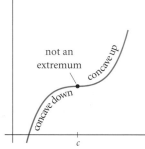

Figure 3.98
$f''(c) = 0$

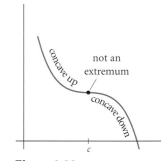

Figure 3.99
$f''(c) = 0$

In all four graphs, it is possible for $f''(c)$ to be zero (remember that graphs that are very flat at their turning points, like $f(x) = x^4$, can have second derivatives of zero at those points even if they are concave up or down). However, the only graph that has a negative second derivative at $x = c$ is the first one, where we have a local maximum at $x = c$. Similarly, the only graph that has a positive second derivative at $x = c$ is the second graph, where there is a local minimum at $x = c$.

When we applied the first derivative test to a critical point, we were interested in the signs of f' to the left and to the right of the critical point. When we use the second derivative to test a critical point, we will look at the sign of f'' *at* the critical point.

THEOREM 3.22

The Second Derivative Test

Suppose that $x = c$ is a critical point of a function f and that f is twice-differentiable in a neighborhood of $x = c$.

(a) If $f''(c)$ is positive, then $x = c$ is a local minimum of f.

(b) If $f''(c)$ is negative, then $x = c$ is a local maximum of f.

(c) If $f''(c) = 0$, then this test says nothing about whether or not $x = c$ is an extremum of f. (You must find another way to test the critical point.)

⬧ *Caution* Make sure you understand part (c) of the theorem; if $x = c$ is a critical point of f and $f''(c) = 0$, the second derivative test tells us nothing. In this case, the function f could have a local maximum, a local minimum, or neither at $x = c$ (see Figures 3.96–3.99). □

PROOF (**THEOREM 3.22**) Suppose $x = c$ is a critical point of f and that f is twice-differentiable in a neighborhood of c. To prove part (a), suppose f'' is positive. Then f'' is positive in a small neighborhood $(c - \delta, c + \delta)$ of c. By Theorem 3.17 in Section 3.7, this means that f' is increasing on $(c - \delta, c + \delta)$ (since f'' is the derivative of f'). Since f is differentiable at the critical point $x = c$, we know that $f'(c) = 0$. Since f' is increasing near c, and zero at c, we must have $f'(x) < 0$ to the immediate left of c and $f'(x) > 0$ to the immediate right of c. By the first derivative test, $x = c$ is a local minimum of f. The proof of part (b) is similar and is left to the exercises.

To prove part (c) we need only exhibit three functions with $f''(c) = 0$ at some point c, where one function has a local maximum at $x = c$, one function has a local minimum at $x = c$, and one function does not have a local extremum at $x = c$. Then we will have shown that any of the three cases can happen when $f''(c) = 0$. Suppose $c = 0$. Three such functions are $f(x) = x^3$ (no extremum at $x = 0$), $f(x) = x^4$ (local minimum at $x = 0$), and $f(x) = -x^4$ (local maximum at $x = 0$). You will show this in the exercises. ▪

One thing that sometimes makes the second derivative test easier to apply than the first derivative test is that if f has n critical points, then the second derivative test involves n computations (one *at* each critical point). In contrast, the first derivative test for such a function would require $n + 1$ computations, since it involves testing the sign of f' on each side of each critical point. Moreover, the expression for f'' is often simpler than the expression for f', so the calculations involved are easier. On the other hand, sometimes the second derivative test yields no information (if $f''(c) = 0$), in which case we would have to apply the first derivative test.

⬧ *Caution* At a critical point $x = c$, a function f has a local maximum if the second derivative is *negative*; this might be the opposite of what you'd expect. Keep in mind the picture of what a concave down function looks like; if a function is concave down at a critical point, then the value at the critical point is *higher* than the value of f at any nearby points. □

ALGORITHM 3.2

Using the Second Derivative to Find Local Extrema

To determine the local extrema of a twice-differentiable function f:

1. Calculate $f'(x)$, and solve for the numbers x at which $f'(x) = 0$ and where $f'(x)$ does not exist. These are the critical points of f and are a list of all the *possible* local extrema of f.

2. Calculate the second derivative of f at each critical point.

3. Apply the second derivative test to each critical point of f to determine whether it is a local maximum, a local minimum, or neither.

EXAMPLE 3.48

Using the second derivative test

Use the second derivative test to determine the local extrema of $f(x) = (x - 1)(x^2 + 2x + 2)$.

Solution The derivative of $f(x) = (x-1)(x^2+2x+2) = x^3 + x^2 - 2$ is $f'(x) = 3x^2 + 2x = x(3x+2)$, which is zero at the points $x = 0$ and $x = -\frac{2}{3}$ and always exists. The only critical points of f are at $x = 0$ and $x = -\frac{2}{3}$. We need to calculate the second derivative of f and evaluate it at these two critical points. Since $f''(x) = 6x + 2$, we have

$$f''(0) = 6(0) + 2 = 2 > 0,$$
$$f''\left(-\tfrac{2}{3}\right) = 6\left(-\tfrac{2}{3}\right) + 2 = -2 < 0.$$

By the second derivative test, since f'' is positive at the critical point $x = 0$, the function f has a local minimum at $x = 0$. Similarly, since f'' is negative at the critical point $x = -\frac{2}{3}$, the f has a local maximum at $x = -\frac{2}{3}$. (Check this with a graphing calculator.) □

3.8.5 Simple curve sketching

In general, to sketch the graph of a function f, we begin by finding the local extrema of f, the intervals where f is increasing, decreasing, concave up, or concave down, and the locations of any inflection points of f. Then we must calculate any "important" values of f, such as the values at critical points and inflection points and (if convenient) any x-intercepts or y-intercepts. With these function values and the first and second derivative information it should be possible to sketch a fairly accurate graph of f. In subsequent chapters, when we work with more complicated types of functions (for example, those with asymptotes, discontinuities, or nondifferentiable points), we will also have to examine various limits.

A function f can change direction only at its critical points, and it can change concavity only at the critical points of f'. On each interval between these critical points, the graph of f is "homogeneous." In other words, for each of these intervals, f is either always increasing or always decreasing and is either always concave up or always concave down. If f is continuous and twice-differentiable, then each segment of the graph of f between critical points of f or f' will look like one of the four arcs shown in Figures 3.100–3.103.

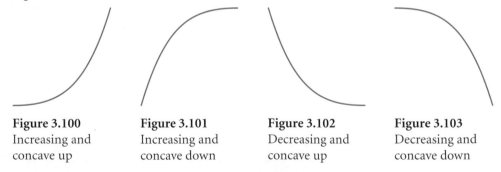

Figure 3.100
Increasing and concave up

Figure 3.101
Increasing and concave down

Figure 3.102
Decreasing and concave up

Figure 3.103
Decreasing and concave down

The algorithm for sketching simple curves (that are continuous and differentiable everywhere) is as follows:

ALGORITHM 3.3

Using the First and Second Derivatives to Sketch Simple Graphs

Suppose f is a continuous and twice-differentiable function. To sketch the graph of f, follow these steps:

1. Find the critical points of f and determine the sign of f' on each interval between these critical points. Record this information on a number line.

2. Find the critical points of f' and determine the sign of f'' on each interval between these critical points. Record this information on a second number line.

3. Locate any local extrema or inflection points of f using the information gathered in Steps 1 and 2. Record this information on the appropriate number lines.

4. Calculate the values of f at each of the critical points of f and f'. Sketch these points on a set of axes with an appropriate scale.

5. Use the information on the number lines to decide which type of arc from the ones shown in Figures 3.100–3.103 should connect the values of f that you plotted.

To sketch more complicated curves, for example, curves with asymptotes or discontinuities or nondifferentiable points or restricted domains, we will need a more sophisticated algorithm. We will investigate curve sketching for more complicated functions in Section 7.5.3.

The following example is written in great detail, with each step explained carefully. However, the process is not difficult; we only provide this much detail to help you understand the procedure and why it works.

EXAMPLE 3.49   **A detailed curve-sketching analysis**

Use Algorithm 3.3 to sketch a graph of the function $f(x) = x^3 - 3x^2 - 9x + 4$.

Solution   We first examine the first derivative information. The derivative of f is $f'(x) - 3x^2 - 6x - 9$. This derivative always exists, so the critical points of f are located where $f'(x) = 0$:

$$f'(x) = 0 \implies 3x^2 - 6x - 9 = 0 \implies 3(x^2 - 2x - 3) = 0$$

$$\implies 3(x - 3)(x + 1) = 0 \implies x = 3 \text{ or } x = -1.$$

Therefore, the only critical points of f are $x = -1$ and $x = 3$.

These critical points divide the real number line into three intervals: $(-\infty, -1)$, $(-1, 3)$, and $(3, \infty)$. We can test the sign of f' on each interval by testing its sign at just one point in each interval, for example,

$$f'(-2) = 3(-2)^2 - 6(-2) - 9 = 12 + 12 - 9 = 15 > 0,$$

$$f'(0) = 3(0)^2 - 6(0) - 9 = 0 - 0 - 9 = -9 < 0,$$

$$f'(4) = 3(4)^2 - 6(4) - 9 = 48 - 24 - 9 = 15 > 0.$$

We record this information on the number line in Figure 3.104. At the same time, we interpret this information in terms of the function f (whether it is increasing or decreasing) and apply the first derivative test to determine whether the critical points are local extrema.

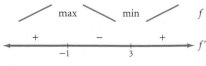

Figure 3.104

Now we look at the second derivative information. The second derivative of f is the derivative of $f'(x) = 3x^2 - 6x - 9$, which is $f''(x) = 6x - 6$. To find the critical points of f' (and thus the possible inflection points of f), we solve the equation $f''(x) = 0$:

$$f''(x) = 0 \implies 6x - 6 = 0 \implies 6(x - 1) = 0 \implies x = 1.$$

Since $f''(x) = 6x - 6$ is continuous and always exists, the only critical point of f' (and thus the only possible inflection point of f) is $x = 1$. This divides the number line into two intervals: $(-\infty, 1)$ and $(1, \infty)$. We test the sign of f'' on each interval by testing the sign of f'' at one point in each interval:

$$f''(0) = 6(0) - 6 = 0 - 6 = -6 < 0,$$
$$f''(2) = 6(2) - 6 = 12 - 6 = 6 > 0.$$

Since f'' changes sign at $x = 1$, the function f changes concavity at that point, and thus $x = 1$ is an inflection point of f. We record the second derivative information (and its implications for the function f) on the number line in Figure 3.105. Note that we have written "IP" above $x = 1$ to indicate that f has an inflection point at $x = 1$ (since f'' changes sign, and thus f changes concavity, at $x = 1$).

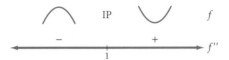

Figure 3.105

We are now ready to sketch the graph of f. The "interesting" points for this function are $x = -1$, $x = 1$, and $x = 3$; the values of $f(x) = x^3 - 3x^2 - 9x + 4$ at these points are

$$f(-1) = (-1)^3 - 3(-1)^2 - 9(-1) + 4 = -1 - 3 + 9 + 4 = 9,$$
$$f(1) = (1)^3 - 3(1)^2 - 9(1) + 4 = 1 - 3 - 9 + 4 = -7,$$
$$f(3) = (3)^3 - 3(3)^2 - 9(3) + 4 = 27 - 27 - 27 + 4 = -23.$$

To make sketching the graph easier, we can combine the information on the number lines for f' and f'' to make a number line describing the behavior of f. The three critical points of f and f' divide this number line into three intervals: $(-\infty, -1)$, $(-1, 3)$, and $(3, \infty)$. (*Note:* On this "combined" number line, the marked points do not necessarily refer to zeros of the function f; instead, they refer to the locations of the zeros, critical points, and inflection points of f.) On each of these intervals, the graph of f looks like one of the arcs shown in Figures 3.100–3.103 (see Figure 3.106).

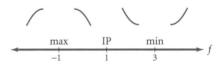

Figure 3.106

Finally, we sketch a set of coordinate axes of an appropriate scale to accommodate the three values of f that we have calculated. After plotting these points, we "connect the dots" according to the arcs in Figure 3.106, which gives us the graph in Figure 3.107.

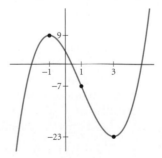

Figure 3.107

✅ *Checking the Answer*   You can use a graphing calculator to check the graph we sketched above. This will produce a graph like the one in Figure 3.108.

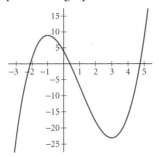

Figure 3.108

Although the calculator graph and our hand-drawn graph have the same shape, each graph provides slightly different information. From our hand-drawn graph we can read off the *exact* locations and values of the local extrema and inflection points of f. On the other hand, from the calculator graph we can get fairly good visual *approximations* for the x- and y-intercepts, extrema, and inflection points of f.   □

Why would we want to spend so much time sketching the graph of a function when we could make essentially the same sketch using a graphing calculator in just a few keystrokes? One reason is that sketching the graph by hand provides more specific information, such as the locations and values of the "important" points of the graph. Moreover, if we do the work by hand, we can easily find "the best window" for the graph of f—one that captures all of the important features of the graph. Furthermore, when we investigate more complicated functions, we will be able to discern features of a graph that are not apparent, or are misleading, on a graphing calculator. Perhaps most important, sketching a graph by hand illustrates the power of the first and second derivatives; together they contain enough information to describe the features of the graph almost completely.

What you should know after reading Section 3.8

▶ What are the definitions of "concave up" and "concave down"? Be able to explain these definitions graphically (in terms of f' being increasing or decreasing).

▶ Be able to find the intervals on which a function is concave up or concave down by testing the sign of f'' between critical points of f'. (What is a critical point of f' and why is it important here?) Be able to record this information on a number line and use it to sketch a rough graph of f.

▶ The definition of an inflection point, and the four possible types of inflection points. Be able to locate inflection points by testing whether the second derivative changes sign at the critical points of f'. If $f''(c) = 0$, is $x = c$ necessarily an inflection point of f?

▶ Be able to state, prove, and give illustrating examples for the second derivative test. What can you say about f at $x = c$ if $f''(c) = 0$? What is the second derivative test for, and how do you use it? What is the algorithm for using f'' to find local extrema of f?

▶ Be able to produce detailed graphs of functions by hand by investigating critical points, extrema, inflection points, and first and second derivative information recorded on number lines.

Exercises 3.8

Concepts

0. Read the section and make your own summary of the material. Include whatever you think will help you review the material later on.

1. Describe intuitively what it means for a function to be *concave down* on an interval I. What is the mathematical definition?

2. Sketch the graph of a function f that is concave down everywhere. Then draw five tangent lines on the graph and explain how you can see that the derivative of f is decreasing.

3. State the converse of part (a) of Theorem 3.21. Is the converse true? If so, explain why; if not, provide a counterexample.

4. State the contrapositive of part (a) of Theorem 3.21. Is the contrapositive true? If so, explain why; if not, provide a counterexample.

5. Fill in the blank: If f is twice-differentiable on an interval I, and if f'' is zero on I, then f must be _____ on the interval I.

■ Determine whether each statement below is true or false. If a statement is true, explain why. If a statement is false, provide a counterexample.

6. True or False: If $f''(2) = 0$, then $x = 2$ is an inflection point of f.

7. True or False: If f'' is continuous, then the only places it can change sign are where it is zero or does not exist.

8. True or False: If $x = -1$ is a maximum of the derivative f' of some function f, then f has an inflection point at $x = -1$.

9. True or False: If f is concave up on an interval I, then f'' is positive on I.

10. True or False: If $f''(2)$ does not exist, then $x = 2$ is a critical point of the function f'.

11. True or False: If $x = 3$ is an inflection point of f, then $x = 3$ is a local minimum or maximum of the derivative f'.

12. True or False: If $x = c$ is a critical point and $f''(c) = 0$, then $x = c$ cannot be a local extremum of f.

13. True or False: If $f'(1) = 0$ and $f''(1) = -2$, then $x = 1$ is a local maximum of f.

■ For each problem below, sketch the graph of a function f with domain all real numbers and the indicated characteristics, if possible. Be especially careful that your function is defined for all real numbers. (In particular, what

happens at the "ends" of your functions?) If it is not possible to graph such a function f, explain why.

14. f is positive everywhere, f' is negative everywhere, and f'' is positive everywhere.

15. f is negative everywhere, f' is negative everywhere, and f'' is negative everywhere.

16. f is negative everywhere, f' is positive everywhere, and f'' is positive everywhere.

■ For each of the three problems below, sketch the graph of a function f that has the given characteristics.

17. f' is positive everywhere, $f''(x) > 0$ for $x < -2$, and $f''(x) < 0$ for $x > -2$.

18. $f(3) = 0$, $f'(3) = 0$, and $f''(3) = 0$.

19. f is zero at $x = -1$, $x = 2$, $x = 4$; f' is zero at $x = -1$, $x = 1$, $x = 3$; and f'' is zero at $x = 0$, $x = 2$.

20. Describe (with supporting graphs) the four possible ways that a differentiable function f can behave near an inflection point $x = c$.

21. The four basic types of inflection points are illustrated in the functions f graphed in Figures 3.92–3.95 in the reading. In which two graphs does the *derivative* of f have a local maximum? In which two graphs does the *derivative* of f have a local minimum? Why?

22. Explain what the second derivative test is for and how it works. Support your explanation with graphs and number lines.

23. If a function f has four critical points, how many calculations (after finding the derivative) will you have to perform to apply the first derivative test? How many calculations (after finding the second derivative) will you have to perform to apply the second derivative test? Explain your answers.

24. Give three reasons why it is worthwhile to know how to sketch graphs of functions by hand, instead of relying on a graphing calculator.

Skills

25. In this problem you will show that the function $f(x) = x^4$ has the property that even though $f''(0) = 0$, the point $x = 0$ is not an inflection point of f.

(a) Show that if $f(x) = x^4$, then $f''(0) = 0$.

(b) Show that $x = 0$ is *not* an inflection point of the function $f(x) = x^4$.

■ For each function f below, find the intervals on which f is concave up and the intervals on which f is concave down. Show your work clearly, and represent your answers in interval notation *and* on a number line. Then support your answer by using a calculator to graph f.

26. $f(x) = x^2 + 2x - 1$　　**27.** $f(x) = 2x^2 - x^3$

28. $f(x) = (x-1)^3 + 2$　　**29.** $f(x) = x^4 - 2$

30. $f(x) = x^4 - 2x^3 - 5$　　**31.** $f(x) = (x-3)^3(x-1)$

■ For each function f below, find any inflection points of f. Show your work clearly. Then support your answer by using a calculator to graph f.

32. $f(x) = x^2 - 3x + 10$　　**33.** $f(x) = x^3 - 5x^2 + 4x$

34. $f(x) = (3x-2)^3 - 4$　　**35.** $f(x) = (x-2)^4$

36. $f(x) = -x^4 + 3x - 1$　　**37.** $f(x) = x^2(2-x)^2$

■ For each function f below, find any local extrema of f by using the second derivative test, if it applies. If the second derivative test does not apply, explain why and then use the first derivative test. Show your work clearly.

38. $f(x) = x^2 + 3x$　　**39.** $f(x) = 2x^3 - 4x^2 + 2x$

40. $f(x) = 2 - (4x+1)^3$　　**41.** $f(x) = (1-x)^4 - 2$

42. $f(x) = 3x^4 + 2x^2 + 4$　　**43.** $f(x) = x^3(x+2) + 1$

■ In each problem below, the derivative f' of some function f is given. Find any local extrema and inflection points of f without finding an equation for the function f. Then use the information you find, and other first and second derivative information, to sketch a possible graph of the function f. Your work should include number lines for f' and f''.

44. $f'(x) = x^2 + x - 2$

45. $f'(x) = x^4 - 1$

46. $f'(x) = 4x^3 - 12x^2 + 12x$

47. Part (c) of Theorem 3.22 says that if f'' is zero at a critical point of f, then the second derivative test gives no information. In this problem we verify this by examining the functions $f(x) = x^3$, $f(x) = x^4$, and $f(x) = -x^4$ at the point $c = 0$.

(a) Show that if $f(x) = x^3$, then $x = 0$ is a critical point of f, $f''(0) = 0$, and $x = 0$ is not a local extremum of f.

(b) Show that if $f(x) = x^4$, then $x = 0$ is a critical point of f, $f''(0) = 0$, and $x = 0$ is a local minimum of f.

(c) Show that if $f(x) = -x^4$, then $x = 0$ is a critical point of f, $f''(0) = 0$, and $x = 0$ is a local maximum of f.

(d) Explain how the three parts above show that the second derivative test does not tell us anything when the second derivative at a critical point is zero.

■ For each problem below, use the given number lines for f, f', and f'' to sketch a possible graph of f. Be sure that all the important features of f are clear, in particular the

intervals where f is positive or negative (if given), increasing or decreasing, and concave up or concave down, as well as any inflection points or extrema.

48.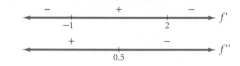

49.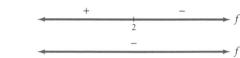

50.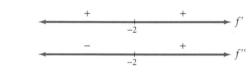

51.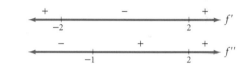

52.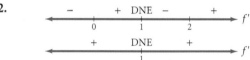

53.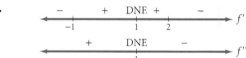

54.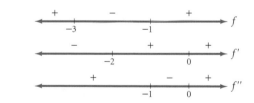

55.

56.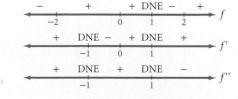

■ For each function f below, follow Algorithm 3.3 to sketch a detailed graph of f. Show and explain your work clearly so it is clear how you arrived at your graph and its features. Then check your answer by using a calculator to graph the function f.

57. $f(x) = (x - 2)(x + 2)$

58. $f(x) = x^3 - 2x^2$

59. $f(x) = x^3 + 6x^2 + 12x + 4$

60. $f(x) = x^4 - 16$

61. $f(x) = 3x^4 - 2x^2 + 4$

62. $f(x) = 1 + 2x^3 + x^4$

■ For each function f graphed below, sketch possible graphs of f' and f''. Be sure that the graphs of f' and f'' reflect the intervals on which f is increasing or decreasing and concave up or concave down, as well as any inflection points or extrema of f. It may help to construct number lines for f' and f''.

63.

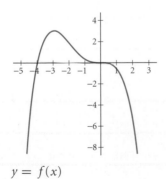

$y = f(x)$

64.

$y = f(x)$

65.

$y = f(x)$

■ Each graph below is the graph of the derivative f' of some function f. Use the graph of f' to sketch possible graphs for f and f''. Be sure that the graphs of f and f'' reflect the important features of the graph of f'.

66.

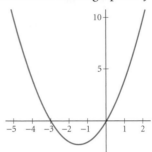

$y = f'(x)$

67.

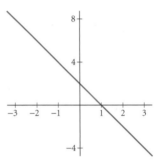

$y = f'(x)$

68.

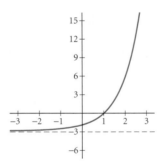

$y = f'(x)$

■ Each graph below is the graph of the second derivative f'' of some function f. Use the graph of f'' to sketch possible graphs for f and f'. Be sure to take into account all the relevant features of the given graph of f''.

69.

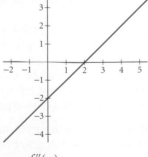

$y = f''(x)$

70.

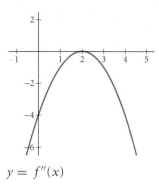

$$y = f''(x)$$

Applications

71. Suppose Rachel drives for 2 hours and that her distance from home (in miles) is given by the function $s(t)$ shown in the accompanying graph. Use this graph to answer the questions below.

   (a) Find the time interval on which Rachel's acceleration is negative. Is her velocity positive or negative on this interval? Describe in real-world terms what is happening during this time interval.

   (b) Find the time interval on which Rachel's acceleration is positive. Is her velocity positive or negative on this interval? Describe in real-world terms what is happening during this time interval.

   (c) The graph $y = s(t)$ of Rachel's position has an inflection point at $t = 1$ hour. What does this inflection point say about Rachel's velocity at $t = 1$? About her acceleration at $t = 1$?

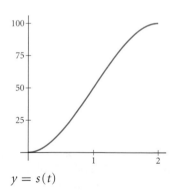

$$y = s(t)$$

72. Alina's distance north from the corner of Main Street and High Street t minutes after noon on Tuesday is given by

the function $s(t)$ (measured in miles) shown below. Use the graph of $s(t)$ to answer the questions below.

   (a) Find a time interval where Alina's velocity is positive and decreasing. Describe what Alina is doing on this time interval.

   (b) Find a time interval where Alina is moving south and her velocity is increasing. Describe what Alina is doing on this time interval.

   (c) Find a time interval where Alina's acceleration and velocity are both negative. Describe what Alina is doing on this time interval.

   (d) At which time is Alina's velocity at a minimum? What is she doing at that moment?

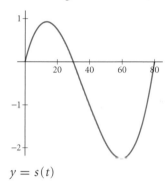

$$y = s(t)$$

Proofs

73. Prove part (b) of Theorem 3.21: If f is twice-differentiable on an interval I, and if f'' is negative on I, then f is concave down on I.

74. Prove the statement in Problem 5 (after filling in the blank). (*Hint:* Use part (c) of Theorem 3.17 in Section 3.7, and remember that f'' is the derivative of the function f'.)

75. Prove part (b) of Theorem 3.22: If $x = c$ is a critical point of f, and f is twice-differentiable near $x = c$, and if $f''(c)$ is negative, then $x = c$ is a local maximum of f.

76. Prove that every quadratic function is either always concave up or always concave down.

77. Prove that every cubic function (that is, every function of the form $f(x) = ax^3 + bx^2 + cx + d$ for some constants a, b, c, and d) has exactly one inflection point. (*Note:* It is not enough just to show that the second derivative of any cubic function has exactly one zero; you must also show that the sign of the second derivative changes.)

PART II

ALGEBRAIC FUNCTIONS

W*e now employ all that we know about functions, limits, and derivatives to investigate algebraic functions. We will discuss power functions, polynomial functions, and rational functions in detail and then investigate general algebraic functions.*

CHAPTER 4

Power Functions

Page from the Book of Kells; Art Resource, NY

P*ower functions are functions of the simple form* $f(x) = Ax^k$. *These functions are the building blocks of all algebraic functions. In this chapter we will define, investigate, and classify power functions, using everything we have learned about functions, limits, and derivatives.*

Power functions naturally fall into categories depending on the type of power involved. By the end of this chapter you will be able to identify, graph, and find limits and derivatives for every type of power function.

CONTENTS

4.1 The Algebra of Power Functions

In this section we will formally develop the notion of "raising a number to a power" (including noninteger powers). We will then investigate the algebraic rules for manipulating expressions involving powers or exponents, and precisely define the class of functions called ***power functions***.

4.1.1 Integer exponents

Multiplication is a way of writing repeated addition; for example, if k is a positive integer, then kx represents the sum of k copies of the number x:

$$(1) \qquad kx := \underbrace{x + x + x + \cdots + x}_{k \text{ times}}.$$

Similarly, exponentiation is a way of writing repeated multiplication; for example, if k is a positive integer, then x^k represents the product of k copies of the number x:

$$(2) \qquad x^k := \underbrace{x \cdot x \cdot x \cdots x}_{k \text{ times}}.$$

In Equation (2), the number x is called the *base* and the number k is called the *exponent*. For example, in the expression 3^4, the base is 3, the exponent is 4, and the expression 3^4 represents the product of four copies of the number 3:

$$3^4 = 3 \cdot 3 \cdot 3 \cdot 3 = 81.$$

What happens when $k = 0$? What would it mean to multiply a number by itself "zero times"? It is unclear what Equation 2 would mean when $k = 0$. We say that *by convention,*

$$(3) \qquad x^0 := \begin{cases} 1, & \text{if } x \neq 0 \\ ?, & \text{if } x = 0. \end{cases}$$

Equation (3) means that we *define* the expression x^0 to be equal to 1 for any nonzero value of x. One reason we do this is that it will make the algebraic rules of power functions work out nicely (see Section 4.1.3). When x and k are both zero, we have the expression 0^0, which is *indeterminate;* in other words, we don't even have a convention for what it should be (this is what the question mark means in Equation (3)). In the exercises, you'll examine why there isn't any number that we could reasonably use to define 0^0.

The definition of x^k in Equation (2) as "repeated multiplication" is easy to understand but not mathematically precise. We can define x^k for positive integer values of k more rigorously using a ***recursive*** definition. This definition specifies the value of x^k for $k = 0$ and then constructs a value for x^k given the value for x^{k-1}.

DEFINITION 4.1

Recursive Definition of Positive Integer Exponents

If k is a positive integer and $x \neq 0$, the expression x^k is defined recursively by

$$x^0 := 1, \qquad x^k := x(x^{k-1}).$$

For $x = 0$, we define $0^k = 0$ for all nonzero values of k.

For example, if $x = 3$, then by Definition 4.1 we have $3^0 = 1$. To find the value of 3^1, we multiply 3^0 by 3; thus $3^1 = 3(3^0) = 3(1) = 3$. Now that we have the value of 3^1, we multiply it by 3 to obtain the value of 3^2, namely $3^2 = 3(3^1) = 3(3) = 9$. In a similar fashion we can calculate $3^3 = 3(3^2) = 3(9) = 27$. By repeating this process, we can use Definition 4.1 to obtain the value of 3^k for any positive integer k.

So far we have discussed the meaning of x^k only when k is a positive integer. What happens if the exponent is a negative integer?

DEFINITION 4.2

Negative Exponents Represent Reciprocals

For any number x and any positive integer k, we define

$$x^{-k} := \frac{1}{x^k}.$$

If $x = 0$, then this expression is undefined.

For example, $3^{-1} = \frac{1}{3^1} = \frac{1}{3}$ and $3^{-2} = \frac{1}{3^2} = \frac{1}{9}$. Probably the most common example of this notation is $x^{-1} = \frac{1}{x}$. Note also that

(4)
$$\left(\frac{1}{x}\right)^{-1} = \frac{1}{1/x} = x.$$

The expression 0^{-k} is undefined for any positive integer k, since

(5)
$$0^{-k} = \frac{1}{0^k} = \frac{1}{0},$$

and division by zero is undefined.

EXAMPLE 4.1 **Working with integer exponents**

Rewrite the following quantities so that they do not involve exponents. Do not use a calculator.

    (a) 2^{-5} **(b)** $\left(\frac{1}{2}\right)^{-3}$ **(c)** 0^{-3} **(d)** 10^{-34}

Solution

    (a) $2^{-5} = \dfrac{1}{2^5} = \dfrac{1}{2 \cdot 2 \cdot 2 \cdot 2 \cdot 2} = \dfrac{1}{32}$ by the first part of Definition 4.2.

    (b) $\left(\frac{1}{2}\right)^{-3} = \dfrac{1}{\left(\frac{1}{2}\right)^3} = \dfrac{1}{\left(\frac{1}{2}\right)\left(\frac{1}{2}\right)\left(\frac{1}{2}\right)} = \dfrac{1}{\left(\frac{1}{8}\right)} = 8.$

    (c) The expression $0^{-3} = \frac{1}{0^3} = \frac{1}{0}$ is undefined.

    (d) $10^{-34} = \frac{1}{10^{34}}$. Note that $10^2 = 100$, $10^3 = 1000$, $10^4 = 10,000$, and so on. Following this pattern, 10^{34} will be a 1 followed by 34 zeros. We want to know what the reciprocal of that very large number will be. To work it out in decimal notation, we will consider powers of 10 and look for a pattern. Notice that $\frac{1}{10^1} = \frac{1}{10} = 0.1$, $\frac{1}{10^2} = \frac{1}{100} = 0.01$, $\frac{1}{10^3} = \frac{1}{1000} = 0.001$, and so on. Following this pattern, the decimal representing $\frac{1}{10^{34}}$ will have 33 zeros after the decimal point. In other words,

$$10^{-34} = 0.0000000000000000000000000000000001.$$

In scientific notation, this number is written as 1×10^{-34}. $\square$

4.1.2 Rational exponents

We now understand the meaning of x^k when k is a positive or negative integer power. The following definition describes what happens when the exponent is the reciprocal of an integer:

DEFINITION 4.3

Fractional Exponents Represent Roots

For any number x and any positive integer k, we define
$$x^{\frac{1}{k}} := \sqrt[k]{x}.$$
If x is negative and k is an even integer, then $\sqrt[k]{x}$ is not a real number.

To understand Definition 4.3 fully, we need to define the expression $\sqrt[k]{x}$, as follows:

DEFINITION 4.4

Roots

For any number x and any positive integer k,

(a) If k is odd, then $\sqrt[k]{x}$ is the unique number whose kth power is x.

(b) If k is even and x is nonnegative, then $\sqrt[k]{x}$ is the unique *nonnegative* number whose kth power is x.

(c) If k is even and x is negative, then $\sqrt[k]{x}$ is not a real number.

The "square root" has two notations, $\sqrt{x}$ and $\sqrt[2]{x}$ (although we generally use the simpler notation, $\sqrt{x}$).

? *Question* If k is a positive odd integer, then the kth root function $\sqrt[k]{x}$ is the inverse of the one-to-one function x^k. If k is a positive even integer, then the kth root function $\sqrt[k]{x}$ is the inverse of the function x^k restricted to the interval $[0, \infty)$. Show this using the definition of inverse functions (Definition 1.19 in Section 1.7.2). □

EXAMPLE 4.2 **Working with fractional exponents**

Rewrite the following quantities so that they do not involve exponents, using Definitions 4.3 and 4.4. Do not use a calculator.

    (a) $8^{\frac{1}{3}}$ (b) $(-8)^{\frac{1}{3}}$ (c) $(16)^{\frac{1}{4}}$ (d) $(-16)^{\frac{1}{4}}$

Solution

(a) Here $k = 3$ is odd, and $8^{\frac{1}{3}} = \sqrt[3]{8}$ is the unique number whose 3rd power is 8. Since $2^3 = (2)(2)(2) = 8$, we have $8^{\frac{1}{3}} = 2$.

(b) Again, $k = 3$ is odd, and since $(-2)(-2)(-2) = -8$, we have $(-8)^{\frac{1}{3}} = \sqrt[3]{-8} = -2$.

(c) $(16)^{\frac{1}{4}} = \sqrt[4]{16}$, so $k = 4$ is even and $x = 16$ is nonnegative. There are two numbers whose fourth power is 16, namely 2 and -2; however, there is only one *nonnegative* number whose fourth power is 16, and that is the number 2. Thus $(16)^{\frac{1}{4}} = 2$.

(d) $(-16)^{\frac{1}{4}} = \sqrt[4]{-16}$ is an even root of a negative number. There is no real number whose fourth power is -16 (since any real number raised to the fourth power must be positive), and therefore $\sqrt[4]{-16}$ is not a real number. □

We now know how to raise a number x to an integer power or to the reciprocal of an integer power. By combining these two concepts, we can define what it means to raise x to a rational power. Recall that by definition, any rational number can be written as the quotient $\frac{p}{q}$ of two integers p and q. We will always assume that $\frac{p}{q}$ is *reduced*, that is, that p and q have no common factors.

DEFINITION 4.5

Rational Exponents

For any number x and any positive rational number $\frac{p}{q}$, we define

$$x^{\frac{p}{q}} := (\sqrt[q]{x})^p.$$

If x is negative and q is even, then $x^{\frac{p}{q}}$ is not a real number.

Definition 4.5 says that the expression $x^{\frac{p}{q}}$ represents the pth power of the number whose qth power is x. Note that if q is even and x is negative, then $\sqrt[q]{x}$ is not a real number, and thus neither is $(\sqrt[q]{x})^p$ (regardless of the value of p).

When we raise a number x to a negative rational power $-\frac{p}{q}$, we will consider the negative sign to "belong" to p; for example,

$$x^{-\frac{2}{3}} = x^{\frac{-2}{3}} = (\sqrt[3]{x})^{-2} = \frac{1}{(\sqrt[3]{x})^2}.$$

EXAMPLE 4.3   **Working with rational exponents**

Calculate the following quantities *by hand*. Do not use a calculator.

  (a) $81^{\frac{3}{4}}$ (b) $(-8)^{-\frac{2}{3}}$ (c) $(-9)^{\frac{3}{2}}$ (d) $1^{-\frac{17}{5}}$

Solution

  (a) By Definition 4.5, $81^{\frac{3}{4}} = (\sqrt[4]{81})^3 = (3)^3 = 27$, since 3 is the number whose 4th power is 81, and the third power of 3 is 27.

  (b) By Definitions 4.2 and 4.5, $(-8)^{-\frac{2}{3}} = (\sqrt[3]{-8})^{-2} = (-2)^{-2} = \frac{1}{(-2)^2} = \frac{1}{4}$.

  (c) By Definition 4.5, $(-9)^{\frac{3}{2}} = (\sqrt{-9})^3$. Since we cannot take an even root of a negative number, $(\sqrt{-9})^3$ is not a real number.

  (d) $1^{-\frac{17}{5}} = \frac{1}{1^{\frac{17}{5}}} = \frac{1}{(\sqrt[5]{1})^{17}} = \frac{1}{1^{17}} = \frac{1}{1} = 1$. In fact, 1 raised to *any* power is equal to 1.   □

4.1.3 Algebraic rules for exponents

Now that we have basic definitions for expressions of the form x^k, we are ready to investigate the algebraic properties of such expressions.

THEOREM 4.1

Algebraic Rules for Exponents

For any numbers x, y, a, and b such that each expression is defined,

(a) $x^a x^b = x^{a+b}$,   (b) $(xy)^a = x^a y^a$,   (c) $(x^a)^b = (x^b)^a = x^{ab}$.

As simple corollaries to these rules we also have

(d) $\dfrac{x^a}{x^b} = x^{a-b}$,   (e) $\left(\dfrac{x}{y}\right)^a = \dfrac{x^a}{y^a}$,   (f) $(\sqrt[b]{x})^a = \sqrt[b]{x^a} = x^{\frac{a}{b}}$.

⊗ *Caution*   It is important to note that the rules in Theorem 4.1 apply only to values of x for which all of the expressions involved are defined. For example, $\sqrt{(-2)^2} = \sqrt{4} = 2$ but $(\sqrt{-2})^2$ is not defined, so Rule (f) does not apply when $a = 2$, $b = 2$, and $x = -2$. As functions, $\sqrt{x^2} = |x|$ is defined for all values of x, whereas $(\sqrt{x})^2 = x$ is defined only for positive values of x.   □

Although the rules in Theorem 4.1 are true for *any* powers a and b, we will usually assume that a and b are rational numbers. Moreover, these rules are difficult to prove when a and b are rational, so we will provide proofs only in the simpler case where a and b are integers. We will prove three of the rules in Theorem 4.1 and leave the rest to you in the exercises.

Rule (a) states that for any numbers x, a, and b, we have $x^a x^b = x^{a+b}$. We will now use Equation (2) to prove Rule (a) in the special case where a and b are positive integers.

PROOF   (**SPECIAL CASE OF PART (A) OF THEOREM 4.1**)   By Equation (2) we have

$$x^a x^b = \underbrace{(x \cdot x \cdot x \cdots x)}_{a \text{ times}}\underbrace{(x \cdot x \cdot x \cdots x)}_{b \text{ times}}$$

$$= \underbrace{(x \cdot x \cdot x \cdots x)}_{a+b \text{ times}}$$

$$= x^{a+b}. \qquad ■$$

The special case of Rule (b) where a is a positive integer can be proved in much the same way using Equation (2). We will instead give a more rigorous proof for this case of Rule (b) by using the recursive definition of x^a given in Definition 4.1 and the method of induction.

PROOF   (**SPECIAL CASE OF PART (B) OF THEOREM 4.1**)   We wish to show that for any positive integer a, we have $x^a y^a = (xy)^a$. We will use induction on the integer a.

True for $a = 1$: $x^1 y^1 = xy = (xy)^1$.
Assume true for $a - 1$, that is, assume that $x^{a-1} y^{a-1} = (xy)^{a-1}$.
Prove true for a (using the Inductive Hypothesis):

$$x^a y^a = (x \cdot x^{a-1})(y \cdot y^{a-1})$$

$$= (xy)(x^{a-1} y^{a-1})$$

$$= (xy)(xy)^{a-1} \qquad \textbf{(by the Inductive Hypothesis)}$$

$$= (xy)^a.$$

Thus by the method of induction, $x^a y^a = (xy)^a$ for any positive integer a.   ■

When a and b are positive integers, Rule (d) can be proved in a manner similar to the proof of Rule (a). However, we can also prove Rule (d) by assuming and using Rule (a), as follows:

PROOF   (**PART (D) OF THEOREM 4.1**)   For any numbers x, y, and a, we have

$$\frac{x^a}{x^b} = (x^a)\left(\frac{1}{x^b}\right)$$

$$= (x^a)(x^{-b}) \qquad \textbf{(Definition 4.2)}$$

$$= x^{a+(-b)} \qquad \textbf{(by Rule (a))}$$

$$= x^{a-b}. \qquad ■$$

4.1.4 Manipulating expressions with exponents

There are many situations in calculus and precalculus where it is necessary to manipulate an expression involving exponents. We can use the rules in Theorem 4.1 to simplify and rewrite such expressions.

| EXAMPLE 4.4 | **Simplifying an expression by using algebraic rules for exponents** |

Rewrite the expression $\dfrac{\sqrt[3]{r^{12}}}{(2s)^4}$ so that only one exponent is involved.

Solution   Using Rules (f), (b), and (e) from Theorem 4.1, we have

$$\frac{\sqrt[3]{r^{12}}}{(2s)^4} = \frac{r^{\frac{12}{3}}}{16s^4} = \frac{1}{16} \cdot \frac{r^4}{s^4} = \frac{1}{16}\left(\frac{r}{s}\right)^4. \qquad \square$$

| EXAMPLE 4.5 | **Simplifying difference quotients** |

Simplify and rewrite the following expressions until you can cancel the h in the denominator. (Computations like this will appear often in Section 4.3.)

$$\textbf{(a)}\ \frac{(x+h)^2 - x^2}{h} \qquad \textbf{(b)}\ \frac{(x+h)^{-1} - x^{-1}}{h} \qquad \textbf{(c)}\ \frac{(x+h)^{\frac{1}{2}} - x^{\frac{1}{2}}}{h}$$

Solution   Even though all of these expressions have the same basic form, in each case the calculation will be quite different.

(a) $\dfrac{(x+h)^2 - x^2}{h} = \dfrac{(x^2 + 2xh + h^2) - x^2}{h} = \dfrac{2xh + h^2}{h} = \dfrac{h(2x+h)}{h} = 2x + h.$

(b) We start by rewriting the expression so that the negative powers are expressed as fractions:

$$\frac{(x+h)^{-1} - x^{-1}}{h} = \frac{\frac{1}{x+h} - \frac{1}{x}}{h} = \frac{\frac{1}{x+h}\left(\frac{x}{x}\right) - \frac{1}{x}\left(\frac{x+h}{x+h}\right)}{h} = \frac{\left(\frac{x-(x+h)}{x(x+h)}\right)}{h}$$

$$= \frac{\left(\frac{-h}{x(x+h)}\right)}{h} = \frac{\left(\frac{-h}{x(x+h)}\right)}{h}\left(\frac{x(x+h)}{x(x+h)}\right)$$

$$= \frac{-h}{xh(x+h)} = \frac{-1}{x(x+h)}.$$

(*Note:* This computation can be done in fewer steps than are shown here. However, you should write out all the steps until you are comfortable with the algebra.)

(c) A common method for rewriting fractions involving a sum or difference of square roots is multiplying the numerator and denominator by the *conjugate*, as follows:

$$\frac{(x+h)^{\frac{1}{2}} - x^{\frac{1}{2}}}{h} = \left(\frac{\sqrt{x+h} - \sqrt{x}}{h}\right)\left(\frac{\sqrt{x+h} + \sqrt{x}}{\sqrt{x+h} + \sqrt{x}}\right)$$

$$= \frac{(\sqrt{x+h})^2 - (\sqrt{x})^2}{h(\sqrt{x+h} + \sqrt{x})} = \frac{(x+h) - x}{h(\sqrt{x+h} + \sqrt{x})}$$

$$= \frac{h}{h(\sqrt{x+h} + \sqrt{x})} = \frac{1}{\sqrt{x+h} + \sqrt{x}}.$$

Notice that we could not combine $\sqrt{x+h}$ and $\sqrt{x}$ in the first line of the calculation; this is one reason why we used the conjugate. $\qquad \square$

? Question List the algebraic rules used in each step of the computations in parts (a), (b), and (c) of Example 4.5. □

4.1.5 Extraneous solutions

When solving equations involving power functions, we occasionally generate *extraneous* (or "extra") solutions. In other words, certain operations concerning power functions may result in an equation with more solutions than the original equation. There are two times that this can happen: when expressions are pulled out of a denominator, and when the expression involves even powers. Both of these situations are illustrated in Example 4.6. When in doubt, check your answer to see whether the solutions you find satisfy the original equation.

EXAMPLE 4.6

Simplifying equations involving exponents

Rewrite each of the following equations so that only positive integer exponents are used. Indicate whether the resulting equation may have more solutions than the original equation.

(a) $a^{-2} = b^{-4}$ (b) $\sqrt[4]{a} = b$ (c) $a^{-1}\sqrt{b} = 0$

(d) $a = b^{\frac{2}{3}}$ (e) $a^{-5} = b^{\frac{1}{3}}$ (f) $a^{-\frac{5}{6}} = (b^7)^{-\frac{1}{4}}$

Solution

(a) First we rewrite the equation, and then we cross-multiply:

$$a^{-2} = b^{-4} \implies \frac{1}{a^2} = \frac{1}{b^4} \implies b^4 = a^2.$$

Since we rewrote the equation such that the a and b are no longer in the denominator, the final equation has $(a, b) = (0, 0)$ as a solution, whereas the original equation did not.

(b) After rewriting, we raise both sides of the equation to the fourth power:

$$\sqrt[4]{a} = b \implies a^{\frac{1}{4}} = b \implies \left(a^{\frac{1}{4}}\right)^4 = b^4 \implies a = b^4.$$

Since we raised both sides of the equation to an even power, the final equation may have more solutions than the original equation. This is in fact the case here: $(a, b) = (16, -2)$ is a solution to the final equation but not to the original equation.

(c) We have

$$a^{-1}\sqrt{b} = 0 \implies \frac{\sqrt{b}}{a} = 0 \implies \sqrt{b} = 0a \implies b = 0.$$

The final equation (considered as a two-variable equation) has a solution that does not work in the original equation, namely $(a, b) = (0, 0)$. The solutions to the original equation are the ordered pairs of the form (a, b), where $a \neq 0$ and $b = 0$.

(d) $a = b^{\frac{2}{3}}$ if and only if $a^3 = b^2$. Raising both sides of an equation to an odd power (in this example, 3) will not generate extraneous solutions.

(e) Depending on how you combine the definitions and rules of exponents, there are various ways that you can solve this problem. Here's one:

$$a^{-5} = b^{\frac{1}{3}} \implies (a^{-5})^3 = \left(b^{\frac{1}{3}}\right)^3 \implies a^{-15} = b$$

$$\implies \frac{1}{a^{15}} = b \implies 1 = a^{15}b.$$

The final equation has the same solution set as the original equation.

(f) One way to rewrite the equation is

$$a^{-\frac{5}{6}} = (b^7)^{-\frac{1}{4}} \implies a^{-\frac{5}{6}} = b^{-\frac{7}{4}}$$
$$\implies \left(a^{-\frac{5}{6}}\right)^{-12} = \left(b^{-\frac{7}{4}}\right)^{-12} \implies a^{10} = b^{21}.$$

The final equation may have extraneous solutions, for two reasons. What are the reasons? Can you find one of these extraneous solutions? Can you find another?   □

4.1.6 What is a power function?

We have already worked with some simple power functions, for example $f(x) = x^2$, $f(x) = x^3$, and $f(x) = x^4$. In general, a power function is defined as follows:

> **DEFINITION 4.6**
>
> **Power Functions**
>
> A ***power function*** is a function that can be written in the form
>
> $$f(x) = Ax^k,$$
>
> for some nonzero real number A and some rational number k.

In Definition 4.6, the constant k is called the *power* or *exponent,* and the constant A is called the *coefficient.* Note that the base of the expression must be the independent *variable,* and the exponent must be a *constant.* Note also that we require k to be a *rational* number; this means that, for example, $f(x) = x^\pi$ is *not* considered to be a power function.

Although power functions all have the simple form $f(x) = Ax^k$, they vary greatly depending on the value of the exponent k. Figures 4.1–4.8 show the graphs of eight of the most often used power functions (with $A = 1$ for each function).

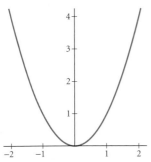

Figure 4.1 $y = x^2$

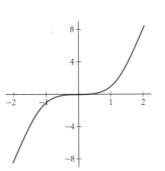

Figure 4.2 $y = x^3$

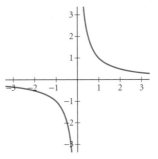

Figure 4.3 $y = x^{-1}$

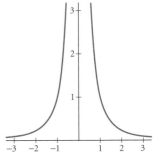

Figure 4.4 $y = x^{-2}$

Figure 4.5 $y = x^{\frac{1}{2}}$

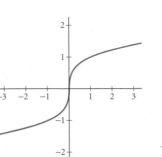

Figure 4.6 $y = x^{\frac{1}{3}}$

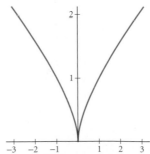

Figure 4.7 $y = x^{\frac{2}{3}}$

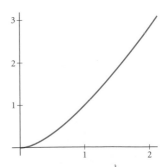

Figure 4.8 $y = x^{\frac{3}{2}}$

Even though all eight graphs represent power functions, they are all very different. They have different domains and ranges, some have asymptotes and some do not, some have discontinuities or cusps and some do not, and so on. Try graphing some other power functions on your calculator, including some examples where $A \neq 1$. We will investigate the graphs of different types of power functions in depth in Sections 4.4 and 4.5.

⑤ **Caution**   Become familiar with the graphs in Figures 4.1–4.8. You will often need to sketch these graphs from memory *without* using your graphing calculator.   □

A function $f(x)$ is a power function if it *can* be written in the form $f(x) = Ax^k$ for some constants A and k. Sometimes it will take a good bit of algebra to determine whether a given function can be written in the form Ax^k. For example, is the function

(6)
$$f(x) = \frac{(2x)^3 + 8x^{\frac{17}{5}}}{x^{\frac{2}{5}} + 1}$$

a power function? If so, what are the numbers A and k? Let's start with some much simpler examples.

EXAMPLE 4.7

Identifying power functions

Which of the following are power functions?

(a) $3x^2$ (b) $4x^{\sqrt{2}}$ (c) $5x^{\frac{3}{2}}$

(d) πx^8 (e) $3(5^x)$ (f) $(10x^3)^2$

(g) $3x^2 + 4$ (h) $\sqrt{x^2 x^3}$ (i) $\dfrac{3x^2}{2\sqrt{x}}$

Solution   All of these functions are power functions except for (b), (e), and (g). The function $4x^{\sqrt{2}}$ is not considered a power function because its exponent is an irrational number. The function $3(5^x)$ cannot be written in the form Ax^k because the variable x is in the exponent, rather than the base, of the expression. The function $3x^2 + 4$ is not a power function; it cannot be written in the form Ax^k because the 4 cannot be combined with the $3x^2$. (It is, however, a *sum* of the power functions $3x^2$ and $4 = 4x^0$.) The functions in parts (a), (c), and (d) are already in the form Ax^k. The functions in parts (f), (h), and (i) can be written as the power functions $100x^6$, $x^{\frac{5}{2}}$, and $\frac{3}{2}x^{\frac{3}{2}}$, respectively.   □

⑤ **Caution**   The function 2^x is *not* a power function, since the variable x is in the exponent rather than the base. Functions like 2^x are called ***exponential functions*** and will be discussed in Part III. For the record, the function x^x is neither a power function nor an exponential function, since neither the exponent nor the base is a constant.   □

Let's go back to our original question: Is the function in Equation (6) a power function? If so, how can we write it in the form $f(x) = Ax^k$? Using the rules in Theorem 4.1, we have

$$f(x) = \frac{(2x)^3 + 8x^{\frac{17}{5}}}{x^{\frac{2}{5}} + 1} = \frac{8x^3 + 8x^{\frac{17}{5}}}{x^{\frac{2}{5}} + 1} \qquad \text{(by Rule (b))}$$

$$= \frac{8\left(x^3 + x^{\frac{17}{5}}\right)}{x^{\frac{2}{5}} + 1} = \frac{8\left(x^3 + x^3 x^{\frac{2}{5}}\right)}{x^{\frac{2}{5}} + 1} \qquad \text{(by Rule (a))}$$

$$= \frac{8x^3\left(1 + x^{\frac{2}{5}}\right)}{x^{\frac{2}{5}} + 1} = 8x^3. \qquad \text{(cancellation)}$$

At the end of the second line, we write $x^{\frac{17}{5}}$ as $x^3 x^{\frac{2}{5}}$ so that we can factor out an x^3 from the expression in the numerator. (Compare this to the more familiar calculation $(x^3 + x^7) = (x^3 + x^3 x^4) = x^3(1 + x^4)$.) Also note that since $1 + x^{\frac{2}{5}}$ is never zero, the cancellation in the last step is valid for all x. Therefore, $f(x)$ is a power function, with $A = 8$ and $k = 3$.

EXAMPLE 4.8   **Writing a function in the form of a power function**

Write the following functions in the form Ax^k. For which values of x are your computations valid?

(a) $\dfrac{x^6 + (x^2)^3}{x^4}$

(b) $\dfrac{x^{\frac{1}{6}} x^{\frac{4}{3}}}{(x^4)^{\frac{1}{3}}}$

(c) $(16x^5 x^7)^{\frac{1}{4}} (\sqrt{x})^{-6}$

Solution

(a) $\dfrac{x^6 + (x^2)^3}{x^4} = \dfrac{x^6 + x^6}{x^4} = \dfrac{2x^6}{x^4} = 2x^2$ (as long as $x \neq 0$), by Rules (c) and (d).

(b) $\dfrac{x^{\frac{1}{6}} x^{\frac{4}{3}}}{(x^4)^{\frac{1}{3}}} = \dfrac{x^{(\frac{1}{6}+\frac{4}{3})}}{x^{\frac{4}{3}}} = \dfrac{x^{\frac{3}{2}}}{x^{\frac{4}{3}}} = x^{(\frac{3}{2}-\frac{4}{3})} = x^{\frac{1}{6}}$ (again, if $x \neq 0$), by Rules (a), (c), and (d).

(c) $(16x^5 x^7)^{\frac{1}{4}} (\sqrt{x})^{-6} = (16x^{12})^{\frac{1}{4}} (x^{\frac{1}{2}})^{-6} = (16)^{\frac{1}{4}} (x^{12})^{\frac{1}{4}} (x^{\frac{1}{2}})^{-6} = 2x^3 x^{-3} = 2x^{3-3} = 2x^0 = 2(1) = 2$, by Rules (a), (b), and (c). This computation holds only for $x > 0$, since if $x \leq 0$, the original expression is undefined. □

☑ *Checking the Answer*   One way to check the answers to Example 4.8 is to verify that the graph of the original expression is the same as the graph of the final expression. Alternatively, choose a value for x and substitute it into the original and final expressions; if you get two different answers, then you've probably made an algebra mistake. Of course neither of these methods absolutely ensures that your answer is correct. However, if your calculation fails either of these tests, then you know you must have done something wrong! □

A natural question to ask about power functions is: When is a combination of two power functions also a power function? The answer to this question depends, of course, on what kind of "combination" you make. For example, a constant multiple of a power function is a power function, and the product or quotient of two power functions is a power function, but the sum or difference of two power functions is not necessarily a power function. These facts can be easily seen by using the definition of power function and the algebraic rules for power functions.

EXAMPLE 4.9   **When is a combination of power functions itself a power function?**

Suppose $f(x)$ and $g(x)$ are power functions.

(a) Prove that the product $f(x)g(x)$ is a power function.

(b) Show by providing a counterexample that the sum $f(x) + g(x)$ is not necessarily a power function. Can you think of an example where a sum of power functions *is* a power function?

Solution

(a) Since $f(x)$ and $g(x)$ are power functions, by definition they can be written in the forms $f(x) = Ax^r$ and $g(x) = Bx^s$ for some constants A, B, r, and s. Their product is

$$f(x)g(x) = (Ax^r)(Bx^s) = ABx^r x^s = (AB)x^{r+s}.$$

This is a power function with exponent $r + s$ and coefficient AB.

(b) If $f(x) = x^2$ and $g(x) = x^3$, then $f(x) + g(x) = x^2 + x^3$. The function $f(x) + g(x)$ cannot be written in the form Ax^r (try a little algebra until you are convinced of this), so it is not a power function. On the other hand, if $f(x) = x^2$ and $g(x) = 3x^2$, then $f(x) + g(x) = x^2 + 3x^2 = 4x^2$ is a power function. $\square$

What you should know after reading Section 4.1

▶ The definitions of expressions x^k where the exponent is a positive integer, zero, a negative integer, the reciprocal of a positive integer, or a rational number.

▶ How to calculate various expressions of the form x^k by hand, including recognizing indeterminate or "undefined" forms such as 0^0, $\frac{1}{0}$, and even roots of negative numbers.

▶ The algebraic rules for power functions, and how to prove them (at least in the integer cases).

▶ The definition of power function, how to determine whether a given function is a power function, and the graphs of the eight most common power functions from Figures 4.1–4.8.

▶ How to use the algebraic rules for power functions to simplify expressions involving power functions, and, if possible, to write complicated expressions in the form Ax^k.

Exercises 4.1

Concepts

0. Read the section and make your own summary of the material. Include whatever you think will help you review the material later on.

■ Which of the following are power functions? Write each power function in the form $f(x) = Ax^k$.

1. $f(x) = x^2 + 1$

2. $f(x) = (x+1)^2$

3. $f(x) = 3(5^x)$

4. $f(x) = 3$

5. $f(x) = \dfrac{3}{2x^5}$

6. $f(x) = 2x^x$

7. $f(x) = \left(\dfrac{1}{\sqrt{x}}\right)^3$

8. $f(x) = (-8)^{\frac{2}{3}}$

9. $f(x) = 0$

10. By definition, a power function is a function that can be written in the form $f(x) = Ax^k$ for some nonzero constant A and rational number k. Why do you think we exclude the case $A = 0$? Use a graph of $f(x) = Ax^k$ when $A = 0$ to explain your answer.

11. Suppose A is any real number and k is any rational number.

(a) Why is $f(x) = Ax^k$ a function? (Think about the definition of function.)

(b) When $k = \frac{1}{2}$, $f(x) = A\sqrt{x}$. What is the domain of this function? Is it really true that for every x in the domain of $f(x)$ we have only *one* output $A\sqrt{x}$? Can you explain why someone might (incorrectly) think otherwise?

12. Recall that, by convention, for nonzero x we set $x^0 = 1$. Why couldn't we instead adopt the convention that $x^0 = 0$? One answer is that the rules in Theorem 4.1 would no longer hold. In this problem you will examine some ways in which these rules fail if we try to set $x^0 = 0$.

(a) Explain why it is not obvious just from Equation (2) what number x^0 should represent. What does it mean to say that setting $x^0 = 1$ is a *convention*?

(b) Suppose we were to adopt the convention that $x^0 = 0$. Try to use Rule (d) and this alternative convention to calculate $\frac{x^2}{x^2}$. What happens? Why is it bad?

(c) Again suppose we try to set $x^0 = 0$. Use the fact that $x^2 = x^{2+0}$ and Rule (a) to show that x^2 is equal to zero for all x. Why is this bad?

(d) You've just seen that if we set $x^0 = 0$, then the rules in Theorem 4.1 no longer work. So what? Why do you think we want those particular rules to work?

13. In this problem you will examine the indeterminate expression 0^0.

(a) We'll first examine what happens when $k = 0$ and x approaches zero. Approximate the value of $\lim_{x \to 0} x^0$ by evaluating at small nonzero values of x.

(b) Now suppose that $x = 0$ and k approaches zero. Use small nonzero values of k to approximate the value of $\lim_{k \to 0} 0^k$.

(c) Explain why your answers above show that there is no reasonable convention or definition for the expression 0^0.

14. Equation (1) defines multiplication as repeated addition. In this problem you will give a more rigorous definition of multiplication using recursion. (Don't know what recursion is? Did you read the section? This problem is analogous to the way we redefined Equation (2) using recursion in Definition 4.1.)

(a) Give a recursive definition of kx for any positive integer k.

(b) Use your definition to calculate $4 \cdot 5$, that is, the value of kx when $k = 4$ and $x = 5$.

15. Rule (f) of Theorem 4.1 states that $x^{\frac{a}{b}} = \sqrt[b]{x^a} = (\sqrt[b]{x})^a$. Suppose we have $a = 6$, $b = 4$, and $x = -2$. Show that Rule (f) does *not* hold in this case, and explain why this is happening.

Skills

■ Evaluate each of the following expressions *by hand*. Do not use a calculator; use the definitions in this section.

16. 20^3 **17.** $\left(\frac{1}{2}\right)^{-1}$ **18.** 4^{-3}

19. $\left(\frac{2}{3}\right)^{-3}$ **20.** $64^{\frac{5}{6}}$ **21.** $(-64)^{\frac{5}{6}}$

22. $\left(\frac{8}{27}\right)^{\frac{2}{3}}$ **23.** $1^{\frac{4}{9}}$ **24.** 5^0

25. 0^5 **26.** 0^0 **27.** 0^{-3}

28. $0^{\frac{1}{100}}$ **29.** 10^{27} **30.** 10^{-27}

31. $\left(\frac{1}{10}\right)^{27}$

■ Find the domains of the following functions.

32. $f(x) = (1 + x)^{-2}$ **33.** $f(x) = (1 + x^2)^{-2}$

34. $f(x) = (1 + x)^{\frac{3}{2}}$ **35.** $f(x) = (1 + x^2)^{\frac{3}{2}}$

36. $f(x) = \sqrt[3]{x + 1}$ **37.** $f(x) = (x + 1)^0$

38. $f(x) = 0^{(x-2)}$ **39.** $f(x) = (x^2 - 1)^{-\frac{1}{4}}$

40. $f(x) = \dfrac{x + 1}{x^2 - 1}$ **41.** $f(x) = \dfrac{x^2 - 1}{x + 1}$

42. $f(x) = \dfrac{1}{\sqrt{x^2 - 5}}$ **43.** $f(x) = \dfrac{1}{\sqrt{5 - x^2}}$

■ Which of the following functions $f(x)$ are power functions? If possible, write $f(x)$ in the form $f(x) = Ax^k$ for some constants A and k, specifying which rules from Theorem 4.1 you use in each step. For which values of x are your calculations valid?

44. $f(x) = \dfrac{1}{\sqrt{x^{\frac{3}{5}}}}$ **45.** $f(x) = x^{\frac{3}{2}} + x^{\frac{5}{3}}$

46. $f(x) = \left(\dfrac{1}{x^{-1}}\right)^{-2}$ **47.** $f(x) = \dfrac{x^3 - x^2}{x - 1}$

48. $f(x) = \dfrac{x^{-3} - x^{-2}}{x^{-1} - 1}$ **49.** $f(x) = 3\sqrt{2^x}$

50. $f(x) = 5\sqrt[3]{x} + 3\sqrt[5]{x}$ **51.** $f(x) = \dfrac{x^{-2}x^3}{x^5(4x^{-3})^2}$

52. $f(x) = 0$

■ All of the following statements are *false* and represent common algebra mistakes. Find a counterexample for each statement (i.e., find a value of x and/or y for which the statement is false).

53. False: $(3x)^2 = 3x^2$

54. False: $x^2 x^3 = x^6$

55. False: $\dfrac{x^{12}}{x^3} = x^4$

56. False: $\sqrt{x^2} = x$

57. False: $x^2 + x^3 = x^5$

58. False: $\sqrt{x^2 + y^2} = x + y$

59. False: $\dfrac{1 + x^{-2}}{y} = \dfrac{1}{y + x^2}$

60. False: $(x + y)^{-1} = \dfrac{1}{x} + \dfrac{1}{y}$

61. False: $\dfrac{x^{-1} + y^{-1}}{z} = \dfrac{1}{(x + y)z}$

■ Use the Intermediate Value Theorem, and what you know about the algebra, signs, and zeros of power functions, to solve each of the following inequalities.

62. $x^2 > x^4$ **63.** $x^3 \leq x^5$

64. $\sqrt{x}(x^2 + 1) < 0$ **65.** $x^{\frac{2}{3}}(x - 2)^{\frac{1}{5}} \geq 0$

66. $\sqrt[3]{x}(x - 1)^{-2} \leq 0$ **67.** $\sqrt[4]{x} + x^8 \geq 0$

■ Simplify and rewrite each expression until you can cancel the h in the denominator. Specify which rules from Theorem 4.1 you use for each step.

68. $\dfrac{(x + h)^3 - x^3}{h}$

69. $\dfrac{(x + h + 3)^2 - (x + 3)^2}{h}$

70. $\dfrac{(2(x + h))^{\frac{1}{2}} - (2x)^{\frac{1}{2}}}{h}$

71. $\dfrac{(x + h + 1)^{-1} - (x + 1)^{-1}}{h}$

72. $\dfrac{(x+h)^{-\frac{1}{2}} - x^{-\frac{1}{2}}}{h}$

73. $\dfrac{(x+h)^{-2} - (x)^{-2}}{h}$

■ Simplify and rewrite each expression until you can cancel a common factor from the numerator and denominator. Specify which rules from Theorem 4.1 you use for each step.

74. $\dfrac{t^3 - x^3}{t - x}$

75. $\dfrac{\sqrt{t} - \sqrt{x}}{t - x}$

76. $\dfrac{t^{-2} - x^{-2}}{t - x}$

■ Rewrite each equation so that only positive integer exponents are involved. Indicate whether the resulting equation may have more solutions than the original equation. If there are extraneous solutions, find one.

77. $b^{-5} = a^3$

78. $b^{-5} = a^{-2}$

79. $\sqrt{a} = b$

80. $a = \sqrt[5]{b}$

81. $a = b^{\frac{1}{3}}$

82. $a^{-1} = b^{\frac{1}{3}}$

83. $a^{-3} = b^{\frac{1}{3}}$

84. $a^{-\frac{1}{6}} = (b^3)^{-\frac{1}{4}}$

■ Use the eight graphs in Figures 4.1–4.8 and your knowledge of the graphs of transformations of functions to give a rough sketch (by hand) of the graph of each function.

85. $f(x) = (x-2)^3 + 1$

86. $f(x) = \dfrac{-3}{x}$

87. $f(x) = \sqrt[3]{x+1}$

88. $f(x) = 2 - \dfrac{3}{x}$

89. $f(x) = (3x)^2 - 5$

90. $f(x) = (x+3)^{\frac{2}{3}} - 2$

91. $f(x) = 2(x+4)^2 - 3$

92. $f(x) = \sqrt{-x}$

93. $f(x) = (\sqrt[2]{x-2})^3$

Proofs

■ In this section, you saw that sometimes a combination of two power functions is itself a power function (and sometimes not). Prove or find counterexamples to the following statements concerning such combinations.

94. If f and g are power functions, is the difference $f - g$ necessarily a power function? Prove or find a counterexample.

95. If f and g are power functions, is the quotient $\dfrac{f}{g}$ necessarily a power function? Prove or find a counterexample.

96. If f is a power function, is the composition $f(3x^2)$ necessarily a power function? Prove or find a counterexample.

97. If f and g are power functions, is the composition $f \circ g$ necessarily a power function? Prove or find a counterexample.

■ Prove each rule from Theorem 4.1 in the manner specified. (*Note:* In the first four problems, you can assume that the exponents involved are positive integers.)

98. Prove Rule (d) using Equation (2).

99. Prove Rule (e) using Equation (2).

100. Prove Rule (c) using Equation (2).

101. Prove Rule (e) using induction.

102. Prove Rule (f) using Rule (c).

103. Prove Rule (e) using Rules (b) and (c).

104. Prove that if Rule (a) of Theorem 4.1 is true, then we must have $x^0 = 1$ for all nonzero values of x. (*Hint:* Look at part (c) of Problem 12.) Be sure to point out where you use the fact that x is nonzero.

4.2 **Limits of Power Functions**

We have already learned how to take limits of power functions with positive integer powers. In this section we investigate the limits of general power functions.

4.2.1 Continuity of power functions

Recall from Section 2.6 that, intuitively, a function $f(x)$ is *continuous at a point $x = c$* if near that point the graph of $f(x)$ is "unbroken." A function $f(x)$ is then said to be a *continuous function* if it is continuous on its domain. The eight graphs of power functions in Figures 4.1–4.8 in Section 4.1.6 do seem to have this property; there are no breaks or jumps in these graphs, except where the functions are undefined. For example, consider Figures 4.9–4.11.

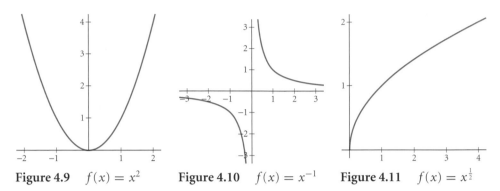

Figure 4.9 $f(x) = x^2$ **Figure 4.10** $f(x) = x^{-1}$ **Figure 4.11** $f(x) = x^{\frac{1}{2}}$

The graph of $f(x) = x^2$ in Figure 4.9 appears to be continuous on its entire domain of $(-\infty, \infty)$; the graph of $f(x) = x^{-1}$ in Figure 4.10 appears to be continuous everywhere except at $x = 0$, which is not in its domain; the graph of $f(x) = x^{\frac{1}{2}}$ in Figure 4.11 seems to be continuous on its entire domain of $[0, \infty)$ (note that at $x = 0$, the function f is only right continuous, since it is not defined to the left of zero).

As the next theorem shows, it is true that these power functions, and in fact *all* power functions, are continuous everywhere they are defined.

THEOREM 4.2

Continuity of Power Functions

Every power function $f(x) = Ax^k$ is continuous on its domain.

✦ *Caution*   We need to be more precise about what we mean when we say a function is continuous "on its domain." The function $f(x) = \sqrt{x}$ is a power function with domain $[0, \infty)$. When we say that this function is continuous "on its domain," we mean it is continuous at every point in the interior $(0, \infty)$ of its domain and is right continuous at the left end of its domain ($x = 0$).   □

Theorem 4.2 is important not only because it shows us that graphs of power functions are "unbroken" along their domains but also because it enables us to calculate limits of power functions. For example, what is $\lim_{x \to 4} \sqrt{x}$? You might guess that this limit is equal to 2, since $\sqrt{4} = 2$. However, in order to say this (i.e., in order to solve this limit by evaluating $\sqrt{x}$ at $x = 2$), we first have to know that $f(x) = \sqrt{x}$ is a continuous function! Recall that a function f is continuous at a point c if

$$\lim_{x \to c} f(x) = f(c).$$

In other words, we can calculate the limit of a function f by "evaluating" at $x = c$ *only if f is continuous at $x = c$.* Theorem 4.2 is equivalent to the following theorem.

THEOREM 4.3

Continuity of Power Functions in Terms of Limits

If a power function $f(x) = Ax^k$ is defined in an interval around $x = c$, then

$$\lim_{x \to c} Ax^k = Ac^k.$$

We already know how to evaluate the type of limit in Theorem 4.3 in the case where k is a positive integer. The point is that now we know how to evaluate limits of power functions for *all* rational powers k.

💲 *Caution*   We have to say "in an interval around $x = c$" rather than just "at $x = c$" in Theorem 4.3. For example, if $f(x) = x^{\frac{1}{2}}$, then f is defined at $x = 0$ (but *not* in an interval around $x = 0$, since $\sqrt{x}$ is not defined for negative numbers); in this case, $\lim_{x \to 0} x^{\frac{1}{2}}$ does not exist, and $\lim_{x \to 0^+} x^{\frac{1}{2}}$ is equal to $0^{\frac{1}{2}} = 0$. This type of example, where only a one-sided limit exists, occurs only for power functions involving even roots; usually, it is sufficient to check that $f(x)$ is defined *at* the point $x = c$.   □

Theorem 4.3 is difficult to prove since power functions are so diverse in their algebraic behavior. For example, x^4, $\frac{1}{x^3}$, and $\sqrt[3]{x^4}$ are very different algebraic expressions but are all power functions! To make matters worse, proving Theorem 4.3 involves a delta–epsilon argument! We have already proved Theorem 4.3 in the case where k is a positive integer (see Section 2.4.6). Since Theorem 4.3 is difficult to prove for general powers k, we will prove it for only a few selected examples. One such proof is presented here; others are included in the exercises.

EXAMPLE 4.10

Proving that the square root function is continuous

Prove that $f(x) = \sqrt{x}$ is a continuous function.

Solution   The domain of $f(x) = \sqrt{x}$ is $[0, \infty)$. We need to show that $\lim_{x \to c} \sqrt{x} = \sqrt{c}$ for all $c \in (0, \infty)$ and that $\lim_{x \to 0^+} \sqrt{x} = \sqrt{0} = 0$. The proof for $x \to 0^+$ is simple and is left to the exercises. We focus here on the proof for $c \in (0, \infty)$.

By the formal definition of limit, we need to show that

(1)
> For all $\varepsilon > 0$, there exists a $\delta > 0$ such that
> if $0 < |x - c| < \delta$, then $|\sqrt{x} - \sqrt{c}| < \varepsilon$.

PROOF   **(THEOREM 4.3)**   Given $\varepsilon > 0$, choose $\delta = \underline{\sqrt{c\varepsilon}}$. The reason why we choose this particular δ will be clear in a moment. Now if $0 < |x - c| < \delta$, we have

$$|\sqrt{x} - \sqrt{c}| = |\sqrt{x} - \sqrt{c}| \left(\frac{|\sqrt{x} + \sqrt{c}|}{|\sqrt{x} + \sqrt{c}|} \right) \qquad \text{(multiply by conjugate)}$$

$$= \frac{|x - c|}{|\sqrt{x} + \sqrt{c}|} \qquad \text{(simplify)}$$

$$< \frac{\delta}{|\sqrt{x} + \sqrt{c}|} \qquad (|x - c| < \delta \text{ by hypothesis})$$

$$= \frac{\delta}{\sqrt{x} + \sqrt{c}} \qquad (\sqrt{x} \text{ and } \sqrt{c} \text{ are} > 0)$$

$$< \frac{\delta}{\sqrt{c}} \qquad (\text{since } \sqrt{x} + \sqrt{c} > \sqrt{c})$$

$$= \frac{\sqrt{c\varepsilon}}{\sqrt{c}} = \varepsilon. \qquad (\text{since we chose } \delta = \sqrt{c\varepsilon})$$

Therefore, for all x such that $0 < |x - c| < \delta$, we have $|\sqrt{x} - \sqrt{c}| < \varepsilon$.   ▪

Now that we know that power functions are continuous, we can easily calculate simple limits involving power functions.

EXAMPLE 4.11   **Using continuity to calculate limits of power functions**

Use Theorem 4.2 to calculate the following limits.

(a) $\lim_{x \to -3} 2x^3$ (b) $\lim_{x \to 4} 5x^{-\frac{3}{2}}$ (c) $\lim_{x \to 0} 5x^{-\frac{3}{2}}$

Solution

(a) Since $f(x) = 2x^3$ is continuous on its entire domain $(-\infty, \infty)$, it is continuous at $x = -3$. Thus we have

$$\lim_{x \to -3} 2x^3 = 2(-3)^3 = 2(-27) = -54.$$

(b) Since $f(x) = 5x^{-\frac{3}{2}} = \frac{5}{(\sqrt{x})^3}$ has domain $(0, \infty)$ and is continuous everywhere on its domain, it is continuous at $x = 4$. Therefore

$$\lim_{x \to 4} 5x^{-\frac{3}{2}} = 5(4)^{-\frac{3}{2}} = \frac{5}{(\sqrt{4})^3} = \frac{5}{8}.$$

(c) The point $x = 0$ is not in the domain of $f(x) = 5x^{-\frac{3}{2}}$, so Theorem 4.2 does not apply here. We will see how to handle limits at nondomain points in the next section.   □

4.2.2 Limits of power functions at zero

We now know that a power function is continuous at each point in the interior of its domain. We can use this fact to compute the limit as $x \to c$ of any power function $f(x) = Ax^k$ as long as c is in the domain of $f(x)$. In this section we investigate what happens when c is not in the domain of $f(x) = Ax^k$.

Given a power function $f(x) = Ax^k$, there are only two ways that c could fail to be in the domain of $f(x)$. The first is if k involves an even root and c is negative. One example is the limit $\lim_{x \to -2} x^{\frac{1}{4}}$. Since $x^{\frac{1}{4}}$ is not defined at -2, or anywhere near -2, this limit does not exist.

The second way that c could fail to be in the domain of Ax^k is if k is negative and $c = 0$. For example, consider the functions graphed in Figures 4.12–4.14.

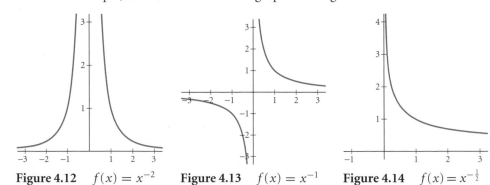

Figure 4.12 $f(x) = x^{-2}$ **Figure 4.13** $f(x) = x^{-1}$ **Figure 4.14** $f(x) = x^{-\frac{1}{2}}$

The limit of each of these functions as $x \to 0$ does not exist (and, in fact, is infinite). This is in fact true in general of the limit $\lim_{x \to 0} x^{-k}$ for any positive integer k.

THEOREM 4.4

Power Functions with Negative Powers Have Infinite Discontinuities

If k is any positive number, then

$$\lim_{x \to 0} x^{-k} \text{ is infinite.}$$

When we say that a limit is "infinite" as $x \to c$, we mean that as $x \to c^-$ or as $x \to c^+$, the limit is ∞ or $-\infty$.

PROOF (THEOREM 4.4) If k is positive, then

$$\lim_{x \to 0} x^{-k} = \lim_{x \to 0} \frac{1}{x^k} \to \frac{1}{0}.$$

This type of limit is infinite (see Section 2.5.3). ∎

Recall from Section 2.5.3 that when we encounter a limit of the form $\frac{1}{0}$, we investigate the left and right limits to see whether they are ∞ or $-\infty$.

EXAMPLE 4.12

Limits of power functions with negative powers

Investigate the following limits.

(a) $\lim_{x \to 0} x^{-2}$ (b) $\lim_{x \to 0} x^{-1}$ (c) $\lim_{x \to 0} x^{-\frac{1}{2}}$

Solution

(a) By Theorem 4.4, this limit does not exist (it is infinite). We examine the left and right limits separately.

$$\lim_{x \to 0^+} x^{-2} = \lim_{x \to 0^+} \frac{1}{x^2} \to \frac{1}{0^+} \to \infty,$$

$$\lim_{x \to 0^-} x^{-2} = \lim_{x \to 0^-} \frac{1}{x^2} \to \frac{1}{0^+} \to \infty.$$

Thus $\lim_{x \to 0} x^{-2} = \infty$. See Figure 4.12.

(b) This limit is also infinite. Therefore, we calculate the left and right limits:

$$\lim_{x \to 0^+} x^{-1} = \lim_{x \to 0^+} \frac{1}{x} \to \frac{1}{0^+} \to \infty,$$

$$\lim_{x \to 0^-} x^{-1} = \lim_{x \to 0^-} \frac{1}{x} \to \frac{1}{0^-} \to -\infty.$$

Thus $f(x) = x^{-1}$ approaches ∞ as x approaches 0 from the right, and it approaches $-\infty$ as x approaches 0 from the left (see Figure 4.13). We conclude that $\lim_{x \to 0} x^{-1}$ does not exist, because the right limit does not equal the left limit.

(c) Once more we investigate what happens as x approaches zero from the right and from the left:

$$\lim_{x \to 0^+} x^{-\frac{1}{2}} = \lim_{x \to 0^+} \frac{1}{\sqrt{x}} \to \frac{1}{0^+} \to \infty,$$

$$\lim_{x \to 0^-} x^{-\frac{1}{2}} = \lim_{x \to 0^-} \frac{1}{\sqrt{x}} \text{ does not exist, since } \sqrt{x} \text{ is not defined for } x < 0.$$

Therefore, $\lim_{x \to 0} x^{-\frac{1}{2}}$ does not exist. The function $f(x) = x^{-\frac{1}{2}}$ approaches ∞ as x approaches zero from the right, and it is not defined for $x < 0$. See Figure 4.14. □

It is interesting to note (but not worth memorizing) that for any positive integers p and q, $\lim_{x \to 0^+} x^{-\frac{p}{q}}$ is always ∞. On the other hand, $\lim_{x \to 0^-} x^{-\frac{p}{q}}$ is equal to ∞ if p is even and q is odd, is equal to $-\infty$ if p and q are both odd, and is "nonexistent" when p is odd and q is even. The proofs of these facts are structured like the calculations in Example 4.12. Look back at the functions in Figures 4.12–4.14 to see whether they fit this profile.

4.2.3 Limits of power functions at infinity

We now investigate the limits of power functions as x increases or decreases without bound. The limit of any power function $f(x) = x^k$ as $x \to \infty$ is either infinite or zero, depending on the sign of the exponent k.

THEOREM 4.5

> ### Global Behavior of Power Functions
>
> For any positive number k,
>
> **(a)** $\displaystyle\lim_{x \to \infty} x^k = \infty$,   **(b)** $\displaystyle\lim_{x \to \infty} x^{-k} = 0$.

Figures 4.15 and 4.16 illustrate this theorem for two common power functions.

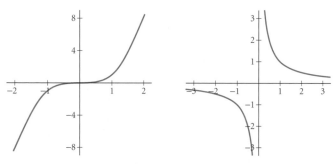

Figure 4.15 $f(x) = x^3$ **Figure 4.16** $f(x) = x^{-1}$

Graphically, we can see that $\displaystyle\lim_{x \to \infty} x^3$ appears to be ∞, whereas $\displaystyle\lim_{x \to \infty} x^{-1}$ appears to be zero.

PROOF (THEOREM 4.5)   The proof of part (a) is a little tricky, because it requires reference to the formal definition of a limit at infinity (see Section 2.2.5). The proof of part (b) will follow easily from part (a).

Suppose k is any positive number. To prove part (a), we must show that

(2)
$$\text{\textit{For all } M > 0\textit{, there exists an } N > 0 \textit{ such that}}$$
$$\text{\textit{if } x > N\textit{, then } x^k > M.}$$

This is like a delta–epsilon proof where the role of ε is played by M and the role of δ is played by N. Given $M > 0$, choose $N = \sqrt[k]{M}$; the reason for this choice of N will be apparent in a moment.

Since $k > 0$, if $x > N$ then $x^k > N^k$ (it is possible, but tedious, to prove this algebraically; we'll accept this as fact for now and prove it later when we know more about the derivatives of power functions). Since we chose N to be $\sqrt[k]{M}$, we have

$$x > N \implies x^k > N^k \implies x^k > (\sqrt[k]{M})^k \implies x^k > M.$$

This completes the proof of part (a).

To prove part (b), we do a little algebra and apply part (a). If k is any positive number, then

$$\lim_{x \to \infty} x^{-k} = \lim_{x \to \infty} \frac{1}{x^k} \to \frac{1}{\infty} \to 0. \qquad \blacksquare$$

The limits of power functions as $x \to -\infty$ can be found by using some algebra and Theorem 4.5. In general, if k is a positive number, then $\displaystyle\lim_{x \to -\infty} x^k$ is ∞, is $-\infty$, or does not exist; on the other hand, $\lim\limits_{x\to-\infty} x^{-k}$ either is zero or does not exist. We will handle each calculation separately rather than attempting to memorize what happens to different kinds of power functions as $x \to -\infty$.

EXAMPLE 4.13

Limits of power functions as $x \to -\infty$

Investigate the following limits.

(a) $\lim\limits_{x\to-\infty} x^2$ (b) $\lim\limits_{x\to-\infty} x^{-3}$ (c) $\lim\limits_{x\to-\infty} x^{\frac{1}{3}}$

Solution

(a) We can use Theorem 4.5 if we first observe that $\lim\limits_{x\to-\infty} f(x) = \lim\limits_{x\to\infty} f(-x)$. Using this fact yields

$$\lim\limits_{x\to-\infty} x^2 = \lim\limits_{x\to\infty} (-x)^2 = \lim\limits_{x\to\infty} x^2 = \infty.$$

You won't usually write out the entire calculation; once you become familiar with such limits, you can do some calculations in your head. For the moment, we will continue to write out all the steps.

(b) By the same method used in part (a),

$$\lim\limits_{x\to-\infty} x^{-3} = \lim\limits_{x\to\infty} \frac{1}{x^3} = \lim\limits_{x\to\infty} \frac{1}{(-x)^3} = \lim\limits_{x\to\infty} \frac{1}{-x^3} \to \frac{1}{-\infty} \to 0.$$

(c) Again using the fact that $\lim\limits_{x\to-\infty} f(x) = \lim\limits_{x\to\infty} f(-x)$, we have

$$\lim\limits_{x\to-\infty} x^{\frac{1}{3}} = \lim\limits_{x\to-\infty} \sqrt[3]{x} = \lim\limits_{x\to\infty} \sqrt[3]{-x} = \lim\limits_{x\to\infty} -\sqrt[3]{x} = -\infty. \qquad \square$$

4.2.4 Limit calculations involving power functions

Now that we know how to calculate limits of general power functions, we can also calculate limits involving sums, products, quotients, and compositions of power functions. The rule for limits of compositions given in Section 2.6.5 can be written to apply specifically to limits of functions of the form $(g(x))^k$, as in Theorem 4.6.

THEOREM 4.6

The Limit of a Composition with a Power Function

If $\lim\limits_{x\to c} g(x) = L$ and $f(x) = x^k$ is continuous at L, then

$$\lim\limits_{x\to c}(g(x))^k = \left(\lim\limits_{x\to c} g(x)\right)^k = L^k.$$

Using this special case of the rule for compositions of limits, we can calculate complicated limits that involve power functions.

EXAMPLE 4.14

Calculating limits that involve power functions

Calculate the following limits. Write each step individually and indicate where you use limit rules or the continuity of power functions.

(a) $\lim\limits_{x\to-2} \left(x^2 + \dfrac{3}{x}\right)$ (b) $\lim\limits_{x\to2}(x^3+1)^{\frac{3}{2}}$ (c) $\lim\limits_{x\to2}(3x^2-12)^{\frac{1}{4}}$

Solution

(a) This limit can be solved using the sum rule for limits and continuity:

$$\lim_{x \to -2} \left(x^2 + \frac{3}{x} \right) = \lim_{x \to -2} x^2 + \lim_{x \to -2} 3x^{-1} \qquad \text{(sum rule for limits)}$$

$$= (-2)^2 + 3(-2)^{-1} \qquad \text{(continuity)}$$

$$= 4 + \frac{3}{-2} = 2.5.$$

Note that $x = -2$ is in the domain of both x^2 and $3x^{-1}$, and thus both of these power functions are indeed continuous at $x = -2$. It is necessary to check this before "plugging in" $x = -2$.

(b) This time we will need Theorem 4.6:

$$\lim_{x \to 2} (x^3 + 1)^{\frac{3}{2}} = \left(\lim_{x \to 2} (x^3 + 1) \right)^{\frac{3}{2}} \qquad \text{(power rule for limits)}$$

$$= \left(\lim_{x \to 2} x^3 + \lim_{x \to 2} 1 \right)^{\frac{3}{2}} \qquad \text{(sum rule for limits)}$$

$$= (2^3 + 1)^{\frac{3}{2}} = 9^{\frac{3}{2}} \qquad \text{(continuity)}$$

$$= (\sqrt{9})^3 = 3^3 = 27.$$

(c) Evaluating the expression at $x = 2$ gives us a real number:

$$(3(2)^2 - 12)^{\frac{1}{4}} = (12 - 12)^{\frac{1}{4}} = 0^{\frac{1}{4}} = 0.$$

However, the limit as $x \to 2$ makes sense only as we approach $x = 2$ from the right; as $x \to 2^-$, the function $f(x) = (3x^2 - 12)^{\frac{1}{4}}$ does not exist. Thus we have

$$\lim_{x \to 2} (3x^2 - 12)^{\frac{1}{4}} \text{ does not exist, but } \lim_{x \to 2^+} (3x^2 - 12)^{\frac{1}{4}} = 0. \qquad \square$$

EXAMPLE 4.15   **Resolving a limit of the indeterminate form $\infty - \infty$**

Calculate $\lim\limits_{x \to \infty} (x^{\frac{3}{2}} - x^5)$.

Solution   By Theorem 4.5, as $x \to \infty$ we have $x^{\frac{3}{2}} \to \infty$ and $x^5 \to \infty$. Therefore, we have

$$\lim_{x \to \infty} (x^{\frac{3}{2}} - x^5) \to \infty - \infty.$$

But what is $\infty - \infty$? It is *not* necessarily zero; a limit of the form $\infty - \infty$ could be infinity, negative infinity, zero, or some other finite number. In other words, $\infty - \infty$ is an indeterminate form. We must use algebra to rewrite the limit. One common method is to turn the difference into a product:

$$\lim_{x \to \infty} (x^{\frac{3}{2}} - x^5) = \lim_{x \to \infty} x^{\frac{3}{2}} \left(1 - x^{\frac{7}{2}} \right).$$

Now by Theorem 4.5 we see that as $x \to \infty$, we have $x^{\frac{3}{2}} \to \infty$ and $(1 - x^{\frac{7}{2}}) \to 1 - \infty = -\infty$. Therefore,

$$\lim_{x \to \infty} (x^{\frac{3}{2}} - x^5) = \lim_{x \to \infty} x^{\frac{3}{2}} \left(1 - x^{\frac{7}{2}} \right) = \infty(-\infty) = -\infty. \qquad \square$$

☑ *Checking the Answer*   Use your calculator to graph $f(x) = x^{\frac{3}{2}} - x^5$ and check that as $x \to \infty$, $f(x)$ decreases without bound. $\qquad \square$

What you should know after reading Section 4.2

▶ How to recognize where power functions are continuous, what this means in terms of limits, and how to investigate continuity of power functions with delta–epsilon arguments. How to use the continuity of power functions to evaluate simple limits of power functions.

▶ How to investigate the limit of a power function at a point that is not in its domain—in particular, limits of power functions as $x \to 0$.

▶ Limits of power functions as $x \to \infty$, and how to use these limits to investigate the limits of power functions as $x \to -\infty$.

▶ How to evaluate limits that involve power functions. Make sure you are able to write out and justify each step in such limit calculations if asked to do so.

Exercises 4.2

Concepts

0. Read the section and make your own summary of the material. Include whatever you think will help you review the material later on.

1. The function $f(x) = \frac{1}{x}$ is a continuous function, since all power functions are continuous. However, $f(x) = \frac{1}{x}$ is clearly not continuous at $x = 0$. Is this a contradiction? Explain.

2. This problem concerns the continuity of power functions that involve even roots. We will concentrate on the function $f(x) = x^{\frac{1}{2}}$.

　(a) For which values of x is the function $f(x) = x^{\frac{1}{2}}$ continuous, and why?

　(b) In what sense is $x^{\frac{1}{2}}$ continuous at $x = 0$? In what sense is it not?

■ Rewrite each of the following statements as (a) a limit statement and (b) a delta–epsilon statement.

3. $f(x) = 3x^{\frac{1}{3}}$ is continuous at $x = 0$.

4. $f(x) = 2x^{-4}$ is continuous at $x = 3$.

5. $f(x) = x^{\frac{1}{4}}$ is continuous at $x = 2$.

6. $f(x) = 7x^{-3.5}$ is continuous at $x = 1$.

7. $f(x) = x^{-\frac{2}{3}}$ is not continuous at $x = 0$.

8. $f(x) = \sqrt[6]{x}$ is not continuous at $x = -1$.

■ Investigate the domains of the following power functions to determine the values of x at which each function is continuous.

9. $f(x) = 5x^{\frac{2}{3}}$

10. $f(x) = 3.28x^{19}$

11. $f(x) = 2x^{-\frac{1}{4}}$

12. $f(x) = -3x^{-3}$

13. $f(x) = -x^{1000}$

14. $f(x) = \frac{1}{3}x^{2.5}$

■ In each problem below, you are asked to find a value of x that satisfies certain conditions. For each problem, (a) use the appropriate part of Theorem 4.4 or Theorem 4.5 to argue why such an x *must* exist. Then (b) find a value of x that satisfies the given conditions. Bonus Question: What do these questions have to do with limits?

15. Find an $x > 0$ such that $0 < x^{-3} < 0.0001$.

16. Find an $x < 0$ such that $-0.0001 < x^{-3} < 0$.

17. Find an $x > 0$ such that $x^{-3} > 1,000,000$.

18. Find an $x < 0$ such that $x^{-3} < -1,000,000$.

Skills

■ For each power function $f(x)$, point $x = c$, and value of ε given below, find the largest value of δ such that $|f(x) - f(c)| < \varepsilon$ whenever $0 < |x - c| < \delta$. Sketch a graph of $f(x)$ with ε and δ intervals that illustrate your answer. Bonus Question: What does this have to do with the continuity of $f(x)$ at $x = c$?

19. $f(x) = x^4$, 　$c = -1$, 　$\varepsilon = 0.5$

20. $f(x) = x^{\frac{1}{4}}$, 　$c = 3$, 　$\varepsilon = 0.25$

21. $f(x) = x^{-2}$, 　$c = 2$, 　$\varepsilon = 0.01$

22. $f(x) = x^5$, 　$c = 0$, 　$\varepsilon = 1$

■ Use the continuity of power functions to calculate the following limits, if possible. Explain in each case how you are using continuity, or why it doesn't apply.

23. $\lim_{x \to -2} 4x^{-3}$

24. $\lim_{x \to 3} 2\sqrt{x}$

25. $\lim_{x \to 0} x^{-\frac{1}{6}}$

26. $\lim_{x \to 0} x^{\frac{3}{8}}$

27. $\lim_{x \to 9} 1.2x^{2.5}$

28. $\lim_{x \to -3} x^{\frac{3}{4}}$

■ Investigate the following limits. If a limit does not exist, examine the left and right limits and describe the behavior of the graph of the function near $x = 0$.

29. $\lim\limits_{x \to 0} 4x^{-2}$

30. $\lim\limits_{x \to 0} 5x^3$

31. $\lim\limits_{x \to 0} -4x^{-3}$

32. $\lim\limits_{x \to 0} x^{\frac{4}{3}}$

33. $\lim\limits_{x \to 0} 2x^{\frac{3}{4}}$

34. $\lim\limits_{x \to 0} 2x^{-\frac{3}{4}}$

■ Investigate the following limits without a calculator, and then check your answer with a graph.

35. $\lim\limits_{x \to \infty} 2x^{-\frac{4}{3}}$

36. $\lim\limits_{x \to -\infty} -3x^{\frac{4}{3}}$

37. $\lim\limits_{x \to \infty} 5x^{\frac{3}{4}}$

38. $\lim\limits_{x \to -\infty} -5x^{\frac{3}{4}}$

39. $\lim\limits_{x \to \infty} -2x^{-\frac{3}{4}}$

40. $\lim\limits_{x \to -\infty} 3x^{-\frac{3}{4}}$

■ Calculate each of the following limits by writing out all steps individually, citing each limit rule you use.

41. $\lim\limits_{x \to -3} \dfrac{x + 3}{x^2 - 9}$

42. $\lim\limits_{x \to 1} \dfrac{x^2}{x - 1}$

43. $\lim\limits_{x \to -8} \left(1 - 2x^{\frac{1}{3}}\right)^2$

44. $\lim\limits_{x \to 0} (x^{-2} - x^{-1})$

45. $\lim\limits_{x \to -8} \left(2x^{\frac{2}{3}} - 3x^2\right)$

46. $\lim\limits_{x \to \infty} (3x - 2x^2)$

■ Calculate each of the following limits.

47. $\lim\limits_{x \to -2} (\sqrt{-3x} + 5x^3)$

48. $\lim\limits_{x \to 1} \dfrac{2x^2 + 3x - 5}{x - 1}$

49. $\lim\limits_{x \to 1} \dfrac{x - 1}{2x^2 + 3x - 5}$

50. $\lim\limits_{x \to 2} (x^2 + 1)^{\frac{1}{4}}$

51. $\lim\limits_{x \to 0} (x^{-2} + 1)$

52. $\lim\limits_{x \to \infty} (\sqrt{x} - x)$

53. $\lim\limits_{x \to -\infty} x^2 - x + 8$

54. $\lim\limits_{x \to \infty} \dfrac{x^{-3}}{x^2 - x^{-1}}$

55. $\lim\limits_{x \to 0} x^0$

■ Determine the values of x for which the functions $f(x)$ below are continuous. Check your answers graphically.

56. $f(x) = \begin{cases} x^3, & \text{if } x \le 1 \\ \sqrt{x}, & \text{if } x > 1 \end{cases}$

57. $f(x) = \begin{cases} x^{\frac{2}{3}}, & \text{if } x \le 1 \\ 2x^{-1}, & \text{if } x > 1 \end{cases}$

■ Calculate each of the following limits. (*Note:* Limits of this form will be important in the next section when we calculate derivatives of power functions.)

58. $\lim\limits_{h \to 0} \dfrac{(1 + h)^3 - 1^3}{h}$

59. $\lim\limits_{h \to 0} \dfrac{\sqrt{3 + h} - \sqrt{3}}{h}$

60. $\lim\limits_{h \to 0} \dfrac{(2 + h)^{-1} - 2^{-1}}{h}$

61. $\lim\limits_{x \to 4} \dfrac{\sqrt{x} - \sqrt{4}}{x - 4}$

62. $\lim\limits_{x \to 2} \dfrac{\frac{1}{x} - \frac{1}{2}}{x - 2}$

63. $\lim\limits_{x \to 1} \dfrac{x^{-\frac{1}{2}} - 1}{x - 1}$

Proofs

64. Prove that the power function $f(x) = \frac{1}{x}$ is continuous on its domain $(-\infty, 0) \cup (0, \infty)$, *without* using a delta–epsilon argument. (*Hint:* If $c \ne 0$, you can compute $\lim\limits_{x \to c} \frac{1}{x}$ using the quotient rule for limits.)

■ Prove each of the following statements about continuity with a delta–epsilon argument.

65. $f(x) = \sqrt{x}$ is right continuous at $x = 0$.

66. $f(x) = \frac{1}{x}$ is continuous at $x = 3$.

67. $f(x) = 4x^{\frac{1}{2}}$ is a continuous function.

68. $f(x) = \frac{3}{x}$ is a continuous function.

69. Prove (with an M, N argument) that $\lim\limits_{x \to \infty} \sqrt[3]{x} = \infty$. You may assume that $b > a \Rightarrow \sqrt[3]{b} > \sqrt[3]{a}$ for all real numbers a and b.

4.3 ## Derivatives of Power Functions

In this section we will find the derivatives of various power functions by using the definition of derivative and then generalize these results into a "rule" that we can use to differentiate power functions quickly.

4.3.1 Differentiating power functions from scratch

We already have a rule that tells us how to differentiate power functions x^k, provided that k is a positive integer. We now wish to find the derivatives of more complicated power functions, like $f(x) = x^{-2}$ and $f(x) = \sqrt{x^3}$. To do this, we must go back to the definition of derivative.

EXAMPLE 4.16

Using the definition of derivative when negative exponents are involved

Find the derivative of $f(x) = x^{-2}$ at $x = -1$. Interpret your answer graphically.

Solution Using the "alternative" definition of derivative, we have

$$f'(-1) = \lim_{x \to -1} \frac{x^{-2} - (-1)^{-2}}{x - (-1)} = \lim_{x \to -1} \frac{\frac{1}{x^2} - \frac{1}{(-1)^2}}{x - (-1)} = \lim_{x \to -1} \frac{\frac{1}{x^2} - 1}{x + 1}$$

$$= \lim_{x \to -1} \frac{\left(\frac{1-x^2}{x^2}\right)}{x + 1} = \lim_{x \to -1} \frac{1 - x^2}{x^2(x + 1)} = \lim_{x \to -1} \frac{(1 - x)(1 + x)}{x^2(x + 1)}$$

$$= \lim_{x \to -1} \frac{1 - x}{x^2} = \frac{1 - (-1)}{(-1)^2} = 2.$$

This means that the line tangent to the graph of $f(x) = x^{-2}$ at the point $(-1, f(-1)) = (-1, 1)$ has a slope of $f'(-1) = 2$ units (see Figure 4.17). □

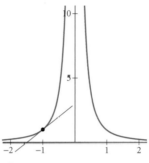

Figure 4.17

❓ *Question* Use the alternative definition of derivative to show that $\frac{d}{dx}(x^{-2}) = \frac{-2}{x^3}$. The calculation will be a more general version of the one in Example 4.16. □

❓ *Question* Repeat the calculation in Example 4.16 using the original definition of derivative (rather than the "alternative" definition). □

As the next example shows, not every power function is differentiable at every point in its domain. Recall that a function $f(x)$ is *differentiable* at $x = c$ if its derivative $f'(c)$ exists; in other words, f is differentiable at c if the limit that defines the derivative $f'(c)$ exists.

EXAMPLE 4.17

Using the definition of derivative to examine a nondifferentiable point

Investigate the derivative of $f(x) = x^{\frac{2}{3}}$ at $x = 0$.

Solution If $f(x) = x^{\frac{2}{3}}$, then by the definition of derivative,

$$f'(0) = \lim_{h \to 0} \frac{(0 + h)^{\frac{2}{3}} - 0^{\frac{2}{3}}}{h} = \lim_{h \to 0} \frac{h^{\frac{2}{3}}}{h} = \lim_{h \to 0} \frac{1}{h^{\frac{1}{3}}} \to \frac{1}{0},$$

which does not exist (i.e., it is not a real number). More precisely,

$$f'_+(0) = \lim_{h \to 0^+} \frac{(0 + h)^{\frac{2}{3}} - 0^{\frac{2}{3}}}{h} = \lim_{h \to 0^+} \frac{1}{h^{\frac{1}{3}}} \to \frac{1}{0^+} \to \infty,$$

and

$$f'_-(0) = \lim_{h \to 0^-} \frac{(0+h)^{\frac{2}{3}} - 0^{\frac{2}{3}}}{h} = \lim_{h \to 0^-} \frac{1}{h^{\frac{1}{3}}} \to \frac{1}{0^-} \to -\infty.$$

In any case, the derivative of $f(x) = x^{\frac{2}{3}}$ at $x = 0$ does not exist, since the limit that would define that derivative does not exist. ☐

✓ *Checking the Answer* Figure 4.18 shows a graph of $f(x) = x^{\frac{2}{3}}$. We clearly cannot have a derivative for $f(x)$ at $x = 0$, because there is not a well-defined tangent line to $f(x)$ at $x = 0$ (there is a "cusp" in the graph). To verify the behavior of the left and right derivatives, look at the slopes of the secant lines from $(0, f(0))$ to $(0 + h, f(0 + h))$ for various small positive and negative values of h. ☐

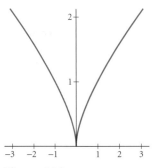

Figure 4.18

The following example involves some particularly nasty algebra, including compound fractions and multiplying by conjugates.

EXAMPLE 4.18

Using the definition of derivative when square roots are involved

Use the definition of derivative to show that the derivative of the function $f(x) = \frac{1}{\sqrt{x}}$ is $f'(x) = -\frac{1}{2\sqrt{x^3}}$.

Solution We first apply the definition of derivative to the function $f(x) = \frac{1}{\sqrt{x}} = x^{-\frac{1}{2}}$:

$$f'(x) = \lim_{h \to 0} \frac{f(x+h) - f(x)}{h} = \lim_{h \to 0} \frac{(x+h)^{-\frac{1}{2}} - x^{-\frac{1}{2}}}{h} = \lim_{h \to 0} \frac{\frac{1}{\sqrt{x+h}} - \frac{1}{\sqrt{x}}}{h}$$

$$= \lim_{h \to 0} \frac{\left(\frac{\sqrt{x} - \sqrt{x+h}}{\sqrt{x}\sqrt{x+h}} \right)}{h} = \lim_{h \to 0} \frac{\sqrt{x} - \sqrt{x+h}}{h\sqrt{x}\sqrt{x+h}}$$

$$= \lim_{h \to 0} \frac{\sqrt{x} - \sqrt{x+h}}{h\sqrt{x}\sqrt{x+h}} \left(\frac{\sqrt{x} + \sqrt{x+h}}{\sqrt{x} + \sqrt{x+h}} \right) = \lim_{h \to 0} \frac{x - (x+h)}{h\sqrt{x}\sqrt{x+h}(\sqrt{x} + \sqrt{x+h})}$$

$$= \lim_{h \to 0} \frac{-h}{h\sqrt{x}\sqrt{x+h}(\sqrt{x} + \sqrt{x+h})} = \lim_{h \to 0} \frac{-1}{\sqrt{x}\sqrt{x+h}(\sqrt{x} + \sqrt{x+h})}$$

$$= \frac{-1}{\sqrt{x}\sqrt{x+0}(\sqrt{x} + \sqrt{x+0})} = \frac{-1}{\sqrt{x}\sqrt{x}(\sqrt{x} + \sqrt{x})} = \frac{-1}{x(2\sqrt{x})}$$

$$= \frac{-1}{2x^{\frac{3}{2}}} = -\frac{1}{2\sqrt{x^3}}.$$
☐

❓ *Question*   Repeat the calculation in Example 4.18 using the alternative definition of derivative. Similar algebra will be involved; in particular, you will have to use a conjugate.   □

❓ *Question*   Show that the derivative of $f(x) = \sqrt{x}$ is $f'(x) = \frac{1}{2\sqrt{x}}$. The limit calculation will be similar to, but simpler than, the calculation in Example 4.18. At some point you will need to multiply by a conjugate.   □

4.3.2 The power rule

By this time you may have noticed that using the definition of derivative to calculate the derivatives of power functions is, well, kind of a pain. Fortunately, after you've done enough of these calculations, you start to see a pattern. We already know that $\frac{d}{dx}(x^k) = kx^{k-1}$ if k is a positive integer. Does this pattern hold in general? Consider the derivatives we have discussed so far in this section:

$$\frac{d}{dx}\left(\frac{1}{x^2}\right) = -\frac{2}{x^3}, \qquad \frac{d}{dx}(\sqrt{x}) = \frac{1}{2\sqrt{x}}, \qquad \frac{d}{dx}\left(\frac{1}{\sqrt{x}}\right) = -\frac{1}{2\sqrt{x^3}}.$$

If we write these results in the form x^k, we have

$$\frac{d}{dx}(x^{-2}) = -2x^{-3}, \qquad \frac{d}{dx}\left(x^{\frac{1}{2}}\right) = \frac{1}{2}x^{-\frac{1}{2}}, \qquad \frac{d}{dx}\left(x^{-\frac{1}{2}}\right) = -\frac{1}{2}x^{-\frac{3}{2}}.$$

These derivatives follow the pattern $\frac{d}{dx}(x^k) = kx^{k-1}$. In general, this rule will tell us the derivative of *any* power function:

THEOREM 4.7

The Power Rule

For any rational number k,

$$\frac{d}{dx}(x^k) = kx^{k-1}.$$

Although in Theorem 4.7 we require that k be a rational number, the power rule is actually true for *any* real number k. However, we consider only rational exponents when we discuss power functions, so Theorem 4.7 will be sufficient for our purposes.

🛇 *Caution*   Be careful that you use the power rule only when it actually applies. The power rule tells you how to differentiate a function that is *exactly* of the form x^k. It does *not* tell you how to differentiate other, similar-looking functions, such as $(4x)^5$ and 2^x. In order for us to apply the power rule, the base must be *only* the variable, and the exponent must be a constant. To differentiate the first function, we would need to do some algebra first (see Example 4.19). The last function is an exponential function, which we won't learn how to differentiate until Part III.   □

🛇 *Caution*   We call the power rule for differentiation simply *the power rule*. When we wish to refer to the power rule for limits, we will write out the entire phrase *the power rule for limits*. Be careful not to confuse these two very different rules, despite their similar names.   □

The proof of the power rule is difficult and must be done in cases. We have already proved that it is true when k is a positive integer; we now prove that it is true when

k is a negative integer. We must postpone the proof of the power rule for noninteger powers k until Section 7.3.4 (we need something called "implicit differentiation"). The proof in the case where k is a negative integer is just a general version of the calculation in Example 4.16.

PROOF   (**THEOREM 4.7** WHEN THE POWER IS A NEGATIVE INTEGER)
Suppose k is a positive integer; we wish to show that $\frac{d}{dx}(x^{-k}) = -kx^{-k-1}$. (Note that we have changed notation so that the negative part of the exponent will be "visible" as a reciprocal.) By the alternative definition of derivative, we have

$$\frac{d}{dx}(x^{-k})$$

$$= \lim_{t \to x} \frac{t^{-k} - x^{-k}}{t - x} \qquad \text{(alternative definition of derivative)}$$

$$= \lim_{t \to x} \frac{\frac{1}{t^k} - \frac{1}{x^k}}{t - x} \qquad \text{(rules of exponents)}$$

$$= \lim_{t \to x} \frac{\left(\frac{x^k - t^k}{t^k x^k}\right)}{t - x} = \lim_{t \to x} \frac{x^k - t^k}{t^k x^k (t - x)} \qquad \text{(algebra)}$$

$$= \lim_{t \to x} \frac{(x - t)(x^{k-1} + x^{k-2}t + \cdots + xt^{k-2} + t^{k-1})}{t^k x^k (t - x)} \qquad \text{(factoring formula)}$$

$$= \lim_{t \to x} \frac{(-1)(x^{k-1} + x^{k-2}t + \cdots + xt^{k-2} + t^{k-1})}{t^k x^k} \qquad \text{(cancel common factor)}$$

$$= \frac{-(x^{k-1} + x^{k-2}x + \cdots + xx^{k-2} + x^{k-1})}{x^k x^k} \qquad \text{(take limit)}$$

$$= \frac{-kx^{k-1}}{x^{2k}} = -kx^{k-1-2k} = -kx^{-k-1}. \qquad \text{(rules of exponents)}$$

If we combine the power rule with the constant multiple rule for differentiation, we can differentiate any power function $f(x) = Ax^k$:

$$\frac{d}{dx}(Ax^k) = A\frac{d}{dx}(x^k) \qquad \text{(constant multiple rule)}$$

$$= A(kx^{k-1}). \qquad \text{(power rule)}$$

EXAMPLE 4.19   **Using the power rule to calculate derivatives**

Use the power rule (and the constant multiple rule) to calculate the derivatives of the following power functions:

(a) $f(x) = 5x^{17}$ (b) $f(x) = \dfrac{5}{x^{\frac{1}{3}}}$ (c) $f(x) = \sqrt{4x}$

Solution

(a) This is a straightforward application of the power and constant multiple rules:

$$\frac{d}{dx}(5x^{17}) = 5\frac{d}{dx}(x^{17}) = 5(17x^{16}) = 85x^{16}.$$

The first equality is due to the constant multiple rule, and the second is an application of the power rule.

(b) Before we can differentiate this function, we must first put it into the form Ax^k (otherwise it will not be clear how to use the power rule):

$$\frac{d}{dx}\left(\frac{5}{x^{\frac{1}{3}}}\right) = \frac{d}{dx}(5x^{-\frac{1}{3}}) = 5\frac{d}{dx}(x^{-\frac{1}{3}}) = 5(-\tfrac{1}{3}x^{-\frac{1}{3}-1}) = -\tfrac{5}{3}x^{-\frac{4}{3}}.$$

(c) Again we must first put $f(x)$ into the form Ax^k. The power rule does not apply directly to the function $\sqrt{4x} = (4x)^{\frac{1}{2}}$. The power rule applies *only* to functions of the form x^k; the only thing that may be in the base of the expression is the independent variable x. We first rewrite $f(x)$ and then apply the power rule:

$$\frac{d}{dx}(\sqrt{4x}) = \frac{d}{dx}(\sqrt{4}\sqrt{x}) = \frac{d}{dx}(2\sqrt{x}) = \frac{d}{dx}(2x^{\frac{1}{2}})$$

$$= 2\frac{d}{dx}(x^{\frac{1}{2}}) = 2(\tfrac{1}{2}x^{-\frac{1}{2}}) = \frac{1}{\sqrt{x}}. \qquad \square$$

☑ *Checking the Answer*   Let's graphically interpret the answer to part (c) of Example 4.19. For each value $x = a$, the *height* of the graph of $f'(x)$ should be the *slope* of the graph of $f(x)$. As illustrated in Figures 4.19 and 4.20, this is true at $x = 1$.

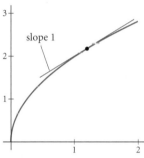

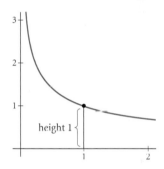

Figure 4.19
Slope of $f(x) = \sqrt{4x}$ at
$x = 1$ is 1

Figure 4.20
Height of $f'(x) = \frac{1}{\sqrt{x}}$ at
$x = 1$ is 1

Take a minute to convince yourself that for *every* point $x = a$, the slope of $f(x)$ in Figure 4.19 is equal to the height of the graph of $f'(x)$ in Figure 4.20. Using graphs like these is a good way to check your differentiation work. $\qquad \square$

4.3.3 Differentiability of power functions

In Example 4.17 we saw that the power function $f(x) = x^{\frac{2}{3}}$ is not differentiable at $x = 0$. Using the power rule, we can easily discover any nondifferentiable points of a power function, as the following example shows.

EXAMPLE 4.20

Determining the differentiability of a power function

Where is the function $f(x) = x^{\frac{2}{3}}$ differentiable?

Solution   By the power rule, we have

$$f'(x) = \frac{d}{dx}(x^{\frac{2}{3}}) = \tfrac{2}{3}x^{-\frac{1}{3}} = \frac{2}{3x^{\frac{1}{3}}}.$$

Clearly this function is defined everywhere except at $x = 0$, so $f(x) = x^{\frac{2}{3}}$ is differentiable on $(-\infty, 0) \cup (0, \infty)$. $\qquad \square$

THEOREM 4.8

Differentiability of Power Functions

Suppose $f(x) = Ax^k$ is a power function.

(a) If $k \geq 1$, then $f(x) = Ax^k$ is differentiable on its domain.

(b) If $k < 1$, then $f(x) = Ax^k$ is differentiable on its domain *except* at $x = 0$.

When we say that a function is differentiable "on its domain," we mean that it is differentiable on the interior of its domain, and differentiable from the left or the right of any closed endpoints of its domain, as appropriate. For example, the domain of the power function $f(x) = x^{\frac{1}{2}}$ is $[0, \infty)$. When we say that $f(x) = x^{\frac{1}{2}}$ is differentiable on its domain, we mean that $f(x)$ is differentiable for all x in $(0, \infty)$ and that $f(x)$ is right differentiable at the left endpoint $x = 0$.

PROOF (**THEOREM 4.8**)   Suppose $f(x) = Ax^k$. Then by the power rule, we have $f'(x) = Akx^{k-1}$.

If $k \geq 1$, then $k - 1 \geq 0$, so $f'(x)$ is defined for all x in the domain of $f(x)$. Note that if k involves an even root, then so will $k - 1$, so sometimes both $f(x)$ and $f'(x)$ are undefined for negative numbers.

If $k < 1$, then $k - 1 < 0$, so $f'(x)$ will not be defined at $x = 0$ (or anywhere not already in the domain of $f(x)$). ∎

Figures 4.21 and 4.22 are examples of Theorem 4.8 when $k < 1$. When $k = \frac{2}{3}$, the graph of $f(x)$ has a cusp at $x = 0$; see Figure 4.21. When $k = \frac{1}{2}$, the derivative is infinite at $x = 0$ (from the right); see Figure 4.22. Figure 4.23 is an example of Theorem 4.8 when $k \geq 1$ and shows the graph for $k = \frac{3}{2}$. The function is not defined for negative values of x, but the derivative at $x = 0$ from the right exists (and is zero).

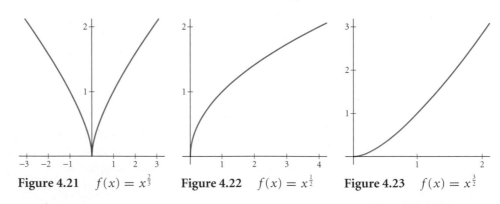

Figure 4.21 $f(x) = x^{\frac{2}{3}}$ **Figure 4.22** $f(x) = x^{\frac{1}{2}}$ **Figure 4.23** $f(x) = x^{\frac{3}{2}}$

As usual, when finding the derivative of a piecewise-defined function, we must be especially careful at any "break" points.

EXAMPLE 4.21   **Finding the derivative of a piecewise function whose pieces are power functions**

Find the derivative of the function $f(x) = \begin{cases} x^2, & \text{if } x \leq 1 \\ \frac{1}{x}, & \text{if } x > 1. \end{cases}$

Solution   For $x < 1$ we have

$$f'(x) = \frac{d}{dx}(x^2) = 2x.$$

On the other hand, when $x > 1$ we have

$$f'(x) = \frac{d}{dx}\left(\frac{1}{x}\right) = \frac{d}{dx}(x^{-1}) = -x^{-2} = \frac{-1}{x^2}.$$

It now remains only to determine what $f'(1)$ is (if it exists). The function $f(x)$ will be differentiable at $x = 1$ if it is continuous at $x = 1$ and the derivatives of each "piece" are equal at $x = 1$. Since $(1)^2 = \frac{1}{1}$, we know that $f(x)$ is continuous at $x = 1$. However, since $2(1) \neq \frac{-1}{(1)^2}$, the derivatives of the left and right pieces are not equal at $x = 1$. Therefore, $f(x)$ is not differentiable at $x = 1$, that is, $f'(1)$ does not exist. Thus the derivative of $f(x)$ is

$$f'(x) = \begin{cases} 2x, & \text{if } x < 1 \\ \text{DNE}, & \text{if } x = 1 \\ \frac{-1}{x^2}, & \text{if } x > 1. \end{cases}$$ □

◈ *Caution*   The derivative $f'(x)$ in Example 4.21 must be defined separately for $x < 1, x = 1$, and $x > 1$ (not for $x \leq 1$ and $x > 1$ as in the original function). You should always carefully examine the derivative at any "break" points of piecewise functions, since it is often the case that the derivative will not exist there.   □

☑ *Checking the Answer*   Figures 4.24 and 4.25 show the graphs of $f(x)$ and $f'(x)$ from Example 4.21. Note that $f(x)$ is indeed continuous, but not differentiable, at $x = 1$. In the graph of $f'(x)$, we have a "hole" at $x = 1$. Make sure you can see in *both* graphs the fact that $f'_- = 2$ but $f'_+ = -1$. Convince yourself that the slope of the graph of $f(x)$ at $x = c$ is equal to the height of the graph of $f'(x)$ at $x = c$ for each $c \neq 1$.

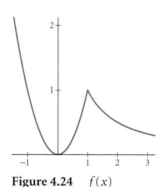

Figure 4.24 $f(x)$

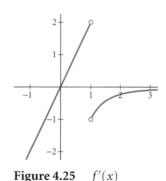

Figure 4.25 $f'(x)$ □

4.3.4 Differentiating with the power rule

Armed with the power rule, the constant multiple rule, and the sum rule, we can differentiate a large variety of algebraic functions.

◈ *Caution*   However, we cannot differentiate *all* algebraic functions using these three differentiation rules, only those that are sums or constant multiples of power functions. For example, at this point we still do not have a rule for differentiating compositions of functions, such as $f(x) = (3x + 1)^{1.5}$.   □

◈ *Caution*   If by some chance you have had calculus already, you may have seen other rules for differentiating. In this section you may use only the rules that we have learned *in this book* up to this point! When a problem asks you to differentiate something "if possible" (see Example 4.22), this of course really means "if possible using what has already been covered in this book."   □

EXAMPLE 4.22

Using the differentiation rules

Use the sum, constant multiple, and power rules to find the derivative of each function, if possible. Write each step individually, and list each differentiation rule you use.

(a) $f(x) = \dfrac{x^2}{x^{-3}\sqrt{x}}$ (b) $f(x) = 5^x$ (c) $f(x) = (2x+3)^2$

(d) $f(x) = 3^4$ (e) $f(x) = \dfrac{(x^2)^5}{7}$ (f) $f(x) = (x^5 - 1)^{\frac{1}{3}}$

Solution

(a) With a little algebra we can write $f(x)$ in the form Ax^k:

$$f(x) = \frac{x^2}{x^{-3}\sqrt{x}} = \frac{x^2 x^3}{x^{\frac{1}{2}}} = x^2 x^3 x^{-\frac{1}{2}} = x^{2+3-\frac{1}{2}} = x^{4.5}.$$

Now we are in a position to differentiate $f(x)$ with the power rule:

$$f'(x) = \frac{d}{dx}\left(\frac{x^2}{x^{-3}\sqrt{x}}\right) = \frac{d}{dx}(x^{4.5}) = 4.5x^{3.5}.$$

(b) This is not a power function (the variable is in the exponent rather than the base), and so the power rule does not apply. In particular, $\frac{d}{dx}(5^x)$ is *not* equal to $x(5^{x-1})$. We will learn how to differentiate functions like 5^x in Chapter 8.

(c) This function cannot be written in the form $f(x) = Ax^k$. You can see this by multiplying it out: $(2x+3)^2 = 4x^2 + 12x + 9$ is not a power function. However, it is a *sum* of power functions. We can use the sum rule for differentiation to split this into three derivative problems:

$$\frac{d}{dx}((2x+3)^2) = \frac{d}{dx}(4x^2 + 12x + 9) \qquad \text{(algebra)}$$

$$= \frac{d}{dx}(4x^2) + \frac{d}{dx}(12x) + \frac{d}{dx}(9) \qquad \text{(sum rule)}$$

$$= 4\frac{d}{dx}(x^2) + 12\frac{d}{dx}(x) + \frac{d}{dx}(9) \qquad \text{(constant multiple rule)}$$

$$= 4(2x^1) + 12(1) + 0 = 8x + 12. \qquad \text{(power rule)}$$

(d) Since $f(x) = 3^4 = 81$ is a constant, we have $\frac{d}{dx}(3^4) = 0$.

(e) With a little algebra we can write $f(x)$ in power function form and use the constant multiple rule and the power rule to find $f'(x)$:

$$\frac{d}{dx}\left(\frac{(x^2)^5}{7}\right) = \frac{d}{dx}\left(\tfrac{1}{7}x^{10}\right) = \tfrac{1}{7}\frac{d}{dx}(x^{10}) = \tfrac{1}{7}(10x^9) = \tfrac{10}{7}x^9.$$

(f) The function $f(x) = (x^5 - 1)^{\frac{1}{3}}$ cannot be written as a power function *or* as a sum of power functions. (We can't multiply this function out as we did in part (c).) There is no way for us to differentiate this function using only the sum, constant multiple, and power rules. We will learn how to differentiate this type of function in Chapter 7 after we develop the chain rule.   □

Usually you won't write out each step of your differentiation calculations as we did here. However, it is important that you are able to do this when asked or when the function you are differentiating is especially complicated.

4.3.5 Using derivatives to calculate limits

Now that we know the power rule for differentiation, we can use what we know about the derivatives of power functions to calculate certain special limits. A limit of the form

$$(1) \qquad \lim_{h \to 0} \frac{A(c + h)^k - Ac^k}{h}$$

represents the derivative $f'(c)$ of the power function $f(x) = Ax^k$ at the point $x = c$. By the power rule we know that $f'(x) = Akx^{k-1}$, so the limit in Expression (1) is equal to $f'(c) = Akc^{k-1}$. The following example illustrates this technique.

EXAMPLE 4.23

Calculating limits by recognizing them as derivatives

Calculate the following limits by recognizing them as the derivatives of power functions.

(a) $\displaystyle\lim_{h \to 0} \frac{(8 + h)^{\frac{2}{3}} - 4}{h}$
(b) $\displaystyle\lim_{x \to 0} \frac{4(1 + x)^{1.7} - 4}{x}$

Solution

(a) This limit represents the derivative of $f(x) = x^{\frac{2}{3}}$ at $x = 8$. Since $\frac{d}{dx}(x^{\frac{2}{3}}) = \frac{2}{3}x^{-\frac{1}{3}}$, we have

$$\lim_{h \to 0} \frac{(8 + h)^{\frac{2}{3}} - 4}{h} = \lim_{h \to 0} \frac{(8 + h)^{\frac{2}{3}} - 8^{\frac{2}{3}}}{h}$$

$$= \frac{d}{dx}\bigg|_{x=8} \left(x^{\frac{2}{3}}\right) = \frac{2}{3}(8)^{-\frac{1}{3}} = \frac{2}{3\sqrt[3]{8}} = \frac{2}{3 \cdot 2} = \frac{1}{3}.$$

(b) Although this limit has $x \to 0$ instead of $h \to 0$, the actual variable used in the limit does not matter. We'll change the x to an h so it is easier to see how this limit represents the derivative of $f(x) = 4x^{1.7}$ at $x = 1$. Since $\frac{d}{dx}(4x^{1.7}) = 4(1.7)x^{0.7}$, we have

$$\lim_{x \to 0} \frac{4(1 + x)^{1.7} - 4}{x} = \lim_{h \to 0} \frac{4(1 + h)^{1.7} - 4}{h} = \lim_{h \to 0} \frac{4(1 + h)^{1.7} - 4(1)^{1.7}}{h}$$

$$= \frac{d}{dx}\bigg|_{x=1} (4x^{1.7}) = 4(1.7)(1)^{0.7} = 4(1.7) = 6.8. \qquad \square$$

By the alternative definition of derivative given in Section 3.1, the limit $\lim_{x \to c} \frac{Ax^k - Ac^k}{x - c}$ also represents the derivative $f'(c)$ of the power function Ax^k at $x = c$. (Compare Example 4.24 to part (a) of Example 4.23.)

EXAMPLE 4.24

Calculating limits by recognizing them as derivatives ("alternative" form)

Calculate $\displaystyle\lim_{x \to 8} \frac{x^{\frac{2}{3}} - 4}{x - 8}$ by recognizing it as the derivative of a power function.

Solution   By the alternative definition of derivative, this limit is the derivative of the function $f(x) = x^{\frac{2}{3}}$ at $x = 8$. Therefore,

$$\lim_{x \to 8} \frac{x^{\frac{2}{3}} - 4}{x - 8} = \lim_{x \to 8} \frac{x^{\frac{2}{3}} - 8^{\frac{2}{3}}}{x - 8} = \frac{d}{dx}\Big|_{x=8} \left(x^{\frac{2}{3}}\right) = \frac{2}{3}(8)^{-\frac{1}{3}} = \frac{2}{3\sqrt[3]{8}} = \frac{2}{3 \cdot 2} = \frac{1}{3}. \quad \square$$

4.3.6 Antidifferentiating power functions

An ***antiderivative*** of a function g is a function f whose derivative is g, that is, a function f with the property that $f' = g$. For example, since the derivative of x^2 is $2x$, an antiderivative of $g(x) = 2x$ is the function $f(x) = x^2$. Since the derivative of a power function is itself a power function, it is sensible to imagine that the *anti*derivatives of a power function will also be power functions. In fact, this is the case:

THEOREM 4.9

Antidifferentiating Power Functions

If $f'(x) = x^k$ and $k \neq -1$, then $f(x) = \frac{1}{k+1}x^{k+1} + C$ for some constant C.

In other words, if $k \neq -1$, then the antiderivatives of x^k are all of the form $\frac{1}{k+1}x^{k+1} + C$, for some constant C.

PROOF (**THEOREM 4.9**)   Since any two functions with the same derivative differ by a constant (see Section 3.7), it suffices to prove that the derivative of $\frac{1}{k+1}x^{k+1}$ is x^k. By the power rule and the constant multiple rule, we have

$$\frac{d}{dx}\left(\frac{1}{k+1}x^{k+1}\right) = \frac{1}{k+1}\frac{d}{dx}(x^{k+1}) = \frac{1}{k+1}(k+1)x^{(k+1)-1} = x^k.$$

Note that if $k = -1$, then there will be a division by zero in the calculation above. Therefore, this proof does not work in the case where $f(x) = x^{-1}$. We will learn how to antidifferentiate x^{-1} in Part III. ∎

We can combine Theorem 4.9 with the constant multiple rule to find the family of antiderivatives of any power function. Knowing the derivative $f'(x)$ and one value of the function $f(x)$, we have enough information to determine the function $f(x)$ completely.

EXAMPLE 4.25

Finding a function with a given derivative and value

Find a function $f(x)$ such that $f'(x) = 8x^3$ and $f(1) = 4$.

Solution   By Theorem 4.9 and the constant multiple rule, if $f'(x) = 8x^3$ then

$$f(x) = 8\left(\tfrac{1}{4}x^4\right) + C = 2x^4 + C.$$

Since $f(1) = 4$, we must have

$$4 = f(1) = 2(1)^4 + C = 2 + C \implies C = 4 - 2 = 2.$$

Thus the function is $f(x) = 2x^4 + 2$. $\square$

☑ *Checking the Answer*   We must check that $f(x) = 2x^4 + 2$ actually has the desired derivative and value. This is straightforward: If $f(x) = 2x^4 + 2$, then $f'(x) = \frac{d}{dx}(2x^4 + 2) = 2(4x^3) + 0 = 8x^3$, and $f(1) = 2(1)^4 + 2 = 2 + 2 = 4$. $\square$

Each function in the family of antiderivatives for $f'(x) = 8x^3$ is of the form $f(x) = 2x^4 + C$ (see Example 4.25). Each graph in this family has the same general shape (since they all have the same derivative), but their graphs are at different heights on the coordinate plane (since they all differ by a constant). Figure 4.26 shows some of the functions $f(x)$ that are in the family of antiderivatives for $f'(x) = 8x^3$. Only *one* of these functions satisfies $f(1) = 4$. This function (namely $f(x) = 2x^4 + 2$) is shown in Figure 4.27.

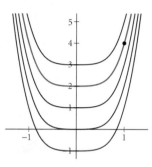

Figure 4.26

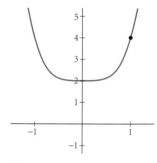

Figure 4.27

What you should know after reading Section 4.3

▶ How to calculate derivatives of power functions using the definition of derivative (both at a point and as a function), and graphical interpretations of these calculations.

▶ The statement of the power rule for differentiation, when it applies, and its proof in the integer case. How to use this rule to calculate the derivatives of expressions involving power functions (including piecewise-defined expressions).

▶ How to test whether or not an expression involving power functions is differentiable at a point. This includes piecewise-defined functions.

▶ How to solve certain limits by recognizing them as the derivatives of power functions. The formula for antidifferentiating power functions, and how to prove and use this formula.

Exercises 4.3

Concepts

0. Read the section and make your own summary of the material. Include whatever you think will help you review the material later on.

1. Let $f(x) = 3\sqrt{x}$. In this problem you will approximate the slope of the tangent line to $f(x)$ at $x = 4$ by calculating the slopes of approaching secant lines.

   (a) Calculate the slopes of the secant lines from $(4, f(4))$ to $(4 + h, f(4 + h))$ for the following values of h: $h = 0.5$, $h = 0.25$, $h = 0.1$, $h = 0.01$.

   (b) What value do these slopes seem to be approaching?

(c) What does this mean about the slope of the tangent line to $f(x) = 3\sqrt{x}$ at $x = 4$?

(d) What does this mean about the value of $f'(4)$?

(e) Sketch a graph of $f(x)$ together with each of the secant lines whose slopes you computed above, and illustrate that the slopes of these secant lines are indeed approaching the slope of the tangent line to $f(x)$ at $x = 4$.

(f) Use the power rule to find the exact value of $f'(4)$, and compare this exact answer with your approximation above.

■ In each problem below, the graph of some power function $f(x)$ is given. Use this graph to sketch a graph of the derivative $f'(x)$.

2.

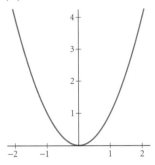

3.

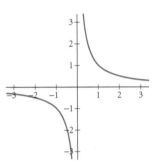

4.

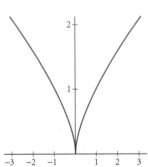

■ For each function, determine whether the function can be differentiated using the differentiation rules we know at this point (and possibly some algebra beforehand). If the function cannot be differentiated using only these rules, explain why the power rule does not apply.

5. $f(x) = (3x^2 + 1)^{\frac{1}{3}}$ 6. $f(x) = (3\sqrt{x} + 1)^4$

7. $f(x) = 3(24)^x$ 8. $f(x) = \dfrac{3}{\sqrt{x}}$

9. $f(x) = (x^5 + 1)^{x+1}$ 10. $f(x) = \dfrac{x^2 + \frac{1}{x}}{2x^3}$

11. Why does the proof of the power rule given in this section work only when the power is a negative integer?

12. The power rule does *not* apply to the function $f(x) = 5^x$. This means that $\frac{d}{dx}(5^x)$ is *not* equal to $x5^{x-1}$. Illustrate this by approximating the derivative of 5^x at $x = 2$ (use the definition of derivative and a small value of h) and showing that this approximation is not close to the value of $x5^{x-1}$ when $x = 2$.

13. Is the derivative of a power function always a power function? Is the antiderivative of a power function always a power function? Justify your answers.

14. Why does Theorem 4.9 hold only when $k \neq -1$? What goes wrong in the proof when $k = -1$?

15. Find a function whose derivative is $4\sqrt{x}$. Then find two more functions whose derivative is $4\sqrt{x}$. Graph all of these functions on the same set of axes. What do you notice about the graphs?

Skills

■ Use the definition of derivative (*not* the power rule!) to calculate the derivative $f'(c)$ of each power function at the indicated point c. Then repeat the calculation using the alternative definition of derivative.

16. $f(x) = 3x^{-1}$, $c = 2$ 17. $f(x) = 3x^{-1}$, $c = 3$

18. $f(x) = x^{-3}$, $c = 2$ 19. $f(x) = \sqrt{x}$, $c = 4$

20. $f(x) = x^{-\frac{1}{2}}$, $c = 4$ 21. $f(x) = x^{\frac{2}{3}}$, $c = 0$

■ Use the definition of derivative (*not* the power rule!) to calculate the derivatives of the following power functions. Then repeat the calculation using the alternative definition of derivative.

22. $f(x) = \dfrac{2}{x}$ 23. $f(x) = \dfrac{1}{x^2}$

24. $f(x) = 3\sqrt{x}$ 25. $f(x) = \sqrt{3x + 1}$

26. $f(x) = \dfrac{1}{\sqrt{x}}$ 27. $f(x) = \dfrac{1}{\sqrt{3x + 1}}$

■ Find the derivatives of the following functions. You may have to do some algebra first so that you can apply the differentiation rules. Write out each step individually, listing each differentiation rule that you use.

28. $f(x) = 5x^{\frac{3}{4}} - 3\sqrt{x} + 1$

29. $f(x) = (3x^5)^{-\frac{1}{3}}$

30. $f(x) = (\sqrt{x} - 1)(\sqrt{x} + 1)$

31. $f(x) = \dfrac{3}{\sqrt[3]{x}}$

32. $f(x) = \dfrac{\sqrt{x} + 3x^5}{x^{-1}}$

33. $f(x) = \dfrac{\left(x^{\frac{1}{3}}\right)^{\frac{5}{4}}}{8x^{\frac{1}{2}}}$

■ Find the derivatives of the following functions. You may have to do some algebra first so that you can apply the differentiation rules. There are some functions below whose derivatives cannot be calculated using the differentiation rules we know at this point; label them as such and explain why the rules do not apply.

34. $f(x) = \sqrt{5}x^2$ 35. $f(x) = (5x^2)^{\frac{2}{3}}$

36. $f(x) = \dfrac{x^2 + 1}{\sqrt{x}}$ 37. $f(x) = 3\sqrt{x} - 2(4x - 1)$

38. $f(x) = (3x^2 + 1)^{\frac{1}{2}}$

39. $f(x) = \left(3x^{\frac{1}{5}} + x^{-1}\right)^2$

40. $f(x) = 2 + 5^x$

41. $f(x) = ((3x+1)\sqrt{x})^3$

42. $f(x) = x(3x+1)^{\frac{5}{4}}$

43. $f(x) = \dfrac{(4x^2)^{\frac{3}{2}}}{5x^{\frac{1}{3}}}$

44. $f(x) = \dfrac{\sqrt{x}}{x^{\frac{3}{4}}+1}$

45. $f(x) = (\sqrt{x}+1)\left(5-x^{\frac{1}{3}}\right)$

46. $f(x) = 1.25x^{2.34}$

47. $f(x) = (x^{1.2} + x^{0.24})^2$

48. $f(x) = \dfrac{x^{0.4} - 1}{x^3}$

■ For each piecewise function $f(x)$, determine whether $f(x)$ is (a) continuous and (b) differentiable at each of its break points. If $f(x)$ fails to be continuous at a point, determine whether it is left or right continuous (or neither) at that point. If $f(x)$ fails to be differentiable at a point, determine whether it is left or right differentiable (or neither) at that point.

49. $f(x) = \begin{cases} -x, & \text{if } x \le -1 \\ \frac{1}{x}, & \text{if } x > -1 \end{cases}$

50. $f(x) - \begin{cases} x^{\frac{1}{3}}, & \text{if } x < 1 \\ 3x - 2, & \text{if } x \ge 1 \end{cases}$

51. $f(x) = \begin{cases} x - 1, & \text{if } x \le 1 \\ 2\sqrt{x}, & \text{if } x > 1 \end{cases}$

■ For each piecewise function $f(x)$, write $f'(x)$ as a piecewise function. (Pay special attention to whether or not $f'(x)$ exists at each of the break points of $f(x)$.)

52. $f(x)$ from Problem 49

53. $f(x)$ from Problem 50

54. $f(x)$ from Problem 51

55. $f(x) = \begin{cases} x^3, & \text{if } x < 1 \\ x, & \text{if } x \ge 1 \end{cases}$

56. $f(x) = \begin{cases} -x^2, & \text{if } x \le 0 \\ x^2, & \text{if } x > 0 \end{cases}$

57. $f(x) = \begin{cases} 1, & \text{if } x \le -1 \\ x^{\frac{2}{3}}, & \text{if } x > -1 \end{cases}$

■ For each piecewise function $f(x)$, find constants a and b such that $f(x)$ is continuous and differentiable everywhere (if possible).

58. $f(x) = \begin{cases} \sqrt{x} + a, & \text{if } x \le 0 \\ bx + 3, & \text{if } x > 0 \end{cases}$

59. $f(x) = \begin{cases} 3x + a, & \text{if } x < 1 \\ x^{\frac{b}{2}}, & \text{if } x \ge 1 \end{cases}$

60. $f(x) = \begin{cases} ax^{-1}, & \text{if } x < 8 \\ x^2 + bx^{\frac{1}{3}}, & \text{if } x \ge 8 \end{cases}$

■ Each of the following limits represents the derivative $f'(c)$ of some power function $f(x)$ at some point c. Determine what $f(x)$ and c are for each problem, use

the power rule to find $f'(c)$, and use this information to calculate the limit.

61. $\lim\limits_{h\to 0} \dfrac{(2+h)^5 - 32}{h}$

62. $\lim\limits_{h\to 0} \dfrac{\frac{4}{(2+h)^2} - 1}{h}$

63. $\lim\limits_{h\to 0} \dfrac{(8+h)^{-\frac{1}{3}} - \frac{1}{2}}{h}$

64. $\lim\limits_{x\to 8} \dfrac{x^{-\frac{1}{3}} - \frac{1}{2}}{x - 8}$

65. $\lim\limits_{x\to 4} \dfrac{3\sqrt{x} - 6}{x - 4}$

66. $\lim\limits_{x\to 0} \dfrac{\left(x^{-\frac{3}{2}} + 4\right) - 4}{x}$

■ For each problem below, find a function $f(x)$ that has the given derivative and value. Check your answer by verifying that your function $f(x)$ indeed has the derivative and value given.

67. $f'(x) = 5\sqrt{x}, \quad f(0) = 1$

68. $f'(x) = 5\sqrt{x}, \quad f(1) = 5$

69. $f'(x) = \dfrac{1}{x^3}, \quad f(1) = 2$

70. $f'(x) = \dfrac{1}{\sqrt{x^3}}, \quad f(9) = 1$

71. $f'(x) - \frac{3}{2}x^{-\frac{4}{5}}, \quad f(1) = 2$

72. $f'(x) = \sqrt{x}\left(1 + x^{\frac{3}{2}}\right), \quad f(0) = 4$

Applications

■ In Problems 73–75 suppose an object moves in a straight line in such a way that its position from a marking flag t seconds after starting is $s(t)$ meters.

73. If the position of the object at time t is $s(t) = \frac{3.2}{x^5}$, find equations for (a) the velocity $v(t)$ and (b) the acceleration $a(t)$.

74. If the velocity of the object at time t is $v(t) = 3t^{-\frac{1}{2}}$ meters per second, and the initial position of the object is 100 meters from the marking flag, find equations for (a) the position $s(t)$ and (b) the acceleration $a(t)$.

75. If the acceleration of the object at time t is $a(t) = 1.3t^{0.24}$, the initial position of the object is 20 meters from the marking flag, and the initial velocity of the object is 2.5 meters per second, find equations for (a) the velocity $v(t)$ and (b) the position $s(t)$.

Proofs

76. Use the alternative definition of derivative to prove that $\frac{d}{dx}(x^{-6}) = -6x^{-7}$.

77. Use the alternative definition of derivative to prove that if k is any negative integer, then $\frac{d}{dx}(x^k) = kx^{k-1}$.

78. Prove that if k is any rational number greater than or equal to 1, then $f(x) = x^k$ is differentiable everywhere on its domain.

79. Prove that if $0 < k < 1$ is any rational number, then $f(x) = x^k$ is differentiable everywhere on its domain except at $x = 0$.

4.4   Graphs of Power Functions with Integer Powers

In this section we investigate the graphs of power functions $f(x) = Ax^k$ where k is an integer. The shape of the graph of such a function depends on the sign of k and its *parity* (i.e., whether k is even or odd).

4.4.1 The four types of integer power functions

Power functions $f(x) = Ax^k$ with integer powers (which we will call *integer power functions*) fall into four categories depending on the sign and parity of the power k. For the moment let's assume that the coefficient A is equal to 1. (We'll examine the effect of A in Section 4.4.4.)

The general shape of the graph of a power function $f(x) = x^k$ with integer power k depends only on whether k is odd or even and on whether it is positive or negative. All power functions with positive even powers have the same shape, all power functions with positive odd powers have the same shape, and so on.

THEOREM 4.10

Graphs of Positive Integer Power Functions

If $f(x) = x^k$ is a power function where $k \neq 1$ is an integer, then the graph of f has one of the four shapes in Figures 4.28–4.31, according to whether k is even or odd, and positive or negative.

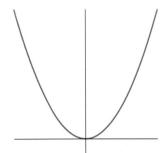 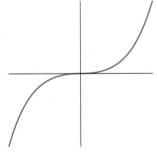 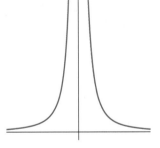 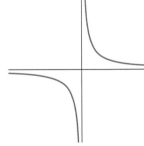

Figure 4.28 $k > 0$ even   **Figure 4.29** $k > 1$ odd   **Figure 4.30** $k < 0$ even   **Figure 4.31** $k < 0$ odd

For example, the first graph could be the graph of $f(x) = x^2$ or $f(x) = x^6$. The second graph could be the graph of $f(x) = x^5$ or $f(x) = x^{13}$. The third and fourth graphs could be the graphs of $f(x) = x^{-4}$ and $f(x) = x^{-7}$, respectively. Notice that in Figure 4.29, we omit the case where $k = 1$, since in that case the graph of $f(x) = x^1 = x$ is a line.

In Section 4.4.2 we will use our curve-sketching abilities, and what we know about limits and derivatives of power functions, to prove Theorem 4.10. Before we attempt to sketch graphs of power functions, let's investigate some basic properties of these four types of integer power functions. You will prove the following two theorems in the exercises.

THEOREM 4.11

Domains of Integer Power Functions

(a) If k is a positive integer, then the domain of $f(x) = x^k$ is $(-\infty, \infty)$.

(b) If k is a negative integer, then the domain of $f(x) = x^k$ is $(-\infty, 0) \cup (0, \infty)$.

THEOREM 4.12

Ranges of Positive Integer Power Functions

(a) If k is a positive even integer, then the range of $f(x) = x^k$ is $[0, \infty)$.

(b) If k is a positive odd integer, then the range of $f(x) = x^k$ is $(-\infty, \infty)$.

❓ Question Can you state and prove a version of Theorem 4.12 for power functions $f(x) = x^k$ where k is a *negative* even or odd integer? Use the graphs in Figures 4.28–4.31 for inspiration, and then prove your theorem algebraically.   □

Check that the graphs in Figures 4.28–4.31 do indeed have the domains and ranges predicted by Theorems 4.11 and 4.12. We prove the first part of Theorem 4.12 here and leave the rest of the proofs to the exercises.

PROOF (THEOREM 4.12) We wish to prove that the range of $f(x) = x^k$ is $[0, \infty)$ if k is a positive even integer. To find the range, we must consider what values an even power function $f(x) = x^k$ can have. Remember that a value y is in the range of $f(x)$ if there exists some x such that $f(x) = y$. Given any nonnegative y, we can choose $x = \sqrt[k]{y}$ and get

$$f(x) = f(\sqrt[k]{y}) = (\sqrt[k]{y})^k = (y^{\frac{1}{k}})^k = y^{\frac{k}{k}} = y^1 = y.$$

This means that any nonnegative value y is equal to $f(x)$ for some value x (namely $x = \sqrt[k]{y}$). This will happen *only* when y is nonnegative, since k is even and we can't take an even root of a negative number. Thus the range of $f(x) = x^k$ is $[0, \infty)$. ▪

It is also worth pointing out that power functions with even integer powers are even functions, and power functions with odd integer powers are odd functions (even for negative powers!). Look back at the graphs in Figures 4.28–4.31. The first and third graphs have even symmetry (and k is even), whereas the second and fourth graphs have odd symmetry (and k is odd). This is in fact one reason why even and odd functions have the names that they do. (See Section 1.6.5 if you need to review odd and even functions.)

THEOREM 4.13

Even and Odd Power Functions

(a) If k is an even integer, then $f(x) = x^k$ is an even function.

(b) If k is an odd integer, then $f(x) = x^k$ is an odd function.

We will prove part (a) and leave the proof of part (b) to the exercises.

PROOF (PART (A) OF THEOREM 4.13) Recall that $f(x)$ is an *even function* if $f(-x) = f(x)$ for all x in the domain of f. If k is an even integer, then $(-1)^k = 1$, and thus

$$f(-x) = (-x)^k = ((-1)x)^k = (-1)^k x^k = (1)x^k = x^k = f(x)$$

for all x in the domain of f. Therefore, $f(x) = x^k$ is an even function. ▪

✆ Caution The converses of parts (a) and (b) of Theorem 4.13 are *not* true. There are plenty of even functions that are not power functions and plenty of odd functions that are not power functions. Power functions simply provide the most basic examples of even and odd functions.   □

4.4.2 Curve sketching

In this section we will use curve-sketching techniques to prove Theorem 4.10. Recall that an incredible amount of information concerning the graph of a function f is given by the sign charts for f' and f''.

We will discuss two of the four types of graphs here (positive even powers and negative odd powers) and leave the other two for the exercises. We start by proving that for *any* positive even integer k, the graph of $f(x) = x^k$ has the basic shape shown in Figure 4.28. The proof is basically a detailed curve-sketching analysis of $f(x) = x^k$ that is valid for any positive even integer k.

PROOF (THEOREM 4.10 WHEN k IS A POSITIVE EVEN INTEGER)
First, suppose that $f(x) = x^k$ where k is a positive even integer. We wish to show that the graph of f must have the upward-pointing "U" shape shown in Figure 4.28. First we have to quantify what we mean when we refer to that "U" shape. One way to describe this shape mathematically is to say that the graph is concave up and positive everywhere and has a minimum at the point $(0, 0)$ where it changes from decreasing to increasing. This is exactly the sort of information we can find using the first and second derivatives of $f(x) = x^k$. We also want to be able to say that the "ends" of the graph of $f(x) = x^k$ both "point up." This we will be able to show by investigating the limit of f as $x \to \infty$ and as $x \to -\infty$.

Assume for the moment that $k > 2$ is an even positive integer. You will investigate the case where $k = 2$ in the exercises. The first and second derivatives of $f(x) = x^k$ are $f'(x) = kx^{k-1}$ and $f''(x) = k(k-1)x^{k-2}$, respectively. Since all three of these functions are power functions with positive integer powers, they are all zero at exactly one point, namely $x = 0$. Moreover, these three functions are all continuous; therefore, by the Intermediate Value Theorem, the only place where any of these three functions can change sign is at $x = 0$.

Consider first the function $f(x) = x^k$. Since k is even, f is positive for all nonzero values of x. Now consider the derivative function $f'(x) = kx^{k-1}$. The power of this function is odd, so x^{k-1} is negative for $x < 0$ and positive for $x > 0$. Since k is positive, this means that $f'(x) = kx^{k-1}$ is also negative for $x < 0$ and positive for $x > 0$. Finally, consider the second derivative $f''(x) = k(k-1)x^{k-2}$. The power $k-2$ is even, and thus x^{k-2} is positive for all nonzero values of x. The numbers k and $k-1$ are also both positive, and thus $f''(x) = k(k-1)x^{k-2}$ is positive for all nonzero x. This sign information about f, f', and f'' is summarized on the number lines in Figure 4.32.

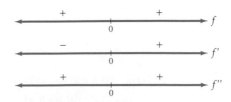

Figure 4.32

These number lines indicate the characteristics of the graph of *any* function $f(x) = x^k$ where k is positive and even. Since $f(x)$ is always positive, $f(x)$ is positive everywhere. Since $f'(x)$ is negative on $(-\infty, 0)$ and positive on $(0, \infty)$, we know that $f(x)$ is decreasing on $(-\infty, 0)$ and increasing on $(0, \infty)$. (Hence $f(x)$ has a minimum at $x = 0$.) Since $f''(x)$ is always positive, $f(x)$ is always concave up.

Note that the domain of $f(x) = x^k$ is all real numbers and that the range of $f(x) = x^k$ is $[0, \infty)$. It is also easy to check that $f(0) = (0)^k = 0$. With this basic information, and the derivative information we have gathered, we are *almost* through describing the graph of f.

The final pieces of information we need about $f(x) = x^k$ are the limits of f as $x \to \infty$ and as $x \to -\infty$. Since k is a positive even integer, we have $\lim\limits_{x \to \infty} x^k = \infty$ and $\lim\limits_{x \to -\infty} x^k = \infty$ (see Section 4.2). This means that both "ends" of the graph point "up." This completes our analysis of $f(x) = x^k$ where k is a positive even integer. ◾

? ***Question***   Go through the analysis in the proof above with $k = 4$ to make sure that you understand the arguments.   ☐

? ***Question***   Why did we exclude the case $k = 2$ in the proof above? What is it we said that is *not* true for $k = 2$? (*Hint:* It has to do with the second derivative.)   ☐

PROOF (**THEOREM 4.10 WHEN k IS A NEGATIVE ODD INTEGER**)
We now give a brief curve-sketching analysis for power functions of the form $f(x) = x^k$ where k is negative and odd. This time we wish to show that f has the general shape shown in Figure 4.31. How can we describe that shape mathematically? First, we want f to be positive, negative, increasing, decreasing, and concave up or down on the correct intervals. Second, the graph of f should have a vertical asymptote at $x = 0$ and a horizontal asymptote at $y = 0$. The first type of information will be found by examining the signs of f, f', and f'', the second type by investigating limits.

Again the first and second derivatives of f are $f'(x) = kx^{k-1}$ and $f''(x) = k(k-1)x^{k-2}$, but this time k and $k-2$ are *odd* and $k-1$ is even (and all of these numbers are negative). The functions f, f', and f'' are all continuous, never zero (why?), and undefined at $x = 0$. Therefore, by the Intermediate Value Theorem, the only place these functions can change sign is at $x = 0$. We obtain the number lines shown in Figure 4.33.

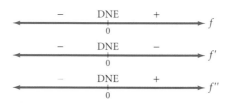

Figure 4.33

The graph of $f(x) = x^k$, where k is a negative odd integer, is negative for $x < 0$ and positive for $x > 0$, always decreasing, and concave down for $x < 0$ and concave up for $x > 0$. We also know that $f(x) = x^k$ has domain $x \neq 0$ and range $y \neq 0$, since k is negative and odd.

Finally, we examine any important limits. The limits at the "ends" (as $x \to \infty$ and as $x \to -\infty$) are always important, as are any limits where the function f does not exist (in this case at $x = 0$). Since k is a negative odd integer, we know that $\lim\limits_{x \to \infty} x^k = 0$ and $\lim\limits_{x \to -\infty} x^k = 0$ (see Section 4.2). Therefore, f has a horizontal asymptote at $y = 0$. As x approaches zero from the right, we have $\lim\limits_{x \to 0^+} x^k = \infty$. From the left, $\lim\limits_{x \to 0^-} x^k = -\infty$. Therefore, f has a vertical asymptote at $x = 0$. ◾

? ***Question***   Go through the analysis in the proof above with $k = -5$ to make sure that you understand the arguments.   ☐

4.4.3 The size of k

We have seen that the sign and the parity of the integer power k determine the general shape of a power function $f(x) = x^k$. Now we see that the *size* of k determines the "steepness" of the curve. For example, $f(x) = x^2$, $f(x) = x^4$, and $f(x) = x^6$ all have the same basic "U" shape, but the graph of $f(x) = x^4$ is steeper on the ends and flatter in the middle than the graph of $f(x) = x^2$, and the graph of $f(x) = x^6$ is steeper and flatter still.

Consider the graphs in Figures 4.34–4.37. In each of the four cases, the largest value of $|k|$ has the steepest and flattest curve.

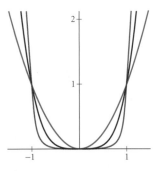

Figure 4.34
$y = x^2$, $y = x^4$, and
$y = x^6$

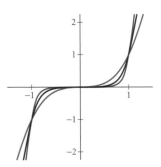

Figure 4.35
$y = x^3$, $y = x^5$, and
$y = x^7$

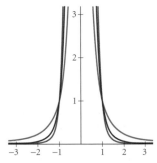

Figure 4.36
$y = x^{-2}$, $y = x^{-4}$, and
$y = x^{-6}$

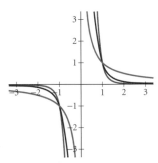

Figure 4.37
$y = x^{-1}$, $y = x^{-3}$, and
$y = x^{-5}$

For example, the blue curve in Figure 4.34 is the graph of $y = x^2$, the maroon curve is the graph of $y = x^4$, and the purple curve is the graph of $y = x^6$.

❓ Question One way to describe the difference between the graphs of, say, $f(x) = x^3$ and $f(x) = x^5$ is to say that $|x^5|$ is *less than* $|x^3|$ for $x \in (-1, 1)$, while $|x^5| > |x^3|$ for $x \in (-\infty, -1) \cup (1, \infty)$. Verify this by examining x^3 and x^5 for various values of x. ☐

It is interesting to note that *each* of the four graphs in Figures 4.34–4.37 goes through the point $(1, 1)$. This is because if $f(x) = x^k$ then $f(1) = 1^k = 1$, regardless of the sign or the parity of the power k. This means that none of the graphs is a vertical or horizontal stretch of another (why not?). For example, the graph of $y = x^4$ is *not* a vertical or horizontal stretch of the graph of $y = x^2$.

4.4.4 Transformations and the coefficient A

So far we have focused on power functions $f(x) = Ax^k$ where the coefficient A is equal to 1. What happens for other values of A? Recall from Section 1.6 that multiplying a function $f(x) = x^k$ by a positive constant A produces a vertical stretch or shrink of the graph of f. If A is negative, then the graph will also be reflected over the x-axis.

Consider the graphs in Figures 4.38–4.41. In each figure we see the graphs of $f(x) = x^k$, $f(x) = 0.5x^k$, and $f(x) = 2x^k$ for some value of k. In each case the graph of $f(x) = x^k$ is the "middle" graph, the graph of $f(x) = 2x^k$ is two times higher at each point x than the value of $f(x) = x^k$, and each point on the graph of $f(x) = 0.5x^k$ is half as high as the corresponding point on the graph of $f(x) = x^k$.

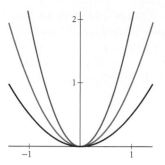

Figure 4.38
$y = x^2$, $y = 0.5x^2$, and $y = 2x^2$

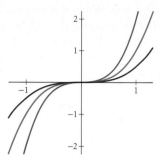

Figure 4.39
$y = x^3$, $y = 0.5x^3$, and $y = 2x^3$

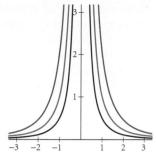

Figure 4.40
$y = x^{-2}$, $y = 0.5x^{-2}$, and $y = 2x^{-2}$

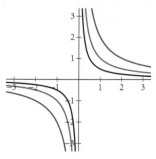

Figure 4.41
$y = x^{-1}$, $y = 0.5x^{-1}$, and $y = 2x^{-1}$

❓ *Question* Use your knowledge of transformations to sketch various graphs $y = Ax^k$ for negative values of A. Specifically, graph $y = -x^k$, $y = -2x^k$, and $y = -0.5x^k$ for $k = 2$, $k = 3$, $k = -2$, and $k = -1$, and compare them to the graphs in Figures 4.28–4.31. □

4.4.5 Graphing reciprocals

If k is a positive integer, how is the graph of $f(x) = x^{-k} = \frac{1}{x^k}$ related to the graph of $g(x) = x^k$? You can tell from the equations; the height of $f(x)$ at any point x is simply the reciprocal of the height of $g(x)$ at that point x. In other words, $f(x) = \frac{1}{g(x)}$. For example, whenever $g(x)$ is equal to 1, $f(x)$ will equal $\frac{1}{1} = 1$. Whenever $g(x)$ is large, $f(x)$ will be small; and whenever $g(x)$ is small, $f(x)$ will be large. Finally, if $g(x)$ is zero, then $f(x)$ will have a vertical asymptote, and vice versa.

Using these facts, we can sketch the graph of any function $f(x) = x^{-k}$ by graphing the reciprocal of the corresponding power function $g(x) = x^k$. This gives us another way to determine what the graph of a negative integer power function looks like.

EXAMPLE 4.26

Graphing power functions with negative integer powers

Use the graphs of $y = x$, $y = x^2$, and $y = x^3$ given in Figures 4.42–4.44 to sketch the graphs of their reciprocals $y = x^{-1}$, $y = x^{-2}$, and $y = x^{-3}$.

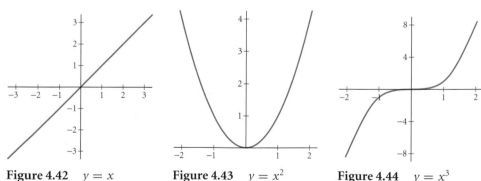

Figure 4.42 $y = x$ **Figure 4.43** $y = x^2$ **Figure 4.44** $y = x^3$

Solution Since all of the graphs in Figures 4.42, 4.43, and 4.44 are zero when $x = 0$, the graph of each of their reciprocals will have a vertical asymptote at $x = 0$. Since all of the graphs increase or decrease without bound as $x \to \infty$ and as $x \to -\infty$, their reciprocals will approach zero as $x \to \infty$ and $x \to -\infty$; therefore, the reciprocals will have horizontal asymptotes at $y = 0$. When the graphs in Figures 4.42–4.44 are large in magnitude, the graphs of their reciprocals will be small in magnitude, and vice versa. Finally, whenever the graphs in Figures 4.42–4.44 have height $y = 1$, the graphs of their reciprocals will also have height $y = 1$. The graphs of the original functions (in maroon), and their reciprocals $y = \frac{1}{x}$, $y = \frac{1}{x^2}$, and $y = \frac{1}{x^3}$ (in blue), are shown in Figures 4.45, 4.46, and 4.47.

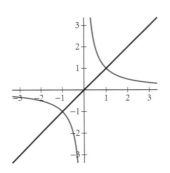 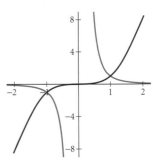

Figure 4.45
$y = x$ and $y = \frac{1}{x}$

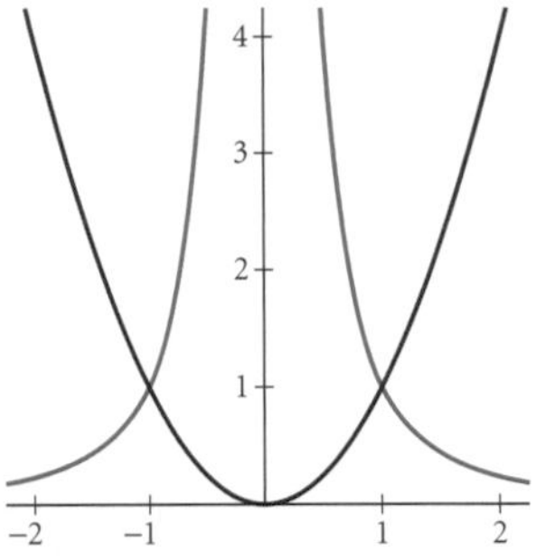

Figure 4.46
$y = x^2$ and $y = \frac{1}{x^2}$

Figure 4.47
$y = x^3$ and $y = \frac{1}{x^3}$ □

4.4.6 Converting between equations and graphs

Suppose $f(x) = Ax^k$ is a power function where k is an integer. Armed with the information in the previous sections, you should be able to sketch a graph of $f(x)$ by considering the effects of A and k. Conversely, given the graph of a function, you should be able to identify it as a power function and determine possible values for A and k. In other words, you should be able to graph a power function given its formula and to find the formula for a power function given its graph.

In general, the power k determines the basic shape of the graph as well as the "steepness" or "flatness" of that shape. Given the graph of a power function, you won't always be able to guess exactly what k is, but you should be able to determine whether k is odd or even and whether k is positive or negative.

Multiplying x^k by a coefficient A causes a vertical stretch or shrink by a factor of $|A|$ (and a reflection over the x-axis if $A < 0$). Note that the graph of $y = x^k$ (for *any* k) will always contain the points $(0, 0)$ and $(1, 1)$, but the graph of $y = Ax^k$ will go through the points $(0, 0)$ and $(1, A)$.

EXAMPLE 4.27

Quickly graphing power functions by hand

Sketch graphs of the following power functions without using a calculator. Explain how you arrived at your graph.

    (a) $f(x) = -2x^5$ **(b)** $f(x) = \dfrac{1}{4x^2}$

Solution

    (a)   Since $k = 5$ is a positive odd power, the graph of $f(x) = -2x^5$ will have a "wave" shape passing through the origin. Since $A = -2$, the graph of $f(x)$ will be the graph of x^5, reflected over the x-axis and vertically stretched by a factor of 2. Since the graph of x^5 passes through the point $(1, 1)$, the graph of $-2x^5$ will pass through the point $(1, -2)$. The graphs of x^5 and $f(x) = -2x^5$ are shown in Figures 4.48 and 4.49; note what happens to the marked point when we multiply x^5 by $A = -2$.

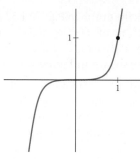

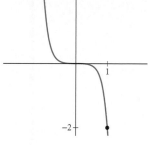

Figure 4.48 $y = x^5$ **Figure 4.49** $y = -2x^5$

(b) Here $f(x) = \frac{1}{4x^2} = \frac{1}{4}x^{-2}$ is a power function with $A = \frac{1}{4}$ and $k = -2$. Thus the graph of $f(x)$ is the graph of $y = x^{-2}$ shrunk vertically by a factor of 4. Figure 4.50 shows the graph of $y = x^{-2}$ with marked point $(1, 1)$. Figure 4.51 shows the graph of $y = \frac{1}{4}x^{-2}$; note that the marked point is now at $\left(1, \frac{1}{4}\right)$.

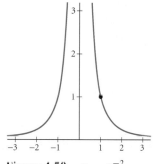

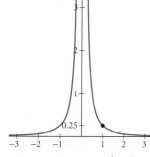

Figure 4.50 $y = x^{-2}$ **Figure 4.51** $y = \frac{1}{4}x^{-2}$

In Example 4.27, we graphed a power function given its equation. The following example illustrates how to find the equation of a power function given its graph.

EXAMPLE 4.28

Finding a power function given its graph

Find power functions $f(x) = Ax^k$ that could have the graphs in **(a)** Figure 4.52 and **(b)** Figure 4.53. (Note that the second graph has an additional marked point.)

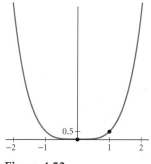

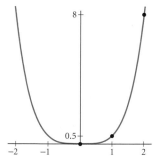

Figure 4.52   **Figure 4.53**

Solution

(a) From the shape of the graph we can tell that if $f(x)$ is a power function $f(x) = Ax^k$, then k must be an even positive integer. From the fact that the graph goes through $(1, 0.5)$ we know that $f(1) = 0.5$. Using this fact, we can solve for A:

$$0.5 = A(1)^k \implies 0.5 = A(1) \implies A = 0.5.$$

Thus the graph in Figure 4.52 could be the graph of $f(x) = 0.5x^2$, or of $f(x) = 0.5x^4$, or of $f(x) = 0.5x^6$, and so on. We don't have enough information to tell exactly which positive even integer k to use; *all* of these functions have the shape shown in Figure 4.52 and pass through the point $(1, 0.5)$.

(b) Figure 4.53 is the same as Figure 4.52 but with one more piece of information: The graph goes through the point $(2, 8)$. By part (a) we know that $f(x) = 0.5x^k$ where k is some positive even integer. Now that we know that $f(2) = 8$, we can determine which positive even integer to choose for k:

$$f(2) = 8 \implies 0.5(2)^k = 8 \implies 2^k = 16 \implies k = 4.$$

Therefore, Figure 4.53 could be the graph of $f(x) = 0.5x^4$.   □

♦ *Caution*   It is important to note that we can't even begin to find equations for the graphs in Figures 4.52 and 4.53 until we have decided what kind of function to use. In this case we are deciding in advance that we wish to find a *power* function that could have the given graph. Then we know the function is of the form $f(x) = Ax^k$, and we can concentrate on finding A and k.   □

The next example shows how to find the equation of a graph without knowing the height of the graph at $x = 1$.

EXAMPLE 4.29   **A more challenging example of finding a power function given its graph**

Find a power function that could have the graph in Figure 4.54.

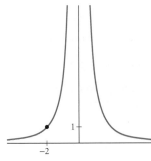

Figure 4.54

Solution   From the shape of the graph we know that if $f(x) = Ax^k$ is a power function, then k must be an even negative integer. Let's try the simplest possible case: $k = -2$. Then $f(x) = Ax^{-2}$ and $f(-2) = 1$ imply that

$$1 = A(-2)^{-2} = \frac{A}{(-2)^2} = \frac{A}{4} \implies A = 4.$$

Thus $f(x) = 4x^{-2}$ is a power function that could have the graph shown in Figure 4.54.   □

❓ *Question*   In Example 4.29 we found *one* possible function that could have the graph in Figure 4.54. Can you think of any others? (*Hint:* Does k really *have* to be equal to -2? What else could k be? Does that change the value of the coefficient A?)   □

> ## What you should know after reading Section 4.4
>
> ▶ The four possible shapes of a power function $f(x) = x^k$ according to the sign and parity of k, and how to describe each shape mathematically.
>
> ▶ The domains and ranges of the four types of integer power functions, and how to prove them algebraically. Which of the four types of integer power functions are odd functions? Even functions? How can you show this algebraically?
>
> ▶ How to use algebra, limits, derivatives, and curve-sketching skills to prove the graphical properties of various types of power functions with integer powers.
>
> ▶ The relationship between power functions with positive integer powers and those with negative integer powers. How to graph the reciprocal of a power function.
>
> ▶ How to make a good sketch of $f(x) = Ax^k$ for any power function with integer power.
>
> ▶ Given a graph, how to get a formula for an integer power function $f(x) = Ax^k$ that could have that graph.

Exercises 4.4

Concepts

0. Read the section and make your own summary of the material. Include whatever you think will help you review the material later on.

■ **Use mathematical language to express all of the graphical properties of each graph below.**

1.

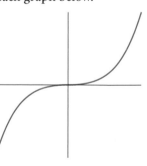

2.

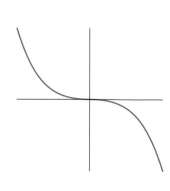

3.

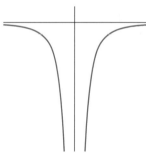

4. Why can't the graph below be the graph of a power function? (*Hint:* Think about the equation $y = Ax^k$.) Although the graph below is not a power function, it is *related* to the graph of some even power function. How?

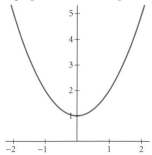

■ **For each type of function f described below, sketch a graph that illustrates the general shape that the graph of f must have.**

5. $f(x) = Ax^k$ where k is a positive even integer and A is a negative number.

6. $f(x) = Ax^k$ where k is any integer and $A = 0$.

7. $f(x) = Ax^k$ where $k = 1$ and A is any positive number.

8. $f(x) = Ax^k$ where k is a positive odd integer and A is a negative number.

9. $f(x) = Ax^{-k}$ where k is a positive even integer and A is a negative number.

10. $f(x) = Ax^{-k}$ where k is a positive odd integer and A is a negative number.

11. If k is an even negative integer, what is the range of $f(x) = x^k$? What if k is an odd negative integer?

12. Suppose that $f(x) = x^k$ where k is a positive integer, and that the graph of $f(x)$ passes through $(0, 0)$, $(1, 2)$, and $(-1.5, 3.375)$. If $g(x) = x^{-k}$ is the reciprocal of $f(x)$, what are $g(0)$, $g(1)$, and $g(-1.5)$?

13. The graphs of $y = x^2$, $y = 2x^2$, and $y = x^4$ are shown below. Determine which graph is which (without using your calculator).

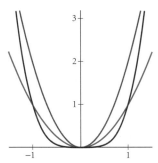

14. The graphs of $y = x^3$, $y = 0.5x^5$, and $y = x^5$ are shown below. Determine which graph is which (without using your calculator).

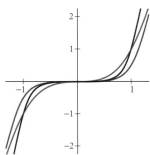

15. The graphs of $y = x^{-2}$, $y = 0.5x^{-2}$, and $y = x^{-4}$ are shown below. Determine which graph is which (without using your calculator).

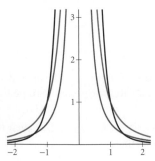

16. If $A > 0$, then $y = Ax^k$ is a vertical stretch or shrink of $y = x^k$ by a factor of A. In the case of power functions, this vertical stretch or shrink can also be thought of as a *horizontal* stretch or shrink. You may need to review Section 1.6 to do this problem.

   (a) If k is a positive integer, then $y = Ax^k$ is a horizontal stretch or shrink of $y = x^k$ by what factor?

   (b) If k is a positive integer, then $y = Ax^{-k}$ is a horizontal stretch or shrink of $y = x^{-k}$ by what factor?

17. Sometimes, knowing the general shape and two points (other than the origin) on the graph of a power function completely determines the power function. However, in certain special cases this is not the case. Sketch graphs with labeled points that satisfy the following criteria:

   (a) The graph of a power function $f(x) = Ax^k$ where k is even and positive, and two points besides the origin are labeled, but there is not enough information to determine both A and k.

   (b) The graph of a power function $f(x) = Ax^k$ where k is odd and negative, and two points besides the origin are labeled, but there is not enough information to determine both A and k.

Skills

■ **Sketch graphs of the following functions by hand. Do not use a calculator. Assume that A is positive and k is a positive integer. Identify and label the coordinates $(1, f(1))$ and $(-1, f(-1))$ on each graph.**

18. $f(x) = -0.2x^{-1}$ 19. $f(x) = \dfrac{2}{x^3}$

20. $f(x) = \dfrac{x^3}{2}$ 21. $f(x) = Ax^{-4}$

22. $f(x) = -Ax^3$ 23. $f(x) = 3x^k$, k odd

24. $f(x) = -2x^{-k}$, k even   25. $f(x) = -Ax^k$, k even

26. $f(x) = -Ax^{-k}$, k odd

■ **Determine algebraically whether the following functions are even functions or odd functions (or neither).**

27. $f(x) = -3x^2$ 28. $f(x) = 5x^{-3}$

29. $f(x) = x^2 + 1$ 30. $f(x) = (x + 1)^5$

31. $f(x) = 1.7x^4$ 32. $f(x) = -2x^{-1}$

■ **For each graph given below, find *two* power functions that could have that graph.**

33.

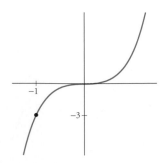

34.

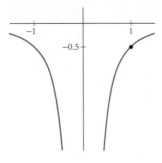

35.

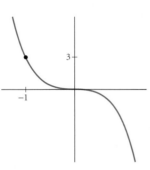

36.

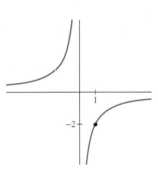

37.

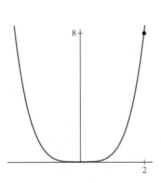

38.

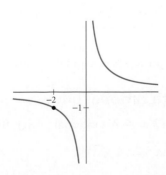

■ For each graph given below, find a power function $f(x) = Ax^k$ that could have that graph. (In these problems there is only *one* power function that will work.)

39.

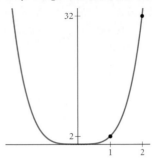

40.

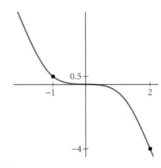

41.

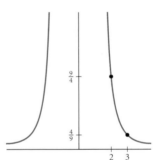

■ Find the reciprocal $\frac{1}{f(x)}$ of each power function $f(x)$. Write the reciprocal as a power function $g(x) = Bx^p$ for some B and p.

42. $f(x) = -2x^3$ **43.** $f(x) = 5x^{-2}$

44. $f(x) = Ax^k$

■ Sketch the reciprocal $\frac{1}{f(x)}$ of each graph $f(x)$ given below. Identify and label the height of $\frac{1}{f(x)}$ at the points $x = -2$, $x = -1$, $x = 0$, $x = 1$, and $x = 2$.

45.

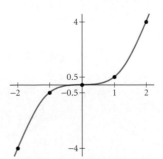

46.

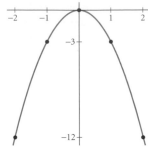

47.

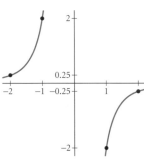

■ Use what you know about the graphs of integer power functions, and what you know about transformations of graphs, to graph each of the following functions quickly (do *not* use a calculator or find any derivatives). Label three points on each graph.

48. $f(x) = (x - 2)^2 + 3$ **49.** $f(x) = 3(x + 12)^5$

50. $f(x) = 2(x + 5)^{-3}$ **51.** $f(x) = \dfrac{1}{x - 2}$

52. $f(x) = \dfrac{3}{(x - 2)^4}$ **53.** $f(x) = 100 - 35(x - 13)^2$

54. $f(x) = \dfrac{1}{3x + 1}$ **55.** $f(x) = (3x + 1)^2 + 2$

56. $f(x) = 3(x + 2)^{-4} + 5$

■ For each function $f(x) = x^k$, do a complete curve-sketching analysis to determine the graph of f. Your analysis should include a set of number lines, a description of the domain and range of the function, and investigation of any important limits. Do *not* produce the graph from memory; show all work algebraically.

57. $f(x) = x^2$ **58.** $f(x) = x^4$

59. $f(x) = x^5$ **60.** $f(x) = x^{-1}$

61. $f(x) = x^{-5}$ **62.** $f(x) = x^{-4}$

Proofs

63. Prove that the domain of $f(x) = x^k$ is $(-\infty, \infty)$ if k is a positive integer, and is $(-\infty, 0) \cup (0, \infty)$ if k is a negative integer.

64. Prove that if k is a positive odd integer, then the range of $f(x) = x^k$ is $(-\infty, \infty)$.

65. Prove that if k is a negative even integer, then the range of $f(x) = x^k$ is $(0, \infty)$.

66. Prove that if k is a negative odd integer, then the range of $f(x) = x^k$ is $(-\infty, 0) \cup (0, \infty)$.

67. Prove the second part of Theorem 4.13: If k is an odd integer, then $f(x) = x^k$ is an odd function.

68. Prove that $x^6 \leq x^4$ for $|x| < 1$, and that $x^6 > x^4$ for $|x| > 1$. What does this say about the graphs of $y = x^4$ and $y = x^6$?

69. Do a curve-sketching analysis of $f(x) = x^k$ when k is a positive odd integer to show that the graph of such a function has the shape of the graph in Figure 4.29. In particular, list the properties you are attempting to prove, and include an investigation of the domain, range, and limits of such a function f and a set of number lines that describe the behavior of the graph of f.

70. Do a curve-sketching analysis of $f(x) = x^k$ when k is a negative even integer to show that the graph of such a function has the shape of the graph in Figure 4.30.

71. Suppose $f(x) = Ax^k$ is a power function where k is a positive or negative integer.

 (a) Show that if k is even, then $f(x)$ goes through the points $(1, A)$ and $(-1, A)$.

 (b) Show that if k is odd, then $f(x)$ goes through the points $(1, A)$ and $(-1, -A)$.

72. Prove that:

 (a) The graph of $y = x^4$ is *not* a vertical or horizontal stretch or shrink of the graph of $y = x^2$.

 (b) The graph of $y = x^{-5}$ is *not* a vertical or horizontal stretch or shrink of the graph of $y = x^{-3}$.

4.5 Graphs of Power Functions with Rational Powers

So far we have investigated power functions $f(x) = Ax^k$ where k is an integer. In this section we consider power functions whose powers are rational numbers $k = \frac{p}{q}$.

4.5.1 Six types of power functions with rational powers

Suppose $f(x) = Ax^k$ is a power function. If $k = \frac{p}{q}$ is a rational power, then

(1)
$$f(x) = Ax^k = Ax^{\frac{p}{q}} = A(\sqrt[q]{x})^p.$$

Throughout this section we assume that $k = \frac{p}{q}$ is a *reduced* rational number, that is, that p and q do not have any common factors. For the moment we will also assume that $\frac{p}{q}$ is positive and that $A = 1$.

THEOREM 4.14

Graphs of Power Functions with Positive Rational Powers

Suppose $\frac{p}{q}$ is a positive reduced rational number. The graph of $f(x) = x^{\frac{p}{q}}$ has one of the six shapes shown in Figures 4.55–4.60, according to the parity of p and q and whether $\frac{p}{q}$ is greater than 1 or less than 1.

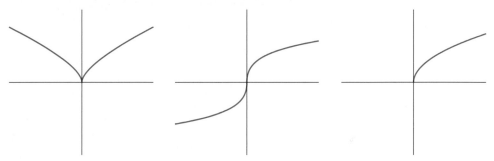

Figure 4.55
p even, q odd
$0 < \frac{p}{q} < 1$

Figure 4.56
p odd, q odd
$0 < \frac{p}{q} < 1$

Figure 4.57
p odd, q even
$0 < \frac{p}{q} < 1$

For example, the graphs of $y = x^{\frac{2}{3}}$, $y = x^{\frac{1}{3}}$, and $y = x^{\frac{1}{2}}$ have the shapes shown in Figures 4.55, 4.56, and 4.57, respectively.

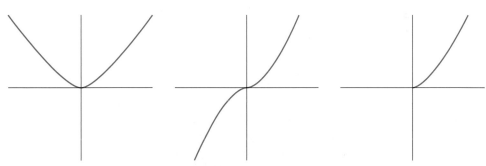

Figure 4.58
p even, q odd
$\frac{p}{q} > 1$

Figure 4.59
p odd, q odd
$\frac{p}{q} > 1$

Figure 4.60
p odd, q even
$\frac{p}{q} > 1$

For example, the graphs of $y = x^{\frac{4}{3}}$, $y = x^{\frac{5}{3}}$, and $y = x^{\frac{3}{2}}$ have the shapes shown in Figures 4.58, 4.59, and 4.60, respectively.

? Question Why don't the graphs above include the case where p and q are both even? (*Hint:* Can $\frac{p}{q}$ be reduced if both p and q are even?) □

? Question Even power functions (such as x^2 and x^4) are special cases of the graph in Figure 4.58, since any even integer k can be written as $\frac{k}{1}$ (where $p = k$ is even, $q = 1$ is odd, and $k = \frac{k}{1} > 1$). Similarly, odd power functions are a special case of the graph in Figure 4.59. Why? □

4.5.2 Graphical properties of power functions with rational powers

The following two theorems mathematically describe the graphs in Theorem 4.14 (except for the limits as $x \to \infty$ and $x \to -\infty$, which we covered in Section 4.2). The first theorem examines the domain, range, and symmetry of functions of the form $f(x) = x^{\frac{p}{q}}$. These properties depend only on the parity of p and q (i.e., on whether p and q are odd or even). In particular, if p is even, then only nonnegative numbers are in the range of $f(x) = x^{\frac{p}{q}}$ (why?). If q is even, then only nonnegative numbers are in the domain of $f(x) = x^{\frac{p}{q}}$ (why?). These facts can be used to prove the following theorem:

THEOREM 4.15

Symmetry, Domain, and Range of Power Functions

If $f(x) = x^{\frac{p}{q}}$ is a power function with positive reduced rational power $\frac{p}{q}$, then

(a) If p is even and q is odd, then $f(x) = x^{\frac{p}{q}}$ is an even function with domain $(-\infty, \infty)$ and range $[0, \infty)$.

(b) If p and q are both odd, then $f(x) = x^{\frac{p}{q}}$ is an odd function with domain $(-\infty, \infty)$ and range $(-\infty, \infty)$.

(c) If p is odd and q is even, then $f(x) = x^{\frac{p}{q}}$ is neither even nor odd and has domain $[0, \infty)$ and range $[0, \infty)$.

The proof of Theorem 4.15 is left to you in the exercises.

If we can determine the shape of the right-hand side of the graph of $f(x) = x^{\frac{p}{q}}$, then we can use the symmetry considerations in Theorem 4.15 to determine the left-hand side of the graph. As we see in Theorem 4.16, the right-hand side of the graph will be determined by whether $\frac{p}{q} < 1$ or $\frac{p}{q} > 1$.

THEOREM 4.16

The Right-Hand Side of the Graph of a Power Function

If $f(x) = x^{\frac{p}{q}}$ is a power function with positive reduced rational power $\frac{p}{q}$, then

(a) If $\frac{p}{q} < 1$, then the graph of $f(x) = x^{\frac{p}{q}}$ is positive, increasing, and concave down on $(0, \infty)$.

(b) If $\frac{p}{q} > 1$, then the graph of $f(x) = x^{\frac{p}{q}}$ is positive, increasing, and concave up on $(0, \infty)$.

The proof of Theorem 4.16 follows from looking at the signs of f, f', and f''. We will prove the first case of the theorem after we do a couple of examples. The proof of the second case is left for the exercises. By using Theorems 4.15 and 4.16, we can quickly determine the shape of the graph of $f(x) = x^{\frac{p}{q}}$ (without memorizing the six cases in Theorem 4.14).

EXAMPLE 4.30

Using Theorems 4.15 and 4.16 to graph a power function quickly

Consider the function $f(x) = x^{\frac{3}{4}}$. Since the exponent $\frac{3}{4}$ is less than 1, the graph of f is concave down in the first quadrant. Since 3 is odd and 4 is even, the function will not be defined for $x < 0$; this means that the graph of $f(x) = x^{\frac{3}{4}}$ must have the general shape of the graph in Figure 4.57. □

PROOF **(PART (A) OF THEOREM 4.16)** Suppose $f(x) = x^{\frac{p}{q}}$ is a power function with positive reduced rational power $\frac{p}{q}$, where $0 < \frac{p}{q} < 1$. Clearly f is positive on $(0, \infty)$. The derivative of f is $f'(x) = \frac{p}{q} x^{\frac{p}{q} - 1}$, and since $\frac{p}{q}$ is assumed to be positive, f' must also be positive on $(0, \infty)$ (and thus f is increasing on that interval). The second derivative of f is $f''(x) = \frac{p}{q}(\frac{p}{q} - 1)x^{\frac{p}{q} - 2}$; since by hypothesis $\frac{p}{q} < 1$, we know that $f''(x)$ is negative on $(0, \infty)$ (and thus f is concave down on that interval). ▪

If $\frac{p}{q}$ is less than 1, then the graph of $f(x) = x^{\frac{p}{q}}$ will fail to be differentiable at $x = 0$ (we showed this in Section 4.3.3, with $k = \frac{p}{q}$). In the case where $0 < \frac{p}{q} < 1$ and p is even while q is odd, the graph will have a "cusp" at $x = 0$ (see Figure 4.55). In the case where $0 < \frac{p}{q} < 1$ and p and q are both odd, the graph will have a vertical tangent line at $x = 0$ (see Figure 4.56). Finally, if $0 < \frac{p}{q} < 1$ and p is odd while q is even, the function $f(x) = x^{\frac{p}{q}}$ will have a right derivative of ∞, but its left derivative will not exist.

❓ *Question*   Use limits of the derivative $f'(x)$ to show that the function $f(x) = x^{\frac{2}{3}}$ has a vertical cusp at $x = 0$, whereas the function $g(x) = x^{\frac{1}{3}}$ has a vertical tangent line at $x = 0$. ☐

The size of the power $\frac{p}{q}$ determines the "steepness" of the graph of $f(x) = x^{\frac{p}{q}}$. The general rule is that the closer $\frac{p}{q}$ is to 1, the closer the graph of $f(x) = x^{\frac{p}{q}}$ will be to the line $y = x$. This means that if $\frac{p}{q}$ is large, the graph of $f(x) = x^{\frac{p}{q}}$ will be relatively steep (and curve rapidly away from the line $y = x$), whereas if $\frac{p}{q}$ is close to 1, then the graph of $f(x) = x^{\frac{p}{q}}$ will be straighter (closer to the graph of $y = x$). Finally, if $\frac{p}{q}$ is very close to zero (and positive), then the graph of $f(x) = x^{\frac{p}{q}}$ will curve downward fairly rapidly away from the graph of $y = x$. This is illustrated in the following example.

EXAMPLE 4.31

Distinguishing between graphs of functions with exponents of different sizes

Consider the four graphs shown in Figures 4.61 and 4.62. These are the graphs of $y = x^{\frac{1}{4}}$, $y = x^{\frac{3}{4}}$, $y = x^{\frac{5}{4}}$, and $y = x^{\frac{11}{4}}$. The dashed line in each figure is the line $y = x$. Determine which of the four graphs is which (without using a calculator).

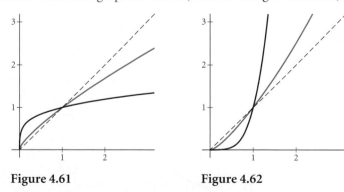

Figure 4.61   **Figure 4.62**

Solution Since the graphs in Figure 4.61 are concave down, they are the graphs of power functions whose exponents are less than 1. Thus $y = x^{\frac{1}{4}}$ and $y = x^{\frac{3}{4}}$ are the two graphs in Figure 4.61. Since $\frac{3}{4}$ is closer to 1 than $\frac{1}{4}$, the graph of $y = x^{\frac{3}{4}}$ is the one that is "straighter" and closer to the graph of $y = x$.

Now we know that the graphs in Figure 4.62 are $y = x^{\frac{5}{4}}$ and $y = x^{\frac{11}{4}}$ (these functions have powers that are greater than 1, since their graphs are concave up). Since $\frac{5}{4}$ is closer to 1 than $\frac{11}{4}$, the "straighter" graph that is closest to the line $y = x$ is the graph of $y = x^{\frac{5}{4}}$. □

So far, we have considered only power functions $f(x) = Ax^{\frac{p}{q}}$ where $A = 1$. If the coefficient A is something other than 1, then the graph of $f(x) = Ax^{\frac{p}{q}}$ will be a vertical stretch or shrink of the graph of $f(x) = x^{\frac{p}{q}}$ (in addition, if A is negative, then the graph will be reflected over the x-axis). For example, the graph of $f(x) = -2x^{\frac{2}{7}}$ will have the cusp shape shown in Figure 4.55 but will be reflected over the x-axis and stretched vertically by a factor of 2. The graph of $f(x) = -2x^{\frac{2}{7}}$ will go through the points $(0, 0)$, $(1, -2)$, and $(-1, -2)$ (why?). Check this using your graphing calculator.

⊗ *Caution*   So far, we have been writing all our rational powers k in fractional form, for example $f(x) = x^{\frac{3}{4}}$. We could also write $f(x)$ using decimal notation: $f(x) = x^{0.75}$. To graph a function written this way, it is often useful to convert the decimal notation to fraction notation. □

4.5.3 Negative rational powers

So far, we have examined power functions $f(x) = x^{\frac{p}{q}}$ where $\frac{p}{q}$ is positive. What happens when the power is negative? By properties of exponents, we have

(2)
$$f(x) = x^{-\frac{p}{q}} = \frac{1}{x^{\frac{p}{q}}}.$$

The function $f(x) = x^{-\frac{p}{q}}$ is the reciprocal of the function $g(x) = x^{\frac{p}{q}}$, because $f(x) = x^{-\frac{p}{q}} = \frac{1}{x^{\frac{p}{q}}} = \frac{1}{g(x)}$. As we saw in Section 4.4.5, we can easily sketch the reciprocal of a function whose graph we know.

EXAMPLE 4.32   **Graphing power functions with negative rational powers**

Use the graphs of $y = x^{\frac{3}{7}}$, $y = x^{\frac{7}{2}}$, and $y = x^{\frac{2}{7}}$ to sketch the graphs of their reciprocals $y = x^{-\frac{3}{7}}$, $y = x^{-\frac{7}{2}}$, and $y = x^{-\frac{2}{7}}$.

Solution   We know the basic shapes of the graphs of $y = x^{\frac{3}{7}}$, $y = x^{\frac{7}{2}}$, and $y = x^{\frac{2}{7}}$. We can sketch the graph of the reciprocal $f(x) = \frac{1}{g(x)} = x^{-\frac{p}{q}}$ of any of these functions by using the techniques of Section 4.4.5 (for example, when f is large, g is small, and vice versa). Figures 4.63–4.65 show each function (in maroon) graphed with its reciprocal (in blue).

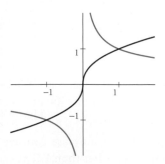

Figure 4.63
$y = x^{\frac{3}{7}}, y = x^{-\frac{3}{7}}$

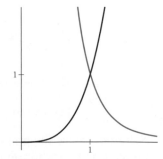

Figure 4.64
$y = x^{\frac{7}{2}}, y = x^{-\frac{7}{2}}$

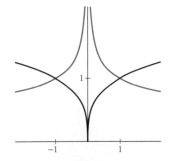

Figure 4.65
$y = x^{\frac{2}{7}}, y = x^{-\frac{2}{7}}$ □

❓ Question   It is possible to categorize the shapes of power functions of the form $x^{-\frac{p}{q}}$ much as we did for power functions with positive rational powers in Theorem 4.14. Do so by graphing the reciprocals of the graphs shown in Figures 4.55–4.60.   □

4.5.4 Inverses of power functions

Recall from Section 1.7 that the **inverse** f^{-1} of a one-to-one function f is the unique function $g(x)$ that satisfies

$$f(f^{-1}(x)) = x \text{ for all } x \in \text{Dom}(f^{-1}) \quad \text{and} \quad f^{-1}(f(x)) = x \text{ for all } x \in \text{Dom}(f).$$

Suppose $f(x) = x^k$ is a power function. If $f(x)$ is one-to-one, then it will have an inverse. Unfortunately, not all power functions are one-to-one.

❓ Question   Give some examples of functions $f(x) = x^k$ that are one-to-one and some examples that are not. What conditions on k will ensure that $f(x) = x^k$ is one-to-one if k is an integer? If $k = \frac{p}{q}$ is a rational power? Think about the shapes of the graphs of power functions for various integers k and rational powers $\frac{p}{q}$.   □

For example, $f(x) = x^2$ is *not* one-to-one (see Figure 4.66). However, if we restrict the domain to $x \geq 0$, then $f(x) = x^2$ *is* one-to-one and thus has an inverse, namely $f^{-1}(x) = \sqrt{x}$. Figure 4.67 shows $f(x) = x^2$ restricted to the domain $x \geq 0$. Figure 4.68 shows this restricted function and its inverse $f^{-1}(x) = \sqrt{x}$ (and the $y = x$ line over which they are symmetric).

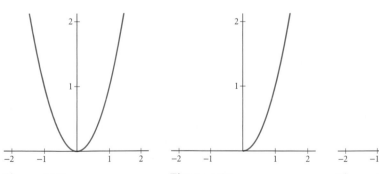

Figure 4.66
$f(x) = x^2$ is not one-to-one

Figure 4.67
Restricted domain

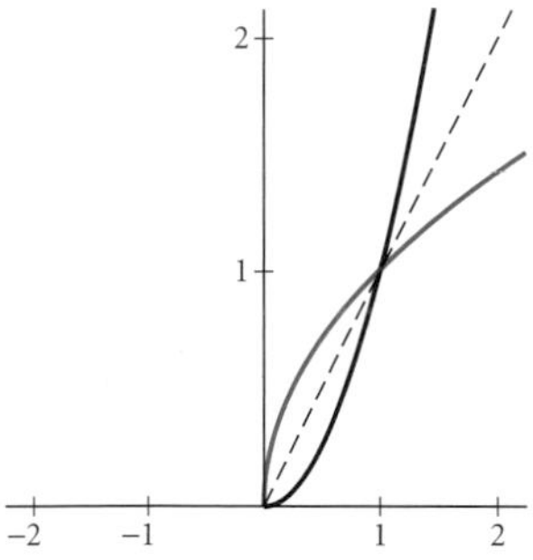

Figure 4.68
Now has inverse

◈ Caution   The inverse $f^{-1}(x)$ of a function $f(x)$ is *not* the same thing as its reciprocal $(f(x))^{-1} = \frac{1}{f(x)}$. For example, the inverse of $f(x) = x^2$ on the restricted domain $[0, \infty)$ is $f^{-1}(x) = x^{\frac{1}{2}} = \sqrt{x}$; note that this is *not* the same as the reciprocal $\frac{1}{f(x)} = \frac{1}{x^2} = x^{-2}$.   □

The following theorem states that the inverse of any power function $f(x) = x^k$ (possibly on a restricted domain) is $f^{-1}(x) = x^{\frac{1}{k}}$.

THEOREM 4.17

Inverses of Power Functions

Suppose $f(x) = x^k$ is a power function.

(a) If k is an odd integer or a reduced rational number $\frac{p}{q}$ where p is odd, then $f(x) = x^k$ is one-to-one and has inverse $f^{-1}(x) = x^{\frac{1}{k}}$.

(b) If k is an even integer or a reduced rational number $\frac{p}{q}$ where p is even, then $f(x) = x^k$ is *not* one-to-one. However, in this case the restriction of $f(x) = x^k$ to the domain $x \geq 0$ (or $x > 0$, if k is negative) *is* one-to-one, and its inverse is $f^{-1}(x) = x^{\frac{1}{k}}$.

❓ *Question*   Theorem 4.17 can be generalized to power functions $f(x) = Ax^k$ where $A \neq 1$. Does the value of A affect whether or not $f(x) = Ax^k$ is one-to-one? What is $f^{-1}(x)$ in this more general case?   □

We will prove that $f^{-1}(x) = x^{\frac{1}{k}}$ (possibly on a restricted domain) when k is an integer. The rest of the proof of Theorem 4.17 is left to the exercises.

PROOF   (**PART (B) OF THEOREM 4.17 WHEN k IS AN INTEGER**)   There are two possible cases, depending on whether k is odd or even. In either case, we will find the inverse of $f(x) = x^k$ by solving the equation $x^k = y$ for x.

If k is an odd integer, then $f(x) = x^k$ is one-to-one, and the calculation of $f^{-1}(x)$ is straightforward:

$$x^k = y \implies \sqrt[k]{x^k} = \sqrt[k]{y} \implies x = \sqrt[k]{y} = y^{\frac{1}{k}}.$$

Therefore, when k is odd, the inverse of $f(x) = x^k$ is $f^{-1}(x) = x^{\frac{1}{k}}$.

If k is an even integer, then the function $f(x) = x^k$ has an inverse only if we restrict the domain. We will restrict the domain to $[0, \infty)$. Solving the equation $x = y^k$ for y yields positive and negative even roots:

$$x^k = y \implies \sqrt[k]{x^k} = \sqrt[k]{y} \implies x = \pm\sqrt[k]{y}.$$

We are restricted to $x \geq 0$, so both x and $y = x^k$ are positive here. Therefore, taking the even kth root in the calculation above is defined. Moreover, since x is assumed to be positive, we need not consider $x = -\sqrt[k]{y}$. Thus we have $x = y^{\frac{1}{k}}$, and $f^{-1}(x) = x^{\frac{1}{k}}$.   ■

EXAMPLE 4.33

Finding the inverse of a power function

For each power function $f(x)$, determine whether $f(x)$ is one-to-one. If $f(x)$ is not one-to-one, then restrict the domain of $f(x)$ such that the restricted function is one-to-one. Then calculate the inverse of the (possibly restricted) function $f(x)$.

(a) $f(x) = -2x^4$ (b) $f(x) = x^{\frac{5}{3}}$

Solution

(a) Since $k = 4$ is even, $f(x)$ is not one-to-one; thus we restrict the domain of $f(x)$ to $x \geq 0$. (Note that we could also choose to restrict the domain to $x \leq 0$ if we wanted to.) For $x \geq 0$ (indeed, for *any* x), the function $y = -2x^4$ is *negative.* To find $f^{-1}(x)$, we solve the equation $-2x^4 = y$ for x:

$$-2x^4 = y \implies x^4 = -\tfrac{1}{2}y \implies x = \pm\sqrt[4]{-\tfrac{1}{2}y}$$

$$\implies x = \sqrt[4]{-\tfrac{1}{2}y} = \left(-\tfrac{1}{2}y\right)^{\frac{1}{4}}.$$

Note that since y is always negative, $-\tfrac{1}{2}y$ is always *positive,* and thus we can take its fourth root. Also note that since x is positive, we need not consider the negative of the fourth root at the end of the first line of the calculation. We have now shown that $f^{-1}(x) = \left(-\tfrac{1}{2}x\right)^{\frac{1}{4}}$. Since $f(x)$ has domain $[0, \infty)$ and range $(-\infty, 0]$, its inverse $f^{-1}(x)$ has domain $(-\infty, 0]$ and range $[0, \infty)$. In particular, $f^{-1}(x) = \left(-\tfrac{1}{2}x\right)^{\frac{1}{4}}$ is defined only for negative values of x. See Figures 4.69–4.71.

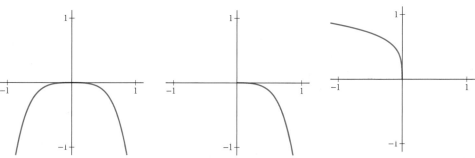

Figure 4.69
$f(x) = -2x^4$

Figure 4.70
$f(x) = -2x^4, \; x \geq 0$

Figure 4.71
$f^{-1}(x) = \left(-\tfrac{1}{2}x\right)^{\frac{1}{4}}$

(b) Since $\frac{p}{q} = \frac{5}{3}$ and $p = 5$ is odd, $f(x)$ is one-to-one. Theorem 4.17 tells us that $f^{-1}(x) = x^{\frac{3}{5}}$, but let's calculate this for ourselves by solving $x^{\frac{5}{3}} = y$ for x:

$$x^{\frac{5}{3}} = y \implies (\sqrt[3]{x})^5 - y \implies \sqrt[3]{x} = y^{\frac{1}{5}} \implies x = \left(y^{\frac{1}{5}}\right)^3 - y^{\frac{3}{5}}.$$

Since this calculation does not involve any even roots, we need not worry about signs. Thus $f^{-1}(x) = x^{\frac{3}{5}}$. Figures 4.72 and 4.73 show the graphs of $f(x)$ and $f^{-1}(x)$, respectively.

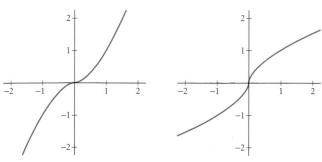

Figure 4.72
$f(x) = x^{\frac{5}{3}}$

Figure 4.73
$f^{-1}(x) = x^{\frac{3}{5}}$

4.5.5 Modeling graphs with general power functions

In Section 4.4.6 we saw how to find an integer power function that could have a given graph. In this section we extend those techniques to find a general power function $f(x) = Ax^{\frac{p}{q}}$ that could have a given graph. As before, the value of $f(x)$ at $x = 1$ will determine the value of A. The shape of the graph will give you a good guess for the value of $k = \frac{p}{q}$, specifically whether p and q are odd or even and whether $\frac{p}{q}$ is greater than or less than 1.

| EXAMPLE 4.34 | **Modeling graphs with power functions** |

Find a power function $f(x) = Ax^k$ that could have the graph in (**a**) Figure 4.74, and (**b**) Figure 4.75.

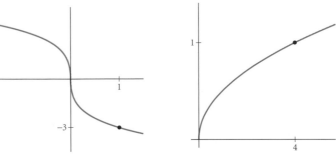

Figure 4.74   **Figure 4.75**

Solution

(**a**) By the shape of the graph we can tell that if $f(x)$ is a power function $f(x) = Ax^k$, then k is probably a rational power $\frac{p}{q}$. Since the graph is defined for all real numbers, q must be odd. Since the graph is odd (has rotational origin symmetry), p must also be odd. Clearly this is the graph of a power function with a *negative* coefficient A. Moreover, by the shape of the graph, we could guess that $\frac{p}{q} < 1$.

The simplest power $k = \frac{p}{q}$ that satisfies these conditions is $\frac{p}{q} = \frac{1}{3}$. Since $f(1) = -3$, we know that $A = -3$ (since $-3 = A(1)^{\frac{1}{3}} = A$). Thus $f(x) = -3x^{\frac{1}{3}}$ is a power function that could have the graph given in Figure 4.74.

(**b**) From the shape of the graph in Figure 4.75, if $f(x) = Ax^k$ is a power function, then k is a rational power $\frac{p}{q}$ where q is even (since the function is defined only for positive x). Since both p and q cannot be even (why?), p must be odd. Since the graph is eventually lower than the line $y = x$, we must have $k < 1$, so $p < q$.

The simplest rational number that satisfies these conditions is $\frac{p}{q} = \frac{1}{2}$. It now remains only to determine the value of the coefficient A. Assuming that $k = \frac{1}{2}$, the fact that $f(4) = 1$ tells us that

$$1 = A(4)^{\frac{1}{2}} = A\sqrt{4} = 2A \implies A = \tfrac{1}{2}.$$

Thus $f(x) = \frac{1}{2}x^{\frac{1}{2}}$ is a function that could have the graph shown in Figure 4.75. ☐

? Question Find another power function that could have the graph shown in Figure 4.74. Does this function have the same coefficient A that we found in Example 4.34? ☐

? Question Find another power function that could have the graph shown in Figure 4.75. ☐

? Question How did we know that p and q couldn't both be even in part (b) of Example 4.34? What would that say about $\frac{p}{q}$? ☐

What you should know after reading Section 4.5

▶ The general shape of a power function $f(x) = x^{\frac{p}{q}}$ with positive, reduced rational power $\frac{p}{q}$, depending on the parity of p and q and on whether $p < q$ or $p > q$. How p and q determine the domain, range, and symmetry of such a power function.

▶ How to make a good sketch of $f(x) = Ax^{\frac{p}{q}}$, including taking into effect the size and sign of the constant A, and how the power $\frac{p}{q}$ compares with the number 1. How to sketch the graph of a power function $f(x) = x^{-\frac{p}{q}}$ with a negative rational power, using the technique of graphing reciprocals.

▶ How to find inverses of power functions algebraically and graphically (including domain considerations). How to find a formula for a power function $f(x) = Ax^{\frac{p}{q}}$ that could have a given graph.

Exercises 4.5

Concepts

0. Read the section and make your own summary of the material. Include whatever you think will help you review the material later on.

1. Why do you think we require that the power $\frac{p}{q}$ of a power function $f(x) = Ax^{\frac{p}{q}}$ be a *reduced* rational number? If $\frac{p}{q}$ is reduced, can p and q both be even? Can they both be odd? Why or why not?

■ Use mathematical language to list all the graphical properties of the following graphs.

2.

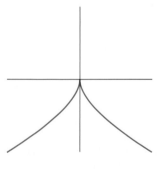

3.

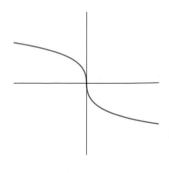

4.

■ For each type of A, p, and q described below, sketch the graph of $f(x) = Ax^{\frac{p}{q}}$.

5. $A > 0$, p odd, q odd, and $0 < \frac{p}{q} < 1$.

6. $A < 0$, p odd, q odd, and $\frac{p}{q} > 1$.

7. $A < 0$, p even, q odd, and $0 < \frac{p}{q} < 1$.

8. $A > 0$, p even, q odd, and $\frac{p}{q} > 1$.

9. $A < 0$, p odd, q even, and $0 < \frac{p}{q} < 1$.

10. $A < 0$, p odd, q even, and $\frac{p}{q} > 1$.

11. Give brief, intuitive explanations for the following statements concerning a power function $f(x) = Ax^{\frac{p}{q}}$, where $\frac{p}{q}$ is a positive reduced rational number.

(a) If q is even, then $f(x)$ will be defined only for $x \geq 0$.

(b) If p is even, then the range of $f(x)$ will have only nonnegative values.

(c) If q is odd and p is even, then $f(x)$ will be an even function.

(d) If q is odd and p is odd, then $f(x)$ will be an odd function.

12. Figures 4.58 and 4.59 in Section 4.5.1 have the same basic shapes as the even and odd power functions with integer powers that we studied in Section 4.4.1. In fact, even and odd power functions with integer powers are special cases of the power functions with rational powers shown in Figures 4.58 and 4.59. Explain how and why.

13. The graphs of $y = x^{\frac{1}{3}}$, $y = 0.5x^{\frac{1}{3}}$, and $y = x^{\frac{1}{5}}$ are shown below. Determine which graph is which (without using your calculator).

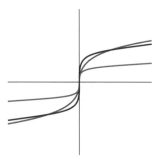

14. The graphs of $y = x^{\frac{2}{3}}$, $y = 2x^{\frac{2}{3}}$, and $y = 2x^{\frac{2}{5}}$ are shown below. Determine which graph is which (without using your calculator).

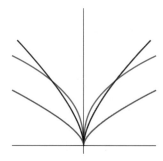

15. The graphs of $y = x^{\frac{3}{8}}$, $y = x^{\frac{5}{8}}$, and $y = x^{\frac{11}{8}}$ are shown below. Determine which graph is which (without using your calculator).

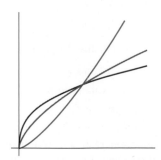

16. When will a power function $f(x) = x^k$ fail to be differentiable at $x = 0$? (*Hint:* Consider what things can happen when $k = \frac{p}{q}$ and what things can happen when $k = -\frac{p}{q}$).

17. Write an analogue of Theorem 4.14 for *negative* reduced rational powers $-\frac{p}{q}$. Include pictures of the six possible shapes that a function $f(x) = x^{-\frac{p}{q}}$ could have.

Skills

■ Sketch quick graphs of the following functions by hand. Do not use a calculator. Identify and label the coordinates $(1, f(1))$ and $(-1, f(-1))$ on each graph.

18. $f(x) = x^{\frac{2}{5}}$ 19. $f(x) = -3x^{\frac{5}{3}}$

20. $f(x) = 0.5x^{\frac{4}{5}}$ 21. $f(x) = 2x^{\frac{4}{3}}$

22. $f(x) = x^{0.7}$ 23. $f(x) = 4.2x^{1.25}$

24. $f(x) = x^{-\frac{2}{3}}$ 25. $f(x) = \frac{1}{3}x^{-\frac{3}{7}}$

26. $f(x) = -1.3x^{-0.25}$

■ Determine algebraically whether the following functions are even functions or odd functions (or neither).

27. $f(x) = 5x^{-\frac{3}{2}}$ 28. $f(x) = \frac{1}{3}x^{\frac{5}{3}}$

29. $f(x) = -3x^{\frac{2}{3}}$ 30. $f(x) = 2x^{-1.25}$

31. $f(x) = -0.25x^{-\frac{5}{7}}$ 32. $f(x) = 4x^{0.81}$

■ Find the reciprocal $\frac{1}{f(x)}$ of each power function $f(x)$. Write the reciprocal as a power function Cx^r for some constants C and r, and list the domains and ranges of $f(x)$ and $\frac{1}{f(x)}$.

33. $f(x) = -2x^{\frac{1}{3}}$ 34. $f(x) = 3x^{\frac{2}{5}}$

35. $f(x) = 0.5x^{1.75}$

■ For each power function below, determine whether $f(x)$ is one-to-one. If $f(x)$ is not one-to-one, then restrict the domain of $f(x)$ such that the restricted function is one-to-one. Then calculate the inverse of the (possibly restricted) function $f(x)$. When possible, write the inverse as a power function Dx^s for some constants D and s. List the domains and ranges of $f(x)$, restricted $f(x)$ (when appropriate), and $f^{-1}(x)$.

36. $f(x) = x^4$ 37. $f(x) = x^{-3}$

38. $f(x) = x^{-\frac{1}{2}}$ 39. $f(x) = -2x^{\frac{1}{3}}$

40. $f(x) = 3x^{\frac{2}{5}}$ 41. $f(x) = 0.5x^{1.75}$

■ For each graph given below, find *two* power functions that could have that graph.

42.

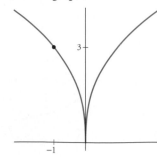

43.

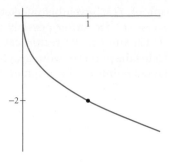

44.

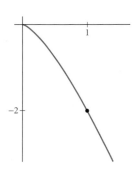

45.

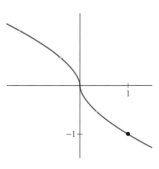

46.

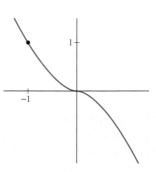

47.

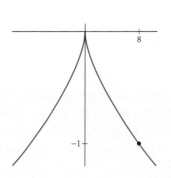

■ For each graph given below, find a power function $f(x) = Ax^k$ that could have that graph. (In these problems there is only *one* power function that will work.)

48.

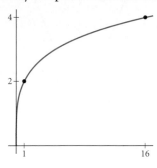

49.

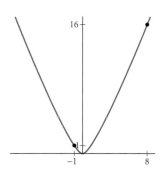

50.

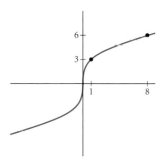

■ Use what you know about the graphs of power functions, and what you know about transformations of graphs, to graph each of the following functions quickly. (Do not use a calculator, and do not take any derivatives.)

51. $f(x) = 2x^{\frac{1}{3}} + 1$

52. $f(x) = 3 - \sqrt[4]{x+1}$

53. $f(x) = 3(x+4)^{\frac{4}{3}} + 2$

54. $f(x) = (x-1)^{-\frac{3}{2}}$

55. $f(x) = (3x+1)^{\frac{5}{3}}$

56. $f(x) = 2x^{-\frac{2}{5}} + 3$

57. In this problem we will investigate the differentiability of $f(x) = x^{\frac{6}{5}}$. Judging on the basis of the graph on the next page, does this function appear to be differentiable at $x = 0$?

   (a) Use your calculator to graph $f(x) = x^{\frac{6}{5}}$, and zoom in a few times centered at the origin. Now do you think the function is differentiable at $x = 0$?

   (b) Use the definition of derivative to show that $f(x) = x^{\frac{6}{5}}$ is differentiable at $x = 0$. What is $f'(0)$?

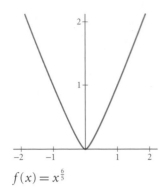

$$f(x) = x^{\frac{6}{5}}$$

58. In this problem we will investigate the differentiability of $f(x) = x^{\frac{2}{3}}$. Judging on the basis of the graph below, does this function appear to be differentiable at $x = 0$?

(a) Use your calculator to graph $f(x) = x^{\frac{2}{3}}$, and zoom in a few times centered at the origin. Now do you think the function is differentiable at $x = 0$?

(b) Use the definition of derivative to show that $f(x) = x^{\frac{2}{3}}$ is *not* differentiable at $x = 0$.

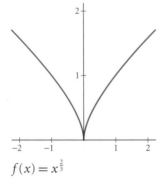

$$f(x) = x^{\frac{2}{3}}$$

59. In this problem we will investigate the differentiability of $f(x) = x^{\frac{1}{3}}$. Judging on the basis of the graph below, does this function appear to be differentiable at $x = 0$? What does the slope at $x = 0$ seem to be?

(a) Use your calculator to graph $f(x) = x^{\frac{1}{3}}$, and zoom in a few times centered at the origin. Now do you think the function is differentiable at $x = 0$? What does the tangent line look like?

(b) Use the definition of derivative to show that $f(x) = x^{\frac{1}{3}}$ is *not* differentiable at $x = 0$. Discuss how this function is nondifferentiable in a different way from the function in Problem 58.

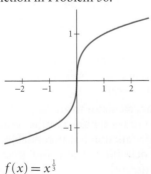

$$f(x) = x^{\frac{1}{3}}$$

■ For each graph of $f(x)$ given below, sketch the graphs of (a) the inverse $f^{-1}(x)$ (after possibly restricting the domain of $f(x)$) and (b) the reciprocal $\frac{1}{f(x)}$ of $f(x)$. Identify and label the points on your graphs corresponding to the marked points on the graph of $f(x)$.

60.

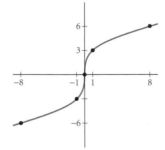

61.

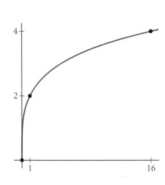

62.

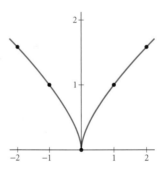

■ Do a complete curve-sketching analysis of each function below. Your analysis should include discussions of domain, range, and symmetry, sign intervals for three number lines, and any important limits.

63. $f(x) = x^{\frac{3}{5}}$ **64.** $f(x) = x^{\frac{3}{4}}$

65. $f(x) = x^{\frac{4}{3}}$ **66.** $f(x) = x^{\frac{5}{2}}$

67. $f(x) = x^{\frac{2}{7}}$ **68.** $f(x) = x^{\frac{7}{3}}$

Proofs

69. Prove that the domain of a power function $f(x) = x^{\frac{p}{q}}$ is $(-\infty, \infty)$ if q is odd, and $[0, \infty)$ if q is even.

70. Prove that the range of a power function $f(x) = x^{\frac{p}{q}}$ is $(-\infty, \infty)$ if p and q are both odd, and $[0, \infty)$ otherwise.

71. Prove that a power function $f(x) = x^{\frac{p}{q}}$ is an even function if p is even and q is odd, is an odd function if p and q are both odd, and is neither odd nor even if q is even.

72. Prove the second part of Theorem 4.16: If $f(x) = x^{\frac{p}{q}}$ is a power function with positive reduced rational power $\frac{p}{q}$, and if $\frac{p}{q} > 1$, then the graph of $f(x) = x^{\frac{p}{q}}$ is positive, increasing, and concave up on $(0, \infty)$.

73. Suppose $\frac{p}{q} > 0$ is a positive reduced rational number. Use the definition of derivative to prove that $f(x) = x^{\frac{p}{q}}$ is differentiable at $x = 0$ (from the right) only if $\frac{p}{q} > 1$.

74. Prove that a power function $f(x) = x^k$ is a one-to-one function only if k is an odd integer or a reduced rational number $k = \frac{p}{q}$ where p is odd.

75. Prove that if a power function $f(x) = x^k$ is not a one-to-one function, then its restriction to $[0, \infty)$ is a one-to-one function.

76. Prove that if $k = \frac{p}{q}$ is a reduced rational power (you may want to handle the positive and negative cases separately), then the inverse of $f(x) = x^{\frac{p}{q}}$ is $f^{-1}(x) = x^{\frac{q}{p}}$, possibly on a restricted domain.

CHAPTER **5**

Polynomial Functions

"Julia set" in a crop circle at Stonehenge. Steve Alexander/Temporary Temple Press Ltd, www. temporarytemples.co.uk

*P*olynomial functions are among the simplest and most common types of functions used in calculus and its applications. Since every polynomial is a sum of power functions with integer powers, we can quickly derive most properties of polynomial functions from what we know of power functions.

In this chapter we will learn how to factor polynomial expressions, including how to factor using synthetic division. We then apply the techniques of precalculus and calculus to investigate the local and global graphical behavior of polynomial functions. We conclude the chapter by using polynomial functions to introduce the concept of optimization.

CONTENTS

5.1   The Algebra of Polynomial Functions

Polynomial functions can often be factored into products of simple functions, and much of this section focuses on factoring techniques (including root-guessing and synthetic division).

5.1.1   What is a polynomial function?

A function f is a **polynomial function** if it is a finite sum of power functions with positive integer powers. In mathematical notation:

DEFINITION 5.1

Polynomial Functions

A **polynomial function** is a function that can be written in the form

$$f(x) = a_n x^n + a_{n-1} x^{n-1} + a_{n-2} x^{n-2} + \cdots + a_2 x^2 + a_1 x + a_0,$$

for some positive integer n and real numbers $a_0, a_1, \ldots, a_n$ with $a_n \neq 0$.

The numbers a_i (for $i = 0, 1, 2, \ldots, n$) are called the **coefficients** of the polynomial. Note that the coefficient belonging to the term containing the power x^i is conveniently named a_i; for example, the coefficient of the x^2 term is called a_2.

EXAMPLE 5.1

The coefficients of a polynomial function

Identify the coefficients $a_0, a_1, \ldots, a_n$ of the polynomial function $f(x) = 4x^2 - 3x + 2x^5 - 2$.

Solution   This is clearly a polynomial function, because it is a sum of power functions whose powers are positive integers. To identify the coefficients of f, we will first write f in the form of Definition 5.1; this means writing the terms in order from highest to lowest power of x, as well as noting that some powers have coefficients of zero:

$$\begin{aligned} f(x) &= 4x^2 - 3x + 2x^5 - 2 \\ &= 2x^5 + 4x^2 - 3x - 2 \\ &= 2x^5 + 0x^4 + 0x^3 + 4x^2 - 3x - 2. \end{aligned}$$

Therefore, the coefficients of f are $a_0 = -2$, $a_1 = -3$, $a_2 = 4$, $a_3 = 0$, $a_4 = 0$, and $a_5 = 2$. $\square$

The coefficient a_n belonging to the highest power of x is called the **leading coefficient,** and the term $a_n x^n$ containing the highest power of x is called the **leading term**. The coefficient a_0 is called the **constant term**. The integer n that represents the highest power of x with a nonzero coefficient is called the **degree** of the polynomial function.

EXAMPLE 5.2

Identifying polynomial functions

Which of the following functions are polynomial functions? For each polynomial function, identify the leading coefficient, leading term, degree, and constant term.

   (a) $f(x) = 3x^5 - 2x^3 + 4$ (b) $f(x) = x^2 + x^{-1} + 2$
   (c) $f(x) = (2x + 1)(x - 3)$ (d) $f(x) = x^2 + 5.2 - 1.7x^4$
   (e) $f(x) = x^7 - x^3 + 4x^7$ (f) $f(x) = 3x^4 + 4x^{1.7} - 5$

Solution   The functions in parts (b) and (f) are not polynomial functions. Although they *are* sums of power functions, those power functions do not all have positive integer powers. No amount of algebra can be used to write these functions in the form of Definition 5.1.

The functions in parts (a), (d), and (e) are clearly sums of power functions whose powers are integers; thus they are polynomial functions. The function in part (a) has leading coefficient 3, leading term $3x^5$, degree 5, and constant term 4. The function in part (d) has leading coefficient -1.7, leading term $-1.7x^4$ (this term has the highest power of x, although it is not the first term written), degree 4, and constant term 5.2.

The function in part (e) clearly has degree 7, but at the moment there are *two* terms involving the power x^7. Before we can identify the leading coefficient of f, we must put it in the standard form: $f(x) = x^7 - x^3 + 4x^7 = 5x^7 - x^3$. Now it is clear that f has leading coefficient 5 and leading term $5x^7$. The constant term of this polynomial is $a_0 = 0$.

The function in part (c) is a polynomial function because it *can* be written in the form in Definition 5.1 (although it is not written in that form at the moment). With a little bit of algebra we see that

$$f(x) = (2x + 1)(x - 3) = 2x^2 - 6x + x - 3 = 2x^2 - 5x - 3.$$

Thus the leading coefficient is 2, the leading term is $2x^2$, the degree is 2, and the constant term is -3.   □

We have special names for some polynomials according to their degrees. For example, a polynomial of degree 2 is called a **quadratic polynomial;** a polynomial of degree 3 is called a **cubic polynomial;** a polynomial of degree 4 is called a **quartic polynomial;** and a polynomial of degree 5 is called a **quintic polynomial.**

Note that if the degree of a polynomial function is 1, then that polynomial function is of the form $f(x) = a_1 x + a_0$ and is a **linear function.** Also note that if the degree of a polynomial function is 0, then that polynomial function is of the form $f(x) = a_0$, which is a **constant function** (this is also a type of linear function).

5.1.2 Roots and factors of polynomial functions

The **roots** or **zeros** of a polynomial function are the values at which $f(x) = 0$ (this is true of all functions, not just polynomial functions). Many of the calculations in this chapter will involve finding roots of polynomial functions. We will soon see that the derivatives of polynomial functions are themselves polynomial functions. This means that finding the local extrema and inflection points of a polynomial function f involves finding roots of the polynomial functions f' and f''.

The following theorem states that every polynomial can be split into linear and irreducible quadratic factors. A quadratic expression is **irreducible** if it cannot be factored with real number coefficients; that is, $ax^2 + bx + c$ is irreducible if it *cannot* be written in the form $A(x - r_1)(x - r_2)$ for some real numbers r_1 and r_2. A quadratic expression $ax^2 + bx + c$ is irreducible if and only if its **discriminant** $b^2 - 4ac$ is negative (think about the quadratic formula to see why). For example, the quadratic expressions $x^2 + 5$ and $x^2 + x + 7$ are irreducible (check this).

THEOREM 5.1

The Splitting Theorem

Every polynomial function can be written as a product

$$f(x) = Ag_1(x)g_2(x) \cdots g_r(x)$$

where A is a real number, and each g_i is either a linear factor of the form $x - c$ or an irreducible quadratic factor of the form $x^2 + bx + c$.

The proof of the Splitting Theorem is beyond the scope of this class, and we will not prove it here. However, we will use this theorem to prove the Fundamental Theorem of Algebra (Theorem 5.3), which states that every polynomial of degree n has at most n real roots.

Note that, for example, the polynomial function $f(x) = (x - 1)(2x + 3)$ is in the form of the Splitting Theorem only after we pull out a factor of 2: $f(x) = 2(x - 1)(x + \frac{3}{2})$. Similarly, the irreducible quadratic $f(x) = 3x^2 + x + 6$ can be written in the form $f(x) = 3(x^2 + \frac{1}{3}x + 2)$.

The Splitting Theorem means that *every* polynomial function of degree higher than 2 can be factored! However, just because a polynomial *can* be factored does not mean that it is easy, or even possible, to do so with the algebraic skills we have.

§ Caution   While it is true that all polynomial functions split into linear and quadratic factors, it is *not* necessarily true that those factors have to be "nice." In particular, the coefficients in these factors are not always going to be integers (although they often will be, since many exercises will be constructed to work out that way). For example, the polynomial function $f(x) = (x - 1.2)(x^2 + \sqrt{2})$ is written as a product of one linear factor and one irreducible quadratic factor, neither of which is particularly "nice." Most of our factoring skills will only help us find "nice"-looking factors, when they exist.   □

The roots of a polynomial function are related to the linear factors of that polynomial function:

THEOREM 5.2

The Factor Theorem

Suppose $f(x)$ is a polynomial function. A value $x = c$ is a root of f if and only if $(x - c)$ is a factor of f.

The Factor Theorem says that $x = c$ is a root of a polynomial function f if and only if f can be written in the form $f(x) = (x - c)g(x)$ for some polynomial function $g(x)$ (of degree one less than the degree of f). The proof of the Factor Theorem follows directly from the Splitting Theorem and is left to the exercises.

EXAMPLE 5.3   **The relationship between roots and factors of a polynomial**

The roots of the polynomial function $f(x) = (x + 1)(2x - 1)$ are $x = -1$ and $x = \frac{1}{2}$. Check the conclusion of the Factor Theorem by showing that $(x - (-1))$ and $(x - \frac{1}{2})$ are factors of f.

Solution   To show that $(x - (-1)) = (x + 1)$ is a factor of f, we need to show that f can be written in the form $f(x) = (x + 1)g(x)$ for some polynomial function $g(x)$. This is already the case, with $g(x) = 2x - 1$.

To show that $(x - \frac{1}{2})$ is a factor of f, we need to show that f can be written as $f(x) = (x - \frac{1}{2})g(x)$ for some polynomial $g(x)$. Since $(2x - 1) = 2(x - \frac{1}{2})$, simply let $g(x) = 2(x + 1)$. Then $f(x) = (x + 1)(2x - 1) = (x + 1)(2(x - \frac{1}{2})) = (x - \frac{1}{2})(2(x + 1)) = (x - \frac{1}{2})g(x)$.   □

The fact that every polynomial function splits into linear and irreducible quadratic factors means that every polynomial function of degree n has at most n real roots. This fact is so important that it is called the Fundamental Theorem of Algebra.

THEOREM 5.3

The Fundamental Theorem of Algebra

Every polynomial function of degree n has at most n real roots.

In fact, the more common statement of the Fundamental Theorem of Algebra says that every polynomial function of degree n has *exactly* n real or complex roots (counting "repeated" roots as separate roots). Since in this text we are concerned only with finding real roots, we will focus on the statement of the Fundamental Theorem of Algebra given in Theorem 5.3.

EXAMPLE 5.4

A cubic polynomial can have at most three real roots

Consider the following third-degree polynomial functions (in factored form). The polynomial $f(x) = (x - 2)(x - 3)(x - 4)$ has three distinct linear factors and three distinct real roots. On the other hand, $f(x) = (x - 2)(x - 3)^2$ has two distinct linear factors (one is repeated) and two distinct roots (one of which is a **repeated root**). Finally, the polynomial function $f(x) = (x - 2)(x^2 + 1)$ has one linear factor and one irreducible quadratic factor and thus has only one real root. (The quadratic factor contributes two complex roots; why?) These functions are graphed in Figures 5.1–5.3.

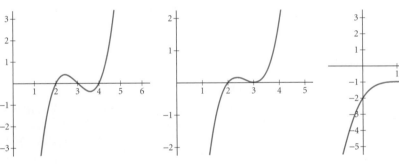

Figure 5.1
$f(x) = (x-2)(x-3)(x-4)$

Figure 5.2
$f(x) = (x-2)(x-3)^2$

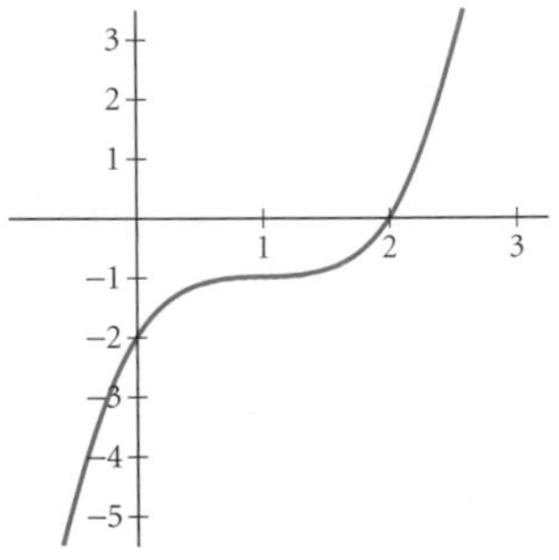

Figure 5.3
$f(x) = (x-2)(x^2+1)$ ☐

? Question   Can you think of a polynomial function that has *no* real roots? What does its graph look like? ☐

The proof of the Fundamental Theorem of Algebra relies on the fact that every polynomial function can be split into linear and irreducible quadratic factors.

PROOF   (THEOREM 5.3)   Let $f(x)$ be a polynomial function of degree n. By the Splitting Theorem, we can write f in the form

$$f(x) = Ag_1(x)g_2(x) \cdots g_r(x)$$

where A is a real number, and each function g_i is either linear (of the form $x - c$) or an irreducible quadratic (of the form $x^2 + bx + c$). How many of these factors g_i can there be? We'll answer this question in cases.

If *every* factor g_i happens to be linear, then there must be n of them, since each linear g_i is of degree 1, and the product of all the g_i's must be a polynomial of degree n. Since each of these linear functions $g_i(x) = x - c$ corresponds to a root $x = c$ of f, this means that in this case, f has *exactly* n real roots (counting repeated roots as separate roots).

On the other hand, if one or more of the g_i's are irreducible quadratics, then there must be fewer than n linear factors among the g_i's (or else the product of all the g_i's would have a degree higher than n). Each irreducible quadratic factor contributes only complex roots (why?), and each linear factor contributes one real root. Since there are fewer than n linear factors in this case, the polynomial f has fewer than n real roots. ∎

5.1.3 Factoring by grouping

Sometimes a polynomial function can be factored by **grouping** its terms together in a clever way. This technique makes use of the **distributive property** $(a + b)c = ac + bc$. For example, with the expressions $a = x^2$, $b = 5$, and $c = x - 3$, we have

$$(x^2 + 5)(x - 3) = x^2(x - 3) + 5(x - 3).$$

The following example uses grouping and the calculation above to factor a cubic polynomial.

EXAMPLE 5.5

Factoring by grouping

Factor the polynomial function $f(x) = x^3 - 3x^2 + 5x - 15$.

Solution If we group the first two terms and the last two terms together, we can factor each group and use the distributive property:

$$
\begin{aligned}
x^3 - 3x^2 + 5x - 15 &= (x^3 - 3x^2) + (5x - 15) &&\text{(grouping)}\\
&= x^2(x - 3) + 5(x - 3)\\
&= (x^2 + 5)(x - 3). &&\text{(distributive property)} \quad\square
\end{aligned}
$$

Factoring by grouping works only for *some* polynomials. The key is to figure out (if possible) how to group the terms of the polynomial so that each group has a common factor. Then, by the distributive property, you can "pull out" that common factor. Sometimes it takes a little trial and error to group the terms of a polynomial the right way; and of course, sometimes it isn't possible at all.

EXAMPLE 5.6

Using multiple factorization techniques

Find the roots of the polynomial function $f(x) = x^4 - 2x^3 - 4x^2 + 8x$.

Solution We need to find the solutions of the equation $x^4 - 2x^3 - 4x^2 + 8x = 0$. Before we attempt to factor by grouping, we can take out the common factor x from all terms. Luckily, we can find a grouping that works:

$$
\begin{aligned}
x^4 - 2x^3 - 4x^2 + 8x &= 0\\
x(x^3 - 2x^2 - 4x + 8) &= 0 &&\text{(common factor)}\\
x((x^3 - 2x^2) + (-4x + 8)) &= 0 &&\text{(group terms together)}\\
x(x^2(x - 2) - 4(x - 2)) &= 0 &&\text{(find common factor in each group)}\\
x(x^2 - 4)(x - 2) &= 0 &&\text{(distributive property)}\\
x(x + 2)(x - 2)(x - 2) &= 0 &&\text{(difference of squares)}\\
x(x + 2)(x - 2)^2 &= 0.
\end{aligned}
$$

The roots of the polynomial function $f(x) = x^4 - 2x^3 - 4x^2 + 8x$ are $x = 0$, $x = -2$, and $x = 2$ (repeated twice). $\square$

☑ *Checking the Answer* You can check these roots algebraically either by multiplying out your factorization (to check that it is indeed equal to f) or by checking that $f(x) = 0$ for $x = 0$, $x = -2$, and $x = 2$. You can also check the roots by graphing $y = f(x)$ and checking that it has roots only at $x = 0$, $x = -2$, and $x = 2$. The graph of $f(x) = x^4 - 2x^3 - 4x^2 + 8x$ is shown in Figure 5.4.

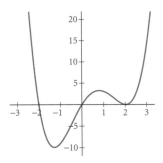

Figure 5.4
$f(x) = x^4 - 2x^3 - 4x^2 + 8x$

Note that the "double" root $x = 2$ looks different from the other roots of f. In Section 5.3.1 we'll discuss how to recognize repeated roots graphically.   □

◈ Caution   Don't forget about the factoring skills you learned in Chapter 0, such as the "reverse FOIL" or "trial-and-error" method for quadratics, the quadratic formula, and factoring formulas for differences of squares and for sums and differences of cubes.   □

5.1.4 Guessing integer roots of a polynomial function

Every polynomial can be split into linear and irreducible quadratic factors, although doing so is not always easy. If the polynomial is a reducible quadratic, it can always be factored by finding roots with the quadratic formula. There are also formulas and algorithms for factoring any polynomial of degree 3 or 4 (although those algorithms are extremely complicated). It is an amazing fact that there is no such formula that will always factor a polynomial of degree 5 or higher. In fact, mathematicians have proved that such a formula *cannot exist*! This doesn't mean that high-degree polynomials can't be factored, but it does mean that we don't have a formula or algorithm that will work every time, as we do for quadratics.

Even without general factoring or root-finding formulas, we can still factor many polynomials by being a little bit clever. Given a polynomial function f, sometimes we can "guess" a root $x = c$ (by finding some c for which $f(c) = 0$). Since every root $x = c$ corresponds to a factor $(x - c)$ of f, this means that $f(x) = (x - c)g(x)$ for some polynomial $g(x)$. Guessing a root gets us one step closer to factoring $f(x)$. Later, we will find integer roots of polynomials by "guessing and checking" and then use a process called synthetic division to find the function $g(x)$ that is left over after the factor $x - c$ is removed.

The following theorem provides a method for making intelligent guesses for those roots of a polynomial function that happen to be integers.

THEOREM 5.4

The Integer Root Theorem

Let $f(x) = a_n x^n + a_{n-1} x^{n-1} + \cdots + a_1 x + a_0$ be a polynomial function where each a_i is an integer and $a_0 \neq 0$. Then every integer root of f is a positive or negative divisor of the constant term a_0.

Given a polynomial function f with integer constant term a_0, the positive and negative divisors of a_0 are the only *possible* integer roots of f. We won't do a formal proof here, but the basic idea behind the proof is that the constant term of f is always the product of the constant terms in the factors of f, and each root $x = c$ of f corresponds to a factor $(x - c)$ of f.

◈ *Caution* Not every positive or negative factor of the constant term of a polynomial f is necessarily a root of f. For example, consider the function $f(x) = x^2 - 16$. The positive and negative factors of -16 are $1, -1, 2, -2, 4, -4, 8, -8, 16$, and -16. However, since $f(x) = x^2 - 16 = (x - 4)(x + 4)$, only $x = 4$ and $x = -4$ are roots of f. ☐

◈ *Caution* Theorem 5.4 helps us to guess the *integer* roots of f, if there are any. Of course, not all roots of a polynomial are necessarily integers. For example, $f(x) = x^2 - 2$ has roots $\sqrt{2}$ and $-\sqrt{2}$, neither of which is an integer; and the function $f(x) = 3x^2 + x - 2$ has roots $x = -1$ and $x = \frac{2}{3}$, only one of which is an integer. Theorem 5.4 won't help us guess noninteger roots of a polynomial function. ☐

EXAMPLE 5.7

Finding the integer roots of a polynomial function

According to Theorem 5.4, what are the possible integer roots of the polynomial function $f(x) = 3x^3 - 5x^2 - 4x + 4$? Test each possible root to see whether it is *actually* a root of this polynomial function.

Solution Theorem 5.4 says that the possible integer roots of $f(x) = 3x^3 - 5x^2 - 4x + 4$ are the positive and negative divisors of the constant term 4, namely $\pm 1, \pm 2$, and ± 4.

To test which, if any, of the six possible integer roots are actually roots of $f(x) = 3x^3 - 5x^2 - 4x + 4$, we evaluate

$$f(1) = 3(1)^3 - 5(1)^2 - 4(1) + 4 = -2,$$
$$f(-1) = 3(-1)^3 - 5(-1)^2 - 4(-1) + 4 = 0,$$
$$f(2) = 3(2)^3 - 5(2)^2 - 4(2) + 4 = 0,$$
$$f(-2) = 3(-2)^3 - 5(-2)^2 - 4(-2) + 4 = -32,$$
$$f(4) = 3(4)^3 - 5(4)^2 - 4(4) + 4 = 100,$$
$$f(-4) = 3(-4)^3 - 5(-4)^2 - 4(-4) + 4 = -252.$$

Therefore, the numbers $x = -1$ and $x = 2$ are roots of f, but the other divisors ($1, -2$, and ± 4) are not. Note that there may be other roots of f that are not integers. In fact, the polynomial function in this example has a non-integer root: It happens that $f(x) = (x + 1)(x - 2)(3x - 2)$, so $x = \frac{2}{3}$ is a root of $f(x)$. ☐

If *all* the roots of a polynomial function f happen to be integers, we can find *every* root of f using Theorem 5.4. In this case we can actually factor f completely using these roots, as the next example shows.

EXAMPLE 5.8

Factoring when all the roots of a polynomial function are known

Every root of the polynomial function $f(x) = 2x^3 + 2x^2 - 8x - 8$ is an integer. Use Theorem 5.4 to find these roots, and then use these roots to write f in factored form.

Solution   Since the constant term of $f(x) = 2x^3 + 2x^2 - 8x - 8$ is $a_0 = -8$, the possible integer roots of f are ± 1, ± 2, ± 4, and ± 8. By evaluating f at each of these values, one can easily see that $x = -1$, $x = 2$, and $x = -2$ are roots of f. Since f is a cubic polynomial, it can have at most three roots, so we have found *all* the roots of f. Since every root $x = c$ of f corresponds to a linear factor $(x - c)$ of f, this means we can write

$$f(x) = A(x + 1)(x - 2)(x + 2)$$

for some constant A. To find A we must multiply out $(x+1)(x-2)(x+2)$ and compare it to the original expression $f(x) = 2x^3 + 2x^2 - 8x - 8$. Since

$$(x + 1)(x - 2)(x + 2) = x^3 + x^2 - 4x - 4,$$

which differs from $f(x) = 2x^3 + 2x^2 - 8x - 8$ by a factor of 2, we know that $A = 2$. This means that $f(x) = 2x^3 + 2x^2 - 8x - 8$ can be written in factored form as

$$f(x) = 2(x + 1)(x - 2)(x + 2).$$

In this particular example, it would not have been hard to factor $f(x)$ by grouping; however, the method used here will be useful for many examples of polynomial functions that would otherwise be difficult to factor. □

? *Question*   Every root of the polynomial function $f(x) = x^3 - 3x^2 + 4$ is an integer. Using Theorem 5.4 and checking each possible root, you can determine that the integer roots of this function are $x = 2$ and $x = -1$ (do this). These are *all* the roots of f, since they are the only integer roots of f, and it was stated that f has only integer roots. How can this be? Write f in factored form using these roots. □

5.1.5 Synthetic division

If $x = c$ is a root of a polynomial function f, then $(x - c)$ is a factor of f. This means that f can be written as $f(x) = (x - c)g(x)$ for some polynomial function g. If we can find that polynomial g, then we have (partially) factored the polynomial f. If $f(x) = (x - c)g(x)$, then

$$g(x) = \frac{f(x)}{x - c}.$$

This means that to find the polynomial g we must divide the polynomial f by the linear factor $(x - c)$. This can be done by a method known as ***synthetic division.*** (It could also be done by "polynomial long division," which we will see in Section 6.1.4; but for division by linear factors, synthetic division can be faster.)

We present the algorithm in general first, and then illustrate what we mean with a few examples. We will see why this algorithm works in Section 6.1.4.

ALGORITHM 5.1

Synthetic Division

Let $f(x) = a_n x^n + a_{n-1} x^{n-1} + \cdots + a_1 x + a_0$ be any polynomial function, and let $x = c$ be any real number. To divide $f(x)$ by $(x - c)$, follow this algorithm:

Make a diagram like the one below with the coefficients a_i of the polynomial f and the number $x = c$. We will find new coefficients b_{n-1}, b_{n-2}, ..., b_1, b_0 as follows:

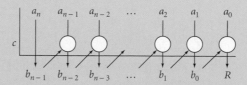

Bring the first coefficient a_n down to become the new coefficient b_{n-1}. Each diagonal arrow indicates multiplication by c, and each vertical arrow represents addition (of the term a_i and the term in the circle). Follow these arrows to find the remaining coefficients b_i and the remainder R. Let $g(x) = b_{n-1}x^{n-1} + b_{n-2}x^{n-2} + \cdots + b_1 x + b_0$.

If $x = c$ is a root of f, then the remainder R will be zero, and we will have

$$\frac{f(x)}{x-c} = g(x), \quad \text{that is,} \quad f(x) = (x-c)g(x).$$

On the other hand, if $x = c$ is not a root of f, then the remainder R will be nonzero, and we will have

$$\frac{f(x)}{x-c} = g(x) + \frac{R}{x-c}, \quad \text{that is,} \quad f(x) = (x-c)g(x) + R.$$

EXAMPLE 5.9

Using synthetic division

Use synthetic division to divide the polynomial $f(x) = x^3 - 3x^2 + 2$ by the linear expression $x - 2$.

Solution We have $a_3 = 1$, $a_2 = -3$, $a_1 = 0$ (since there is no "x" term), and $a_0 = 2$. We will use the synthetic-division algorithm with these coefficients and $c = 2$ (since we are dividing by the linear expression $x - 2$). When we start, our diagram looks like the one in Figure 5.5. We will follow the algorithm to fill in the blanks and circles in this diagram.

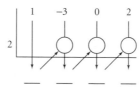   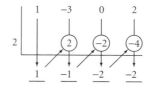

Figure 5.5   **Figure 5.6**

The result of following the synthetic-division algorithm is shown in Figure 5.6. We began by bringing down the coefficient 1 into the first blank. Then we multiplied this 1 by $c = 2$ and put the result in the first circle (following along the first diagonal line). Third, we added the two numbers in the second column (the coefficient -3 and the number 2 in the circle) to get -1 in the second blank. Repeating this algorithm, we can fill in the remaining blanks and circles.

What do we do with this calculation? By Algorithm 5.1, we now know that $f(x) = (x - c)g(x) + R$, where $g(x) = x^2 - x - 2$ and $R = -2$. Notice that in particular, $(x - 2)$ is *not* a factor of f (since there is a nonzero remainder), and thus $x = 2$ is *not* a root of f. In summary, we have just shown that

$$\frac{x^3 - 3x^2 + 2}{x - 2} = (x^2 - x - 2) + \frac{-2}{x - 2}.$$

We can rewrite this result by multiplying both sides of the equation by $x - 2$:

$$x^3 - 3x^2 + 2 = (x - 2)(x^2 - x - 2) - 2.$$

In this case $x = 2$ does not happen to be a root of $f(x)$. Most of the time we will apply synthetic division when we *know* that $x = c$ is a root of f (for example, after applying the Integer Root Theorem), and in those cases the remainder will be zero.   □

☑ *Checking the Answer*   To check the synthetic-division calculation in Example 5.9, simply multiply out $(x - c)g(x) + R$ and check that it is indeed equal to $f(x)$:

$$\begin{aligned}(x - c)g(x) + R &= (x - 2)(x^2 - x - 2) - 2 \\ &= x^3 - x^2 - 2x - 2x^2 + 2x + 4 - 2 \\ &= x^3 - 3x^2 + 2 = f(x).\end{aligned}$$

   □

◈ *Caution*   When using synthetic division, don't forget to include the coefficients that are equal to zero. Example 5.9 dealt with $f(x) = x^3 - 3x^2 + 2$, which has standard form $f(x) = x^3 - 3x^2 + 0x + 2$. If we leave out the $a_1 = 0$ coefficient in the synthetic-division algorithm, we will not get the correct answer.   □

By "guessing" roots and applying synthetic division, we can often factor complicated polynomials, as shown in the next example.

EXAMPLE 5.10

Factoring using the Integer Root Theorem and synthetic division

Factor the polynomial function $f(x) = 2x^4 + 7x^3 + x^2 - 7x - 3$ as much as possible.

Solution   By the Integer Root Theorem, we know that the only possible integer roots of this function are ± 1 and ± 3 (the integer divisors of the constant term -3). Evaluating $f(x)$ at these values, we see that the values $x = 1$, $x = -1$, and $x = -3$ are roots of f (since $f(1) = 0$, $f(-1) = 0$, and $f(-3) = 0$). Note that $f(3) = 336 \neq 0$, so $x = 3$ is not a root.

Since $x = \pm 1$ and $x = -3$ are roots of f, we know that $(x - 1)$, $(x + 1)$, and $(x + 3)$ are factors of f. We now use synthetic division with the root $x = -3$ (it would work just as well to choose one of the other roots at this point). We leave out the arrows and circles and include just the numbers; this is the most typical way to write synthetic-division computations (see Figure 5.7).

$$\begin{array}{r|rrrrr} & 2 & 7 & 1 & -7 & -3 \\ -3 & & -6 & -3 & 6 & 3 \\ \hline & 2 & 1 & -2 & -1 & 0 \end{array}$$

Figure 5.7

We now know that $f(x) = (x - c)g(x)$, where $c = -3$ and $g(x) = 2x^3 + x^2 - 2x - 1$. (Note that the remainder is zero here, since $x = -3$ is a root of f, and thus $x + 3$ is a factor of $f(x)$.) We have shown that

$$f(x) = 2x^4 + 7x^3 + x^2 - 7x - 3 = (x + 3)(2x^3 + x^2 - 2x - 1).$$

Now that we have managed to extract one linear factor from f, we concentrate on what is left, namely $2x^3 + x^2 - 2x - 1$. As shown in Figure 5.8, we can do another synthetic-division calculation to divide $2x^3 + x^2 - 2x - 1$ by $x - 1$ (since $x = 1$ is one of the other roots of f, and therefore also not a root of $2x^3 + x^2 - 2x - 1$).

$$
\begin{array}{r|rrrr}
 & 2 & 1 & -2 & -1 \\
1 & & 2 & 3 & 1 \\
\hline
 & 2 & 3 & 1 & 0
\end{array}
$$

Figure 5.8

Now we know that $2x^3 + x^2 - 2x - 1 = (x-1)(2x^2 + 3x + 1)$ (alternatively, we could have factored $2x^3 + x^2 - 2x - 1$ by grouping). Therefore, we have

$$
\begin{aligned}
f(x) &= 2x^4 + 7x^3 + x^2 - 7x - 3 \\
&= (x+3)(2x^3 + x^2 - 2x - 1) && \text{(the first synthetic division)} \\
&= (x+3)(x-1)(2x^2 + 3x + 1) && \text{(the second synthetic division)} \\
&= (x+3)(x-1)(2x+1)(x+1). && \text{(factoring a quadratic)}
\end{aligned}
$$

Note that not all of the roots of f are integers; that is why at the start of this problem we could identify the roots $x = \pm 1$ and $x = 3$ but not the root $x = -\frac{1}{2}$. □

☑ ***Checking the Answer*** You can check the answer simply by multiplying out the factorization $(x+3)(x-1)(2x+1)(x+1)$ to see that it is equal to the original expression $f(x) = 2x^4 + 7x^3 + x^2 - 7x - 3$. Alternatively, you could graph $f(x) = 2x^4 + 7x^3 + x^2 - 7x - 3$ and check that it has roots at $x = -3$, $x = 1$, $x = -\frac{1}{2}$, and $x = -1$. □

What you should know after reading Section 5.1

▶ The definition of a polynomial function, including notation for and definitions of the coefficients a_i, leading coefficient, leading term, degree, and constant term. The definitions of quadratic, cubic, quartic, and quintic polynomials.

▶ The relationship between roots and linear factors of polynomials, and the fact that every polynomial function can be factored into linear and irreducible quadratic factors. How can you tell whether a quadratic function is irreducible? What is the Fundamental Theorem of Algebra?

▶ How to use synthetic division and a combination of other factoring techniques to factor polynomials completely (when possible). How to use synthetic division to divide a polynomial by a linear expression $x - c$. What does it mean if the remainder is zero? Nonzero?

Exercises 5.1

Concepts

0. Read the section and make your own summary of the material. Include whatever you think will help you review the material later on.

■ For each function f, determine whether f is a polynomial function. If f is not a polynomial function, explain why not. If f is a polynomial function, determine the leading coefficient, the leading term, the degree, the constant term, and the coefficients a_1 and a_3. Indicate any

polynomial functions that are quadratic, cubic, quartic, or quintic.

1. $f(x) = 3x^4 - x^3 + 2x^2 - x + 1$

2. $f(x) = 1 - x^4$

3. $f(x) = 5x^2 - 3x^{1.7} + 2$

4. $f(x) = 3x + 1$

5. $f(x) = \pi x^{17} - x(x - 3)$

6. $f(x) = 4$

7. $f(x) = -3x^4 - 4x^{-1} - 2$

8. $f(x) = x(x+1)(x-2)^2$

9. $f(x) = 1.7x + \sqrt{2}x^3 - 44.1$

10. Consider the polynomial function
    $f(x) = 4x^3 + 11x^2 - 3x$.

    (a) Find the roots of this polynomial function.
    (b) For each root $x = c$ found above, verify that $(x - c)$ is a factor of f by writing f in the form $f(x) = (x - c)g(x)$ for some polynomial function $g(x)$.
    (c) Write $f(x)$ in the form $f(x) = A(x - b)(x - c) \cdot (x - d)$ for some real numbers A, b, c, and d.

11. Consider the polynomial function
    $f(x) = 2x^3 - 3x^2 + 8x$.

    (a) Factor this polynomial function as much as possible. You should find that there is one linear factor and one irreducible quadratic factor for this polynomial. Explain how you know that the quadratic factor is irreducible.
    (b) Write $f(x)$ in the form $f(x) = A(x - b)(x^2 + cx + d)$ for some real numbers A, b, c, and d.

■ Describe the meanings of the following terms. Be as precise and concise as possible.

12. Leading coefficient   13. Leading term

14. Constant term

15. Degree

16. Root

17. Factor

18. Quartic polynomial

19. Repeated root

20. Irreducible quadratic

21. State the Fundamental Theorem of Algebra.

22. What is the distributive property and what does it have to do with the method of factoring by grouping?

23. Verify that $x = -1$, $x = 1$, and $x = \frac{1}{2}$ are roots of the polynomial function $f(x) = 2x^4 - 3x^3 - x^2 + 3x - 1$. On the other hand, show that $x = 2$ and $x = -3$ are *not* roots of this function.

24. Suppose f is a polynomial with constant term $a_0 = -12$. List all the possible integer roots of f.

25. Suppose f is a polynomial with constant term a_0. If the only possible integer roots of f are ± 1, ± 2, ± 3, and ± 6, what is a_0? Is there only one possible answer?

■ For each problem below, give an example (or examples) of a polynomial function f that has the stated properties.

26. f is a quartic polynomial with no constant term.

27. f has leading coefficient -2, degree 7, and exactly four terms.

28. f has roots at $x = 0$, $x = 2$, and $x = -5$. Give three examples.

29. f is an irreducible quadratic. Give three examples.

30. f is a cubic polynomial with three real roots.

31. f is a quartic polynomial with no real roots.

32. f is a quintic polynomial with only one real root.

33. f has two distinct real roots, one of which is a repeated root.

34. f has two irreducible quadratic factors and three linear factors.

35. f has four real roots, only two of which are integers.

36. f has four real roots, only two of which are integers, but all of its coefficients are integers.

37. All of the roots of f are rational (but noninteger) numbers.

38. All of the roots of f are irrational numbers.

39. The only roots of the polynomial function $f(x) = 2x^3 - 6x^2 + 8$ are $x = -1$ and $x = 2$. Write $f(x)$ in the form of the Splitting Theorem.

40. Explain why a cubic polynomial function f can never have exactly two nonrepeated real roots. (*Hint:* Think about the factored form of f in the Splitting Theorem.)

41. What is synthetic division? Specifically, what is synthetic division used for, and how is it done?

42. Fill in the blank: If there is no remainder after synthetic division is used to divide a polynomial function f by a linear expression $x - c$, then $f(x) = $ _____ for some polynomial function g.

43. Fill in the blank: If there is a nonzero remainder R after synthetic division is used to divide a polynomial function f by a linear expression $x - c$, then $f(x) = $ _____ for some polynomial function g.

44. This problem concerns the proof of the Fundamental Theorem of Algebra in the text. In the problems below you will justify some of the steps in that proof. Suppose f is a polynomial function of degree n that has been written in the form $f(x) = Ag_1(x)g_2(x) \cdots g_r(x)$ guaranteed by the Splitting Theorem.

    (a) Suppose all of the factors g_i are linear. Explain why there must be n factors in this case (i.e., explain why $r = n$ here).
    (b) Now suppose that at least one of the factors g_i is an irreducible quadratic. Explain why there must therefore be less than n linear factors in the factorization of f.
    (c) Suppose one of the factors g_i is an irreducible quadratic. Explain why this factor will not contribute any real roots to f.

45. Write down a polynomial function that cannot be factored with the techniques we have covered up to this point. (*Hint:* Start with the factored form of the polynomial and then multiply it out.) Why can't we factor this polynomial? Does this mean that it is irreducible?

46. Suppose f is the polynomial function graphed below. Approximate values of k such that the translation $f(x) + k$ has the indicated number of roots. Sketch the graph of $f(x) + k$ for each value of k.

(a) One root   (b) Two roots   (c) Three roots
(d) Four roots   (e) Five roots

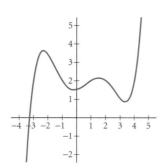

47. Suppose f is the polynomial function graphed below. Approximate values of k such that the translation $f(x) + k$ has the indicated number of roots. Sketch the graph of $f(x) + k$ for each value of k.

(a) No roots   (b) One root   (c) Two roots
(d) Three roots   (e) Four roots

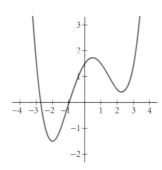

■ Determine whether each of the following statements is true or false. If a statement is true, explain why. If a statement is false, provide a counterexample.

48. True or False: Every power function is a polynomial function.

49. True or False: Every polynomial function is a power function.

50. True or False: Every polynomial function is a sum of power functions.

51. True or False: For any real numbers a and b, the polynomial $x^4 - ax^3 + bx + 7$ has four real roots.

52. True or False: If f is a polynomial function with constant term $a_0 = 0$, then $x = 0$ is a root of f.

53. True or False: If f is a polynomial function with constant term $a_0 = 3$, then $x = 3$ is a root of f.

54. True or False: Theorem 5.4 can be used to find all the roots of any polynomial function f.

Skills

■ For each polynomial function f, list all the possible integer roots of f. Then determine which of these possible roots are actually roots of the polynomial.

55. $f(x) = 2x^5 - 3x^2 + 3$

56. $f(x) = x^6 - 2x - 1$

57. $f(x) = 3x^4 - 2x^3 + 5x$

58. $f(x) = 28 - 3x^2 + 2x^3$

59. $f(x) = 3(x + 1)(x - 2) + 8$

60. $f(x) = x(x - 1) - x^2(x + 1)$

■ Each polynomial function f below has *only* integer roots. Find all these roots by "guessing and checking" using Theorem 5.4. Then use these roots to write f in factored form. (*Note:* Some roots may be repeated roots; in that case, you can determine which roots are repeated by trial and error, and by multiplying out your guesses.)

61. $f(x) = x^3 - 3x^2 - x + 3$

62. $f(x) = x^3 - 5x^2 - 2x + 24$

63. $f(x) = x^4 + 7x^3 + 9x^2 - 7x - 10$

64. $f(x) = x(x + 3) + 2$

65. $f(x) = x^3 - 5x^2 + 3x + 9$

66. $f(x) = x^4 - 6x^3 + 13x^2 - 12x + 4$

■ Factor each of the following polynomial functions as much as possible using factoring by grouping (and other factoring methods).

67. $f(x) = x^3 - 2x^2 + x - 2$

68. $f(x) = -8x^2 + 14x - 3$

69. $f(x) = x^3 + 8x^2 - 4x - 32$

70. $f(x) = 5 - x + 5x^4 - x^5$

71. $f(x) = 2x^3 + x^2 - 32x - 16$

72. $f(x) = 3(x^4 + 2) - x(3 + 6x^2)$

■ Write each polynomial function in the form $f(x) = Ag_1(x) g_2(x) \cdots g_r(x)$ for some real number A and some functions g_i that are either linear (of the form $x - c$) or irreducible quadratics (of the form $x^2 + bx + c$).

73. $f(x) = 2x^3 + 4x^2 - 6x$

74. $f(x) = 2x^2 - x + 1$

75. $f(x) = 2x^3 - 2x^2 + x$

76. $f(x) = 5 - 7x - 6x^2$

77. $f(x) = -6x^4 + 9x^3 + 6x^2$

78. $f(x) = 3x^3 - 2x^2 + 12x - 8$

■ Use synthetic division (and other factoring methods) to factor each of the following polynomial functions as much as possible. Check your answers by multiplying out your factorization.

79. $f(x) = x^3 - 2x^2 - 5x + 6$

80. $f(x) = 3x^3 - 8x^2 + 5x - 2$

81. $f(x) = x^3 + 4x^2 - 11x + 6$

82. $f(x) = x^4 - 8x^3 + 24x^2 - 32x + 16$

83. $f(x) = 3x^4 + 11x^3 + 15x^2 + 9x + 2$

84. $f(x) = 2x^4 + 6x^2 - 8$

85. $f(x) = 3x^4 + 7x^3 - 21x^2 - 6x - 8$

86. $f(x) = 2x^6 - 2x^5 - 10x^4 + 2x^3 + 16x^2 + 8x$

■ For each polynomial $f(x)$ and linear expression $l(x) = x - c$, **(a)** use synthetic division to find $\frac{f(x)}{l(x)}$. Then **(b)** use your answer to write $f(x)$ in terms of $x - c$ and the "remainder" (see Algorithm 5.1). Check your answers by multiplying out this expression for $f(x)$.

87. $f(x) = 2x^3 + 2x^2 - x - 5$, $l(x) = x - 3$

88. $f(x) = 2x^3 - 3x^2 + 8$, $l(x) = x - 2$

89. $f(x) = 1 - x^4$, $l(x) = x + 3$

90. $f(x) = 5x^4 - 3x + 2$, $l(x) = x + 1$

91. $f(x) = (x - 2)(x^5 - 3x^2) + 4$, $l(x) = x - 2$

92. $f(x) = x^6 + x^5 - 3x^3 - 2x^2 - 7x - 8$, $l(x) = x + 1$

■ There is an analogue of the Integer Root Theorem that finds *rational* roots of polynomials (sensibly called the

Rational Root Theorem):

(1) *If $f(x) = a_n x^n + a_{n-1}x^{n-1} + \cdots + a_1 x + a_0$, where each a_i is an integer and $a_0 \neq 0$, then every (reduced) rational root $\frac{p}{q}$ has the property that p is a positive or negative divisor of the constant term a_0, and q is a positive or negative divisor of the leading coefficient a_n.*

Use this more general theorem to guess the rational roots of each of the following polynomials, and then use synthetic division (which works for dividing a polynomial expression by $x - c$ even if c is not an integer) to factor these polynomials completely.

93. $f(x) = 2x^3 + x^2 + 8x + 4$

94. $f(x) = 6x^4 - 7x^3 + 8x^2 - 7x + 2$

95. $f(x) = 16x^3 - 12x^2 + 1$

96. $f(x) = 27x^4 - 18x^2 - 8x - 1$

Proofs

97. Prove the two implications of the Factor Theorem: If f is a polynomial function, then **(a)** if $x - c$ is a factor of f, then $x = c$ is a root of f, and **(b)** if $x = c$ is a root of f, then $x - c$ is a factor of f.

98. Suppose f is a polynomial function that factors as $f(x) = 3(x - a)(x^2 + cx + d)$, where $c^2 < 4d$. Prove that f has only one real root.

99. Prove that if f is a polynomial function whose constant term is $a_0 = 0$, then $x = 0$ is a root of f.

5.2   Limits and Derivatives of Polynomial Functions

In this section we will show that polynomial functions are always continuous, differentiable, and defined for all real numbers. Every property of polynomial functions will follow directly from the fact that polynomials are sums of power functions with positive integer powers.

5.2.1 Continuity of polynomial functions

We begin with a simple theorem that describes the domain and continuity of polynomial functions.

THEOREM 5.5

Continuity of Polynomial Functions
Every polynomial function is continuous and defined on all of $(-\infty, \infty)$.

This theorem is easy to prove using the fact that polynomials are sums of power functions with integer powers.

PROOF (**THEOREM 5.5**)   Let f be a polynomial function of the form

$$f(x) = a_n x^n + a_{n-1} x^{n-1} + \cdots + a_1 x + a_0.$$

By definition, f is a sum of power functions whose powers are nonnegative integers (namely, the power functions $a_n x^n$, $a_{n-1} x^{n-1}$, ..., $a_1 x$, and a_0). From Chapter 4 we know that power functions with nonnegative integer powers are defined for all real numbers. Since f is a sum of such power functions, it is also defined for all real numbers.

We know from Section 4.2.1 that every power function is continuous (on its entire domain). Moreover, we know that sums of continuous functions are themselves continuous functions. Therefore, f is a sum of continuous functions and is thus continuous everywhere on its domain $(-\infty, \infty)$. ∎

Recall that a function f is continuous at a point $x = c$ if and only if its limit as $x \to c$ is its value $f(c)$. We can always calculate limits of polynomial functions by evaluating at $x = c$ (since every polynomial function is continuous). If n is any positive integer and $a_0, a_1, \ldots, a_n$ are any real numbers, then for all real numbers c,

$$\lim_{x \to c}(a_n x^n + a_{n-1} x^{n-1} + \cdots + a_1 x + a_0) = a_n c^n + a_{n-1} c^{n-1} + \cdots + a_1 c + a_0.$$

For example, $\lim_{x \to 3}(2x^4 - 5x^2 - 8) = 2(3)^4 - 5(3)^2 - 8 = 109$. In general, if f is a polynomial function, then for any real number c, we have $\lim_{x \to c} f(x) = f(c)$.

5.2.2 Global behavior of polynomial functions

Since every polynomial function is continuous on $(-\infty, \infty)$, we know how to compute $\lim_{x \to c} f(x)$ for any polynomial function $f(x)$ and real number c. We now focus on the limits of polynomial functions as $x \to \infty$ and as $x \to -\infty$. These limits will determine part of the *global behavior* of polynomial functions, namely how polynomial functions behave at their "ends."

As the following theorem shows, the limit of a polynomial function f as $x \to \infty$ or as $x \to -\infty$ depends only on the sign of the leading coefficient and the parity of the degree of f.

THEOREM 5.6

Global Behavior of Polynomial Functions

Suppose $f(x) = a_n x^n + a_{n-1} x^{n-1} + \cdots + a_1 x + a_0$ is a polynomial function of degree n. If the leading coefficient a_n is positive, then

(a) If n is even, then $\lim_{x \to \infty} f(x) = \infty$ and $\lim_{x \to -\infty} f(x) = \infty$.

(b) If n is odd, then $\lim_{x \to \infty} f(x) = \infty$ and $\lim_{x \to -\infty} f(x) = -\infty$.

If the leading coefficient a_n is negative, then the signs above are reversed.

Figures 5.9–5.12 illustrate Theorem 5.6 graphically. The dashed part in the middle of each graph indicates that this theorem does not tell us the behavior of the graph in the "middle," only at the "ends" (the dashed part is just an example of what *might* happen in the middle part of the graph).

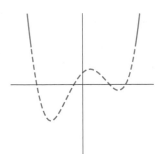

Figure 5.9
n even, $a_n > 0$

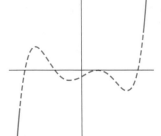

Figure 5.10
n odd, $a_n > 0$

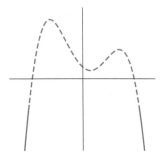

Figure 5.11
n even, $a_n < 0$

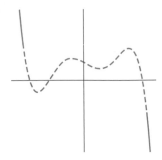

Figure 5.12
n odd, $a_n < 0$

The "ends" of the graph of an even-degree polynomial either both point up or both point down, whereas the "ends" of the graph of an odd-degree polynomial point in opposite directions. In particular, a polynomial function always increases or decreases without bound as $x \to \infty$ and as $x \to -\infty$; therefore, the graph of a polynomial function never has a horizontal asymptote.

EXAMPLE 5.11

Limits at the "ends" of a polynomial function

The polynomial function $f(x) = -3x^5 + 3x^2 + 10$ has leading coefficient $a_n = -3$ and degree $n = 5$. Since the degree is odd, the "ends" of the graph of f will point in opposite directions. Since the leading coefficient is negative, the height of the graph will approach infinity ("go up") at the left end and negative infinity ("go down") at the right end. In the language of limits,

$$\lim_{x \to -\infty} (-3x^5 + 3x^2 + 10) = \infty \quad \text{and} \quad \lim_{x \to \infty} (-3x^5 + 3x^2 + 10) = -\infty.$$

Figure 5.13 shows the graph of $f(x) = -3x^5 + 3x^2 + 10$. Verify that this graph does indeed have the global behavior predicted by Theorem 5.6.

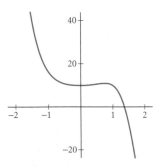

Figure 5.13
$f(x) = -3x^5 + 3x^2 + 10$

The behavior at the "ends" of a polynomial function f depends entirely on the leading term of that polynomial (the remaining terms determine the local behavior of f, as we will see in the next section). This is because for values of x with very large (positive or negative) magnitude, the leading term of a polynomial f is the "most powerful." In fact, as $x \to \infty$ and as $x \to -\infty$, a polynomial will behave more and more like its leading term. This will be the main argument in our proof of Theorem 5.6. Before we get to the proof, let's look at an example that illustrates the "power" of the leading term of a polynomial.

EXAMPLE 5.12

The global behavior of a polynomial is determined by its leading term

Consider the polynomial $f(x) = x^4 - x^3 - 11x^2 + 9x + 18$. The shape of the graph of f at the "ends" is determined by the leading term, x^4. Figures 5.14–5.16 show graphs of the function $y = x^4 - x^3 - 11x^2 + 9x + 18$ and its leading term $y = x^4$ (the dashed curve) in different viewing windows. The more we zoom out (i.e., as $x \to \infty$ or as $x \to -\infty$), the more the graph of the function f looks like the graph of $y = x^4$.

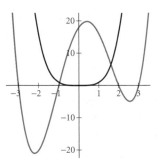

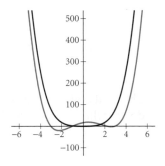

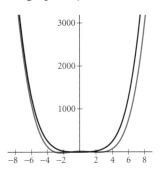

Figure 5.14
$y = x^4 - x^3 - 11x^2 + 9x + 18$ and $y = x^4$ on $[-3, 3]$

Figure 5.15
$y = x^4 - x^3 - 11x^2 + 9x + 18$ and $y = x^4$ on $[-6, 6]$

Figure 5.16
$y = x^4 - x^3 - 11x^2 + 9x + 18$ and $y = x^4$ on $[-8, 8]$

Why would the limit of a polynomial as $x \to \infty$ or as $x \to -\infty$ be difficult to compute? After all, a polynomial is just a sum of very simple power functions! The problem arises when we are forced to add or subtract "infinities"; for example, what is the limit as $x \to \infty$ of the function $f(x) = x^3 - 12x^2$? We know that as $x \to \infty$ we have $x^3 \to \infty$ and $-12x^2 \to -\infty$. However, we cannot simply add these two limits together to find the limit of $f(x)$, since limits of the form $\infty - \infty$ are indeterminate.

❓ ***Question*** Find the limits of $f(x) = x^2 - 2x^2$, $g(x) = 2x^2 - x^2$, and $h(x) = x^2 - x^2$ as $x \to \infty$ (each of these limits starts out in the form "$\infty - \infty$"; simplify first to find the limit). Use the results to show that limits of the form "$\infty - \infty$" are indeterminate. □

We will now prove Theorem 5.6 by showing that the limit of a polynomial function as $x \to \infty$ or as $x \to -\infty$ is determined entirely by the limit of its leading term.

PROOF (THEOREM 5.6) Suppose $f(x) = a_n x^n + a_{n-1} x^{n-1} + \cdots + a_1 x + a_0$ is any polynomial function. We will first show that the limit of this polynomial function as $x \to \infty$ is equal to the limit of its leading term as $x \to \infty$:

$$\lim_{x \to \infty} (a_n x^n + a_{n-1} x^{n-1} + \cdots + a_1 x + a_0) = \lim_{x \to \infty} (a_n x^n).$$

This is just a simple calculation; the key is to factor out the leading term:

$\lim_{x \to \infty} (a_n x^n + a_{n-1} x^{n-1} + \cdots + a_1 x + a_0)$

$= \lim_{x \to \infty} \left(a_n x^n \left(1 + \dfrac{a_{n-1} x^{n-1}}{a_n x^n} + \cdots + \dfrac{a_1 x}{a_n x^n} + \dfrac{a_0}{a_n x^n} \right) \right)$ (factor out leading term)

$= \lim_{x \to \infty} \left(a_n x^n \left(1 + \dfrac{a_{n-1}}{a_n x} + \cdots + \dfrac{a_1}{a_n x^{n-1}} + \dfrac{a_0}{a_n x^n} \right) \right)$ (algebra)

$= \left(\lim_{x \to \infty} a_n x^n \right) \left(\lim_{x \to \infty} \left(1 + \dfrac{a_{n-1}}{a_n x} + \cdots + \dfrac{a_1}{a_n x^{n-1}} + \dfrac{a_0}{a_n x^n} \right) \right).$ (limit rule)

Each of the quotients in the second limit approaches zero as $x \to \infty$ (why?), and thus the second limit is equal to 1. Continuing our calculation, the original limit is equal to

$$\left(\lim_{x \to \infty} a_n x^n \right)(1) = \lim_{x \to \infty} a_n x^n.$$

Technical Note: We applied the product rule for limits in this calculation, which we have shown to be valid only when all limits involved exist. In this example, some limits are infinite (and thus "do not exist"), so technically our product rule for limits does not apply. However, we *can* use the product rule for limits even when the limits involved "do not exist" if we are careful with how we handle "infinity" and indeterminate forms, although we will not prove this here.

We now continue with the proof of Theorem 5.6. We have shown that the limit at infinity of any polynomial function is equal to the limit at infinity of its leading term. Since that leading term is a power function (with a positive integer power), we can use what we know about the limits of power functions at infinity to find the limits of polynomial functions at infinity. We know that if n is an even or odd positive integer, then $\lim_{x \to \infty} x^n = \infty$. Therefore, we have

$$\lim_{x \to \infty} a_n x^n = \infty \text{ if } a_n > 0, \quad \text{and} \quad \lim_{x \to \infty} a_n x^n = -\infty \text{ if } a_n < 0.$$

By our previous calculation, this means that whether n is even or odd,

$$\lim_{x \to \infty} (a_n x^n + a_{n-1} x^{n-1} + \cdots + a_1 x + a_0) = \lim_{x \to \infty} a_n x^n = \infty \text{ if } a_n > 0, \quad \text{and}$$

$$\lim_{x \to \infty} (a_n x^n + a_{n-1} x^{n-1} + \cdots + a_1 x + a_0) = \lim_{x \to \infty} a_n x^n = -\infty \text{ if } a_n < 0.$$

The proof for the limit of a polynomial function as $x \to -\infty$ is similar; it is left to the exercises. The major difference will be in the last couple of steps, where it will be important whether n is even or odd, since we know that if n is a positive even integer, then $\lim_{x \to -\infty} x^n = \infty$, whereas if n is a positive odd integer, then $\lim_{x \to -\infty} x^n = -\infty$. ■

◈ **Caution** The statement that a polynomial and its leading term have the same limit applies only as $x \to \infty$ or as $x \to -\infty$. A polynomial and its leading term usually have different limits as $x \to c$. For example, $\lim_{x \to 2}(2x^3 - 3x + 1) = 16 - 6 + 1 = 11$, whereas $\lim_{x \to 2}(2x^3) = 16$. □

5.2.3 Derivatives of polynomial functions

At this point we already know how to differentiate a polynomial function using the power rule, sum rule, and constant multiple rule, since every polynomial function is a sum of constant multiples of power functions x^k. In general, the derivative of a polynomial is given by the following formula.

THEOREM 5.7

Differentiating Polynomial Functions

If $f(x) = a_n x^n + a_{n-1} x^{n-1} + a_{n-2} x^{n-2} + \cdots + a_2 x^2 + a_1 x + a_0$ is any polynomial function, then

$$f'(x) = n a_n x^{n-1} + (n-1) a_{n-1} x^{n-2} + (n-2) a_{n-2} x^{n-3} + \cdots + 2 a_2 x + a_1.$$

You don't need to memorize this formula; you just need to be able to use the power, sum, and constant multiple rules to differentiate any polynomial function quickly. For example, $\frac{d}{dx}(2x^5 - 5x^3 - 4x + 12) = 2(5x^4) - 5(3x^2) - 4(1) + 0 = 10x^4 - 15x^2 - 4$. You will use these same differentiation rules to prove Theorem 5.7 in the exercises.

Since every polynomial function has a well-defined derivative (as given in Theorem 5.7) at every point in its domain $(-\infty, \infty)$, polynomial functions are always differentiable.

THEOREM 5.8

Differentiability of Polynomial Functions
Every polynomial function is differentiable on all of $(-\infty, \infty)$.

Since polynomials are defined, continuous, and differentiable for all real numbers, their graphs are unbroken and smooth everywhere. We will use these nice properties to identify and graph polynomial functions in Section 5.3.

The derivative of a polynomial function is itself a polynomial function. One might then expect that the *anti*derivative of a polynomial function would itself be a polynomial function, and this is in fact the case. Antiderivatives of polynomial functions can be calculated by using the sum and constant multiple rules and the antiderivatives of power functions.

EXAMPLE 5.13   **Finding an antiderivative of a polynomial function**

Find a function $f(x)$ whose derivative is $f'(x) = 5x^4 - 2x^3 + 3x - 4$ that has value $f(0) = 3$.

Solution   From Section 4.3.6 we know that for every positive integer k, if $f'(x) = x^k$ then $f(x) = \frac{1}{k+1}x^{k+1} + C$ for some positive integer C. Using this rule (and the sum and constant multiple rules), we thus calculate that every antiderivative of $f'(x) = 5x^4 - 2x^3 + 3x - 4$ has the form

$$f(x) = 5\left(\tfrac{1}{5}x^5\right) - 2\left(\tfrac{1}{4}x^4\right) + 3\left(\tfrac{1}{2}x^2\right) - 4\left(\tfrac{1}{1}x^1\right) + C$$

$$= x^5 - \tfrac{1}{2}x^4 + \tfrac{3}{2}x^2 - 4x + C.$$

Given that $f(0) = 3$, we can solve for C:

$$f(0) = 3 \implies 0^5 - \tfrac{1}{2}0^4 + \tfrac{3}{2}0^2 - 4(0) + C = 3 \implies C = 3.$$

Therefore, we must have $f(x) = x^5 - \tfrac{1}{2}x^4 + \tfrac{3}{2}x^2 - 4x + 3$.   □

☑ *Checking the Answer*   To check the answer, verify that the function $f(x) = x^5 - \tfrac{1}{2}x^4 + \tfrac{3}{2}x^2 - 4x + 3$ has derivative $f'(x) = 5x^4 - 2x^3 + 3x - 4$ and value $f(0) = 3$.   □

Recall that the position, velocity, and acceleration functions ($s(t)$, $v(t)$, and $a(t)$, respectively) of a moving object are related by the equations

$$v(t) = s'(t), \qquad a(t) = v'(t), \qquad a(t) = s''(t).$$

If any one of these three functions is a polynomial, then we can find the other two functions by differentiating or antidifferentiating, as appropriate. For example, if $v(t)$ is a given polynomial, then we can differentiate $v(t)$ to get $a(t) = v'(t)$ and antidifferentiate $v(t)$ to get $s(t)$ (since $v(t) = s'(t)$). Remember that the initial conditions $s_0 = s(0)$, $v_0 = v(0)$, and $a_0 = a(0)$ represent the initial position, velocity, and acceleration of the moving object, respectively. These conditions can be used to solve for the constant "C" that arises in any antidifferentiation calculations.

EXAMPLE 5.14

Obtaining position and acceleration functions from velocity

A spaceship is moving along a straight path from Venus straight into the heart of the sun. The velocity of the spaceship t hours after leaving Venus is $v(t) = 0.015t^2 + 200$ thousands of miles per hour. Find equations for the position and acceleration of the spaceship t hours after leaving Venus. What can you say about the initial values s_0, v_0, and a_0?

Solution   From the equation $v(t) = 0.015t^2 + 200$, we can see that the initial velocity of the spaceship is $v_0 = v(0) = 0.015(0)^2 + 200 = 200$ thousand miles per hour.

Since $a(t) = v'(t)$, we know that

$$a(t) = \frac{d}{dt}(0.015t^2 + 200) = 2(0.015)t + 0 = 0.03t.$$

In particular, the initial acceleration of the spaceship is $a_0 = a(0) = 0.03(0) = 0$ thousands of miles per hour per hour.

Finally, since $v(t) = s'(t)$, we know that $s'(t) = 0.015t^2 + 200$. Therefore, we can antidifferentiate $0.015t^2 + 200$ to get a formula for $s(t)$:

$$s(t) = 0.015\left(\tfrac{1}{3}t^3\right) + 200(t) + C = 0.005t^3 + 200t + C.$$

By assumption, the spaceship starts out at Venus at position $s_0 = 0$; we can use this initial position to solve for C:

$$s(0) = 0 \implies 0.005(0)^3 + 200(0) + C = 0 \implies C = 0.$$

Therefore, t hours after leaving, the spaceship is $s(t) = 0.005t^3 + 200t$ thousand miles from Venus.   ☐

5.2.4 Local behavior of polynomial functions

The global behavior of a polynomial function is determined solely by its leading term. On the other hand, *all* of the terms of a polynomial function help determine the **local behavior** of the function. Examples of the local behavior of a function include roots, maxima, minima, and inflection points.

Recall that finding the extrema and inflection points of a function involves determining the zeros and signs of the first and second derivatives of that function. Since the derivatives of polynomial functions are always themselves polynomial functions, much of the local behavior of a polynomial function f can be described by examining roots and signs of polynomial functions (namely, f, f', and f''). Here is where all our factoring skills from Section 5.1 will come in handy!

EXAMPLE 5.15

Using derivatives to determine the local behavior of a polynomial function

Find all the local extrema and inflection points of the function $f(x) = x^4 + 4x^3 - 16x$.

Solution   To find the local extrema of f, we must first find its critical points. The derivative of f is $f'(x) = 4x^3 + 12x^2 - 16$, which is a polynomial and thus is always defined; we need only find its zeros. First we pull out the common factor of 4:

$$f'(x) = 4x^3 + 12x^2 - 16 = 4(x^3 + 3x^2 - 4).$$

The polynomial $x^3 + 3x^2 - 4$ looks difficult to factor at first glance (no grouping or common factors), so we'll guess some roots. The constant term is -4, so by the Integer Root Theorem, the only possible integer roots are ± 1, ± 2, and ± 4. We only need to find one root to use synthetic division and reduce the problem of factoring a cubic to a problem of factoring a quadratic. We test the easiest root possibility, $x = 1$, and find that it is in fact a root, since $(1)^3 + 3(1)^2 - 4 = 1 + 3 - 4 = 0$. Now a quick synthetic-division calculation (see Figure 5.17) will enable us to factor $x - 1$ from $x^3 + 3x^2 - 4$.

$$
\begin{array}{r|rrrr}
 & 1 & 3 & 0 & -4 \\
1 & & 1 & 4 & 4 \\
\hline
 & 1 & 4 & 4 & 0
\end{array}
$$

Figure 5.17

Thus $x^3 + 3x^2 - 4 = (x - 1)(x^2 + 4x + 4)$ (you can check this by multiplying out), so

$$
\begin{aligned}
f'(x) &= 4x^3 + 12x^2 - 16 \\
&= 4(x^3 + 3x^2 - 4) \\
&= 4(x - 1)(x^2 + 4x + 4) \qquad \text{(synthetic division)} \\
&= 4(x - 1)(x + 2)(x + 2). \qquad \text{(factoring a quadratic)}
\end{aligned}
$$

The only zeros of f' (and thus the only critical points of f) are $x = 1$ and $x = -2$.

To find the local extrema of f, we test each of these critical points to determine whether it is a local maximum, a local minimum, or neither. A sign analysis of f' gives us the number line in Figure 5.18 (check this by evaluating $f'(-3)$, $f'(0)$, and $f'(2)$).

Figure 5.18

Since the sign of f' changes only at $x = 1$ (and not at $x = -2$), the only local extremum of f is at $x = 1$. The sign of f' changes from negative to positive at $x = 1$, so by the first derivative test, f has a local minimum at $x = 1$.

To find the inflection points of $f(x) = x^4 + 4x^3 - 16x$, we must do a similar analysis for the second derivative of f. The second derivative is the derivative of $f'(x) = 4x^3 + 12x^2 - 16$, which is $f''(x) = 12x^2 + 24x$. Luckily this is very easy to factor:

$$
f''(x) = 12x^2 + 24x = 12x(x + 2).
$$

The only possible inflection points of f are at $x = 0$ and $x = -2$. By evaluating, for example, $f''(-3)$, $f''(-1)$, and $f''(1)$, we can quickly get a sign analysis for f'' (see Figure 5.19).

Figure 5.19

Since the sign of f'' changes both at $x = -2$ and at $x = 0$, the function f changes concavity at both of these values. The function f has two inflection points, at $x = -2$ and at $x = 0$. $\qquad \square$

✓ ***Checking the Answer*** The graph of the function $f(x) = x^4 + 4x^3 - 16x$ is shown in Figure 5.20. Note that $x = -2$ is a critical point that is not a local extremum, and $x = 1$ is a critical point that is a local minimum. Moreover, both $x = -2$ and $x = 0$ are inflection points of f.

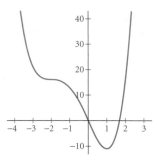

Figure 5.20
$$f(x) = x^4 + 4x^3 - 16x$$ □

A point $x = c$ is a ***turning point*** of the graph of a function f if the graph "turns around" at that point. More precisely, the turning points of a function f are the local maxima and minima of f. The following theorem tells us how many turning points a polynomial function of degree n can have.

THEOREM 5.9

The Maximum Number of Turning Points for a Polynomial Function

A polynomial function of degree n can have at most $n - 1$ turning points.

🔷 *Caution*   Note the words "at most" in Theorem 5.9; a polynomial function of degree n can have $n - 1$ *or fewer* turning points.   □

EXAMPLE 5.16   **A polynomial of degree 5 has at most four turning points**

Consider the three quintic polynomials shown in Figures 5.21–5.23.

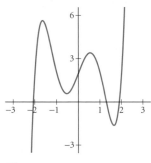   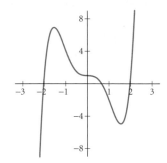   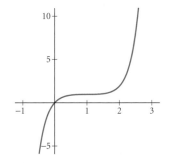

Figure 5.21   **Figure 5.22**   **Figure 5.23**
$f(x) = x^5 - 5x^3 + 4x + 2$ $g(x) = x^5 - 4x^3 + 1$ $h(x) = x^5 - 5x^4 + 10x^3$
$- 10x^2 + 5x$

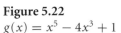

The function f in Figure 5.21 has four turning points (the maximum amount, since $5 - 1 = 4$). The function g has only two turning points, and the function h has no turning points at all. All of these polynomial functions are of degree 5 and have *at most* four turning points.   □

The proof of Theorem 5.9 follows directly from the Fundamental Theorem of Algebra, since every turning point of a polynomial function f is a critical point of f and thus a zero of the polynomial function f'.

PROOF (**THEOREM 5.9**)   Suppose f is an nth-degree polynomial function. Then by Theorem 5.7, its derivative f' is a polynomial of degree $n - 1$. Since the polynomial f' is defined for all real numbers (every polynomial is), the only critical points of f are the zeros of f'. By the Fundamental Theorem of Algebra, the derivative f' can have at most $n-1$ roots (since it is of degree $n-1$), and thus the function f can have at most $n-1$ critical points. Moreover, some of these critical points may be turning points of f, and some may not. Thus there are at the very *most n − 1* turning points for f.

Given the graph of a polynomial function, Theorems 5.6 and 5.9 can help us identify the parity and size of the degree of that polynomial function. This information will be especially useful in the next section when we try to find polynomial functions to model graphs.

❓ Question Can an even-degree polynomial have an even number of turning points? Why or why not? What can you say about whether the number of turning points of an odd-degree polynomial must be odd or even? □

What you should know after reading Section 5.2

▶ How to prove that polynomial functions are defined, continuous, and differentiable on all of $(-\infty, \infty)$.

▶ How to calculate limits, derivatives, and antiderivatives of polynomial functions (and why these calculations work). Also, be able to use derivatives and antiderivatives to solve problems involving position, velocity, and acceleration.

▶ Be able to tell from the global behavior of the graph of a polynomial function f whether the degree of f must be odd or even, and be able to say something about the size of the degree of f by looking at the number of turning points in the graph.

▶ Be able to find any local extrema and inflection points of a polynomial function f. Note that this may involve some fancy polynomial factoring.

Exercises 5.2

Concepts

0. Read the section and make your own summary of the material. Include whatever you think will help you review the material later on.

1. Fill in the blanks: Every polynomial function is _____, _____, and _____ on $(-\infty, \infty)$.

■ Explain why each graph cannot possibly be the graph of a polynomial function.

2.

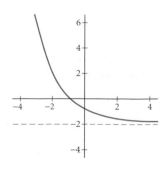

3.

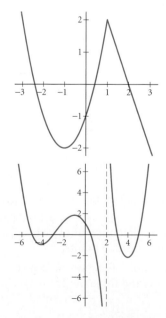

4.

■ Given the polynomial functions graphed below, what can you say about the parity and size of the degree of f, and why? What can you say about the leading coefficient of f, and why?

5.

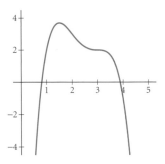

6.

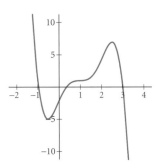

7.

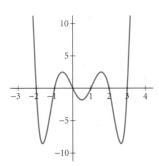

■ For each problem, either sketch a graph representing a polynomial function with the given conditions, or explain why such a polynomial function cannot exist.

8. f is a polynomial function of odd degree and negative leading coefficient.

9. f is a polynomial function of degree 4 with five turning points.

10. f is a polynomial function of even degree, three turning points, and one root.

11. f is a polynomial function of degree 5 with three turning points.

12. f is a polynomial function of even degree that is never positive.

13. f is a polynomial function of degree 7 that has seven roots and four turning points.

14. f is a polynomial function of degree 6 that has five turning points and three roots.

15. Use a graphing calculator to sketch the graphs of $f(x) = x^5 - 2x^4 - 3x^3 + 8x^2 - 4x$ and $g(x) = x^5$ in two different graphing windows: first, a relatively small window where the local behavior of these two functions is clear (the two graphs should look very different); second, a graphing window large enough that the graphs look almost identical. Why is it possible to find a window large enough so that the two graphs look almost the same?

16. Consider the functions $f(x) = x^2 - 2x^2$, $g(x) = 2x^2 - x^2$, and $h(x) = x^2 - x^2$. The limit of each of these functions as $x \to \infty$ is initially of the form $\infty - \infty$. Simplify these functions so that you can compute their limits as $x \to \infty$, and use your answers to show that limits of the form $\infty - \infty$ are indeterminate.

17. Can an even-degree polynomial have an even number of turning points? An odd number? Why or why not?

18. Can an odd-degree polynomial have an even number of turning points? An odd number? Why or why not?

■ Determine whether each of the following statements is true or false. If a statement is true, explain why. If a statement is false, provide a counterexample.

19. True or False: If f is a function whose graph has a horizontal asymptote, then f cannot be a polynomial function.

20. True or False: If f is a polynomial function of degree n, then the graph of f must have $n - 1$ turning points.

21. True or False: If f is a polynomial function with k roots, then f must have at least $k - 1$ turning points.

22. True or False: If f is a polynomial function of odd degree, then f must have at least one real root.

23. True or False: If f is a polynomial function with leading term $a_n x^n$, then $\lim_{x \to 3} f(x) = \lim_{x \to 3} a_n x^n$.

24. True or False: If f is a polynomial function with leading term $a_n x^n$, then $\lim_{x \to \infty} f(x) = \lim_{x \to \infty} a_n x^n$.

25. True or False: If f is a polynomial function of even degree, then f is an even function.

26. True or False: If f is a polynomial function of odd degree, then f is an odd function.

Skills

■ Calculate the following limits.

27. $\lim\limits_{x \to -1} (3 - 5x^3 + 2x^7)$

28. $\lim\limits_{x \to 3} (2x^4 - 12x^2 + 2)$

29. $\lim\limits_{x \to 0} (2x^6 - 5)(3x^4 - 2x^3)$

30. $\lim\limits_{x \to \infty} (-5x^4 + 3x^2 - 10)$

31. $\lim\limits_{x \to \infty} (-2x^5 + 8x^4 - 6)$

32. $\lim\limits_{x \to -\infty} (2x^7 + x^4 - x^3 + 16)$

33. $\lim\limits_{x \to -\infty} (3x^3 - 1)(5 - 7x^4)$

34. $\lim\limits_{x \to \infty} \dfrac{1}{1 - 2x^7}$

35. $\lim\limits_{x \to -\infty} \dfrac{-2}{3x^5 - 2x^2 + 17}$

36. $\lim\limits_{x \to -2} \dfrac{x^2 - 4x + 4}{x^5 - 6x^4 + 12x^3 - 8x^2}$

37. $\lim\limits_{x \to 3} \dfrac{x^4 - 4x^3 + 2x^2 + 4x - 3}{x^2 - 5x + 6}$

38. $\lim\limits_{x \to -\infty} \dfrac{x^3 - 3x + 2}{x^3 + 2x^2 - x - 2}$

■ Calculate the following derivatives using the sum, constant multiple, and power rules. Some algebra may be needed before differentiating.

39. $\dfrac{d}{dx}(5x^6 - 3x^4 + 2x^3 - 3x + 6)$

40. $\dfrac{d}{dx}(2 - 3x^3 + x^8 - 4x^{11})$

41. $\dfrac{d^2}{dx^2}(3x^5 - 2x^4 + 3x^2 - 5x + 7)$

42. $\dfrac{d}{dx}(3x + 1)(4 - x^7)$

43. $\dfrac{d}{dx}((2x - 1)^4)$

44. $\dfrac{d^3}{dx^3}(3x^5 + 5(1 - 2x^4))$

45. $\dfrac{d^5}{dx^5}(3x^5 - 2x^4 + 7x^2 - 1)$

46. $\dfrac{d^{11}}{dx^{11}}(-2x^8 + 3x^5 - 2x^2 + 6)$

47. $\dfrac{d^7}{dx^7}(4x^7 + 3x^2 - 5)$

■ For each problem, find a function f that has the given derivative and value. Check your answer by checking that your function f does indeed have the given derivative and value.

48. $f'(x) = 3x^5 - 2x^2 + 4, \quad f(0) = 1$

49. $f'(x) = 7x^2 + 8x^{11} - 18, \quad f(0) = -2$

50. $f'(x) = 1 - 4x^6, \quad f(1) = 3$

51. $f'(x) = x(4 - 2x), \quad f(0) = 0$

52. $f'(x) = (3x + 1)^3, \quad f(2) = 1$

53. $f'(x) = (x^4 - 8)(1 - 3x^5), \quad f(0) = 2$

■ Find all local extrema and inflection points of each polynomial function f. Do all calculations by hand, and show and explain your work carefully. Check

your work with a graphing calculator when you are finished.

54. $f(x) = 2x^3 - 3x^2 - 12x$

55. $f(x) = x^4 - 6x^2 + 8x$

56. $f(x) = x^{11} + 4$

57. $f(x) = x^4 - 4x^3 + 6x^2 - 36x$

58. $f(x) = x^4 + 4x^3 + 6x^2 + 8x$

59. $f(x) = 2x^5 - 15x^4 - 30x^3 + 70x^2$

60. $f(x) = x^4 - 2x^3 - 8x^2 - 6x$

61. $f(x) = x^3(1 + 3x)$

62. $f(x) = (x^2 - 2x)(x^2 - 2x - 6)$

Applications

■ In each of the next three problems, use the information given to find functions and initial values for position, velocity, and acceleration. In other words, find equations for $s(t)$, $v(t)$, and $a(t)$, and find the values of s_0, v_0, and a_0.

63. A mouse is running back and forth on a straight track. He runs along the track for 6 minutes, and his position after t minutes is $s(t) = 0.1t^4 + 1.15t^3 - 4.1t^2 + 4.4t + 3$ feet from the left end of the track.

64. Alina is driving from New York to San Francisco. At 12:00 noon she is 550 miles into her trip. Her velocity t hours after noon is $v(t) = 40t^3 - 100t^2 + 50t + 45$ miles per hour.

65. On the planet XV-37, gravity acts differently than it does here on Earth. An object dropped from a 1000-foot building on planet XV-37 will have a downward gravitational acceleration of $a(t) = -6t$ feet per second per second after falling for t seconds.

66. A spaceship is moving along a straight path from Venus straight into the heart of the sun. The velocity of the spaceship t hours after leaving Venus is $v(t) = 0.012t^2 + 400$ thousands of miles per hour.

   (a) Find equations for the position and acceleration of the spaceship (make sure to include the appropriate units). Say what you can about the initial values s_0, v_0, and a_0.

   (b) Is the spaceship always moving toward the sun? How can you tell?

   (c) Is the spaceship traveling at a constant acceleration? If not, is it *always* speeding up or slowing down as it approaches the sun? How can you tell?

   (d) How long will it take the spaceship to reach the sun? How fast will it be going when it gets there? (Assume that the distance between Venus and the sun is 67 million miles.)

67. After you drink a cup of coffee, the concentration of caffeine in your bloodstream increases at first and then

decreases steadily. Let $C(t)$ represent the amount of caffeine (in grams) in your bloodstream t hours after you drink a cup of coffee. Explain why $C(t)$ cannot possibly be a polynomial function. (*Hint:* Try drawing a possible graph of $C(t)$.)

68. On Earth, a falling object has a downward acceleration of -32 feet per second per second because of gravity. Suppose an object falls from an initial height of s_0 feet, with an initial velocity of v_0 feet per second. Show that the equations for the position and velocity of the falling object after t seconds are

$$s(t) = -16t^2 + v_0 t + s_0 \quad \text{and} \quad v(t) = -32t + v_0.$$

Proofs

69. Prove in your own words that every polynomial function is defined on all of $(-\infty, \infty)$. (*Hint:* Use what you know about power functions.)

70. Prove in your own words that every polynomial function is continuous on all of $(-\infty, \infty)$. (*Hint:* Use what you know about power functions.)

71. Prove in your own words that every polynomial function is differentiable on all of $(-\infty, \infty)$. (*Hint:* Use what you know about power functions.)

72. Prove that if $f(x) = a_n x^n + a_{n-1} x^{n-1} + a_{n-2} x^{n-2} + \cdots + a_2 x^2 + a_1 x + a_0$ is any polynomial function, then its derivative is

$$f'(x) = n a_n x^{n-1} + (n-1) a_{n-1} x^{n-2}$$
$$+ (n-2) a_{n-2} x^{n-3} + \cdots + 2a_2 x + a_1.$$

73. Suppose $g(x) = a_n x^n + a_{n-1} x^{n-1} + a_{n-2} x^{n-2} + \cdots + a_2 x^2 + a_1 x + a_0$ is any polynomial function. Prove that an antiderivative of g is

$$f(x) = \frac{a_n}{n+1} x^{n+1} + \frac{a_{n-1}}{n} x^n + \frac{a_{n-2}}{n-1} x^{n-1}$$
$$+ \cdots + \frac{a_2}{3} x^3 + \frac{a_1}{2} x^2 + a_0 x + C.$$

(*Hint:* It suffices to show that $f'(x) = g(x)$. Why?)

74. Suppose f is a polynomial function of odd degree. Use the global behavior of f and the Intermediate Value Theorem to prove that f must have at least one real root.

75. Suppose f is a polynomial function with k roots. Use Rolle's Theorem to prove that f must have at least $k - 1$ turning points.

76. Prove the parts of Theorem 5.6 that were not proved in the text. In other words, prove that if f is a polynomial function of degree n with leading coefficient a_n, then

    (a) If n is even and $a_n > 0$, then $\lim\limits_{x \to -\infty} f(x) = \infty$.

    (b) If n is even and $a_n < 0$, then $\lim\limits_{x \to -\infty} f(x) = -\infty$.

    (c) If n is odd and $a_n > 0$, then $\lim\limits_{x \to -\infty} f(x) = -\infty$.

    (d) If n is odd and $a_n < 0$, then $\lim\limits_{x \to -\infty} f(x) = \infty$.

77. Prove that the $(n+1)$st derivative of any nth-degree polynomial function is zero.

78. Prove that the nth derivative of an nth-degree polynomial function with leading coefficient a_n is equal to $n! a_n$. (Recall that $n!$ stands for "n factorial," which is equal to the product $n(n-1)(n-2)\cdots(2)(1)$ of all integers from 1 to n.)

5.3   Graphing Polynomial Functions

In this section we use the information from the previous sections to construct graphs (quick sketches as well as precise, accurate graphs) of polynomial functions. We will also see how to model graphs with polynomial functions.

5.3.1 Repeated roots

As we have seen, every polynomial function can be split into linear and irreducible quadratic factors, and each linear factor corresponds to a root of the function. Sometimes a linear factor appears more than once in the factorization; for example, consider the (factored) polynomial function

$$f(x) = (x + 1)(x - 2)(x - 2)(x - 6),$$

or, equivalently,

$$f(x) = (x + 1)(x - 2)^2 (x - 6).$$

This function has four linear factors, namely $x+1$, $x-2$, $x-2$ (again), and $x-6$. On the other hand, f has three *distinct* (i.e., different) linear factors, namely $x+1$, $x-2$, and $x-6$. Therefore, this function has four roots ($x=-1$, $x=2$, $x=2$ (again), and $x=6$) but only *three* distinct roots ($x=-1$, $x=2$, and $x=6$). The root $x=2$ occurs more than once and thus is called a *repeated root.*

Repeated roots have interesting graphical properties. Consider the graph of the function $f(x) = (x+1)(x-2)^2(x-6)$ shown in Figure 5.24.

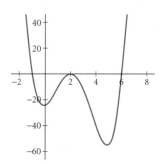

Figure 5.24
$$f(x) = (x+1)(x-2)^2(x-6)$$

The graph of f behaves differently near the *double root* $x=2$ than it does at its other two roots. At the *single roots* $x=-1$ and $x=6$ the graph of f passes through the x-axis; but at the double root $x=2$ the graph "bounces off" the x-axis. In fact, near $x=2$ the graph of f has a "quadratic" type of shape. Near $x=2$ the graph of f looks similar to the way the graph of $y=x^2$ does near the origin (but flipped over).

What happens if a root is repeated more than twice? Consider the polynomial function $g(x) = (x+1)^3(x-2)^4$ graphed in Figure 5.25.

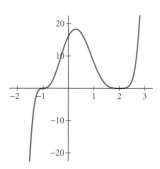

Figure 5.25
$$g(x) = (x+1)^3(x-2)^4$$

The function g has two distinct roots: $x=-1$ (repeated three times, so a *triple root*) and $x=2$ (repeated four times, so a *quadruple root*). Notice that near the triple root $x=-1$, the graph behaves as $y=x^3$ does at the origin. Similarly, the graph behaves as $y=x^4$ does at the quadruple root $x=2$. Note that the graph is very flat at $x=2$, much flatter than it would be for a double root.

In general, if a function f has a root $x=c$ that is repeated n times, then the graph of f will look something like the curve of x^n near that repeated root $x=c$. We cannot *prove* that repeated roots always behave this way graphically until after we learn the product and chain rules in Section 7.2, so for now we will just take it for granted. The idea of the proof will be that if $x=c$ is a root of f repeated n times, then the first $n-1$ derivatives of f will also have $x=c$ as a root. We can, of course, use derivatives to show that this happens in particular examples, and we will do so in the exercises.

5.3.2 Graphing polynomial functions quickly

If a polynomial function f of degree n factors completely into n linear factors (with no irreducible quadratic factors), then the graph of f will have n roots. In this case we can sketch a rough graph of f very quickly just by using the roots of f (paying attention to repeated roots) and the sign of its leading coefficient. The following theorem tells us that a polynomial that splits entirely into linear factors will have only *one* turning point between adjacent roots (although it may also have some turning points *at* some of those roots).

THEOREM 5.10

The Graph of a Polynomial That "Splits" into Linear Factors

Suppose a polynomial function f of degree n splits entirely into linear factors (i.e., has n real roots, counting repeated roots separately). Then the graph of f has exactly one turning point between every adjacent pair of distinct real roots.

We won't attempt to prove this theorem until Chapter 7 (when we know the product and chain rules). The main idea of the proof is that a repeated root of f is also a root of the derivative f'; this cuts down on the number of possible turning points of f. Then Rolle's Theorem guarantees that between each two distinct real roots of f, the graph of f has a turning point. The following example illustrates why this theorem is useful for quickly graphing polynomial functions.

EXAMPLE 5.17   **Quickly graphing a polynomial function that "splits"**

Sketch a rough graph of the polynomial function $f(x) = -2(x + 1)^2(x - 2)(x - 3)$.

Solution   The polynomial function f is completely split into linear factors (no irreducible quadratic factors) and thus has the maximum number of roots. More precisely, f is a fourth-degree polynomial (why?), and we have four roots if we count repeated roots separately: $x = -1$, $x = -1$ (again), $x = 2$, and $x = 3$. In terms of *distinct* roots, we have only three: $x = -1$, $x = 2$, and $x = 3$.

By Theorem 5.10, we know that there is exactly one turning point between the roots $x = -1$ and $x = 2$ and exactly one turning point between the roots $x = 2$ and $x = 3$. Using this fact, the fact that $x = -1$ is a double root, and the fact that f is a degree-4 polynomial with a negative leading coefficient, we can quickly sketch a graph of f (see Figure 5.26). To make this sketch, we use the fact that the degree of f is even to determine that the "ends" of the graph of f will either both point up or both point down. Since the leading coefficient $a_n = -2$ is negative, we know that these "ends" must

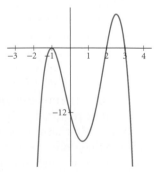

Figure 5.26

point down. Because f is the product of linear factors, there is exactly one turning point between $x = -1$ and $x = 2$ and exactly one turning point between $x = 2$ and $x = 3$. Finally, we evaluate $f(0) = -2(0 + 1)^2(0 - 2)(0 - 3) = -2(1)(-2)(-3) = -12$ so that we can mark the height of one point on the graph.   □

◈ *Caution*   The method used in Example 5.17 does not work if the polynomial function f does not split entirely into linear factors (i.e., if f has any irreducible quadratic factors). Irreducible quadratic factors of f can cause f to have "extra" turning points. For example, consider the function

$$f(x) = (x + 2)(x^2 - 5x + 10),$$

which has one linear and one irreducible quadratic factor. The graph of this function is given in Figure 5.27. We could not have used Theorem 5.10 to graph this function quickly.

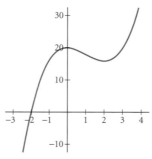

Figure 5.27   □

5.3.3 Graphing polynomial functions accurately

In general, we can make a very accurate graph of a polynomial function by using information from the first and second derivatives. We have already seen how to find roots, local extrema, and inflection points of polynomial functions, as well as how to determine the behavior of a polynomial function at its "ends." We also know that polynomial functions (and their derivatives, which are also polynomial functions) are defined, continuous, and differentiable on all of $(-\infty, \infty)$. If we put all of this information together, we can make a very accurate graph of any polynomial function. The only possible problem is that the function f and/or its derivatives might be difficult to factor (especially if they have irrational or complicated rational roots).

EXAMPLE 5.18

A detailed curve-sketching analysis of a polynomial

Sketch the graph of the polynomial function $f(x) = x^5 - 15x^3$. Identify the coordinates of each root, local extremum, and inflection point.

Solution   We'll start by factoring the polynomial functions f, f', and f''. Note the use of the factoring formula $(a^2 - b^2) = (a - b)(a + b)$ in each case.

$$f(x) = x^5 - 15x^3 = x^3(x^2 - 15) = x^3(x - \sqrt{15})(x + \sqrt{15})$$
$$f'(x) = 5x^4 - 45x^2 = 5x^2(x^2 - 9) = 5x^2(x - 3)(x + 3)$$
$$f''(x) = 20x^3 - 90x = 10x(2x^2 - 9) = 10x(\sqrt{2}x - 3)(\sqrt{2}x + 3)$$

From these factorizations we know that f has roots at $x = 0$ and $x = \pm\sqrt{15} \approx \pm 3.87$, that the only critical points (and thus the only possible local extrema) of f are $x = 0$

and $x = \pm 3$, and that the only possible inflection points of f are $x = 0$ and $x = \pm \frac{3}{\sqrt{2}} \approx \pm 2.12$. By checking the signs of f, f', and f'' at appropriate values, we obtain the number lines shown in Figure 5.28.

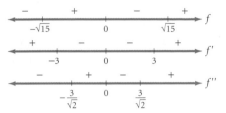

Figure 5.28

From this information we know that f has a maximum at $x = -3$ and a minimum at $x = 3$ and has inflection points at $x = 0$ and at $x = \pm \frac{3}{\sqrt{2}}$. We also know the intervals on which f is positive or negative, increasing or decreasing, and concave up or concave down.

Since f is a fifth-degree polynomial with a positive leading coefficient ($a_5 = 1$), we know that the behavior of f at the "ends" is as follows:

$$\lim_{x \to \infty} f(x) = \infty \quad \text{and} \quad \lim_{x \to -\infty} f(x) = -\infty.$$

We need to know the values of $f(x) = x^5 - 15x^3$ at all the "important" points, that is, at the roots of f, f', and f''. Since $x = 0$ and $x = \pm\sqrt{15}$ are roots of f, we know that $f(x) = 0$ at these points. The other values are

$$f(3) = -162, \quad f(-3) = 162, \quad f\left(\frac{3}{\sqrt{2}}\right) \approx -100.232, \quad f\left(-\frac{3}{\sqrt{2}}\right) \approx 100.232.$$

Now we need only plot these points and "connect the dots" using the sign information in the number lines. Remember to be careful that on each interval between the seven "important" points, you sketch a piece of a graph that accurately represents the signs of f, f', and f''. For example, according to the number lines, between $x = -3$ and $x = -\frac{3}{\sqrt{2}}$ the graph should be positive, decreasing, and concave down. The resulting sketch is shown in Figure 5.29.

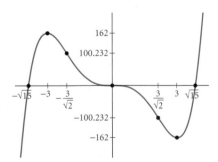

Figure 5.29 $f(x) = x^5 - 15x^3$

In Example 5.18, the function

$$f(x) = x^5 - 15x^3 = x^3(x - \sqrt{15})(x + \sqrt{15})$$

was split into all linear factors, namely x (repeated three times), $x - \sqrt{15}$, and $x + \sqrt{15}$. Therefore, we could have sketched a rough graph of f using the techniques of the last section. However, without examining the roots and signs of f' and f'', we would not have known the precise locations of the extrema and inflection points of f. Moreover, the more precise technique used in Example 5.18 will work for *any* polynomial, whether or not it splits into linear factors.

> ◈ *Caution*   If f is a polynomial function, then so are f' and f''; therefore, we don't need to worry about finding values at which f, f', or f'' do not exist (polynomials are defined everywhere). In the future, however, when we deal with more complicated types of functions, we *will* have to worry about such values. Don't forget that the only reason we don't look for them here is that we know polynomial functions are always "nice."   □

5.3.4 Modeling graphs with polynomial functions

In the last couple of sections we saw how to graph a polynomial function given its formula. In this section we learn how to find a polynomial function that could have a given graph.

Given a graph, the first step is to determine whether or not that graph could be the graph of a polynomial function: It must be defined, continuous, and differentiable everywhere, and it must increase or decrease without bound as $x \to \infty$ and as $x \to -\infty$.

Once we determine that the graph could be modeled with a polynomial function, we can make educated guesses about the degree and leading coefficient of the polynomial. The global behavior of the function tells us whether the degree is even or odd, as well as the sign of the leading coefficient. The magnitude of the degree depends on the number of turning points in the graph; if the graph has k turning points, then the degree of the polynomial must be *at least* $k + 1$.

Finally, the locations and multiplicities of the roots of the graph can help us write the polynomial in factored form, and any additional given points on the graph can help us solve for coefficients or constants. Remember that we are looking for a polynomial function that *could* have the given graph; this means we want to find a polynomial function that has all of the easily identifiable properties (roots, number of turning points, global behavior, etc.) of the graph. Sometimes more than one polynomial function will have these properties.

EXAMPLE 5.19   **Modeling a graph with a polynomial**

Find a polynomial function that could have the graph in Figure 5.30.

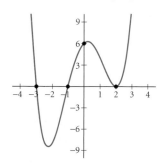

Figure 5.30

Solution   This graph is defined, continuous, and differentiable everywhere and increases without bound as $x \to \infty$ and as $x \to -\infty$. This could be the graph of a polynomial function, and if so, that polynomial function must have an even degree and a positive leading coefficient.

The roots of f are $x = -3$, $x = -1$, and $x = 2$. The root $x = 2$ is repeated an even number of times; for simplicity we will assume that $x = 2$ is a double root. Since there are four roots (counting $x = 2$ twice) and three turning points, the degree of f is at least 4. Since we also know the degree must be even, the simplest assumption is that f is a degree-4 polynomial. (Although it *could* be a higher-degree polynomial, we will always make the simplest choices possible.) Given these assumptions, f would have the form

$$f(x) = A(x + 3)(x + 1)(x - 2)^2,$$

where A is some positive constant. The y-intercept $(0, 6)$ marked on the graph can help us determine A. Since $f(0) = 6$, we have

$$6 = A(0 + 3)(0 + 1)(0 - 2)^2 \implies 6 = A(3)(1)(4) \implies 6 = 12A \implies A = \tfrac{1}{2}.$$

Therefore, a function that could have the graph shown in Figure 5.30 is the quartic polynomial function $f(x) = \tfrac{1}{2}(x + 3)(x + 1)(x - 2)^2$. ☐

✓ *Checking the Answer* You can check the answer by graphing the function $f(x) = \tfrac{1}{2}(x + 3)(x + 1)(x - 2)^2$ and checking that it has the roots, y-intercept, and general shape of the graph in Figure 5.30. ☐

If a given graph must be that of at least an nth-degree polynomial but has fewer than n roots (counting multiple roots separately), or if not all of the roots are marked, then it can be difficult to find a polynomial function that could have that graph. In this case it can be useful to write f in factored form with constants to be filled in later, as in the next example.

EXAMPLE 5.20   **Using a polynomial to model a graph with an unknown root**

Find a polynomial function that could have the graph shown in Figure 5.31.

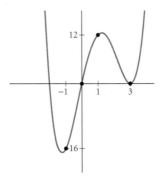

Figure 5.31

Solution   If we were given the coordinates of the leftmost root, this problem would be easy; unfortunately, we are not given this root value. We can tell from the graph that f could be a fourth-degree polynomial with positive leading coefficient and that f has three distinct roots: $x = 0$, and $x = 3$ (twice), and one that we do not know. Thus f could have the form

$$f(x) = Ax(x - 3)^2(x - c)$$

for some real number A (which should be positive) and some value c (which is the location of the unidentified root).

Given that f goes through the points $(-1, -16)$ and $(1, 12)$, we also know that $f(-1) = -16$ and that $f(1) = 12$. We can use these two datapoints to write two equations involving the unknown constants A and c:

$$f(-1) = -16 \implies A(-1)(-1-3)^2(-1-c) = -16$$
$$\implies -16A(-1-c) = -16$$
$$\implies A(1+c) = -1,$$

and

$$f(1) = 12 \implies A(1)(1-3)^2(1-c) = 12$$
$$\implies 4A(1-c) = 12$$
$$\implies A(1-c) = 3.$$

We now solve this system of equations for A and c. We have $A = \frac{-1}{1+c}$ from the first equation and $A = \frac{3}{1-c}$ from the second equation (as long as $c \neq \pm 1$, which we know is the case because $x = 1$ and $x = -1$ are not roots of f). We can set these equations equal to each other and solve for c:

$$\frac{-1}{1+c} = \frac{3}{1-c} \implies -1+c = 3(1+c) \implies -1+c = 3+3c$$
$$\implies -4 = 2c \implies c = -2.$$

If $c = -2$, then $A = \frac{-1}{1+c} = \frac{-1}{1-2} = 1$. Therefore, the graph in Figure 5.31 could be the graph of the polynomial function

$$f(x) = x(x-3)^2(x+2).$$

Use a graphing calculator to check that the graph of f has the features of the graph in Figure 5.31. $\square$

Sometimes we can get enough information from points on a graph that are not roots to help us find a polynomial function f to model the graph. This can be information about values of f, f', or f''. Recall that we know that $f'(c) = 0$ if the graph of f has a horizontal tangent line at $x = c$ and that $f''(c) = 0$ if the graph of f has an inflection point at $x = c$. Sometimes it is useful to determine the degree of f, write f in expanded form, and use known values of f, f', and f'' to solve for the coefficients of f.

EXAMPLE 5.21

Using derivative information to find a polynomial model for a graph

Find a polynomial function that could have the graph shown in Figure 5.32.

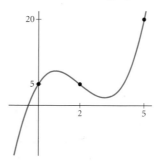

Figure 5.32

Solution   First, note that this graph is defined, continuous, and differentiable for all real numbers and has no asymptotes. Therefore, this graph could possibly be modeled by a polynomial function f. We can see that f has one root just to the left of the origin, but we are not given that value. What we know from the given graph is that

$$f(0) = 5, \quad f(2) = 5, \quad \text{and} \quad f(5) = 20.$$

Since $x = 2$ seems to be an inflection point of f, we assume that

$$f''(2) = 0.$$

We can also see from the global behavior of the graph that f must be an odd-degree polynomial function with a positive leading coefficient. Since there are two turning points in the graph, we also know that the degree of f must be at least 3; that's the lowest (and thus simplest) possible degree, so let's assume that f is a cubic polynomial.

If f is a cubic polynomial function, then

$$f(x) = ax^3 + bx^2 + cx + d$$

for some real numbers a, b, c, and d. Our problem is now reduced to using the values of f (and f'') that we know to find these coefficients. If f has the cubic form above, then its first and second derivatives are

$$f'(x) = 3ax^2 + 2bx + c,$$
$$f''(x) = 6ax + 2b.$$

Therefore, the values of f and f'' that we are given result in the following system of equations:

$$\begin{cases} a(0)^3 + b(0)^2 + c(0) + d = 5 \\ a(2)^3 + b(2)^2 + c(2) + d = 5 \\ a(5)^3 + b(5)^2 + c(5) + d = 20 \\ 6a(2) + 2b = 0 \end{cases} \implies \begin{cases} d = 5 \\ 8a + 4b + 2c + d = 5 \\ 125a + 25b + 5c + d = 20 \\ 12a + 2b = 0. \end{cases}$$

If we can solve this system of equations for a, b, c, and d, then we will have found a polynomial f that has the features of the graph in Figure 5.32. We'll give a few steps in the solution process and leave the details to you:

$$\implies \begin{cases} d = 5 \\ 8a + 4b + 2c = 0 \\ 25a + 5b + c = 3 \\ b = -6a \end{cases} \implies \begin{cases} d = 5 \\ c = 8a \\ c = 3 + 5a \\ b = -6a. \end{cases}$$

Now we can use the second and third equations to solve for a; we find that $8a = 3 + 5a$, and thus $a = 1$. Since all the other coefficients are in terms of a, we have found a solution:

$$\begin{cases} a = 1 \\ b = -6 \\ c = 8 \\ d = 5. \end{cases}$$

Therefore, we have shown that the polynomial

$$f(x) = x^3 - 6x^2 + 8x + 5$$

has a graph with the features shown in Figure 5.32. Of course, after all that work it would be a good idea to graph this function f and make sure (do this)!   □

We can use the techniques used in Example 5.21 to find polynomial functions f for which certain values of f and its derivatives are given. You will see such problems in the exercises. It is an interesting fact that each degree-n polynomial is completely determined by its value and first through nth derivatives at *one* point $x = c$. For example, this means that if f is a fourth-degree polynomial function, and you know the values of $f(2)$, $f'(2)$, $f''(2)$, $f'''(2)$, and $f''''(2)$, then you can completely determine the coefficients of f. We'll discuss this more in the exercises.

What you should know after reading Section 5.3

▶ The graphical properties of polynomial functions with repeated roots, and how to recognize repeated roots given the graph of a polynomial function.

▶ How to sketch quickly the graph of a polynomial function with "enough" real roots. What conditions on the factors of a polynomial function must we have in order to be able to sketch relatively accurate graphs of polynomial functions using only the roots? Be sure that you take into account the value and sign of the leading coefficient when making these quick sketches.

▶ How to sketch an accurate, labeled graph of a polynomial function f by finding the zeros and sign intervals of f, f', and f'' and identifying the global behavior of the graph of f.

▶ Given a graph, be able to determine whether the graph could be the graph of a polynomial function f, and if so, what must be true of the degree and leading coefficient of f. If "enough" roots are given in the graph, you should be able to find a polynomial model very quickly.

▶ Given a graph of a polynomial function of degree at least n where fewer than n roots are given, be able to find a polynomial function that has the general features of the given graph by solving a system of equations. Don't forget that values of f, f', and f'' can contribute to the system of equations.

Exercises 5.3

Concepts

0. Read the section and make your own summary of the material. Include whatever you think will help you review the material later on.

1. If $x = c$ is a double root of a polynomial function $f(x)$, then $f(x)$ can be written in the form _____ for some polynomial function $g(x)$.

2. If $x = c$ is a triple root of a polynomial function $f(x)$, then $f(x)$ can be written in the form _____ for some polynomial function $g(x)$.

■ For each polynomial function f graphed below, identify the roots of f. Make an educated guess as to the multiplicity of each root.

3.
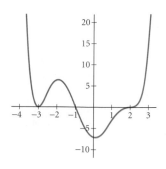

4.
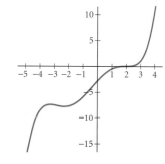

5.
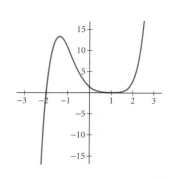

■ For each problem, write down an equation for a polynomial function f that has the properties described. Then use a graphing calculator to confirm that your function has the properties it should.

6. f has five real roots but only three *distinct* real roots.

7. f has six real roots but only three *distinct* real roots.

8. f is a seventh-degree polynomial with five real roots but only four *distinct* real roots.

9. f is a sixth-degree polynomial with two distinct real roots.

10. f has four distinct real roots, one of which is a triple root, and $f(0) = 4$.

11. f is a degree-4 polynomial with three distinct real roots.

12. f has one real root and two turning points.

13. The only root of f is a double root at $x = 0$, but f has five turning points.

■ Determine whether each of the following statements is true or false. If a statement is true, explain why. If a statement is false, provide a counterexample.

14. True or False: The critical points of a polynomial function f are the values at which f' is zero.

15. True or False: If a polynomial function has k real roots, then it has at most k distinct real roots.

16. True or False: A polynomial function of degree 5 cannot have three double roots.

17. True or False: If f is a quintic polynomial with three roots, two of which are double roots, then f must have exactly four turning points.

18. True or False: There is no sixth-degree polynomial with exactly three distinct roots, one of which is a triple root and two of which are single roots.

19. True or False: If f is a seventh-degree polynomial with four distinct roots, one of which is a double root, then f must have exactly four turning points.

20. True or False: Every graph is the graph of some polynomial function.

Skills

■ If possible, make a quick sketch of each of the following functions using Theorem 5.10. If it is not possible to make a quick sketch of f by using its roots and Theorem 5.10, explain why not.

21. $f(x) = (x + 1)(x - 2)(x - 3)^2$

22. $f(x) = (x - 2)^3(x + 1)^2$

23. $f(x) = (x - 3)^2(x + 2)(x^2 + 1)^2$

24. $f(x) = -2(x + 2)^4(x - 1)(x - 3)^2$

25. $f(x) = x^5 + x^4 - 2x^3$

26. $f(x) = 2x^4 - x^3 - x^2$

27. $f(x) = 3x^3 - 2x^2 + 3x - 2$

28. $f(x) = x^3 - 2x^2 - 4x + 8$

29. $f(x) = x^4 + x^3 - 6x^2 - 4x - 8$

30. $f(x) = x^4 + x^3 - 4x^2 + 4x + 16$

■ For each polynomial function f, use the zeros and signs of f, f', and f'' to sketch an accurate graph of f. Identify the coordinates of each root, local extremum, and inflection point. Show all work, including factorizations of f, f', and f'' and appropriate number lines.

31. $f(x) = x^3 - 12x^2 - 45x$

32. $f(x) = x^3 - 2x^2 + x - 2$

33. $f(x) = x^4 + 4x^2 + 4$

34. $f(x) = (x - 2)^2(1 + x)$

35. $f(x) = x^2(x - 1)(x + 1)$

36. $f(x) = (1 + x^2)(x - 3)$

37. $f(x) = 8x^3 + 3x^4$

38. $f(x) = 6x^5 - 15x^4 + 10x^3$

39. $f(x) = x^4 + 4x(8 - 3x - x^2)$

■ For each graph given, determine whether a polynomial function f could have that graph, and if so, find such a function f. When you are finished, use a graphing calculator to check that your function f has the properties and features of the given graph.

40.

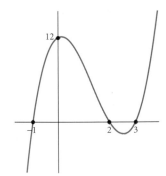

41.

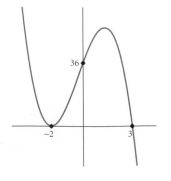

42.

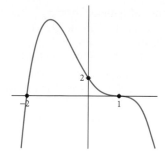

43.

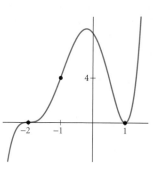

44.

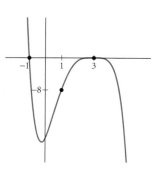

45.

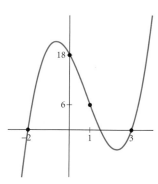

46.

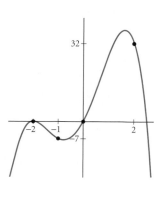

47.

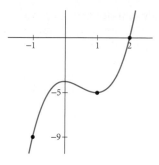

48.

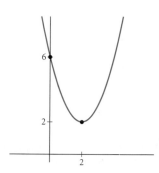

■ For each problem, find a polynomial function f with the given characteristics. Do not use your calculator (except to check your answers); instead, use known values of f and its derivatives to determine conditions on the coefficients of f, and then choose some coefficients for f that satisfy these conditions. There may be more than one possible answer for each problem.

49. f is a quartic polynomial function with two double roots and a y-intercept of 5.

50. f is a cubic polynomial function with an inflection point at $x = 2$.

51. f is a quadratic polynomial function with no roots and a local maximum at $x = -1$.

52. f is a cubic polynomial function with $f(0) = 2$, $f'(0) = 3$, $f''(0) = -4$, and $f'''(0) = 12$.

53. f is a cubic polynomial function with $f(0) = -5$, $f'(0) = -3$, $f''(0) = -2$, and $f'''(0) = 6$.

54. f is a quadratic polynomial function with $f(2) = 5$, $f'(2) = 7$, and $f''(2) = 2$.

55. f is a quartic polynomial function with $f(1) = 4$, $f'(1) = 0$, $f''(1) = 0$, $f'''(1) = -24$, and $f''''(1) = -48$.

56. In this problem you will find a polynomial function f that could have the graph below, by finding f' and then antidifferentiating, as follows:

(a) List all of the zeros of the *derivative* f' of the function f graphed below. Assuming that f' is a polynomial function, determine a possible value for the degree of f' and the sign of its leading coefficient.

(b) Use the information above to write down a possible polynomial expression for f'.

(c) Find an antiderivative f of f' (you should multiply out the expression for f' before antidifferentiating).

(d) Use the one known value of f to determine the value of the antidifferentiation constant "C".

(e) Use a graphing calculator to graph the function f that you found, and check that it does have the properties of the graph of f given below.

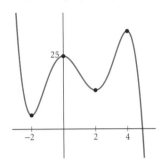

57. Find a polynomial function that could have the graph below, by the method used in Problem 56. Be aware that the derivative f' has a repeated root; you'll have to determine the location of this repeated root and how many times it is repeated.

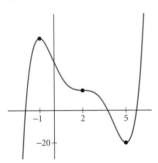

Applications

58. A watermelon is thrown down from a 100-foot building at an initial velocity of -30 feet per second. If the height $s(t)$ of the watermelon t seconds after it is thrown is a quadratic polynomial function, use $s(0)$, $s'(0)$, and $s''(0)$ to find an equation for $s(t)$. (Assume that the only force acting on the watermelon is the downward acceleration of -32 feet per second per second due to gravity.)

59. A bowling ball is thrown down from a twentieth-story window. After 3 seconds, the bowling ball is 26 feet from the ground and falling at a rate of -106 feet per second (downward).

(a) If the height $s(t)$ of the bowling ball t seconds after being thrown is a quadratic polynomial function, use $s(3)$, $s'(3)$, and $s''(3)$ to find an equation for $s(t)$.

(b) How high is the twentieth-story window from which the bowling ball was thrown?

(c) How fast was the bowling ball initially thrown?

60. On the planet of Xillian, "gravity" acts very differently than it does on Earth. The height $s(t)$ of a falling object on Xillian is always a cubic polynomial function. Suppose a kiwi fruit is dropped (with initial velocity of zero) from the top of a Xillian radio tower. After 5 seconds, the kiwi fruit is 100 feet from the ground and falling at a rate of -200 feet per second. The acceleration of the kiwi fruit at that moment is -46 feet per second per second.

(a) Use the four values of s, s', and s'' given in the description above to find a formula for the height $s(t)$ of the kiwi fruit t seconds after it is dropped.

(b) Verify that the function $s(t)$ you just found produces the correct values for $s(5)$, $s'(5)$, $s''(5)$, and $s'(0)$.

(c) How high is the radio tower from which the kiwi fruit was dropped?

(d) On Earth, acceleration due to gravity is given by a constant -32 feet per second per second; that is, on Earth we always have $a(t) = -32$ for falling objects. What is the function $a(t)$ for acceleration due to Xillian "gravity"? Is this acceleration constant?

61. The table below describes the number of cars that were on a particular 1-mile stretch of Route 97 at t hours after 6 a.m. Monday morning.

(a) Make a plot of the data in the table below.

(b) Suppose you wanted to find a polynomial function that models these data; in particular, suppose you wanted to find a polynomial function $N(t)$ that passes through each of the four datapoints. What is the minimum degree that this polynomial function could be, and why?

(c) Find a polynomial function $N(t)$ that goes through all four datapoints. (*Hint:* Use the data to solve for the coefficients of $N(t)$.)

(d) Overlay a plot of the function $N(t)$ you found above with the plot of the data. Verify that your function $N(t)$ goes through all four datapoints. Does $N(t)$ look the way you expected it to?

(e) Use the function $N(t)$ that you found to predict the number of cars that were on that 1-mile stretch of Route 97 at 7:30 a.m.

(f) How many cars does your model function predict will be on the road at 3 p.m. that day? Does that make sense? On what time interval do you think your model function makes practical sense?

t (in hours after 6 a.m.)	0	1	2	3
N (number of cars on one mile of road)	0	28	12	18

Proofs

62. Prove that if f is any quadratic polynomial function $f(x) = ax^2 + bx + c$, then the coefficients of f are completely determined by the values of f and its derivatives at $x = 0$, as follows:

$$a = \frac{f''(0)}{2}, \quad b = f'(0), \quad \text{and} \quad c = f(0).$$

In particular, this means that if f is any quadratic polynomial function, then $f(x) = \frac{f''(0)}{2}x^2 + f'(0)x + f(0)$.

63. Prove that if f is any cubic polynomial function $f(x) = ax^3 + bx^2 + cx + d$, then the coefficients of f are completely determined by the values of f and its derivatives at $x = 0$, as follows:

$$a = \frac{f'''(0)}{6}, \quad b = \frac{f''(0)}{2}, \quad c = f'(0), \quad \text{and} \quad d = f(0).$$

This means that every cubic polynomial function can be written $f(x) = \frac{f'''(0)}{6}x^3 + \frac{f''(0)}{2}x^2 + f'(0)x + f(0)$.

64. Suppose f is any cubic polynomial function $f(x) = ax^3 + bx^2 + cx + d$. Prove that the coefficients a, b, c, and d of f can be expressed in terms of the values of f and its derivatives at the point $x = 2$. (In other words, show that you can write the coefficients a, b, c, and d in terms of $f(2)$, $f'(2)$, $f''(2)$, and $f'''(2)$.)

65. Suppose f is a polynomial of degree n, and let k be some integer with $0 \leq k \leq n$. Prove that if f is of the form

$$f(x) = a_nx^n + a_{n-1}x^{n-1} + \cdots + a_kx^k + \cdots + a_1x + a_0,$$

then $a_k = \frac{f^{[k]}(0)}{k!}$, where $f^{[k]}(x)$ is the kth derivative of $f(x)$, and $k! = k(k-1)\cdots(2)(1)$. (*Hint:* Find $f^{[k]}(x)$ and use it to show that $f^{[k]}(0) = k!a_k$.) (*Note:* This problem, and the three before it, are a brief glimpse of **Taylor polynomials,** which you may see in your next calculus course.)

5.4 Optimization with Polynomial Functions

In this section we will introduce the type of word problems known as **optimization** problems. Many real-world problems involve optimization, that is, finding the global maximum and minimum values of a function on an interval. For example, a company might be interested in the maximum amount of profit it can make or in the minimum cost of producing a product.

In this section we will focus on optimizing polynomial functions. In later chapters we will apply these same techniques to optimize more complicated functions.

5.4.1 Strategies for word problems

We'll begin by saying a few words about word problems in general. What makes a word problem more difficult than a calculational problem is that there is **translating** involved. A word problem is written in full sentences (hence its name). Often variables, constants, functions, and relationships are described in these sentences, but with a minimum of mathematical notation.

To solve a word problem, we must first translate it into mathematical language. Then it is nothing more than a calculational problem. Once the mathematical problem has been solved, the answer needs to be translated back into real-world terms, to answer the original question. This process is summarized in Figure 5.33.

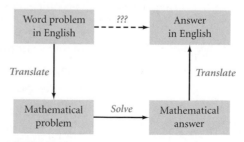

Figure 5.33

In the diagram in Figure 5.33 we start in the upper left-hand corner. We would like to go across the top of the diagram and get from the written word problem to some kind of written, real-world answer. However, it is not usually immediately clear how to get from the problem to the answer. Most word problems are not solvable with just "common sense" or just by reading the problem; a word problem usually describes a real-world situation that can, and must, be analyzed mathematically.

Our first step is to translate the word problem into a mathematical problem (the arrow down the left-hand side of the diagram). Then we can solve this mathematical problem (the arrow along the bottom of the diagram) and translate the solution as a written answer to the original word problem (the arrow going up the right-hand side).

The following algorithm describes strategies for translating word problems into mathematical problems.

ALGORITHM 5.2

Translating Word Problems

When translating a written word problem into mathematical language, the following steps are often helpful:

1. If applicable, make a picture or sketch that illustrates the problem. Label this sketch with the variable names you choose in step 2.

2. Identify any variables or functions in the problem, and give them names (such as x, V, t, $f(t)$, $C(s)$, and so on, as appropriate). Keep track of any units associated with these variables or functions.

3. Determine what is *given* in the problem and what it is that you want to *find*. Write this information in mathematical notation using the variable and function names.

4. Identify any functions or rates of change (derivatives) described in the word problem. If possible, construct equations for them on the basis of the descriptions given. Again, be sure to keep track of any units.

5. Determine the mathematical problem that needs to be solved to reach an answer to the original word problem. Write this mathematical problem entirely in mathematical notation.

After translating the word problem and setting up the mathematical problem to be solved, you can use your calculus, precalculus, and algebra skills to solve the problem. Then you need to translate your mathematical solution into an answer to the original word problem.

As a simple example of "translating", consider the following word problem. (*Note:* This is not an "optimization" problem; it is a word problem that we already know how to solve from Chapter 3.) The parts of the solution in italics are the main steps of the problem and are the steps that your grader or instructor might look for to verify that you have solved the problem correctly.

EXAMPLE 5.22   **Translating and solving a simple word problem**

A car is traveling at 75 mph at 1:00 p.m. and at 40 mph at 3:00 p.m. Determine the average acceleration of the car over this 2-hour time period.

Solution   The only information we are given in this problem is the velocity of the car at two different times. We'll start by putting this information in mathematical notation. The velocity of the car is a function (it changes over time); let's call it something sensible, like $v(t)$. To be entirely precise, we must write down what we mean by "$v(t)$":

$$v(t) = \textit{the velocity of the car, in miles per hour, } t \textit{ hours after 1:00 p.m.}$$

Note that the units of $v(t)$ are miles per hour and that the units of t are hours. In this notation, 1:00 is represented by $t = 0$, and 3:00 is represented by $t = 2$.

Given: $v(0) = 75$ and $v(2) = 40$.

Sometimes a picture or diagram is useful in solving a word problem, but it is not necessary in this particular example (although you can draw a picture of a car on a road and record the time and velocity information on the picture if you like).

Now we have to figure out what mathematical problem will help us find the average acceleration of the car. Since acceleration is the derivative of velocity, the average acceleration of the car is equal to the average rate of change of the velocity of the car over the 2-hour time period. Using the formula for the average rate of change of $v(t)$ from $t = 2$ to $t = 0$, we discover that the mathematical problem is

$$\textit{Find: } \frac{v(2) - v(0)}{2 - 0}.$$

We have now "translated" the word problem and can solve the mathematical problem:

$$\frac{v(2) - v(0)}{2 - 0} = \frac{40 - 75}{2 - 0} = \frac{-35}{2} = -17.5.$$

Finally, we need to use this mathematical answer to get an answer to the original word problem. We translate the answer "-17.5" (which represents a *deceleration* of 17.5 miles per hour per hour) into the following real-world answer:

Over the 2-hour time period from 1:00 to 3:00 p.m.,
the car had an average deceleration of 17.5 miles per hour per hour.

This last step is very important. One of the mistakes that students most commonly make when solving a word problem is that they forget to answer the original question! Also, be careful to include the correct units in the final answer. □

5.4.2 Optimizing a function on an interval

An optimization word problem is a word problem whose underlying mathematical problem involves finding the global maximum and minimum locations and values for a function on an interval. Before we get to the word problems, we must review and expand our knowledge of how to find such extrema.

Recall that a function f on an interval I has a ***global maximum of f on I*** at a number $c \in I$ if $f(c) \geq f(x)$ for all $x \in I$. The ***value*** of the global maximum is the value $f(c)$ of the function at the point $x = c$. A ***global minimum of f on I*** is defined similarly. Don't forget that the global maximum (or minimum) of a function on an interval may occur in more than one location. (Can you think of an example?) Also, a function might not have a global maximum (or minimum) value on a given interval.

If f is a continuous function on a *closed* interval, then by the Extreme Value Theorem (Section 2.7.1), f must have *at least one* global maximum and *at least one* global minimum on that closed interval. Since polynomial functions are always continuous, this means that every polynomial function f has a global maximum and a global minimum on any *closed* interval I. On an open or half-closed interval, however, not even continuous functions necessarily have global maximum and minimum values. (Can you think of examples?)

So how can we find the global extrema of a function f on an interval I? Recall that we can use the first or second derivative test to determine whether any of the critical points of f are *local* extrema. If the interval I has any *closed* endpoints, then f may also have local extrema at those endpoints; these local extrema are called ***endpoint extrema*** of f. *Open* endpoints of I cannot be local extrema of f. The following example illustrates this graphically.

EXAMPLE 5.23   **Endpoint extrema at closed endpoints, no extrema at open endpoints**

Consider the function $f(x) = 2x^3 - 15x^2 + 24x + 20$ on the closed interval $[0, 6]$ and also on the open interval $(0, 6)$. See Figures 5.34 and 5.35.

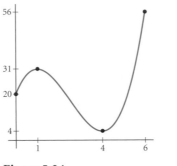

Figure 5.34   **Figure 5.35**

The function f has two critical points, both of which are local extrema, on the closed interval $[0, 6]$. In addition, the endpoints $x = 0$ and $x = 6$ are also locations of local extrema of f; however, the first derivative information of f does not detect these extrema. The *global* maximum of f on $[0, 6]$ is at the endpoint $x = 6$, and the *global* minimum of f on $[0, 6]$ is at $x = 4$.

On the open interval $(0, 6)$, we still have the "interior" local extrema at $x = 1$ and at $x = 4$. However, in this case we do *not* have any endpoint extrema, since the interval $(0, 6)$ is open and thus does not include the endpoints $x = 0$ and $x = 6$. In this case the global minimum of f on $(0, 6)$ is at $x = 4$, but there is *no* global maximum of f on $(0, 6)$.   □

Given a list of all the *local* interior and endpoint extrema of f on I, how can we determine whether f has a *global* maximum (or minimum) on I, and if so, where this global extremum is?

◈ *Caution*   It is not enough just to find all the local extrema of f on I and then compare them to see which one has the largest (or smallest) value. For example, in Figure 5.35, this incorrect process would lead us to believe that $x = 1$ is a global maximum of f on $(0, 6)$, which is not the case.   □

In general, to find a global maximum (or minimum) of a function f on an interval I, we must take into account the values of the local interior extrema as well as the behavior of f at the "ends" of the interval I. At closed endpoints we will have more local extrema, and at open endpoints we will have to consider the *limit* of f.

ALGORITHM 5.3

Optimizing a Function on an Interval

To find a global maximum of a continuous function f on an interval I (if a global maximum exists), we must compare the following quantities:

(a) The values $f(c)$ for each interior local extremum $c \in I$;

(b) The values $f(c)$ for each closed endpoint $x = c$ of I; and

(c) The limits $\lim_{x \to c} f(x)$ for each open endpoint $x = c$ of I.

If the limit of f at either open endpoint of I is larger than all of the values $f(c)$ in parts (a) and (b), then f does not have a global maximum on I.

Otherwise, the value $x = c$ from parts (a) and (b) that has the highest value $f(c)$ is the location of the global maximum of f on I.

The global minimum of f on I can be determined in a similar fashion.

Since in Algorithm 5.3 we are considering only the values of f in the interval I, we are really looking only at *one-sided* limits at the open endpoints of the interval. For example, if $I = (a, b)$ is an open interval, then we will be interested in the limit of f as $x \to a^+$ and as $x \to b^-$.

In Example 5.23 we graphically determined the global maximum and minimum values of $f(x) = 2x^3 - 15x^2 + 24x + 20$ on the interval $(0, 6)$. In the following example we do this analysis algebraically.

EXAMPLE 5.24 **Finding the global extrema of a function on an interval**

Find the global maximum and minimum values (if any) of $f(x) = 2x^3 - 15x^2 + 24x + 20$ on the interval $(0, 6)$.

Solution Since $f'(x) = 6x^2 - 30x + 24 = 6(x^2 - 5x + 4) = 6(x - 1)(x - 4)$, the only critical points of f are $x = 1$ and $x = 4$. The second derivative of f is $f''(x) = 12x - 30$. Since $f''(1) = -18 < 0$ and $f''(4) = 18 > 0$, we know that $x = 1$ is a local maximum of f and $x = 4$ is a local minimum of f (by the second derivative test). These local extrema have values $f(1) = 31$ and $f(4) = 4$.

Since the endpoints of the interval $(0, 6)$ are open, we must examine the limit of f as $x \to 0^+$ and as $x \to 6^-$. We have

$$\lim_{x \to 0^+} f(x) = \lim_{x \to 0^+} (2x^3 - 15x^2 + 24x + 20) = 20,$$

$$\lim_{x \to 6^-} f(x) = \lim_{x \to 6^-} (2x^3 - 15x^2 + 24x + 20) = 56.$$

Since the limit of f at the right-hand open endpoint is higher than all of the values of the local maxima of f, there is no global maximum of f on $(0, 6)$. However, the local minimum $x = 4$ has value $f(4) = 4$, which is less than the limit of f at either end of the interval $(0, 6)$; thus f has a global minimum on $(0, 6)$ at $x = 4$. See Figure 5.35. □

◈ *Caution* Suppose we are interested in finding the global extrema of a function f on an open interval (a, b), as in Example 5.24. If f is a polynomial function, and thus necessarily continuous, then the limit of f as $x \to a^+$ must be equal to the *value* $f(a)$; similarly, the limit of f as $x \to b^-$ is equal to $f(b)$. This means that we could have calculated $f(0)$ and $f(6)$, instead of calculating the limits. However, the techniques in this section work for *any* function f, not just polynomial or continuous functions; when f is not continuous, its limit can be different from its value at an open endpoint. In later sections the function f may be sufficiently complicated that we *have* to look at the limit rather than the value of f; thus we will compute limits (rather than values) now so that we won't forget to do it later. □

5.4.3 Optimization word problems

Many real-world problems involve a function that is to be maximized or minimized on an interval, such as the problem of finding the cheapest way to construct a soda can with a particular volume. When we are translating such an optimization problem into mathematical notation, the key step is to determine the function to be optimized and the relevant interval.

EXAMPLE 5.25

Translating a real-world problem into an optimization problem

A baseball is thrown straight up from a height of 5 feet with an initial velocity of 50 feet per second. What is the greatest height of the baseball in the first second of flight? When is the baseball at its greatest height over its entire flight?

Solution A quick look at this word problem tells us that we will need to maximize the function that describes the height of the baseball on two different intervals: first on the interval $[0, 1]$ and then on the interval $[0, b]$, where b is the time at which the baseball hits the ground. First, we must write a function that represents the position of the baseball.

Let's use the standard notation $s(t)$ for the position of the baseball (in feet from the ground) t seconds after it is thrown. The position $s(t)$ of a falling object under the influence of a constant gravity g is given by the formula

$$s(t) = -\frac{g}{2}t^2 + v_0 t + s_0.$$

You can derive this equation by starting from a constant acceleration $-g$ in the downward direction and antidifferentiating twice. We are given that $s_0 = 5$ feet and that the initial velocity of the baseball is $v_0 = 50$ feet per second (in the positive, or upward, direction). We also know that the acceleration due to gravity is given by $g = -32$ feet per second per second (in the negative, or downward, direction). Therefore, we have

$$s(t) = -16t^2 + 50t + 5.$$

The first part of the problem thus translates into the mathematical optimization problem

Find the maximum value of $s(t) = -16t^2 + 50t + 5$ on the interval $[0, 1]$.

The derivative of s is $s'(t) = -32t + 50$, so the only critical point of s is $t = \frac{50}{32} = 1.5625$. However, this critical point is not in the interval $[0, 1]$; thus the function s has no critical points in the interval $[0, 1]$. The only possible place that s could have a maximum on $[0, 1]$ is at one of the endpoints $t = 0$ or $t = 1$. Since $s(0) = 5$ and $s(1) = 39$, the maximum value of $s(t) = -16t^2 + 50t + 5$ on $[0, 1]$ is at time $t = 1$. The answer to the first part of the problem is

The greatest height of the baseball in the first second of flight is 39 feet above the ground.

The second part of the problem asks us to find the time at which the baseball is at its greatest height. We wish to maximize the function $s(t) = -16t^2 + 50t + 5$ over the entire time interval that the ball is in the air. We can solve the equation $s(t) = 0$ to find the time that the ball hits the ground; a quick application of the quadratic formula shows that the only positive time at which the height of the ball is zero is at $t = \frac{1}{16}(25 + \sqrt{705}) \approx 3.22$ seconds (we will use the decimal approximation here). The second part of the problem asks us to solve the mathematical problem

Find the maximum of $s(t) = -16t^2 + 50t + 5$ on the interval $[0, 3.22]$.

Again the only critical point of s is at $t = \frac{50}{32} = 1.5625$. Since $s''(t) = -32$ is always negative, the second derivative test tells us that this critical point is a local maximum of s. Moreover, since $s(1.5625) \approx 44$, while the height of the baseball at the endpoints is $s(0) = 0$ and $s(3.22) \approx 0$, it must be that $t = 1.5625$ is the global maximum of s on the interval $[0, 3.22]$. (Alternatively, note that $s(t)$ is a quadratic with negative leading coefficient, and thus its graph is a downward-pointing parabola. This means that the critical point at $t = 1.5625$ must be a local maximum, and in fact the global maximum, of $s(t)$.) The answer to the second part of the problem is

The baseball is at its greatest height 1.5625 seconds into its flight.

Notice that the first part of the problem asked us "how high," whereas the second part of the problem asked us "when." Be sure, when giving your final answer, that you are actually answering the questions that were asked. □

⊗ Caution It is very important to give your final answer in the context of the original problem, with units. If someone asked you when the baseball was at its highest, the answer "1.5625" would not be adequate. Is that in seconds? Seconds after what? Be as specific as possible when giving final answers to word problems. □

☑ Checking the Answer Checking that you have correctly translated a word problem is not always possible, but you can at least check that your answer makes sense. Of course you can always check your mathematical work to see whether you solved the associated mathematical problem correctly (but if you translated incorrectly, that won't make much difference). In this example, our answers do seem reasonable (why?). We can check our mathematical work by inspecting a graph of $s(t) = -16t^2 + 50t + 5$; see Figure 5.36.

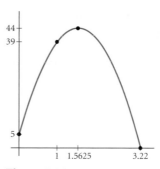

Figure 5.36

From the graph, we can see that the highest value of this function on $[0, 1]$ is indeed 39 feet and that the highest value overall occurs at approximately $t = 1.5625$. □

5.4.4 Optimization with constraints

Optimization problems often involve **constraints**. For example, you might want to maximize the area of a fenced enclosure built with a fixed amount of fencing material. The amount of fencing is the *constraint*. As another example, you might want to minimize the cost of the material used to make an oil drum, with a desired volume as the constraint. In general, such problems involve optimizing a function of two (or more) variables, and the constraints in the problem allow you to solve for one variable in terms of the other(s). We'll start with an easy, but classic, example.

EXAMPLE 5.26

Optimizing a function on an interval, given a constraint

Farmer Joe wants to build a rectangular chicken pen for his chickens. He wants to build the pen so that it has the largest area possible, and he only has 100 feet of chickenwire fencing. What dimensions should he use for the chicken pen?

Solution   Before we do anything else, we need to set up a picture and name any relevant variables. Let l and w represent the length and width of the rectangular pen, in feet. We know that the perimeter of the pen must be 100 feet, since that is how much chickenwire fencing Farmer Joe has. We are interested in maximizing the area of the pen.

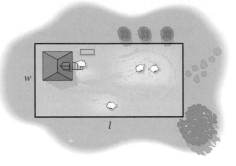

Constraint: $P = 2w + 2l = 100$

Maximize: $A = lw$

What makes this problem difficult is that the function A that we wish to maximize is written in terms of *two* variables, and our calculus techniques work only for functions of *one* variable. Luckily, we can use the constraint $P = 2w + 2l = 100$ to solve for one variable in terms of the other; this will reduce A to a one-variable function that we can optimize. We'll solve for l in terms of w (we could solve for w in terms of l if we preferred to, or if the calculations involved looked easier):

$$P = 2w + 2l = 100 \implies 2l = 100 - 2w \implies l = 50 - w.$$

Using the equation $l = 50 - w$, we can write the area function entirely in terms of w:

$$A = lw = (50 - w)w = 50w - w^2.$$

We now know the function we need to maximize, but what is the interval? We need to know which values of w we are willing to consider in this problem. Clearly we must have $w \geq 0$, since the width of the pen cannot be negative. (Of course, if we had $w = 0$, then the resulting fenced region would have an area of zero, but Farmer Joe *could* use his fencing that way.) What is the upper bound on w? The key is to realize that w is the largest when l is the smallest (why?). The smallest l could be is zero, and when $l = 0$ we have $0 = 50 - w$, or $w = 50$. Therefore, we must have $w \leq 50$. (This makes physical sense; why?) We have now determined that the underlying mathematical problem for this word problem is

Find the maximum of $A(w) = 50w - w^2$ on the interval $[0, 50]$.

Now we have just a simple mathematical problem to solve. Since $A'(w) = 50 - 2w$, the only critical point of A is $w = 25$ (which is indeed in the interval $[0, 50]$). Since $A''(w) = -2$ is always negative, in particular $A(25)$ is negative, and thus $w = 25$ is a local maximum of A. We now need only to compare the value of this local maximum and the values of A at the endpoints $w = 0$ and $w = 50$:

$$A(25) = 50(25) - (25)^2 = 625, \quad A(0) = 50(0) - (0)^2 = 0,$$
$$A(50) = 50(50) - (50)^2 = 0.$$

Therefore, $w = 25$ is indeed the global maximum of A on the interval $[0, 50]$. (Again, in this particular case the function we were optimizing was a downward-pointing parabola, and thus its only critical point must be its global maximum.)

We were asked to find the dimensions of the "optimal" chicken pen, and we now know that the width of the pen should be $w = 25$ feet. Since $l = 50 - w$, the length of the pen should also be $l = 25$ feet, and the pen is square! The final answer to the word problem is

Farmer Joe should build a square chicken pen
each side of which is 25 feet long.   □

⊛ *Caution*   It is true in general that, given a fixed perimeter, the rectangular enclosure with the maximum area will always be a square. (This should make sense to you. Think about it until it does.) However, it is not generally a good idea to *assume* something like this when solving an optimization problem; some problems that look similar have solutions that are different. Do the math first, and then check that your answer makes sense.   □

Most optimization problems with constraints follow exactly the same pattern as Example 5.26. In general, optimization word problems can be solved using the following algorithm.

ALGORITHM 5.4

Solving Optimization Problems

The following steps form a general outline for solving optimization word problems.

1. Follow Algorithm 5.2 to translate the word problem into a mathematical problem. In particular, identify the function to be optimized and any constraints.

2. Use the constraints to write the function to be optimized in terms of one variable. Find the domain of this function that is relevant to the problem. You have now translated the word problem into a mathematical problem of finding a global maximum or minimum of a function on an interval. Be sure to identify the function and the interval clearly.

3. Find any critical points of f in the interior of the interval, and use the first or second derivative test to determine whether these critical points are local maxima, local minima, or neither. Then apply Algorithm 5.3 to compare these interior local extrema to the behavior of f at the "ends" of the interval, and determine the *global* maximum or minimum of the function on the interval.

4. Once you have determined the global maximum or minimum of the given function on the given interval, find the answer to the original problem. Sometimes this involves finding the *value* of the maximum or minimum that you found, and sometimes this involves finding the dimensions or quantities that produce that maximum or minimum value.

The following example is similar in character to Example 5.26 but involves a more complicated translation process.

EXAMPLE 5.27

A more challenging real-world optimization problem

You work for a company that makes jewelry boxes. Your boss tells you that each jewelry box must have a square base and an open top and that you can spend $3.75 on the materials for each box. The people in production tell you that the material for the sides of the box costs two cents per square inch and that the reinforced material for the base of the box costs five cents per square inch. What is the largest jewelry box you can make (in terms of volume) and still stay within budget?

Solution   From a quick review of the problem, we can see that we wish to maximize the volume of a jewelry box given certain monetary constraints. Let's begin with a simple picture, assign variable names, and list what we know in mathematical language. Suppose x is the length (and width) of the sides of the square base, in inches, and let y

be the height (also in inches) of the jewelry box. The material to make each of the four sides of the box will cost $0.02(xy)$ dollars, and the material to make the base of the box will cost $0.05(x^2)$ dollars.

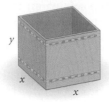

Constraint: $0.02(4xy) + 0.05(x^2) = 3.75$

Maximize: $V = x^2 y$

Again, the function to be optimized is written in terms of *two* variables at this point, but we can use the constraint to solve for one variable in terms of the other. It is easiest to solve for y in terms of x:

$$0.02(4xy) + 0.05(x^2) = 3.75 \implies 0.08xy + 0.05x^2 = 3.75$$
$$\implies 0.08xy = 3.75 - 0.05x^2$$
$$\implies y = \frac{3.75 - 0.05x^2}{0.08x}.$$

Now we write the volume function V in terms of one variable (and simplify it as much as possible):

$$V = x^2 y = x^2\left(\frac{3.75 - 0.05x^2}{0.08x}\right) = \tfrac{1}{0.08}x(3.75 - 0.05x^2)$$

$$= 12.5x(3.75 - 0.05x^2) = 46.875x - 0.625x^3.$$

Now we must determine the appropriate domain for V (in terms of the problem; as an abstract function, V has domain $(-\infty, \infty)$). Clearly the smallest that x can be is zero (why?). To find the largest that the base side length x can be, consider the smallest possible value of the height: $y = 0$. If $y = 0$, then we have

$$0 = \frac{3.75 - 0.05x^2}{0.08x} \implies 0 = 3.75 - 0.05x^2 \implies x = \sqrt{\tfrac{3.75}{0.05}} = \sqrt{75} \approx 8.66.$$

Note that we do not consider the negative square root of 75, since we know that the length x must be nonnegative.

We have now completely translated the word problem into the mathematical problem

Find the maximum value of $V(x) = 46.875x - 0.625x^3$ *on the interval* $[0, \sqrt{75}]$.

To solve this problem, we will find all the local interior extrema of $V(x)$ in the interval $[0, \sqrt{75}]$ and compare their values to the values of V at the endpoints of the interval. The derivative of V is $V'(x) = 46.875 - 3(0.625)x^2$, which is zero when

$$46.875 - 3(0.625)x^2 = 0 \implies x = \sqrt{\tfrac{46.875}{3(0.625)}} = \sqrt{25} = \pm 5.$$

Thus there are two critical points of V, one at $x = 5$ and one at $x = -5$. The only critical point in the interval $[0, \sqrt{75}]$ is at $x = 5$. Since $V''(5) = -2(3)(0.625)(5)$ is negative, we know by the second derivative test that this critical point $x = 5$ is a local maximum. To see whether $x = 5$ is the *global* maximum of V on $[0, \sqrt{75}]$ we must compare the value of V at $x = 5$ with the values of V at the endpoints $x = 0$ and $x = \sqrt{75}$:

$$V(5) = 46.875(5) - 0.625(5)^3 = 156.25,$$
$$V(0) = 46.875(0) - 0.625(0)^3 = 0,$$
$$V(\sqrt{75}) = 46.875(\sqrt{75}) - 0.625(\sqrt{75})^3 = 0.$$

(Of course, since $x = 0$ and $x = 75$ correspond to "flat" boxes, we expected that the volume would be zero at these endpoint values.) Therefore, $x = 5$ is the global maximum of $V(x)$ on $[0, \sqrt{75}]$.

The original word problem asked us to find the largest possible volume of a jewelry box with the given conditions. Since $V(5) = 156.25$, the answer to this question is

> *The largest possible jewelry box you could make, and still stay within your budget, is a box with volume* 156.25 *cubic inches.*

We weren't asked to find the dimensions of the box, but we could do so if we wanted to. We know that the largest possible jewelry box will have base side length of $x = 5$ inches and thus a height of

$$y = \frac{3.75 - 0.05(5)^2}{0.08(5)} = 6.25 \text{ inches.} \qquad \square$$

☑ *Checking the Answer* You can check some of the work in Example 5.27 for algebraic errors by checking that with the dimensions $x = 5$ and $y = 6.25$ inches, the volume of the jewelry box is indeed $V = x^2 y = 156.25$ inches, and the cost of producing the jewelry box is indeed $0.02(4xy) + 0.05(x^2) = \$3.75$. You can check your work regarding the optimization of V (and the choice of endpoints for your interval) by graphing $V(x) = 46.875x - 0.625x^3$. This graph is shown in Figure 5.37.

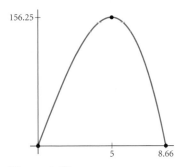

Figure 5.37 $\square$

We will see more optimization problems, with more complicated setups and functions, in Chapter 7. Right now we are doing optimization problems only for polynomial functions. In particular, this means that the functions we are optimizing in this section are always defined, continuous, and differentiable everywhere. In Chapter 7 the functions we optimize will not necessarily be so nice.

What you should know after reading Section 5.4

▶ How to translate a word problem into a mathematical problem. This includes naming variables and constructing functions on the basis of the information given in the word problem. Also, know how to translate the answer to this mathematical problem into real-world terms.

▶ How to find the global maximum or minimum of a function on an interval. This includes closed, half-closed, and open intervals. When will a function have a local extremum at an endpoint of an interval?

▶ How to translate, solve, and answer optimization word problems. This can include using constraints given in the problem to write the function to be optimized in terms of one variable.

Exercises 5.4

Concepts

0. Read the section and make your own summary of the material. Include whatever you think will help you review the material later on.

1. Explain the meaning of the diagram in Figure 5.33 in this section.

2. Use mathematical notation (in particular, inequalities) to express each of the following statements.

(a) $x = 2$ is a local maximum of the function f on the interval $[-3, 5]$.

(b) $x = -1$ is a global maximum of the function f on the interval $[-3, 5]$.

(c) $x = c$ is a global minimum of the function f on the interval I.

3. Explain the difference between an *endpoint extremum* and an *interior extremum* of a function on an interval I.

4. Suppose f is a function that is defined and continuous on an *open* interval I. Will the endpoints of I be local extrema of f? Will f necessarily have a global maximum or minimum in the interval I?

5. Suppose f is a function that is defined and continuous on a *closed* interval I. Will the endpoints of I be local extrema of f? Will f necessarily have a global maximum or minimum in the interval I?

6. When you try to find the local extrema of a function f on an interval I, one of the first steps is to find the critical points of f. Explain why these critical points of f won't help you locate "endpoint" extrema.

7. Explain why you *can't* find the global maximum of a function f on an interval I just by finding all the local extrema of f and then checking to see which one has the highest value $f(c)$.

8. Write down in mathematical notation what Algorithm 5.3 says about finding the global minimum of a function f on an interval I.

Skills

■ **Use Algorithm 5.3 to solve each of the problems below. Then check your answers with appropriate graphs.**

9. Find the locations and values of any global extrema of the function $f(x) = 2x^3 - 3x^2 - 12x$ on each of the following intervals:

(a) $[-3, 3]$ (b) $[0, 3]$
(c) $(-1, 2]$ (d) $(-2, 1)$
(e) $[-2, \infty)$

10. Find the locations and values of any global extrema of the function $f(x) = 3x^4 + 4x^3 - 36x^2$ on each of the following intervals:

(a) $[-5, 5]$ (b) $[-2, 2]$
(c) $[-4, 2]$ (d) $(-3, 1]$
(e) $(-1, 3)$

11. Find the locations and values of any global extrema of the function $f(x) = -12x + 6x^2 + 4x^3 - 3x^4$ on each of the following intervals:

(a) $[-1, 1]$ (b) $(-1, 1)$
(c) $(-3, 0]$ (d) $[0, 3]$
(e) $(-\infty, \infty)$

12. Find two real numbers x and y whose sum is 35 and whose product is as large as possible.

13. Find two real numbers x and y whose sum is 35 and whose product is as small as possible.

14. Find real numbers a and b for which the sum of a and b is 100 and the sum of the squares of a and b is as small as possible.

■ **Find dimensions for each of the following three shapes such that the total area enclosed is as large as possible, assuming that the total edge length of each shape is 120 inches. (You may assume that the rounded shapes are half-circles and that the triangular shapes are equilateral.)**

15.

16.

17.

■ **For each problem, find the point on the graph of the given function f that is closest to the given point (a, b), by minimizing the *square* of the distance. (*Hint:* Remember that a point (x, y) is on the graph of a function f if and only if $f(x) = y$.)**

18. $f(x) = 3x + 1$, $(-2, 1)$

19. $f(x) = x^2$, $(0, 3)$

20. $f(x) = x^2 - 2x + 1$, $(1, 2)$

Applications

- A toy car is set down on a straight track and moves back and forth for 4 minutes. The function $s(t) = 96t - 84t^2 + 28t^3 - 3t^4$ describes how far the car is to the right of the starting point after t minutes. Optimize an appropriate function on an appropriate interval to solve each of the following problems (don't forget to check endpoints!). When you have finished each problem, check your answer with a graph.

21. When is the toy car closest to the left end of the track? When is the toy car closest to the right end?

22. When is the toy car moving fastest to the right? When is the toy car moving fastest to the left?

23. When is the toy car accelerating fastest to the right? When is the toy car accelerating fastest to the left?

- Farmer Jill wants to build four fenced enclosures on her farmland for her free-range ostriches. To keep costs down, she is always interested in enclosing as much area as possible with a given amount of fence. For each of the four fencing projects, a specific type of ostrich pen is described, for which Jill has allotted a certain length of fencing. Determine how to set up each ostrich pen so that the maximum possible amount of area is enclosed, and find this maximum area. As with all optimization problems, make sure you show your work clearly, following the general outline of Algorithm 5.4.

24. A rectangular ostrich pen built with 350 feet of fencing material.

25. A rectangular ostrich pen built along the side of a river (so only three sides of fence are needed), with 540 feet of fencing material.

26. A rectangular ostrich pen that is divided into three equal sections by two interior fences that run parallel to the exterior side fences, with 1000 feet of fencing material.

27. A rectangular ostrich pen that is divided into six equal sections by two interior fences that run parallel to the east and west fences, and another interior fence running parallel to the north and south fences. Jill has allotted 2400 feet of fencing material for this important project.

28. Your family makes and sells velvet Elvis paintings. After many years of research, you have found a function that predicts how many paintings you will sell in a year on the basis of the price that you charge per painting. You always charge between 5 and 55 dollars per painting. Specifically, if you charge c dollars per painting, you have found that you can sell $N(c) = 0.6c^2 - 54c + 1230$ paintings in a year.

    (a) What price should you charge to sell the *greatest* number of velvet Elvis paintings, and how many could you sell at that price? For what price would you sell the *least* number of paintings, and how many would you sell? Find your answers algebraically using Algorithm 5.3, and then check them graphically.

    (b) Write down a function that predicts the revenue $R(c)$, in dollars, that you will earn in a year if you charge c dollars per painting. (*Hint:* Try some examples first; for instance, what would your yearly revenue be if you charged \$10.00 per painting? What about \$50.00? Then write down a function that works for all values of c.)

    (c) What price should you charge to earn the *most* money in revenue, and how much money would you earn? What price per painting would cause you to make the *least* amount of revenue in a year, and how much revenue would you make in that case? Find your answers algebraically using Algorithm 5.3 (you may need the quadratic formula at some point), and then check them graphically.

    (d) Explain why you do not make the most revenue at the same price per painting at which you sell the most paintings.

29. Suppose you want to make an open-topped box out of a 4-by-6-inch index card by cutting a square out of each corner and then folding up the edges (see the figure below). How large a square should you cut out of each corner in order to maximize the volume of the resulting box? Find your answer algebraically using Algorithm 5.3 (what is the interval here?), and then check your answer graphically.

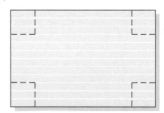

- Alina wants to make keepsake boxes for her two best friends. She doesn't have a lot of money, so she wants to make each box as large as possible (so that the boxes will hold the greatest amount possible) with a limited amount of material.

30. For Eliza, Alina wants to make a rectangular box with a top that is lined (on the inside *and* outside) with velvet, and whose base is twice as long as it is wide. If Alina has 240 square inches of velvet, how can she make Eliza's box so that it holds as many keepsakes as possible?

31. For Jen, Alina wants to make a box with a square base that is made of wood except for a metal top. The wood she wants to use costs 5 cents per square inch, and the material for the metal top costs 12 cents per square inch. What is the largest possible box that Alina can make for Jen if she has only \$20.00 to spend on materials?

- The U.S. Postal Service ships a package under "large package" rates if the sum of the length and the girth of the package is greater than 84 inches and less than or equal to 108 inches. The *length* of a package is considered to be the length of its longest side, and the *girth* of the package is the distance around the package perpendicular to its length. Linda wants to ship three packages under the USPS "large package" rates, as described below.

32. Linda's first package must be rectangular and 40 inches in length. What is the largest volume that her package can have? What is the largest surface area that her package can have?

33. Linda also needs to mail a rectangular package with square ends (in other words, with equal width and height). What is the largest volume that her package can have? What is the largest surface area that her package can have?

34. Linda also needs to mail some architectural plans, which must be shipped in a cylindrical container. What is the largest volume that her package can have? What is the largest surface area that her package can have? (*Hints:* The volume of a right circular cylinder with radius r and height h is $V = \pi r^2 h$; the total surface area of such a cylinder is $SA = 2\pi r h + 2\pi r^2$.)

■ **Suppose you have a 10-inch length of wire that you wish to cut and form into shapes. In each of the following three problems you will determine how to cut the** wire to minimize or maximize the area of the resulting shapes.

35. Suppose you wish to make one cut in the wire and use the two pieces to form a square and a circle. Determine how to cut the wire so that the combined area enclosed by the square and the circle is (**a**) as small as possible, and (**b**) as large as possible.

36. Suppose you wish to make one cut in the wire and use the two pieces to form a square and an equilateral triangle. Determine how to cut the wire so that the combined area is (**a**) as small as possible, and (**b**) as large as possible.

37. Suppose you wish to make two cuts in the wire and use the three pieces to form a square, a circle, and an equilateral triangle. Determine how to cut the wire so that the combined area of these three shapes is (**a**) as small as possible, and (**b**) as large as possible.

Proofs

38. Prove that the rectangle with the largest possible area, given a fixed perimeter P, is always a square.

39. Prove that the most efficient way to build a rectangular fenced area along a river (so only three sides of fencing are needed) is to make the side parallel to the river twice as long as the other sides. (Assume that you have a fixed amount of fencing material.)

CHAPTER 6

Rational Functions

A rational function is a function that can be
written as a quotient of two polynomials. In this
section we will use our understanding of polynomial
functions to investigate rational functions. Rational
functions are significantly more complicated than
power or polynomial functions; their graphs can
have many kinds of asymptotes and can even have
"holes."

Rational functions can be simplified by
factorization and by a process called polynomial
long division that enables us to divide one
polynomial by another. The roots of the numerator
and denominator of a rational function reveal much
about the behavior of its graph. Together with first
and second derivative information, these roots can be
used to produce very accurate graphs of rational
functions.

*Mirror maze at Wookey by
Adrian Fisher Mazes Ltd,
www.mazemaker.com*

CONTENTS

The Algebra of Rational Functions

In this section we learn how to simplify rational functions algebraically, that is, functions that are quotients of polynomials. In Section 5.1.5 we saw how to use synthetic division to divide a polynomial by a linear expression. We will now learn the more general algorithm of *polynomial long division,* which enables us to divide one polynomial by another polynomial.

6.1.1 What is a rational function?

A rational number is a number that can be written as a quotient of the simplest possible numbers you can imagine, namely integers. Similarly, a rational *function* is a quotient of very simple functions, namely polynomials.

DEFINITION 6.1

Rational Functions Are Quotients of Polynomials

A function f is a *rational function* if it can be written in the form

$$f(x) = \frac{p(x)}{q(x)} = \frac{a_n x^n + a_{n-1} x^{n-1} + \cdots + a_1 x + a_0}{b_m x^m + b_{m-1} x^{m-1} + \cdots + b_1 x + b_0},$$

where $p(x)$ and $q(x)$ are polynomial functions.

This is analogous to our definition of a rational *number,* which is a quotient of integers. A rational *function* is a quotient of polynomials. Just as every integer is also a rational number (with denominator 1), note that every polynomial function is also a rational function (whose denominator $q(x)$ is 1).

EXAMPLE 6.1   **Identifying rational functions**

Which of the following are rational functions?

(a) $f(x) = \dfrac{1 + 2x - 3x^3}{x^2 - 4}$ (b) $f(x) = \dfrac{\sqrt{1 + x^2}}{x^3 - x + 1}$

(c) $f(x) = \dfrac{x(x-1)(x+2)}{(x+1)(x-3)}$

Solution   Although all of these functions involve quotients, only (a) and (c) are rational functions. The numerator in (b) is not a polynomial, since it involves a square root.   □

A rational function is called *proper* if the degree of its numerator is less than the degree of its denominator and *improper* if the degree of its numerator is greater than or equal to the degree of its denominator. As we will see in Section 6.1.4, an improper rational function can be written as the sum of a polynomial and a proper rational function. This is analogous to what happens with improper rational *numbers* (also known as "improper fractions"), where the numerator is greater than or equal to the denominator. Every improper rational number can be written as the sum of an integer and a proper rational number (for example, $\frac{3}{2} = 1 + \frac{1}{2}$).

6.1.2 Factoring, canceling, and "holes"

Since rational functions are quotients of polynomial functions, much of their behavior can be described using what we know about polynomial functions. To understand the polynomial functions that make up a rational function, it is usually a good idea to write them in factored form. Moreover, we will often be concerned with what happens when the numerator and denominator of a rational function f have **common factors,** and to find these common factors we must be able to factor the numerator and denominator of f.

In general you *cannot* simply cancel common factors in the numerator and denominator of a rational function. If the numerator and denominator of a rational function f have a common factor $x = c$, then f will be *undefined* at $x = c$ (why?). Everywhere away from $x = c$ the original function f and the "cancelled" function will be equal, but at $x = c$ they may differ. The following theorem is a special case of a theorem from Chapter 2.

THEOREM 6.1

Cancellation of a Common Factor

Given any polynomial functions $r(x)$ and $s(x)$, and any real number c,

$$\frac{(x-c)r(x)}{(x-c)s(x)} = \begin{cases} \dfrac{r(x)}{s(x)}, & \text{if } x \neq c \\ \text{undefined}, & \text{if } x = c. \end{cases}$$

This theorem states that the functions $\frac{(x-c)r(x)}{(x-c)s(x)}$ and $\frac{r(x)}{s(x)}$ are identical everywhere except at $x = c$. The graphs of these functions will be identical everywhere except at $x = c$. At $x = c$ the graph of $\frac{(x-c)r(x)}{(x-c)s(x)}$ will have a **hole** (assuming that there are no more factors of $x - c$ in the denominator; see the "Caution" before Example 6.4).

EXAMPLE 6.2

Two rational functions that are equal except at one point

Consider the rational functions

$$f(x) = \frac{(x-2)(x+1)}{(x-1)(x+1)} \quad \text{and} \quad g(x) = \frac{x-2}{x-1}.$$

These functions are equal for all values of x that are not equal to -1 (try some values, like $x = 0$ and $x = 5$, to see why). However, at $x = -1$ we have $f(-1) = \frac{(-3)(0)}{(-2)(0)}$, which is undefined, whereas $g(-1) = \frac{-1-2}{-1-1} = \frac{-3}{-2} = \frac{3}{2}$. The graphs of f and g are shown in Figures 6.1 and 6.2.

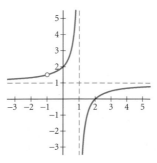

Figure 6.1
$$f(x) = \frac{(x-2)(x+1)}{(x-1)(x+1)}$$

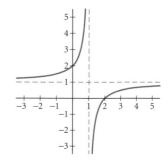

Figure 6.2
$$g(x) = \frac{x-2}{x-1}$$

Note that these two graphs are identical except at $x = -1$, where f is undefined and has a "hole." ☐

Writing a rational function in factored form simply involves writing its numerator and denominator polynomials in factored form. Remember that we have a lot of techniques for factoring, including synthetic division and factoring by grouping.

EXAMPLE 6.3

Finding common factors by factoring

Factor the following rational function $f(x)$ as much as possible, and identify any common factors in the numerator and the denominator.

$$f(x) = \frac{x^3 - x^2 - 4x + 4}{x^2 - 3x + 2}$$

Solution   The numerator polynomial $p(x) = x^3 - x^2 - 4x + 4$ can be factored by grouping:

$$p(x) = x^3 - x^2 - 4x + 4 = x^2(x - 1) - 4(x - 1)$$
$$= (x^2 - 4)(x - 1) = (x - 2)(x + 2)(x - 1).$$

The denominator polynomial $q(x) = x^2 - 3x + 2$ is an easy quadratic to factor:

$$q(x) = x^2 - 3x + 2 = (x - 2)(x - 1).$$

Therefore, the given rational function $f(x)$ can be written in factored form as

$$f(x) = \frac{x^3 - x^2 - 4x + 4}{x^2 - 3x + 2} = \frac{(x - 2)(x + 2)(x - 1)}{(x - 2)(x - 1)}.$$

This rational function has two factors common to the numerator and denominator, namely $x - 2$ and $x - 1$. This means that if $x \neq 2$ and $x \neq 1$, then $f(x)$ is actually equal to the linear function (after cancellations) $x + 2$. However, at $x = 2$ and at $x = 1$, the original function $f(x)$ is not defined (why?).   □

? *Question*   Use the information in the solution to Example 6.3, and what you know about "holes" of rational functions, to sketch graphs of $f(x)$ and of $y = x + 2$.   □

$ *Caution*   If a rational function f has a common factor $x - c$ in its numerator and its denominator, then it *may or may not* have a hole at $x = c$. If the factor $x - c$ appears with greater multiplicity in the denominator than in the numerator, then the graph of f will have an *asymptote* at $x = c$ instead of a hole.   □

In the following example we examine three situations that occur when a rational function has a common factor in the numerator and the denominator. The common factor can appear with a higher multiplicity in the numerator, or with the same multiplicity in the numerator and the denominator, or with a higher multiplicity in the denominator.

EXAMPLE 6.4

Common factors sometimes cause "holes" and sometimes do not

Consider the graphs of $f(x) = \frac{(x-1)^2}{x-1}$, $g(x) = \frac{x-1}{x-1}$, and $h(x) = \frac{x-1}{(x-1)^2}$ shown in Figures 6.3–6.5.

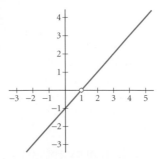

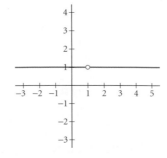

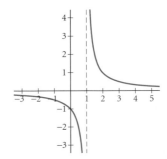

Figure 6.3

$$f(x) = \frac{(x - 1)^2}{x - 1}$$

Figure 6.4

$$g(x) = \frac{x - 1}{x - 1}$$

Figure 6.5

$$h(x) = \frac{x - 1}{(x - 1)^2}$$

The graph of f is identical to the graph of $y = x - 1$, except that it has a hole at $x = 1$. The graph of g is identical to the graph of $y = 1$, except that it has a hole at $x = 1$. On the other hand, the graph of h is identical to the graph of $y = \frac{1}{x-1}$ everywhere; why? The reason is that the function $h(x)$ is equal to the function $\frac{1}{x-1}$ (after canceling) everywhere except at $x = 1$, where *both* of these functions are undefined. Note in particular that in this last case, h does *not* have a hole at $x = 1$, even though it has a common factor of $x - 1$ in the numerator and the denominator. ☐

6.1.3 Domains and roots of rational functions

The roots and domain of a rational function are completely determined by the roots of its numerator and denominator polynomials. Specifically, we have the following theorem.

THEOREM 6.2

Domain and Roots of a Rational Function

Let $f(x) = \frac{p(x)}{q(x)}$ be a rational function. Then:

(a) $x = c$ is a root of $f(x)$ if and only if $p(c) = 0$ but $q(c) \neq 0$.

(b) The domain of f is the set $\{x \mid q(x) \neq 0\}$.

In other words, the zeros of the numerator polynomial $p(x)$ are the roots of f that are *not* also roots of $q(x)$, and the zeros of the denominator polynomial $q(x)$ are the values at which f is *not* defined. Recall that if a value $x = c$ is a root of *both* $p(x)$ and $q(x)$, then f is not defined at $x = c$ and thus cannot have a root at $x = c$. The proof of this theorem is simple (involving only properties of polynomials and quotients) and is left to the exercises.

EXAMPLE 6.5

Finding the domain and roots of a rational function

Find the domain and roots of the rational function

$$f(x) = \frac{x^2 - 1}{x^3 - 2x^2 - 5x + 6}.$$

Solution We must first factor the numerator and denominator. Clearly the numerator factors as $p(x) = x^2 - 1 = (x - 1)(x + 1)$. To factor the denominator $q(x) = x^3 - 2x^2 - 5x + 6$, we need synthetic division. By the method of guessing integer roots, we know that the only integers that *could* be roots of $q(x)$ are ± 1, ± 2, ± 3, and ± 6. The "easiest" of these four possible roots is $x = 1$, and a quick check ($q(1) = (1)^3 - 2(1)^2 - 5(1) + 6 = 1 - 2 - 5 + 6 = 0$) shows us that $x = 1$ is indeed a root of $q(x)$. We can now use synthetic division to pull out a factor of $x - 1$ from $q(x)$. See Figure 6.6.

$$
\begin{array}{r|rrrr}
 & 1 & -2 & -5 & 6 \\
1 & & 1 & -1 & -6 \\
\hline
 & 1 & -1 & -6 & 0
\end{array}
$$

Figure 6.6

Therefore, we have

$$q(x) = x^3 - 2x^2 - 5x + 6$$
$$= (x-1)(x^2 - x - 6) \qquad \text{(synthetic division)}$$
$$= (x-1)(x+2)(x-3). \qquad \text{(factoring a quadratic)}$$

The rational function $f(x)$ can be written in factored form as

$$f(x) = \frac{x^2 - 1}{x^3 - 2x^2 - 5x + 6} = \frac{(x-1)(x+1)}{(x-1)(x+2)(x-3)}.$$

The roots of the numerator are $x = 1$ and $x = -1$, and the roots of the denominator are $x = 1$, $x = -2$, and $x = 3$. Since the common factor $(x - 1)$ occurs with the same multiplicity in the numerator and the denominator, we know that $x = 1$ is a "hole" of f (and thus cannot be a root). The only root of f is the value $x = -1$. Moreover, f is defined everywhere except at $x = 1$, at $x = -2$, and at $x = 3$. The domain of f is therefore

$$\text{Domain}(f) = \{x \,|\, (x-1)(x+2)(x-3) \neq 0\}$$
$$= \{x \,|\, x \neq 1, x \neq -2, x \neq 3\}$$
$$= (-\infty, -2) \cup (-2, 1) \cup (1, 3) \cup (3, \infty). \qquad \square$$

⬧ *Caution*   If you use a graphing calculator to check the answers to Example 6.5, you will not be able to see from the graph that f is undefined at $x = 1$; your calculator will not draw the "hole" in the graph. However, if you use your calculator to trace along the function to $x = 1$, you will see that the function is not defined at $x = 1$ and thus has a "hole" there. $\square$

✓ *Checking the Answer*   The graph of $f(x) = \frac{x^2-1}{x^3-2x^2-5x+6}$ is shown in Figure 6.7. Note that f has only one root, at $x = -1$, and that f is undefined at $x = -2$ and $x = 3$ (where there are asymptotes) and at $x = 1$ (where there is a hole).

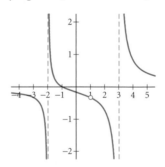

Figure 6.7 $\square$

6.1.4 Polynomial long division

As we mentioned previously, every improper rational function can be written as the sum of a polynomial function and a proper rational function. To write an improper rational function in this form, we must divide the numerator polynomial by the denominator polynomial. In this section we will present an algorithm called *polynomial long division* that will enable us to divide one polynomial by another.

Polynomial long division works almost like the long-division algorithm that we use to divide one integer by another, except that it involves polynomials instead of positive integers. Let's quickly review the long-division algorithm for dividing positive integers.

| EXAMPLE 6.6 | **A review of the long-division algorithm** |

The long-division algorithm for dividing 131 by 4 is shown in Figure 6.8.

$$
\begin{array}{r}
3\,2 \\
4\overline{)1\,3\,1} \\
\underline{1\,2} \\
1\,1 \\
\underline{8} \\
3
\end{array}
$$

Figure 6.8

The work illustrated in Figure 6.8 can be briefly explained as follows: The first step is to figure out how many times 4 goes into 13, which is three times. We then multiply the number 3 by 4 and subtract this product (12) from 13. This gives us a remainder of 1. We bring down the remaining 1 from the number 131. Now we repeat the process and determine that 4 goes into 11 two times (with a remainder). This gives us the "2" in the "32" at the top of the diagram. If we multiply 2 by 4 and subtract the product (8) from 11, we are left with a remainder of 3. Since 4 does not go into 3 (because $4 > 3$), we are finished, and we have shown that 4 goes into 131 thirty-two times, with a remainder of 3. Therefore, $4(32) + 3 = 128 + 3 = 131$ (check this). We can rewrite this equality by dividing both sides by 4:

$$
\frac{131}{4} = 32 + \frac{3}{4}.
$$

We'll introduce the algorithm of *polynomial* long division by example. Keep in mind that we are following the same algorithm as in Example 6.6, with the exception that we are working with polynomials instead of integers.

| EXAMPLE 6.7 | **Using long division to divide one polynomial by another** |

Suppose we wish to divide the polynomial $4x^2 + 2x + 3$ by the polynomial $x - 2$. The work for using polynomial long division to do this is given in Figure 6.9 and explained below.

$$
\begin{array}{r}
4x + 10 \\
x - 2\overline{)\,4x^2 + 2x + 3} \\
\underline{4x^2 - 8x} \\
10x + 3 \\
\underline{10x - 20} \\
23
\end{array}
$$

Figure 6.9

The leading term of $x - 2$ is x, and the leading term of $4x^2 + 2x + 3$ is $4x^2$. The first step in the algorithm is to determine how many times x goes into $4x^2$, which is $4x$. Therefore, the first entry on top of the line is $4x$. Now we multiply $x - 2$ by $4x$ to get $4x^2 - 8x$ and then *subtract* this from the polynomial $4x^2 + 2x + 3$. Note that the first terms (both equal to $4x^2$) cancel each other out; this should *always* happen when doing polynomial long division. In this example we have $(4x^2 + 2x + 3) - (4x^2 - 8x) = 4x^2 + 2x + 3 - 4x^2 + 8x = 10x + 3$, and we write that below the first horizontal line. (See the "Caution" below.)

Now we repeat the process. The leading term of $x - 2$ goes into the leading term of $10x + 3$ ten times, since $10(x) = 10x$. Thus the second entry above the top line is 10. Now we multiply 10 by $x - 2$ and subtract this product $(10x - 20)$ from $10x + 3$. When we do so, we get a remainder of 23. Since 23 has a degree less than $x - 2$, the process is over. (*Note:* If $x - 2$ had been a factor of $4x^2 + 2x + 3$, then the remainder would have been zero.) We have now shown that

$$(x - 2)(4x + 10) + 23 = 4x^2 + 2x + 3.$$

We can rewrite this equality by dividing both sides by $x - 2$:

$$\frac{4x^2 + 2x + 3}{x - 2} = (4x + 10) + \frac{23}{x - 2}. \qquad \square$$

⑂ Caution   Be sure that you *subtract* in the polynomial long-division algorithm (for example, where we subtracted $4x^2 - 8x$ from $4x^2 + 2x + 3$ in Example 6.7). A common mistake is to add the polynomials instead of subtracting them. Some people find it useful to change all the signs of the second polynomial (effectively multiplying it by -1) and then add this altered polynomial to the first one; this is equivalent to subtracting the original two polynomials. Whatever you do, be consistent, and check your work. $\square$

Synthetic division is a shorthand notation for polynomial long division. However, synthetic division works only for problems where we wish to divide by a *linear* function, such as $x - 2$, whereas polynomial long division can be used to divide a polynomial by another polynomial of *any* degree. To divide $4x^2 + 2x + 3$ by $x - 2$ using synthetic division, we would proceed as shown in Figure 6.10.

$$
\begin{array}{r|rrr}
 & 4 & 2 & 3 \\
2 & & 8 & 20 \\
\hline
 & 4 & 10 & 23 \\
\end{array}
$$

Figure 6.10

Compare this to the polynomial long division done in Figure 6.9. In particular, notice where the numbers 4, 10, and 23 appear in both calculations and how they are found. Where do the numbers 8 and 20 appear in the polynomial long-division algorithm?

The general algorithm for polynomial long division is given in Algorithm 6.1. You can use this algorithm to divide any polynomial $p(x)$ by any polynomial $q(x)$, provided that the degree of $q(x)$ is less than or equal to the degree of $p(x)$ (in other words, provided that $\frac{p(x)}{q(x)}$ is an *improper* rational function).

ALGORITHM 6.1

Polynomial Long Division

To use polynomial long division to divide a polynomial $p(x)$ by a polynomial $q(x)$ (of equal or lower degree), start by writing the diagram

$$q(x)\overline{)\,p(x)}$$

Then follow these steps (compare with Figure 6.9).

1. Find an expression ax^k whose product with the leading term of $q(x)$ is equal to the leading term of $p(x)$. Record this expression above the line.

2. Multiply $q(x)$ by the expression ax^k you found in step 1, and *subtract* the product from $p(x)$. For the purposes of the next step, call the resulting polynomial the "remainder polynomial."

3. Repeat steps 1 and 2 with $p(x)$ replaced by the remainder polynomial. Repeat until the degree of the remainder polynomial is less than the degree of $q(x)$.

When the algorithm terminates, the "remainder polynomial" from the last step is the **remainder.** At the end of the algorithm, if the expression above the top line is $m(x)$ and the remainder is $R(x)$, then

$$p(x) = q(x)m(x) + R(x).$$

In other words,

$$\frac{p(x)}{q(x)} = m(x) + \frac{R(x)}{q(x)}.$$

In this text, we won't try to prove that this algorithm always works. However, given that this algorithm successfully allows us to divide one polynomial by another, we can use it to explain why *synthetic* division works; you will do so in the exercises.

Algorithm 6.1 can always be used to write an improper rational function $\frac{p(x)}{q(x)}$ as the sum of a polynomial function $m(x)$ and a proper rational function $\frac{R(x)}{q(x)}$.

EXAMPLE 6.8   **Using polynomial long division to simplify an improper rational function**

Write the following improper rational function $f(x)$ as a sum of proper rational functions.

$$f(x) = \frac{x^4 - 3x^3 + x - 1}{x^2 + x - 3}$$

Solution   Figure 6.11 illustrates the polynomial long-division algorithm.

$$
\begin{array}{r}
x^2 - 4x + 7 \\
x^2 + x - 3 \overline{\smash{\big)}\ x^4 - 3x^3 - 0x^2 + x - 1} \\
\underline{x^4 + x^3 - 3x^2} \\
-4x^3 + 3x^2 + x - 1 \\
\underline{-4x^3 - 4x^2 + 12x} \\
7x^2 - 11x - 1 \\
\underline{7x^2 + 7x - 21} \\
-18x + 20
\end{array}
$$

Figure 6.11

Note that we added "$0x^2$" to the function $p(x)$ so that the terms of all our polynomials would line up by degree. Note also that we stopped when the degree of the "remainder polynomial" $(-18x + 20)$ became smaller than the degree of $q(x) = x^2 + x - 3$. We now know that

(1) $x^4 - 3x^3 + x - 1 = (x^2 + x - 3)(x^2 - 4x + 7) + (-18x + 20).$

We can rewrite this result by dividing both sides of the equality by $x^2 + x - 3$:

(2) $\dfrac{x^4 - 3x^3 + x - 1}{x^2 + x - 3} = (x^2 - 4x + 7) + \dfrac{-18x + 20}{x^2 + x - 3}.$

We have now written the original improper rational function as a sum of a polynomial and a proper rational function, as desired.   □

☑ *Checking the Answer*   There are two ways to check the answer to Example 6.8. First, we could multiply out the right-hand side of Equation (1) and check that it is equal to $x^4 - 3x^3 + x - 1$. Second, we could combine the terms on the right-hand side of Equation (2) by finding a common denominator and then show that this is equal to the original improper fraction on the left-hand side.   □

✦ *Caution*   If you don't add "placeholders" for any "invisible" terms (like the $0x^2$ in Example 6.8), then your polynomials won't line up correctly. If your polynomials aren't lined up with each other, you are likely to make errors in the subtraction step.   □

What you should know after reading Section 6.1

▶ The definition of and notation for rational functions. When is a rational function "improper"? How to write a rational function in factored form. What happens if f has a common factor in the numerator and the denominator?

▶ How to identify the roots and domain of a rational function $f(x) = \frac{p(x)}{q(x)}$. What happens if $x = c$ is a root of *both* $p(x)$ and $q(x)$?

▶ How to divide one polynomial by another using the method of polynomial long division. Be able to use the results of this algorithm to write an improper rational function as a sum of a polynomial and a proper rational function.

Exercises 6.1

Concepts

0. Read the section and make your own summary of the material. Include whatever you think will help you review the material later on.

1. What is a rational function? Give three examples of rational functions and three examples of functions that involve quotients but are not rational.

2. Explain what it means for a rational function to be *improper* and what it means for a rational function to be *proper*. Give an example of each type of rational function.

■ **Determine which of the following functions are rational functions. For each rational function f, write the function in the form $f(x) = \frac{p(x)}{q(x)}$ for some polynomials $p(x)$ and $q(x)$.**

3. $f(x) = 1 + \dfrac{x+1}{4x-3}$

4. $f(x) = \dfrac{1}{x} - \dfrac{x^2-1}{x+3}$

5. $f(x) = \dfrac{\left(\frac{3x+1}{x-1}\right)}{\left(\frac{x^2-4}{x+5}\right)}$

6. $f(x) = x^2 - 3x^{\frac{1}{2}}$

7. $f(x) = \dfrac{x^{-1}}{3x^2+1}$

8. $f(x) = 3x^{-1} - x^{-2} + 2x^{-3}$

9. Explain what it means to say that a rational function f has a *common factor* in its numerator and its denominator.

10. How can you find the roots of a polynomial function $f(x) = \frac{p(x)}{q(x)}$ by looking at the polynomials $p(x)$ and $q(x)$?

11. How can you determine the domain of a rational function $f(x) = \frac{p(x)}{q(x)}$ by looking at the polynomials $p(x)$ and $q(x)$?

12. Suppose that $f(x)$ is a rational function of the form
$$f(x) = \frac{(x-2)\,p(x)}{(x-2)q(x)}$$
for some polynomial functions $p(x)$ and $q(x)$. What can you say about $f(4)$? About $f(2)$?

13. Consider the functions
$$f(x) = \frac{x^2-1}{x^2+4x-5} \quad \text{and} \quad g(x) = \frac{x+1}{x+5}.$$
For which values of x are these two functions equal? For which values of x are these two functions different?

14. Use what you know about "holes" of rational functions to sketch a graph of the following rational function $f(x)$ without using your calculator.
$$f(x) = \frac{(x+1)(x-3)(x+2)}{(x+2)(x-3)}$$

15. Suppose $f(x) = \frac{p(x)}{q(x)}$ is a rational function and that $p(x)$ has roots only at $x = 1$, $x = 3$, and $x = -3$, whereas $q(x)$ has roots only at $x = 2$, $x = 3$, and $x = 4$. What are the roots of f? What is the domain of f?

16. Use a graphing calculator to make a graph of the function $f(x) = \frac{x^2+x-2}{x^2-4}$. Somewhere on this graph is a "hole" where the function is not defined; determine the location of this "hole," and trace along the graph on your calculator until you find it there.

17. Explain why the graph of $f(x) = \frac{(x-1)(x+2)}{(x+2)^2}$ does *not* have a "hole" in it at $x = -2$.

■ **For each of the seven problems below, write the equation of a rational function f that has the properties described.**

18. f has roots at $x = 1$ and $x = 3$, is defined everywhere except at $x = -2$ and $x = 4$, and has no "holes."

19. f is defined everywhere except at $x = 3$ and $x = 0$, has roots at $x = -2$ and $x = 2$, and has a "hole" at $x = 3$.

20. f is defined everywhere except at $x = -1$, $x = 1$, and $x = 2$, has roots at $x = 0$ and at $x = -3$, and has "holes" at $x = -1$ and at $x = 1$. In addition, the denominator polynomial of f is a quartic polynomial.

21. f is defined everywhere except at $x = 3$ and $x = -2$, has only one root at $x = 1$, and has a "hole" at $x = 3$. In addition, f is an *improper* rational function.

22. f has a common factor of $x - 4$ in the numerator and the denominator, but the graph of f does *not* have a hole at $x = 4$.

23. The graph of f has a hole at the coordinates $(3, 0)$.

24. The graph of f is a line with three holes in it.

25. We have seen that a common linear factor $x - c$ of a rational function f can cause the graph of f to have a "hole" at $x = c$. What happens if a rational function f has an *irreducible quadratic* factor that is common to the numerator and the denominator, and why? Give an example to support your argument.

26. What is polynomial long division for? Give three examples of ways in which polynomial long division is like the long-division algorithm for numbers, and give two examples of ways in which it is different.

27. What would happen if you tried to use polynomial long division to divide a second-degree polynomial by a third-degree polynomial? Identify the part of Algorithm 6.1 that justifies your answer. (It may help to try an example first.)

28. Suppose you use polynomial long division to divide the polynomial $p(x)$ by the polynomial $q(x)$, and after doing your calculations you end up with the polynomial $x^2 - x + 3$ above the top line and the polynomial $3x - 1$ as a remainder. Then $p(x) = $ _____ and $\frac{p(x)}{q(x)} = $ _____ .

29. Use polynomial long division to divide the polynomial $x^3 - 2x + 3$ by the polynomial $x - 2$. Then use synthetic division to perform the same calculation. Compare the two calculations, and determine where each number that appears in the synthetic-division algorithm appears in the polynomial long-division algorithm. Explain how the synthetic-division algorithm is really the same as the polynomial long-division algorithm, but with abbreviated notation.

■ **Determine whether each of the following statements is true or false. If a statement is true, explain why. If a statement is false, provide a counterexample.**

30. True or False: Every polynomial function is also a rational function.

31. True or False: If $f(x) = \frac{p(x)}{q(x)}$ is a rational function, and $p(3) = 0$, then $x = 3$ is a root of f.

32. True or False: If $f(x) = \frac{p(x)}{q(x)}$ is a rational function, and $q(3) = 0$, then $x = 3$ is not in the domain of f.

33. True or False: If $f(x) = \frac{p(x)}{q(x)}$ is a rational function, and $q(3) = 0$, then $x = 3$ is a "hole" of f.

34. True or False: If $f(x) = \frac{p(x)}{q(x)}$ is a rational function where $p(x)$ and $q(x)$ are both cubic polynomials, then $f(x)$ is an improper rational function.

35. True or False: Synthetic division can always be used in the place of polynomial long division.

Skills

■ **For each rational function, factor the numerator and denominator as much as possible (if the function is not written that way already). Then determine the domain of f, and locate any roots or "holes" in the graph of f.**

36. $f(x) = \dfrac{(x-1)(x+5)(x-3)}{x(x-2)}$

37. $f(x) = \dfrac{(x-1)(x+5)(x-3)}{x(x-1)(x-2)}$

38. $f(x) = \dfrac{x^2(x+2)(x^2+1)}{(x+1)(x^2+3)}$

39. $f(x) = \dfrac{x^2+x-2}{2x^2-x-6}$

40. $f(x) = \dfrac{x^2-6x+9}{x+x^3}$

41. $f(x) = \dfrac{x^2+4}{x^2-4}$

42. $f(x) = \dfrac{x^3-6x^2+12x-8}{-x^3+3x^2+4x-12}$

43. $f(x) = \dfrac{x^3+6x^2+3x-10}{x^5+3x^4-x-3}$

44. $f(x) = \dfrac{x^3-x^2-16x+16}{x^4-16}$

45. $f(x) = \dfrac{x^4-4x^2+4}{x^3+x^2-2x-2}$

■ Use polynomial long division to write each of the following improper rational functions as the sum of a polynomial function and a proper rational function. Check some of your answers by multiplying out $q(x)m(x) + R(x)$ and the rest of your answers by simplifying $m(x) + \frac{R(x)}{q(x)}$ (see Algorithm 6.1).

46. $f(x) = \dfrac{x^3 - 2x^2 + 5}{x - 2}$

47. $f(x) = \dfrac{x^2 + 3x - 2}{x^2 + x - 1}$

48. $f(x) = \dfrac{x^5 - 5x^3 - 4x^2 + 6x + 12}{x^2 - 3}$

49. $f(x) = \dfrac{1 - x^4}{x^2 + 2x - 1}$

50. $f(x) = \dfrac{x^4 - 2x^3 - 3x^2 + 8x - 4}{x^2 - 2x + 1}$

51. $f(x) = \dfrac{x^4 + 12}{x^3 - 3}$

52. $f(x) = \dfrac{x^5 - 3x^2 + x - 1}{2x + 5}$

53. $f(x) = \dfrac{2 - 3x^2 + x^4}{2x^2 + 1}$

54. $f(x) = \dfrac{x^5 - 3x^4 - 3x^3 - 3x^2 - 4x}{x^2 - 3x - 4}$

55. $f(x) = \dfrac{x^7 - 1}{1 - 2x + x^2}$

Proofs

56. Prove the first part of Theorem 6.2: If $f(x) = \frac{p(x)}{q(x)}$ is a rational function, then $x = c$ is a root of $f(x)$ if and only if $p(c) = 0$ but $q(c) \neq 0$. (*Hint:* When is a quotient of real numbers equal to zero?)

57. Prove the second part of Theorem 6.2: If $f(x) = \frac{p(x)}{q(x)}$ is a rational function, then the domain of f is the set $\{x \mid q(x) \neq 0\}$. (*Hint:* When are polynomial functions defined? When is a quotient defined?)

6.2   Limits and Asymptotes of Rational Functions

As you may have noticed, the graphs of rational functions tend to have asymptotes. In this section we use our knowledge of limits to find these asymptotes. Rational functions can have vertical or horizontal asymptotes as well as "slant" asymptotes (linear asymptotes that are neither vertical nor horizontal) and asymptotes that are themselves curves. Limits will also enable us to quantify what happens near a "hole" of a rational function.

6.2.1 Limits of rational functions

Since every polynomial function is continuous, and the quotient of two continuous functions is continuous, we know that every rational function is continuous everywhere it is defined (that is, on its domain). Therefore, if $x = c$ is in the domain of a rational function $f(x)$, we have $\lim\limits_{x \to c} f(x) = f(c)$.

What happens if $x = c$ is not in the domain of f? One way this can happen is for $x - c$ to be a common factor of the numerator and denominator of f. In this case we can find the limit of f as $x \to c$ by using a special case of a theorem from Chapter 2:

THEOREM 6.3

Cancellation in a Limit

If $r(x)$ and $s(x)$ are polynomial functions, and $x = c$ is any real number, then
$$\lim_{x \to c} \frac{(x - c)r(x)}{(x - c)s(x)} = \lim_{x \to c} \frac{r(x)}{s(x)}.$$

This theorem says that to find the limit as $x \to c$ of a rational function $f(x)$ having a common factor $x - c$ in the numerator and denominator, we can cancel the common factor and take the limit of the new function. Note that if we do not cancel the common factor of $x - c$, we cannot evaluate the limit (it would be in the indeterminate form $\frac{0}{0}$).

EXAMPLE 6.9

The limit of a rational function at a "hole"

Calculate $\lim\limits_{x \to 2} \dfrac{x^2 + 4x - 12}{x^3 - 4x}$.

Solution This limit is initially of the form $\frac{0}{0}$ (evaluate at $x = 2$ to see why). To find this limit, we will factor the numerator and denominator and see whether there are any common factors.

$$\lim_{x \to 2} \frac{x^2 + 4x - 12}{x^3 - 4x} = \lim_{x \to 2} \frac{(x - 2)(x + 6)}{x(x - 2)(x + 2)}$$

$$= \lim_{x \to 2} \frac{x + 6}{x(x + 2)} \qquad \text{(Theorem 6.3)}$$

$$= \frac{2 + 6}{2(2 + 2)} = \frac{8}{8} = 1.$$

The graph of $f(x) = \frac{x^2 + 4x - 12}{x^3 - 4x} = \frac{(x-2)(x+6)}{x(x-2)(x+2)}$ has a "hole" at $x = 2$ but otherwise is exactly the same as the graph of the "canceled" function $g(x) = \frac{x+6}{x(x+2)}$. □

The graph of a function $f(x)$ will have a "hole" in it at $x = c$ if $\lim\limits_{x \to c} f(x)$ exists, but it is not equal to $f(c)$ (possibly because $f(c)$ does not exist). The following theorem describes a quick way to determine whether a rational function will have a "hole."

THEOREM 6.4

Hole Theorem for Rational Functions

Suppose $f(x) = \frac{p(x)}{q(x)}$ is a rational function, and suppose $x = c$ is a root of both $p(x)$ and $q(x)$. The graph of $f(x) = \frac{p(x)}{q(x)}$ will have a "hole" at $x = c$ if and only if the multiplicity of $x = c$ as a root of $p(x)$ is greater than or equal to the multiplicity of $x = c$ as a root of $q(x)$.

For example, the graphs of $f(x) = \frac{(x-3)}{(x-3)(x+1)}$ and $g(x) = \frac{(x-3)^4}{(x-3)(x+1)}$ will have "holes" at $x = 3$, but the graph of $h(x) = \frac{(x-3)^2}{(x-3)^3(x+1)}$ will not (it has a vertical asymptote at $x = 3$). The proof of Theorem 6.4 follows from Theorem 6.3 and is left for the exercises.

6.2.2 Vertical asymptotes

The results in this section will concern "reduced" rational functions. A rational function f is said to be **reduced** if it does not have any factors common to the numerator and the denominator. Remember that if a rational function f is not reduced (i.e., there *is* a common factor in the numerator and the denominator), and if g is the rational function obtained by canceling all the common factors, then the functions f and g will be identical except at the roots of those common factors. Thus we can examine the asymptotes of a nonreduced rational function by first canceling common factors and examining the asymptotes of the "reduced" function and then taking into account any holes produced by the common factors.

We have already seen that a rational function $f(x) = \frac{p(x)}{q(x)}$ is undefined everywhere that the denominator $q(x)$ is zero and that at these points, the graph of f has either a hole or a vertical asymptote. If f is a *reduced* rational function whose denominator has a root at $x = c$, then the graph of f will have a vertical asymptote at $x = c$.

THEOREM 6.5

Vertical Asymptote Theorem for Reduced Rational Functions

Suppose $f(x) = \frac{p(x)}{q(x)}$ is a *reduced* rational function. If the denominator $q(x)$ has a root at $x = c$, then the graph of f has a vertical asymptote at $x = c$.

Before proving this theorem, let's look at an example.

EXAMPLE 6.10   **Investigating a rational function near its vertical asymptote**

Show that the rational function $f(x) = \dfrac{x-1}{x-3}$ has a vertical asymptote at $x = 3$.

Solution   Recall that by definition, a function f has a vertical asymptote at $x = c$ if the limit as $x \to c$ (from either side) of $f(x)$ is infinite. Thus we need only compute the limit:

$$\lim_{x \to 3} f(x) = \lim_{x \to 3} \frac{x-1}{x-3} \to \frac{2}{0}.$$

Since this limit has the form "constant over zero," it is infinite; thus the graph of f must have a vertical asymptote at $x = 3$. We can also be more precise and examine the limit from either side of $x = 3$:

$$\lim_{x \to 3^+} f(x) = \lim_{x \to 3^+} \frac{x-1}{x-3} \to \frac{2}{0^+} = \infty,$$

$$\lim_{x \to 3^-} f(x) = \lim_{x \to 3^-} \frac{x-1}{x-3} \to \frac{2}{0^-} = -\infty.$$

From these calculations, we know that the graph of f increases without bound as it approaches $x = 3$ from the right and that it decreases without bound as it approaches $x = 3$ from the left (see Figure 6.12).   □

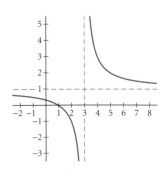

Figure 6.12

$$f(x) = \frac{x-1}{x-3}$$

The proof of Theorem 6.5 is just a more general version of the calculation we did in Example 6.10.

PROOF (**THEOREM 6.5**)   Suppose $f(x) = \frac{p(x)}{q(x)}$ is a reduced rational function whose denominator $q(x)$ has a root at $x = c$. If this root $x = c$ has multiplicity k, then we can write $q(x) = (x-c)^k r(x)$ for some polynomial function $r(x)$ that does not have a factor of $x - c$. We also know from the assumption that f is reduced that the numerator polynomial $p(x)$ has no factors of $x - c$. We then have

$$\lim_{x \to c} f(x) = \lim_{x \to c} \frac{p(x)}{q(x)} = \lim_{x \to c} \frac{p(x)}{(x - c)^k r(x)} \to \frac{p(c)}{0^k(r(c))}.$$

Since $p(c)$ and $r(c)$ are nonzero (why?), this limit is infinite. We'd have to examine the limit from the left and from the right to determine whether the limit is ∞ or $-\infty$ on either side. In any case, we know that the graph of the function f increases or decreases without bound as $x \to c$, and therefore that there is a vertical asymptote at $x = c$. ■

◈ *Caution*   The Vertical Asymptote Theorem does *not* apply to rational functions that are not reduced. For example, the rational function

$$f(x) = \frac{(x - 1)(x + 2)(x + 3)}{(x + 1)(x + 2)}$$

does *not* have a vertical asymptote at $x = -2$, even though the denominator of f has a root at $x = -2$. Since the factor $x + 2$ in the denominator also appears in the numerator, this function will have a *hole*, and not an asymptote, at $x = -2$. □

6.2.3 Horizontal asymptotes

The horizontal asymptotes of a function are determined by the global behavior of the function as $x \to \infty$ and as $x \to -\infty$. Since rational functions are by definition quotients of polynomial functions, we can use what we know about the global behavior of polynomials to determine the global behavior of rational functions.

Recall from Section 5.2.2 that the global behavior of a polynomial function is determined by the leading term of the polynomial function. Therefore, the long-term behavior of a rational function is determined by the *quotient* of the leading terms of its numerator and denominator.

THEOREM 6.6

Horizontal Asymptote Theorem for Rational Functions

Suppose that

$$f(x) = \frac{p(x)}{q(x)} = \frac{a_n x^n + a_{n-1} x^{n-1} + \cdots + a_1 x + a_0}{b_m x^m + b_{m-1} x^{m-1} + \cdots + b_1 x + b_0}$$

is a rational function. Then

(a) If $n < m$, then the graph of $y = f(x)$ has a horizontal asymptote at $y = 0$.

(b) If $n = m$, then the graph of $y = f(x)$ has a horizontal asymptote at $y = \frac{a_n}{b_m}$.

(c) If $n > m$, then the graph of $y = f(x)$ does not have a horizontal asymptote.

Unlike the Vertical Asymptote Theorem, this theorem does not require that the rational function be reduced. The proof of this theorem will be a more general version of the calculations discussed in Algorithm 6.2 and Example 6.11; it is left for the exercises.

To find the horizontal asymptotes of a rational function, we must look at the limit of the function as $x \to \infty$ and as $x \to -\infty$. Unfortunately, the limit $\lim_{x \to \infty} f(x)$ of a rational function f is often initially of the indeterminate form $\frac{\infty}{\infty}$. To find such a limit, we must first do some algebra. A useful algebraic method for finding limits of rational functions at infinity is described in Algorithm 6.2 and illustrated in Example 6.11.

ALGORITHM 6.2

Calculating Limits of Rational Functions at Infinity

Suppose f is a rational function of the form

$$f(x) = \frac{p(x)}{q(x)} = \frac{a_n x^n + a_{n-1} x^{n-1} + \cdots + a_1 x + a_0}{b_m x^m + b_{m-1} x^{m-1} + \cdots + b_1 x + b_0}.$$

To find $\lim\limits_{x \to \infty} f(x)$ or $\lim\limits_{x \to -\infty} f(x)$, it is often useful to divide the numerator and denominator of f by the highest power of x that appears in f (i.e., by x^n or by x^m, depending on which of n and m is larger).

EXAMPLE 6.11

Using limits to find horizontal asymptotes

Find any horizontal asymptotes of the following rational functions:

(a) $f(x) = \dfrac{x^2 - 1}{x^3 - 4x}$ (b) $g(x) = \dfrac{4x^2 - 1}{2x^2 - 2}$ (c) $h(x) = \dfrac{x^3 - 9x}{x^2 - 4}$

Solution

(a) In this case the degree of the numerator is less than the degree of the denominator, and thus Theorem 6.6 predicts that f will have a horizontal asymptote at $y = 0$. Let's see whether that is indeed the case.

To determine whether $f(x)$ has a horizontal asymptote (and if so, where), we must find the limit of $f(x)$ as $x \to \infty$ and as $x \to -\infty$. If either of these limits is a real number b, then f will have a horizontal asymptote at $y = b$. We can use Algorithm 6.2 to find the limit of $f(x)$ as $x \to \infty$:

$$\lim_{x \to \infty} \frac{x^2 - 1}{x^3 - 4x} = \lim_{x \to \infty} \frac{x^2 - 1}{x^3 - 4x} \left(\frac{1/x^3}{1/x^3} \right) \qquad \text{(Algorithm 6.2)}$$

$$= \lim_{x \to \infty} \frac{\frac{1}{x} - \frac{1}{x^3}}{1 - \frac{4}{x^2}} \to \frac{0 - 0}{1 - 0} = \frac{0}{1} = 0.$$

In a similar way we can show that $f(x) \to 0$ as $x \to -\infty$. Therefore, $f(x)$ has a horizontal asymptote at $y = 0$.

(b) Again we examine the limit as $x \to \infty$:

$$\lim_{x \to \infty} \frac{4x^2 - 1}{2x^2 - 2} = \lim_{x \to \infty} \frac{4x^2 - 1}{2x^2 - 2} \left(\frac{1/x^2}{1/x^2} \right)$$

$$= \lim_{x \to \infty} \frac{4 - \frac{1}{x^2}}{2 - \frac{2}{x^2}} \to \frac{4 - 0}{2 - 0} = \frac{4}{2} = 2.$$

In a similar fashion, we can show that $g(x) \to 2$ as $x \to -\infty$. Therefore $g(x)$ has a horizontal asymptote (on both sides) at $y = 2$. Note that, as predicted by Theorem 6.6, this is equal to the ratio of the leading coefficients $\frac{a_2}{b_2} = \frac{4}{2} = 2$.

(c) In this case the degree of the numerator is larger than the degree of the denominator, so Theorem 6.6 predicts that the function h will not have a horizontal asymptote. We can verify this by examining the limit of h as $x \to \infty$:

$$\lim_{x \to \infty} \frac{x^3 - 9x}{x^2 - 4} = \lim_{x \to \infty} \frac{x^3 - 9x}{x^2 - 4} \left(\frac{1/x^3}{1/x^3} \right)$$

$$= \lim_{x \to \infty} \frac{1 - \frac{9}{x^2}}{\frac{1}{x} - \frac{4}{x^3}} \to \frac{1 - 0}{0 - 0} \to \frac{1}{0},$$

which is infinite. In a similar way we can show that the limit of $h(x)$ as $x \to -\infty$ is infinite. (In fact, $\lim\limits_{x \to \infty} h(x) = \infty$ and $\lim\limits_{x \to -\infty} h(x) = -\infty$.) Therefore, the graph of h does not have a horizontal asymptote, since it increases or decreases without bound at either "end." ☐

☑ ***Checking the Answer*** The graphs of the functions f, g, and h from Example 6.11 are shown in Figures 6.13–6.15.

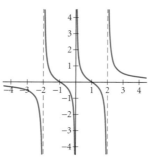

Figure 6.13
$$f(x) = \frac{x^2 - 1}{x^3 - 4x}$$

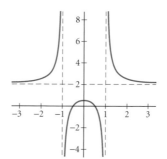

Figure 6.14
$$g(x) = \frac{4x^2 - 1}{2x^2 - 2}$$

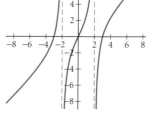

Figure 6.15
$$h(x) = \frac{x^3 - 9x}{x^2 - 4}$$ ☐

6.2.4 Slant asymptotes

Some rational functions have linear asymptotes that are neither horizontal nor vertical; such asymptotes are called ***slant asymptotes.*** For example, the function h in Figure 6.15 has a slant asymptote of $y = x$; the line $y = x$ is shown with the graph of h in Figure 6.16.

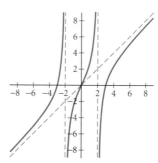

Figure 6.16
$$h(x) = \frac{x^3 - 9x}{x^2 - 4} \text{ and the line}$$
$$y = x$$

As $x \to \infty$ and as $x \to -\infty$, the graph of $y = h(x)$ looks more and more like the graph of $y = x$. How can we tell whether a rational function has a slant asymptote, and if so, how do we find the equation for this slant asymptote? The key is polynomial long division. Consider the following example.

EXAMPLE 6.12

A rational function with a slant asymptote

Use polynomial long division to write the improper rational function $h(x) = \frac{x^3 - 9x}{x^2 - 4}$ as the sum of a polynomial and a proper rational function.

Solution   The polynomial long-division algorithm in this case looks as shown in Figure 6.17.

$$x^2 - 4\overline{)\,x^3 + 0x^2 - 9x + 0\,}$$

with quotient x and steps
$$x^3 + 0x^2 - 4x$$
$$-5x$$

Figure 6.17

Therefore, $x(x^2 - 4) - 5x = x^3 - 9x$, and

$$\frac{x^3 - 9x}{x^2 - 4} = x + \frac{-5x}{x^2 - 4}.$$

We have now written the improper rational function $h(x) = \frac{x^3 - 9x}{x^2 - 4}$ as the sum of the polynomial function $y = x$ and the proper rational function $\frac{-5x}{x^2 - 4}$.   □

So what does this mean in terms of slant asymptotes? Note that we wrote $h(x)$ in terms of the polynomial function $y = x$ and a proper rational function, and note that $y = x$ is the slant asymptote of $h(x)$. Observe what happens when we look at the difference of $h(x)$ and $y = x$ as $x \to \infty$:

$$\lim_{x \to \infty} (h(x) - x) = \lim_{x \to \infty} \left(\frac{x^3 - 9x}{x^2 - 4} - x \right)$$

$$= \lim_{x \to \infty} \left(\left(x + \frac{-5x}{x^2 - 4} \right) - x \right) \qquad \text{(by the long division above)}$$

$$= \lim_{x \to \infty} \frac{-5x}{x^2 - 4} = \lim_{x \to \infty} \frac{-5x}{x^2 - 4} \left(\frac{1/x^2}{1/x^2} \right)$$

$$= \lim_{x \to \infty} \frac{\frac{-5}{x}}{1 - \frac{4}{x^2}} \to \frac{0}{1 - 0} = \frac{0}{1} = 0.$$

As $x \to \infty$, the difference between $h(x)$ and x gets smaller and smaller and approaches zero. This is why the graph of $y = h(x)$ approaches the graph of the line $y = x$ as $x \to \infty$. Similar arguments can be made to show that the graph of $y = h(x)$ approaches $y = x$ as $x \to -\infty$.

In general, a rational function f has a slant asymptote when the degree of its numerator is *exactly* 1 *higher* than the degree of its denominator. When this is the case, we will always be able to write f in terms of a *linear* function and a proper rational function.

THEOREM 6.7

Using Polynomial Long Division to Find Slant Asymptotes

Suppose $f(x) = \frac{p(x)}{q(x)}$ is a rational function where the degree of $p(x)$ is n and the degree of $q(x)$ is m. If $n = m + 1$, then we can use polynomial long division to write f as a sum:

$$\frac{p(x)}{q(x)} = l(x) + \frac{R(x)}{q(x)},$$

where $l(x)$ is a linear function and $\frac{R(x)}{q(x)}$ is a proper rational function.

Moreover, the line $y = l(x)$ will be a slant asymptote of the rational function $y = f(x)$.

The proof of Theorem 6.7 involves an analysis of the polynomial long-division algorithm, a little algebra, and a general version of the limit argument given above. We leave the details to you in the exercises.

6.2.5 Curve asymptotes

We can actually say something much more general than we did in Theorem 6.7. Whenever we have an improper rational function f, we can write f as the sum of a polynomial function $c(x)$ and a proper rational function. In Theorem 6.7, the polynomial function $c(x)$ happens to be linear, but in general it may be a quadratic, or a cubic, or some polynomial of higher degree. If its degree is greater than 1, then this polynomial function $c(x)$ is called a ***curve asymptote*** of f. What this means is that the graph of $y = f(x)$ will get closer and closer to the graph of $y = c(x)$ as $x \to \infty$ and as $x \to -\infty$. Consider the following example.

EXAMPLE 6.13

A rational function with a curve asymptote

Consider the improper rational function $f(x) = \frac{x^3 - 4x}{x - 1}$. Using polynomial long division, we can write this function as a sum: $\dfrac{x^3 - 4x}{x - 1} = (x^2 + x - 3) + \dfrac{-3}{x - 1}$.

As $x \to \infty$ and as $x \to -\infty$, the graph of the function f will get closer and closer to the graph of the polynomial function $x^2 + x - 3$, since the proper rational function $\frac{-3}{x-1}$ will approach zero. The graph of f is shown, together with the curve $y = x^2 + x - 3$, in Figure 6.18.

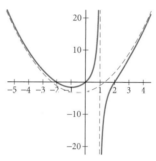

Figure 6.18
$$f(x) = \frac{x^3 - 4x}{x - 1} \text{ and}$$
$$y = x^2 + x - 3$$

As $x \to \infty$ and $x \to -\infty$, f does not have a horizontal asymptote or a slant asymptote; instead, the graph of f approaches the graph of a *curve*. Up until now we would have simply said that the ends of the graph of such a function f go "to infinity" or "to negative infinity." Now we know *how* the ends go to infinity: As $x \to \infty$ and as $x \to -\infty$, $f(x)$ behaves like the (simpler) curve $y = x^2 + x - 3$.

In general, if f is an improper rational function for which the degree of the numerator is k more than the degree of the denominator, then the graph of f will have a curve asymptote that is a polynomial of degree k. This is just a more general version of Theorem 6.7.

THEOREM 6.8

Using Polynomial Long Division to Find Curve Asymptotes

If $f(x) = \frac{p(x)}{q(x)}$ is a rational function where $p(x)$ has degree n, $q(x)$ has degree m, and $n \geq m$, then we can use polynomial long division to write f as a sum

$$\frac{p(x)}{q(x)} = c(x) + \frac{R(x)}{q(x)},$$

where $c(x)$ is a polynomial of degree $n - m$, and $\frac{R(x)}{q(x)}$ is a proper rational function. Moreover, the curve $y = c(x)$ will be a curve asymptote of $y = f(x)$.

The proof of this theorem works just like the proof of Theorem 6.7 and is left for the exercises. After working out the degree information, it is just a question of showing that the difference between $\frac{p(x)}{q(x)}$ and $c(x)$ goes to zero as $x \to \infty$ and as $x \to -\infty$.

6.2.6 Finding asymptotes of nonreduced rational functions

In all of the examples so far, we were working with *reduced* rational functions. To find the asymptotes of a nonreduced rational function, we will cancel the common factors and work with the corresponding reduced rational function. Since the nonreduced rational function and its "canceled" reduced rational function are the same everywhere except at a finite number of points (any "holes"), their asymptotes will be the same. After we find the asymptotes of the reduced rational function, we will take into account any holes caused by the common factors in the original rational function.

EXAMPLE 6.14

Finding asymptotes of a nonreduced rational function

Find all asymptotes (vertical, horizontal, slant, or curve), zeros, and holes of the rational function

$$f(x) = \frac{2x^3 - 8x}{x^3 + 2x^2 - x - 2}.$$

Solution   We need to know the roots of the numerator and the denominator to determine the vertical asymptotes, zeros, and "holes" of f, so we begin by factoring:

$$f(x) = \frac{2x^3 - 8x}{x^3 + 2x^2 - x - 2} = \frac{2x(x^2 - 4)}{x^2(x + 2) - 1(x + 2)}$$

$$= \frac{2x(x - 2)(x + 2)}{(x^2 - 1)(x + 2)} = \frac{2x(x - 2)(x + 2)}{(x - 1)(x + 1)(x + 2)}.$$

Everywhere except at $x = -2$ (where f will have a "hole"), the function f is identical to the reduced rational function

$$g(x) = \frac{2x(x - 2)}{(x - 1)(x + 1)}.$$

The reduced function g has roots at $x = 0$ and $x = 2$ and vertical asymptotes at $x = 1$ and $x = -1$, and thus so does the original function f. Since the degrees of the numerator and denominator of g are equal, the Horizontal Asymptote Theorem tells us that g will have a horizontal asymptote at $y = \frac{2}{1} = 2$ (the ratio of the leading coefficients of the numerator and denominator). Since f is identical to g everywhere except at $x = -2$, we know that f will also have a horizontal asymptote at $y = 2$.

In summary, the graph of f will have a hole at $x = -2$, roots at $x = 0$ and $x = 2$, vertical asymptotes at $x = 1$ and $x = -1$, and a horizontal asymptote at $y = 2$.   □

☑ *Checking the Answer*   The graph of f is shown in Figure 6.19 (you'll have to check the "hole" with the "trace" feature on your calculator). Note in particular that this graph has the asymptotes, zeros, and "holes" predicted.

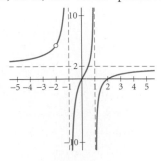

Figure 6.19

✦ *Caution* Don't forget that a rational function can have irreducible quadratic factors in its numerator or denominator. Since these irreducible quadratic factors do not have real roots, they will not contribute to the roots, holes, or vertical asymptotes of a rational function. However, the presence of an irreducible quadratic factor *can* affect the horizontal, slant, or curve asymptotes of the function. □

Working with a reduced rational function can also save you work when applying polynomial long division or synthetic division, as shown in the following example.

EXAMPLE 6.15

Reducing a rational function before finding slant or curve asymptotes

Find all asymptotes (vertical, horizontal, slant, or curve), zeros, and holes of the rational function

$$f(x) = \frac{x^3 - x}{x^2 - 3x + 2}.$$

Solution We begin by factoring:

$$f(x) = \frac{x^3 - x}{x^2 - 3x + 2} = \frac{x(x+1)(x-1)}{(x-1)(x-2)}.$$

Therefore, everywhere but at $x = 1$ the function f is identical to the reduced function

$$g(x) = \frac{x(x+1)}{x-2}.$$

The function g has roots at $x = 0$ and $x = -1$ and a vertical asymptote at $x = 2$. The function f has these same roots and the same asymptote. In addition, the function f has a hole at the point $x = 1$.

Since the degree of the numerator of g is 1 higher than the degree of the denominator, the graph of g (and thus also the graph of f) will have a slant asymptote. To find this slant asymptote, we need to divide the polynomial $x(x+1) = x^2 + x$ by the polynomial $x - 2$. We could use polynomial long division, but it is easier to use synthetic division (since the divisor is linear). Note that the coefficients of $x(x + 1) = x^2 + x + 0$ are 1, 1, and 0 and that we are careful to include the zero coefficient in the top line of the synthetic division algorithm (see Figure 6.20).

$$
\begin{array}{r|rrr}
 & 1 & 1 & 0 \\
2 & & 2 & 6 \\
\hline
 & 1 & 3 & 6
\end{array}
$$

Figure 6.20

Therefore, we know that $x^2 + x = (x - 2)(x + 3) + 6$, which means that

$$g(x) = \frac{x(x+1)}{x-2} = (x+3) + \frac{6}{x-2}.$$

Therefore, the function g (and also the function f) has a slant asymptote whose equation is $y = x + 3$.

In summary, we have determined that f has a hole at $x = 1$, roots at $x = 0$ and $x = -1$, a vertical asymptote at $x = 2$, and a slant asymptote with equation $y = x + 3$. □

Note that we could have used polynomial long division to find the slant asymptote of f, even if we had not reduced the function first. However, this calculation would have been more complex.

✓ *Checking the Answer*   The graph of the function f in Example 6.15 is shown in Figure 6.21.

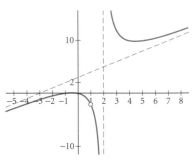

Figure 6.21   ☐

What you should know after reading Section 6.2

► How to find the limit of a rational function at domain points and nondomain points, including at holes, at vertical asymptotes, and at infinity.

► The Vertical Asymptote Theorem and the Horizontal Asymptote Theorem, when they apply, how to use them, and why they work. Be able to find the vertical and horizontal asymptotes of a rational function both by applying these theorems and by doing the relevant limit calculations by hand.

► What kind of rational function will have a "slant" asymptote, and how you can find the equation of that slant asymptote. How can you use the results of polynomial long division to prove that a given rational function has a given slant asymptote?

► When a rational function will have a "curve" asymptote. How can you tell from the degrees of the numerator and denominator of a rational function what kind of polynomial will be a curve asymptote for that function? How can you find the equation of that curve asymptote? Be able to provide a limit argument using the results of polynomial long division to prove that a given rational function has a given slant asymptote.

► How to find the graph of a nonreduced function (including its asymptotes) by examining the corresponding reduced rational function.

Exercises 6.2

Concepts

0. Read the section and make your own summary of the material. Include whatever you think will help you review the material later on.

1. Theorem 6.3 says that we can cancel a common factor $x - c$ from a rational function f while taking the limit as $x \to c$. However, we know that the "canceled" function is *not* equal to the original function f at the point $x = c$. Why is this not a problem?

2. What does it mean for a rational function to be *reduced*? Give an example of a rational function that is reduced and an example of a rational function that is not.

3. Can a rational function have two different horizontal asymptotes? Why or why not? Can *any* function have two different horizontal asymptotes?

4. Find a rational function with a horizontal asymptote whose graph crosses this horizontal asymptote. Explain how you arrived at your function.

5. Can a rational function cross one of its vertical asymptotes? Can *any* function? Why or why not?

6. Suppose that $f(x) = \frac{p(x)}{q(x)}$ is a rational function and that $p(x)$ has degree n while $q(x)$ has degree m. Fill in the blanks below.

   (a) The graph of f will have a horizontal asymptote if _____ .

   (b) The graph of f will have a horizontal asymptote at $y = 0$ if _____ .

   (c) The graph of f will have a slant asymptote if _____ .

   (d) The graph of f will have a curve asymptote of degree greater than or equal to 2 if _____ .

   (e) The graph of f will have a curve asymptote of degree 4 if _____ .

■ **For each problem, construct an equation for a rational function f that has the properties listed.**

7. The graph of f has one hole at $x = 1$, three roots at $x = 0$, $x = 3$, and $x = -1$, and no vertical asymptote.

8. f has a common factor of $x - 3$ in its numerator and denominator but does not have a hole at $x = 3$.

9. f is a *proper* rational function whose graph has three roots at $x = 0$, $x = 3$, and $x = -2$ and one vertical asymptote at $x = 1$.

10. The graph of f has a hole at the coordinates $(-2, 0)$, two vertical asymptotes at $x = 1$ and $x = -3$, and a horizontal asymptote at $y = -5$.

11. The graph of f has no holes, no vertical asymptotes, two roots at $x = 3$ and $x = 0$, and a horizontal asymptote at $y = 0$.

12. The graph of f has no roots, no holes, two vertical asymptotes at $x = -2$ and $x = 2$, and a horizontal asymptote at $y = 3$.

13. The graph of f has a hole at the coordinates $(-2, 0)$, two vertical asymptotes at $x = 1$ and $x = -3$, and a horizontal asymptote at $y = -5$.

14. The graph of f has a hole at $x = 0$, one vertical asymptote at $x = 2$, and a slant asymptote.

15. The graph of f has a curve asymptote of $y = x^3 + 2$ and a vertical asymptote at $x = 1$.

■ **Determine whether each of the following statements is true or false. If a statement is true, explain why. If a statement is false, provide a counterexample.**

16. True or False: Every rational function is continuous on its domain.

17. True or False: Every rational function is continuous on $(-\infty, \infty)$.

18. True or False: Every rational function has at least one vertical asymptote.

19. True or False: Every rational function has a horizontal asymptote, a slant asymptote, or a curve asymptote.

20. True or False: A rational function f has a horizontal asymptote if and only if it is a proper rational function.

Skills

■ **Calculate the following limits by hand. Then check your answers using a graphing calculator.**

21. $\displaystyle\lim_{x \to 2} \frac{1 - x^3}{x^5 - 3x^2 + 1}$

22. $\displaystyle\lim_{x \to -1} \frac{(x - 1)(x + 2)(x + 5)}{(x + 2)(x + 1)}$

23. $\displaystyle\lim_{x \to -2} \frac{(x - 1)(x + 2)(x + 5)}{(x + 2)(x + 1)}$

24. $\displaystyle\lim_{x \to 2} \frac{(x + 1)(x - 2)}{(x - 2)(x + 2)}$

25. $\displaystyle\lim_{x \to 2} \frac{(x + 1)(x - 2)^2}{(x - 2)(x + 2)}$ $\frac{3(0)}{4} = 0$

26. $\displaystyle\lim_{x \to 2} \frac{(x + 1)(x - 2)}{(x - 2)^2(x + 2)}$ 3

27. $\displaystyle\lim_{x \to 3} \frac{x^2 - 9}{2x^2 - 5x - 3}$

28. $\displaystyle\lim_{x \to 2} \frac{x^2 - 4x + 4}{x^3 - 2x^2 - x - 2}$ -8

29. $\displaystyle\lim_{x \to 2} \frac{x^3 - 2x^2 - x - 2}{x^2 - 4x + 4}$

30. $\displaystyle\lim_{x \to 0} \frac{x^4 - 4x^3 + 3x^2}{x^5 - x}$

31. $\displaystyle\lim_{x \to 1} \frac{x^4 - 4x^3 + 3x^2}{x^5 - x}$

32. $\displaystyle\lim_{x \to \infty} \frac{x^2 - 100x + 5}{2x^3 + 1}$

33. $\displaystyle\lim_{x \to \infty} \frac{1 - x^4}{x^2 + x + 2}$

34. $\displaystyle\lim_{x \to -\infty} \frac{1 - x^4}{x^2 + x + 2}$

35. $\displaystyle\lim_{x \to -\infty} \frac{(x + 1)(2x - 3)(x + 5)}{(1 - x)(3x + 2)}$

36. $\displaystyle\lim_{x \to \infty} \frac{(x + 1)(x - 2)}{(x - 2)(x + 2)}$

37. $\displaystyle\lim_{x \to \infty} \frac{(x + 1)(x - 2)^2}{(x - 2)(x + 2)}$

38. $\displaystyle\lim_{x \to \infty} \frac{(x + 1)(x - 2)}{(x - 2)^2(x + 2)}$

■ Find all roots, holes, and vertical or horizontal asymptotes of the following rational functions. When you are finished, use a graphing calculator to check your answers.

39. $f(x) = \dfrac{(x-2)(x+5)(x-1)}{(x-3)(x-1)}$

40. $f(x) = \dfrac{x^2(x+3)}{(x-1)(x+1)(x^2+5)}$

41. $f(x) = \dfrac{(x-1)^2(x-2)}{(x-1)(3x+2)}$

42. $f(x) = \dfrac{(x-1)(x-2)}{(x-1)^2(3x+2)}$

43. $f(x) = \dfrac{6x^2+7x-5}{x^3+x^2-2x}$

44. $f(x) = \dfrac{16-x^4}{x^4-4x^2}$

45. $f(x) = \dfrac{x^3-x^2-4x+4}{x^3-7x-6}$

46. $f(x) = \dfrac{x^3+2x^2-4x-8}{2x^3-x^2-1}$

47. $f(x) = \dfrac{x^4+2x^2+1}{3x^3-5x^2+5x-3}$

48. $f(x) = \dfrac{9x^4+12x^2+4}{x^4+6x^2+9}$

■ Find all roots, holes, and any asymptotes (vertical, horizontal, slant, or curve) of the following rational functions. When you are finished, use a graphing calculator to check your answers.

49. $f(x) = \dfrac{2x^2+x-3}{x+1}$

50. $f(x) = \dfrac{x^3+x^2-6x}{x^2-4x+4}$

51. $f(x) = \dfrac{(x+3)(x-2)^2}{(x-2)(x+1)}$

52. $f(x) = \dfrac{(x-1)(3x+2)(x+1)}{x^2+x+1}$

53. $f(x) = \dfrac{16x^4-81}{3-2x}$

54. $f(x) = \dfrac{x^4+2x^2+1}{x^3-1}$

55. $f(x) = \dfrac{x^4-2x^3-3x^2+4x+4}{x^2-x-2}$

56. $f(x) = \dfrac{x^5-2x^4+x^3}{x^2-1}$

Proofs

57. Prove the "Hole Theorem" for rational functions by showing that a rational function of the form

$$f(x) = \frac{(x-c)^k\, p(x)}{(x-c)^l q(x)},$$

where $(x-c)$ is *not* a factor of $p(x)$ or $q(x)$, has a horizontal asymptote at $x=c$ if $k < l$ but has a hole at $x=c$ if $k \geq l$.

58. Use Algorithm 6.2 to prove Theorem 6.6: If $f(x) = \frac{p(x)}{q(x)}$ is a rational function where $p(x)$ has degree n and $q(x)$ has degree m, then:

(a) If $n < m$, then the graph of $y = f(x)$ has a horizontal asymptote at $y = 0$.

(b) If $n = m$, then the graph of $y = f(x)$ has a horizontal asymptote at $y = \frac{a_n}{b_m}$.

(c) If $n > m$, then the graph of $y = f(x)$ does not have a horizontal asymptote.

59. In this problem you will prove Theorem 6.7 in steps. Suppose that $f(x) = \frac{p(x)}{q(x)}$ is a rational function and that the degree of $p(x)$ is exactly 1 higher than the degree of $q(x)$.

(a) We know from the polynomial long-division algorithm that we can write

$$\frac{p(x)}{q(x)} = l(x) + \frac{R(x)}{q(x)}$$

for some polynomial functions $l(x)$ and $R(x)$. Show that the degree of $l(x)$ must be 1. (*Hint:* Combine the terms on the right-hand side of the equality, and then compare the numerator $p(x)$ with the numerator on the right.)

(b) Explain how you know that the degree of $R(x)$ is less than the degree of $q(x)$; this will show that $\frac{R(x)}{q(x)}$ is a *proper* rational function. (*Hint:* Look back at the algorithm for polynomial long division and how it terminates.)

(c) Prove that the limits $\displaystyle\lim_{x\to\infty}\left(\frac{p(x)}{q(x)} - l(x)\right)$ and $\displaystyle\lim_{x\to-\infty}\left(\frac{p(x)}{q(x)} - l(x)\right)$ are both equal to zero.

(d) Explain why the limits in the part above mean that $y = l(x)$ is a slant asymptote of the function $f(x) = \frac{p(x)}{q(x)}$.

60. Mimic the proof of Theorem 6.7 in the exercise above to prove Theorem 6.8. The main difference will be that the linear polynomial $l(x)$ will be replaced with a polynomial $c(x)$ whose degree is the difference of the degrees of $p(x)$ and $q(x)$.

| 6.3 | **Derivatives of Rational Functions** |

In this section we will investigate the derivatives of rational functions and even develop a formula for calculating these derivatives called the **quotient rule.** The derivatives of rational functions will be used in Section 6.4 when we investigate graphs of rational functions, and at the end of this section when we do application problems that involve rational functions.

6.3.1 Calculating derivatives of rational functions from scratch

At this point the only functions that we know how to differentiate with "rules" are sums of power functions. We do not yet have a differentiation rule that deals with quotients. At this point we will have to go back to the definition of derivative when we want to calculate the derivative of a rational function.

◈ *Caution* The derivative of a quotient of two functions is *not* in general the quotient of the derivatives of those functions. In other words, $(\frac{f}{g})' \neq \frac{f'}{g'}$. The formula for the derivative of a quotient is more complicated, and we will learn it in Section 6.3.2. □

| EXAMPLE 6.16 | **Applying the definition of the derivative at a point to a rational function** |

Find the slope of the line tangent to the graph of the function $f(x) = \frac{3x^2+4}{x+2}$ at the point $x = 2$.

Solution We need to find $f'(2)$, and since we don't know any differentiation rules that apply to the rational function f, we will have to use the definition of derivative:

$$f'(2) = \lim_{h \to 0} \frac{f(2+h) - f(2)}{h} = \lim_{h \to 0} \frac{\frac{3(2+h)^2+4}{(2+h)+2} - \frac{3(2)^2+4}{2+2}}{h} \qquad \text{(definition of derivative)}$$

$$= \lim_{h \to 0} \frac{\frac{3(4+4h+h^2)+4}{h+4} - \frac{16}{4}}{h} = \lim_{h \to 0} \frac{\frac{3h^2+12h+16}{h+4} - 4}{h} \qquad \text{(algebra)}$$

$$= \lim_{h \to 0} \frac{\left(\frac{(3h^2+12h+16)-4(h+4)}{h+4}\right)}{h} = \lim_{h \to 0} \frac{3h^2 + 8h}{h(h+4)} \qquad \text{(algebra)}$$

$$= \lim_{h \to 0} \frac{h(3h+8)}{h(h+4)} = \lim_{h \to 0} \frac{3h+8}{h+4} = \frac{3(0)+8}{0+4} = \frac{8}{4} = 2. \qquad \text{(cancellation, limit rules)}$$

Therefore, the slope of the line tangent to the graph of $f(x) = \frac{3x^2+4}{x+2}$ at $x = 2$ is $f'(2) = 2$. □

Note that, as usual, we could not evaluate the limit without doing some algebra first. (If we try to "plug in" $h = 0$ at the start, we will get the indeterminate form $\frac{0}{0}$.) Most definitions of derivative calculations for rational functions involve the kind of algebra in Example 6.16, where the goal is to get one or more "h's" to cancel so that the limit can be evaluated. This is true even if we are finding the derivative *function* (rather than the derivative at a *point*), as illustrated in the following example.

| EXAMPLE 6.17 | **Applying the definition of the derivative to a rational function** |

Find the derivative of $f(x) = \frac{x^2 + 1}{x^2 - 4}$.

Solution   The calculation of the derivative of f is a simple application of the definition of derivative, although the algebra required is quite involved.

$$f'(x) = \lim_{h \to 0} \frac{f(x+h) - f(x)}{h} \qquad \text{(definition of derivative)}$$

$$= \lim_{h \to 0} \frac{\frac{(x+h)^2+1}{(x+h)^2-4} - \frac{x^2+1}{x^2-4}}{h} \qquad \text{(definition of } f(x)\text{)}$$

$$= \lim_{h \to 0} \frac{\left(\frac{((x+h)^2+1)(x^2-4)-(x^2+1)((x+h)^2-4)}{((x+h)^2-4)(x^2-4)}\right)}{h} \qquad \text{(combine fractions)}$$

$$= \lim_{h \to 0} \frac{((x+h)^2+1)(x^2-4)-(x^2+1)((x+h)^2-4)}{h((x+h)^2-4)(x^2-4)} \qquad \text{(simplify complex fraction)}$$

$$= \lim_{h \to 0} \frac{-5h^2 - 10hx}{h((x+h)^2-4)(x^2-4)} \qquad \text{(multiply out numerator)}$$

$$= \lim_{h \to 0} \frac{-5h - 10x}{((x+h)^2-4)(x^2-4)} \qquad \text{(cancellation)}$$

$$= \frac{-5(0) - 10x}{((x+0)^2-4)(x^2-4)} = \frac{-10x}{(x^2-4)^2}. \qquad \text{(limit rules)}$$

Thus the derivative of $f(x) = \dfrac{x^2+1}{x^2-4}$ is the function $f'(x) = \dfrac{-10x}{(x^2-4)^2}$.   □

As you can see, calculating the derivative of a rational function using the definition of derivative is a tedious, lengthy process, even for relatively simple rational functions. Luckily, we are about to develop a rule for quickly differentiating rational functions.

6.3.2 The quotient rule

Since we will often need to differentiate rational functions, we need to develop a differentiation rule that works for quotients. Theorem 6.9 actually works for *all* quotients, not just for quotients of polynomials. However, for now we will be using the quotient rule to differentiate only rational functions, so we will use the notation of rational functions.

THEOREM 6.9

The Quotient Rule

If p and q are differentiable functions, then the derivative of their quotient is given by

$$\left(\frac{p(x)}{q(x)}\right)' = \frac{p'(x)q(x) - p(x)q'(x)}{(q(x))^2}.$$

We'll prove the quotient rule at the end of this section. First we will do some examples.

◈ *Caution*   As we pointed out earlier, the derivative of a quotient $f(x) = \frac{p(x)}{q(x)}$ is *not* equal to the quotient of derivatives $\frac{p'(x)}{q'(x)}$. One good example is when $p(x) = 1$ and $q(x) = x$. Then $\frac{d}{dx}(\frac{p(x)}{q(x)}) = \frac{d}{dx}(\frac{1}{x}) = -\frac{1}{x^2}$, which is not equal to $\frac{p'(x)}{q'(x)} = \frac{0}{1} = 0$.   □

EXAMPLE 6.18   **Using the quotient rule to differentiate a rational function**

Find the derivative of $f(x) = \dfrac{x^2+1}{x^2-4}$.

Solution This is the function whose derivative we found using the definition of derivative in Example 6.17. Now we can use the quotient rule, with $p(x) = x^2 + 1$ and $q(x) = x^2 - 4$. Then $p'(x) = 2x$ and $q'(x) = 2x$, so by the quotient rule

$$\frac{d}{dx}\left(\frac{x^2+1}{x^2-4}\right) = \frac{p'(x)q(x) - p(x)q'(x)}{(q(x))^2} = \frac{(2x)(x^2-4) - (x^2+1)(2x)}{(x^2-4)^2}.$$

The numerator simplifies as $(2x)(x^2-4) - (x^2+1)(2x) = 2x^3 - 8x - 2x^3 - 2x = -10x$, and thus

$$\frac{d}{dx}\left(\frac{x^2+1}{x^2-4}\right) = \frac{-10x}{(x^2-4)^2},$$

which is exactly the answer we found using the lengthy definition of derivative calculation in Example 6.17. ☐

EXAMPLE 6.19

A more challenging quotient rule calculation

Find the derivative of the function $f(x) = \dfrac{3x^6 - 4x^2 - 7}{x^{17} - 2x^8 + 9x^3 - 1}$.

Solution Using the quotient rule, this is a very simple calculation (even though the degrees of the numerator and denominator are large):

$$\frac{d}{dx}\left(\frac{3x^6 - 4x^2 - 7}{x^{17} - 2x^8 + 9x^3 - 1}\right)$$

$$= \frac{\left(\frac{d}{dx}(3x^6 - 4x^2 - 7)\right)(x^{17} - 2x^8 + 9x^3 - 1) - (3x^6 - 4x^2 - 7)\left(\frac{d}{dx}(x^{17} - 2x^8 + 9x^3 - 1)\right)}{(x^{17} - 2x^8 + 9x^3 - 1)^2}$$

$$= \frac{(18x^5 - 8x)(x^{17} - 2x^8 + 9x^3 - 1) - (3x^6 - 4x^2 - 7)(17x^{16} - 16x^7 + 27x^2)}{(x^{17} - 2x^8 + 9x^3 - 1)^2}.$$

We could simplify this expression further, but the differentiation part of the calculation is done, so we will stop here. Stop for a minute and imagine the calculations that would be involved if you tried to find the derivative using the definition of derivative! ☐

❧ *Caution* In some cases you will need to multiply out, simplify, or factor the derivative you obtain so that you can use it (to find its zeros, for example). However, if you are only looking for an expression for the derivative, it is enough just to use the quotient rule and leave the expression unsimplified. ☐

The proof of Theorem 6.9 is just a general application of the definition of derivative. The proof will not require that $p(x)$ and $q(x)$ be polynomials, only that they be differentiable.

PROOF (THEOREM 6.9) Suppose p and q are differentiable functions. We'll work from the left-hand side to the right-hand side of the equality. The first part of the calculation is exactly like the first part of the calculation in Example 6.17.

$$\left(\frac{p(x)}{q(x)}\right)' = \lim_{h \to 0} \frac{\frac{p(x+h)}{q(x+h)} - \frac{p(x)}{q(x)}}{h} \qquad \text{(definition of derivative)}$$

$$= \lim_{h \to 0} \frac{\left(\frac{p(x+h)q(x) - p(x)q(x+h)}{q(x+h)q(x)}\right)}{h} \qquad \text{(combine fractions)}$$

$$= \lim_{h \to 0} \frac{p(x+h)q(x) - p(x)q(x+h)}{hq(x+h)q(x)} \qquad \text{(simplify)}$$

At this point in our calculation we would usually multiply out the numerator and attempt to cancel the "h" in the denominator. However, this time we don't have specific equations for p and q, and so we can't do that. Instead, we will use some clever algebra to help us simplify the expression. We will add and subtract the quantity $p(x)q(x)$ from the numerator of the expression (essentially adding zero, which does not change the expression). Remember that the goal is to write the expression as $\frac{p'q - pq'}{q^2}$. Continuing our calculation, we have

$$= \lim_{h \to 0} \frac{p(x+h)q(x) - p(x)q(x) + p(x)q(x) - p(x)q(x+h)}{hq(x+h)q(x)} \qquad \text{(see above)}$$

$$= \lim_{h \to 0} \frac{q(x)(p(x+h) - p(x)) - p(x)(q(x+h) - q(x))}{hq(x+h)q(x)} \qquad \text{(grouping)}$$

$$= \lim_{h \to 0} \frac{\left(q(x)\frac{p(x+h) - p(x)}{h} - p(x)\frac{q(x+h) - q(x)}{h} \right)}{q(x+h)q(x)} \qquad \text{(algebra)}$$

$$= \frac{q(x)\left(\lim\limits_{h \to 0} \frac{p(x+h) - p(x)}{h} \right) - p(x)\left(\lim\limits_{h \to 0} \frac{q(x+h) - q(x)}{h} \right)}{\lim\limits_{h \to 0}(q(x+h)q(x))} \qquad \text{(limit rules)}$$

$$= \frac{q(x)\,p'(x) - p(x)q'(x)}{\lim\limits_{h \to 0}(q(x+h)q(x))} \qquad \begin{array}{c}\text{(definition of}\\ \text{derivative)}\end{array}$$

$$= \frac{p'(x)q(x) - p(x)q'(x)}{(q(x))^2}. \qquad \begin{array}{c}\text{(evaluate last}\\ \text{limit)}\end{array}$$

From the equation for the quotient rule, we can see that a rational function $f(x) = \frac{p(x)}{q(x)}$ will be differentiable everywhere except where its denominator $q(x)$ is zero (why?). This means that all rational functions are differentiable everywhere on their domains. In particular, this means that a rational function will be differentiable everywhere that it does not have a hole or a vertical asymptote. Note also that the derivative of a rational function is also a rational function.

6.3.3 Local extrema of rational functions

To find the local extrema of a function, we must first find its critical points (where the derivative is zero or does not exist). Up until now, we have always been dealing with functions whose derivatives exist everywhere, and thus all of our critical points have been zeros of the derivative. Now, since the derivatives of rational functions are not always defined everywhere, some critical points will be places where the derivative does not exist.

EXAMPLE 6.20

Using derivatives to find the local extrema of a rational function

Find the local extrema of the function $f(x) = \dfrac{x^2}{x^2 - 2x + 1}$.

Solution First we must find the derivative of f. Using the quotient rule, we have

$$f'(x) = \frac{(2x)(x^2 - 2x + 1) - (x^2)(2x - 2)}{(x^2 - 2x + 1)^2}.$$

Because we need to find the critical points of f, we must find the values at which the derivative f' either is zero or does not exist. The function f' is a quotient; therefore, f' will be zero whenever its numerator is zero (and its denominator isn't) and will not exist whenever its denominator is zero. We start by simplifying and factoring f':

$$f'(x) = \frac{(2x)(x^2 - 2x + 1) - (x^2)(2x - 2)}{(x^2 - 2x + 1)^2}$$

$$= \frac{(2x^3 - 4x^2 + 2x) - (2x^3 - 2x^2)}{(x^2 - 2x + 1)^2} \qquad \text{(multiply out)}$$

$$= \frac{-2x^2 + 2x}{((x - 1)^2)^2} = \frac{-2x(x - 1)}{(x - 1)^4} \qquad \text{(collect terms and factor)}$$

It is now easy to see that f' is zero when $x = 0$ and fails to exist at $x = 1$. (In particular, notice that f' is *not* zero at $x = 1$ even though a factor of $x - 1$ is in the numerator; why?) It makes sense that f' does not exist at $x = 1$, since the original function f does not exist at $x = 1$. The critical points of f are $x = 0$ and $x = 1$; these are the only places at which the sign of f' can change.

Now we make a number line representing the sign of the derivative f' between each pair of adjacent critical points. Notice that the denominator of f' is *always positive* (why?). Therefore, the sign of the derivative f' depends only on the sign of its numerator. We will check the sign of f' at $x = -1$, $x = 0.5$, and $x = 2$. In the first case, we have

$$f'(-1) = \frac{-2(-1)(-1 - 1)}{(-1 - 1)^4},$$

which is negative, since its numerator is the product of three negative numbers (and so is negative), while its denominator is positive. In a similar fashion we can find that $f'(0.5) > 0$ and $f'(2) < 0$. This gives us the number line shown in Figure 6.22.

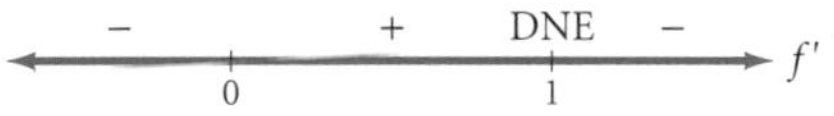

$$\xleftarrow{\qquad - \qquad \underset{0}{|} \qquad + \qquad \underset{1}{\overset{\text{DNE}}{|}} \qquad - \qquad} f'$$

Figure 6.22

By the first derivative test, f has a local minimum at $x = 0$ (since f is decreasing to the left of zero and increasing to the right of zero). At first glance you might be tempted to say that f has a local maximum at $x = 1$; however, the original function is *not defined* at $x = 1$ and thus cannot have a maximum value there. In fact, since $x = 1$ is a root of the denominator of f (and f is reduced), the graph of f has an asymptote at $x = 1$. Therefore, the only local extremum of f is at $x = 0$, where we have a local minimum. $\square$

☑ *Checking the Answer*   The graph of $f(x) = \dfrac{x^2}{x^2 - 2x + 1}$ is shown in Figure 6.23. Notice that f does indeed have a local minimum at $x = 0$ and a vertical asymptote (not an extremum) at $x = 1$.

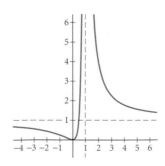

Figure 6.23 $\square$

6.3.4 Global extrema of rational functions

To find the global extrema of a function f on an interval I, we must compare the values of the local extrema of f in I with the behavior of f at the "ends" of the interval I. If I is a closed interval $[a, b]$, then the behavior of f at the "ends" of the interval can be determined by examining the values $f(a)$ and $f(b)$. On the other hand, if I is an open interval (a, b), then we must look at the limits $\lim_{x \to a^+} f(x)$ and $\lim_{x \to b^-} f(x)$. For polynomial functions, the limits of f at the ends of an open interval are equal to the values of f at those endpoints. For rational functions, this is not necessarily the case.

EXAMPLE 6.21

Using limits and derivatives to find global extrema of rational functions

Find the global extrema of $f(x) = \dfrac{1}{-x^2 + 6x - 5}$ on $(1, 4)$.

Solution   We must first find the local extrema of f that are in the interval $(1, 4)$. The derivative of f is

$$f'(x) = \frac{(0)(-x^2 + 6x - 5) - (1)(-2x + 6)}{(-x^2 + 6x - 5)^2} \qquad \text{(quotient rule)}$$

$$= \frac{2x - 6}{((x - 1)(-x + 5))^2} = \frac{2(x - 3)}{(x - 1)^2(x - 5)^2}. \qquad \text{(factor)}$$

Therefore, f' is zero at $x = 3$ and does not exist at $x = 1$ and $x = 5$. Of these critical points, the only one in $(1, 4)$ is the point $x = 3$. Construct a number line representing the signs of f' on the interval $(1, 4)$. See Figure 6.24.

Figure 6.24

By the first derivative test, the function f has a local minimum at $x = 3$, with value

$$f(3) = \frac{1}{-(3)^2 + 6(3) - 5} = \frac{1}{-9 + 18 - 5} = \frac{1}{4}.$$

It now remains only to examine the behavior of f at the ends of the interval $(1, 4)$. We compute

$$\lim_{x \to 1^+} f(x) = \lim_{x \to 1^+} \frac{1}{-x^2 + 6x - 5} = \lim_{x \to 1^+} \frac{1}{(x - 1)(-x + 5)} \to \frac{1}{(0^+)(4)} \to \infty.$$

$$\lim_{x \to 4^-} f(x) = \lim_{x \to 4^-} \frac{1}{-x^2 + 6x - 5} = \frac{1}{-16 + 24 - 5} = \frac{1}{3}.$$

(We did not have to factor in the second calculation, since we did not need to know whether the denominator approached 3^+ or 3^-.) On the interval $(1, 4)$, the function f has no global maximum (since $\lim_{x \to 1^+} f(x) = \infty$) and has a global minimum at $x = 3$ with value $f(3) = \frac{1}{4}$.   □

✓ *Checking the Answer*   The graph of $f(x) = \frac{1}{-x^2 + 6x - 5}$ is shown in Figure 6.25. Figure 6.26 shows the function f on the interval $(1, 4)$. Note that on the interval $(1, 4)$, the function f does indeed have a global minimum at $x = 3$ and no global maximum.

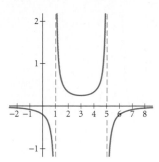

Figure 6.25

$$f(x) = \frac{1}{-x^2 + 6x - 5}$$

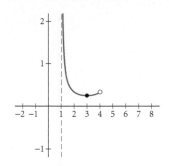

Figure 6.26

$$f(x) = \frac{1}{-x^2 + 6x - 5} \text{ on } (1, 4) \qquad \Box$$

❓ *Question* Consider $f(x) = \frac{1}{-x^2+6x-5}$ from Example 6.21. What are the global extrema of f on $[2, 4]$? On $(2, 4)$? On $[2, 6]$? On $(5, 7)$? You should be able to answer these questions algebraically (by examining limits or values at the endpoints) as well as graphically (using the graph in Figure 6.25). $\qquad \Box$

Suppose we wish to find the global extrema of a rational function f on an interval I. If f has a critical point $x = c$ in I where f' does not exist, then f may have a vertical asymptote at $x = c$. This is important to consider when looking for the global extrema.

EXAMPLE 6.22

Considering vertical asymptotes when looking for global extrema

Find the global extrema of the function $f(x) = \frac{1}{x^2 - 2x + 1}$ on the interval $[0, 3]$.

Solution   The derivative of f is

$$f'(x) = \frac{(0)(x^2 - 2x + 1) - (1)(2x - 2)}{(x^2 - 2x + 1)^2} = \frac{-2x + 2}{((x-1)^2)^2} = \frac{-2(x-1)}{(x-1)^4}.$$

This derivative is never zero but is undefined at $x = 1$. Since $x = 1$ is a critical point in the interval $I = [0, 3]$, we must take into account the behavior of f near $x = 1$. We cannot evaluate $f(1)$, but we can examine the limit of f as x approaches 1 from the right and from the left:

$$\lim_{x \to 1^+} f(x) = \lim_{x \to 1^+} \frac{1}{x^2 - 2x + 1} = \lim_{x \to 1^+} \frac{1}{(x-1)^2} \to \frac{1}{0^+} \to \infty,$$

$$\lim_{x \to 1^-} f(x) = \lim_{x \to 1^-} \frac{1}{x^2 - 2x + 1} = \lim_{x \to 1^-} \frac{1}{(x-1)^2} \to \frac{1}{0^+} \to \infty.$$

Since there are no other critical points of f, it remains only to examine the behavior of f at the endpoints of $[0, 3]$. We have

$$f(0) = \frac{1}{0^2 - 2(0) + 1} = \frac{1}{1} = 1 \quad \text{and} \quad f(3) = \frac{1}{3^2 - 2(3) + 1} = \frac{1}{9 - 6 + 1} = \frac{1}{4}.$$

Since the only possible places a global extremum of f could occur are at the critical points or the endpoints, we now know that the global minimum of f on $[0, 3]$ is at $x = 3$ (with a value of $f(3) = \frac{1}{4}$) and that there is no global maximum of f (the graph of f increases without bound on either side of $x = 1$). The graph of f is shown in Figure 6.27 and is restricted to the interval $[0, 3]$ in Figure 6.28.

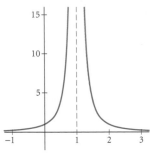

Figure 6.27

$$f(x) = \frac{1}{x^2 - 2x + 1}$$

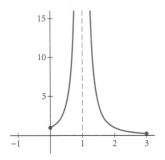

Figure 6.28

$$f(x) = \frac{1}{x^2 - 2x + 1} \text{ on } [0, 3]$$

□

What you should know after reading Section 6.3

▶ How to find the derivative of a rational function using the definition of derivative.

▶ The statement and proof of the quotient rule, and how to use it to find derivatives of rational functions.

▶ How to find local extrema of rational functions. Don't forget to consider any critical points where the derivative does not exist.

▶ How to find global extrema of rational functions. Note that when finding global extrema of a rational function f on an open interval $I = (a, b)$, it is necessary to examine the limit of f as $x \to a^+$ and as $x \to b^-$. Moreover, f might have a vertical asymptote at some point $x = c$ inside the interval I, in which case you must examine the limit of f as $x \to c^+$ and as $x \to c^-$.

Exercises 6.3

Concepts

0. Read the section and make your own summary of the material. Include whatever you think will help you review the material later on.

1. Find functions f and g that illustrate the fact that the derivative of a quotient of two functions is *not* equal to the quotient of their derivatives. In other words, find functions f and g for which $\left(\frac{f}{g}\right)' \neq \frac{f'}{g'}$.

2. Although in this section we have been discussing the quotient rule only in terms of rational functions, the quotient rule actually applies to *any* quotient of functions that we know how to differentiate. Use the quotient rule to differentiate the (nonrational) function
$$f(x) = \frac{\sqrt{x} + 3x^{-2}}{x^{\frac{3}{2}} + 1}.$$

3. Suppose $f(x) = \frac{p(x)}{q(x)}$ is a rational function whose numerator $p(x)$ is degree n and whose denominator $q(x)$ is degree m. When we use the quotient rule to

differentiate f, the resulting function f' will also be a rational function.

(a) What can you say about the degree of the numerator of f'? What can you say about the degree of the denominator of f'? (Be careful when examining the numerator of f'. Is it possible that there will be some cancelation that lowers the degree of the numerator of f' more than you would expect?)

(b) If f is a *proper* rational function, is f' necessarily a proper rational function? Use your answer to part (a) to explain your answer.

(c) Suppose f has a horizontal asymptote at $y = 2$. Does f' necessarily have a horizontal asymptote, and if so, can you tell where it is? Explain your answer two ways: first, in terms of the degrees of the numerators and denominators of f and f'; second, in terms of the graph of f and its behavior at the horizontal asymptote $y = 2$.

4. Find a function whose *derivative* is

$$f'(x) = \frac{(3x^2 - 1)(x^2 + 3x - 1) - (x^3 - x + 5)(2x + 3)}{(x^2 + 3x - 1)^2}.$$

5. Suppose $f(x)$ is a function with the following properties:

▶ $f(x)$ has domain $(-\infty, 4) \cup (4, \infty)$, and $f(-1) = 2$, $f(1) = 3$, $f(2) = 0$, $f(3) = -2$, and $f(6) = 3$.

▶ $f'(x)$ is zero only at $x = 1$ and $x = 3$, and $f'(x)$ does not exist at $x = 4$.

▶ $\lim\limits_{x \to \infty} f(x) = 1$ and $\lim\limits_{x \to -\infty} f(x) = 1$.

▶ $\lim\limits_{x \to 4^+} f(x) = \infty$ and $\lim\limits_{x \to 4^-} f(x) = \infty$.

Given the information above, find the global extrema of $f(x)$ on each of the intervals below.

(a) $[-1, 4)$ (b) $[-1, 2]$ (c) $(-1, 2)$
(d) $[2, 6]$ (e) $(2, 6)$

6. Let f be the rational function graphed below. Determine graphically the global extrema of f (if any) on each of the following intervals:

(a) $[-1, 1]$ (b) $(-1, 1)$ (c) $[-1, 3)$
(d) $[-3, 1]$ (e) $(3, 5]$

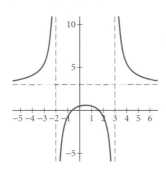

Skills

■ For each rational function f and value c below, use the definition of derivative (or the "alternative" definition) to calculate $f'(c)$.

7. $f(x) = \dfrac{x - 1}{x + 3}$, $c = 2$

8. $f(x) = \dfrac{1}{x^2 - 1}$, $c = -2$

9. $f(x) = \dfrac{x^2 - 3x}{x^2 - 2x + 1}$, $c = 0$

10. $f(x) = \dfrac{x - 1}{(x + 1)(x + 2)}$, $c = 1$

■ Calculate the derivatives of the rational functions below by using the definition of derivative (or the "alternative" definition of derivative).

11. $f(x) = \dfrac{x - 1}{x + 3}$

12. $f(x) = \dfrac{1}{x^2 - 1}$

13. $f(x) = \dfrac{x^2 - 3x}{x^2 - 2x + 1}$

14. $f(x) = \dfrac{x - 1}{(x + 1)(x + 2)}$

15. $f(x) = \dfrac{x^3}{x + 1}$

16. $f(x) = \dfrac{x^2 - 1}{x^2 - x - 2}$

■ Use the quotient rule to calculate the derivatives of the following rational functions.

17. $f(x) = \dfrac{2x - 3}{5x + 4}$

18. $f(x) = \dfrac{x^3}{x + 1}$

19. $f(x) = \dfrac{x^2 - 3x}{x^2 - 2x + 1}$

20. $f(x) = \dfrac{1}{x^3 - 2x^2 + x - 3}$

21. $f(x) = \dfrac{x^7 - 3x^5 + 4}{1 - 3x^4}$

22. $f(x) = \dfrac{x^2}{x^3 + 5x^2 - 3x}$

23. $f(x) = \dfrac{1}{(x + 1)^3}$

24. $f(x) = \dfrac{x - 1}{(x + 1)(x + 2)}$

25. $f(x) = \dfrac{(x - 2)^2}{(x^2 + 1)(x - 3)}$

■ Find the critical points of each rational function f.

26. $f(x) = \dfrac{1 + x + x^2}{x^2 + x - 2}$

27. $f(x) = \dfrac{(x - 1)^2}{x + 2}$

28. $f(x) = \dfrac{x^3}{x^2 - 3x + 2}$

29. $f(x) = \dfrac{x^2 - 2x + 1}{x^2 - 1}$

30. $f(x) = \dfrac{x^2(x - 1)}{(x - 2)^2}$

31. $f(x) = \dfrac{1}{x^3 - x}$

■ Find the local extrema of the following rational functions. Do all work algebraically (by hand), and then check your answers with a graphing calculator. Notice that the first six problems involve the same functions you investigated in the block of problems above.

32. $f(x) = \dfrac{1 + x + x^2}{x^2 + x - 2}$

33. $f(x) = \dfrac{(x - 1)^2}{x + 2}$

34. $f(x) = \dfrac{x^3}{x^2 - 3x + 2}$

35. $f(x) = \dfrac{x^2 - 2x + 1}{x^2 - 1}$

36. $f(x) = \dfrac{x^2(x - 1)}{(x - 2)^2}$

37. $f(x) = \dfrac{1}{x^3 - x}$

38. $f(x) = \dfrac{x^2 - x - 2}{x^3}$

39. $f(x) = \dfrac{1}{(x - 2)^2}$

40. $f(x) = \dfrac{x^2 - 4x + 4}{x - 2}$

■ Find the global extrema of each of the following rational functions on the given intervals. Do all work algebraically (by hand), and then check your answers with a graphing calculator. Note that all of the functions in this block are also in the previous block of problems.

41. $f(x) = \dfrac{1 + x + x^2}{x^2 + x - 2}, \quad (-1, 1)$

42. $f(x) = \dfrac{1 + x + x^2}{x^2 + x - 2}, \quad (1, 5]$

43. $f(x) = \dfrac{(x-1)^2}{x+2}, \quad [-1, 3]$

44. $f(x) = \dfrac{(x-1)^2}{x+2}, \quad (-4, -2)$

45. $f(x) = \dfrac{1}{(x-2)^2}, \quad [0, 2)$

46. $f(x) = \dfrac{1}{(x-2)^2}, \quad [1, 4]$

47. $f(x) = \dfrac{x^2 - 2x + 1}{x^2 - 1}, \quad [-2, 1]$

48. $f(x) = \dfrac{x^2 - 2x + 1}{x^2 - 1}, \quad (-1, 2)$

Proofs

49. Use the quotient rule to prove that a rational function f is differentiable everywhere on its domain.

50. Prove that if f is a rational function with a horizontal asymptote, then its derivative f' also has a horizontal asymptote.

51. Prove that if f is a rational function with a slant asymptote, then its derivative f' has a horizontal asymptote.

52. Use the definition of derivative to prove the quotient rule. Justify each step in your calculation.

6.4 | Graphs of Rational Functions

In this section we will graph rational functions two ways. First, we will learn how to make a quick sketch of the graph of a rational function *without* using the derivative at all. Second, we will use the zeros and signs of the first and second derivatives to make an accurate graph of a rational function, including the locations of its local extrema and inflection points. In the last part of this section, we will reverse the problem and construct equations for rational functions that will model given graphs.

6.4.1 Graphing rational functions quickly

Given a factored rational function f, it is easy to identify any holes, vertical asymptotes, or horizontal asymptotes. If we also know selected values of f, or the intervals on which f is positive or negative, we can use this information to sketch a rough graph of f.

EXAMPLE 6.23   **Making a rough graph of a rational function**

Sketch a rough graph of the function $f(x) = \dfrac{2(x-1)^2(x+2)}{(x-1)(x+1)(x-2)}$.

Solution   We can see immediately from the factors in the numerator and denominator of the equation for f that the graph of f will have

▶ A hole at $x = 1$

▶ A root at $x = -2$

▶ Vertical asymptotes at $x = -1$ and $x = 2$

▶ A horizontal asymptote at $y = 2$.

(Note that the leading term of the numerator is $2x^3$ (after multiplying out) and the leading term of the denominator is x^3; this is why there is a horizontal asymptote at

$y = \frac{2}{1} = 2$.) We also know that the height of the hole at $x = 1$ will be the value $g(1)$, where $g(x) = \frac{2(x-1)(x+2)}{(x+1)(x-2)}$ is the function obtained by canceling the common factors in f. The height of the hole at $x = 1$ will be

$$g(1) = \frac{2(1-1)(1+2)}{(1+1)(1-2)} = \frac{2(0)(3)}{(2)(-1)} = 0.$$

At this point we know a lot about the graph of f (even without calculating any derivatives!). Figure 6.29 shows the hole, root, and asymptotes described above.

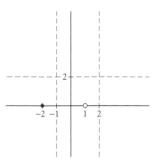

Figure 6.29

To get a better idea of what the graph of f looks like, we can calculate the y-intercept:

$$f(0) = \frac{2(0-1)^2(0+2)}{(0-1)(0+1)(0-2)} = \frac{2(1)(2)}{(-1)(1)(-2)} = \frac{4}{2} = 2.$$

We now have some idea what the function might look like to the left of $x = -1$, and for x between -1 and 2. Since it is not yet clear what the graph of f might look like to the right of $x = 2$, we'll also calculate $f(3)$:

$$f(3) = \frac{2(3-1)^2(3+2)}{(3-1)(3+1)(3-2)} = \frac{2(4)(5)}{(2)(4)(1)} = 5.$$

Figure 6.30 shows the information from Figure 6.29 together with the two new coordinates at $(0, 2)$ and $(3, 5)$. Combining this information, we might guess that the function f looks like the graph in Figure 6.31 (this is at least one of the simplest possibilities for the graph). Remember that a function cannot cross its vertical asymptotes, and that f must approach $y = 2$ as $x \to \infty$ and as $x \to -\infty$.

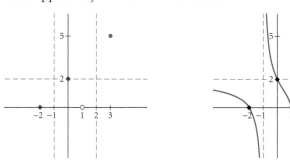

Figure 6.30   **Figure 6.31**

Of course, this is a very rough sketch of the graph of f. To check the accuracy of our graph, we can make a quick sign analysis of f. (In fact, it is often a good idea to make a number line for f *before* attempting to sketch its graph, especially if the asymptotes and zeros don't provide enough information to guess the shape of the graph.) Since f is in factored form, we know that it can change sign only at $x = -2$, $x = -1$, $x = 1$, or $x = 2$. By testing some values between these points, we can see that f has the signs given in the number line in Figure 6.32.

Figure 6.32

Notice that the graph we drew in Figure 6.31 does indeed have the signs indicated by this number line. □

🔷 *Caution*   The graph of f in Example 6.23 may actually have features not recorded in the rough sketch (although in this example the graph is accurate). The method used in Example 6.23 produces only a rough guess for the graph of a rational function. □

❓ *Question*   Sketch two other graphs that have the holes, values, and asymptotes shown in Figure 6.30. Make one of the graphs so that it has the sign intervals shown in Figure 6.32. □

Calculators are notoriously bad at graphing rational functions. Sometimes they "connect" a graph over its vertical asymptotes, and most times the "holes" of a rational function are not immediately clear from a calculator graph. The techniques in Example 6.23 can be used to augment a calculator graph of a rational function by precisely locating any intercepts or asymptotes of the graph. To get a more accurate sketch of the graph of f, including the locations of its extrema and inflection points, we can do a complete first and second derivative analysis.

6.4.2 Graphing rational functions accurately

If we wish to have an accurate graph of f that correctly reflects the increasing or decreasing behavior, local extrema, concavity, and inflection points, then we must do a first and second derivative analysis of f. We will also use the hole, root, and asymptote information that can be obtained using the techniques in Section 6.4.1.

EXAMPLE 6.24

A detailed curve-sketching analysis of a rational function

Sketch a labeled, accurate graph of $f(x) = \dfrac{2(x-3)(x+3)}{x^2+1}$.

Solution   From the factors and degrees of the numerator and denominator of f we can immediately see that the graph of f will have

▶ Roots at $x = -3$ and $x = 3$

▶ No holes

▶ No vertical asymptotes

▶ A horizontal asymptote at $y = 2$.

To make an accurate, labeled graph, begin by examining the zeros and sign intervals of the first derivative. Before we can differentiate f we have to multiply it out (we don't know how to differentiate products yet—although we will soon!):

$$f(x) = \frac{2(x-3)(x+3)}{x^2+1} = \frac{2(x^2-9)}{x^2+1} = \frac{2x^2-18}{x^2+1}.$$

Given a graph that we wish to model with a rational function, we first look for the following information: the locations of any roots, holes, or other coordinates on the graph; the locations of any vertical, horizontal, slant, or curve asymptotes; and any known locations of extrema or inflection points. Given this information, we need to find the equation of a rational function f that will have the features that we identified.

EXAMPLE 6.25 **Modeling a graph with a rational function**

Find a rational function f that could have the graph shown in Figure 6.37.

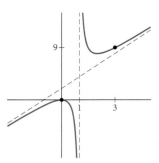

Figure 6.37

Solution We know two points on the graph of this function: $(0, 0)$ and $(3, 9)$. We also see a vertical asymptote at $x = 1$ and a slant asymptote (whose equation we cannot determine at this point). The graph does not have any holes and has no zeros other than the one at $x = 0$.

If Figure 6.37 is the graph of a rational function f, then we know that the numerator of f will have only one root, at $x = 0$, at that the denominator of f will have a zero only at $x = 1$. Since the graph has a slant asymptote, the degree of the numerator of f must be exactly 1 higher than the degree of the denominator. There are a number of ways we could get an equation for f with this information, using irreducible quadratic factors and repeated linear factors to get the degrees and zeros we desire. Perhaps the simplest such function having all these characteristics is

$$f(x) = \frac{Ax^2}{x - 1}.$$

Notice that we have a repeated linear factor x^2 in the numerator. We squared the x because we needed the degree of the numerator to be 1 higher than the degree of the denominator. We included a constant A in the numerator because we have not yet used the fact that the graph passes through the point $(3, 9)$. The value of A will determine the extent to which the graph of f is stretched vertically. We can use the fact that $f(3) = 9$ to solve for A:

$$f(3) = 9 \implies \frac{A(3)^2}{3 - 1} = 9 \implies \frac{9A}{2} = 9 \implies A = 2.$$

One function that might have the graph in Figure 6.37 is therefore

$$f(x) = \frac{2x^2}{x - 1}.$$

☐

☑ *Checking the Answer* Use a calculator to graph the function $f(x) = \frac{2x^2}{x-1}$, and check that this graph has the features shown in Figure 6.37. By synthetic division, we can show that the equation of the slant asymptote of $f(x) = \frac{2x^2}{x-1}$ is $y = 2x + 2$, which seems to match the slant asymptote shown in Figure 6.37. ☐

In Example 6.25 we used repeated linear factors to get the desired degrees in the numerator and denominator of our rational function. In the next example we do the same thing with irreducible quadratic factors.

EXAMPLE 6.26

Using irreducible quadratic factors when modeling with a rational function

Find a rational function f that could have the graph shown in Figure 6.38.

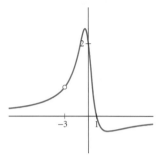

Figure 6.38

Solution   Suppose f is a rational function whose graph is the one given in Figure 6.38. Since the graph has a hole at $x = -3$, the numerator and denominator of f must have a common factor of $x + 3$. The graph has only one root, at $x = 1$, and thus the only other linear factor of the numerator of f is $x - 1$. Since the graph has no vertical asymptotes, the only linear factor in the denominator of f is the factor $x + 3$. At this point you might be tempted to guess that f could be the function

$$f(x) = \frac{A(x - 1)(x + 3)}{x + 3},$$

but this *cannot* be the case! The graph in Figure 6.38 has a horizontal asymptote at $y = 0$, so the degree of the denominator of f must be *larger* than the degree of the numerator, and in the equation this is not so. However, we aren't allowed to add any other linear factors to the denominator of f (this would create vertical asymptotes), and if we make $x + 3$ a repeated linear factor of the denominator, then the graph of f will have an asymptote at $x = -3$ (why?).

If we include an irreducible quadratic factor in the denominator, the roots of the denominator will not be changed, but the degree of the denominator will be increased by 2. One of the simplest irreducible quadratics is $x^2 + 1$, so we'll use that. (Remember that modeling a graph by a function often involves some guesswork and that there is often more than one possible model function for a given graph.) Our guess for the function f is now

$$f(x) = \frac{A(x - 1)(x + 3)}{(x + 3)(x^2 + 1)}.$$

Once again, we include a "stretching" constant A so that we can use the information that $f(0) = 2$. Solving for A, we have

$$f(0) = 2 \ \Rightarrow \ \frac{A(0 - 1)(0 + 3)}{(0 + 3)(0^2 + 1)} = 2 \ \Rightarrow \ \frac{-3A}{3} = 2 \ \Rightarrow \ A = -2.$$

Finally, our best guess for a rational function whose graph has the features shown in Figure 6.38 is the function

$$f(x) = \frac{-2(x - 1)(x + 3)}{(x + 3)(x^2 + 1)}.$$

Obviously, after all that work it is a good idea to graph the function f and check that it does look like the graph shown in Figure 6.38.   □

What you should know after reading Section 6.4

▶ How to sketch a rough graph of a rational function quickly using the degrees and factors of its numerator and denominator polynomials. This includes identifying roots, holes, and asymptotes. To make a slightly more accurate graph, it is advisable to plot a few points on the graph, by hand, and/or examine the sign intervals of the function.

▶ How to sketch an accurate, labeled graph of a rational function by doing a sign analysis of its first and second derivatives, as well as using the degree and factor information for the original function. If the function has slant or curve asymptotes, then the equations for these asymptotes should also be found.

▶ How to find a rational function that has the properties of a given graph. This includes constructing a rational function that has the correct roots, holes, and vertical asymptotes and then using repeated roots and/or irreducible quadratic factors so that the function will have the correct horizontal, slant, or curve asymptotes. If additional datapoints are given on the graph, then make sure that your equation has those datapoints as solutions!

Exercises 6.4

Concepts

0. Read the section and make your own summary of the material. Include whatever you think will help you review the material later on.

1. The rational function $f(x) = \frac{1}{x-1}$ has a vertical asymptote at $x = 1$. Determine the behavior of the graph of f on either side of $x = 1$ in two ways:

  (a) By examining $\lim\limits_{x \to 1^+} f(x)$ and $\lim\limits_{x \to 1^-} f(x)$.

  (b) By creating and using a number line describing the sign intervals of the derivative f'.

2. Use the graph of $y = \frac{1}{x}$ and what you know about transformations of graphs to sketch graphs of each of the following rational functions. Label the coordinates of each graph at $x = 1$ and $x = -2$.

  (a) $f(x) = \dfrac{-3}{x}$

  (b) $f(x) = \dfrac{1}{x-3}$

  (c) $f(x) = \dfrac{2}{1-4x}$

3. Which graphical features of a rational function f can be determined by examining the roots and degrees of the numerator and denominator of f, and how?

4. Which graphical features of a rational function f can be determined by a sign analysis of the first and second derivatives of f, and how?

5. Which graphical features of a rational function f can be determined by calculating limits, and how? (*Note:* There may also be other ways to determine these graphical features; however, list as many features as possible that can be determined with limits.)

6. Suppose f is a rational function with roots at $x = 1$ and $x = 3$, a hole at $x = -1$, a vertical asymptote at $x = 2$, and a horizontal asymptote at $y = -1$.

  (a) Sketch three different possible graphs of f. Make the graphs as different as possible while still having the characteristics given above.

  (b) Write down the equations of three different functions f that have the properties given above. (*Note:* The functions you write down need not be related to the graphs you drew above.)

7. Suppose f is a rational function with a slant asymptote of $y = 3 - x$ and one vertical asymptote at $x = 0$. Sketch six different graphs with these properties, making your graphs as different as possible.

8. The graph on the next page shows asymptotes, roots, and holes that a rational function f could have. Assuming that the only asymptotes, roots, and holes are those shown in the graph, sketch three possible graphs with the given characteristics. What extra pieces of information would you need to know in order to determine completely the graph of f?

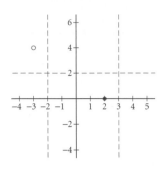

9. Given the number lines below, sketch a possible graph of f. Make sure that your graph reflects all of the information given in the number lines. Identify all roots, holes, and asymptotes of your graph. What additional information would you need in order to sketch a completely accurate graph of f?

$$\begin{array}{ccccccccc} - & & - & \text{DNE} & + & & - & \text{DNE} & + & \text{DNE} & + \\ \hline & -3 & & -1 & & 2 & & 3 & & 5 & \end{array} f$$

$$\begin{array}{ccccccccc} + & & - & \text{DNE} & - & & - & \text{DNE} & - & \text{DNE} & + \\ \hline & -3 & & -1 & & 1 & & 3 & & 5 & \end{array} f'$$

$$\begin{array}{cccccccc} & - & \text{DNE} & + & & - & \text{DNE} & + & \text{DNE} & + \\ \hline & -1 & & 1 & & 3 & & 5 & \end{array} f''$$

10. Find an equation for a rational function that would have the number lines from Problem 9. You may have to do some educated guessing (and use a graphing calculator to check and refine your guesses).

11. Many of the rational functions in the graphing exercises are chosen precisely because they (and/or their derivatives) can be factored using the techniques in this course. In general, however, rational functions are not always this nice to work with! Think of a "random" rational function that isn't too simple or too complicated. Can you factor its numerator and denominator easily? What about the numerators and denominators of its first and second derivatives?

12. If you can factor the denominator of a rational function $f(x) = \frac{p(x)}{q(x)}$, can you also factor the denominators of f' and f''? Why or why not? What if you can factor the numerator of f; could you then factor the numerators of f' and f''?

Skills

■ Make rough sketches of the following rational functions without using any derivative information. Be as accurate as you can. Begin by using the factors and degrees of the numerator and denominator, and then, as necessary, plot points of the function by hand and/or do a sign analysis of the function. Find the equations of any horizontal, vertical, slant, or curve asymptotes.

13. $f(x) = \dfrac{(x-1)(x+2)}{(x+1)(x-1)}$

14. $f(x) = \dfrac{x^2 + x - 6}{x - 2}$

15. $f(x) = \dfrac{x^2 + 1}{x - 1}$

16. $f(x) = \dfrac{1}{16 - x^4}$

17. $f(x) = \dfrac{3x(x+1)}{(x-5)^2(x+2)}$

18. $f(x) = \dfrac{2x^3 + 4x^2 - 6x}{x^2 - 4}$

19. $f(x) = \dfrac{2x^3 + 3x^2 - 2x - 3}{x^2 - 2x - 3}$

20. $f(x) = \dfrac{(x^2 - 1)^2}{2x^2 - 3x - 2}$

21. $f(x) = \dfrac{(x^2 - 4)^2}{2x^2 - 3x - 2}$

■ Sketch accurate graphs of the following rational functions. Use first and second derivative information as well as root, hole, and asymptote information. If possible, label the coordinates of any roots, extrema, or inflection points. Give the equations of any asymptotes (vertical, horizontal, slant, or curve).

22. $f(x) = \dfrac{x^2 - 1}{x^2 - x - 2}$

23. $f(x) = \dfrac{x - 2}{(x - 1)^2}$

24. $f(x) = \dfrac{x^2}{x - 1}$

25. $f(x) = \dfrac{3x^2 - 2x + 4}{x^2 - 4x + 4}$

26. $f(x) = \dfrac{x(x+1)(x-1)}{(x-2)(x+1)}$

27. $f(x) = \dfrac{x^3 - 5x}{x - 1}$

28. $f(x) = \dfrac{(x^2 - 1)^2}{3x^2 - x - 2}$

29. $f(x) = \dfrac{x^4}{x - 2}$

30. $f(x) = \dfrac{x^3 - 3x^2 - x + 3}{x^3 - 6x^2 + 12x - 8}$

■ For each of the graphs given below, find a rational function f that has the properties shown in the graph. Explain how you arrived at your function. Then check your answer (revising if necessary) with a graphing calculator.

31.

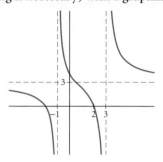

32.

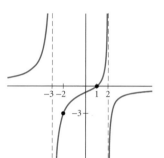

33.

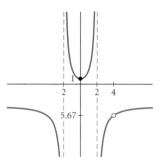

34.

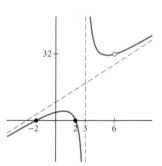

35.

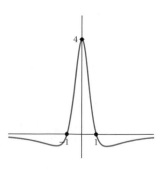

36.

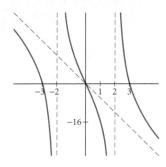

37.

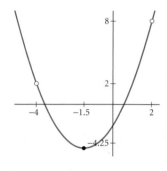

38.

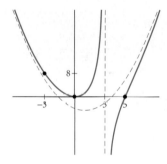

39.

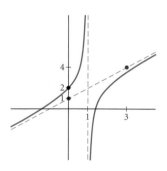

General Algebraic Functions

*A*ll of the functions we have examined so far have been special types of algebraic functions. Power functions, polynomial functions, and rational functions are all algebraic. Sums, products, compositions, and powers of these types of functions are also algebraic. In this chapter we will apply familiar techniques and theorems to the general class of algebraic functions.

"The Orb," a modular structure by Gregg Fleishman, www. greggfleishman.com

CONTENTS

7.1 | Working with Algebraic Functions

Recall that a function f is **algebraic** if it can be expressed in terms of a variable x using only arithmetic operations $(+, -, \times, \text{ and } \div)$ and rational powers. For example, the function $f(x) = \sqrt{5-x} + x^{\frac{3}{4}}$ is algebraic, but the function $f(x) = 2^x$ is not (since the independent variable is in the exponent). Power functions, polynomial functions, and rational functions are all algebraic functions, as are their sums, products, quotients, and compositions.

In this section we will review algebraic, limit, and differentiation techniques in the context of general algebraic functions. We have already learned most of these techniques while studying power, polynomial, and rational functions; here we put everything together to study algebraic functions in general.

7.1.1 Domains of algebraic functions

The domain of a function is the set of values for which it is defined. In general, there are only two things to look out for when determining the domain of an algebraic function: division by zero, and even roots of negative numbers.

EXAMPLE 7.1 **Finding domains of functions with square roots and denominators**

Find the domain of each of the following algebraic functions:

(a) $f(x) = \sqrt{5-x} + x^{\frac{3}{4}}$ (b) $g(x) = \dfrac{3}{\sqrt{x^2 - 4}}$

Solution

(a) The function f is a sum and will be defined whenever *both* $\sqrt{5-x}$ and $x^{\frac{3}{4}}$ are defined. The expression $\sqrt{5-x}$ is defined only when $5 - x \geq 0$, that is, when $x \leq 5$. On the other hand, the expression $x^{\frac{3}{4}}$ is defined for $x \geq 0$. Thus the domain of f is

$$(-\infty, 5] \cap [0, \infty) = [0, 5].$$

(b) The function g will be defined if and only if its denominator $\sqrt{x^2 - 4}$ is nonzero and the expression $x^2 - 4$ is nonnegative (why?). Since $\sqrt{x^2 - 4}$ is zero if and only if $x^2 - 4$ is zero, g is defined whenever $x^2 - 4 > 0$. Since $x^2 - 4 = (x-2)(x+2)$ is continuous, the Intermediate Value Theorem states that the only places it can change sign are at $x = 2$ and at $x = -2$. Testing the sign of $x^2 - 4$ at $x = -3$, $x = 0$, and $x = 3$ (for example), we obtain the number line shown in Figure 7.1.

Figure 7.1

Therefore $x^2 - 4$ is strictly greater than zero when $x < -2$ or when $x > 2$, and thus the domain of g is $(-\infty, -2) \cup (2, \infty)$. □

☑ *Checking the Answer* One way to check Example 7.1 is to try to evaluate each function at points inside and outside the proposed domain. For example, try to evaluate g at $x = 3$ (supposedly in the domain) and at $x = 0$ (supposedly not in the domain). Clearly $g(3) = \frac{3}{\sqrt{5}}$ exists but $g(0) = \frac{3}{\sqrt{-4}}$ does not, so our proposed domain is valid at least for those two values. We could also use a graphing calculator to graph the functions f and g and see where they appear to be defined. Neither of these methods will tell you for certain that your answers are correct, but you will know that you have made an error if either of these tests fails. □

7.1.2 Limits of algebraic functions

We have already seen that power functions, polynomial functions, and rational functions are continuous everywhere on their domains. Moreover, we know that sums, products, and quotients of continuous functions are also continuous everywhere they are defined. These facts enable us to solve a limit as $x \to c$ of an algebraic function f by evaluating f at $x = c$ (if f is defined at that point). To solve limits of algebraic functions at nondomain points and as $x \to \infty$ and $x \to -\infty$, we will use the techniques we have developed in previous sections.

EXAMPLE 7.2

Finding limits of algebraic functions

Calculate the following limits.

(a) $\displaystyle \lim_{x \to -2} \frac{\sqrt{3+x}}{x+1}$ **(b)** $\displaystyle \lim_{x \to 0} \frac{x^2}{x^{\frac{7}{2}} - x^{\frac{8}{3}}}$ **(c)** $\displaystyle \lim_{x \to \infty} \frac{x^{\frac{1}{3}}}{1 - x^{\frac{1}{3}}}$

Solution

(a) Since $\frac{\sqrt{3+x}}{x+1}$ is defined and continuous at $x = -2$, its limit is simply its value at $x = -2$:

$$\lim_{x \to -2} \frac{\sqrt{3+x}}{x+1} = \frac{\sqrt{3-2}}{-2+1} = \frac{\sqrt{1}}{-1} = -1.$$

(b) If we attempt to evaluate the expression $\frac{x^2}{x^{\frac{7}{2}} - x^{\frac{8}{3}}}$ at $x = 0$, we get the indeterminate form $\frac{0}{0}$. We need to do some algebra before we can evaluate the limit:

$$\lim_{x \to 0} \frac{x^2}{x^{\frac{7}{2}} - x^{\frac{8}{3}}} = \lim_{x \to 0} \frac{x^2}{x^2 \left(x^{\frac{3}{2}} - x^{\frac{2}{3}} \right)} = \lim_{x \to 0} \frac{1}{x^{\frac{3}{2}} - x^{\frac{2}{3}}} \to \frac{1}{0}.$$

We now know that the limit is infinite. To be completely precise, as $x \to 0^-$ this limit does not exist (since $x^{\frac{3}{2}}$ is not defined for negative values of x), and as $x \to 0^+$ this limit approaches positive infinity.

(c) As $x \to \infty$, both the numerator and the denominator of the expression $\frac{x^{\frac{1}{3}}}{1 - x^{\frac{1}{3}}}$ become infinite (in particular, we have the indeterminate form $\frac{\infty}{-\infty}$). For a rational function (a quotient of two polynomials) we would divide the numerator and denominator by the highest power of x, and although this is not a rational function, a similar method works here. In this case, we will divide the numerator and denominator by $x^{\frac{1}{3}}$:

$$\lim_{x \to \infty} \frac{x^{\frac{1}{3}}}{1 - x^{\frac{1}{3}}} = \lim_{x \to \infty} \frac{x^{\frac{1}{3}}}{1 - x^{\frac{1}{3}}} \left(\frac{1/x^{\frac{1}{3}}}{1/x^{\frac{1}{3}}} \right) = \lim_{x \to \infty} \frac{1}{\frac{1}{x^{\frac{1}{3}}} - 1} \to \frac{1}{0 - 1} = -1. \qquad \square$$

7.1.3 Derivatives of algebraic functions

We already know how to differentiate power functions, polynomial functions, and rational functions using the power, sum, constant multiple, and quotient rules for differentiation. In fact, we can differentiate a wider variety of algebraic functions using these rules, including any sum, constant multiple, or quotient of power functions. In Section 7.2 we will learn rules for differentiating products and compositions, and with those rules we will be able to differentiate *any* algebraic function. For now we restrict our attention to the power, sum, constant multiple, and quotient rules.

EXAMPLE 7.3	**Finding derivatives of algebraic functions using the "rules" we have developed**

Find the derivatives of the following functions using the power, sum, constant multiple, and quotient rules, if possible.

(a) $f(x) = \dfrac{\sqrt{x}}{3x^{\frac{2}{3}} - 4x^2}$
  (b) $g(x) = \sqrt{x}\,(1 - 3\sqrt[4]{x})^2$

(c) $h(x) = (x^3 + 4x - 1)^{\frac{1}{5}}$

Solution

(a) The function f is a quotient of two functions that we know how to differentiate using the power, sum, and constant multiple rules, so we apply the quotient rule:

$$f'(x) = \frac{d}{dx}\left(\frac{\sqrt{x}}{3x^{\frac{2}{3}} - 4x^2}\right) = \frac{d}{dx}\left(\frac{x^{\frac{1}{2}}}{3x^{\frac{2}{3}} - 4x^2}\right)$$

$$= \frac{\left(\frac{d}{dx}\left(x^{\frac{1}{2}}\right)\right)\left(3x^{\frac{2}{3}} - 4x^2\right) - \left(x^{\frac{1}{2}}\right)\left(\frac{d}{dx}\left(3x^{\frac{2}{3}} - 4x^2\right)\right)}{\left(3x^{\frac{2}{3}} - 4x^2\right)^2}$$

$$= \frac{\left(\frac{1}{2}x^{-\frac{1}{2}}\right)\left(3x^{\frac{2}{3}} - 4x^2\right) - x^{\frac{1}{2}}\left(3\left(\frac{2}{3}\right)x^{-\frac{1}{3}} - 8x\right)}{\left(3x^{\frac{2}{3}} - 4x^2\right)^2}.$$

We could simplify this expression if we needed to, but we are finished with the differentiation steps and so we will stop here.

(b) If we knew how to differentiate products and compositions of functions, we could differentiate the function g as it is currently written. However, we don't yet have such rules (we will soon!), so we'll have to do some algebra first:

$$g'(x) = \frac{d}{dx}\left(\sqrt{x}\,(1 - 3\sqrt[4]{x})^2\right) = \frac{d}{dx}\left(x^{\frac{1}{2}}\left(1 - 3x^{\frac{1}{4}}\right)^2\right)$$

$$= \frac{d}{dx}\left(x^{\frac{1}{2}}\left(1 - 2\left(3x^{\frac{1}{4}}\right) + \left(3x^{\frac{1}{4}}\right)^2\right)\right) = \frac{d}{dx}\left(x^{\frac{1}{2}}\left(1 - 6x^{\frac{1}{4}} + 9x^{\frac{1}{2}}\right)\right)$$

$$= \frac{d}{dx}\left(x^{\frac{1}{2}} - 6x^{\frac{1}{2}}x^{\frac{1}{4}} + 9x^{\frac{1}{2}}x^{\frac{1}{2}}\right) = \frac{d}{dx}\left(x^{\frac{1}{2}} - 6x^{\frac{3}{4}} + 9x\right)$$

$$= \tfrac{1}{2}x^{-\frac{1}{2}} - 6\left(\tfrac{3}{4}\right)x^{-\frac{1}{4}} + 9 = \tfrac{1}{2}x^{-\frac{1}{2}} - \tfrac{9}{2}x^{-\frac{1}{4}} + 9.$$

Notice that most of the work involved algebra, and not differentiation. After we learn the product and chain rules in Section 7.2, we will be able to differentiate g directly, without doing algebra first.

(c) The function h is a composition that cannot be written as a sum or quotient of functions we currently know how to differentiate. Since we don't have a rule for differentiating compositions yet, we cannot differentiate this function using "rules" at this point. We will be able to differentiate compositions after we learn the chain rule in Section 7.2. □

Working with general algebraic functions is often messy. For example, suppose we wanted to find the critical points of the function $g(x) = \sqrt{x}\,(1 - 3\sqrt[4]{x})^2$ from Example 7.3(b). To do this we would need to find the places at which the derivative $g'(x) = \frac{1}{2}x^{-\frac{1}{2}} - \frac{9}{2}x^{-\frac{1}{4}} + 9$ is zero or does not exist. This means we will have to rewrite $g'(x)$ in the form of a simple quotient, as in Example 7.4.

EXAMPLE 7.4

Finding the critical points of an algebraic function

Find the critical points of the function $g(x) = \sqrt{x}\,(1 - 3\sqrt[4]{x})^2$.

Solution   As we saw in part (b) of Example 7.3, the derivative of $g(x) = \sqrt{x}\,(1 - 3\sqrt[4]{x})^2$ is the function

$$g'(x) = \tfrac{1}{2}x^{-\frac{1}{2}} - \tfrac{9}{2}x^{-\frac{1}{4}} + 9.$$

We are interested in finding the values at which g' is zero or does not exist. Note that since the equation for g' involves negative powers, the function g' is really made up of *fractions* (for example, $x^{-\frac{1}{2}} = \frac{1}{x^{\frac{1}{2}}}$). It will make our job easier if we first write g' as a quotient:

$$g'(x) = \tfrac{1}{2}x^{-\frac{1}{2}} - \tfrac{9}{2}x^{-\frac{1}{4}} + 9$$

$$= \frac{1}{2x^{\frac{1}{2}}} - \frac{9}{2x^{\frac{1}{4}}} + 9 \qquad \text{(rewrite as fractions)}$$

$$= \frac{1}{2x^{\frac{1}{2}}} - \frac{9}{2x^{\frac{1}{4}}}\left(\frac{x^{\frac{1}{4}}}{x^{\frac{1}{4}}}\right) + 9\left(\frac{2x^{\frac{1}{2}}}{2x^{\frac{1}{2}}}\right) \qquad \text{(get common denominator)}$$

$$= \frac{1 - 9x^{\frac{1}{4}} + 18x^{\frac{1}{2}}}{2x^{\frac{1}{2}}}. \qquad \text{(combine fractions)}$$

The zeros of g' will be the values of x that are roots of the numerator (but not the denominator), and the values at which g' is not defined will be the roots of the denominator. Although the numerator involves fractional powers, in this case we can actually factor it, since it can be thought of as a quadratic expression in $x^{\frac{1}{4}}$ (because $(x^{\frac{1}{4}})^2 = x^{\frac{1}{2}}$). We have

$$g'(x) = \frac{1 - 9x^{\frac{1}{4}} + 18x^{\frac{1}{2}}}{2x^{\frac{1}{2}}} = \frac{\left(1 - 3x^{\frac{1}{4}}\right)\left(1 - 6x^{\frac{1}{4}}\right)}{2x^{\frac{1}{2}}}.$$

Thus $g'(x)$ is zero when $1 - 3x^{\frac{1}{4}} = 0$ and when $1 - 6x^{\frac{1}{4}} = 0$, that is, when $x = \frac{1}{3^4} = \frac{1}{81}$ and when $x = \frac{1}{6^4} = \frac{1}{1296}$. Moreover, $g'(x)$ is undefined when $2x^{\frac{1}{2}} = 0$, that is, when $x = 0$. Therefore, there are critical points of $g(x)$, namely $x = \frac{1}{81}$, $x = \frac{1}{1296}$, and $x = 0$.   □

☑ *Checking the Answer*   One way to check the answer is to graph the function $g(x) = \sqrt{x}\,(1 - 3\sqrt[4]{x})^2$. The critical points of g should be evident as places where the graph has a horizontal tangent line (the zeros of g') or an undefined tangent line (the places where g' does not exist).

However, you need to pay attention to your graphing window in this problem! If you graph g without thinking about the appropriate window, you might first end up with a graph like the one in Figure 7.2. From this graph, it erroneously appears that g is always increasing (and, in particular, has no horizontal tangent lines), contrary to our answer in Example 7.4.

With a little thought, it is clear that we are interested in very small values of x (like $x = \frac{1}{81} \approx 0.0123457$). The function g is shown with a more appropriate graphing window in Figure 7.3. The graph of g *does* have some interesting behavior, but the graphing window in Figure 7.2 is too large for us to see it! On the other hand, the graph in Figure 7.3 shows the critical points at $x = 0$, $x = \frac{1}{1296}$, and $x = \frac{1}{81}$. (In addition, not only is the function undefined to the left of zero, its graph appears to have a vertical tangent line from the right at $x = 0$.)

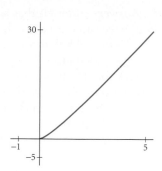

Figure 7.2
$y = g(x)$ in a bad window

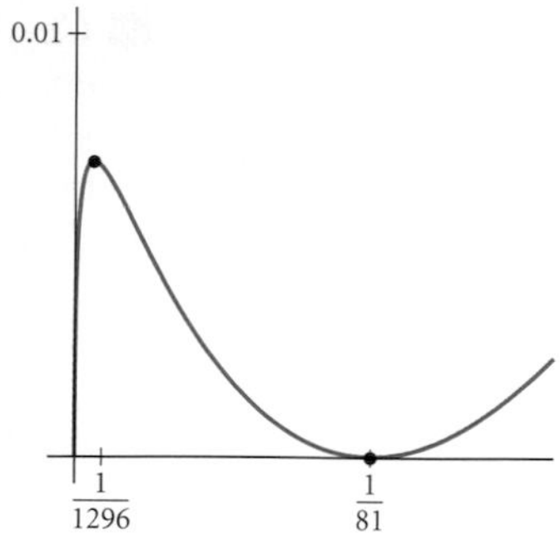

Figure 7.3
$y = g(x)$ in a good window ☐

7.1.4 Differentiating algebraic functions by hand

As we mentioned previously, we do not yet have rules for differentiating *every* type of algebraic function. If we really wanted to, we could go back to the definition of derivative to calculate such derivatives. In general this can involve a lot of nasty algebra, as in the following example.

EXAMPLE 7.5

Using the definition of derivative

If $f(x) = \sqrt{x^2 - 5}$, use the definition of derivative to find $f'(x)$.

Solution   First notice that we cannot differentiate f with any of the differentiation rules we currently know (why not?). We will use the definition of derivative to find $f'(x)$. Note the use of the conjugate in the third line of the calculation.

$$f'(x) = \lim_{h \to 0} \frac{f(x+h) - f(x)}{h} \qquad \text{(definition of derivative)}$$

$$= \lim_{h \to 0} \frac{\sqrt{(x+h)^2 - 5} - \sqrt{x^2 - 5}}{h}$$

$$= \lim_{h \to 0} \frac{\sqrt{(x+h)^2 - 5} - \sqrt{x^2 - 5}}{h} \left(\frac{\sqrt{(x+h)^2 - 5} + \sqrt{x^2 - 5}}{\sqrt{(x+h)^2 - 5} + \sqrt{x^2 - 5}} \right)$$

$$= \lim_{h \to 0} \frac{((x+h)^2 - 5) - (x^2 - 5)}{h(\sqrt{(x+h)^2 - 5} + \sqrt{x^2 - 5})}$$

$$= \lim_{h \to 0} \frac{x^2 + 2xh + h^2 - 5 - x^2 + 5}{h(\sqrt{(x+h)^2 - 5} + \sqrt{x^2 - 5})}$$

$$= \lim_{h \to 0} \frac{2xh + h^2}{h(\sqrt{(x+h)^2 - 5} + \sqrt{x^2 - 5})}$$

$$= \lim_{h \to 0} \frac{2x + h}{\sqrt{(x+h)^2 - 5} + \sqrt{x^2 - 5}} \qquad \text{(cancel } h\text{)}$$

$$= \frac{2x + 0}{\sqrt{(x+0)^2 - 5} + \sqrt{x^2 - 5}} \qquad \text{(limit rules)}$$

$$= \frac{2x}{\sqrt{x^2 - 5} + \sqrt{x^2 - 5}} = \frac{2x}{2\sqrt{x^2 - 5}} = \frac{x}{\sqrt{x^2 - 5}} \qquad \text{(algebra)}$$

Therefore, the derivative of $f(x) = \sqrt{x^2 - 5}$ is the function $f'(x) = \frac{x}{\sqrt{x^2-5}}$. ☐

✓ *Checking the Answer*   After all that algebra, it's a very good idea to check your answer. Although we can't use a graph as a sure check that the answer is correct, we *can* use graphs to check that our answer makes sense (and is at least likely to be correct). One way to do this is to plot graphs of f and its proposed derivative f' and then check that the height of the graph of f' seems to be the slope of the graph of f. At the very least, you should be able to see that the graph of f' has zeros where the graph of f has horizontal tangent lines and that the graph of f' is positive when the graph of f is increasing (and negative when the graph of f is decreasing). Take a minute to visualize these relationships in Figures 7.4 and 7.5.

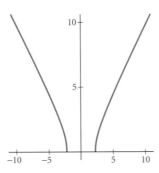

Figure 7.4
$$f(x) = \sqrt{x^2 - 5}$$

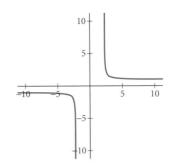

Figure 7.5
$$f'(x) = \frac{x}{\sqrt{x^2 - 5}}$$ □

What you should know after reading Section 7.1

▸ How to find the domain of an algebraic function. Note that this includes solving equations and inequalities involving algebraic functions and pieces of algebraic functions.

▸ How to calculate limits of algebraic functions. When is an algebraic function continuous?

▸ How to calculate derivatives using the power, sum, constant multiple, and quotient rules. How to find the critical points of an algebraic function (these calculations can be algebraically tricky).

▸ How to differentiate algebraic functions for which we do not currently have differentiation rules. How can you check your answer after doing such a calculation?

Exercises 7.1

Concepts

0. Read the section and make your own summary of the material. Include whatever you think will help you review the material later on.

1. What is an algebraic function? Give three examples of algebraic functions and three examples of functions that are not algebraic.

2. Explain why the function $f(x) = 3^x$ is *not* an algebraic function.

3. Suppose you go through a lengthy calculation using the definition of derivative in order to find the derivative of a given function. How can you check your answer with graphs?

4. Explain why we cannot differentiate the function $f(x) = \sqrt{x^2 - 5}$ using only the power, sum, constant multiple, and quotient rules.

5. The function $f(x) = |x|$ doesn't appear at first glance to be algebraic, but it is! In other words, $f(x) = |x|$ can be written in terms of arithmetic operations, integer powers, and roots of the independent variable x. How?

■ Determine whether each of the following statements is true or false. If a statement is true, explain why. If a statement is false, provide a counterexample.

6. True or False: The sum of two algebraic functions is itself an algebraic function.

7. True or False: The product of two algebraic functions is itself an algebraic function.

8. True or False: The quotient of two algebraic functions is itself an algebraic function.

9. True or False: Every algebraic function is a power function, a polynomial function, or a rational function.

10. True or False: Every algebraic function is continuous everywhere on its domain.

11. True or False: We can differentiate every algebraic function by using just the power, sum, constant multiple, and quotient rules.

Skills

■ Find the domains of the following functions. Check your answers by attempting to evaluate each function at points inside and outside of your proposed domain, as well as by using a graphing calculator to make a graph of the function.

12. $f(x) = \dfrac{1}{x^{\frac{2}{3}} - x^{\frac{1}{2}}}$

13. $f(x) = (-2x^2 + 11x - 5)^{\frac{1}{4}}$

14. $f(x) = \dfrac{\sqrt{3 - x}}{x^2 - 3x - 4}$

15. $f(x) = (x^2 - 9)^{\frac{3}{4}}$

16. $f(x) = (x^2 - 9)^{-\frac{3}{4}}$

17. $f(x) = |x^2 - 9|^{-\frac{3}{4}}$

18. $f(x) = \dfrac{\sqrt{3x^3 - 2x^2 + 3x - 2}}{(x - 3)^{-2}}$

19. $f(x) = \sqrt{\dfrac{x^2 - 1}{x^3 - 7x + 6}}$

20. $f(x) = \dfrac{3x^{-3}}{x^{-\frac{1}{4}} - x^{\frac{3}{4}}}$

■ Calculate the following limits. Check your answers using graphs.

21. $\displaystyle\lim_{x \to -1} \dfrac{x^2 - 4}{\sqrt{x + 3}}$

22. $\displaystyle\lim_{x \to 2} (x - 2)^{-\frac{1}{3}}$

23. $\displaystyle\lim_{x \to 1^+} \dfrac{1}{\sqrt{x - 1}}$

24. $\displaystyle\lim_{x \to 1} \dfrac{x^{\frac{7}{2}} - x^{\frac{8}{3}}}{x^2}$

25. $\displaystyle\lim_{x \to 0^+} \dfrac{x^{\frac{7}{2}} - x^{\frac{8}{3}}}{x^2}$

26. $\displaystyle\lim_{x \to \infty} \dfrac{x^{\frac{7}{2}} - x^{\frac{8}{3}}}{x^2}$

27. $\displaystyle\lim_{x \to 64} \left(x^{-\frac{1}{3}} - x^{-\frac{1}{2}}\right)$

28. $\displaystyle\lim_{x \to 0^+} \left(x^{-\frac{1}{3}} - x^{-\frac{1}{2}}\right)$

29. $\displaystyle\lim_{x \to \infty} \left(x^{-\frac{1}{3}} - x^{-\frac{1}{2}}\right)$

30. $\displaystyle\lim_{x \to 0^+} \left(x^{\frac{1}{3}} - x^{\frac{1}{2}}\right)$

31. $\displaystyle\lim_{x \to \infty} \left(x^{\frac{1}{3}} - x^{\frac{1}{2}}\right)$

32. $\displaystyle\lim_{x \to \infty} \left(x^{\frac{1}{2}} - x^{\frac{1}{3}}\right)$

33. $\displaystyle\lim_{h \to 0} \dfrac{\sqrt{(3 + h)^2 - 4} - \sqrt{5}}{h}$

34. $\displaystyle\lim_{h \to 0} \dfrac{\sqrt{(x + h)^2 - 4} - \sqrt{x^2 - 4}}{h}$

35. $\displaystyle\lim_{h \to 0} \dfrac{\frac{1}{1 + \sqrt{1 + h}} - \frac{1}{2}}{h}$

■ Calculate the derivative of each of the following functions using the power, sum, constant multiple, and quotient rules, if possible. If it is not possible to differentiate a given function with these rules, explain why.

36. $f(x) = 3x^{\frac{7}{3}} - 2x^{\frac{2}{3}} + 5x^{-3}$

37. $f(x) = \sqrt{x} - \dfrac{1}{\sqrt{x}}$

38. $f(x) = \dfrac{3x^2 - 5}{\sqrt{x} + 1}$

39. $f(x) = (\sqrt{x} - 5)^2$

40. $f(x) = (\sqrt{x} - 5)^{-2}$

41. $f(x) = (\sqrt{x} - 5)^{\frac{1}{3}}$

42. $f(x) = x^{-\frac{2}{3}}(x^3 - \sqrt{x})$

43. $f(x) = x^{\frac{2}{3}}(x^3 - \sqrt{x})$

44. $f(x) = x^{-\frac{2}{3}}(x^3 - \sqrt{x})^2$

45. $f(x) = \dfrac{x^2 - 3x^5 + \sqrt{x}}{\sqrt[3]{x}}$

46. $f(x) = \dfrac{\sqrt{4 + x^2}}{3 - 2x}$

47. $f(x) = \dfrac{3(1 - 2\sqrt{x})}{1 + x^{-\frac{1}{2}}}$

■ Calculate the following derivatives using the power, sum, constant multiple, and quotient rules.

48. $\dfrac{d}{dx}\Big|_{x=4} \left((1 + \sqrt{x})^{-1}\right)$

49. $\dfrac{d^2}{dx^2}\left((1 + \sqrt{x})^{-1}\right)$

50. $\dfrac{d^2}{dx^2}\Big|_{x=4} \left((1 + \sqrt{x})^{-1}\right)$

51. $\dfrac{d^2}{dx^2}\left(\dfrac{x}{x^{\frac{3}{2}} - x^3}\right)$

52. $\dfrac{d^2}{dx^2}\Big|_{x=2} \left(\dfrac{x}{x^{\frac{3}{2}} - x^3}\right)$

53. $\dfrac{d^4}{dt^4}\left(t^{\frac{3}{2}}(1 - t)\right)$

■ For each function f and value c, calculate $f'(c)$. Use the power, sum, constant multiple, and/or quotient rules if possible; if those rules do not apply, then use the definition of derivative.

54. $f(x) = (4 - \sqrt{x})^2, \quad c = 0$

55. $f(x) = (4 - \sqrt{x})^2, \quad c = 4$

56. $f(x) = \sqrt{4 - x^2}, \quad c = -1$

57. $f(x) = (x - 1)^{-\frac{1}{2}}, \quad c = 5$

58. $f(x) = x^{-\frac{1}{3}}(1 - \sqrt{x}), \quad c = 1$

59. $f(x) = \dfrac{1}{x - \sqrt{x}}, \quad c = 2$

60. $f(x) = \dfrac{1}{\sqrt{x}} - \dfrac{\sqrt[3]{x}}{\sqrt[4]{x}}, \quad c = 4$

61. $f(x) = \dfrac{1}{\sqrt{x^2 - 8x}}, \quad c = -1$

62. $f(x) = \dfrac{1}{\sqrt{5x - x^2}}, \quad c = 2$

■ Calculate the derivatives of the following functions by using the definition of derivative.

63. $f(x) = \sqrt{x^2 - 4}$ **64.** $f(x) = (x^2 - 4)^{-\frac{1}{2}}$

65. $f(x) = \dfrac{1}{\sqrt{x^2 - 10x + 9}}$

■ Find the critical points of each of the following functions. When you are finished, check your answer with a graph of the function; explain how your graph verifies your answer.

66. $f(x) = \sqrt{x} - \dfrac{1}{\sqrt{x}}$

67. $f(x) = (\sqrt{x} - 5)^2$

68. $f(x) = \sqrt{x}(3\sqrt[4]{x} - 1)^2$

69. $f(x) = x^{\frac{1}{4}} - 2x^{-\frac{1}{2}}$

70. $f(x) = \dfrac{x^2}{x^{\frac{1}{2}} - 1}$

71. $f(x) = (\sqrt{x} - 5x + 1)^{-2}$

72. $f(x) = (\sqrt{x} - 5)^{-2}$

73. $f(x) = x^{-\frac{2}{3}}(x^2 - \sqrt{x})$

74. $f(x) = x^{\frac{2}{3}}(x^2 - \sqrt{x})$

7.2   The Product Rule and the Chain Rule

In this section we (finally) learn rules for differentiating products and compositions of functions. With these new rules, we will be able to differentiate *any* algebraic function quickly!

7.2.1 The product rule

We know that the derivative of a sum of functions is the sum of the derivatives of those functions (by the sum rule, $(f + g)' = f' + g'$). Is the same sort of thing true for products of functions? In other words, is the derivative of a product of functions simply equal to the product of the derivatives of those functions? The answer is no: In general, $(f g)' \neq f' g'$.

> ❓ **Question**   Find functions f and g for which $(f g)'$ and $f' g'$ are different.   □

We cannot find the derivative of a product $f g$ simply by multiplying f' and g'. However, we *can* write the derivative of $f g$ in terms of f, g, f', and g', as follows:

THEOREM 7.1

The Product Rule

Given any two differentiable functions f and g, the derivative of their product is given by the formula

$$(f g)' = f' g + f g'.$$

Using slightly different notation, we can also write the product rule as

$$(f(x)g(x))' = f'(x)g(x) + f(x)g'(x).$$

In Leibnitz notation, the product rule is written as

$$\frac{d}{dx}(f(x)g(x)) = \frac{df}{dx}g(x) + f(x)\frac{dg}{dx}.$$

EXAMPLE 7.6   **Using the product rule**

Use the product rule (and other differentiation rules) to differentiate the following functions.

(a) $h(x) = x^{\frac{2}{3}}(1 - x^3)$ (b) $f(x) = (5x^3 - 3x + 1)(3 - \sqrt{x})$

Solution

(a) Previously, we had to expand the function $h(x) = x^{\frac{2}{3}}(1 - x^3)$ before differentiating. Now that we know the product rule, we can start differentiating immediately. The function $h(x)$ is a product of two functions, namely $f(x) = x^{\frac{2}{3}}$ and $g(x) = 1 - x^3$. Thus we have

$$h'(x) = (f(x)g(x))' = f'(x)g(x) + f(x)g'(x) \qquad \text{(product rule)}$$
$$= \left(x^{\frac{2}{3}}\right)'(1 - x^3) + \left(x^{\frac{2}{3}}\right)(1 - x^3)'$$
$$= \left(\tfrac{2}{3}x^{-\frac{1}{3}}\right)(1 - x^3) + \left(x^{\frac{2}{3}}\right)(0 - 3x^2) \qquad \text{(other differentiation rules)}$$
$$= \tfrac{2}{3}x^{-\frac{1}{3}}(1 - x^3) - 3x^{\frac{2}{3}}(x^2). \qquad \text{(simplify)}$$

We could simplify further, but for this problem there is no need to do so. Usually we will not actually give names to the factors in the product before applying the product rule; we only did so here to make the process of the product rule more clear. (*Note:* In this example, it is actually easier to use algebra first than to use the product rule; try it!)

(b) This function f is the product of two functions, $5x^3 - 3x + 1$ and $3 - \sqrt{x}$, both of which we know how to differentiate. We apply the product rule to find the derivative of f:

$$f'(x) = (5x^3 - 3x + 1)'(3 - \sqrt{x})$$
$$\qquad\quad + (5x^3 - 3x + 1)(3 - \sqrt{x})' \qquad \text{(product rule)}$$
$$= (15x^2 - 3)(3 - \sqrt{x})$$
$$\qquad\quad + (5x^3 - 3x + 1)\left(-\tfrac{1}{2}x^{-\frac{1}{2}}\right). \qquad \text{(more differentiation)} \quad \square$$

When we are differentiating complicated functions, it may be advisable to apply the product rule separately from other differentiation steps (as we did in Example 7.6). However, we can often differentiate simple products in one step (Example 7.7).

EXAMPLE 7.7   **Applying multiple differentiation rules in one step**

Suppose we wish to differentiate the function $f(x) = (1 - x^2)(x^{-2} + x^{\frac{1}{4}})$. This function is the product of the functions $1 - x^2$ and $x^{-2} + x^{\frac{1}{4}}$. By the product rule, we have

$$f'(x) = (-2x)\left(x^{-2} + x^{\frac{1}{4}}\right) + (1 - x^2)\left(-2x^{-3} + \tfrac{1}{4}x^{-\frac{3}{4}}\right). \qquad \square$$

The proof of the product rule is quite similar to the proof of the quotient rule and will consist of a general calculation with the definition of derivative.

PROOF (**THEOREM 7.1**)   Suppose f and g are differentiable functions. We wish to write the derivative of the product fg in terms of f', g', f, and g. Specifically, we wish to show that $(fg)' = f'g + fg'$. By the definition of derivative, we know that

$$\frac{d}{dx}(f(x)g(x)) = \lim_{h \to 0} \frac{f(x+h)g(x+h) - f(x)g(x)}{h}$$

and that

$$\frac{d}{dx}(f(x)) = \lim_{h \to 0} \frac{f(x+h) - f(x)}{h} \quad \text{and} \quad \frac{d}{dx}(g(x)) = \lim_{h \to 0} \frac{g(x+h) - g(x)}{h}.$$

We want to write the first expression in terms of the last two expressions. To do this we will begin with the expression for $\frac{d}{dx}(f(x)g(x))$ and add and subtract the quantity $f(x)g(x+h)$ from the numerator. Then we will simplify and use limit rules until we have extracted the limits that define the derivatives of f and g.

$$\frac{d}{dx}(f(x)g(x))$$

$$= \lim_{h \to 0} \frac{f(x+h)g(x+h) - f(x)g(x)}{h} \qquad \text{(definition of derivative)}$$

$$= \lim_{h \to 0} \frac{f(x+h)g(x+h) - f(x)g(x+h) + f(x)g(x+h) - f(x)g(x)}{h} \qquad \text{(see above)}$$

$$= \lim_{h \to 0} \frac{(f(x+h) - f(x))g(x+h) + f(x)(g(x+h) - g(x))}{h} \qquad \text{(factoring)}$$

$$= \lim_{h \to 0} \left(\left(\frac{f(x+h) - f(x)}{h} \right) g(x+h) + f(x) \left(\frac{g(x+h) - g(x)}{h} \right) \right) \qquad \text{(algebra)}$$

$$= \left(\lim_{h \to 0} \frac{f(x+h) - f(x)}{h} \right) \left(\lim_{h \to 0} g(x+h) \right)$$

$$+ \left(\lim_{h \to 0} f(x) \right) \left(\lim_{h \to 0} \frac{g(x+h) - g(x)}{h} \right) \qquad \text{(limit rules)}$$

$$= \left(\lim_{h \to 0} \frac{f(x+h) - f(x)}{h} \right) g(x) + f(x) \left(\lim_{h \to 0} \frac{g(x+h) - g(x)}{h} \right) \qquad \text{(limit rules)}$$

$$= f'(x)g(x) + f(x)g'(x). \qquad \text{(definition of derivative)}$$

Notice that throughout the entire calculation, our goal was to extract the expressions that represent the derivatives of f and g. The algebra steps were not meant to "simplify"; they were meant to get us closer to the final form of $f'g + fg'$.

7.2.2 The chain rule

We now know how to differentiate functions that are sums, constant multiples, quotients, and products of power functions. However, we still don't know how to differentiate a function as simple as $y = (x^2 + 1)^{\frac{1}{2}}$ (since it involves a composition). We will now finally learn the **chain rule,** which allows us to differentiate a composition $f(g(x))$ of functions f and g whose derivatives we know.

THEOREM 7.2

The Chain Rule

If f and g are differentiable functions, then the derivative of their composition $f(g(x))$ is:

$$(f(g(x)))' = f'(g(x))g'(x).$$

? *Question* In Theorem 7.2, $f'(g(x))$ is the derivative of f evaluated at $g(x)$. How is this different from $(f(g(x)))'$? □

Before we get into the proof of the chain rule or its expression in other notations, let's do a simple example.

EXAMPLE 7.8

Using the chain rule

Suppose that f and g are the functions $f(x) = x^{\frac{1}{2}}$ and $g(x) = x^2 + 1$. Then the composition $f \circ g$ is the function

$$f(g(x)) = (x^2 + 1)^{\frac{1}{2}}.$$

Since $f(g(x))$ is a composition, we can use the chain rule to evaluate its derivative. To do so we first need to know the derivatives of f and g. Using the power and sum rules, we know that

$$f'(x) = \tfrac{1}{2}x^{-\frac{1}{2}} \quad \text{and} \quad g'(x) = 2x.$$

Moreover, when we evaluate the derivative f' at $g(x)$, we have

$$f'(g(x)) = \tfrac{1}{2}(g(x))^{-\frac{1}{2}} = \tfrac{1}{2}(x^2 + 1)^{-\frac{1}{2}}.$$

Therefore, the derivative of the composition $f(g(x)) = (x^2 + 1)^{\frac{1}{2}}$ is, by the chain rule,

$$\frac{d}{dx}(f(g(x))) = f'(g(x))g'(x) = \tfrac{1}{2}(x^2 + 1)^{-\frac{1}{2}}(2x). \qquad \square$$

---

⑤ *Caution* The chain rule formula $f'(g(x))g'(x)$ is *not* the same as the product of the derivatives of $f(x)$ and $g(x)$. The difference is that we use function $g(x)$, not x, as the "input" for the derivative of $f(x)$. □

The process of performing the chain rule can be verbalized as follows: To differentiate a composition $f(g(x))$ of functions, first take the derivative of the "outside" function f, and evaluate it at the "inside" function g (that is, replace all the x's in the expression for f' with the expression $g(x)$). Then multiply this by the derivative of the "inside" function g. Notice that to use the chain rule, we must first identify the "outside" and "inside" functions f and g.

EXAMPLE 7.9

Decomposing a function for the chain rule

Use the chain rule to differentiate $h(x) = \left(\dfrac{x}{1 - 3x^2}\right)^2$.

Solution The function h is the composition $h(x) = f(g(x))$, where the "outside" function is $f(x) = x^2$ and the "inside" function is $g(x) = \frac{x}{1-3x^2}$. By the chain rule, we have

$$
\begin{aligned}
h'(x) &= f'(g(x))g'(x) & \text{(chain rule)} \\
&= 2(g(x))^1 g'(x) & \text{(derivative of "outside" function)} \\
&= 2\left(\frac{x}{1 - 3x^2}\right)g'(x) & \text{(evaluated at "inside" function)} \\
&= 2\left(\frac{x}{1 - 3x^2}\right)\left(\frac{1(1 - 3x^2) - x(-6x)}{(1 - 3x^2)^2}\right). & \text{(derivative of "inside" function)}
\end{aligned}
$$

If we needed to, we could simplify this expression. Usually you won't write out each step separately as we did here; instead, you will simply write

$$\frac{d}{dx}\left(\left(\frac{x}{1-3x^2}\right)^2\right) = 2\left(\frac{x}{1-3x^2}\right)\left(\frac{1(1-3x^2)-x(-6x)}{(1-3x^2)^2}\right).$$ □

If f is a composition that is an *algebraic* function, then the "outside" function of the composition will always be a power function (why?). This means that at this point, the chain rule will not be very difficult to apply, and most chain rule problems will resemble the examples we have done so far. We will use the chain rule for functions of the form $(u(x))^k$ so often that we state it as a rule.

THEOREM 7.3

Combining the Chain Rule and the Power Rule
If $u(x)$ is a differentiable function and k is a rational number, then

$$\frac{d}{dx}((u(x))^k) = k(u(x))^{k-1}u'(x).$$

For example, $\frac{d}{dx}(3x+1)^{100} = 100(3x+1)^{99}(3)$. The proof of Theorem 7.3 is an application of the chain and power rules and is left for the exercises. In later chapters, after we introduce more complicated, nonalgebraic functions, we will have more complicated compositions to deal with, and applying the chain rule will be more difficult.

The following theorem restates the chain rule in Leibnitz notation. It turns out that Leibnitz notation provides a very natural way to express the chain rule.

THEOREM 7.4

The Chain Rule in Leibnitz Notation
If f and u are differentiable functions, then the derivative of their composition $f(u(x))$ is

$$\frac{df}{dx} = \frac{df}{du}\frac{du}{dx}.$$

Note that we are using the notation $u(x)$ instead of the notation $g(x)$ for the "inside" function. We did not have to do this, but it is traditional to use a letter like "u" or "v" to represent the "inside" function so that it will look like a variable (although "u" is actually a function of the independent variable x).

The Leibnitz notation for the chain rule makes sense if we think of derivatives as rates of change. Consider the following simple example.

EXAMPLE 7.10 **Using Leibnitz notation in a real-world application of the chain rule**

Suppose you can produce 30 widgets per hour, and you make a profit of $10.00 per widget. How much profit do you make per hour?

Solution The common-sense answer is that since you can make 30 widgets every hour, and you make $10.00 per widget, you will make a profit of $300.00 per hour. We arrived at this answer by multiplying the number of dollars per widget by the rate of widgets per hour:

$$\left(10 \ \frac{\text{dollars}}{\text{widget}}\right)\left(30 \ \frac{\text{widgets}}{\text{hour}}\right) = 300 \ \frac{\text{dollars}}{\text{hour}}.$$

What does this have to do with the derivative of a composition? Let $w(t)$ be the number of widgets you have t hours after starting production, and let $p(w)$ be the profit made from producing w widgets. We know that you can make $\frac{dw}{dt} = 30$ widgets per hour and that you make a profit of $\frac{dp}{dw} = 10$ dollars per widget produced. The profit is a function of the number of widgets, which in turn is a function of time. More precisely, the profit made t hours after production starts is given by the composition $p(w(t))$. We are interested in the rate $\frac{dp}{dt}$ of profit per hour. By the chain rule, we have

$$\frac{dp}{dt} = \frac{dp}{dw}\frac{dw}{dt} = (10)(30) = 300.$$

□

◈ **Caution** The chain rule is easy to remember when written in Leibnitz notation $\frac{df}{dx} = \frac{df}{du}\frac{du}{dx}$; simply think of "canceling du." However, keep in mind that "canceling du" is *not* sufficient mathematical justification for the chain rule; it is just notationally convenient (in particular, $\frac{df}{du}$ and $\frac{du}{dx}$ are not fractions).

□

The Leibnitz notation version of the chain rule reveals where the "chain rule" gets its name. It turns out that if you have a composition of several functions, then the derivative of that composition is the product of derivatives that make a "chain" from the outermost function to the innermost independent variable. For example, consider a composition $y(u(v(x)))$. The derivative of y with respect to x is given by the formula

$$\frac{dy}{dx} = \frac{dy}{du}\frac{du}{dv}\frac{dv}{dx}.$$

Can you see the "chain" involved? We take the derivative first of y with respect to u, then of u with respect to v, and then of v with respect to x. We are working our way from the "outside" to the "inside" of the function.

The proof of the chain rule is too involved to treat properly here. Instead, we will give an argument for the chain rule using approximations rather than true limits. The "proof" that follows is not rigorous, but it should give you some idea why the chain rule should be true.

PROOF (**USING APPROXIMATIONS, TO SUPPORT THEOREM 7.2**)
We want to show that if f and g are differentiable functions, then $\frac{d}{dx}(f(g(x))) = f'(g(x))g'(x)$. Showing equality is difficult; we will settle for showing an approximation.

We start by stating some approximations, which will prove to be useful later in the proof. By the definition of derivative, we know that

$$g'(x) = \lim_{h \to 0} \frac{g(x+h) - g(x)}{h}.$$

Therefore, for very small values of h, we have

$$g'(x) \approx \frac{g(x+h) - g(x)}{h}.$$

(This is the same as saying that the slope of the tangent line to g at x is very close to the slope of the secant line from $(x, g(x))$ to $(x+h, g(x+h))$ when h is small.) Moving things around a bit, this means that

(1) $$g(x+h) \approx g(x) + g'(x)h.$$

We can say the same kind of thing about the function f; we'll use the letters z and k in the place of x and h to avoid confusion later on:

$$(2) \qquad\qquad f(z + k) \approx f(z) + f'(z)k.$$

Moreover, the approximations above get better as $h \to 0$ and as $k \to 0$. In fact, we can make the error in the approximations as small as we like (this is due to the fact that f and g are differentiable functions).

We now use these approximations to show that $\frac{d}{dx}(f(g(x))) \approx f'(g(x))g'(x)$. By the definition of derivative, we have

$$\frac{d}{dx}(f(g(x))) = \lim_{h \to 0} \frac{f(g(x+h)) - f(g(x))}{h} \qquad \text{(definition of derivative)}$$

$$\approx \frac{f(g(x+h)) - f(g(x))}{h}. \qquad \text{(for very small } h)$$

Since $g(x+h) \approx g(x) + g'(x)h$ by Equation (1), we have $f(g(x+h)) \approx f(g(x) + g'(x)h)$, and

$$\approx \frac{f(g(x) + g'(x)h) - f(g(x))}{h}.$$

Now we can apply Equation (2) with $z = g(x)$ and $k = g'(x)h$. Note that when h is very small, so is $k = g'(x)h$ (since $g'(x)$ must be finite, due to the fact that g is differentiable). We have $f(g(x) + g'(x)k) = f(z+k) \approx f(z) + f'(z)k = f(g(x)) + f'(g(x))g'(x)h$. Continuing our calculation, we have

$$\approx \frac{f(g(x)) + f'(g(x))g'(x)h - f(g(x))}{h}$$

$$= \frac{f'(g(x))g'(x)h}{h} \qquad \text{(cancellation)}$$

$$= f'(g(x))g'(x). \qquad \text{(cancellation)}$$

As you can see, even an "approximate" proof of the chain rule is difficult. In the following example, we prove that the chain rule works in the special case where the "outside" function is x^3. You will prove other special cases in the exercises.

EXAMPLE 7.11   Use the definition of derivative (*not* the chain rule) to prove that if $g(x)$ is a differentiable function, then $\frac{d}{dx}((g(x))^3) = 3(g(x))^2 g'(x)$.

PROOF   This proof will use the definition of derivative and the factoring formula $a^3 - b^3 = (a - b)(a^2 + ab + b^2)$ (see Section 0.2.2). Once we use the definition of derivative, our main goal will be to try to extract the limit that defines $g'(x)$ from our expression. We have

$$\frac{d}{dx}((g(x))^3) = \lim_{h \to 0} \frac{(g(x+h))^3 - (g(x))^3}{h}$$

$$= \lim_{h \to 0} \frac{(g(x+h) - g(x))((g(x+h))^2 + g(x+h)g(x) + (g(x))^2)}{h}$$

$$= \left(\lim_{h \to 0} \frac{g(x+h) - g(x)}{h} \right) \left(\lim_{h \to 0} (g(x+h))^2 + g(x+h)g(x) + (g(x))^2 \right)$$

$$= g'(x)((g(x))^2 + g(x)g(x) + (g(x))^2)$$

$$= 3(g(x))^2 g'(x).$$

The first step is the definition of derivative, the second step uses the factoring formula we mentioned earlier, and the third step uses algebra and the product rule for limits. In the fourth and fifth steps, we evaluate the limits (using the definition of derivative again) and then rewrite the expression.

7.2.3 Differentiating using the "rules"

To use differentiation rules accurately, you must know the rules and when to apply them. At this point we can differentiate *any* algebraic function with the rules that are summarized in Theorem 7.5.

THEOREM 7.5

Summary of Differentiation Rules

Let f and g be differentiable functions, and let k be any real number.

$$\text{Sum Rule} \quad (f(x) + g(x))' = f'(x) + g'(x)$$

$$\text{Constant Multiple Rule} \quad (kf(x))' = kf'(x)$$

$$\text{Power Rule} \quad (x^k)' = kx^{k-1}$$

$$\text{Quotient Rule} \quad \left(\frac{f(x)}{g(x)}\right)' = \frac{f'(x)g(x) - f(x)g'(x)}{(g(x))^2}$$

$$\text{Product Rule} \quad (f(x)g(x))' = f'(x)g(x) + f(x)g'(x)$$

$$\text{Chain Rule} \quad (f(g(x)))' = f'(g(x))g'(x)$$

? Question Can you write each of these differentiation rules using Leibnitz notation? ☐

When faced with a differentiation problem, the first question to ask is: Which differentiation rule applies here? Or, more precisely: Which differentiation rule applies *first*? Try to figure out whether the function you are attempting to differentiate is, at its outermost level, a sum, product, quotient, or composition; then apply the appropriate differentiation rule. Work your way from the "outside" to the "inside" until you have done all of the differentiation steps.

EXAMPLE 7.12

Using the differentiation rules, one step at a time

Differentiate the function $f(x) = 6(1 - \sqrt{x})^7 + 5x^{\frac{3}{2}}$. Write out each step individually, indicating which differentiation rules you use.

Solution At its outermost level this function is a sum of two functions, $6(1 - \sqrt{x})^7$ and $5x^{\frac{3}{2}}$, so the first rule we will need to use is the sum rule. To differentiate the first function we will need to use the chain rule (as well as the constant multiple, sum, and power rules), and to differentiate the second function we will need the power and constant multiple rules.

$$\frac{d}{dx}\left(6(1 - \sqrt{x})^7 + 5x^{\frac{3}{2}}\right)$$

$$= \frac{d}{dx}\left(6(1 - \sqrt{x})^7\right) + \frac{d}{dx}\left(5x^{\frac{3}{2}}\right) \qquad \text{(sum rule)}$$

$$= 6\left(\frac{d}{dx}(1 - \sqrt{x})^7\right) + 5\frac{d}{dx}\left(x^{\frac{3}{2}}\right) \qquad \text{(constant multiple rule)}$$

$$= 6\left(7(1 - \sqrt{x})^6\left(\frac{d}{dx}(1 - \sqrt{x})\right)\right) + 5\frac{d}{dx}\left(x^{\frac{3}{2}}\right) \qquad \text{(chain and power rules)}$$

$$= 6\left(7(1 - \sqrt{x})^6\left(0 - \tfrac{1}{2}x^{-\frac{1}{2}}\right)\right) + 5\left(\tfrac{3}{2}\right)x^{\frac{1}{2}} \qquad \text{(sum and power rules)}$$

$$= -21x^{-\frac{1}{2}}(1 - \sqrt{x})^6 + \tfrac{15}{2}x^{\frac{1}{2}}. \qquad \text{(algebra)} \quad ☐$$

✧ *Caution*   As you become more familiar with differentiating, you will be able to do almost all of these steps in your head. However, always be sure that you can explain which rules you are using, and in which order.   □

EXAMPLE 7.13   **Determining which differentiation rule to apply first**

For each function f, find the derivative f'. What is the *first* differentiation rule that you must use in each case?

(a) $f(x) = \left(\sqrt[3]{x} - \frac{1}{x}\right)^{-2}$

(b) $f(x) = \dfrac{\sqrt{x^2 - 1}}{1 - \frac{1}{x}}$

(c) $f(x) = (x^2 - 1)^5(3x - 1)^2$

Solution

(a) This function is, at its outermost level, a composition. Specifically, this function is of the form $(u(x))^{-2}$, where $u(x) = \sqrt[3]{x} - \frac{1}{x}$. The first rule to use is the chain rule. (*Note:* The power rule is the *second* differentiation rule you will use, since you need to use the chain rule even to *begin* "differentiating the outside function.")

Before differentiating, we will use algebra to rewrite the function f in a more easily differentiable form:

$$\frac{d}{dx}\left(\left(\sqrt[3]{x} - \frac{1}{x}\right)^{-2}\right) = \frac{d}{dx}\left((x^{\frac{1}{3}} - x^{-1})^{-2}\right)$$

$$= -2(x^{\frac{1}{3}} - x^{-1})^{-3}\left(\tfrac{1}{3}x^{-\frac{2}{3}} - (-1)x^{-2}\right).$$

(b) This function is, at its outermost level, a quotient, so we begin by using the quotient rule. Again we write roots and fractions as exponents first so that the differentiation steps will be easier.

$$\frac{d}{dx}\left(\frac{\sqrt{x^2 - 1}}{1 - \frac{1}{x}}\right) = \frac{d}{dx}\left(\frac{(x^2 - 1)^{\frac{1}{2}}}{1 - x^{-1}}\right)$$

$$= \frac{\frac{d}{dx}\left((x^2 - 1)^{\frac{1}{2}}\right)(1 - x^{-1}) - (x^2 - 1)^{\frac{1}{2}}\frac{d}{dx}(1 - x^{-1})}{(1 - x^{-1})^2}$$

$$= \frac{\frac{1}{2}(x^2 - 1)^{-\frac{1}{2}}(2x)(1 - x^{-1}) - (x^2 - 1)^{\frac{1}{2}}(0 - (-1)x^{-2})}{(1 - x^{-1})^2}.$$

(c) This function is, at its outermost level, a product, so we begin by applying the product rule.

$$\frac{d}{dx}((x^2 - 1)^5(3x - 1)^2)$$

$$= \frac{d}{dx}((x^2 - 1)^5)(3x - 1)^2 + (x^2 - 1)^5\frac{d}{dx}((3x - 1)^2)$$

$$= 5(x^2 - 1)^4(2x)(3x - 1)^2 + (x^2 - 1)^5(2)(3x - 1)(3).$$ □

✧ *Caution*   Don't forget that a little algebra *before* differentiating can often make your life a lot easier! For example, would you rather differentiate the function $f(x) = \frac{(\sqrt{x})^3 - 3x^4}{\sqrt{x}}$ or do some algebra first and then differentiate the function $f(x) = x - 3x^{\frac{7}{2}}$? What about the function $g(x) = \left(\frac{1}{\sqrt[3]{x}}\right)^2$ and its simplified version $g(x) = x^{-\frac{2}{3}}$? There will be many problems in the exercises where simplifying first will result in a *much* simpler differentiation problem.   □

> ## What you should know after reading Section 7.2
>
> ▶ The product and chain rules, what they are for, how to state them in different notations, and how to use them to differentiate functions.
>
> ▶ The proof of the product rule, and the approximation argument for the chain rule. You should also know how the chain rule gets its name and how it is related to products of rates in application problems.
>
> ▶ How to differentiate complicated combinations of functions both quickly and step by step. This involves identifying whether a function is a sum, constant multiple, product, quotient, or composition of functions, and it may also involve some preliminary algebra. What functions are we able to differentiate at this point? Be able to state all of the differentiation rules.

Exercises 7.2

Concepts

0. Read the section and make your own summary of the material. Include whatever you think will help you review the material later on.

■ State each of the following differentiation rules in "prime" notation and in Leibnitz notation. Then give one example calculation for each rule.

1. The sum rule

2. The constant multiple rule

3. The power rule

4. The quotient rule

5. The product rule

6. The chain rule

7. Give an example of functions f and g that show that the derivative $(fg)'$ of a product of functions is *not* equal to the product $f'g'$ of their derivatives.

8. Differentiate $f(x) = (3x+1)(2-5x^2)$ two ways: first, with the product rule; second, without the product rule. Show that the answers you get are the same.

9. Differentiate $f(x) = (3x + \sqrt{x})^2$ three ways: first, with the chain rule; second, with the product rule (but not the chain rule); third, without the chain or product rule. Show that the answers you get are the same.

10. Differentiate $f(x) = \frac{1}{3x^2+1}$ two ways: first, by using the quotient rule; second, without using the quotient rule, by using algebra first and then the chain rule. Show that the answers you get are the same.

11. Differentiate $f(x) = \frac{\sqrt{x}}{1-x^7}$ two ways: first, by using the quotient rule; second, without using the quotient rule (use algebra and the product and chain rules). Show that the answers you get are the same.

12. Use the product rule to find and state the rule for differentiating a product of *three* functions f, g, and h. In other words, fill in the blank: $(f(x)g(x)h(x))' = $ _____. Then use your rule to differentiate the function $y = (2x-1)(x^2+x+1)(1-3x^4)$. Check your answer by differentiating the function $y(x)$ another way.

13. Suppose $u(x) = \sqrt{3x^2+1}$ and $f(u) = \frac{u^2+3u^5}{1-u}$. Find $\frac{d}{dx}(f(u(x)))$ without first finding the formula for $f(u(x))$.

14. In this section we noted that if $f(u(v(x)))$ is a composition of three functions, then its derivative is given by $\frac{df}{dx} = \frac{df}{du}\frac{du}{dv}\frac{dv}{dx}$. Write this rule in "prime" (rather than Leibnitz) notation.

■ Suppose you wanted to differentiate each of the following functions. Determine which differentiation rule you would have to apply *first* in each case (assuming that you don't perform any algebra first). Then list any other differentiation rules you would have to use to complete the problem. Don't actually differentiate the functions; just list the first rule and the rest of the rules that you would use.

15. $f(x) = \dfrac{(1+\sqrt{x})^2}{3x^2-4x+1}$

16. $f(x) = (5x^4 - 3x^2)^7(2x^3+1)$

17. $f(x) = \left(\dfrac{1}{\sqrt{x}} + 3\sqrt{x} - 4\right)^{-3}$

18. $f(x) = 7\left(\dfrac{2}{x} + 3x - \dfrac{5}{x+1}\right)$

19. $f(x) = \dfrac{(x+1)(3x-4)}{\sqrt{x^3-27}}$

20. $f(x) = (x^4 - \sqrt{3-4x})^8 + 5x$

■ Suppose f, g, and h are differentiable functions. Find formulas for the derivatives of the following combinations of functions (your answers may involve the letters f, g, h, and x).

21. $f(g(h(x)))$

22. $f(x^{-2})$

23. $(f(x))^{-2}$

24. $(f(x) + g(x))^7$

25. $\dfrac{1}{f(x)}$

26. $\dfrac{f(x)}{g(x)h(x)}$

27. $\dfrac{f(x)g(x)}{h(x)}$

28. $f(g(x)h(x))$

29. $x^2 f(x)$

30. $x^2 f(g(x))$

31. $(f \circ g)(x)(h(x))^2$

32. $xf(x) + g(x^3) - \sqrt{h(x)}$

■ Suppose $g(x)$ is some function with values $g(0) = -2$, $g(1) = 0$, $g(2) = 3$, $g(3) = 4$, and $g(4) = 5$ whose derivative has values $g'(0) = 1$, $g'(1) = 4$, $g'(2) = 5$, $g'(3) = 3$, and $g'(4) = 2$. For each function $f(x)$ below, find $f'(2)$.

33. $f(x) = g(x^2 - 1)$

34. $f(x) = g\left(\frac{2}{x}\right)$

35. $f(x) = xg(x^2)$

■ Suppose g, h, and j are differentiable functions with the function and derivative values given in the table below. Use the table to calculate each of the derivatives in Problems 36–45.

x	$g(x)$	$h(x)$	$j(x)$	$g'(x)$	$h'(x)$	$j'(x)$
-3	0	3	1	1	0	2
-2	1	2	3	2	-3	0
-1	3	0	1	-1	-2	-2
0	2	3	0	-2	3	-2
1	0	-1	-2	-2	-2	-1
2	-2	-2	-3	-1	0	2
3	-3	0	1	0	1	2

36. If $f(x) = g(x)h(x)$, find $f'(0)$.

37. If $f(x) = g(x)(h(x) + j(x))$, find $f'(2)$.

38. If $f(x) = g(h(x))$, find $f'(3)$.

39. If $f(x) = h(g(x))$, find $f'(3)$.

40. If $f(x) = (g(x))^3$, find $f'(-2)$.

41. If $f(x) = g(x^3 - 6)$, find $f'(-2)$.

42. If $f(x) = j(2x)$, find $f'(-1)$.

43. If $f(x) = h(g(j(x)))$, find $f'(1)$.

44. If $f(x) = h(x)g(x)j(x)$, find $f'(1)$.

45. If $f(x) = h(g(x)j(x))$, find $f'(0)$.

■ Determine whether each of the following statements is true or false. If a statement is true, explain why. If a statement is false, provide a counterexample.

46. True or False: Using the sum, constant multiple, power, quotient, product, and chain rules, we can differentiate *any* algebraic function.

47. True or False: The chain rule is used to differentiate compositions of functions.

48. True or False: If f and g are differentiable functions, then $(f(x)g(x))' = g'(x)f(x) + f'(x)g(x)$.

49. True or False: If f and g are differentiable functions, then $\frac{d}{dx}(f(g(x))) = f'(x)g'(x)$.

50. True or False: If f is a differentiable function, then for small values of h it is true that $f'(x + h) \approx f(x) + f'(x)h$.

Skills

■ Differentiate each of the following functions, writing out each step individually. State the differentiation rules that you use in each step.

51. $f(x) = x^3 \sqrt{3x + 1}$

52. $f(x) = (1 - x^4)^{87}$

53. $f(x) = \sqrt{3 - 4x^6}$

54. $f(x) = (x^2 + 3x - 1)(x - 2)^{\frac{3}{2}}$

55. $f(x) = 3x\left(x + \frac{1}{x}\right)$

56. $f(x) = \dfrac{\sqrt{x}}{x\sqrt{x - 1}}$

■ Differentiate each of the following functions. Keep in mind that preliminary algebraic manipulations can often (but not always) make a problem easier.

57. $f(x) = x(3x^2 + 1)^9$

58. $f(x) = (1 - 4x)^2(3x^2 + 1)^9$

59. $f(x) = (x\sqrt{x + 1})^{-2}$

60. $f(x) = \dfrac{3}{x^{-\frac{3}{2}}\sqrt{x}}$

61. $f(x) = 3((x^2 + 1)^8 - 7x)^{-\frac{2}{3}}$

62. $f(x) = \dfrac{x^2 + 1}{(x^2 + 4)(3x - 2)}$

63. $f(x) = \sqrt{3x - 4(2x + 1)^6}$

64. $f(x) = \dfrac{\frac{1}{x} - 3x^2}{x^5 - \frac{1}{\sqrt{x}}}$

65. $f(x) = \dfrac{x^2 - \sqrt[3]{x} + 5x^9}{\sqrt{x^{-1}}}$

66. $f(x) = \dfrac{(1 + \sqrt{x})^2}{3x^2 - 4x + 1}$

67. $f(x) = (5x^4 - 3x^2)^7(2x^3 + 1)$

68. $f(x) = \left(\dfrac{1}{\sqrt{x}} + 3\sqrt{x} - 4 \right)^{-3}$

69. $f(x) = 7 \left(\dfrac{2}{x} + 3x - \dfrac{5}{x+1} \right)$

70. $f(x) = \dfrac{(x+1)(3x-4)}{\sqrt{x^3 - 27}}$

71. $f(x) = (x^4 - \sqrt{3 - 4x})^8 + 5x$

72. $f(x) = x(x^2)(\sqrt{x})\left(x^{\frac{2}{3}} \right)$

73. $f(x) = x\sqrt{3x^2 + 1}\sqrt[3]{2x + 5}$

74. $f(x) = \dfrac{(3x+1)(x^4 - 3)^4}{(x+5)^{-2}(1 + x^2)^5}$

■ Calculate each of the following derivatives.

75. $\dfrac{d^2}{dx^2} \left((x\sqrt{x+1})^{-2} \right)$

76. $\dfrac{d^2}{dx^2} \left(\dfrac{\frac{1}{x} - 3x^2}{x^5 - \frac{1}{\sqrt{x}}} \right)$

77. $\dfrac{d^2}{dx^2} \left((5x^4 - 3x^2)^7 (2x^3 + 1) \right)$

78. $\dfrac{d}{dx} \bigg|_{x=-1} \left(\dfrac{x^2 + 1}{(x^2 + 4)(3x - 2)} \right)$

79. $\dfrac{d^2}{dx^2} \bigg|_{x=2} \left(\dfrac{3}{x^{-\frac{3}{2}}\sqrt{x}} \right)$

80. $\dfrac{d^3}{dx^3} \bigg|_{x=0} \left(x(3x^2 + 1)^9 \right)$

■ Find the critical points of each of the following functions. Then check your answers using a graph.

81. $f(x) = x^3 \sqrt{3x + 1}$

82. $f(x) = (1 - x^4)^7$

83. $f(x) = (x^2 + 3x - 1)(x - 2)^{\frac{3}{2}}$

84. $f(x) = 3x\left(x + \dfrac{1}{x} \right)$

85. $f(x) = \dfrac{\sqrt{x}}{x\sqrt{x} - 1}$

86. $f(x) = (x\sqrt{x + 1})^{-2}$

Applications

87. Linda can sell 12 magazine subscriptions a week and makes $4.00 for each magazine subscription she sells. Obviously this means that Linda will make 12($4.00) = $48.00 per week from magazine subscriptions. Explain this result mathematically, using mathematical notation and the chain rule (see Example 7.10).

88. The area of a circle can be written in terms of its radius as $A = \pi r^2$. Suppose a circular area from a spotlight on a stage floor is slowly expanding.

(a) Find $\frac{dA}{dr}$ and explain its meaning in practical terms.

(b) Does the rate $\frac{dA}{dr}$ depend on how fast the radius of the circle is increasing? Does it depend on the radius of the circle? Why or why not? (*Hint:* What are the mathematical notations for the radius of the circle and for the rate at which the radius of the circle increases? Does the formula for $\frac{dA}{dr}$ that you found above depend on either of these quantities?)

(c) Find $\frac{dA}{dt}$ and explain its meaning in practical terms. (*Hint:* As the circle expands, both the area A and the radius r are functions of time.)

(d) Does the rate $\frac{dA}{dt}$ depend on how fast the radius of the circle is increasing? Does it depend on the radius of the circle?

(e) If the radius of the circle of light is increasing at a constant rate of 2 inches per second, how fast is the area of the circle of light increasing at the moment that the spotlight has a radius of 24 inches?

Proofs

89. Use the power and chain rules to prove Theorem 7.3: If $u(x)$ is a differentiable function and k is a rational number, then $\frac{d}{dx}((u(x))^k) = k(u(x))^{k-1} u'(x)$.

90. In Section 6.3.2 we proved the quotient rule using the definition of derivative. Prove it now another way: by writing a quotient $\frac{f}{g}$ as a product and applying the product, power, and chain rules. Point out where you use each rule.

91. Use the product rule to prove the "triple" product rule: If f, g, and h are differentiable functions, then the derivative of their triple product is $(fgh)' = f'gh + fg'h + fgh'$. (Bonus Question: What do you think the "quadruple" product rule is? Can you prove it?)

92. Use the chain rule (twice) to prove that $\frac{d}{dx}(f(u(v(x)))) = f'(u(v(x)))u'(v(x))v'(x)$. (Bonus Questions: What do you think the formula is for the derivative of the composition of *four* functions? Can you prove it?)

93. Prove the product rule. You may use the proof in this section for inspiration, but be sure to write the proof in your own words and explain your work. Justify each step of your proof.

94. Explain the approximation argument used in this section to "prove" the chain rule. (We put "prove" in quotation marks because it was not a rigorous proof.) Go through each step of the proof and explain why we are doing that step and why that step is true.

■ Use the definition of derivative (*not* the chain rule) to prove the following special cases of the chain rule.

95. $\frac{d}{dx}((g(x))^2) = 2g(x)g'(x)$

96. $\frac{d}{dx}((g(x))^4) = 4(g(x))^3 g'(x)$

97. $\frac{d}{dx}((g(x))^{-1}) = -(g(x))^{-2} g'(x)$

98. $\frac{d}{dx}\left((g(x))^{\frac{1}{2}} \right) = \frac{1}{2}(g(x))^{-\frac{1}{2}} g'(x)$

99. Prove that if $f(x)$ has a double root at $x = c$, then the graph of $f(x)$ has a local minimum or maximum at $x = c$. (*Hint:* Since $x = c$ is a double root, $f(x)$ can be written in the form $(x - c)^2 g(x)$ for some polynomial $g(x)$, where $x = c$ is *not* a root of $g(x)$.)

100. Prove that if $f(x)$ has a triple root at $x = c$, then the first and second derivatives of $f(x)$ are zero at $x = c$. What does this mean about the graph of $f(x)$ at $x = c$?

7.3   Implicit Differentiation

So far we have always worked with equations that define functions. Of course, there are many equations that do not define functions; for example, the equation $x^2 + y^2 = 1$ (the graph of a circle) does not explicitly define y as a function of x (why not?). However, we can still investigate the graphs and tangent lines of such equations using a more general form of differentiation called implicit differentiation. Implicit differentiation will help us solve word problems that involve relationships between rates ("related-rates" problems; see Section 7.4), and will finally enable us to prove the general version of the power rule.

7.3.1   Implicit functions

Consider the equation $x^2 + y^2 = 1$. The solutions (x, y) to this equation describe a circle of radius 1 centered at the origin (see Figure 7.6). Clearly, the graph is not of a function (it does not pass the Vertical Line Test). However, *locally*, that is, in small pieces, the graph defines y as a function of x. Consider, for example, the "top" piece of the graph shown in Figure 7.7. This graph *is* a function. So is the graph of the "bottom" half of the circle shown in Figure 7.8.

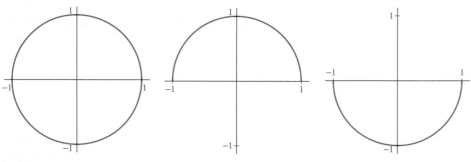

Figure 7.6
$x^2 + y^2 = 1$

Figure 7.7
$y = \sqrt{1 - x^2}$

Figure 7.8
$y = -\sqrt{1 - x^2}$

The equation of the top part of the circle is $y = \sqrt{1 - x^2}$, and the equation of the bottom part is $y = -\sqrt{1 - x^2}$. We can obtain these equations by attempting to solve the original equation $x^2 + y^2 = 1$ for y:

$$x^2 + y^2 = 1 \quad \Rightarrow \quad y^2 = 1 - x^2 \quad \Rightarrow \quad y = \pm\sqrt{1 - x^2}.$$

Even though the equation $x^2 + y^2 = 1$ does not define y as a function of x, the equation *implicitly* defines y in terms of x. Although we cannot solve the equation for y and obtain a function, we can still think of the x-values as "inputs" and of the y-values as "outputs"; the only difference is that there may be more than one y-value for each x-value. We say that y is an ***implicit function*** of x.

Of course, some equations cannot be solved for y at all; for example, consider the equation

$$y^3 + xy + 2 = 0.$$

This equation cannot be solved for y (try it!). However, we *can* make a graph of the solutions (x, y) to this equation (using a graphing calculator or computer algebra system, or by plotting lots of points; see Figure 7.9). Locally, this graph implicitly defines y as a function of x.

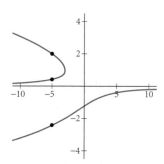

Figure 7.9
$y^3 + xy + 2 = 0$

Note that the graph of $y^3 + xy + 2 = 0$ has *three* values corresponding to $x = -5$. Even though we cannot solve this equation for y, we can sometimes solve for y given a particular value of x.

EXAMPLE 7.14   **An implicit function can have more than one y-value for a given x-value**

If $y^3 + xy + 2 = 0$ and $x = -5$, what are the possible values for y?

Solution   We substitute $x = -5$ into the equation $y^3 + xy + 2 = 0$ and solve for y:

$y^3 - 5y + 2 = 0$

$\quad \Rightarrow \quad (y - 2)(y^2 + 2y - 1) = 0$ (synthetic division)

$\quad \Rightarrow \quad y = 2, \ y = -1 + \sqrt{2}, \ \text{or} \ y = -1 - \sqrt{2}.$ (quadratic formula)

Note that $-1 + \sqrt{2} \approx 0.414$ and $-1 - \sqrt{2} \approx -2.414$. The three marked points in Figure 7.9 have coordinates $(-5, 2)$, $(-5, -1 + \sqrt{2})$, and $(-5, -1 - \sqrt{2})$.   □

7.3.2 Implicit differentiation

Given an equation that defines y as an implicit function of x, we can talk about the derivative $\frac{dy}{dx}$, or the rate of change of y with respect to x, even if we cannot solve the equation for y as a function of x. Since the graph of an implicit function will *locally* look like the graph of a function, the graph of an implicit function will *locally* have well-defined tangent lines. In this section we learn how to calculate the slopes of these tangent lines with a more general type of differentiation called ***implicit differentiation.***

Consider the equation $y^3 + xy + 2 = 0$ discussed in the previous section. At $x = -5$ the graph of this equation has three possible y-values, and at each of these y-values the graph has a tangent line (see Figure 7.10).

To find the slope of each tangent line, we must calculate $\frac{dy}{dx}$ without first solving for y in terms of x. This can be done using Algorithm 7.1.

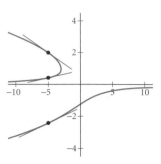

Figure 7.10
$$y^3 + xy + 2 = 0$$

ALGORITHM 7.1

Implicit Differentiation

Given an equation that defines y as an implicit function of x, the following process will find $\frac{dy}{dx}$ in terms of x and y:

1. Differentiate both sides of the equation. Keep in mind that we are thinking of y as a function of x, and thus we may need the chain rule to differentiate parts of the equation that involve y.
2. Solve for $\frac{dy}{dx}$ in terms of x and y.

⊗ *Caution*   Suppose y is an (implicit) function of x, and we want to differentiate the expression y^3 with respect to x. Note that the variable we are differentiating with respect to (x) is *not* the same as the variable in our expression (y). This means that

$$\frac{d}{dx}(y^3) \neq 3y^2.$$

Instead, since y is an (implicit) function of x, we must use the chain rule when differentiating y^3 with respect to x. It helps to write $y(x)$ for y to see how the chain rule works in this case (although most of the time we will simply write y and remember that it is a function of x):

$$\frac{d}{dx}(y^3) = \frac{d}{dx}((y(x))^3) = 3(y(x))^2 y'(x) = 3y^2 y'. \qquad \square$$

EXAMPLE 7.15   **Using implicit differentiation**

Given that $y^3 + xy + 2 = 0$, find an equation for $\frac{dy}{dx}$ in terms of x and y. Interpret this result in terms of the graph of $y^3 + xy + 2 = 0$, and use it to find the slope of the line tangent to the graph of $y^3 + xy + 2 = 0$ at the point $(-5, 2)$.

Solution   We will differentiate both sides (using the chain rule as necessary) and then attempt to solve for y'.

$$y^3 + xy + 2 = 0$$

$$\frac{d}{dx}(y^3 + xy + 2) = \frac{d}{dx}(0) \qquad \text{(differentiate both sides)}$$

$$3y^2 y' + (1)(y) + (x)(y') + 0 = 0 \qquad \text{(chain and product rules)}$$

$$3y^2 y' + xy' = -y \qquad \text{(start solving for } y')$$

$$y'(3y^2 + x) = -y$$

$$y' = \frac{-y}{3y^2 + x}.$$

This means that if a point (a, b) is on the graph of $y^3 + xy + 2 = 0$, then the line tangent to the graph of $y^3 + xy + 2 = 0$ at (a, b) has slope

$$\left.\frac{dy}{dx}\right|_{\substack{x=a \\ y=b}} = \frac{-b}{3b^2 + a}.$$

The slope of the tangent line to the graph of $y^3 + xy + 2 = 0$ at the point $(a, b) = (-5, 2)$ is

$$\left.\frac{dy}{dx}\right|_{\substack{x=-5 \\ y=2}} = \frac{-2}{3(2)^2 + (-5)} = \frac{-2}{12 - 5} = \frac{-2}{7}.$$

Look back to the graph in Figure 7.10 to check that this answer is reasonable. ☐

※ **Caution**   In Example 7.15, the expression $y'(-5)$ would not be well defined, since there are three possible y-values when $x = -5$, and thus three possible tangent lines. This means that we can't specify which derivative value we mean just by specifying a value for x; we must specify both x and y. This is why we use the Leibnitz notation $\left.\frac{dy}{dx}\right|_{\substack{x=-5 \\ y=2}}$ to refer to the slope of the line tangent to the equation at $(-5, 2)$. ☐

7.3.3 Finding vertical and horizontal tangent lines

Given an equation that defines y as an implicit function of x, how can we find the places at which the graph has a horizontal or vertical tangent line? By implicit differentiation we can find $\frac{dy}{dx}$; then the values (x, y) on the graph of the equation for which $\frac{dy}{dx} = 0$ are the values where the equation has a horizontal tangent line, and the values (x, y) on the graph for which $\frac{dy}{dx}$ is infinite are the values where the tangent line is vertical.

EXAMPLE 7.16

Finding the horizontal and vertical tangent lines of an implicit function

Find the coordinates on the graph of $x^2 - y^2 = 1$ where the tangent line is horizontal, and the coordinates where the tangent line is vertical.

Solution   We begin by using implicit differentiation to find $\frac{dy}{dx}$.

$$x^2 - y^2 = 1$$
$$\frac{d}{dx}(x^2 - y^2) = \frac{d}{dx}(1) \qquad \text{(differentiate both sides)}$$
$$2x - 2yy' = 0 \qquad \text{(remember the chain rule!)}$$
$$-2yy' = -2x \qquad \text{(begin solving for } y')$$
$$y' = \frac{-2x}{-2y}$$
$$y' = \frac{x}{y}.$$

Since $\frac{d}{dx} = \frac{x}{y}$, the tangent line to the graph of $x^2 - y^2 = 1$ will be horizontal at any point (x, y) on the graph for which $x = 0$. We must now find out whether there are any such points. We will use the equation $x^2 - y^2 = 1$ to solve for y when $x = 0$. Because

$$(0)^2 - y^2 = 1 \quad \Rightarrow \quad -y^2 = 1 \quad \Rightarrow \quad y^2 = -1,$$

there are no such values. In other words, there is no point of the form $(0, y)$ on the graph of $x^2 - y^2 = 1$. Since those are the only possible points at which the graph of $x^2 - y^2 = 1$ could have a horizontal tangent line, we know that there are *no* points on the graph of the equation $x^2 - y^2 = 1$ where the tangent line is horizontal.

On the other hand, the tangent line to the graph of $x^2 - y^2 = 1$ will be vertical at any point (x, y) where the derivative $\frac{dy}{dx} = \frac{x}{y}$ is infinite. This derivative will be infinite whenever $y = 0$ (but $x \neq 0$). Are there any points on the graph of $x^2 - y^2 = 1$ of the form $(x, 0)$? We solve the equation for x with $y = 0$ to find out:

$$x^2 - (0)^2 = 1 \implies x^2 = 1 \implies x = \pm 1.$$

Therefore, the graph of $x^2 - y^2 = 1$ has vertical tangent lines at $(-1, 0)$ and at $(1, 0)$. The graph of $x^2 - y^2 = 1$ is shown in Figure 7.11. Check that it has no horizontal tangent lines and has vertical tangent lines at the coordinates we predicted.

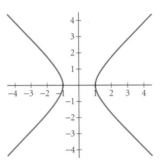

Figure 7.11 $x^2 - y^2 = 1$

☐

❓ Question   Why do we say that $\frac{dy}{dx} = \frac{x}{y}$ is infinite whenever $y = 0$ and x is not equal to zero? What would happen if both x and y were zero?   ☐

✎ Caution   Don't forget that finding the x- or y-values at which $\frac{dy}{dx}$ is zero or infinite is only half of the story. We must also ensure that the coordinates (x, y) are actually on the graph of the equation! In Example 7.16, the derivative is infinite whenever $y = 0$ but $x \neq 0$. Does this imply that the graph has a vertical tangent line at the point $(x, y) = (2, 0)$? No, since $(2, 0)$ is not on the graph of the equation!   ☐

7.3.4 Proving the general power rule

If your memory is very good, you may recall that we never proved the general version of the power rule (although we use it all the time). We did prove that if k is an *integer*, then $\frac{d}{dx}(x^k) = kx^{k-1}$ (we proved the case when k is a positive integer in Section 3.5 and the case where k is a negative integer in Section 4.3.2). However, we didn't prove that the power rule works for any *rational* power k! All this time we have been trusting that it is true, but so far we have not actually *proved* it.

To prove that the power rule holds for any *rational* power, we will use implicit differentiation and the power rule for integer powers. Theorem 7.6 simply restates the power rule while explicitly writing the power as a rational number $\frac{p}{q}$.

THEOREM 7.6

The Power Rule (for Rational Powers)

If $\frac{p}{q}$ is any rational number, then

$$\frac{d}{dx}\left(x^{\frac{p}{q}}\right) = \frac{p}{q}x^{\frac{p}{q}-1}.$$

Recall that power functions are by definition functions Ax^k where the exponent k is rational (for example, we don't consider x^π to be a power function). Therefore, the power rule in Theorem 7.6 is the most general one we will need.

PROOF (**THEOREM 7.6**)   Suppose $y = x^{\frac{p}{q}}$, where $\frac{p}{q}$ is a rational number (so p and q are integers, and $q \neq 0$). We wish to show that $y' = \frac{p}{q} x^{\frac{p}{q}-1}$. To do this, we will use algebra to write the equation $y = x^{\frac{p}{q}}$ as an equation that involves only integer powers (since we know we can differentiate integer power functions). Then we will use implicit differentiation to calculate y'.

$$y = x^{\frac{p}{q}}$$

$$(y)^q = \left(x^{\frac{p}{q}}\right)^q \qquad\qquad\qquad \text{(raise both sides to the } q\text{th power)}$$

$$y^q = x^p \qquad\qquad\qquad\qquad \text{(algebra)}$$

$$\frac{d}{dx}(y^q) = \frac{d}{dx}(x^p) \qquad\qquad\qquad \text{(differentiate both sides)}$$

$$qy^{q-1}y' = px^{p-1} \qquad\qquad \text{(chain rule and power rule for \textit{integer} powers)}$$

$$y' = \frac{px^{p-1}}{qy^{q-1}} \qquad\qquad\qquad\qquad \text{(solve for } y')$$

If we were unable to solve for y as a function of x, as was the case in our previous implicit differentiation problems, then we would stop at this point. However, in this proof we know that $y = x^{\frac{p}{q}}$, and thus we can write y' as an expression entirely in terms of x. We can then use algebra to show that $y' = \frac{p}{q} x^{\frac{p}{q}-1}$.

$$y' = \frac{px^{p-1}}{qy^{q-1}} = \frac{px^{p-1}}{q\left(x^{\frac{p}{q}}\right)^{q-1}} \qquad\qquad \text{(since } y = x^{\frac{p}{q}})$$

$$= \frac{px^{p-1}}{qx^{\frac{p(q-1)}{q}}} = \frac{p}{q}x^{\left(p-1-\frac{p(q-1)}{q}\right)} = \frac{p}{q}x^{\frac{p}{q}-1}. \qquad \text{(lots of algebra)}$$

What you should know after reading Section 7.3

▶ What an implicit function is. Is an implicit function really a function? Be able to find points (x, y) that lie on the graph of an implicit function (by solving appropriate equations).

▶ How to use implicit differentiation to differentiate an implicit function. This includes using the chain rule as appropriate. How to find the slope of the tangent line to the graph of an implicit function at a given point (x, y).

▶ How to find the points where the graph of an implicit function has horizontal or vertical tangent lines.

▶ How to use implicit differentiation to prove the power rule in the case where the power is a rational number.

Exercises 7.3

Concepts

0. Read the section and make your own summary of the material. Include whatever you think will help you review the material later on.

■ Each equation below defines y as an implicit function of x. Solve for y and use the answer (and your graphing calculator) to sketch a graph of the solutions (x, y) to this equation. (*Hint:* After solving for y, you should be able to determine two functions that, when graphed together on the same set of axes, form the graph of the solutions to the original equation.)

1. $\frac{1}{4}x^2 + y^2 = 9$

2. $(x - 1)^2 + (y + 2)^2 = 4$

3. $x^2 - 3y^2 = 16$

■ The equations in Problems 4–6 are graphed below, in some random order. For each equation, determine (without using a calculator) which graph is the graph of that equation's solutions.

4. $(x + 1)(y^2 + y - 1) = 1$

5. $xy^2 + y = 1$

6. $xy^2 + x = 1$

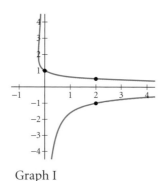

Graph I

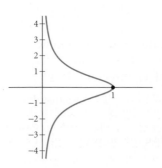

Graph II

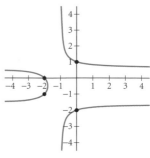

Graph III

■ Suppose that r is an independent variable, s is a function of r, and q is a constant. Calculate the following derivatives. (Your answers may involve r, s, q, and/or s'.)

7. $\frac{d}{dr}(r^3)$ **8.** $\frac{d}{dr}(s^3)$ **9.** $\frac{d}{dr}(q^3)$

10. $\frac{d}{dr}(sr^2)$ **11.** $\frac{d}{dr}(rs^2)$ **12.** $\frac{d}{dr}(qs^2)$

13. Show that, for any integers p and q (with $q \neq 0$), it is true that $(p - 1) - \frac{p(q-1)}{q} = \frac{p}{q} - 1$. What does this have to do with this section?

Skills

■ Determine which of the following equations define y as a function of x. Justify your answers. When possible (whether or not the equation defines an explicit function), solve each equation for y.

14. $y^4 - 2x + 3 = 0$ **15.** $y^5 - 2x + 3 = 0$

16. $y^5 - 3y^2 = x - 1$ **17.** $xy + 3x^2y - 2 = x + 1$

18. $\frac{y - 1}{y + 1} = x$ **19.** $xy^2 + 5x^3 - 3y^2 = 4$

■ Each of the equations below defines y as an implicit function of x. Use implicit differentiation to find $\frac{dy}{dx}$.

20. $x^2 + y^2 = 4$ **21.** $4x^2 - y^2 = 9$

22. $y^6 - 3x + 4 = 0$ **23.** $xy^2 + 3x^2 = 4$

24. $x^2y - y^2x = x^2 + 3$ **25.** $(3x + 1)(y^2 - y + 6) = 0$

26. $\sqrt{3y - 1} = 5xy$ **27.** $\frac{y^2 + 1}{3y - 1} = x$

28. $(3y^2 + 5xy - 2)^4 = 2$ **29.** $3y = 5x^2 + \sqrt[3]{y - 2}$

30. $\frac{1}{y} - \frac{1}{x} = \frac{x^2}{y + 1}$ **31.** $\frac{x + 1}{y^2 - 3} = \frac{1}{xy}$

■ The following six problems concern the circle of radius 1 centered at the origin, which is given by the equation $x^2 + y^2 = 1$.

32. Find all points on the graph with an x-coordinate of $x = \frac{1}{2}$.

33. Find the slope of the tangent line at each of the points on the graph whose x-coordinate is $x = \frac{1}{2}$.

34. Find all points on the graph with a y-coordinate of $y = \frac{\sqrt{2}}{2}$.

35. Find the slope of the tangent line at each of the points on the graph whose y-coordinate is $y = \frac{\sqrt{2}}{2}$.

36. Find all points on the graph where the tangent line is vertical.

37. Find all points on the graph where the tangent line has a slope of -1.

■ The following six problems concern the graph of $4y^2 - x^2 + 2x = 2$.

38. Find all points on the graph with an x-coordinate of $x = 3$.

39. Find the slope of the tangent line at each of the points on the graph whose x-coordinate is $x = 3$.

40. Find all points on the graph with a y-coordinate of $y = 3$.

41. Find the slope of the tangent line at each of the points on the graph whose y-coordinate is $y = 3$.

42. Find all points where the graph has a horizontal tangent line.

43. Find all points where the graph has a vertical tangent line.

■ The following six problems concern the graph of $y^3 + xy + 2 = 0$.

44. Find all points on the graph with an x-coordinate of $x = 1$.

45. Find the slope of the tangent line at each of the points on the graph whose x-coordinate is $x = 1$.

46. Find all points on the graph with a y-coordinate of $y = 1$.

47. Find the slope of the tangent line at each of the points on the graph whose y-coordinate is $y = 1$.

48. Find all points where the graph has a horizontal tangent line.

49. Find all points where the graph has a vertical tangent line.

■ The following six problems concern the graph of $y^3 - 3y - x = 1$.

50. Find all points on the graph with an x-coordinate of $x = -1$.

51. Find the slope of the tangent line at each of the points on the graph whose x-coordinate is $x = -1$.

52. Find all points on the graph with a y-coordinate of $y = 2$.

53. Find the slope of the tangent line at each of the points on the graph whose y-coordinate is $y = 2$.

54. Find all points where the graph has a horizontal tangent line.

55. Find all points where the graph has a vertical tangent line.

Proofs

56. Use implicit differentiation and the fact that $\frac{d}{dx}(x^4) = 4x^3$ to prove that $\frac{d}{dx}(x^{-4}) = -4x^{-5}$.

57. Use implicit differentiation and the fact that $\frac{d}{dx}(x^3) = 3x^2$ and $\frac{d}{dx}(x^5) = 5x^4$ to prove that $\frac{d}{dx}(x^{\frac{3}{5}}) = \frac{3}{5}x^{-\frac{2}{5}}$. (Mimic the proof of Theorem 7.6.)

58. Use implicit differentiation and the power rule for positive integer powers to prove (in your own words) the power rule for negative integer powers.

59. Use implicit differentiation and the power rule for integer powers to prove (in your own words) the power rule for rational powers (Theorem 7.6).

7.4 Related Rates

One important application of implicit differentiation is related-rates problems. Many real-life problems involve two or more rates that are related to each other, and implicit differentiation can help us write down explicitly how two or more rates are related.

7.4.1 What is a related-rates problem?

A *related-rates* problem is a word problem involving two rates that are related in some way. For example, consider an expanding circle. As the circle expands over time, the rate at which the radius of the circle is increasing is related to the rate at which the area of the circle is increasing. It turns out that if we know one of these rates, we can find the other. The key to relating these two rates is implicit differentiation.

Let's put some mathematical notation into this example. Suppose a circle of radius r is expanding in such a way that the radius increases at a constant rate of 3 inches per minute. The radius $r = r(t)$ is a function of time, and its rate of change with respect to time (its derivative with respect to time) is $\frac{dr}{dt} = 3$ inches per minute. The area $A = A(t)$ of the circle is also a function of time, since it changes over time. The rate at which the area is increasing (the derivative of the area with respect to time) is $\frac{dA}{dt}$. How is this rate related to the rate at which the radius is increasing?

The rates $\frac{dr}{dt}$ and $\frac{dA}{dt}$ are related because the quantities r and A are related. The area and radius of a circle are related by the equation $A = \pi r^2$. To find the relationship between the rates $\frac{dA}{dt}$ and $\frac{dr}{dt}$, we use implicit differentiation:

$$A(t) = \pi(r(t))^2 \qquad \text{(relationship between } A \text{ and } r\text{)}$$

$$\frac{d}{dt}(A(t)) = \frac{d}{dt}(\pi(r(t))^2) \qquad \text{(differentiate both sides with respect to } t\text{)}$$

$$\frac{dA}{dt} = \pi\left(2r(t)\frac{dr}{dt}\right) \qquad \text{(remember the chain rule!)}$$

$$\frac{dA}{dt} = 2\pi r(t)\frac{dr}{dt}. \qquad \text{(relationship between } \frac{dA}{dt} \text{ and } \frac{dr}{dt}\text{)}$$

We used the notation $A(t)$ and $r(t)$ to remind us that the area and radius of the circle are functions of t. We can write simply A and r, but we will have to remember that A and r are functions of t. Using this simpler notation, the computation looks like this:

$$A = \pi r^2 \qquad \text{(relationship between } A \text{ and } r\text{)}$$

$$\frac{d}{dt}(A) = \frac{d}{dt}(\pi r^2) \qquad \text{(differentiate both sides with respect to } t\text{)}$$

$$\frac{dA}{dt} = \pi\left(2r\frac{dr}{dt}\right) \qquad \text{(remember the chain rule!)}$$

$$\frac{dA}{dt} = 2\pi r\frac{dr}{dt}. \qquad \text{(relationship between } \frac{dA}{dt} \text{ and } \frac{dr}{dt}\text{)}$$

The rate $\frac{dA}{dt}$ at which the area is increasing at time t depends on both the size of the radius r at time t and the rate $\frac{dr}{dt}$ at which the radius is increasing. For example, if the radius is increasing at a constant rate of $\frac{dr}{dt} = 3$ inches per minute, then at the instant that the circle has a radius of 12 inches, the area of the circle is increasing at a rate of

$$\frac{dA}{dt}\bigg|_{r=12} = 2\pi(12)(3) = 72\pi \text{ square inches per minute.}$$

On the other hand, when the radius of the circle is 20 inches, the area of the circle is increasing at a rate of

$$\frac{dA}{dt}\bigg|_{r=20} = 2\pi(20)(3) = 120\pi \text{ square inches per minute.}$$

Although the rate of change of the radius of the circle is constant, the rate at which the area is changing is not constant! Note also that we evaluated the derivative of the area with respect to time at a certain *radius*, not at a *time*. If we wanted to know the rate of change of the area of the circle after 6 seconds, we would first need to know the size of the radius at that time (why?).

◈ *Caution*   It is important to remember that the expression $\frac{dA}{dt}$ represents a *function*, not a number. If you want to know the rate of change of the area of the circle after $t = 6$ seconds, then you are interested in finding $\frac{dA}{dt}\big|_{t=6}$. On the other hand, if you want to find the rate of change of the area of the circle at the instant that the circle has a radius of 20 inches (as in the example above), then you need to find $\frac{dA}{dt}\big|_{r=20}$.

Be sure you understand the difference between the *function* $\frac{dA}{dt}$ and *numerical values* like $\frac{dA}{dt}\big|_{t=6}$ and $\frac{dA}{dt}\big|_{r=20}$. Be sure that you use Leibnitz notation correctly to reflect the quantity you wish to evaluate. □

Most related-rates problems involve two rates, one of which is known and one of which you are asked to find. These rates will be related in some way that is determined by the way the corresponding quantities are related. You will implicitly differentiate the equation relating the quantities to get an equation relating the rates.

7.4.2 Important formulas and equations

Many word problems, including many related-rates problems, involve geometric quantities such as volume, area, and surface area. The following formulas may come in handy for such word problems. You should know these formulas by memory.

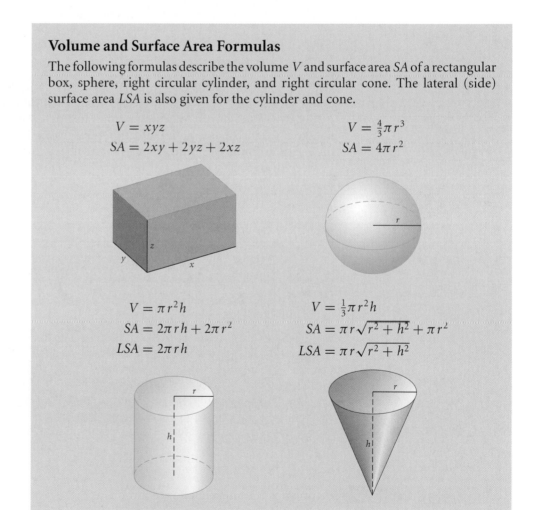

Volume and Surface Area Formulas

The following formulas describe the volume V and surface area SA of a rectangular box, sphere, right circular cylinder, and right circular cone. The lateral (side) surface area LSA is also given for the cylinder and cone.

$$V = xyz$$
$$SA = 2xy + 2yz + 2xz$$

$$V = \tfrac{4}{3}\pi r^3$$
$$SA = 4\pi r^2$$

$$V = \pi r^2 h$$
$$SA = 2\pi r h + 2\pi r^2$$
$$LSA = 2\pi r h$$

$$V = \tfrac{1}{3}\pi r^2 h$$
$$SA = \pi r \sqrt{r^2 + h^2} + \pi r^2$$
$$LSA = \pi r \sqrt{r^2 + h^2}$$

Also, recall that a circle with radius r has area $A = \pi r^2$ and circumference $C = 2\pi r$, that a rectangle with length l and width w has area $A = lw$ and perimeter $P = 2l + 2w$, and that any triangle with base b and height h has area $A = \tfrac{1}{2}bh$.

It is also common for word problems to involve the following two well-known theorems concerning right triangles. You should memorize these theorems and be able to apply them when appropriate.

Two Theorems About Right Triangles

The two theorems below describe well-known relationships among the side lengths of right triangles.

The Pythagorean Theorem

$$a^2 + b^2 = c^2$$

Similar Triangles

$$\frac{h}{b} = \frac{H}{B}, \quad \frac{d}{b} = \frac{D}{B}, \quad \frac{d}{h} = \frac{D}{H}$$

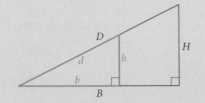

7.4.3 Setting up related-rates problems

The first step in solving a word problem is translating the written problem into a mathematical problem. For a related-rates problem, this means identifying the rate or rates that are known and the rate or rates that you wish to find. You will also have to introduce mathematical notation (i.e., give names to variables) and determine the relationship between the variables that will eventually determine the relationship between the rates. The process is summarized in the following algorithm.

ALGORITHM 7.2

Setting up Related-Rates Problems

Given a related-rates problem, use the following steps to help you translate the word problem into a mathematical problem.

1. Sketch a diagram describing the problem. Introduce variable names for the important variables in the problem, and use these variable names to label your diagram.

2. Determine the known rates given in the problem and the rate or rates that you wish to find. Clearly state, in mathematical language, the information you are given and the rate you wish to find.

3. Find a relationship between the variable(s) whose rate(s) you know and the variable(s) whose rate(s) you wish to find. This is the equation that you will differentiate to find the relationship between the rates in the problem.

EXAMPLE 7.17   **Setting up a simple related-rates problem**

Set up the following related-rates problem: Suppose a circle is expanding in such a way that its radius is growing at a constant rate of 3 inches per minute. How fast is the area of the circle changing at the instant that the radius of the circle is 12 inches?

Solution   Clearly the problem involves a circle, so we will draw a picture of a circle. We'll call the radius of the circle r (in inches) and then use A to represent the area of the circle (in square inches). Note that both $r = r(t)$ and $A = A(t)$ are functions of t.

Using this notation, the problem tells us that $\frac{dr}{dt} = 3$ inches per minute and that we are interested in $\frac{dA}{dt}\big|_{r=12}$. These are the rates we wish to relate, so we need to find an equation relating the variables A and r. Of course, the equation $A = \pi r^2$ does just that.

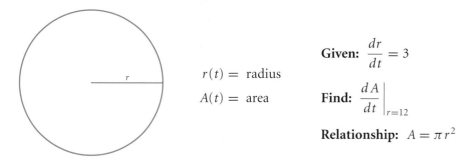

$r(t) = $ radius

$A(t) = $ area

Given: $\dfrac{dr}{dt} = 3$

Find: $\dfrac{dA}{dt}\bigg|_{r=12}$

Relationship: $A = \pi r^2$

Now the word problem has been completely translated into a mathematical problem. We must now use implicit differentiation to find an equation relating $\frac{dA}{dt}$ and $\frac{dr}{dt}$ and then evaluate this equation using the rate we know ($\frac{dr}{dt} = 3$) at the indicated value of the radius ($r = 12$). We did this earlier in this section and found that $\frac{dA}{dt}\big|_{r=12}$ was equal to 72π square inches per minute.   □

⊗ Caution   Note that we did *not* label the radius of the circle in the diagram with the number 12. This is because the radius of the circle is not always 12. As the circle expands, the radius of the circle is a *variable* that depends on time. It is true that at the final step of the problem we will see what happens when $r = 12$, but r is not equal to 12 throughout the problem. Be sure to distinguish between values that are variable and values that are constant.   □

7.4.4 Related-rates word problems

We now work through two common examples of related-rates problems.

EXAMPLE 7.18

Relating the rates of change of the volume and radius of an inflating balloon

A pink spherical party balloon is being inflated at the rate of 44 cubic inches per second. How fast is the radius of the balloon increasing at the instant that the balloon has a radius of 4 inches? How fast is the radius of the balloon increasing at the instant that the balloon contains 100 cubic inches of air?

Solution   The first thing to notice about this problem is that it involves two rates, namely the rate at which the balloon is being inflated (the rate of change of the *volume* of the balloon) and the rate of change of the radius of the balloon. We know something about one rate and wish to say something about the other; this is a related-rates problem.

Let's establish some notation. Let $r = r(t)$ stand for the radius of the balloon (in inches, after t seconds), and let $V = V(t)$ denote the volume of the balloon (in cubic inches, after t seconds). Note that both the radius and the volume are functions of time. The rate at which the balloon is being inflated is $\frac{dV}{dt} = 44$. We wish to find various values of $\frac{dr}{dt}$. This means that we need to find an equation relating the variables V and r; this equation is $V = \frac{4}{3}\pi r^3$.

Given: $\dfrac{dV}{dt} = 44$

$r(t) =$ radius

$V(t) =$ volume

Find: $\dfrac{dr}{dt}\Big|_{r=4}$ and $\dfrac{dr}{dt}\Big|_{V=100}$

Relationship: $V = \frac{4}{3}\pi r^3$

We are finished translating the word problem into a mathematical problem. We must now use the equation relating the variables V and r to get an equation involving the rates $\frac{dV}{dt}$ and $\frac{dr}{dt}$. Differentiate both sides of the equation $V = \frac{4}{3}\pi r^3$ with respect to t (this will involve the chain rule since both r and V are themselves functions of t):

$$V = \tfrac{4}{3}\pi r^3 \qquad\qquad \text{(relationship between } V \text{ and } r)$$

$$\frac{d}{dt}(V) = \frac{d}{dt}\left(\tfrac{4}{3}\pi r^3\right) \qquad\qquad \text{(differentiate with respect to } t)$$

$$\frac{dV}{dt} = \tfrac{4}{3}\pi (3r^2)\frac{dr}{dt} \qquad\qquad \text{(chain rule)}$$

$$\frac{dV}{dt} = 4\pi r^2 \frac{dr}{dt}. \qquad\qquad \text{(relationship between } \tfrac{dV}{dt} \text{ and } \tfrac{dr}{dt})$$

We know from the given information that $\frac{dV}{dt}$ is always equal to 44, thus we can say that

(1) $$44 = 4\pi r^2 \frac{dr}{dt}.$$

To answer the first part of the word problem, we must evaluate the expression at $r = 4$ and then solve for $\frac{dr}{dt}\big|_{r=4}$:

$$44 = 4\pi (4^2) \frac{dr}{dt}\Big|_{r=4} \quad \Longrightarrow \quad \frac{dr}{dt}\Big|_{r=4} = \frac{44}{4\pi(16)} \approx 0.2188.$$

(Note that in the equations above we write $\frac{dr}{dt}\big|_{r=4}$ and not $\frac{dr}{dt}$, since we are evaluating the function $\frac{dr}{dt}$ at $r = 4$.) Thus, at the instant that the radius of the balloon is 4 inches, the radius is increasing at a rate of 0.2188 inch per second.

To answer the second part of the word problem, we must find $\frac{dr}{dt}\big|_{V=100}$. However, Equation (1) involves r and not V. We can use the equation $V = \frac{4}{3}\pi r^3$ to determine the value of r when $V = 100$:

$$100 = \frac{4}{3}\pi r^3 \quad \Longrightarrow \quad r^3 = \frac{100}{\frac{4}{3}\pi} \quad \Longrightarrow \quad r = \left(\frac{100}{\frac{4}{3}\pi}\right)^{\frac{1}{3}} \approx 2.879.$$

We have now shown that when $V = 100$, the radius of the balloon is approximately $r = 2.879$; therefore, $\frac{dr}{dt}\big|_{V=100}$ is equal to $\frac{dr}{dt}\big|_{r=2.879}$. Now we can evaluate Equation (1) with this value of r:

$$44 = 4\pi (2.879)^2 \frac{dr}{dt}\Big|_{V=100} \quad \Longrightarrow \quad \frac{dr}{dt}\Big|_{V=100} = \frac{44}{4\pi(2.879)^2} \approx 0.422.$$

Therefore, at the instant that the volume of the balloon is 100 cubic inches, the radius is increasing at a rate of 0.422 inch per second. □

❓ *Question* In Example 7.18, the radius of the balloon is increasing at a greater rate when the balloon is *smaller*. In particular, when the radius of the balloon is 4 inches, the radius is increasing at 0.2188 inch per second; when the balloon has a smaller radius of 2.879 inches, the radius is increasing faster, at 0.422 inch per second. Explain why this makes sense in real-world terms. ☐

The following example is a bit more complicated. However, the methods we will use to solve the problem are identical to those we used in Examples 7.17 and 7.18. Related-rates problems can be tricky, but the good news is that they all work the same way. Once you have established that a word problem is a related-rates problem, the path to the solution should be clear.

EXAMPLE 7.19

The shadow of a person walking away from a streetlight

Matt is 6 feet tall and is walking away from a 10-foot streetlight at a rate of 3 feet per second. As he walks away from the streetlight, his shadow gets longer. How fast is the length of Matt's shadow increasing when he is 8 feet from the streetlight?

Solution Probably the hardest thing about this problem is setting up the picture and determining the relevant variables. We are given the rate at which Matt walks away from the streetlight, and we wish to find the rate of change of the length of Matt's shadow. Thus we will have two associated variables, the distance between Matt and the streetlight (we'll call this s) and the length of Matt's shadow (we'll call this l). By the Similar Triangles Theorem given in Section 7.4.2, s and l are related by the equation $\frac{10}{s+l} = \frac{6}{l}$.

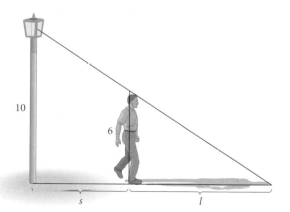

Given: $\dfrac{ds}{dt} = 3$

Find: $\dfrac{dl}{dt}\Big|_{s=8}$

Relationship: $\dfrac{10}{s+l} = \dfrac{6}{l}$

To find the relationship between $\frac{ds}{dt}$ and $\frac{dl}{dt}$, we must implicitly differentiate the equation relating s and l. We will simplify the equation first to make our lives easier:

$$\frac{10}{s+l} = \frac{6}{l} \qquad \text{(from above)}$$

$$10l = 6(s+l)$$

$$4l = 6s \qquad \text{(algebra)}$$

$$\frac{d}{dt}(4l) = \frac{d}{dt}(6s) \qquad \text{(differentiate with respect to } t\text{)}$$

$$4\frac{dl}{dt} = 6\frac{ds}{dt}.$$

Since we are given that $\frac{ds}{dt} = 3$, we now know that

$$4\frac{dl}{dt} = 6(3) \implies \frac{dl}{dt} = \frac{18}{4} = 4.5.$$

We have just discovered that Matt's shadow is increasing at a *constant* rate of 4.5 feet per second! In particular, when $s = 8$ we have

$$\left.\frac{dl}{dt}\right|_{s=8} = 4.5.$$

When Matt is 8 feet from the streetlight (and, in fact, at any time), the length of his shadow is increasing at a rate of 4.5 feet per second.   □

What you should know after reading Section 7.4

▶ How to recognize a related-rates problem and how to set up related-rates problems (how to translate them into mathematical problems). This includes sketching and labeling any relevant pictures and writing all of the information in mathematical notation.

▶ Know from memory the formulas for the areas, perimeters, volumes, and surface areas of various geometric objects. Also know the Pythagorean Theorem and the Similar Triangles Theorem.

▶ The mathematical steps involved in solving related-rates problems, including using implicit differentiation to find an equation relating the rates in the problem. Be sure that you use the chain rule correctly!

Exercises 7.4

Concepts

0. Read the section and make your own summary of the material. Include whatever you think will help you review the material later on.

1. State the Pythagorean Theorem.

2. When are two right triangles "similar," and what can you say about similar triangles?

3. Give formulas for the volumes and surface areas of:

(a) A cylinder with radius y and height s.
(b) A cone with radius u and height w.
(c) A sphere with diameter x.
(d) A cylinder whose radius r is always half of its height h.
(e) A cone whose height h is always three times its radius r.

4. Explain what a "related-rates" word problem is and the general method for setting up and solving such a word problem.

5. If the volume and radius of a sphere are functions of time, what is the relationship between the rate of change of the volume of the sphere and the rate of change of the radius of the sphere?

6. If the volume and radius of a cone are functions of time, what is the relationship between the rate of change of the volume of the cone and the rate of change of the radius of the cone? (Your answer will also involve the radius and the height of the cone.)

7. If the volume and circumference of a sphere are functions of time, what is the relationship between the rate of change of the volume of the sphere and the rate of change of the circumference of the sphere?

8. Suppose the side lengths x, y, and z of a rectangular box are all functions of time.

(a) How is the rate of change of the volume of the box related to the rates of change of x, y, and z?
(b) How is the rate of change of the surface area of the box related to the rates of change of x, y, and z?

9. Suppose the radius r, volume V, and surface area S of a sphere are functions of time t.

(a) How are $\frac{dV}{dt}$ and $\frac{dr}{dt}$ related?
(b) How are $\frac{dS}{dt}$ and $\frac{dr}{dt}$ related?
(c) How are $\frac{dV}{dt}$ and $\frac{dS}{dt}$ related?

10. Suppose the radius r, height h, and volume V of a cylinder are functions of time t.

(a) Express $\frac{dV}{dt}$ in terms of $\frac{dr}{dt}$, $\frac{dh}{dt}$, and r.
(b) How is $\frac{dV}{dt}$ related to $\frac{dr}{dt}$ if the height of the cylinder is constant?

(c) How is $\frac{dV}{dt}$ related to $\frac{dh}{dt}$ if the radius of the cylinder is constant?

(d) Suppose the height of the cylinder is always twice its radius. Write $\frac{dV}{dt}$ in terms of h and $\frac{dh}{dt}$ (and no other variables or derivatives).

Skills

■ Given that $u = u(t)$, $v = v(t)$, and $w = w(t)$ are functions of t, and that k is a constant, calculate the derivative $\frac{df}{dt}$ of each function $f(t)$ below. (Your answers may involve u, v, w, $\frac{du}{dt}$, $\frac{dv}{dt}$, $\frac{dw}{dt}$, k, and/or t.)

11. $f(t) = u + v + w$
12. $f(t) = u^2 + kv$
13. $f(t) = tv + kv$
14. $f(t) = 2v\sqrt{u+w}$
15. $f(t) = kuvw$
16. $f(t) = 3u^2v + vt$
17. $f(t) = \dfrac{w}{uv}$
18. $f(t) = w(u+t)^2$
19. $f(t) = \dfrac{ut + w}{k}$

Applications

20. A rock dropped into a pond causes a circular wavefront of ripples whose radius increases at 4 inches per second.

    (a) How fast is the area of the circle of ripples expanding at the instant that the radius of the circle is 12 inches? 24 inches? 100 inches? Explain why it makes sense that the rate of change of the area increases as the radius increases.

    (b) How fast is the area of the circle of ripples expanding at the instant that the area of the circle is 100 square inches? 200 square inches? 1000 square inches? Explain why it makes sense that the rate of change of the area increases as the area increases.

21. The sides of a cube are expanding at a rate of 2 inches per minute.

    (a) How fast is the volume of the cube changing when it has a side length of 8 inches?

    (b) How fast is the volume of the cube changing when its volume is 55 cubic inches?

22. Suppose the width w of a rectangle is decreasing at a rate of 3 inches per second, while the height h of the rectangle is increasing at a rate of 3 inches per second. The rectangle initially has a width of 100 inches and a height of 75 inches.

    (a) Find the rate of change of the area of the rectangle (in terms of the width and height of the rectangle).

    (b) On what intervals do the variables w and h make sense in this problem?

    (c) When will the area of the rectangle be increasing, and when will it be decreasing? Will there be a time at which the rate of change of the area is zero, and, if so, when? (Answer these questions in terms of the width and height of the rectangle, not in terms of time.)

23. A large helium balloon is being inflated at the rate of 120 cubic inches per second.

    (a) How fast is the radius of the balloon increasing at the instant that the balloon has a radius of 12 inches?

    (b) How fast is the radius of the balloon increasing at the instant that the balloon contains 300 cubic inches of air?

    (c) How fast is the surface area of the balloon increasing at the instant that its radius is 15 inches?

24. Alison is 6 feet tall and is walking toward a 20-foot streetlight at a rate of 4 feet per second. As she walks toward the streetlight, her shadow gets shorter.

    (a) How fast is the length of Alison's shadow changing? Does it depend on how far Alison is from the streetlight?

    (b) How fast is the tip of Alison's shadow moving? Does it depend on how far Alison is from the streetlight?

25. Alina is trying to sneak into her upstairs bedroom through the window. She props a 12-foot ladder up against the side of the house and starts to climb it. Unfortunately, the ground is muddy, because of a recent rainstorm, and the base of the ladder starts to slide away from the house at a rate of half a foot per second.

    (a) How fast is the top of the ladder moving down the side of the house when the base of the ladder is 4 feet from the house?

    (b) How fast is the area formed by the ladder, the house, and the ground changing when the top of the ladder is 6 feet from the ground?

26. Linda is bored one afternoon and decides to pour an entire container of salt into a pile on the kitchen floor. She pours 3 cubic inches of salt per second into a conical pile (like an upside-down ice cream cone) whose height is always two-thirds of its radius.

    (a) How fast is the radius of the conical salt pile changing when the radius of the pile is 2 inches?

    (b) How fast is the radius of the conical salt pile changing when the height of the pile is 4 inches?

    (c) How fast is the height of the conical salt pile changing when the radius of the pile is 2 inches?

    (d) How fast is the height of the conical salt pile changing when the height of the pile is 4 inches?

27. Linda is still bored and is now pouring a container of sugar onto the floor. The poured-out sugar forms a conical pile whose height is three-quarters of its radius, and the height of the pile of sugar is growing at a rate of 1.5 inches per second. How fast is Linda pouring the sugar at the instant that the pile of sugar is 3 inches high?

28. Alina is holding an ice cream cone on a hot summer day. As usual, the cone has a small hole at the tip, and Alina's ice cream is melting and dripping through the hole at a rate of half a cubic inch per minute. The cone holding the ice cream has a radius of 2 inches and a height of 5 inches. How fast is the height of the ice cream in the cone changing when the height of the ice cream in the cone is 3 inches?

| 7.5 | Optimization and Curve Sketching |

In this section we review optimization and curve-sketching techniques and apply these techniques to general algebraic functions. In some sense this section is a review, but the functions we will be working with will be much more complicated than the ones we used previously.

7.5.1 Summary of graphical features of algebraic functions

In previous chapters, we have worked mostly with specific types of functions: power functions, polynomial functions, and rational functions. When we are working only with these specialized types of functions, we can make many generalizations about their graphical properties. For example, we can predict the shape of the graph of a power function simply by looking at the value of its power (whether it is an odd or even negative or positive integer, or a certain type of rational number). The global behavior of a polynomial function is determined by its leading term; in particular, the degree of a polynomial function tells us how many roots and turning points its graph can have. The asymptotes, roots, and holes of a rational function are easy to determine from the roots of its numerator and denominator.

On the other hand, the category of general algebraic functions is very broad (in particular, all power, polynomial, and rational functions are algebraic, and many algebraic functions are much more complicated), and thus it is more difficult to classify the graphical properties that algebraic functions can have. In general, the graph of an algebraic function f could have roots, holes, asymptotes, nondifferentiable points, and other features that cannot be immediately determined by just looking at the equation for f.

We can always use limits and an analysis of the roots and sign intervals of f, f', and f'' to determine the features of the graph of a given algebraic function f. Table 7.1 briefly summarizes some of the graphical features that can be determined by examining roots and sign intervals.

Feature	How to Find
Roots of f	Solutions to $f(x) = 0$
Local extrema of f	Find places where $f'(x)$ changes sign
Inflection points of f	Find places where $f''(x)$ changes sign
Positive/negative intervals of f	Sign analysis of $f(x)$
Increasing/decreasing intervals of f	Sign analysis of $f'(x)$
Concave up/down intervals of f	Sign analysis of $f''(x)$

Table 7.1 Roots and Signs of f, f', and f'' Determine Basic Features of a Graph

Table 7.2 summarizes the techniques for examining the asymptotes and discontinuities of a function. Each of these features can be detected by calculating a limit involving the function f.

Suppose a function f is continuous, but not differentiable, at a point $x = c$. We can use the techniques in Table 7.3 to determine how differentiability fails. Notice that corners, cusps, vertical cusps, and vertical tangent lines can all be detected by examining limits of the *derivative* function f'.

Feature	How to Find
f has a vertical asymptote at $x = c$.	$\lim\limits_{x \to c^-} f(x)$ and/or $\lim\limits_{x \to c^+} f(x)$ are infinite.
f has a horizontal asymptote.	$\lim\limits_{x \to \infty} f(x)$ and/or $\lim\limits_{x \to -\infty} f(x)$ are finite.
f has a hole at $x = c$.	$\lim\limits_{x \to c} f(x)$ exists but does not equal $f(c)$.
f has a jump discontinuity at $x = c$.	$\lim\limits_{x \to c^-} f(x)$ and $\lim\limits_{x \to c^+} f(x)$ exist but are not equal.

Table 7.2 Examining Limits of $f(x)$ to Investigate Discontinuities

Feature	How to Find
f has a corner or cusp at $x = c$.	$\lim\limits_{x \to c^-} f'(x)$ and $\lim\limits_{x \to c^+} f'(x)$ exist but are not equal.
f has a vertical cusp at $x = c$.	$\lim\limits_{x \to c^-} f'(x)$ and $\lim\limits_{x \to c^+} f'(x)$ are infinite, with opposite signs.
f has a vertical tangent line at $x = c$.	$\lim\limits_{x \to c^-} f'(x)$ and $\lim\limits_{x \to c^+} f'(x)$ are infinite, with the same sign.

Table 7.3 Examining Limits of $f'(x)$ to Investigate Nondifferentiable Points

? Question You need to know and understand the information in Tables 7.1–7.3 so that you know what to look for when you are investigating the graphical behavior of a function. Think of these tables as a "toolbox" for investigating function behavior. □

7.5.2 Optimization with general algebraic functions

Recall that an "optimization" problem is a word problem where we want to find the global maximum or minimum value (if any) of a function on a given interval. In Section 5.4 we discussed optimization of polynomial functions; we now examine optimization problems involving general algebraic functions.

All of the optimization techniques we learned in the context of polynomials are applicable to general algebraic functions, with one exception. Since polynomial functions are always continuous, our previous optimization problems always involved functions that were continuous everywhere. To optimize a function f that is continuous on an interval I, we simply find all the local extrema of f in the interior of I and then compare these values with the behavior of the function at the "ends" of the interval. This method is guaranteed to find the global maximum or minimum (if any) of f on the interval I, provided that f is continuous on I. However, not all algebraic functions are continuous everywhere!

If a function f is discontinuous somewhere on an interval I, then it is possible that there is a global maximum or minimum of f on I that will not be detected by the algorithm described in the previous paragraph. For example, suppose we wanted to find the global minimum (if any) on the interval $[0, 5]$ of the function f shown in Figure 7.12.

Clearly f has one local extremum in $[0, 5]$, which is located at $x = 1$ and has a maximum of approximately $f(1) \approx 3$. According to our old algorithm, we would compare the height at this extremum with the height of f at the ends of the interval $[0, 5]$. (If the ends of the interval had been open, we would have instead considered the *limit* of $f(x)$ as x approached the ends of the interval.) From the graph we can see that $f(0) \approx 2.2$ and $f(5) \approx 1.5$. If this is the only information we were given,

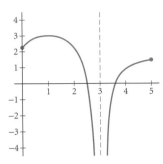

Figure 7.12

and we forgot to check for discontinuities in the interior of [0, 5], we would *incorrectly* conclude that the minimum of f is at the right-hand endpoint $x = 5$ (since that point had the lowest value among the local extrema and the endpoints). However, clearly the function f shown in Figure 7.12 does not have a global minimum value on [0, 5], because f decreases without bound as x approaches the discontinuous point $x = 3$.

To find the global extremum of a general algebraic function on a given interval, we must be careful to examine not only the local extrema and endpoints but also the behavior of the function near any discontinuities, especially at vertical asymptotes. In general we can follow Algorithm 7.3.

ALGORITHM 7.3

Finding the Global Extrema of a Function on an Interval

To find a global maximum or minimum of a function f on an interval I (if such a global extremum exists), we must consider the following:

(a) The values $f(c)$ for each interior local extremum $c \in I$;

(b) The behavior of f at the "ends" of the interval I (specifically, the *value* of f at any closed endpoint and the *limit* of f at any open endpoint); and

(c) The behavior of f at any discontinuous points in the interval I (in particular, the behavior near any vertical asymptotes, jumps, or holes).

This algorithm actually works for *any* function, whether or not it is algebraic, but for the moment we will be concerned only with algebraic functions.

Although you may not have noticed it at the time, when we restricted our attention in Section 5.4 to optimization of *polynomial* functions, we could solve only certain types of optimization problems. Now we can optimize any (algebraic) function we like!

EXAMPLE 7.20   **The fastest way to put out a flaming tent**

While you are on a camping trip, your tent accidentally catches on fire. Luckily, you happen to be standing right at the edge of a stream with a bucket in your hand. The stream runs east-west, and the tent is 40 feet north of the stream and 100 feet farther east than you are (see Figure 7.13). You can run only half as fast while carrying the full bucket, and thus any distance traveled with the full bucket is effectively twice as long. What is the fastest way for you to get water to the tent?

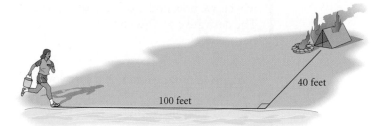

Figure 7.13

Solution   First notice that if you were to get water immediately and run diagonally to the tent, you would run the entire distance with a full bucket of water (and thus at half your normal speed). On the other hand, if you were to run along the side of the stream until you were directly south of the tent, then get water and run north to the tent, you would have a lot of total distance to run. Clearly, it would be more efficient for you to run along the side of the stream for a while, fill the bucket, and then run diagonally to the tent. The question is: How far should you run along the side of the stream before you stop to fill up the bucket?

 We need some mathematical notation; let x be the distance you will run along the stream before filling your bucket. The distance you will run with the full bucket is then the hypotenuse of a right triangle with legs of lengths 40 feet and $100 - x$ feet. By the Pythagorean Theorem, you will have to run $\sqrt{40^2 + (100 - x)^2}$ feet with a full bucket of water (see Figure 7.14).

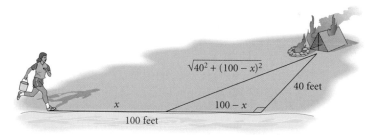

Figure 7.14

Since it takes you twice as long to run with a full bucket of water, the total *effective* distance you will have to run is

$$D(x) = x + 2\sqrt{40^2 + (100 - x)^2}.$$

This is the function we want to minimize. What is the interval of x-values that we are interested in? From the diagram we can see that x must be between 0 and 100 feet (you wouldn't want to run away from the tent or past the tent!). The endpoint cases $x = 0$ and $x = 100$ correspond to the case when you get water immediately and the case when you run 100 feet along the stream before getting water and then run directly north to the tent. We have now translated the word problem into the following mathematical optimization problem:

Find the minimum value of $D(x) = x + 2\sqrt{40^2 + (100 - x)^2}$ on the interval $[0, 100]$.

 The function $D(x)$ is continuous, and in particular has no vertical asymptotes, on $[0, 100]$. $D(x)$ is defined on the entire interval $[0, 100]$ (and in fact for all real numbers x;

why?). We now find any local extrema of $D(x)$ by finding and testing its critical points. The derivative of $D(x)$ is

$$D'(x) = 1 + 2(\tfrac{1}{2})(40^2 + (100 - x)^2)^{-\frac{1}{2}}(2(100 - x)(-1)).$$

(Stop for a minute and make sure you followed that computation; notice that we used the chain rule twice.) We need to simplify this expression before we attempt to find its zeros or the values where it does not exist; remember that it will be easier to find such values if we first write $D'(x)$ in the form of a quotient:

$$D'(x) = 1 + 2(\tfrac{1}{2})(40^2 + (100 - x)^2)^{-\frac{1}{2}}(2(100 - x)(-1))$$

$$= 1 + \frac{-2(100 - x)}{\sqrt{40^2 + (100 - x)^2}}$$

$$= \frac{\sqrt{40^2 + (100 - x)^2} - 2(100 - x)}{\sqrt{40^2 + (100 - x)^2}}.$$

The denominator of $D'(x)$ is never zero, and thus $D'(x)$ always exists. We must find the values of x for which $D'(x) = 0$, that is, the values of x for which the numerator is zero:

$$\sqrt{40^2 + (100 - x)^2} - 2(100 - x) = 0$$

$$\sqrt{40^2 + (100 - x)^2} = 2(100 - x)$$

$$40^2 + (100 - x)^2 = 4(100 - x)^2 \qquad \text{(square both sides)}$$

$$1600 + 10{,}000 - 200x + x^2 = 40{,}000 - 800x + 4x^2$$

$$0 = 3x^2 - 600x + 28{,}400$$

$$x = 100 \pm \tfrac{40}{3}\sqrt{3}. \qquad \text{(quadratic formula)}$$

Note that $x = 100 + \tfrac{40}{3}\sqrt{3} \approx 123.1$ is not in the interval $[0, 100]$, but $x = 100 - \tfrac{40}{3}\sqrt{3} \approx 76.9$ is. (In addition, recall that squaring both sides of an equation sometimes leads to extraneous, or false, solutions; $x = 100 + \tfrac{40}{3}\sqrt{3}$ happens to be one of those false solutions.) The first derivative test can be applied to show that $x = 100 - \tfrac{40}{3}\sqrt{3}$ is a local minimum of $D(x)$. But is it the *global* minimum of $D(x)$ on $[0, 100]$?

Since $D(x)$ is continuous, we need only compare $D(100 - \tfrac{40}{3}\sqrt{3})$ with the endpoint values $D(0)$ and $D(100)$:

$$D\left(100 - \frac{40}{3}\sqrt{3}\right) \approx 169.282,$$

$$D(0) \approx 215.407,$$

$$D(100) \approx 180.$$

The *global* minimum of $D(x)$ on $[0, 100]$ is at $x = 100 - \tfrac{40}{3}\sqrt{3}$. The quickest way to bring water to the burning tent is to run for $100 - \tfrac{40}{3}\sqrt{3} \approx 76.9$ feet along the side of the stream, fill the bucket with water, and then run diagonally to the tent.   □

7.5.3 Curve sketching with general algebraic functions

At this point you have the skills to graph any algebraic function by hand (provided that the algebra involved is not too difficult). Remember to use the tools summarized in Tables 7.1–7.3! It's usually easy to remember to do a sign analysis of the first and second derivatives, but don't forget that you can't get *all* the information you need just from number lines for f' and f''.

To detect asymptotes, holes, cusps, and so on, and the behavior of the graph near these features, you must consider some limits. In particular, you will always need to look at the limits of the function as $x \to \infty$ and as $x \to -\infty$ to determine the global behavior of the graph. It is a good idea to start by finding the domain of the function you are attempting to graph.

EXAMPLE 7.21

A detailed curve-sketching analysis

Sketch a careful, labeled graph of the function $f(x) = x^{\frac{2}{3}}(5x^2 - 32x + 80)$.

Solution The function f is defined for all real numbers, and moreover it is a continuous function (it is a product of continuous functions). Therefore, the graph of f will not have any holes, jumps, or vertical asymptotes. We now calculate (and simplify) the first and second derivatives of f. Our calculations will be simpler if we first expand $f(x)$:

$$f(x) = x^{\frac{2}{3}}(5x^2 - 32x + 80) = 5x^{\frac{8}{3}} - 32x^{\frac{5}{3}} + 80x^{\frac{2}{3}}.$$

We first find $f'(x)$, and then use algebra to write it as a simple quotient (this will help us find its zeros and undefined points):

$$f'(x) = \frac{40}{3}x^{\frac{5}{3}} - \frac{160}{3}x^{\frac{2}{3}} + \frac{160}{3}x^{-\frac{1}{3}} \qquad \text{(differentiate)}$$

$$= \frac{40}{3}\left(x^{\frac{5}{3}} - 4x^{\frac{2}{3}} + \frac{4}{x^{\frac{1}{3}}}\right) = \frac{40}{3}\left(\frac{x^{\frac{5}{3}}x^{\frac{1}{3}} - 4x^{\frac{2}{3}}x^{\frac{1}{3}} + 4}{x^{\frac{1}{3}}}\right) \qquad \text{(combine fractions)}$$

$$= \frac{40(x^2 - 4x + 4)}{3x^{\frac{1}{3}}} = \frac{40(x - 2)^2}{3x^{\frac{1}{3}}}. \qquad \text{(factor)}$$

Using the multiplied-out version of $f'(x)$, we can easily find $f''(x)$ (and then simplify as a simple quotient):

$$f''(x) = \frac{40}{3}\left(\frac{5}{3}\right)x^{\frac{2}{3}} - \frac{160}{3}\frac{2}{3}x^{-\frac{1}{3}} + \frac{160}{3}\left(-\frac{1}{3}\right)x^{-\frac{4}{3}} \qquad \text{(differentiate)}$$

$$= \frac{40}{9}\left(5x^{\frac{2}{3}} - \frac{8}{x^{\frac{1}{3}}} - \frac{4}{x^{\frac{4}{3}}}\right) = \frac{40}{9}\left(\frac{5x^{\frac{2}{3}}x^{\frac{4}{3}} - 8x - 4}{x^{\frac{4}{3}}}\right) \qquad \text{(combine fractions)}$$

$$= \frac{40(5x^2 - 8x - 4)}{9x^{\frac{4}{3}}} = \frac{40(x - 2)(5x + 2)}{9x^{\frac{4}{3}}}. \qquad \text{(factor)}$$

From the work above we can see that f' is undefined at $x = 0$ and is zero at $x = 2$, and that f'' is undefined at $x = 0$ and zero at $x = 2$ and $x = -\frac{2}{5} = -0.4$. We now know that f is continuous but not differentiable at $x = 0$, so the graph of f could have a corner, cusp, vertical cusp, or vertical tangent line at $x = 0$. A sign analysis of f' and f'' between the critical points gives us the number lines in Figure 7.15. (For example, you could check the signs of f' at $x = -1$, $x = 1$, and $x = 3$, and check the signs of f'' at $x = -1$, $x = -\frac{1}{5}$, $x = 1$, and $x = 3$.)

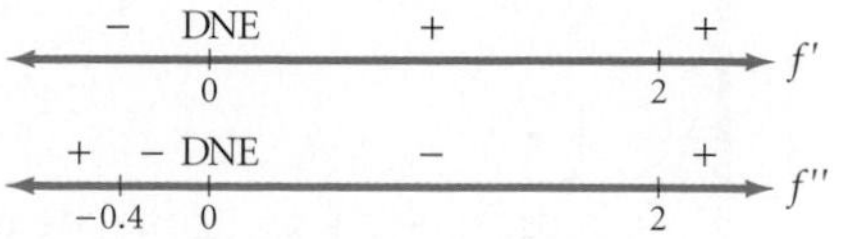

Figure 7.15

From these number lines we can see that f has a local minimum at $x = 0$ (although the derivative doesn't exist there, the function does). We can also see that the graph of f has a horizontal tangent line at $x = 2$ that is not a local extremum, and inflection points at $x = -\frac{2}{5}$ and at $x = 2$.

Let's examine carefully what is happening near $x = 0$. We know that f is continuous but not differentiable at that point and that $x = 0$ is a local minimum of f. This means the graph of f could have a corner, cusp, or vertical cusp at $x = 0$ (note that we have now excluded the possibility of a vertical tangent line; why?). The graph would have a *corner* at $x = 0$ only if the graph of f was a straight line to the left of $x = 0$ and to the right of $x = 0$; this clearly is not happening, because the second derivative would be zero to either side of $x = 0$, which is not the case. So is the cusp at $x = 0$ a vertical or a nonvertical cusp? Let's look at the appropriate limits (see Table 7.3). Using the expanded form of $f'(x)$, we have

$$\lim_{x \to 0^-} f'(x) = \lim_{x \to 0^-} \left(\frac{40}{3} x^{\frac{5}{3}} - \frac{160}{3} x^{\frac{2}{3}} + \frac{160}{3 x^{\frac{1}{3}}} \right) \to 0 - 0 + \frac{1}{0^-} = -\infty.$$

A similar calculation shows that $\lim_{x \to 0^+} f'(x) = \infty$. Therefore, the graph of f has a vertical cusp (point down) at $x = 0$.

We must also investigate the global behavior of f (as $x \to \infty$ and as $x \to -\infty$). So far we don't know whether the graph of f increases or decreases without bound at the "ends" or whether it has a horizontal asymptote. We will use our knowledge of the limits of power functions and polynomial functions to calculate the limits:

$$\lim_{x \to \infty} f(x) = \lim_{x \to \infty} x^{\frac{2}{3}}(5x^2 - 32x + 80) \to (\infty)(\infty) = \infty,$$

$$\lim_{x \to -\infty} f(x) = \lim_{x \to -\infty} x^{\frac{2}{3}}(5x^2 - 32x + 80) \to (\infty)(\infty) = \infty.$$

Therefore, the graph of f increases without bound both as $x \to \infty$ and as $x \to -\infty$.

To put all this information together in a graph, we just need to know the values of f at any "important" points. We have

$$f(-0.4) \approx 50.8, \quad f(0) = 0, \quad f(2) \approx 57.1.$$

Now we know the coordinates of the inflection points and extrema of f, and we can (finally) make a graph of f from all the information we have collected. Note in particular that it is important to make sure that the graph increases or decreases, and is concave up or concave down, in the appropriate places (based on the number lines). See Figure 7.16.

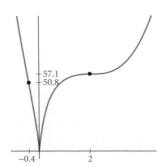

Figure 7.16
$f(x) = x^{\frac{2}{3}}(5x^2 - 32x + 80)$

It is difficult to see from the graph, but note that the function f is concave up to the left of $x = -0.4$ and concave down between $x = -0.4$ and $x = 0$.   □

⊗ *Caution* Example 7.21 showed how to detect a vertical cusp. Be sure that you also know how to detect a vertical tangent line or the behavior of a graph near a vertical asymptote. Also be on the lookout for "holes" in the graph! ☐

⊗ *Caution* Don't forget that if you are asked to graph a function that happens to be a power, polynomial, or rational function, then you have a good idea what the graph should look like (and what features to look for) even before you start differentiating. Use that knowledge to your advantage! ☐

⊗ *Caution* Not every function factors easily or nicely. When doing root and sign analyses of a function f and its derivatives f' and f'', you may not be able to factor all three of f, f', and f''. If this is the case, don't worry about it (of course, don't forget to try root-guessing and synthetic division!). Use whatever information you *can* find to make your graph. ☐

What you should know after reading Section 7.5

▶ The types of features that the graph of an algebraic function can have, and how to determine algebraically whether the graph of a given function f will have these features. This includes root and sign analyses of f, f', and f'' as well as an investigation of various limits involving f and f'.

▶ How to find the global maximum or minimum (if any) of a function on an interval. This includes functions that are continuous as well as those that are discontinuous. How to solve optimization word problems where the function to be optimized is a general algebraic function.

▶ How to sketch an accurate, labeled graph of an algebraic function by performing a sign analysis of f, f', and f'' as well as examining any relevant limits.

Exercises 7.5

Concepts

0. Read the section and make your own summary of the material. Include whatever you think will help you review the material later on.

■ Various graphical features are listed below. For each graphical feature listed, explain in detail how such a feature could be detected algebraically (that is, by hand). Give more detail than is given in the tables in the reading.

1. f has a root at $x = c$.

2. f has a critical point at $x = c$.

3. f has a local extremum at $x = c$.

4. f has an inflection point at $x = c$.

5. f is negative on an interval I.

6. f is increasing on an interval I.

7. f is concave down on an interval I.

8. f has a nonvertical cusp at $x = c$.

9. f has a vertical cusp at $x = c$.

10. f has a vertical tangent line at $x = c$.

11. f has a vertical asymptote at $x = c$.

12. f has a horizontal asymptote at $y = k$.

13. f has a hole at $x = c$.

14. f has a jump discontinuity at $x = c$.

15. Suppose f is a function that is discontinuous somewhere on an interval I. Explain why comparing the values of any local extrema of f on I and the values or limits of f at the endpoints of I is not in general sufficient to determine any *global* maximum or minimum of f on I.

16. Explain the basic steps involved in solving an optimization word problem.

17. Give a basic outline of a method for sketching the graph of an algebraic function by hand. What would you do first? What would you do last? What would you do in between? There is no one right answer to this question; just develop a strategy for discovering the graphical information about a function.

Skills

■ Find the domains of the following functions. Do all work by hand, and check your answers with a graphing calculator.

18. $f(x) = 1 - \dfrac{3x}{\sqrt{9x^2 + 6x + 1}}$

19. $f(x) = 5x^{\frac{1}{3}} - 3x^{\frac{4}{3}} + x^2$

20. $f(x) = (x^2 - 4x + 3)^{-\frac{1}{2}}$

■ Find all points (if any) at which the following functions fail to be continuous. Determine the graphical behavior of each function near any points of discontinuity. Do all work by hand, and check your answers with a graphing calculator.

21. $f(x) = 1 - \dfrac{3x}{\sqrt{9x^2 + 6x + 1}}$

22. $f(x) = \dfrac{(x^2 - 1)^{\frac{2}{3}}}{(x - 1)^{\frac{2}{3}}}$

23. $f(x) = \dfrac{(x^2 - 1)^{\frac{2}{3}}}{(x - 1)^2}$

■ Find all points (if any) at which the following functions fail to be differentiable. Determine the graphical behavior of each function near any nondifferentiable points. Do all work by hand, and check your answers with a graphing calculator.

24. $f(x) = \sqrt{1 + x^{\frac{2}{3}}}$

25. $f(x) = (x^2 - 4x + 3)^{-\frac{1}{2}}$

26. $f(x) = 5x^{\frac{1}{3}} - 3x^{\frac{4}{3}} + x^2$

■ Find all vertical and horizontal asymptotes (if any) of the following functions. Do all work by hand, and check your answers with a graphing calculator.

27. $f(x) = \dfrac{4}{\sqrt{x^2 + 1}} + 3$

28. $f(x) = 1 - \dfrac{3x}{\sqrt{9x^2 + 6x + 1}}$

29. $f(x) = (x^2 - 4x + 3)^{-\frac{1}{2}}$

■ Find any roots of each of the following functions. Do all work by hand, and check your answers with a graphing calculator.

30. $f(x) = x^{-\frac{1}{3}}(x + 3)^2$

31. $f(x) = x^{\frac{3}{2}} + 2\sqrt{x} - \dfrac{3}{\sqrt{x}}$

32. $f(x) = (-x^3 + x^2 + 5x - 5)^{\frac{1}{3}}$

■ Find the intervals (if any) on which the following functions are positive or negative. Do all work by hand, and check your answers with a graphing calculator.

33. $f(x) = x^{-\frac{1}{3}}(x + 3)^2$

34. $f(x) = \dfrac{(x^2 - 1)^{\frac{2}{3}}}{(x - 1)^{\frac{2}{3}}}$

35. $f(x) = (-x^3 + x^2 + 5x - 5)^{\frac{1}{3}}$

■ Find the intervals (if any) on which the following functions are increasing or decreasing. Do all work by hand, and check your answers with a graphing calculator.

36. $f(x) = 3(3 + x)(x - 1)^{\frac{1}{3}}$

37. $f(x) = \dfrac{x\sqrt{x^2 + 1}}{x^2 - 4}$

38. $f(x) = \dfrac{(x^2 - 1)^{\frac{2}{3}}}{(x - 1)^2}$

■ Find the intervals (if any) on which the following functions are concave up or concave down. Do all work by hand, and check your answers with a graphing calculator.

39. $f(x) = (4x - 2x^2)^{-1}$

40. $f(x) = \dfrac{4}{\sqrt{x^2 + 1}} + 3$

41. $f(x) = 3(3 + x)(x - 1)^{\frac{1}{3}}$

■ Find any local extrema of each of the following functions. Do all work by hand, and check your answers with a graphing calculator.

42. $f(x) = 1 - \dfrac{3x}{\sqrt{9x^2 + 6x + 1}}$

43. $f(x) = \dfrac{1}{1 + \sqrt{x}}$

44. $f(x) = 2\sqrt{1 + x}(3x^2 - 4x - 7)$

■ Find any inflection points of each of the following functions. Do all work by hand, and check your answers with a graphing calculator.

45. $f(x) = 3(3 + x)(x - 1)^{\frac{1}{3}}$

46. $f(x) = \dfrac{4}{\sqrt{x^2 + 1}} + 3$

47. $f(x) = 2\sqrt{1 + x}(3x^2 - 4x - 7)$

■ For each function f and interval I given below, determine (by hand) any global extrema of f on I. When you are finished, check your answer using a graphing calculator.

48. $f(x) = x^{\frac{3}{2}}(3x - 5), \quad I = (0, 3)$

49. $f(x) = \dfrac{1}{1 + \sqrt{x}}, \quad I = [0, 3]$

50. $f(x) = \dfrac{4}{\sqrt{x^2 + 1}} + 3, \quad I = (-5, 2]$

51. $f(x) = (x^2 - 4x + 3)^{-\frac{1}{2}}, \quad I = [0, 10]$

■ Make a careful sketch (by hand) of each of the functions below. Label any important features of each graph. Your work should include number lines, domains, selected values, and limit calculations. When you are finished, check your answer using a graphing calculator.

52. $f(x) = (4x - 2x^2)^{-1}$

53. $f(x) = 6x^{-\frac{2}{3}}$

54. $f(x) = (x^2 - 1)^{\frac{4}{5}}$

55. $f(x) = \dfrac{1}{1 + \sqrt{x}}$

56. $f(x) = x^{\frac{3}{2}}(3x - 5)$

57. $f(x) = 2\sqrt{1 + x}(3x^2 - 4x - 7)$

58. $f(x) = (x^2 + x - 2)(x^2 - x - 6)^{-1}$

59. $f(x) = \dfrac{4}{\sqrt{x^2 + 1}} + 3$

60. $f(x) = (x^2 - 4x + 3)^{-\frac{1}{2}}$

61. $f(x) = 3(3 + x)(x - 1)^{\frac{1}{3}}$

62. $f(x) = x(3x - 10)^{\frac{2}{3}}$

63. $f(x) = \dfrac{x\sqrt{x^2 + 1}}{x^2 - 4}$

Applications

64. Your company produces cylindrical metal oil drums each of which must hold 40 cubic feet of oil. How should the oil drums be constructed so that they use as little metal as possible? Can they be constructed to use as *much* metal as possible?

65. The cost of the material for the top and bottom of a cylindrical can is 5 cents per square inch. The material for the "side" of the can (the curvy part) costs only 2 cents per square inch. If the can must hold 400 cubic inches of liquid, what is the cheapest way to make the can? What is the most expensive way?

66. Consider the can-making situation in the previous problem, but suppose that the cans are made with open tops. If each can must hold 400 cubic inches of liquid, what is the cheapest way to make the cans? What is the most expensive way?

67. Find the point on the graph of the function $f(x) = \sqrt{x^2 + 1}$ that is closest to the point $(2, 0)$.

68. Find the area of the largest rectangle that fits inside a circle of radius 10.

69. Find the volume of the largest cylinder that fits inside a sphere of radius 10.

70. A steam pipe must be buried underground to reach from one corner of a rectangular parking lot to the diagonally opposite corner. The dimensions of the parking lot are 500 feet by 800 feet. It costs 5 dollars per foot to lay steam pipe under the pavement but only 3 dollars per foot to lay the pipe along one of the long edges of the parking lot. Because of nearby sidewalks, the pipe cannot be laid along the 500-foot sides of the parking lot. How should the steam pipe be buried so as to cost as little as possible?

71. Alina needs to make a flier for her band's concert. The flier must contain 20 square inches of printed material and, for design purposes, should have side margins of 1 inch and top and bottom margins of 2 inches. What size paper should Alina use in order to use as little paper per flier as possible?

72. An airplane leaves Chicago at noon and travels south at 500 miles per hour. Another airplane is traveling east toward Chicago at 650 miles per hour and arrives at 2:00 p.m. When are these two airplanes closest to each other, and how far apart are they at that time?

73. While you are on a camping trip, your tent accidentally catches on fire. At the time, you and the tent are both 50 feet from a stream, and you are 200 feet away from the tent, via a straight line parallel to the stream (see the figure below). You have a bucket with you, and you need to run to the stream, fill the bucket, and

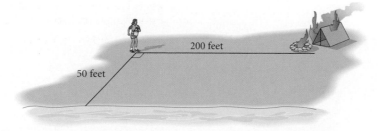

Figure for Problem 73

run to the tent as fast as possible. You can run only half as fast while carrying the full bucket, so any distance traveled with the full bucket is effectively twice as long. How should you run from your position, to the stream, and to the tent in order to get there as soon as possible?

(a) Let x represent the distance from the point on the stream directly "below" you to the point on the stream that you run to. Label x on the diagram on the preceding page, and sketch the path that you would follow to run from your starting position, to the point x along the stream, to the tent.

(b) Let $D(x)$ represent the effective distance (counting twice any distance traveled while carrying a full bucket) you have to run in order to collect water and get to the tent. Write an equation for $D(x)$.

(c) Determine the interval I of x-values on which $D(x)$ should be minimized. Explain in practical terms what happens at the endpoints of this interval, and calculate the value of $D(x)$ at these endpoints.

(d) Find $D'(x)$, and simplify as much as possible. Are there any points (in the interval I) at which $D'(x)$ is undefined?

(e) It is difficult to find the zeros of $D'(x)$ by hand. Use a graphing calculator to approximate any zeros of $D'(x)$ in the interval I, and test these zeros by evaluating D' at each one.

(f) Use the information above to determine the minimum value of $D(x)$, and then use this value to answer the original word problem.

Proofs

74. Use limits of f to prove that the graph of $f(x) = \dfrac{(x^2 - 4)^{\frac{2}{5}}}{(x+2)^{\frac{2}{5}}}$ has a hole at $x = -2$.

75. Use limits of f' to prove that the graph of $f(x) = (x-3)^{\frac{4}{5}} + 2$ has a vertical cusp at $x = 3$.

76. Use limits of f' to prove that the graph of $f(x) = x(x-2)^{\frac{1}{3}}$ has a vertical tangent line at $x = 2$.

PART III

TRANSCENDENTAL FUNCTIONS

We now use the tools we have developed to investigate exponential functions, logarithmic functions, trigonometric functions, and inverse trigonometric functions. Specifically, we examine the algebra, limits, derivatives, and graphs of each of these types of nonalgebraic, or transcendental, functions.

CHAPTER 8

Exponential Functions

"Double Knot," a sculpture by Charles Perry,
www.charlesperry.com

*E*xponential functions are the most basic type of transcendental (nonalgebraic) functions. In this chapter we define exponential functions and examine their algebraic and graphical properties, their limits, and their derivatives. As we will see, exponential functions have a number of interesting properties that make them particularly suited for certain types of real-world applications.

At the end of the chapter we will introduce "L'Hôpital's Rule," a new technique for calculating limits. This rule will be important throughout Part III.

CONTENTS

530

8.1	**The Algebra of Exponential Functions**

In this section we examine some basic properties of **exponential functions.** Up to this point, we have dealt only with "algebraic" functions; exponential functions are our first example of nonalgebraic, or **transcendental** functions. After defining what an exponential function is, we will discuss the rules for manipulating exponential functions and some simple properties of exponential functions.

8.1.1 What is an exponential function?

We have already studied *power functions,* functions of the form $f(x) = Ax^k$ where the base is the independent variable x and the exponent is a rational number k. If instead the base is constant while the exponent is the independent variable, the function is said to be **exponential.**

DEFINITION 8.1	

The Form of an Exponential Function

An **exponential function** is a function that can be written in the form

$$f(x) = Ab^x$$

for some real numbers A and b such that $A \neq 0$, $b > 0$, and $b \neq 1$.

§ *Caution* Be careful not to confuse exponential functions with power functions! Remember that power functions have the independent variable in the *base* and a constant in the exponent (e.g., $f(x) = x^2$), whereas exponential functions have the independent variable in the *exponent* and a constant for the base (e.g., $f(x) = 2^x$). □

? *Question* Is the function $f(x) = x^x$ an exponential function? Is it a power function? Why or why not? □

Note that we require that the base b of an exponential function $f(x) = Ab^x$ is greater than zero and not equal to 1. The reason we do not want to consider the case $b = 1$ is because the function $f(x) = 1^x$ is not very interesting (in fact, it is a constant function, since $1^x = 1$ for all values of x). We also do not want to consider $f(x) = 0^x$ to be an exponential function, since it is constantly zero for all x, except at $x = 0$, where it is undefined (because 0^0 is undefined). Moreover, the base b must be positive so that we don't encounter any negative numbers under even roots; for example, the function $f(x) = (-2)^x$ is not defined for many values of x, including $x = \frac{1}{2}$, $x = \frac{3}{4}$, and $x = 7.5$.

In Chapter 4, we defined what it means to raise a number to a rational power: $a^{\frac{p}{q}} := (\sqrt[q]{a})^p$. This tells us how to compute the value of an exponential function

$f(x) = Ab^x$ at any rational value of x. But what does Ab^x mean when x is an *irrational* number? For example, what is $f(\pi)$ if $f(x) = 2^x$? (In other words, what is 2^π?) One way to define b^x where x is irrational is as a limit:

(1)   *If x is an irrational number, then* $b^x := \lim\limits_{\substack{r \to x \\ r \text{ rational}}} b^r$.

For example, 2^π can be approximated by 2^r for some rational number r that is very close to π (like $r = 3.14$). As we consider rational numbers r that are closer and closer to π, the expression 2^r will get closer and closer to 2^π.

Technical Note: This method of defining irrational powers is not completely rigorous. To define irrational powers properly, we must use something called an "integral" to define logarithms and then define exponential functions (and, in particular, irrational powers) from that. At this point we do not know what an "integral" is, so for now we will use the idea in Statement (1).

Exponential functions are our first example of **transcendental** (i.e., nonalgebraic) functions and have behavior that is much different from that of power functions or other algebraic functions. The graphs of a few common exponential functions are shown in Figures 8.1–8.3 (note that a different graphing window is used for each function). We will investigate the graphs of exponential functions in detail in Section 8.5.

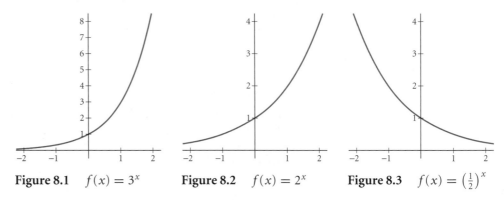

Figure 8.1 $f(x) = 3^x$ **Figure 8.2** $f(x) = 2^x$ **Figure 8.3** $f(x) = \left(\frac{1}{2}\right)^x$

These graphs are representative of the graphs of exponential functions; many of the properties shown in these graphs are shared by all exponential functions. Notice in particular that each of these exponential functions has a horizontal asymptote at $y = 0$; as we will see in Section 8.5, all exponential functions have this property.

8.1.2 Rules for the algebra of exponential functions

Consider an exponential function $f(x) = b^x$ for some $b > 0$ and $b \neq 1$. (We'll set the coefficient A to be 1 for now so we can focus on the exponentiation.) By two of the most basic algebraic properties of exponents, we know that

(2) $b^0 = 1$ and $b^{-x} = \frac{1}{b^x}$.

In terms of exponential functions, this means that if $f(x) = b^x$, then $f(0) = b^0 = 1$ (so the y-intercept of $f(x) = b^x$ is $y = 1$) and that if $f(x) = b^x$, then $f(-x) = b^{-x} = \frac{1}{b^x} = \frac{1}{f(x)}$.

Theorem 8.1 is identical to the theorem in Section 4.1 that describes the algebra of exponents, except that it is written in notation that suggests exponential functions (where the bases are constants and the variable is in the exponent).

THEOREM 8.1

Algebraic Rules for Exponential Functions

For any real numbers b, c, x, and y for which the following are defined, we have

(a) $b^{x+y} = b^x b^y$, **(b)** $(bc)^x = b^x c^x$, **(c)** $b^{xy} = (b^x)^y = (b^y)^x$.

As simple corollaries to these rules, we also have

(d) $b^{x-y} = \dfrac{b^x}{b^y}$, **(e)** $\left(\dfrac{b}{c}\right)^x = \dfrac{b^x}{c^x}$, **(f)** $b^{\frac{x}{y}} = (\sqrt[y]{b})^x = \sqrt[y]{b^x}$.

❧ *Caution* Don't forget that Ab^x is *not* in general equal to $(Ab)^x$. For example, $2(3^x)$ is not equal to 6^x. (If you don't believe this, try evaluating the two expressions at $x = 2$.) □

Using the algebraic rules for exponents, we can easily determine whether a given function is exponential, as shown in the following example.

EXAMPLE 8.1

Identifying exponential functions

Determine which of the following functions is (are) exponential. Write each exponential function in the form $f(x) = Ab^x$ for some real numbers A and b.

(a) $f(x) = 4(1.2)^x - 5$ **(b)** $f(x) = 2^x 3^{1-x}$ **(c)** $f(x) = 3.1x^{7.4}$

Solution The functions in parts (a) and (c) are *not* exponential functions, because they cannot be written in the form $f(x) = Ab^x$. The function in part (a) is a translation of an exponential function, but it is not itself exponential. The function in part (c) is not an exponential function because the variable x appears in the *base* of the expression $x^{7.4}$.

The function in part (b) *can* be written in the form $f(x) = Ab^x$ by using the rules in Theorem 8.1:

$$f(x) = 2^x 3^{1-x} = 2^x 3^1 3^{-x} = 2^x (3)\frac{1}{3^x} = 3(2)^x \left(\tfrac{1}{3}\right)^x = 3\left((2)\left(\tfrac{1}{3}\right)\right)^x = 3\left(\tfrac{2}{3}\right)^x.$$

Thus $f(x) = Ab^x$ with $A = 3$ and $b = \tfrac{2}{3}$. □

8.1.3 Basic properties of exponential functions

The following theorem describes some simple properties of exponential functions, including their domains and ranges.

THEOREM 8.2

Basic Properties of Exponential Functions

Suppose b is a real number with $b > 0$ and $b \neq 1$.

(a) The function $f(x) = b^x$ has domain $\mathbb{R}$ and range $(0, \infty)$.

(b) In particular, $b^x \neq 0$ for any value of x.

(c) The function $f(x) = b^x$ is one-to-one.

The first part of this theorem tells us that the range of an exponential function of the form b^x is all positive numbers. For example, the range of $f(x) = 2^x$ is $(0, \infty)$. This means that $y = 5$ is in the range of $f(x) = 2^x$ and thus that there is some number x for which $5 = 2^x$. However, Theorem 8.2 does not tell us how to *find* this value of x.

Notice that Theorem 8.2 refers only to exponential functions of the form $f(x) = b^x$. Most of the theorem is also true for general exponential functions $f(x) = Ab^x$, except that if $A < 0$, then the range of $f(x) = Ab^x$ will be $(-\infty, 0)$ instead of $(0, \infty)$.

Rigorously proving part (a) of Theorem 8.2 is difficult because of the way that we have defined exponential functions (specifically, because of our definition of b^x for irrational values of x). To define exponential functions properly, we would have to know about something called "integration," which we won't see until Part IV. For now, we will prove only parts (b) and (c) of Theorem 8.2.

PROOF **(PARTS (B) AND (C) OF THEOREM 8.2)**   Part (b) is actually a consequence of the range statement in part (a), but we will prove it separately. We wish to show that if $b > 0$ and $b \neq 1$, then b^x is never zero (no matter what x is). Seeking a contradiction, suppose that $b^x = 0$ for some value of x. Clearly we could not have $x = 0$, because in that case $b^x = b^0 = 1$, which is not equal to 0. If $x \neq 0$, then

$$b^x = 0 \quad \Rightarrow \quad (b^x)^{\frac{1}{x}} = 0^{\frac{1}{x}} \quad \Rightarrow \quad b = 0,$$

but by assumption, $b \neq 0$. We now have a contradiction, and thus there is no number x such that $b^x = 0$.

To prove part (c) we have to show that $f(x) = b^x$ is one-to-one. By the definition of one-to-one, this means we have to show that if $x \neq y$, then $b^x \neq b^y$. This is equivalent to proving the contrapositive of this statement:

$$\text{If } b^x = b^y, \text{ then } x = y.$$

Before proving this statement, notice that if $b > 0$ and $b \neq 1$, then $b^x = 1$ only when $x = 0$, for the following reason. If $b^x = 1$ but x were some number *other* than zero, then we could write $(b^x)^{\frac{1}{x}} = 1^{\frac{1}{x}}$, which simplifies to $b = 1$. However, we are assuming that $b \neq 1$. We use the fact that $b^x = 1$ only when $x = 0$ in our argument.

$$
\begin{aligned}
b^x = b^y \quad &\Rightarrow \quad \frac{b^x}{b^y} = 1 && \text{(note } b^y \neq 0) \\
&\Rightarrow \quad b^{x-y} = 1 && \text{(properties of exponents)} \\
&\Rightarrow \quad x - y = 0 && (b^z = 1 \text{ only if } z = 0) \\
&\Rightarrow \quad x = y.
\end{aligned}
$$

❓ *Question*   Theorem 8.2 concerns exponential functions of the form $f(x) = b^x$. What can you say about the domain and range of a general exponential function $f(x) = Ab^x$? Can $f(x)$ ever be zero? Is $f(x)$ one-to-one? Can you prove your answers?   □

Part (c) of Theorem 8.2 is particularly useful for solving equations that involve exponential expressions. For example, since the exponential function $f(x) = 2^x$ is a function and is one-to-one, we know that $2^x = 2^3$ if and only if $x = 3$ (the "if" comes from the fact that $f(x) = 2^x$ is a function; the "only if" comes from the fact that $f(x) = 2^x$ is one-to-one). In general, for all $b > 0$ with $b \neq 1$,

$$(3) \qquad\qquad b^x = b^y \text{ if and only if } x = y.$$

As we will see in the next subsection, many equations that involve exponential expressions can be solved by "making the bases equal" and then using Statement (3).

8.1.4 Equations involving exponential functions

Many equations that involve exponential expressions can be solved using the algebraic and functional properties of exponential functions. More complicated exponential functions must be solved using "logarithms," which we will introduce in the next section.

The equations in the following example can be solved by factoring and then using the fact that an exponential function $f(x) = b^x$ is never zero.

EXAMPLE 8.2

Factoring expressions that involve exponential functions

Solve the following equations.

(a) $x^2 2^x = x 2^{x+1}$

(b) $\dfrac{x 3^x - 27x}{x - 2} = 0$

Solution

(a) Remember that we can't just divide both sides of this equation by x (or we will lose the solution $x = 0$). Instead, we move everything to one side of the equation and factor:

$$x^2 2^x = x 2^{x+1}$$
$$x^2 2^x - x 2^{x+1} = 0$$
$$x 2^x (x - 2) = 0 \qquad \text{(because } 2^{x+1} = 2^x 2^1 = 2(2^x))$$

Since 2^x is never zero, the only solutions to this equation are $x = 0$ and $x = 2$.

(b) Recall that a quotient is equal to zero only when its numerator is zero but its denominator is not.

$$\frac{x 3^x - 27x}{x - 2} = 0$$
$$x 3^x - 27x = 0, \quad (x \neq 2)$$
$$x(3^x - 27) = 0, \quad (x \neq 2).$$

Clearly the left-hand side of this equation is zero when $x = 0$ or when $3^x - 27 = 0$. Solving, we get

$$3^x - 27 = 0$$
$$3^x = 27$$
$$3^x = 3^3 \qquad\qquad (27 = 3^3)$$
$$x = 3. \qquad \text{(since } 3^x \text{ is one-to-one)}$$

The solutions to the original equation are $x = 0$ and $x = 3$.   □

EXAMPLE 8.3

Writing all terms in an equation in terms of the same base

Solve the equation $3^x = 9^{x-2}$.

Solution We will try to write each side of the equation with the same base and then use the fact that exponential functions are one-to-one:

$$3^x = 9^{x-2}$$
$$3^x = (3^2)^{x-2} \qquad\qquad (9 = 3^2)$$
$$3^x = 3^{2(x-2)} \qquad \text{(since } (a^b)^c = a^{bc})$$
$$x = 2(x - 2) \qquad\quad (3^x \text{ is one-to-one})$$
$$x = 2x - 4$$
$$4 = x. \qquad\qquad\qquad\qquad\qquad\qquad □$$

✅ *Checking the Answer* To check the answer, simply check to see whether $x = 4$ satisfies the original equation. It does, since $3^4 = 81$ and $9^{4-2} = 9^2 = 81$. Of course, this won't tell you if you missed any solutions, but it does tell you that $x = 4$ is a valid solution. ☐

Sometimes we can apply the factoring techniques we have for quadratics to solve equations that involve exponential functions. In particular, an equation of the form

$$A(b^x)^2 + B(b^x) + C = 0$$

is a quadratic equation where the role of x is played by b^x. We say that this equation is quadratic "in b^x." (Keep in mind that $(b^x)^2$ can also be written as b^{2x} or $(b^2)^x$.) If we are lucky, we can factor such equations by inspection.

EXAMPLE 8.4	**An equation that is quadratic in 3^x**

Solve the equation $9(9^x) = 10(3^x) - 1$.

Solution Once we put everything on one side of the equation and apply a little algebra, we'll see that this equation has a quadratic form:

$$9(9^x) = 10(3^x) - 1$$
$$9(9^x) - 10(3^x) + 1 = 0$$
$$9(3^2)^x - 10(3^x) + 1 = 0$$
$$9(3^x)^2 - 10(3^x) + 1 = 0.$$

This equation is quadratic "in 3^x." We can factor it as follows:

$$(9(3^x) - 1)(3^x - 1) = 0.$$

(If you don't see how we factored that, try to factor $9z^2 - 10z + 1$ and compare it to the answer above.) To find the solutions, we set each factor equal to zero and solve:

$$9(3^x) - 1 = 0 \qquad\qquad 3^x - 1 = 0$$
$$3^x = \frac{1}{9} \qquad\qquad 3^x = 1$$
$$x = -2 \qquad\qquad x = 0.$$

We know that $3^x = \frac{1}{9}$ if and only if $x = -2$ (because $3^{-2} = \frac{1}{3^2} = \frac{1}{9}$ and 3^x is one-to-one). On the other hand, $3^x = 1$ if and only if $x = 0$. Therefore, the solutions are $x = -2$ and $x = 0$. ☐

We can solve equations like $3^x = \frac{1}{9}$ or $3^x = 1$ by "guessing" solutions, since the solutions are integers that we can easily spot (and we know that there is only one possible solution since exponential functions are one-to-one). However, consider an equation like $3^x = 4$. Although there is some x that is a solution to this equation (since $y = 4$ is in the range $(0, \infty)$ of the exponential function $y = 3^x$), we can't guess what it is. Using a calculator, we can approximate the solution to $3^x = 4$ as $x \approx 1.2618595$; surely you could not have guessed that! The exact solution requires a new kind of notation that we don't know yet. In Section 8.2, we will learn how to write down the exact solutions to equations like $3^x = 4$ using "logarithms."

What you should know after reading Section 8.1

▶ The definition of an exponential function. In particular, which bases are allowed in an exponential function, and why? How does an exponential function differ from a power function? Basic properties of exponential functions, including the domain and range of an exponential function and the fact that an exponential function is one-to-one.

▶ The algebraic rules for manipulating exponential functions, and how to use these rules to simplify expressions and, if possible, write a function in the form of an exponential function.

▶ How to solve equations involving exponential functions by using the fact that exponential functions are one-to-one and never zero.

Exercises 8.1

Concepts

0. Read the section and make your own summary of the material. Include whatever you think will help you review the material later on.

1. Define what it means for a function to be an exponential function. Be sure to specify which values are allowed to be in the "base."

2. What is the difference between an exponential function and a power function? Give an example of each.

3. Is the function $f(x) = x^x$ an exponential function? A power function? Explain why or why not.

4. Explain why we require that $b > 0$ and $b \neq 1$ for $f(x) = Ab^x$ to be considered an exponential function.

5. In Chapter 4 we defined what we mean by the expression a^b when b is a rational number. Now we need to have a definition for a^b that works when b is an *irrational* number (why?). In this problem we investigate two different ways to think about what 2^π means.

   (a) One way to define 2^π is to think of it as a limit. If we take a sequence $a_1, a_2, a_3, \ldots$ of rational numbers that approach π, then the sequence $2^{a_1}, 2^{a_2}, 2^{a_3}, \ldots$ should approach 2^π. Said in terms of limits, this means that
   $$2^\pi = \lim_{a \to \pi} 2^a,$$
   where each a is assumed to be a rational number. Can you think of a sequence of rational numbers that get closer and closer to π? (*Hint:* Think about the decimal expansion of π.)

   (b) Another way to consider 2^π is to write it as an infinite product:
   $$2^\pi = 2^3 \, 2^{\frac{1}{10}} \, 2^{\frac{4}{100}} \, 2^{\frac{1}{1000}} \, 2^{\frac{5}{10,000}} \, 2^{\frac{9}{100,000}} \cdots .$$

What will the next term in the product be? How could 2^π equal the product of infinitely many numbers? Wouldn't that make 2^π infinitely large? Calculate some of the later terms in the product (for example, $2^{\frac{5}{10,000}}$ or $2^{\frac{9}{100,000}}$) and use these calculations to argue that even though 2^π can be written as a product of infinitely many numbers, it is not necessarily infinitely large.

6. Approximate $2^{\sqrt{3}}$ by calculating 2^r for rational values r that get closer and closer to $\sqrt{3}$. (*Hint:* You can use the decimal expansion of $\sqrt{3}$ to get a sequence of rational numbers that approach $\sqrt{3}$.)

7. The functions $f(x) = 2^{(x^2)}$ and $g(x) = (2^x)^2$ look similar but are very different functions. (This is why we try to avoid using the ambiguous notation 2^{x^2}.) Calculate $f(3)$ and $g(3)$ and show that they are not the same (and thus that $f(x)$ and $g(x)$ are not the same function). Then find all the values for which $f(x) = g(x)$.

8. Every exponential function of the form $f(x) = b^x$ (with $b > 0$ and $b \neq 1$) is one-to-one. Explain why this fact implies that $b^x = b^y$ if and only if $x = y$.

9. Explain why the solutions to the equation $4^{2x} + 5(4^x) + 3 = 0$ are the same as the solutions to
   $$4^x = \frac{-5 + \sqrt{13}}{2} \quad \text{and} \quad 4^x = \frac{-5 - \sqrt{13}}{2}.$$
   (Don't try to solve for the solution set; just explain why the solutions to the original equation are the numbers that satisfy one of the two equations displayed above.)

■ **Determine whether each of the following statements is true or false. If a statement is true, explain why. If a statement is false, provide a counterexample.**

10. True or False: $3(2^x)$ is always equal to 6^x.

11. True or False: The range of any exponential function of the form $f(x) = b^x$ is $(0, \infty)$.

12. True or False: The range of any general exponential function $f(x) = Ab^x$ is $(0, \infty)$.

13. True or False: If $b > 0$ and $b \neq 1$, and if x and y are any real numbers, then $b^x = b^y$ if and only if $x = y$.

Skills

■ Determine which of the following are exponential functions. Write each exponential function in the form $f(x) = Ab^x$ for some real numbers A and b (with $b > 0$ and $b \neq 1$).

14. $f(x) = \dfrac{3}{2^x}$ 15. $f(x) = \dfrac{3^x}{2^x}$

16. $f(x) = (2 + x)^x$ 17. $f(x) = 3(5^x) + 1$

18. $f(x) = 5^x 2^{3-x}$ 19. $f(x) = 3(-2)^x$

20. $f(x) = 3^{2x+1}$ 21. $f(x) = 3^{x^2+1}$

22. $f(x) = 4(3^x)^2$

■ Find the domain of each of the following functions without using a calculator. Then use a calculator to graph each function and check your answers.

23. $f(x) = \dfrac{1}{3}(3^{-2x+1})$ 24. $f(x) = \dfrac{x-1}{2(4^{3x})}$

25. $f(x) = \dfrac{1}{3^x - 1}$ 26. $f(x) = \sqrt{\left(\tfrac{1}{2}\right)^x}$

27. $f(x) = \dfrac{1}{9^x - 2(3^x) + 1}$ 28. $f(x) = \dfrac{\sqrt{x+1}}{3^x - 2^x}$

■ Solve the following equations by hand. When you are finished, check your answers either by testing your solutions or by graphing an appropriate function.

29. $2^x = \dfrac{1}{8}$

30. $4^{x-3} = 2^x$

31. $10^x = 0.001$

32. $x3^{x+1} = x^2 3^x$

33. $\dfrac{2^x + 1}{3^x - 5} = 0$

34. $3^x \left(\dfrac{1}{2^x} - 4 \right) = 0$

35. $4(2^{2x}) = 1 - 5(2^x)$

36. $25^x - 26(5^x) + 25 = 0$

37. $\dfrac{1}{4^x} - \dfrac{5}{2^x} + 4 = 0$

Proofs

38. Prove that every exponential function $f(x) = Ab^x$ has the property that $f(x)$ is never zero. (*Hint:* Mimic the appropriate part of the proof of Theorem 8.2.)

39. Prove that every exponential function $f(x) = Ab^x$ is one-to-one. (*Hint:* Mimic the appropriate part of the proof of Theorem 8.2.)

8.2 The Natural Exponential Function

We can use any positive number b (except for $b = 1$) for the base of an exponential function $f(x) = Ab^x$. For reasons that will become clear once we learn how to differentiate exponential functions, there is one base that is particularly "natural" to use.

8.2.1 The natural base for exponential functions

Interestingly, the most "natural" base to use for an exponential function isn't a simple number like $b = 2$ or $b = 3$. Instead, the most "natural" base for an exponential function is the irrational number known as e.

DEFINITION 8.2

The Number e

The number e is defined as the following limit:

$$e := \lim_{h \to 0} (1 + h)^{\frac{1}{h}}.$$

Technical Note: It is not obvious that this limit even exists, but proving that it does exist is beyond the scope of this course. (The limit is of the form "1^∞," which is indeterminate.)

Assuming that the limit exists, and that we use "e" to denote the number it is equal to, it is not clear why the number e is irrational rather than rational; this is another fact that we do not have the time or tools to prove at this point.

We can approximate the limit in Definition 8.2 (and thus the value of e) by using a table of approximate values of $(1 + h)^{\frac{1}{h}}$ for small values of h. See Table 8.1.

h	-0.1	-0.01	-0.001	0	0.001	0.01	0.1
$(1 + h)^{\frac{1}{h}}$	2.867972	2.731999	2.719642	-	2.716924	2.704814	2.593742

Table 8.1

Although $(1+h)^{\frac{1}{h}}$ does not seem to approach a very "nice" number as $h \to 0$, it appears that the limit does exist and that it is somewhere between 2.719642 and 2.716924. If we evaluate $(1 + h)^{\frac{1}{h}}$ at an extremely small value of h, say $h = 0.000001$, we can get a relatively accurate approximation for the limit in Definition 8.2 and thus for the number e:

$$e \approx (1 + 0.000001)^{\frac{1}{0.000001}} \approx 2.718280.$$

As it turns out, e is an *irrational* number, which means that its decimal expansion will never terminate or repeat. This means that no decimal expansion will ever tell you what e "really is"; the exact value of e is defined by limit in Definition 8.2, but we can't write e down in decimal notation without approximating. The first 74 decimal places of e are

2.71828182845904523536028747135266249775724709369995957496696762772407663035...

As a decimal approximation for e, we commonly use 2.71828 or even just 2.718.

The following theorem gives us an alternative characterization of the number e.

THEOREM 8.3

An Alternative Way of Defining e

The number "e" is the unique number such that

$$\lim_{h \to 0} \frac{e^h - 1}{h} = 1.$$

This is equivalent to the definition of e given in Definition 8.2.

Theorem 8.3 is equivalent to Definition 8.2; in fact, some books use Theorem 8.3 as the definition of the number e. We won't prove rigorously that Theorem 8.3 and Definition 8.2 define the same number, but we can use approximations to argue the fact:

PROOF (USING APPROXIMATIONS TO SUPPORT THEOREM 8.3)

Suppose $e = \lim_{h \to 0}(1 + h)^{\frac{1}{h}}$, as in Definition 8.2. Then for sufficiently small values of h, we have

$$e \approx (1 + h)^{\frac{1}{h}} \iff e^h \approx 1 + h \iff e^h - 1 \approx h \iff \frac{e^h - 1}{h} \approx 1.$$

This approximation gets better as $h \to 0$, so we have $\lim_{h \to 0} \frac{e^h - 1}{h} = 1$, as in Theorem 8.3.

When we use e as the base of an exponential function, we have what is known as the "natural" exponential function.

The Natural Exponential Function

The exponential function with base e and coefficient 1 is called the ***natural exponential function*** and is denoted

$$e^x.$$

The function $f(x) = e^x$ has many interesting properties, as we will see in upcoming sections. We will also be interested in functions of the form

$$f(x) = e^{kx}, \text{ where } k \text{ is a nonzero real number.}$$

Figures 8.4–8.6 show the graphs of the functions $f(x) = e^x$, $f(x) = e^{2x}$, and $f(x) = e^{-2x}$. We will investigate graphs of exponential functions of the form e^{kx} in Section 8.5.

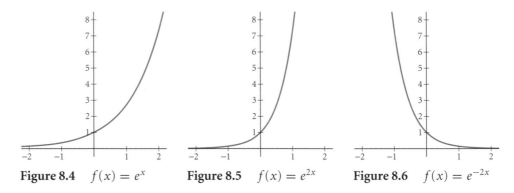

Figure 8.4 $f(x) = e^x$ **Figure 8.5** $f(x) = e^{2x}$ **Figure 8.6** $f(x) = e^{-2x}$

In fact, it turns out that every exponential function can be written in the form $f(x) = Ae^{kx}$ for some nonzero number k:

Two Ways of Expressing Exponential Functions

Every exponential function $f(x) = Ab^x$ can also be written in the form $f(x) = Ae^{kx}$. In other words, for all real numbers $b > 0$ with $b \neq 1$, there is some real number k such that

$$b^x = e^{kx}.$$

To prove this theorem, we need to define the "inverse" of the function $f(x) = e^x$. We will now define this inverse and then use it to prove Theorem 8.4.

8.2.2 The natural logarithmic function

We will study logarithmic functions in detail in Chapter 9, but we introduce the natural logarithmic function now so that we can prove Theorem 8.4.

Since $f(x) = e^x$ is an exponential function (where the base b is equal to the number e), it is one-to-one. Thus $f(x) = e^x$ is invertible, that is, has a well-defined inverse function $f^{-1}(x)$. Since the natural exponential function $f(x) = e^x$ will be so important to us, its inverse deserves a special name. We call it the "natural logarithmic function" and use the notation "ln x." (Despite how it is spelled, many mathematicians pronounce ln x as "log" x.)

DEFINITION 8.4

The Natural Logarithmic Function

The inverse of the function e^x is called the **natural logarithmic function** and is denoted

$$\ln x.$$

We don't have a way to calculate most values of ln x by hand. However, we can use a calculator to approximate values of ln x as necessary (your calculator should have an "ln" key).

◈ *Caution* Don't forget that the symbol "ln" represents a *function,* and not a quantity. This means that the expression ln x represents "the natural logarithm of x," and *not* "ln times x." It also means that you can't just write "ln" by itself (without an input "x") and that you can't just cancel "ln" symbols as though they were quantities. □

Using the fact that the natural exponential function e^x has domain $(-\infty, \infty)$ and range $(0, \infty)$, and the fact that the natural logarithmic function ln x is the inverse of e^x, we obtain the following theorem:

THEOREM 8.5

Domain and Range of ln x

The natural logarithmic function $f(x) = \ln x$ has domain $(0, \infty)$ and range $(-\infty, \infty)$.

We can obtain the graph of $y = \ln x$ by reflecting the graph of $y = e^x$ over the line $y = x$ (see Figures 8.7 and 8.8). Notice that since $y = e^x$ has a horizontal asymptote at $y = 0$, the graph of its inverse $y = \ln x$ has a vertical asymptote at $x = 0$.

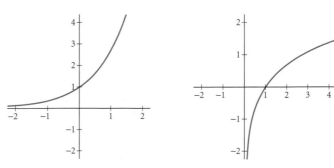

Figure 8.7 $y = e^x$ **Figure 8.8** $y = \ln x$

Since the natural logarithmic function is the inverse of the natural exponential function, we can use properties of inverse functions to write down some relationships between ln x and e^x.

THEOREM 8.6

Relationships Between ln x and e^x

Since the natural exponential function and the natural logarithmic function are inverses of each other,

(a) For all x and all $y > 0$, $y = e^x$ if and only if $\ln y = x$.

(b) If $x > 0$, then $e^{\ln x} = x$.

(c) For all x, $\ln(e^x) = x$.

The proof of this theorem follows directly from the properties of inverse functions.

PROOF (**THEOREM 8.6**) If $f(x) = e^x$, then by definition, $f^{-1}(x) = \ln x$. Since $y = f(x)$ if and only if $f^{-1}(y) = x$, we know that $y = e^x$ if and only if $\ln y = x$. Moreover, since $f(f^{-1}(x)) = x$ for all x in the domain of $f^{-1}(x)$, we know that $e^{\ln x} = x$ for all $x > 0$. Finally, since $f^{-1}(f(x)) = x$ for all x in the domain of $f(x)$, we know that $\ln(e^x) = x$ for all x. ∎

Given that the natural logarithmic function and the natural exponential function are inverses, we can calculate two simple values of $\ln x$. First, since $e^0 = 1$, we know from part (a) of Theorem 8.6 that $\ln 1 = 0$. Second, since $e^1 = e$, we know that $\ln e = 1$. We'll put those identities in a box since they are important to know from memory:

$$(1) \qquad\qquad \ln 1 = 0 \quad \text{and} \quad \ln e = 1.$$

Theorem 8.6 basically says that the natural logarithmic function "undoes" the natural exponential function (and vice versa). This is an important tool to use in simplifying expressions involving exponential and logarithmic functions.

EXAMPLE 8.5

Using Theorem 8.6 to simplify expressions

Use the fact that $y = e^x$ and $y = \ln x$ are inverses to simplify the following expressions.

(a) $e^{\ln 2}$ (b) $e^{3\ln x}$ (c) $\ln(e^{3x})$

Solution

(a) Since $e^{\ln x} = x$ for all $x > 0$, we know that $e^{\ln 2} = 2$.

(b) We know that $e^{\ln x} = x$ for all $x > 0$, but the expression $e^{3\ln x}$ is not quite in this form. We must first do some algebra to take care of the extra "3" in the expression:

$$e^{3\ln x} = (e^{\ln x})^3 \qquad \text{(since } b^{xy} = (b^x)^y = (b^y)^x\text{)}$$
$$= x^3. \qquad \text{(by part (b) of Theorem 8.6)}$$

(c) Since $\ln(e^x) = x$ for any value of x, we know that $\ln(e^{3x}) = 3x$ for all values of x. ☐

✎ *Caution* In part (b) of Example 8.5, we cannot "cancel" the e and the ln before using algebra to put the expression in the form $e^{\ln x}$. In other words,

$$e^{3\ln x} \neq 3x.$$

The same is true for expressions like $\ln(3e^x)$; we will see how to handle such expressions in Chapter 9 when we discuss the algebra of logarithmic functions. ☐

The fact that the natural exponential function and the natural logarithmic function are inverses of each other can be used to solve equations. In the following example we use this fact (and Theorem 8.6) to solve two very simple equations. In Section 8.2.4 we will learn more about the natural logarithmic function and its properties; then we will be able to solve more complicated equations.

EXAMPLE 8.6   **Using Theorem 8.6 to solve equations**

Use the fact that $y = e^x$ and $y = \ln x$ are inverses to solve the following equations for x. Your answers will involve ln and e; after finding exact solutions, use a calculator to approximate the numerical values of the solutions.

(a) $\ln x = 3$ (b) $e^{2x} = 4$

Solution

(a) By part (a) of Theorem 8.6, we know that $\ln a = b$ if and only if $e^b = a$. Therefore, $\ln x = 3$ if and only if $e^3 = x$. Using a calculator, we can approximate that $x = e^3 \approx 20.0855$.

(b) Since $\ln a = b$ if and only if $e^b = a$, we know that $e^{2x} = 4$ if and only if $\ln 4 = 2x$. The solution is $x = \frac{\ln 4}{2}$. Using a calculator, we can approximate $x = \frac{\ln 4}{2} \approx 0.693147$.

8.2.3 Two ways to express exponential functions

Now that we have defined the natural logarithmic function, we are in a position to prove Theorem 8.4 and thus show that every exponential function $f(x) = Ab^x$ can be written such that its base is the "natural" base e. Specifically, we will use the natural logarithmic function to show that if $b > 0$ and $b \neq 1$, then b^x is equal to e^{kx} for some real number k.

PROOF (**THEOREM 8.4**)   Suppose b is greater than zero and not equal to 1. We want to show that $b^x = e^{kx}$ (for all values of x) for some real number k. Since b is positive, it is in the domain of $\ln x$. Moreover, since $\ln x$ is the inverse of e^x, we know that

$$e^{\ln b} = b.$$

Therefore, for all x, we have

$$b^x = (e^{\ln b})^x = e^{(\ln b)x}.$$

In other words, if we define $k := \ln b$, then $b^x = e^{kx}$.   ▪

Now that we know that every exponential function $f(x) = b^x$ can be written with the "natural" base e as $f(x) = e^{kx}$, we will work almost entirely with exponential functions in this "natural" form. As you will see in the next few sections, the natural exponential form $f(x) = e^{kx}$ is usually much easier to work with than the form $f(x) = b^x$.

Of course we can also go in the other direction; that is, we can write any exponential function of the form $f(x) = e^{kx}$ in the form $f(x) = b^x$ for some $b > 0$, $b \neq 1$. This direction is easy, since by properties of exponents, we have

$$e^{kx} = (e^k)^x.$$

Therefore, if we set $b := e^k$, then $e^{kx} = b^x$. We summarize the method for translating between the two different forms for exponential functions as follows:

$$(2) \qquad b^x = e^{(\ln b)x} \quad \text{and} \quad e^{kx} = (e^k)^x.$$

In other words

$$(3) \qquad \text{If } b^x = e^{kx}, \text{ then } k = \ln b \text{ and } b = e^k.$$

EXAMPLE 8.7   **Converting from b^x form to e^{kx} form**

Write the function $f(x) = 2^x$ in the form $f(x) = e^{kx}$.

Solution   We need to find a number k such that $2^x = e^{kx}$. By Statement (3), we know that we must have $k = \ln 2 \approx 0.693147$:

$$2^x = (e^{\ln 2})^x = e^{(\ln 2)x} \approx e^{0.693147x}. \qquad \square$$

EXAMPLE 8.8   **Converting from e^{kx} form to b^x form**

Write the function $f(x) = e^{2x}$ in the form $f(x) = b^x$.

Solution   We need to find a number b such that $e^{2x} = b^x$. By Statement (3), we know that we must have $b = e^2 \approx 7.38906$:

$$e^{2x} = (e^2)^x \approx (7.38906)^x. \qquad \square$$

Since every exponential function of the form $f(x) = Ab^x$ can be written in the form $f(x) = Ae^{kx}$ for some nonzero number k (and vice versa), we have the following alternative definition of an exponential function:

DEFINITION 8.5

Another Form for Exponential Functions

An *exponential function* is a function that can be written in the form

$$f(x) = Ae^{kx}$$

for some real number A and some nonzero real number k.

It is useful to know what types of values of b correspond to what types of values of k if $b^x = e^{kx}$ (although we will not prove this fact right now). The following theorem explains the two possible cases.

THEOREM 8.7

Exponential Growth and Exponential Decay

If $e^{kx} = b^x$, then $k > 0$ if and only if $b > 1$, and $k < 0$ if and only if $0 < b < 1$.

Theorem 8.7 splits exponential functions into two types, one that represents "growth" and one that represents "decay" (as we will see in Section 8.6). Figures 8.9 and 8.10 illustrate these two types of exponential functions.

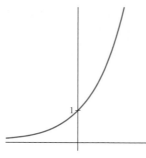
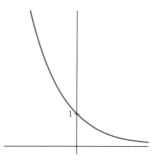

Figure 8.9
$f(x) = e^{kx}$ with $k > 0$
$f(x) = b^x$ with $b > 1$

Figure 8.10
$f(x) = e^{kx}$ with $k < 0$
$f(x) = b^x$ with $0 < b < 1$

In Section 8.5 we will prove that all exponential functions have one of the graphs shown in Figures 8.9 and 8.10.

8.2.4 Using the natural logarithm to solve equations

We will learn about all of the algebraic properties of logarithmic functions in Chapter 9. For now, we concentrate on one important algebraic property of logarithmic functions that will enable us to solve equations involving exponential functions.

THEOREM 8.8

Logarithms Turn Powers into Constant Multiples
For any real number x and any positive real number b,

$$\ln(b^x) = x \ln b.$$

The proof of this theorem follows directly from the fact that the natural logarithmic function is the inverse of the natural exponential function.

PROOF (THEOREM 8.8) Given any real number x and any positive real number b,

$$
\begin{aligned}
\ln(b^x) &= \ln((e^{\ln b})^x) && \text{(since } b = e^{\ln b}) \\
&= \ln(e^{(\ln b)x}) && \text{(algebra of exponential functions)} \\
&= (\ln b)x && \text{(property of inverse functions)} \\
&= x \ln b.
\end{aligned}
$$

What makes Theorem 8.8 so useful is that it basically says that the natural logarithmic function has the power to turn exponentiation into multiplication. This is a very powerful tool for solving equations involving exponential expressions.

EXAMPLE 8.9 **Using logarithms to solve an equation**

Use Theorem 8.8 to solve the equation

$$3.25(1.72)^x = 1000.$$

After finding exact expressions for the solutions, use a calculator to approximate their numerical values.

Solution To solve for x, we will isolate the expression $(1.72)^x$ and then take the natural logarithm of both sides of the equation. (Since the natural logarithmic function is one-to-one, the solution set of an equation is not changed by taking the natural logarithm of both sides.) This will "get the x out of the exponent," as follows:

$$3.25(1.72)^x = 1000$$

$$(1.72)^x = \frac{1000}{3.25}$$

$$\ln((1.72)^x) = \ln\left(\frac{1000}{3.25}\right) \qquad \text{(take the natural logarithm of both sides)}$$

$$x\ln(1.72) = \ln\left(\frac{1000}{3.25}\right) \qquad \text{(Theorem 8.8)}$$

$$x = \frac{\ln\left(\frac{1000}{3.25}\right)}{\ln(1.72)} \qquad \text{(exact answer)}$$

$$x \approx 10.564. \qquad \text{(approximate answer)}$$

Notice that we did not use any calculator approximations until the last step. This way we can obtain an exact answer before finding the approximate answer (and there is less rounding error because we are approximating only once). □

❖ *Caution* If we were to take the natural logarithm of both sides of the equation in the first step, then the constant multiple 3.25 would be *inside* the logarithm. Then we would have to use some more algebra (which we won't see until Chapter 9) before we would be able to use Theorem 8.8. It is a good rule always to get one side in the form b^x (as opposed to, say, Ab^x or $b^x + c$) *before* taking the natural logarithm of both sides. Otherwise, Theorem 8.8 will not apply! □

We can also apply the natural exponential function to both sides of an equation to "cancel" the natural logarithmic function, as in the following example.

EXAMPLE 8.10

Using exponential functions to solve an equation

Find the exact solutions of the equation

$$2\ln(3x^2 - 1) = 4,$$

and then use a calculator to obtain numerical approximations of these solutions.

Solution Using the fact that $y = e^x$ is the inverse of $y = \ln x$, we have

$$2\ln(3x^2 - 1) = 4$$

$$\ln(3x^2 - 1) = 2 \qquad \text{(isolate the logarithmic function)}$$

$$e^{\ln(3x^2-1)} = e^2 \qquad \text{(apply the natural exponential function to both sides)}$$

$$3x^2 - 1 = e^2 \qquad \text{(since } e^{\ln A} = A\text{)}$$

$$3x^2 = e^2 + 1$$

$$x^2 = \frac{e^2 + 1}{3}$$

$$x = \pm\sqrt{\frac{e^2 + 1}{3}} \qquad \text{(using algebra to get the exact answer)}$$

$$x \approx \pm 1.67223. \qquad \text{(approximate answer)}$$

Notice that before applying the natural exponential function to both sides, we divided both sides of the equation by 2 to isolate the natural logarithmic function on the left-hand side. If we hadn't done this preliminary algebra step, we would have been forced to do more complicated algebra later on (try it and see!).   □

What you should know after reading Section 8.2

▶ The definition of e as a limit, and the equivalent characterization of e given in Theorem 8.3. Why are these characterizations of e equivalent? How could you approximate the value of e using a table?

▶ The definition of the natural exponential function, and its domain, range, and invertibility properties.

▶ The definition of the natural logarithmic function as the inverse of the natural exponential function. What properties can you determine about the natural logarithmic function using the fact that it is the inverse of the natural exponential function? What are the values of $\ln 1$ and $\ln e$ and why?

▶ How to write any exponential function of the form b^x in the form e^{kx} (and vice versa). How can you *prove* that every exponential function can be expressed in these two ways? What can you say about b if $k > 0$? If $k < 0$?

▶ How to use the natural logarithmic function to solve equations involving exponential functions (both exactly and approximately). In particular, know that $\ln(b^x) = x \ln b$ and be able to prove this fact.

Exercises 8.2

Concepts

0. Read the section and make your own summary of the material. Include whatever you think will help you review the material later on.

1. What is the definition of the number e? Use the definition to approximate e to five decimal places.

2. What is the "alternative" definition of the number e? Use this alternative definition to approximate e to six decimal places.

3. Argue that the characterizations of the number e in Definition 8.2 and Theorem 8.3 are equivalent by showing that $e \approx (1 + h)^{\frac{1}{h}}$ for small h if and only if $\frac{e^h - 1}{h} \approx 1$ for small h.

4. What is the "natural exponential function"? What are its domain and range? Is it invertible?

5. What is the definition of the "natural logarithmic function"? What are its domain and range? Is it invertible?

6. Sketch graphs of $y = e^x$ and $y = \ln x$ on the same set of axes. Label two points on each graph.

7. Fill in the blanks in each statement below.
   (a) $y = e^x$ if and only if $x = $ _____, for all $x \in$ _____ and $y \in$ _____.
   (b) $e^{\ln x} = $ _____ for all $x \in$ _____.
   (c) $\ln(e^x) = $ _____ for all $x \in$ _____.

8. Suppose that $e^{kx} = b^x$ for some real numbers k and b. Given that fact, fill in the blanks below.
   (a) $k \in (0, \infty)$ if and only if $b \in$ _____.
   (b) $k \in (-\infty, 0)$ if and only if $b \in$ _____.

9. If $e^{kx} = b^x$ for all values of x, how must the constants k and b be related? Express your answer in two ways: with k in terms of b and with b in terms of k.

10. If $e^{kx} = b^x$ and $k > 0$, what can you say about b? If $e^{kx} = b^x$ and $k < 0$, what can you say about b?

■ **Determine whether each of the following statements is true or false. If a statement is true, explain why. If a statement is false, explain why or provide a counterexample.**

11. True or False: $\ln(9) = 2 \ln 3$

12. True or False: $(\ln 3)^2 = 2 \ln 3$

13. True or False: $e^{3 \ln x} = 3x$

14. True or False: $\ln(x^2 + 1) = 2 \ln x + 1$

Skills

15. Calculate the following values by hand (if defined), using the definitions of the natural exponential function and the natural logarithmic function.

(a) e^0 (b) e^1 (c) $\ln 0$ (d) $\ln 1$ (e) $\ln e$

■ Use algebra and the fact that $y = e^x$ and $y = \ln x$ are inverses to simplify the following expressions.

16. $e^{\ln 3}$ **17.** $e^{2 \ln 3}$ **18.** $\ln(e^2)$

19. $\ln(e^{2x})$ **20.** $\frac{1}{2} \ln(x^2)$ **21.** $e^{\ln 3 + \ln 2}$

22. $e^{2 \ln(x+1)}$ **23.** $e^{2 \ln x + 1}$

■ Find the domain of each function without using a calculator. (Use logarithms as necessary to get *exact* answers.) Then use a graphing calculator to check your answers.

24. $f(x) = -2e^{1-3x^2}$

25. $f(x) = e^{(2 - \frac{1}{x})}$

26. $f(x) = \dfrac{1}{\sqrt{e^x + 1}}$

27. $f(x) = \dfrac{1}{2e^{3x-1}(x-2)(x+1)}$

28. $f(x) = \dfrac{\sqrt{2x + 1}}{e^{3x} - 5}$

29. $f(x) = \dfrac{1}{3e^{2x} - e^x - 2}$

■ Use properties of the natural exponential and logarithmic functions (and the fact that they are inverses) to solve the following equations. Justify each step as appropriate. Give exact answers and then use a calculator to make numerical approximations of your answers.

30. $\ln x = 2$ **31.** $e^{3x} = 5$

32. $2e^{3x+1} = 4$ **33.** $2(3^x) = 4$

34. $3\left(\frac{1}{2}\right)^x = 2$ **35.** $1.76(0.24)^x = 500$

36. $\ln(x^2 + x - 5) = 0$ **37.** $4\ln(2x^2 - 3) = 7$

38. $e^{2 \ln x + 4} = 6$

■ Write each of the functions below in the form $f(x) = Ae^{kx}$ for some numbers A and k (you may approximate k).

39. $f(x) = 5^x$ **40.** $f(x) = 2(3^x)$

41. $f(x) = 1.2(0.72)^x$ **42.** $f(x) = -\frac{1}{3}(3^{x+2})$

■ Write each of the functions below in the form $f(x) = Ab^x$ for some numbers A and b (you may approximate b).

43. $f(x) = e^{3x}$ **44.** $f(x) = 1.2e^{-7.2x}$

45. $f(x) = 3e^{x+1}$ **46.** $f(x) = 2e^{3x-4}$

■ For each pair of values $f(a)$ and $f(b)$ given below, find an exponential function $f(x) = Ae^{kx}$ that has those given values. (*Note:* This will involve solving for the constants A and k; give both an exact and an approximate value for k.)

47. $f(0) = 9$, $f(2) = 6$

48. $f(-4) = 10$, $f(0) = 2$

49. $f(0) = -3$, $f(5) = -7$

50. $f(2) = 18$, $f(3) = 54$

51. $f(-2) = 8$, $f(3) = \frac{1}{4}$

52. $f(-1) = 1$, $f(1) = 3$

Applications

53. In January you borrowed $500 from LoanSharks, Incorporated. LoanSharks charges interest on your debt in such a way that after t months, you will owe $L(t) = 500(1.17)^t$ dollars. How much will you owe LoanSharks in April? How much will you owe them in November?

54. The population of fish in a stream t years after 1990 is given by the exponential function $P(t) = 750(1.08)^t$. In what year will the fish population have doubled in size?

55. Ten years ago, Jenny deposited $10,000 in an investment account. Her investment account now holds $2260.98. Her accountant tells her that her investment account balance $I(t)$ is an exponential function.

(a) Find an exponential function of the form $I(t) = Ae^{kt}$ to model Jenny's investment account balance.

(b) Find an exponential function of the form $I(t) = Ab^t$ to model Jenny's investment account balance.

Proofs

56. Use the fact that the natural logarithmic function is the inverse of the natural exponential function to prove that for all x and all $y > 0$, we have $y = e^x$ if and only if $\ln y = x$.

57. Use the fact that the natural logarithmic function is the inverse of the natural exponential function to prove that for all $x > 0$, we have $e^{\ln x} = x$.

58. Use the fact that the natural logarithmic function is the inverse of the natural exponential function to prove that for all values of x, we have $\ln(e^x) = x$.

59. Prove that every exponential function of the form $f(x) = Ab^x$ can be written in the form $f(x) = Ae^{kx}$ for some nonzero real number k.

60. Prove that every exponential function of the form $f(x) = Ae^{kx}$ can be written in the form $f(x) = Ab^x$ for some positive real number b with $b \neq 1$.

61. Use the fact that the natural logarithmic function is the inverse of the natural exponential function to prove that for all real numbers x and $b > 0$, we have $\ln(b^x) = x \ln b$.

8.3 Limits of Exponential Functions

In this section we investigate limits of exponential functions. We do so by first examining exponential functions of the form $f(x) = e^x$ and $f(x) = e^{kx}$ and then interpreting our results for exponential functions of the form $f(x) = b^x$.

8.3.1 Continuity of the natural exponential function

From their graphs, exponential functions certainly *appear* to be continuous. In the next few pages, we will *prove* that exponential functions are continuous. This will mean that the limit of an exponential function $f(x)$ as $x \to c$ can be found simply by evaluating $f(c)$. Recall that the definition of continuity is as follows:

(1) $\qquad\qquad f(x) \text{ is continuous at } x = c \quad \Longleftrightarrow \quad \lim_{x \to c} f(x) = f(c).$

We begin by showing that the natural exponential function $f(x) = e^x$ is continuous at $x = 0$; by Statement (1), and the fact that $e^0 = 1$, this is equivalent to the following theorem.

THEOREM 8.9

The Natural Exponential Function Is Continuous at One Point

The natural exponential function $f(x) = e^x$ is continuous at $x = 0$; in other words,

$$\lim_{x \to 0} e^x = e^0 = 1.$$

The proof of Theorem 8.9 uses the characterization of the number e given in Theorem 8.3 in Section 8.2.

PROOF (THEOREM 8.9) By Theorem 8.3 in Section 8.2, the number e is the unique number with the property that

(2) $\qquad\qquad\qquad\qquad \lim_{x \to 0} \dfrac{e^x - 1}{x} = 1.$

(*Note:* We have replaced the letter h that appears in Theorem 8.3 in Section 8.2 with the letter x.) The key is to manipulate $\lim_{x \to 0} e^x$ so that we can use the limit in Equation (2).

$$\lim_{x \to 0} e^x = \lim_{x \to 0} (e^x - 1 + 1) \qquad\qquad (-1 + 1 = 0)$$

$$= \lim_{x \to 0} \left(\frac{e^x - 1}{x} \cdot x + 1 \right) \qquad\qquad (\tfrac{x}{x} = 1)$$

$$= \left(\lim_{x \to 0} \frac{e^x - 1}{x} \right) \left(\lim_{x \to 0} x \right) + \left(\lim_{x \to 0} 1 \right) \qquad\qquad \textbf{(limit rules)}$$

$$= (1)(0) + (1) = 1. \qquad\qquad \textbf{(Equation (2), limit rules)}$$

Amazingly, the fact that the natural exponential function is continuous at *one point* $x = 0$ is all we need to know to prove that the natural exponential function is continuous *everywhere!* In other words, the proof of the following theorem is mostly an application of Theorem 8.9.

THEOREM 8.10

Continuity of the Natural Exponential Function

The natural exponential function is continuous everywhere; in other words, for all real numbers c, we have

$$\lim_{x \to c} e^x = e^c.$$

❓ Question   Usually when we prove a limit statement, we have to go back to the original "delta–epsilon" definition of a limit. What is the delta–epsilon expression of the limit in Theorem 8.10? Luckily, we won't have to prove Theorem 8.10 using this delta–epsilon definition; instead, we will use the fact that the natural exponential function is continuous at zero.   □

PROOF (THEOREM 8.10)   Suppose c is any real number. We will use the substitution $h = x - c$ to "change variables" in our limit. Note that if $h = x - c$, then $x = c + h$. Moreover, as $x \to c$, the new variable $h = x - c$ approaches 0. This change of variables will enable us to apply Theorem 8.9.

$$\lim_{x \to c} e^x = \lim_{h \to 0} e^{c+h} \qquad \text{(let } h = x - c, \text{ so } x = c + h)$$

$$= \lim_{h \to 0} e^c e^h \qquad \text{(algebra)}$$

$$= e^c \lim_{h \to 0} e^h \qquad (e^c \text{ is a constant})$$

$$= e^c(1) = e^c. \qquad \text{(Theorem 8.9)}$$

Using limit rules and the fact that the natural exponential function is continuous, we can now calculate limits that involve the natural exponential function.

EXAMPLE 8.11

A limit that involves the natural exponential function

Calculate $\lim_{x \to 3} \dfrac{3 - 2e^x}{1 + x}$.

Solution   Although we don't write out each step individually, the calculation involves the sum, constant multiple, and quotient rules for limits, as well as the fact that the natural exponential function is continuous.

$$\lim_{x \to 3} \frac{3 - 2e^x}{1 + x} = \frac{3 - 2e^3}{1 + 3} = \frac{3 - 2e^3}{4}.$$

The answer given is the *exact* value of the limit. We can use a calculator to give a numerical approximation of this answer; since $\frac{3-2e^3}{4} \approx -9.29277$, we have

$$\lim_{x \to 3} \frac{3 - 2e^x}{1 + x} \approx -9.29277.$$

□

8.3.2 Continuity of exponential functions

We have proved that the natural exponential function $f(x) = e^x$ is continuous. In fact, *every* exponential function $f(x) = Ae^{kx}$ is continuous.

THEOREM 8.11

Continuity of Exponential Functions

Every exponential function $f(x) = Ae^{kx}$ is continuous; that is, if A, k, and c are any real numbers with $k \neq 0$, then

$$\lim_{x \to c} Ae^{kx} = Ae^{kc}.$$

The proof of this theorem follows directly from the fact that the natural exponential function is continuous.

PROOF (**THEOREM 8.11**)   Given any real numbers A, k, and c, we have

$$\lim_{x \to c} Ae^{kx} = \lim_{x \to c} A(e^x)^k \qquad \text{(algebra)}$$

$$= A\left(\lim_{x \to c} e^x\right)^k \qquad \text{(limit rules)}$$

$$= A(e^c)^k \qquad \text{(since } e^x \text{ is continuous)}$$

$$= Ae^{kc}. \qquad \text{(algebra)}$$

We can use the fact that every exponential function $f(x) = Ae^{kx}$ is continuous (at every point in its domain) to calculate many limits that involve exponential functions.

EXAMPLE 8.12

A limit involving an exponential function of the form e^{kx}

We can now quickly calculate

$$\lim_{x \to -2} 3(e^{2x} - 1) = 3\left(e^{2(-2)} - 1\right) = 3(e^{-4} - 1) \approx -2.945.$$

Notice that we have given both an exact and an approximate answer.   □

Since every exponential function of the form $f(x) = Ab^x$ can be written in the form $f(x) = Ae^{kx}$, Theorem 8.11 is equivalent to the following theorem.

THEOREM 8.12

Continuity of Exponential Functions in the Form b^x

Every exponential function $f(x) = Ab^x$ is continuous; that is, if A and b are any real numbers (with $b > 0$ and $b \neq 1$), and c is any real number, then

$$\lim_{x \to c} Ab^x = Ab^c.$$

We can use Theorem 8.12 to calculate even more limits.

EXAMPLE 8.13

A limit involving an exponential function of the form b^x

We can now quickly calculate

$$\lim_{x \to 2} \frac{3^x - 1}{2^x} = \frac{3^2 - 1}{2^2} = \frac{9 - 1}{4} = \frac{8}{4} = 2.$$

□

8.3.3 Limits of exponential functions at infinity

So far we have learned how to calculate limits of exponential functions as $x \to c$ for some real number c. We now investigate limits of exponential function "at infinity," that is, as $x \to \infty$ and as $x \to -\infty$.

Consider first the natural exponential function $f(x) = e^x$. Looking at a graph of $f(x) = e^x$, we can easily "see" that the limit of $f(x) = e^x$ as $x \to \infty$ is infinity, while the limit of $f(x) = e^x$ as $x \to -\infty$ is zero (see Figure 8.11).

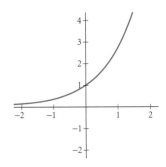

Figure 8.11 $f(x) = e^x$

Of course, guessing limits by looking at a graph does not always produce the right answers. How do we know that the graph of $f(x) = e^x$ doesn't do something funny outside of the graphing window? (It doesn't, but how can we be *sure* that it doesn't?) Luckily, we can prove that the natural exponential function has these limits at infinity. We begin by stating our guess as a theorem.

THEOREM 8.13

Limits at Infinity for the Natural Exponential Function

Let $f(x) = e^x$ be the natural exponential function. The limits of this function as $x \to \infty$ and as $x \to -\infty$ are

$$\lim_{x \to \infty} e^x = \infty, \qquad \lim_{x \to -\infty} e^x = 0.$$

PROOF (**THEOREM 8.13**)   Proving the first limit will take some work (in fact, we'll have to go back to the definition of a limit at infinity). Given the first limit, the second limit will be easy to prove.

To prove that $\lim_{x \to \infty} e^x = \infty$, we must show the following (see Section 2.1.5):

(3)   *For all $M > 0$, there exists an $N > 0$*
              such that if $x > N$, then $e^x > M$.

The proof of Statement (3) is like a "delta–epsilon" proof, except that the role of δ will be played by N and the role of ε will be played by M (since we are working with an infinite limit "at" infinity).

Given $M > 0$, choose $N = \underline{\ \ln M\ }$. (The reason for this choice will be clear soon.) Now if $x > N = \ln M$, then we know that $x = \ln M + c$ for some positive number c (why?). Therefore,

$$
\begin{aligned}
e^x &= e^{\ln M + c} && \text{(since } x = \ln M + c\text{)} \\
&= e^{\ln M} e^c && \text{(algebra)} \\
&= M(e^c) && \text{(since } e^{\ln M} = M\text{)} \\
&> M. && \text{(since } c > 0\text{, we know that } e^c > 1\text{)}
\end{aligned}
$$

The last step, where we claim that $e^c > 1$ if $c > 0$, actually requires a little bit of proof, but we will assume it here (look at the graph of $f(x) = e^x$ in Figure 8.11 to see why it makes sense graphically). The argument above proves that $\lim_{x \to \infty} e^x = \infty$.

Now that we know that $\lim_{x \to \infty} e^x = \infty$, we can use this fact to show that $\lim_{x \to -\infty} e^x = 0$, as follows:

$$\lim_{x \to -\infty} e^x = \lim_{u \to \infty} e^{-u} \qquad \text{(change variables to } u = -x \text{, so } x = -u)$$

$$= \lim_{u \to \infty} \frac{1}{e^u} \qquad \text{(algebra)}$$

$$\to \frac{1}{\infty} \to 0. \qquad \text{(since } \lim_{u \to \infty} e^u = \infty)$$

The limits at infinity of more general exponential functions are described in the following theorem.

THEOREM 8.14

Limits at Infinity for Exponential Functions of the Form e^{kx}

If $k > 0$, then

(a) $\lim_{x \to \infty} e^{kx} = \infty$ and $\lim_{x \to -\infty} e^{kx} = 0$.

(b) $\lim_{x \to \infty} e^{-kx} = 0$ and $\lim_{x \to -\infty} e^{-kx} = \infty$.

Of course, since exponential functions can also be expressed in the form $f(x) = Ab^x$, we also need to state a version of Theorem 8.14 for limits of expressions of the form b^x:

THEOREM 8.15

Limits at Infinity for Exponential Functions of the Form b^x

Given $b > 0$ and $b \neq 1$,

(a) if $b > 1$, then $\lim_{x \to \infty} b^x = \infty$ and $\lim_{x \to -\infty} b^x = 0$.

(b) if $0 < b < 1$, then $\lim_{x \to \infty} b^x = 0$ and $\lim_{x \to -\infty} b^x = \infty$.

We leave the proofs of Theorems 8.14 and 8.15 as exercises. The proof of Theorem 8.14 is an application of Theorem 8.13 (using the fact that $e^{kx} = (e^x)^k$). The proof of Theorem 8.15 is a rewording of Theorem 8.14 using the fact that if $e^{kx} = b^x$, then either $k > 0$ and $b > 1$, or $k < 0$ and $0 < b < 1$.

⊛ *Caution*   Do not simply memorize Theorems 8.14 and 8.15. You can recover the information in those theorems by remembering the graphs of exponential functions for $k > 0$ (or $b > 1$) and $k < 0$ (or $0 < b < 1$). See Figures 8.9 and 8.10 on page 545. □

8.3.4 Calculating limits involving exponential functions

We can now calculate many different types of limits that involve exponential functions.

EXAMPLE 8.14

Factoring to find a limit that involves exponential functions

Calculate $\lim\limits_{x \to 0} \dfrac{3^{2x} + 3^x - 2}{1 - 3^x}$.

Solution We can *try* to calculate this limit by evaluating $\frac{3^{2x}+3^x-2}{1-3^x}$ at $x = 0$, but this process does not always work. In fact, if we try to evaluate $\frac{3^{2(0)}+3^0-2}{1-3^0} = \frac{1+1-2}{1-1}$, we get the indeterminate form $\frac{0}{0}$. More algebra is required before we can attempt to evaluate this limit. Fortunately, the numerator of the function is quadratic in 3^x and can be factored:

$$\lim_{x \to 0} \frac{3^{2x} + 3^x - 2}{1 - 3^x} = \lim_{x \to 0} \frac{(3^x - 1)(3^x + 2)}{1 - 3^x} \qquad \text{(factor the numerator)}$$

$$= \lim_{x \to 0} \frac{-(1 - 3^x)(3^x + 2)}{1 - 3^x} \qquad \text{(rewrite numerator)}$$

$$= \lim_{x \to 0} (-(3^x + 2)) \qquad \text{(cancel common factor)}$$

$$= -(3^0 + 2) \qquad \text{(continuity of } 3^x \text{, limit rules)}$$

$$= -(1 + 2) = -3. \qquad \square$$

EXAMPLE 8.15

A function with two different horizontal asymptotes

Find any horizontal asymptotes of the function $f(x) = \dfrac{3}{2 + e^{-4x}}$.

Solution To see whether $f(x)$ has any horizontal asymptotes, we must examine the limit of $f(x)$ as $x \to \infty$ and as $x \to -\infty$.

$$\lim_{x \to \infty} f(x) = \lim_{x \to \infty} \frac{3}{2 + e^{-4x}}$$

$$= \lim_{x \to \infty} \frac{3}{2 + \frac{1}{e^{4x}}} \qquad \text{(algebra)}$$

$$\to \frac{3}{2 + \frac{1}{\infty}} \to \frac{3}{2 + 0} = \frac{3}{2}. \qquad (e^{4x} \to \infty \text{ as } x \to \infty)$$

$$\lim_{x \to -\infty} f(x) = \lim_{x \to -\infty} \frac{3}{2 + e^{-4x}}$$

$$\to \frac{3}{2 + \infty} \qquad (e^{-4x} \to \infty \text{ as } x \to -\infty)$$

$$\to \frac{3}{\infty} \to 0.$$

The function $f(x) = \dfrac{3}{2 + e^{-4x}}$ has *two* horizontal asymptotes: one at $y = 0$ (to the left, as $x \to -\infty$), and one at $y = \frac{3}{2}$ (to the right, as $x \to \infty$). $\square$

☑ *Checking the Answer* We can check the answers by using a graph of the function $f(x) = \dfrac{3}{2 + e^{-4x}}$ to spot any horizontal asymptotes (see Figure 8.12).

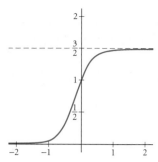

Figure 8.12

$$f(x) = \frac{3}{2 + e^{-4x}}$$

Note that the graph does have horizontal asymptotes at $y = 0$ and at $y = \frac{3}{2}$. ☐

When we studied limits of rational functions (quotients of polynomials), we used a technique of "dividing numerator and denominator by the highest power of x" that enabled us to solve infinite limits of rational functions quickly. This same technique can be modified to apply to infinite limits of quotients that involve exponential functions. The following example illustrates this technique.

EXAMPLE 8.16

Finding a limit by dividing by the "largest" exponential function

Calculate $\displaystyle\lim_{x \to \infty} \frac{3^x - 2^x}{1 + 2(3^x)}$.

Solution   At first inspection we see that

$$\lim_{x \to \infty} \frac{3^x - 2^x}{1 + 2(3^x)} \to \frac{\infty - \infty}{\infty}.$$

We have no idea what the limit might be, because "$\infty - \infty$" is indeterminate (in some cases a limit of this form exists, and in some cases it does not).

Instead of dividing numerator and denominator by the "highest power of x," we will divide by the exponential function b^x with the largest base b. In this example, we divide by 3^x.

$$\lim_{x \to \infty} \frac{3^x - 2^x}{1 + 2(3^x)} = \lim_{x \to \infty} \frac{3^x - 2^x}{1 + 2(3^x)} \cdot \frac{1/3^x}{1/3^x} \qquad (\tfrac{1/3^x}{1/3^x} = 1)$$

$$= \lim_{x \to \infty} \frac{3^x/3^x - 2^x/3^x}{1/3^x + 2(3^x)/3^x} \qquad \text{(distribute)}$$

$$= \lim_{x \to \infty} \frac{1 - \left(\frac{2}{3}\right)^x}{\left(\frac{1}{3}\right)^x + 2} \qquad \text{(algebra)}$$

$$\to \frac{1 - 0}{0 + 2} = \frac{1}{2}. \qquad \text{(limit of } b^x \text{ as } x \to \infty \text{ if } 0 < b < 1)$$

☐

A similar technique can be used for infinite limits of quotients that involve expressions of the form e^{kx}. In that case you would divide the numerator and the denominator of the quotient by the expression e^{kx} with the highest value of k.

What you should know after reading Section 8.3

▶ Be able to show that the natural exponential function $f(x) = e^x$ is continuous at $x = 0$, and be able to use this to show that $f(x) = e^x$ is continuous everywhere. Why is this useful information to know, in terms of limits?

▶ How to use the fact that the natural exponential function is continuous to prove that *every* exponential function is continuous (whether in the form Ae^{kx} or Ab^x). What does this mean in terms of limits?

▶ How do various types of exponential functions behave as $x \to \infty$ and as $x \to -\infty$? How can you *prove* that the limit of e^x as $x \to \infty$ is infinite, while the limit of e^x as $x \to -\infty$ is zero?

▶ How to calculate limits that involve exponential functions. This includes using the fact that exponential functions are continuous, and the limits of exponential functions "at" infinity as well as algebraic properties of exponential functions and various limit-solving techniques.

Exercises 8.3

Concepts

0. Read the section and make your own summary of the material. Include whatever you think will help you review the material later on.

■ Write each of the following statements in terms of a limit.

1. The natural exponential function is continuous at $x = 0$.

2. The natural exponential function is continuous.

3. Every exponential function is continuous (use the form $f(x) = Ae^{kx}$ for your exponential function).

4. Every exponential function is continuous (use the form $f(x) = Ab^x$ for your exponential function).

5. The natural exponential function grows without bound as x approaches infinity.

6. The natural exponential function has a horizontal asymptote at $y = 0$ on the left.

■ Express each of the following limits using the appropriate formal definition of limit. (According to the type of limit, the formal definition will involve ε and δ, or ε and N, or M and δ, or M and N.)

7. $\lim\limits_{x \to 0} e^x = 1$

8. $\lim\limits_{x \to c} e^x = e^c$

9. $\lim\limits_{x \to 1} 2(e^{3x}) = 2e^3$

10. $\lim\limits_{x \to 3} 5(2^x) = 40$

11. $\lim\limits_{x \to \infty} e^{3x} = \infty$

12. $\lim\limits_{x \to -\infty} e^{3x} = 0$

13. $\lim\limits_{x \to \infty} 2^x = \infty$

14. $\lim\limits_{x \to -\infty} 2^x = 0$

15. $\lim\limits_{x \to \infty} e^{-3x} = 0$

16. Explain why you know immediately that the function $f(x) = 3e^{2x}(x^2 + 1)$ is continuous. (*Hint:* Is the product of two continuous functions continuous?)

■ Fill in the blanks in each statement below (with as general a statement as possible).

17. If $k > 0$, then $\lim\limits_{x \to \infty} e^{kx} = $ _____.

18. If $k < 0$, then $\lim\limits_{x \to \infty} e^{kx} = $ _____.

19. If $k \in $ _____, then $\lim\limits_{x \to -\infty} e^{kx} = \infty$.

20. If $k \in $ _____, then $\lim\limits_{x \to -\infty} e^{kx} = 0$.

21. If $b \in $ _____, then $\lim\limits_{x \to \infty} b^x = 0$.

22. If $b \in $ _____, then $\lim\limits_{x \to \infty} b^x = \infty$.

23. If $b > 1$, then $\lim\limits_{x \to -\infty} b^x = $ _____.

24. If $0 < b < 1$, then $\lim\limits_{x \to -\infty} b^x = $ _____.

Skills

25. Consider the limit $\lim\limits_{x \to \infty} \dfrac{2^x - 4^x}{3^x}$.

(a) Solve this limit by using the method of dividing numerator and denominator by the "largest" exponential function that appears in the expression.

(b) Solve this limit by a different method (use algebra).

26. Consider the limit $\lim\limits_{x \to -\infty} \dfrac{2^x - 4^x}{3^x}$.

(a) Explain why the method of dividing numerator and denominator by the "largest" exponential function in the expression does *not* work for this limit. (*Hints:* Try it; what goes wrong? Note that "$\frac{\infty}{\infty}$" is an indeterminate form.)

(b) Solve this limit by using algebra to simplify $\frac{2^x - 4^x}{3^x}$.

■ Calculate the following limits. Find exact answers and then use your calculator to get a numerical approximation of your exact answer. Check each answer with a calculator graph.

27. $\lim\limits_{x \to 2} 4e^{-3x}$

28. $\lim\limits_{x \to -2} 3\left(\frac{1}{4}\right)^x$

29. $\lim\limits_{x \to 0} \dfrac{2 - 3e^x}{x + 4}$

30. $\lim\limits_{x \to -3} 2(4 - 3e^{4x})$

31. $\lim\limits_{x \to 3} \dfrac{2^x - 1}{3^x}$

32. $\lim\limits_{x \to \infty} -2e^{4x}$

33. $\lim\limits_{x \to -\infty} 1.5(0.21)^x$

34. $\lim\limits_{x \to \infty} \dfrac{1}{2 + 3^x}$

35. $\lim\limits_{x \to -\infty} \dfrac{1}{2 + 3^x}$

36. $\lim\limits_{x \to 0} \dfrac{2}{4 + e^{-2x}}$

37. $\lim\limits_{x \to \infty} \dfrac{2}{4 + e^{-2x}}$

38. $\lim\limits_{x \to -\infty} \dfrac{2}{4 + e^{-2x}}$

39. $\lim\limits_{x \to 0} \dfrac{4^x - 6(2^x) + 5}{1 - 2^x}$

40. $\lim\limits_{x \to 2} \dfrac{4^x - 6(2^x) + 5}{1 - 2^x}$

41. $\lim\limits_{x \to \infty} \dfrac{4^x - 6(2^x) + 5}{1 - 2^x}$

42. $\lim\limits_{x \to 2} \dfrac{4^x - 3(2^x) - 4}{2^x - 4}$

43. $\lim\limits_{x \to \infty} \dfrac{4^x - 3^x}{5^x}$

44. $\lim\limits_{x \to \infty} \dfrac{2^x - 4^{-x}}{3^x}$

45. $\lim\limits_{x \to -\infty} \dfrac{3^x - 5^x}{4^x}$

46. $\lim\limits_{x \to \infty} \dfrac{e^{2x} - e^{4x}}{e^{3x}}$

47. $\lim\limits_{x \to \infty} \dfrac{4(3^x)}{3 + 2^x}$

48. $\lim\limits_{x \to \infty} \dfrac{4(3^x)}{2 + 3^x}$

49. $\lim\limits_{x \to \infty} \dfrac{2e^{1.5x}}{3e^{2x} + e^{1.5x}}$

50. $\lim\limits_{x \to \infty} \dfrac{2e^{2x}}{3e^{1.5x} + e^{2x}}$

■ Determine any horizontal asymptotes of the following functions by calculating limits. Then check your answers by using a calculator to graph each function.

51. $f(x) = 25 - (0.5)^x$

52. $f(x) = 3e^{-2x} + 4$

53. $f(x) = \dfrac{2}{4 + e^{-2x}}$

54. $f(x) = \dfrac{1}{2 + 3^x}$

55. $f(x) = \dfrac{2^x - 4^x}{3^x}$

56. $f(x) = \dfrac{4^x - 6(2^x) + 5}{1 - 2^x}$

Proofs

57. Use Theorem 8.3 in Section 8.2 to prove that the natural exponential function is continuous at $x = 0$.

58. Use the fact that the natural exponential function is continuous at $x = 0$ to prove that the natural exponential function is continuous *everywhere*.

59. Use the fact that the natural exponential function is continuous to prove that *every* exponential function $f(x) = Ae^{kx}$ is continuous.

60. Use the formal definition of a limit at infinity to prove that $\lim\limits_{x \to \infty} e^x = \infty$.

61. Use the fact that $\lim\limits_{x \to \infty} e^x = \infty$ to prove that $\lim\limits_{x \to -\infty} e^x = 0$.

62. Use Theorem 8.13 to prove that if $k > 0$, then $\lim\limits_{x \to \infty} e^{kx} = \infty$.

63. Use Theorem 8.13 to prove that if $k > 0$, then $\lim\limits_{x \to -\infty} e^{kx} = 0$.

8.4 Derivatives of Exponential Functions

When we investigated the limits of exponential functions in Section 8.3, we began by determining the limits of the natural exponential function. All limits of exponential functions in general were based on the limits of the natural exponential function. In a similar way, we begin our investigation of the derivatives of exponential functions by examining the derivative of the natural exponential function.

8.4.1 The derivative of the natural exponential function

In this section we find the derivative of the natural exponential function $f(x) = e^x$. Unfortunately, none of the differentiation rules we currently know applies to this function, so we will have to use the definition of derivative to determine the derivative of $f(x) = e^x$.

✦ *Caution*   The power rule does *not* apply to exponential functions! In other words,

$$\frac{d}{dx}(e^x) \neq xe^{x-1}.$$

Remember that the power rule, which says that $\frac{d}{dx}(x^k) = kx^{k-1}$, applies only if the variable x is in the *base* and the exponent k is a constant. This is *not* the case with the natural exponential function $f(x) = e^x$, which has the constant e in the base, and the variable x in the *exponent*.   □

As the next theorem shows, the natural exponential function $f(x) = e^x$ has the amazing property that it is its own derivative! In fact, this is one of the main reasons why the special irrational number e is considered the "natural" base for an exponential function.

THEOREM 8.16

The Derivative of the Natural Exponential Function

The natural exponential function is its own derivative; that is,

$$\frac{d}{dx}(e^x) = e^x.$$

PROOF (THEOREM 8.16)   The proof is just a definition of derivative calculation, using one of the definitions of the number e from Section 8.2.

$$\frac{d}{dx}(e^x) = \lim_{h \to 0} \frac{e^{x+h} - e^x}{h} \qquad \text{(definition of derivative)}$$

$$= \lim_{h \to 0} \frac{e^x e^h - e^x}{h} \qquad \text{(algebra)}$$

$$= \lim_{h \to 0} \frac{e^x(e^h - 1)}{h} \qquad \text{(factor out } e^x)$$

$$= e^x \lim_{h \to 0} \frac{e^h - 1}{h} \qquad (e^x \text{ does not involve } h)$$

$$= e^x(1) = e^x. \qquad \text{(Theorem 8.3 in Section 8.2)}$$ ■

Since the natural exponential function $f(x) = e^x$ is its own derivative, the height of its graph at every point is equal to the slope of the graph at that point. More precisely, for any real number c, the height $f(c)$ of the graph of $f(x) = e^x$ is equal to the slope $f'(c)$ of the tangent line to the graph at $x = c$. Figure 8.13 illustrates this amazing property of $f(x) = e^x$ at the point $x = 1$.

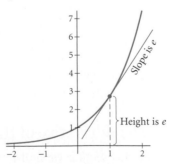

Figure 8.13
$f(x) = e^x$ Height at $x = 1$
equals slope at $x = 1$

The height of the graph of $f(x) = e^x$ in Figure 8.13 at $x = 1$ (the length of the dashed line) is $e \approx 2.718$, and the slope of the tangent line to $f(x) = e^x$ at $x = 1$ (the slope of the purple line) is also equal to $e \approx 2.718$. Since $f(x) = e^x$ is its own derivative, this is true at *every* value: The slope of $f(x) = e^x$ at $x = c$ is equal to the height of $f(x) = e^x$ at $x = c$. Note in particular that for larger values of x, both the height and the slope of $f(x) = e^x$ are larger, and for smaller values of x, both the height and the slope of $f(x) = e^x$ are smaller. Draw more heights and slopes on the graph in Figure 8.13 to convince yourself that this is the case.

Of course, now that we have a differentiation rule for the natural exponential function, we can differentiate functions that involve e^x.

EXAMPLE 8.17

A simple differentiation problem

We can now quickly differentiate the function $f(x) = e^x(x^2 + 3x - 1)$, as follows:

$$\frac{d}{dx}(e^x(x^2 + 3x - 1))$$
$$= \frac{d}{dx}(e^x) \cdot (x^2 + 3x - 1) + e^x \cdot \frac{d}{dx}(x^2 + 3x - 1) \quad \text{(product rule)}$$
$$= e^x(x^2 + 3x - 1) + e^x(2x + 3).$$

Of course, we could simplify this answer further, but we are finished with the differentiation steps and will stop here. $\square$

8.4.2 Differentiating exponential functions

We can use the fact that the natural exponential function is its own derivative to develop a rule for differentiating *any* exponential function. We call this rule the "exponential rule" for differentiation; in particular, notice that this new rule is very different from the power rule! Since we have two ways of writing every exponential function, the "exponential rule" has two forms.

THEOREM 8.17

The Exponential Rule for Differentiation

(a) If k is any real number, then $\frac{d}{dx}(e^{kx}) = ke^{kx}$.

(b) If b is any positive real number, then $\frac{d}{dx}(b^x) = (\ln b)b^x$.

The second rule is just a generalization of the first: When $b = e$, the second rule says that $\frac{d}{dx}(e^x) = (\ln e)e^x = e^x$ (since $\ln e = 1$).

EXAMPLE 8.18

Using Theorem 8.17

By Theorem 8.17, we have $\frac{d}{dx}(e^{3x}) = 3e^{3x}$, and $\frac{d}{dx}(2^x) = (\ln 2)2^x$. $\square$

Although the two forms of the exponential rule look very different, they are in fact the same rule written two different ways (as we will see in the following proof).

PROOF (**THEOREM 8.17**) To prove part (a), we need only use the chain rule and the fact that the natural exponential function is its own derivative. The key is to notice that e^{kx} is really a composition $f(u(x))$ of the functions $f(x) = e^x$ and $u(x) = kx$:

$$f(u(x)) = f(kx) = e^{kx}.$$

By the chain rule, the derivative of e^x is given by $f'(u(x))u'(x)$; since $f'(x) = e^x$ and $u'(x) = k$, we have

$$\frac{d}{dx}(e^{kx}) = \frac{d}{dx}(f(u(x))) = f'(u(x))u'(x) = e^{u(x)}(k) = ke^{kx}.$$

Part (b) actually follows from part (a) with a bit of algebra:

$$\frac{d}{dx}(b^x) = \frac{d}{dx}((e^{\ln b})^x) \qquad \text{(since } b = e^{\ln b})$$

$$= \frac{d}{dx}\left(e^{(\ln b)x}\right) \qquad \text{(algebra)}$$

$$= (\ln b)e^{(\ln b)x} \qquad \text{(part (a), with } k = \ln b)$$

$$= (\ln b)(e^{\ln b})^x \qquad \text{(algebra)}$$

$$= (\ln b)b^x. \qquad \text{(since } b = e^{\ln b})$$

Now that we can differentiate expressions of the form e^{kx} and b^x, we can differentiate any exponential function. Using the constant multiple rule, we have

$$(1) \qquad \frac{d}{dx}(Ae^{kx}) = Ake^{kx}$$

$$(2) \qquad \frac{d}{dx}(Ab^x) = A(\ln b)b^x$$

| EXAMPLE 8.19 | **Differentiating exponential functions** |

Using Equations (1) and (2), $\frac{d}{dx}(2e^{3x}) = 2(3)e^{3x} = 6e^{3x}$ and $\frac{d}{dx}(5(2^x)) = 5(\ln 2)2^x$. ☐

We can also differentiate many more complicated functions that involve exponential expressions. Before doing some examples, we formally state as a theorem the rule for differentiating a composition whose "outside" function is exponential. This rule is nothing more than a special case of the chain rule.

THEOREM 8.18

Combining the Exponential Rule and the Chain Rule

If $u(x)$ is a function of x, then

(a) If k is any real number, then $\dfrac{d}{dx}(e^{u(x)}) = e^{u(x)}u'(x)$.

(b) If b is any positive real number, then $\dfrac{d}{dx}(b^{u(x)}) = (\ln b)b^{u(x)}u'(x)$.

The proof of Theorem 8.18 is simply an application of the chain rule and the rule for differentiating exponential functions (similar to the proof of Theorem 8.17) and is left to the exercises. Note that since $\ln e = 1$, part (a) is just a special case of part (b).

EXAMPLE 8.20

A simple differentiation problem using the chain and exponential rules

Consider the function $f(x) = e^{x^2+1}$ (here the "inside" function is $u(x) = x^2 + 1$). The derivative of $f(x)$ is

$$\frac{d}{dx}(e^{x^2+1}) = e^{x^2+1}\frac{d}{dx}(x^2 + 1) = e^{x^2+1}(2x).$$

Now consider the function $g(x) = 2^{x^2+1}$ (again we have $u(x) = x^2 + 1$). The derivative of $g(x)$ is

$$\frac{d}{dx}(2^{x^2+1}) = (\ln 2)2^{x^2+1}\frac{d}{dx}(x^2 + 1) = (\ln 2)2^{x^2+1}(2x). \qquad \square$$

⑤ **Caution**   Don't forget that the power rule does not apply to exponential functions! The power rule applies only to power functions, where the base is x and the exponent is a constant. On the other hand, the exponential rule applies only to exponential functions, where the base is a constant and the exponent is the independent variable. $\qquad \square$

❓ **Question**   What is the derivative of the function $f(x) = x^x$? If the power rule applied to this function, then the derivative would be $x(x^{x-1})$; on the other hand, if the exponential rule applied to this function, then the derivative would be $(\ln x)x^x$. Does either of these rules apply? Why or why not? $\qquad \square$

The following example illustrates how to differentiate functions that involve exponential expressions.

EXAMPLE 8.21

Using the differentiation rules

Differentiate the following functions.

(a) $f(x) = xe^{\frac{1}{x}}$ $\qquad$ (b) $f(x) = \dfrac{5^{3x}}{1+x}$ $\qquad$ (c) $f(x) = 3\sqrt{e^x} - x^2$

Solution

(a) At its outermost level, this function is a product (of x and $e^{\frac{1}{x}}$). Therefore, we first apply the product rule.

$$\frac{d}{dx}\left(xe^{\frac{1}{x}}\right) = \frac{d}{dx}(x) \cdot e^{\frac{1}{x}} + x \cdot \frac{d}{dx}\left(e^{\frac{1}{x}}\right) \qquad \text{(product rule)}$$

$$= (1)e^{\frac{1}{x}} + x\left(e^{\frac{1}{x}}\right)(-x^{-2}) \qquad \text{(chain, exponential rules)}$$

$$= e^{\frac{1}{x}} - \frac{e^{\frac{1}{x}}}{x}. \qquad \text{(a little algebra)}$$

We didn't have to do that last bit of algebra, but it's nice to clean up our answers when we have a chance.

(b) Starting with the quotient rule, we have

$$\frac{d}{dx}\left(\frac{5^{3x}}{1+x}\right) = \frac{\frac{d}{dx}(5^{3x}) \cdot (1+x) - 5^{3x} \cdot \frac{d}{dx}(1+x)}{(1+x)^2} \qquad \text{(quotient rule)}$$

$$= \frac{(\ln 5)5^{3x}(3)(1+x) - 5^{3x}(1)}{(1+x)^2} \qquad \text{(chain, exponential rules)}$$

$$= \frac{5^{3x}(3(\ln 5)(1+x) - 1)}{(1+x)^2}. \qquad \text{(simplify a bit)}$$

(c) In this example, an exponential function is on the *inside* of a composition (note that e^x appears inside the square root).

$$\frac{d}{dx}(3\sqrt{e^x} - x^2) = \frac{d}{dx}\left(3(e^x)^{\frac{1}{2}} - x^2\right) \qquad \text{(rewrite so differentiation will be easier)}$$

$$= 3\left(\tfrac{1}{2}\right)(e^x)^{-\frac{1}{2}}(e^x) - 2x \qquad \text{(don't forget the chain rule!)}$$

$$= \tfrac{3}{2}e^x(e^x)^{-\frac{1}{2}} - 2x. \qquad \text{(some algebra)}$$

We could simplify the answer further, but there's no need to do that right now. One interesting thing about this particular example is that we could have saved ourselves some differentiation work by doing some algebra first (this is quite often the case). In particular, we could have done the calculation this way:

$$\frac{d}{dx}(3\sqrt{e^x} - x^2) = \frac{d}{dx}\left(3(e^x)^{\frac{1}{2}} - x^2\right) \qquad \text{(rewrite so differentiation will be easier)}$$

$$= \frac{d}{dx}\left(3e^{\frac{1}{2}x} - x^2\right) \qquad \text{(algebra makes differentiation even easier!)}$$

$$= 3\left(\tfrac{1}{2}\right)e^{\frac{1}{2}x} - 2x. \qquad \text{(differentiation is easy now)}$$

This answer is equivalent to the answer we got in the first calculation (do some algebra to see why). Notice that the differentiation steps used in the second calculation were much easier than the differentiation steps used in the first calculation!   □

8.4.3 The rate of change of an exponential function

We have seen that the natural exponential function e^x (or in fact any function Ae^x) is its own derivative, and that the derivative of a general exponential function Ae^{kx} is a constant multiple of Ae^{kx} (namely $k \cdot Ae^{kx}$). In other words, the rate of change of an exponential function is proportional to the function itself! If $f(x) = Ae^{kx}$ is an exponential function, then its derivative $f'(x)$ (the "rate of change") is proportional to the function $f(x)$:

$$f'(x) = kf(x).$$

(Remember that "proportional to" means "a constant multiple of.") In fact, it turns out that the *only* functions whose rates of change are proportional to themselves are exponential functions. That is, if $f'(x) = kf(x)$, then $f(x)$ *must* be an exponential function. This gives us a nice characterization of exponential functions: *All* exponential functions, and *only* exponential functions, have derivatives that are proportional to themselves.

THEOREM 8.19

Identifying Exponential Functions Using Rate-of-Change Information

If $f(x)$ is any function, then
(a) If $f(x) = Ae^{kx}$ is exponential, then $f'(x) = kf(x)$.
(b) If $f'(x) = kf(x)$ for some real number $k \neq 0$, then $f(x)$ is an exponential function of the form $f(x) = Ae^{kx}$.

Note in particular that not only is the derivative of an exponential function proportional to the function itself, but also the proportionality constant involved is exactly the "k" from the equation for that exponential function. This fact will be very useful to us later on, especially when we look at real-life applications of exponential functions.

Proving part (a) of Theorem 8.19 is an easy application of the differentiation rules for exponential functions. Proving part (b) is more difficult and requires a knowledge of "differential equations," which we will not go into here.

PROOF (**PART (A) OF THEOREM 8.19**)   Suppose $f(x)$ is an exponential function. Then $f(x)$ can be written in the form $f(x) = Ae^{kx}$ for some A and k. Therefore, the derivative of $f(x)$ is

$$f'(x) = \frac{d}{dx}(Ae^{kx}) = Ake^{kx} = k(Ae^{kx}) = kf(x).$$

Hence if $f(x)$ is exponential, then its derivative $f'(x)$ is proportional to the function $f(x)$ itself, that is, $f'(x) = kf(x)$. ◼

❷ *Question*   We could also have used the form $f(x) = Ab^x$ for an exponential function in the proof of part (a) of Theorem 8.19. Try re-doing the proof using this form of exponential function. ☐

◈ *Caution*   Theorem 8.19 applies only to functions that are truly "exponential," and not to *any* function that involves an exponential expression. A function is said to be exponential only if it can be written in the form Ae^{kx} (or, equivalently, Ab^x). For example, the function $f(x) = 3e^{2x} + 1$ is not exponential and does not have the property that its derivative is proportional to the function itself. ☐

Theorem 8.19 is important because so many real-life functions have the property that their derivatives are proportional to themselves. This means that many real-life functions are exponential! In Section 8.6 we will see that populations, radioactive decay, and investment balances can all be modeled by exponential functions.

If a word problem states that the rate of change of a function is constant, we immediately know that the function is linear. On the other hand, considering Theorem 8.19, if a word problem states that the rate of change of a function is proportional to the function itself, we immediately know that that function is exponential. Then it is just a question of finding values A and k to determine the function $f(x) = Ae^{kx}$ that models the situation.

EXAMPLE 8.22

A real-world problem that can be modeled with an exponential function

Suppose a population of wombats on a small island is growing at a rate proportional to the number of wombats on the island. If there were 12 wombats on the island in 1990 and 37 wombats on the island in 1998, how many wombats will be on the island in the year 2010?

Solution   The rate of change of the population of wombats is proportional to the population of wombats. If we let $W(t)$ be the number of wombats on the island at time t, this means that the rate of change $W'(t)$ is proportional to the number $W(t)$ of wombats on the island; in other words, $W'(t) = kW(t)$ for some real number k. Now, by Theorem 8.19, the population $W(t)$ of wombats on the island at time t must be an exponential function $W(t) = Ae^{kt}$. We need only to find constants A and k that match the information given in the problem.

If we let $t = 0$ represent 1990 (so $t = 8$ will represent 1998), then from the information given in the problem, we have $W(0) = 12$ and $W(8) = 37$. We will use these two datapoints to find A and k, as follows: Using the first datapoint, we have

$$W(0) = 12 \quad \Rightarrow \quad Ae^{k(0)} = 12 \quad \Rightarrow \quad Ae^0 = 12 \quad \Rightarrow \quad A(1) = 12 \quad \Rightarrow \quad A = 12,$$

so we know that $W(t) = 12e^{kt}$ for some k. Using the second datapoint, we can now solve for the value of k:

$$W(8) = 37 \quad \Rightarrow \quad 12e^{k(8)} = 37 \quad \Rightarrow \quad e^{8k} = \tfrac{37}{12} \quad \Rightarrow \quad 8k = \ln\left(\tfrac{37}{12}\right)$$

$$\Rightarrow \quad k = \frac{\ln\left(\tfrac{37}{12}\right)}{8} \approx 0.14.$$

We now know that $W(t) = 12e^{0.14t}$. Using this function, we can now easily calculate the number of wombats that will be on the island in the year 2010. Since 2010 is 20 years after 1990, we need to find

$$W(20) = 12e^{0.14(20)} \approx 197.34.$$

In the year 2010, there will be approximately 197 wombats on the island (assuming that we can't have "parts" of wombats). □

What you should know after reading Section 8.4

▶ How to prove that the natural exponential function is its own derivative, and how to interpret this fact graphically.

▶ The exponential rule for differentiation (in both of its forms), and how to prove it. Be sure you can tell when the exponential rule applies and when it does not.

▶ How to differentiate functions of the form $e^{u(x)}$ and $b^{u(x)}$, and how to prove the formulas for these derivatives. Be able to differentiate any function that involves exponential expressions using these rules and other differentiation rules.

▶ Know that a function is exponential if and only if its rate of change is proportional to the function itself. Be able to use this fact to identify exponential functions in real-world problems and to find exponential functions that model a given real-world situation.

Exercises 8.4

Concepts

0. Read the section and make your own summary of the material. Include whatever you think will help you review the material later on.

1. Explain why the power rule does not apply to the function $f(x) = 2^x$. What is the *incorrect* derivative that you would get if you mistakenly applied the power rule to this function? What is the correct derivative?

2. Compare the derivatives of $f(x) = x^5$ and $g(x) = 5^x$. Which rule did you use in each case and why?

3. Does the exponential rule apply to the function $f(x) = x^x$? What about the power rule? Explain your answers.

4. The natural exponential function is its own derivative. Explain what this means graphically (use words like "height" and "slope").

5. Given that A and k are constants, what is the derivative of the exponential function $f(x) = Ae^{kx}$? Given that A and b are constants, what is the derivative of the exponential function $f(x) = Ab^x$?

6. Given that $u(x)$ is a function of x, what is the derivative of the function $f(x) = e^{u(x)}$? What is the derivative of the function $f(x) = b^{u(x)}$?

7. Explain why part (a) of Theorem 8.18 is really just a special case of part (b) of Theorem 8.18.

■ Differentiate each function below in two ways: first, by differentiating the function without doing any algebra first, and second, by doing as much algebraic simplification as possible before differentiating. Which way was easier? Show that the two answers you get are equivalent.

8. $f(x) = \dfrac{1}{\sqrt{3^x}}$

9. $f(x) = 3e^{-2x} - x^2 e^{-2x}$

10. $f(x) = (e^{2x+1})^3$

11. Differentiate the function $f(x) = (2^x)^x$. Note that you cannot differentiate this function without doing some algebra first; as it stands before simplifying, no differentiation rule applies (why?).

12. Find a function $f(x)$ whose *derivative* is $f'(x) = 2e^{3x}$.

13. Find a function $f(x)$ whose *derivative* is $f'(x) = 2^x$.

14. Fill in the blank: Given a function $y(x)$, $y'(x)$ is proportional to $y(x)$ if and only if $y(x)$ is a _____.

Skills

■ Differentiate the following functions.

15. $f(x) = e^x$

16. $f(x) = e^{-2x}$

17. $f(x) = \left(\frac{1}{3}\right)^x$

18. $f(x) = (3 - e^x)^5$

19. $f(x) = e^{(x^2)}$

20. $f(x) = (e^x)^2$

21. $f(x) = x^3 3^x$

22. $f(x) = \dfrac{1}{2 - e^{5x}}$

23. $f(x) = \dfrac{2^{3x}}{x^2 - 1}$

24. $f(x) = 2\sqrt{e^{3x+1}}$

25. $f(x) = 2e^{3\sqrt{x}+1}$

26. $f(x) = 3x\left(\frac{2}{3}\right)^x$

27. $f(x) = \dfrac{4^x - 6^x}{2^x}$

28. $f(x) = 2^{1-3^x}$

29. $f(x) = e^{(e^x)}$

30. $f(x) = e^{(x^e)}$

31. $f(x) = (e^x)^e$

32. $f(x) = (x^e)^e$

■ Find the critical points of the following functions. (Recall that $x = c$ is a *critical point* of a function $f(x)$ if $f'(c) = 0$ or $f'(c)$ does not exist.)

33. $f(x) = e^{-3x} + 2x$

34. $f(x) = x^2 3^x$

35. $f(x) = 3e^{x^2-4x+1}$

36. $f(x) = \dfrac{1}{3 - 2e^x}$

37. $f(x) = \dfrac{2^x}{1 - 2^x}$

38. $f(x) = (e^{3x+1} - 2)^{\frac{1}{3}}$

Applications

39. A fruit fly population is being cultivated in Dr. Drosophila's laboratory. Dr. Drosophila thinks that the population $P(t)$ of fruit flies after t days could be modeled by an exponential function. His laboratory assistant agrees and correctly says, "The population of fruit flies is an exponential function if the rate of change of the population of fruit flies is proportional to _____." Fill in the blank.

40. An abandoned building had a population of 45 rats on the first day of the year and a population of 53 rats 30 days later. Let $r(t)$ be the function that describes the number of rats in the building t days after the first of the year.

 (a) Find a formula for $r(t)$ given that the rate of change of the rat population is constant, and use this formula to predict the number of rats in the building on the 100th day of the year ($t = 99$).

 (b) Find a formula for $r(t)$ given that the rate of change of the rat population is proportional to the number of rats in the building, and use this formula to predict the number of rats in the building on the 100th day of the year ($t = 99$).

41. Alina started an investment account with an initial deposit of $1000. After the initial deposit, the amount of money increased at a rate proportional to her investment account balance. After 3 years the investment account had a balance of $1260.

 (a) Write down a function $A(t)$ that describes the amount of money in Alina's investment account t years after her initial deposit.

 (b) How much money will Alina have in her investment account after 30 years?

 (c) How long will it be before Alina's initial investment quadruples?

Proofs

42. Use the definition of derivative and the characterization of e from Theorem 8.3 in Section 8.2 to prove that the natural exponential function is its own derivative.

43. Use the chain rule and the fact that $\frac{d}{dx}(e^x) = e^x$ to prove that $\frac{d}{dx}(e^{kx}) = ke^{kx}$. Be sure that your argument is clear; in particular, point out exactly how you are using the chain rule.

44. Use the fact that $\frac{d}{dx}(e^{kx}) = ke^{kx}$ to prove that $\frac{d}{dx}(b^x) = (\ln b)b^x$. (*Hint:* First write b^x in the form e^{kx} for some k that depends on b.)

45. Prove part (b) of Theorem 8.18: If b is any positive real number, and $u(x)$ is any function of x, then $\frac{d}{dx}(b^{u(x)}) = (\ln b)b^{u(x)}u'(x)$. (*Hint:* Use the chain rule and the exponential rule for differentiation.)

46. Use part (b) of Theorem 8.18 to prove part (a): If k is any real number, and $u(x)$ is any function of x, then $\frac{d}{dx}(e^{u(x)}) = e^{u(x)}u'(x)$. (*Hint:* Use the fact that $\ln e = 1$.)

| **8.5** | **Graphs of Exponential Functions** |

In this section we will see that graphs of exponential functions always have the same basic shape. We will then investigate graphs of transformations of exponential functions and graphs of functions that involve exponential expressions. At the end of this section, we will do some word problems that involve exponential functions and their graphs.

8.5.1 Exponential growth and decay

Let's first examine the graphs of exponential functions of the form $f(x) = e^{kx}$ (or, equivalently, $f(x) = b^x$). In Section 8.2 we claimed that functions of this form have one of two types of graphs; we now state and prove this fact as a theorem.

THEOREM 8.20

Graphs of Basic Exponential Functions

Each exponential function of the form $f(x) = e^{kx}$ (or, equivalently, $f(x) = b^x$) has a graph with one of the following two general shapes:

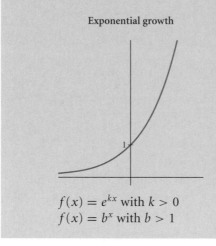

Exponential growth

$f(x) = e^{kx}$ with $k > 0$
$f(x) = b^x$ with $b > 1$

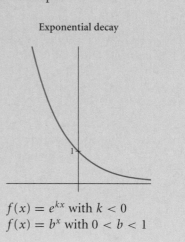

Exponential decay

$f(x) = e^{kx}$ with $k < 0$
$f(x) = b^x$ with $0 < b < 1$

To prove this theorem, we need only do sign analyses of f, f', and f'' and calculate the limits of f as x approaches positive and negative infinity. In particular, we need to show that (in the first case) $f(x) = e^{kx}$ has a y-intercept of $y = 1$; is always positive, increasing, and concave up; increases without bound as $x \to \infty$; and has a horizontal asymptote at $y = 0$ as $x \to -\infty$.

? *Question*   What features does the second graph in Theorem 8.20 have? In other words, what would you have to show in order to prove Theorem 8.20 in the case where $k < 0$ (or, equivalently, $0 < b < 1$)?   □

We will prove that every exponential function $f(x) = e^{kx}$ with $k > 0$ has the shape shown in the first graph in Theorem 8.20 and will leave the case $k < 0$ to the exercises.

PROOF   (**THEOREM 8.20**)   Suppose $f(x) = e^{kx}$ is an exponential function with $k > 0$. First, it is easy to see that the graph of f has a y-intercept of $y = 1$, since $f(0) = e^{k(0)} = e^0 = 1$. Also, we have already proved (in Section 8.3) that

$$\lim_{x \to \infty} e^{kx} = \infty \quad \text{and} \quad \lim_{x \to -\infty} e^{kx} = 0$$

when $k > 0$. Now it remains only to show that the graph of f is always positive, increasing, and concave up. We know from Section 8.1 that the range of every exponential function $f(x) = e^{kx}$ is $(0, \infty)$ and thus that the graph of f is always positive (regardless of whether k is positive or negative).

To show that $f(x) = e^{kx}$ is increasing and concave up when $k > 0$, we need to calculate its first and second derivatives. By the exponential rule for differentiation, we have

$$f'(x) = ke^{kx} \quad \text{and} \quad f''(x) = k^2 e^{kx}.$$

Since e^{kx} is always positive, $k > 0$ (by hypothesis), and $k^2 > 0$, we know that

$$f'(x) > 0 \quad \text{and} \quad f''(x) > 0$$

for all values of x. The graph of f is indeed always increasing and always concave up.

Since a function $f(x) = b^x$ with $b > 1$ can always be written in the form $f(x) = e^{kx}$ for some $k > 0$, this also proves that every exponential function of the form $f(x) = b^x$, with $b > 1$, has a graph like the first graph shown in Theorem 8.20. ∎

8.5.2 Comparing graphs of exponential functions

In this section we compare the graphs of exponential functions of the forms $f(x) = e^{kx}$ and $f(x) = b^x$ for various values of k and b. We have already seen that if k is positive (or $b > 1$), then $f(x)$ will be an exponential *growth* function, whereas if k is negative (or $0 < b < 1$), then $f(x)$ will be an exponential *decay* function. The rate at which an exponential function grows or decays depends on the size of k (or the size of b).

In general, an exponential growth function of the form $f(x) = e^{kx}$ (with $k > 0$) will grow quickly if k is large and slowly if k is small. On the other hand, an exponential decay function of the form $f(x) = e^{-kx}$ will decay quickly for large k and slowly for small k. The following example illustrates how the value of k affects the steepness of the graph of an exponential function.

EXAMPLE 8.23

The size of k affects the graph of e^{kx}

Consider the graphs of $f(x) = e^x$, $g(x) = e^{2x}$, and $h(x) = e^{3x}$ shown on the same set of axes in Figure 8.14. The function $f(x)$ has the smallest value of k (namely $k = 1$) and thus grows the slowest; the function $g(x)$ is the graph in the middle, and the function $h(x)$ has the highest value of k (namely $k = 3$) and thus grows the fastest. Notice that all three graphs have the same y-intercept (and therefore are *not* related by vertical stretches; why?).

Similarly, the size of k affects the steepness of the graph of an exponential decay function e^{kx}; Figure 8.15 shows the graphs of $f(x) = e^{-x}$, $g(x) = e^{-2x}$, and $h(x) = e^{-3x}$ on the same set of axes. Once again, the function with the highest value of k (namely $h(x) = e^{-3x}$) decays the fastest (and is the steepest graph). Again notice that all three graphs have the same y-intercept.

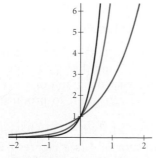

Figure 8.14
$y = e^{3x}$, $y = e^{2x}$, and $y = e^x$

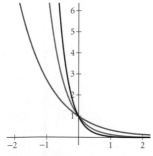

Figure 8.15
$y = e^{-x}$, $y = e^{-2x}$, and $y = e^{-3x}$ □

If an exponential function of the form $f(x) = b^x$ represents exponential growth (so $b > 1$), the function will grow quickly when b is large and slowly when b is close to 1; on the other hand, an exponential decay function of the form $f(x) = b^x$ (with $0 < b < 1$) decays quickly if b is close to zero and slowly if b is close to 1.

EXAMPLE 8.24

The size of b affects the graph of b^x

Figure 8.16 shows the graphs of $f(x) = 2^x$, $g(x) = 3^x$, and $h(x) = 4^x$. All three graphs have the same y-intercept, but the graph of $h(x) = 4^x$ (in purple) is the steepest, and the graph of $f(x) = 2^x$ (in maroon) grows at the slowest rate. Figure 8.17 shows the graphs of $f(x) = (\frac{1}{2})^x$, $g(x) = (\frac{1}{3})^x$, and $h(x) = (\frac{1}{4})^x$. The steepest graph (in maroon) is the one where b is the smallest, namely $h(x) = (\frac{1}{4})^x$.

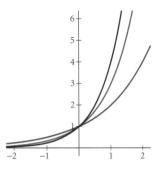   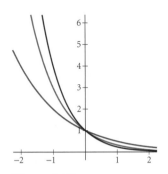

Figure 8.16
$y = 4^x$, $y = 3^x$, and $y = 2^x$

Figure 8.17
$y = (\frac{1}{2})^x$, $y = (\frac{1}{3})^x$, and
$y = (\frac{1}{4})^x$

One way to remember which types of exponential functions grow or decay the fastest is to compare them with the constant function $f(x) = 1$. An exponential function $f(x) = e^{kx}$ is "flattest" if k is very close to zero (so that $f(x) = e^{kx}$ is very close to $f(x) = e^{0x} = e^0 = 1$). On the other hand, an exponential function $f(x) = b^x$ is "flattest" if b is very close to 1 (so that $f(x) = b^x$ is very close to $f(x) = 1^x = 1$).

8.5.3 Graphing general exponential functions

We now know the shapes of the graphs of functions of the form $f(x) = e^{kx}$ (or $f(x) = b^x$). Using what we know about transformations, we can now completely describe the graph of any exponential function $f(x) = Ae^{kx}$ or $f(x) = A(b^x)$.

EXAMPLE 8.25

Using transformations to graph exponential functions

Sketch the graphs of the functions $f(x) = -2e^{0.5x}$ and $g(x) = 0.25(0.5)^x$.

Solution   The graph of $f(x) = -2e^{0.5x}$ can be obtained by reflecting the graph of $y = e^{0.5x}$ over the y-axis and then stretching the resulting graph vertically by a factor of 2. Since we know what the graph of $y = e^{0.5x}$ looks like (see Theorem 8.20 in the case where $k > 0$), we can use it to produce the graph of $f(x) = -2e^{0.5x}$ (see Figures 8.18–8.20).

The graph of $g(x) = 0.25(0.5)^x$ can be obtained by vertically contracting the graph of $y = (0.5)^x$ by a factor of 4 (see Figures 8.21 and 8.22). Note that $y = (0.5)^x$ is an exponential decay function because $b = 0.5 < 1$.

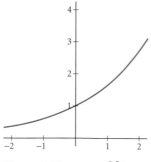

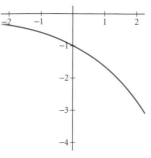

Figure 8.18 $y = e^{0.5x}$ **Figure 8.19** $y = -e^{0.5x}$

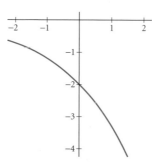

Figure 8.20 $y = -2e^{0.5x}$

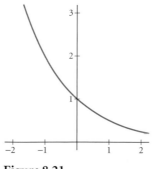

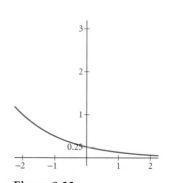

Figure 8.21
$y = (0.5)^x$

Figure 8.22
$y = 0.25(0.5)^x$ □

8.5.4 Transformations of exponential functions

Given the graph of a function $f(x)$, we can easily sketch the graphs of any of the transformations $f(x) + C$, $f(x + C)$, $Af(x)$, and $f(Ax)$. We can use what we know about transformations to graph transformations of exponential functions. In particular, we should always be able to mark at least one point on the graph and label the location of any horizontal asymptote.

❓ *Question* Do you remember the graphical effects of the transformations $f(x) + C$, $f(x + C)$, $Af(x)$, and $f(Ax)$? How are the graphs of these transformations related to the graph of $f(x)$? If you can't answer this question with confidence you might want to review Section 1.6.   □

| EXAMPLE 8.26 | **Graphing a transformation of an exponential function** |

Sketch the graph of the function $f(x) = 5 - 3e^{1.7x}$.

Solution   The function $f(x) = 5 - 3e^{1.7x}$ is a transformation of the exponential function $y = e^{1.7x}$. We know that $y = e^{1.7x}$ is an "exponential growth" graph and that $y = -3e^{1.7x}$ is that same graph, reflected over the y-axis and stretched vertically by a factor of 3. The graph of $f(x) = 5 - 3e^{1.7x}$ can then be obtained by shifting the graph up five units. The graph of $f(x)$ will have a y-intercept of $(0, 2)$ and a horizontal asymptote at $y = 5$; moreover, the shape of the graph will be an upside-down version of the "exponential growth" shape (with an asymptote on the left). See Figure 8.23.

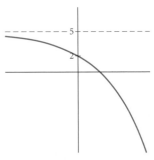

Figure 8.23
$f(x) = 5 - 3e^{1.7x}$

8.5.5   More complicated graphing problems

If a function $f(x)$ involves exponential expressions but is not a simple transformation of an exponential function, we need to perform a limit and derivative analysis in order to sketch its graph. The process of sketching such a function by hand involves everything we have learned so far about exponential functions, including algebra, limits, and derivatives. We now take the time to do one such example in depth.

| EXAMPLE 8.27 | **A complete curve-sketching analysis** |

Sketch an accurate, labeled graph of the function $f(x) = \dfrac{6}{4 - 2^x}$. Include complete sign analyses of f, f', and f'', and calculate any relevant limits.

Solution   Our general strategy will follow the methods developed in Section 7.5: We will find the domain of the function, make sign analyses of f, f', and f'', locate any local extrema or inflection points, and examine any interesting limits. The domain of $f(x)$ is all values of x for which the denominator $4 - 2^x \neq 0$; in other words, the values for which $4 \neq 2^x$, or $x \neq 2$. The graph of $f(x)$ is defined everywhere except at $x = 2$. Later on we will examine the limit of the function as $x \to 2$.

The first derivative of f is

$$f'(x) = \frac{d}{dx}\left(\frac{6}{4 - 2^x}\right) = \frac{d}{dx}(6(4 - 2^x)^{-1}) \qquad \text{(algebra)}$$

$$= 6(-1)(4 - 2^x)^{-2}(-(\ln 2)2^x) \qquad \text{(differentiate)}$$

$$= \frac{6(\ln 2)2^x}{(4 - 2^x)^2}. \qquad \text{(algebra)}$$

The second derivative of f is

$$f''(x)$$

$$= \frac{d}{dx}\left(\frac{6(\ln 2)2^x}{(4-2^x)^2}\right)$$

$$= \frac{\frac{d}{dx}(6(\ln 2)2^x)\cdot(4-2^x)^2 - (6(\ln 2)2^x)\cdot\frac{d}{dx}((4-2^x)^2)}{((4-2^x)^2)^2} \qquad \text{(quotient rule)}$$

$$= \frac{6(\ln 2)(\ln 2)2^x(4-2^x)^2 - (6(\ln 2)2^x)(2)(4-2^x)^1(-(\ln 2)2^x)}{(4-2^x)^4} \qquad \text{(differentiate)}$$

$$= \frac{6(\ln 2)^2 2^x(4-2^x)^2 + 12(\ln 2)^2(2^x)^2(4-2^x)}{(4-2^x)^4} \qquad \text{(simplify)}$$

$$= \frac{6(\ln 2)^2 2^x(4-2^x)((4-2^x)+2(2^x))}{(4-2^x)^4} \qquad \text{(factor)}$$

$$= \frac{6(\ln 2)^2 2^x(4-2^x)(4+2^x)}{(4-2^x)^4} = \frac{6(\ln 2)^2 2^x(4+2^x)}{(4-2^x)^3}. \qquad \text{(algebra)}$$

Now we can determine the intervals on which f, f', and f'' are positive and those on which they are negative. We do this by locating the values of x for which these functions are zero or do not exist, and then testing the sign of each function in the intervals between these points.

The function $f(x) = \frac{6}{4-2^x}$ is never zero, but it is undefined at $x = 2$ (since its denominator is zero at $x = 2$). We can determine the sign intervals for f by checking the sign of f at one point to the left of $x = 2$ and one point to the right of $x = 2$, say at $x = 1$ and $x = 3$:

$$f(1) = \frac{6}{4-2^1} = \frac{6}{2} > 0$$

$$f(3) = \frac{6}{4-2^3} = \frac{6}{-4} < 0.$$

Therefore, the function f is positive for $x < 2$ and negative for $x > 2$. This information is recorded in the number line for f in Figure 8.24.

Looking at the formula for $f'(x)$, we can easily see that $f'(x)$ is never zero (since 2^x is never zero) but is undefined if $x = 2$ (since then the denominator $4 - 2^x$ is zero). Thus $x = 2$ is the only critical point of f, and we can make a sign analysis of f' by testing the sign of f' in the intervals $(-\infty, 2)$ and $(2, \infty)$—for example, at $x = 1$ and at $x = 3$:

$$f'(1) = \frac{6(\ln 2)2^1}{(4-2^1)^2} = \frac{6(\ln 2)(2)}{(2)^2} > 0$$

$$f'(3) = \frac{6(\ln 2)2^3}{(4-2^3)^2} = \frac{6(\ln 2)(8)}{(-4)^2} > 0.$$

(Note that we used the fact that $\ln 2$ is positive in these calculations; in general, $\ln x$ is positive if $x > 1$ and negative if $0 < x < 1$.) We now know that f' is positive everywhere that it is defined (see the number line in Figure 8.24). This means that the graph of f will be increasing everywhere (except at $x = 2$, where it is not defined).

Similarly, from the formula for $f''(x)$, it is clear that $f''(x)$ is never zero (since neither 2^x nor $4 + 2^x$ can ever be zero; why?) but is undefined at $x = 2$. To make a sign analysis of f'' we only have to check the sign of f'' at one point to the left, and one point to the right, of $x = 2$; for example, we could check at $x = 1$ and at $x = 3$:

$$f''(1) = \frac{6(\ln 2)^2 2^1(4 + 2^1)}{(4 - 2^1)^3} = \frac{6(\ln 2)^2(2)(6)}{(2)^3} > 0$$

$$f''(3) = \frac{6(\ln 2)^2 2^3(4 + 2^3)}{(4 - 2^3)^3} = \frac{6(\ln 2)^2(8)(12)}{(-4)^3} < 0.$$

Therefore, f'' is positive for $x < 2$ and negative for $x > 2$. The graph of f is concave up on $(-\infty, 2)$ and concave down on $(2, \infty)$.

Figure 8.24

It now remains to calculate any interesting limits. As always, we must calculate the limits at the "ends" (for this example, as $x \to \infty$ and as $x \to -\infty$):

$$\lim_{x \to \infty} f(x) = \lim_{x \to \infty} \frac{6}{4 - 2^x} \to \frac{6}{-\infty} \to 0, \qquad \text{(since } 2^x \to \infty \text{ as } x \to \infty\text{)}$$

$$\lim_{x \to -\infty} f(x) = \lim_{x \to -\infty} \frac{6}{4 - 2^x} = \frac{6}{4 - 0} = \frac{3}{2}. \qquad \text{(since } 2^x \to 0 \text{ as } x \to -\infty\text{)}$$

We now know that the graph of f has *two* horizontal asymptotes: one at $y = 0$ (as $x \to \infty$) and one at $y = \frac{3}{2}$ (as $x \to -\infty$). The only other place where it is worth calculating a limit is at the point $x = 2$ (where the function and its first and second derivatives are undefined). We calculate the left and right limits separately:

$$\lim_{x \to 2^-} f(x) = \lim_{x \to 2^-} \frac{6}{4 - 2^x} \to \frac{6}{0^+} \to \infty, \qquad \begin{array}{l}\text{(if } x < 2 \text{ then } 2^x < 4,\\ \text{so } 4 - 2^x > 0\text{)}\end{array}$$

$$\lim_{x \to 2^+} f(x) = \lim_{x \to 2^+} \frac{6}{4 - 2^x} \to \frac{6}{0^-} \to -\infty. \qquad \begin{array}{l}\text{(if } x > 2 \text{ then } 2^x > 4,\\ \text{so } 4 - 2^x < 0\text{)}\end{array}$$

This means that the graph of f has a vertical asymptote at $x = 2$ (and that the graph approaches ∞ as we approach $x = 2$ from the left, but $-\infty$ as $x \to 2^+$).

Putting all this information together, we can now sketch a very accurate graph of $f(x) = \frac{6}{4 - 2^x}$. We know that the graph must have a domain of $(-\infty, 2) \cup (2, \infty)$; must be positive, increasing, and concave up on $(-\infty, 2)$; must be negative, increasing, and concave down on $(2, \infty)$; must have horizontal asymptotes at $y = \frac{3}{2} = 1.5$ and at $y = 0$ (on the left and the right, respectively); and must have a vertical asymptote at $x = 2$. See Figure 8.25.

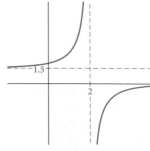

Figure 8.25 $f(x) = \dfrac{6}{2 - 4^x}$

Of course, you should use a graphing calculator to check this answer.   □

What you should know after reading Section 8.5

▶ The graphs of exponential growth and decay functions. If $f(x) = e^{kx}$, which values of k would indicate growth? Which would indicate decay? Similarly, if $f(x) = b^x$, which values of b indicate growth, and which indicate decay?

▶ How to prove (by doing sign analyses and investigating limits) that the graphs of exponential functions always have one of two basic shapes. How do the numbers A and k (or b) affect the graph of an exponential function Ae^{kx} or $A(b^x)$?

▶ How to graph transformations of exponential functions quickly. This includes being able to identify quickly the shapes, y-intercepts, and asymptotes of any such transformations.

▶ How to sketch an accurate graph of a function that involves exponential functions. This includes finding the domain, performing sign analyses of the function and its first and second derivatives, and investigating any interesting limits.

Exercises 8.5

Concepts

0. Read the section and make your own summary of the material. Include whatever you think will help you review the material later on.

1. If $f(x) = Ae^{kx}$ is an exponential function, what is its y-intercept?

2. Consider the exponential function $f(x) = 2(3^x)$. Is the function $f(-x)$ also an exponential function? If not, explain why not. If so, write the function $f(-x)$ in the form Ab^x for some real numbers A and b.

3. Argue that if you reflect the graph of $y = 2(3^x)$ over the y-axis, you get the graph of $y = 2(\frac{1}{3})^x$. (*Hint:* What transformation reflects a graph over the y-axis? Are the two functions in this problem related by that transformation?)

4. List all the graphical features of an exponential function $f(x) = e^{kx}$ with $k < 0$ (or, equivalently, of an exponential function $f(x) = b^x$ with $0 < b < 1$). In other words, what would you have to show in order to prove the second part of Theorem 8.20?

■ Fill in the blanks in each statement below with an interval of values.

5. An exponential function $f(x) = Ae^{kx}$ represents exponential growth if $k \in$ _____.

6. An exponential function $f(x) = Ae^{kx}$ represents exponential decay if $k \in$ _____.

7. An exponential function $f(x) = Ab^x$ represents exponential growth if $b \in$ _____.

8. An exponential function $f(x) = Ab^x$ represents exponential decay if $b \in$ _____.

■ For each graph below, use your knowledge of transformations and what you know about the graphs of exponential functions to find the equation of a function $f(x)$ that could have the given graph. Check that your answer is valid by using a calculator to graph your function.

9.

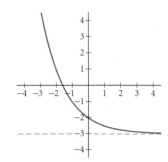

10.

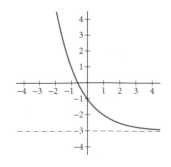

11.

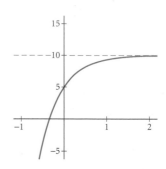

■ Determine whether each of the following statements is true or false. If a statement is true, explain why. If a statement is false, explain why or provide a counterexample.

12. True or False: Every exponential function has a horizontal asymptote at $y = 0$.

13. True or False: Every exponential function of the form $f(x) = e^{kx}$ is positive on its entire domain $(-\infty, \infty)$.

14. True or False: Every exponential function $f(x) = Ae^{kx}$ is positive on its entire domain $(-\infty, \infty)$.

15. True or False: The function $f(x) = 3e^{0.5x} - 2$ is an exponential function.

16. True or False: If $(1, 3)$ is on the graph of $y = Ae^{kx}$, then $(-2, 7)$ is on the graph of $y = Ae^{k(x+3)} + 4$.

17. True or False: The function $f(x) = 50 - 2^x$ has a horizontal asymptote at $y = 50$.

18. True or False: The graph of $y = (\frac{1}{2})^x$ can be obtained by reflecting the graph of $y = 2^x$ over the x-axis.

19. True or False: The graph of $y = 2(3^x)$ is the same as the graph of $y = 6^x$.

20. The exponential functions $f(x) = e^x$, $g(x) = e^{3x}$, and $h(x) = e^{-2x}$ are shown in the graph below. Determine which graph is which, without using a graphing calculator.

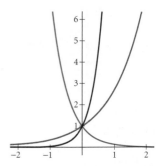

21. The exponential functions $f(x) = 2^x$, $g(x) = (\frac{1}{2})^x$, and $h(x) = (\frac{1}{5})^x$ are shown in the following graph. Determine which graph is which, without using a graphing calculator.

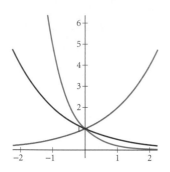

22. The exponential functions $f(x) = e^x$, $g(x) = e^{3x}$, and $h(x) = 3e^x$ are shown in the graph below. Determine which graph is which, without using a graphing calculator.

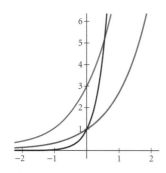

23. The exponential functions $f(x) = 2^x$, $g(x) = 4^x$, and $h(x) = 2(2^x)$ are shown in the graph below. Determine which graph is which, without using a graphing calculator.

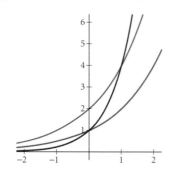

Skills

■ Sketch a rough graph of each of the following exponential functions by hand, without using a calculator. Be sure to indicate the y-intercept of each graph and any asymptotes.

24. $f(x) = 3^x$ **25.** $f(x) = (0.25)^x$

26. $f(x) = e^{2x}$ **27.** $f(x) = e^{0.3x}$

28. $f(x) = e^{-2x}$ **29.** $f(x) = -2e^{3x}$

30. $f(x) = -0.5(0.75)^x$ **31.** $f(x) = -1.5e^{-4.2x}$

■ Each of the following functions is a transformation of an exponential function. Sketch a rough graph of each function by hand (that means *without* using a calculator), using your knowledge of transformations and the shapes of the graphs of exponential functions. Label the y-intercept and any asymptotes of each graph.

32. $f(x) = 2^{-x} + 3$

33. $f(x) = 20 - 5e^{-2x}$

34. $f(x) = 5 + 4e^{3x}$

35. $f(x) = 4(e^{3x} + 2)$

36. $f(x) = -0.25(3^{x-2})$

37. $f(x) = -\left(\frac{1}{2}\right)^x + 10$

■ Sketch an accurate, labeled graph of each function $f(x)$ below. Your work should include finding the domain, doing complete sign analyses of f, f', and f'', and calculating any relevant limits.

38. $f(x) = 400 - 30e^{-2x}$

39. $f(x) = \dfrac{3}{9 - 3^x}$

40. $f(x) = \dfrac{10}{2e^{-3x} + 5}$

41. $f(x) = e^{\sqrt{x}}$

42. $f(x) = \sqrt{1 - 2^x}$

43. $f(x) = \dfrac{e^{3x}}{1 - e^{3x}}$

Proofs

44. Prove the second part of Theorem 8.20. Specifically, if $f(x) = e^{kx}$ where $k < 0$, prove each of the following (remember that you can use results from previous sections).

(a) $f(x)$ has y-intercept $(0, 1)$.

(b) $f(x)$ is always positive.

(c) $f(x)$ is always increasing.

(d) $f(x)$ is always concave up.

(e) $f(x)$ increases without bound as $x \to \infty$.

(f) $f(x)$ has a horizontal asymptote at $y = 0$.

8.6 | Applications of Exponential Functions

In this section we investigate various real-world applications that involve exponential functions. As we will see, exponential functions have a number of very nice properties that make them particularly suited for modeling quantities like investment balances, biological populations, and radioactive decay.

8.6.1 Percentage growth

Recall that a quantity $Q(t)$ that varies over time can be written as a linear function of time if the quantity changes by a fixed amount each time period. As we will see in a moment, if instead the quantity changes by a fixed *percentage* each time period, then that quantity is an *exponential* function of time.

EXAMPLE 8.28

An account balance that grows by a fixed percentage each month

Consider a savings account balance with an initial balance of \$1000.00 that earns 3% interest each month. Let $B(t)$ represent the balance of the savings account t months after the initial deposit. After 1 month, the amount in the savings account will be $B(1) = 1000 + 0.03(1000) = \$1030.00$. Since $1000 + 0.03(1000) = 1.03(1000)$, we could also write this as

$$B(1) = 1000(1.03) = \$1030.00.$$

To find the balance after 2 months, we need to multiply $B(1)$ by 1.03 (since the balance should grow by 3% more). The balance of the savings account after 2 months will be $B(1)(1.03) = (1000(1.03))(1.03) = 1000(1.03)^2$:

$$B(2) = 1000(1.03)^2 = \$1060.90.$$

Similarly, after 3 months the savings account balance will be

$$B(3) = 1000(1.03)^3 = \$1092.73.$$

Each month the balance in the savings account will be 1.03 times the amount in the savings account the previous month. The savings account balance $B(t)$ after t months is the function $B(t) = 1000(1.03)(1.03) \cdots (1.03)(1.03)$ (with t copies of 1.03). This is an exponential function:

$$B(t) = 1000(1.03)^t. \qquad \square$$

In general, if a quantity $Q(t)$ increases by a fixed percentage r each time period (where t is measured in terms of that time period and r is written in decimal form), then $Q(t)$ is an exponential function with base $1 + r$, exponent t, and coefficient Q_0 (the initial amount of the quantity). This is summarized in Theorem 8.21.

THEOREM 8.21

Percentage Growth Formula

If $Q(t)$ grows at a fixed percentage r (in decimal form) every time period t and has an initial quantity of $Q(0) = Q_0$, then $Q(t)$ is the exponential function

$$Q(t) = Q_0(1 + r)^t.$$

This formula works for *any* value of t, not just integer values. For instance, in Example 8.28, this formula says that you will have $B(3.5) = 1000(1.03)^{3.5} \approx \$1109.00$ in your savings account after three and a half months.

Technical Note: Of course, in real life, this is not precisely the case; your savings account balance will grow only at the beginning of each month (not in the middle of the month). Therefore, since $B(3) = \$1092.73$, you will still have $\$1092.73$ in your savings account when $t = 3.5$. The formula in Theorem 8.21 is what is called a "continuous" model (it works for the continuum of real numbers) of a "discrete" phenomenon (which makes sense only for integer values of t).

As the next example illustrates, the percentage growth formula also works for *decay* (i.e., where the percentage growth is *negative*).

EXAMPLE 8.29

A population that is reduced by a fixed percentage each week

A termite colony under your house is slowly dying off. Three weeks ago there were 2500 termites in the colony; now there are only 1250. Assuming that the population of termites is decreasing by a fixed percentage each week, determine when there will be only 300 termites in the colony.

Solution   Let $P(t)$ be the number of termites under your house after t weeks, where $t = 0$ represents 3 weeks ago. Then $P_0 = P(0) = 2500$, and $P(3) = 1250$. Assuming that the population of termites decreases by a fixed percentage each week, we know that $P(t)$ is a function of the form

$$P(t) = 2500(1 - r)^t.$$

(We write $1 - r$ instead of $1 + r$ because we know that the percentage rate will be negative, since the size of the colony is decreasing. This way the fact that the rate is negative will be "visible" in our equation.) We use the fact that $P(3) = 1250$ to solve for the percentage decay rate r:

$$P(3) = 1250 \;\Rightarrow\; 2500(1-r)^3 = 1250 \;\Rightarrow\; (1-r)^3 = \frac{1}{2}$$

$$\Rightarrow\; 1 - r = \left(\frac{1}{2}\right)^{\frac{1}{3}} \;\Rightarrow\; -r = \left(\frac{1}{2}\right)^{\frac{1}{3}} - 1$$

$$\Rightarrow\; r = 1 - \left(\frac{1}{2}\right)^{\frac{1}{3}} \approx 0.206.$$

This calculation tells us that the population of termites must be decreasing by approximately 20.6% each week and that the number of termites after t weeks is given by the formula

$$P(t) = 2500(1 - 0.206)^t = 2500(0.794)^t.$$

Finally, to solve the problem, we must find the time t at which $P(t) = 300$. Now that we have a formula for $P(t)$, we can find this t by solving the equation

$$P(t) = 300 \;\Rightarrow\; 2500(0.794)^t = 300 \;\Rightarrow\; (0.794)^t = \frac{300}{2500}$$

$$\Rightarrow\; \ln((0.794)^t) = \ln\left(\frac{300}{2500}\right) \;\Rightarrow\; t\ln(0.794) = \ln\left(\frac{300}{2500}\right)$$

$$\Rightarrow\; t = \frac{\ln\left(\frac{300}{2500}\right)}{\ln(0.794)} \approx 9.19.$$

It will be just over 9 weeks since $t = 0$ (just over 6 weeks from now) before there are exactly 300 termites left in the colony. □

8.6.2 Multiple compounding

Sometimes a quantity increases by a fixed percentage each time period but we wish to measure the time in terms of a larger time period. For example, an outstanding credit card balance might increase (because of interest) monthly, whereas we wish to measure the balance in terms of years. Credit card companies that offer an annual percentage rate (say, 14.99%) actually "compound" the interest monthly. This does not mean that the outstanding balance is increased by 14.99% each month but, rather, that the balance increases by $\frac{14.99}{12}$% each month. The formula for such "multiple compounding" of percentage growth is given in the following theorem.

THEOREM 8.22

Percentage Growth with Multiple Compounding

If $Q(t)$ grows at r percent, compounded n times per time period t, and has an initial value of $Q(0) = Q_0$, then $Q(t)$ is given by the function

$$Q(t) = Q_0 \left(1 + \frac{r}{n}\right)^{nt}.$$

◈ *Caution* A credit card balance with a 14.99% annual percentage rate, compounded monthly, actually grows more than 14.99% over the course of a year! This is because each month there is interest not only on the original balance but also on the interest accrued during previous months. □

578 CHAPTER 8 • Exponential Functions

EXAMPLE 8.30 | **Credit card companies use percentage growth with multiple compounding**

If you have a \$3000.00 debt on a credit card with a 14.99% annual percentage rate, and you leave that debt on the card for 2 years without making any payments (and somehow without getting any late fees or making any more purchases), how much will you owe the credit card company after 2 years? Assume that the credit card company compounds interest monthly.

Solution Let $D(t)$ be the size of your debt (in dollars) after t years. Here $r = 0.1499$ and $n = 12$ (since the interest is compounded 12 times per year). We also know that $D_0 = D(0) = 3000$. Using the multiple compounding formula, we can calculate the amount of your debt after t years with the function

$$D(t) = 3000 \left(1 + \frac{0.1499}{12}\right)^{12t}.$$

After 2 years, your debt will be

$$D(2) = 3000 \left(1 + \frac{0.1499}{12}\right)^{12(2)} \approx \$4041.25. \qquad \square$$

❓ *Question* Compare the answer in Example 8.30 to what would happen if the credit card company compounded interest only *once* a year. Then compare these answers to what would happen if the credit card company compounded interest every *day* (this is in fact what most credit card companies do now). What if they compounded interest every *minute*? $\qquad \square$

8.6.3 Continuous growth

As we have seen, an account balance where interest is compounded monthly will grow more quickly than one where the interest is compounded only once a year. If the interest is compounded *daily*, then the balance will grow even faster; and if the interest is compounded every *minute*, or every *second*, then the balance will grow even more quickly.

EXAMPLE 8.31 | **A balance grows faster if the interest is compounded more often**

Consider an account that is opened with a \$1000.00 deposit and earns 8% interest per year. Table 8.2 describes the balance of the account after 5 years if interest is calculated yearly, monthly, daily, hourly, and every minute.

Compounding	Times Compounded	Balance After Five Years
Yearly	1	$1000(1 + 0.08)^5 = \$1469.33$
Monthly	12	$1000\left(1 + \frac{0.08}{12}\right)^{12(5)} = \$1489.85$
Daily	365	$1000\left(1 + \frac{0.08}{365}\right)^{365(5)} = \$1491.76$
Hourly	8760	$1000\left(1 + \frac{0.08}{8760}\right)^{8760(5)} = \$1491.82$
Every minute	525,600	$1000\left(1 + \frac{0.08}{525,600}\right)^{525,600(5)} = \$1491.82$

Table 8.2

The last two balances are not actually the same, but they differ by only a fraction of a cent. The balance with hourly compounding is approximately $1491.821973, and the balance with compounding every minute is approximately $1491.824652. □

Notice that you earn more money when the interest is compounded more often. However, the amount of money you earn over the 5-year period seems to level off as the number of times the interest is compounded gets very large. In particular, there is not much difference in the amount of money you would earn if your interest was compounded every minute instead of hourly or even daily. There seems to be some limit to the amount of money you can earn, even if interest is compounded very frequently.

Wait, did we say *limit?* If n is the number of times that the interest is compounded, what happens as $n \to \infty$? Let's try to calculate the limit

$$(1) \qquad \lim_{n \to \infty} 1000\left(1 + \tfrac{0.08}{n}\right)^{n(5)}.$$

As $n \to \infty$, this limit approaches $100(1+0)^{\infty}$, which is an indeterminate form (since 1^{∞} is an indeterminate form). At this point we cannot be sure whether or not this limit exists. However, remember the definition of the number e from Section 8.2:

$$e = \lim_{h \to 0} (1 + h)^{\frac{1}{h}}.$$

If we substitute $n = \tfrac{1}{h}$, then $n \to \infty$ as $h \to 0$, and this limit is equivalent to

$$(2) \qquad e = \lim_{n \to \infty} \left(1 + \tfrac{1}{n}\right)^{n}.$$

This expression for e has some similarities to the limit from Expression (1). The following theorem will help us make the connection.

THEOREM 8.23

Expressing e^r as a Limit

For any real number r,

$$e^r = \lim_{n \to \infty} \left(1 + \frac{r}{n}\right)^{n}.$$

PROOF (**THEOREM 8.23**)   The key is to write $\lim_{n \to \infty} (1 + \tfrac{r}{n})^n$ in terms of the characterization of e given in Equation (2). This is easily done with a bit of algebra and a change of variables:

$$\lim_{n \to \infty} \left(1 + \frac{r}{n}\right)^{n} = \lim_{n \to \infty} \left(1 + \frac{1}{n/r}\right)^{n} \qquad \left(\tfrac{r}{n} = \tfrac{1}{n/r}\right)$$

$$= \lim_{n \to \infty} \left(\left(1 + \frac{1}{n/r}\right)^{n/r}\right)^{r} \qquad \left(\text{since } \tfrac{n}{r} \cdot r = n\right)$$

$$= \lim_{m \to \infty} \left(\left(1 + \frac{1}{m}\right)^{m}\right)^{r} \qquad \begin{array}{l}\left(\text{let } m = n/r; \text{ note } m \to \infty\right.\\ \left.\text{as } n \to \infty\right)\end{array}$$

$$= \left(\lim_{m \to \infty} \left(1 + \frac{1}{m}\right)^{m}\right)^{r} \qquad (\text{limit of a composition})$$

$$= e^r. \qquad (\text{Equation (2)})$$

Using Theorem 8.23, we can now calculate the limit in Expression (1):

$$\lim_{n \to \infty} 1000 \left(1 + \tfrac{0.08}{n}\right)^{n(5)} = 1000 \left(\lim_{n \to \infty} \left(1 + \tfrac{0.08}{n}\right)^{n}\right)^{5} \qquad \text{(limit rules)}$$

$$= 1000(e^{0.08})^{5} \qquad \text{(Theorem 8.23)}$$

$$= 1000 e^{0.08(5)} \qquad \text{(rules of exponents)}$$

$$\approx 1491.824698.$$

This limit is what we might have guessed from the information in Table 8.2. It represents the amount of money you would make if you deposited \$1000 into a bank account where it earned 8% yearly interest, compounded "infinitely many" times.

In general, we call compounding "infinitely many" times **continuous compounding**. Continuous compounding is seen often in biological growth, where growth happens all the time in small increments, rather than all at once at the end of the month. In the example we have been discussing, the constant 0.08 represents the **continuous growth rate** of your account balance. The general formula for continuous compounding is given in Theorem 8.24.

❖ **Caution**   A *continuous* growth rate of 0.08 (which represents a yearly growth rate of 8 percent, compounded *continuously*) is very different from a *percentage* growth rate of 8% (compounded once a year).   □

THEOREM 8.24

Continuous Growth Formula

If $Q(t)$ has a *continuous growth rate* of k and an initial value of $Q_0 = Q(0)$, then $Q(t)$ is given by the exponential function

$$Q(t) = Q_0 e^{kt}.$$

PROOF (**THEOREM 8.24**)   We want to prove that the continuous growth formula is the limit, as $n \to \infty$, of the formula for percentage growth with multiple compounding. The key is Theorem 8.23:

$$\lim_{n \to \infty} Q_0 \left(1 + \frac{r}{n}\right)^{nt} = \lim_{n \to \infty} Q_0 \left(\left(1 + \frac{r}{n}\right)^{n}\right)^{t} \qquad \text{(algebra)}$$

$$= Q_0 \left(\lim_{n \to \infty} \left(1 + \frac{r}{n}\right)^{n}\right)^{t} \qquad \text{(limit rules)}$$

$$= Q_0 (e^{r})^{t} \qquad \text{(Theorem 8.23)}$$

$$= Q_0 e^{rt}. \qquad \text{(algebra)}$$

We will soon see several real-world examples that have continuous growth rates. First we give an example of translating between continuous and percentage growth rates. We will use the fact that every exponential function can be written in the form Ae^{kt} or in the form $Ab^{t} = A(1 + r)^{t}$.

EXAMPLE 8.32   **Converting from a percentage growth rate to a continuous growth rate**

The number of worms in a backyard garden increases every year by 7%. What is the *continuous* growth rate of the worm population in the garden?

Solution   The information given tells us that the number of worms in the garden after t years will be

$$W(t) = W_0(1 + 0.07)^t.$$

We are interested in finding some number k such that $W(t) = W_0 e^{kt}$ (this k will be the continuous growth rate). Thus we must solve

$$W_0(1 + 0.07)^t = W_0 e^{kt} \implies (1.07)^t = (e^k)^t \implies 1.07 = e^k$$

$$\implies k = \ln(1.07) \approx 0.06766.$$

The continuous growth rate of the worms is 0.06766. Note that this is close, but not quite equal to, the yearly percentage growth rate of 0.07. Notice also that our computation did not depend on the initial value W_0, nor did it depend on the time t.   □

Most of the time, we will use exponential functions of the form $Q(t) = Q_0 e^{kt}$ (with *continuous* growth rates) to describe real-world examples of exponential growth or decay. Continuous growth or decay rates make sense for human population growth or radioactive decay, which change continuously, whereas percentage growth or decay rates work better for account balances and population growth for animals that have fixed breeding seasons.

◈ *Caution*   You can always translate between a continuous growth rate k (an exponential function of the form $Q(t) = Q_0 e^{kt}$) and a percentage growth rate r (an exponential function of the form $Q(t) = Q_0(1 + r)^t$); just remember that the continuous growth rate k will *not* be equal to the percentage growth rate r.   □

8.6.4 Doubling time and half-life

We have already seen two situations in which a real-life quantity can be represented by an exponential function. In Section 8.4 we saw that if a quantity $Q(t)$ grows at a rate proportional to the quantity itself, then $Q(t)$ is an exponential function of time. Earlier in this chapter, we saw that a quantity $Q(t)$ that grows by a fixed percentage each time period is an exponential function. Now we examine a third type of situation where the quantity to be represented must be an exponential function.

THEOREM 8.25

Exponential Functions Have a Constant Doubling Time or Half-Life

Suppose $Q(t)$ is a quantity that increases with time. The function $Q(t)$ has a constant doubling time if and only if $Q(t)$ is an exponential function.

Similarly, suppose $Q(t)$ is a quantity that decreases with time. The function $Q(t)$ has a constant half-life if and only if $Q(t)$ is an exponential function.

What we mean by ***doubling time*** is the period of time that it takes the quantity to double. Theorem 8.25 states that if the time it takes for an increasing quantity Q to double is the same (i.e., constant) regardless of when you start measuring the quantity, then Q must be exponential. Similarly, the ***half-life*** of a decreasing quantity Q is the amount of time that it takes for the quantity to decrease by half. If this half-life is always the same, then Q can be represented by an exponential decay function. (In fact, Theorem 8.25 also holds for quantities that have a constant tripling time, or a constant quadrupling time, or a constant time to decrease by a third, and so on.)

| EXAMPLE 8.33 | **A function without, and a function with, a constant doubling time** |

Let $P(t) = 2t$ (a linear function) and let $Q(t) = 2^t$ (an exponential function). Table 8.3 shows values of $P(t)$ and $Q(t)$ at various points in time.

t	0	1	2	3	4	5	6
$P(t) = 2t$	0	2	4	6	8	10	12
$Q(t) = 2^t$	1	2	4	8	16	32	64

Table 8.3

By looking at the second row of the table, we can see that the exponential function $Q(t)$ doubles for every unit increase in time. Its doubling time is a constant value of one time unit.

On the other hand, this is not the case for the linear function $P(t)$ shown in the first row of the table. For example, it takes one time unit for $P(t)$ to double from $P(1) = 2$ to $P(2) = 4$. However it then takes *two* time units for $P(t)$ to double from $P(2) = 4$ to $P(4) = 8$. The doubling time of the linear function $P(t)$ is *not* constant; it changes depending on the starting point. □

It is difficult to prove that *only* exponential functions have constant doubling times. However, it is not too difficult to prove the reverse direction of Theorem 8.25—that *every* exponential function has a constant doubling time. We will prove the latter fact in the case where $Q(t)$ is an exponential *growth* function and leave the proof of the exponential decay case for the exercises.

PROOF **("REVERSE" DIRECTION OF THEOREM 8.25)**   Suppose $Q(t) = Q_0 e^{kt}$ is any exponential function representing exponential growth (so $k > 0$). Let α be the time it takes for the quantity $Q(t)$ to double from its initial value Q_0 to twice its initial value $2Q_0$ (since $Q(t) \to \infty$ as $t \to \infty$, we know there must be such a value α; in other words, we know that the quantity must eventually double). In function notation, we have $Q(0) = Q_0$ and $Q(\alpha) = 2Q_0$.

We want to show that regardless of our starting point, the quantity $Q(t)$ will *always* take a time of α to double. In other words, we must show that $Q(t + \alpha) = 2Q(t)$ for all values of t. (Make sure that you understand why this is what we need to prove.) Since $Q(\alpha) = 2Q_0$, we know that $Q_0 e^{k\alpha} = 2Q_0$ and thus that $e^{k\alpha} = 2$. Using this, we can calculate

$$Q(t + \alpha) = Q_0 e^{k(t+\alpha)} \qquad \text{(definition of } Q(t)\text{)}$$

$$= Q_0 e^{kt + k\alpha} = Q_0 e^{kt} e^{k\alpha} \qquad \text{(algebra)}$$

$$= Q_0 e^{kt}(2) \qquad \text{(since } Q(\alpha) = 2Q_0\text{)}$$

$$= 2Q_0 e^{kt} = 2Q(t). \qquad \text{(definition of } Q(t)\text{)}$$

Knowing the doubling time of an exponential function is equivalent to knowing the continuous growth value k; If you know one of these values, you can always find the other. This is illustrated in the following example.

| EXAMPLE 8.34 | **Using doubling time to determine the continuous growth rate** |

Suppose a population of bacteria is growing in a petri dish in such a way that the number of bacteria doubles every 3 hours. Given that there are initially 1000 bacteria in the petri dish, how long will it take for there to be 1,000,000 bacteria?

Solution   Let $B(t)$ be the number of bacteria in the dish after 3 hours. Since $B(t)$ has a constant doubling time, it is an exponential function. Moreover, we know that $B(0) = 1000$; therefore, $B(t) = 1000e^{kt}$ for some value of k. We can use the doubling time to solve for this value of k.

Because the bacteria population doubles every 3 hours, we know that $B(3) = 2B(0) = 2(1000) = 2000$. This means that

$$B(3) = 2000 \implies 1000e^{k(3)} = 2000 \implies e^{3k} = 2 \implies 3k = \ln 2 \implies k = \frac{\ln 2}{3}.$$

We could make a numerical approximation of k at this point, but we will wait until the end of the problem to make any approximations; this will make our final answer much more accurate.

The number of bacteria in the petri dish after t hours is given by the function $B(t) = 1000e^{\left(\frac{\ln 2}{3}\right)t}$. It now remains only to solve the equation $B(t) = 1,000,000$ for t:

$$B(t) = 1,000,000 \implies 1000e^{\left(\frac{\ln 2}{3}\right)t} = 1,000,000 \implies e^{\left(\frac{\ln 2}{3}\right)t} = 1000$$

$$\implies \left(\frac{\ln 2}{3}\right)t = \ln(1000) \implies t = \frac{\ln(1000)}{\frac{\ln 2}{3}} \implies t \approx 29.9.$$

It will be approximately 29.9 hours before there are 1,000,000 bacteria in the petri dish.   □

⊗ *Caution*   You *cannot* solve the problem in Example 8.34 simply by using "common sense" (i.e., by doubling the number of bacteria until the population is greater than 1,000,000). You *may* do this to determine whether your answer is reasonable. (In Example 8.34, you need to double the population just less than 10 times, which means that it will take just under $10(3) = 30$ hours to have 1,000,000 bacteria.) This method will not give you an exact answer. Remember that if a quantity has a constant doubling time (or half-life), then that quantity is represented by an exponential function. You will have to find an equation for that exponential function to solve the problem.   □

❓ *Question*   Can you find the general relationship between the doubling time α of an exponential growth function $Q(t) = Q_0 e^{kt}$ and the continuous growth rate k of that function? (*Hint:* The computations are similar to the ones in Example 8.34, where we used the fact that $Q(\alpha) = 2Q_0$ to solve for the value of k.)   □

Many biological populations tend to have a constant doubling time or half-life. Half-life is also seen in radioactive decay; the amount of a radioactive substance over time always has a constant half-life. This means that a quantity of a radioactive substance always takes the same amount of time to decrease by half, no matter how much or how little of that substance you start with.

EXAMPLE 8.35

Using half-life to determine the percentage growth rate

The half-life of the radioactive substance strontium-90 is 29 years. A sample of rock is found with 80 grams of strontium-90. What is the percentage by which the amount of strontium-90 in the rock decreases each year?

Solution   Let $S(t)$ represent the number of grams of strontium-90 that are left after t years. Since $S(t)$ has a constant half-life, we know that $S(t)$ is an exponential function. We also know that its initial value is $S_0 = S(0) = 80$ grams. Therefore, $S(t) = 80e^{kt}$ for some value k.

We begin by using the given half-life to solve for the value of k. Since $S(t)$ has a half-life of 29 years, we know that $S(29) = \frac{S_0}{2} = \frac{80}{2} = 40$ grams (after 29 years there should be only half of the substance left). Therefore,

$$S(29) = 40 \;\Rightarrow\; 80e^{k(29)} = 40 \;\Rightarrow\; e^{29k} = \tfrac{1}{2} \;\Rightarrow\; k = \frac{\ln\left(\frac{1}{2}\right)}{29}.$$

This means that the amount (in grams) of strontium-90 in the rock after t years is given by the exponential function $S(t) = 80e^{\left(\frac{\ln(0.5)}{29}\right)t}$.

Since we were asked by what percentage the amount of strontium-90 decreased each year, we need to convert $S(t)$ into an exponential function of the form $S(t) = S_0(1-r)^t$ (we are looking for the value of r). We have

$$S(t) = 80e^{\left(\frac{\ln(0.5)}{29}\right)t} = 80\left(e^{\frac{\ln(0.5)}{29}}\right)^t \approx 80(0.97638)^t = 80(1-0.02362)^t.$$

The amount of strontium-90 in the rock decreases by approximately 2.362% each year. □

What you should know after reading Section 8.6

▶ The percentage growth formula, why it makes sense, and how to use it in word problems. Any quantity that undergoes a fixed percentage growth each time period is an exponential function of time.

▶ The formula for percentage growth with multiple compounding, why it makes sense, and how to use it to solve word problems. How does compounding more often affect the yearly total of an account balance, and why?

▶ The formula for continuous growth and how it is obtained using a limit of the multiple compounding formula. Also know that $e^r = \lim\limits_{n \to \infty} \left(1 + \frac{r}{n}\right)^n$, and how to use this expression in computational problems. Know the difference between a continuous growth rate and a percentage growth rate, and how to translate between the two.

▶ The definitions of doubling time and half-life, and what it means for the doubling time or half-life of a quantity to be constant. Any quantity with a constant doubling time or half-life is an exponential function. Be able to solve word problems involving doubling time and half-life.

Exercises 8.6

Concepts

0. Read the section and make your own summary of the material. Include whatever you think will help you review the material later on.

1. If a quantity $Q(t)$ grows or decreases by a fixed percentage each time period, then $Q(t)$ is a _____ function.

2. If a quantity $Q(t)$ grows at a rate proportional to _____, then $Q(t)$ is an exponential function.

3. If a quantity $Q(t)$ has a _____ doubling time or half-life, then $Q(t)$ is a _____ function.

4. Show that the percentage growth formula is just a special case of the formula for percentage growth with multiple compounding (where the compounding occurs only once per time period).

5. Suppose you owe money to a loan company that charges 15% interest annually. Would you rather have that company compound your interest monthly, daily, or continuously, and why?

6. If a credit card company charges an annual percentage rate of 13.99% and compounds interest monthly, what percentage is charged each month as interest on any outstanding balances?

7. Suppose you wish to put some money in an investment account. Which is better, to have 16% annual interest compounded yearly or to have 15% annual interest compounded daily? (*Hint:* In each case, think about the factor by which your balance would increase each year.)

8. Suppose a quantity $Q(t)$ is an exponential function that increases with time (measured in days). If the quantity takes 7 days to double in size, how long will it take for that quantity to increase eightfold?

9. Suppose a quantity $Q(t) = Q_0 e^{kt}$, with time measured in days, has a doubling time of 5 days. Without finding a value for k, fill in the blanks: $e^{5k} = $ _____ and $e^{10k} = $ _____ .

10. Suppose a quantity $Q(t)$, with time measured in years, has a half-life of 17 years and an initial amount of Q_0. Without finding an equation for $Q(t)$, fill in the blanks: $Q(17) = $ _____ , $Q(51) = $ _____ , and $Q(-34) = $ _____ .

11. Suppose a growing population $W(t)$ of warthogs on an island grows at a rate proportional to the number of warthogs on the island. Will the function $W(t)$ have a constant doubling time? Why or why not?

12. Show that the function $P(t) = t^2 + 1$ does not have a constant doubling time. (*Hint:* Show that $P(t)$ takes different amounts of time to double if you choose two different starting points.)

Skills

■ Calculate each of the following limits by using the fact that $\lim_{x \to \infty} \left(1 + \frac{r}{x}\right)^x = e^r$.

13. $\lim_{x \to \infty} \left(1 + \frac{3}{x}\right)^{2x}$

14. $\lim_{x \to \infty} \left(1 - \frac{1}{x}\right)^x$

15. $\lim_{x \to \infty} (1 + 7x^{-1})^{-3x}$

16. $\lim_{x \to \infty} \left(1 + \frac{1}{3x}\right)^{4x+1}$

17. $\lim_{x \to 0^+} (1 + 5x)^{\frac{1}{x}}$

18. $\lim_{x \to 0^+} \left(1 - \frac{x}{2}\right)^{\frac{3}{x}}$

19. Suppose a quantity $Q(t)$, with t measured in years, has a continuous growth rate of 0.17. What is the yearly percentage growth rate of the quantity?

20. Suppose a quantity $Q(t)$, with t measured in years, has a yearly percentage growth rate of 24%. What is the continuous growth rate of the quantity?

21. Suppose a quantity $Q(t)$, with t measured in years, has a monthly percentage growth rate of 2.3%. What is the yearly percentage growth rate of the quantity?

22. Suppose a quantity $Q(t)$, with t measured in years, has a monthly percentage growth rate of 3.1%. What is the continuous growth rate of the quantity?

23. Suppose a quantity $Q(t)$, with t measured in years, has a half-life of 32 years. What is the continuous growth rate of the quantity?

24. Suppose a quantity $Q(t)$, with t measured in years, triples every 17 years. What is the yearly percentage growth rate of the quantity?

25. Suppose a quantity $Q(t)$, with t measured in years, has a yearly percentage growth rate of 13%. What is the doubling time of the quantity?

26. Suppose a quantity $Q(t)$, with t measured in years, decreases by 4% each year.

(a) How long will it take for the quantity to decrease by half?

(b) How long will it take for the quantity to decrease by a third?

(c) What is the continuous decay rate of the quantity $Q(t)$?

27. Suppose a quantity $Q(t)$, with t measured in years, has a doubling time of 6.5 years.

(a) What is the percentage by which $Q(t)$ increases each year?

(b) How long will it take for the quantity $Q(t)$ to triple?

(c) What is the continuous growth rate of the quantity $Q(t)$?

Applications

28. Suppose you earn 4% annual interest, compounded monthly, on the balance in your savings account. If you wanted to change to an account where interest is compounded yearly, what annual percentage rate would earn you the same total amount of interest over the course of the year?

29. Suppose you owe 18% annual interest, compounded monthly, on the balance of your credit card account. What is the percentage by which your balance would increase each year? What would that yearly percentage be if the credit card company decided to compound your interest daily instead?

30. Suppose you owe $5000 to Transcam, a loan company that charges 22% interest annually. Answer the following questions, assuming that you do not make any payments on, or add to, this loan.

(a) How much would you owe Transcam in 15 years if the company compounded the interest only once a year? What if they compounded the interest monthly? Daily? Four times each day?

(b) How much would you owe Transcam in 15 years if it compounded the interest *continuously*?

(c) By what percentage would your balance with Transcam increase each year if it compounded the interest yearly? Monthly? Daily? Continuously?

(d) How many years will it take before you owe Transcam $1,000,000, assuming that it compounds interest yearly? Monthly? Daily? Continuously?

31. Dr. Drosophila is breeding fruit flies in his laboratory. He started with 600 fruit flies, and after 3 days there were already 840 fruit flies in Dr. Drosophila's laboratory. Answer the following questions, assuming that the fruit fly population grows at a rate proportional to the number of fruit flies in the laboratory.

   (a) By what percentage does the fruit fly population increase each day?
   (b) What is the continuous growth rate of the fruit fly population?
   (c) How long will it take for the fruit fly population to quadruple?
   (d) How long will it be before there are 1,000,000 fruit flies packed into Dr. Drosophila's laboratory?

32. Linda and Alina start a rumor that peanut butter makes your hair fall out. The number of people that have heard this rumor triples every 4 hours.

   (a) How long will it be before 700 people have heard this rumor?
   (b) How long will it take for the number of people who have heard this rumor to double?
   (c) By what percentage does the number of people who have heard this rumor increase each hour?
   (d) By what percentage does the number of people who have heard this rumor increase each day? (Careful; this one is harder!)

33. The half-life of the radioactive substance strontium-90 is 29 years. Suppose a sample of rock contains 250 grams of strontium-90.

   (a) What percentage of strontium-90 decays each year?
   (b) What is the continuous decay rate of strontium-90?
   (c) How long will it be before the rock sample contains only 6 grams of strontium-90?
   (d) At one point the rock sample contained 900 grams of strontium-90. How long ago was that?
   (e) What percentage of the strontium-90 will be left in 300 years?

(f) How long will it be before 95% of the strontium-90 has decayed?

Proofs

34. Given that $\lim_{n \to \infty} (1 + \frac{r}{n})^n = e^r$, prove that the limit of the multiple compounding formula $Q_0(1 + \frac{r}{n})^{nt}$ as $n \to \infty$ is the continuous growth formula $Q_0 e^{rt}$.

35. Suppose that $Q(t)$ is an exponential growth function and that it takes 12 time units for the quantity $Q(t)$ to increase from its initial amount, Q_0, to twice that amount, $2Q_0$. Prove that $Q(t)$ has a *constant* doubling time of 12 time units. (In other words, prove that for *any* time t, we have $Q(t + 12) = 2Q(t)$.)

36. Prove that every decreasing exponential function has a constant half-life, as follows: Suppose that $Q(t)$ is a decreasing exponential function with $Q_0 = Q(0)$ and that it takes α time units for the function $Q(t)$ to decrease by half from Q_0 to $\frac{Q_0}{2}$. Prove that this half-life α for $Q(t)$ is constant; other words, prove that it *always* takes $Q(t)$ a period of α time units to decrease by half (regardless of your starting point).

■ Suppose a quantity $Q(t)$, with t measured in years, has a yearly percentage growth rate of r, a continuous growth rate of k, a doubling time of α, and a tripling time of β.

37. Find a formula that expresses the continuous growth rate k in terms of the percentage growth rate r, and prove this formula.

38. Find a formula that expresses the doubling time α in terms of the continuous growth rate k, and prove this formula.

39. Find a formula that expresses the percentage growth rate r in terms of the doubling time α, and prove this formula.

40. Find a formula that expresses the tripling time β in terms of the doubling time α, and prove this formula.

8.7 | **L'Hôpital's Rule**

Although we currently have many powerful techniques for calculating limits, there are still some limits that we don't know how to calculate. In this section we discuss three of the most common indeterminate forms of limits and introduce a new computational technique that will help us solve these types of limits when algebraic simplification is not possible. At the conclusion of this section, we will use this new technique to discuss the "dominance" of exponential functions over power functions.

8.7.1 The indeterminate forms $\frac{0}{0}$, $\frac{\infty}{\infty}$, and $0 \cdot \infty$

We have already seen many limits of the form $\frac{0}{0}$ and $\frac{\infty}{\infty}$; both of these types of limits are **indeterminate.** This means that we cannot say whether or not they exist, or, if they exist, what they are equal to—at least not before doing some algebra. For example, the four limits in Example 8.36 are all initially of the form $\frac{0}{0}$, but some exist and some do not. Moreover, the ones that do exist are equal to different real numbers.

EXAMPLE 8.36

Four limits that illustrate that limits of the form $\frac{0}{0}$ are indeterminate

Calculate each of the following limits.

(a) $\displaystyle\lim_{x \to 1} \frac{x-1}{(x-1)^2}$

(b) $\displaystyle\lim_{x \to 1} \frac{(x-1)^2}{x-1}$

(c) $\displaystyle\lim_{x \to 1} \frac{x-1}{x-1}$

(d) $\displaystyle\lim_{x \to 1} \frac{x-1}{3(x-1)}$

Solution   Each of the four limits initially has the indeterminate form $\frac{0}{0}$ but can be evaluated after some algebraic simplification.

(a) $\displaystyle\lim_{x \to 1} \frac{x-1}{(x-1)^2} = \lim_{x \to 1} \frac{1}{x-1} \to \frac{1}{0}$ does not exist.

(b) $\displaystyle\lim_{x \to 1} \frac{(x-1)^2}{x-1} = \lim_{x \to 1} \frac{x-1}{1} = \frac{1-1}{1} = \frac{0}{1} = 0.$

(c) $\displaystyle\lim_{x \to 1} \frac{x-1}{x-1} = \lim_{x \to 1} 1 = 1.$

(d) $\displaystyle\lim_{x \to 1} \frac{x-1}{3(x-1)} = \lim_{x \to 1} \frac{1}{3} = \frac{1}{3}.$

? *Question*   Calculate the limits of the functions in Example 8.36 as $x \to \infty$ to illustrate that limits of the form $\frac{\infty}{\infty}$ are indeterminate.

? *Question*   Example 8.36 did not involve any exponential functions (we chose to use quotients of polynomial functions because they are so simple). Can you find other examples of limits of the forms $\frac{0}{0}$ and $\frac{\infty}{\infty}$ that can be solved using algebraic simplification? Try to find examples where the limit does not exist as well as various examples where the limit exists.

Limits of the form $0 \cdot \infty$ are also indeterminate and require algebraic simplification before they can be solved. In fact, limits of the form $0 \cdot \infty$ are actually limits of the form $\frac{0}{0}$ or $\frac{\infty}{\infty}$ in disguise. They can always be rewritten in the form $\frac{0}{0}$ or $\frac{\infty}{\infty}$. Example 8.37 illustrates this fact. Since all limits of the form $0 \cdot \infty$ can be rewritten in the form $\frac{0}{0}$ or $\frac{\infty}{\infty}$, we will focus primarily on these latter two indeterminate forms.

EXAMPLE 8.37

Converting a limit of the form $0 \cdot \infty$ into a limit of the form $\frac{0}{0}$ or $\frac{\infty}{\infty}$

Rewrite the following limit so it is in the form $\frac{0}{0}$. Then rewrite the limit in the form $\frac{\infty}{\infty}$.

(1) $$\lim_{x \to \infty} x^2 e^{-3x}.$$

Solution   This limit is of the form $0 \cdot \infty$, since as $x \to \infty$, we have $x^2 \to \infty$ and $e^{-3x} \to 0$. This limit can be written in the form $\frac{0}{0}$, as follows:

(2) $$\lim_{x \to \infty} \frac{x^2}{e^{3x}}.$$

We could also write the original limit in the form $\frac{\infty}{\infty}$, using the fact that $x^2 = \frac{1}{1/x^2}$:

(3)
$$\lim_{x\to\infty} \frac{e^{-3x}}{\frac{1}{x^2}}.$$

We will solve the limit in Expression (1) in Example 8.42 by writing it in the form of Expression (2) and using L'Hôpital's Rule. □

⑤ **Caution** Note that $0 \cdot \infty$ is *not* always equal to zero, even though zero times any number is zero. Since "∞" is not a real number, we cannot say that its product with zero is necessarily equal to zero. □

⑤ **Caution** Although $\frac{0}{0}$, $\frac{\infty}{\infty}$, and $0 \cdot \infty$ are indeterminate forms, the forms $\frac{0}{\infty}$ and $\frac{\infty}{0}$ are *not* indeterminate. For example, consider the limit $\lim_{x\to\infty} \frac{2^{-x}}{x^2+1}$. This limit is of the form $\frac{0}{\infty}$, which can always be rewritten in the form $0 \cdot 0$, which is equal to zero:

$$\lim_{x\to\infty} \frac{2^{-x}}{x^2+1} = \lim_{x\to\infty} 2^{-x} \frac{1}{x^2+1} \to 0 \cdot 0 = 0.$$

Similarly, the form $\frac{\infty}{0}$ can always be rewritten in the form $\infty \cdot \infty$, which is infinite. □

⑤ **Caution** Don't forget that limits of the form $\infty + \infty$ are infinite but limits of the form $\infty - \infty$ are indeterminate. To solve a limit of the form $\infty - \infty$, it is often a good idea to do some algebra first. For example, you can evaluate the limit $\lim_{x\to 0}\left(\frac{1}{x} - \frac{1}{2^x-1}\right)$ by rewriting it as

$$\lim_{x\to 0}\left(\frac{1}{x} - \frac{1}{2^x-1}\right) = \lim_{x\to 0} \frac{2^x - 1 - x}{x(2^x-1)}.$$

We will soon be able to solve this (rewritten) limit using L'Hôpital's Rule (since it is of the form $\frac{0}{0}$). □

The following theorem summarizes the indeterminate forms we have discussed, as well as some *non*indeterminate forms and what they are equal to. We will be a little loose with our mathematical notation in this theorem. For example, a limit of the form $\frac{1}{0}$ could equal ∞, $-\infty$, or "DNE," depending on what the limit approaches from the right-hand and left-hand sides; however, here we simply state that "$\frac{1}{0} \to \infty$," to indicate that $\frac{1}{0}$ is infinite.

THEOREM 8.26

Basic Indeterminate and Nonindeterminate Forms for Limits

The following forms are *indeterminate*:

$$\frac{0}{0}, \qquad \frac{\infty}{\infty}, \qquad 0 \cdot \infty, \qquad \infty - \infty.$$

The following forms are *nonindeterminate* (and approach the indicated limit):

$$\frac{1}{0} \to \infty, \qquad \frac{0}{1} \to 0, \qquad \frac{1}{\infty} \to 0, \qquad \frac{\infty}{1} \to \infty,$$

$$\frac{0}{\infty} \to 0, \qquad \frac{\infty}{0} \to \infty, \qquad \infty + \infty \to \infty, \qquad \infty \cdot \infty \to \infty.$$

Technical Note: We will use the expression "$\frac{\infty}{\infty}$" to represent any limit where the numerator and denominator become infinite; this means that the indeterminate form $\frac{\infty}{\infty}$ also includes limits that approach $\frac{-\infty}{\infty}$, $\frac{\infty}{-\infty}$, and $\frac{-\infty}{-\infty}$.

8.7.2 Statement of L'Hôpital's Rule

Theorem 8.27 presents a calculational tool that we can use to solve some limits of the form $\frac{0}{0}$ or $\frac{\infty}{\infty}$ (as well as limits of the form $0 \cdot \infty$, which can be rewritten in either of the first two indeterminate forms).

THEOREM 8.27

L'Hôpital's Rule

Suppose f and g are differentiable functions on some punctured interval $(c - \delta, c) \cup (c, c + \delta)$ around $x = c$ and that $g(x)$ is nonzero on that punctured interval. If $\lim_{x \to c} f(x)$ and $\lim_{x \to c} g(x)$ are both zero, or if $f(x)$ and $g(x)$ both become infinite as $x \to c$, then

$$\lim_{x \to c} \frac{f(x)}{g(x)} = \lim_{x \to c} \frac{f'(x)}{g'(x)},$$

as long as the second limit exists or is infinite.

The conclusion also holds if $x \to \infty$ or $x \to -\infty$ as long as f and g are differentiable on some interval (N, ∞) and g is nonzero on that interval.

There are a few technical points to notice about the hypotheses of this theorem. First, notice that the functions f and g must be differentiable near, but not necessarily at, $x = c$. Second, the function g in the denominator must be nonzero everywhere near, but not necessarily at, $x = c$. Finally, notice that the conclusion does *not* necessarily hold if the limit on the right-hand side fails to be a real number or some kind of infinity; in other words, the conclusion might not hold if the limit on the right does not exist for some reason *other* than being infinite (for example, the function $\frac{f'(x)}{g'(x)}$ might oscillate as $x \to c$).

EXAMPLE 8.38

Applying L'Hôpital's Rule to a limit of the form $\frac{0}{0}$

Consider the limit

$$\lim_{x \to 0} \frac{x^2}{2^x - 1}.$$

We cannot solve this limit using algebraic simplification. Let's see whether L'Hôpital's Rule applies to this limit. Both $f(x) = x^2$ and $g(x) = 2^x - 1$ are differentiable everywhere (so in particular they are differentiable near $x = 0$). It is also true that the function $g(x)$ is nonzero as $x \to 0$ *except* at the actual point $x = 0$. Most important, both $f(x)$ and $g(x)$ approach zero as $x \to 0$ (this is what causes the limit to be in the indeterminate form $\frac{0}{0}$).

Now that we know it applies, we can use L'Hôpital's Rule to solve this limit by replacing f by f' and g by g':

$$\lim_{x \to 0} \frac{x^2}{2^x - 1} \overset{\text{L'H}}{=} \lim_{x \to 0} \frac{\frac{d}{dx}(x^2)}{\frac{d}{dx}(2^x - 1)} \qquad \text{(L'Hôpital's Rule)}$$

$$= \lim_{x \to 0} \frac{2x}{(\ln 2)2^x} \qquad \text{(differentiation)}$$

$$= \frac{2(0)}{(\ln 2)2^0} = \frac{0}{\ln 2} = 0. \qquad \text{(limit rules)}$$

Notice that we write "L'H" to indicate the application of L'Hôpital's Rule. □

✿ *Caution*   When we apply L'Hôpital's Rule, we do *not* differentiate the quotient $\frac{f(x)}{g(x)}$ (which would require the quotient rule). Instead, we differentiate the numerator and denominator individually. □

✿ *Caution*   L'Hôpital's Rule does *not* allow us to replace $\lim_{x \to c} f(x)$ with $\lim_{x \to c} f'(x)$. A quotient must be present in order for us to apply L'Hôpital's Rule. □

The proof of L'Hôpital's Rule requires a more general version of the Mean Value Theorem called the ***Cauchy Mean Value Theorem***. This theorem (and the resulting proof of L'Hôpital's Rule) are too involved to treat properly here. However, although the general proof of L'Hôpital's Rule is somewhat difficult, it is easy to prove L'Hôpital's Rule in a simple special case. We include the proof of the following special case so that you can have some idea of why L'Hôpital's Rule is reasonable.

THEOREM 8.28

A Simple Case of L'Hôpital's Rule

Suppose that f and g are differentiable (and thus continuous) on an interval around $x = c$. If $f(c) = g(c) = 0$ but $g'(c) \neq 0$, then

$$\lim_{x \to c} \frac{f(x)}{g(x)} = \frac{f'(c)}{g'(c)}.$$

PROOF   (**THEOREM 8.28**)   We will start from the right-hand side of the equality in Theorem 8.28 and work toward the left-hand side.

$$\frac{f'(c)}{g'(c)} = \frac{\left(\lim\limits_{x \to c} \dfrac{f(x) - f(c)}{x - c} \right)}{\left(\lim\limits_{x \to c} \dfrac{g(x) - g(c)}{x - c} \right)} \qquad \text{(alternative definition of derivative)}$$

$$= \lim_{x \to c} \frac{\left(\frac{f(x) - f(c)}{x - c} \right)}{\left(\frac{g(x) - g(c)}{x - c} \right)} \qquad \text{(quotient rule for limits)}$$

$$= \lim_{x \to c} \frac{f(x) - f(c)}{g(x) - g(c)} \qquad \text{(cancellation)}$$

$$= \lim_{x \to c} \frac{f(x) - 0}{g(x) - 0} = \lim_{x \to c} \frac{f(x)}{g(x)}. \qquad \text{(since } f(c) = 0 \text{ and } g(c) = 0\text{)}$$

■

8.7.3 Using L'Hôpital's Rule

L'Hôpital's Rule can be used to solve limits that have the indeterminate forms $\frac{0}{0}$, $\frac{\infty}{\infty}$, and $0 \cdot \infty$ (recall that this last type of indeterminate form can always be converted into either of the first two types). In Chapter 9 we will look at more complicated types of limits that can be written in such a way that L'Hôpital's Rule applies.

✿ *Caution*   Be careful not to use L'Hôpital's Rule unless it applies! Make sure, before applying L'Hôpital's Rule, that the limit you are attempting to find is of the form $\frac{0}{0}$ or $\frac{\infty}{\infty}$. □

EXAMPLE 8.39

Checking to see whether L'Hôpital's Rule applies

Determine whether or not L'Hôpital's Rule applies to each of the following limits (as they are currently written).

(a) $\displaystyle\lim_{x \to 1} \frac{(x-1)^2}{x-1}$

(b) $\displaystyle\lim_{x \to 0} \frac{x^2}{1+2^x}$

(c) $\displaystyle\lim_{x \to \infty} (2^x - x^3)$

(d) $\displaystyle\lim_{x \to \infty} \frac{2^x}{1-3^x}$

Solution

(a) L'Hôpital's Rule applies in this case, since $\frac{(x-1)^2}{x-1} \to \frac{0}{0}$ as $x \to 1$. The other hypotheses of Theorem 8.27 are also satisfied by the numerator and denominator, as they are in most common examples (check this!). Usually we will focus mostly on whether or not the limit is of the form $\frac{0}{0}$ or $\frac{\infty}{\infty}$. Although L'Hôpital's Rule does apply here, it is actually easier to use algebra to solve this limit than it is to use L'Hôpital's Rule! Try it both ways and see!

(b) L'Hôpital's Rule does *not* apply in this case, since this limit approaches $\frac{0}{2} = 0$ as $x \to 0$.

(c) L'Hôpital's Rule does *not* apply here, since $(2^x - x^3)$ is not a quotient! Remember that L'Hôpital's Rule applies only to quotients $\frac{f(x)}{g(x)}$ that approach $\frac{0}{0}$ or $\frac{\infty}{\infty}$.

(d) L'Hôpital's Rule applies in this case, since $\frac{2^x}{1-3^x} \to \frac{\infty}{-\infty}$ as $x \to \infty$. We will solve this limit using L'Hôpital's Rule in Example 8.40. □

Examples 8.40–8.42 illustrate various limits for which L'Hôpital's Rule *does* apply.

EXAMPLE 8.40

Applying L'Hôpital's Rule to a limit of the form $\frac{\infty}{\infty}$

Calculate $\displaystyle\lim_{x \to \infty} \frac{2^x}{1-3^x}$.

Solution As $x \to \infty$, we have $2^x \to \infty$ and $1 - 3^x \to 1 - \infty = -\infty$. Therefore, this limit is of the indeterminate form $\frac{\infty}{\infty}$ (to be more precise, we could say that it is of the form $\frac{\infty}{-\infty}$). L'Hôpital's rule applies to this limit.

L'Hôpital's Rule says that we can take the derivatives of the numerator and the denominator without changing the value of the limit:

$$\lim_{x \to \infty} \frac{2^x}{1-3^x} \overset{\text{L'H}}{=} \lim_{x \to \infty} \frac{(\ln 2)2^x}{0 - (\ln 3)3^x} \qquad \text{(L'Hôpital's Rule)}$$

$$= \lim_{x \to \infty} \frac{(\ln 2)2^x}{-(\ln 3)3^x} \qquad \text{(simplify)}$$

$$= -\frac{\ln 2}{\ln 3} \lim_{x \to \infty} \left(\frac{2}{3}\right)^x \qquad \text{(algebra and limit rules)}$$

$$\to -\frac{\ln 2}{\ln 3}(0) = 0. \qquad \text{(limit of } b^x \text{ as } x \to \infty \text{ for } 0 < b < 1)$$

In the second line of the calculation, even though L'Hôpital's Rule applies, it would not be of any help (try it to see why!). □

⚜ **Caution** Although L'Hôpital's Rule is an important and powerful computational tool, it does not do all of our work for us! First of all, after applying L'Hôpital's Rule, we still have to find a limit. Second, sometimes L'Hôpital's Rule applies but is not helpful (in which case we might try algebra instead). □

The next example illustrates that sometimes it is necessary to apply L'Hôpital's Rule more than once in order to change a limit into one that is simple enough to solve.

EXAMPLE 8.41

Applying L'Hôpital's Rule twice

Calculate $\lim\limits_{x \to \infty} \dfrac{2^x - 1}{x^2 + 3x + 5}$.

Solution   We cannot simplify this limit with algebra. However, it is a limit of the form $\frac{\infty}{\infty}$, so we can try L'Hôpital's Rule:

$$\lim_{x \to \infty} \frac{2^x - 1}{x^2 + 3x + 5} \overset{\text{L'H}}{=} \lim_{x \to \infty} \frac{(\ln 2)2^x}{2x + 3} \qquad \text{(L'Hôpital's Rule)}$$

Although L'Hôpital's Rule did seem to simplify our limit, we are still left with a limit that we cannot compute using just algebra and limit rules. This new limit is also of the form $\frac{\infty}{\infty}$, so it might be worth trying L'Hôpital's Rule again (especially since we can't think of any other method for calculating this limit). Continuing the calculation, we have

$$\overset{\text{L'H}}{=} \lim_{x \to \infty} \frac{(\ln 2)(\ln 2)2^x}{2} \qquad \text{(L'Hôpital's Rule)}$$

$$\to \frac{\infty}{2} = \infty. \qquad \text{(solve limit)} \qquad \square$$

? *Question*   Why was it a good idea to apply L'Hôpital's Rule twice in Example 8.41 but not in Example 8.40? $\qquad \square$

Example 8.42 illustrates how to compute a limit of the form $0 \cdot \infty$ using L'Hôpital's Rule. The key step in the computation is rewriting the limit in the form $\frac{0}{0}$ or $\frac{\infty}{\infty}$ so that we can apply L'Hôpital's Rule.

EXAMPLE 8.42

Rewriting a limit of the form $0 \cdot \infty$ so that we can apply L'Hôpital's Rule

Calculate the limit $\lim\limits_{x \to \infty} x^2 e^{-3x}$.

Solution   Since $x^2 \to \infty$ and $e^{-3x} \to 0$ as x approaches ∞, this limit is of the indeterminate form $\infty \cdot 0$. By Example 8.37, we can write this limit in the form $\frac{0}{0}$ or $\frac{\infty}{\infty}$; then we will be able to apply L'Hôpital's Rule. Since the form $\frac{\infty}{\infty}$ turns out to be simpler here, we will use that form:

$$\lim_{x \to \infty} x^2 e^{-3x} = \lim_{x \to \infty} \frac{x^2}{e^{3x}}.$$

Whenever you rewrite a function of the form $0 \cdot \infty$, you will have a choice between two ways to do so. Usually one way is easier to work with than the other, and in many cases one way will work (after applying L'Hôpital's Rule) whereas the other will not! After a bit of practice with problems of this type, you should be able to determine which way of rewriting the function is more likely to be useful.

We can now use L'Hôpital's Rule to calculate the limit:

$$\lim_{x \to \infty} x^2 e^{-3x} = \lim_{x \to \infty} \frac{x^2}{e^{3x}} \qquad \left(\text{rewrite; now of form } \tfrac{\infty}{\infty}\right)$$

$$\overset{\text{L'H}}{=} \lim_{x \to \infty} \frac{2x}{3e^{3x}} \qquad \left(\text{L'Hôpital's Rule; still of form } \tfrac{\infty}{\infty}\right)$$

$$\overset{\text{L'H}}{=} \lim_{x \to \infty} \frac{2}{9e^{3x}} \qquad \text{(L'Hôpital's Rule)}$$

$$\to \frac{2}{\infty} \to 0. \qquad \text{(solve limit)}$$

Notice that we used L'Hôpital's Rule twice. We decided to apply it for a second time for two reasons. First, there were no possible algebraic simplifications at that point that would help us solve the limit. Second, the first application of L'Hôpital's Rule simplified the limit by decreasing the power of the numerator by 1, and thus it was reasonable to anticipate that a second application of L'Hôpital's Rule would further simplify the limit. □

❓ *Question* The examples of L'Hôpital's Rule we have seen so far have all involved $x \to \infty$. L'Hôpital's Rule also applies if x approaches a real number c, as long as the limit is in the form $\frac{0}{0}$ or $\frac{\infty}{\infty}$ as $x \to c$. For example, we can calculate the limit $\lim_{x \to 1} \frac{2 - 2^x}{x - 1}$ using L'Hôpital's Rule. Try it! □

✸ *Caution* Don't forget that when a limit is in an "indeterminate form" (like $\frac{0}{0}$, $\frac{\infty}{\infty}$, or $0 \cdot \infty$), it *may or may not* exist. This means that you can't assume the limit "DNE" when you find one of these indeterminate forms; you must first use algebra or L'Hôpital's Rule (if it applies). An indeterminate form is never a good "final answer" to a limit. □

8.7.4 Dominance of exponential functions over power functions

A function f is said to **dominate** another function g if it grows much faster than g in the long run (i.e., as $x \to \infty$). (Sometimes this is rather loosely written as $x \gg 0 \Rightarrow f(x) \gg g(x)$, meaning that if x is *much* greater than 0, then $f(x)$ is *much* bigger than $g(x)$.) As we will soon see, exponential functions grow so quickly (at least in the long run) that they dominate all power and polynomial functions.

DEFINITION 8.6

Dominance of Functions

Suppose f and g are functions that approach ∞ as $x \to \infty$. We say that f **dominates** g if

$$\lim_{x \to \infty} \frac{f(x)}{g(x)} = \infty.$$

If a function f dominates a function g, then eventually the values of $f(x)$ are larger (in fact *much* larger) than the values of $g(x)$. However, be careful not to equate the concept of dominance with the concept of being "larger." Dominance is a condition related to the rates of growth of functions, not just to the function values. If a function f grows significantly faster than a function g, then f will eventually become larger than g. However, if f is a function that is eventually larger than a function g, it is *not* necessarily true that f dominates g. This is summarized in the following theorem.

THEOREM 8.29

Dominance Is a Stronger Condition Than "Greater Than"

If f dominates g, then there exists some positive number N such that $f(x) > g(x)$ for all $x > N$. The converse of the previous sentence is false.

The following example proves (by giving a counterexample) that the converse of the first sentence in Theorem 8.29 is false.

EXAMPLE 8.43

A function with larger values is not necessarily dominant

If $f(x) = x + 1$ and $g(x) = x$, then

$$\lim_{x \to \infty} \frac{f(x)}{g(x)} = \lim_{x \to \infty} \frac{x+1}{x} = \lim_{x \to \infty} \frac{x+1}{x} \left(\frac{1/x}{1/x} \right) = \lim_{x \to \infty} \frac{1 + \frac{1}{x}}{1} \to \frac{1+0}{1} = 1.$$

The function $f(x) = x + 1$ does *not* dominate the function $g(x) = x$, even though $f(x)$ is always larger than $g(x)$. (Note that we could also have used L'Hôpital's Rule to solve this limit.) Intuitively, f does not dominate g because f and g are growing at essentially the same rate. One function dominates another as $x \to \infty$ only if it is growing at a much faster rate.   □

🐧 *Caution*   Given two functions f and g, it can be that *neither* function dominates the other; this is in fact the case in Example 8.43.   □

PROOF (**THEOREM 8.29**)   Suppose that f and g are functions such that $\lim_{x \to \infty} \frac{f(x)}{g(x)} = \infty$. By definition (see Section 2.2), this means that

For all $M > 0$, there exists some $N > 0$ such that if $x > N$, then $\frac{f(x)}{g(x)} > M$.

In particular, if we choose $M = 1$, then there must be some $N > 0$ such that $\frac{f(x)}{g(x)} > 1$ for all $x > N$. Since both $f(x)$ and $g(x)$ approach infinity as $x \to \infty$ (by hypothesis), we can assume that we have chosen N large enough so that $g(x)$ is positive. Assuming that $g(x)$ is positive, we can multiply both sides of the inequality $\frac{f(x)}{g(x)} > 1$ by $g(x)$ and preserve the inequality to get $f(x) > g(x)$. This means that

There is some $N > 0$ such that if $x > N$, then $f(x) > g(x)$.

Therefore, if f dominates g, then $f(x) > g(x)$ for all sufficiently large values of x.   ■

One interesting fact about exponential functions Ab^x or Ae^{kx} is that they always dominate power functions Cx^r. (We are comparing only those exponential functions and power functions that approach infinity as $x \to \infty$.) Even if the constant k (or b) is very small and the power r is very large, and even if the leading coefficient A of the exponential function is small while the leading coefficient C of the power function is large, the exponential function will still eventually overtake the power function!

THEOREM 8.30

Exponential Functions Always Dominate Power and Polynomial Functions

If $f(x)$ is *any* exponential growth function $f(x) = Ae^{kx}$ (with $k > 0$) or $f(x) = Ab^x$ (with $b > 1$), and $g(x)$ is *any* power function or polynomial function, then f dominates g.

We won't prove Theorem 8.30, but its proof has the same type of format as the calculation we will do in Example 8.45. The following example graphically illustrates Theorem 8.30.

EXAMPLE 8.44

An exponential function will eventually dominate a power function

Consider the functions $f(x) = 0.1e^{0.2x}$ and $g(x) = 30x^{10}$. Figures 8.26 and 8.27 show graphs of these two functions (note that vastly different scales are used for the y-axis!).

Although $f(x)$ initially grows more slowly than $g(x)$, Theorem 8.30 guarantees that in the long run, the exponential function $f(x)$ will overtake, and grow faster than,

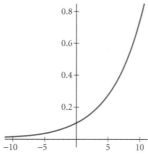

Figure 8.26 $f(x) = 0.1e^{0.2x}$

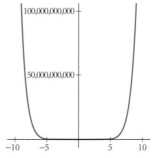

Figure 8.27 $g(x) = 30x^{10}$

the power function $g(x)$. Figure 8.28 shows the two graphs on the same set of axes, with a graphing window large enough so that we can see that f eventually becomes larger than g (and in fact grows at a much faster rate than g). Note that f is smaller than g until $x \approx 303$, at which time f finally overtakes g and grows much more quickly (note how steep the graph of f is at that point, compared to the steepness of the graph of g).

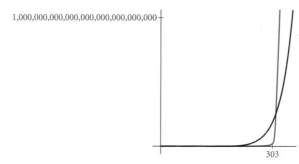

Figure 8.28 $f(x) = 0.1e^{0.2x}$ and $g(x) = 30x^{10}$

In the following example we use L'Hôpital's Rule to verify the dominance of an exponential function over a power function (using the same functions we graphed in Example 8.44).

EXAMPLE 8.45

Showing that an exponential function dominates a power function

Show that the exponential function $f(x) = 0.1e^{0.2x}$ dominates the power function $g(x) = 30x^{10}$ as x approaches ∞.

Solution   By Definition 8.6, we must calculate

$$\lim_{x \to \infty} \frac{f(x)}{g(x)} = \lim_{x \to \infty} \frac{0.1e^{0.2x}}{30x^{10}} \qquad \text{(note this limit is of the form } \tfrac{\infty}{\infty}\text{)}$$

$$\overset{\text{L'H}}{=} \lim_{x \to \infty} \frac{0.1(0.2)e^{0.2x}}{30(10)x^9} \qquad \text{(L'Hôpital's Rule; still } \tfrac{\infty}{\infty}\text{)}$$

$$\overset{\text{L'H}}{=} \lim_{x \to \infty} \frac{0.1(0.2)^2 e^{0.2x}}{30(10)(9)x^8} \qquad \text{(L'Hôpital's Rule; still } \tfrac{\infty}{\infty}\text{)}$$

$$\overset{\text{L'H}}{=} \lim_{x \to \infty} \frac{0.1(0.2)^3 e^{0.2x}}{30(10)(9)(8)x^7} \qquad \text{(L'Hôpital's Rule; still } \tfrac{\infty}{\infty}\text{)}$$

$$= \cdots = \qquad \text{(continue pattern)}$$

$$\overset{\text{L'H}}{=} \lim_{x \to \infty} \frac{0.1(0.2)^9 e^{0.2x}}{30(10)(9)(8)(7)(6)(5)(4)(3)(2)x} \qquad \text{(L'Hôpital's Rule; still } \tfrac{\infty}{\infty}\text{)}$$

$$\overset{\text{L'H}}{=} \lim_{x \to \infty} \frac{0.1(0.2)^{10} e^{0.2x}}{30(10)(9)(8)(7)(6)(5)(4)(3)(2)(1)} \qquad \text{(L'Hôpital's Rule)}$$

$$\to \frac{\infty}{30(10)(9)(8)(7)(6)(5)(4)(3)(2)(1)} \to \infty. \qquad \text{(solve limit)}$$

Therefore, the exponential function $f(x) = 0.1e^{0.2x}$ (despite its small leading coefficient 0.1 and its small continuous growth rate 0.2) dominates the power function $g(x) = 30x^{10}$. ☐

Theorem 8.30 can help us quickly anticipate the values of certain limits even before we apply any calculational techniques, as illustrated in the following example.

EXAMPLE 8.46   **Using dominance to calculate a limit**

Using Theorem 8.30, we know immediately that $\lim\limits_{x\to\infty} \dfrac{e^{3x}}{x^{100}+1} = \infty$ and that $\lim\limits_{x\to\infty} \dfrac{x^{100}+1}{e^{3x}} = 0$ (since $\frac{1}{\infty} \to 0$). Note that these particular limits would be difficult to solve calculationally; we would have to apply L'Hôpital's Rule one hundred times! ☐

What you should know after reading Section 8.7

▶ Why are limits of the form $\frac{0}{0}$, $\frac{\infty}{\infty}$, and $0 \cdot \infty$ "indeterminate"? Be able to construct various examples of limits with these indeterminate forms, and to calculate limits that have these forms using algebra, if possible. How can a limit of the form $0 \cdot \infty$ be rewritten as a limit of the form $\frac{0}{0}$ or $\frac{\infty}{\infty}$?

▶ The statement of L'Hôpital's Rule (including all the hypotheses that determine when the rule applies), and how to prove it in the simplest case.

▶ How to use L'Hôpital's Rule to calculate limits of the forms $\frac{0}{0}$, $\frac{\infty}{\infty}$, and $0 \cdot \infty$ (after rewriting). In particular, be able to determine whether L'Hôpital's Rule applies to a given limit, and if so, whether it might be useful to apply it (as opposed to doing some algebra).

▶ What does it mean for one function to "dominate" another as $x \to \infty$? Is the concept of dominance equivalent to the concept of "greater than"? Why or why not? Be able to use the definition of dominance to determine whether one function dominates another. Also be aware that exponential growth functions always dominate any power and polynomial functions.

Exercises 8.7

Concepts

0. Read the section and make your own summary of the material. Include whatever you think will help you review the material later on.

1. Explain what we mean when we say that a limit has an "indeterminate form."

2. Which of the following forms are indeterminate? For each form that is not indeterminate, what limit does it approach?

(a) $\dfrac{1}{0}$ (b) $\dfrac{0}{1}$ (c) $\dfrac{0}{0}$

(d) $\dfrac{\infty}{0}$ (e) $\dfrac{0}{\infty}$ (f) $\dfrac{\infty}{\infty}$

(g) $0 \cdot \infty$ (h) $\infty \cdot \infty$ (i) $\dfrac{1}{\infty}$

(j) $\dfrac{\infty}{1}$ (k) $\infty + \infty$ (l) $\infty - \infty$

3. Why does a limit of the form $\frac{0}{\infty}$ approach zero? (*Hint:* Think of the quotient as a product.)

4. For each part (a)–(d) below, find functions $f(x)$ and $g(x)$ such that the limit $\lim\limits_{x \to 2} \frac{f(x)}{g(x)}$ is initially of the indeterminate form $\frac{0}{0}$ but after algebraic simplification equals (a) ∞; (b) 0; (c) 1; and (d) 3. Explain how this shows that limits of the form $\frac{0}{0}$ are "indeterminate."

5. For each part (a)–(d) below, find functions $f(x)$ and $g(x)$ such that the limit $\lim\limits_{x \to 2} \frac{f(x)}{g(x)}$ is initially of the indeterminate form $\frac{\infty}{\infty}$ but after algebraic simplification equals (a) ∞; (b) 0; (c) 1; and (d) 3. Explain how this shows that limits of the form $\frac{\infty}{\infty}$ are "indeterminate."

6. The limit $\lim\limits_{x \to \infty} 2^{-x} x$ is of the form $0 \cdot \infty$. Rewrite this limit so that it is (a) in the form $\frac{0}{0}$, and (b) in the form $\frac{\infty}{\infty}$. Although L'Hôpital's Rule applies in both cases, it makes the limit simpler in only one case. Which one? (Apply L'Hôpital's Rule in both cases to find out.)

7. The limit $\lim\limits_{x \to 0} (2^x - 1)x^{-2}$ is of the form $0 \cdot \infty$. Rewrite this limit so that it is (a) in the form $\frac{0}{0}$, and (b) in the form $\frac{\infty}{\infty}$. Although L'Hôpital's Rule applies in both cases, it makes the limit simpler in only one case. Which one? (Apply L'Hôpital's Rule in both cases to find out.)

8. Give an example of a limit of the form $\lim\limits_{x \to c} \frac{f(x)}{g(x)}$ that is initially of the form $\frac{0}{0}$ and *cannot* be solved using algebraic manipulation. If possible, solve this limit using L'Hôpital's Rule.

9. Give an example of a limit of the form $\lim\limits_{x \to c} \frac{f(x)}{g(x)}$ that is initially of the form $\frac{0}{0}$ and *cannot* be solved using L'Hôpital's Rule (i.e., even though L'Hôpital's Rule applies, it does not simplify the limit). If possible, solve this limit using algebraic simplification.

10. Give an example of a limit of the form $\lim\limits_{x \to c} \frac{f(x)}{g(x)}$ that is initially of the form $\frac{0}{0}$ and can be solved using algebraic manipulation, as well as by using L'Hôpital's Rule. Then solve this limit (a) using algebraic manipulation, and (b) using L'Hôpital's Rule.

11. Suppose a function f "dominates" a function g as $x \to \infty$; what does this mean? How is it different from just saying that eventually $f(x) > g(x)$ for large values of x? Give an example where $f(x) > g(x)$ for large values of x, although f does *not* dominate g.

12. Find *all* of the errors in the following *incorrect* calculation. Then calculate the limit correctly.

$$\lim\limits_{x \to 0} \frac{x^2 - x}{2^x - 1} \overset{\text{L'H}}{=} \lim\limits_{x \to 0} \frac{2x - 1}{(\ln 2)2^x} \overset{\text{L'H}}{=} \lim\limits_{x \to 0} \frac{2}{(\ln 2)^2 2^x}$$
$$= \frac{2}{(\ln 2)^2 2^0} = \frac{2}{(\ln 2)^2}.$$

■ **Determine whether each of the following statements is true or false. If a statement is true, explain why; if a statement is false, provide a counterexample.**

13. True or False: If a limit has an indeterminate form, then that limit does not exist.

14. True or False: L'Hôpital's Rule can be used to find the limit of *any* quotient $\frac{f(x)}{g(x)}$ as $x \to c$.

15. True or False: When using L'Hôpital's Rule, we need to apply the quotient rule.

16. True or False: L'Hôpital's Rule applies only to limits where $x \to 0$ or where $x \to \infty$.

17. True or False: f dominates g if and only if the values $f(x)$ are eventually larger than the values $g(x)$ as $x \to \infty$.

18. True or False: If f does not dominate g, then g must dominate f.

19. True or False: The function $f(x) = 0.001(1.003)^x$ dominates the function $g(x) = 500x^{1500}$ as $x \to \infty$.

20. True or False: There is some large value of x for which $0.001(1.003)^x > 500x^{1500}$.

21. True or False: If f dominates g, then $\lim\limits_{x \to 3} \frac{f(x)}{g(x)} = \infty$.

Skills

■ **Calculate each of the following limits (a) using L'Hôpital's Rule and (b) without using L'Hôpital's Rule (for example, by using algebra). Compare your answers to parts (a) and (b).**

22. $\lim\limits_{x \to 1} \dfrac{x^2 + x - 2}{x - 1}$

23. $\lim\limits_{x \to \infty} \dfrac{x - 1}{2 - 3x^2}$

24. $\lim\limits_{x \to 2} \dfrac{x^2 - 4x + 4}{x - 2}$

25. $\lim\limits_{x \to \infty} \dfrac{3^x}{1 - 4^x}$

26. $\lim\limits_{x \to \infty} \dfrac{e^{3x}}{1 - e^{2x}}$

27. $\lim\limits_{x \to 0} \dfrac{2^x - 1}{4^x - 1}$

■ **Calculate the following limits, using L'Hôpital's Rule if it applies and if it is useful. (*Hints:* Some problems are easier to do *without* L'Hôpital's Rule, even if it applies; remember to use algebraic simplification whenever possible.)**

28. $\lim\limits_{x \to 3} \dfrac{2^x - 8}{3 - x}$

29. $\lim\limits_{x \to 1} \dfrac{x - 1}{e^{x-1} - 1}$

30. $\lim\limits_{x \to 2} \dfrac{e^{2x-4} - 1}{x^2 - 4}$

31. $\lim\limits_{x \to \infty} \dfrac{e^x - 7}{8x^2 + 12x + 5}$

32. $\lim\limits_{x \to \infty} \dfrac{2^{-x}}{x^2 + 1}$

33. $\lim\limits_{x \to \infty} \dfrac{x + 3^x}{3 - 2^x}$

34. $\lim\limits_{x \to 0} \dfrac{x^{-2}}{x^{-3} + 1}$

35. $\lim\limits_{x \to \infty} x^2 e^{-x}$

36. $\lim\limits_{x \to \infty} \left(e^x - \dfrac{e^x}{x + 1} \right)$

37. $\lim\limits_{x \to 0^+} \left(x^{-1} - \dfrac{1}{2^x - 1} \right)$

38. $\lim\limits_{x \to \infty} x \left(\dfrac{1}{2} \right)^x$

39. $\lim\limits_{x \to \infty} \dfrac{xe^x}{e^{2x} + 1}$

40. $\lim\limits_{x \to \infty} \dfrac{e^{-3x}}{x^2 + 3x + 1}$

41. $\lim\limits_{x \to \infty} \dfrac{xe^{-3x}}{x^2 + 3x + 1}$

42. $\lim\limits_{x \to 0} \left(\dfrac{2^x - 1}{x^2} - \dfrac{1}{1 - e^x} \right)$

■ For each function f and g below, use limits to determine whether f dominates g, or g dominates f, or neither, as $x \to \infty$. (*Note:* Do not just use Theorem 8.30; use the definition of dominance given in Definition 8.6.)

43. $f(x) = x^2$, $g(x) = 2^x$

44. $f(x) = 100x^2$, $g(x) = 2x^{100}$

45. $f(x) = 50x^3$, $g(x) = (1.1)^x$

46. $f(x) = 2^x$, $g(x) = 3^x$

47. $f(x) = 0.001e^{0.025x}$, $g(x) = 5000x^4$

48. $f(x) = e^{3x}$, $g(x) = 1000e^{3x}$

■ Determine the values of each of the following limits by using the fact that exponential growth functions always dominate power or polynomial functions (Theorem 8.30).

49. $\displaystyle\lim_{x\to\infty} \frac{x^{101} + 500}{e^x}$

50. $\displaystyle\lim_{x\to\infty} \frac{e^x}{x^{101} + 500}$

51. $\displaystyle\lim_{x\to -\infty} 2^x x^{100}$

Proofs

52. Prove Theorem 8.28 in your own words. Provide supporting reasons for each step of your proof.

53. Prove Theorem 8.29 in your own words. Provide supporting reasons for each step of your proof.

CHAPTER 9

Logarithmic Functions

*I*n Chapter 8 we introduced the "natural" logarithmic function $f(x) = \ln x$. In this chapter we introduce general logarithmic functions $f(x) = \log_b x$. Just as $\ln x$ is, by definition, the inverse of the natural exponential function e^x, the function $\log_b x$ will be defined as the inverse of the exponential functions b^x.

All of the algebraic, limit, and derivative properties of logarithmic functions will be derived from the fact that they are inverses of exponential functions. After investigating these properties, we will learn how to use logarithms as a calculational tool.

Decorated Lithuanian eggs. © *James A. Sugar/ Corbis*

CONTENTS

| 9.1 | The Algebra of Logarithmic Functions |

In this section we define general logarithmic functions as the inverses of exponential functions and investigate their algebraic properties. It is worth noting that there is more than one way to define logarithmic functions. Logarithmic functions can also be defined in terms of something called "definite integrals," which we will see in Part IV.

9.1.1 What is a logarithmic function?

In Section 8.2.2 we saw that the natural exponential function $y = e^x$ is one-to-one, and thus has an inverse, which we call the natural logarithmic function $y = \ln x$. In fact, since *every* exponential function $y = b^x$ is one-to-one, *every* exponential function has an inverse; these inverses are called the **logarithmic functions.** The inverse of the function $y = b^x$ is written in the notation $y = \log_b x$.

DEFINITION 9.1

Logarithmic Functions Are the Inverses of Exponential Functions
For any $b > 0$ with $b \neq 1$, the **logarithmic function with base b**, denoted by

$$y = \log_b x,$$

is defined to be the inverse of the exponential function $y = b^x$.

The "b" in the expression $y = \log_b x$ is called the **base** of the logarithmic function. We require that $b > 0$ and $b \neq 1$, because these are exactly the conditions we must have for $y = b^x$ to be an exponential function. Using function notation, the exponential function $f(x) = b^x$ has an inverse $f^{-1}(x)$, which we denote using the logarithmic notation $f^{-1}(x) = \log_b x$.

EXAMPLE 9.1   **Logarithmic and exponential functions are inverses**

The function $y = \log_2 x$ is by definition the inverse of the function $y = 2^x$, and the function $y = \log_3 x$ is the inverse of the exponential function $y = 3^x$.   □

   We have two different notations that represent the logarithmic function with base e; the logarithmic function $y = \log_e x$ with base e is by definition the inverse of the natural exponential function $y = e^x$, and thus

$$(1) \qquad \qquad \ln x = \log_e x.$$

There are also two notations for the base 10 logarithm (which is used often in the sciences):

$$(2) \qquad \qquad \log x = \log_{10} x.$$

9.1.2 Basic properties of logarithmic functions

Because logarithmic functions are by definition the inverses of exponential functions, they inherit all of their functional and algebraic properties from exponential functions. For example, since we know that every exponential function of the form $y = b^x$ has domain $(-\infty, \infty)$ and range $(0, \infty)$, we know that the domain and range of $y = \log_b x$ are as in Theorem 9.1.

THEOREM 9.1

Domain and Range of Logarithmic Functions

Every logarithmic function $y = \log_b x$ has domain $(0, \infty)$ and range $(-\infty, \infty)$.

Moreover, since every logarithmic function $y = \log_b x$ is by definition the inverse of the exponential function $y = b^x$, logarithmic and exponential functions have the following relationships:

THEOREM 9.2

Relationships Between Logarithmic and Exponential Functions

Suppose $b > 0$ and $b \neq 1$. Since the logarithmic function $y = \log_b x$ is the inverse of the exponential function $y = b^x$, we have

(a) For all $x > 0$ and all y, $\log_b x = y$ if and only if $b^y = x$.

(b) If $x > 0$, then $b^{\log_b x} = x$.

(c) For all x, $\log_b(b^x) = x$.

The proof of this theorem involves applying the basic properties of inverse functions. If $f(x)$ and $f^{-1}(x)$ are inverses, then:

(a) For all $x \in \text{Domain}(f^{-1})$ and $y \in \text{Domain}(f)$, $f^{-1}(x) = y$ if and only if $f(y) = x$.

(b) If x is in the domain of f^{-1}, then $f(f^{-1}(x)) = x$.

(c) If x is in the domain of f, then $f^{-1}(f(x)) = x$.

We leave the details of the proof of Theorem 9.2 to the exercises. Another way of thinking of part (b) of Theorem 9.2 is as the following statement:

(3) $\log_b x$ is the exponent to which you have to raise b in order to get x.

In other words, suppose you want to fill in the box representing the exponent in the equation $b^\square = x$. Our name for the quantity that must go in this box is $\log_b x$. Of course, having a name for this exponent is not the same as being able to *calculate* this exponent! As we will soon see, logarithms can help us solve equations where the variable is in an exponent (like $2^x = 7$).

EXAMPLE 9.2 | **Relating $\log_3 9 = 2$ and $3^2 = 9$**

Because the exponent to which you have to raise 3 in order to get 9 is 2 (that is, $3^2 = 9$), we say that $\log_3 9 = 2$. In the notation of Theorem 9.2, we can also say that

(a) $\log_3 9 = 2 \iff 3^2 = 9$;

(b) $3^{\log_3 9} = 9$ (note that $\log_3 9 = 2$ and that $3^2 = 9$);

(c) $\log_3(3^2) = 2$ (note that $3^2 = 9$ and that $\log_3 9 = 2$).   □

Since $y = \log_b(x)$ is the inverse of $y = b^x$ for all $b > 0$ with $b \neq 1$, we can obtain the graph of a logarithmic function $y = \log_b x$ by reflecting the graph of the exponential function $y = b^x$ over the line $y = x$. For example, Figures 9.1 and 9.2 show the graphs of $y = 2^x$ and $y = \log_2 x$, respectively.

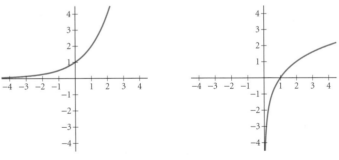

Figure 9.1 $y = 2^x$ **Figure 9.2** $y = \log_2 x$

Since all exponential functions $y = b^x$ where $b > 1$ have the same general shape as the graph in Figure 9.1, all logarithmic functions $y = \log_b x$ with $b > 1$ have the same general shape as the graph in Figure 9.2. Logarithmic functions with $b > 1$ are called **logarithmic growth** functions. Notice in particular that (if $b > 1$) the function $\log_b x$ is positive for $x > 1$ and negative for $0 < x < 1$ (see Figure 9.2). We will use this fact often in the next few sections—in particular, in the case where $b = e$ (i.e., when we are discussing the natural logarithmic function $\ln x$).

On the other hand, if $0 < b < 1$, then the exponential function $y = b^x$ will have a graph with the shape shown in Figure 9.3 (which is the graph of $y = \left(\frac{1}{2}\right)^x$). In this case, the graph of the logarithmic function $y = \log_b x$ will have the shape shown in Figure 9.4 (which is the graph of $y = \log_{\frac{1}{2}} x$). Logarithmic functions with base $0 < b < 1$ are called **logarithmic decay** functions.

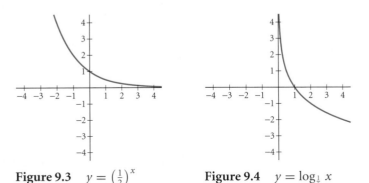

Figure 9.3 $y = \left(\frac{1}{2}\right)^x$ **Figure 9.4** $y = \log_{\frac{1}{2}} x$

✥ **Caution**   You should know the shapes and important features of the graphs of logarithmic functions $y = \log_b x$ (both for $b > 1$ and for $0 < b < 1$). With our knowledge of graphical transformations, we can also quickly sketch graphs of functions like $y = \log_2(x - 1) + 3$ and $y = -3 \log_{\frac{1}{2}} x$. For example, the graph of $y = \log_2(x - 1) + 3$ can be obtained by shifting the graph of $y = \log_2 x$ three units up and one unit to the right.   □

Note that for any $b > 0$ with $b \neq 1$, the graph of $y = \log_b x$ has a vertical asymptote at $x = 0$ (since the graph of $y = b^x$ has a horizontal asymptote at $x = 0$), and the graph of $y = \log_b x$ passes through the point $(1, 0)$ (since the graph of $y = b^x$ passes through the point $(0, 1)$). The latter fact means that for any $b > 0$ with $b \neq 1$, we have

(4) $$\log_b 1 = 0.$$

It is important to note that all logarithmic functions $y = \log_b x$ are one-to-one, since they are invertible (their inverses are the exponential functions $y = b^x$).

9.1.3 Algebraic rules for logarithmic functions

We already have a set of rules for algebraically manipulating expressions that involve exponential functions. For example, we have $b^{x+y} = b^x b^y$; an exponential function of a sum is a product of exponential functions. Since logarithmic functions are the inverses of exponential functions, the algebraic rules for manipulating logarithmic functions are in some sense the "opposites" of the rules for manipulating exponential functions. For example, as we will see in Theorem 9.3, we have $\log_b(xy) = \log_b x + \log_b y$; the logarithm of a product is a sum of logarithms.

THEOREM 9.3

Algebraic Rules for Logarithmic Functions

Suppose that $b > 0$ and $b \neq 1$. Then for all values of x, y, and a for which the expressions below are defined, we have

(a) $\log_b(x^a) = a \log_b x$,

(b) $\log_b(xy) = \log_b x + \log_b y$.

As simple corollaries to Rules (a) and (b), we have

(c) $\log_b \left(\frac{1}{x}\right) = -\log_b x$,

(d) $\log_b \left(\frac{x}{y}\right) = \log_b x - \log_b y$.

We have already seen that Rule (a) applies when the base b is equal to that special number e. In Section 8.2 we saw that $\ln(x^a) = a \ln x$, and in "base e" notation this means that $\log_e(x^a) = a \log_e x$. Rule (a) now tells us that this rule applies for *any* base b.

❓ *Question* Write out the four algebraic rules for logarithmic functions in the special case where $b = e$ (i.e., for the function $\ln x$). ☐

We'll prove Theorem 9.3 in a moment; first we will do a couple of examples to illustrate how the rules work and why they are reasonable.

EXAMPLE 9.3

Logarithms turn exponents into constant multiples

Using Rule (a), we have

$$\log_2(64) = \log_2(4^3) = 3\log_2 4.$$

This makes sense in light of Statement (3) on page 601, since the exponent to which you must raise 2 to get 64 is 6 (that is, $\log_2(64) = 6$), and the exponent to which you must raise 2 to get 4 is 2 (that is, $\log_2 4 = 2$, and thus $3\log_2 4 = 3(2) = 6$). ☐

⚜ *Caution*   Be careful not to misuse Rule (a) of Theorem 9.3. In particular, note that

$$(\log_b x)^a \neq a \log_b x.$$

In order for us to use Rule (a), the entire expression x^a must be *inside* the logarithm.

□

| EXAMPLE 9.4 |

A difference of logarithms is the logarithm of a quotient

Using Rule (d), we know that

$$\log_3 9 - \log_3 27 = \log_3 \left(\frac{9}{27}\right) = \log_3 \left(\frac{1}{3}\right).$$

Again, this makes sense in the context of Statement (3). The exponent to which you have to raise 3 to get 9 is 2 (that is, $\log_3 9 = 2$), and the exponent to which you have to raise 3 to get 27 is 3 (that is, $\log_3 27 = 3$). Therefore, the left-hand side of the equality above is equal to $2 - 3 = -1$. On the other hand, the exponent to which you have to raise 3 to get $\frac{1}{3}$ is -1 (that is, $3^{-1} = \frac{1}{3}$), and thus the right-hand side of the equality is also equal to -1.

□

⚜ *Caution*   The following statements illustrate some common errors that students make when working with logarithms. Be careful not to make these mistakes!

$$\frac{\log_2 5}{\log_2 4} \neq \log_2 \tfrac{5}{4} \qquad\qquad \frac{\log_2 5}{\log_2 4} \neq \frac{5}{4} \qquad\qquad \frac{\log_2 3}{\log_5 3} \neq \log_{\frac{2}{5}} 3$$

□

The proof of Rule (a) of Theorem 9.3 will be similar to the proof of the corresponding rule from Theorem 8.1 in Section 8.2. Rule (b) will follow from the fact that $b^{x+y} = b^x b^y$. We will prove Rules (a) and (b). Rules (c) and (d) follow easily from the first two rules, and we leave those proofs for the exercises.

PROOF   (RULES (A) AND (B) OF THEOREM 9.3)   Suppose $b > 0$ and $b \neq 1$, and let a and $x > 0$ be real numbers. Note that this implies that x^a is positive and thus that $\log_b(x^a)$ will be defined. Our proof of Rule (a) will use the fact that $x = b^{\log_b x}$ and the algebraic rule $(m^u)^v = m^{uv}$:

$$\begin{aligned}
\log_b(x^a) &= \log_b((b^{\log_b x})^a) &\text{(Theorem 9.2)}\\
&= \log_b(b^{a \log_b x}) &\text{(properties of exponents)}\\
&= a \log_b x. &\text{(Theorem 9.2)}
\end{aligned}$$

To prove Rule (b), let x and y be any positive real numbers. The proof of Rule (b) will use the algebraic rule for exponents that says that $m^{u+v} = m^u m^v$. We will manipulate $\log_b(xy)$ until we arrive at $\log_b x + \log_b y$. The key will be to write $x = b^{\log_b x}$ and $y = b^{\log_b y}$. We have

$$\begin{aligned}
\log_b(xy) &= \log_b(b^{\log_b x} b^{\log_b y}) &\text{(Theorem 9.2)}\\
&= \log_b(b^{\log_b x + \log_b y}) &\text{(property of exponents)}\\
&= \log_b x + \log_b y &\text{(Theorem 9.2)}
\end{aligned}$$

■

9.1.4 Exact and approximate values of logarithmic expressions

Sometimes we can calculate the value of a logarithmic function at a particular value of x by hand, and sometimes we cannot. For example, it is easy to calculate that the value of $\log_5 125$ is 3, since 3 is the exponent to which we have to raise 5 to get 125 (i.e., $5^3 = 125$). However, we would have a lot of trouble trying to calculate $\log_5 2$ by hand; to do so, we would need to know the exponent to which we would have to raise 5 to obtain the number 2 (we would have to find a number a such that $5^a = 2$). Since this number a isn't an integer, it is difficult for us to calculate it by hand. A calculator can be used to show that $\log_5 2 \approx 0.43067656$.

So what can we do if we wish to calculate a logarithmic expression whose answer cannot be easily guessed? One solution is to try to rewrite the expression in such a way that it involves "easy" logarithms, as in the following example.

EXAMPLE 9.5

Rewriting a logarithmic expression so that we can calculate its value

Find the exact value of $2\log_3 6 - \log_3 4$.

Solution Calculating the value of either $\log_3 6$ or $\log_3 4$ would be very difficult (what would you put in the boxes in the expressions $3^\square = 6$ and $3^\square = 4$?). However, if we apply the algebraic rules for logarithms, it turns out that this is an "easy" logarithm in disguise:

$$2\log_3 6 - \log_3 4 = \log_3(6^2) - \log_3 4 = \log_3\left(\frac{6^2}{4}\right) = \log_3\left(\frac{36}{4}\right) = \log_3 9 = 2. \quad \square$$

Of course, not every logarithmic expression can be rewritten in terms of logarithms that are easy to calculate; this example is a rarity. We know that $\log_b x$ is the exponent to which we would have to raise b in order to get x, but except in very special cases it is not easy to *calculate* this exponent mentally. Your calculator should have keys for computing values of $\ln x$ and $\log x$.

However, most calculators are not programmed to handle logarithm bases other than e and 10. How can we use a calculator to calculate a logarithm in some other base, such as $\log_3 5$? For logarithmic expressions with other bases, the following conversion formula is helpful.

THEOREM 9.4

Base Conversion Formula for Logarithms

For any positive numbers a, b, and x (with $a \neq 1$ and $b \neq 1$),

$$\log_b x = \frac{\log_a x}{\log_a b}.$$

Theorem 9.4 can be used to convert a logarithm with base b into an expression involving logarithms in *any* other base a. The base conversion formula is most commonly used to convert logarithms with arbitrary bases into expressions involving logarithms with base e or base 10. In these special cases, the formula says that

(5) $$\log_b x = \frac{\ln x}{\ln b} \quad \text{and} \quad \log_b x = \frac{\log x}{\log b}.$$

With these formulas, we can now use a calculator to approximate the value of $\log_b x$, regardless of the base b (although how the calculator makes these approximations is another story). We will prove Theorem 9.4 at the end of this section; first we do an example.

EXAMPLE 9.6

Converting to base *e* or base 10

Use a calculator to approximate $\log_3 5$.

Solution   First notice that it would be difficult to find the exact value of $\log_3 5$ by hand, since to do so we would have to find the exponent y such that $3^y = 5$. However, with the base conversion formula, we can write $\log_3 5$ in terms of the natural logarithm and then use a calculator to approximate its value:

$$\log_3 5 = \frac{\ln 5}{\ln 3} \approx 1.46497.$$

Alternatively, we could convert $\log_3 5$ into an expression involving the base 10 logarithm:

$$\log_3 5 = \frac{\log 5}{\log 3} \approx 1.46497.$$

(Note that we get the same answer regardless of whether we use the base 10 logarithm or the base *e* logarithm.) ☐

Keep in mind what it *means* to say that $\log_3 5 \approx 1.46497$; it means that the exponent to which we must raise 3 to get 5 is approximately 1.46497. In other words, $3^{1.46497} \approx 5$ (check this on your calculator).

⚙ *Caution*   Rounding error can be significant when you work with an expression involving logarithms. For this reason it is usually a good idea not to make any approximations until the *end* of a problem (as opposed to making and then working with multiple approximations, which introduces more rounding error). For example, when computing $\frac{\ln 5}{\ln 3}$, don't find ln 5 and ln 3 and then divide these numbers. Instead, put $\frac{\ln 5}{\ln 3}$ into your calculator in one line, perhaps by entering (ln 5)/(ln 3). ☐

We now prove Theorem 9.4. We will use Theorem 9.2 to rewrite the formula in Theorem 9.4 and then prove using direct calculation.

PROOF (**THEOREM 9.4**)   We want to show that for any positive numbers a, b, and x, we have $\log_b x = \frac{\log_a x}{\log_a b}$. This expression is somewhat difficult to work with, especially since none of our algebraic rules for manipulating logarithms tells us what to do when we have a quotient of two logarithms. To simplify things a bit, we will set $y = \log_b x$ and then show that $y = \frac{\log_a x}{\log_a b}$:

$$y = \log_b x \implies b^y = x \qquad\qquad \text{(Theorem 9.2)}$$
$$\implies \log_a(b^y) = \log_a(x) \qquad \text{(we can do this for \emph{any} positive } a)$$
$$\implies y\log_a b = \log_a x \qquad\qquad \text{(Theorem 9.3)}$$
$$\implies y = \frac{\log_a x}{\log_a b}. \qquad\qquad\qquad \text{(algebra)}$$

What you should know after reading Section 9.1

▶ The definition of the function $y = \log_b x$ (including what bases b are allowed), as well as the two different notations for base e and base 10 logarithms.

▶ The basic properties of logarithmic functions, and how they follow from the fact that logarithmic functions are by definition the inverses of exponential functions. This includes domain and range as well as the algebraic properties that immediately follow from the fact that $y = \log_b x$ and $y = b^x$ are inverses.

▶ The algebraic rules for logarithmic functions. This includes knowing how to rewrite the logarithm of a power function, a product, a reciprocal, or a quotient in terms of simpler logarithmic expressions.

▶ How to find the *exact* value of a logarithmic expression, if possible (this may involve doing some algebra first). Also, how to use the logarithmic base conversion formula to write logarithmic expressions in terms of logarithms of base e or base 10 so that you can use a calculator to make approximations.

▶ Be able to *prove* the basic properties of logarithmic functions in Theorems 9.1 and 9.2, the algebraic properties of logarithmic functions in Theorem 9.3, and the base conversion formula in Theorem 9.4.

Exercises 9.1

Concepts

0. Read the section and make your own summary of the material. Include whatever you think will help you review the material later on.

1. What is the definition of the logarithmic function $f(x) = \log_b x$?

2. What bases b are permitted in a logarithmic function $f(x) = \log_b x$, and why?

3. Use the domain and range of logarithmic functions to find the domains and ranges of the functions (**a**) $f(x) = 3\log_4 x - 8$ and (**b**) $f(x) = \log_2(x - 3)$.

■ Fill in the blanks in each statement below.

4. For all $x \in$ _____, $\log_2 x = y$ if and only if $x =$ _____.

5. For all $x \in$ _____, $3^{\log_3 x} =$ _____.

6. For all $x \in$ _____, $\log_4(4^x) =$ _____.

7. $\log_2 3$ is the exponent to which you have to raise _____ to get _____.

8. For any logarithmic base b, $\log_b 1 =$ _____.

■ Complete each of the algebraic rules for logarithmic functions given below.

9. $\log_b x - \log_b y =$ _____.

10. $\log_b x + \log_b y =$ _____.

11. $\log_b(x^a) =$ _____.

12. $\log_b\left(\frac{1}{x}\right) =$ _____.

13. The graphs of $y = \log_2 x$ and $y = \log_4 x$ are shown below. Determine which graph is which, without using a calculator. (*Hint:* Think about the graphs $y = 2^x$ and $y = 4^x$, and then reflect those graphs over the line $y = x$.)

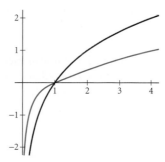

$y = \log_2 x$ and $y = \log_4 x$

14. The graphs of $y = \log_{\frac{1}{2}} x$ and $y = \log_{\frac{1}{4}} x$ are shown on the following page. Determine which graph is which, without using a calculator. (*Hint:* Think about the graphs $y = (\frac{1}{2})^x$ and $y = (\frac{1}{4})^x$, and then reflect those graphs over the line $y = x$.)

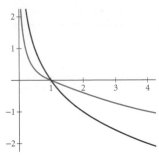

$$y = \log_{\frac{1}{2}} x \text{ and } y = \log_{\frac{1}{4}} x$$

15. Use algebra, the graph of $y = \ln x$, and your knowledge of transformations to sketch graphs of $f(x) = \ln(x^2)$ and $g(x) = \ln(\frac{1}{x})$.

■ Determine whether each of the following statements is true or false. Explain your answers.

16. True or False: $(\log_2 5)^3 = 3 \log_2 5$

17. True or False: $\dfrac{\log_3 7}{\log_3 2} = \log_3 5$

18. True or False: $\dfrac{\log_5 2}{\log_5 7} = \dfrac{2}{7}$

19. True or False: $(\log_{10} 3)(\log_{10} 4) = \log_{10} 7$

20. True or False: $(\log_{10} 3)(\log_{10} 4) = \log_{10} 12$

21. True or False: $\log_4 5 + \log_3 5 = \log_7 5$

22. Solve the inequality $\ln\left(\dfrac{x+1}{x-1}\right) \geq 0$. Be careful; some of the algebra is tricky.

23. State the base conversion formula for logarithms, and explain why it is helpful, when using a calculator, to approximate the numerical value of an expression involving logarithms.

24. Use the base conversion formula for logarithms to show that the function $f(x) = \log_2 x$ is equal to the function $g(x) = \log_3 x$ only when $x = 1$.

Skills

■ Calculate the *exact* numerical value of each of the following expressions. Do not use a calculator. You may have to do some algebra before you can evaluate the logarithms.

25. $\log_{10} 1000$

26. $\log_{10} 0.01$

27. $\log_4 \frac{1}{2}$

28. $\log_{\frac{1}{2}} 4$

29. $\ln 1$

30. $\ln e$

31. $\ln(e^2)$

32. $\ln\left(\dfrac{1}{e^2}\right)$

33. $\log_3 1$

34. $\log_{1.729} 1$

35. $\log_{64} \frac{1}{2}$

36. $\dfrac{2 \log_4 8}{\log_4 2}$

37. $\log_6 2 + \log_6 3$

38. $\log_2 6 - \log_2 3$

39. $\log_{10} 4 + 2 \log_{10} 5$

40. $4 \log_2 6 - 2 \log_2 9$

41. $\dfrac{\ln 1}{\ln e}$

42. $\dfrac{\ln 4}{\ln 16}$

43. $\dfrac{\log_7 9}{\log_7 \frac{1}{3}}$

44. $\dfrac{\log_{10} 49}{\log_{10} 7}$

■ Use a calculator to approximate the numerical value of each of the following expressions.

45. $\log_{10} 5{,}000{,}000$

46. $\log_4 7$

47. $\log_{\frac{1}{2}} 9$

48. $\log_{34} 3$

■ Find the domain of each of the following functions.

49. $f(x) = \dfrac{2}{\ln(x+2)}$

50. $f(x) = \dfrac{\log_2 \sqrt{x}}{3-x}$

51. $f(x) = \dfrac{1}{x \ln(3x+2)}$

52. $f(x) = \log_3(x^2 - 1)$

53. $f(x) = \ln\left(\dfrac{x+1}{x-2}\right)$

54. $f(x) = \dfrac{\ln(x+1)}{\ln(x-2)}$

55. $f(x) = \sqrt{\ln(x-2)}$

56. $f(x) = \dfrac{1}{\sqrt{\ln(x+1)}}$

57. $f(x) = \dfrac{\ln(x^2 - 1)}{\ln(\sqrt{3x+1})}$

■ Solve each of the following equations exactly; then find numerical approximations for your solutions using a calculator.

58. $\log_4 x = -2$

59. $\log_x 8 = 3$

60. $\log_2\left(\dfrac{x-1}{x+1}\right) = 4$

61. $3^x = 10$

62. $2^{x+1} - 4 = 6$

63. $10^{x^2+1} = 4$

64. $3(7^{2x-1}) + 4 = 12$

65. $2^x = 3^{x-1}$

66. $10^x = e^x$

67. $2(4^x) = 5^x$

68. $\dfrac{3^x - 2}{x - 1} = 0$

69. $\dfrac{4^{x-1}}{2^x} = 6$

■ Using what you know about transformations and the shape of logarithmic graphs, sketch a rough graph of each of the following functions by hand. When you are finished, check your answer with a graphing calculator (you may need to use the base conversion formula to enter these functions into your calculator).

70. $f(x) = 3 \log_4 x - 8$

71. $f(x) = \log_2(x-3)$

72. $f(x) = \log_{\frac{1}{2}}(3x)$

Proofs

73. Use the definition of a logarithmic function $y = \log_b x$ to prove that for any $b > 0$ with $b \neq 1$, the quantity $\log_b 1$ is equal to zero.

74. Prove Theorem 9.1 as follows: Use the domains of exponential functions and the relationship between logarithmic and exponential functions to prove that every logarithmic function $y = \log_b x$ has domain $(0, \infty)$ and range $(-\infty, \infty)$.

75. Prove Theorem 9.2. (*Hint:* Use properties of inverse functions and the fact that if $f(x) = b^x$ is any exponential function, then $f^{-1}(x) = \log_b x$.)

76. Rule (a) from Theorem 9.3 was proved in the reading. Write the proof in your own words, justifying and explaining all steps.

77. Rule (b) from Theorem 9.3 was proved in the reading. Write the proof in your own words, justifying and explaining all steps.

78. Prove rule (c) from Theorem 9.3. (*Hint:* Use part (a) of the same theorem.)

79. Prove rule (d) from Theorem 9.3. (*Hint:* Use rules (a) and (b) from the same theorem.)

80. Prove the base conversion formula for logarithms (Theorem 9.4). Use the proof in the reading as a guide, but write the proof in your own words.

9.2 Limits and Derivatives of Logarithmic Functions

In this section we learn how to calculate limits and derivatives of functions that involve logarithms. We begin by examining the limits and derivatives of functions of the form $y = \log_b x$ and then move on to more complicated functions. Remember that $y = \log_b x$ is the inverse of the exponential function $y = b^x$. This means that we can use what we know about the limits and derivatives of exponential functions to investigate the limits and derivatives of logarithmic functions.

9.2.1 Continuity of logarithmic functions

As usual, the first thing we would like to know about the limits of logarithmic functions is whether we can solve them by "plugging in." For example, is it true that $\lim_{x \to 8} \log_2 x = \log_2 8$? In other words, is the limit of the function $f(x) = \log_2 x$ as x approaches 8 equal to the value $f(8) = \log_2 8$? In *other* words, is $f(x) = \log_2 x$ continuous at $x = 8$?

THEOREM 9.5

Continuity of Logarithmic Functions

If $b > 0$ and $b \neq 1$, then the logarithmic function $f(x) = \log_b x$ is continuous at each point in its domain $(0, \infty)$.

In the language of limits, Theorem 9.5 says that for all c in the domain $(0, \infty)$ of the function $f(x) = \log_b x$, we have

(1)
$$\lim_{x \to c} \log_b x = \log_b c.$$

In previous chapters, when we wanted to prove the continuity of a function, we had to use the delta–epsilon definition of limit. In this case (to show Equation (1), and thus Theorem 9.5), we would have to prove the delta–epsilon statement

For all $\varepsilon > 0$, there exists a $\delta > 0$ such that
if $0 < |x - c| < \delta$, then $|\log_b x - \log_b c| < \varepsilon$.

Luckily, we won't have to use deltas and epsilons to prove Theorem 9.5. Instead, we will use the fact that exponential functions are continuous and the rule for the limit of a composition.

PROOF (**THEOREM 9.5**)   Let $L := \lim_{x \to c} \log_b x$. We need to show that, for $c > 0$, $L = \log_b c$. This is equivalent to showing that $b^L = c$:

$$b^L = b^{\lim_{x \to c} \log_b x} \qquad \text{(definition of } L)$$
$$= \lim_{x \to c} b^{\log_b x} \qquad \text{(limit of a composition; note that } b^x \text{ is continuous)}$$
$$= \lim_{x \to c} x \qquad \text{(since } b^{\log_b x} = x)$$
$$= c. \qquad \text{(evaluate limit)}$$

Technical Point: The second equality holds only if $\lim_{x \to c} \log_b x$ exists, that is, is a real number L. As we will see, the technique used in the proof of Theorem 9.5 does not apply as $x \to 0$, since in that case the limit of $\log_b x$ does not exist (it is infinite).

Since $b^L = c$, we know that $L = \log_b c$, and therefore that $\lim_{x \to c} \log_b x = \log_b c$, for all positive values of x. Hence $f(x) = \log_b x$ is continuous everywhere on its domain $(0, \infty)$. ∎

Now that we know that logarithmic functions are continuous, we can easily calculate the limit of any logarithmic function $f(x) = \log_b x$ as x approaches a positive number c:

EXAMPLE 9.7

Using continuity of logarithms to calculate limits

Since $\log_2 x$ is continuous at $x = 8$, we can calculate $\lim_{x \to 8} \log_2 x$ by evaluating at $x = 8$. Since $2^3 = 8$, we know that $\log_2 8 = 3$ and therefore that

$$\lim_{x \to 8} \log_2 x = \log_2 8 = 3. \qquad \square$$

✦ **Caution**   Note that $\lim_{x \to c} \log_b x = \log_b c$ only if c is a positive number. If c is negative, then $\log_b x$ is not defined at $x = c$, so the limit of $\log_b x$ as $x \to c$ does not exist. For example, $\lim_{x \to -2} \log x$ does not exist, since $f(x) = \log x$ does not exist anywhere near $x = -2$. $\square$

9.2.2 Limits of logarithmic functions at 0 and ∞

We now know how to find the limit of a logarithmic function $f(x) = \log_b x$ as x approaches any positive number c. What happens to $f(x) = \log_b x$ as $x \to 0$? To be more precise, we know that $\log_b x$ is not defined for negative values of x, so its limit as $x \to 0^-$ will not exist; but what about the limit as $x \to 0^+$? We also want to know the limit of $f(x) = \log_b x$ as $x \to \infty$. Theorem 9.6 describes both of these limits.

THEOREM 9.6

Limits of Logarithmic Functions at Zero and Infinity

If $b > 1$, then $\lim_{x \to 0^+} \log_b x = -\infty$ and $\lim_{x \to \infty} \log_b x = \infty$.

If $0 < b < 1$, then $\lim_{x \to 0^+} \log_b x = \infty$ and $\lim_{x \to \infty} \log_b x = -\infty$.

You need not memorize this theorem; instead, you should know the shape of the graph of $f(x) = \log_b x$ (for $b > 1$ and for $0 < b < 1$) and determine the limits from the graph (see Figures 9.5 and 9.6). Theorem 9.6 simply describes the behavior of these graphs near $x = 0$ and as $x \to \infty$ in terms of limits. In particular, Theorem 9.6 is just a precise and detailed way of saying that, whether $b > 1$ or $0 < b < 1$, the graph of a logarithmic function $f(x) = \log_b x$ always has a vertical asymptote at $x = 0$ and always becomes infinite (either $+\infty$ or $-\infty$) as x approaches infinity.

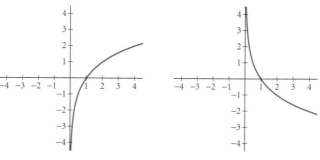

Figure 9.5
$f(x) = \log_b x, b > 1$

Figure 9.6
$f(x) = \log_b x, 0 < b < 1$

We won't prove Theorem 9.6, but it is "clear" if you remember that the graph of $y = \log_b x$ can be obtained by reflecting the graph of its inverse $y = b^x$ over the line $y = x$. For $b > 1$, the fact that $b^x \to 0^+$ as $x \to -\infty$ means that $\log_b x \to -\infty$ as $x \to 0^+$ (the roles of x and y are interchanged, since $y = \log_b x$ is the inverse of $y = b^x$). We have already proved that exponential functions $y = b^x$ have the shapes that they do; the graphs of logarithmic functions are obtained from these exponential graphs by reflecting them over the line $y = x$. Therefore, we know that the graphs of $y = \log_b x$ (for $b > 1$ and for $0 < b < 1$) *must* look like those in Figures 9.5 and 9.6.

EXAMPLE 9.8   **Limits of logarithmic functions at infinity**

By Theorem 9.6, we have $\lim\limits_{x \to \infty} \log_2 x = \infty$, $\lim\limits_{x \to \infty} \ln x = \infty$, and $\lim\limits_{x \to \infty} \log x = \infty$. On the other hand, $\lim\limits_{x \to \infty} \log_{0.5} x = -\infty$, and $\lim\limits_{x \to \infty} \log_{\frac{1}{3}} x = -\infty$. Although a logarithmic function $f(x) = \log_b x$ with $b > 1$ grows slowly, it does keep getting larger and larger (approaches infinity) as $x \to \infty$. ☐

EXAMPLE 9.9   **Limits of logarithmic functions at zero**

By Theorem 9.6, we have $\lim\limits_{x \to 0^+} \log_2 x = -\infty$, $\lim\limits_{x \to 0^+} \ln x = -\infty$, and $\lim\limits_{x \to 0^+} \log_{10} x = -\infty$. On the other hand, $\lim\limits_{x \to 0^+} \log_{0.5} x = \infty$, and $\lim\limits_{x \to 0^+} \log_{\frac{1}{3}} x = \infty$. ☐

We can now calculate fairly complicated limits involving logarithmic functions. However, we save these calculations until after the next subsection, so we can first investigate the derivatives of logarithmic functions. We will then be able to use the additional tool of L'Hôpital's Rule.

9.2.3 Derivatives of logarithmic functions

At this point we would usually go back to the definition of derivative to develop a "rule" for differentiating the type of function we are investigating. However, the fact that logarithmic functions are the inverses of exponential functions (whose derivatives we know already) will save us from having to do that. The following rule for differentiating logarithmic functions follows directly from what we know about the derivatives of exponential functions.

THEOREM 9.7

The Logarithmic Rule for Differentiation

For any positive real number b with $b \neq 1$,

$$\frac{d}{dx}(\log_b x) = \frac{1}{(\ln b)x}.$$

Technical Note: Since the domain of any logarithmic function $f(x) = \log_b x$ is $(0, \infty)$, we should really say that the derivative of $\frac{d}{dx}(\log_b x)$ is the function $\frac{1}{(\ln b)x}$ restricted to the domain $(0, \infty)$. We will revisit this idea in Section 9.3.3.

EXAMPLE 9.10

Differentiating logarithmic functions

By Theorem 9.7, we have $\dfrac{d}{dx}(\log_2 x) = \dfrac{1}{(\ln 2)x}$ and $\dfrac{d}{dx}(\log_{10} x) = \dfrac{1}{(\ln 10)x}$. □

PROOF (**THEOREM 9.7**)   Suppose b is any positive real number with $b \neq 1$. We wish to show that $\frac{d}{dx}(\log_b x) = \frac{1}{(\ln b)x}$; we will do so using implicit differentiation. Since $y = \log_b x$ and $y = b^x$ are inverses, we know that

$$b^{\log_b x} = x$$

for all x in the domain $(0, \infty)$ of $y = \log_b x$. The "outside part" of this equation is something we know how to differentiate (since it is an exponential function). We will now differentiate both sides of the equation (using the chain rule as appropriate) and then solve for the derivative we are interested in.

$$b^{\log_b x} = x \qquad \text{(property of inverses)}$$

$$\frac{d}{dx}(b^{\log_b x}) = \frac{d}{dx}(x) \qquad \text{(differentiate both sides)}$$

$$(\ln b)b^{\log_b x}\frac{d}{dx}(\log_b x) = 1 \qquad \text{(note use of chain rule here)}$$

$$\frac{d}{dx}(\log_b x) = \frac{1}{(\ln b)b^{\log_b x}} \qquad \text{(solve for } \frac{d}{dx}(\log_b x))$$

$$\frac{d}{dx}(\log_b x) = \frac{1}{(\ln b)x}. \qquad \text{(since } b^{\log_b x} = x)$$

Make sure that you understand the differentiation step in the proof of Theorem 9.7. It may help to think of $b^{\log_b x}$ as $b^{u(x)}$; then the derivative of $b^{u(x)}$ (by the chain rule) is $(\ln b)b^{u(x)}u'(x)$. In this case we did not know how to find the derivative $u'(x) = \frac{d}{dx}(\log_b x)$, so we just wrote "$\frac{d}{dx}(\log_b x)$"; since this was exactly the quantity we were trying to find, we then solved for this quantity.

❓ *Question*   In the proof of Theorem 9.7, we used the fact that $b^{\log_b x} = x$. It is also true that $\log_b(b^x) = x$; could we have started with this equality instead? In other words, can we differentiate both sides of the equality $\log_b(b^x) = x$ to solve for $\frac{d}{dx}(\log_b x)$? It turns out that the answer is no. Try it and explain what goes wrong. □

In particular, the differentiation rule in Theorem 9.7 tells us how to find the derivative of the natural logarithmic function $f(x) = \ln x$, since

$$\frac{d}{dx}(\ln x) = \frac{d}{dx}(\log_e x) = \frac{1}{(\ln e)x} = \frac{1}{(1)x} = \frac{1}{x}.$$

This is the most common usage of Theorem 9.7 (since the natural logarithmic function is the most commonly used logarithmic function). We state it as a separate theorem, even though it is merely a special case of Theorem 9.7.

THEOREM 9.8

The Derivative of the Natural Logarithmic Function

$$\frac{d}{dx}(\ln x) = \frac{1}{x}.$$

We can combine the chain rule with the rules in Theorems 9.7 and 9.8 as follows: If u is a function of x, then

$$\frac{d}{dx}(\ln u) = \frac{u'}{u} \quad \text{and} \quad \frac{d}{dx}(\log_a u) = \frac{u'}{(\ln a)u}.$$

For example, $\frac{d}{dx}(\ln(x^2 + 1)) = \frac{2x}{x^2+1}$. The following example provides some more complicated examples of how these differentiation rules work in practice. Keep in mind that many differentiation problems can be greatly simplified if some algebra is done first!

EXAMPLE 9.11

Differentiating expressions that involve logarithms

Find the derivatives of the following functions.

(a) $f(x) = (x^4 - \log_2(3x + 1))^3$ **(b)** $f(x) = \ln\left(\left(\frac{x^2 - 1}{1 - 2x}\right)^2\right)$

Solution

(a) Starting with the chain and power rules, and then using some other differentiation rules to differentiate the "inside," we have

$$f'(x) = 3(x^4 - \log_2(3x + 1))^2\left(4x^3 - \frac{1}{(\ln 2)(3x + 1)}(3)\right).$$

(b) There are a number of ways we could find $f'(x)$. One way is to jump right in and start differentiating; although it is possible to do this all in "one step," we'll do the calculation in a few separate steps to be as clear as possible:

$$\frac{d}{dx}\left(\ln\left(\left(\frac{x^2 - 1}{1 - 2x}\right)^2\right)\right)$$

$$= \frac{1}{\left(\frac{x^2-1}{1-2x}\right)^2}\frac{d}{dx}\left(\left(\frac{x^2 - 1}{1 - 2x}\right)^2\right)$$

$$= \frac{1}{\left(\frac{x^2-1}{1-2x}\right)^2}(2)\left(\frac{x^2 - 1}{1 - 2x}\right)^1\frac{d}{dx}\left(\frac{x^2 - 1}{1 - 2x}\right)$$

$$= \frac{1}{\left(\frac{x^2-1}{1-2x}\right)^2}(2)\left(\frac{x^2 - 1}{1 - 2x}\right)^1\left(\frac{(2x)(1 - 2x) - (x^2 - 1)(-2)}{(1 - 2x)^2}\right).$$

The first step used the chain rule and the logarithmic rule, the second step used the chain rule and the power rule, and the last step used the quotient rule.

If we do a little bit of algebra first, the differentiation step becomes much easier:

$$\frac{d}{dx}\left(\ln\left(\left(\frac{x^2-1}{1-2x}\right)^2\right)\right) = \frac{d}{dx}\left(2\ln\left(\frac{x^2-1}{1-2x}\right)\right)$$

$$= \frac{d}{dx}(2(\ln(x^2-1)-\ln(1-2x)))$$

$$= 2\left(\frac{1}{x^2-1}(2x)-\frac{1}{1-2x}(-2)\right).$$

In the calculation above, the first two steps were algebra; only the final step involved differentiation. Although the two answers obtained look much different, they are in fact the same (use algebra to show this!). □

9.2.4 Limits involving logarithmic functions

Now that we know the derivatives of logarithmic functions, we can use L'Hôpital's Rule to calculate some limits that involve logarithmic functions. As we have seen, L'Hôpital's Rule does not always apply, and even when it does, it is not always useful. Example 9.12 involves a limit where L'Hôpital's Rule applies and is useful; Example 9.13 involves a limit where L'Hôpital's Rule does not apply.

EXAMPLE 9.12   **Using L'Hôpital's Rule to find a limit involving logarithms**

Find the exact value of $\displaystyle\lim_{x\to 2}\frac{\ln(2x-3)}{x^2-4}$.

Solution   As $x\to 2$, we have $\ln(2x-3)\to\ln(1)=0$ and $x^2-4\to 0$. This limit is of the indeterminate form $\frac{0}{0}$. Since there is no possible algebraic simplification to be done, we'll try using L'Hôpital's Rule:

$$\lim_{x\to 2}\frac{\ln(2x-3)}{x^2-4} \overset{\text{L'H}}{=} \lim_{x\to 2}\frac{\frac{1}{2x-3}(2)}{2x} \qquad \text{(L'Hôpital's Rule)}$$

$$= \lim_{x\to 2}\frac{2}{2x(2x-3)} \qquad \text{(algebra)}$$

$$= \frac{2}{4(4-3)} = \frac{2}{4} = \frac{1}{2}. \qquad \text{(solve limit)} \qquad □$$

✓ *Checking the Answer*   Remember that to check your answer you could inspect a graph of $f(x)=\frac{\ln(2x-3)}{x^2-4}$ near $x=2$ (to see whether its height seems to approach $\frac{1}{2}$), or you could try evaluating $f(x)$ at values of x very close to 2, like $x=2.001$ or $x=1.999$. □

EXAMPLE 9.13   **A limit involving logarithms where L'Hôpital's Rule does not apply**

Calculate $\displaystyle\lim_{x\to\infty}(\log_2(x+1)-\log_2(x^2+1))$.

Solution   As $x\to\infty$, we have $\log_2(x+1)\to\infty$ and $\log_2(x^2+1)\to\infty$. This limit is of the indeterminate form $\infty-\infty$. Using the algebraic rules of logarithms, we can rewrite this limit in an alternative form:

$$\lim_{x\to\infty}(\log_2(x+1)-\log_2(x^2+1)) = \lim_{x\to\infty}\log_2\left(\frac{x+1}{x^2+1}\right) \qquad \text{(algebra)}$$

$$\to \log_2(0)\to -\infty. \qquad \text{(since } \frac{x+1}{x^2+1}\to 0 \\ \text{as } x\to\infty) \qquad □$$

Recall that exponential growth functions **dominate** any power or polynomial function as $x \to \infty$. Intuitively, this means that exponential growth functions grow much faster than any power or polynomial function. Logarithmic growth functions have the opposite property; they grow much *more slowly* than any power or polynomial function.

THEOREM 9.9

> **Power and Polynomial Functions Dominate Logarithmic Functions**
> If $f(x) = \log_b x$ is any logarithmic growth function (with $b > 1$), and $g(x)$ is any power function or polynomial function, then g dominates f.

This theorem can be proved with one application of L'Hôpital's Rule; the proof is left for the exercises. Recall that when we say that "g dominates f," we mean that the limit $\lim\limits_{x \to \infty} \frac{g(x)}{f(x)}$ approaches ∞.

EXAMPLE 9.14 **Showing that a polynomial function dominates a logarithmic function**

Show that as $x \to \infty$, the polynomial function $g(x) = 0.1x^2 - 300x$ dominates the logarithmic function $f(x) = \log_3 x$.

Solution We must show that $\lim\limits_{x \to \infty} \dfrac{g(x)}{f(x)} = \infty$. To do so we will need L'Hôpital's Rule:

$$\lim_{x \to \infty} \frac{g(x)}{f(x)} = \lim_{x \to \infty} \frac{0.1x^2 - 300x}{\log_3 x} \qquad \text{(note that this is of the form } \tfrac{\infty}{\infty}\text{)}$$

$$\overset{\text{L'H}}{=} \lim_{x \to \infty} \frac{0.1(2x) - 300}{\frac{1}{(\ln 3)x}} \qquad \text{(L'Hôpital's Rule)}$$

$$= \lim_{x \to \infty} ((0.2x - 300)(\ln 3)x) \qquad \text{(algebra)}$$

$$\to \infty \cdot \infty \to \infty. \qquad \text{(take limit)}$$

Notice that the logarithmic function $f(x) = \log_3 x$ is actually greater than the polynomial function $g(x) = 0.1x^2 - 300x$ until x is a little bit greater than 1000. However, as $x \to \infty$, $g(x)$ grows *much* faster than $f(x)$ and eventually overtakes the slow-growing logarithmic function $f(x)$ (to see this, graph $f(x)$ and $g(x)$ in a large viewing window). $\square$

9.2.5 Graphing functions that involve logarithms

Now that we understand the algebra, limits, and derivatives of logarithmic functions, we can use these computational skills to sketch graphs of functions that involve logarithms. To review all of the skills we have learned, we now present a curve-sketching example in complete detail.

EXAMPLE 9.15 **A detailed curve-sketching analysis of a function that involves a logarithm**

Sketch a careful, labeled graph of the function $f(x) = \dfrac{\ln 2x}{x}$.

Solution As usual, we begin by finding the domain of $f(x)$. Since the domain of the natural logarithmic function is $(0, \infty)$, we must have $2x > 0$. Moreover, since we cannot divide by zero, we must also have $x \neq 0$. These two requirements combine to give us a domain of $x > 0$ for $f(x)$.

We now investigate the signs of f, f', and f'', beginning with the function f. The function $f(x)$ is zero only when $\ln 2x = 0$, that is, when $2x = 1$ (since $\ln a = 0$ if and only if $a = 1$). Thus $f(x) = 0$ when $x = \frac{1}{2}$, and we know already that f is defined only on $(0, \infty)$. We must test the sign of f for one point in the interval $(0, \frac{1}{2})$ and for one point in $(\frac{1}{2}, \infty)$. For example, we could test at $x = \frac{1}{4}$ and $x = 1$:

$$f\left(\tfrac{1}{4}\right) = \frac{\ln\left(2\left(\tfrac{1}{4}\right)\right)}{\tfrac{1}{4}} = \frac{\ln\left(\tfrac{1}{2}\right)}{\tfrac{1}{4}} < 0, \qquad\qquad (\ln \tfrac{1}{2} < 0 \text{ since } \tfrac{1}{2} < 1)$$

$$f(1) = \frac{\ln(2 \cdot 1)}{1} = \frac{\ln 2}{1} > 0. \qquad\qquad (\ln 2 > 0 \text{ since } 2 > 1)$$

Therefore, the graph of $f(x)$ is negative for $x \in (0, \frac{1}{2})$ and positive for $x \in (\frac{1}{2}, \infty)$. This information is recorded on the number line for f in Figure 9.7 on the next page.

To investigate the signs of the derivative $f'(x)$, we must first differentiate $f(x)$:

$$f'(x) = \frac{d}{dx}\left(\frac{\ln 2x}{x}\right) = \frac{\frac{d}{dx}(\ln 2x) \cdot x - \ln 2x \cdot \frac{d}{dx}(x)}{x^2} \qquad \text{(quotient rule)}$$

$$= \frac{\frac{1}{2x}(2)(x) - (\ln 2x)(1)}{x^2} = \frac{\frac{2x}{2x} - \ln 2x}{x^2} = \frac{1 - \ln 2x}{x^2}.$$

We can now see that $f'(x)$ is zero only when $1 - \ln 2x = 0$, which is easy to solve if we remember that $\ln e = 1$:

$$1 - \ln 2x = 0 \;\Rightarrow\; 1 = \ln 2x \;\Rightarrow\; 2x = e \;\Rightarrow\; x = \frac{e}{2} \approx 1.359.$$

The derivative $f'(x)$ is defined only on $(0, \infty)$, which also happens to be the domain of $f(x)$. The only critical point of $f(x)$ is $x = \frac{e}{2} \approx 1.359$. We need to check the sign of $f'(x)$ on the intervals $(0, \frac{e}{2})$ and $(\frac{e}{2}, \infty)$. For example, look at the signs of $f'(x)$ at $x = 1$ and $x = 2$:

$$f'(1) = \frac{1 - \ln(2 \cdot 1)}{1^2} = \frac{1 - \ln 2}{1} \approx 0.307 > 0,$$

$$f'(2) = \frac{1 - \ln(2 \cdot 2)}{2^2} = \frac{1 - \ln 4}{4} \approx -0.097 < 0.$$

We now know that $f'(x)$ is positive (and thus f is increasing) on $(0, \frac{e}{2})$ and that $f'(x)$ is negative (and thus f is decreasing) on $(\frac{e}{2}, \infty)$. See the number line for $f'(x)$ in Figure 9.7 on the next page.

Finally, we calculate and investigate the signs of the second derivative:

$$f''(x) = \frac{d}{dx}(f'(x)) = \frac{d}{dx}\left(\frac{1 - \ln 2x}{x^2}\right)$$

$$= \frac{\left(-\frac{1}{2x}(2)\right)(x^2) - (1 - \ln 2x)(2x)}{(x^2)^2} \qquad \text{(quotient rule)}$$

$$= \frac{-x - 2x + 2x\ln 2x}{x^4} = \frac{x(-3 + 2\ln 2x)}{x^4} = \frac{-3 + 2\ln 2x}{x^3}.$$

Clearly $f''(x)$ is zero when $-3 + 2\ln 2x = 0$. To solve this equation, we need to remember that $\ln x = y$ if and only if $x = e^y$:

$$-3 + 2\ln 2x = 0 \;\Rightarrow\; 2\ln 2x = 3 \;\Rightarrow\; \ln 2x = \tfrac{3}{2}$$

$$\Rightarrow\; 2x = e^{\frac{3}{2}} \;\Rightarrow\; x = \tfrac{1}{2}e^{\frac{3}{2}} \approx 2.241.$$

The second derivative $f''(x)$ has only one zero, at $x = \frac{1}{2}e^{\frac{3}{2}}$, and is defined everywhere on $(0, \infty)$. We now check the signs of $f''(x)$ to the left and the right of

$x = \frac{1}{2}e^{\frac{3}{2}} \approx 2.241$:

$$f''(1) = \frac{-3 + 2\ln(2 \cdot 1)}{1^3} = \frac{-3 + 2\ln 2}{1} \approx -1.614 < 0,$$

$$f''(3) = \frac{-3 + 2\ln(2 \cdot 3)}{3^3} = \frac{-3 + 2\ln 6}{27} \approx 0.022 > 0.$$

Therefore, $f''(x)$ is negative (and thus $f(x)$ is concave down) on $(0, \frac{1}{2}e^{\frac{3}{2}})$, and $f''(x)$ is positive (and thus $f(x)$ is concave up) on $(\frac{1}{2}e^{\frac{3}{2}}, \infty)$. This information is recorded on the third number line in Figure 9.7.

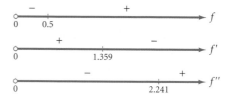

Figure 9.7

It now remains only to find the values of f at the "interesting points" $x = \frac{1}{2}$, $x = \frac{e}{2} \approx 1.359$, and $x = \frac{1}{2}e^{\frac{3}{2}} \approx 2.241$ and to find the limits of the function $f(x)$ as $x \to 0^+$ and as $x \to \infty$ (the "ends" of the domain). We calculate the limits first:

$$\lim_{x \to 0^+} f(x) = \lim_{x \to 0^+} \frac{\ln 2x}{x} \qquad \text{(note that this approaches } \frac{-\infty}{0} \text{)}$$

$$= \lim_{x \to 0^+} \left(\ln 2x \cdot \frac{1}{x} \right) \qquad \text{(rewrite limit)}$$

$$\to -\infty \cdot \infty \to -\infty.$$

$$\lim_{x \to \infty} f(x) = \lim_{x \to \infty} \frac{\ln 2x}{x} \qquad \text{(note that this is of the form } \frac{\infty}{\infty} \text{)}$$

$$\overset{\text{L'H}}{=} \lim_{x \to \infty} \frac{\frac{1}{2x}(2)}{1} = \lim_{x \to \infty} \frac{2}{2x}$$

$$= \lim_{x \to \infty} \frac{1}{x} \to 0.$$

This limit information tells us that the graph of $f(x)$ has a vertical asymptote (approaching $-\infty$) as x approaches zero from the right, and a horizontal asymptote (at $y = 0$) as $x \to \infty$. It is simple to calculate that the "interesting" values of f are

$$f\left(\tfrac{1}{2}\right) = 0, \quad f\left(\tfrac{e}{2}\right) \approx 0.736, \quad \text{and} \quad f\left(\tfrac{1}{2}e^{\frac{3}{2}}\right) \approx 0.669.$$

Putting all this information together, we obtain the graph in Figure 9.8. The dots on the graph are the "interesting" values, where f, f', and f'' change sign.

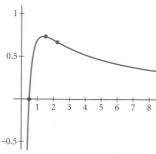

Figure 9.8 $\quad f(x) = \dfrac{\ln 2x}{x}$

(It is difficult to see the horizontal asymptote at $y = 0$ in this graph, but it is there. If you graph $f(x)$ in a "longer" window, you will be able to see it, but then the locations of the maximum and the inflection point will be harder to see.)   □

What you should know after reading Section 9.2

▶ The continuity of logarithmic functions (and how to *prove* it), and what this means about calculating limits of logarithmic functions.

▶ The rule for differentiating logarithmic functions, how to prove it, and how to use it in differentiation problems. What is the rule for differentiating the *natural* logarithmic function, and why?

▶ How to use L'Hôpital's Rule and other techniques to calculate limits of functions that involve logarithms. How to take limits of logarithmic functions as $x \to 0^+$ and as $x \to \infty$.

▶ How to use algebra, limits, and derivatives to sketch accurate graphs (by hand) of functions that involve logarithms. This includes identifying local extrema, inflection points, and asymptotes.

Exercises 9.2

Concepts

0. Read the section and make your own summary of the material. Include whatever you think will help you review the material later on.

1. What is $\lim\limits_{x \to 1} \ln(x - 1)$? Be sure to consider both the left and right limits.

2. What is $\lim\limits_{x \to 2} \log_3(x - 5)$? Why?

3. Consider the function $f(x) = \frac{\ln(x^5)}{\ln(x^4)}$. Calculate $\lim\limits_{x \to \infty} f(x)$ in two ways: **(a)** using L'Hôpital's Rule, and **(b)** by doing algebra first. Verify that your two answers are the same.

4. Consider the function $f(x) = \frac{\ln(x^5)}{\ln(x^4)}$. Find the derivative $f'(x)$ in two ways: **(a)** without doing any algebra first, and **(b)** by doing algebra before differentiating. Verify that your two answers are the same.

5. Every logarithmic function is continuous (on its domain). Express this fact **(a)** using a limit, and **(b)** using the delta–epsilon definition of limit. Why did we *not* have to use the delta–epsilon definition of limit to prove that logarithmic functions are continuous? What did we use instead?

6. Every logarithmic function is differentiable (on its domain). Express this fact using a limit (with the definition of differentiability).

7. The derivative of any logarithmic function $f(x) = \log_b x$ is $f'(x) = \frac{1}{(\ln b)x}$. Express this fact using the formal "limit" definition of derivative. Why did we *not* have to use the definition of derivative to prove this differentiation rule? What did we use instead?

8. Fill in the blanks below.

   (a) $\lim\limits_{x \to 0^+} \log_3 x = $ _____

   (b) $\lim\limits_{x \to 0^+} \log_{\frac{1}{2}} x = $ _____

   (c) $\lim\limits_{x \to \infty} \log_{0.2} x = $ _____

   (d) $\lim\limits_{x \to \infty} \log_4 x = $ _____

9. Find the error in the following limit calculation:

$$\lim_{x \to \infty} \frac{x}{\ln x} \stackrel{\text{L'H}}{=} \lim_{x \to \infty} \frac{1}{\frac{1}{x}} = \lim_{x \to \infty} x = \infty.$$

10. Explain why Theorem 9.8 is just a special case of Theorem 9.7.

11. In this section we proved that $\frac{d}{dx}(\log_b x) = \frac{1}{(\ln b)x}$ by differentiating both sides of the equation $b^{\log_b x}$. Explain why the proof would *not* work if we instead differentiated both sides of the equation $\log_b(b^x) = x$.

12. Use the fact that logarithmic functions are always dominated by power functions to find $\lim\limits_{x \to \infty} \frac{500 \ln x}{x^{\frac{1}{5}}}$.

13. Find a function $f(x)$ with derivative $f'(x) = \frac{1}{x}$ and value $f(1) = 4$.

14. Find a function $f(x)$ with derivative $f'(x) = \frac{3}{2x+1}$ and value $f(2) = 6$.

15. Find a function $f(x)$ with derivative $f'(x) = \frac{2x}{x^2+1}$ and value $f(0) = -2$.

Skills

■ Calculate the following limits.

16. $\lim\limits_{x \to 5} \log_3(x^2 - 4)$

17. $\lim\limits_{x \to 2^+} \log_3(x^2 - 4)$

18. $\lim\limits_{x \to 0^+} \log_{\frac{1}{2}}\left(\frac{1}{x}\right)$

19. $\lim\limits_{x \to 1} \dfrac{\ln x}{x - 1}$

20. $\lim\limits_{x \to 0^+} \dfrac{x}{\log_2 x}$

21. $\lim\limits_{x \to 0^+} x \log_2 x$

22. $\lim\limits_{x \to \infty} (\ln x^2 - \ln(2x + 1))$

23. $\lim\limits_{x \to \infty} (\ln 3x - \ln 2x)$

24. $\lim\limits_{x \to \infty} (\ln(3x - 1) + \ln(x^{-2}))$

25. $\lim\limits_{x \to 0^+} \dfrac{\log_2 x}{\log_2 3x}$

26. $\lim\limits_{x \to 2^+} \dfrac{\ln(x - 2)}{\ln(x^2 - 4)}$

27. $\lim\limits_{x \to 2^+} \ln\left(\dfrac{2^x - 4}{x - 2}\right)$

28. $\lim\limits_{x \to \infty} \log_3\left(\dfrac{x^2 - 1}{2 - 3x^2}\right)$

29. $\lim\limits_{x \to \infty} \dfrac{\ln x}{\ln(2x + 1)}$

30. $\lim\limits_{x \to \infty} \ln\left(\dfrac{e^x}{1 + e^x}\right)$

■ Calculate the derivative of each of the following functions $f(x)$. (Don't forget that sometimes you can make a differentiation problem easier by doing some algebra first!) Then determine the critical points of f (i.e., the values where $f'(x)$ is zero or undefined).

31. $f(x) = \log_2(3x^2 - 2x - 5)$

32. $f(x) = \ln \sqrt{x}$

33. $f(x) = \sqrt{\ln x}$

34. $f(x) = \log_2(3^x)$

35. $f(x) = \log_2(2^x)$

36. $f(x) = e^{(\ln 2)x}$

37. $f(x) = 2^{1 - \ln x}$

38. $f(x) = \ln\left(\dfrac{x^3}{x^2 + x + 1}\right)$

39. $f(x) = \ln(x\sqrt{2x + 1})$

■ Calculate the derivative of each of the following functions $f(x)$.

40. $f(x) = x^2 \log_2(x 2^x)$

41. $f(x) = \dfrac{1 - \log_3 x}{x^2 + 5}$

42. $f(x) = \dfrac{e^x \ln x}{x^2 - 1}$

43. $f(x) = x^2 \ln(\ln x)$

44. $f(x) = \log_3(\sqrt{x^5 + 1} - \sqrt{5x})$

45. $f(x) = \sqrt{\ln(3x^2 - 2x - 5)}$

■ For each pair of functions $f(x)$ and $g(x)$ given below, determine whether f dominates g, or g dominates f, or neither. Support your answer with a limit calculation.

46. $f(x) = \log_2 x,\ g(x) = 0.002x^{0.001}$

47. $f(x) = \ln(x^2 + 1),\ g(x) = x^2 + 1$

48. $f(x) = \log_3(1000x^{50}),\ g(x) = x + 2$

49. $f(x) = \ln x,\ g(x) = \ln(x^{100} + x^{50} + 200)$

50. $f(x) = \log_{\frac{1}{2}} x,\ g(x) = x^2 - 50$

51. $f(x) = \log_2 x,\ g(x) = \log_{30} x$

■ Sketch an accurate graph of each function $f(x)$ below, *without* using a graphing calculator. Your work should include finding the domain of f; constructing number lines for f, f', and f''; and showing how you found the locations of any roots, local extrema, inflection points, and asymptotes.

52. $f(x) = x \log_2 x$

53. $f(x) = \dfrac{\ln 3x}{x}$

54. $f(x) = \ln(\ln x)$

55. $f(x) = \ln\left(\dfrac{1}{1 + x^2}\right)$

56. $f(x) = \ln\left(\dfrac{1}{x^2 - 1}\right)$

57. $f(x) = \dfrac{x}{\ln x - 1}$

Proofs

58. Prove that the natural logarithmic function $f(x) = \ln x$ is continuous at every point in its domain (following the method of proof used for Theorem 9.5).

59. Prove that the function $f(x) = \log_2 x$ is continuous at every point in its domain (following the method of proof used for Theorem 9.5).

60. Prove that the derivative of $f(x) = \ln x$ is $f'(x) = \frac{1}{x}$ (follow the method of proof for Theorem 9.7).

61. Prove that the derivative of $f(x) = \log_2 x$ is $f'(x) = \frac{1}{(\ln 2)x}$ (follow the method of proof for Theorem 9.7).

9.3   Using Logarithms as a Calculational Tool

In this section we examine three ways in which logarithms can be used as a calculational tool: to solve equations, to calculate limits, and to calculate derivatives. The first way we have seen already; many equations can be solved if we first "take logs of both sides." The other two types of calculations involve calculating a limit or derivative of $\ln(f(x))$ in order to determine a limit or derivative of some function $f(x)$.

9.3.1   Using logarithms to solve equations

In this section we use logarithms to solve equations where the variable x is in the exponent of an expression—for example, the equation $2^x = 3$. In such examples, logarithms can be used to "cancel" the exponentiation and isolate the variable x.

There are two ways that we could use logarithms to solve the equation $2^x = 3$: by using the natural logarithm (and algebraic properties of logarithms) or by using the base 2 logarithm and the fact that $\log_2 x$ is the inverse of the function 2^x. We usually use the natural logarithm for solving such equations (regardless of the base in the exponential function), but using a logarithmic function with a well-chosen base can often make the problem easier. We have seen such calculations already, but we repeat them in the following example.

EXAMPLE 9.16

Solving an equation using logarithms of different bases

Solve the equation $2^x = 3$ in two ways:

(a) Using the natural logarithmic function.

(b) Using the base 2 logarithmic function.

Solution

(a) Applying the natural logarithm to both sides and then using a little algebra will enable us to move the variable x out of the exponent:

$$2^x = 3 \quad\Rightarrow\quad \ln(2^x) = \ln 3 \quad\Rightarrow\quad x \ln 2 = \ln 3 \quad\Rightarrow\quad x = \frac{\ln 3}{\ln 2}.$$

(b) Applying the base 2 logarithm to both sides will "cancel" the exponential function with base 2 (since $\log_2(2^x) = x$):

$$2^x = 3 \quad\Rightarrow\quad \log_2(2^x) = \log_2 3 \quad\Rightarrow\quad x = \log_2 3.$$

It is important to note that the two answers we found are the same (by the base conversion formula for logarithms). Now that we have the exact solution to the equation $2^x = 3$, we can use a calculator to obtain a numerical approximation for the solution: $x = \frac{\ln 3}{\ln 2} \approx 1.58496$. ☐

When we add or multiply an equation by a real number, we do not change the solution set of the equation. Similarly, the process of "taking the logarithms of both sides" of an equation does not change the solution set of the equation as long as both sides are always positive. For example, the solution set of the equation $2^x = 3$ is the same as the solution set of the equation $\ln(2^x) = \ln 3$. Why? The reason is that all logarithmic functions are one-to-one functions, and thus $A = B$ if and only if $\log_b A = \log_b B$.

❓ *Question*   How do we know that all logarithmic functions are one-to-one? Why does this mean that $A = B$ if and only if $\log_b A = \log_b B$? ☐

As illustrated in the next example, sometimes we need to do some preliminary algebra before "taking the logarithm" of each side of an equation. In particular, if you wish to isolate or solve for a variable x, you should first isolate the exponential expression involving x and *only then* take the logarithm of each side. This is similar to the fact that in the equation $2(x+1) = 3$, you should first isolate the expression $x + 1$ (by dividing both sides by 2) before attempting to subtract 1 from both sides.

EXAMPLE 9.17

Using logarithms to solve an equation

Solve the equation $(5^x + 1)^2 = 9$.

Solution   Before using logarithms, we must isolate the exponential expression 5^x:

$$(5^x + 1)^2 = 9 \implies 5^x + 1 = \pm 3 \qquad \text{(take square roots of both sides)}$$

$$\implies 5^x + 1 = 3 \qquad \text{($5^x + 1$ is always positive, so it can't equal -3)}$$

$$\implies 5^x = 2 \qquad \text{(subtract 1 from both sides)}$$

$$\implies \ln(5^x) = \ln 2 \qquad \text{(take the natural logarithms of both sides)}$$

$$\implies x \ln 5 = \ln 2 \qquad \text{(algebraic property of logarithms)}$$

$$\implies x = \frac{\ln 2}{\ln 5} \approx 0.4307. \qquad \text{(divide both sides by $\ln 5$)}$$

$\square$

You don't always have to isolate an expression of the form b^x before taking logarithms. For example, consider the equation $4(2^x) = 3$. We could apply the natural logarithm to both sides of this equation as it stands and then use the fact that $\ln(4(2^x)) = \ln 4 + \ln(2^x) = \ln 4 + x \ln 2$. On the other hand, we could start by dividing both sides of the equation by 4 and *then* taking logarithms. You'll get the same answer either way; try it! In either case you will have to do some algebra, although many times the algebra is simpler if you do it *before* applying logarithms.

9.3.2 Using logarithms to calculate limits (0^0, 1^∞, and ∞^0)

Logarithms are a useful tool for calculating limits that involve variables in both the exponent and the base of an expression. Some limits of this type can be calculated using methods that we already know. For example,

$$(1) \qquad \lim_{x \to \infty} (x^2 - 1)^x \to \infty^\infty \to \infty.$$

However, limits of the forms 0^0, 1^∞, and ∞^0 are ***indeterminate;*** limits of these forms may or may not exist, depending on the functions involved.

THEOREM 9.10

Limit Forms Involving a Variable Raised to a Variable

The following forms are ***indeterminate:***

$$0^0, \qquad 1^\infty, \qquad \infty^0.$$

The following forms are *not* indeterminate (and approach the indicated limits):

$$\infty^\infty \to \infty, \qquad 0^\infty \to 0, \qquad \infty^1 \to \infty, \qquad 1^0 \to 1, \qquad 0^1 \to 0.$$

You may be surprised that a limit of the form 1^∞ is indeterminate; why doesn't such a limit always approach 1? Although all three of the limits in Statement (2) are of the form 1^∞, only *one* of them approaches 1 (we will examine the first two limits in the exercises; the third limit should already be familiar to you).

$$(2) \qquad \lim_{x \to \infty} \left(\frac{x}{x-1} \right)^x = 1, \quad \text{but} \quad \lim_{x \to 1^+} x^{\frac{1}{x-1}} = \infty \quad \text{and} \quad \lim_{x \to \infty} \left(1 + \tfrac{1}{x} \right)^x = e.$$

Loosely speaking, if the base approaches 1 somehow "faster" than the exponent approaches ∞, then a limit of the form 1^∞ will approach 1 (since raising a number very close to 1 to a large exponent results in a number that is also very close to 1). This happens in the first limit in Statement (2). On the other hand, if the exponent approaches ∞ somehow "faster" than the base approaches 1, then a limit of the form 1^∞ will approach ∞. This is what happens in the second limit in Statement (2). If the race between the base (which is approaching 1) and the exponent (which is approaching ∞) is somehow "balanced," then we can get a limit like the third one in Statement (2).

◈ *Caution*   Consider a limit where the base *is* the number 1 while the exponent approaches ∞, like $\lim_{x \to \infty} 1^{x+1}$. Such a limit is *not* of the indeterminate form 1^∞, since the base is not *approaching* 1 as $x \to \infty$; it is *equal* to 1 all the time (regardless of the value of x). In other words, we have $\lim_{x \to \infty} 1^{x+1} = \lim_{x \to \infty} 1 = 1$; the expression 1^{x+1} is equal to 1 even before we take the limit.   □

You may also wonder why a limit of the form 0^∞ must approach zero (and thus is *not* indeterminate). Remember that a small number (less than 1) raised to a large positive exponent is very *small*; as the base approaches 0 and the exponent approaches ∞, a limit of the form 0^∞ will very rapidly approach zero.

How can we calculate limits that are initially in the indeterminate forms 0^0, 1^∞, or ∞^0? Usually there will not be any algebraic simplification that can "fix" such indeterminate limits. However, remember that logarithms have the power to change exponentiation into multiplication, in the sense that $\ln(a^b) = b \ln a$. The problem with limits of the indeterminate forms 0^0, 1^∞, and ∞^0 is that there is a variable in both the base and the exponent; if we could somehow apply logarithms to these limits, then we could take care of this problem by moving the exponent. The key to using logarithms to calculate limits is the following theorem.

THEOREM 9.11

Relating the Limit of a Function to the Limit of Its Logarithm

If $\lim_{x \to c} \ln(f(x)) = L$, then $\lim_{x \to c} f(x) = e^L$.

This theorem applies even if $\lim_{x \to c} \ln(f(x))$ is infinite (although this case is not covered in the proof we will give). In this case the theorem says that

$$\lim_{x \to c} \ln(f(x)) \to \infty \implies \lim_{x \to c} f(x) \to e^\infty \to \infty,$$

and

$$\lim_{x \to c} \ln(f(x)) \to -\infty \implies \lim_{x \to c} f(x) \to e^{-\infty} \to 0.$$

We'll prove Theorem 9.11 in a moment; first let's see how it can be useful for calculating limits of the indeterminate forms discussed above. To calculate a limit $\lim_{x \to c} f(x)$ (where the variable x is involved both in the base and in the exponent of the function $f(x)$), we will first calculate the limit $\lim_{x \to c} \ln(f(x))$. This limit will be easier to calculate because the logarithm will enable us to "get rid of" the exponent. Then we can use the answer to this new limit to find the value of the original limit, using Theorem 9.11.

EXAMPLE 9.18

Using logarithms to calculate a limit

Calculate $\lim_{x \to \infty} x^{\frac{1}{x}}$.

Solution Since $x \to \infty$ and $\frac{1}{x} \to 0$ as x approaches ∞, this limit is of the indeterminate form ∞^0. Let's calculate the related (but *different*) limit $\lim_{x \to \infty} \ln(x^{\frac{1}{x}})$ and see what we get:

$$
\begin{aligned}
\lim_{x \to \infty} \ln\left(x^{\frac{1}{x}}\right) &= \lim_{x \to \infty} \left(\frac{1}{x}\right) \ln x && \text{(algebra; note that this limit is of the form } 0 \cdot \infty) \\
&= \lim_{x \to \infty} \frac{\ln x}{x} && \text{(algebra; this limit is in the form } \tfrac{\infty}{\infty}) \\
&\overset{\text{L'H}}{=} \lim_{x \to \infty} \frac{\frac{1}{x}}{1} && \text{(L'Hôpital's Rule)} \\
&= \lim_{x \to \infty} \frac{1}{x} = 0. && \text{(solve limit)}
\end{aligned}
$$

But wait! This is not the answer to our *original* limit; however, by Theorem 9.11 we know that our original limit is equal to

$$
\lim_{x \to \infty} x^{\frac{1}{x}} = e^{\lim_{x \to \infty} \ln\left(x^{\frac{1}{x}}\right)} = e^0 = 1.
$$

(*Note:* You don't have to write down the second step; we included it here only to make it clear why you use e^0 to calculate the value of the original limit.) □

In fact, the method of calculating $\lim_{x \to c} \ln(f(x))$ to find the value of $\lim_{x \to c} f(x)$ illustrated in Example 9.18 can be used to solve lots of limits that involve exponents; it is not necessary for those limits to be of one of the indeterminate forms 0^0, 1^∞, and ∞^0. For example, we could use this logarithm technique to find the limit in Equation (1); however, it is much easier to use the second part of Theorem 9.10 to determine a limit if it is not indeterminate.

We now prove Theorem 9.11. The proof follows directly from the "composition" rule for limits of continuous functions (see Section 2.6).

PROOF (**THEOREM 9.11**) Suppose $f(x)$ is a function that is positive as x approaches c (so $\ln(f(x))$ is defined near $x = c$). Since the function $f(x) = \ln x$ is continuous on $(0, \infty)$, the rule for limits of compositions of continuous functions tells us that

$$
L = \lim_{x \to c} \ln(f(x)) = \ln\left(\lim_{x \to c} f(x)\right);
$$

Since $L = \ln(A)$ if and only if $A = e^L$, this equation implies that

$$
\lim_{x \to c} f(x) = e^L.
$$

9.3.3 Using logarithms to calculate derivatives

Logarithms can sometimes be a useful tool for calculating derivatives of very complicated functions. Because logarithms can in some sense turn exponentiation into multiplication (since $\ln(a^b) = b \ln a$) as well as turn multiplication into addition (since $\ln(ab) = \ln a + \ln b$), it is often easier to calculate the derivative of $\ln(f(x))$ than the derivative of $f(x)$. We can then use the derivative of $\ln(f(x))$ to find the derivative of $f(x)$ (as we will see in Example 9.19). Calculating $\frac{d}{dx}(\ln f(x))$ and then using it to find the derivative of $f(x)$ is often called **logarithmic differentiation.** In general, logarithmic differentiation is a very useful tool for differentiating functions that involve products or quotients with a lot of factors.

There is one possible problem: Given a function $f(x)$, the function $\ln(f(x))$ will be defined only for those values of x where $f(x)$ is positive. To solve this problem, we will consider the function $\ln|f(x)|$ rather than the function $\ln(f(x))$. It is often a complicated task to differentiate a function that involves an absolute value, but luckily it is not difficult in this case. In fact, the derivative of $\ln|x|$ is "the same" as the derivative of $\ln x$, but on a larger domain:

THEOREM 9.12

The Derivative of $\ln|x|$

For all $x \neq 0$, $\dfrac{d}{dx}(\ln|x|) = \dfrac{1}{x}$.

You will prove this fact in the exercises. Recall that the derivative of $\ln x$ is the function $\frac{1}{x}$ restricted to the domain $(0, \infty)$. Theorem 9.12 simply states that the derivative of $\ln|x|$ is the function $\frac{1}{x}$ on its usual domain.

❓ Question   If y is a function of x, what is $\frac{d}{dx}(\ln|y|)$? (Don't forget the chain rule!) For which values of x is your answer valid?   □

The following example illustrates how logarithmic differentiation can simplify the task of differentiating a function that involves a product or quotient with many factors. The general process is to set $y = f(x)$, take the natural logarithms of both sides to get $\ln y = \ln(f(x))$, use implicit differentiation, and then solve for y'.

EXAMPLE 9.19

Using logarithmic differentiation when products and quotients are involved

Use logarithmic differentiation to calculate the derivative of the function

$$f(x) = \frac{\sqrt{x}(x^2 - 1)^5}{(x + 2)(x - 4)^3}.$$

Solution   Obviously, it would take a great deal of work to differentiate this function as it is currently written. We would have to apply the quotient rule once and the product rule twice (among other things). We could do some algebra and multiply out some of the factors in the numerator and the denominator of $f(x)$ to make our job easier, but that is in itself a pretty nasty calculation. As we will soon see, it is not nearly so difficult to differentiate the related but *different* function $\ln|f(x)|$.

$$y = \frac{\sqrt{x}(x^2 - 1)^5}{(x+2)(x-4)^3}$$

$$\ln|y| = \ln\left|\frac{\sqrt{x}(x^2 - 1)^5}{(x+2)(x-4)^3}\right| \qquad \text{(apply } \ln|x|\text{)}$$

$$\ln|y| = \ln|\sqrt{x}| + \ln|(x^2 - 1)^5| - \ln|x+2| - \ln|(x-4)^3| \qquad \text{(algebra)}$$

$$\ln|y| = \tfrac{1}{2}\ln|x| + 5\ln|x^2 - 1| - \ln|x+2| - 3\ln|x-4| \qquad \text{(algebra)}$$

$$\frac{d}{dx}(\ln|y|) = \frac{d}{dx}\left(\tfrac{1}{2}\ln|x| + 5\ln|x^2 - 1| - \ln|x+2| - 3\ln|x-4|\right) \qquad \text{(differentiate)}$$

$$\frac{1}{y}y' = \frac{1}{2x} + \frac{5(2x)}{x^2 - 1} - \frac{1}{x+2} - \frac{3}{x-4} \qquad \text{(derivative rules)}$$

$$y' = y\left(\frac{1}{2x} + \frac{10x}{x^2 - 1} - \frac{1}{x+2} - \frac{3}{x-4}\right) \qquad \text{(solve for } y'\text{)}$$

$$y' = \frac{\sqrt{x}(x^2 - 1)^5}{(x+2)(x-4)^3}\left(\frac{1}{2x} + \frac{10x}{x^2 - 1} - \frac{1}{x+2} - \frac{3}{x-4}\right). \qquad \text{(definition of } y\text{)}$$

Although this looks very complicated, the differentiation step is very simple. However, notice the use of the chain rule when we say that $\frac{d}{dx}(\ln y) = \frac{1}{y}y'$. Note also that we use the fact that $|ab| = |a||b|$ and the fact that $|a^b| = |a|^b$. ☐

❓ Question Although there are absolute values in the fifth line of the calculation in Example 9.19, there are no absolute values in the sixth line; what happened to the absolute values and why? ☐

❓ Question Try to differentiate the function $f(x)$ from Example 9.19 using the quotient and product rules (as it is written initially). Then differentiate $f(x)$ by doing algebra (multiplying out expressions) and then differentiating. You will find that both of these methods take considerably more work, and leave more room for error, than the calculation in Example 9.19. ☐

Logarithmic differentiation is also useful for calculating the derivatives of functions that have variables in both the base *and* the exponent. For example, suppose we wish to differentiate the function $f(x) = (x^2 + 3)^x$. The power rule does not apply to this function, since the exponent x is not a constant. On the other hand, the exponential rule for differentiation does not apply either, since the base $(x^2 + 3)$ is not a constant. Luckily, we will not have to go back to the definition of derivative (it wouldn't be very helpful for this function anyway); instead we will use logarithmic differentiation.

EXAMPLE 9.20

Using logarithms to differentiate a function with variables in both base and exponent

Find $\dfrac{d}{dx}((x^2 - 3)^x)$.

Solution In contrast to Example 9.19, using logarithmic differentiation here is not a choice; it is a necessity. By setting $y = (x^2 - 3)^x$ and then applying the natural logarithm to both sides, we will be able to "remove" the variable x from the exponent. Since our whole problem with differentiating $y = (x^2 - 3)^x$ stems from the fact that the variable x appears in both the base and the exponent, "getting rid of" the exponent should prove very helpful. We have

$$y = (x^2 - 3)^x$$

$$\ln y = \ln((x^2 - 3)^x) \hspace{3cm} \text{(apply } \ln(x) \text{ to both sides)}$$

$$\ln y = x \ln(x^2 - 3) \hspace{3cm} \text{(algebra)}$$

$$\frac{d}{dx}(\ln y) = \frac{d}{dx}(x \ln(x^2 - 3)) \hspace{2cm} \text{(differentiate both sides)}$$

$$\frac{1}{y} y' = (1) \ln(x^2 - 3) + (x)\frac{2x}{x^2 - 3} \hspace{2cm} \text{(chain and product rules)}$$

$$y' = y\left(\ln(x^2 - 3) + \frac{2x^2}{x^2 - 3}\right) \hspace{2cm} \text{(solve for } y)$$

$$y' = (x^2 - 3)^x \left(\ln(x^2 - 3) + \frac{2x^2}{x^2 - 3}\right). \hspace{1cm} \text{(since } y = (x^2 - 3)^x)$$

Note that we did not need any absolute values in this calculation (usually we would apply $\ln|x|$ to both sides, rather than just $\ln x$). This is because $(x^2 - 3)^x$ is always positive, so the absolute value symbols would have been redundant. This is always the case when we are dealing with a function that involves a variable in both the exponent and the base, since such functions are well defined only when the base is positive. □

Logarithmic differentiation can also be used to prove the product, quotient, power, and exponential rules for differentiation. In fact, the proofs of these rules via logarithmic differentiation are easier than the proofs we used when we learned these rules. However, at that time we didn't know about logarithmic functions! In the exercises you'll prove the quotient, power, and exponential rules using logarithmic differentiation. Example 9.21 uses logarithmic differentiation to prove the product rule.

EXAMPLE 9.21

Using implicit differentiation to prove a differentiation rule

Use implicit differentiation to prove that the derivative of a product $f(x)g(x)$ of functions is given by the formula $f'(x)g(x) + f(x)g'(x)$. □

PROOF (EXAMPLE 9.21)   Let $y(x) = f(x)g(x)$; we have to show that $y'(x) = f'(x)g(x) + f(x)g'(x)$. Using logarithmic differentiation, we have

$$y(x) = f(x)g(x)$$

$$\ln|y(x)| = \ln|f(x)g(x)| \hspace{3cm} \text{(apply } \ln|x| \text{ to both sides)}$$

$$\ln|y(x)| = \ln(|f(x)||g(x)|) \hspace{3cm} \text{(algebra of absolute values)}$$

$$\ln|y(x)| = \ln|f(x)| + \ln|g(x)| \hspace{3cm} \text{(algebra of logarithms)}$$

$$\frac{d}{dx}(\ln|y(x)|) = \frac{d}{dx}(\ln|f(x)| + \ln|g(x)|) \hspace{2cm} \text{(differentiate both sides)}$$

$$\frac{1}{y(x)}y'(x) = \frac{f'(x)}{f(x)} + \frac{g'(x)}{g(x)} \hspace{2cm} \text{(chain and logarithmic rules)}$$

$$y'(x) = y(x)\left(\frac{f'(x)}{f(x)} + \frac{g'(x)}{g(x)}\right) \hspace{2cm} \text{(algebra)}$$

$$y'(x) = f(x)g(x)\left(\frac{f'(x)g(x) + g'(x)f(x)}{f(x)g(x)}\right) \hspace{1.5cm} \text{(since } y(x) = f(x)g(x))$$

$$y'(x) = f'(x)g(x) + g'(x)f(x). \hspace{3cm} \text{(algebra)}$$

Although this calculation may look long and complicated, it is mostly very simple algebra steps. Of course, we proved the product rule long ago, but the proof in Example 9.21 illustrates the power of logarithmic differentiation. ◼

What you should know after reading Section 9.3

▶ How to use logarithms as a tool for solving equations. This includes knowing the algebraic properties of logarithms and how (and when) to use them to your advantage.

▶ Be able to identify indeterminate forms that involve exponents (as well as those forms that are *not* indeterminate).

▶ How to solve a limit of a function $f(x)$ by first finding the limit of $\ln(f(x))$. How is the limit of $\ln(f(x))$ related to the limit of $f(x)$, and why? In what situations is this method of solving limits useful?

▶ How to calculate the derivative of a function $f(x)$ by first finding the derivative of $\ln | f(x)|$. How is the derivative of $\ln | f(x)|$ related to the derivative of the original function $f(x)$, and why? What is the derivative of $\ln |x|$? When is it useful (or necessary) to use this method of calculating derivatives?

Exercises 9.3

Concepts

0. Read the section and make your own summary of the material. Include whatever you think will help you review the material later on.

1. Solve the equation $3^{x-1} = 2$ in two ways: **(a)** by using the natural logarithmic function, and **(b)** by using the base 3 logarithmic function. Show that the answers you get are the same.

2. Solve the equation $4(2^x) = 3$ in two ways: **(a)** by immediately applying the natural logarithmic function to both sides and then solving, and **(b)** by isolating the expression 2^x and *then* applying the natural logarithmic function to both sides. Do you get the same answer? Is one way easier than the other?

3. Explain how we know that logarithmic functions are one-to-one. Why does this mean that $A = B$ if and only if $\log_b A = \log_b B$ (assuming A and B are positive)?

4. Which of the following forms are indeterminate? For those that are not indeterminate, what limits do they approach?

  (a) 0^∞ **(b)** 1^∞ **(c)** ∞^∞ **(d)** ∞^1
  (e) ∞^0 **(f)** 1^0 **(g)** 0^1 **(h)** 0^0
  (i) 0^{-1} **(j)** $0^{-\infty}$ **(k)** $\infty^{-\infty}$ **(l)** $1^{-\infty}$

5. Explain why the limit $\lim_{x \to \infty} 1^{x^2 - 4}$ is *not* of the indeterminate form 1^∞. What is the value of this limit?

6. Suppose $f(x)$ is a positive function.

  (a) If $\lim_{x \to 3} \ln(f(x)) = 0$, what can you say about $\lim_{x \to 3} f(x)$?

  (b) If $\lim_{x \to 1} \ln(f(x)) = \infty$, what can you say about $\lim_{x \to 1} f(x)$?

  (c) If $\lim_{x \to 0} \ln(f(x)) = -\infty$, what can you say about $\lim_{x \to 0} f(x)$?

7. Sketch the graphs of $y = \ln |x|$ and $y = \frac{1}{x}$. Use the graphs to argue that $\frac{1}{x}$ could be the derivative of $\ln |x|$.

8. Suppose $f(x)$ is a function with $f(2) = 4$ and $\frac{d}{dx}|_{x=2}(\ln | f(x)|) = -3$. What can you say about $f'(2)$, and why?

9. What types of functions are best differentiated using logarithmic differentiation? (There are two types of such functions.)

10. Suppose we wish to find the derivative of $y = (x + 1)^x$ using logarithmic differentiation. Why is it enough to consider $\ln y = \ln((x + 1)^x)$, rather than $\ln |y| = \ln |(x + 1)^x|$?

11. Consider the function $f(x) = x^x$.

  (a) Alina says that the derivative of this function is $f'(x) = x(x^{x-1})$. What is she thinking and why is she wrong?

(b) Linda says that the derivative of this function is $f'(x) = (\ln x)x^x$. What is she thinking and why is she wrong?

(c) Use logarithmic differentiation to find the derivative of $f(x) = x^x$.

Skills

■ Find the exact solution of each of the following equations. Then use a calculator to find a numerical approximation for your exact solution.

12. $2^x = 3^{x+1}$

13. $(2^x + 1)^2 = 4$

14. $3^x = 4\sqrt{2^x}$

15. $2e^{3x+1} - 5 = 0$

16. $\dfrac{3^x + 1}{3^x - 1} = 4$

17. $\dfrac{3^{x-1}}{3^{2x+1}} = 4$

■ Solve the following limits. You may or may not need to use logarithms.

18. $\lim\limits_{x \to 0^+} x^{2x}$

19. $\lim\limits_{x \to 0^+} x^{\ln x}$

20. $\lim\limits_{x \to 0^+} (x^2 + 1)^x$

21. $\lim\limits_{x \to \infty} x^{\ln x}$

22. $\lim\limits_{x \to 2} (x^2 - 4)^{x-2}$

23. $\lim\limits_{x \to 2^+} (x - 2)^{x^2 - 4}$

24. $\lim\limits_{x \to 1^+} x^{\frac{1}{x-1}}$

25. $\lim\limits_{x \to \infty} x^{\frac{1}{x}}$

26. $\lim\limits_{x \to \infty} \left(\dfrac{1}{x}\right)^x$

27. $\lim\limits_{x \to \infty} \left(\dfrac{1}{x+1}\right)^x$

28. $\lim\limits_{x \to \infty} \left(\dfrac{x}{x-1}\right)^x$

29. $\lim\limits_{x \to 1} (\ln x)^{x-1}$

30. Find $\lim\limits_{x \to \infty} (1 + \frac{3}{x})^{2x}$ in two ways: **(a)** by using the fact that $\lim\limits_{x \to \infty} (1 + \frac{r}{x})^x = e^r$, and **(b)** by using logarithms and the methods of this section.

31. Consider the following especially nasty function:

$$f(x) = \frac{x^2 \sqrt{3x + 1}(x^3 + x + 1)}{(2x - 1)(3x^2 + 8)}.$$

Find the derivative of this function in two ways: **(a)** by using logarithmic differentiation, and **(b)** without using logarithmic differentiation.

■ Calculate the derivatives of the following functions. You may or may not need to use logarithmic differentiation.

32. $f(x) = \ln |3x^2 + 4|$

33. $f(x) = 3(\ln |x - 1|)^2 + 4$

34. $f(x) = \sqrt{x \ln |2^x + 1|}$

35. $f(x) = x(2 - x)(3 - x)(4 - x)$

36. $f(x) = 12x\sqrt[3]{1 - x}\sqrt{x + 1}$

37. $f(x) = \dfrac{e^{2x}(x^3 - 2)^4}{x(3e^{5x} + 1)}$

38. $f(x) = \dfrac{2^x \sqrt{x^3 - 1}}{\sqrt{x}(2x - 1)}$

39. $f(x) = \sqrt{\dfrac{x^2(3x + 1)^{99}}{4x^2 - 3x + 2}}$

40. $f(x) = x^3 \ln(\ln x)\sqrt{3x + 1}$

41. $f(x) = (2x + 1)^{3x}$

42. $f(x) = x^{\ln x}$

43. $f(x) = (\ln x)^x$

44. $f(x) = \left(\dfrac{1}{x + 1}\right)^x$

45. $f(x) = \left(\dfrac{x}{x - 1}\right)^x$

46. $f(x) = (\ln x)^{\ln x}$

Proofs

47. Prove that for all $x \neq 0$, the derivative of $f(x) = \ln |x|$ is $f'(x) = \frac{1}{x}$ by considering the cases $x > 0$ and $x < 0$ separately. (*Hints:* If x is positive, then $|x| = x$ and this is just the usual rule for differentiating the natural logarithmic function. If x is negative, then $|x| = -x$; in this case use the chain rule.)

48. Prove that for any function $f(x)$, $\frac{d}{dx}(\ln |f(x)|) = \frac{f'(x)}{f(x)}$. Provide a reason for each step of your proof.

49. Use logarithms to prove that for any real number r, we have $\lim\limits_{x \to \infty} (1 + \frac{r}{x})^x = e^r$.

■ In the following five problems, you will use logarithmic differentiation to prove various differentiation rules.

50. Use logarithmic differentiation to prove the rule for differentiating power functions: If k is any constant, then $\frac{d}{dx}(Ax^k) = kAx^{k-1}$.

51. Use logarithmic differentiation to prove the following rule for differentiating exponential functions: If k is any real number, then $\frac{d}{dx}(e^{kx}) = ke^{kx}$.

52. Use logarithmic differentiation to prove the following rule for differentiating exponential functions: For any positive real number b with $b \neq 1$, $\frac{d}{dx}(b^x) = (\ln b)b^x$.

53. Use logarithmic differentiation to prove the product rule.

54. Use logarithmic differentiation to prove the quotient rule.

CHAPTER 10

Trigonometric Functions

*I*n this chapter we investigate the six trigonometric functions: sine, cosine, tangent, secant, cosecant, and cotangent. We begin by defining these functions both in terms of the ratios of side lengths of right triangles and as the coordinates of points on the unit circle. The trigonometric functions have very interesting algebraic properties (better known as "identities") and a periodic behavior. After establishing the definitions and algebraic properties of these functions, we investigate their limits, derivatives, and graphs.

"Cassini," a sculpture by Charles Perry, www. charlesperry.com

CONTENTS

10.1   Right Triangle Trigonometry

In this section we will use right triangles to define the six ***trigonometric functions.*** The definitions in this section hold only for *acute* angles; in the next section we will develop a more general definition for the trigonometric functions that works for *all* angles. After defining the trigonometric functions as the ratios of side lengths of right triangles, we will learn to calculate certain values of trigonometric functions by hand. We will finish the section by investigating some word problems that can be solved with the help of trigonometric functions.

10.1.1 Definitions of the trigonometric functions for acute angles

The ***Law of Similar Triangles*** says that if two triangles have the same angles, then the ratios of their side lengths are the same. More precisely, if a triangle abc and a triangle ABC have the same three angles (see Figure 10.1), then the ratio of the lengths of any two sides of triangle abc is equal to the ratio of the lengths of the corresponding two sides of triangle ABC:

$$\frac{a}{b} = \frac{A}{B}, \qquad \frac{a}{c} = \frac{A}{C}, \qquad \frac{b}{c} = \frac{B}{C}.$$

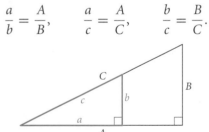

Figure 10.1

Of course, there are six possible ratios of side lengths for any given triangle: the three listed in Figure 10.1 and their reciprocals. Another way to express the Law of Similar Triangles is to say that the ratio of any two side lengths of a triangle is determined only by the angles of that triangle. Although the Law of Similar Triangles actually applies to all triangles, in this section we are interested in focusing on ***right triangles***—that is, triangles that contain a 90-degree angle.

Recall that an angle of 1 "degree" is an angle that makes up one 360th of a circle. Thus an angle of 90 degrees is one-fourth of the way around a circle and is a "right angle" (a perpendicular corner). In the next section we will use a more intrinsic measure of angles called ***radians,*** but for the moment we will continue to use degrees.

An angle θ is called an ***acute*** angle if $0° < \theta < 90°$. (*Note:* θ is the Greek letter ***theta*** and is commonly used to represent angles.) Since the sum of the three angles in any triangle is always $180°$, if one of the acute angles of a right triangle is θ, then the other acute angle of the triangle must be $90° - \theta$ (why?). This means that if you know *one* acute angle of a right triangle, you know the remaining angles (one of which is, of course, 90 degrees). Any acute angle determines a unique type of right triangle (all right triangles containing that angle will be similar), which in turn determines the six side length ratios.

Keeping in mind the previous discussion, we can construct six functions for each of which the "input" is an acute angle and the "output" is one of the six ratios of side lengths for the right triangle determined by that angle. Given an acute angle θ in a right triangle, we will call the leg of the triangle across from θ the **opposite side,** and we will call the leg of the triangle next to θ the **adjacent side** of the right triangle. The longest side of the right triangle will always be the side of the triangle across from the right angle of the triangle and is called the **hypotenuse** of the triangle. See Figure 10.2.

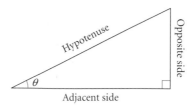

Figure 10.2

Each of the ratios of side lengths of such a right triangle is determined only by the angle θ (and not by the size of the triangle); thus each ratio of side lengths is a function of θ. These six ratios are what we use to define the six **trigonometric functions** of the angle θ. In the following definition, "opp," "adj," and "hyp" denote the lengths of the opposite side, adjacent side, and hypotenuse, respectively.

DEFINITION 10.1

Trigonometric Functions for Acute Angles

Given an acute angle θ, the six **trigonometric functions** are defined to be

$$\sin\theta := \frac{\text{opp}}{\text{hyp}}, \qquad \cos\theta := \frac{\text{adj}}{\text{hyp}}, \qquad \tan\theta := \frac{\text{opp}}{\text{adj}},$$

$$\csc\theta := \frac{\text{hyp}}{\text{opp}}, \qquad \sec\theta := \frac{\text{hyp}}{\text{adj}}, \qquad \cot\theta := \frac{\text{adj}}{\text{opp}}.$$

The function $\sin\theta$ is pronounced as "sine theta," but its full name is the **sine function.** The full names of the remaining trigonometric functions are (reading across the first row and then across the second row) **cosine, tangent, cosecant, secant,** and **cotangent.**

EXAMPLE 10.1

Finding values of trigonometric functions given side lengths

Given the angle θ shown in the right triangle in Figure 10.3, find the values of the six trigonometric functions.

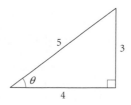

Figure 10.3

Solution   Here opp $= 3$, adj $= 4$, and hyp $= 5$. Therefore,

$$\sin\theta = \frac{3}{5}, \quad \cos\theta = \frac{4}{5}, \quad \tan\theta = \frac{3}{4}, \quad \csc\theta = \frac{5}{3}, \quad \sec\theta = \frac{5}{4}, \quad \text{and} \quad \cot\theta = \frac{4}{3}.$$

$\square$

Some people like to remember the definitions of the trigonometric functions by remembering the spelling of the following "word" (which is usually pronounced "so-ka-toe-ahh"):

SOHCAHTOA.

The letters of this word spell out the definitions of the sine, cosine, and tangent functions, as follows: Sine is Opposite over Hypotenuse, Cosine is Adjacent over Hypotenuse, and Tangent is Opposite over Adjacent. The remaining three trigonometric functions are the reciprocals of the sine, cosine, and tangent functions. In fact, the sine and cosine functions can be used to define the other four trigonometric functions:

THEOREM 10.1

Sine and Cosine Determine All of the Trigonometric Functions

Given any acute angle θ,

(a) $\csc \theta = \dfrac{1}{\sin \theta}$,

(b) $\sec \theta = \dfrac{1}{\cos \theta}$,

(c) $\tan \theta = \dfrac{\sin \theta}{\cos \theta}$,

(d) $\cot \theta = \dfrac{1}{\tan \theta} = \dfrac{\cos \theta}{\sin \theta}$.

The proof of this theorem follows directly from the definitions of the six trigonometric functions and is left to you in the exercises.

⬥ *Caution*   Note that $\csc \theta$ is the reciprocal of the *sine* function, whereas $\sec \theta$ is the reciprocal of the *cosine* function. This might be the opposite of what you would expect, since "sine" and "secant" both begin with the letter "s," and "cosine" and "cosecant" both begin with "co." However, sine and cosecant are reciprocals, and cosine and secant are reciprocals.   □

⬥ *Caution*   It is also a common mistake to think that a prefix of "co" means that a reciprocal is taken; that's not true! The sine and cosine functions are not reciprocals of each other, and neither are the secant and cosecant functions (although it *is* true that tangent and cotangent are reciprocals).   □

10.1.2 Two special triangles

In order to calculate the trigonometric functions of an acute angle θ, we must know the (ratios of the) lengths of the sides of a right triangle that contains the angle θ. Usually this is difficult; for example, it is difficult to determine the possible side lengths of a right triangle that contains an angle of 12 degrees.

However, there are two special right triangles for which we can determine (by hand) the ratios of side lengths. These are the 30–60–90 triangle and the 45–45–90 triangle (the numbers refer to the degree measures of the angles). The following theorem describes the lengths of the legs of these triangles when we assume that the hypotenuse is 1 unit in length.

THEOREM 10.2

Side Lengths of Two Important Right Triangles

The 45–45–90 right triangle and the 30–60–90 right triangle with hypotenuse of length 1 unit have side lengths as shown in the following figures:

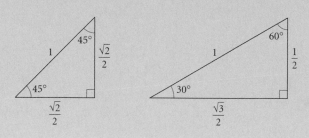

If you have seen something like Theorem 10.2 before, you may remember it a bit differently. Because increasing or decreasing the size of a right triangle while preserving its angles does not change the ratios of its side lengths, Theorem 10.2 is equivalent to the following:

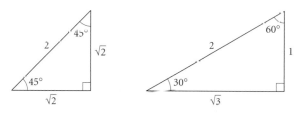

All of the side lengths above are obtained from the side lengths in Theorem 10.2 by multiplication by 2. We will generally use the side lengths given in Theorem 10.2, because in the future we will be working with triangles where the hypotenuse is of length 1.

In general, if two triangles are similar, then the side lengths of the larger triangle are common multiples of the corresponding side lengths of the smaller triangle.

EXAMPLE 10.2

Scaling up all side lengths to obtain a similar triangle

Find the lengths of the sides of a 30–60–90 triangle whose shortest side is 5 units long.

Solution   The shortest side of a 30–60–90 triangle is the side across from its smallest angle (30 degrees). The shortest side of the 30–60–90 triangle in question is ten times larger than the shortest side of the triangle in Theorem 10.2 (since $5 = 10(\frac{1}{2})$). Therefore, *each* side of our triangle is ten times as long as the corresponding side of the 30–60–90 triangle with hypotenuse of length 1. Thus the length of the hypotenuse of our triangle is $10(1) = 10$, and the length of the other leg of the triangle (across from the 60-degree angle) is $10(\frac{\sqrt{3}}{2}) = 5\sqrt{3}$.   □

The proof of Theorem 10.2 is an application of the ***Pythagorean Theorem,*** which states that if a right triangle has legs of length a and b and a hypotenuse of length c, then $a^2 + b^2 = c^2$.

PROOF (**THEOREM 10.2**)   Let's start with the 45–45–90 triangle. Suppose a right triangle with a hypotenuse of length 1 has two angles of 45 degrees. Then the triangle is an isosceles triangle; that is, its two legs both have the same length a (see

Figure 10.4). It remains to show that $a = \frac{\sqrt{2}}{2}$. By the Pythagorean Theorem,

$$a^2 + a^2 = 1 \quad \Rightarrow \quad 2a^2 = 1 \quad \Rightarrow \quad a^2 = \frac{1}{2} \quad \Rightarrow \quad a = \sqrt{\frac{1}{2}} = \frac{1}{\sqrt{2}} = \frac{1}{\sqrt{2}}\left(\frac{\sqrt{2}}{\sqrt{2}}\right) = \frac{\sqrt{2}}{2}.$$

Notice that we considered only the positive square root of $\frac{1}{2}$, since we know that a is a length and thus must be positive.

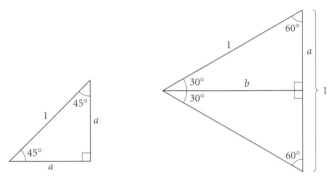

Figure 10.4   **Figure 10.5**

Now suppose that a right triangle has a hypotenuse of length 1, and angles of 30, 60, and 90 degrees. Let a be the length of the leg across from the 30-degree angle, and let b be the length of the leg across from the 60-degree angle. Two of these triangles together make the equilateral triangle (since all the angles are 60 degrees) shown in Figure 10.5. Since the triangle in Figure 10.5 is equilateral, we know that $2a = 1$ and thus $a = \frac{1}{2}$. It now remains to show that $b = \frac{\sqrt{3}}{2}$. Again we use the Pythagorean Theorem:

$$a^2 + b^2 = 1^2 \quad \Rightarrow \quad \frac{1}{4} + b^2 = 1 \quad \Rightarrow \quad b^2 = \frac{3}{4} \quad \Rightarrow \quad b = \sqrt{\frac{3}{4}} = \frac{\sqrt{3}}{2}.$$

Note that we considered only the positive square root in this calculation, since b is a length and must be positive.   ■

10.1.3 Calculating values of trigonometric functions

Using the two "special triangles" described in Theorem 10.2, we can calculate the values of the trigonometric functions for acute angles of 30, 60, and 45 degrees.

EXAMPLE 10.3   **Using known side lengths to calculate trigonometric values**

Find $\sin 60°$ and $\cot 60°$.

Solution   We begin by making a right triangle with an angle of $60°$ (the other angles must be $90°$ and $30°$; why?). Then we label the lengths of the sides of the triangle, assuming that the hypotenuse of the triangle is of length 1. See Figure 10.6.

Now we can see that the "opposite" side of the triangle (with respect to the angle 60 degrees) has length $\frac{\sqrt{3}}{2}$, the "adjacent" side has length $\frac{1}{2}$, and the hypotenuse has length 1. Therefore,

$$\sin 60° = \frac{\text{opp}}{\text{hyp}} = \frac{\frac{\sqrt{3}}{2}}{1} = \frac{\sqrt{3}}{2},$$

$$\cot 60° = \frac{\text{adj}}{\text{opp}} = \frac{\frac{1}{2}}{\frac{\sqrt{3}}{2}} = \frac{1}{\sqrt{3}}.$$

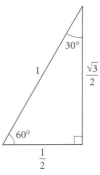

Figure 10.6

⬅ *Caution* When you label the sides of a 30–60–90 triangle, keep in mind that the shortest side is always across from the smallest angle. Therefore, the side across from the 30-degree angle is the side of length $\frac{1}{2}$ (since that is smaller than $\frac{\sqrt{3}}{2}$).

Table 10.1 lists the values of the six trigonometric functions that can be obtained using 45–45–90 and 30–60–90 triangles.

θ	$\sin\theta$	$\cos\theta$	$\tan\theta$	$\csc\theta$	$\sec\theta$	$\cot\theta$
30°	$\frac{1}{2}$	$\frac{\sqrt{3}}{2}$	$\frac{1}{\sqrt{3}}$	2	$\frac{2}{\sqrt{3}}$	$\sqrt{3}$
60°	$\frac{\sqrt{3}}{2}$	$\frac{1}{2}$	$\sqrt{3}$	$\frac{2}{\sqrt{3}}$	2	$\frac{1}{\sqrt{3}}$
45°	$\frac{\sqrt{2}}{2}$	$\frac{\sqrt{2}}{2}$	1	$\sqrt{2}$	$\sqrt{2}$	1

Table 10.1

Table 10.1 follows directly from the side lengths of the "special triangles" and the definitions of the trigonometric functions as the ratios of side lengths.

⬅ *Caution* Do not just blindly memorize the values of the trigonometric functions given in Table 10.1! Instead, remember the definitions of the trigonometric functions and the side lengths of the two "special triangles," and work out the entries in the table from those definitions and side lengths, as we did in Example 10.3.

We now know how to calculate the values of any of the six trigonometric functions of the angles 30 degrees, 45 degrees, and 60 degrees. What about the values of the trigonometric functions of other acute angles? For example, what is cos 12°? Or csc 75°? If we don't know the lengths of the sides of the right triangle containing θ, then we cannot calculate the values of the trigonometric functions of θ.

For angles other than 30, 45, and 60 degrees, we will use a calculator to approximate the values of the trigonometric functions. Your calculator should have keys for the sine, cosine, and tangent functions, but it may not have keys for the cosecant, secant, and cotangent functions. Remember that the latter three trigonometric functions are the reciprocals of the first three.

EXAMPLE 10.4

Using a calculator to approximate trigonometric values

Using a calculator, we can approximate that $\cos 12° \approx 0.978$ and that $\csc 75° = \frac{1}{\sin 75°} \approx \frac{1}{0.966} \approx 1.035$.

💲 *Caution*   To calculate the value of a trigonometric function of an angle measured in degrees, make sure that your calculator is set to "degree mode." (In later sections we will use a different angle measure, and then we will need "radian mode.")   □

💲 *Caution*   Your calculator should also have keys labeled "$\sin^{-1}$," "$\cos^{-1}$," and "$\tan^{-1}$." These do *not* represent the reciprocals of the sine, cosine, and tangent functions (despite the notation). You *cannot* calculate the values of cosecant, secant, or cotangent using these keys. The functions $\sin^{-1} x$, $\cos^{-1} x$, and $\tan^{-1} x$ are "inverse trigonometric functions," and we will learn about them in Chapter 11.   □

10.1.4 Real-world problems involving triangles

Given an angle and the length of any side of a right triangle, we can use what we've learned about trigonometric functions to find the length of any other side of the triangle. Many real-world problems involve this type of situation. For example, if you know your distance from a building and the angle of your line of sight to the top of the building, you can determine the height of the building.

EXAMPLE 10.5   **Using trigonometry to find a side length given an angle and another side length**

Suppose that you are 500 feet from the base of an office building and that the angle of elevation of your line of sight to the top of the building (the angle between the horizontal ground and your diagonal line of sight) is 36 degrees. (Note: To make the problem easier, we assume that your eye is at ground level.) How tall is the building?

Solution   We start, as always, with a labeled picture of the situation (see Figure 10.7). Let h be the height of the building.

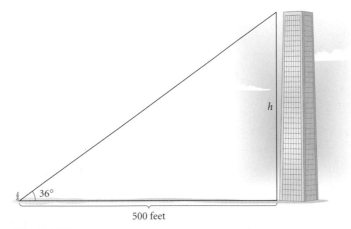

Figure 10.7

We know the length of the side of the triangle adjacent to the 36-degree angle, and we would like to find the length of the side that is opposite that angle. One trigonometric function that involves the adjacent and opposite sides is the tangent function (cotangent would also work). Specifically, we know that $\tan 36°$ is the ratio of the height of the building to the distance you are from the building:

$$\tan 36° = \frac{h}{500} \quad \Rightarrow \quad h = 500 \tan 36° \quad \Rightarrow \quad h \approx 500(0.7265) = 363.25 \text{ feet.}$$

(Of course, we used a calculator to approximate $\tan 36°$.)   □

What you should know after reading Section 10.1

▶ The basics: What is a "degree"? An "acute angle"? A "right triangle"? What is the sum of the degrees in any triangle? What are the Law of Similar Triangles and the Pythagorean Theorem?

▶ The definitions of the six trigonometric functions of an acute angle, and how these six functions are related to each other. Why do these functions depend only on the acute angle θ and not on the "size" of the triangle? Given the lengths of the sides of a right triangle, be able to find the values of the trigonometric functions of the acute angles of that triangle.

▶ The side lengths of 45–45–90 and 30–60–90 triangles with hypotenuse of length 1. Be able to *prove* that these special triangles have these side lengths. What does this say about the lengths of the sides of *any* 45–45–90 or 30–60–90 triangle?

▶ How to calculate the exact values of the six trigonometric functions, *without* a calculator, for 30-, 45-, and 60-degree angles. How to use your calculator to approximate the values of the six trigonometric functions of other acute angles.

▶ How to set up word problems that involve right triangles and solve them by using trigonometric functions.

Exercises 10.1

Concepts

0. Read the section and make your own summary of the material. Include whatever you think will help you review the material later on.

1. What is the "Law of Similar Triangles," and why is it important for the definitions of the trigonometric functions of an acute angle?

2. Determine whether each of the following is equal to $\cot \theta$, and why.

(a) $\dfrac{\cos \theta}{\sin \theta}$ (b) $\csc \theta \cos \theta$ (c) $\dfrac{1}{\tan \theta}$

(d) $\dfrac{\csc \theta}{\sec \theta}$ (e) $\dfrac{1}{\sec \theta \sin \theta}$

3. Use the definitions of the trigonometric functions to argue that if θ is an acute angle, then all six trigonometric functions of θ are positive.

4. Use a triangle to argue that for any acute angle, $\sin \theta$ is always less than 1. (*Hint:* Use the fact that the longest side of a right triangle is its hypotenuse.) Is $\tan \theta$ always less than 1? Why or why not?

5. Suppose θ is an acute angle. As θ increases, does $\sin \theta$ increase or decrease, and why? What about $\cos \theta$? What about $\tan \theta$? Use pictures of triangles to support your arguments.

6. Show that the side lengths of the 30–60–90 and 45–45–90 triangles in Theorem 10.2 satisfy the Pythagorean Theorem.

7. Given the angle θ in the right triangle below, which of the following are true, which are false, and why?

(a) $\sin \theta < \cos \theta$ (b) $\tan \theta > 1$

(c) $\sec \theta < \csc \theta$

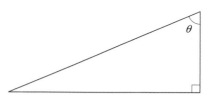

Skills

8. Given the angle α shown in the right triangle below, find $\sin \alpha$, $\cos \alpha$, $\tan \alpha$, $\sec \alpha$, $\csc \alpha$, and $\cot \alpha$.

9. Given the angle β shown in the right triangle below, find $\sin\beta$, $\cos\beta$, $\tan\beta$, $\sec\beta$, $\csc\beta$, and $\cot\beta$.

■ Use a calculator to determine approximate numerical value of each of the expressions below.

10. $\cos 11°$ **11.** $\sin 89°$ **12.** $\tan 20°$

13. $\sec 72°$ **14.** $\cos 1°$ **15.** $\tan 1°$

16. $\csc 44°$ **17.** $\cot 65°$

■ Find the *exact* numerical value of each of the expressions below. Support your work with a labeled 30–60–90 or 45–45–90 triangle.

18. $\sin 60°$ **19.** $\cos 45°$ **20.** $\tan 30°$

21. $\sec 45°$ **22.** $\csc 30°$ **23.** $\cot 45°$

24. $\sec 60°$ **25.** $\csc 45°$

■ For each triangle below, use similar triangles to find the *exact* values of the remaining side lengths (without a calculator, and without using trigonometric functions). Then use a calculator to find numerical approximations for these exact values.

26.

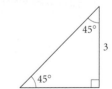

27.

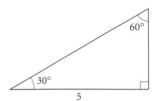

28.

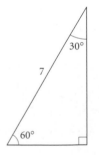

■ For each triangle below, use trigonometric functions (and your calculator) to approximate the side length *x*.

29.

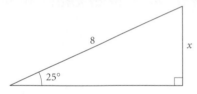

30.

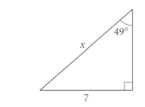

31.

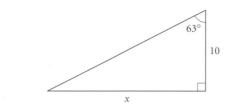

Applications

32. Find the area of an equilateral triangle with three sides of length 5. (Recall that the area of any triangle is $A = \frac{1}{2}bh$, where b is the length of the "base" (any side), and h is the length of the "height," that is, the perpendicular distance from the base to the opposite vertex of the triangle.)

33. Find the area of an isosceles triangle with two 30-degree angles and two sides of length 3.

34. Suppose you are standing on the ground looking up to the top of a 2500-foot building and that the angle of elevation of your line of sight is 18 degrees. How far away are you from the building? (You may assume that your eyes are level with the ground.)

35. Alina is flying a kite, and she has managed to get her kite so high in the air that she has let out 400 feet of kite string. If the angle made by the ground and the line of kite string is 32 degrees, how high is the kite?

36. Suppose two stars are each 60 light-years away from Earth. The angle between the line of sight to the first star and the line of sight to the second star is 2 degrees. (In other words, if you look at the first star and then turn your head to look at the second star, your head will move through an angle of 2 degrees.) How far apart are the stars? (*Hint:* The initial triangle you draw will *not* be a right triangle, but you can split it into two right triangles.)

Proofs

37. Use the definitions of the trigonometric functions of an acute angle θ to prove Theorem 10.1.

38. Prove the first part of Theorem 10.2: A 45–45–90 triangle with a hypotenuse of length 1 must have two legs of length $\frac{\sqrt{2}}{2}$.

39. Prove the second part of Theorem 10.2: A 30–60–90 triangle with a hypotenuse of length 1 must have legs of lengths $\frac{\sqrt{3}}{2}$ and $\frac{1}{2}$.

40. Use the Pythagorean Theorem to prove that for any acute angle θ,

$$(\sin\theta)^2 + (\cos\theta)^2 = 1.$$

(*Hint:* Suppose θ is an angle in a right triangle whose hypotenuse is of length 1. Label the remaining sides of the triangle a and b, and then use the definitions of the sine and cosine functions and the Pythagorean Theorem.)

10.2 Unit Circle Trigonometry

In the previous section we used right triangles to define the six trigonometric functions for any acute angle θ. In this section we will use the unit circle to define the six trigonometric functions for *any* angle θ.

10.2.1 Radians and standard position

So far we have been measuring angles in terms of degrees. Unfortunately, degrees are not a very intrinsic (i.e., "natural") way of measuring angles. Why would we divide a circle into 360 pieces? Why not 100 pieces? Or 1000 pieces? There is no intrinsic mathematical reason why we should measure angles in terms of units that represent $\frac{1}{360}$ of a circle. In this section we develop another method for describing the size of an angle—a method that *is* mathematically intrinsic.

Before we discuss this new way of measuring angles, we'll establish a fixed way to orient these angles. We consider an angle to be in **standard position** in the xy-plane if it satisfies the following three conditions: (1) Its vertex is at the origin of the xy-plane; (2) its **initial edge** (the edge from which we begin measuring the angle) lies along the positive side of the x-axis; and (3) the angle opens up in a counterclockwise or clockwise direction until it reaches its **terminal edge.** Figure 10.8 shows such an angle θ in standard position. We can now think of any angle as having a "direction": A **positive angle** travels counterclockwise from its initial edge, whereas a **negative angle** travels clockwise from its initial edge.

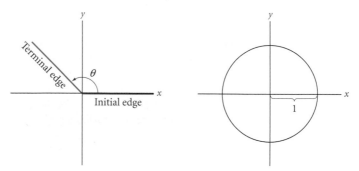

Figure 10.8   **Figure 10.9**

When we measured the size of an angle in degrees, we thought of that angle as a fraction of the angle that makes a complete circle. Although that is a sensible way to measure an angle, it involves an arbitrary division of the circle into 360 pieces. We now present a more mathematical way to divide the circle. Consider the **unit circle,** that is, the circle with radius 1 unit, centered at the origin of the xy-plane (see Figure 10.9). What is the circumference of this circle?

Since the unit circle has radius $r = 1$ unit, its circumference is $C = 2\pi r = 2\pi(1) = 2\pi$ units (that's approximately 6.283185 units). If the unit circle had some "nice" circumference like 10 units, then we might divide the circle up into 10 pieces, but the unit circle has an *irrational* circumference! However, we can still use this circumference to develop a new way to measure angles.

Suppose you start at the point $(1, 0)$ in the xy-plane (this is the rightmost point on the unit circle) and travel in a counterclockwise direction for 1 unit; your stopping point will be just under one-sixth of the way around the circle (why?) and is marked on the unit circle in Figure 10.10. The angle in standard position that starts at the x-axis and then opens up in the counterclockwise direction until its terminal edge intersects this stopping point is said to measure 1 **radian**; see Figure 10.11.

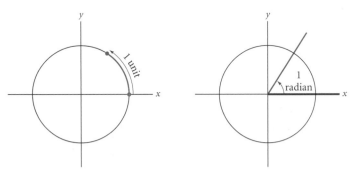

Figure 10.10   **Figure 10.11**

Since the distance all the way around the circle is 2π units, the distance halfway around is π units, and the distance one-quarter of the way around the circle is $\frac{\pi}{2}$ units. This means that an angle of $90°$ measures $\frac{\pi}{2}$ radians, an angle of $180°$ measures π radians, and an angle of $360°$ measures, of course, 2π radians.

◈ *Caution*   By convention, angles measured in radians are usually not followed by unit names. When we talk about the angle $\frac{\pi}{2}$, we mean an angle measuring $\frac{\pi}{2}$ radians. On the other hand, to refer to an angle in degrees, you *must* use units (the degree symbol). An angle of $3°$ is an angle of 3 degrees, whereas an angle of 3 is assumed to be an angle of 3 *radians*.   □

❓ *Question*   Which is bigger, an angle of 3 degrees or an angle of 3 radians? Show both angles in standard position.   □

Since an angle of $180°$ is the same as an angle that measures π radians, we can convert from degrees to radians (or vice versa) by multiplying by $\frac{\pi}{180°}$ (or $\frac{180°}{\pi}$).

EXAMPLE 10.6

Converting from radians to degrees

An angle of 3 radians measures approximately $172°$, since

$$3 \cdot \frac{180°}{\pi} \approx 171.887°.$$

This answer makes sense, because 3 radians is almost π radians, which would be halfway around the unit circle, and 172 degrees is nearly $180°$.   □

EXAMPLE 10.7

Converting from degrees to radians

An angle of $60°$ measures $\frac{\pi}{3}$ radians, since

$$60° \cdot \frac{\pi}{180°} = \frac{\pi}{3}.$$

This answer makes sense because a $60°$ angle is one-third of halfway around the unit circle, that is, one-third of π radians. ☐

We can now discuss negative angles, as well as angles that measure more than 2π radians (i.e., $360°$). Negative angles open in the clockwise direction, and angles that measure more than 2π radians go around the unit circle more than once.

EXAMPLE 10.8

A negative angle that goes around the unit circle more than once

An angle of $-\frac{11\pi}{4}$ radians is pictured in Figure 10.12.

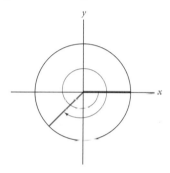

Figure 10.12

Note that $-\frac{11\pi}{4} = -\frac{11}{4}\pi = -2\pi - \frac{3}{4}\pi$, and recall that 2π radians is one full revolution around the unit circle. Thus the angle $-\frac{11\pi}{4}$ opens up in the clockwise direction for one full revolution plus three-quarters of halfway (i.e., $\frac{3}{4}$ of π) around the unit circle. ☐

From this point on, we will work almost exclusively in radians, rather than degrees. Be sure that you can quickly identify the radian measures of angles, or sketch angles measured in radians, *without* first converting to degrees. Figures 10.13–10.15 show the radian measure of various important angles between 0 and 2π radians.

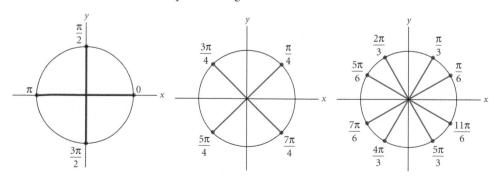

Figure 10.13   **Figure 10.14**   **Figure 10.15**

10.2.2 Definitions of the trigonometric functions for any angle

In Section 10.1.1 we defined the six trigonometric functions for any acute angle θ (any angle between 0 and 90 degrees, or between 0 and $\frac{\pi}{2}$ radians). We now construct more general definitions of the trigonometric functions that works for *any* angle θ.

Given any angle θ in standard position, the terminal edge of θ intersects the unit circle at some point (x, y) in the xy-plane. We will define the height y of that point to be the *sine* of θ, and the *cosine* of θ will be defined as the x-coordinate of that point.

DEFINITION 10.2

Trigonometric Functions for Any Angle

Given any angle θ in standard position, let (x, y) be the point where the terminal edge of θ intersects the unit circle. The six *trigonometric functions* are

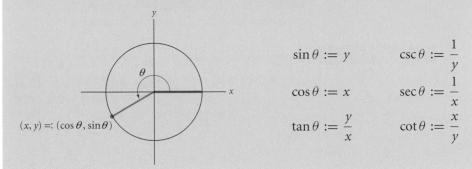

$$\sin \theta := y \qquad\qquad \csc \theta := \frac{1}{y}$$

$$\cos \theta := x \qquad\qquad \sec \theta := \frac{1}{x}$$

$$\tan \theta := \frac{y}{x} \qquad\qquad \cot \theta := \frac{x}{y}$$

$(x, y) =: (\cos \theta, \sin \theta)$

According to Definition 10.2, the trigonometric functions depend only on the terminal edge of θ (in standard position), not on how many times (or in what direction) the angle rotates around the unit circle. Therefore, angles like $\frac{4\pi}{3}$ and $2\pi + \frac{4\pi}{3} = \frac{10\pi}{3}$, which have the same terminal edge (i.e., end in the same place on the unit circle), will have the same trigonometric function values. Note that Definition 10.2 implies the same relationships between the trigonometric functions that we saw in Theorem 10.1 in Section 10.1.1.

THEOREM 10.3

Sine and Cosine Determine All of the Trigonometric Functions

Given *any* angle θ where the following quantities are defined,

(a) $\csc \theta = \dfrac{1}{\sin \theta}$,

(b) $\sec \theta = \dfrac{1}{\cos \theta}$,

(c) $\tan \theta = \dfrac{\sin \theta}{\cos \theta}$,

(d) $\cot \theta = \dfrac{1}{\tan \theta} = \dfrac{\cos \theta}{\sin \theta}$.

How does Definition 10.2 fit our previous definitions of trigonometric functions for acute angles? In particular, do the two definitions agree in the case where θ is an acute angle? (They should, or we would get different values for the trigonometric functions depending on which definition we used!) The two definitions do, in fact, agree for acute angles. Consider the acute angle shown in standard position in Figure 10.16.

Any acute angle θ in standard position will terminate in the first quadrant (why?). Moreover, if the point at which the terminal edge intersects the unit circle has coordinates (x, y), then the right triangle shown in Figure 10.16 will have adjacent side of length x, opposite side of length y, and hypotenuse of length 1 (since we are on the *unit* circle). Thus, according to the right triangle definition of sine, $\sin \theta = \frac{\text{opp}}{\text{hyp}} = \frac{y}{1} = y$, which is equal to the unit circle definition of $\sin \theta$ given in Definition 10.2. In a similar way, we can see that the two definitions of $\cos \theta$ agree if θ is an acute angle, and of course the remaining four trigonometric functions can be determined from the sine and cosine functions.

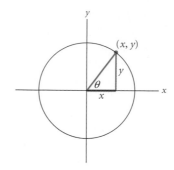

Figure 10.16

As we will see in the following example, it is possible to calculate the trigonometric functions of an angle in standard position given only the coordinates of the point where the terminal edge of that angle intersects the unit circle. Note that the angle θ in the following example is *not* an acute angle, so our original "right triangle" definitions of the trigonometric functions will not apply.

EXAMPLE 10.9   **Using the unit circle to calculate trigonometric values**

Suppose θ is an angle in standard position whose terminal edge intersects the unit circle in the third quadrant at the point (x, y), where $x = -\frac{1}{5}$. Find $\sin\theta$, $\tan\theta$, and $\sec\theta$.

Solution   In order to find the values of $\sin\theta$, $\tan\theta$, and $\sec\theta$, we have to know both coordinates of the point where the angle θ intersects the unit circle. We are given that the x-coordinate is $-\frac{1}{5}$ and that the point is in the third quadrant. What is the y-coordinate? Since we are on the unit circle, we can use the equation $x^2 + y^2 = 1$ to relate x and y. If $x = -\frac{1}{5}$, then

$$\left(-\tfrac{1}{5}\right)^2 + y^2 = 1 \quad \Rightarrow \quad y^2 = 1 - \tfrac{1}{25} \quad \Rightarrow \quad y = \pm\sqrt{1 - \tfrac{1}{25}} = \pm\sqrt{\tfrac{24}{25}} = \pm\tfrac{2\sqrt{6}}{5}.$$

Since the point (x, y) is in the third quadrant, we know that the y-coordinate must be negative and thus that $y = -\frac{2\sqrt{6}}{5}$. Therefore, $(x, y) = (-\frac{1}{5}, -\sqrt{1 - \frac{1}{25}})$ is the point where the terminal edge of θ intersects the unit circle. By Definition 10.2, we have

$$\sin\theta = y = -\frac{2\sqrt{6}}{5} \approx -0.9798,$$

$$\tan\theta = \frac{y}{x} = \frac{-\frac{2\sqrt{6}}{5}}{-\frac{1}{5}} = 2\sqrt{6} \approx 4.899,$$

$$\sec\theta = \frac{1}{x} = \frac{1}{-\frac{1}{5}} = -5.$$
◻

10.2.3 Graphing $\sin\theta$ and $\cos\theta$

The sine and cosine functions have inputs that are *angles* and outputs that are the y- and x-coordinates, respectively, of the points (x, y) corresponding to these angles. By watching how these x- and y-coordinates change as the angle θ changes, we can get an idea of how the graphs of $\sin\theta$ and $\cos\theta$ should look. We'll investigate these graphs (and the graphs of the other four trigonometric functions) in detail in Section 10.6.

The sine function measures the height of the y-coordinate of the point (x, y) corresponding to an angle θ. Let's look at what happens to the height y as θ makes one complete (counterclockwise) revolution around the unit circle. When $\theta = 0$ this height is zero; as θ increases from 0 to $\frac{\pi}{2}$, the height y increases, until at $\frac{\pi}{2}$ the height is $y = 1$. Then, as θ moves from $\frac{\pi}{2}$ to π, the height y decreases back to zero. As θ moves from π to $\frac{3\pi}{2}$, the height y continues to decrease, until it reaches its lowest height, $y = -1$. Finally, as θ changes from $\frac{3\pi}{2}$ to 2π, the height y increases, until at 2π the height of the point is zero. If θ were to make a second revolution around the unit circle, this process would repeat. If θ were to make a clockwise revolution from $\theta = 0$ to $\theta = -2\pi$, we would see the same process in reverse. The function $\sin\theta$ oscillates between -1 and 1 as θ changes.

Figure 10.17 shows this oscillating behavior of $\sin\theta$ in a graph. In the graph, θ runs along the horizontal axis, and $\sin\theta$ is graphed against the vertical axis. Notice that the horizontal axis of the graph is marked in intervals of π units (rather than in intervals of 1 unit, for example, which is more common for general graphs). This is because the zeros of the graph of $\sin\theta$ naturally fall in intervals of π units.

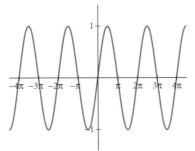 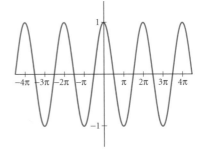

Figure 10.17   Graph of $\sin\theta$ **Figure 10.18**   Graph of $\cos\theta$

A similar analysis can be used to argue that the graph of $\cos\theta$ should look like the graph in Figure 10.18. In this case, the height of the graph at an angle θ represents the x-coordinate of the point (x, y) corresponding to θ. Notice that the graph of $\cos\theta$ also oscillates between -1 and 1 but that it does so shifted over from the graph of $\sin\theta$; why?

10.2.4 Calculating values of trigonometric functions

If the terminal edge of an angle θ intersects the unit circle at quarters or sixths of π radians, then we can calculate the trigonometric functions of θ *exactly*, using the two "special triangles" from Section 10.1.2 to determine the x- and y-coordinates of the point where the terminal edge of θ intersects the unit circle. We illustrate this process with the following example.

EXAMPLE 10.10   **Using reference triangles and the unit circle to calculate trigonometric values**

Find the exact values of $\cos\frac{5\pi}{6}$ and $\cot\frac{5\pi}{6}$.

Solution   First of all, notice that $\frac{5\pi}{6}$ is not an acute angle, so we can't use the right triangle definition of the cosine function to find $\cos\frac{5\pi}{6}$. However, we *can* use a right triangle to determine the coordinates (x, y) at which the angle $\frac{5\pi}{6}$ intersects the unit circle, and after doing that, we can use the unit circle definition of cosine to find $\cos\frac{5\pi}{6}$. Figure 10.19 shows where this angle lies on the unit circle. If we draw a line from the point (x, y) where the angle meets the unit circle to the x-axis, we obtain a right triangle (called the ***reference triangle*** for $\frac{5\pi}{6}$) that has an angle of $30°$. (We use degrees, and not radians, to measure the reference angle because it is not in standard position.)

Since we know the side lengths of a 30–60–90 triangle with hypotenuse of length 1, we know the lengths of the sides of the reference triangle (see Figure 10.20). This in turn means that we know the coordinates (x, y) of the point at which the terminal edge

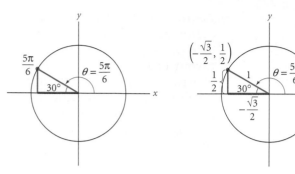

Figure 10.19   **Figure 10.20**

of θ intersects the unit circle; we have $x = -\frac{\sqrt{3}}{2}$ and $y = \frac{1}{2}$. Using the definitions of the trigonometric functions, we can now say that

$$\cos \frac{5\pi}{6} = x = -\frac{\sqrt{3}}{2},$$

$$\cot \frac{5\pi}{6} = \frac{x}{y} = \frac{-\frac{\sqrt{3}}{2}}{\frac{1}{2}} = -\sqrt{3}.$$

☑ *Checking the Answer*   To check these answers, we could use a calculator (in *radian* mode) to approximate $\cos \frac{5\pi}{6}$ and $\cot \frac{5\pi}{6}$ and then compare these approximations to $-\frac{\sqrt{3}}{2}$ and $-\sqrt{3}$, respectively.   ☐

In general, to calculate the *exact* values of the trigonometric functions of an angle θ, we have to know the coordinates (x, y) at which the terminal edge of θ intersects the unit circle. We can find these coordinates if we can determine the lengths of the sides of the reference triangle for θ. The **reference triangle** for θ is the triangle whose vertices are the origin, the point (x, y) on the unit circle, and the point on the x-axis directly above or below the point (x, y). The reference triangle will always look like one of the four pieces of the "bowtie" shape shown in Figure 10.21.

The reference triangle will always be a right triangle and will have a hypotenuse of length 1 (since its hypotenuse is a radius of the unit circle). If the reference triangle is a 30–60–90 or a 45–45–90 triangle, then we can determine its side lengths. If the reference triangle is *not* one of these special triangles, then we do not know (at this point) how to determine the *exact* values of the trigonometric functions of θ.

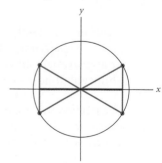

Figure 10.21

Of course, if θ is an angle whose terminal edge lies on the x-axis or the y-axis (like π or $\frac{\pi}{2}$; see Figure 10.13 on page 641), then there won't be a "reference triangle" for θ; why not? In this case, however, it is a simple task to find the values of the trigonometric functions of θ, because such angles θ intersect the unit circle at the point $(1, 0)$, $(0, 1)$, $(-1, 0)$, or $(0, -1)$.

§ *Caution*   If you have had some trigonometry before, then you may know how to determine the trigonometric functions of an angle θ by finding the *reference angle* for θ (the angle at the origin of the reference triangle), calculating the values of the trigonometric functions of the reference angle, and then inserting a sign (positive or negative) depending on which quadrant contains the terminal edge of θ. This is a suitable method for calculating trigonometric values, but it obscures the *definitions* of the trigonometric functions. Don't forget that the sine of an angle in standard position is the *y*-coordinate of the point at which its terminal side intersects the unit circle and that the cosine of such an angle is the *x*-coordinate of that point.   □

If θ is an angle whose reference triangle is not a 30–60–90 or a 45–45–90 triangle, then we currently do not have the tools to calculate the exact values of the trigonometric functions of θ. You can approximate the values of the trigonometric functions of such angles using a calculator.

§ *Caution*   Your calculator must be set to **radian mode** in order to calculate correctly the trigonometric functions of angles that are measured in radians. Be sure that you know how to change the mode (radian or degree) in which your calculator is operating.   □

EXAMPLE 10.11

Using a calculator in radian mode to approximate a trigonometric value

Using a calculator in radian mode, we can approximate $\cos(-\frac{\pi}{5}) \approx 0.809$. This makes sense because $-\frac{\pi}{5}$ is an angle in the fourth quadrant, and the point where the terminal edge of this angle meets the unit circle has an *x*-coordinate of 0.809 (why?). See Figure 10.22.

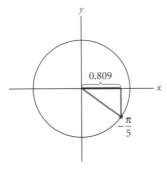

Figure 10.22   □

§ *Caution*   Remember that even if an angle measure does not involve π, it is assumed to be in radians unless the degree symbol is shown. For example, using radian mode we approximate $\sin 4 \approx -0.7568$. (Does this make sense in terms of the unit circle? Think about how 4 compares to π.)   □

The next example shows that the value of one of the trigonometric functions of an angle θ can be used to determine the values of the other trigonometric functions of θ (as long as we know what quadrant θ is in).

EXAMPLE 10.12

Using one trigonometric value to find another

Given that $\cos\theta = -\frac{1}{3}$ and that θ is an angle that terminates in the third quadrant, find $\tan\theta$.

Solution   Since $\cos\theta = -\frac{1}{3}$, the angle θ must have one of the two terminal edges shown in Figure 10.23 (since those are the terminal edges that intersect the unit circle at a point with x-coordinate $-\frac{1}{3}$). Since θ is in the third quadrant, we know which terminal edge θ must have (we have marked that terminal edge with θ in Figure 10.23).

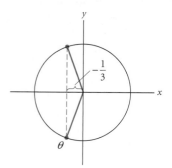

Figure 10.23

We know the lengths of two of the sides of the reference triangle for θ (the leg on the x-axis has length $\frac{1}{3}$, and the length of the hypotenuse is 1). We can use the Pythagorean Theorem to find the length a of the remaining leg of the reference triangle:

$$a^2 + \left(\tfrac{1}{3}\right)^2 = 1^2 \;\Rightarrow\; a^2 + \tfrac{1}{9} = 1 \;\Rightarrow\; a = \pm\sqrt{1 - \tfrac{1}{9}} = \pm\sqrt{\tfrac{8}{9}} = \pm\frac{2\sqrt{2}}{3} \approx \pm 0.9428.$$

Since θ terminates in the third quadrant, the y-coordinate corresponding to θ is $-\frac{2\sqrt{2}}{3} \approx -0.9428$. (We choose the *negative* height, since the point is below the x-axis. Also notice that -0.9428 is a reasonable answer, since the point in question has a y-coordinate very close to -1.) Therefore, the tangent of θ is

$$\tan\theta = \frac{y}{x} = \frac{-\frac{2\sqrt{2}}{3}}{-\frac{1}{3}} \approx \frac{-0.9428}{-\frac{1}{3}} = 3(0.9428) = 2.8284.$$

It is interesting to note that we never actually determined the angle measure of θ. We know the (x, y) coordinates of the corresponding point on the unit circle, but we never had to find the radian measure of θ (nor could we have!) ☐

What you should know after reading Section 10.2

► What is a "radian" and how is it related to a "degree"? Be able to sketch positive and negative angles. In particular, be able to identify angles that are quarters or sixths of π radians, and be able to convert between radians and degrees.

► The definitions of the six trigonometric functions for *any* angle θ, in terms of the unit circle. When θ is an acute angle, the unit circle definitions of the trigonometric functions are equivalent to the right triangle definitions of the trigonometric functions; why?

► Be able to calculate the exact values of the trigonometric functions at angles whose reference triangles are 30–60–90 or 45–45–90 triangles. Be able to calculate trigonometric values given the (x, y) coordinates of the point where the terminal edge of an angle intersects the unit circle. Be able to use a calculator to approximate the values of the trigonometric functions for angles in radian measure.

Exercises 10.2

Concepts

0. Read the section and make your own summary of the material. Include whatever you think will help you review the material later on.

1. Why is the distance once around the unit circle equal to 2π units?

2. What is a radian? Is it larger or smaller than a degree? Show an angle of 1 degree and an angle of 1 radian in standard position.

3. List four angles (two positive and two negative) with the same terminal edge as the angle $\frac{4\pi}{3}$.

4. List four angles (two positive and two negative) whose sine is $-\frac{1}{2}$.

5. Sketch four angles in standard position whose cosine is $-\frac{4}{5}$.

6. Why do angles measured in radians so often involve the number π? Do they *have* to involve the number π?

7. Give a formal definition of $\sin\theta$, for any angle θ. Your definition should include the words "unit circle," "standard position," "terminal," and "coordinate."

8. Give a formal definition of $\tan\theta$, for any angle θ. Your definition should include the words "unit circle," "standard position," "terminal," and "coordinate."

9. Suppose θ is an angle in standard position whose terminal edge intersects the unit circle at the point $(-\frac{1}{4}, \frac{\sqrt{15}}{4})$. Find $\sin\theta$, $\cos\theta$, and $\tan\theta$.

10. Suppose θ is an angle in standard position whose terminal edge intersects the unit circle at the point (x, y). If $y = -\frac{1}{3}$, what are the possible values of $\cos\theta$? If you know that the terminal edge of θ is in the third quadrant, what can you say about $\cos\theta$? What if the terminal edge of θ is in the fourth quadrant? Could the terminal edge of θ be in the first or second quadrant?

11. Explain in your own words why the graph of $\cos\theta$ shown in Figure 10.18 looks the way that it does. In particular, what does this graph have to do with the unit circle definition of the cosine function? Why are the zeros of the graph where they are? Why is the height of the graph always between -1 and 1? Why does the graph keep "repeating" itself?

12. Use the definition of the sine function to explain why $\sin(\frac{\pi}{4})$ is equal to $\sin(\frac{9\pi}{4})$ and $\sin(-\frac{7\pi}{4})$.

13. Suppose θ is an angle in standard position whose terminal edge meets the circle (centered at the origin) of radius 5 at the point $(\frac{3}{2}, -\frac{\sqrt{91}}{2})$. Find $\sin\theta$, $\cos\theta$, and $\tan\theta$.

14. Suppose θ is an angle in standard position whose terminal edge meets the circle (centered at the origin) of radius 7 at the point $(-2, 3\sqrt{5})$. Find $\sin\theta$, $\cos\theta$, and $\tan\theta$.

■ Fill in the blanks in each statement below.

15. The sine of an angle θ is the _____ of the point where the terminal edge of θ intersects the _____.

16. The angle $-\frac{5\pi}{6}$ has a reference angle of _____ degrees.

17. For any angle θ, the cosecant of θ is the _____ of the sine of θ.

18. The tangent of an angle θ is negative if the terminal edge of θ is in the _____ quadrant.

19. If the sine of θ is $-\frac{1}{3}$, then the cosecant of θ is _____.

■ Determine whether each of the following statements is true or false. If a statement is true, explain why. If a statement is false, provide a counterexample.

20. True or False: If (x, y) is the point on the unit circle corresponding to the angle $-\frac{7\pi}{3}$, then x is positive and y is negative.

21. True or False: If θ is any angle, then θ and $2\pi\theta$ have the same terminal edge.

22. True or False: If θ is any angle, then θ and $2\pi + \theta$ have the same terminal edge.

23. True or False: The sine of an angle θ is always equal to the sine of the reference angle for θ.

24. True or False: If the terminal edge of θ is in the first quadrant, then the values of all six trigonometric functions of θ are positive.

25. True or False: If the terminal edge of θ is in the third quadrant, then the values of all six trigonometric functions of θ are negative.

26. True or False: There are only two angles whose sine is $-\frac{1}{4}$.

27. True or False: If θ is an integer multiple of π, then $\cos\theta = -1$.

28. True or False: If $\tan\theta = 1$, then the sine and cosine of θ must be equal.

29. True or False: For any angle θ, $-1 \le \cos\theta \le 1$.

30. True or False: For any angle θ, $|\sec\theta| \ge 1$.

31. True or False: If $\theta = -\frac{6\pi}{7}$, then the sine of θ is between $-\frac{1}{2}$ and 0. (Do not use a calculator.)

32. True or False: If $\theta = \frac{4\pi}{5}$, then the tangent of θ is greater than 1. (Do not use a calculator.)

Skills

■ Convert each of the following angles from degree to radian measure.

33. $60°$ **34.** $-30°$ **35.** $-135°$ **36.** $720°$ **37.** $-12°$

■ Convert each of the following angles from radian to degree measure.

38. $\dfrac{5\pi}{6}$ **39.** $-\pi$ **40.** $-\dfrac{3\pi}{4}$ **41.** $\dfrac{13\pi}{12}$ **42.** 4

■ For each angle θ given below, sketch θ in standard position on a picture of the unit circle. (*Note:* Since no degree symbols are used, each of these angles is in radian measure.)

43. $\dfrac{12\pi}{13}$ **44.** $-\dfrac{14\pi}{5}$ **45.** 201π

46. 5   **47.** -1

■ For each angle θ given below, sketch θ in standard position on a picture of the unit circle, and find the reference triangle for θ (if one exists).

48. $\dfrac{\pi}{4}$ **49.** $-\dfrac{3\pi}{2}$ **50.** $\dfrac{7\pi}{6}$

51. $-\pi$ **52.** $-\dfrac{2\pi}{3}$ **53.** $-\dfrac{11\pi}{4}$

54. $\dfrac{37\pi}{3}$ **55.** $\dfrac{201\pi}{2}$

■ Find the *exact* value of each of the following trigonometric functions for the angles given. Show your work by sketching the angle in standard position with the unit circle, finding the reference triangle for that angle (if applicable), and finding the lengths of the sides of that reference triangle.

56. $\sin 0$ **57.** $\cos \pi$ **58.** $\cot \dfrac{\pi}{2}$

59. $\sin\left(-\dfrac{\pi}{3}\right)$ **60.** $\cos \dfrac{3\pi}{2}$ **61.** $\tan\left(-\dfrac{\pi}{4}\right)$

62. $\sec\left(-\dfrac{\pi}{6}\right)$ **63.** $\csc\left(-\dfrac{5\pi}{4}\right)$ **64.** $\tan \dfrac{11\pi}{6}$

65. $\sin(201\pi)$ **66.** $\sec(201\pi)$ **67.** $\cot(201\pi)$

68. $\cos \dfrac{48\pi}{3}$ **69.** $\tan \dfrac{48\pi}{3}$ **70.** $\csc \dfrac{48\pi}{3}$

■ For each angle θ given below, find the coordinates (x, y) where the terminal edge of θ meets the unit circle.

71. $\theta = \dfrac{\pi}{3}$ **72.** $\theta = -\dfrac{5\pi}{4}$ **73.** $\theta = \dfrac{7\pi}{2}$

■ For each angle θ given below, find the coordinates (x, y) where the terminal edge of θ meets the circle centered at the origin with a radius of 3.

74. $\theta = \dfrac{2\pi}{3}$ **75.** $\theta = -\dfrac{3\pi}{4}$ **76.** $\theta = \dfrac{11\pi}{6}$

■ Use a calculator to approximate each of the following trigonometric functions at the angles given. Check that the answers are reasonable with a sketch of the angle in standard position and what you know about the definitions of the trigonometric functions. (*Note:* As usual, since no degree symbols are shown, these angles are measured in radians.)

77. $\tan \dfrac{4\pi}{17}$ **78.** $\sec\left(-\dfrac{\pi}{5}\right)$ **79.** $\csc \dfrac{11\pi}{7}$

80. $\sin 1$ **81.** $\cos(-6)$ **82.** $\csc 40$

■ For each problem below, use the information given to calculate the *exact* values of the given trigonometric functions of θ. Show your work using a sketch of the unit circle and a triangle.

83. Given that $\cos \theta = -\frac{1}{4}$ and that θ is in the second quadrant, find $\sin \theta$.

84. Given that $\sin \theta = \frac{1}{5}$ and that $\frac{\pi}{2} < \theta < \pi$, find $\cos \theta$.

85. Given that $\tan \theta = 1$ and that $\sin \theta$ is negative, find $\sec \theta$.

86. Given that $\csc \theta = -3$ and that $\sec \theta$ is positive, find $\cot \theta$.

Proofs

87. Prove that if θ is an acute angle, then the unit circle definition of $\cos \theta$ is equivalent to the right triangle definition of $\cos \theta$.

88. Use the unit circle definition of the trigonometric functions (Definition 10.2) to prove Theorem 10.3.

89. Prove that $\sin \theta = 0$ if and only if $\theta = \pi k$ for some integer k. (Be sure to prove both directions: "if" and "only if.")

90. Prove that $\cos \theta = 0$ if and only if $\theta = \frac{\pi}{2} + \pi k$ for some integer k. (Be sure to prove both directions, "if" and "only if.")

10.3   The Algebra of Trigonometric Functions

In this section we examine the basic algebraic properties of trigonometric functions. We begin by examining the domain and range of each trigonometric function and then develop trigonometric "identities" that tell us how to algebraically manipulate expressions that involve trigonometric functions.

10.3.1 Domains and ranges of trigonometric functions

The trigonometric functions $\sin \theta$, $\cos \theta$, $\tan \theta$, $\sec \theta$, $\csc \theta$, and $\cot \theta$ are all truly *functions* because for each input θ (in the domain) there is exactly one output, uniquely determined by the coordinates (x, y) on the unit circle that correspond to the angle θ. The domain of each trigonometric function is the set of *angles* for which that function is defined. The range of each trigonometric function is determined by the possible coordinates (x, y) on the unit circle corresponding to the angles in the domain.

The domains and ranges of sine and cosine can easily be determined by considering the unit circle definitions of $\sin \theta$ and $\cos \theta$. For example, we can find the sine of *any* angle θ because *every* angle θ corresponds to a point (x, y) on the unit circle with a well-defined height y (which is what we define to be the sine of θ). Therefore, the domain of the sine function is $\theta \in (-\infty, \infty)$. The range of the sine function is the set of possible heights y that can occur at a point (x, y) on the unit circle. As θ sweeps around the unit circle, the corresponding point (x, y) on the unit circle passes through all heights between, and including, -1 and 1. Thus the range of the sine function is $[-1, 1]$.

THEOREM 10.4

Domains and Ranges of the Trigonometric Functions
In the notation below, we assume that k represents all positive or negative integers.
- **(a)** $\sin \theta$ has domain $(-\infty, \infty)$ and range $[-1, 1]$.
- **(b)** $\cos \theta$ has domain $(-\infty, \infty)$ and range $[-1, 1]$.
- **(c)** $\tan \theta$ has domain $\{\theta \mid \theta \neq \frac{\pi}{2} + \pi k\}$ and range $(-\infty, \infty)$.
- **(d)** $\sec \theta$ has domain $\{\theta \mid \theta \neq \frac{\pi}{2} + \pi k\}$ and range $(-\infty, -1] \cup [1, \infty)$.
- **(e)** $\csc \theta$ has domain $\{\theta \mid \theta \neq \pi k\}$ and range $(-\infty, -1] \cup [1, \infty)$.
- **(f)** $\cot \theta$ has domain $\{\theta \mid \theta \neq \pi k\}$ and range $(-\infty, \infty)$.

PROOF **(THEOREM 10.4)**   The paragraph above Theorem 10.4 proves part (a). The proof of part (b) is similar and is left for the exercises. We will prove parts (c) and (e) and leave parts (d) and (f), which are similar, for the exercises.

To prove the domain portion of part (c), recall that $\tan \theta = \frac{\sin \theta}{\cos \theta}$. Since $\sin \theta$ and $\cos \theta$ are defined for all angles, the tangent function is defined for all angles except the ones that make $\cos \theta$ equal to zero. By looking at a picture of the unit circle, we can quickly see that $\cos \theta$ is zero when the terminal edge of θ ends at $\frac{\pi}{2}$ or $-\frac{\pi}{2}$ on the unit circle (see Figure 10.24). Thus the angles $\frac{\pi}{2}, \frac{3\pi}{2}, \frac{5\pi}{2}, \ldots$, as well as the angles $-\frac{\pi}{2}, -\frac{3\pi}{2}, -\frac{5\pi}{2}, \ldots$, all have a cosine of zero. In general, any angle of the form $\frac{\pi}{2} + \pi k$, where k is an integer, has a cosine of zero. Thus the domain of $\tan \theta$ is the set $\{\theta \mid \theta \neq \frac{\pi}{2} + \pi k\}$.

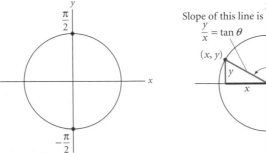

Figure 10.24
$\cos \theta = 0$ at these points

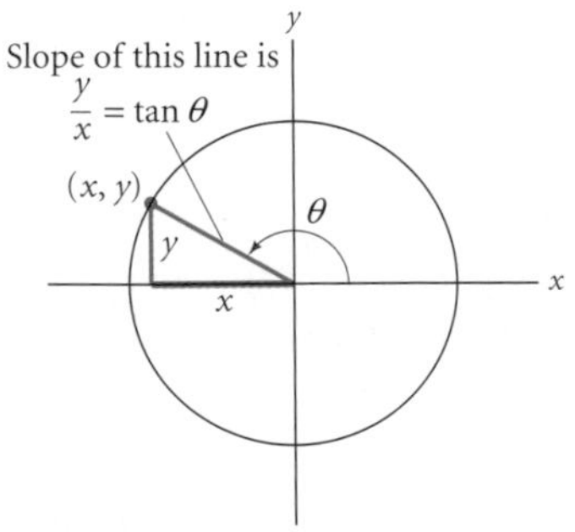

Figure 10.25
$\tan \theta$ seen as a slope

To prove the range portion of part (c), notice that for any angle θ, the tangent of θ is the slope $\frac{y}{x}$ of the line from the origin to the point (x, y) on the unit circle that corresponds to θ (see Figure 10.25). Given any slope (i.e., any real number), we can find an angle θ

for which the tangent of θ is equal to that slope. Therefore, the range of $\tan \theta$ is all real numbers.

To prove part (e), first notice that $\csc \theta = \frac{1}{\sin \theta}$ is defined everywhere except at the values of θ for which $\sin \theta$ is zero. We have $\sin \theta = 0$ for any angle whose terminal edge is at 0 or π radians (why?); thus $\sin \theta = 0$ if θ is any (positive or negative) integer multiple of π, that is, if $\theta = \pi k$ for any positive or negative integer k. Thus the domain of $\csc \theta$ is $\{\theta \mid \theta \neq \pi k\}$. The range of cosecant can be determined using the range of sine (its reciprocal). Since $-1 \leq \sin \theta \leq 1$ for all θ, we know that $\csc \theta = \frac{1}{\sin \theta}$ must always be either greater than or equal to 1 or less than or equal to -1. In fact, if r is any number in $(-\infty, -1] \cup [1, \infty)$, then $\frac{1}{r} \in [-1, 1]$, and thus there is some angle θ for which $\sin \theta = \frac{1}{r}$, that is, for which $\csc \theta = r$. Therefore, the range of $\csc \theta$ is the entire set $(-\infty, -1] \cup [1, \infty)$. ∎

> **✖ Caution**   You should know the domains and ranges of the six trigonometric functions from memory, but please don't simply memorize Theorem 10.4. Instead, think about the unit circle and what possible angles can be used as inputs for each trigonometric function, as well as what possible outputs can be obtained from each trigonometric function. □

10.3.2 Basic trigonometric identities

An equation that is always true (for values in the domains of both sides) is called an *identity.* For example, the equation $3^{2x} = 9^x$ is true for *all* values of x and thus is an identity. In this section we present some basic *trigonometric identities.* These identities will be our basic algebraic rules for simplifying and rewriting trigonometric expressions. We will prove each of these identities either by using the unit circle definitions of trigonometric functions or by rewriting identities we have already proved.

Perhaps the most obvious trigonometric identity is the *Pythagorean identity:*

$$(1) \qquad\qquad (\sin \theta)^2 + (\cos \theta)^2 = 1.$$

Equation (1) is true for *all* angles θ (that's why we call it an "identity") and follows from the unit circle definitions of sine and cosine. For any angle θ, if (x, y) is the point on the unit circle corresponding to θ, then $\sin \theta = y$ and $\cos \theta = x$. Since (x, y) is on the unit circle, we know that $y^2 + x^2 = 1$ (this is the equation for the circle of radius 1 centered at the origin) and thus $(\sin \theta)^2 + (\cos \theta)^2 = 1$. We could also obtain this identity by noticing that $y = \sin \theta$ and $x = \cos \theta$ are the legs of a right triangle with hypotenuse of length 1 unit and then applying the Pythagorean Theorem.

At this point we need some new notation. We will write $\sin^2 \theta$ to represent the quantity $(\sin \theta)^2$. In fact, for any positive integer k, we will write $\sin^k \theta$ to represent the kth power of $\sin \theta$:

NOTATION 10.1

Shorthand Notation for Powers of Trigonometric Functions

For any positive integer k,

$$\sin^k \theta := (\sin \theta)^k.$$

We use similar notation to represent the kth power of any trigonometric function.

> **✖ Caution**   This notation is used *only* when the exponent k is a positive integer. For example, we do *not* write $\sin^{-1} x$ to represent $(\sin x)^{-1}$ (in Chapter 11 we will use the notation "$\sin^{-1} x$" to represent an entirely different function). Also, we do *not* write $\sin^{-3} x$ to represent $(\sin x)^{-3}$, and we do *not* write $\sin^{\frac{1}{2}} x$ to represent $(\sin x)^{\frac{1}{2}}$. □

Part (a) of the following theorem simply rewrites Equation (1) in this new notation. Parts (b) and (c) of Theorem 10.5 follow directly from part (a).

THEOREM 10.5

Pythagorean Identities

(a) $\sin^2\theta + \cos^2\theta = 1$ (b) $\tan^2\theta + 1 = \sec^2\theta$ (c) $1 + \cot^2\theta = \csc^2\theta$

PROOF   (**THEOREM 10.5**)   We have already proved part (a). To prove part (b), we simply divide both sides of the identity in part (a) by $\cos^2\theta$ and then simplify.

$$\sin^2\theta + \cos^2\theta = 1 \qquad\qquad \text{(by the original Pythagorean identity)}$$

$$\frac{\sin^2\theta + \cos^2\theta}{\cos^2\theta} = \frac{1}{\cos^2\theta} \qquad\qquad \text{(divide both sides by } \cos^2\theta\text{)}$$

$$\frac{\sin^2\theta}{\cos^2\theta} + \frac{\cos^2\theta}{\cos^2\theta} = \frac{1}{\cos^2\theta} \qquad\qquad \text{(algebra)}$$

$$\frac{(\sin\theta)^2}{(\cos\theta)^2} + \frac{(\cos\theta)^2}{(\cos\theta)^2} = \frac{1}{(\cos\theta)^2} \qquad\qquad \text{(notation)}$$

$$(\tan\theta)^2 + 1 = (\sec\theta)^2 \qquad \text{(relationships between trigonometric functions)}$$

$$\tan^2\theta + 1 = \sec^2\theta. \qquad\qquad \text{(notation)}$$

We did not actually have to write the fourth and fifth lines of this calculation; usually we won't translate out of the $\sin^2\theta$ notation. We did so only because it was the first time we had used the $\sin^2\theta$ notation, and we wanted the steps to be perfectly clear. Part (c) of the theorem is proved in a similar fashion (except this time dividing both sides of the identity in part (a) by $\sin^2\theta$) and is left for the exercises. ∎

We can determine from the unit circle definitions that each of the six trigonometric functions is either an even function or an odd function. Recall that a function $f(x)$ is **even** if $f(-x) = f(x)$ for all x in its domain and is **odd** if $f(-x) = -f(x)$. The following theorem tells us that sine and tangent are odd functions and that cosine is an even function.

THEOREM 10.6

Even/Odd Identities

(a) $\sin(-\theta) = -\sin\theta$ (b) $\cos(-\theta) = \cos\theta$ (c) $\tan(-\theta) = -\tan\theta$

Since the cosecant, secant, and cotangent functions are the reciprocals of the sine, cosine, and tangent functions, respectively, Theorem 10.6 implies that cosecant and cotangent are odd functions and that secant is an even function (why?). The even/odd identities are often useful for simplifying or rewriting trigonometric expressions.

❓ *Question*   Look back at the graphs of $\sin\theta$ and $\cos\theta$ in Figures 10.17 and 10.18 in Section 10.2.3. Can you tell from these graphs that $\sin\theta$ is an odd function and $\cos\theta$ is an even function? ☐

PROOF   (**THEOREM 10.6**)   We'll prove part (a) first. Given any angle θ in standard position, the angle $-\theta$ is the same angle, opening in the "opposite" direction. This means that if (x, y) is the point on the unit circle corresponding to θ, then the point $(x, -y)$ is the point on the unit circle corresponding to $-\theta$ (see Figure 10.26).

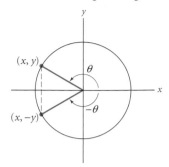

Figure 10.26

Since the y-coordinate for $-\theta$ is the negative of the y-coordinate for θ, the sine of $-\theta$ is the negative of the sine of θ; in other words, $\sin(-\theta) = -\sin\theta$. The proof of part (b) is similar and is left for the exercises.

Part (c) of Theorem 10.6 follows directly from the first two parts of the theorem and the fact that tangent is the quotient of sine and cosine:

$$\tan(-\theta) = \frac{\sin(-\theta)}{\cos(-\theta)} \qquad \text{(tangent is the quotient of sine and cosine)}$$

$$= \frac{-\sin\theta}{\cos\theta} \qquad \text{(sine is an odd function and cosine is even)}$$

$$= -\frac{\sin\theta}{\cos\theta} \qquad \text{(algebra)}$$

$$= -\tan\theta. \qquad \text{(tangent is the quotient of sine and cosine)}$$

Since $\tan(-\theta) = -\tan(\theta)$, $\tan\theta$ is an odd function.

◈ Caution There is *not* a typographical error in Figure 10.26! If the uppermost dot has coordinates (x, y) (x happens to be a negative number and y happens to be positive), then the lowermost dot has coordinates $(x, -y)$ (the new x-coordinate is the same, but the new y-coordinate is the negative of the old y-coordinate; x is negative and $-y$ is also negative). □

10.3.3 Sum and difference identities

One might initially hope that the sine function is "additive" in the sense that for any two angles α and β, $\sin(\alpha + \beta) = \sin\alpha + \sin\beta$. However, this is in general *false*! Try it for two angles whose sines you know, such as $\alpha = \frac{\pi}{4}$ and $\beta = \frac{\pi}{2}$. We have $\sin\frac{\pi}{4} = \frac{\sqrt{2}}{2}$ and $\sin\frac{\pi}{2} = 1$, but $\sin(\frac{\pi}{4} + \frac{\pi}{2}) = \sin(\frac{3\pi}{4}) = \frac{\sqrt{2}}{2}$ is *not* equal to the sum of $\frac{\sqrt{2}}{2}$ and 1. Although the sine function is not additive, we *can* say something about the sine of a sum of angles. The following theorem describes how to rewrite sines and cosines of sums and differences of angles.

THEOREM 10.7

Sum and Difference Identities

(a) $\sin(\alpha + \beta) = \sin\alpha\cos\beta + \sin\beta\cos\alpha$

(b) $\cos(\alpha + \beta) = \cos\alpha\cos\beta - \sin\alpha\sin\beta$

(c) $\sin(\alpha - \beta) = \sin\alpha\cos\beta - \sin\beta\cos\alpha$

(d) $\cos(\alpha - \beta) = \cos\alpha\cos\beta + \sin\alpha\sin\beta$

The proofs of parts (a) and (b) are lengthy, and we do not have time to include them here. The proofs of parts (c) and (d) follow easily from the first two identities, using the fact that sine is an odd function and cosine is an even function. We prove part (c) and leave the proof of part (d) for the exercises.

PROOF (**PART (C) OF THEOREM 10.7**)   By writing the difference $\alpha - \beta$ as a sum, we can apply the sum identity for sine:

$$
\begin{aligned}
\sin(\alpha - \beta) & \\
&= \sin(\alpha + (-\beta)) && \text{(rewrite difference as a sum)} \\
&= \sin(\alpha)\cos(-\beta) + \sin(-\beta)\cos(\alpha) && \text{(part (a) with } \alpha \text{ and } -\beta \text{ as the angles)} \\
&= \sin\alpha \cos\beta - \sin\beta \cos\alpha. && (\cos(-\beta) = \cos\beta, \sin(-\alpha) = -\sin\alpha)
\end{aligned}
$$

∎

The following example illustrates that the sum identity for the sine function is true for the angles $\alpha = \frac{\pi}{4}$ and $\beta = \frac{\pi}{2}$.

EXAMPLE 10.13   **An example verifying the sum identity for sine**

Suppose $\alpha = \frac{\pi}{4}$ and $\beta = \frac{\pi}{2}$. In this case we have

$$
\sin(\alpha + \beta) = \sin\left(\frac{\pi}{4} + \frac{\pi}{2}\right) = \sin\left(\frac{3\pi}{4}\right) = \frac{\sqrt{2}}{2},
$$

whereas

$$
\sin\alpha \cos\beta + \sin\beta \cos\alpha = \sin\left(\frac{\pi}{4}\right)\cos\left(\frac{\pi}{2}\right) + \sin\left(\frac{\pi}{2}\right)\cos\left(\frac{\pi}{4}\right)
$$

$$
= \frac{\sqrt{2}}{2}(0) + (1)\frac{\sqrt{2}}{2} = \frac{\sqrt{2}}{2}.
$$

□

Previously we could calculate the exact values of trigonometric functions only for angles that were quarters or sixths of π. Now we can use the sum and difference identities to calculate the exact values of trigonometric functions for angles that are sums or differences of these angles. For example, we know how to calculate $\sin 45°$ and $\sin 30°$ by hand, so we can use the difference identity for sine to find $\sin(45° - 30°) = \sin(15°)$ by hand. In the language of radians, this means we can use $\sin(\frac{\pi}{4})$ and $\sin(\frac{\pi}{6})$ to calculate $\sin(\frac{\pi}{4} - \frac{\pi}{6}) = \sin(\frac{\pi}{12})$; see Example 10.14.

EXAMPLE 10.14   **Using sum or difference identities to calculate a trigonometric value**

Calculate the *exact* value of $\sin\dfrac{\pi}{12}$.

Solution   The reference triangle for the angle $\frac{\pi}{12}$ is not a 30–60–90 or a 45–45–90 triangle, so we cannot calculate the value of $\sin\frac{\pi}{12}$ directly. However, notice that

$$
\frac{\pi}{4} - \frac{\pi}{6} = \frac{\pi}{12}.
$$

By the difference identity for sine, we have

$$\sin\frac{\pi}{12} = \sin\left(\frac{\pi}{4} - \frac{\pi}{6}\right)$$

$$= \sin\left(\frac{\pi}{4}\right)\cos\left(\frac{\pi}{6}\right) - \sin\left(\frac{\pi}{6}\right)\cos\left(\frac{\pi}{4}\right) \qquad \text{(by the difference identity for sine)}$$

$$= \frac{\sqrt{2}}{2} \cdot \frac{\sqrt{3}}{2} - \frac{1}{2} \cdot \frac{1}{\sqrt{2}} \qquad \text{(using reference triangles)}$$

$$= \frac{\sqrt{6}}{4} - \frac{\sqrt{2}}{4} = \frac{\sqrt{6} - \sqrt{2}}{4}. \qquad \text{(algebra)}$$

$\square$

☑ *Checking the Answer*   You can check the answer to Example 10.14 by using a calculator (in radian mode) to approximate $\sin\frac{\pi}{12}$ and then comparing this to an approximation of $\frac{\sqrt{6}-\sqrt{2}}{4}$. $\qquad\square$

10.3.4 More trigonometric identities

Many trigonometric identities can be obtained from the sum and difference identities. For example, we can obtain identities for the sines and cosines of "double angles"—that is, identities that tell us how to rewrite the sine and cosine of 2θ in terms of the sine and cosine of θ.

THEOREM 10.8

Double-Angle Identities

(a) $\sin 2\theta = 2\sin\theta\cos\theta$ (b) $\cos 2\theta = \cos^2\theta - \sin^2\theta$

(c) $\cos 2\theta = 1 - 2\sin^2\theta$ (d) $\cos 2\theta = 2\cos^2\theta - 1$

These double-angle identities show that constants do *not* factor out of the sine and cosine functions; in other words, $\sin 2\theta$ is *not* equal to $2\sin\theta$. Note that we have three different ways to express the cosine of a double angle. The identities in Theorem 10.8 follow directly from the sum identities for sine and cosine.

PROOF (**THEOREM 10.8**)   To prove part (a), we will write 2θ as the sum $\theta + \theta$ of two angles and apply the sum identity for sine:

$$\sin 2\theta = \sin(\theta + \theta) \qquad \text{(rewrite 2θ as a sum)}$$

$$= \sin\theta\cos\theta + \sin\theta\cos\theta \qquad \text{(sum identity for sine)}$$

$$= 2\sin\theta\cos\theta. \qquad \text{(algebra)}$$

The proof of part (b) is similar and is left for the exercises.

We can prove parts (c) and (d) from part (b) and the Pythagorean identity for sine and cosine. We will prove part (c) and leave part (d) for the exercises. Since $\sin^2\theta + \cos^2\theta = 1$, we know that $\cos^2\theta = 1 - \sin^2\theta$. Now we rewrite the identity in part (b) to get the identity in part (c):

$$\cos 2\theta = \cos^2\theta - \sin^2\theta \qquad \text{(by part (b))}$$

$$\cos 2\theta = (1 - \sin^2\theta) - \sin^2\theta \qquad \text{(by the Pythagorean identity)}$$

$$\cos 2\theta = 1 - 2\sin^2\theta. \qquad \text{(algebra)}$$

◈ *Caution*   Remember that we *never* prove an identity by starting with what we are trying to prove and then working on both sides until we arrive at an "obvious" identity (see Section 0.5). For example, we cannot prove that $\sin 2\theta = 2\sin\theta\cos\theta$ by working on both sides of the equality until we arrive at something we know; instead, we start with one side and work toward the other. However, we *can* start with an identity that we already know and modify it until we get the identity we are trying to prove (as in the proof of part (c) of Theorem 10.8).   □

The double-angle identities can be rewritten in two different ways to obtain some new identities that are traditionally called the "half-angle" identities. These identities (especially the first two forms) will be particularly useful in Part IV when we study integration. The proofs are left for the exercises.

THEOREM 10.9

Half-Angle Identities

(a) $\sin^2\theta = \dfrac{1 - \cos 2\theta}{2}$

(b) $\cos^2\theta = \dfrac{1 + \cos 2\theta}{2}$

(c) $\sin\left(\dfrac{\theta}{2}\right) = \pm\sqrt{\dfrac{1 - \cos\theta}{2}}$

(d) $\cos\left(\dfrac{\theta}{2}\right) = \pm\sqrt{\dfrac{1 + \cos\theta}{2}}$

The first two identities in Theorem 10.10 are special cases of the sum identity for sine and the difference identity for cosine. They can be useful when we want to write sines in terms of cosines, or vice versa. The second two identities in Theorem 10.10 follow from the unit circle definitions of the trigonometric functions and mean that the sine and cosine functions are "periodic" (see Section 10.6). The proofs are left for the exercises.

THEOREM 10.10

Shift Identities

(a) $\cos(\theta - \frac{\pi}{2}) = \sin\theta$ 　　　　　(b) $\sin(\theta + \frac{\pi}{2}) = \cos\theta$

(c) $\sin(\theta + 2\pi) = \sin\theta$ 　　　　　(d) $\cos(\theta + 2\pi) = \cos\theta$

If you look back at the graphs of $\sin\theta$ and $\cos\theta$ in Section 10.2.3, you should be able to see that the graph of the sine function is a horizontal translation of the graph of the cosine function (you shift the cosine function right $\frac{\pi}{2}$ units to get the sine function). Similarly, the cosine function is a horizontal translation of the sine function (in the opposite direction).

There are many other trigonometric identities. In particular, there are sum and difference identities for tangent and the other trigonometric functions; you will see even more trigonometric identities in the exercises. You should memorize the basic trigonometric identities and should know how to derive or where to find formulas for more complicated trigonometric identities when you need them.

What you should know after reading Section 10.3

▶ The domains and ranges of the six trigonometric functions (and how to explain these domains and ranges using the unit circle definitions of the trigonometric functions).

▶ The new notation $\sin^k \theta := (\sin \theta)^k$ for positive integers k. Be able to manipulate algebraically expressions written in this notation.

▶ What is an identity and how is it different from an equation? What are some possible ways of proving an identity?

▶ Know the Pythagorean, even/odd, sum and difference, double-angle, half-angle, and shift identities. Be able to prove these identities (except for the sum identities). Which identities follow from the unit circle definitions of the trigonometric functions, and which identities follow from previous identities?

▶ Be able to use trigonometric identities to simplify trigonometric expressions and to calculate exact values of trigonometric functions at angles that are sums or differences of angles with 30–60–90 or 45–45–90 reference triangles.

Exercises 10.3

Concepts

0. Read the section and make your own summary of the material. Include whatever you think will help you review the material later on.

1. Is the angle $\theta = -\frac{3\pi}{2}$ in the domain of $\sec \theta$? Why or why not? What about the angle $\theta = \frac{6\pi}{2}$?

2. Show that 2 is in the range of secant by finding an angle θ for which $\sec \theta = 2$.

3. Show that $-\sqrt{3}$ is in the range of tangent by finding an angle θ for which $\tan \theta = \sqrt{3}$.

■ **Fill in the blanks below.**

4. The cosine function has domain _____ and range _____.

5. The secant function has domain _____ and range _____.

6. The cotangent function has domain _____ and range _____.

7. Use the unit circle definition of sine to explain why the sine function has domain $(-\infty, \infty)$ and range $[-1, 1]$.

8. Use the unit circle definition of tangent to explain why the tangent function has domain $\{\theta \mid \theta \neq \frac{\pi}{2} + \pi k\}$ and range $(-\infty, \infty)$.

9. Use the domain and range of the sine function to explain why the cosecant function has domain $\{\theta \mid \theta \neq \pi k\}$ and range $(-\infty, -1] \cup [1, \infty)$.

10. Show on a unit circle the types of angles for which $\cos \theta = 0$. What is the solution set of the equation $\cos \theta = 0$? (*Note:* There are *infinitely* many angles in the solution set; be sure that you completely describe the *entire* solution set.)

11. Show on a unit circle the types of angles for which $\sin \theta = \cos \theta$. What is the solution set of the equation $\sin \theta = \cos \theta$? (*Note:* There are *infinitely* many angles in the solution set; be sure that you completely describe the *entire* solution set.)

12. Show that $\sin \theta \cos \theta + \sin \theta = 1 + \cos \theta$ is *not* a trigonometric identity.

13. Use the equation for the unit circle to explain why $\sin^2 \theta + \cos^2 \theta = 1$ for all angles θ. Then use the Pythagorean Theorem to explain this identity.

14. Use the unit circle definition of sine to explain why, for any angle θ, we have $\sin(-\theta) = -\sin \theta$.

15. Which of the six trigonometric functions are even functions? Which of the six trigonometric functions are odd functions?

16. Find angles α and β that illustrate that $\cos(\alpha + \beta) \neq \cos \alpha + \cos \beta$.

17. Find an angle θ that illustrates that $\sin 2\theta \neq 2 \sin \theta$.

18. Show that the difference formula for cosine works for the angles $\alpha = \frac{5\pi}{6}$ and $\beta = \frac{\pi}{3}$.

19. Show that for the angle $\theta = -\frac{\pi}{3}$, the quantities $\cos 2\theta$, $\cos^2 \theta - \sin^2 \theta$, $1 - 2\sin^2 \theta$, and $2\cos^2 \theta - 1$ are all equal. What does this have to do with trigonometric identities?

20. Show that for the angle $\theta = \frac{4\pi}{3}$, we have $\sin \theta = \cos(\theta - \frac{\pi}{2})$ and $\cos \theta = \sin(\theta + \frac{\pi}{2})$.

■ Fill in the blanks below to complete the trigonometric identities.

21. $\tan^2 \theta + 1 = $ _____

22. $\sin(-\theta) = $ _____

23. $\sin(\alpha - \beta) = $ _____

24. $\cos(\alpha + \beta) = $ _____

25. $2\cos^2 \theta - 1 = $ _____

26. $\sin(\frac{\theta}{2}) = $ _____

27. $\dfrac{1 + \cos 2\theta}{2} = $ _____

28. $\sin(\theta + \frac{\pi}{2}) = $ _____

29. List two trigonometric identities that we prove by using the unit circle definitions of the trigonometric functions. Then list two trigonometric identities that follow from previous identities (and list the previous identities as well).

30. The following "proof" of the trigonometric identity $\sin 2\theta = 2\sin\theta\cos\theta$ uses bad style and bad logic. What is wrong with the way this proof is written? Rewrite the proof so that it is logically and stylistically correct.

PROOF

$$\sin 2\theta = 2\sin\theta\cos\theta$$
$$\sin(\theta + \theta) = 2\sin\theta\cos\theta \quad \text{(rewrite } 2\theta \text{ as a sum)}$$
$$\sin\theta\cos\theta + \sin\theta\cos\theta = 2\sin\theta\cos\theta \quad \text{(sum identity for sine)}$$
$$2\sin\theta\cos\theta = 2\sin\theta\cos\theta.$$

∎

Skills

■ For each equation below, (a) find all $\theta \in [0, 2\pi]$ that make the equation true, and (b) find *all* angles θ that make the equation true. Note that in part (b) the solution set will include *infinitely many* values; make sure that you clearly describe the values that are in the solution set.

31. $\sin\theta = 1$

32. $\tan\theta = -1$

33. $\sec\theta = 2$

34. $\cos\theta = \frac{1}{2}$

35. $\csc\theta = \frac{2}{\sqrt{3}}$

36. $\cot\theta = -\sqrt{3}$

37. $\cos 2\theta = 0$

38. $\sin(\theta - \frac{\pi}{3}) = \frac{1}{2}$

39. $\tan(\pi - \theta) = 1$

■ Find the domains of the following functions.

40. $f(\theta) = \sin\theta + \cos\theta$

41. $f(\theta) = \csc\theta\cot\theta$

42. $f(\theta) = \dfrac{\sec\theta}{\tan\theta}$

43. $f(\theta) = \dfrac{1}{1 - \tan\theta}$

44. $f(\theta) = \cos(\theta - \frac{2\pi}{3})$

45. $f(\theta) = \sqrt{\sec\theta}$

■ Determine algebraically whether the following functions are odd or even (or neither).

46. $f(\theta) = \sin\theta + \cos\theta$

47. $f(\theta) = \dfrac{\sin\theta}{1 + \cos\theta}$

48. $f(\theta) = \dfrac{\cos\theta}{1 + \sin\theta}$

49. $f(\theta) = \sin^2\theta - \sec\theta$

50. $f(\theta) = \cos 2\theta \sin 2\theta$

51. $f(\theta) = \dfrac{\tan\theta}{\csc\theta + \cot\theta}$

■ Calculate the *exact* values of each of the trigonometric expressions below. (*Hint:* You will need to use some trigonometric identities to write each expression in terms of trigonometric functions of angles that you know how to deal with.) Is your answer reasonable in terms of the unit circle? Also use a calculator to check your answers.

52. $\sin(-\frac{\pi}{12})$

53. $\cos(-\frac{\pi}{12})$

54. $\tan(-\frac{\pi}{12})$

55. $\cos(\frac{7\pi}{12})$

56. $\sec(\frac{7\pi}{12})$

57. $\cot(\frac{7\pi}{12})$

Proofs

58. Use the unit circle definition of cosine to prove that the cosine function has domain $(-\infty, \infty)$ and range $[-1, 1]$.

59. Use the domain and range of the cosine function to prove that the secant function has domain $\{\theta \mid \theta \neq \frac{\pi}{2} + \pi k\}$ and range $(-\infty, -1] \cup [1, \infty)$.

60. Prove that the cotangent function has domain $\{\theta \mid \theta \neq \pi k\}$ and range $(-\infty, \infty)$.

61. Use the fact that $\sin^2\theta + \cos^2\theta = 1$ to show that $1 + \cot^2\theta = \csc^2\theta$ for all angles θ.

62. Use the unit circle definition of cosine to prove that cosine is an even function.

63. Use the fact that sine is an odd function and cosine is an even function to prove that cotangent is an odd function.

64. Use the sum identity for cosine to prove the difference identity for cosine: $\cos(\alpha - \beta) = \cos\alpha\cos\beta + \sin\alpha\sin\beta$.

65. Use the sum identity for cosine to prove the identity $\cos 2\theta = \cos^2\theta - \sin^2\theta$.

66. Use the identity in Problem 65 and a Pythagorean identity to prove the identity $\cos 2\theta = 2\cos^2\theta - 1$.

67. Use Theorem 10.8 (and some algebra) to prove the identities in Theorem 10.9. In other words, prove that $\sin^2\theta = \frac{1 - \cos 2\theta}{2}$ and $\cos^2\theta = \frac{1 + \cos 2\theta}{2}$.

68. Use the double-angle identities to prove the half-angle identities $\sin(\frac{\theta}{2}) = \pm\sqrt{\frac{1 - \cos\theta}{2}}$ and $\cos(\frac{\theta}{2}) = \pm\sqrt{\frac{1 + \cos\theta}{2}}$.

69. Use the sum and difference formulas for sine and cosine to prove the shift identities $\sin\theta = \cos(\theta - \frac{\pi}{2})$ and $\cos\theta = \sin(\theta + \frac{\pi}{2})$.

■ Prove the following trigonometric identities (by using the trigonometric identities in this section).

70. $\tan 2\theta = \dfrac{2 \tan \theta}{1 - \tan^2 \theta}$

71. $\csc 2\theta = \frac{1}{2} \sec \theta \csc \theta$

72. $(\sin \theta + \cos \theta)^2 = \sin 2\theta + 1$

73. $\sec \theta \cot \theta = \dfrac{\sin \theta}{1 - \cos^2 \theta}$

74. $\tan(\alpha + \beta) = \dfrac{\tan \alpha + \tan \beta}{1 - \tan \alpha \tan \beta}$

75. $\tan(\alpha - \beta) = \dfrac{\tan \alpha - \tan \beta}{1 + \tan \alpha \tan \beta}$

10.4   Limits of Trigonometric Functions

In this section we examine limits of trigonometric expressions. We begin with two important limits involving sine and cosine functions, which can be used to prove that all six of the trigonometric functions are continuous on their domains. We will then investigate the long-term behavior of trigonometric functions (all of which are periodic).

10.4.1   Two important trigonometric limits

Two limits lie at the heart of the continuity and differentiability of the trigonometric functions. Armed with these two limits alone (and our other mathematical skills), we can prove that the sine and cosine functions are continuous and differentiable everywhere.

THEOREM 10.11

Two Important Trigonometric Limits

(a) $\displaystyle\lim_{\theta \to 0} \frac{\sin \theta}{\theta} = 1$ **(b)** $\displaystyle\lim_{\theta \to 0} \frac{1 - \cos \theta}{\theta} = 0$

If we knew how to differentiate the sine and cosine functions, then we could use L'Hôpital's Rule to solve these limits (since both of them are of the indeterminate form $\frac{0}{0}$). However, we won't know the derivatives of the trigonometric functions until Section 10.5, and besides, these limits will be an integral part of finding those derivatives! Rigorous proof of the limits in Theorem 10.11 requires fairly technical geometric arguments, and we do not have the time to treat those arguments properly here. However, for the remainder of this subsection we will present some reasons why these limits make sense.

The first limit basically says that as θ gets very small, the quantities $\sin \theta$ and θ are very close to each other (since their ratio approaches 1). Suppose θ is a very small angle, and let (x, y) be the point on the unit circle corresponding to θ (see Figure 10.27). By

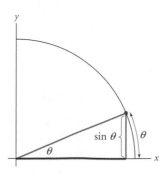

Figure 10.27

definition, $\sin\theta$ is the y-coordinate of the point (x, y) (i.e., the small vertical distance from the x-axis to the point (x, y)), and θ is the distance along the arc of the unit circle from $(0, 0)$ to (x, y). (Recall that if θ is measured in radians, then the point (x, y) corresponding to θ will be a distance of θ from $(0, 0)$ along the circumference of the unit circle.)

If θ is a very small angle, then the small vertical distance and the small arc distance will be very similar; in other words, we will have $\sin\theta \approx \theta$. As $\theta \to 0$, these two distances become even more similar, so their ratio $\frac{\sin\theta}{\theta}$ approaches 1.

We can also see that the limits in Theorem 10.11 are reasonable by looking at tables of values for $\frac{\sin\theta}{\theta}$ and $\frac{1-\cos\theta}{\theta}$ as θ approaches zero. Table 10.2 can be obtained using a calculator (in radian mode) to approximate these values.

θ	-0.5	-0.1	-0.01	0	0.01	0.1	0.5
$\frac{\sin\theta}{\theta}$	0.95885	0.99833	0.99998	-	0.99998	0.99833	0.95885
$\frac{1-\cos\theta}{\theta}$	-0.24483	-0.04996	-0.00500	-	0.00500	0.04996	0.24483

Table 10.2

Neither $\frac{\sin\theta}{\theta}$ nor $\frac{1-\cos\theta}{\theta}$ is defined at $\theta = 0$. However, as θ approaches zero from the left and from the right, the values of $\frac{\sin\theta}{\theta}$ approach 1 and the values of $\frac{1-\cos\theta}{\theta}$ approach 0.

We could also look at the graphs of $y = \frac{\sin\theta}{\theta}$ and $y = \frac{1-\cos\theta}{\theta}$ near $\theta = 0$ and check that the height of the first graph approaches 1, while the height of the second graph approaches 0. This is shown in the graphs in Figures 10.28 and 10.29. (Be sure that your calculator is in radian mode.)

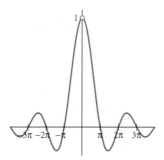

Figure 10.28
$$y = \frac{\sin\theta}{\theta}$$

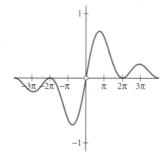

Figure 10.29
$$y = \frac{1 - \cos\theta}{\theta}$$

Again, although neither function is defined near $\theta = 0$ (note the "holes" in the graphs), the first graph approaches a height of 1 as $\theta \to 0$, while the second graph approaches a height of 0.

In Section 10.4.3 we will use the special limits in Theorem 10.11 to prove that the sine and cosine functions are continuous. In Section 10.5 we will use these limits to find the derivatives of sine and cosine. These limits also come in handy when we try to compute limits of certain types of trigonometric expressions, as we will see in Section 10.4.5.

10.4.2 From the unit circle to the xy-plane

It is time for us to change notation. So far we have been using θ as the independent variable when discussing trigonometric functions; this has been convenient because the variable θ suggests an angle and because the unit circle definitions of the trigonometric functions already involve our usual independent variable name "x" (as the x-coordinate of a point on the unit circle). Of course, we can use whatever letters we like for the variables; for example, we could write $w = \sin t$ or $L = \cos\phi$. However, we will now start to use the familiar "*x*" as the independent variable (and "*y*" as the dependent variable).

Although we will still often refer to the unit circle definitions of the trigonometric functions, we will now be thinking of $y = \sin x$ and $y = \cos x$ the way that we have thought of all the previous types of functions that we have studied: as functions graphed in the *xy*-plane. This will make it easier for us to discuss the limits and derivatives of the trigonometric functions using the tools that we used to investigate other types of functions that we have studied. It will also enable us to observe the behavior of each of the six trigonometric functions.

◈ ***Caution***   Keep in mind that $y = \sin x$ and $y = \cos x$ are still functions whose inputs should be thought of as *angles* and that the "*x*" we are using for the independent variable should not be confused with the "*x*" in the unit circle definitions of trigonometric functions (similarly for the dependent variable "*y*"). Of course, everything we have proved about the trigonometric functions in terms of the variable θ is true when we change notation from θ to *x*.   ☐

If you use your calculator to graph the six trigonometric functions (in radian mode, of course), you will get the graphs in Figures 10.30–10.35. All six graphs are drawn in the same "graphing window" so that you can easily compare them. Each of the six trigonometric functions has ***periodic*** behavior, which means that the shape of the graph repeats over and over again as $x \to \infty$ and as $x \to -\infty$. For example, the sine function repeats every 2π units, and thus we say that $\sin x$ has a ***period*** of 2π (we will discuss periodic behavior in depth in Section 10.6). Note that for these graphs, we do *not* assume that the function "keeps going" in the directions indicated at the ends of the graphing window; instead, we assume that the graph repeats it shape over and over again.

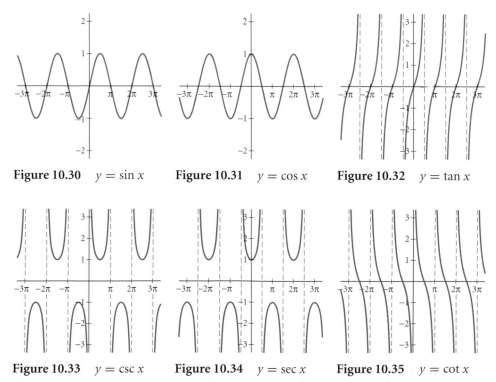

Figure 10.30 $y = \sin x$ **Figure 10.31** $y = \cos x$ **Figure 10.32** $y = \tan x$

Figure 10.33 $y = \csc x$ **Figure 10.34** $y = \sec x$ **Figure 10.35** $y = \cot x$

We will investigate graphs of trigonometric functions in depth (and prove that they have the shapes that they do) in Section 10.6. For the rest of this section, we will focus on the graphical properties that can be detected with limits (for example, continuity and vertical asymptotes).

10.4.3 Continuity of trigonometric functions

In this section we will prove that all six trigonometric functions are continuous on their domains. Amazingly enough, the only new piece of information we need in order to do this is the two limits from Section 10.4.1.

Consider the graphs in Figures 10.30–10.35. All six of the trigonometric functions appear to be continuous everywhere they are defined (since there are no "jumps" in the graphs except at nondomain points). It also makes sense that sine and cosine would be continuous because the *x*- and *y*-coordinates of a point traveling around the unit circle cannot "jump over" any values. Our goal is to prove this fact using the two special trigonometric limits from Section 10.4.1. We'll start by looking at the sine and cosine functions near $x = 0$.

THEOREM 10.12

Continuity of Sine and Cosine at One Point

The functions $\sin x$ and $\cos x$ are continuous at $x = 0$.

PROOF (**THEOREM 10.12**)   To prove that the sine function is continuous at $x = 0$, we need to show that

$$\lim_{x \to 0} \sin x = \sin 0.$$

This follows directly from the first trigonometric limit in Theorem 10.11:

$$\lim_{x \to 0} \sin x = \lim_{x \to 0} \left(\frac{\sin x}{x} \cdot x \right) \qquad \text{(multiply by } \tfrac{x}{x}\text{)}$$

$$= \left(\lim_{x \to 0} \frac{\sin x}{x} \right) \left(\lim_{x \to 0} x \right) \qquad \text{(product rule for limits)}$$

$$= (1)(0) = 0 \qquad \text{(Theorem 10.11)}$$

$$= \sin 0. \qquad \text{(since } \sin 0 = 0\text{)}$$

Notice that when we used the product rule for limits, we *did* know that both of the limits involved existed.

To prove that cosine is continuous at $x = 0$, we must show that $\lim_{x \to 0} \cos x = \cos 0 = 1$. This is equivalent to showing that $\lim_{x \to 0}(\cos x - 1) = 0$. Proving this second limit uses the second limit in Theorem 10.11, and a calculation analogous to the one above (the details are left to the exercises). ∎

We can use the fact that the sine and cosine functions are continuous at $x = 0$ to prove that they are continuous everywhere. Think for a moment about how amazing that is; just by showing continuity at zero and knowing the identities and algebra of sine and cosine, we can determine continuity *everywhere*! Recall that we went through a similar process with exponential functions in Section 8.3: First we showed that e^x is continuous at $x = 0$, and then we used that fact to prove that e^x is continuous everywhere.

THEOREM 10.13

Continuity of Sine and Cosine

The functions $\sin x$ and $\cos x$ are continuous everywhere.

PROOF (**THEOREM 10.13**)   To show that $\sin x$ is continuous everywhere, we must show that for all values c, we have

$$\lim_{x \to c} \sin x = \sin c.$$

We'd like to be able to use the properties of the sine function that we know: the continuity of the sine function at zero and the trigonometric identities that involve the sine function. If we change variables by setting $h := x - c$, then $h \to 0$ as $x \to c$, and $x = c + h$, so the limit above is equivalent to the limit

$$\lim_{h \to 0} \sin(c + h) = \sin c.$$

This new limit involves the sine of a sum; let's apply the sum identity for sine and see where it gets us.

$$
\begin{aligned}
\lim_{h \to 0} & \sin(c + h) \\
&= \lim_{h \to 0}(\sin c \cos h + \sin h \cos c) && \text{(sum identity for sine)} \\
&= \left(\lim_{h \to 0} \sin c \cos h \right) + \left(\lim_{h \to 0} \sin h \cos c \right) && \text{(sum rule for limits)} \\
&= \sin c \left(\lim_{h \to 0} \cos h \right) + \cos c \left(\lim_{h \to 0} \sin h \right) && \text{(constant multiple rule for limits)} \\
&= (\sin c)(\cos 0) + (\cos c)(\sin 0) && \text{(continuity at zero)} \\
&= (\sin c)(1) + (\cos c)(0) = \sin c. && \text{(since } \cos 0 = 1 \text{ and } \sin 0 = 0)
\end{aligned}
$$

All the limits involved in this calculation do exist, and therefore the sum and constant multiple rules for limits do apply here. In the third line we used the fact that $\sin c$ and $\cos c$ are constants with respect to the variable h that is in the limit. We now know that $\sin x$ is continuous for all values $x = c$.

The proof that cosine is continuous everywhere is entirely similar (use the sum identity for cosine) and is left for the exercises.

Now that we know that $\sin x$ and $\cos x$ are continuous functions, it is simple to prove that the remaining four trigonometric functions are continuous (on their domains).

THEOREM 10.14

Continuity of the Trigonometric Functions

$\sin x$, $\cos x$, $\tan x$, $\sec x$, $\csc x$, and $\cot x$ are all continuous on their domains.

PROOF (**THEOREM 10.14**)   We have already shown that sine and cosine are continuous. The other four trigonometric functions are reciprocals or quotients of the sine and cosine functions. Since the reciprocal of a continuous function is continuous (everywhere that it is defined), and since the quotient of two continuous functions is continuous (everywhere that it is defined), we know that $\tan x$, $\sec x$, $\csc x$, and $\cot x$ are continuous on their entire domains.

10.4.4 Vertical asymptotes and long-term behavior

As you can see from the graphs in Figures 10.30–10.35, four of the trigonometric functions have vertical asymptotes. For example, the graph of $y = \tan x$ has vertical asymptotes at the points where the function $y = \cos x$ is zero (why?), that is, at the points $x = \frac{\pi}{2} + \pi k$, where k is any positive or negative integer.

? *Question*   List all the vertical asymptotes of $y = \csc x$, $y = \sec x$, and $y = \cot x$. How are the vertical asymptotes of these functions related to the roots of the functions $y = \sin x$, $y = \cos x$, and $y = \tan x$, and why?   □

When you are calculating limits of trigonometric functions, it is often a good idea to write the functions in terms of sine and cosine. Don't forget that a limit that approaches $\frac{c}{0}$ for some nonzero constant c is always infinite, although you must examine the left and right limits separately.

EXAMPLE 10.15   **Using sine and cosine to investigate a trigonometric limit**

Investigate $\lim\limits_{x \to -\pi} \csc x$.

Solution   Since cosecant is the reciprocal of sine, and the sine of $-\pi$ is zero, we have

$$\lim\limits_{x \to -\pi} \csc x = \lim\limits_{x \to -\pi} \frac{1}{\sin x} \to \frac{1}{0}.$$

The function $y = \csc x$ will have a vertical asymptote at $x = -\pi$, but we don't yet know whether the function approaches $-\infty$ or ∞ from the left and right. We examine these limits separately:

$$\lim\limits_{x \to -\pi^{-}} \csc x = \lim\limits_{x \to -\pi^{-}} \frac{1}{\sin x} \to \frac{1}{0^{+}} \to \infty,$$

$$\lim\limits_{x \to -\pi^{+}} \csc x = \lim\limits_{x \to -\pi^{+}} \frac{1}{\sin x} \to \frac{1}{0^{-}} \to -\infty.$$

We used the unit circle definition of sine in these calculations, as follows. For an angle x slightly less than $-\pi$ (for example, $x = -\pi - 0.1$), the point on the unit circle corresponding to the angle x is in the second quadrant (why?), where the sine function is positive. On the other hand, an angle x slightly more than $-\pi$ (for example, $x = -\pi + 0.1$) terminates in the third quadrant and thus has a negative sine.   □

Since the six trigonometric functions are periodic, their graphical behavior repeats over and over again as $x \to \infty$ and as $x \to -\infty$. For example, the function $y = \sin x$ oscillates between 1 and -1 over and over again as $x \to \infty$ (see Figure 10.30). In general, limits of trigonometric expressions as $x \to \infty$ can be complicated.

EXAMPLE 10.16   **A limit that does not exist but is not ∞ or $-\infty$**

Investigate $\lim\limits_{x \to \infty} x \sin x$.

Solution   As $x \to \infty$, $\sin x$ oscillates between -1 and 1. Your first thought might be that since x is approaching ∞, the function $x \sin x$ will also approach infinity. However, since $\sin x$ oscillates between negative and positive numbers, so will the function $x \sin x$. Thus the limit as $x \to \infty$ of $x \sin x$ does not exist. The right side of the graph of $y = x \sin x$ (with a wide graphing window) is shown in Figure 10.36.

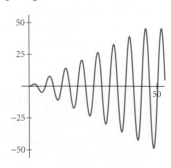

Figure 10.36 $y = x \sin x$ □

EXAMPLE 10.17

A limit that approaches "bounded over infinity"

Investigate $\lim\limits_{x\to\infty} \dfrac{\sin x}{x}$.

Solution   As $x \to \infty$, $\sin x$ oscillates between -1 and 1. Therefore, the numerator $\sin x$ stays bounded while the denominator x becomes infinitely large. A bounded quantity divided by an infinite quantity that increases without bound is zero; in other words,

$$(1) \qquad\qquad \frac{\text{bounded}}{\infty} \to 0.$$

Thus $\lim\limits_{x\to\infty} \dfrac{\sin x}{x} = 0$. The right side of the graph of $y = \frac{\sin x}{x}$ is shown in Figure 10.37.

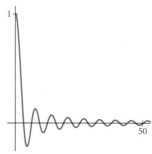

Figure 10.37 $y = \dfrac{\sin x}{x}$

Statement (1) can also be phrased in the following way:

$$(2) \qquad\qquad (\text{bounded}) \cdot (0) \to 0.$$

In other words, if a limit approaches the product of a bounded quantity and a quantity that is approaching zero, then the entire limit also approaches zero.   □

10.4.5 Calculating limits of trigonometric expressions

We can use the trigonometric limits presented at the beginning of Section 10.4.1 to calculate a number of trigonometric limits that are initially in an indeterminate form. Remember that since we don't know how to differentiate trigonometric functions at this point, L'Hôpital's Rule is not an option yet; limits of the form $\frac{0}{0}$ must be solved using algebraic simplification.

⊗ *Caution*   You may already know the derivatives of the sine and cosine functions. However, at this point in our development of trigonometric functions we do not yet know these derivatives. You may *not* use the derivatives of the trigonometric functions (and thus you cannot use L'Hôpital's Rule) until Section 10.5.   □

EXAMPLE 10.18

Limits related to the special trigonometric limits from Theorem 10.11

Calculate the following limits.

  (a) $\lim\limits_{x\to 0} \dfrac{\sin 3x}{3x}$ 	 (b) $\lim\limits_{x\to 0} \dfrac{\sin 3x}{2x}$

Solution

  (a) If we change variables by defining $u = 3x$, then u approaches zero as x approaches 0 (why?), and $x = \frac{u}{3}$. We have

$$\lim_{x\to 0} \frac{\sin 3x}{3x} = \lim_{u\to 0} \frac{\sin 3\left(\frac{u}{3}\right)}{3\left(\frac{u}{3}\right)} \qquad \text{(change variables to } u := 3x)$$

$$= \lim_{u\to 0} \frac{\sin u}{u} = 1. \qquad \text{(by Theorem 10.11)}$$

The last step of our calculation uses Theorem 10.11 with the variable x replaced by u. In general, the quotient $\frac{\sin u}{u}$ approaches 1 as u approaches 0, where u can be any expression that approaches 0; in this example, u is the expression $3x$. Most of the time we do not explicitly change variables when solving limits of this type; we will simply recognize that $\frac{\sin 3x}{3x}$ approaches 0 as x approaches 0 (since then $3x$ also approaches 0).

(b) If we can write this limit in the form $\frac{\sin u}{u}$, then we will be able to evaluate the limit. We need the expression inside the sine function to match the expression in the denominator. We can rewrite the $2x$ in the denominator as $2x = \frac{2}{3}(3x)$. We then have

$$\lim_{x \to 0} \frac{\sin 3x}{2x} = \lim_{x \to 0} \frac{\sin 3x}{\frac{2}{3}(3x)} \qquad \text{(rewrite } 2x \text{ in terms of } 3x\text{)}$$

$$= \frac{3}{2} \lim_{x \to 0} \frac{\sin 3x}{3x} \qquad \text{(algebra, limit rules)}$$

$$= \frac{3}{2}(1) = \frac{3}{2}. \qquad \text{(Theorem 10.11, since } 3x \to 0 \text{ as } x \to 0\text{)} \qquad \square$$

We can also use the techniques used in Example 10.18 to evaluate limits such as

$$\lim_{x \to 1} \frac{\sin(x - 1)}{x - 1}, \qquad \lim_{x \to 0} \frac{1 - \cos 4x}{4x}, \qquad \text{and} \qquad \lim_{x \to -2} \frac{1 - \cos(x + 2)}{x + 2}.$$

Each of these limits is equivalent to one of the limits in Theorem 10.11; the first limit is equal to 1, and the last two limits are equal to 0.

? Question   Why doesn't Theorem 10.11 apply to the limit $\lim_{x \to 0} \frac{\sin(x-1)}{x-1}$? What is the value of this limit? What about the limit $\lim_{x \to 1} \frac{\sin x}{x}$? $\qquad \square$

The two special trigonometric limits in Theorem 10.11 can also be used to solve more complicated limits, as shown in the next example. The key is to rewrite the limits in terms of the limits in Theorem 10.11.

EXAMPLE 10.19   **Writing limits in terms of special trigonometric limits**

Calculate the following limits.

(a) $\displaystyle\lim_{x \to 3} \frac{1 - \cos(x - 3)}{x^2 - x - 6}$

(b) $\displaystyle\lim_{x \to 0} \frac{2x}{\tan^2 x}$

Solution

(a) First of all, notice that this limit has the indeterminate form $\frac{0}{0}$ (try evaluating at $x = 3$ to see why). We can rewrite this limit so that it involves a limit of the form $\frac{1 - \cos u}{u}$:

$$\lim_{x \to 3} \frac{1 - \cos(x - 3)}{x^2 - x - 6} = \lim_{x \to 3} \frac{1 - \cos(x - 3)}{(x - 3)(x + 2)} \qquad \text{(factor denominator)}$$

$$= \lim_{x \to 3} \left(\frac{1 - \cos(x - 3)}{x - 3} \cdot \frac{1}{x + 2} \right) \qquad \text{(algebra)}$$

$$= \left(\lim_{x \to 3} \frac{1 - \cos(x - 3)}{x - 3} \right) \left(\lim_{x \to 3} \frac{1}{x + 2} \right) \qquad \text{(limit rule)}$$

$$= (0)\left(\tfrac{1}{5}\right) = 0. \qquad \text{(Theorem 10.11)}$$

(b) Since $\tan^2 0 = 0$, this limit is of the indeterminate form $\frac{0}{0}$. Since we don't know the derivative of tangent, we can't use L'Hôpital's Rule; thus we must try doing some algebra to evaluate this limit. We will write the limit in terms of sines and cosines and then apply Theorem 10.11:

$$\lim_{x \to 0} \frac{2x}{\tan^2 x} = \lim_{x \to 0} \frac{2x \cos^2 x}{\sin^2 x} \qquad \text{(since } \tan x = \tfrac{\sin x}{\cos x}\text{)}$$

$$= \lim_{x \to 0} \left(\frac{x}{\sin x} \cdot \frac{2 \cos^2 x}{\sin x} \right) \qquad \text{(algebra)}$$

$$= \left(\lim_{x \to 0} \frac{x}{\sin x} \right) \left(\lim_{x \to 0} \frac{2 \cos^2 x}{\sin x} \right) \qquad \text{(limit rule)}$$

$$\to (1)\left(\tfrac{2(1)^2}{0} \right) \to \tfrac{2}{0}. \qquad \text{(Theorem 10.11)}$$

Note that we used the fact that since $\frac{\sin x}{x}$ approaches 1 as $x \to 0$, its reciprocal $\frac{x}{\sin x}$ approaches $\frac{1}{1} = 1$ as $x \to 0$. We aren't finished with this limit yet; we know that the limit is infinite, but we need to examine both sides of the limit to see whether they approach ∞ or $-\infty$.

$$\lim_{x \to 0^+} \frac{2 \cos^2 x}{\sin x} \to \tfrac{2}{0^+} \to \infty, \qquad \text{(since } \sin x \text{ is positive for } x > 0\text{)}$$

$$\lim_{x \to 0^-} \frac{2 \cos^2 x}{\sin x} \to \tfrac{2}{0^-} \to -\infty. \qquad \text{(since } \sin x \text{ is negative for } x < 0\text{)}$$

Therefore, the original limit does not exist as $x \to 0$ (since it approaches ∞ from the right but $-\infty$ from the left). $\qquad \square$

✓ *Checking the Answer* To check the answers to Example 10.19, we can use graphs or approximations. For example, to check that $\lim\limits_{x \to 0^+} \frac{2x}{\tan^2 x}$ is ∞, look at the graph of $f(x) = \frac{2x}{\tan^2 x}$ near $x = 0$ and check that the height of the graph approaches ∞ as x approaches 0 from the right. Or evaluate $f(x) = \frac{2x}{\tan^2 x}$ at various small positive values of x (say, at $x = 0.1$, $x = 0.01$, and $x = 0.001$), and check that the values of $f(x)$ seem to grow without bound as $x \to 0^+$. $\qquad \square$

❓ *Question* What does the quotient $\frac{x}{1 - \cos x}$ approach as $x \to 0$, and why? $\qquad \square$

❓ *Question* Why don't you need Theorem 10.11 to determine that $\lim\limits_{x \to 0} \frac{\sin x}{x + 1}$ is equal to zero? $\qquad \square$

What you should know after reading Section 10.4

▶ The two important trigonometric limits from Theorem 10.11, and how to use them to calculate limits of other trigonometric expressions. Be able to check these limits graphically or by numerical approximation. How to investigate limits of trigonometric expressions, including limits near vertical asymptotes and limits as x approaches $\pm\infty$.

▶ The graphs of the six trigonometric functions and their basic features.

▶ The fact that the sine and cosine functions are continuous everywhere, and how to prove it. Why does this imply that the remaining four trigonometric functions are continuous on their domains?

Exercises 10.4

Concepts

0. Read the section and make your own summary of the material. Include whatever you think will help you review the material later on.

1. Use a diagram of the unit circle to explain why $\sin \theta \approx \theta$ when θ is a small positive or negative angle.

2. Use a table of values to estimate $\lim\limits_{\theta \to 0} \frac{\sin 2\theta}{\theta}$. Then check your approximation by examining a graph of $f(\theta) = \frac{\sin 2\theta}{\theta}$ near $\theta = 0$.

3. Use a table of values to estimate $\lim\limits_{\theta \to 0} \frac{1-\cos 2\theta}{\theta}$. Then check your approximation by examining a graph of $f(\theta) = \frac{1-\cos 2\theta}{\theta}$ near $\theta = 0$.

4. What type of discontinuity does the function $f(x) = \frac{\sin x}{x}$ have at $x = 0$, and why?

5. Sketch a graph of $y = \sin x$ from memory. (*Hint:* Thinking about the unit circle may help you remember the features of the graph.) Using what you know about transformations of functions, use this graph to sketch the graphs of (**a**) $y = \sin x + 2$, (**b**) $y = \sin(x + 2)$, (**c**) $y = 2 \sin x$, and (**d**) $y = \sin 2x$.

6. Sketch a graph of $y = \tan x$ from memory. (*Hint:* Thinking about the unit circle may help you remember the features of the graph.) Using what you know about transformations of functions, use this graph to sketch the graphs of (**a**) $y = \tan x + 2$, (**b**) $y = \tan(x + 2)$, (**c**) $y = 2 \tan x$, and (**d**) $y = \tan 2x$.

7. Sketch a graph of $y = \sec x$ from memory. (*Hint:* Thinking about the unit circle may help you remember the features of the graph.) Using what you know about transformations of functions, use this graph to sketch the graphs of (**a**) $y = \sec x + 2$, (**b**) $y = \sec(x + 2)$, (**c**) $y = 2 \sec x$, and (**d**) $y = \sec 2x$.

8. List the periods of all of the six trigonometric functions.

9. Suppose you know that $\lim\limits_{x \to 0} \sin x = 0$; explain why this means that $f(x) = \sin x$ is continuous at $x = 0$.

10. Suppose you know that $\lim\limits_{x \to 0} (\cos x - 1) = 0$; explain why this means that $f(x) = \cos x$ is continuous at $x = 0$.

11. Explain why the limit $\lim\limits_{x \to c} \cos x$ is equivalent to the limit $\lim\limits_{h \to 0} \cos(c + h)$.

12. Suppose $x = c$ is a real number in the domain of $f(x) = \sec x$. How can we calculate $\lim\limits_{x \to c} \sec x$, and why?

13. Guess the value of the limit $\lim\limits_{x \to \infty} e^x \sin x$. Then check your answer by sketching a *good* graph. You should use both your calculator and your knowledge of the sine and exponential functions to sketch your graph.

14. Guess the value of the limit $\lim\limits_{x \to \infty} \frac{\sin x}{e^x}$. Then check your answer by sketching a *good* graph. You should use both your calculator and your knowledge of the sine and exponential functions to sketch your graph.

15. Guess the value of the limit $\lim\limits_{x \to \infty} \frac{e^x}{\sin x}$. Then check your answer by sketching a *good* graph. You should use both your calculator and your knowledge of the sine and exponential functions to sketch your graph.

■ Determine whether each of the following statements is true or false. If a statement is true, explain why. If a statement is false, explain why or provide a counterexample.

16. True or False: When θ is a small angle, the quantities $\sin \theta$ and θ are equal.

17. True or False: $y = \tan x$ is continuous at $x = 0$.

18. True or False: $y = \tan x$ is continuous at $x = \frac{\pi}{2}$.

19. True or False: The graph of $y = \csc x$ has vertical asymptotes at $x = k\pi$, for any integer k.

20. True or False: The graph of $y = \tan x$ intersects the x-axis at every integer multiple of π.

21. True or False: For any real number x, $\sin(x + 2\pi) = \sin x$.

Skills

■ Calculate the following limits (note that you will not be able to apply L'Hôpital's Rule, since we do not yet know how to differentiate trigonometric functions). Show all work, and then check your answers with graphs.

22. $\lim\limits_{x \to \pi} \dfrac{\cos x}{\sin x + 2}$

23. $\lim\limits_{x \to -\frac{\pi}{4}} \dfrac{1}{\sin x + \cos x}$

24. $\lim\limits_{x \to 0} \dfrac{1 + \sin x}{1 - \cos x}$

25. $\lim\limits_{x \to 0} \dfrac{x}{\sin x}$

26. $\lim\limits_{x \to 0} \dfrac{x}{1 - \cos x}$

27. $\lim\limits_{x \to 0} \dfrac{\sin 2x}{2x}$

28. $\lim\limits_{x \to 0} \dfrac{\sin 3x}{5x}$

29. $\lim\limits_{x \to 0} \dfrac{1 - \cos 2x}{7x}$

30. $\lim\limits_{x \to 0} \dfrac{3 \sin x + x}{x}$

31. $\lim\limits_{x \to 0} \dfrac{\sin^2 3x}{x^3 - x}$

32. $\lim\limits_{x \to 0} \dfrac{\sin(3x^2)}{x^3 - x}$

33. $\lim\limits_{x \to 0} \left(\dfrac{1}{2x} - \dfrac{2 \cos 5x}{4x} \right)$

34. $\lim\limits_{x \to 1^+} \dfrac{1 - \cos \sqrt{x - 1}}{\sqrt{x - 1}}$

35. $\lim\limits_{x \to 0^+} \dfrac{x^2 \csc 3x}{1 - \cos 2x}$

36. $\lim\limits_{x \to \frac{\pi}{2}} \dfrac{\sin\left(x - \frac{\pi}{2}\right)}{\frac{\pi}{2} - x}$

37. $\lim\limits_{x \to 0} \dfrac{x^2 \cot x}{\sin x}$

38. $\lim\limits_{x \to 0} \dfrac{\sec x \tan x}{x}$

39. $\lim\limits_{x \to 0} 3x^2 \cot^2 x$

40. $\lim\limits_{x \to 2} \dfrac{x^2 - 4}{\sin(x - 2)}$

41. $\lim\limits_{x \to 1} \dfrac{\sin^2(x - 1)}{x^2 - 1}$

42. $\lim\limits_{x \to 2} \dfrac{1 - \cos(x - 2)}{x^2 + x - 6}$

43. $\lim\limits_{x \to \pi} \tan x$

44. $\lim\limits_{x \to \frac{\pi}{2}} \tan x$

45. $\lim\limits_{x \to \pi} \csc^2 x$

46. $\lim\limits_{x \to \infty} \cos x$

47. $\lim\limits_{x \to \infty} x \cos x$

48. $\lim\limits_{x \to \infty} \dfrac{\cos x}{x}$

49. $\lim\limits_{x \to 0} \sin \frac{1}{x}$

50. $\lim\limits_{x \to 0} x \sin \frac{1}{x}$

51. $\lim\limits_{x \to 0} x^2 \sin \frac{1}{x}$

■ **Use limits and the definition of continuity to determine whether each of the following functions is continuous at $x = 0$. Then use a graph to support your answer.**

52. $f(x) = \begin{cases} \sin\left(\frac{1}{x}\right), & \text{if } x \neq 0 \\ 0, & \text{if } x = 0 \end{cases}$

53. $f(x) = \begin{cases} x \sin\left(\frac{1}{x}\right), & \text{if } x \neq 0 \\ 0, & \text{if } x = 0 \end{cases}$

54. $f(x) = \begin{cases} x^2 \sin\left(\frac{1}{x}\right), & \text{if } x \neq 0 \\ 0, & \text{if } x = 0 \end{cases}$

Applications

55. Consider a mass hanging from the ceiling at the end of a spring. If you pull down on the mass and let go, it will oscillate up and down according to the equation

$$s(t) = A \sin\left(\sqrt{\tfrac{k}{m}}\, t\right) + B \cos\left(\sqrt{\tfrac{k}{m}}\, t\right),$$

where $s(t)$ is the distance of the mass from its equilibrium position, m is the mass of the bob on the end of the spring, and k is a "spring coefficient" that measures how tight or stiff the spring is. The constants A and B depend on initial conditions—specifically, how far you pull down the mass (s_0) and the velocity at which you release the mass (v_0). This equation does *not* take into account any friction due to air resistance.

(a) Determine whether or not the limit of $s(t)$ as $t \to \infty$ exists. What does this say about the long-term behavior of the mass on the end of the spring?

(b) Explain how this limit is related to the fact that the equation above does *not* take friction due to air resistance into account.

(c) Suppose the bob at the end of the spring has a mass of 2 grams and that the coefficient for the spring is $k = 9$. Suppose also that the spring is released in such a way that $A = \sqrt{2}$ and $B = 2$. Use a calculator to graph the function $s(t)$ that describes

the distance of the mass from its equilibrium position. Use your graph to support your answer to part (a).

56. In Problem 55 we gave an equation describing spring motion without air resistance. If we take into account friction due to air resistance, the mass will oscillate up and down according to the equation

$$s(t) = e^{\frac{-f}{2m} t} \left(A \sin\left(\tfrac{\sqrt{4km - f^2}}{2m}\, t \right) + B \cos\left(\tfrac{\sqrt{4km - f^2}}{2m}\, t \right) \right),$$

where m, k, A, and B are the constants described in Problem 55, and f is a positive "friction coefficient" that measures the amount of friction due to air resistance.

(a) Find the limit of $s(t)$ as $t \to \infty$. What does this say about the long-term behavior of the mass on the end of the spring?

(b) Explain how this limit is related to the fact that the equation above *does* take friction due to air resistance into account.

(c) Suppose the bob at the end of the spring has a mass of 2 grams, the coefficient for the spring is $k = 9$, and the friction coefficient is $f = 6$. Suppose also that the spring is released in such a way that $A = 4$ and $B = 2$. Use a calculator to graph the function $s(t)$ that describes the distance of the mass from its equilibrium position. Use your graph to support your answer to part (a).

Proofs

57. Use the fact that $\lim\limits_{x \to 0} \frac{\sin x}{x} = 1$ to prove that $\lim\limits_{x \to 0} \sin x = \sin 0$. Explain why this means that $f(x) = \sin x$ is continuous at $x = 0$.

58. Use the fact that $\lim\limits_{x \to 0} \frac{1 - \cos x}{x} = 0$ to prove that $\lim\limits_{x \to 0} (\cos x - 1) = 0$. Explain why this means that $f(x) = \cos x$ is continuous at $x = 0$.

59. Use Theorem 10.12 and the sum identity for the sine function to prove that $\lim\limits_{h \to 0} \sin(c + h) = \sin c$ for all real numbers c. Explain why this means that $f(x) = \sin x$ is continuous everywhere.

60. Use Theorem 10.12 and the sum identity for the cosine function to prove that $\lim\limits_{h \to 0} \cos(c + h) = \cos c$ for all real numbers c. Explain why this means that $f(x) = \cos x$ is continuous everywhere.

61. Use the fact that the cosine function is continuous to prove that $f(x) = \sec x$ is continuous everywhere on its domain.

62. Use the fact that the sine and cosine functions are continuous to prove that $f(x) = \tan x$ is continuous everywhere on its domain.

10.5   Derivatives of Trigonometric Functions

In this section we develop formulas for the derivatives of the six trigonometric functions. At the end of the section we use these derivatives to do some limits with L'Hôpital's Rule.

10.5.1   Derivatives of sine and cosine

We do not yet have any differentiation rules that apply to the sine and cosine functions, since they aren't sums, products, quotients, or compositions of functions whose derivatives we already know. We will need to go back to the definition of derivative in order to determine their derivatives. As we will soon see, this produces a limit that can be simplified using the sum identities for sine and cosine and the limits from Theorem 10.11 in Section 10.4.1.

The graph of $f(x) = \sin x$ has "periodic" behavior; it repeats the same shape over and over again. Clearly, the derivative $f'(x)$ of this function will also have periodic behavior, since the slope of the graph of $f(x) = \sin x$ repeats its behavior over and over again. We might expect the derivative of the sine function to be another trigonometric function (since it is periodic). As the following theorem shows, this is in fact the case:

THEOREM 10.15

> **Derivatives of Sine and Cosine**
>
> **(a)** $\frac{d}{dx}(\sin x) = \cos x$ **(b)** $\frac{d}{dx}(\cos x) = -\sin x$

Before we prove this theorem, let's check that it is reasonable. Figures 10.38 and 10.39 show the graphs of $f(x) = \sin x$ and its derivative $f'(x) = \cos x$.

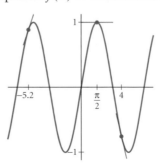

Figure 10.38
$f(x) = \sin x$ with three marked slopes

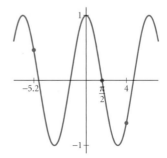

Figure 10.39
Corresponding heights of $f'(x) = \cos x$

Notice that at each value of x, the slope of the graph of $f(x) = \sin x$ is given by the height of the graph of $f'(x) = \cos x$. For example, at $x = -5.2$ the graph of $f(x) = \sin x$ has a positive slope equal to the the positive height on the graph of $f'(x) = \cos x$ at $x = -5.2$. At $x = \frac{\pi}{2}$ the graph of $f(x) = \sin x$ has a horizontal tangent line (thus a slope of 0), and the graph of $f'(x) = \cos x$ has a zero at $x = \frac{\pi}{2}$.

❓ Question   Describe the relationship between the slope of the graph of $f(x) = \sin x$ in Figure 10.38 at the point $x = 4$ and the height of the graph of its derivative $f'(x) = \cos x$ in Figure 10.39 at the point $x = 4$.   ☐

❓ Question   Sketch graphs of $f(x) = \cos x$ and $f(x) = -\sin x$, and use them to illustrate part (b) of Theorem 10.15.   ☐

We now prove Theorem 10.15. We will prove part (a) and leave part (b), which is similar, for the exercises. Note that the only pieces of information we need are the sum identities for sine and cosine and the important trigonometric limits from Section 10.4.1.

PROOF (**PART (A) OF THEOREM 10.15**)   The proof is nothing more than an annotated calculation using the definition of derivative:

$$\frac{d}{dx}(\sin x) = \lim_{h \to 0} \frac{\sin(x+h) - \sin x}{h} \qquad \text{(definition of derivative)}$$

At this point we would get the indeterminate form $\frac{0}{0}$ if we tried to evaluate this limit at $h = 0$; thus we must do some algebra. There is no "simplification" that can be done here, but we *can* rewrite $\sin(x+h)$ using the sum identity for sine:

$$= \lim_{h \to 0} \frac{(\sin x \cos h + \sin h \cos x) - \sin x}{h} \qquad \text{(sum identity for sine)}$$

At this point it looks like we have made the problem more complicated, but using algebra and limit rules, we can rewrite this limit in terms of the two trigonometric limits from Section 10.4.1:

$$= \lim_{h \to 0} \frac{\sin x \cos h - \sin x + \sin h \cos x}{h} \qquad \text{(reorder terms)}$$

$$= \lim_{h \to 0} \frac{\sin x(\cos h - 1) + \sin h \cos x}{h} \qquad \text{(algebra)}$$

$$= \lim_{h \to 0} \left(\sin x \frac{\cos h - 1}{h} + \cos x \frac{\sin h}{h} \right) \qquad \text{(algebra)}$$

$$= \sin x \left(\lim_{h \to 0} \frac{\cos h - 1}{h} \right) + \cos x \left(\lim_{h \to 0} \frac{\sin h}{h} \right) \qquad \text{(limit rules)}$$

$$= (\sin x)(0) + (\cos x)(1) = \cos x. \qquad \text{(trigonometric limits)}$$

We can pull $\sin x$ and $\cos x$ out of the limits above because they do not involve the variable h. In a sense, $\sin x$ and $\cos x$ are constants with respect to h.

❖ *Caution* The derivative of the sine function is the cosine function *only* if we are measuring in radians. In degree measure, the derivative of the sine function is a constant multiple of cosine; this is one reason why we always choose to work in radian measure. Graphically, this makes sense because the graph of $y = \sin x$ is more stretched out if x is measured in degrees (for example, it has zeros at multiples of $180°$ rather than at multiples of $\pi \approx 3.14$ radians). This means that the "bumps" of the degree sine function will be shallower, which affects the slopes of the tangent lines to the graph. □

10.5.2 Derivatives of the trigonometric functions

Since the tangent, cotangent, secant, and cosecant functions can be written in terms of the sine and cosine functions, we can use the derivatives of sine and cosine to find the derivatives of the other four trigonometric functions.

THEOREM 10.16

Derivatives of the Trigonometric Functions

$$\frac{d}{dx}(\sin x) = \cos x \qquad \frac{d}{dx}(\tan x) = \sec^2 x \qquad \frac{d}{dx}(\sec x) = \sec x \tan x$$

$$\frac{d}{dx}(\cos x) = -\sin x \qquad \frac{d}{dx}(\cot x) = -\csc^2 x \qquad \frac{d}{dx}(\csc x) = -\csc x \cot x$$

The following observation might make it easier for you to remember these derivatives. Consider the six trigonometric functions as three pairs of "partners," with $\sin x$ and $\cos x$ as partners, $\tan x$ and $\cot x$ as partners, and $\sec x$ and $\csc x$ as partners. Given the derivative of any of the six trigonometric functions, the derivative of its "partner" can be obtained by switching sign and changing all the trigonometric functions in the derivative to their "partners." For example, the derivative of $\sec x$ is $\sec x \tan x$. The "partner" of $\sec x$ is $\csc x$, and the derivative of $\csc x$ is the negative of the product of the "partner" of $\sec x$ and the "partner" of $\tan x$—namely, $\csc x \cot x$. If you use this memory device, then all you have to remember is the top row of derivatives.

❖ *Caution*   When writing down the derivatives from Theorem 10.16, be careful *not* to write $\sin x = \cos x$ when you really mean that $\frac{d}{dx}(\sin x) = \cos x$.   □

Since the tangent, secant, cotangent, and cosecant functions are reciprocals and quotients of $\sin x$ and $\cos x$, the proof of Theorem 10.16 is simply an application of differentiation rules that we already know. In particular, notice that we do *not* need to use the definition of derivative in any of the calculations.

PROOF (**THEOREM 10.16**)   We already know that $\frac{d}{dx}(\sin x) = \cos x$ and that $\frac{d}{dx}(\cos x) = -\sin x$. Since $\tan x$ can be written as the quotient $\frac{\sin x}{\cos x}$, the quotient rule tells us that

$$
\begin{aligned}
\frac{d}{dx}(\tan x) &= \frac{d}{dx}\left(\frac{\sin x}{\cos x}\right) && \text{(trigonometric identity)} \\[2mm]
&= \frac{\frac{d}{dx}(\sin x) \cdot (\cos x) - (\sin x) \cdot \frac{d}{dx}(\cos x)}{(\cos x)^2} && \text{(quotient rule)} \\[2mm]
&= \frac{(\cos x)(\cos x) - (\sin x)(-\sin x)}{\cos^2 x} && \text{(derivatives of sine and cosine)} \\[2mm]
&= \frac{\cos^2 x + \sin^2 x}{\cos^2 x} && \text{(algebra)} \\[2mm]
&= \frac{1}{\cos^2 x} = \sec^2 x. && \text{(trigonometric identities)}
\end{aligned}
$$

We actually finished differentiating in the third line of the calculation. To write the derivative in the simplest form possible, we simplified using some algebra and trigonometric identities. (After all, wouldn't you rather memorize that the derivative of tangent is $\sec^2 x$ than have to memorize the messy third line in the calculation?)

To prove that $\frac{d}{dx}(\csc x) = -\csc x \cot x$, we use a similar strategy and the fact that $\csc x$ is the reciprocal of $\sin x$:

$$
\begin{aligned}
\frac{d}{dx}(\csc x) &= \frac{d}{dx}\left(\frac{1}{\sin x}\right) && \text{(trigonometric identity)} \\[2mm]
&= \frac{d}{dx}((\sin x)^{-1}) && \text{(rewrite with algebra)} \\[2mm]
&= -(\sin x)^{-2} \cdot \frac{d}{dx}(\sin x) && \text{(chain and power rules)} \\[2mm]
&= -\csc^2 x (\cos x) && \text{(derivative of sine)}
\end{aligned}
$$

At this point we have finished differentiating; it remains only to use algebra to write the expression as $-\csc x \cot x$. Continuing our calculation, we have

$$= -\csc x \csc x \cos x \qquad\qquad \text{(algebra)}$$

$$= -\csc x \frac{1}{\sin x} \cos x \qquad\qquad \text{(trigonometric identity)}$$

$$= -\csc x \frac{\cos x}{\sin x} \qquad\qquad \text{(algebra)}$$

$$= -\csc x \cot x. \qquad\qquad \text{(trigonometric identity)}$$

The proofs for the derivatives of $\cot x$ and $\sec x$ are similar to the two calculations above, respectively, and are left for the exercises. ∎

10.5.3 Derivative calculations

We now know how to differentiate an extremely large variety of functions. Specifically, we can differentiate any combination (sum, product, quotient, or composition) of algebraic functions, exponential functions, logarithmic functions, and trigonometric functions. When differentiating complicated functions, remember to begin by identifying which differentiation rule should be used *first* (by identifying whether the function you are differentiating is a sum, product, quotient, or composition at its outermost level).

EXAMPLE 10.20

Differentiating functions that involve trigonometric expressions

Find the derivatives of the following functions.

(a) $f(x) = x \sin(x^3)$

(b) $g(x) = \sec^2(e^x)$

Solution

(a) Although you can do this calculation in one line, we will split it up into steps so it will be clear what we are doing. The function $f(x) = x \sin(x^3)$ is a product of two functions, x and $\sin(x^3)$, so we begin by using the product rule:

$$\frac{d}{dx}(x \sin(x^3)) = \frac{d}{dx}(x) \cdot \sin(x^3) + x \cdot \frac{d}{dx}(\sin(x^3)) \qquad \textbf{(product rule)}$$

$$= 1 \cdot \sin(x^3) + x \cdot \cos(x^3) \frac{d}{dx}(x^3) \qquad \textbf{(chain rule)}$$

$$= \sin(x^3) + x \cos(x^3)(3x^2) \qquad \textbf{(power rule)}$$

Because $\sin(x^3)$ is a composition of x^3 (on the "inside") and $\sin x$ (on the "outside"), we needed to use the chain rule. The first part of the derivative of $\sin(x^3)$ is the derivative of the "outside" (i.e., the derivative of the sine function, which is the cosine function) with the "inside" (x^3) plugged in: $\cos(x^3)$. Then we need to multiply by the derivative of the "inside" ($\frac{d}{dx}(x^3) = 3x^2$) to complete the use of the chain rule.

(b) The function $g(x) = \sec^2(e^x)$ is, at its outermost level, a composition of three functions, since we can rewrite $g(x)$ as

$$g(x) = \sec^2(e^x) = (\sec(e^x))^2.$$

We need to use the chain rule *twice*:

$$\frac{d}{dx}(\sec^2(e^x)) = \frac{d}{dx}((\sec(e^x))^2) \qquad \text{(rewrite so notation is clear)}$$

$$= 2(\sec(e^x))^1 \frac{d}{dx}(\sec(e^x)) \qquad \text{(first application of chain rule)}$$

$$= 2\sec(e^x)\sec(e^x)\tan(e^x)\frac{d}{dx}(e^x) \qquad \text{(second application of chain rule)}$$

$$= 2\sec(e^x)\sec(e^x)\tan(e^x)e^x \qquad \text{(derivative of e^x is e^x)}$$

Perhaps the trickiest part of this calculation is that the derivative of $\sec x$ has *two* instances of the independent variable: $\frac{d}{dx}(\sec x) = \sec\underline{x}\tan\underline{x}$. This means that in the calculation above, we needed to put the "inside" function e^x into *both* of these variable slots. In general, the derivative of $\sec(u(x))$, for any function $u(x)$, is

$$\frac{d}{dx}(\sec(u(x))) = \sec(u(x))\tan(u(x))u'(x).$$

In this calculation, $u(x)$ is the function e^x (so $u'(x)$ also happens to be equal to e^x). $\qquad\square$

10.5.4 L'Hôpital's Rule problems

Now that we know how to differentiate trigonometric functions, we can use L'Hôpital's Rule to evaluate certain trigonometric limits. Recall that L'Hôpital's Rule applies only if the limit is written in the form $\frac{0}{0}$ or $\frac{\infty}{\infty}$ and that even when L'Hôpital's Rule *does* apply, it is not always useful.

EXAMPLE 10.21   **Using L'Hôpital's Rule to calculate limits involving trigonometric expressions**

Find the following limits.

(a) $\displaystyle\lim_{x\to 0}\frac{x\cos x}{\sin x}$ (b) $\displaystyle\lim_{x\to 0^+}\frac{\csc x}{\ln x}$ (c) $\displaystyle\lim_{x\to\frac{\pi}{2}}(\tan x - \sec x)$

Solution

(a) This limit is of the form $\frac{0}{0}$, since as $x\to 0$ both $x\cos x$ and $\sin x$ approach 0. Therefore, L'Hôpital's Rule applies:

$$\lim_{x\to 0}\frac{x\cos x}{\sin x} \overset{\text{L'H}}{=} \lim_{x\to 0}\frac{(1)\cos x + x(-\sin x)}{\cos x} \qquad \text{(L'Hôpital's Rule and the product rule)}$$

$$= \lim_{x\to 0}\frac{\cos x - x\sin x}{\cos x} \qquad \text{(simplify)}$$

$$= \frac{\cos 0 - 0(\sin 0)}{\cos 0} = \frac{1 - 0(0)}{1} = 1. \qquad \text{(evaluate limit)}$$

(Don't forget that L'Hôpital's Rule requires that we take the derivative of the numerator and the derivative of the denominator *separately*. We do *not* use the quotient rule when applying L'Hôpital's Rule!) We could have solved this limit another way, using the fact that $\displaystyle\lim_{x\to 0}\frac{\sin x}{x} = 1$ (see Section 10.4.5). Make sure you can solve this limit *both* ways!

(b) As $x \to 0^+$, csc x approaches ∞ and ln x approaches $-\infty$ (why?); therefore, this is a limit of the form $\frac{\infty}{\infty}$, and L'Hôpital's Rule applies:

$$\lim_{x \to 0^+} \frac{\csc x}{\ln x} \overset{\text{L'H}}{=} \lim_{x \to 0^+} \frac{-\csc x \cot x}{\frac{1}{x}} \qquad \text{(L'Hôpital's Rule)}$$

$$= \lim_{x \to 0^+} (-x \csc x \cot x) \qquad \text{(note indeterminate form } 0(\infty)(\infty))$$

This limit is still in an indeterminate form, so we will try simplifying the function. It is often helpful to convert a trigonometric expression into sines and cosines, so we try that:

$$= \lim_{x \to 0^+} (-x) \left(\frac{1}{\sin x} \right) \left(\frac{\cos x}{\sin x} \right) \qquad \text{(trigonometric identities)}$$

$$= \lim_{x \to 0^+} \frac{-x \cos x}{\sin^2 x} \qquad \text{(algebra)}$$

Now we have a limit of the form $\frac{0}{0}$, so we can again apply L'Hôpital's Rule:

$$\overset{\text{L'H}}{=} \lim_{x \to 0^+} \frac{(-1) \cos x + (-x)(-\sin x)}{2 \sin x \cos x} \qquad \text{(note use of product and chain rules)}$$

$$= \lim_{x \to 0^+} \frac{-\cos x + x \sin x}{2 \sin x \cos x} \qquad \text{(simplify)}$$

$$\to \frac{-\cos 0 + 0(\sin 0)}{2 \sin 0 \cos 0} \to \frac{-1}{0^+} \to -\infty. \qquad \text{(evaluate limit)}$$

We know that the denominator approaches 0 from the right because $\sin x$ approaches 0 from the right as $x \to 0^+$ (since $\sin x$ is positive for small positive numbers). After all that work, it would be a very good idea to check the answer; see Figure 10.41.

(c) At first glance this does not look like a L'Hôpital's Rule problem, since the function $\tan x - \sec x$ is not in the form of a quotient. However, this is a limit of the indeterminate form $\infty - \infty$ (why?), so we can't solve it simply by evaluating the function at $x = \frac{\pi}{2}$. After using a bit of algebra, we will see that L'Hôpital's Rule does apply here:

$$\lim_{x \to \frac{\pi}{2}}(\tan x - \sec x) = \lim_{x \to \frac{\pi}{2}} \left(\frac{\sin x}{\cos x} - \frac{1}{\cos x} \right) \qquad \text{(trigonometric identities)}$$

$$= \lim_{x \to \frac{\pi}{2}} \frac{\sin x - 1}{\cos x} \qquad \text{(this is now in the form } \frac{0}{0})$$

$$\overset{\text{L'H}}{=} \lim_{x \to \frac{\pi}{2}} \frac{\cos x}{-\sin x} \qquad \text{(L'Hôpital's Rule)}$$

$$= \frac{\cos \frac{\pi}{2}}{-\sin \frac{\pi}{2}} = \frac{0}{-1} = 0. \qquad \text{(evaluate limit)}$$

Note that we had to calculate values of trigonometric functions (here, $\cos \frac{\pi}{2}$ and $\sin \frac{\pi}{2}$) in order to evaluate this limit. Be sure that you can do this *without* your calculator! □

✓ *Checking the Answer* You can use graphs to check these answers. Figure 10.40 shows the graph of $f(x) = \frac{x \cos x}{\sin x}$. This function does approach a height of 1 as $x \to 0$ (although it is not defined at $x = 0$). Figures 10.41 and 10.42 show the graphs of the functions from parts (b) and (c) of Example 10.21; use these graphs to verify the limits we found.

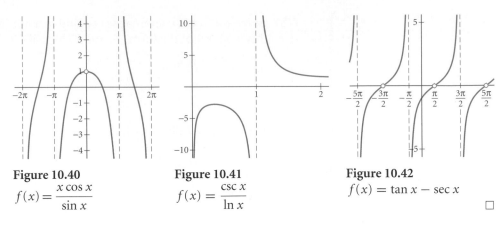

Figure 10.40

$$f(x) = \frac{x \cos x}{\sin x}$$

Figure 10.41

$$f(x) = \frac{\csc x}{\ln x}$$

Figure 10.42

$$f(x) = \tan x - \sec x$$

As a final note, let's return to the two important trigonometric limits from Theorem 10.11 in Section 10.4.1. For example, consider the limit $\lim\limits_{x \to 0} \frac{\sin x}{x}$. Now that we know the derivative of $\sin x$, we can use L'Hôpital's Rule to evaluate this limit (try it). Why couldn't we have proved this limit using L'Hôpital's Rule? Think about that for a moment and then read the next paragraph.

The reason why we can't use L'Hôpital's Rule to prove the limits from Theorem 10.11 is that we needed those limits to prove the differentiation rules for sine and cosine. If we had tried to prove those two limits using L'Hôpital's Rule, we would have ended up in a logical circle! Of course, now that we have established these limits, as well as the derivatives of sine and cosine, we can use whatever method we choose to evaluate $\lim\limits_{x \to 0} \frac{\sin x}{x}$.

What you should know after reading Section 10.5

▶ The derivatives of the six trigonometric functions, and how to prove them. Note that proving the rules for differentiating sine and cosine involves the definition of derivative and the two important trigonometric limits from the previous section.

▶ How to differentiate complicated functions that involve trigonometric expressions.

▶ How to use L'Hôpital's Rule to evaluate limits of trigonometric expressions. Be able to use graphs to check your answers to such problems.

Exercises 10.5

Concepts

0. Read the section and make your own summary of the material. Include whatever you think will help you review the material later on.

1. State the derivatives of the six trigonometric functions.

2. What does the height of the graph of $y = \cos x$ at $x = \frac{\pi}{2}$ tell you about the graph of $y = \sin x$ at $x = \frac{\pi}{2}$? What does the height of the graph of $y = \sin x$ at $x = \frac{\pi}{2}$ tell you about the graph of $y = \cos x$ at $x = \frac{\pi}{2}$?

3. Sketch graphs of the sine and cosine functions, and use them to illustrate the fact that the derivative of $\cos x$ is $-\sin x$.

4. Find *all* of the errors in the following calculation, and then calculate the limit correctly.

$$\lim_{x \to \frac{\pi}{2}} \frac{\cos x}{1 - \sin 2x} \overset{\text{L'H}}{=} \lim_{x \to \frac{\pi}{2}} \frac{-\sin x}{-\cos 2x}$$

$$= \frac{-\sin \frac{\pi}{2}}{-\cos \pi} = 1$$

5. Find *all* of the errors in the following calculation, and then calculate the limit correctly.

$$\lim_{x\to 0} \frac{\sin x}{x^2 - x} \overset{L'H}{=} \lim_{x\to 0} \frac{\cos x}{2x - 1} \overset{L'H}{=} \lim_{x\to 0} \frac{\sin x}{2} = \frac{\sin 0}{2} = 0$$

6. We mentioned in the text that $\frac{d}{dx}(\sin x) = \cos x$ *only* if the independent variable x is measured in radians. In this problem you will explore why this derivative relationship does not hold if x is measured in degrees. Set your calculator to degree mode (don't forget to change it back to radian mode after this problem!) and find a graph of $\sin x$ that shows at least two periods of the graph. If the derivative of sine were cosine (in degrees), then the slope of your graph at $x = 0$ would be equal to $\cos 0 = 1$ (why?). Use your graph to explain why this is not the case. (*Hint:* Think about your graphing window.)

Skills

■ Differentiate each of the following functions. For some problems, you may find it useful to simplify first and/or to do the differentiation in more than one step.

7. $f(x) = \dfrac{x^2 + 1}{\cos x}$

8. $f(x) = 2\tan(x^3)$

9. $f(x) = 3x - 2\csc x$

10. $f(x) = e^x \sin x$

11. $f(x) = x\sec(x^3)$

12. $f(x) = x^3 \sec x$

13. $f(x) = \sin x \ln x$

14. $f(x) = \sin(\ln x)$

15. $f(x) = \ln(\sin x)$

16. $f(x) = \dfrac{\sin^2 x}{\cos x}$

17. $f(x) = x\sqrt{\sin x}$

18. $f(x) = \dfrac{x\sin x + 2}{\cot x}$

19. $f(x) = \sin^2 x + \cos^2 x$

20. $f(x) = \ln(x\sin x)$

21. $f(x) = \sin(\cos(\sec(x)))$

22. $f(x) = \csc^2(e^x)$

23. $f(x) = e^{\csc^2 x}$

24. $f(x) = e^x \csc^2 x$

25. $f(x) = x\sqrt{\sin x \cos x}$

26. $f(x) = \dfrac{\sin x \csc x}{\cot x \cos x}$

27. $f(x) = \dfrac{x^{\frac{2}{3}} \sin\left(x^{\frac{1}{3}}\right)}{\csc(x^2)}$

28. $f(x) = x^{\sin x}$

29. $f(x) = (\sec x)^x$

30. $f(x) = (\sin x)^{\cos x}$

■ Find each of the following limits in two ways: (a) without using L'Hôpital's Rule, and (b) by using L'Hôpital's Rule.

31. $\displaystyle\lim_{x\to 0} \frac{\sin(x^2)}{x}$

32. $\displaystyle\lim_{x\to 0} \frac{\sin^2 x}{x}$

33. $\displaystyle\lim_{x\to 0} \frac{\sin^2 3x}{2x^2}$

34. $\displaystyle\lim_{x\to 0} \frac{\sin x}{x\cos x}$

35. $\displaystyle\lim_{x\to 0} \frac{1 - \cos 3x}{x}$

36. $\displaystyle\lim_{x\to 0} x\cot x$

37. $\displaystyle\lim_{x\to 0} \frac{x\sin x}{\sin x^2}$

38. $\displaystyle\lim_{x\to 0} x^2 \csc 3x$

39. $\displaystyle\lim_{x\to 2} \frac{x^2 - x - 2}{\sin(x - 2)}$

■ Find each of the following limits. L'Hôpital's Rule will be useful for some, but not all, of these limits. Check your answers using graphs.

40. $\displaystyle\lim_{x\to 0} \frac{\sin x}{e^x}$

41. $\displaystyle\lim_{x\to 0} \frac{e^x - \cos x}{\sin x}$

42. $\displaystyle\lim_{x\to 0} \frac{x\cos x}{1 - e^x}$

43. $\displaystyle\lim_{x\to 0} \frac{x\sin x}{1 - e^x}$

44. $\displaystyle\lim_{x\to 0} \frac{e^x - 1}{\cos x}$

45. $\displaystyle\lim_{x\to \frac{\pi}{2}} (\sec x - 2\tan x)$

46. $\displaystyle\lim_{x\to 0^+} x\ln(\sin x)$

47. $\displaystyle\lim_{x\to 1} \frac{\sin(\ln x)}{x - 1}$

48. $\displaystyle\lim_{x\to \frac{\pi}{2}} \frac{\sin(\cos x)}{\cos x}$

49. $\displaystyle\lim_{x\to 0} \frac{\sin x \cos x}{\sec x - 1}$

50. $\displaystyle\lim_{x\to 0} \frac{\tan^2 x}{\cos x - 1}$

51. $\displaystyle\lim_{x\to \frac{\pi}{2}} \frac{x - \frac{\pi}{2}}{\csc 2x}$

52. $\displaystyle\lim_{x\to 0^+} x^{\sin x}$

53. $\displaystyle\lim_{x\to 0^+} (\sin x)^x$

54. $\displaystyle\lim_{x\to 0^+} (\cos x)^{\frac{1}{x}}$

■ For each problem below, suppose $f(x)$ is a function with the given derivative $f'(x)$ and value $f(c)$. Find an equation for $f(x)$.

55. $f'(x) = \sin x, f(0) = 0$

56. $f'(x) = \sec^2 x, f\left(\frac{\pi}{2}\right) = 2$

57. $f'(x) = \csc x \cot x, f\left(\frac{\pi}{2}\right) = 1$

58. $f'(x) = 2\cos(3x), f(0) = 4$

59. $f'(x) = 2x\cos(x^2), f(0) = 1$

60. $f'(x) = \sec 2x \tan 2x, f(\pi) = 0$

■ Determine whether each of the following functions is differentiable at $x = 0$. Then use a graph to support your answer.

61. $f(x) = \begin{cases} \sin\left(\frac{1}{x}\right), & \text{if } x \neq 0 \\ 0, & \text{if } x = 0 \end{cases}$

62. $f(x) = \begin{cases} x\sin\left(\frac{1}{x}\right), & \text{if } x \neq 0 \\ 0, & \text{if } x = 0 \end{cases}$

63. $f(x) = \begin{cases} x^3 \sin\left(\frac{1}{x}\right), & \text{if } x \neq 0 \\ 0, & \text{if } x = 0 \end{cases}$

Applications

64. In Problem 55 in Section 10.4, we learned that the oscillating position of a mass hanging from the end of a spring (neglecting air resistance) is given by the equation below. (Recall that A, B, k, and m are all constants.)

$$s(t) = A\sin\left(\sqrt{\frac{k}{m}}\, t\right) + B\cos\left(\sqrt{\frac{k}{m}}\, t\right)$$

(a) Show that this function $s(t)$ has the property that $s''(t) + \frac{k}{m}s(t) = 0$. This is the differential equation for the spring motion (meaning it is an equation, involving derivatives, that describes the motion of the bob on the end of the spring). (*Hint:* Find the second derivative of $s(t)$, multiply $s(t)$ by the constant $\frac{k}{m}$, and add these two equations together; you should get zero.)

(b) Suppose the spring is released from an initial position of s_0 and with an initial velocity of v_0. Show that $A = v_0 \sqrt{\frac{m}{k}}$ and $B = s_0$.

65. In Problem 56 in Section 10.4, we learned that the oscillating position of a mass hanging from the end of a spring, taking air resistance into account, is given by the equation below. (Recall that A, B, k, and m are all constants.)

$$s(t) = e^{\frac{-f}{2m}t}\left(A \sin\left(\frac{\sqrt{4km - f^2}}{2m}\, t \right) + B \cos\left(\frac{\sqrt{4km - f^2}}{2m}\, t \right) \right)$$

(a) Show that this function $s(t)$ has the property that $s''(t) + \frac{a}{m}s'(t) + \frac{k}{m}s(t) = 0$. This is the differential equation for the spring motion, taking air resistance into account. (*Hint:* Find the first and second derivatives of $s(t)$ first, and then show that the equations for $s(t)$, $s'(t)$, and $s''(t)$ have the relationship given by the differential equation.)

(b) Suppose the spring is released from an initial position of s_0 and with an initial velocity of v_0. Show that $A = \frac{2mv_0 + fs_0}{\sqrt{4km - f^2}}$ and $B = s_0$.

Proofs

66. Use the definition of derivative (and some trigonometric limits) to prove that $\frac{d}{dx}(\sin x) = \cos x$.

67. Use the definition of derivative (and some trigonometric limits) to prove that $\frac{d}{dx}(\cos x) = -\sin x$.

68. Use the derivatives of sine and/or cosine to prove that $\frac{d}{dx}(\tan x) = \sec^2 x$.

69. Use the derivatives of sine and/or cosine to prove that $\frac{d}{dx}(\cot x) = -\csc^2 x$.

70. Use the derivatives of sine and/or cosine to prove that $\frac{d}{dx}(\sec x) = \sec x \tan x$.

71. Use the derivatives of sine and/or cosine to prove that $\frac{d}{dx}(\csc x) = -\csc x \cot x$.

72. In Problem 67 we used the definition of derivative to find the derivative of the cosine function. Give an alternative proof for the fact that $\frac{d}{dx}(\cos x) = -\sin x$. This proof should *not* use the definition of derivative but instead should use the derivative of the sine function and the fact that $\cos x = \sin(x + \frac{\pi}{2})$.

10.6   Graphs of Trigonometric Functions

Now that we understand the algebra, limits, and derivatives of trigonometric functions, we bring all this information together to investigate their graphs. We also examine transformations of trigonometric functions (especially sine and cosine functions).

10.6.1   The six basic trigonometric functions

In Section 10.2.3 we saw how to sketch rough graphs of the sine and cosine functions by considering their definitions on the unit circle (it may be a good idea to review that section before continuing with this one). In Section 10.4.2 we saw graphs of all six trigonometric functions. Now that we know how to differentiate trigonometric functions, we can investigate these graphs more completely.

The graphs of the six basic trigonometric functions are recorded in Theorem 10.17. Notice that the graph of $y = \csc x$ is the reciprocal of the graph of $y = \sin x$ (because $\csc x = \frac{1}{\sin x}$). In fact, each of the graphs in the second column is the reciprocal of the graph immediately to its left. Remember that you can use the graph of a function $f(x)$ to sketch the graph of its reciprocal $\frac{1}{f(x)}$. In particular, the zeros of $f(x)$ will be vertical asymptotes of $\frac{1}{f(x)}$ (and vice versa), and large heights on the graph of $f(x)$ will become small heights on the graph of $\frac{1}{f(x)}$ (and vice versa).

❖ *Caution*   You should know the graphs in Theorem 10.17 from memory and should be able to describe their domains, ranges, zeros, and vertical asymptotes. Thinking about the unit circle definitions of the trigonometric functions (and the method of graphing reciprocals) can help you remember these graphs.   □

THEOREM 10.17

Graphs of the Six Trigonometric Functions

The graphs of the sine, cosine, tangent, cosecant, secant, and cotangent functions are as shown below.

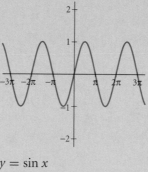

$y = \sin x$

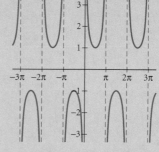

$y = \csc x$

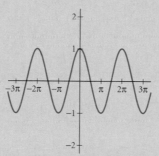

$y = \cos x$

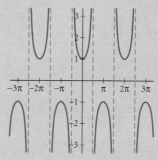

$y = \sec x$

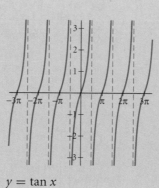

$y = \tan x$

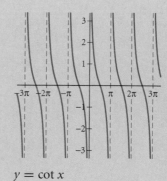

$y = \cot x$

To prove Theorem 10.17, we must do a curve-sketching analysis of each of the trigonometric functions. We will give the full curve-sketching analysis for the sine function and the major calculational steps for the tangent function. We leave the rest of the proof for the exercises. Each curve-sketching analysis will rely mostly on root and sign information for the sine and cosine functions (which will require thinking about the unit circle definitions of these functions).

PROOF (**THEOREM 10.17**) Consider the function $f(x) = \sin x$. The first and second derivatives of this function are $f'(x) = \cos x$ and $f''(x) = -\sin x$. From the unit circle definition of sine (as the height of the point on the unit circle corresponding to an angle of x radians), we know that $\sin x = 0$ if x is an angle that terminates at 0 or π radians (see Figure 10.43). This means that $\sin x = 0$ if $x = k\pi$ for any integer k. Therefore, both f and f'' have zeros at $x = k\pi$ for any integer k. Similarly, from the unit circle definition of cosine, we know that f' has zeros at $x = \frac{\pi}{2} + k\pi$ for all integers k (see Figure 10.44).

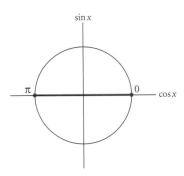

Figure 10.43

$\sin x = 0$ if x is an angle terminating at 0 or π radians

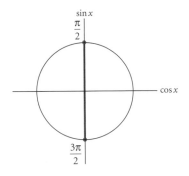

Figure 10.44

$\cos x = 0$ if x is an angle terminating at $\frac{\pi}{2}$ or $\frac{3\pi}{2}$ radians

Note that we are using "x" as an angle name here; thus we name the axes on the picture of the unit circle the "$\cos x$" axis and the "$\sin x$" axis, rather than the "x" and "y" axes (otherwise "x" would represent both an angle and a horizontal coordinate, and we want it to represent only an angle). It might be a good idea to review the unit circle definitions of sine and cosine at this point if you do not understand how we found the roots of f, f', and f''.

Again using the unit circle definition of sine, we know that $\sin x$ is positive if the angle x terminates in the first or second quadrant and negative if x terminates in the third or fourth quadrant (why?). Similarly, $\cos x$ is positive if x terminates in the first or fourth quadrant and negative if the angle x terminates in the second or third quadrant. Thus the number lines for f, f', and f'' (which is just the negative of the function f) are as shown in Figure 10.45.

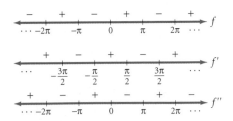

Figure 10.45

These number lines continue this pattern over and over as $x \to \infty$ and as $x \to -\infty$. From the number lines, we can see the intervals on which $f(x) = \sin x$ is positive or negative, the intervals on which it is increasing or decreasing, and the intervals on which it is concave up or concave down. Notice in particular (by the first derivative test) that f has local maxima at $x = \frac{\pi}{2} + 2\pi k$ (for example, at $x = \frac{\pi}{2}$ and at $x = \frac{\pi}{2} + 2\pi(-1) = -\frac{3\pi}{2}$) and local minima at $x = -\frac{\pi}{2} + 2\pi k$. Also, f has inflection points at $x = k\pi$ for every integer k. Furthermore, the unit circle definition of the sine function implies that $f(x) = \sin x$ has domain $(-\infty, \infty)$ and range $[-1, 1]$.

It now remains only to find the coordinates on the graph of $f(x) = \sin x$ at the "interesting" points. From the unit circle we see that $f(k\pi) = \sin(k\pi) = 0$, $f(\frac{\pi}{2} + k(2\pi)) = \sin(\frac{\pi}{2} + k(2\pi)) = 1$, and $f(-\frac{\pi}{2} + k(2\pi)) = \sin(-\frac{\pi}{2} + k(2\pi)) = 1$. Putting all this information together, we obtain the familiar graph of $f(x) = \sin x$ (see the first graph shown in Theorem 10.17).

The curve-sketching analyses for the other five trigonometric functions are similar, except that they have slightly more computational complexity. For example, consider the function $f(x) = \tan x$. The first and second derivatives of this function are

$$f'(x) = \frac{d}{dx}(\tan x) = \sec^2 x = \frac{1}{\cos^2 x},$$

$$f''(x) = \frac{d}{dx}(\sec^2 x) = 2(\sec x)^1(\sec x \tan x) = 2\sec^2 x \tan x$$

$$= 2\frac{1}{\cos^2 x}\frac{\sin x}{\cos x} = \frac{2\sin x}{\cos^3 x}.$$

The first derivative is never zero, but it fails to exist at values of x that make $\cos^3 x = 0$, so the number line for f' will be marked at $x = \frac{\pi}{2} + k\pi$ for each integer k. The domain of $f(x) = \tan x$ is all real numbers except for these points. The second derivative is 0 when $\sin x = 0$ and does not exist when $\cos^3 x = 0$, so its number line will be marked at the points $x = k\pi$ and the points $x = \frac{\pi}{2} + k\pi$. The signs on the number lines can be determined by considering the signs of sine and cosine in each interval. The behavior of f near the nondomain points $x = \frac{\pi}{2} + k\pi$ can be determined using limits. For example, at $\frac{\pi}{2}$ we have

$$\lim_{x \to \frac{\pi}{2}} f(x) = \lim_{x \to \frac{\pi}{2}} \tan x = \lim_{x \to \frac{\pi}{2}} \frac{\sin x}{\cos x} \to \frac{1}{0}.$$

A more careful inspection shows that the limit from the left is ∞ (since $\cos x$ is positive for values just a bit less than $\frac{\pi}{2}$) and the limit from the right is $-\infty$. This tells us that f has a vertical asymptote (going "up" on the left and "down" on the right) at $x = \frac{\pi}{2}$. A similar analysis shows that $f(x) = \tan x$ has this same kind of vertical asymptote at each point $x = \frac{\pi}{2} + k\pi$. We leave the construction of the number lines and the synthesis of this information to the reader. ∎

❓ Question Complete the details of the curve-sketching analysis of the tangent function in the proof of Theorem 10.17. □

You should be able to use the unit circle definitions to determine whether each trigonometric function is zero, positive, or negative at a given point. This will enable you to make number lines for each trigonometric function and its first and second derivatives. Four of the trigonometric functions have vertical asymptotes, which can be detected using limit calculations. The graphs of more complicated functions—for example, $f(x) = e^x \sin x$—can be obtained using a similar curve-sketching analysis.

❓ Question Do a curve-sketching analysis of $f(x) = e^x \sin x$. Include number lines for f, f', and f'', an analysis of the domain and range of f, and any important values or limits. □

10.6.2 General trigonometric functions

A *general sine function* is a function that can be written in the form

(1) $$f(x) = A\sin(B(x + C)) + D,$$

for some real numbers A, B, C, and D. Similarly, a *general cosine function* is a function of the form

(2) $$f(x) = A\cos(B(x + C)) + D.$$

Sometimes these general trigonometric functions are called simply "sine functions" or "cosine functions." For example, the function $f(x) = 2\sin(3x) - 5$ is a (general) sine function, and the function $g(x) = -\cos(1 - 2x) + 4$ is a (general) cosine function.

As we will see in Section 10.6.3, many real-world phenomena are described with such general trigonometric functions. The constants A, B, C, and D affect the size, period, and location of the graph. We will use our knowledge of transformations to investigate the effect of each of these constants. We will work with sine functions only, although everything we discuss will also hold for cosine functions.

Let's begin with the constant A. Depending on the value of A, the graph of $f(x) = A \sin x$ will differ from the graph of $y = \sin x$ by a vertical stretch or compression (and possibly by a reflection over the x-axis, if A is negative). The absolute value $|A|$ is called the *amplitude* of the function. The amplitude measures the size of the oscillation of the function $f(x) = A \sin x$; in particular, it is the vertical distance from the horizontal "center axis" of the sine function to the top of its peaks. Notice that the *total* vertical distance from the lower to the upper peaks is *twice* the amplitude.

EXAMPLE 10.22   **Vertical stretches and compressions**

Figures 10.46, 10.47, and 10.48 are examples of vertical stretches and compressions of $y = \sin x$. In each figure the graph of $y = \sin x$ is shown in maroon. The amplitude of the first and third graphs is 2, and the amplitude of the second graph is $\frac{1}{2}$.

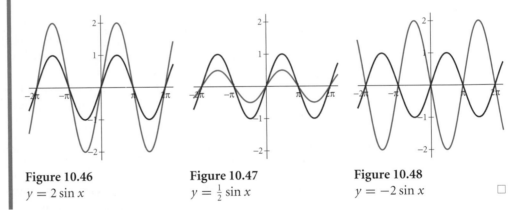

Figure 10.46   **Figure 10.47**   **Figure 10.48**
$y = 2 \sin x$ $y = \frac{1}{2} \sin x$ $y = -2 \sin x$ □

The constant B affects the *period* of the function. Before we describe how the constant B is related to the period, we need to have a more precise definition of "period." Many functions, including all trigonometric functions, are periodic.

DEFINITION 10.3

Periodicity

A function $f(x)$ has a *period* of p if p is the smallest positive real number such that, for all x in the domain of $f(x)$,

$$f(x + p) = f(x).$$

A function with a period is said to be *periodic*.

EXAMPLE 10.23   **The sine function is periodic**

The function $f(x) = \sin x$ has a period of 2π, since for all real numbers x, we have $\sin(x + 2\pi) = \sin x$. (Can you explain why, using the unit circle definition of sine?) Graphically, we can see that $f(x) = \sin x$ has a period of 2π because the graph of $f(x) = \sin x$ "repeats" every 2π units.   □

💲 *Caution*   Note that sine, cosine, secant, and cosecant have periods of 2π, whereas tangent and cotangent have periods of π units. See the graphs in Theorem 10.17 on page 679.   □

💲 *Caution*   The definition of period requires that p be the *smallest* number with the property that $f(x + p) = f(x)$ for all x in the domain of f. Thus, although the function $f(x) = \sin x$ has the property that $\sin(x + 4\pi) = \sin x$ for all real numbers x, we do *not* say that $f(x) = \sin x$ has a period of 4π. There is a smaller number (namely $p = 2\pi$) with the property that $\sin(x + p) = \sin x$ for all x.   □

The graph of $f(x) = \sin(Bx)$ will differ from the graph of $y = \sin x$ by a horizontal stretch or shrink, depending on the value of B (and by a reflection over the y-axis, if B happens to be negative).

EXAMPLE 10.24

Horizontal stretches and compressions

Figures 10.49–10.51 show the graph of $y = \sin x$ (in maroon) and three of its transformations. The function $f(x) = \sin(2x)$ is a horizontal *compression* of $y = \sin x$ by a factor of 2 (see Section 1.6 if you need to review this). Thus the graph of the function $f(x) = \sin(2x)$ repeats its behavior twice as fast as the function $y = \sin x$. In other words, $f(x) = \sin(2x)$ has a period of $\frac{2\pi}{2} = \pi$ (see Figure 10.49). On the other hand, the function $f(x) = \sin(\frac{1}{2}x)$ is a horizontal *stretch* of the function $y = \sin x$ by a factor of 2; its period is *twice* the period of $y = \sin x$, or $2(2\pi) = 4\pi$ units (see Figure 10.50). Figure 10.51 shows the graph of $f(x) = \sin(-2x)$, which is a horizontal compression, *and* a reflection over the y-axis, of the graph of $y = \sin x$.

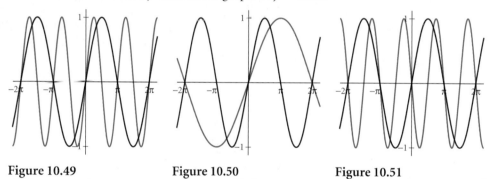

Figure 10.49	**Figure 10.50**	**Figure 10.51**
$f(x) = \sin(2x)$	$f(x) = \sin(\frac{1}{2}x)$	$f(x) = \sin(-2x)$

The period of $f(x) = \sin(Bx)$ is obtained by dividing 2π by the constant B. For example, the function $f(x) = \sin(2x)$ shown in Figure 10.49 has a period of $\frac{2\pi}{2} = \pi$. A more general version of this fact is stated in the following theorem (which you will prove in the exercises).

THEOREM 10.18

The Period of a Transformation

If $g(x)$ is a function with period p, and $f(x) = Ag(B(x + C)) + D$, then the period of $f(x)$ is $\frac{p}{B}$.

EXAMPLE 10.25

Finding the period of a trigonometric transformation

Since $\sin x$ has period 2π, the transformation $f(x) = 2\sin(3(x + 1)) + 5$ has period $\frac{2\pi}{3}$. Since $\tan x$ has period π, the transformation $f(x) = 2\tan(3(x + 1)) + 5$ has period $\frac{\pi}{3}$.   □

The constants C and D determine the "location" of the sine or cosine function. For example, $f(x) = \sin x + D$ is a vertical shift (up or down) of $y = \sin x$, and $f(x) = \sin(x + C)$ is a horizontal shift (left or right). These vertical and horizontal shifts do not change the shape or size of the graph, just its location.

EXAMPLE 10.26   **Vertical and horizontal translations**

The graph of $f(x) = \sin(x + \pi) - 1$ will be 1 unit lower, and π units farther to the *left*, than the graph of $y = \sin x$. Figure 10.52 shows a graph of this transformation (in blue).

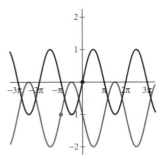

Figure 10.52
$f(x) = \sin(x + \pi) - 1$

It is easy to sketch the graph of a transformation that involves only translations. If stretches or compressions are also involved, things are more complicated. A general sine function $f(x) = A\sin(B(x + C)) + D$ is a transformation of $y = \sin x$ *first* by stretches and compressions and *then* by translations. This means that the graph of $y = \sin x$ gets stretched or compressed *before* the translation, which can make the translation hard to identify. The only point on the graph of $y = \sin x$ that we *know* will not be moved by a stretch or compression is the origin $(0, 0)$; we will call this the ***center point*** of $y = \sin x$.

This center point $(0, 0)$ of $y = \sin x$ will move to the center point $(-C, D)$ on the graph of the transformation $f(x) = A\sin(B(x + C)) + D$. (Most other points on the graph of $y = \sin x$ will be moved by stretches or compressions *first,* and *then* be translated C units left or right and D units up or down.) In Example 10.26, the center point of the transformation $f(x) = \sin(x + \pi) - 1$ is $(-\pi, 1)$; locate this point on the graph in Figure 10.52.

When dealing with cosine functions, we must be more careful. We will consider the center of $y = \cos x$ to be the origin $(0, 0)$, even though that point is not on the graph of $y = \cos x$. The point $(-C, D)$ will be the corresponding center point for the transformation $f(x) = A\cos(B(x + C)) + D$. Although the point $(-C, D)$ is not on the graph of $f(x)$, it will help us keep track of the translation parts of the transformation. The point $(0, 0)$ is the only point in the plane that we know will move *only* by translation (not by stretches or compressions).

EXAMPLE 10.27   **The center point of a general cosine function**

The function $f(x) = \cos(x + \pi) - 4$ is a translation of $y = \cos x$ by π units to the left and 4 units down, so a center of this function is at $(-\pi, -4)$. See Figures 10.53 and 10.54; in these graphs we use an asterisk to represent the center point so that we will not mistakenly think that the center point is part of the graph.

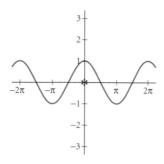

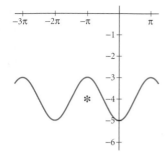

Figure 10.53

$y = \cos x$ and center point

Figure 10.54

$f(x) = \cos(x + \pi) - 4$

and center point ☐

⊛ *Caution* Don't forget that the center point of a cosine graph is *not* actually a point on the graph; it's just a point we use for reference to determine the translation part of the transformation. ☐

⊛ *Caution* In Examples 10.26 and 10.27, no stretches or compressions were involved. Therefore, for these examples, *every* point on the graph of $y = \sin x$ or $y = \cos x$ was moved only by translation. We didn't really need to talk about center points to graph the translations. In future examples where stretches or compressions are involved (like Example 10.28), the center point will be more important. ☐

The following algorithm summarizes the information we have developed concerning the constants A, B, C, and D.

ALGORITHM 10.1

The Effects of the Constants in a General Trigonometric Function

If $f(x)$ is a general sine function $f(x) = A\sin(B(x + C)) + D$, or a general cosine function $f(x) = A\cos(B(x + C)) + D$, then

(a) The amplitude of f is $|A|$.

(b) The period of f is $\dfrac{2\pi}{B}$.

(c) $(-C, D)$ is a center point of f.

(d) If $A < 0$, then the graph of f is "upside down" when compared with the graph of $y = \sin x$ or $y = \cos x$ (viewed near the center point).

With Algorithm 10.1, we can quickly sketch graphs of (general) sine or cosine functions. A good method for sketching such graphs is to determine the amplitude and period of the graph, find the new center of the graph, and then take into account any vertical or horizontal reflections (if A or B is negative). Also keep in mind that the sine or cosine function should be at a high or low point one-quarter of a period on either side of the center point.

EXAMPLE 10.28

Sketching the graph of a general sine function

Use your knowledge of transformations to make a quick, labeled sketch of

$$f(x) = -2\sin\left(\tfrac{1}{4}(x - \pi)\right) + 3.$$

Solution   The graph of $f(x)$ has an amplitude of $|-2| = 2$ and a period of $\frac{2\pi}{1/4} = 8\pi$. Its center will be at the point $(\pi, 3)$. Finally, the graph will have the same basic shape as the graph of $y = \sin x$, except "upside down," since the constant A is negative. This means that the function will have a range of $[1, 5]$ (2 units above and below the height of the center point) and will repeat itself every 8π units. The graph is shown in Figure 10.55.

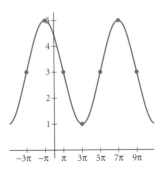

Figure 10.55
$f(x) = -2\sin\left(\frac{1}{4}(x-\pi)\right) + 3$

Notice that near the center $(\pi, 3)$, the graph looks like the graph of $y = \sin x$ does near the origin, except "upside down."   □

? Question   Use the same techniques to make a quick, labeled graph of the general cosine function $f(x) = -2\cos(\frac{1}{4}(x - \pi)) + 3$.   □

◈ Caution   It is important that you always use the standard form $f(x) = A\sin(B(x + C)) + D$ when investigating general sine graphs. In particular, beware of functions like $f(x) = \sin(2x + \pi)$. This function *does* have a period of π, but its center point is at $(-\frac{\pi}{2}, 0)$, not at $(-\pi, 0)$ as you might expect. This is because in standard form this function is written

$$f(x) = \sin(2x + \pi) = \sin\left(2\left(x + \frac{\pi}{2}\right)\right),$$

so $B = 2$ and $C = \frac{\pi}{2}$ (not π).   □

A derivative analysis is not really necessary when we are graphing general sine and cosine functions. The locations of the local extrema and inflection points of the graph will always be clear after use of Algorithm 10.1. For example, the function $f(x) = -2\sin(\frac{1}{4}(x - \pi)) + 3$ in Example 10.28 has local maxima at $x = -\pi + 8\pi k$ for all positive or negative integers k, since there is a local maximum at one-quarter of a period before the center point (i.e., at $\pi - \frac{1}{4}(8\pi) = -\pi$) and the function has a period of 8π. Similarly, $f(x)$ has local minima at $x = 3\pi + 8\pi k$ and inflection points at $x = \pi + 4\pi k$ for all (positive or negative) integers k.

10.6.3 Modeling with trigonometric functions

Many real-world phenomena can be modeled by trigonometric functions. For example, cycling populations and high/low tide cycles are periodic—and thus good candidates for trigonometric models. Data that seem to be oscillating regularly can often be represented by a general sine or cosine function. Given just a few datapoints (such as a local maximum value followed by a local minimum value), we can come up with a good trigonometric model.

We will start with a purely graphical problem and then deal with a set of data. When finding a general sine or cosine function to fit a given graph, we must find a center point, determine the period and amplitude of the graph, and take into account any reflections.

EXAMPLE 10.29

Modeling a graph with a general cosine function

Find a general cosine function that could have the graph shown in Figure 10.56.

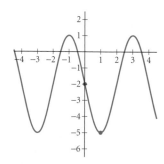

Figure 10.56

Solution The goal is to find a function of the form $f(x) = A\cos(B(x + C)) + D$ that has the graph in Figure 10.56. We need to find appropriate values for the constants A, B, C, and D.

The horizontal distance between the two marked points is 1 unit. This represents one-quarter of a period, so this function has a period of 4 units. Since the period of a general cosine function is $\frac{2\pi}{B}$, we have

$$4 = \frac{2\pi}{B} \quad \Rightarrow \quad 4B = 2\pi \quad \Rightarrow \quad B = \frac{\pi}{2}.$$

The amplitude $|A|$ is half of the vertical distance between the local maximum values and the local minimum values:

$$|A| = \frac{1 - (-5)}{2} = \frac{6}{2} = 3.$$

The sign of A (positive or negative) depends on the center point that we choose. Suppose we choose $(1, -2)$ as our center point (this point is on the horizontal "center axis" of the graph and at an x-coordinate where the function has a minimum). With this center point we are thinking of this function as a stretch or shrink, and translation, of $y = -\cos x$, so A is negative. Therefore, $A = -3$. (We could have chosen another center point; see the "Question" following this solution.) Moreover, with a center point of $(1, -2) = (-C, D)$, we know that $C = -1$ and $D = -2$. Putting this all together, we get the function

$$f(x) = -3\cos\left(\tfrac{\pi}{2}(x - 1)\right) - 2. \qquad \square$$

❓ *Question* The solution given in Example 10.29 is only one of many possibilities for the constants A, B, C, and D. If we start with a different center point, then we will get different values for these constants. The *function* would be the same, just written in a different form. Find a cosine function that models the graph in Figure 10.56 by using a center point that will make the constant A a positive number. $\qquad \square$

? *Question*   Can you find a general *sine* function that has the graph in Figure 10.56? You'll have to choose a different center point, but the period and amplitude information will be the same.   □

The methods used in Example 10.29 can be used to find a trigonometric function that models a set of cyclical data. We need to know only a few key pieces of data to produce a general sine or cosine function that models the data.

EXAMPLE 10.30

Using a general sine function to model periodic data

On a certain island, a population of wombats cycles each year. There are a minimum of 400 wombats in the winter (January 1) and a maximum of 900 wombats in the summer (midway through the year). Find a sine function that models the population $P(t)$ of wombats on the island t days after January 1.

Solution   Although we are given only two pieces of data, the maximum and the minimum populations of wombats, we can find a function to model the population if we assume that that function is a general sine function $P(t) = A \sin(B(t + C)) + D$. Specifically, we are given that $P(0) = 400$ is the minimum population, followed by a maximum population of $P(182.5) = 900$. These two datapoints are plotted in Figure 10.57.

The horizontal distance between a minimum and a successive maximum of a sine function is half of a period, so the period of this function is $2(182.5 - 0) = 365$ (as you might expect). Therefore $365 = \frac{2\pi}{B}$, and thus $B = \frac{2\pi}{365}$. The amplitude is half of the vertical distance between the maximum and minimum heights, so $|A| = \frac{1}{2}(900 - 400) = 250$. One center point of the general sine function has an x-coordinate halfway between 0 and 182.5 and a y-coordinate halfway between 400 and 900; thus we can take $(91.25, 650) = (-C, D)$ as our center point. This gives $C = -91.25$ and $D = 650$. With this center point the constant A is positive, so $A = 250$. This gives us the function

$$P(t) = 250 \sin\left(\tfrac{2\pi}{365}(t - 91.25)\right) + 650.$$

Notice that we found *four* constants A, B, C, and D with only *two* datapoints! Note also that this is not the only possible answer; if we had chosen a different center point, we would have found different values for C and D, as well as possibly a different sign for A (although the resulting graph would have been the same). The graph of $P(t)$ is shown in Figure 10.58; note that it passes though our two datapoints as well as the center point that we chose.

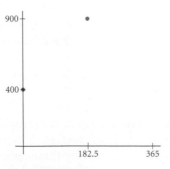

Figure 10.57
Successive minimum and maximum

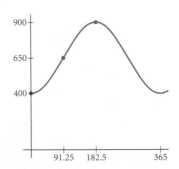

Figure 10.58
$P(t) = 250 \sin\left(\tfrac{2\pi}{365}(t-91.25)\right)$
$+ 650$ □

> ### What you should know after reading Section 10.6

> ► The graphs of the six trigonometric functions, their domains and ranges, and the locations of their roots and vertical asymptotes. Be able to use notation like "$x = \frac{\pi}{2} + k\pi$ for any integer k" to indicate the locations of zeros, extrema, inflection points, and vertical asymptotes.

> ► How to use a curve-sketching analysis to make a careful, labeled graph of a trigonometric function. In particular, this includes how to use the unit circle definitions to determine the zeros and signs of sine and cosine.

> ► The definition of a general trigonometric function, and the effects of the constants A, B, C, and D for such functions. This includes understanding the definitions of the amplitude and period of a function, and how to determine a "center point" of a general sine or cosine function.

> ► How to sketch quickly an accurate, labeled graph of any general sine or cosine function, and how to find a general sine or cosine function that models a given graph or set of data.

Exercises 10.6

Concepts

0. Read the section and make your own summary of the material. Include whatever you think will help you review the material later on.

1. Sketch labeled graphs of $y = \sin x$, $y = \cos x$, and $y = \tan x$ from memory. Show at least three periods of each graph.

2. Use the graphs of $y = \sin x$, $y = \cos x$, and $y = \tan x$ to sketch labeled graphs of their reciprocals $y = \csc x$, $y = \sec x$, and $y = \cot x$.

3. Explain with limits why $f(x) = \csc x$ has a vertical asymptote at each value of x for which $\sin x = 0$.

4. Explain how the shape of the graph of $f(x) = \sin x$ is related to the unit circle definition of the sine function.

5. Explain how the shape of the graph of $f(x) = \tan x$ is related to the unit circle definition of the tangent function.

6. Explain how the unit circle definitions of sine and cosine lead to the number lines shown in Figure 10.45 in the proof of Theorem 10.17.

7. Fill in the missing details in the curve-sketching analysis of tangent given in the proof of Theorem 10.17.

8. List all the graphical features of $f(x) = \csc x$ that you can think of, both verbally and mathematically (i.e., in terms of limits and derivatives).

9. Write down the equations of three general sine functions that have amplitude 4 and period 3π.

10. Write down the equations of three general cosine functions that have amplitude $\frac{1}{2}$ and period 4.

11. Sketch graphs (by hand) of $f(x) = A \sin x$ and $g(x) = A \cos x$ for **(a)** $A = \frac{1}{3}$, **(b)** $A = -3$, and **(c)** $A = \pi$.

12. Sketch graphs (by hand) of $f(x) = \sin(Bx)$ and $g(x) = \cos(Bx)$ for **(a)** $B = \frac{1}{4}$, **(b)** $B = 2\pi$, and **(c)** $B = -\frac{\pi}{3}$.

13. Sketch graphs (by hand) of $f(x) = \sin(x + C) + D$ and $g(x) = \cos(x + C) + D$ for **(a)** $C = \frac{\pi}{2}$, $D = -5$, **(b)** $C = -3\pi$, $D = 2$, and **(c)** $C = -2$, $D = \frac{1}{2}$.

14. Find three center points for the function $f(x) = -2 \sin(\frac{\pi}{4}(x - 2)) + 8$.

15. Find three center points for the function $f(x) = \frac{1}{3} \cos(2(x - \frac{\pi}{2})) - 2$.

16. Without using derivatives or using a calculator, find the coordinates of the inflection points of the function $f(x) = 3 \sin(4(x - \frac{\pi}{2})) + 1$.

17. Without using derivatives or using a calculator, find the coordinates of the inflection points of the function $f(x) = -2 \cos(\pi(x + 1)) - 3$.

18. In Example 10.29 we found that $f(x) = -3 \cos(\frac{\pi}{2}(x - 1)) - 2$ had the graph in Figure 10.56. We obtained this function by choosing $(1, -2)$ as the "center point" of the graph. Show that if we instead choose $(-1, -2)$ as the "center point," we obtain the function $g(x) = 3 \cos(\frac{\pi}{2}(x + 1)) - 2$. Then use trigonometric identities and algebra to show that $f(x) = g(x)$.

■ Determine whether each of the following statements is true or false. If a statement is true, explain why. If a statement is false, explain why or provide a counterexample.

19. True or False: If $\csc x = 5$, then $\sec x = \frac{1}{5}$.

20. True or False: If $\tan x = 2$, then $\cot x = \frac{1}{2}$.

21. True or False: If $\cos x = \frac{1}{3}$ and $\sin x = \frac{2}{5}$, then $\tan x = \frac{6}{5}$.

22. True or False: If $\sin x \geq 0$, then $\csc x \geq 1$.

23. True or False: Only trigonometric functions can be periodic.

24. True or False: If $f(x+5) = f(x)$ for all x in the domain of f, then f is periodic with a period of 5.

25. True or False: The period of $\sin(4x)$ is 4.

26. True or False: The period of $\tan(4x)$ is $\frac{\pi}{2}$.

Skills

■ Identify the period of each of the following functions. Check your answers with calculator graphs.

27. $f(x) = 2\sin(\pi x) + 8$

28. $f(x) = \cos\left(-\frac{\pi}{2}(x+1)\right)$

29. $f(x) = \sin(5 - 4x) - 1$

30. $f(x) = \tan(\pi(x-1))$

31. $f(x) = \sec(2(x-\pi))$

32. $f(x) = -2\cot(1-x) + 3$

■ Write each of the following in compact notation, using k to represent any integer. (For example, the set of angles that terminate at $\frac{\pi}{4}$ on the unit circle can be compactly represented as $\frac{\pi}{4} + 2\pi k$, for any integer k.)

33. The set of angles that terminate at $\frac{\pi}{2}$

34. The set of angles that terminate at $\frac{\pi}{2}$ or $-\frac{\pi}{2}$

35. $\ldots, -3\pi, -2\pi, -\pi, \pi, 2\pi, 3\pi, \ldots$

36. $\ldots, -\frac{5\pi}{4}, -\frac{3\pi}{4}, -\frac{\pi}{4}, \frac{\pi}{4}, \frac{3\pi}{4}, \frac{5\pi}{4}, \ldots$

37. The solutions to $\sin x = 0$

38. The solutions to $\tan x = 1$

39. The values where $\cot x$ is zero or does not exist

40. The locations of the vertical asymptotes of $\csc x$

41. The critical points of $f(x) = \sec x$

42. The inflection points of $f(x) = \cos x$

■ Sketch quick but labeled graphs (by hand) of each of the following general sine and cosine functions. For each function, identify the period and amplitude, and find a "center point." Check your answers with calculator graphs.

43. $f(x) = \sin(2x) + \frac{\pi}{2}$ **44.** $f(x) = \sin\left(2\left(x + \frac{\pi}{2}\right)\right)$

45. $f(x) = \sin\left(2x + \frac{\pi}{2}\right)$ **46.** $f(x) = 2\cos\left(\frac{\pi}{2}x\right)$

47. $f(x) = -\cos(2x) + 3$ **48.** $f(x) = \cos(\pi(x+2)) - 1$

49. $f(x) = \frac{1}{2}\sin(3(x-\pi)) + 2$

50. $f(x) = -\cos(\pi(x-1)) - 3$

51. $f(x) = 2\sin\left(\frac{1}{3}(x-\pi)\right) + 1$

52. $f(x) = \pi\cos\left(\frac{\pi}{2}(x-2)\right) - \pi$

53. $f(x) = 4\sin(2x+\pi) - 3$

54. $f(x) = 3\cos\left(\pi - \frac{\pi}{2}x\right) + 1$

■ Sketch quick but labeled graphs (by hand) of each of the following general trigonometric functions. Identify the period of each function.

55. $f(x) = \tan\left(\frac{1}{3}x\right) + 5$ **56.** $f(x) = \sec(\pi(x-1))$

57. $f(x) = 2\csc(x+3) - 5$

■ For each graph below, find a general sine function *and* a general cosine function that have that graph.

58.

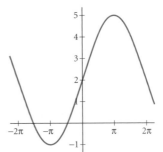

59.

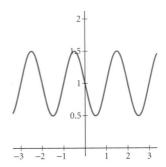

60.

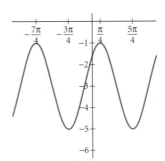

61.

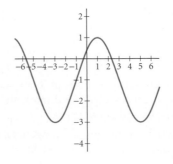

62.

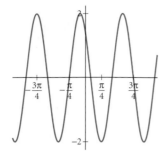

63.

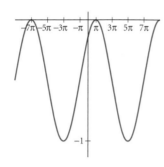

■ For each function $f(x)$ below, create number lines for f, f', and f''.

64. $f(x) = \cos\left(\frac{\pi}{2} + \pi x\right)$

65. $f(x) = \sec 2x$

66. $f(x) = 3\sin\left(4\left(x - \frac{\pi}{2}\right)\right) - 1$

67. $f(x) = \sin x - \tan x$

68. $f(x) = e^x \sin x$

69. $f(x) = \dfrac{\sin x}{e^x}$

■ Do a complete curve-sketching analysis of each of the following functions.

70. $f(x) = \cos\left(\frac{\pi}{2} + \pi x\right)$

71. $f(x) = \sec 2x$

72. $f(x) = 3\sin\left(4\left(x - \frac{\pi}{2}\right)\right) - 1$

73. $f(x) = x - \tan x$

74. $f(x) = e^x \sin x$

75. $f(x) = \dfrac{\sin x}{e^x}$

Applications

76. Suppose a population of wombats on an island cycles each year, between a minimum of 200 wombats in the winter (January 1) and a maximum of 500 wombats in the summer (midway through the year).

(a) Find a general sine function that models the population $P(t)$ of wombats on the island t days after January 1. Make a labeled sketch of the graph of your function.

(b) Find a general cosine function that models the population $P(t)$ of wombats on the island t days after January 1.

77. The following table describes the heights (in feet) of the high and low tides at Paradise Cove on September 3 and 4 last year.

Day	Time	Tide	Height
9/3	12:06 a.m.	High	7.2
9/3	6:20 a.m.	Low	0.3
9/3	12:27 p.m.	High	7.1
9/3	6:37 p.m.	Low	0.5
9/4	12:43 a.m.	High	7.2
9/4	6:54 a.m.	Low	0.3
9/4	1:03 p.m.	High	7.2
9/4	7:14 p.m.	Low	0.5

(a) Plot the data above, with time t in minutes, where $t = 0$ represents 12:00 a.m. on September 3. Why might you suspect that the heights $h(t)$ of the tides could be modeled by a general sine function?

(b) Find the average height M of a high tide, the average height N of a low tide, the average length of time P between two high tides, and the average length of time Q between two low tides.

(c) Find a general sine function that models data (i.e., whose graph will closely follow the datapoints). Use M as the maximum value of the function, and use N as the minimum value. Use the average $\frac{P+Q}{2}$ as the period of your function.

(d) Graph the general sine function you found above, and compare it to the plot of the original data. Is the general sine function a good model for the data?

Proofs

78. Prove that the graph of $y = \cos x$ has the shape and features shown in Theorem 10.17. Use as a model the proof for the graph of $y = \sin x$ that is given in the chapter.

79. Prove that the graph of $y = \cot x$ has the shape and features shown in Theorem 10.17. Use as a model the proof for the graph of $y = \tan x$ that is given in the chapter.

80. Prove that the graph of $y = \sec x$ has the shape and features shown in Theorem 10.17. Identify the location of and behavior near each asymptote, and identify the locations of any inflection points of the graph.

81. Prove Theorem 10.18 by following the steps below. Suppose $g(x)$ is a function with period p, and define $f(x) = Ag(B(x+C)) + D$. We wish to show that $f(x)$ has a period of $\frac{p}{B}$.

(a) Show that $f\left(x + \frac{p}{B}\right) = f(x)$ for all x in the domain of f.

(b) Argue that $\frac{p}{B}$ is the *smallest* number with the property you showed in part (a).

CHAPTER 11

Inverse Trigonometric Functions

Egyptian spiral circuit for playing the Serpent Game.
© *Gianni Dagli Orti/Corbis*

*A*lthough the trigonometric functions are not one-to-one, on restricted domains we can define their inverses. The definitions and derivatives of the inverse trigonometric functions follow from what we know of the trigonometric functions. The inverse sine, inverse cosine, inverse tangent, and inverse secant functions will be useful in Part IV when we investigate integrals.

CONTENTS

11.1   Defining the Inverse Trigonometric Functions

In this section we define four new functions: the inverses of the sine, cosine, tangent, and secant functions (on restricted domains). These ***inverse trigonometric functions*** will be a useful tool in integral calculus in Part IV.

11.1.1   Restricted domains of the trigonometric functions

None of the trigonometric functions is one-to-one (their graphs do not pass the Horizontal Line Test), and only one-to-one functions are invertible. Thus it may seem odd that we have an entire chapter (even a short one) devoted to the inverses of trigonometric functions. However, we can restrict the domains of the trigonometric functions so that they *are* one-to-one and then consider their inverses.

Consider first the function $f(x) = \sin x$. There are many ways that we could restrict the domain of this function to make it one-to-one; all we have to do is restrict our attention to a piece of the graph that passes the Horizontal Line Test.

EXAMPLE 11.1

Three restricted domains of the sine function

We could consider the function $f(x) = \sin x$ restricted to the domain $[\frac{\pi}{2}, \frac{3\pi}{2}]$ or restricted to the domain $[-\frac{\pi}{2}, 0]$ (see Figures 11.1–11.3).

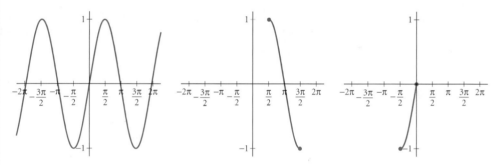

Figure 11.1
$f(x) = \sin x$

Figure 11.2
$f(x) = \sin x$ with restricted domain $[\frac{\pi}{2}, \frac{3\pi}{2}]$

Figure 11.3
$f(x) = \sin x$ with restricted domain $[-\frac{\pi}{2}, 0]$

Both of the restricted functions shown in Figures 11.2 and 11.3 are one-to-one and thus invertible. The graphs of their inverses can be obtained by reflecting each graph over the line $y = x$. Notice that we would get different "inverses" depending on which restricted domain we use.   ☐

? *Question* Sketch the graphs of the inverses of the restricted functions shown in Figures 11.2 and 11.3. Are your graphs the same? What other restricted domains could we use for the sine function? Which one do you think we should call "the" inverse of the sine function?   ☐

For the purposes of constructing inverses, we have a convention for restricting the domains of the trigonometric functions. Starting at the origin, we move to the right as far as possible, taking as much of the function as we can, while ensuring that our entire segment passes the Horizontal Line Test. Then we do the same thing moving to the left from the origin. This gives us a way of choosing a unique "piece" of the function that is one-to-one and thus invertible.

DEFINITION 11.1

The Restricted Sine and Restricted Cosine Functions

The restricted sine and cosine functions have the graphs, domains, and ranges given below.

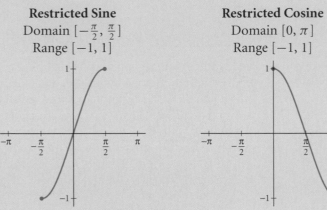

Restricted Sine
Domain $[-\frac{\pi}{2}, \frac{\pi}{2}]$
Range $[-1, 1]$

Restricted Cosine
Domain $[0, \pi]$
Range $[-1, 1]$

Similarly, we can restrict the domains of the tangent and secant functions according to our "convention":

DEFINITION 11.2

The Restricted Tangent and Restricted Secant Functions

The restricted tangent and secant functions have the graphs, domains, and ranges given below.

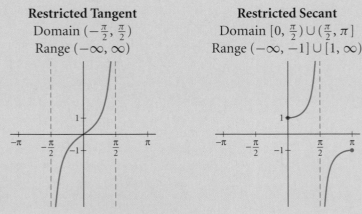

Restricted Tangent
Domain $(-\frac{\pi}{2}, \frac{\pi}{2})$
Range $(-\infty, \infty)$

Restricted Secant
Domain $[0, \frac{\pi}{2}) \cup (\frac{\pi}{2}, \pi]$
Range $(-\infty, -1] \cup [1, \infty)$

Note that the ranges of the restricted sine, cosine, tangent, and secant functions are the same as the ranges of the complete sine, cosine, tangent, and secant functions.

❓ Question We could also define restricted domains for the cosecant and cotangent functions, although the inverses of cosecant and cotangent are seldom used. For each of the functions $f(x) = \csc x$ and $f(x) = \cot x$, find a restricted domain on which the function is one-to-one, sketch the graph of the restricted function, and list the domain and range. ☐

✎ Caution You should be able to reproduce the information in Definitions 11.1 and 11.2. You already know the graphs of the six trigonometric functions; each of the restricted graphs is obtained by using the "convention" for restricting domains. Be sure you can picture the "complete" graph for each of these trigonometric functions, as well as the portion that corresponds to the restricted domain. ☐

11.1.2 Definitions of the inverse trigonometric functions

Now that we have restricted sine, cosine, tangent, and secant so that they are one-to-one, we can consider their inverses. These inverse functions are so important that we give them special names (remember that this was also the case with the inverse of the natural exponential function, called $\ln x$). In fact, each of these inverse functions has *two* commonly used names. For example, the inverse of the (restricted) sine function is called both $\sin^{-1} x$ (pronounced "sine inverse of x") and $\arcsin x$ (pronounced "arcsine of x").

DEFINITION 11.3

The Inverse Trigonometric Functions

We use the following notation to denote the inverses of the *restricted* trigonometric functions $\sin x$, $\cos x$, $\tan x$, and $\sec x$, respectively:

$$\sin^{-1} x, \quad \cos^{-1} x, \quad \tan^{-1} x, \quad \text{and} \quad \sec^{-1} x.$$

We also use the following notation for these same functions:

$$\arcsin x, \quad \arccos x, \quad \arctan x, \quad \text{and} \quad \arcsec x.$$

We also define functions $\csc^{-1} x$ and $\cot^{-1} x$ (also known as $\arccsc x$ and $\arccot x$) as the inverses of the cosecant and cotangent functions, but they are not used very often and thus are not included here.

⬥ *Caution*   The notation $\sin^{-1} x$ is supposed to remind you of the inverse function notation $f^{-1}(x)$. Just as $f^{-1}(x)$ is *not* equal to $\frac{1}{f(x)}$, remember that $\sin^{-1} x$ is *not* equal to $(\sin x)^{-1}$ or $\frac{1}{\sin x}$ (even though $\sin^2 x$ *is* equal to $(\sin x)^2$). By definition, $\sin^{-1} x$ is the *inverse*, and not the reciprocal, of the function $\sin x$.   □

⬥ *Caution*   We often say simply "the inverse of sine" when what we mean is the inverse of the *restricted* sine function.   □

We can graph the inverse trigonometric functions simply by reflecting the graphs of the (restricted) trigonometric functions over the line $y = x$. We begin with the inverses of the sine and cosine functions (compare with Definition 11.1).

THEOREM 11.1

Basic Properties of Inverse Sine and Inverse Cosine

The graphs, domains, and ranges of $\sin^{-1} x$ and $\cos^{-1} x$ are as follows:

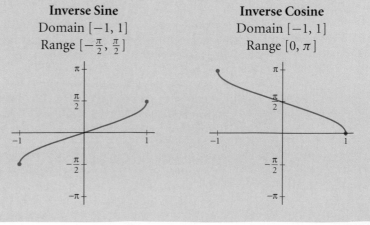

Inverse Sine
Domain $[-1, 1]$
Range $\left[-\frac{\pi}{2}, \frac{\pi}{2}\right]$

Inverse Cosine
Domain $[-1, 1]$
Range $[0, \pi]$

Similarly, the graphs of $\tan^{-1} x$ and $\sec^{-1} x$ are obtained by reflecting the graphs of $\tan x$ and $\sec x$ across the line $y = x$ (compare with Definition 11.2):

THEOREM 11.2

Basic Properties of Inverse Tangent and Inverse Secant

The graphs, domains, and ranges of $\tan^{-1} x$ and $\sec^{-1} x$ are as follows:

Inverse Tangent	Inverse Secant
Domain $(-\infty, \infty)$	Domain $(-\infty, -1] \cup [1, \infty)$
Range $\left(-\frac{\pi}{2}, \frac{\pi}{2}\right)$	Range $\left[0, \frac{\pi}{2}\right) \cup \left(\frac{\pi}{2}, \pi\right]$

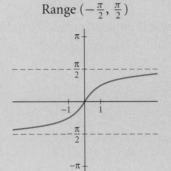
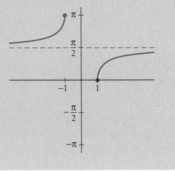

⊗ *Caution*   You should know the graphs, domains, and ranges of the four inverse trigonometric functions without looking them up. As usual, memorization isn't the best option here; a good knowledge of the properties of inverse functions and the graphs of (restricted) sine, cosine, tangent, and secant is enough to re-create Theorems 11.1 and 11.2.   □

❓ *Question*   What would the graphs of $\csc^{-1} x$ and $\cot^{-1} x$ look like? What are their domains and ranges?   □

11.1.3 Calculating values of inverse trigonometric functions

To understand inverse trigonometric functions fully, we need to consider the unit circle. Let's begin by focusing on the (restricted) sine function and its inverse. Remember that if x is an angle (in radian measure), then $\sin x$ is the height of the point on the unit circle where the terminal edge of the angle x meets the unit circle. This means that the inputs of $y = \sin x$ are angles, and the outputs are heights of points on the unit circle. Since $y = \sin^{-1} x$ is the inverse of $y = \sin x$, the *inputs* of $y = \sin^{-1} x$ are heights of points on the unit circle, and the *outputs* of $y = \sin^{-1} x$ are angles.

A more precise way to say this is as follows:

(1) $\sin^{-1} x$ is the angle in $\left[-\frac{\pi}{2}, \frac{\pi}{2}\right]$ whose sine is x.

The reason we must restrict our attention to the angles in the interval $\left[-\frac{\pi}{2}, \frac{\pi}{2}\right]$ is that this is the range of the inverse sine function (since it is the domain of the restricted sine function). Given any height between -1 and 1, there are *two* points on the unit circle with that height, and each of these points corresponds to infinitely many angles. If we want $\sin^{-1} x$ to be a function, then we must choose *one* of these angles to be $\sin^{-1} x$.

⊗ *Caution*   Don't forget that $\sin^{-1} x$ represents an *angle*. Since the "inputs" of trigonometric functions are angles, the "outputs" of the inverse trigonometric functions are angles.   □

You can estimate values of inverse trigonometric functions using a calculator (using the "$\sin^{-1}$" or "arcsin" button). Be sure that your calculator is in radian mode, and keep in mind that most calculators can give only *approximate* answers, not exact answers. The following example illustrates how to compute an *exact* inverse trigonometric value by hand.

EXAMPLE 11.2

Determining an inverse trigonometric value

Find $\sin^{-1}(\frac{1}{2})$. Interpret this quantity in terms of the unit circle and in terms of the graph of the inverse sine function.

Solution Here the input is the height $\frac{1}{2}$, and $\sin^{-1}(\frac{1}{2})$ is the unique angle in the restricted domain $[-\frac{\pi}{2}, \frac{\pi}{2}]$ whose sine (i.e., whose height on the unit circle) is equal to $\frac{1}{2}$. There are two points on the unit circle with a height of $\frac{1}{2}$ (one in the first quadrant and one in the second quadrant). Since we want to consider only the angles in the interval $[-\frac{\pi}{2}, \frac{\pi}{2}]$, we can restrict our attention to the point in the first quadrant (see Figures 11.4 and 11.5). There are infinitely many angles that terminate at this point. However, there is only *one* angle in the restricted domain $[-\frac{\pi}{2}, \frac{\pi}{2}]$ that does. To find it we draw a reference triangle (see Figure 11.6).

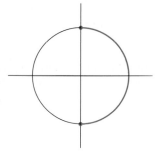 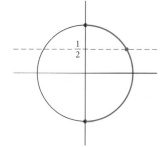 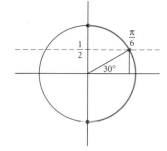

Figure 11.4
Restricted domain of sine (range of inverse sine) is $[-\frac{\pi}{2}, \frac{\pi}{2}]$

Figure 11.5
The only point in the restricted domain with a height of $\frac{1}{2}$

Figure 11.6
$\sin^{-1}(\frac{1}{2}) = \frac{\pi}{6}$ is the angle corresponding to the point on the unit circle

The height of the reference triangle is $\frac{1}{2}$ (because the point on the unit circle has a height of $\frac{1}{2}$), and the hypotenuse of the triangle has length 1 (since we are on the *unit* circle). Therefore, the reference triangle must be a 30–60–90 triangle (why?), and the interior angle is 30°. In radian measure, this means that the only angle in $[-\frac{\pi}{2}, \frac{\pi}{2}]$ that terminates at our marked point is the angle $\frac{\pi}{6}$. In other words, $\sin^{-1}(\frac{1}{2}) = \frac{\pi}{6}$ (you can check this answer on your calculator). This fact is illustrated graphically in Figure 11.7.

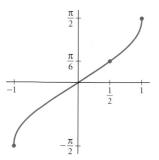

Figure 11.7
$y = \sin^{-1} x$ when $x = \frac{1}{2}, y = \frac{\pi}{6}$

In the language of inverse functions, we are using the fact that if f^{-1} is the inverse of a function f, then

(2)
$$y = f^{-1}(x) \text{ if and only if } f(y) = x,$$
$$\text{for all } x \in \text{Domain}(f^{-1}) \text{ and } y \in \text{Domain}(f).$$

Here we are considering the function $f(x) = \sin x$ (on a restricted domain) and its inverse $f^{-1}(x) = \sin^{-1}(x)$. Since in this case the domain of $f(x) = \sin x$ is restricted to $[-\frac{\pi}{2}, \frac{\pi}{2}]$, and the domain of $f^{-1}(x) = \sin^{-1}(x)$ is $[-1, 1]$, we have

(3)
$$y = \sin^{-1}(x) \text{ if and only if } \sin(y) = x,$$
$$\text{for all } x \in [-1, 1] \text{ and } y \in [-\tfrac{\pi}{2}, \tfrac{\pi}{2}].$$

? Question   Similar statements can be made about the inverse cosine, tangent, and secant functions. Can you state them?   □

◈ Caution   The domain considerations in Statement (3) are vital. For example, although it is true that $\sin(\frac{5\pi}{6}) = \frac{1}{2}$, it is *not* true that $\sin^{-1}(\frac{1}{2}) = \frac{5\pi}{6}$. This is because $\frac{5\pi}{6}$ is not in the domain $[-\frac{\pi}{2}, \frac{\pi}{2}]$ of the restricted sine function.   □

In general, we can calculate the exact value of $\sin^{-1}(x)$ (or $\cos^{-1} x$) for a particular value of x if x is recognizable as the side length of a 30–60–90 or a 45–45–90 triangle. Given a value of x, say $x = \frac{1}{2}$ or $x = -\frac{\sqrt{2}}{2}$, we can calculate $\sin^{-1}(x)$ (or $\cos^{-1}(x)$) by finding an angle in the appropriate restricted domain whose sine (or cosine) is x. Some calculations of inverse trigonometric values do not even require a reference triangle, as illustrated in the following example.

EXAMPLE 11.3   **The arccosine of zero is one of the angles whose cosine is zero**

Find all of the angles whose cosine is zero, and then find arccos(0).

Solution   An angle θ has a cosine of zero if it terminates at $\frac{\pi}{2}$ or $-\frac{\pi}{2}$ radians (see Figure 11.8), so $\cos\theta = 0$ if $\theta = \frac{\pi}{2}, \frac{3\pi}{2}, \frac{5\pi}{2}, \ldots$ or if $\theta = -\frac{\pi}{2}, -\frac{3\pi}{2}, -\frac{5\pi}{2}, \ldots$. In other words, $\cos\theta = 0$ if $\theta = \frac{\pi}{2} + k\pi$ for any positive or negative integer k.

Although there are infinitely many angles with a cosine of zero, there is only *one* such angle in the restricted domain of cosine ($[0, \pi]$). The only angle in $[0, \pi]$ whose cosine is zero is $\theta = \frac{\pi}{2}$; therefore, arccos $0 = \frac{\pi}{2}$ (see Figure 11.9).

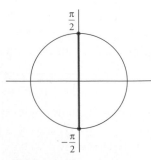

Figure 11.8
$\cos\theta = 0$ if θ terminates
at $\frac{\pi}{2}$ or $-\frac{\pi}{2}$

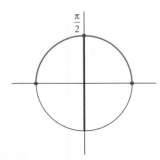

Figure 11.9
$\theta = \frac{\pi}{2}$ is the only angle in
$[0, \pi]$ whose cosine is 0

□

Finding angles with a particular tangent or secant requires a little more work than finding angles with a particular sine or cosine. We illustrate both types of calculations in the following example.

EXAMPLE 11.4 **Calculating values of inverse tangent and inverse secant by hand**

Find the exact values of:

(a) $\tan^{-1}(-\sqrt{3})$ (b) $\sec^{-1}\left(-\frac{2}{\sqrt{3}}\right)$

Solution

(a) Since $\tan^{-1} x$ is the inverse of the tangent function, we know that $\theta = \tan^{-1}(-\sqrt{3})$ if and only if $\tan(\theta) = -\sqrt{3}$. In other words, the quantity $\tan^{-1}(-\sqrt{3})$ is the angle θ (in the restricted domain $(-\frac{\pi}{2}, \frac{\pi}{2})$ of tangent) whose tangent is $-\sqrt{3}$. In terms of sine and cosine, this means that

$$\tan\theta = -\sqrt{3} \implies \frac{\sin\theta}{\cos\theta} = -\sqrt{3} \implies \sin\theta = -\sqrt{3}\cos\theta.$$

We now have to think of an angle θ in the first or fourth quadrant of the unit circle whose sine is $-\sqrt{3}$ times its cosine. Since we know that $\tan^{-1} x$ is a *function*, there must be only one such angle; we will try to "guess" it. Since $\tan\theta$ is negative, we know θ must be in the fourth quadrant. We also know that the reference triangle must have a vertical side length that is $\sqrt{3}$ times its horizontal side length. A 30–60–90 triangle has this property; see Figures 11.10–11.12.

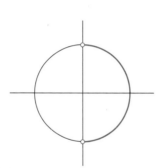
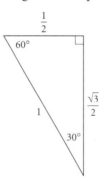
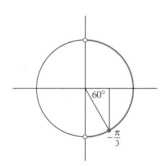

Figure 11.10
The restricted domain of tangent

Figure 11.11
Guess for a reference triangle

Figure 11.12
The angle θ must be $-\frac{\pi}{3}$

The angle in $(-\frac{\pi}{2}, \frac{\pi}{2})$ that has this reference triangle is the angle $-\frac{\pi}{3}$. We have now shown that $\tan^{-1}(-\sqrt{3}) = -\frac{\pi}{3}$.

(b) First notice that $x = -\frac{2}{\sqrt{3}}$ is less than -1 (since 2 is greater than $\sqrt{3}$) and thus is in the domain $(-\infty, -1] \cup [1, \infty)$ of $\sec^{-1}(x)$. We are looking for an angle $\theta = \sec^{-1}(-\frac{2}{\sqrt{3}})$, that is, an angle θ such that $\sec\theta = -\frac{2}{\sqrt{3}}$. Since secant is the reciprocal of cosine, this means that

$$\sec\theta = -\frac{2}{\sqrt{3}} \implies \frac{1}{\cos\theta} = -\frac{2}{\sqrt{3}} \implies \cos\theta = -\frac{\sqrt{3}}{2}.$$

There are infinitely many angles θ with a cosine of $-\frac{\sqrt{3}}{2}$, but we want the *one* angle with this property that is in the range $[0, \frac{\pi}{2}) \cup (\frac{\pi}{2}, \pi]$ of inverse secant (i.e., the restricted domain of secant). This means θ must be in the first or second quadrant. Since $\cos\theta$ is negative, we know θ must be in the second quadrant. Since $\frac{\sqrt{3}}{2}$ is the length of the long leg of a 30–60–90 triangle, we know that the reference triangle for θ is a 30–60–90 triangle. Figures 11.13–11.15 show how to put this information together to determine that $\theta = \frac{5\pi}{6}$.

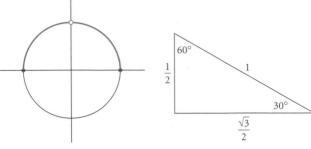

Figure 11.13
The restricted domain
of secant

Figure 11.14
Guess for a reference
triangle

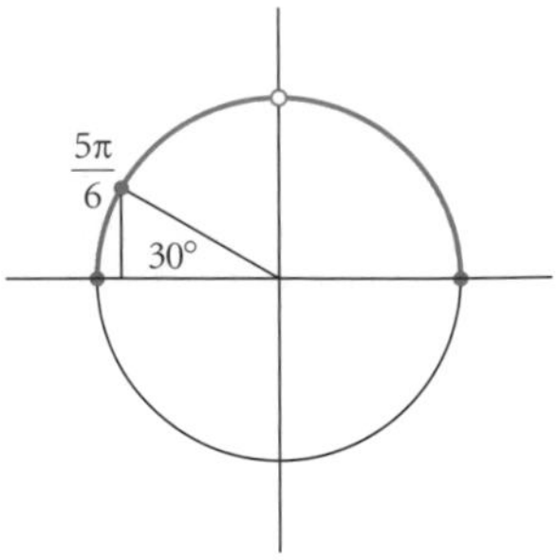

Figure 11.15
The angle θ must be $\frac{5\pi}{6}$

Since $\cos\frac{5\pi}{6} = -\frac{\sqrt{3}}{2}$, we have $\sec\frac{5\pi}{6} = \frac{1}{-\sqrt{3}/2} = -\frac{2}{\sqrt{3}}$ and thus $\sec^{-1}(-\frac{2}{\sqrt{3}}) = \frac{5\pi}{6}$. $\qquad\square$

⊛ Caution A good understanding of the domain and range of each inverse trigonometric function is vital for the calculations in Example 11.4. An understanding of the unit circle definitions of the trigonometric functions is also needed. $\qquad\square$

The inverse trigonometric values asked for in Example 11.4 were carefully chosen. It is possible to calculate *exact* values of the inverse trigonometric functions in only a few cases: those that involve the side lengths of 45–45–90 or 30–60–90 triangles, or the angles that terminate at the top, bottom, leftmost, or rightmost points of the unit circle. For example, we can calculate $\tan^{-1}(-\sqrt{3})$ by hand but cannot calculate the seemingly simpler expression $\tan^{-1}(3)$ without a calculator. You can probably find keys for inverse sine, inverse cosine, and inverse tangent on your calculator, but most calculators don't have buttons for the other inverse trigonometric functions. The following example illustrates how to use a calculator to find values of these other inverse trigonometric functions.

EXAMPLE 11.5 **Approximating a value of inverse secant with a calculator**

Use a calculator to find $\sec^{-1}(5)$.

Solution If $\theta = \sec^{-1}(5)$, then $\sec\theta = 5$, so $\frac{1}{\cos\theta} = 5$. This means that $\cos\theta = \frac{1}{5}$ and thus that $\theta = \cos^{-1}(\frac{1}{5})$. Using a calculator we can approximate $\sec^{-1}(5) = \cos^{-1}(\frac{1}{5}) \approx 1.36944$. $\qquad\square$

❓ Question Prove that $\sec^{-1}x = \cos^{-1}(\frac{1}{x})$ for all x in the domain of inverse secant. The argument is a more general version of the computation in Example 11.5. $\qquad\square$

⊛ Caution Don't make the mistake of thinking that $\sec^{-1}(5)$ is the reciprocal of $\cos^{-1}(5)$. In other words, do not confuse $\sec^{-1}(x)$ with $\frac{1}{\cos^{-1}x}$. Also, note that you cannot write $\sec^{-1}x$ as $(\frac{1}{\cos})^{-1}(x)$; this notation makes no mathematical sense, since you can't write "cos" by itself without an "input" variable or value. $\qquad\square$

11.1.4 Composing a trigonometric function and its inverse

Recall that if $f^{-1}(x)$ is the inverse of a function $f(x)$, then by definition we have

(4)
$$f^{-1}(f(x)) = x, \quad \text{for all } x \in \text{Domain}(f), \quad \text{and}$$
$$f(f^{-1}(x)) = x, \quad \text{for all } x \in \text{Domain}(f^{-1}).$$

Because $\sin^{-1} x$ is by definition the inverse of (restricted) $\sin x$, we know that

(5)
$$\sin^{-1}(\sin(x)) = x, \quad \text{for all } x \in \left[-\tfrac{\pi}{2}, \tfrac{\pi}{2}\right],$$
$$\sin(\sin^{-1}(x)) = x, \quad \text{for all } x \in [-1, 1].$$

❓ *Question*   You can say similar things about the inverse cosine, inverse tangent, and inverse secant functions. Write out what Statement (4) means for these three functions. Note that you will need to know the domains of the restricted and inverse functions in order to write out the complete version of Statement (4).   ☐

🛇 *Caution*   The domain restrictions in Statement (5) are essential. The inverse sine function "undoes" only the *restricted* sine function, not the general sine function (which is not one-to-one and thus is not invertible).   ☐

The next example illustrates what happens when Statement (5) applies and what happens when it does not.

EXAMPLE 11.6

Compositions of sine and inverse sine

Calculate the *exact* values of:

(a) $\sin^{-1}\!\left(\sin\!\left(-\tfrac{\pi}{4}\right)\right)$

(b) $\sin^{-1}\!\left(\sin\!\left(\tfrac{3\pi}{4}\right)\right)$

Solution

(a) Since $-\tfrac{\pi}{4}$ is in the domain $\left[-\tfrac{\pi}{2}, \tfrac{\pi}{2}\right]$ of the restricted sine function, we can use Statement (5). Therefore, $\sin^{-1}(\sin(-\tfrac{\pi}{4})) = -\tfrac{\pi}{4}$.

(b) You may be tempted to say that $\sin^{-1}(\sin(\tfrac{3\pi}{4}))$ is equal to $\tfrac{3\pi}{4}$. However, $\tfrac{3\pi}{4}$ is not in the domain of the restricted sine function, so we *cannot* use Statement (5) here. Instead we must calculate $\sin(\tfrac{3\pi}{4})$ and then evaluate the inverse sine function at that value. We know that $\sin(\tfrac{3\pi}{4}) = \tfrac{\sqrt{2}}{2}$ (see Figure 11.16). To find the value of $\sin^{-1}(\tfrac{\sqrt{2}}{2})$, we must find the angle in $\left[-\tfrac{\pi}{2}, \tfrac{\pi}{2}\right]$ whose sine is $\tfrac{\sqrt{2}}{2}$. This angle is $\tfrac{\pi}{4}$ (see Figure 11.17). Therefore, we have $\sin^{-1}(\sin(\tfrac{3\pi}{4})) = \sin^{-1}(\tfrac{\sqrt{2}}{2}) = \tfrac{\pi}{4}$.

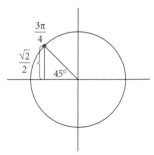

Figure 11.16
$\sin\left(\tfrac{3\pi}{4}\right) = \tfrac{\sqrt{2}}{2}$

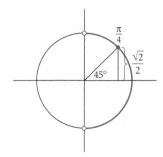

Figure 11.17
$\sin^{-1}\left(\tfrac{\sqrt{2}}{2}\right) = \tfrac{\pi}{4}$ ☐

> ### *What you should know after reading Section 11.1*
>
> ▶ The domains, ranges, and graphs of the restricted sine, cosine, tangent, and secant functions.
>
> ▶ How to use the restricted trigonometric functions to define the inverse trigonometric functions. What are the domains, ranges, and graphs of the arcsine, arccosine, arctangent, and arcsecant functions?
>
> ▶ How to calculate exact values of inverse trigonometric functions by hand, by considering the appropriate domains and ranges and using the unit circle definitions of the trigonometric functions.
>
> ▶ When is $\sin^{-1}(\sin(x))$ equal to x, and when is it not equal to x? What does this have to do with inverse functions?

Exercises 11.1

Concepts

0. Read the section and make your own summary of the material. Include whatever you think will help you review the material later on.

1. Find three restricted domains on which the function $f(x) = \cos x$ is invertible. Sketch the graphs of these three restricted cosine functions, and then sketch the graphs of their inverses.

2. List the domains and ranges of the restricted sine and cosine functions, and sketch their graphs.

3. List the domains and ranges of the restricted tangent and secant functions, and sketch their graphs.

4. Illustrate the domains of the restricted sine and cosine functions (i.e., the ranges of the inverse sine and cosine functions) on a picture of the unit circle.

5. Illustrate the domains of the restricted tangent and secant functions (i.e., the ranges of the inverse tangent and secant functions) on a picture of the unit circle.

6. What is the definition of the function "$\sin^{-1} x$"? What is the definition of the function "$\arctan x$"?

7. List the domains and ranges of the inverse sine and inverse cosine functions, and sketch their graphs.

8. List the domains and ranges of the inverse tangent and inverse secant functions, and sketch their graphs.

9. Following the convention for restricting domains of trigonometric functions, find "the" restricted domain of the cosecant function. Sketch the graph of this restricted function and that of its inverse.

10. Suppose $\arcsin x = y$. Which variable (x or y) represents an angle, and why? What does the other variable represent (in terms of the unit circle)?

11. Suppose $\cos^{-1} x = y$. Which variable (x or y) represents an angle, and why? What does the other variable represent (in terms of the unit circle)?

12. Suppose $\tan^{-1} x = y$. Which variable (x or y) represents an angle, and why? What does the other variable represent (in terms of the unit circle)?

13. Suppose $\operatorname{arcsec} x = y$. Which variable ($x$ or y) represents an angle, and why? What does the other variable represent (in terms of the unit circle)?

14. Explain why $\sin^{-1}(\frac{1}{2})$ is *not* equal to $\frac{5\pi}{6}$, even though $\sin(\frac{5\pi}{6}) = \frac{1}{2}$.

15. List nine values of x for which you would be able to calculate the exact value of $\sin^{-1} x$ by hand. (*Hint:* Think about the two right triangles whose side lengths we know, and think about the unit circle.) Could you also calculate $\cos^{-1} x$, $\tan^{-1} x$, and $\sec^{-1} x$ by hand for these values?

16. Which of the following expressions are defined? Why or why not?

(a) $\arcsin\left(-\frac{1}{25}\right)$ (b) $\cos^{-1}\left(\frac{3}{2}\right)$

(c) $\tan^{-1}(100)$ (d) $\operatorname{arcsec}\left(\frac{\pi}{4}\right)$

17. Without calculating the exact or approximate values of the following expressions, use the unit circle to determine whether each of the following quantities is positive or negative.

(a) $\sin^{-1}\left(-\frac{1}{5}\right)$ (b) $\arccos\left(-\frac{2}{3}\right)$

(c) $\arctan(2)$ (d) $\sec^{-1}(-5)$

■ **Fill in the blanks:**

18. "$\cos^{-1} x$" is the angle in the interval _____ whose _____ is x.

19. _____ is the angle in the interval _____ whose tangent is x.

20. $y = \arcsin x$ if and only if $\sin y =$ _____, for all $x \in$ _____ and $y \in$ _____.

21. $y =$ _____ if and only if $\cos y = x$, for all $x \in$ _____ and $y \in$ _____.

22. $\cos^{-1}(\cos(x)) =$ _____, for all $x \in$ _____.

23. $\text{arcsec}(\sec(x)) =$ _____, for all $x \in$ _____.

24. If $\sin^{-1} x = \theta$ and $\sin \theta$ is negative, then θ is in the _____ quadrant.

25. If $\tan^{-1} x = \theta$ and $\tan \theta$ is positive, then θ is in the _____ quadrant.

26. If $\sec^{-1} x = \theta$, then $\cos \theta =$ _____.

27. If $\arctan x = \theta$ and $\sin \theta = \frac{1}{3}$, then $\cos \theta =$ _____.

28. If $1 + 3 \arcsin x = \theta$, then $x =$ _____.

29. If $\arcsin(3x + 1) = \theta$, then $x =$ _____.

30. Find all angles whose sine is zero, and then find $\sin^{-1}(0)$.

31. Find all angles whose cosine is $-\frac{1}{2}$, and then find $\cos^{-1}(-\frac{1}{2})$.

32. Find all angles whose secant is 2, and then find $\sec^{-1}(2)$.

33. Find all angles whose tangent is 1, and then find $\tan^{-1}(1)$.

■ **Determine whether each of the following statements is true or false. If a statement is true, explain why. If a statement is false, provide a counterexample.**

34. True or False: $\tan^{-1} x = \text{arctan } x$ for all values of x.

35. True or False: $\sec^{-1}(\sec x) = x$ for all values of x.

36. True or False: If $\cos \theta = \frac{1}{3}$, then $\arccos(\frac{1}{3}) = \theta$.

37. True or False: If $\arccos(\frac{1}{3}) = \theta$, then $\cos \theta = \frac{1}{3}$.

38. True or False: $\tan^{-1} x = \frac{1}{\tan x}$.

39. True or False: $\sec^{-1} x = \frac{1}{\cos^{-1} x}$.

40. True or False: If $\arccos x = \theta$, then $\sin \theta$ is greater than or equal to zero.

Skills

■ **Calculate the *exact* values of the following expressions by hand. Include the unit circle and/or a triangle as part of your work.**

41. $\sin^{-1}(0)$ 42. $\arccos(0)$ 43. $\arcsin(1)$

44. $\cos^{-1}(1)$ 45. $\sin^{-1}(-1)$ 46. $\cos^{-1}(-1)$

47. $\sin^{-1}(\frac{1}{2})$ 48. $\arccos(-\frac{1}{2})$ 49. $\sin^{-1}(-\frac{\sqrt{2}}{2})$

50. $\arccos(\frac{\sqrt{2}}{2})$ 51. $\arcsin(-\frac{\sqrt{3}}{2})$ 52. $\cos^{-1}(\frac{\sqrt{3}}{2})$

53. $\text{arcsec}(-2)$ 54. $\sec^{-1}(\frac{2}{\sqrt{3}})$ 55. $\sec^{-1}(-1)$

56. $\text{arcsec}(-\frac{2}{\sqrt{2}})$ 57. $\tan^{-1}(0)$ 58. $\arctan(-1)$

59. $\arctan(\sqrt{3})$ 60. $\tan^{-1}(-\frac{1}{\sqrt{3}})$

■ **Use a calculator to find an approximate numerical value for each of the following expressions, if it exists. If an expression does not exist, use the unit circle to explain why.**

61. $\sin^{-1}(\frac{1}{5})$ 62. $\arccos(-\frac{2}{3})$

63. $\cos^{-1}(\frac{3}{2})$ 64. $\arcsin(-2)$

65. $\text{arcsec}(0)$ 66. $\sec^{-1}(-5)$

67. $\text{arcsec}(\frac{\sqrt{2}}{2})$ 68. $\tan^{-1}(3)$

69. $\csc^{-1}(\frac{5}{4})$ 70. $\cot^{-1}(3)$

71. $\sin^{-1}(\sin \frac{5\pi}{4})$ 72. $\sec^{-1}(\sec(-\frac{\pi}{5}))$

Proofs

73. Prove that, for all x in the domain of the inverse secant function, we have $\sec^{-1} x = \cos^{-1}(\frac{1}{x})$.

74. Prove that, for all x in the domain of the inverse cosecant function, we have $\csc^{-1} x = \sin^{-1}(\frac{1}{x})$.

75. Prove that, for all x in the domain of the inverse tangent function, we have $\cot^{-1} x = \tan^{-1}(\frac{1}{x})$.

11.2 Derivatives of Inverse Trigonometric Functions

In this section we will develop formulas for the derivatives of the inverse trigonometric functions. These derivatives will have useful applications to certain calculations in integral calculus in Part IV.

11.2.1 Rules for differentiating the inverse trigonometric functions

We can determine the derivatives of the inverse trigonometric functions using implicit differentiation and the derivatives of the trigonometric functions. The derivatives of the inverse sine, inverse cosine, inverse tangent, and inverse secant functions are listed in Theorem 11.3. We will need to consider the unit circle definitions of the trigonometric and inverse trigonometric functions to prove these differentiation formulas.

THEOREM 11.3

Derivatives of the Inverse Trigonometric Functions

(a) $\frac{d}{dx}(\sin^{-1} x) = \dfrac{1}{\sqrt{1 - x^2}}$

(b) $\frac{d}{dx}(\cos^{-1} x) = \dfrac{-1}{\sqrt{1 - x^2}}$

(c) $\frac{d}{dx}(\tan^{-1} x) = \dfrac{1}{1 + x^2}$

(d) $\frac{d}{dx}(\sec^{-1} x) = \dfrac{1}{|x|\sqrt{x^2 - 1}}$

You might be surprised that the derivatives of the inverse trigonometric functions are *algebraic*. So far, many types of functions we have seen have had derivatives of the same "type." For example, the derivatives of exponential functions are themselves exponential functions, and the derivatives of the trigonometric functions are also made up of trigonometric functions. However, as we see in Theorem 11.3, the derivatives of inverse trigonometric functions are algebraic. That is, they are built out of sums, differences, products, and power functions with rational powers. This interesting property of the inverse trigonometric functions will be useful in certain "integration" problems in Part IV.

◈ *Caution*   Be sure to memorize the differentiation formulas given in Theorem 11.3. It may help to notice that the derivatives of inverse sine and inverse cosine differ only by a minus sign.   □

Before we prove the differentiation rules in Theorem 11.3, let's try some simple computational examples.

EXAMPLE 11.7

Differentiating functions that involve inverse trigonometric expressions

Find the derivatives of the following functions:

(a) $f(x) = \sin^{-1}(2x)$

(b) $g(x) = (\sec^{-1} x)^2$

Solution

(a) The function $f(x)$ is a composition of $2x$ (the "inside" function) with the inverse trigonometric function $\sin^{-1} x$ (the "outside" function). By the chain rule, we have

$$\frac{d}{dx}(\sin^{-1}(2x)) = \frac{1}{\sqrt{1 - (2x)^2}}(2) \qquad \text{(chain rule and Theorem 11.3)}$$

$$= \frac{2}{\sqrt{1 - 4x^2}}. \qquad \text{(algebra)}$$

The multiplication by 2 in the first line of the calculation is the chain rule step where we multiply by the derivative of the "inside" function $2x$.

(b) The function $g(x)$ is also a composition, but this time the inverse trigonometric function is on the "inside." By the chain rule, we have

$$\frac{d}{dx}((\sec^{-1} x)^2) = 2(\sec^{-1} x)^1 \left(\frac{1}{|x|\sqrt{x^2 - 1}} \right) \qquad \text{(chain rule and Theorem 11.3)}$$

$$= \frac{2 \sec^{-1} x}{|x|\sqrt{x^2 - 1}}. \qquad \text{(algebra)}$$

□

❧ **Caution**   Be careful that you do not confuse the reciprocal $(\sin x)^{-1} = \frac{1}{\sin x}$ of $\sin x$ with the inverse function $\sin^{-1} x$ when finding derivatives. For example,

$$\frac{d}{dx}(\sin^{-1} x) = \frac{1}{\sqrt{1 - x^2}}$$

by Theorem 11.3, whereas

$$\frac{d}{dx}((\sin x)^{-1}) = -(\sin x)^{-2}(\cos x) = -\frac{\cos x}{\sin^2 x}.$$ □

❧ **Caution**   Theorem 11.3 says that the derivative of $\sin^{-1} x$ is $\frac{1}{\sqrt{1-x^2}}$. It does *not* say that $\sin^{-1} x$ and $\frac{1}{\sqrt{1-x^2}}$ are equal to each other. Avoid this common error.   □

11.2.2 Proving the derivative formulas for inverse sine and inverse cosine

The proofs of all of the differentiation rules in Theorem 11.3 are similar (although proving the differentiation rule for the inverse secant function is slightly more complicated, as you might guess by the presence of an absolute value in the formula). We present a careful, detailed proof for the derivative of the inverse sine function. The proof for the derivative of the inverse cosine function is entirely similar and is left for the exercises.

> **PROOF**   **(PART (A) OF THEOREM 11.3)**   We wish to show that the derivative of $\sin^{-1} x$ is $\frac{1}{\sqrt{1-x^2}}$. However, at this point we do not have any differentiation rules for the inverse trigonometric functions (in fact, that is what we are trying to prove). Fortunately, it will not be necessary to go all the way back to the definition of derivative; in fact, if we did, we would quickly get stuck (try it to see why). Instead, we can use the fact that $\sin^{-1} x$ is the inverse of $\sin x$, whose derivative we know. The method we use is similar to the proof we used in Section 9.2 to find the derivative of a logarithmic function.
>
> Since $\sin^{-1} x$ is by definition the inverse of the function $\sin x$, the composition of these two functions is the identity x:

(1) $$\sin(\sin^{-1} x) = x$$

By implicitly differentiating both sides of Equation (1) with respect to the variable x, we will be able to solve for the derivative of $\sin^{-1} x$, as follows:

$$\sin(\sin^{-1} x) = x \qquad \text{($\sin^{-1} x$ is the inverse of $\sin x$)}$$

$$\frac{d}{dx}(\sin(\sin^{-1} x)) = \frac{d}{dx}(x) \qquad \text{(differentiate both sides)}$$

$$\cos(\sin^{-1} x) \cdot \frac{d}{dx}(\sin^{-1} x) = 1 \qquad \text{(chain rule)}$$

Note that we do not know the derivative of the "inside" function $\sin^{-1} x$, and thus we simply write "$\frac{d}{dx}(\sin^{-1} x)$" to represent that derivative. Luckily, this derivative is exactly what we are trying to find, and we can now use the result above to solve for it:

$$\frac{d}{dx}(\sin^{-1} x) = \frac{1}{\cos(\sin^{-1} x)}. \qquad \text{(algebra)}$$

At this point we have an expression for the derivative of $\sin^{-1} x$. It now remains only to show that $\frac{1}{\cos(\sin^{-1} x)}$ is equal to $\frac{1}{\sqrt{1-x^2}}$, for all values of x in the domain $[-1, 1]$ of $\sin^{-1} x$. At first glance you might not even believe that these two expressions could possibly be equal; after all, there is no apparent algebraic manipulation that could lead from one to the other. However, notice that the two expressions *do* have almost the same domain (check this). The key is to write the expression $\cos(\sin^{-1} x)$ in a simpler way.

Recall that the outputs of the function $\sin^{-1} x$ are angles; specifically, $\sin^{-1} x$ is the angle (in $[-\frac{\pi}{2}, \frac{\pi}{2}]$) whose sine is x. If we define $\theta := \sin^{-1} x$, then $\sin \theta = x$. With this notation we are interested in finding the cosine of θ, since $\cos \theta = \cos(\sin^{-1} x)$ is the expression we are attempting to rewrite. We know from the range of the inverse sine function that θ must be in the interval $[-\frac{\pi}{2}, \frac{\pi}{2}]$.

Let's first consider the case where θ is in the interval $[0, \frac{\pi}{2}]$, that is, the case where θ is in the first quadrant. (In that case x must be positive; why?) The reference triangle for such a θ is shown in Figure 11.18. If we wish θ to have a sine of x, then the length of the vertical leg of the triangle must be x. The hypotenuse of the triangle is of length 1 (since we are on the *unit* circle). We could also have considered that the sine of θ is "opposite over hypotenuse"; thus one triangle involving our angle θ could have an opposite side of length x and a hypotenuse of length 1 (since then $\sin \theta = \frac{\text{opp}}{\text{hyp}} = \frac{x}{1} = x$, as desired).

Using the Pythagorean Theorem, we find that the length of the remaining leg of the triangle is $\sqrt{1 - x^2}$; see Figure 11.19.

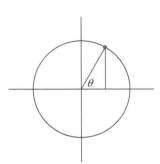

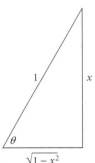

Figure 11.18
Reference triangle for an angle θ in $[0, \frac{\pi}{2}]$

Figure 11.19
Use Pythagorean Theorem to determine length of remaining leg

Finally, we now see that the cosine of θ (the horizontal coordinate of the point on the unit circle corresponding to θ) is $\sqrt{1 - x^2}$. The case where θ is in the fourth quadrant (i.e., where $\theta \in [-\frac{\pi}{2}, 0]$) is similar and also shows that $\cos \theta = \sqrt{1 - x^2}$. We now know that

$$\cos(\sin^{-1} x) = \cos \theta = \sqrt{1 - x^2},$$

and thus

$$\frac{d}{dx}(\sin^{-1} x) = \frac{1}{\cos(\sin^{-1} x)} = \frac{1}{\sqrt{1 - x^2}}. \qquad \blacksquare$$

? *Question*   In the proof of part (a) of Theorem 11.3, we used the fact that $\sin(\sin^{-1} x) = x$ for all x in the domain of the inverse sine function. It is also true that $\sin^{-1}(\sin x) = x$ for all x in the domain of the restricted sine function; why did we not use this equation instead? Try reworking the proof with this second equation; what goes wrong?   □

? *Question*   Although it is true that the derivative of $\sin^{-1} x$ is $\frac{1}{\cos(\sin^{-1} x)}$ (as we saw in the proof of part (a) of Theorem 11.3), we took the time to rewrite this expression in the algebraic form $\frac{1}{\sqrt{1-x^2}}$. Why did we do this? Which formulation of the derivative do you think it would be easier to work with?   □

? *Question* Notice that the derivative of $\cos^{-1} x$ is the negative of the derivative of $\sin^{-1} x$. Does this make sense in terms of the graphs of these functions? □

11.2.3 Proving the derivative formulas for inverse secant and inverse tangent

The proofs of the last two parts of Theorem 11.3 are similar to the proof of part (a) of Theorem 11.3. For the most part, we will show only the major steps and leave some of the details for the exercises. However, we will explain in detail why the absolute value appears in the differentiation formula for inverse secant.

We begin with a proof showing that the derivative of the inverse tangent function is $\frac{1}{1+x^2}$.

PROOF (**PART (C) OF THEOREM 11.3**) We know that $\tan(\tan^{-1} x) = x$ for all x in the domain $(-\infty, \infty)$ of inverse tangent. Using this fact, implicit differentiation, and algebra, we have

$$\frac{d}{dx}(\tan^{-1} x) = \frac{1}{\sec^2(\tan^{-1} x)}.$$

It now suffices to show that $\sec^2(\tan^{-1} x) = 1 + x^2$ for all x in the domain of $\tan^{-1} x$. If we define $\theta := \tan^{-1} x$, then $\tan \theta = x$. We are interested in finding $\sec^2 \theta$ in terms of x. We can construct a triangle with an angle θ so that $\tan \theta = x$ by letting the length of the side of the triangle "opposite" to θ be x and letting the length of the side "adjacent" to θ be 1. By the Pythagorean Theorem, the remaining side has a length of $\sqrt{1 + x^2}$ (see Figure 11.20).

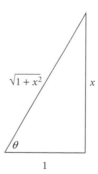

Figure 11.20

Note that the hypotenuse of this triangle is *not* of length 1, so technically it is not a reference triangle in the unit circle (although a scaled version of it would be and would have the same sine and cosine values). Since θ is in the range $(-\frac{\pi}{2}, \frac{\pi}{2})$ of inverse tangent, this (scaled) reference triangle would be in the first or fourth quadrant.

Using the triangle in Figure 11.20, we can see that the cosine of θ is $\frac{1}{\sqrt{1+x^2}}$. Using this information, we have $\sec^2(\tan^{-1} x) = \sec^2 \theta = 1 + x^2$. Now we can rewrite the derivative of tangent as the algebraic function $\frac{d}{dx}(\tan^{-1} x) = \frac{1}{1+x^2}$. You will work out these arguments in more detail in the exercises. ■

In the proofs of parts (a) and (c) of Theorem 11.3, the quadrant we worked in did not make a difference. This is not the case in the proof of part (d). The range of $\sec^{-1} x$ is $[0, \frac{\pi}{2}) \cup (\frac{\pi}{2}, \pi]$, so we will be considering angles terminating in the first and second quadrants. The derivative of inverse secant involves an absolute value because of a sign difference that will occur depending on whether the angle is in the first or second quadrant.

PROOF (PART (D) OF THEOREM 11.3) Differentiating both sides of $\sec(\sec^{-1} x) = x$ and solving for the derivative of inverse secant, we find

$$\frac{d}{dx}(\sec^{-1} x) = \frac{1}{\sec(\sec^{-1} x)\tan(\sec^{-1} x)} = \frac{1}{x\tan(\sec^{-1} x)}.$$

We need to rewrite $\tan(\sec^{-1} x)$. Define $\theta := \sec^{-1} x$, so that $\sec\theta = x$. From the range of the inverse secant function we know that θ is in $[0, \frac{\pi}{2}) \cup (\frac{\pi}{2}, \pi]$, that is, in the first or second quadrant in terms of the unit circle. We examine these two cases separately, since in this particular proof the cases are somewhat different.

First, suppose that θ is in the interval $[0, \frac{\pi}{2})$, in which case x is positive (in fact, $x \geq 1$; think about the graph of $\theta = \sec^{-1} x$). See Figure 11.21. If $\sec\theta = x$, then a triangle with angle θ will have $\frac{\text{hyp}}{\text{adj}} = x$. Using the Pythagorean Theorem, we can determine the length of the remaining side (see Figure 11.22). We can then use this triangle to calculate $\tan(\sec^{-1} x)$:

$$\tan(\sec^{-1} x) = \tan\theta = \frac{\sqrt{x^2 - 1}}{1} = \sqrt{x^2 - 1}.$$

Therefore, for $x \geq 1$ (since $\theta \in [0, \frac{\pi}{2})$), we have

(2)
$$\frac{d}{dx}(\sec^{-1} x) = \frac{1}{x\tan(\sec^{-1} x)} = \frac{1}{x\sqrt{x^2 - 1}}.$$

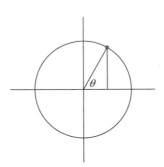

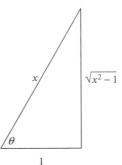

Figure 11.21
Reference triangle for
an angle θ in $[0, \frac{\pi}{2})$

Figure 11.22
Scaled reference triangle
with $\sec\theta = x$

On the other hand, suppose θ is in the interval $(\frac{\pi}{2}, \pi]$. In this case x is negative (in fact, $x \leq -1$). The reference triangle in this case has the same side lengths as the one in Figure 11.22, but since we are in the second quadrant, where tangent is negative, we must consider signs. Note that the angle θ is different from the reference angle ϕ shown in Figures 11.23 and 11.24.

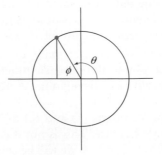

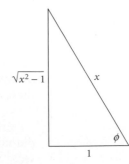

Figure 11.23
Reference triangle for
an angle θ in $(\frac{\pi}{2}, \pi]$

Figure 11.24
Scaled reference triangle
with $\sec\theta = x$

Taking into account the fact that the horizontal coordinate of this reference triangle is negative, we can now calculate $\tan(\sec^{-1} x)$:

$$\tan(\sec^{-1} x) = \tan\theta = \frac{\sqrt{x^2-1}}{-1} = -\sqrt{x^2-1}.$$

Therefore, for $x \le -1$ (since $\theta \in (\frac{\pi}{2}, \pi]$), we have

(3) $$\frac{d}{dx}(\sec^{-1} x) = \frac{1}{x\tan(\sec^{-1} x)} = \frac{1}{x(-\sqrt{x^2-1})} = \frac{1}{-x\sqrt{x^2-1}}.$$

We have now found an algebraic representation for the derivative of $\sec^{-1} x$, but unfortunately, it depends on whether x is positive or negative. This is where the absolute value comes in; recall that the absolute value of x is equal to x if x is positive and equal to $-x$ if x is negative. Thus in Equation (2) we can replace x by $|x|$ (since x is positive in that case), and in Equation (3) we can replace $-x$ with $|x|$ (since in that case, x is negative). Therefore, for all values of x in the domain $(-\infty, -1] \cup [1, \infty)$ of $\sec^{-1} x$, we have

$$\frac{d}{dx}(\sec^{-1} x) = \frac{1}{|x|\sqrt{x^2-1}}.$$ ∎

11.2.4 Derivative and antiderivative calculations

Now that we know the derivatives of the inverse sine, inverse cosine, inverse tangent, and inverse secant functions, we can differentiate an even wider variety of functions.

EXAMPLE 11.8

Differentiating when inverse trigonometric functions are involved

Differentiate the following functions.

(a) $f(x) = x\cos^{-1}(3x+1)$
(b) $f(x) = \sec^{-1}(\sin^{-1} x)$
(c) $f(x) = \sin^2(\tan^{-1} e^{3x})$

Solution

(a) This is a product of functions, and thus we begin with the product rule. We will also need the chain rule to differentiate the composition $\cos^{-1}(3x+1)$:

$$f'(x) = (1)(\cos^{-1}(3x+1)) + x\left(\frac{-1}{\sqrt{1-(3x+1)^2}}\right)(3)$$

$$= \cos^{-1}(3x+1) + \frac{-3x}{\sqrt{1-(3x+1)^2}}.$$

(b) This function is the composition of two inverse trigonometric functions, where the "inside" function is $\sin^{-1} x$ and the "outside" function is $\sec^{-1} x$. By the chain rule, we have

$$f'(x) = \frac{1}{|\sin^{-1}x|\sqrt{(\sin^{-1}x)^2-1}} \cdot \frac{d}{dx}(\sin^{-1} x) \qquad \text{(chain rule, derivative of } \sec^{-1} x)$$

$$= \frac{1}{|\sin^{-1}x|\sqrt{(\sin^{-1}x)^2-1}}\left(\frac{1}{\sqrt{1-x^2}}\right). \qquad \text{(derivative of } \sin^{-1} x)$$

(c) We'll begin by rewriting $\sin^2 u$ as $(\sin u)^2$ so we can see the correct order for the differentiation rules. The function $f(x)$ is actually a composition of *five* functions: x^2, $\sin x$, $\tan^{-1} x$, e^x, and $3x$. We will have to apply the chain rule quite a few times. We will proceed carefully, writing several intermediate steps:

$$f'(x)$$

$$= \frac{d}{dx}(\sin^2(\tan^{-1} e^{3x}))$$

$$= \frac{d}{dx}((\sin(\tan^{-1} e^{3x}))^2) \qquad \text{(rewrite function)}$$

$$= 2(\sin(\tan^{-1} e^{3x})) \frac{d}{dx}(\sin(\tan^{-1} e^{3x})) \qquad \text{(chain and power rules)}$$

$$= 2(\sin(\tan^{-1} e^{3x})) \cos(\tan^{-1} e^{3x}) \frac{d}{dx}(\tan^{-1} e^{3x}) \qquad \text{(chain rule again)}$$

$$= 2(\sin(\tan^{-1} e^{3x})) \cos(\tan^{-1} e^{3x}) \left(\frac{1}{1+(e^{3x})^2}\right) \frac{d}{dx}(e^{3x}) \qquad \text{(chain rule again)}$$

$$= 2(\sin(\tan^{-1} e^{3x})) \cos(\tan^{-1} e^{3x}) \left(\frac{1}{1+(e^{3x})^2}\right) (3e^{3x}). \qquad \text{(the last chain rule)}$$

☐

We can also *anti*differentiate some new functions. Remember that the derivatives of the inverse trigonometric functions are algebraic (rather than trigonometric); therefore, the antiderivatives of some algebraic functions involve inverse trigonometric functions.

EXAMPLE 11.9

Using inverse trigonometric functions to antidifferentiate

Find a function $f(x)$ whose *derivative* is $f'(x) = \dfrac{1}{1+4x^2}$.

Solution   Since the derivative of $\tan^{-1} x$ is $\frac{1}{1+x^2}$, we might suspect that the function $f(x)$ we are looking for is related to inverse tangent. We will use an intelligent guess-and-check method to find $f(x)$. Clearly $f(x) = \tan^{-1} x$ isn't exactly right (its derivative is missing the "4"). A good guess might be $f(x) = \tan^{-1}(4x)$. Let's try that:

$$\frac{d}{dx}(\tan^{-1}(4x)) = \frac{1}{1+(4x)^2}(4) = \frac{4}{1+16x^2}.$$

Obviously that wasn't quite right either, but by examining the results, we can make a new guess. We might try $\tan^{-1}(2x)$, since the "$2x$" will be squared in the derivative and become the "$4x^2$" we are looking for in the denominator:

$$\frac{d}{dx}(\tan^{-1}(2x)) = \frac{1}{1+(2x)^2}(2) = \frac{2}{1+4x^2}.$$

Now we are getting somewhere; this differs by a multiplicative constant from the derivative $f'(x)$ we are looking for, and that is easy to fix. We need only divide our guess by that constant. Try the function $f(x) = \frac{1}{2}\tan^{-1}(2x)$:

$$\frac{d}{dx}\left(\frac{1}{2}\tan^{-1}(2x)\right) = \left(\frac{1}{2}\right)\frac{1}{1+(2x)^2}(2) = \frac{1}{1+4x^2}.$$

We now know that $f(x) = \frac{1}{2}\tan^{-1}(2x)$ is a function whose derivative is $f'(x) = \frac{1}{1+4x^2}$. Of course, we could also add any constant to this function $f(x)$ and still obtain the same derivative; for example, $f(x) = \frac{1}{2}\tan^{-1}(2x) + 5$ would also work. In fact, any function of the form $f(x) = \frac{1}{2}\tan^{-1}(2x) + C$ will have $f'(x) = \frac{1}{1+4x^2}$.   ☐

> **What you should know after reading Section 11.2**
>
> ▶ The formulas for the derivatives of the inverse sine, inverse cosine, inverse tangent, and inverse secant functions.
>
> ▶ How to prove these derivative formulas. Note that this will require a strong understanding of the unit circle definitions of the trigonometric functions, as well as the domains and ranges of the inverse trigonometric functions.
>
> ▶ How to calculate the derivatives of functions that involve inverse sine, inverse cosine, inverse tangent, or inverse secant. How to find antiderivatives that involve inverse trigonometric functions.

Exercises 11.2

Concepts

0. Read the section and make your own summary of the material. Include whatever you think will help you review the material later on.

1. List all of the derivative formulas we have learned so far in this course (for example, but not limited to, the derivatives of the inverse trigonometric functions and the derivatives of exponential and logarithmic functions).

2. Use the graphs of inverse sine and inverse cosine to illustrate that (where their domains overlap) the derivative of inverse cosine is the negative of the derivative of inverse sine.

3. Find the domains of the functions $\sin^{-1} x$ and $\frac{1}{\sqrt{1-x^2}}$. Are they the same? Why does it make sense that the two domains are so similar?

4. Find the domains of the functions $\tan^{-1} x$ and $\frac{1}{1+x^2}$. Are they the same? Why does it make sense that the two domains are so similar?

5. Find the domains of the functions $\sec^{-1} x$ and $\frac{1}{|x|\sqrt{x^2-1}}$. Are they the same? Why does it make sense that the two domains are so similar?

6. The function $\sin^{-1} x$ is defined on $[-1, 1]$, but its derivative $\frac{1}{\sqrt{1-x^2}}$ is defined only on $(-1, 1)$. Explain why the tangent lines to the graph of $y = \sin^{-1} x$ at $x = 1$ and $x = -1$ do not exist. (*Hint:* Think about the corresponding tangent lines on the graph of the restricted sine function.)

7. How can the derivative of $\sin^{-1} x$ be equal to *both* $\frac{1}{\sqrt{1-x^2}}$ and $\frac{1}{\cos(\sin^{-1} x)}$? Discuss why it is more convenient to use the algebraic formulation of the derivative of inverse sine when calculating derivatives.

8. In this section we proved that the derivative of $\sin^{-1} x$ is $\frac{1}{\sqrt{1-x^2}}$. During this proof we defined $\theta = \sin^{-1} x$.

Explain how we know that the reference triangle for θ must be in the first or fourth quadrant of the unit circle.

9. In this section we proved that the derivative of $\sin^{-1} x$ is $\frac{1}{\sqrt{1-x^2}}$. During this proof we used the fact that $\sin(\sin^{-1} x) = x$ for all x in the domain of the inverse sine function. It is also true that $\sin^{-1}(\sin x) = x$ for all x in the domain of the restricted sine function. Explain why using this equation instead would not have worked.

10. In the proof for the derivative of inverse tangent given in the text, we used a reference triangle whose hypotenuse was *not* of length 1. Why is this okay?

11. Briefly explain why the formula for the derivative of $\sec^{-1} x$ has an absolute value in it. It may help to consult the proof of this derivative formula given in the text.

Skills

■ Write each of the following as an algebraic expression. Use the unit circle and a reference triangle to explain your answers.

12. $\cos(\sin^{-1} x)$ **13.** $\sin(\cos^{-1} x)$

14. $\sin(\sin^{-1} x)$ **15.** $\sec^2(\tan^{-1} x)$

16. $\tan(\sec^{-1} x)$ **17.** $\sec(\cos^{-1} x)$

18. $\sin(\sec^{-1} x)$ **19.** $\cot(\tan^{-1} x)$

20. $\sin^2(\tan^{-1} x)$

■ Calculate the derivatives of the following functions.

21. $f(x) = \cos^{-1}(2x)$ **22.** $f(x) = (\cos(2x))^{-1}$

23. $f(x) = \tan^{-1}(\ln x)$ **24.** $f(x) = 3(\sin^{-1} x)^2$

25. $f(x) = \sin^{-1}(3x^2)$ **26.** $f(x) = \dfrac{\sin^{-1} x}{\cos^{-1} x}$

27. $f(x) = \sec^{-1}(e^x)$ **28.** $f(x) = x^2 \arctan x^2$

29. $f(x) = \sin^2(\sec^{-1} x)$ **30.** $f(x) = e^{\arccos x}$

31. $f(x) = 3\tan^{-1}(\sin 2^x)$

32. $f(x) = \ln(\text{arcsec}(\sin^2 x))$

33. $f(x) = \dfrac{\sin(\arcsin x)}{\tan^{-1} x}$

34. $f(x) = \sec(1 + \tan^{-1} x)$

35. $f(x) = \sin^{-1} x \cos^{-1}(x^2)$

■ **For each problem below, find a function $f(x)$ that has the given derivative $f'(x)$. (*Note:* The function $f(x)$ may or may not involve inverse trigonometric functions. You may have to guess and check a number of times before you discover a function that works.)**

36. $f'(x) = \dfrac{2}{1 + 9x^2}$

37. $f'(x) = \dfrac{1}{\sqrt{1 - 3x^2}}$

38. $f'(x) = x(1 - 16x^2)^{-1}$

39. $f'(x) = \dfrac{1}{4 + x^2}$

40. $f'(x) = \dfrac{2x}{1 + x^2}$

41. $f'(x) = \dfrac{2x}{1 + x^4}$

42. $f'(x) = \dfrac{3}{\sqrt{4 - 36x^2}}$

43. $f'(x) = \dfrac{e^x}{1 + e^x}$

44. $f'(x) = \dfrac{e^x}{1 + e^{2x}}$

Proofs

45. Prove that the derivative of the inverse sine function is $\dfrac{1}{\sqrt{1-x^2}}$. Use the proof given in this section as a guide, but write the proof in your own words.

46. Prove that the derivative of the inverse cosine function is $\dfrac{-1}{\sqrt{1-x^2}}$.

47. Prove that the derivative of the inverse tangent function is $\dfrac{1}{1+x^2}$. Use the proof outline given in this section as a guide, but write the proof in your own words.

48. Prove that the derivative of the inverse secant function is $\dfrac{1}{|x|\sqrt{x^2-1}}$. Use the proof outline given in this section as a guide, but write the proof in your own words.

PART IV

INTEGRATION

So far we have investigated the differential calculus of various types of functions. In this part of the book, we examine the integral calculus of each of these types of functions.

CHAPTER 12

Definite Integrals

*I*n this chapter we investigate methods of calculating areas. To represent such areas, we will develop a formal notation called the definite integral, which will be defined in terms of limits of sums. We begin with a discussion of sigma notation, which will enable us to express sums in a compact way. Through this notation we will develop the concept of the definite integral and find approximations and exact values for various areas. As we will see at the end of the chapter, definite integrals are useful for more than just areas; they also enable us to define the average value of a function on an interval.

CONTENTS

12.1 # Geometric Approximation and Sigma Notation

In this section we introduce the concept of approximating the area of a region by dividing up the region into pieces and then approximating the areas of the pieces. This will be the key idea behind the development of the definite integral in Section 12.3. Approximating area in this manner will require adding up many small quantities, and thus we also introduce the idea of *sigma notation,* a notation for compactly representing sums.

12.1.1 Geometric approximation

As you know, a circle with radius r has area $A = \pi r^2$. In particular, a circle of radius 2 units has an area of $A = \pi 2^2 = 4\pi$. But wait a moment. Where does the "area formula" $A = \pi r^2$ come from? Why is it true? Suppose for a moment that we didn't know this area formula. How could we find, or at least approximate, the area of a circle with radius 2? Example 12.1 suggests an answer.

EXAMPLE 12.1

Approximating the area of a circle

Use a grid to approximate the area of a circle of radius 2.

Solution   We will give three approximations, each better than the one that precedes it. In Figure 12.1, a circle of radius 2 units is shown with a grid of squares, each 1 unit in length. In Figure 12.2, each square has a side length of $\frac{1}{2}$ unit, and in Figure 12.3, the length of each square is $\frac{1}{4}$ unit.

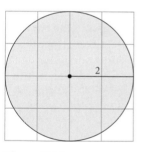 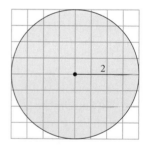 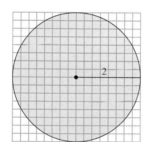

Figure 12.1   **Figure 12.2**   **Figure 12.3**

The area of each square in Figure 12.1 is 1 square unit. We will approximate the area of the circle by counting the squares that intersect the circle. The circle includes four full squares (in the center), and 12 partial squares. We will approximate by counting each partial square as *half* of a square. (There are other reasonable ways to count the partial squares; we just chose this way because it was simple.) This method produces an approximation of

$$A \approx 4(1) + 12\left(\tfrac{1}{2}\right) = 10 \text{ square units.}$$

As we "know," the actual area of the circle is $4\pi \approx 12.5664$, so the approximation we found is not very accurate. If we use the grid in Figure 12.2, we can obtain a better approximation. In this case, the circle covers 32 full squares (each of area $\left(\tfrac{1}{2}\right)^2 = \tfrac{1}{4}$) and 28 "partial" squares (which we will count as "half" squares, or $\tfrac{1}{8}$ square unit each). We now have the approximation

$$A \approx 32\left(\tfrac{1}{4}\right) + 28\left(\tfrac{1}{8}\right) = 11.5 \text{ square units.}$$

This approximation is closer to the true area since the squares we used were smaller. An even better approximation can be obtained using Figure 12.3, where there are 164 "full" squares of area $(\frac{1}{4})^2 = \frac{1}{16}$ and 52 "half" squares, which we'll count as having an area of $\frac{1}{32}$ unit each. This gives us an even better approximation:

$$A \approx 164\left(\tfrac{1}{16}\right) + 52\left(\tfrac{1}{32}\right) = 11.875 \text{ square units.} \qquad \square$$

❓**Question**   Can you think of a different way of approximating the area of a circle of radius 2? Specifically, did we have to count each of the "partial" squares as half a square? Can you think of a method that would be more accurate?   □

The same kind of technique can be used to approximate the area of the "blob" in Figure 12.4. One possible grid (suppose each square has length $\frac{1}{2}$ unit) is shown in Figure 12.5.

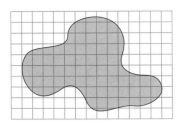

Figure 12.4   **Figure 12.5**

❓**Question**   Given that each of the squares in Figure 12.5 has a side length of $\frac{1}{2}$ unit, approximate the area of the "blob." How could you get a better approximation?   □

In Example 12.1, we effectively cut the circle into a number of small pieces, approximated the area of each of those pieces (by counting each piece as either a "full" square or a "half" square), and then added all of the small approximated areas. This process of "subdividing, approximating, and adding" is the cornerstone of the **definite integral,** which we will introduce in Section 12.3. When we have learned the theory of integrals, we will actually be able to *prove* that the area formula $A = \pi r^2$ for a circle of radius r is correct (we will do so in Section 14.4).

In the rest of this chapter we will be concerned primarily with finding or approximating the area "under" a curve, that is, the area between the graph of some function and the x-axis. For example, we might be interested in calculating the shaded area shown in Figure 12.6.

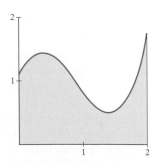

Figure 12.6

The area shown in Figure 12.6 is the area "under" the graph of some function $f(x)$ from $x = 0$ to $x = 2$. We can approximate the area under this curve using a method similar to the one we used in Example 12.1. Figures 12.7–12.9 show some possible grids we could use.

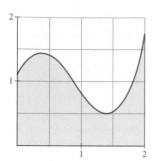

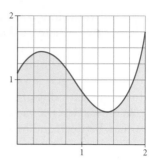

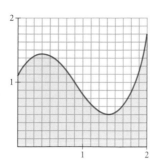

Figure 12.7   **Figure 12.8**   **Figure 12.9**

? *Question*   Approximate the area of the shaded region using the grid shown in Figure 12.7. You may use whatever method you like to determine which squares to "count," or how much of each square to "count," but be consistent. Then use the grids in Figures 12.8 and 12.9, and the same method, to approximate the area of the region.   □

In the exercises you will see another way that we could approximate the area under the graph of a function $f(x)$, using rectangles whose heights depend on the height of the graph of $y = f(x)$. This method will be the key idea in Section 12.2.

12.1.2 Sigma notation

In Section 12.2.3, we will use sums of numbers to approximate the area under a curve. It will be convenient to have a compact notation to represent the sum of a sequence of numbers—in particular, a sequence of numbers that have a recognizable pattern. This notation is called *sigma notation,* since it uses the letter "sigma" (written $\sum$), which is the Greek equivalent of the letter "S" (for "sum").

EXAMPLE 12.2

Writing a sum with "sigma" notation

Consider the following sum:

(1)
$$1 + \frac{1}{2} + \frac{1}{3} + \frac{1}{4} + \frac{1}{5} + \frac{1}{6} + \frac{1}{7} + \frac{1}{8} + \frac{1}{9} + \frac{1}{10}.$$

Notice that each of the numbers being summed is of the form $\frac{1}{k}$ for some integer k. The first number in the sum is $\frac{1}{k}$ with $k = 1$, the second number is $\frac{1}{k}$ with $k = 2$, and so on, until we reach the final number in the sum, which is $\frac{1}{k}$ with $k = 10$. Another way to describe this sum is as follows:

(2)   *The sum of all numbers of the form $\frac{1}{k}$, where k is*
        an integer greater than or equal to 1 and less than or equal to 10.

At this point we are not interested in the *value* of the sum, but we want to find a way to rewrite the sum in a compact, efficient way. We can do this with the following "sigma notation":

(3)
$$\sum_{k=1}^{10} \frac{1}{k}$$

This notation is usually pronounced "the sum from 1 to 10 of $\frac{1}{k}$," but it means exactly the same thing as Statement (2) (and thus the same thing as Statement (1)).   □

In Expression (3) in Example 12.2, "$\sum$" stands for "sum," the "$k = 1$" below the sigma shows where we should "begin" the sum (at $k = 1$), and it also specifies the ***index*** k for the sum. The "10" above the sigma shows us where to "end" the sum, and the $\frac{1}{k}$ tells us the form of each of the numbers that we plan to add. With sigma notation, the "step" from one value of k to the next is always equal to 1. If we start with $k = 1$, then the next value of k will be $k = 2$, the next will be $k = 3$, and so on. In particular, this means that the variable in the sigma notation (in this case k) can have only integer values, between and including the starting and ending values.

In general, sigma notation has the form shown in Definition 12.1. It is traditional to use a_k (rather than $a(k)$) to denote the function of k that describes the numbers that make up the sum. This is because a_k is a function only of *integers* k, not of real numbers. This is why we use k (rather than x) to represent the "variable" in the sum; the letter k suggests an integer value, whereas the letter x would usually suggest a real number.

DEFINITION 12.1

Sigma Notation

If a_k is a function of k, and m and n are nonnegative integers with $m \leq n$, then

$$\sum_{k=m}^{n} a_k := a_m + a_{m+1} + a_{m+2} + \cdots + a_{n-1} + a_n.$$

§ *Caution* When it appears in a line of text, the notation

$$\sum_{k=m}^{n} a_k$$

is often written as $\sum_{k=m}^{n} a_k$ (otherwise it is too big for the line of text). These two notations represent the same sum. □

Definition 12.1 requires a bit of explanation. The symbol a_k is a function whose "inputs" are nonnegative integers and whose outputs are real numbers. In Example 12.2, a_k was equal to the function $\frac{1}{k}$. In that example, $a_1 = \frac{1}{1} = 1$, $a_2 = \frac{1}{2}$, and so on, and we had a starting value of $m = 1$ and an ending value of $n = 10$. Sigma notation always follows the general pattern

$$\sum_{k=\text{starting value}}^{\text{ending value}} (\text{function of } k).$$

Remember that $\sum_{k=m}^{n} a_k$ is just fancy notation for "the sum of all numbers of the form a_k, where k is an integer greater than or equal to m and less than or equal to n." In other words, to find the value of $\sum_{k=m}^{n} a_k$, you would evaluate the function a_k at $k = m$, $k = m + 1$, $k = m + 2$, and so on until $k = n$, and then add up all of these values.

Sigma notation takes a little while to get used to, so be sure to practice with a lot of examples until you are sure that you understand how it works. You should be able to convert between sigma notations and "expanded" sums, as illustrated in the following two examples.

EXAMPLE 12.3   **Converting from sigma notation to an expanded sum**

Write the sum represented by the sigma notation $\sum_{k=2}^{8} k^2$ in "expanded" form, and find the value of this sum.

Solution This sigma notation is a compact way of writing "the sum of all numbers of the form k^2, where k is an integer greater than or equal to 2 and less than or equal to 8." This means we must find the values of k^2 for $k = 2$, $k = 3$, and so on until $k = 8$, and then add all of these values:

$$\sum_{k=2}^{8} k^2 = 2^2 + 3^2 + 4^2 + 5^2 + 6^2 + 7^2 + 8^2 = 203.$$

☐

EXAMPLE 12.4

Converting from an expanded sum to sigma notation

Write the following sum in sigma notation:

$$\frac{3}{4} + \frac{4}{5} + \frac{5}{6} + \frac{6}{7} + \frac{7}{8} + \frac{8}{9} + \frac{9}{10}$$

Solution First, we look for a pattern in the numbers of the sum, so we can determine the function a_k. You might notice that each number in the sum is of the form $\frac{k}{k+1}$ for some nonnegative integer k. (You could also find a different pattern; see the "Question" that follows.) Therefore, we can set $a_k := \frac{k}{k+1}$. The first number, $\frac{3}{4}$, corresponds to the value $k = 3$. The last number, $\frac{9}{10}$, corresponds to the value $k = 9$ (when $k = 9$, we have $\frac{k}{k+1} = \frac{9}{10}$). The sum can be written in sigma notation as

$$\frac{3}{4} + \frac{4}{5} + \frac{5}{6} + \frac{6}{7} + \frac{7}{8} + \frac{8}{9} + \frac{9}{10} = \sum_{k=3}^{9} \frac{k}{k+1}.$$

☐

❓ *Question* There can be more than one way to write a sum in sigma notation. Verify that the sum in Example 12.4 can also be expressed in the following sigma notation:

$$\sum_{k=4}^{10} \frac{k-1}{k}.$$

☐

12.1.3 The algebra of sigma notation

The following theorem expresses two common properties of sums in sigma notation. For the moment we will restrict our attention to sums whose index k starts at $k = 1$. However, Theorem 12.1 is true for sums that begin at any nonnegative integer $k = m$.

THEOREM 12.1

Constant Multiple and Sum Rules for Sums

Given any functions a_k and b_k defined for nonnegative integers k, and any real number c,

(a) $\displaystyle\sum_{k=1}^{n} c a_k = c \sum_{k=1}^{n} a_k.$ (b) $\displaystyle\sum_{k=1}^{n} (a_k + b_k) = \sum_{k=1}^{n} a_k + \sum_{k=1}^{n} b_k.$

Part (a) of this theorem is a general version of the distributive property $c(x + y) = cx + cy$. Part (b) is a general version of the associative and commutative properties of addition (i.e., the fact that we can reorder and regroup the numbers in a sum). A simple example is the fact that $(a_1 + b_1) + (a_2 + b_2) = (a_1 + a_2) + (b_1 + b_2)$. The proof of Theorem 12.1 consists mostly of translating the sigma notation.

PROOF **(THEOREM 12.1)**   To prove part (a), we translate the sigma notation and then apply the distributive property:

$$\sum_{k=1}^{n} ca_k = ca_1 + ca_2 + ca_3 + \cdots + ca_{n-1} + ca_n \qquad \text{(write out the sum)}$$

$$= c(a_1 + a_2 + a_3 + \cdots + a_{n-1} + a_n) \qquad \text{(factor out } c\text{)}$$

$$= c \sum_{k=1}^{n} a_k. \qquad \text{(write in sigma notation)}$$

The proof of part (b) is similar; we will write the sigma notation in "expanded" form, reorder and regroup the terms in the sum, and then write the reordered sum in sigma notation:

$$\sum_{k=1}^{n}(a_k + b_k) = (a_1 + b_1) + (a_2 + b_2) + (a_3 + b_3) + \cdots + (a_{n-1} + b_{n-1}) + (a_n + b_n)$$

$$= (a_1 + a_2 + a_3 + \cdots + a_{n-1} + a_n) + (b_1 + b_2 + b_3 + \cdots + b_{n-1} + b_n)$$

$$= \sum_{k=1}^{n} a_k + \sum_{k=1}^{n} b_k. \qquad ■$$

🔹 **Caution**   It is *not* true that the sum of a product is the product of sums! In simple algebra, this means that $(a_1 b_1) + (a_2 b_2)$ is *not* in general equal to $(a_1 + a_2)(b_1 + b_2)$. In sigma notation, it means that

$$\sum_{k=1}^{n}(a_k b_k) \neq \left(\sum_{k=1}^{n} a_k \right) \left(\sum_{k=1}^{n} b_k \right).$$

☐

Theorem 12.1 allows us to manipulate algebraically sums that are written in sigma notation without first expanding the sigma notation. In fact, as the next example shows, given the value of two sums, we can find the value of some combination of those sums even without knowing the values of the terms in each sum.

EXAMPLE 12.5	**Algebraically manipulating sums in sigma notation**

Given that $\displaystyle\sum_{k=1}^{4} a_k = 7$ and $\displaystyle\sum_{k=1}^{4} b_k = 10$, find $\displaystyle\sum_{k=1}^{4}(a_k + 3b_k)$.

Solution   Notice that we do not know what the functions a_k and b_k are in this problem, and thus we cannot directly compute $\sum_{k=1}^{4}(a_k + 3b_k)$. However, using Theorem 12.1 we have

$$\sum_{k=1}^{4}(a_k + 3b_k) = \sum_{k=1}^{4} a_k + \sum_{k=1}^{4} 3b_k \qquad \text{(part (a) of Theorem 12.1)}$$

$$= \sum_{k=1}^{4} a_k + 3 \sum_{k=1}^{4} b_k \qquad \text{(part (b) of Theorem 12.1)}$$

$$= 7 + 3(10) = 37. \qquad \text{(using the values of the sums given)} \qquad ☐$$

🔹 **Caution**   It is very important that both sums given in Example 12.5 started and ended with the same index numbers. We cannot apply Theorem 12.1 to sums that start or end at different integers (at least not without doing some work first; see Example 12.6).

☐

The following example illustrates how to strip terms off the beginning or end of a sum to change the starting or ending index. This is necessary if you wish to combine two sums with different starting or ending indices.

EXAMPLE 12.6   **Sums can be combined only if they start and end at the same value**

Given that $\displaystyle\sum_{k=2}^{20} \frac{1}{k} \approx 2.5977$ and $\displaystyle\sum_{k=0}^{19} \sqrt{k} \approx 57.1938$, calculate $\displaystyle\sum_{k=2}^{19} \left(\frac{2}{k} + \sqrt{k}\right)$.

Solution   If we are clever, we can calculate the value of this sum without writing out all the terms and adding them. (This would be especially important if the sums ran from $k = 0$ to $k = 5000$, for example.) The key is to use the given information to find $\sum_{k=2}^{19} \frac{1}{k}$ and $\sum_{k=2}^{19} \sqrt{k}$; then we can use Theorem 12.1. We are given that $\sum_{k=2}^{20} \frac{1}{k} = \frac{1}{2} + \frac{1}{3} + \cdots + \frac{1}{19} + \frac{1}{20}$ is approximately 2.5977. To get the sum of the $k = 2$ through $k = 19$ terms, all we have to do is subtract the "extra" term $\frac{1}{20}$.

$$\sum_{k=2}^{19} \frac{1}{k} = \left(\sum_{k=2}^{20} \frac{1}{k}\right) - \frac{1}{20} \approx 2.5977 - \frac{1}{20} = 2.5477.$$

Similarly, we can find the sum from $k = 2$ to $k = 19$ of $\sqrt{k}$ by subtracting the two extra terms (at $k = 0$ and $k = 1$) at the beginning of the sum $\sum_{k=0}^{19} \sqrt{k}$:

$$\sum_{k=2}^{19} \sqrt{k} = \left(\sum_{k=0}^{19} \sqrt{k}\right) - \sqrt{0} - \sqrt{1} \approx 57.1938 - 0 - 1 = 56.1938.$$

We can now use Theorem 12.1 to combine these sums and find the value of the desired sum:

$$\sum_{k=2}^{19} \left(\frac{2}{k} + \sqrt{k}\right) = \sum_{k=2}^{19} \frac{2}{k} + \sum_{k=2}^{19} \sqrt{k} \qquad \text{(sum of sums)}$$

$$= 2\sum_{k=2}^{19} \frac{1}{k} + \sum_{k=2}^{19} \sqrt{k} \qquad \text{(constant times a sum)}$$

$$\approx 2(2.5477) + 56.1938 = 61.2892. \qquad \text{(values computed above)}$$

$\square$

The fact that we can "strip off" terms from the beginning or end of a sum, or split a sum into two pieces, is expressed in general notation in the following theorem:

THEOREM 12.2

Splitting a Sum

Given any function a_k defined for nonnegative integers k, and any integers m, n, and p such that $0 \leq m < p < n$,

$$\sum_{k=m}^{n} a_k = \sum_{k=m}^{p-1} a_k + \sum_{k=p}^{n} a_k.$$

In Example 12.6, we had $m = 0$, $p = 1$, and $n = 19$, and $a_k = \sqrt{k}$, and we used the fact that

$$\sum_{k=0}^{19} \sqrt{k} = \sum_{k=0}^{1} \sqrt{k} + \sum_{k=2}^{19} \sqrt{k}.$$

The proof of this theorem is fairly simple; we need only write out the terms of the sum.

PROOF   (THEOREM 12.2)   If $m < p < n$, we have

$$\sum_{k=m}^{n} a_k = a_m + a_{m+1} + \cdots + a_{p-1} + a_p + a_{p+1} + \cdots + a_{n-1} + a_n \qquad \text{(expand)}$$

$$= (a_m + a_{m+1} + \cdots + a_{p-1}) + (a_p + a_{p+1} + \cdots + a_{n-1} + a_n) \qquad \text{(regroup)}$$

$$= \sum_{k=m}^{p-1} a_k + \sum_{k=p}^{n} a_k. \qquad \text{(sigma notation)}$$

■

12.1.4 Sum formulas and limits of sums

It can be tedious to calculate a long sum (for example, $\sum_{k=1}^{1000} k^2$) by hand; to do so, you would have to write out all the terms of the sum and then add them together. Luckily, we can prove formulas for the values of a few simple sums:

THEOREM 12.3

Sum Formulas

If n is a positive integer, then

(a) $\displaystyle\sum_{k=1}^{n} 1 = n$ (b) $\displaystyle\sum_{k=1}^{n} k = \frac{n(n+1)}{2}$

(c) $\displaystyle\sum_{k=1}^{n} k^2 = \frac{n(n+1)(2n+1)}{6}$ (d) $\displaystyle\sum_{k=1}^{n} k^3 = \frac{n^2(n+1)^2}{4}$

◈ *Caution*   The sum formulas in Theorem 12.3 work only for sums starting at $k = 1$. To use these formulas on a sum with a different starting index (for example, $k = 0$ or $k = 3$), we would first have to write that sum so that it started at $k = 1$.   □

To prove part (a) of Theorem 12.3, we only have to write out the sum

$$\sum_{k=1}^{n} 1 = \underbrace{1 + 1 + 1 + \cdots + 1}_{n \text{ times}} = n.$$

The remaining three formulas can be proved by induction. For example, the formula in part (b) is the sigma notation version of the statement

$$1 + 2 + 3 + \cdots + n = \frac{n(n+1)}{2},$$

which we proved in Section 0.5.5 using induction. You used induction to prove the sum formula in part (c) in the exercises for Section 0.5.5. In the exercises for this section, you will use a similar proof to show the sum formula in part (d).

Using these sums and the algebraic properties of sums, we can quickly calculate any sum whose general term is a cubic polynomial in k, as shown in the following examples.

EXAMPLE 12.7

Using sum formulas to calculate the value of a sum

Find the value of the sum $\displaystyle\sum_{k=1}^{300} (k^3 - 4k^2 + 2)$.

Solution   By writing the sum above as three simple sums and using the formulas in Theorem 12.3, we have

$$\sum_{k=1}^{300}(k^3 - 4k^2 + 2) = \left(\sum_{k=1}^{300} k^3\right) - 4\left(\sum_{k=1}^{300} k^2\right) + 2\left(\sum_{k=1}^{300} 1\right) \qquad \text{(properties of sums)}$$

$$= \frac{300^2(301)^2}{4} - 4\frac{300(301)(601)}{6} + 2(300) \qquad \text{(Theorem 12.3)}$$

$$= 2{,}002{,}342{,}900.$$

Of course that was a whole lot easier than expanding the sum and then adding all 300 terms! □

EXAMPLE 12.8

A sum involving both *k* and *n*

Find the values of the sum $\displaystyle\sum_{k=1}^{n} \frac{k^2}{n^3}$ for $n = 3$, $n = 4$, $n = 100$, and $n = 1000$.

Solution   For $n = 3$ and $n = 4$, we might as well write out the sum

$$\sum_{k=1}^{3} \frac{k^2}{n^3} = \frac{1^2}{3^3} + \frac{2^2}{3^3} + \frac{3^2}{3^3} = \frac{1}{27} + \frac{4}{27} + \frac{9}{27} = \frac{14}{27}.$$

$$\sum_{k=1}^{4} \frac{k^2}{n^3} = \frac{1^2}{4^3} + \frac{2^2}{4^3} + \frac{3^2}{4^3} + \frac{4^2}{4^3} = \frac{1}{64} + \frac{4}{64} + \frac{9}{64} + \frac{16}{64} = \frac{30}{64} = \frac{15}{32}.$$

Notice that the denominator in each sum is constantly equal to n^3 and does not change as k changes. Notice also that the sum from $n = 1$ to $n = 4$ *cannot* be obtained from the sum from $n = 1$ to $n = 3$ just by "adding a fourth term"; *all* the terms in the second sum are different from the terms in the first sum. In fact, the second sum is actually *smaller* than the first sum, since $\frac{14}{27} \approx 0.5185$ and $\frac{15}{32} = 0.46875$. Although we are adding more terms in the second sum, each of those terms is smaller than the terms in the first sum (since their denominators are larger).

To find the sum for $n = 100$, we will use properties of sums and one of the sum formulas from Theorem 12.3:

$$\sum_{k=1}^{100} \frac{k^2}{n^3} = \sum_{k=1}^{100} \frac{k^2}{100^3} = \frac{1}{100^3} \sum_{k=1}^{100} k^2 \qquad \left(\text{pull out the constant } \tfrac{1}{100^3}\right)$$

$$= \frac{1}{100^3} \frac{100(101)(201)}{6} \qquad \text{(sum formula from Theorem 12.3)}$$

$$= \frac{2{,}030{,}100}{6{,}000{,}000} = 0.33835.$$

Similarly, the sum from $k = 1$ to $k = 1000$ is

$$\sum_{k=1}^{1000} \frac{k^2}{n^3} = \sum_{k=1}^{1000} \frac{k^2}{1000^3} = \frac{1}{1000^3} \sum_{k=1}^{1000} k^2$$

$$= \frac{1}{1000^3}\left(\frac{1000(1001)(2001)}{6}\right) = \frac{2{,}003{,}001{,}000}{6{,}000{,}000{,}000} \approx 0.33383. \qquad □$$

Even when we sum up a thousand terms, the sum in Example 12.8 is a very small number. In fact, the more terms we sum up, the smaller that number seems to get! You may have noticed that the sum for $n = 1000$ was very close to the sum for $n = 100$

in Example 12.8. In fact, the sum for $n = 10,000,000$ is also very close to these values. What happens if we sum up "infinitely many" terms? Can the sum somehow be finite? If we can find a formula for the sum of $\frac{k^2}{n^3}$ from $k = 1$ to $k = n$ (without specifying a value for n), we can use a limit to investigate what happens as $n \to \infty$.

EXAMPLE 12.9   **An "infinite" sum**

Calculate $\displaystyle\lim_{n \to \infty} \sum_{k=1}^{n} \frac{k^2}{n^3}$.

Solution   We first find a formula for the sum in exactly the same way we did in the previous example. After we have found this formula, we will take the limit as $n \to \infty$.

$$\lim_{n \to \infty} \sum_{k=1}^{n} \frac{k^2}{n^3} = \lim_{n \to \infty} \frac{1}{n^3} \sum_{k=1}^{n} k^2 \qquad (n \text{ is constant with respect to } k)$$

$$= \lim_{n \to \infty} \frac{1}{n^3} \left(\frac{n(n+1)(2n+1)}{6} \right) \qquad (\text{sum formula from Theorem 12.3})$$

$$= \lim_{n \to \infty} \frac{(n+1)(2n+1)}{6n^2} \qquad (\text{algebra})$$

$$= \lim_{n \to \infty} \frac{2n^2 + 3n + 1}{6n^2} = \frac{2}{6} = \frac{1}{3}. \qquad (\text{take the limit})$$

This answer makes sense because the sums from $k = 1$ to $k = n$ in Example 12.8 were approaching $\frac{1}{3}$ as n got larger; for example, the sum for $n = 100$ was 0.33835, and the sum for $n = 1000$ was 0.33383.   □

What you should know after reading Section 12.1

▶ How to approximate the area of a region by dividing the region into small pieces and approximating the areas of the pieces (for example, by using a grid). How can you make your approximation better?

▶ What is meant by "the area under a curve"? How could you approximate such an area?

▶ How to write sums in sigma notation, and how to find the value of a sum written in sigma notation. What do all the parts of sigma notation stand for? In particular, what is "a_k"?

▶ How to manipulate algebraically sums written in sigma notation. For example, what can you say about the sum of two sums or the constant multiple of a sum? What can you say about the product of two sums?

▶ How to strip terms off the beginning or end of a sum to change the indices in the sigma notation, and how to split a sum into two sums, in sigma notation.

▶ The formulas for the sum from $k = 1$ to $k = n$ of a_k when $a_k = 1$, $a_k = k$, $a_k = k^2$, and $a_k = k^3$, and how to prove these formulas. What does it mean to take the limit as $n \to \infty$ of a sum from $k = 1$ to $k = n$, and how can you calculate such limits of sums (if they exist)?

Exercises 12.1

Concepts

0. Read the section and make your own summary of the material. Include whatever you think will help you review the material later on.

1. In this section we approximated the area of a circle of radius 2 by dividing the circle into pieces (with a grid), approximating the areas of the pieces, and then adding up these approximate areas (see Figure 12.1, for example).

  (a) What method did we use to approximate the area of each of the pieces?

  (b) Think of another way to approximate the area of each of the pieces, and use this method (and the grid in Figure 12.1) to approximate the area of a circle of radius 2. There are many possible ways that you could do this.

  (c) Approximate the area of a circle of radius 2 using your method from part (b), but this time use the grid in Figure 12.2. Is this approximation better? Why or why not?

  (d) Is your method of approximating more accurate than the one used in this section? Why or why not? Compare each approximation to the actual area of the circle.

2. Approximate the area of the "blob" in Figure 12.4 by using the grid in Figure 12.5. Explain the method that you use to approximate the area of each of the pieces of the "blob." Then describe a method that would make an even better approximation of the area (using the same grid), and explain why this new method should be more accurate.

3. Figure 12.6 shows the area between the graph of some function $f(x)$ and the x-axis between $x = 0$ and $x = 2$.

  (a) Approximate this area by using the grid in Figure 12.7, using whatever method you like.

  (b) Use the same method to approximate the area by using the grid in Figure 12.8. Then use the grid in Figure 12.9.

  (c) Make a guess for the area under the curve on the basis of your three approximations.

4. Consider the area in Figure 12.6 between the graph of some function $f(x)$ and the x-axis between $x = 0$ and $x = 2$.

  (a) Use the grid in Figure 12.8 to get an upper bound on the area of the region. In other words, make an approximation for the area of the region that you *know* is greater than the actual area.

  (b) Use the grid in Figure 12.8 to get a lower bound on the area of the region. In other words, make an approximation for the area of the region that you *know* is less than the actual area.

  (c) Make a guess for the area under the curve on the basis of your upper and lower bounds for the area.

5. Explain why it would be difficult to write the sum $\frac{1}{3} + \frac{1}{4} + \frac{1}{5} + \frac{1}{8} + \frac{1}{11} + \frac{1}{12} + \frac{1}{13}$ in sigma notation.

6. Why do we use a_k (rather than $a(k)$, the usual function notation) to represent the general term in a sum written in sigma notation? How is a_k different from the functions we usually work with in this course? Why do we use the letter k for the index, or variable, of the sum instead of the letter x that is most often used for a variable?

7. Use a sentence to describe what the notation $\sum\limits_{k=3}^{87} k^2$ means. (*Hint:* Start with "The sum of....")

8. Use a sentence to describe what the notation $\sum\limits_{k=m}^{n} a_k$ means. (*Hint:* Start with "The sum of....")

9. Consider the sum $\sum\limits_{i=p}^{q} b_i$.

  (a) Write out this sum in expanded form (i.e., without sigma notation).

  (b) What is the "index" of the sum? What is the "starting value"? What is the "ending value"? Which part of the notation describes the form of each of the terms in the sum?

  (c) Do p and q have to be integers? Can they be negative? What about b_i? What else can you say about p and q?

10. Consider the sum $\sum\limits_{k=m}^{n} a_k = 9 + 16 + 25 + 36 + 49$. What is a_k? What is m? What is n?

11. Consider the sum $\sum\limits_{k=2}^{5} \frac{k}{1-k}$. Identify $a_2, a_3, a_4,$ and a_5.

12. Show that $\sum\limits_{k=3}^{9} \frac{k}{k+1}$ is equal to $\sum\limits_{k=4}^{10} \frac{k-1}{k}$ by writing out the terms in each sum.

13. Show that $\sum\limits_{k=0}^{8} \frac{1}{k^2+1}$ is equal to $2\sum\limits_{k=0}^{8} \frac{1}{2k^2+2}$ by writing out the terms in each sum.

14. Write the sum $\frac{4}{7} + \frac{5}{8} + \frac{6}{9} + \frac{7}{10} + \frac{8}{11}$ in sigma notation in three ways: **(a)** with a starting value of $k = 4$; **(b)** with a starting value of $k = 7$; and **(c)** with a starting value of $k = 5$.

15. Write the sum $2 + \frac{2}{4} + \frac{2}{9} + \frac{2}{16} + \frac{2}{25}$ in sigma notation in three ways: **(a)** with a starting value of $k = 1$; **(b)** with a starting value of $k = 2$; and **(c)** with a starting value of $k = 0$.

■ Determine whether each of the following statements is true or false. If a statement is true, explain why. If a statement is false, explain why or provide a counterexample.

16. True or False: $\displaystyle\sum_{k=0}^{n}\frac{1}{k+1}+\sum_{k=1}^{n}k^2=\sum_{k=0}^{n}\frac{k^3+k+1}{k+1}$.

17. True or False: $\displaystyle\sum_{k=1}^{n}\frac{1}{k+1}+\sum_{k=0}^{n}k^2=\sum_{k=1}^{n}\frac{k^3+k+1}{k+1}$.

18. True or False: $\displaystyle\left(\sum_{k=1}^{n}\frac{1}{k+1}\right)\left(\sum_{k=1}^{n}k^2\right)=\sum_{k=1}^{n}\frac{k^2}{k+1}$.

19. True or False: $\displaystyle\sum_{k=0}^{n}a_k=-a_0-a_n+\sum_{k=1}^{n-1}a_k$.

20. True or False: $\displaystyle\sum_{k=0}^{m}\sqrt{k}+\sum_{k=m}^{n}\sqrt{k}=\sum_{k=0}^{n}\sqrt{k}$.

21. True or False: $\displaystyle\left(\sum_{k=1}^{10}a_k\right)^2=\sum_{k=1}^{10}a_k^2$.

22. State algebraic formulas that express the following sums, where n is any positive integer:

(a) $\displaystyle\sum_{k=1}^{n}1$ (b) $\displaystyle\sum_{k=1}^{n}k$ (c) $\displaystyle\sum_{k=1}^{n}k^2$ (d) $\displaystyle\sum_{k=1}^{n}k^3$

23. Explain why terms in the sum in Example 12.8 with $n=4$ are completely different from the terms in the sum when $n=3$. How can the sum from $k=1$ to $k=4$ possibly be *smaller* than the sum from $k=1$ to $k=3$?

24. Discuss the fact that the infinite sum in Example 12.9 is finite. Why might you *not* expect it to be finite? Why might it make sense that it is finite after all? Why does it make sense that the infinite sum is equal to $\frac{1}{3}$?

Skills

■ Consider the area under the graph of $f(x)=x^2-2x+2$ from $x=1$ to $x=3$ (see the graph). For each problem below, approximate this area by calculating the sum of the areas of the rectangles shown. Note that each rectangle has a height of $f(x)$ for some value of x; for example, in Problem 25, the first rectangle has a height of $f(1)=1^2-2(1)+2=1$ unit, and the second rectangle has a height of $f(1.5)=(1.5)^2-2(1.5)+2=1.25$ units.

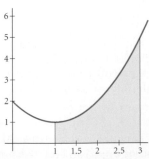

Area under $f(x)=x^2-2x+2$ from $x=1$ to $x=3$

25.

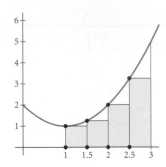

26.

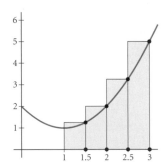

27.

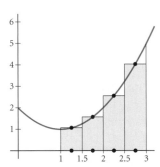

■ Write out each sum below in expanded form, and then calculate the value of the sum.

28. $\displaystyle\sum_{k=4}^{9}k^2$

29. $\displaystyle\sum_{k=0}^{6}\frac{2}{k+1}$

30. $\displaystyle\sum_{k=3}^{10}\ln k$

31. $\displaystyle\sum_{k=0}^{5}\left(\tfrac{1}{2}k\right)^2\left(\tfrac{1}{2}\right)$

32. $\displaystyle\sum_{k=1}^{4}((2+k)^2+1)$

33. $\displaystyle\sum_{k=0}^{9}\sqrt{3+\tfrac{1}{10}k}\left(\tfrac{1}{10}\right)$

■ Write the following sums in sigma notation. Identify m, n, and a_k in each problem.

34. $\frac{4}{3}+\frac{5}{4}+\frac{6}{5}+\frac{7}{6}+\frac{8}{7}+\frac{9}{8}+\frac{10}{9}+\frac{11}{10}$

35. $3+3+3+3+3+3+3+3$

36. $\frac{1}{4}+\frac{1}{9}+\frac{1}{16}+\frac{1}{25}+\frac{1}{36}+\frac{1}{49}+\frac{1}{64}$

37. $3+\frac{4}{8}+\frac{5}{27}+\frac{6}{64}+\frac{7}{125}$

38. $5+10+17+26+37+50+65+82+101$

39. $9+12+15+18+21+24+27$

■ Write each of the following expressions in *one* sigma notation (with some "extra" terms added to or subtracted from the sum, as necessary).

40. $2\sum_{k=0}^{100} a_k - \sum_{k=3}^{101} a_k$

41. $\sum_{k=1}^{40} \frac{1}{k} - \sum_{k=0}^{39} \frac{1}{k+1}$

42. $3\sum_{k=2}^{25} k^2 + 2\sum_{k=2}^{24} k - \sum_{k=0}^{25} 1$

■ For each problem below, find the sum or quantity *without* completely expanding and calculating any sums.

43. Given that $\sum_{k=3}^{10} a_k = 12$ and $\sum_{k=2}^{10} a_k = 23$, find a_2.

44. Given that $\sum_{k=0}^{25} k = 325$ and $\sum_{k=3}^{28} (k-3)^2 = 14{,}910$, find $\sum_{k=3}^{25} (k^2 - 5k + 9)$.

45. Given that $\sum_{k=1}^{4} a_k = 7$, $\sum_{k=0}^{4} b_k = 10$, and $a_0 = 2$, find $\sum_{k=0}^{4} (2a_k + 3b_k)$.

■ Find a formula for each of the following sums, and then use these formulas to calculate each sum for $n = 100$, $n = 500$, and $n = 1000$.

46. $\sum_{k=1}^{n} (3 - k)$

47. $\sum_{k=1}^{n} (k^3 - 10k^2 + 2)$

48. $\sum_{k=3}^{n} (k + 1)^2$

49. $\sum_{k=1}^{n} \frac{k^3 - 1}{4}$

50. $\sum_{k=1}^{n} \frac{k^3 - 1}{n^4}$

51. $\sum_{k=1}^{n} \frac{k^2 + k + 1}{n^3}$

■ Determine which of the following infinite sums are infinite and which are finite. For each sum that is finite, compute the value of the infinite sum. (*Hint:* Find a formula for the sum, and then take the limit as *n* approaches ∞.)

52. $\lim_{n\to\infty} \sum_{k=1}^{n} (k^2 + k + 1)$

53. $\lim_{n\to\infty} \sum_{k=1}^{n} \frac{k^2 + k + 1}{n^3}$

54. $\lim_{n\to\infty} \sum_{k=1}^{n} \frac{k^2 + k + 1}{n^2}$

55. $\lim_{n\to\infty} \sum_{k=1}^{n} \frac{(k+1)^2}{n^3 - 1}$

56. $\lim_{n\to\infty} \sum_{k=1}^{n} \frac{k^3}{n^4 + n + 1}$

57. $\lim_{n\to\infty} \sum_{k=1}^{n} \left(1 + \frac{k}{n}\right)^2 \cdot \frac{1}{n}$

Proofs

58. Prove that $\sum_{k=0}^{n} 3a_k = 3\sum_{k=0}^{n} a_k$.

59. Prove that $\sum_{k=5}^{n} (a_k + b_k) = \sum_{k=5}^{n} a_k + \sum_{k=5}^{n} b_k$.

60. Prove that if *n* is a positive integer, then $\sum_{k=1}^{n} 1 = n$.

61. Prove that if *n* is a positive integer and *c* is any real number, then $\sum_{k=1}^{n} c = cn$.

12.2 Approximating Area with Riemann Sums

In this section we will develop a method (and a lot of notation) for approximating the area between the graph of a nonnegative function $f(x)$ and the *x*-axis on an interval. We'll start by developing some notation in the context of a particular example.

12.2.1 An introduction to approximating the area under a curve

In this section we will formalize the process of finding (and defining) the area under a curve. More precisely, given a function $f(x)$ and an interval $[a, b]$, we are interested in finding the area bounded by the graph of $f(x)$, the *x*-axis, and the vertical lines $x = a$ and $x = b$. For the moment we will assume in all of our examples that the function $f(x)$ is nonnegative on the entire interval $[a, b]$. Figure 12.10 shows one such area.

In Section 12.1.1, we discussed one way of approximating this area. We subdivided the area by "dicing" it, using a grid of squares. We approximated the area of each of the small pieces by counting its area as one square (if the piece was a full square) or as half a square (if the piece was only part of a square). Then we added all these approximations

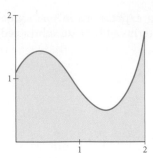

Figure 12.10
Area between graph of $f(x)$ and x-axis from $x = 0$ to $x = 2$

(the areas of all the full and partial squares) to get an overall approximation for the entire area under the curve. There are many other ways that we could have approximated this area, and in fact the way that we will approximate the area from now on will be different.

We want to have a method for approximating area (and, eventually, finding the *actual* area) that can be written using mathematical notation. To do this, we must make use of the height information provided by $f(x)$. We will "slice" the region into four vertical slices of equal width. This is equivalent to dividing the interval $[0, 2]$ into four subintervals, $[0, \frac{1}{2}]$, $[\frac{1}{2}, 1]$, $[1, \frac{3}{2}]$, and $[\frac{3}{2}, 2]$. Since the total length of the interval $[0, 2]$ is 2 units, and we are dividing it into four pieces, the length of each subinterval is $\frac{2}{4} = \frac{1}{2}$. We will label the kth dividing point of the interval $[0, 2]$ with an x_k; this means that we have $x_0 = 0$, $x_1 = \frac{1}{2}$, $x_2 = 1$, $x_3 = \frac{3}{2}$, and $x_4 = 2$. (We call x_0 the "zeroth point," x_1 the "first point," x_2 the "second point," and so on.) The four slices and the five subdivision points x_0, x_1, x_2, x_3, and x_4 are shown in Figure 12.11.

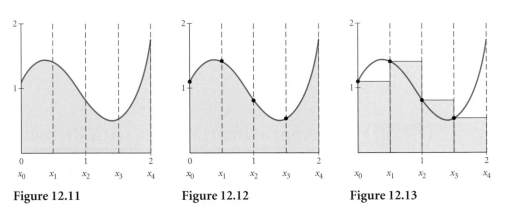

Figure 12.11   **Figure 12.12**   **Figure 12.13**

Now we will approximate the area of each "slice" with a rectangle. We now need to decide how tall each rectangle should be. The height of the rectangle should be related somehow to the height of the function (why?). One way to accomplish this is to make the height of the kth rectangle equal to the height $f(x)$ of the function at some point x in the kth interval. We only have to choose which point in each subinterval to use; for now we will choose the leftmost point in each subinterval. This means that $x_0 = 0$ will determine the height $f(x_0) = f(0)$ of the rectangle over the subinterval $[0, \frac{1}{2}]$, $x_1 = \frac{1}{2}$ will determine the height $f(x_1) = f(\frac{1}{2})$ of the rectangle over the subinterval $[\frac{1}{2}, 1]$, and so on. Notice that we will not need to use the last marked point x_4, since x_4 is not the leftmost point of any subinterval. The points $(x_0, f(x_0))$, $(x_1, f(x_1))$, $(x_2, f(x_2))$, and $(x_3, f(x_3))$ are marked on the graphs in Figures 12.12 and 12.13.

? *Question*   We could also use the height of the function at the midpoint of each subinterval to determine the heights of the rectangles. What are some other ways that we could determine the height of each rectangle? There are many possible answers, but each should somehow involve the height of the function (otherwise, we wouldn't be using the function information at all!).   □

Now we know the height and width of each of the four rectangles. The area of each rectangle (height times width) will be used to approximate the area of each respective slice. The width of each rectangle is $\frac{1}{2}$, and the height of the first rectangle is $f(x_0)$. Therefore, the area of the first rectangle is $f(x_0)(\frac{1}{2})$. Similarly, the area of the second rectangle is $f(x_1)(\frac{1}{2})$, and so on. The total area under the curve is approximately the sum of the areas of the four rectangles. Notice that the total area of the rectangles is not a very good approximation for the area under the curve (do you think it is an underestimate or an overestimate?). However, if we were to consider more rectangles (using thinner "slices"), we could get a better approximation.

Here's where the sigma notation comes in. The sum of the areas of the four rectangles is

$$f(x_0)(\tfrac{1}{2}) + f(x_1)(\tfrac{1}{2}) + f(x_2)(\tfrac{1}{2}) + f(x_3)(\tfrac{1}{2}).$$

When the terms in a sum have a recognizable pattern, we can often write the sum in sigma notation. Here, we have terms for $k = 0$, $k = 1$, $k = 2$, and $k = 3$, where the general form of the kth term is $f(x_k)(\frac{1}{2})$. Therefore, the sum of the areas of the four rectangles (our approximation for the area under the curve) is

(1)
$$\sum_{k=0}^{3} f(x_k)(\tfrac{1}{2}).$$

We could factor the constant $\frac{1}{2}$ out of the sum, but we will leave it inside the sum for such area computations (the reason for this will become apparent later in the chapter). Although in this particular example there isn't much need to write the sum in sigma notation, this notation will allow us to develop a very general theory for approximating the area under a curve (see Section 12.2.2).

To find the areas of the rectangles, we first need to know the equation of the function $f(x)$. The equation of the function shown in Figures 12.11–12.13 happens to be

$$f(x) = \tfrac{1}{80}(125x^3 - 325x^2 + 175x + 89).$$

Therefore, the sum of the areas of the four rectangles (our approximation for the area under the curve) is

$$\sum_{k=0}^{3} f(x_k)(\tfrac{1}{2}) = f(x_0)(\tfrac{1}{2}) + f(x_1)(\tfrac{1}{2}) + f(x_2)(\tfrac{1}{2}) + f(x_3)(\tfrac{1}{2})$$
$$= f(0)(\tfrac{1}{2}) + f(\tfrac{1}{2})(\tfrac{1}{2}) + f(1)(\tfrac{1}{2}) + f(\tfrac{3}{2})(\tfrac{1}{2})$$
$$\approx (1.11)(\tfrac{1}{2}) + (1.39)(\tfrac{1}{2}) + (0.80)(\tfrac{1}{2}) + (0.53)(\tfrac{1}{2}) \approx 1.915.$$

The area under the graph of $f(x)$ from $x = 0$ to $x = 2$ is approximately 1.915 square units. Notice that the sigma and x_k notation did not make our calculation easier; in fact, we had to spend some time translating the notation to do the calculation. However, as we will soon see, having a general notation for such area approximations will be very useful.

The sigma notation in Equation (1) could also be written as the sum

(2)
$$\sum_{k=1}^{4} f(x_{k-1})(\tfrac{1}{2}).$$

To see that this sum is equivalent to the one in Equation (1), write out the terms in the sum; for example, the first term ($k = 1$) in Statement (2) is $f(x_{1-1})(\frac{1}{2}) = f(x_0)(\frac{1}{2})$, which is also the first term in the sum in Equation (1).

12.2.2 Riemann sums

We now generalize the discussion in Section 12.2.1 to define the area under a curve. Suppose $f(x)$ is a function that is nonnegative on an interval $[a, b]$. Let A be the area "under" the graph of f on this interval, that is, the area of the region bounded above by the graph of f, below by the x-axis, and to the left and right by the lines $x = a$ and $x = b$. We now use the "subdivide, approximate, and add up" process to define a ***Riemann sum*** that will help us to approximate this area.

Subdivide the interval $[a, b]$ into n subintervals $[x_{k-1}, x_k]$ of equal length, where $x_0 := a$ and $x_n := b$. We will call the width of each subinterval "Δx" (for "change in x"). Since there are n subintervals and the total length of $[a, b]$ is $b - a$, we have

$$(3) \qquad \Delta x = \frac{b - a}{n}.$$

The subdivision points $x_0, x_1, x_2, \ldots, x_n$ are evenly spaced at intervals of Δx units. The "0th" point x_0 is equal to a, the "first" point x_1 is Δx units further to the right, the "second" point x_2 is Δx units further still, and so on. This means that $x_1 = a + \Delta x$ and $x_2 = x_1 + \Delta x = a + 2\,\Delta x$. Similarly, $x_3 = a + 3\,\Delta x$ (see Figure 12.14), and so on.

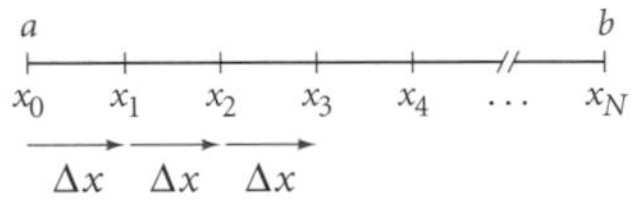

Figure 12.14
$x_3 = a + 3\,\Delta x$

In general, the kth partition point in the subdivision is

$$(4) \qquad x_k = a + k\,\Delta x.$$

We have now subdivided the interval $[a, b]$. The vertical lines passing through each of the partition points x_k will "slice" the area we are interested in. Now we can approximate the area of each slice with a rectangle. The height of the kth rectangle will be the value of $f(x)$ at some point that we will call x_k^* in the kth subinterval $[x_{k-1}, x_k]$.

❖ ***Caution***   Notice that the kth subinterval is $[x_{k-1}, x_k]$, and not $[x_k, x_{k+1}]$. This is because we want the "first" rectangle to correspond to $k = 1$, and this rectangle sits over the subinterval $[x_0, x_1]$. □

We now need to choose, from each subinterval, one point that will determine the height of the rectangle for that subinterval. In the previous examples we chose the leftmost point of each subinterval, but in general we can choose *any* point x_k^* in $[x_{k-1}, x_k]$. The "star" in the x_k^* is supposed to suggest that we are *choosing* any point x_k^* from the kth interval $[x_{k-1}, x_k]$. The most we can say about x_k^* right now is that

$$(5) \qquad x_k^* \text{ is some point in } [x_{k-1}, x_k].$$

Now we will take the value of $f(x)$ at our chosen point x_k^* to be the height of the kth rectangle. In other words,

(6) $\qquad\qquad\qquad\qquad$ $f(x_k^*)$ is the height of the kth rectangle.

Of course, since we started with subintervals of equal width Δx, the width of the kth rectangle is always Δx:

(7) $\qquad\qquad\qquad\qquad$ Δx is the width of the kth rectangle.

Figure 12.15 shows one possible kth rectangle in such a subdivision. In Figure 12.16 we zoom in on this rectangle to illustrate the meanings of the notation we have just developed.

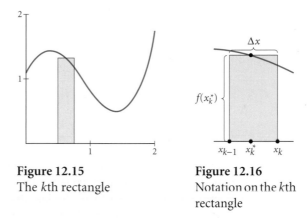

Figure 12.15
The kth rectangle

Figure 12.16
Notation on the kth rectangle

From Figure 12.16 it is clear that the area of the kth rectangle is $f(x_k^*)\,\Delta x$ (the product of the height $f(x_k^*)$ and the width Δx of the rectangle):

(8) $\qquad\qquad\qquad\qquad$ $f(x_k^*)\,\Delta x$ is the area of the kth rectangle.

If the width Δx of each subinterval is small, then the area of the kth rectangle will be close to the area under the graph of $f(x)$ on the interval $[x_{k-1}, x_k]$. We have now approximated the area of the kth slice of the region; see Figures 12.17 and 12.18.

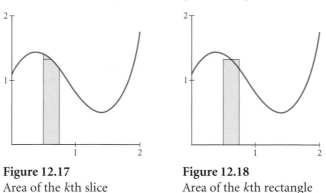

Figure 12.17
Area of the kth slice

Figure 12.18
Area of the kth rectangle

To approximate the entire area under the graph of $f(x)$ from $x = a$ to $x = b$, we must add the approximate areas of the slices. There are n slices (approximated by n rectangles), from the $k = 1$ slice to the $k = n$ slice. Adding up the n approximate areas (in sigma notation) gives us

$$(9) \qquad \sum_{k=1}^{n} f(x_k^*) \, \Delta x = \begin{array}{c} \text{area of all} \\ \text{the rectangles} \end{array} \approx \begin{array}{c} \text{area under the} \\ \text{graph of } f \end{array}$$

Since the area of each slice is approximately equal to the area of the corresponding rectangle, the total area under the graph of $f(x)$ from $x = a$ to $x = b$ is approximately equal to the sum of the areas of these rectangles.

Sums of the form in Equation (9) are called **Riemann sums.** The precise definition of a Riemann sum is as follows.

DEFINITION 12.2

A Riemann Sum for a Function on an Interval

A **Riemann sum** for a function $f(x)$ on an interval $[a, b]$ is a sum of the form

$$\sum_{k=0}^{n} f(x_k^*) \, \Delta x,$$

where $\Delta x := \frac{b-a}{n}$, $x_k := a + k \Delta x$, and x_k^* is some point in the interval $[x_{k-1}, x_k]$.

Our definition of a Riemann sum assumes that each subinterval $[x_{k-1}, x_k]$ has the same width Δx. In general this need not be the case; we could have divided the interval $[a, b]$ into n subintervals of varying lengths. However, for our purposes we will not need to consider anything other than subintervals of equal width.

As we have seen, a Riemann sum for a function $f(x)$ on an interval $[a, b]$ approximates the area under the graph of $f(x)$ between $x = a$ and $x = b$. In Section 12.3.1 we will make the previous sentence more mathematically precise; for now, we will concentrate on the method of approximating areas with various Riemann sums.

12.2.3 Methods for approximating areas with Riemann sums

Suppose $f(x)$ is a function defined on $[a, b]$. For now, we will continue to assume that $f(x)$ is nonnegative on the interval $[a, b]$. Suppose we wish to approximate the area under the graph of $f(x)$ from $x = a$ to $x = b$ with a Riemann sum using n rectangles. The values a, b, and n determine Δx and the values of the subdivision points $x_0, x_1, x_2, \ldots, x_n$. We need only one more piece of information to construct the Riemann sum: We must decide how we are going to determine the heights of the rectangles. This means that we must decide how to choose the point x_k^* in each subinterval $[x_{k-1}, x_k]$.

EXAMPLE 12.10

Different choices for x_k^* result in different Riemann sums

Consider the area shown in Figure 12.10 on page 728, and suppose we wish to construct a Riemann sum with eight rectangles (so $n = 8$). Since our interval is $[a, b] = [0, 2]$, we have $\Delta x = \frac{2-0}{8} = \frac{1}{4}$, and $x_0 = 0$, $x_1 = \frac{1}{4}$, $x_2 = \frac{1}{2}$, $x_3 = \frac{3}{4} \ldots$, $x_8 = 2$. Figures 12.19–12.21 illustrate three of the ways that we could select the points x_k^* that determine the heights of the rectangles.

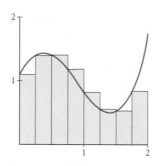

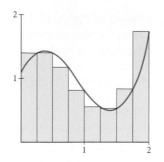

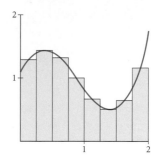

Figure 12.19
Leftmost point of each
subinterval determines
rectangle heights

Figure 12.20
Rightmost point of each
subinterval determines
rectangle heights

Figure 12.21
Midpoint of each
subinterval determines
rectangle heights ☐

Figure 12.19 illustrates what is known as the ***Left-Hand Sum,*** where x_k^* is the leftmost point of the subinterval $[x_{k-1}, x_k]$; in other words, where x_k^* is the point x_{k-1}. For example, x_1^* is the point $x_{1-1} = x_0 = 0$, and this point determines the height $f(x_1^*) = f(x_0) = f(0)$ of the first rectangle. Similarly, x_2^* is the leftmost point x_1 of the second subinterval $[x_1, x_2]$, and the second rectangle has a height of $f(x_2^*) = f(x_1) = f(\frac{1}{4})$. Figure 12.20 illustrates the ***Right-Hand Sum,*** where in each subinterval $[x_{k-1}, x_k]$ we choose the rightmost point $x_k^* = x_k$. In Figure 12.21 we see the ***Midpoint Sum,*** where x_k^* is the center of the subinterval $[x_{k-1}, x_k]$, that is, where $x_k^* = \frac{x_{k-1}+x_k}{2}$. These special Riemann sums are described in sigma notation in the following definition.

DEFINITION 12.3

Left-Hand, Right-Hand, and Midpoint Sums

Suppose $f(x)$ is a function defined on the interval $[a, b]$. The following Riemann sums provide three ways to approximate the area under the graph of $f(x)$ from $x = a$ to $x = b$ with n rectangles (where $x_k := a + k \Delta x$ and $\Delta x := \frac{b-a}{n}$):

(a) The ***Left-Hand Sum*** is $\displaystyle\sum_{k=0}^{n-1} f(x_k) \Delta x = \sum_{k=1}^{n} f(x_{k-1}) \Delta x.$

(b) The ***Right-Hand Sum*** is $\displaystyle\sum_{k=1}^{n} f(x_k) \Delta x.$

(c) The ***Midpoint Sum*** is $\displaystyle\sum_{k=1}^{n} f(\frac{x_{k-1}+x_k}{2}) \Delta x.$

The Riemann sums in Definition 12.3 are not the only ones we can use to approximate the area under a curve. For example, we could use a "one-third" sum, which would define x_k^* to be the point one-third of the way across the interval $[x_{k-1}, x_k]$.

❓ ***Question*** How would you write x_k^* in terms of x_{k-1} and x_k for the "one-third" sum? What other methods for choosing x_k^* can you think of? ☐

Three more common Riemann sums are shown in Figures 12.22–12.24. Figure 12.22 shows the ***Upper Sum,*** where each x_k^* is chosen to be a point M_k in $[x_{k-1}, x_k]$ that produces the maximum possible value of $f(x)$ on that interval. The Upper Sum is always greater than or equal to the actual area under the curve. The ***Lower Sum*** is shown in Figure 12.23. This time the x_k^* is chosen to be a point $m_k \in [x_{k-1}, x_k]$ so that $f(m_k)$ is the minimum value of $f(x)$ on that subinterval; this Riemann sum always produces an approximation that is less than or equal to the actual area under the curve. Finally, Figure 12.24 shows what might happen if you chose x_k^* to be a *random* point in the interval $[x_{k-1}, x_k]$; we'll call this the ***Random Sum.***

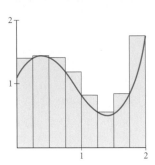
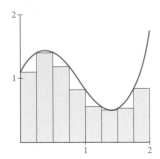
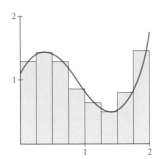

Figure 12.22
Maximum value of f on each subinterval determines its height

Figure 12.23
Minimum value of f on each subinterval determines its height

Figure 12.24
Random value of f on each subinterval determines its height

The Upper Sum and the Lower Sum can often be difficult to compute. However, for any given slice, the global extrema of $f(x)$ often, but not always, occur at the leftmost or rightmost point of the subinterval. When a global extremum occurs in the *interior* of the subinterval, it can sometimes take a bit of work to determine exactly where it occurs. The following definition expresses Upper, Lower, and Random Sums in sigma notation.

DEFINITION 12.4

Upper, Lower, and Random Sums

Suppose $f(x)$ is a function defined on the interval $[a, b]$. The following Riemann sums provide three ways to approximate the area under the graph of $f(x)$ from $x = a$ to $x = b$ with n rectangles (where $x_k := a + k\Delta x$ and $\Delta x := \frac{b-a}{n}$).

(a) The ***Upper Sum*** is $\displaystyle\sum_{k=1}^{n} f(M_k)\,\Delta x$,

where for all k, $f(x)$ has a global maximum on $[x_{k-1}, x_k]$ at M_k.

(b) The ***Lower Sum*** is $\displaystyle\sum_{k=1}^{n} f(m_k)\,\Delta x$,

where for all k, $f(x)$ has a global minimum on $[x_{k-1}, x_k]$ at m_k.

(c) The ***Random Sum*** is $\displaystyle\sum_{k=1}^{n} f(r_k)\,\Delta x$,

where for all k, r_k is a random value in the subinterval $[x_{k-1}, x_k]$.

❓ ***Question*** Can you label each of the chosen points x_k^* in Figures 12.19–12.24? Show in each case how x_k^* determines the height of the kth rectangle. ☐

We don't have to approximate the area under a curve with rectangles; for example, we could use trapezoids instead. The ***Trapezoid Sum*** uses a trapezoid with width Δx, left height of $f(x_{k-1})$, and right height of $f(x_k)$ to approximate the slice of area in the subinterval $[x_{k-1}, x_k]$. Figure 12.25 shows one such trapezoid; Figure 12.26 shows the Trapezoid Sum with $n = 8$ trapezoids. The Trapezoid Sum is often a very good approximation, as you can see from Figure 12.26.

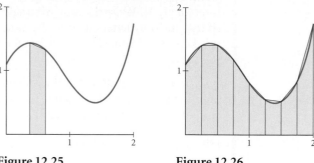

| **Figure 12.25** | **Figure 12.26** |

The area of a trapezoid with base b and heights h_1 and h_2 is $\left(\frac{h_1 + h_2}{2}\right)b$ (the average of the heights times the base). This leads to the following definition for the Trapezoid Sum.

DEFINITION 12.5

Trapezoid Sum

Suppose $f(x)$ is a function defined on the interval $[a, b]$. We can approximate the area under the graph of $f(x)$ from $x = a$ to $x = b$ with n trapezoids by the *Trapezoid Sum:*

$$\sum_{k=1}^{n} \frac{f(x_{k-1}) + f(x_k)}{2} \Delta x.$$

Notice that the Trapezoid Sum averages the heights of f at the endpoints of each subinterval, whereas the Midpoint Sum uses the height of f at the average of the endpoints of each subinterval. According to our definition of Riemann sums in Definition 12.2, the Trapezoid Sum is not technically a Riemann sum for f. In fact, the Trapezoid Sum is the *average* of two Riemann sums: the Left-Hand Sum and the Right-Hand Sum. For this reason we will often refer to the Trapezoid Sum as a Riemann sum.

12.2.4 Examples of approximating areas with Riemann sums

We will now approximate the area under the graph of $f(x) = x^2 - 2x + 2$ from $x = 1$ to $x = 3$ (shown in Figure 12.27) using all seven of the methods described in Section 12.2.3. We begin with the Left-Hand Sum.

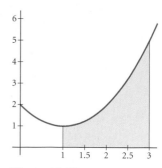

Figure 12.27
Area under $f(x) = x^2 - 2x + 2$ from $x = 1$ to $x = 3$

EXAMPLE 12.11

Writing a Left-Hand Sum in sigma notation

Consider the area between the graph of $f(x) = x^2 - 2x + 2$ and the x-axis on the interval $[1, 3]$. Write the four-rectangle Left-Hand Sum for this area in sigma notation so that the only letters used in the notation are f and k. Identify n, Δx, x_k, and x_k^* for this Left-Hand Sum. Then calculate the value of the sum to approximate the area under the curve.

Solution   First of all, notice that calculating the areas of the rectangles in the Left-Hand Sum is an easy task. By sketching the picture in Figure 12.28, we can easily see that the sum of the areas of the rectangles for the Left-Hand Sum is

(10) $$f(1)\left(\tfrac{1}{2}\right) + f(1.5)\left(\tfrac{1}{2}\right) + f(2)\left(\tfrac{1}{2}\right) + f(2.5)\left(\tfrac{1}{2}\right),$$

which is easy to calculate using the fact that $f(x) = x^2 - 2x + 2$. However, we are asked to write each Riemann sum in sigma notation. The notation is the difficult part of the problem. Let's start over at the beginning.

First we must subdivide the interval $[1, 3]$ into $n = 4$ rectangles; this means that

$$\Delta x = \frac{b - a}{n} = \frac{3 - 1}{4} = \frac{1}{2}.$$

This creates five subdivision points called x_0, x_1, x_2, x_3, and x_4. Since the distance between successive subdivision points is $\tfrac{1}{2}$, we have

$$x_k = a + k\,\Delta x = 1 + k\left(\tfrac{1}{2}\right) = 1 + \tfrac{k}{2}.$$

Since we are using the Left-Hand Sum, the point x_k^* that we choose from each subinterval $[x_{k-1}, x_k]$ will be the leftmost point x_{k-1} of the subinterval. Putting this together with the fact that $x_k = 1 + \tfrac{k}{2}$, we have

$$x_k^* = x_{k-1} = 1 + \tfrac{k-1}{2}.$$

Finally, putting all this into the sigma notation for the Left-Hand Sum, we have the following (notice that the final sigma notation involves only the letters f and k).

$$\sum_{k=1}^{4} f(x_k^*)\,\Delta x = \sum_{k=1}^{4} f\left(1 + \tfrac{k-1}{2}\right)\left(\tfrac{1}{2}\right).$$

Now, to make the area approximation, we need to calculate this sum, using the fact that $f(x) = x^2 - 2x + 2$.

$$\begin{aligned}
\sum_{k=1}^{4} f\left(1 + \tfrac{k-1}{2}\right)\left(\tfrac{1}{2}\right) &= f\left(1 + \tfrac{1-1}{2}\right)\left(\tfrac{1}{2}\right) + f\left(1 + \tfrac{2-1}{2}\right)\left(\tfrac{1}{2}\right) \\
&\quad + f\left(1 + \tfrac{3-1}{2}\right)\left(\tfrac{1}{2}\right) + f\left(1 + \tfrac{4-1}{2}\right)\left(\tfrac{1}{2}\right) \\
&= f(1)\left(\tfrac{1}{2}\right) + f(1.5)\left(\tfrac{1}{2}\right) + f(2)\left(\tfrac{1}{2}\right) + f(2.5)\left(\tfrac{1}{2}\right) \\
&= (1)\left(\tfrac{1}{2}\right) + (1.25)\left(\tfrac{1}{2}\right) + (2)\left(\tfrac{1}{2}\right) + 3.25\left(\tfrac{1}{2}\right) \\
&= 3.75
\end{aligned}$$

In the third line of this calculation, we finally arrived at the sum we found in Expression (10). The sigma notation does not in general make it easier to calculate the areas of the rectangles. However, it will be very important when we discuss the theory of definite integrals in the next section. It would also be very useful if you wanted to write a computer program to approximate the area under a curve, especially if you were using a large number of rectangles.

Figure 12.28 shows the four rectangles used in our Left-Hand Sum. The four chosen points x_k^* and the corresponding heights $f(x_k^*)$ are marked with dots on the graph. From the graph it is clear that our approximation of 3.75 is an underestimate, since the rectangles are always at or below the height of the graph of $f(x)$.

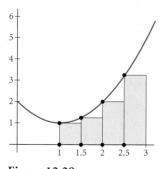

Figure 12.28
Left-Hand Sum   □

❓ *Question*   Write the Left-Hand Sum approximation in the alternative sigma notation shown in Definition 12.3.   □

❓ *Question*   In general, the Left-Hand Sum will always underestimate an increasing function, and overestimate a decreasing function; why? What about the Right-Hand Sum?   □

What about the other types of Riemann sum approximations? We could repeat Example 12.11 with $x_k^* = x_k$ (the rightmost point of each subinterval) to get the Right-Hand Sum approximation for the area under the curve; see Figure 12.29. Or we could take the midpoint of each subinterval as x_k^* to get the Midpoint Sum approximation of the area; see Figure 12.30. The Upper Sum is actually the same as the Right-Hand Sum in this particular example (why)?, and the Lower Sum happens in this case to be the same as the Left-Hand Sum.

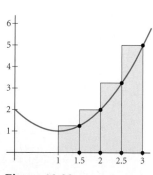
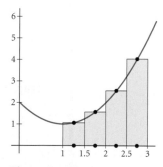

Figure 12.29   **Figure 12.30**
Right-Hand Sum   Midpoint Sum

We could also do a Random Sum approximation by choosing x_k^* as we please in each subinterval; for example, we could choose $x_1^* = 1.35$, $x_2^* = 1.7$, $x_3^* = 2.5$, and $x_4^* = 2.9$ (note that each of these choices is within the appropriate subinterval). Figure 12.31 shows this Random Sum approximation, and Figure 12.32 shows a Trapezoid Sum approximation.

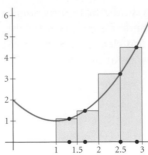

Figure 12.31
Random Sum

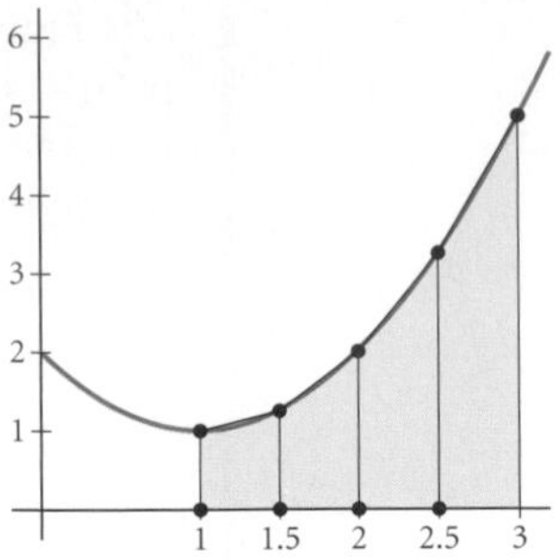

Figure 12.32
Trapezoid Sum

EXAMPLE 12.12

Using each type of Riemann sum to approximate area

Consider the area under the graph of $f(x) = x^2 - 2x + 2$ from $x = 1$ to $x = 3$. Use sigma notation to express approximations for this area, using each type of Riemann sum in Definitions 12.3, 12.4, and 12.5, with $\Delta x = \frac{1}{2}$. Then find a numerical approximation for the area with each of these Riemann sums.

Solution   We have the following approximations for the area in question:

$$\textbf{Left-Hand Sum} = \sum_{k=1}^{4} f\left(1 + \tfrac{k-1}{2}\right)\left(\tfrac{1}{2}\right)$$

$$= f(1)\left(\tfrac{1}{2}\right) + f(1.5)\left(\tfrac{1}{2}\right) + f(2)\left(\tfrac{1}{2}\right) + f(2.5)\left(\tfrac{1}{2}\right) = 3.75$$

$$\textbf{Right-Hand Sum} = \sum_{k=1}^{4} f\left(1 + \tfrac{k}{2}\right)\left(\tfrac{1}{2}\right)$$

$$= f(1.5)\left(\tfrac{1}{2}\right) + f(2)\left(\tfrac{1}{2}\right) + f(2.5)\left(\tfrac{1}{2}\right) + f(3)\left(\tfrac{1}{2}\right) = 5.75$$

$$\textbf{Midpoint Sum} = \sum_{k=1}^{4} f\left(\tfrac{3+2k}{4}\right)\left(\tfrac{1}{2}\right)$$

$$= f(1.25)\left(\tfrac{1}{2}\right) + f(1.75)\left(\tfrac{1}{2}\right) + f(2.25)\left(\tfrac{1}{2}\right) + f(2.75)\left(\tfrac{1}{2}\right) = 4.625$$

$$\textbf{Random Sum} = \sum_{k=1}^{4} f(x_k^*)\left(\tfrac{1}{2}\right)$$

$$= f(1.35)\left(\tfrac{1}{2}\right) + f(1.7)\left(\tfrac{1}{2}\right) + f(2.5)\left(\tfrac{1}{2}\right) + f(2.9)\left(\tfrac{1}{2}\right) \approx 5.236$$

$$\textbf{Trapezoid Sum} = \sum_{k=1}^{4} \frac{f\left(1 + \tfrac{k-1}{2}\right) + f\left(1 + \tfrac{k}{2}\right)}{2}\left(\tfrac{1}{2}\right)$$

$$= \frac{f(1) + f(1.5)}{2}\left(\tfrac{1}{2}\right) + \frac{f(1.5) + f(2)}{2}\left(\tfrac{1}{2}\right) + \frac{f(2) + f(2.5)}{2}\left(\tfrac{1}{2}\right)$$

$$+ \frac{f(2.5) + f(3)}{2}\left(\tfrac{1}{2}\right) = 4.75$$

$$\textbf{Upper Sum} = \textbf{Right-Hand Sum} = 5.75$$

$$\textbf{Lower Sum} = \textbf{Left-Hand Sum} = 3.75$$

The x_k^* for the Midpoint Sum was found by simplifying the expression $\frac{x_{k-1}+x_k}{2}$. The Random Sum was done with a particular choice of x_k^* values, namely $x_1^* = 1.35$, $x_2^* = 1.7$, $x_3^* = 2.5$, and $x_4^* = 2.9$. Of course if you choose different values for x_k^*, then you will get a different answer for the Random Sum.   □

⚠ **Caution**   The Upper Sum is *not* always equal to the Right-Hand Sum, although it is in Example 12.12. (Why?) For example, consider the Upper Sum shown in Figure 12.22 on page 734; the sum in that figure is neither a Right-Hand Sum nor a Left-Hand Sum.   □

❓ **Question**   The actual area (which we won't know how to calculate until Chapter 13) under the graph of $f(x) = x^2 - 2x + 2$ on the interval $[1, 3]$ is $\frac{14}{3} \approx 4.6667$. Which approximation worked better in this case? Is it the sum that you would have guessed from looking at Figures 12.28–12.32? Do you think that certain approximations are *always* better than others?   □

What you should know after reading Section 12.2

▶ How to approximate the area under the graph of a function $f(x)$ on an interval $[a, b]$ by the "subdivide, approximate, and add" method.

▶ The definition of a Riemann sum, and the meanings of the associated notation. For example, what are x_k, x_k^*, Δx, and n? What is the notation for the kth subinterval? What are the geometric meanings of $f(x_k^*)$ and $f(x_k^*)\,\Delta x$?

▶ The definitions of the various methods for constructing Riemann sums. Be able to write these Riemann sums in sigma notation in general as well as in particular examples, and be able to find the numerical values of these sums in particular examples.

Exercises 12.2

Concepts

0. Read the section and make your own summary of the material. Include whatever you think will help you review the material later on.

■ Suppose $f(x)$ is a function that is nonnegative on $[a, b]$, and we are interested in approximating the area under the graph of $f(x)$ from $x = a$ to $x = b$. Determine whether each of the following statements is true or false, and provide graphical examples or counterexamples supporting your answers.

1. True or False: If $f(x)$ is increasing on $[a, b]$, then any Left-Hand Sum will be an underapproximation.

2. True or False: If $f(a) > f(b)$, then any Right-Hand Sum approximation will be an underapproximation.

3. True or False: A Midpoint Sum is always a better approximation than a Left-Hand Sum.

4. True or False: If $f(x)$ is concave up on all of $[a, b]$, then any Left-Hand Sum will be an underapproximation.

5. True or False: If $f(x)$ is concave up on all of $[a, b]$, then any Trapezoid Sum will be an overapproximation.

6. Sketch a function $f(x)$ on an interval $[a, b]$ for which the Left-Hand Sum, with $n = 4$, finds the *exact* area under the graph of $f(x)$ from $x = a$ to $x = b$.

7. Sketch a function $f(x)$ on an interval $[a, b]$ for which *all* Trapezoid Sums (regardless of the size of n) will find the *exact* area under the graph of $f(x)$ from $x = a$ to $x = b$.

8. In this section we mentioned that the Trapezoid Sum is the average of the Left-Hand Sum and the Right-Hand Sum. Use Example 12.12 to show that for $f(x) = x^2 - 2x + 2$, $[a, b] = [1, 3]$, and $n = 4$, the Trapezoid Sum is the average of the Left-Hand Sum and the Right-Hand Sum.

9. Suppose that the Left-Hand Sum approximation with eight rectangles for the area under the graph of a function $f(x)$ from $x = a$ to $x = b$ is equal to 8.2 and that the corresponding Right-Hand Sum approximation is equal to 7.5.

   (a) What is the corresponding Trapezoid Sum approximation for this area?

   (b) Is the corresponding Midpoint Sum for this area necessarily between 7.5 and 8.2? If so, explain why.

If not, sketch an example of a function $f(x)$ on an interval $[a, b]$ whose Midpoint Sum is *not* between the Left-Hand Sum and the Right-Hand Sum.

(c) What can you say about the corresponding Upper Sum for this area? The corresponding Lower Sum?

(d) Is it necessarily true that $f(x)$ is decreasing on the entire interval $[a, b]$? If so, explain why. If not, sketch a possible counterexample (where the Left-Hand Sum is greater than the Right-Hand Sum, but $f(x)$ is *not* decreasing on all of $[a, b]$).

(e) Could the function $f(x)$ possibly be increasing on the entire interval $[a, b]$? If not, explain why not. If so, sketch a possible example (where the Left-Hand Sum is greater than the Right-Hand Sum, and $f(x)$ is increasing on all of $[a, b]$).

10. Explain why the Upper Sum approximation for the area under the graph of a function $f(x)$ on $[a, b]$ must always be larger than or equal to any other Riemann sum approximation for this area (with the same number n of rectangles).

11. Explain why the Upper Sum approximation for the area under the graph of a function $f(x)$ on $[a, b]$ must always be larger than or equal to the actual area of the region (for *any* number n of rectangles).

12. Consider the area under the graph of a nonnegative function $f(x)$ on an interval $[a, b]$. Explain why the Upper Sum approximation for this area with $n = 8$ boxes must be smaller than or equal to the Upper Sum approximation with $n = 4$ boxes. It may help to draw an example picture.

■ Each of the sums in Problems 13–16 approximates the area under the graph of some function $f(x)$ from $x = a$ to $x = b$. Determine the type of approximation (Left-Hand Sum, Midpoint Sum, etc.), identify $f(x)$, a, b, n, Δx, and x_k, and sketch the approximation described.

13. $\displaystyle\sum_{k=1}^{4} \left(1 + \tfrac{k}{2}\right)^2 \left(\tfrac{1}{2}\right)$

14. $\displaystyle\sum_{k=0}^{2} \ln\left(2 + \tfrac{k}{3}\right)\left(\tfrac{1}{3}\right)$

15. $\displaystyle\sum_{k=1}^{100} \frac{\sin(0.05(k-1)) + \sin(0.05k)}{2}(0.05)$

16. $\displaystyle\sum_{k=1}^{9} \sqrt{\frac{\left(1 + \tfrac{k-1}{3}\right) + \left(1 + \tfrac{k}{3}\right)}{2}}\left(\tfrac{1}{3}\right)$

17. Explain why the sum $\displaystyle\sum_{k=1}^{20} f(-3 + \tfrac{k}{2})(0.25)$ *can't* be a Right-Hand Sum for f on $[a, b] = [-3, 2]$.

18. Explain why the sum $\displaystyle\sum_{k=1}^{100} f(2 + 0.1(k-1))(0.1)$ *can't* be a Left-Hand Sum for f on $[a, b] = [2, 5]$.

19. Suppose you wanted to calculate the Upper Sum approximation for the area under the graph of $f(x) = (x-1)^2$ from $x = 0$ to $x = 2$. List all of the values M_k if you use (a) $n = 2$ rectangles, (b) $n = 3$ rectangles, and (c) $n = 4$ rectangles. Sketch graphs of your rectangles to illustrate your answers.

20. Repeat Problem 19 using the Lower Sum approximation and the values m_k.

21. Repeat Problem 19 using the Random Sum approximation and some possible values for r_k.

Skills

■ Your calculator should be able to approximate the area under the graph of a function $f(x)$ from $x = a$ to $x = b$. Determine how to do this on your particular calculator, and then use this method to approximate the area under the graph of each function $f(x)$ below on the given interval $[a, b]$. Include a graph of the function $f(x)$, and shade in the area you are approximating.

22. $f(x) = x^2$, $[a, b] = [0, 3]$

23. $f(x) = \sqrt{x - 1}$, $[a, b] = [2, 3]$

24. $f(x) = e^x$, $[a, b] = [1, 4]$

25. $f(x) = (x - 2)^2 + 1$, $[a, b] = [1, 3]$

26. $f(x) = \sin x$, $[a, b] = [0, \pi]$

27. $f(x) = 2^x$, $[a, b] = [-3, 1]$

■ For each function $f(x)$ and interval $[a, b]$ given below, approximate the area under the graph of $f(x)$ from $x = a$ to $x = b$ using the given approximation method with n rectangles (or trapezoids, if using the Trapezoid Sum). Sketch the graph of $f(x)$ together with the rectangles or trapezoids being used in each approximation. Determine whether each of your approximations is likely to be an overapproximation or an underapproximation of the actual area.

28. $f(x) = x^2$, $[a, b] = [0, 3]$, Left-Hand Sum with (a) $n = 3$ and (b) $n = 6$.

29. $f(x) = \sqrt{x - 1}$, $[a, b] = [2, 3]$, $n = 4$, with (a) Left-Hand Sum and (b) Right-Hand Sum.

30. $f(x) = e^x$, $[a, b] = [1, 4]$, $n = 6$, with (a) Midpoint Sum and (b) Trapezoid Sum.

31. $f(x) = (x - 2)^2 + 1$, $[a, b] = [1, 3]$, Lower Sum with (a) $n = 2$, (b) $n = 3$, and (c) $n = 4$.

32. $f(x) = \sin x$, $[a, b] = [0, \pi]$, $n = 3$, with (a) Trapezoid Sum and (b) Upper Sum.

33. $f(x) = 2^x$, $[a, b] = [-3, 1]$, $n = 8$, with (a) Right-Hand Sum and (b) Random Sum.

34. $f(x) = 4 - x^2$, $[a, b] = [-1, 9]$, $n = 5$, with (a) Midpoint Sum and (b) Lower Sum.

■ Write out the sigma notation for the Riemann sums described below in such a way that the only letter that appears in the general term of the sum is **k**. (Don't calculate the value of the sum; just write it down in sigma notation.)

35. $f(x) = x^2$, $[a, b] = [0, 3]$, Left-Hand Sum with $n = 3$.

36. $f(x) = \sqrt{x - 1}$, $[a, b] = [2, 3]$, Right-Hand Sum with $n = 4$.

37. $f(x) = e^x$, $[a, b] = [1, 4]$, Midpoint Sum with $n = 6$.

38. $f(x) = \sin x$, $[a, b] = [0, \pi]$, Trapezoid Sum with $n = 4$.

39. $f(x) = \ln x$, $[a, b] = [2, 5]$, Left-Hand Sum with $n = 100$.

40. $f(x) = \sqrt{1 - x^2}$, $[a, b] = [-1, 1]$, Midpoint Sum with $n = 20$.

■ For each function $f(x)$ and interval $[a, b]$ below, it is possible to find the *exact* area under the graph of $f(x)$ from $x = a$ to $x = b$ geometrically using the areas of circles, triangles, and rectangles. Find this exact area, and then calculate the Left-Hand Sum, Right-Hand Sum, Midpoint Sum, Upper Sum, Lower Sum, and Trapezoid Sum approximations for this area with **n = 4**. Which approximation rule is most accurate in each case?

41. $f(x) = 4 - x$, $[a, b] = [0, 4]$

42. $f(x) = 3x + 1$, $[a, b] = [3, 5]$

43. $f(x) = \sqrt{1 - x^2}$, $[a, b] = [-1, 1]$

44. $f(x) = 3 + \sqrt{4 - x^2}$, $[a, b] = [-2, 2]$

Proof

45. Use sigma notation to prove that, for a given function $f(x)$, a given interval $[a, b]$, and a given value of n, the Trapezoid Sum is the average of the Left-Hand Sum and the Right-Hand Sum.

12.3 The Definite Integral

In this section we consider limits of Riemann sums and use these limits to define the exact area under a curve, which we call the **definite integral.** We will also discuss the properties of definite integrals and use them to do a few exact area calculations.

12.3.1 Defining the exact area under a curve

What is the area under the graph of a nonnegative function $f(x)$ on an interval $[a, b]$? Think about that for a minute; what do we *mean* by "area"? In a loose sense, we mean the "size" of a shaded region, like the one in Figure 12.33.

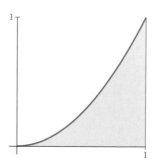

Figure 12.33
The "area" under the graph
of $f(x) = x^2$ from $x = 0$
to $x = 1$

However, we don't yet have any rigorous mathematical description of "area." On the other hand, we do have a definition for the sums of areas of rectangles (Riemann sums). Consider the Upper Sum for $f(x) = x^2$ on $[0, 1]$ with $n = 10$, $n = 50$, and $n = 100$ rectangles. (In this case, since $f(x) = x^2$ is increasing on $[0, 1]$, the Upper Sum is the same as the Right-Hand Sum.) See Figures 12.34–12.36.

As the number n of rectangles gets larger, the Upper Sum approximation gets closer and closer to our intuitive notion of the "area" under the graph of $f(x) = x^2$ from $x = 0$

to $x = 1$. The same thing happens if we consider any other Riemann sum approximation, say the Lower Sum (see Figures 12.37–12.39).

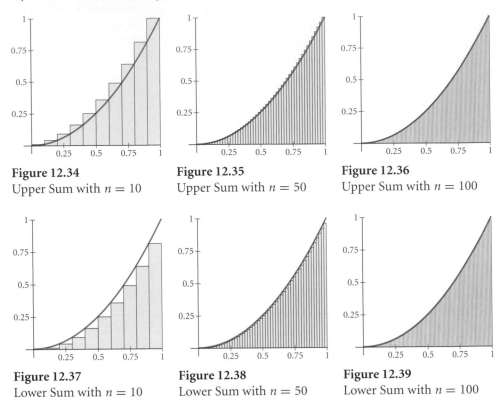

Figure 12.34
Upper Sum with $n = 10$

Figure 12.35
Upper Sum with $n = 50$

Figure 12.36
Upper Sum with $n = 100$

Figure 12.37
Lower Sum with $n = 10$

Figure 12.38
Lower Sum with $n = 50$

Figure 12.39
Lower Sum with $n = 100$

In general, we will *define* the "area" under the graph of a continuous function $f(x)$ from $x = a$ to $x = b$ to be the limit, as the number n of rectangles approaches infinity, of any Riemann sum for $f(x)$ on $[a, b]$. To denote this area, we use the symbol

$$\int_a^b f(x)\, dx.$$

This notation is pronounced "the ***definite integral*** of $f(x)$ from a to b" or sometimes just "the ***integral*** of $f(x)$ from a to b." The letters a and b are called the ***limits of integration.*** In this case, the word "limits" means "ends" (rather than our usual mathematical notion of "limit").

Definition 12.6 defines area (and thus the definite integral) as a limit of Riemann sums. This definition does *not* require that f be a nonnegative function, but for the present we will limit our examples to functions that are nonnegative in the interval $[a, b]$. In Section 12.4 we will expand our experience to include functions that are negative.

DEFINITION 12.6

The Definite Integral of a Function on an Interval

Let $f(x)$ be any continuous function defined on an interval $[a, b]$. The ***definite integral*** of $f(x)$ from $x = a$ to $x = b$ is defined to be

$$\int_a^b f(x)\, dx := \lim_{n \to \infty} \sum_{k=1}^n f(x_k^*)\, \Delta x,$$

if this limit exists, where $\Delta x := \frac{b-a}{n}$, $x_k := a + k\,\Delta x$, and $x_k^* \in [x_{k-1}, x_k]$.

The "$\int$" symbol may remind you of the letter "S" for "Sum." The symbol $\sum$ represents a discrete sum (of areas of rectangles, for example), whereas in some sense the integral symbol $\int$ represents a continuous sum (of the areas of "infinitely many" rectangles, each of which is "infinitely thin"). It may help you remember Definition 12.6 if you think of it in the following way: As $n \to \infty$, the finite sum ($\sum$) becomes an integral ($\int$). In addition, the discrete list of values $f(x_k^*)$ becomes the continuous function $f(x)$, and the small change Δx becomes an "infinitely small" change dx. The dx indicates that we are integrating along the x-axis, and the a and b at the ends of the integral symbol indicate that we are considering the area from $x = a$ to $x = b$.

> ◈ **Caution**   Don't forget that since the symbol $\int_a^b f(x)\, dx$ represents an area, it stands for a *number*. Make sure you understand the notation *and* the geometrical interpretation of the definite integral as a limit of Riemann sums.   □

We can use *any* Riemann sum in Definition 12.6. This may surprise you, since different Riemann sums can give different area approximations. The limit in Definition 12.6 does not specify the choices for the partition points x_k^*, and it is not obvious that choosing x_k^* differently will not change the limit. However, it is true that *in the limit*, all Riemann sums are the same, and each Riemann sum approaches the "area" under the curve (provided that its limit as $n \to \infty$ exists). For example, consider the Upper and Lower Sums shown earlier in Figures 12.34–12.39. For any given n, the Upper and Lower Sums are different. However, as $n \to \infty$, both of these sums approach the same area.

If the limit defining the integral exists for a given function $f(x)$ and interval $[a, b]$, then we say that $f(x)$ is **integrable** on $[a, b]$. It turns out that every continuous function is integrable, although the proof of this theorem is difficult, and will not be presented here.

12.3.2 Calculating a definite integral exactly

We can now calculate the *exact* area under the graph of a function on an interval, provided that we can find the limit as $n \to \infty$ of the corresponding Riemann sums. In the following example we will approximate, and then find *exactly*, the area shown in Figure 12.33 (on page 741). The key to doing this will be writing a Riemann sum for this area as a formula rather than a sum. This will enable us to make very good approximations (with very large n), as well as make it possible for us to evaluate the limit as $n \to \infty$.

EXAMPLE 12.13

Finding approximate and exact area

Consider the area under the graph of $f(x) = x^2$ from $x = 0$ to $x = 1$. Use the Right-Hand Sum to approximate this area with $n = 100$ and with $n = 1000$ rectangles. Then find the *exact* area (in other words, calculate $\int_0^1 x^2\, dx$).

Solution   We will begin by finding an expression for the Right-Hand Sum with n rectangles, and then use this expression to see what happens when $n = 100$, when $n = 1000$, and when $n \to \infty$. The Right-Hand Sum approximation with n rectangles is

(1) $$\sum_{k=1}^{n} f(x_k)\, \Delta x.$$

Since $a = 0$ and $b = 1$, we have

$$\Delta x = \tfrac{b-a}{n} = \tfrac{1-0}{n} = \tfrac{1}{n} \quad \text{and} \quad x_k = a + k\,\Delta x = 0 + k\left(\tfrac{1}{n}\right) = \tfrac{k}{n}.$$

Finally, since $f(x) = x^2$, we have

$$f(x_k) = (x_k)^2 = \left(\tfrac{k}{n}\right)^2.$$

This means that Statement (1) can be written

$$\sum_{k=1}^{n} \left(\tfrac{k}{n}\right)^2 \left(\tfrac{1}{n}\right).$$

Using algebra and a sum formula from Theorem 12.3 in Section 12.1.4, we can write this sum as

$$\sum_{k=1}^{n} \left(\tfrac{k}{n}\right)^2 \left(\tfrac{1}{n}\right) = \sum_{k=1}^{n} \frac{k^2}{n^3} \qquad \text{(algebra)}$$

$$= \frac{1}{n^3} \sum_{k=1}^{n} k^2 \qquad \text{(\textit{n} is constant with respect to \textit{k})}$$

$$= \frac{1}{n^3} \frac{n(n+1)(2n+1)}{6} \qquad \text{(sum formula)}$$

$$= \frac{(n+1)(2n+1)}{6n^2}. \qquad \text{(algebra)}$$

We did this same calculation in Example 12.9, before we even knew about Riemann sums.

We can use the result of the previous calculation to find the values of the Right-Hand Sum approximations for $n = 100$ and $n = 1000$ (this is much easier than adding a hundred or a thousand terms):

The Right-Hand Sum with $n = 100$ is $\quad \dfrac{(101)(201)}{60{,}000} = \dfrac{20{,}301}{60{,}000} = 0.33835.$

The Right-Hand Sum with $n = 1000$ is $\quad \dfrac{(1001)(2001)}{6{,}000{,}000} = \dfrac{2{,}003{,}001}{6{,}000{,}000} \approx 0.33383.$

To find the *exact* area of the region, we take the limit as $n \to \infty$. We will use Definition 12.6, with the Right-Hand Sum as our Riemann sum.

$$\int_0^1 x^2 \, dx = \lim_{n \to \infty} \sum_{k=1}^{n} f(x_k) \, \Delta x \qquad \text{(Definition 12.6)}$$

$$= \lim_{n \to \infty} \frac{(n+1)(2n+1)}{6n^2} \qquad \text{(by the work above)}$$

$$= \lim_{n \to \infty} \frac{2n^2 + 3n + 1}{6n^2} \qquad \text{(algebra)}$$

$$= \frac{2}{6} = \frac{1}{3}. \qquad \text{(ratio of leading coefficients)}$$

(In the step where we evaluated the limit, we used the fact that $\frac{2n^2+3n+1}{6n^2}$ is a rational function with the same degree in the numerator and denominator. See Section 6.2.3 if you need to review limits of rational functions.) Notice that we would not have been able to calculate this limit if we did not have a *formula* for the Riemann sum (in terms of n, but with no "Σ"). Notice also that the approximation with $n = 1000$ was quite accurate; it was only approximately 0.0005 square unit larger than the true area under the curve.   □

12.3.3 Properties of definite integrals

Now that we have a rigorous definition for the definite integral, we will investigate its properties. The following theorem illustrates two simple properties.

THEOREM 12.4

The Definite Integral on $[a, a]$ and $[b, a]$

For any function $f(x)$ and real numbers a and b, we have:

(a) $\displaystyle\int_a^a f(x)\, dx = 0.$ (b) $\displaystyle\int_b^a f(x)\, dx = -\int_a^b f(x)\, dx.$

Part (a) of Theorem 12.4 makes sense graphically, since the area under the graph of any function $f(x)$ from $x = a$ to $x = a$ should be zero (the width of the region would be zero). Part (b) follows directly from the definition of the definite integral as a limit of Riemann sums, where in this case if $a < b$ then the right-hand side involves a *negative* change $\Delta x = \frac{a-b}{n}$. This enables us to consider definite integrals like $\int_4^1 x^2\, dx$, where the "starting" x-value is greater than the "ending" x-value. You can think of this part of the theorem as saying that if we integrate "backwards," say from $x = 4$ to $x = 1$, then we count the area given by the definite integral negatively. You will prove both parts of this theorem in the exercises.

Like limits and derivatives, definite integrals behave well with sums and constant multiples. In other words, definite integrals commute with sums and constant multiples.

THEOREM 12.5

Sum and Constant Multiple Rules for Definite Integrals

For any continuous functions $f(x)$ and $g(x)$ on $[a, b]$, and any real number k,

(a) $\displaystyle\int_a^b (f(x) + g(x))\, dx = \int_a^b f(x)\, dx + \int_a^b g(x)\, dx.$

(b) $\displaystyle\int_a^b kf(x)\, dx = k\int_a^b f(x)\, dx.$

Before we present the formal proof of Theorem 12.5, let's think about the situation graphically. Figure 12.40 shows a Left-Hand Sum for the area under the graph of $f(x) = x^2$ from $x = 0$ to $x = 2$. Figure 12.41 shows a Left-Hand Sum (with the same n) for the area under $g(x) = x$ on the same interval. What is the Left-Hand Sum (with the same n) for the area under the graph of the sum $f(x) + g(x) = x^2 + x$? Figure 12.42 illustrates that this Left-Hand Sum is the sum of the two Left-Hand Sums in Figures 12.40 and 12.41. For example, the fourth rectangle in Figure 12.42 is shown as a stack of the fourth rectangle in Figure 12.40 (on the top) and the fourth rectangle in Figure 12.41 (on the bottom).

❓ *Question*   Can you explain the second part of Theorem 12.5 graphically? Start by arguing that each rectangle in a Riemann sum for a function $kf(x)$ is k times the height of the corresponding rectangle in a Riemann sum for $f(x)$.   □

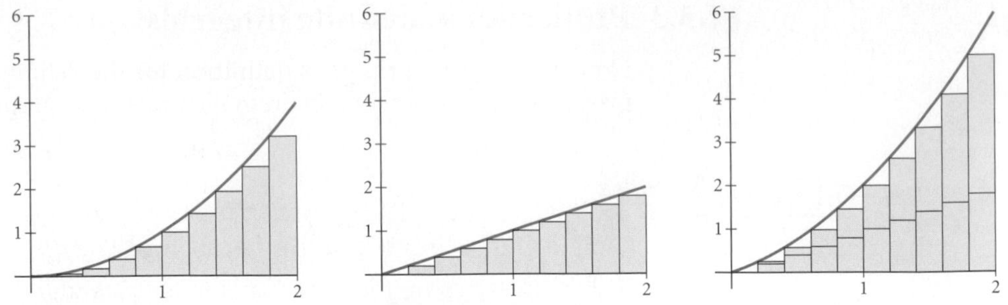

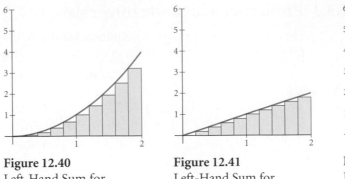

Figure 12.40
Left-Hand Sum for
$f(x) = x^2$

Figure 12.41
Left-Hand Sum for
$g(x) = x$

Figure 12.42
Left-Hand Sum for
$f(x) + g(x) = x^2 + x$

The key to the proof of Theorem 12.5 is that sums and constant multiples commute with limits and with Riemann sums. Since the definite integral is a limit of Riemann sums, it must be that sums and constant multiples commute with definite integrals.

PROOF (**THEOREM 12.5**)   Given a positive integer n, define $\Delta x := \frac{b-a}{n}$ and $x_k := a + k\,\Delta x$, and let x_k^* be any point in the subinterval $[x_{k-1}, x_k]$. Then, by the definition of the definite integral, we have

$$\int_a^b (f(x) + g(x))\, dx$$

$$= \lim_{n \to \infty} \sum_{k=1}^n (f(x_k^*) + g(x_k^*))\, \Delta x \qquad \text{(definition of definite integral)}$$

$$= \lim_{n \to \infty} \sum_{k=1}^n (f(x_k^*)\, \Delta x + g(x_k^*)\, \Delta x) \qquad \text{(algebra)}$$

$$= \lim_{n \to \infty} \left(\sum_{k=1}^n f(x_k^*)\, \Delta x + \sum_{k=1}^n g(x_k^*)\, \Delta x \right) \qquad \text{(split into two sums)}$$

$$= \lim_{n \to \infty} \sum_{k=1}^n f(x_k^*)\, \Delta x + \lim_{n \to \infty} \sum_{k=1}^n g(x_k^*)\, \Delta x \qquad \text{(sum rule for limits)}$$

$$= \int_a^b f(x)\, dx + \int_a^b g(x)\, dx. \qquad \text{(definition of definite integral)}$$

The proof of part (b) is similar: Write the definite integral of $kf(x)$ as a limit of Riemann sums, use properties of sums and limits to "pull out" the constant k, and then use the definition of the definite integral again to get k times the definite integral of $f(x)$. You will write out the details in the exercises.   ■

⊗ *Caution*   Although the definite integral of a sum of functions is equal to the sum of the definite integrals of those functions, it is *not* true that the definite integral of a product is a product of definite integrals. In other words,

$$\int_a^b f(x)g(x)\, dx \neq \left(\int_a^b f(x)\, dx \right) \left(\int_a^b g(x)\, dx \right).$$

One counterexample is $f(x) = x$, $g(x) = x$, and $[a, b] = [0, 1]$. (Work out the details to show that this is a counterexample.)   □

The following theorem states that we can split a definite integral on an interval $[a, b]$ at any point c between a and b.

THEOREM 12.6

Splitting a Definite Integral

For any $c \in [a, b]$,

$$\int_a^b f(x)\, dx = \int_a^c f(x)\, dx + \int_c^b f(x)\, dx.$$

This theorem can be proved using the definition of the definite integral as a limit of Riemann sums, but the proof is tedious (it requires very technical and careful manipulation of notation). In Figures 12.43–12.45 we present a "proof by picture," which illustrates Theorem 12.6 with $a = 0$, $b = 3$, and $c = 1 \in [0, 3]$. The area from $x = 0$ to $x = 3$ is the sum of the area from $x = 0$ to $x = 1$ and the area from $x = 1$ to $x = 3$.

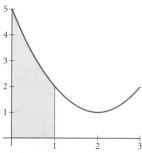

Figure 12.43
Area from $x = 0$ to $x = 1$

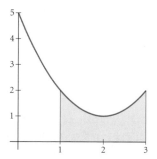

Figure 12.44
Area from $x = 1$ to $x = 3$

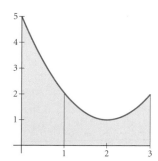

Figure 12.45
Area from $x = 0$ to $x = 3$

EXAMPLE 12.14

Using the algebraic properties of definite integrals

If $\int_1^3 f(x)\, dx = 4$ and $\int_5^3 2f(x)\, dx = -3$, find $\int_1^5 f(x)\, dx$.

Solution   First of all, by Theorem 12.4,

$$\int_5^3 2f(x)\, dx = -3 \quad \Rightarrow \quad \int_3^5 2f(x)\, dx = -(-3) = 3.$$

Now, using part (b) of Theorem 12.5, we know that

$$\int_3^5 2f(x)\, dx = 2\int_3^5 f(x)\, dx = 3 \quad \Rightarrow \quad \int_3^5 f(x) = \frac{3}{2}.$$

Finally, by Theorem 12.6, we have

$$\int_1^5 f(x)\, dx = \int_1^3 f(x)\, dx + \int_3^5 f(x)\, dx = 4 + \frac{3}{2} = \frac{11}{2}.$$

The following theorem describes the definite integrals of constant functions $f(x) = c$, the identity function $f(x) = x$, and the squaring function $f(x) = x^2$ on an arbitrary interval $[a, b]$.

THEOREM 12.7

Definite Integral Formulas

For any real numbers a, b, and c,

(a) $\displaystyle\int_a^b c\,dx = c(b-a)$ 　　　　 (b) $\displaystyle\int_a^b x\,dx = \tfrac{1}{2}(b^2 - a^2)$

(c) $\displaystyle\int_a^b x^2\,dx = \tfrac{1}{3}(b^3 - a^3)$

Theorem 12.7 is true for all real numbers a, b, and c, even those for which the definite integral in question turns out to be a *negative* number. For example,

$$\int_{-2}^{1} x\,dx = \tfrac{1}{2}(1^2 - (-2)^2) = \tfrac{1}{2}(1 - 4) = -\tfrac{3}{2}.$$

In Section 12.4 we will discuss what it means when a definite integral is negative, and we will discuss "negative area." For now, we will treat Theorem 12.7 as a calculational theorem and not worry about what it means when the definite integral is negative. Our proof of Theorem 12.7 will assume conditions on a, b, and c so that the functions involved are nonnegative (although the theorem is true even for negative functions).

The first two parts of Theorem 12.7 can be proved geometrically, since the region under the graph of a constant function $f(x) = c$ from $x = a$ to $x = b$ is always a rectangle, and the region under the graph of the identity function $f(x) = x$ is a trapezoid. The proof of part (c) is left for the exercises.

PROOF (**GEOMETRICAL ARGUMENTS FOR PARTS (A) AND (B) OF THEOREM 12.7**) To prove part (a), consider the constant function $f(x) = c$ on an interval $[a, b]$. We will assume that c is positive and that $a < b$. The area $\int_a^b c\,dx$ under this curve is shown in Figure 12.46.

Clearly this area is a rectangle, so we can easily compute its area. The length of the base of this rectangle is $|b - a| = b - a$ (since $b > a$). Since $f(x) = c$ for all x, the height of the rectangle is c. Using the formula for the area of a rectangle, we have $\int_a^b c\,dx = c(b - a)$.

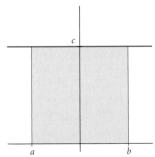

Figure 12.46
Area under $f(x) = c$ on $[a, b]$

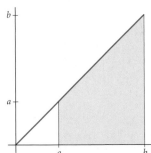

Figure 12.47
Area under $f(x) = x$ on $[a, b]$

Now, to prove part (b), consider the identity function $f(x) = x$ on some interval $[a, b]$. We will assume that $0 \le a < b$. Figure 12.47 shows one possible picture of the area $\int_a^b x\,dx$.

The area in question is a trapezoid with base of length $b - a$, a height of length a on the left, and a height of length b on the right. Using the formula for the area of a trapezoid, we have $\int_a^b x\,dx = \frac{a+b}{2}(b-a) = \frac{1}{2}(ab + b^2 - a^2 - ab) = \frac{1}{2}(b^2 - a^2)$. ∎

It is also possible to prove Theorem 12.7 algebraically, using the definition of the definite integral as a limit of Riemann sums. We do so for part (a) of the theorem. (Although we have already proved this part of the theorem, the algebraic proof is a good illustration of how the definition of the definite integral works.)

PROOF (**ALGEBRAIC ARGUMENT FOR PART (A) OF THEOREM 12.7**)
Suppose $f(x) = c$ is a constant function on $[a, b]$. By the definition of the definite integral, we have

$$\int_a^b f(x)\,dx = \lim_{n\to\infty} \sum_{k=1}^n f(x_k^*)\,\Delta x \qquad \text{(definition of definite integral)}$$

$$= \lim_{n\to\infty} \sum_{k=1}^n c\,\Delta x \qquad (f(x) = c \text{ is a constant function})$$

$$= \lim_{n\to\infty} \sum_{k=1}^n c\left(\tfrac{b-a}{n}\right) \qquad (\Delta x = \tfrac{b-a}{n})$$

$$= \lim_{n\to\infty} c\left(\tfrac{b-a}{n}\right) \sum_{k=1}^n 1 \qquad \text{(pull constants out of sum)}$$

$$= \lim_{n\to\infty} c\left(\tfrac{b-a}{n}\right)(n) \qquad \text{(sum formula)}$$

$$= \lim_{n\to\infty} c(b-a) \qquad \text{(algebra)}$$

$$= c(b-a). \qquad \text{(limit of a constant)}$$

The proof of part (c) of Theorem 12.7 must be done algebraically (like the proof above), since the region under the graph of $f(x) = x^2$ on $[a, b]$ is not a geometrical shape whose area we know (like a rectangle or a trapezoid). This proof is algebra-intensive and is left to you in the exercises.

Now that we know various properties of definite integrals, there are many definite integrals that we can calculate *exactly* without using any Riemann sums or limits.

EXAMPLE 12.15   **Using definite integral formulas**

Find the exact value of $\int_2^3 (5x^2 - 3x + 2)\,dx$.

Solution   Using the sum and constant multiple rules for definite integrals (Theorem 12.5), we have

$$\int_2^3 (5x^2 - 3x + 2)\,dx$$

$$= \int_2^3 5x^2\,dx + \int_2^3 (-3x)\,dx + \int_2^3 2\,dx \qquad \text{(part (a) of Theorem 12.5)}$$

$$= 5\int_2^3 x^2\,dx - 3\int_2^3 x\,dx + \int_2^3 2\,dx \qquad \text{(part (b) of Theorem 12.5)}$$

$$= 5\left(\tfrac{1}{3}(3^3 - 2^3)\right) - 3\left(\tfrac{1}{2}(3^2 - 2^2)\right) + 2(3 - 2) \qquad \text{(Theorem 12.7)}$$

$$= \frac{95}{3} - \frac{15}{2} + 2 = \frac{157}{6}.$$

Notice that we did not have to calculate any Riemann sums (or any limits) to find the exact value of the definite integral.

🔷 *Caution*   In Example 12.15, we wanted to find the *exact* area; that is why we did not write down a decimal approximation for our exact answer of $\frac{157}{6}$. Be careful that you do not accidentally turn an exact answer into an approximate answer by using your calculator to simplify fractions.   □

☑ *Checking the Answer*   You can use your graphing calculator to approximate the answer, and then check that your approximation is close to the true area you found by hand. Graph $f(x) = 5x^2 - 3x + 2$ so that the entire portion of the graph between $x = 2$ and $x = 3$ can be seen, and have your calculator approximate the area under the graph.   □

What you should know after reading Section 12.3

▶ The definition of the definite integral as a limit of Riemann sums. Be able to write down this definition algebraically (with the proper notation), and be able to describe what this definition means graphically (in terms of area.)

▶ How to compute the *exact* value of a definite integral by calculating a limit of Riemann sums. (This will involve the sum formulas from Section 12.1.4.)

▶ Be able to state, prove, and use the various properties of definite integrals presented in this section. You should also be able to sketch graphs that illustrate why these properties hold (both in terms of exact area and in terms of Riemann sums).

Exercises 12.3

Concepts

0. Read the section and make your own summary of the material. Include whatever you think will help you review the material later on.

1. Explain why it makes sense that *every* Riemann sum for a continuous function $f(x)$ on an interval $[a, b]$ approaches the same number as the number n of rectangles approaches infinity. Illustrate your argument with graphs.

2. Fill in the blanks: The area under the graph of a continuous function $f(x)$ from $x = a$ to $x = b$ is represented by the notation _____ and is called the _____.

3. Fill in the blanks: The definite integral of a continuous function $f(x)$ from $x = a$ to $x = b$ is defined to be

$$\int_a^b f(x)\, dx := \lim_{\boxed{}} \sum_{\boxed{}}^{\boxed{}} \boxed{},$$

where $\Delta x :=$ _____, $x_k :=$ _____, and $x_k^* \in$ _____.

4. Explain geometrically what the definition of the definite integral as a limit of Riemann sums represents. Include

a labeled picture of a Riemann sum (for a particular n) that illustrates the roles of n, Δx, x_k, x_k^*, and $f(x_k^*)$. What happens in the picture as $n \to \infty$?

5. Draw pictures illustrating the fact that if $a \le c \le b$, then $\int_a^c f(x)\, dx + \int_c^b f(x)\, dx = \int_a^b f(x)\, dx$.

6. If $f(x)$ is defined at $x = a$, then $\int_a^a f(x)\, dx$ is defined to be zero. Explain why this makes sense in terms of area.

7. Suppose $f(x)$ is a continuous, nonnegative function on an interval $[a, b]$, and let k be any positive real number. Use pictures of Riemann sums to illustrate that the Right-Hand Sum for the function $kf(x)$ on $[a, b]$ is k times the value of the Right-Hand Sum (with the same n) for $f(x)$ on $[a, b]$. What does this mean when $n \to \infty$? In other words, what does this say about the definite integrals $\int_a^b f(x)\, dx$ and $\int_a^b kf(x)\, dx$?

■ **Determine whether each of the following statements is true or false. If a statement is true, explain why. If a statement is false, explain why or provide a counterexample.**

8. True or False: The Left-Hand Sum and Right-Hand Sum approximations for the definite integral of a function are the same if the number n of rectangles is very large.

9. True or False: $\displaystyle\int_{-2}^{5} (x+2)^3 \, dx$ is a real number.

10. True or False: $\displaystyle\int_{-3}^{-2} f(x) \, dx = -\int_{2}^{3} f(x) \, dx.$

11. True or False: If $\displaystyle\int_{0}^{2} f(x) \, dx = 3$ and $\displaystyle\int_{0}^{2} g(x) \, dx = 2$, then $\displaystyle\int_{0}^{2} f(g(x)) = 6.$

12. True or False: If $\displaystyle\int_{0}^{2} f(x) \, dx = 3$ and $\displaystyle\int_{0}^{2} g(x) \, dx = 2$, then $\displaystyle\int_{0}^{2} f(x)g(x) \, dx = 6.$

13. True or False: If $\displaystyle\int_{0}^{1} f(x) \, dx = 3$ and $\displaystyle\int_{0}^{-1} f(x) \, dx = 4$, then $\displaystyle\int_{-1}^{1} f(x) \, dx = 7.$

14. True or False: If $\displaystyle\int_{0}^{2} f(x) \, dx = 3$ and $\displaystyle\int_{2}^{4} g(x) \, dx = 4$, then $\displaystyle\int_{0}^{4} (f(x) + g(x)) \, dx = 7.$

15. Although the definite integral of a sum of functions is equal to the sum of the definite integrals of those functions, the definite integral of a product of functions is *not* the product of two definite integrals.

    (a) Write the sentence above in mathematical notation by filling in the blanks:

$$\underline{\hspace{1.5cm}} = \underline{\hspace{1.5cm}},$$

but $\underline{\hspace{1.5cm}} \neq \underline{\hspace{1.5cm}}.$

    (b) Choose two simple functions $f(x)$ and $g(x)$ so that you can calculate the definite integrals of $f(x)$, $g(x)$, and $f(x) + g(x)$ on $[0, 1]$ and show that the sum of the first two definite integrals is equal to the third.

    (c) Find two simple functions $f(x)$ and $g(x)$ such that $\int_{0}^{1} f(x)g(x) \, dx$ is not equal to the product of $\int_{0}^{1} f(x) \, dx$ and $\int_{0}^{1} g(x) \, dx$. (*Hint:* Choose $f(x)$ and $g(x)$ so that you can calculate the definite integrals involved.)

16. Consider the definite integral $\int_{0}^{3} x^2 \, dx$. Clearly the Right-Hand Sum and Left-Hand Sum, for a particular number of rectangles n, will be different (draw a picture to show why). However, as $n \to \infty$, these two Riemann sums approach the same number. Show that this is true, as follows:

    (a) Write down a general Right-Hand Sum for $\int_{0}^{3} x^2 \, dx$ (for an arbitrary n), and use algebra and a sum formula to write this sum as a formula in terms of n.

    (b) Write down a general Left-Hand Sum for $\int_{0}^{3} x^2 \, dx$ (for an arbitrary n), and use algebra and a sum formula to write this sum as a formula in terms of n.

    (c) Use your answers to parts (a) and (b) to show that the Right-Hand Sum and the Left-Hand

Sum for $\int_{0}^{3} x^2 \, dx$ are different for $n = 100$ and $n = 1000$.

    (d) Use your answers to parts (a) and (b) to show that the Right-Hand Sum and the Left-Hand Sum for $\int_{0}^{3} x^2 \, dx$ approach the same quantity as $n \to \infty$. What does this quantity represent geometrically?

17. The definite integral of a function $f(x)$ on an interval $[a, b]$ is defined as a limit of Riemann sums. How can it be that the sum of the areas of "infinitely many" rectangles each of which is "infinitely thin" is a finite number? Shouldn't it be infinite, since we are adding up infinitely many rectangles? On the other hand, shouldn't it always be zero, since the width of each of the rectangles is approaching zero as $n \to \infty$?

Skills

▪ Use geometry (i.e., areas of triangles, rectangles, and circles) to find the exact value of each of the following definite integrals.

18. $\displaystyle\int_{0}^{2} (2 - x) \, dx$ **19.** $\displaystyle\int_{2}^{3} (4x - 3) \, dx$

20. $\displaystyle\int_{-3}^{8} 24 \, dx$ **21.** $\displaystyle\int_{-1}^{1} \sqrt{1 - x^2} \, dx$

22. $\displaystyle\int_{-r}^{r} \sqrt{r^2 - x^2} \, dx$ **23.** $\displaystyle\int_{-2}^{2} (3 + \sqrt{4 - x^2}) \, dx$

▪ Write down the general Right-Hand Sum for each definite integral below (with an arbitrary n). Using algebra and the sum formulas from Section 12.1.4, write the Right-Hand Sum as a formula in terms of n (without a sum). Use this formula to approximate the area of the given definite integral with $n = 100$ and $n = 1000$. Then take the limit as $n \to \infty$ to find the exact area.

24. $\displaystyle\int_{0}^{3} (2x + 1) \, dx$ **25.** $\displaystyle\int_{2}^{5} (5 - x) \, dx$

26. $\displaystyle\int_{0}^{1} 2x^2 \, dx$ **27.** $\displaystyle\int_{-3}^{2} x^2 \, dx$

28. $\displaystyle\int_{-1}^{1} (1 - x^2) \, dx$ **29.** $\displaystyle\int_{2}^{3} (x + 1)^2 \, dx$

▪ Use the integration formulas in this section to calculate the exact values of the following integrals.

30. $\displaystyle\int_{2}^{4} (x^2 + 1) \, dx$ **31.** $\displaystyle\int_{0}^{6} (3x + 2) \, dx$

32. $\displaystyle\int_{5}^{2} (9 + 10x - x^2) \, dx$ **33.** $\displaystyle\int_{1}^{3} (x + 1)^2 \, dx$

34. $\displaystyle\int_{0}^{4} ((2x - 3)^2 + 5) \, dx$ **35.** $\displaystyle\int_{6}^{1} (3(1 - 2x)^2 + 4x) \, dx$

■ Given that $f(x)$ and $g(x)$ are continuous functions with $\int_{-2}^{3} f(x)\, dx = 4$, $\int_{-2}^{6} f(x)\, dx = 9$, $\int_{-2}^{3} g(x)\, dx = 2$, and $\int_{3}^{6} g(x)\, dx = 3$, find the values of the definite integrals in the problems below. (If there is not enough information, explain why.)

36. $\displaystyle\int_{-2}^{6} g(x)\, dx$ 37. $\displaystyle\int_{3}^{6} f(x)\, dx$

38. $\displaystyle\int_{-2}^{6} (f(x) + g(x))\, dx$ 39. $\displaystyle\int_{-2}^{3} f(x)g(x)\, dx$

40. $\displaystyle\int_{3}^{6} (2f(x) - g(x))\, dx$ 41. $\displaystyle\int_{-2}^{6} (g(x) + x)\, dx$

42. $\displaystyle\int_{-2}^{3} xf(x)\, dx$ 43. $\displaystyle\int_{3}^{6} (g(x))^2\, dx$

44. $\displaystyle\int_{-2}^{6} (4f(x) - 2)\, dx$ 45. $\displaystyle\int_{6}^{3} (f(x) + g(x))\, dx$

46. $\displaystyle\int_{-2}^{-2} x(f(x) + 3)^2\, dx$

47. $\displaystyle\int_{3}^{-2} (2x^2 - 3g(x))\, dx$

Proofs

48. Prove that $\displaystyle\int_{1}^{3} (3x+4)\, dx = 20$ in three ways, as follows:

   (a) Using the definition of the definite integral; that is, write the area as a limit of Riemann sums, find a formula for the Riemann sum, and take the limit as $n \to \infty$.

   (b) Geometrically; that is, recognize the region in question as the trapezoid, and calculate its area.

   (c) Using the properties of definite integrals and the formulas for the definite integrals of identity and constant functions.

49. Prove that if c is any real number, then $\int_{a}^{b} cf(x)\, dx = c\int_{a}^{b} f(x)\, dx$. Use the definition of the definite integral as a limit of Riemann sums. (The outline of this proof is given in the proof of Theorem 12.5 in the reading.)

50. Prove part (a) of Theorem 12.4 by using the definition of the definite integral as a limit of Riemann sums.

51. Prove part (b) of Theorem 12.4 by using the definition of the definite integral as a limit of Riemann sums.

52. Give an algebraic proof of part (b) of Theorem 12.7. Use the proof of part (a) of that theorem as a guide.

53. Prove part (c) of Theorem 12.7. Use the algebraic proof of part (a) of that theorem as an guide. (*Warning:* This proof involves a great deal of algebra!)

12.4   Area and Average Value

In this section we use definite integrals to describe positive and negative area, the area between two curves, and the average value of a function.

12.4.1 Positive and negative area

So far we have considered the area under a graph only for a *nonnegative* function $f(x)$ on an interval $[a, b]$. However, the definition and properties of Riemann sums and the definite integral work just as well if $f(x)$ is negative.

EXAMPLE 12.16   **The definite integral of a negative function**

Consider the Left-Hand Sum with $n = 4$ rectangles for $f(x) = x^2 - 1$ on $[0, 1]$. This sum is shown in Figure 12.48. Note that the height of each of the rectangles in this Riemann sum is *negative*. For example, the first rectangle has a height of $f(0) = 0^2 - 1 = -1$. Since $n = 4$ and $b - a = 1 - 0 = 1$, each rectangle has a width of $\Delta x = \frac{1}{4} = 0.25$. The Left-Hand Sum is equal to

$$f(0)\,\Delta x + f(0.25)\,\Delta x + f(0.5)\,\Delta x + f(0.75)\,\Delta x$$

$$= (0^2 - 1)(0.25) + ((0.25)^2 - 1)(0.25) + ((0.5)^2 - 1)(0.25) + ((0.75)^2 - 1)(0.25)$$

$$= (-1)(0.25) + (-0.9375)(0.25) + (-0.75)(0.25) + (-0.4375)(0.25)$$

$$= -0.78125.$$

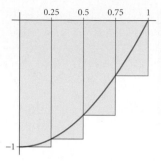

Figure 12.48
Left-Hand Sum with $n = 4$
for $f(x) = x^2 - 1$ on $[0, 1]$ □

In general, as in Example 12.16, the definite integral counts the area between a negative function and the x-axis *negatively*. We call this a "signed" area. The "true" area of the same region is 0.78125, whereas the definite integral of $f(x) = x^2 - 1$ on $[0, 1]$ is -0.78125 (the "signed" area). The following definition makes precise this distinction between "signed" and "true" area.

DEFINITION 12.7

> ### Signed Area and True Area
> Given any continuous function $f(x)$ on an interval $[a, b]$,
> **(a)** The *signed area* between the graph of $f(x)$ and the x-axis from $x = a$ to $x = b$ is defined to be the definite integral $\int_a^b f(x)\, dx$. The signed area is often simply called the *area.*
> **(b)** The *true area* between the graph of $f(x)$ and the x-axis from $x = a$ to $x = b$ is defined to be $\int_a^b |f(x)|\, dx$.

Signed area counts the area between the graph of $f(x)$ and the x-axis differently, depending on whether the graph of $f(x)$ is above or below the x-axis. When signed area is used, regions above the x-axis will have positive area and regions below the x-axis will have "negative" area. The absolute value in the definition of true area will count area positively regardless of whether it is above or below the x-axis.

◈ *Caution* When we refer to the "area" between the graph of a function $f(x)$ and the x-axis on an interval $[a, b]$, we mean the *signed* area (where regions above the x-axis have positive area, and regions below the x-axis have negative area). Also, be warned that it is common to refer to the signed area as the area "under the graph," even though the negative regions of area are not "under" the graph of the function.
□

EXAMPLE 12.17 **Using definite integrals to find signed area and true area**

Let $f(x)$ be the function shown in Figure 12.49. Express **(a)** the signed area and **(b)** the true area between the graph of $f(x)$ and the x-axis from $x = -2$ to $x = 4$ in terms of definite integrals of the function $f(x)$. It may help to look at the graph in Figure 12.50.

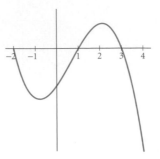

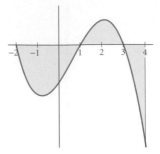

Figure 12.49

$f(x)$ on $[-2, 4]$

Figure 12.50

Region between $f(x)$ and
x-axis

Solution   The signed area of this region is simply the definite integral of $f(x)$ from
$x = -2$ to $x = 4$; in other words,

$$\text{signed area} = \int_{-2}^{4} f(x)\,dx.$$

The true area of the region is $\int_{-2}^{4} |f(x)|\,dx$. We can express this area in terms of
definite integrals of $f(x)$ (i.e., without using absolute value) by using three definite
integrals. From $x = -2$ to $x = 1$ the function $f(x)$ is negative, and thus on the interval
$[-2, 1]$ we have $|f(x)| = -f(x)$. From $x = 1$ to $x = 3$ the function is positive, and so
$|f(x)| = f(x)$ on this interval. Finally, from $x = 3$ to $x = 4$ the function is negative,
so $|f(x)| = -f(x)$ on $[3, 4]$. Therefore, the true area of the region is

$$\text{true area} = \int_{-2}^{4} |f(x)|\,dx = -\int_{-2}^{1} f(x)\,dx + \int_{1}^{3} f(x)\,dx - \int_{3}^{4} f(x)\,dx. \qquad \square$$

? *Question*   There is only *one* function $f(x)$ on $[0, 4]$ for which the true area
between the graph of $f(x)$ and the x-axis is zero. What is that function? On the
other hand, there are *many* functions $f(x)$ on $[0, 4]$ for which the signed area between the graph of $f(x)$ and the x-axis is zero. Draw graphs of three such functions.
$\qquad \square$

? *Question*   Absolute values do *not* commute with definite integrals. In other
words,

$$\int_{a}^{b} |f(x)|\,dx \neq \left| \int_{a}^{b} f(x)\,dx \right|.$$

Give an example of a function $f(x)$ on an interval $[a, b]$ that illustrates that the
two quantities above are not the same. $\qquad \square$

12.4.2 Areas between curves

We now know how to find the area between the graph of a function $f(x)$ and the
x-axis. We could think of this as the area between the graph of $y = f(x)$ and the graph
of $y = 0$ on an interval $[a, b]$. What about the area between the graph of $f(x)$ and
the graph of some other function $g(x)$ on $[a, b]$? For example, consider the functions
$f(x) = x + 2$ and $g(x) = x^2$ (shown in Figure 12.51) and the region between these
curves on the interval $[-1, 2]$ (shown in Figure 12.52). How can we find the area of the
shaded region?

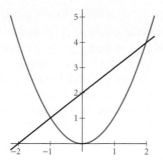

Figure 12.51
$f(x) = x + 2$ and $g(x) = x^2$

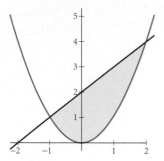

Figure 12.52
Region between f and g on $[-1, 2]$

One way to find the area of the region shown in Figure 12.52 would be to calculate the area under the upper graph ($f(x) = x + 2$) on $[-1, 2]$, and subtract from this the area under the lower graph ($f(x) = x^2$) on $[-1, 2]$. These two areas are shown in Figures 12.53 and 12.54.

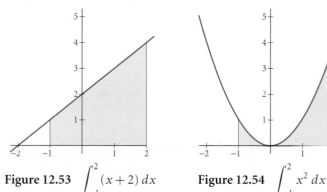

Figure 12.53 $\displaystyle\int_{-1}^{2} (x + 2)\, dx$ **Figure 12.54** $\displaystyle\int_{-1}^{2} x^2\, dx$

From these pictures and the properties of definite integrals, we have

$$\text{Area between } f(x) = x + 2 \atop \text{and } g(x) = x^2 \text{ on } [-1, 2] = \int_{-1}^{2} (x + 2)\, dx - \int_{-1}^{2} x^2\, dx = \int_{-1}^{2} ((x + 2) - x^2)\, dx.$$

? Question   Find the exact area between the graphs of $f(x) = x + 2$ and $g(x) = x^2$ from $x = -1$ to $x = 2$ by finding the exact value of the definite integral $\int_{-1}^{2} ((x + 2) - x^2)\, dx$. (*Hint:* Use the definite integral formulas in Theorem 12.7.)   □

The following definition generalizes this method of finding the (true) area between two curves.

DEFINITION 12.8

The Area Between Two Graphs

If $f(x)$ and $g(x)$ are continuous functions, then the ***area between the graphs of*** $f(x)$ ***and*** $g(x)$ ***from*** $x = a$ ***to*** $x = b$ is defined to be

$$\int_{a}^{b} |f(x) - g(x)|\, dx.$$

⊗ Caution   The phrase "area between the graphs of $f(x)$ and $g(x)$" will always refer to the *true* area between the graphs. In this context we will *not* consider the area to be positive or negative depending on whether it is above or below the x-axis.   □

On any interval $[c, d]$ where $f(x) \geq g(x)$, we will have $|f(x) - g(x)| = f(x) - g(x)$. The area between the graphs of $f(x)$ and $g(x)$ on such an interval is $\int_c^d (f(x) - g(x)) \, dx$. On any interval $[r, s]$ where $g(x) \geq f(x)$, we will have $|f(x) - g(x)| = g(x) - f(x)$. On such an interval, the area between the graphs of $f(x)$ and $g(x)$ is $\int_r^s (g(x) - f(x)) \, dx$.

When calculating the area between two curves, it will often be necessary to split the interval $[a, b]$ into smaller intervals where $f(x)$ is always greater than $g(x)$ (or vice versa) so that we can express the area without using any absolute values. Example 12.18 illustrates this method. The first step is to determine the intervals for which $f(x) \geq g(x)$ and the intervals for which $f(x) \leq g(x)$. We can then express the area between the graphs as a sum (or difference) of definite integrals.

EXAMPLE 12.18

Using definite integrals to find the area between two graphs

Find the area of the region between the graphs of $f(x) = x^2 - x - 2$ and $g(x) = 4 - x^2$ on the interval $[-2, 3]$.

Solution   Our first goal is to split the interval $[-2, 3]$ so that we can express the area without using absolute values. We must determine when the graph of $f(x) = x^2 - x - 2$ is above the graph of $g(x) = 4 - x^2$ and vice versa (i.e., we have to determine the intervals on which $f(x) \geq g(x)$ and the intervals on which $f(x) \leq g(x)$). The first step is to determine the values of x at which the graphs of $f(x)$ and $g(x)$ intersect:

$$f(x) = g(x) \implies x^2 - x - 2 = 4 - x^2$$
$$\implies 2x^2 - x - 6 = 0$$
$$\implies (2x + 3)(x - 2) = 0$$
$$\implies x = -\tfrac{3}{2} \text{ or } x = 2.$$

These points of intersection are the only places where the graphs of $f(x)$ and $g(x)$ could "switch places" (why?). Notice that both $x = -\frac{3}{2} = -1.5$ and $x = 2$ are in the interval $[-2, 3]$. By testing points, we can determine which graph is larger on each subinterval $[-2, -1.5]$, $[-1.5, 2]$, and $[2, 3]$. For example, since $f(0) = 0^2 - 0 - 2 = -2$ is less than $g(0) = 4 - 0^2 = 4$, we must have $f(x) \leq g(x)$ on the entire interval $[-1.5, 2]$. Thus the area of the center piece of the region is given by $\int_{-1.5}^{2} (g(x) - f(x)) \, dx$. Similarly, $f(x) \geq g(x)$ on the intervals $[-2, -1.5]$ and $[2, 3]$. This is confirmed in the graphs of $f(x)$ and $g(x)$ in Figures 12.55 and 12.56.

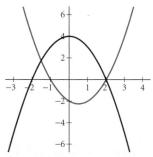

Figure 12.55
$f(x) = x^2 - x - 2$ and
$g(x) = 4 - x^2$

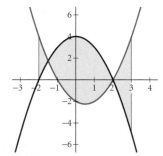

Figure 12.56
The region between the
graphs on $[-2, 3]$

Therefore, the area $\int_{-2}^{3} |f(x) - g(x)| \, dx$ of the region shown in Figure 12.56 can be written as the sum of three definite integrals:

$$\int_{-2}^{-1.5} (f(x) - g(x)) \, dx + \int_{-1.5}^{2} (g(x) - f(x)) \, dx + \int_{2}^{3} (f(x) - g(x)) \, dx.$$

Since we know how to calculate definite integrals involving constants, the identity function, and x^2, we can calculate the exact value of this area. The first integral in the expression for the area (corresponding to the area of the triangular wedge on the left-hand side of the region in Figure 12.56) is

$$\int_{-2}^{-1.5} (f(x) - g(x))\, dx$$

$$= \int_{-2}^{-1.5} ((x^2 - x - 2) - (4 - x^2))\, dx$$

$$= \int_{-2}^{-1.5} (2x^2 - x - 6)\, dx$$

$$= 2\int_{-2}^{-1.5} x^2\, dx - \int_{-2}^{-1.5} x\, dx - \int_{-2}^{-1.5} 6\, dx$$

$$= 2\left(\tfrac{1}{3}((-1.5)^3 - (-2)^3)\right) - \tfrac{1}{2}((-1.5)^2 - (-2)^2)) - 6((-1.5) - (-2))$$

$$= \frac{23}{24} \approx 0.9583333.$$

Note that we used the definite integral formulas from Section 12.3.3. Note also that to find the *exact* answer $\frac{23}{24}$, we must take the trouble to write out and simplify the second-to-last line in fractional form; to find the decimal approximation of the answer, we can just put the second-to-last line into a calculator.

In a similar fashion, we can calculate

$$\int_{-1.5}^{2} (g(x) - f(x))\, dx = \int_{-1.5}^{2} (-2x^2 + x + 6)\, dx = \frac{343}{24} \approx 14.2916667,$$

$$\int_{2}^{3} (f(x) - g(x))\, dx = \int_{-1.5}^{2} (2x^2 - x - 6)\, dx = \frac{25}{6} \approx 4.1666667.$$

The total area of the region between the graphs of $f(x) = x^2 - x - 2$ and $g(x) = 4 - x^2$ from $x = -2$ to $x = 3$ is

$$\int_{-2}^{3} |f(x) - g(x)|\, dx$$

$$= \int_{-2}^{-1.5} (f(x) - g(x))\, dx + \int_{-1.5}^{2} (g(x) - f(x))\, dx + \int_{2}^{3} (f(x) - g(x))\, dx$$

$$= \frac{23}{24} + \frac{343}{24} + \frac{25}{6} = \frac{466}{24} \approx 19.416667.$$

Although that was a long process, it was certainly better than working with limits of Riemann sums! This is why it was so vital for us to describe the properties of definite integrals. □

Although we *defined* the area between two curves in Definition 12.8, we should check that this definition makes sense in terms of the Riemann sum definition of the definite integral. Given continuous functions $f(x)$ and $g(x)$, the definite integral of the function $|f(x) - g(x)|$ on an interval $[a, b]$ is defined to be

(1)
$$\int_{a}^{b} |f(x) - g(x)|\, dx = \lim_{n \to \infty} \sum_{k=1}^{n} |f(x_k^*) - g(x_k^*)|\, \Delta x.$$

In other words, the definite integral of $|f(x) - g(x)|$ is the limit as $n \to \infty$ of the sum of n rectangles, each of which has width Δx and height $|f(x_k^*) - g(x_k^*)|$ for some x_k^* in the subinterval $[x_{k-1}, x_k]$. One example of such a Riemann sum is shown in Figure 12.57. In this figure, x_k^* is always taken to be the right-hand side of the subinterval $[x_{k-1}, x_k]$, and the height of the kth rectangle is $|f(x_k^*) - g(x_k^*)| = f(x_k) - g(x_k) = (x_k + 2) - (x_k)^2$.

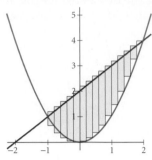

Figure 12.57
Height of the kth rectangle
is $|f(x_k^*) - g(x_k^*)|$

12.4.3 Average value

Definite integrals can be used to find the **average value** of a function on an interval. We will begin our development of the definition of average value with a discrete approximation.

EXAMPLE 12.19

Approximating the average height of a growing plant

Suppose $f(x) = 0.375x^2$ describes the height, in centimeters, of a growing plant x days after it breaks through the soil (see Figure 12.58). Approximate the average height of the plant during the first four days of its growth.

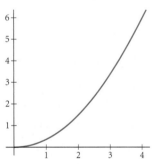

Figure 12.58
Height of the plant after
x days is $f(x) = 0.375x^2$

Solution   We can approximate the average height of the plant by choosing a discrete number of x-values (for example, $x = 1$, $x = 2$, $x = 3$, and $x = 4$) and averaging the heights $f(x)$ of the plant at these times. We will add up $f(1)$, $f(2)$, $f(3)$, and $f(4)$ and divide the result by 4:

$$\begin{aligned}\text{Average height of the plant} & \approx \frac{f(1) + f(2) + f(3) + f(4)}{4} \\ \text{in the first four days} & \\ &= \frac{0.375(1)^2 + 0.375(2)^2 + 0.375(3)^2 + 0.375(4)^2}{4} \\ &= 2.8125 \text{ centimeters.}\end{aligned}$$

Of course, during the first four days the plant has more than just four heights. The height of the plant changes *continuously* as x changes, and we took into account only a *discrete* number of plant heights.

To make a better approximation of the average height of the plant, we could average over a larger number of heights; for example, we could consider the height of the plant every quarter day (at $x = 0.25$, $x = 0.5$, $x = 0.75$, and so on until $x = 4$) and average those 16 heights:

$$\text{Average height of the plant in the first four days} \approx \frac{f(0.25) + f(0.5) + f(0.75) + \cdots + f(3.75) + f(4)}{16}$$

$$= \frac{0.375(0.25)^2 + 0.375(0.5)^2 + \cdots + 0.375(3.75)^2 + 0.375(4)^2}{16}$$

$$\approx 2.1914 \text{ centimeters.} \qquad \square$$

Considering the plant growth function in Example 12.19, suppose we measure n plant heights, evenly spaced over the four-day period. This means that the time between measurings will be $\frac{4-0}{n} = \frac{4}{n}$ days and thus that we will be measuring at times $x_k := 0 + k(\frac{4}{n})$ for $k = 1$ to $k = n$. (For example, when $n = 16$ as above, we would measure the height of the plant at $x_1 = 0 + 1(\frac{4}{16}) = 0.25$, $x_2 = 0 + 2(\frac{4}{16}) = 0.5$, and so on through $x_{16} = 0 + 16(\frac{4}{16}) = 4$ days.) The average height of the plant during the first four days is approximated by the average of these n heights:

$$(2) \quad \text{Average height of the plant in the first four days} \approx \frac{f(x_1) + f(x_2) + f(x_3) + \cdots + f(x_n)}{n} = \frac{\sum_{k=1}^{n} f(x_k)}{n}.$$

We can make this approximation for the average plant height as accurate as we like by choosing a large enough sample size n.

In general, the sigma notation in Equation (2) gives an approximation for the average value of a function $f(x)$ from $x = a$ to $x = b$, using n evenly spaced values called $x_k := a + k\,\Delta x$, where $\Delta x := \frac{b-a}{n}$. This should look very familiar! As we consider larger and larger values of n, that is, as $n \to \infty$, this approximation approaches what we will define to be the **average value** of the function $f(x)$ on $[a, b]$:

$$\text{Average value of } f(x) \text{ from } x = a \text{ to } x = b := \lim_{n\to\infty} \left(\frac{\sum_{k=1}^{n} f(x_k)}{n} \right) = \lim_{n\to\infty} \sum_{k=1}^{n} f(x_k)\tfrac{1}{n}.$$

This is *almost* identical to the definition of the definite integral as a limit of Riemann sums. The only difference is that the terms we are summing are of the form $f(x_k)(\frac{1}{n})$, rather than $f(x_k)\,\Delta x$. However, note that

$$\Delta x = \frac{b-a}{n} \quad \Rightarrow \quad \frac{1}{n} = \frac{\Delta x}{b-a}.$$

This means that

$$\text{Average value of } f(x) \text{ from } x = a \text{ to } x = b = \lim_{n\to\infty} \sum_{k=1}^{n} f(x_k)\left(\tfrac{1}{n}\right) \qquad \text{(above)}$$

$$= \lim_{n\to\infty} \sum_{k=1}^{n} f(x_k)\left(\frac{\Delta x}{b-a}\right) \qquad \text{(definition of } \Delta x)$$

$$= \frac{1}{b-a} \lim_{n\to\infty} \sum_{k=1}^{n} f(x_k)\,\Delta x \qquad (b-a \text{ is a constant})$$

$$= \frac{1}{b-a} \int_a^b f(x)\,dx. \qquad \begin{array}{l}\text{(definition of the}\\\text{definite integral)}\end{array}$$

The last line implicitly used the facts that $x_k := a + k\,\Delta x$ and $\Delta x = \frac{b-a}{n}$ (this is what tells us that the limits of integration should be a and b). The work we have done leads us to make the following definition for average value.

DEFINITION 12.9

The Average Value of a Function on an Interval

The *average value* of a function $f(x)$ on an interval $[a, b]$ is defined to be $\frac{1}{b-a}$ times the definite integral of $f(x)$ on $[a, b]$:

$$\begin{array}{c}\text{Average value of } f(x) \\ \text{from } x = a \text{ to } x = b\end{array} = \frac{1}{b-a}\int_a^b f(x)\,dx.$$

✿ *Caution*   Do not confuse the average *value* of a function $f(x)$ on an interval $[a, b]$ with the average *rate of change*. For the plant height function $f(x)$ from Example 12.19, the average value of $f(x)$ on $[0, 4]$ is the average height, in centimeters, of the plant from time $x = 0$ to time $x = 4$. On the other hand, the average rate of change of $f(x)$ on $[0, 4]$ would be the average rate of growth of the plant, measured in centimeters per day (see Example 12.20). ☐

In the following example, we use Definition 12.9 to calculate the *exact* value of the average height of the growing plant from Example 12.19 (as well as its average rate of growth).

EXAMPLE 12.20   **Using a definite integral to find the average value of a function**

Suppose the function $f(x) = 0.375x^2$ describes the height, in centimeters, of a growing plant after x days. Find (**a**) the average height of the plant during the first four days of its growth, and (**b**) the average rate of growth of the plant over those four days.

Solution   The average height of the plant during the first four days is the average value of the function $f(x) = 0.375x^2$ from $x = 0$ to $x = 4$, which by definition is

$$\begin{array}{c}\text{Average height of the plant} \\ \text{in the first four days}\end{array} = \frac{1}{4-0}\int_0^4 0.375x^2\,dx \qquad \text{(definition of average value)}$$

$$= \frac{0.375}{4}\int_0^4 x^2\,dx \qquad \text{(property of definite integrals)}$$

$$= \frac{0.375}{4}\left(\tfrac{1}{3}(4^3 - 0^3)\right) \qquad \text{(formula for definite integral of } x^2)$$

$$= 2 \text{ centimeters.}$$

Compare this result with our earlier approximations for the average height of the plant.
On the other hand, the average rate of growth of the plant over the first four days is the average rate of change of $f(x) = 0.375x^2$ on $[0, 4]$, which is equal to

$$\begin{array}{c}\text{Average rate of growth} \\ \text{in the first four days}\end{array} = \frac{f(4) - f(0)}{4 - 0} \qquad \text{(definition of average rate of change)}$$

$$= \frac{0.375(4)^2 - 0.375(0)^2}{4}$$

$$= 1.5 \tfrac{\text{cm}}{\text{day}}.$$

Note that this answer makes sense, because the plant grew six centimeters in four days, and thus its average rate of growth was $\frac{6}{4} = 1.5$ centimeters per day. ☐

❓ Question How can it be that the height of the plant in Example 12.20 changes from zero centimeters to six centimeters and yet has an average height of *two* centimeters, rather than *three*? Why might you naively guess that the average was three centimeters? Use the shape of the graph of $f(x) = 0.375x^2$ to explain why it is reasonable that the average height is only two centimeters. □

The average value of a function on an interval is related to the *area* under the function on that interval. Specifically, if K is the average value of a function $f(x)$ on an interval $[a, b]$, then we have

$$K = \frac{1}{b-a} \int_a^b f(x)\, dx \implies K(b-a) = \int_a^b f(x)\, dx.$$

This means the area under the graph of $f(x)$ from $x = a$ to $x = b$ is equal to $K(b-a)$, which is the area under the graph of the constant function $y = K$ on $[a, b]$ (since $\int_a^b K\, dx = K(b-a)$). One way to think of this is to imagine the region between the graph of $f(x)$ and the x-axis as the cross section of a wave of water. Suppose you are looking at water sloshing in a glass tank from the side. When the water settles, it will have a height of K, i.e., its height will be the average value of the function $f(x)$ on the interval. See Figures 12.59–12.61.

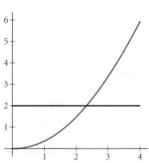

Figure 12.59
$y = f(x)$ and $y = K$
on $[0, 4]$

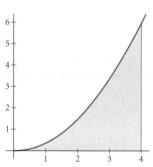

Figure 12.60
Area under the graph of
$y = f(x)$ on $[0, 4]$

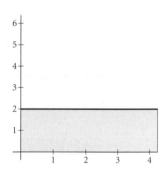

Figure 12.61
Area under the graph of
$y = K$ on $[0, 4]$

What you should know after reading Section 12.4

▶ The difference between the "signed" area and the "true" area under a curve, how to write both of these quantities in terms of definite integrals, and how to calculate these areas. Be able to describe how a Riemann sum could be negative.

▶ How to write the area of the region between two curves in terms of definite integrals, and how to calculate this area.

▶ The definition of the average value of a function on an interval, and how to calculate this average value (exactly and approximately). Be able to explain how the definite integral that defines average value is related to the intuitive notion of an "average."

Exercises 12.4

Concepts

0. Read the section and make your own summary of the material. Include whatever you think will help you review the material later on.

1. Consider the function $f(x) = x^2 - 1$ on the interval $[0, 2]$. If you were to make a Right-Hand Sum for this function with four rectangles, how many rectangles would have negative heights? What if you used seven rectangles? In each case, would your estimate for the signed area between $f(x)$ and $[0, 2]$ be positive or negative? Do you think the exact value of the signed area is positive or negative?

2. Suppose $f(x)$ is positive on $[-\infty, -1]$ and $[2, \infty]$ and negative on the interval $[-1, 2]$. Write **(a)** the signed area and **(b)** the true area between the graph of $f(x)$ and the x-axis from $x = -3$ to $x = 4$ in terms of definite integrals.

3. Suppose $f(x) \geq g(x)$ on the interval $[1, 3]$, whereas $f(x) \leq g(x)$ on $[-\infty, 1]$ and $[3, \infty]$. Write the area of the region between the graphs of $f(x)$ and $g(x)$ from $x = -2$ to $x = 5$ in terms of definite integrals.

4. How many definite integrals would you have to calculate to find the true area between the graph of $f(x) = \sin x$ and the x-axis on the interval $[-\frac{\pi}{2}, 2\pi]$? Do you think the true area will be positive or negative, and why? Do you think the signed area will be positive or negative, and why?

5. How many definite integrals would you have to calculate to find the area of the region between the graphs of $f(x) = \sin x$ and $g(x) = \frac{1}{2}$ on the interval $[-\frac{\pi}{2}, 2\pi]$? Sketch graphs of $f(x)$ and $g(x)$ and shade in the region described.

6. There is only *one* function $f(x)$ on $[0, 4]$ for which the true area between the graph of $f(x)$ and the x-axis is zero; what is that function? On the other hand, there are *many* functions $f(x)$ on $[0, 4]$ for which the signed area between the graph of $f(x)$ and the x-axis is zero; draw graphs of three such functions.

7. Sketch the graph of a function $f(x)$ whose signed area on the interval $[-5, 2]$ is **(a)** positive; **(b)** negative; **(c)** less than 20; and **(d)** zero.

8. Sketch the graph of a function $f(x)$ whose average value on the interval $[-1, 6]$ is **(a)** positive; **(b)** negative; **(c)** more than 10; and **(d)** zero.

9. Explain in terms of Riemann sums how you could approximate the area between the graphs of two functions $f(x)$ and $g(x)$ on an interval $[a, b]$. You may assume that $f(x) \geq g(x)$ on the entire interval $[a, b]$.

10. Write the area of the region between the graphs of two functions $f(x)$ and $g(x)$ on an interval $[a, b]$ as a limit of Riemann sums. You may assume that $f(x) \geq g(x)$ on the entire interval $[a, b]$.

11. Explain the difference between the average value of a function $f(x)$ on an interval $[a, b]$ and the average rate of change of the function on that interval. It may help to illustrate your answer with a real-world example (different from the one described in this section).

12. Suppose $f(x)$ is a function whose average value on $[-3, 1]$ is -2 and whose average rate of change on the same interval is 4. Sketch a possible graph for $f(x)$. Illustrate the average value and the average rate of change graphically on your graph of $f(x)$.

13. Suppose $f(x)$ is a function whose average value on $[-2, 5]$ is 10 and whose average rate of change on the same interval is -3. Sketch a possible graph for $f(x)$. Illustrate the average value and the average rate of change graphically on your graph of $f(x)$.

14. If $f(x)$ is negative on $[-3, 2]$, is the definite integral $\int_{-3}^{2} f(x)\,dx$ positive or negative? What about the definite integral $-\int_{-3}^{2} f(x)\,dx$?

15. Consider the function $f(x)$ graphed below. Shade in the regions between $f(x)$ and the x-axis on the intervals **(a)** $[-2, 6]$ and **(b)** $[-4, 2]$. Do you think the signed areas on these intervals will be positive or negative, and why?

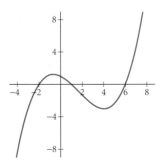

16. Consider the function $y = f(x)$ and the line $y = g(x)$ shown below. Shade in the regions between $f(x)$ and $g(x)$ on the intervals **(a)** $[-2, 3]$; **(b)** $[-1, 2]$; and **(c)** $[-2.5, 3.5]$.

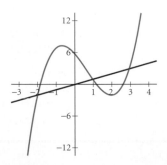

17. Consider the function $f(x)$ shown in the graph below. Use the graph to make a rough estimate of the average value of $f(x)$ on $[-4, 4]$, and show this average value as a height on the same set of axes. (*Hint:* Think about how the area under the graph is related to the average value of the function.)

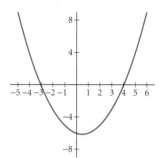

■ For each function $f(x)$ and interval $[a, b]$ described below, draw the rectangles that would be used to approximate the signed area between the graph of $f(x)$ and the x-axis on $[a, b]$ with a Left-Hand Sum and $n = 8$.

18.

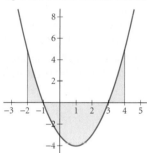

$[a, b] = [-2, 4]$

19.

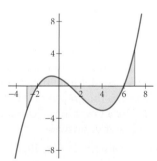

$[a, b] = [-3, 7]$

20.

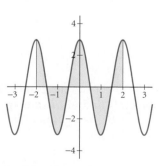

$[a, b] = [-2, 2]$

■ For each function $f(x)$ and interval $[a, b]$ described below, express (a) the signed area and (b) the true area between the graph of $f(x)$ and the x-axis on $[a, b]$ in terms of definite integrals. (Do not use any absolute values.)

21. See graph in Problem 18.

22. See graph in Problem 19.

23. See graph in Problem 20.

■ For each pair of functions $f(x)$ and $g(x)$ and interval $[a, b]$ described below, draw the rectangles that would be used to approximate the signed area between the graphs of $f(x)$ and $g(x)$ on $[a, b]$ with a Left-Hand Sum and $n = 8$.

24.

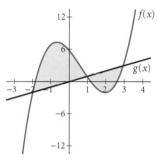

$[a, b] = [-2, 3]$

25.

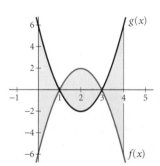

$[a, b] = [0, 4]$

26.

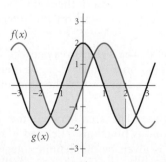

$[a, b] = [-2.5, 2]$

■ For each pair of functions $f(x)$ and $g(x)$ and interval $[a, b]$ described, express the area of the region between the graphs of $f(x)$ and $g(x)$ on $[a, b]$ in terms of definite integrals. (Do not use any absolute values.)

27. See graphs of $f(x)$ and $g(x)$ in Problem 24.

28. See graphs of $f(x)$ and $g(x)$ in Problem 25.

29. See graphs of $f(x)$ and $g(x)$ in Problem 26.

30. Use the definition of absolute value to explain why the true area $\int_a^b |f(x)|\, dx$ counts area positively regardless of whether it is above or below the x-axis. Include graphs that support your explanation.

31. Give an example of a function $f(x)$ and an interval $[a, b]$ that illustrates that $\int_a^b |f(x)|\, dx$ is *not* equal to $|\int_a^b f(x)\, dx|$.

32. Explain why a person might naively think that the average height of a plant that grew six centimeters in four days would be three centimeters. Then explain why it makes sense that a plant whose height at day x in centimeters is $f(x) = 0.375x^2$ would have an average height of *less* than three centimeters over the first four days. (*Hint:* Think about the graph of $f(x)$. What feature does this graph have that would make the average value of $f(x)$ on $[0, 4]$ less than 3?)

■ Without calculating any sums or definite integrals, determine the values of the quantities described below.

33. The (signed) area between the graph of $f(x) = \cos x$ and the x-axis from $x = -\pi$ to $x = \pi$.

34. The area of the region between the graphs of $f(x) = \sqrt{4 - x^2}$ and $g(x) = -\sqrt{4 - x^2}$ on the interval $[-2, 2]$.

35. The average value of $f(x) = \cos x$ from $x = 0$ to $x = 2\pi$.

■ Determine whether each of the following statements is true or false. If a statement is true, explain why. If a statement is false, explain why or provide a counterexample.

36. True or False: The "true area" between the graph of a function $f(x)$ on an interval $[a, b]$ is equal to $\int_a^b f(x)\, dx$.

37. True or False: The area of the region between $f(x) = x - 4$ and $g(x) = -x^2$ on the interval $[-3, 3]$ is negative.

38. True or False: The signed area between the graph of a function $f(x)$ on an interval $[a, b]$ is always less than or equal to the true area of the function on the same interval.

39. True or False: The area between the graphs of any two continuous functions $f(x)$ and $g(x)$ on an interval $[a, b]$ is given by the definite integral $\int_a^b (f(x) - g(x))\, dx$.

40. True or False: If $f(x) = x^2 - 3$, then the average value of $f(x)$ on the interval $[2, 6]$ is $\frac{f(6) + f(2)}{2} = \frac{33+1}{2} = 17$.

41. True or False: If $f(x) = x^2 - 3$, then the average value of $f(x)$ on the interval $[2, 6]$ is $\frac{f(6) - f(2)}{4} = \frac{33-1}{4} = 8$.

Skills

■ Approximate each of the following quantities using at least eight rectangles. Sketch a picture in each case that includes your rectangles.

42. The (**a**) signed area and (**b**) true area between the graph of $f(x) = 1 - e^x$ and the x-axis from $x = -1$ to $x = 3$.

43. The (**a**) signed area and (**b**) true area between the graph of $g(x) = \cos x$ and the x-axis on the interval $[-2\pi, 2\pi]$.

44. The area between the graphs of $f(x) = e^x$ and $g(x) = \ln x$ from $x = 0.5$ to $x = 2.5$.

45. The area between the graphs of $f(x) = \sin x$ and $f(x) = \cos x$ on the interval $[0, \pi]$.

■ For each function $f(x)$ and interval $[a, b]$ below, approximate the average value of $f(x)$ from $x = a$ to $x = b$ using a sample size of at least eight.

46. $f(x) = \ln x$, $[1, 5]$

47. $f(x) = 2^x$, $[-1, 1]$

48. $f(x) = \sin x$, $[0, \frac{3\pi}{2}]$

■ For each function $f(x)$ and interval $[a, b]$ given below, use definite integrals to find (**a**) the signed area and (**b**) the true area of the region between the graph of $f(x)$ and the x-axis from $x = a$ to $x = b$. Use a graphing calculator to sketch each region and to check your answers.

49. $f(x) = 3 - x$, $[0, 5]$

50. $f(x) = x^2$, $[-2, 2]$

51. $f(x) = x^2 - 1$, $[-1, 3]$

52. $f(x) = 1 - x^2$, $[0, 3]$

53. $f(x) = 2x^2 - 7x + 3$, $[0, 4]$

54. $f(x) = 3x^2 + 5x - 2$, $[-3, 2]$

■ For each pair of functions $f(x)$ and $g(x)$ and interval $[a, b]$ given below, use definite integrals to find the area of the region between the graphs of $f(x)$ and $g(x)$ from $x = a$ to $x = b$. Use a graphing calculator to sketch each region and to check your answers.

55. $f(x) = 1 + x$, $g(x) = 2 + x$, $[0, 3]$

56. $f(x) = 1 + x$, $g(x) = 2 - x$, $[0, 3]$

57. $f(x) = x^2$, $g(x) = x - 2$, $[-2, 2]$

58. $f(x) = x^2$, $g(x) = x + 2$, $[-2, 2]$

59. $f(x) = x^2$, $g(x) = x + 2$, $[-3, 3]$

60. $f(x) = x^2 - x - 1$, $g(x) = 5 - x^2$, $[-2, 3]$

■ For each function $f(x)$ and interval $[a, b]$ given below, use definite integrals to find the average value of $f(x)$ from $x = a$ to $x = b$. Use a graph of $f(x)$ to check that your answer is reasonable.

61. $f(x) = 3x + 1$, $[0, 4]$

62. $f(x) = 4$, $[-37.2, 103.75]$

63. $f(x) = x - 1$, $[-1, 3]$

64. $f(x) = 4 - x^2$, $[-2, 2]$

65. $f(x) = x^2 - 2x - 1$, $[0, 3]$

66. $f(x) = (x + 2)^2 - 5$, $[-5, 0]$

Applications

67. The Worldcom Burger Shack wants to put a giant sign up by Interstate 81. According to local sign ordinances, any sign visible from the interstate must have a frontal square footage of 529 feet or less. The entire sign will be a gigantic "W" cut out from billboard material. The lower boundaries of the "W" are described by the functions $f(x) = 0.5(x - 12)^2$ and $g(x) = 0.5(x - 24)^2$, and the upper boundaries are described by the functions $r(x) = (x - 12)^2 + 10$ and $s(x) = (x - 24)^2 + 10$. The top edges of the "W" are at a height of 55 feet. See the accompanying graph.

   (a) Write the total area of the front of the "W" sign in terms of definite integrals. Be careful about how you divide the region to compute the area. (*Note:* You will have to determine, and use, the solutions to $f(x) = 55$, $r(x) = 55$, $s(x) = 55$, and $g(x) = 55$.)

   (b) Use your answer to part (a) to calculate the exact frontal square footage of the "W" sign. Will Worldcom's sign meet the local square footage requirements?

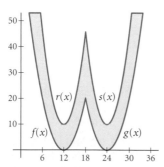

68. Just as you are driving past the big oak tree on Main Street, you notice a new stop sign 50 feet ahead of you. You slam on your brakes and end up coming to a full stop exactly at the stop sign. Your distance from the stop sign (in feet) t seconds after stepping on the brakes is given by the function $s(t) = 50 - 9t^2$.

   (a) How long did it take you to come to a full stop?

   (b) What was your average distance from the stop sign from the time that you first saw it to the time that you came to a stop?

   (c) Use an average rate of change to find your average velocity during the time that you were trying to stop the car. (Assume that the average velocity is the average rate of change of position.)

   (d) Find your average velocity from the oak tree to the stop sign in another way, as follows: Differentiate the formula for position to get a formula $v(t)$ for your velocity, in feet per second, t seconds after hitting the brakes. Then use a definite integral to find the average velocity during the time that you were trying to stop the car. (Your final answer should of course be the same as the average velocity you found in part (c)!)

   (e) You have just calculated your average velocity in two ways: once using the formula $\frac{s(b)-s(a)}{b-a}$ for the average rate of change of position, and once using the definition $\frac{1}{b-a}\int_a^b v(t)\,dt$ of the average value of $v(t)$ on $[a, b]$, where $a = 0$ and b is the amount of time it took you to stop the car. Use the fact that these two quantities are equal to discuss the relationship between the area under the graph of your velocity $v(t)$ on $[a, b]$ and the total distance that you traveled while trying to stop the car.

69. Suppose the height h, in feet, of a growing tree t years after it is planted is $h(t) = 0.25t^2 + 1$.

   (a) Approximate the average height of the tree during the first five years of its growth, using a sample size of $n = 5$ times.

   (b) Approximate the average height of the tree during the first five years of its growth, using a sample size of $n = 10$ times.

   (c) Write the average height of the tree during the first five years of its growth in terms of a definite integral.

   (d) Use your answer to part (c) to find the exact value of the average height of the tree during the first five years of its growth.

Proof

70. Given that $x_k := a + k(\frac{b-a}{n})$, show that $\lim_{n\to\infty} \frac{\sum_{k=1}^n f(x_k)}{n} = \frac{1}{b-a}\int_a^b f(x)\,dx$. What do the left-hand and right-hand sides of this equation have to do with average value?

CHAPTER 13

The Fundamental Theorem of Calculus

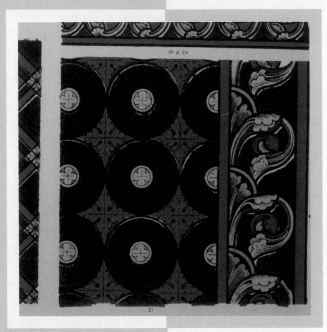

Stained glass at the Cathedral of Bourges, Middle Ages. From The Grammar of Ornament *by Owen Jones, 1856. Reproduced by permission of Octavo, www.octavo.com.*

In this chapter we finally make the connection between areas and antiderivatives. This connection is so fundamental to the study of calculus that it is called the Fundamental Theorem of Calculus. After developing a compact notation for antiderivatives called the indefinite integral, we will use the Fundamental Theorem of Calculus to relate indefinite integrals to definite integrals (which represent areas). This will allow us to calculate definite integrals quickly without having to set up Riemann sums. The Second Fundamental Theorem of Calculus will enable us to construct antiderivatives for all continuous functions.

CONTENTS

13.1   Indefinite Integrals

In the last chapter we defined the ***definite integral*** of a function $f(x)$ on an interval $[a, b]$ to be the number that represents the area between the graph of $f(x)$ and the x-axis from $x = a$ to $x = b$. We now define a completely different object called the ***indefinite integral*** of a function $f(x)$. This new object will be a family of functions, not a number, but we will see in Section 13.2 that these two types of "integrals" are related. In fact, that's why we call them by similar names.

13.1.1   The definition of the indefinite integral

Previously, whenever we wanted to find an "antiderivative" of a function, we used a guess-and-check method. In this section we formalize the process of "undoing" differentiation. More precisely, consider the following definition:

DEFINITION 13.1

Antiderivatives

A function $F(x)$ is an ***antiderivative*** of a function $f(x)$ if the derivative of $F(x)$ is $f(x)$, that is, if $F'(x) = f(x)$.

Notice that we say *an* antiderivative of $f(x)$, not "the" antiderivative of $f(x)$. Although a function has only *one* derivative, it has *many* antiderivatives. Any two of these antiderivatives will differ by a constant.

THEOREM 13.1

Functions with the Same Derivative Must Differ by a Constant

Suppose $F(x)$ and $G(x)$ are functions that are differentiable on an interval (a, b). Then $F'(x) = G'(x)$ on (a, b) if, and only if, for all $x \in (a, b)$ we have $F(x) = G(x) + C$ for some constant C.

Theorem 13.1 means that if $F(x)$ is any antiderivative of a function $f(x)$, then *every* other antiderivative of $f(x)$ differs from $F(x)$ by a constant. We proved this theorem in Chapter 3.

Since all of the antiderivatives of a function $f(x)$ are "related" to each other, we call them a "family" of functions. The "family" of antiderivatives of a function $f(x)$ consists of all the functions $F(x)$ whose derivatives are $F'(x) = f(x)$. All the functions in this family differ from each other by constants. We say that any particular antiderivative $F(x)$ of a function $f(x)$ is a ***representative*** of the family of antiderivatives and can then write $F(x) + C$ to represent the ***family of antiderivatives*** of $f(x)$.

EXAMPLE 13.1   **A family of antiderivatives**

Both $F(x) = x^2$ and $G(x) = x^2 - 3$ are antiderivatives of $f(x) = 2x$, since $F'(x) = \frac{d}{dx}(x^2) = 2x$ and $G'(x) = \frac{d}{dx}(x^2 - 3) = 2x$. Note that these two functions differ by a constant. Theorem 13.1 tells us that *every* antiderivative of $f(x) = 2x$ is of the form $x^2 + C$ for some constant C. In this case we say that $x^2 + C$ is the family of antiderivatives of $f(x) = 2x$, and that $F(x) = x^2$ is a representative of this family.   □

We have a compact notational way of expressing the fact that the derivative of x^2 is $2x$ (by writing "$\frac{d}{dx}(x^2) = 2x$"). We would also like to have a compact way of saying that "$x^2 + C$ is the family of antiderivatives of $2x$." To do this, we need to introduce some new notation.

DEFINITION 13.2

The Indefinite Integral of a Function

The *indefinite integral* of a function $f(x)$ is defined to be the family of antiderivatives

$$\int f(x)\,dx := F(x) + C,$$

where $F(x)$ is any antiderivative of $f(x)$ (i.e., any function with $F'(x) = f(x)$).

The "dx" in the notation represents the fact that we are antidifferentiating with respect to the variable x. The function $f(x)$ inside the integral notation is called the *integrand,* and when we find $\int f(x)\,dx$, we say that we are *integrating* the function $f(x)$.

◈ *Caution*   You have probably noticed that our notation $\int f(x)\,dx$ for the indefinite integral is strikingly similar to the notation $\int_a^b f(x)\,dx$ used for the definite integral. However, at this point we don't have much reason to believe that the indefinite integral (which is a family of antiderivatives) is related to the definite integral (which represents an area). The amazing fact that these two very different objects are related will be discussed in Section 13.2.   □

We can also state Definition 13.2 without specifically giving a name to the representative antiderivative, as follows:

$$\int f(x)\,dx := \begin{pmatrix} \text{any} \\ \text{antiderivative} \\ \text{of } f(x) \end{pmatrix} + C.$$

The indefinite integral of $f(x)$, often called simply the *integral* of $f(x)$, can be obtained by finding *any* antiderivative of $f(x)$ and then adding an arbitrary constant C. The following theorem describes the relationship among derivatives, antiderivatives, and integrals.

THEOREM 13.2

Relating Indefinite Integrals, Antiderivatives, and Derivatives

The following statements are equivalent:

(a) $\int f(x)\,dx = F(x) + C$.

(b) $F(x)$ is an antiderivative of $f(x)$.

(c) The derivative of $F(x)$ is $f(x)$ (i.e., $F'(x) = f(x)$).

The proof of Theorem 13.2 follows directly from the definitions of indefinite integrals and antiderivatives.

EXAMPLE 13.2

Using different antiderivatives to represent the same indefinite integral

Consider the function $f(x) = 2x$. The function x^2 is an antiderivative of $f(x)$, since $\frac{d}{dx}(x^2) = 2x$, and thus

$$\int 2x\, dx = x^2 + C.$$

On the other hand, the function $x^2 - 3$ is *also* an antiderivative of $f(x) = 2x$, since $\frac{d}{dx}(x^2 - 3) = 2x - 0 = 2x$. Therefore, we can *also* write

$$\int 2x\, dx = (x^2 - 3) + C.$$

We usually pick the easiest antiderivative of $f(x)$ we can find to represent the family of antiderivatives, so most people would write the indefinite integral of $f(x) = 2x$ the first way.

As an illustration of Theorem 13.2, in this example the three equivalent statements are (a) $\int 2x\, dx = x^2 + C$; (b) x^2 is an antiderivative of $2x$; and (c) the derivative of x^2 is $2x$ (i.e., $\frac{d}{dx}(x^2) = 2x$). ☐

❖ Caution It is important to recognize that C represents an arbitrary constant, not a particular constant that we care about (unless we have some reason to choose a *particular* antiderivative of the integrand). ☐

13.1.2 Antidifferentiation formulas

All of the rules we developed for differentiating various types of functions can be interpreted as *anti*differentiation rules. For example, the rule for differentiating power functions says that for any constant k, we have $\frac{d}{dx}(x^k) = kx^{k-1}$. The rule for antidifferentiating a power function should "undo" this process and is given in the following theorem.

THEOREM 13.3

Integrals of Power Functions

(a) $\displaystyle\int x^k\, dx = \frac{1}{k+1}x^{k+1} + C,$ for all $k \neq -1$

(b) $\displaystyle\int \frac{1}{x}\, dx = \ln|x| + C$

The first formula applies only for $k \neq -1$, because $\frac{1}{k+1}$ is not defined for $k = -1$. We know from Chapter 9 that $\frac{d}{dx}(\ln|x|) = \frac{1}{x}$, and this fact gives us part (b) of the theorem.

EXAMPLE 13.3

Integrating power functions

By Theorem 13.3, we have

$$\int x^5\, dx = \tfrac{1}{6}x^6 + C.$$

This is equivalent to saying that $\frac{d}{dx}(\frac{1}{6}x^6) = x^5$. In Theorem 13.3, the k in the exponent can be a negative or rational number. Sometimes we have to do a little algebra to write the integrand in the form x^k, as in the following calculations:

$$\int \frac{1}{x^2}\, dx = \int x^{-2}\, dx = \frac{1}{-2+1}x^{-2+1} = \frac{1}{-1}x^{-1} + C = -\frac{1}{x} + C$$

$$\int \sqrt{x^3}\, dx = \int x^{\frac{3}{2}}\, dx = \frac{1}{\frac{3}{2}+1}x^{\frac{3}{2}+1} = \frac{1}{\frac{5}{2}}x^{\frac{5}{2}} + C = \frac{2}{5}\sqrt{x^5} + C.$$

☐

PROOF (**THEOREM 13.3**)   Suppose $k \neq -1$. To prove part (a), we need to show that $\frac{1}{k+1}x^{k+1}$ is an antiderivative of x^k for all $k \neq -1$. Then, since every antiderivative of x^k differs by a constant, we will know that *all* the antiderivatives of x^k are of the form $\frac{1}{k+1}x^{k+1} + C$ for some constant C.

$$\frac{d}{dx}\left(\frac{1}{k+1}x^{k+1}\right) = \frac{1}{k+1}(k+1)x^k = x^k.$$

Since the derivative of $\frac{1}{k+1}x^{k+1}$ is x^k, all of the antiderivatives of x^k are of the form $\frac{1}{k+1}x^{k+1} + C$, and therefore $\int x^k\, dx = \frac{1}{k+1}x^{k+1} + C$.

Now consider the case where $k = -1$ (part (b) of the theorem). Since $\frac{d}{dx}(\ln|x|) = \frac{1}{x}$ (see Section 9.2), the function $\ln|x|$ is an antiderivative of the function $x^{-1} = \frac{1}{x}$. Therefore, we have $\int \frac{1}{x}\, dx = \ln|x| + C$, and we have proved part (b). ∎

In general, to show that $\int f(x)\, dx = F(x) + C$, it is sufficient to show that $F'(x) = f(x)$, since these two statements are equivalent. Theorems 13.4, 13.5, and 13.6 describe more antidifferentiation rules ("integration rules"). Each of these rules can be proved by differentiating, and you will do so in the exercises.

THEOREM 13.4

Integrals of Exponential Functions

(a) $\displaystyle\int e^{kx}\, dx = \frac{1}{k}e^{kx} + C,$ for all $k \neq 0$

(b) $\displaystyle\int b^x\, dx = \frac{1}{\ln b}b^x + C,$ for all $b > 0,\, b \neq 1$

For example, $\int 2^x\, dx = \frac{1}{\ln 2}2^x$, which is equivalent to the fact that $\frac{d}{dx}\left(\frac{1}{\ln 2}2^x\right) = \frac{1}{\ln 2}(\ln 2)2^x = 2^x$. Notice that both of the rules in Theorem 13.4 imply that the integral of e^x is itself, that is, that $\int e^x\, dx = e^x + C$.

Theorem 13.5 describes some trigonometric integrals. These integrals are just re-wordings of the rules for differentiating the six trigonometric functions. For example, we have $\int \sec^2 x\, dx = \tan x + C$ because $\frac{d}{dx}(\tan x) = \sec^2 x$.

THEOREM 13.5

Integrals of Certain Trigonometric Expressions

(a) $\displaystyle\int \sin x\, dx = -\cos x + C$ (b) $\displaystyle\int \cos x\, dx = \sin x + C$

(c) $\displaystyle\int \sec^2 x\, dx = \tan x + C$ (d) $\displaystyle\int \csc^2 x\, dx = -\cot x + C$

(e) $\displaystyle\int \sec x \tan x\, dx = \sec x + C$ (f) $\displaystyle\int \csc x \cot x\, dx = -\csc x + C$

✪ **Caution**   At this point we do *not* know how to integrate all six of the trigonometric functions. Theorem 13.5 tells us how to integrate the sine and cosine functions and how to integrate some other more complicated trigonometric functions. However, we do not yet know how to integrate $\sec x$, $\tan x$, $\csc x$, or $\cot x$. (We will learn those integrals in Chapter 14.)   □

Finally, Theorem 13.6 gives rules for antidifferentiating three special functions, namely the functions that are the derivatives of the inverse sine, inverse cosine, and inverse tangent functions. In other words, like all of the integration rules we have seen so far, these are just the differentiation rules we know and love, written "backwards."

THEOREM 13.6

Integrals Defined by Inverse Trigonometric Functions

(a) $\displaystyle\int \frac{1}{\sqrt{1-x^2}}\,dx = \sin^{-1}x + C$

(b) $\displaystyle\int \frac{1}{1+x^2}\,dx = \tan^{-1}x + C$

(c) $\displaystyle\int \frac{1}{|x|\sqrt{x^2-1}}\,dx = \sec^{-1}x + C$

❓ **Question**   Why doesn't Theorem 13.6 include a rule that has to do with the inverse cosine function? Think about the derivatives of $\cos^{-1}x$ and $\sin^{-1}x$.   □

13.1.3   Integrals of sums and constant multiples

Using the differentiation rules, we know how to differentiate any function that is a constant multiple, sum, product, quotient, or composition of functions whose derivatives we know. It would also be convenient to be able to integrate combinations of functions, like $f(x) = 5x^3 - \sin x$ or $f(x) = xe^x$. Unfortunately, antidifferentiation is a much more complicated process than differentiation, and only the constant multiple rule and the sum rule translate directly into integration rules:

THEOREM 13.7

Constant Multiple and Sum Rules for Indefinite Integrals

(a) $\displaystyle\int kf(x)\,dx = k\int f(x)\,dx$

(b) $\displaystyle\int (f(x) + g(x))\,dx = \int f(x)\,dx + \int g(x)\,dx$

Theorem 13.7 says that indefinite integrals commute with sums and constant multiples. We can use these two integration rules to integrate any sum or constant multiple of functions whose integrals we already know. Before we prove Theorem 13.7, let's do a simple example.

EXAMPLE 13.4   **Using the sum and constant multiple rules for definite integrals**

Find $\displaystyle\int (5x^3 - \sin x)\,dx$.

Solution   The function $f(x) = 5x^3 - \sin x$ is a sum of constant multiples of functions whose integrals we know (the functions x^3 and $\sin x$). By Theorem 13.7, we have

$$\int (5x^3 - \sin x)\, dx = \int (5x^3 + (-\sin x))\, dx \qquad \text{(write the difference as a sum)}$$

$$= \int 5x^3\, dx + \int (-\sin x)\, dx \qquad \text{(part (b) of Theorem 13.7)}$$

$$= 5\int x^3\, dx - \int \sin x\, dx \qquad \text{(part (a) of Theorem 13.7)}$$

$$= 5\left(\tfrac{1}{4}x^4\right) - (-\cos x) + C \qquad \text{(Theorems 13.3 and 13.5)}$$

$$= \tfrac{5}{4}x^4 + \cos x + C. \qquad \text{(simplify)}$$

Notice that we added only *one* constant C in this calculation. Technically, there are *two* such constants, since the family of antiderivatives for x^3 is $\tfrac{1}{4}x^4 + C_1$, and the family of antiderivatives for $\sin x$ is $-\cos x + C_2$. The C we are using is really the sum $C_1 + C_2$. There is no need to write the two constants C_1 and C_2 separately, since the sum of two arbitrary constants is also an arbitrary constant.   □

> ❓ *Question*   Is the integral of a difference of functions always equal to the difference of the integrals of those two functions? Can you prove it?   □

The proof of Theorem 13.7 follows directly from the sum and constant multiple rules for differentiation. In some sense, Theorem 13.7 is just a "backwards" version of those differentiation rules.

PROOF **(THEOREM 13.7)**   To prove the first part of Theorem 13.7, we have only to show that the integral of a function $kf(x)$ is k times the integral of $f(x)$. This means we must show that if $F(x)$ is an antiderivative of $f(x)$, then $kF(x)$ is an antiderivative of $kf(x)$. (Stop right now and make sure that you really understand why that is what we have to prove!). Suppose $\int f(x)\, dx = F(x) + C$; that is, suppose $F(x)$ is an antiderivative of $f(x)$. This means that $F'(x) = f(x)$. Then, by the constant multiple rule, we know that for any constant k,

$$\frac{d}{dx}(kF(x)) = k\frac{d}{dx}(F(x)) = kF'(x) = kf(x).$$

This means that $kF(x)$ is an antiderivative of $kf(x)$, so we have

$$\int kf(x)\, dx = kF(x) + C = k\int f(x)\, dx.$$

If we wanted to be very technical, we could point out that $kF(x) + C = k(F(x) + \tfrac{C}{k})$ and that $F(x) + \tfrac{C}{k}$ is also an antiderivative of $f(x)$ since $\tfrac{C}{k}$ is a constant.

To prove the second part of the theorem, we must show that the integral of a sum of functions $f(x) + g(x)$ is equal to the sum of the integrals of $f(x)$ and $g(x)$. Suppose $\int f(x)\, dx = F(x) + C$ and $\int g(x)\, dx = G(x) + C$. (In other words, suppose that $F(x)$ is an antiderivative of $f(x)$ and that $G(x)$ is an antiderivative of $g(x)$.) Therefore, $F'(x) = f(x)$ and $G'(x) = g(x)$. We need to show that $F(x) + G(x)$ is an antiderivative of $f(x) + g(x)$. Using the sum rule for differentiation, we have

$$\frac{d}{dx}(F(x) + G(x)) = F'(x) + G'(x) = f(x) + g(x),$$

and thus $F(x) + G(x)$ is indeed an antiderivative of $f(x) + g(x)$. Therefore,

$$\int (f(x) + g(x))\, dx = F(x) + G(x) + C = \int f(x)\, dx + \int g(x)\, dx.$$

Technically, the integral on the right is equal to $(F(x) + C_1) + (G(x) + C_2)$ for some arbitrary constants C_1 and C_2; our "C" represents their sum.

13.1.4 Antidifferentiating other combinations of functions

Unfortunately, at this point we do not have formulas that tell us how to integrate products or quotients of functions. In fact, there *isn't* a general formula for these types of integrals (as there was for derivatives).

✦ *Caution* Although the integral of a sum is the sum of integrals, the integral of a product is *not* in general equal to the product of integrals. In other words,

$$\int (f(x)g(x))\, dx \neq \int f(x)\, dx \int g(x)\, dx$$

For example, we *cannot* find $\int xe^x\, dx$ by multiplying together the integrals $\int x\, dx = \frac{1}{2}x^2 + C$ and $\int e^x\, dx = e^x + C$. □

❓ *Question* Show that $\int x^2\, dx$ is *not* equal to $(\int x\, dx)^2$. What does this have to do with the previous "Caution"? □

The reason why we don't have an easy "product rule" for antidifferentiation is that the derivative of a product of functions is *not* the product of the derivatives of those functions; instead, $(fg)' = f'g + fg'$ (by the product rule). However, if we think of the product rule "backwards," we see that the integral of a function of the form $f'g + fg'$ is equal to $fg + C$:

(1) $$\int (f'(x)g(x) + f(x)g'(x))\, dx = f(x)g(x) + C.$$

Of course, the first step is recognizing when the integrand is of this special form. Equation (1) will be the basis for the method of integration known as "integration by parts," which we will study in Section 14.2. If we are lucky enough to recognize an integrand as the "result" of a product rule calculation, then we can make an educated guess for the integral and can check our answer by differentiating.

EXAMPLE 13.5

Recognizing an integrand as the "result" of a product rule

Although we cannot integrate the function $f(x) = xe^x$ at this point, by the product rule we do know that $\frac{d}{dx}(xe^x) = e^x + xe^x$. Therefore, we know that

$$\int (e^x + xe^x)\, dx = xe^x + C.$$

However, if we had begun with the integral on the left, without thinking about the function xe^x at the start, it would have been difficult to recognize the integrand $e^x + xe^x$ as the derivative of xe^x (although we could do so with some educated guessing and checking). □

We can also write the chain rule "backwards" to get an integration formula. Since the derivative of a composition $f(g(x))$ is $\frac{d}{dx}(f(g(x))) = f'(g(x))g'(x)$, we can say that

$$\int f'(g(x))g'(x)\, dx = f(g(x)) + C.$$

This means that if an integrand is in the form $f'(g(x))g'(x)$ (the "result" of a chain rule computation), then we can use the chain rule "backwards" to integrate it. We will formalize the process of antidifferentiating integrands of this form in Section 14.1 when we learn how to integrate "by substitution." For now, if we recognize an integrand as the "result" of a chain rule calculation, then we can guess its antiderivative and check by differentiating.

EXAMPLE 13.6

Recognizing an integrand as the "result" of a chain rule

Calculate $\int 5(x^2 + 1)^4 (2x)\, dx$.

Solution   We want to recognize the integrand $5(x^2 + 1)^4(2x)$ as a function of the form $f'(g(x))g'(x)$ for some functions $f(x)$ and $g(x)$. One way to start is to notice that $5(x^2 + 1)^4$ is a composition of functions with an "inside" function of $x^2 + 1$. If we set $g(x) = x^2 + 1$, then $g'(x) = 2x$, which fits the form we are looking for, since the integrand has a "$2x$" in it. Now we have to make a guess for the "outside" function. We might try $f'(x) = 5x^4$; then the integrand is indeed of the form $f'(g(x))g'(x)$. We now know that the integral given is equal to $f(g(x)) + C$, and we know what $g(x)$ is, but we still do not know $f(x)$. To find $f(x)$ we must find a function whose derivative is $f'(x) = 5x^4$; one such function is $f(x) = x^5$. Therefore, we might guess that the integral given is equal to $f(g(x)) + C = (x^2 + 1)^5 + C$. Finally, we check by differentiating our guess:

$$\frac{d}{dx}((x^2 + 1)^5) = 5(x^2 + 1)^4(2x).$$

Therefore, $\int 5(x^2 + 1)^4 (2x)\, dx = (x^2 + 1)^5 + C$.   □

Sometimes an integrand isn't *exactly* in the form $f'(g(x))g'(x)$, but we can still recognize it as the result of a chain rule calculation, with some constant multiples "missing." Example 13.7 illustrates how to solve two such integrals by guessing and checking.

EXAMPLE 13.7

Integrating by educated guess-and-check

Calculate the following integrals.

(a) $\int \sin 3x\, dx$
    (b) $\int x e^{x^2 + 1}\, dx$

Solution

(a) At first glance the function $\sin(3x)$ does not appear to be of the form $f'(g(x))g'(x)$; specifically, if we choose $f'(x)$ to be $\sin x$, so that the "inside" function is $g(x) = 3x$, then the derivative $g'(x) = 3$ does not appear in the integrand (the 3 is "missing"). However, since an antiderivative of $\sin x$ is $-\cos x$, we might try $F(x) = -\cos 3x$ as a first guess for an antiderivative of $\sin 3x$. Testing this guess, we find that $\frac{d}{dx}(-\cos 3x) = -(-\sin 3x)(3) = 3 \sin 3x$. This is *almost* the integrand we are looking for, but with an "extra" constant multiple of 3. This leads us to update our guess to $F(x) = -\frac{1}{3} \cos 3x$, so that the $\frac{1}{3}$ at the front will cancel out the factor of 3 that appears when we apply the chain rule. We now have

$$\frac{d}{dx}\left(-\tfrac{1}{3} \cos 3x\right) = -\tfrac{1}{3}(-\sin 3x)(3) = \sin 3x,$$

and therefore $\int \sin 3x\, dx = -\frac{1}{3} \cos 3x + C$.

(b) The first thing to recognize about this problem is that the function $x^2 + 1$ is an "inside" function in the integrand, and its derivative $2x$ is *almost* part of the integrand (we have an x instead of a $2x$, so we're missing only a constant multiple). A good first guess for an antiderivative of xe^{x^2+1} might be $F(x) = e^{x^2+1}$ (since the function e^x is its own antiderivative). Checking this guess, we find that $\frac{d}{dx}(e^{x^2+1}) = e^{x^2+1}(2x)$, which is *almost* what we want. To get rid of the extra 2, we update our guess to $F(x) = \frac{1}{2}e^{x^2+1}$; with this guess we have

$$\frac{d}{dx}\left(\tfrac{1}{2}e^{x^2+1}\right) = \tfrac{1}{2}e^{x^2+1}(2x) = xe^{x^2+1}.$$

This means that $\int xe^{x^2+1}\, dx = \frac{1}{2}e^{x^2+1} + C$. It is important to note that the method used in this example, where we update our guess by dividing by what is "missing," works only if what is missing is a *constant*. ☐

☑ Checking the Answer You should always check your answers when you integrate! If you decide that $\int f(x)\, dx = F(x) + C$ for some function $F(x)$, then you should check that $F'(x) = f(x)$, especially if you found the function $F(x)$ by "guessing." ☐

In general, antidifferentiation is more difficult than differentiation. We can differentiate *every* function that we know how to write down; however, at this point we cannot integrate even very simple functions like $\ln x$ and $\sec x$. You should think of integration as a puzzle, not a procedure. Unlike differentiation, where it is always clear which rules to apply, and in which order, it is *not* always immediately clear how to find a given integral. We will learn some more methods for calculating integrals in Chapter 14, but even then we will not be able to calculate all integrals. In fact, some functions have no "elementary" antiderivatives, which means that their antiderivatives cannot be written in terms of the functions we now know.

What you should know after reading Section 13.1

▶ The definitions of and notations for indefinite integrals, antiderivatives, families of functions, arbitrary constants, and integrands. What is the meaning of the "dx" in the integral notation? How are two antiderivatives of a function related? How can you phrase an antidifferentiation statement as a differentiation statement?

▶ The integration formulas that come from the differentiation formulas we already know, and how to prove them. The formulas for integrals of sums and constant multiples of functions, and how to use these formulas. (You may have to do some algebra before you can apply these formulas.)

▶ Do we have rules for integrating products, quotients, and compositions of functions? What do the product, quotient, and chain rules for differentiation mean in terms of integrals?

▶ Be able to solve integrals either by educated "guessing and checking" or by recognizing the integrand as the "result" of a product, quotient, or chain rule calculation.

Exercises 13.1

Concepts

0. Read the section and make your own summary of the material. Include whatever you think will help you review the material later on.

1. What is the difference between an antiderivative of a function and the indefinite integral of a function?

2. Explain why we call the collection of antiderivatives of a function $f(x)$ a "family." How are the antiderivatives of a function "related"?

3. Write out the three parts of Theorem 13.2 in the case where the integrand is the function x^6.

4. Write out the three parts of Theorem 13.2 in the case where the antiderivative is the function x^6.

5. Write out all the integration formulas and rules that we know at this point.

6. Explain why the formula for the integral of x^k does not apply when $k = -1$. What *is* the integral of x^{-1}?

7. Explain why at this point we don't have an integration formula for the function $f(x) = \sec x$, when we *do* have an integration formula for $f(x) = \sin x$.

8. Why don't we bother to state an integration formula that has to do with $\cos^{-1} x$? (*Hint:* Think about the derivatives of $\cos^{-1} x$ and $\sin^{-1} x$.) What would the integration formula be? Why is it "redundant," given the integration formula that has to do with $\sin^{-1} x$?

■ Determine whether each of the following statements is true or false. If a statement is true, explain why. If a statement is false, explain why or provide a counterexample.

9. True or False: If $f(x) - g(x) = 2$, then $f(x)$ and $g(x)$ have the same derivative.

10. True or False: If $f(x) - g(x) = 2$, then $f(x)$ and $g(x)$ have the same antiderivative.

11. True or False: If $f(x) = 2x$, then $F(x) = x^2$.

12. True or False: $\displaystyle\int (x^3 + 1)\, dx = (\frac{1}{4}x^4 + x + 2) + C$.

13. True or False: $\displaystyle\int \sin(x^2)\, dx = -\cos(x^2) + C$.

14. True or False: $\displaystyle\int e^x \cos x\, dx = e^x \sin x + C$.

15. True or False: $\displaystyle\int \frac{1}{x^2 + 1}\, dx = \ln|x^2 + 1| + C$.

16. True or False: $\displaystyle\int \frac{1}{x^2 + 1}\, dx = \frac{1}{2x}\ln|x^2 + 1| + C$.

17. Show that $F(x) = \sin x - x\cos x + 2$ is an antiderivative of $f(x) = x \sin x$.

18. Show that $F(x) = \frac{1}{2}e^{(x^2)}(x^2 - 1)$ is an antiderivative of $f(x) = x^3 e^{(x^2)}$.

19. Show that $\displaystyle\int \ln x\, dx = x(\ln x - 1) + C$. (*Hint:* Don't try to integrate!)

20. Show that $\displaystyle\int \cot x\, dx = \ln(\sin x) + C$. (*Hint:* Don't try to integrate!)

21. Show that $\displaystyle\int xe^x(x + 2)\, dx = x^2 e^x + C$. (*Hint:* Don't try to integrate!)

22. Show that in general $\int f(x)g(x)\, dx \neq (\int f(x)\, dx) \cdot (\int g(x)\, dx)$, by exhibiting a counterexample. In other words, find two functions $f(x)$ and $g(x)$ such that the integral of their product is not equal to the product of their integrals.

23. Show that in general $\int \frac{f(x)}{g(x)}\, dx \neq \frac{\int f(x)\, dx}{\int g(x)\, dx}$, by exhibiting a counterexample. In other words, find two functions $f(x)$ and $g(x)$ such that the integral of their quotient is not equal to the quotient of their integrals.

24. Write down six functions that we do not currently know how to integrate, and explain what it is about each function that makes it impossible to integrate given only what we currently know.

Skills

■ For each *derivative* $f'(x)$ and value $f(c)$ given below, find a function $f(x)$ that has that derivative and value. Be sure to check your answers!

25. $f'(x) = x^8 - x^2 + 1$, $f(0) = -1$

26. $f'(x) = \dfrac{2}{\sqrt[3]{x}}$, $f(1) = 2$

27. $f'(x) = 3e^{2x} + 1$, $f(0) = 2$

28. $f'(x) = 3\left(\frac{4}{5}\right)^x$, $f(1) = 0$

29. $f'(x) = \dfrac{5}{3x - 2}$, $f(1) = 8$

30. $f'(x) = 3\sin(-2x) - 5\cos(3x)$, $f(0) = 0$

■ Use integration formulas to calculate each of the following integrals. You may have to use algebra before any integration rules apply, and you may have to use a small amount of educated "guessing and checking." Check each of your answers by differentiating.

31. $\displaystyle\int (x^2 - 3x^5 - 7)\, dx$ **32.** $\displaystyle\int (x^2 - 1)(3x + 5)\, dx$

33. $\displaystyle\int (x^3 + 4)^2\, dx$ **34.** $\displaystyle\int \left(\frac{2}{3x} + 1\right) dx$

35. $\displaystyle\int \left(\frac{3}{x^2} - 4\right) dx$ **36.** $\displaystyle\int \frac{x + 1}{\sqrt{x}}\, dx$

37. $\int (3.2(1.43)^x - 50)\, dx$ **38.** $\int 4e^{2x-3}\, dx$

39. $\int 4(e^{x-3})^2\, dx$ **40.** $\int 3\csc(\pi x)\cot(\pi x)\, dx$

41. $\int 3\tan^2 x\, dx$ **42.** $\int (\sin^2(3x) + \cos^2(3x))\, dx$

43. $\int \dfrac{-3}{1+x^2}\, dx$ **44.** $\int \dfrac{7}{\sqrt{1-(2x)^2}}\, dx$

45. $\int \dfrac{1}{|3x|\sqrt{9x^2-1}}\, dx$

■ Solve each of the following integrals, where *a*, *b*, and *c* are any constants with the given restrictions.

46. $\int \dfrac{a}{bx^c}\, dx \quad (b \neq 0,\ c \neq 1)$

47. $\int ae^{bx+c}\, dx \quad (b \neq 0)$

48. $\int b(c^x) + a\, dx \quad (c > 0,\ c \neq 1)$

49. $\int a\sin(bx) - c\, dx \quad (b \neq 0)$

50. $\int \dfrac{\sec(ax)\tan(ax)}{b}\, dx \quad (a \neq 0,\ b \neq 0)$

51. $\int \dfrac{a}{1+(bx)^2}\, dx \quad (b \neq 0)$

■ Calculate each of the following integrals by recognizing the integrand as the "result" of a product, quotient, or chain rule calculation, or by educated guessing and checking. Check each of your answers by differentiating. (*Hint for Problem 60:* $\tan x = \frac{\sin x}{\cos x}$.)

52. $\int (x^2 e^x + 2xe^x)\, dx$

53. $\int 3x^2 \sin(x^3 + 1)\, dx$

54. $\int \dfrac{2x\sin x - (x^2+1)\cos x}{\sin^2 x}\, dx$

55. $\int \dfrac{6x}{3x^2+1}\, dx$

56. $\int x^3 e^{3x^4-2}\, dx$

57. $\int \left(\dfrac{1}{x}\cos x - \sin x \ln x\right) dx$

58. $\int x^2(x^3+1)^5\, dx$

59. $\int \dfrac{2x\ln x - x}{(\ln x)^2}\, dx$

60. $\int \tan x\, dx$

Proofs

■ Use Theorem 13.2 to prove each of the following integration formulas.

61. $\int x^k\, dx = \frac{1}{k+1}x^{k+1} + C, \quad$ for all $k \neq -1$

62. $\int \dfrac{1}{x}\, dx = \ln|x| + C$

63. $\int b^x\, dx = \frac{1}{\ln b}b^x + C, \quad$ for all $b > 0,\ b \neq 1$

64. $\int e^{kx}\, dx = \frac{1}{k}e^{kx} + C, \quad$ for all $k \neq 0$

65. $\int \sin x\, dx = -\cos x + C$

66. $\int \cos x\, dx = \sin x + C$

67. $\int \sec^2 x\, dx = \tan x + C$

68. $\int \csc^2 x\, dx = -\cot x + C$

69. $\int \sec x\tan x\, dx = \sec x + C$

70. $\int \csc x\cot x\, dx = -\csc x + C$

71. $\int a\sin(bx)\, dx = -\frac{a}{b}\cos(bx) + C$

72. $\int \dfrac{1}{\sqrt{1-x^2}}\, dx = \sin^{-1} x + C$

73. $\int \dfrac{1}{1+x^2}\, dx = \tan^{-1} x + C$

74. $\int \dfrac{1}{|x|\sqrt{x^2-1}}\, dx = \sec^{-1} x + C$

■ Prove the two integration rules below in your own words. Justify each step of your proofs.

75. $\int kf(x)\, dx = k\int f(x)\, dx$

76. $\int (f(x) + g(x))\, dx = \int f(x)\, dx + \int g(x)\, dx$

77. In this problem you will prove that if $F(x)$ and $G(x)$ are any differentiable functions, then $F'(x) = G'(x)$ if, and only if, $F(x) = G(x) + C$ for some constant C.

  (a) Prove that if $F(x) = G(x) + C$ for some constant C then $F'(x) = G'(x)$, by differentiating both sides of the equality $F(x) = G(x) + C$.

  (b) Prove that if $F'(x) = G'(x)$ then $F(x) = G(x) + C$ for some constant C, by showing that if $F'(x) = G'(x)$, then the function $F(x) - G(x)$ is constant. (*Hint:* Recall that if the derivative of a function is zero, that function must be a constant function.)

  (c) What does this problem have to do with integrals?

13.2   The Fundamental Theorem of Calculus

So far we have two very different mathematical objects that we call "integrals," namely the definite integral $\int_a^b f(x)\,dx$ and the indefinite integral $\int f(x)\,dx$. The reason why these objects have such similar names and notation is that they are related by a very important theorem called the ***Fundamental Theorem of Calculus.***

13.2.1   The Fundamental Theorem

Consider for a moment a position function $s(t)$ representing the position of an object moving in a straight line. As we know, the derivative of the position function $s(t)$ is the velocity function $v(t) = s'(t)$. What is the average velocity of the object over a particular time interval $[a, b]$? We have two ways to answer this question. First of all, the average velocity on $[a, b]$ is the average value of the velocity function $v(t)$ on the interval $[a, b]$, and by the definition of the average value of a function (see Section 12.4), this is

$$(1) \qquad \begin{matrix} \text{Average velocity} \\ \text{on } [a,\,b] \end{matrix} = \begin{matrix} \text{Average value of} \\ v(t) \text{ on } [a,\,b] \end{matrix} = \frac{1}{b-a} \int_a^b v(t)\,dt.$$

On the other hand, it makes intuitive sense that the average velocity of the object should be equal to the average rate of change of the position of the object. We would hope that the mathematics would make sense as well! From Section 3.2.2 we know that the average rate of change of the position of the object is given by

$$(2) \qquad \begin{matrix} \text{Average rate of change} \\ \text{of } s(t) \text{ on } [a,\,b] \end{matrix} = \frac{s(b) - s(a)}{b-a}.$$

Physical intuition tells us that the quantities in Equations (1) and (2) should be equal. The ***Fundamental Theorem of Calculus*** (Theorem 13.8) will tell us that these two expressions are in fact the same:

$$\frac{1}{b-a} \int_a^b v(t)\,dt = \frac{s(b) - s(a)}{b-a}.$$

This is equivalent to the following equation (which we obtain by multiplying both sides of the equation above by $b - a$):

$$(3) \qquad \int_a^b v(t)\,dt = s(b) - s(a).$$

Equation (3) states that the area under the graph of the velocity function $v(t)$ from $t = a$ to $t = b$ is equal to the change in position $s(b) - s(a)$ on that time interval! If we wanted to find the area $\int_a^b v(t)\,dt$ under the graph of the velocity function on $[a, b]$, we wouldn't have to do any Riemann sums or limits! We could simply determine the initial position $s(a)$ and the final position $s(b)$ of the object and calculate $s(b) - s(a)$.

EXAMPLE 13.8   **Distance traveled is a definite integral of velocity**

Suppose a car traveled 100 miles in 2 hours. If $v(t)$ describes the car's velocity in miles per hour after t hours, find the area between the graph of $v(t)$ and the x-axis from $t = 0$ to $t = 2$.

Solution   Suppose the position of the car after t hours, in miles, is given by the function $s(t)$. If $s(0) = 0$ miles, then $s(2) = 100$ miles. By Equation (3), we know that the area under the graph of the velocity function $v(t)$ from $t = 0$ to $t = 2$ is

$$\int_0^2 v(t)\,dt = s(2) - s(0) = 100 - 0 = 100.$$

We didn't need to know anything about the velocity function to do this; all we had to know was the change in the position of the car over the time interval [0, 2]. Neither did we have to calculate any Riemann sums to find the definite integral $\int_0^2 v(t)\,dt$.   □

As we will soon see, the reason we get the amazing relationship in Equation (3) is that the position function $s(t)$ is an antiderivative of the velocity function $v(t)$. In fact, Equation (3) is true in a more general sense: The area under the graph of a function $f(x)$ on $[a, b]$ is equal to the change $F(b) - F(a)$ in its antiderivative $F(x)$ on that same interval. This incredible relationship is so important, in fact so *fundamental*, to the study of calculus that it is called the ***Fundamental Theorem of Calculus.***

THEOREM 13.8

The Fundamental Theorem of Calculus

If $f(x)$ is continuous on $[a, b]$, and $F(x)$ is any antiderivative of $f(x)$, then

$$\int_a^b f(x)\,dx = F(b) - F(a).$$

Instead of saying that $F(x)$ is any antiderivative of $f(x)$, we could equivalently say that $F(x)$ is any function whose derivative is $f(x)$, that is, any function such that $F'(x) = f(x)$. We will prove the Fundamental Theorem of Calculus in Section 13.2.2. It is important to note that the previous discussion concerning the velocity and position functions is not a valid *proof* of Equation (3) or of the Fundamental Theorem.

Caution   If $f(x)$ is not continuous on $[a, b]$, then the Fundamental Theorem does not apply to $f(x)$ on that interval. Remember that if f is discontinuous on $[a, b]$, then we can't even be sure that its "area" on the interval $[a, b]$ exists. For example, graph the function $\frac{1}{x-1}$ on the interval [0, 3], and try to shade in the area between the graph and the x-axis.   □

Caution   The fact that $F(x)$ is an antiderivative of $f(x)$ is a *major* part of the Fundamental Theorem. If the Fundamental Theorem is stated without this assumption, then it is meaningless; it would say that the area between the graph of a function $f(x)$ and the x-axis from $x = a$ to $x = b$ is equal to the difference $F(b) - F(a)$ of values of some unspecified function $F(x)$ on that interval. The entire key to the Fundamental Theorem is that $F(x)$ is an antiderivative of $f(x)$, and thus $F'(x) = f(x)$.   □

We could also rephrase the Fundamental Theorem in terms of a function and its derivative, rather than a function and its antiderivative. In the following alternative form of the Fundamental Theorem, the function $f(x)$ plays the role of the antiderivative "$F(x)$" and its derivative $f'(x)$ plays the role of the integrand, called "$f(x)$" in the original Fundamental Theorem.

THEOREM 13.9

Alternative Form of the Fundamental Theorem

If $f(x)$ is differentiable on $[a, b]$, and its derivative $f'(x)$ is continuous on $[a, b]$, then

$$\int_a^b f'(x)\,dx = f(b) - f(a).$$

Notice that $f(x)$ is the antiderivative of $f'(x)$, since the derivative of $f(x)$ is $f'(x)$. In this form, the Fundamental Theorem says that a definite integral "undoes" the process of differentiation, in the sense that if you integrate the derivative of a function from $x = a$ to $x = b$, you get something that has to do with the original function (namely $f(b) - f(a)$). It also says that if you "accumulate" all the instantaneous rates of change of $f(x)$ by integrating its derivative $f'(x)$ on the interval $[a, b]$, you get the *total* change $f(b) - f(a)$ of the function $f(x)$ on that interval.

Try to spend a few minutes thinking about how absolutely amazing the Fundamental Theorem of Calculus is. Who would have thought that the area under the graph of a function on an interval would have anything to do with antiderivatives? As we will see in Section 13.2.4, the Fundamental Theorem of Calculus is an extraordinarily useful theorem; it will enable us to calculate the *exact* values of some definite integrals without having to do any Riemann sums or limits!

13.2.2 Proving the Fundamental Theorem

To prove the Fundamental Theorem of Calculus, we must show that the Riemann sums that define the definite integral of a function $f(x)$ on an interval can somehow be written in terms of an antiderivative $F(x)$ of $f(x)$. The key to doing this will be applying the Mean Value Theorem to the antiderivative $F(x)$. This will allow us to relate the average rate of change of $F(x)$ to the instantaneous rate of change (i.e., the derivative) of $F(x)$. Before reading this proof, you may want to review the definition of and notation for the definite integral of a function (Section 12.3), as well as the Mean Value Theorem (Section 3.6.3).

PROOF (**THEOREM 13.8**)   Suppose $f(x)$ is a continuous function on $[a, b]$, and let $F(x)$ be any antiderivative of $f(x)$ (therefore, $F'(x) = f(x)$). Since $f(x)$ is continuous, its definite integral from $x = a$ to $x = b$ is well defined by a limit of Riemann sums, and we can write

$$(4) \qquad \int_a^b f(x)\, dx = \lim_{n \to \infty} \sum_{k=1}^n f(x_k^*)\, \Delta x,$$

where for each n we define $\Delta x := \frac{b-a}{n}$ and $x_k := a + k\,\Delta x$, and where each x_k^* is a point in the subinterval $[x_{k-1}, x_k]$. Since by hypothesis we know that $F'(x) = f(x)$, we know that $f(x_k^*) = F'(x_k^*)$, and thus

$$(5) \qquad \int_a^b f(x)\, dx = \lim_{n \to \infty} \sum_{k=1}^n F'(x_k^*)\, \Delta x.$$

We can choose each x_k^* to be *any* point we like in the subinterval $[x_{k-1}, x_k]$. The key to this proof is to choose the points x_k^* in a very special way.

Consider for a moment the function $F(x)$ on the subinterval $[x_{k-1}, x_k]$. Since $F(x)$ is differentiable on this subinterval by hypothesis, it is also continuous. Therefore, the Mean Value Theorem applies to $F(x)$ on this subinterval and says that

$$(6) \qquad \textit{There exists some } x_k^* \in [x_{k-1}, x_k] \textit{ such that } F'(x_k^*) = \frac{F(x_k) - F(x_{k-1})}{x_k - x_{k-1}}.$$

In Statement (6), x_k^* is playing the role of "c" from the Mean Value Theorem, and x_{k-1} and x_k are playing the roles of "a" and "b," respectively. The denominator $x_k - x_{k-1}$ is equal to Δx (the width of the subinterval $[x_{k-1}, x_k]$). We now choose the points x_k^* according to the Mean Value Theorem. For each subinterval $[x_{k-1}, x_k]$, we choose x_k^* to be a point satisfying Statement (6). In other words, for each x_k^* we will have

$$F'(x_k^*) = \frac{F(x_k) - F(x_{k-1})}{\Delta x}.$$

With these choices of points x_k^*, Equation (5) becomes

(7) $\displaystyle\int_a^b f(x)\,dx = \lim_{n\to\infty}\sum_{k=1}^{n}\frac{F(x_k)-F(x_{k-1})}{\Delta x}\,\Delta x = \lim_{n\to\infty}\sum_{k=1}^{n}(F(x_k)-F(x_{k-1})).$

Luckily, for every n the sum $\sum_{k=1}^{n}(F(x_k)-F(x_{k-1}))$ is what is known as a ***telescoping sum;*** this means that if we write out the terms of this sum, most of the terms of the sum will cancel each other out:

$$\sum_{k=1}^{n}(F(x_k)-F(x_{k-1})) = (F(x_1)-F(x_0))+(F(x_2)-F(x_1))+(F(x_3)-F(x_2))$$

$$+\cdots+(F(x_{n-1})-F(x_{n-2}))+(F(x_n)-F(x_{n-1}))$$

$$= -F(x_0)+(F(x_1)-F(x_1))+(F(x_2)-F(x_2))$$

$$+\cdots+(F(x_{n-1})-F(x_{n-1}))+F(x_n)$$

$$= -F(x_0)+0+0+\cdots+0+F(x_n)$$

$$= F(x_n)-F(x_0)$$

$$= F(b)-F(a).$$

The last line follows from the fact that x_0 and x_n are by definition equal to the endpoints a and b of the original interval.

Putting this computation together with what we had in Equation (7), we have

$$\int_a^b f(x)\,dx = \lim_{n\to\infty}\sum_{k=1}^{n}(F(x_k)-F(x_{k-1})) \qquad \text{(Equation (7))}$$

$$= \lim_{n\to\infty}(F(b)-F(a)) \qquad \text{(telescoping sum)}$$

$$= F(b)-F(a). \qquad (F(b)-F(a) \text{ does not depend on } n)$$

13.2.3 Evaluation notation

Now that we know the Fundamental Theorem of Calculus, we can calculate the *exact* value of a definite integral $\int_a^b f(x)\,dx$ very easily, provided that we can find an antiderivative of $f(x)$.

EXAMPLE 13.9

Applying the Fundamental Theorem of Calculus

Use the Fundamental Theorem of Calculus to find the exact value of $\displaystyle\int_0^1 x^2\,dx.$

Solution   In the notation of the Fundamental Theorem, we have $f(x)=x^2$, $a=0$, and $b=1$. We know that the definite integral $\int_0^1 x^2\,dx$ is equal to $F(b)-F(a)=F(1)-F(0)$ for any antiderivative $F(x)$ of $f(x)$. Now we must find an antiderivative $F(x)$. Since $\frac{d}{dx}(\frac{1}{3}x^3)=\frac{1}{3}(3x^2)=x^2$, the function $F(x)=\frac{1}{3}x^3$ is an antiderivative of $f(x)=x^2$. By the Fundamental Theorem of Calculus, we have

$$\int_0^1 x^2\,dx = F(1)-F(0) = \tfrac{1}{3}(1)^3 - \tfrac{1}{3}(0)^3 = \tfrac{1}{3}.$$

Compare the calculation we just did with the calculation of the same quantity in Example 12.13 in Section 12.3.2. Before we knew the Fundamental Theorem, we had to write out a general Riemann sum for x^2 on $[0, 1]$, find a formula for this sum that works for every n, and then take the limit as $n \to \infty$ of this formula!   □

In Example 13.9, we had to give a name to the antiderivative of x^2 (we called it $F(x)$) so that we could see how to apply the Fundamental Theorem of Calculus. The following *evaluation notation* will enable us to use the Fundamental Theorem without specifically naming an antiderivative.

DEFINITION 13.3

Evaluation Notation

For any function $F(x)$ on an interval $[a, b]$, the difference $F(b) - F(a)$ will be called *the evaluation of $F(x)$ on $[a,\ b]$* and will be denoted by

$$[F(x)]_a^b := F(b) - F(a).$$

The notation $[F(x)]_a^b$ is pronounced "$F(x)$ evaluated on $[a, b]$" or "$F(x)$ evaluated from $x = a$ to $x = b$." This notation is called the "evaluation" of $F(x)$ on $[a, b]$ because $[F(x)]_a^b$ is the difference of $F(x)$ *evaluated* at $x = b$ and $F(x)$ *evaluated* at $x = a$.

EXAMPLE 13.10

Using evaluation notation when applying the Fundamental Theorem of Calculus

With evaluation notation, we can do the calculation in Example 13.9 more efficiently:

$$\int_0^1 x^2\, dx = \left[\tfrac{1}{3}x^3\right]_0^1 = \tfrac{1}{3}(1)^3 - \tfrac{1}{3}(0)^3 = \tfrac{1}{3}.$$

It does not matter *which* antiderivative we choose for $f(x) = x^2$. For example, the function $\frac{1}{3}x^3 + 2$ is also an antiderivative of x^2 (because its derivative is x^2); thus we could also write

$$\int_0^1 x^2\, dx = \left[\tfrac{1}{3}x^3 + 2\right]_0^1 = \left(\tfrac{1}{3}(1)^3 + 2\right) - \left(\tfrac{1}{3}(0)^3 + 2\right) = \tfrac{1}{3} + 2 - \tfrac{0}{3} - 2 = \tfrac{1}{3}.$$

Notice that the 2 and -2 cancel out; choosing a different antiderivative of x^2 does not change the final result.   □

The following rewording of the Fundamental Theorem of Calculus shows how the definite integral and the indefinite integral are related.

THEOREM 13.10

The Fundamental Theorem of Calculus in Evaluation Notation

For any continuous function $f(x)$ on an interval $[a, b]$,

$$\int_a^b f(x)\, dx = \left[\int f(x)\, dx\right]_a^b.$$

In other words, the definite integral of a function $f(x)$ on an interval $[a, b]$ is equal to the indefinite integral of $f(x)$ evaluated on that interval. The proof of this theorem follows directly from the Fundamental Theorem, evaluation notation, and the definition of the indefinite integral. It is left to the exercises.

13.2.4 Using the Fundamental Theorem

We can use the Fundamental Theorem of Calculus to solve a variety of problems that involve definite integrals. For example, we can now quickly calculate the exact signed area under a curve, the exact "true" area under a curve, the area between two curves, or the average value of a function, since all of these quantities are defined in terms of definite integrals. Moreover, we will be able to find the *exact* values of these quantities even for relatively complicated functions; recall that before we knew the Fundamental Theorem, we could calculate definite integrals exactly only for very simple functions (like x^2).

EXAMPLE 13.11

Using the Fundamental Theorem of Calculus to find signed and "true" area

Find (**a**) the signed area and (**b**) the true area between the graph of $f(x) = 3e^{2x-5} - 3$ and the x-axis from $x = 1$ to $x = 3$.

Solution By definition, the signed area is the definite integral of $f(x)$ on $[1, 3]$. Using the integration formula for exponential functions (and a little guess-and-check), we find that $\frac{3}{2}e^{2x-5} - 3x$ is an antiderivative of $f(x) = 3e^{2x-5} - 3$. By the Fundamental Theorem of Calculus, we have

$$\int_1^3 (3e^{2x-5} - 3)\, dx = \left[\frac{3}{2}e^{2x-5} - 3x\right]_1^3$$

$$= \left(\frac{3}{2}e^{2(3)-5} - 3(3)\right) - \left(\frac{3}{2}e^{2(1)-5} - 3(1)\right)$$

$$= \frac{3}{2}e - \frac{3}{2}e^{-3} - 6 \approx -1.99726.$$

To find the "true" area, we must first know where the function $f(x) = 3e^{2x-5} - 3$ is positive and where it is negative. The function $f(x)$ crosses the x-axis when $f(x) = 0$, that is, when

$$0 = 3e^{2x-5} - 3 \implies e^{2x-5} = 1 \implies 2x - 5 = 0 \implies x = \frac{5}{2} = 2.5.$$

As you can see in the graph of $f(x)$ shown in Figure 13.1, the function $f(x)$ is negative on $[1, 2.5]$ and positive on $[2.5, 3]$.

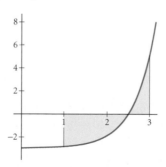

Figure 13.1

Therefore, the "true" area between the graph of $f(x)$ and the x-axis on $[1, 3]$ is

$$-\int_1^{2.5} (3e^{2x-5} - 3)\, dx + \int_{2.5}^3 (3e^{2x-5} - 3)\, dx$$

$$= -\left[\frac{3}{2}e^{2x-5} - 3x\right]_1^{2.5} + \left[\frac{3}{2}e^{2x-5} - 3x\right]_{2.5}^3$$

$$= -\left(\left(\frac{3}{2}e^{2(2.5)-5} - 3(2.5)\right) - \left(\frac{3}{2}e^{2(1)-5} - 3(1)\right)\right)$$

$$\quad + \left(\left(\frac{3}{2}e^{2(3)-5} - 3(3)\right) - \left(\frac{3}{2}e^{2(2.5)-5} - 3(2.5)\right)\right)$$

$$= \frac{3}{2}e^{-3} + \frac{3}{2}e^1 \approx 4.1521.$$

(Notice that a simplified, *exact* answer is given before the approximation.) The true area is larger than the signed area, since for the true area we are counting the area under the x-axis as positive area.   □

✓ *Checking the Answer*   There are two ways that you can check Fundamental Theorem of Calculus calculations. First, you can check your antidifferentiation step by differentiating. Second, you can check your numerical answer by approximating the value of the definite integral with a graphing calculator. You can also quickly check that the size and sign of your numerical answer make sense in terms of the graph of the function.   □

In the exercises, you will use the Fundamental Theorem of Calculus to calculate areas between curves and average values; see Section 12.4 if you need to review how to write these quantities as definite integrals. The following example uses the fact that the Fundamental Theorem of Calculus relates the area under a function to values of its antiderivative. In the language of the "alternative" version of the Fundamental Theorem, given the value of a function $f(x)$ at a point $x = a$, and the area under the graph of its derivative $f'(x)$ on $[a, b]$, we can determine the value of $f(x)$ at $x = b$.

EXAMPLE 13.12   **Given the area under f' and a value of f, computing other values of f**

Suppose $f(x)$ is a function whose *derivative* $f'(x)$ is shown in Figure 13.2. Given that $f(1) = 3$, use the Fundamental Theorem of Calculus and the graph in Figure 13.3 to approximate $f(4)$.

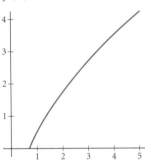

Figure 13.2
Graph of the derivative $f'(x)$

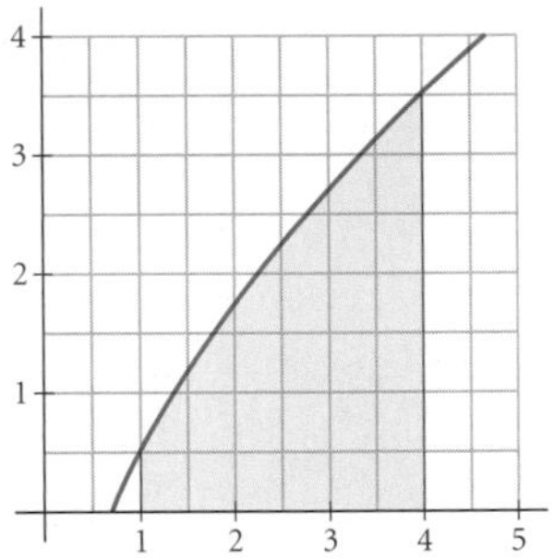

Figure 13.3 $\displaystyle\int_1^4 f'(x)\,dx$

Solution   By the Fundamental Theorem of Calculus, we know that the value $f(1)$, the value $f(4)$, and the area under the graph of $f'(x)$ on $[1, 4]$ are related. Specifically, we know that

$$\int_1^4 f'(x)\,dx = f(4) - f(1).$$

Solving for $f(4)$ gives us

$$f(4) = f(1) + \int_1^4 f'(x)\,dx.$$

We are given that $f(1) = 3$, so all that remains is to approximate the value of the definite integral. We can do this using the grid in Figure 13.3. Each box in the grid has a length and height of half a unit, and therefore each box has an area of $(\frac{1}{2})(\frac{1}{2}) = \frac{1}{4}$ square units. By counting boxes (and estimating "partial" boxes), we find that the area under the graph of $f'(x)$ from $x = 1$ to $x = 4$ covers approximately 26.5 boxes. Therefore, the area in question is approximately $26.5(\frac{1}{4}) = 6.625$ square units. This means that the value of $f(4)$ must be approximately

$$f(4) = f(1) + \int_1^4 f'(x)\,dx \approx 3 + 6.625 = 9.625.$$

Notice that we were able to find the approximate value of $f(4)$ even though we were never given a graph or equation for the function $f(x)$.   □

What you should know after reading Section 13.2

▶ The statement, significance, and proof of the Fundamental Theorem of Calculus (and its "alternative" form). What does the Fundamental Theorem say about velocity and position functions?

▶ The meaning of evaluation notation, and how to use it. Be able to use this notation to describe how definite and indefinite integrals are related.

▶ How to use the Fundamental Theorem of Calculus to find exact values of definite integrals, and thus exact areas or average values. Also be able to use the Fundamental Theorem to relate two values of a function to a definite integral of the derivative of that function.

Exercises 13.2

Concepts

0. Read the section and make your own summary of the material. Include whatever you think will help you review the material later on.

1. Intuitively, the average velocity of an object (the average value of the velocity function) should be the same as the average rate of change of the position function. Explain why this would suggest that the area under the velocity graph on an interval is equal to the difference in the position function on that interval.

2. State the Fundamental Theorem of Calculus (**a**) in its original form, (**b**) in its "alternative" form, and (**c**) using an indefinite integral and evaluation notation. Why are these three forms equivalent?

3. Why do we require that the integrand be a continuous function in the Fundamental Theorem of Calculus?

4. Why is the Fundamental Theorem of Calculus such an incredible theorem?

5. What important theorem is the key to proving the Fundamental Theorem of Calculus?

6. Explain what "evaluation notation" means and what it is used for.

7. Why is it not necessary to write down an antiderivative "family" when using the Fundamental Theorem of Calculus to calculate definite integrals? In other words, why don't we have to write "$+C$" as part of the antiderivative in this case?

8. State the Mean Value Theorem.

9. Determine whether or not each statement below is equivalent to the Fundamental Theorem of Calculus. (You may assume that all functions named are continuous.)

(**a**) $\int_a^b f(x)\,dx = g(b) - g(a)$, where $g'(x) = f(x)$

(**b**) If $f'(x) = F(x)$ then $\int_a^b f(x)\,dx = [F(x)]_a^b$.

(**c**) If $f'(x) = F(x)$ then $\int_a^b F(x)\,dx = [f(x)]_a^b$.

(**d**) If $g(x)$ is any antiderivative of $h(x)$, then $\int_a^b h(x)\,dx = g(a) - g(b)$.

(**e**) $\int_a^b G(x)\,dx = [G'(x)]_a^b$

(**f**) $\int_a^b h''(x)\,dx = [h'(x)]_a^b$

(**g**) $[\int p(x)\,dx]_a^b = \int_a^b p(x)\,dx$

Skills

■ Calculate each definite integral below in four ways, as follows: (**a**) Approximately, using a Riemann sum with ten rectangles. (**b**) Exactly, using the definition of the definite integral as a limit of Riemann sums. (**c**) Exactly, using the definite integral formulas from Theorem 12.7 (from the previous chapter). (**d**) Exactly, by using the Fundamental Theorem of Calculus.

10. $\displaystyle\int_1^2 3x^2\,dx$ **11.** $\displaystyle\int_{-3}^5 (x^2 - 3x + 2)\,dx$

12. $\displaystyle\int_0^4 (3x+2)^2\,dx$

■ Use the Fundamental Theorem of Calculus to find the exact value of each of the following definite integrals. (*Hint:* Some of the integrands in these problems are the "result" of a product, quotient, or chain rule calculation; you may have to use an educated "guess-and-check" method to find the antiderivatives.)

13. $\displaystyle\int_0^3 (x^2-4)^2\,dx$ **14.** $\displaystyle\int_1^4 3(2^x)\,dx$

15. $\displaystyle\int_{-\pi}^{\pi} \sin(3x)\,dx$ **16.** $\displaystyle\int_2^5 \frac{1}{\sqrt{x^5}}\,dx$

17. $\displaystyle\int_0^1 \frac{1}{2e^x}\, dx$ **18.** $\displaystyle\int_2^4 \frac{1}{4-3x}\, dx$

19. $\displaystyle\int_0^1 \frac{1}{1+x^2}\, dx$ **20.** $\displaystyle\int_{-\frac{\pi}{2}}^0 (\sin^2 4x + \cos^2 4x)\, dx$

21. $\displaystyle\int_{\frac{\pi}{4}}^{\frac{\pi}{2}} \csc x \cot x\, dx$ **22.** $\displaystyle\int_{-3}^3 2\cos(\pi x)\, dx$

23. $\displaystyle\int_1^4 \frac{2x+3}{x^2+3x+4}\, dx$

24. $\displaystyle\int_0^{\frac{3}{2}} \frac{1}{\sqrt{9-x^2}}\, dx$

25. $\displaystyle\int_{-\pi}^{\pi} (e^x \sin x - e^x \cos x)\, dx$

26. $\displaystyle\int_{-\frac{\pi}{4}}^{\frac{\pi}{4}} x^2 \sec^2(x^3)\, dx$

27. $\displaystyle\int_{-1}^1 \frac{e^x - xe^x}{e^{2x}}\, dx$

■ Calculate the exact value of each of the following definite integrals by using the Fundamental Theorem of Calculus. Because the integrands involve absolute values (which we do not know how to antidifferentiate), you will have to rewrite them as piecewise functions and, accordingly, split the definite integral into pieces.

28. $\displaystyle\int_{-2}^5 |x-2|\, dx$

29. $\displaystyle\int_{-1}^3 |4-x^2|\, dx$

30. $\displaystyle\int_{-2}^4 |2x^2 - 5x - 3|\, dx$

■ For Problems 31–34, suppose $f(x)$ is a function whose *derivative* $f'(x)$ is given by the accompanying graph. For each given value of f, use an approximation of a definite integral and the Fundamental Theorem of Calculus to approximate the requested value of f.

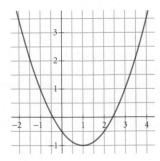

Graph of the derivative $f'(x)$

31. Given that $f(3) = 2$, approximate $f(4)$.
32. Given that $f(0) = -1$, approximate $f(2)$.
33. Given that $f(-1) = 2$, approximate $f(1)$.
34. Given that $f(2) = 3$, approximate $f(-2)$.

■ Use the Fundamental Theorem of Calculus to calculate **(a)** the exact signed area and **(b)** the exact "true" area between the graph of $f(x)$ and the x-axis on the interval $[a, b]$. Sketch a graph of $f(x)$, and shade in the region in question.

35. $f(x) = \cos x$, $[a, b] = [-\pi, \pi]$
36. $f(x) = (2x-1)^2 - 4$, $[a, b] = [-2, 4]$
37. $f(x) = \dfrac{1}{2x^3}$, $[a, b] = [1, 3]$
38. $f(x) = \dfrac{1}{1+x^2}$, $[a, b] = [-1, 1]$
39. $f(x) = 2 - e^{-x}$, $[a, b] = [-1, 0]$
40. $f(x) = \dfrac{3 - 2x}{x^2}$, $[a, b] = [1, 4]$

■ Use the Fundamental Theorem of Calculus to calculate the exact area between the graphs of $f(x)$ and $g(x)$ on the interval $[a, b]$. Sketch graphs of $f(x)$ and $g(x)$, and shade in the region in question.

41. $f(x) = \sin x$, $g(x) = \cos x$, $[a, b] = [-\frac{\pi}{2}, \frac{\pi}{2}]$
42. $f(x) = x-1$, $g(x) = x^2 - 2x - 1$, $[a, b] = [-1, 3]$
43. $f(x) = x^3$, $g(x) = (x-2)^2$, $[a, b] = [-1, 2]$
44. $f(x) = 2^x$, $g(x) = x^2$, $[a, b] = [0, 3]$
45. $f(x) = \dfrac{3}{x-2}$, $g(x) = 6 - x$, $[a, b] = [3, 7]$
46. $f(x) = \dfrac{2}{1+x^2}$, $g(x) = 1$, $[a, b] = [0, \sqrt{3}]$

■ Use the Fundamental Theorem of Calculus to find the exact average value of $f(x)$ on the interval $[a, b]$. Sketch a graph of $f(x)$, and use it to determine whether your answer is reasonable.

47. $f(x) = 4x^{\frac{3}{2}}$, $[a, b] = [0, 2]$
48. $f(x) = (e^x)^2$, $[a, b] = [-1, 1]$
49. $f(x) = 2 - \sqrt[3]{x}$, $[a, b] = [-1, 8]$
50. $f(x) = \dfrac{1}{3x+1}$, $[a, b] = [2, 5]$
51. $f(x) = \sin x + x \cos x$, $[a, b] = [-\pi, \pi]$
52. $f(x) = x^2 \sin(x^3 + 1)$, $[a, b] = [-1, 2]$

Proofs

53. Prove the Fundamental Theorem of Calculus in your own words (you can use the proof in this section as a guide).

54. Prove that if $F(x)$ and $G(x)$ differ by a constant, then $[F(x)]_a^b = [G(x)]_a^b$.

55. Use the Fundamental Theorem of Calculus to prove that on any interval $[a, b]$, we have $\int_a^b x^2\, dx = \frac{1}{3}(b^3 - a^3)$.

56. Use the Fundamental Theorem of Calculus to prove that for any continuous functions $f(x)$ and $g(x)$ on an interval $[a, b]$, we have $\int_a^b (f(x) + g(x))\, dx = \int_a^b f(x)\, dx + \int_a^b g(x)\, dx$. (*Hint:* Suppose $F(x)$ is an

antiderivative of $f(x)$, and $G(x)$ is an antiderivative of $g(x)$.)

57. Use the Fundamental Theorem of Calculus to prove that for any continuous function $f(x)$ on an interval $[a, b]$, we have $\int_a^b kf(x)\, dx = k \int_a^b f(x)\, dx$. (*Hint:* Suppose $F(x)$ is an antiderivative of $f(x)$.)

13.3 Functions Defined by Integrals

In this section we examine functions that are defined in terms of definite integrals. An extension of the Fundamental Theorem of Calculus will show that these "area functions" can be used to construct antiderivatives.

13.3.1 Area functions

Up to this point, all the functions we have considered could be neatly expressed using algebraic, exponential, logarithmic, trigonometric, or inverse trigonometric functions. We can write a great variety of functions using these building blocks—for example, $f(x) = \sin x \sqrt{\ln(x^2 + 1)}$ and $f(x) = \cot^{-1}(e^{3x})$. Functions that can be written in terms of these basic building blocks are called **elementary.** In this section we examine a **nonelementary** type of function called an **area function.**

Suppose f is a continuous function defined on an interval $[a, b]$. For any real number x in $[a, b]$, we can consider the area under the graph of f from a to x; note that this area is also a real number. Since *every* number $x \in [a, b]$ has a *unique* area associated with it, this relationship is a function. The inputs of the function are the values $x \in [a, b]$, and the outputs of the function are the areas under the graph of f from a to x. This function is called the "area accumulation function" (or simply the "area function") for f on $[a, b]$. Definition 13.4 defines this function using mathematical notation.

DEFINITION 13.4

Area Accumulation Functions

Suppose f is a continuous function on $[a, b]$. Then for any $x \in [a, b]$, we can define the **area accumulation function** for f to be the function

$$A(x) := \int_a^x f(t)\, dt.$$

Note that the variable we use in the integrand is called t and that the variable we use for the input of the area accumulation function is called x. The variable t in the integrand is called a "dummy variable"; in Definition 13.4, t varies from a to x. There will be no "t" left after we integrate. For example, $\int_1^x t\, dt = [\frac{1}{2}t^2]_1^x = \frac{1}{2}x^2 - \frac{1}{2}(1)^2$ does not involve the variable t. We can use any variable name we like in the place of t (except for x) and still get the same function $A(x)$; for example, we could define $A(x) := \int_a^x f(w)\, dw$. We cannot use x for the dummy variable because we have to keep the dummy variable separate from the "real" variable x of our area accumulation function. You can think of t as the variable for the function f and of x as the variable for the area accumulation function $A(x)$ of f; see Figure 13.4.

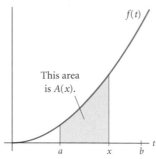

Figure 13.4
The value of the area
function $A(x) = \int_a^x f(t)\,dt$
for a particular x

🛇 *Caution*   Make sure you understand the difference between the variable x and the variable t in Definition 13.4. Which is the input variable for the area accumulation function? What role does the other variable play?   □

EXAMPLE 13.13

Calculating values of an area accumulation function

Define $A(x) := \int_0^x \sqrt{t}\,dt$. Then, given any input x, the value of $A(x)$ is the area under the graph of $f(t) = \sqrt{t}$ from $t = 0$ to $t = x$. For example, the value of $A(1)$ is the area under the graph of $f(x)$ on the interval $[0, 1]$ shown in Figure 13.5. We can calculate this area (and thus the value of $A(1)$) with the Fundamental Theorem of Calculus:

$$(1) \qquad A(1) = \int_0^1 \sqrt{t}\,dt = \left[\tfrac{2}{3}t^{\frac{3}{2}}\right]_0^1 = \tfrac{2}{3}\left(1^{\frac{3}{2}}\right) - \tfrac{2}{3}\left(0^{\frac{3}{2}}\right) = \tfrac{2}{3} \approx 0.67.$$

Figures 13.6 and 13.7 show the areas corresponding to the values $A(2)$ and $A(3)$.

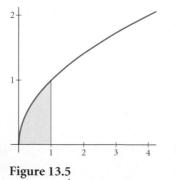
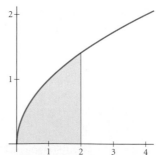
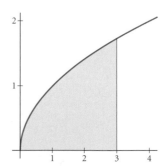

Figure 13.5
$A(1) = \int_0^1 \sqrt{t}\,dt \approx 0.67$

Figure 13.6
$A(2) = \int_0^2 \sqrt{t}\,dt \approx 1.89$

Figure 13.7
$A(3) = \int_0^3 \sqrt{t}\,dt \approx 3.46$

□

Using the techniques in Example 13.13, we can calculate $A(x)$ for any nonnegative value of x we like. In fact, using a calculation similar to the one in Equation (1), we can even get an *equation* for $A(x)$.

EXAMPLE 13.14

Finding an elementary equation for an area accumulation function

Use the Fundamental Theorem of Calculus to find an elementary equation for the area accumulation function $A(x) = \int_0^x \sqrt{t}\,dt$.

Solution Since $\frac{2}{3}t^{\frac{3}{2}}$ is an antiderivative of $\sqrt{t} = t^{\frac{1}{2}}$, we have

$$A(x) = \int_0^x \sqrt{t}\, dt = \left[\tfrac{2}{3}t^{\frac{3}{2}}\right]_0^x = \tfrac{2}{3}\left(x^{\frac{3}{2}}\right) - \tfrac{2}{3}\left(0^{\frac{3}{2}}\right) = \tfrac{2}{3}x^{\frac{3}{2}}.$$

You might notice that the function $A(x) = \frac{2}{3}x^{\frac{3}{2}}$ is the antiderivative of the function $f(x) = \sqrt{x}$. In fact, every area function for a given function f will *always* be an antiderivative of f, as we will see in Section 13.3.2. ☐

Since the definite integral of a function f on an interval $[a, x]$ represents a *signed* area, the area accumulation function $A(x) = \int_a^x f(t)\, dt$ is actually a *signed* area accumulation function. Example 13.15 illustrates the behavior of the area accumulation function of a function that is sometimes positive and sometimes negative.

EXAMPLE 13.15

Graphically interpreting an area accumulation function

Suppose f is the function shown in Figure 13.8, and define the area accumulation function $A(x) = \int_1^x f(t)\, dt$. Use the graph of f to answer the following questions:

(a) Which is larger, $A(2)$ or $A(3)$? Which is larger, $A(3)$ or $A(6)$?

(b) List the intervals on which $A(x)$ is increasing and those on which it is decreasing.

(c) Sketch a rough graph of the function $A(x)$ on $[1, 8]$.

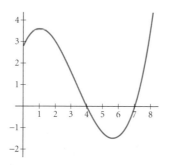

Figure 13.8
The function $f(t)$

Solution $A(2)$ is the signed area under the graph of f from $t = 1$ to $t = 2$, and $A(3)$ is the signed area from $t = 1$ to $t = 3$. Since the function f is positive on $[2, 3]$, the area under the graph of f on $[2, 3]$ is positive. Note that $A(3)$ is the sum of $A(2)$ and the area under f on $[2, 3]$ (since $\int_1^3 f(t)\, dt = \int_1^2 f(t)\, dt + \int_2^3 f(t)\, dt$). Since the "extra" area from $t = 2$ to $t = 3$ is positive, we have $A(3) > A(2)$; see Figures 13.9 and 13.10. On the other hand, as we move from $t = 3$ to $t = 6$, we accumulate a small amount of positive area (until $t = 4$) and then accumulate *negative* area until $t = 6$. Since there is more negative area than positive area on $[3, 6]$, we know that $A(6) < A(3)$; see Figures 13.10 and 13.11.

By similar arguments, the area accumulation function $A(x)$ will be increasing if the function f is positive, decreasing if the function f is negative. Think of the function $A(x)$ as an "accumulation" of area as x moves from the left to the right along the t-axis. When f is positive, the area accumulated is positive, and thus the area accumulation function $A(x)$ increases. When f is negative, the area accumulated is negative, so the area accumulation function $A(x)$ decreases. From the graph of f in Figure 13.8 we see that f is positive on $(1, 4)$, negative on $(4, 7)$, and positive on $(7, 8)$. Therefore, $A(x)$ is increasing on $(1, 4) \cup (7, 8)$ and decreasing on $(4, 7)$.

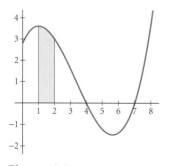 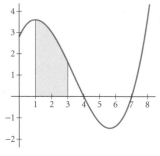 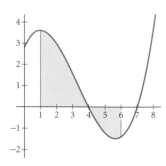

Figure 13.9   **Figure 13.10**   **Figure 13.11**

$$A(2) = \int_1^2 f(t)\,dt \qquad A(3) = \int_1^3 f(t)\,dt \qquad A(6) = \int_1^6 f(t)\,dt$$

Using what we know about where $A(x)$ is increasing and decreasing, and the fact that $A(1) = 0$ (why?), we can sketch a rough graph of $A(x)$; see Figure 13.12. The height of the graph of A at each point x is the signed area under the graph of f from $t = 1$ to $t = x$.

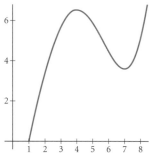

Figure 13.12
The area accumulation
function $A(x)$ □

❓ *Question*   The following questions concern Example 13.15. In this example, why is $A(x)$ always positive, even though f is sometimes negative? Why is $A(1) = 0$? How could you estimate the scale on the y-axis in Figure 13.12? (For example, how could you estimate that $A(4)$ is a bit more than 6?) Why do you think $A(x)$ is concave down until about $x = 5.8$? (*Hint:* What can you say about the graph of f on $[1, 5.8]$? Why would this make the area accumulation function concave down?)   □

13.3.2 The Second Fundamental Theorem

In Example 13.13 we saw that the area accumulation function $A(x) = \int_0^x \sqrt{t}\,dt$ is an antiderivative of the function $f(t) = \sqrt{t}$. This is true in general; the area function for any continuous function f is an antiderivative of f. This fact is so important that it is called the ***Second Fundamental Theorem of Calculus.***

THEOREM 13.11

The Second Fundamental Theorem of Calculus
If f is continuous on $[a, b]$ and we define

$$F(x) := \int_a^x f(t)\,dt$$

for all $x \in [a, b]$, then $F(x)$ is continuous on $[a, b]$ and differentiable on (a, b), with a derivative of $F'(x) = f(x)$.

In other words, given any continuous function f, if we define $F(x) := \int_a^x f(t)\, dt$ to be the area accumulation function for f, then F is an antiderivative of f. This is true regardless of the choice of "starting point" a for the area accumulation function. Theorem 13.11 shows that every function $f(x)$ has an antiderivative given by the "area function" $F(x) := \int_a^x f(t)\, dt$ (where a is any constant).

PROOF **(THEOREM 13.11)** Suppose f is continuous on $[a, b]$, and for all $x \in [a, b]$ define $F(x) = \int_a^x f(t)\, dt$. We'll leave the proof that $F(x)$ is continuous on $[a, b]$ to the exercises. Here we will show that for all $x \in (a, b)$, we have $F'(x) = f(x)$. We will calculate the right derivative of $F(x)$; the calculation for the left derivative is similar. By the definition of the derivative, we have

$$F'(x) = \lim_{h \to 0^+} \frac{F(x+h) - F(x)}{h} \qquad \text{(definition of derivative)}$$

$$= \lim_{h \to 0^+} \frac{\int_a^{x+h} f(t)\, dt - \int_a^x f(t)\, dt}{h} \qquad \text{(definition of } F(x) \text{)}$$

$$= \lim_{h \to 0^+} \frac{\int_x^{x+h} f(t)\, dt}{h}. \qquad \text{(see below)}$$

Note that since we are considering only the right derivative here, h is positive. The last step in the calculation follows from the fact that $a < x < x + h$, and thus

$$\int_a^{x+h} f(t)\, dt = \int_a^x f(t)\, dt + \int_x^{x+h} f(t)\, dt.$$

Let's examine the quantity $\int_x^{x+h} f(t)\, dt$ for a small value h. This definite integral represents the area under the graph of $f(t)$ from $t = x$ to $t = x + h$; see Figure 13.13. We can approximate this definite integral using a rectangle; Figures 13.14 and 13.15 show the one-rectangle Minimum Sum and Maximum Sum, respectively.

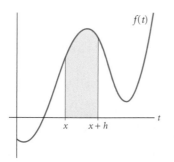

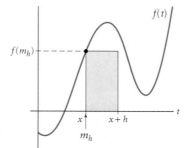

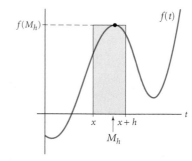

Figure 13.13
$$\int_x^{x+h} f(t)\, dt$$

Figure 13.14
Minimum Sum with one rectangle

Figure 13.15
Maximum Sum with one rectangle

As shown in Figures 13.14 and 13.15, we define m_h to be a point in $[x, x + h]$ for which $f(t)$ has a minimum value and define M_h to be a point in $[x, x + h]$ for which $f(t)$ has a maximum value. Notice that these values depend on h. For example, suppose h was one-tenth the size shown in the figures; then m_h would remain the same, but M_h would change (since the M_h shown in Figure 13.15 would not be in the new, smaller interval).

The area of the rectangle in Figure 13.14 is $f(m_h) \cdot h$ (length times width), and the area of the rectangle in Figure 13.15 is $f(M_h) \cdot h$. Since any Minimum Sum is always less than or equal to the actual value of the definite integral it approximates, and the Maximum Sum is always greater than or equal to the actual area, we have

$$f(m_h) \cdot h \leq \int_x^{x+h} f(t)\, dt \leq f(M_h) \cdot h.$$

When we divide by h (which in this case is positive), this becomes

$$f(m_h) \leq \frac{\int_x^{x+h} f(x)\, dt}{h} \leq f(M_h).$$

Note that the quantity in the middle is exactly the quantity whose limit we need to find to calculate $F'(x)$. Since our results hold for all $h > 0$, we have

$$\lim_{h \to 0} f(m_h) \leq \lim_{h \to 0} \frac{\int_x^{x+h} f(t)\, dt}{h} \leq \lim_{h \to 0} f(M_h).$$

As $h \to 0$, both $f(m_h)$ and $f(M_h)$ approach the value $f(x)$, since f is continuous, and as the interval $[x, x+h]$ shrinks, both m_h and M_h approach x. Therefore, the limit we are interested in is both less than or equal to, *and* greater than or equal to, $f(x)$; thus it is *equal* to $f(x)$. Combining this with our initial work, we have

$$F'(x) = \lim_{h \to 0} \frac{\int_x^{x+h} f(t)\, dt}{h} = f(x).$$

Therefore, $F(x)$ is an antiderivative of $f(x)$, and we are done. ∎

13.3.3 Using the Second Fundamental Theorem

Using the Second Fundamental Theorem, we can construct an antiderivative for *any* function f. For example, consider the function $f(x) = \sin(e^x)$. We do not know how to antidifferentiate this function; that is, we do not know an "elementary" antiderivative for this function. However, the area accumulation function $F(x) := \int_0^x \sin(e^t)\, dt$ is an antiderivative of $f(x) = \sin(e^x)$ by the Second Fundamental Theorem. So is the function $G(x) := \int_3^x \sin(e^t)\, dt$, since we can make the starting limit of integration anything we like.

? Question   If $F(x) = \int_0^x \sin(e^t)\, dt$ and $G(x) = \int_3^x \sin(e^t)\, dt$ are *both* antiderivatives of $f(x) = \sin(e^x)$, then they must differ by a constant. What is that constant? Can you illustrate that constant as an area under the graph of $f(x) = \sin(e^x)$?   □

Another way to phrase the Second Fundamental Theorem is as follows:

THEOREM 13.12

A Rewording of the Second Fundamental Theorem of Calculus
If f is continuous on $[a, b]$, then for all $x \in [a, b]$ we have

$$\frac{d}{dx} \int_a^x f(t)\, dt = f(x).$$

This theorem illustrates another way that integration and differentiation can be thought of as "inverse" operations: Given a function f and a real number a, we can find the definite integral of f from a to x, differentiate that function with respect to x, and arrive back at the function $f(x)$. To put it another way, the derivative of an area accumulation function for any function f is equal to the original function f.

⊗ Caution   Be careful when using Theorem 13.12. In particular, in order for us to apply the theorem, the upper limit of the integral *must* be x, and the lower limit of the integral must be a constant (although it can be *any* constant we like). If the upper limit is a function of x, then the chain rule will also be needed (see part (b) of Example 13.16).   □

The following examples illustrate how we can use Theorem 13.12 to differentiate functions that are defined in terms of area accumulation functions.

EXAMPLE 13.16   **Differentiating area accumulation functions**

Find the derivatives of each of the following functions.

(a) $F(x) = \displaystyle\int_x^2 \ln t\, dt$ (b) $F(x) = \displaystyle\int_0^{x^2} \sec t\, dt$ (c) $F(x) = \displaystyle\int_0^{3x} \frac{x}{t^2+1}\, dt$

Solution

(a) We cannot apply Theorem 13.12 immediately because here the *lower* limit of integration (rather than the upper limit of integration) is the variable x. However, we can easily fix this before differentiating:

$$F'(x) = \frac{d}{dx}\left(\int_x^2 \ln t\, dt\right) = \frac{d}{dx}\left(-\int_2^x \ln t\, dt\right)$$
$$= -\frac{d}{dx}\left(\int_2^x \ln t\, dt\right) = -\ln x.$$

(b) We cannot apply Theorem 13.12 directly to this function because the upper limit of integration is x^2, not x. In particular, the derivative of this function is *not* $\sec x$, nor is it $\sec x^2$. To differentiate this function we will need the chain rule, because this function is a *composition*. The "outside" function of the composition is the area accumulation function $A(x) = \int_0^x \sec t\, dt$, and the "inside" function is x^2. In other words, $F(x) = A(x^2)$. By the chain rule, the derivative of the composition $F(x) = A(x^2)$ is $\frac{d}{dx}(A(x^2)) = A'(x^2)(2x)$. By Theorem 13.12, $A'(x) = \sec x$; therefore, $A'(x^2)$ is equal to $\sec x^2$. The derivative of $F(x)$ is

$$F'(x) = \frac{d}{dx}(A(x^2)) = A'(x^2)(2x) = \sec(x^2)(2x).$$

(It may have occurred to you that we could solve the integral $\int_0^x \sec t\, dt$ first and then find the derivative. In this example, we cannot do that, because we do not know how to integrate the function $\sec t$. Our only option is to use Theorem 13.12 and the chain rule.)

(c) Notice that the integrand $\frac{x}{t^2+1}$ involves *both* the variable x and the "dummy" variable t. Since the integral is in terms of t (as indicated by the presence of a "dt"), the x can be factored out of the integral:

$$F(x) = \int_0^{3x} \frac{x}{t^2+1}\, dt = x\int_0^{3x} \frac{1}{t^2+1}\, dt.$$

Now it is clear that $F(x)$ is really the *product* of two functions, namely x and $\int_0^{3x} \frac{1}{t^2+1}\, dt$. We must use the product rule to differentiate $F(x)$. Notice also that the function $\int_0^{3x} \frac{1}{t^2+1}\, dt$ is the composition of the area accumulation function $A(x) = \int_0^x \frac{1}{t^2+1}$ with the function $3x$, so we will also need the chain rule:

$$F'(x) = \frac{d}{dx}\left(\int_0^{3x} \frac{x}{t^2+1}\, dt\right)$$
$$= \frac{d}{dx}\left(x\int_0^{3x} \frac{1}{t^2+1}\, dt\right) \qquad \text{(factor out the } x\text{)}$$
$$= (1)\int_0^{3x} \frac{1}{t^2+1}\, dt + (x)\frac{d}{dx}\left(\int_0^{3x} \frac{1}{t^2+1}\, dt\right) \qquad \text{(product rule)}$$
$$= \int_0^{3x} \frac{1}{t^2+1}\, dt + x\left(\frac{1}{9x^2+1}\right)(3). \qquad \text{(chain rule and Theorem 13.12)}$$

$\square$

The applications of the chain rule in Example 13.16 hint at the following formula for finding the derivative of an area accumulation function whose upper limit of integration is itself a function:

THEOREM 13.13

Differentiating a Composition That Involves an Area Accumulation Function

If f is continuous on $[a, b]$ and $u(x)$ is a differentiable function, then for all $x \in [a, b]$ we have

$$\frac{d}{dx}\left(\int_a^{u(x)} f(t)\, dt\right) = f(u(x))u'(x).$$

The right-hand side of the equation in Theorem 13.13 looks like the chain rule, with the important exception that it begins with $f(u(x))$ rather than $f'(u(x))$. In fact, it *is* the chain rule, and $f(x)$ is the derivative of the area accumulation function $F(x) = \int_a^x f(t)\, dt$. The proof of Theorem 13.13 simply involves recognizing $\int_a^{u(x)} f(t)\, dt$ as a composition and then applying the chain rule (much as we did in Example 13.16).

PROOF (**THEOREM 13.13**)   If $F(x) = \int_a^x f(t)\, dt$, then $\int_a^{u(x)} f(t)\, dt$ is the composition $F(u(x))$. Thus, by the chain rule, we have

$$\frac{d}{dx}\left(\int_a^{u(x)} f(t)\, dt\right) = \frac{d}{dx}(F(u(x))) \qquad \text{(write as a composition)}$$

$$= F'(u(x))u'(x) \qquad \text{(chain rule)}$$

$$= f(u(x))u'(x). \qquad \left(F'(x) = \frac{d}{dx}\left(\int_a^x f(t)\, dt\right) = f(x)\right)$$

13.3.4 The Mean Value Theorem for Integrals

If f is a continuous, differentiable function on an interval $[a, b]$, the Mean Value Theorem (Section 3.6.3) tells us that there is some point $c \in (a, b)$ where the instantaneous rate of change of f is equal to the average rate of change of f. By combining the Mean Value Theorem with the Second Fundamental Theorem, we get the *Mean Value Theorem for Integrals:*

THEOREM 13.14

The Mean Value Theorem for Integrals

If $f(x)$ is continuous on a closed interval $[a, b]$, then there exists some $c \in (a, b)$ such that

$$\int_a^b f(x)\, dx = f(c)(b - a).$$

Graphically, this theorem means that the area under the graph of a continuous function f on an interval $[a, b]$ is equal to the area of a rectangle of height $f(c)$ and width $b - a$ for some point $c \in (a, b)$. Consider the area under the graph of f as the side view of a "wave" of water sloshing in a tank; when the water settles (so that its surface is horizontal), this side view should have the same area (see Section 12.4.3). The Mean

Value Theorem for Integrals tells us that the height of the settled water is equal to $f(c)$ for some $c \in (a, b)$; see Figures 13.16 and 13.17.

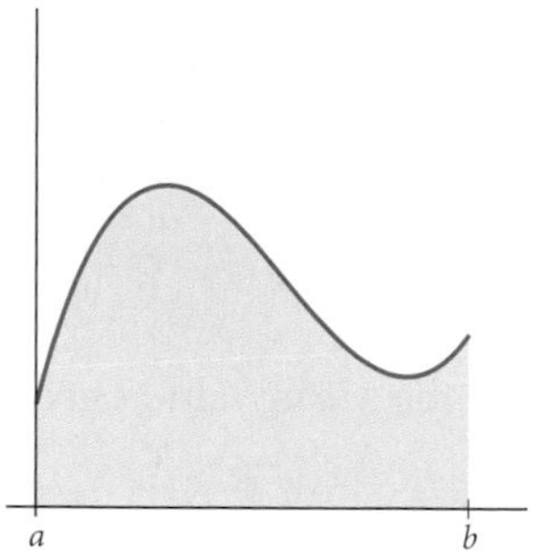

Figure 13.16

The area $\displaystyle\int_a^b f(x)\,dx$

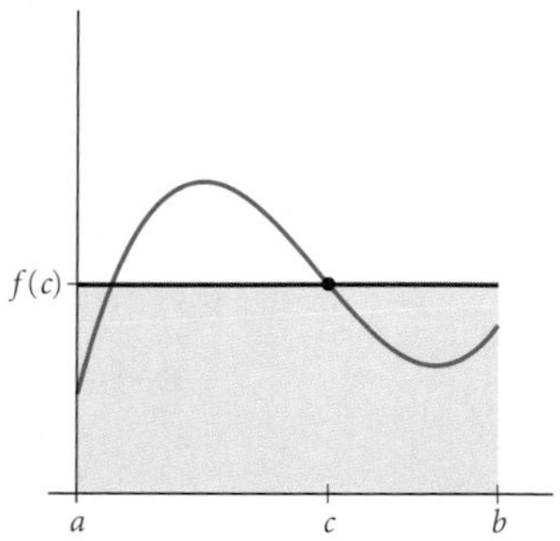

Figure 13.17

The area $f(c)(b - a)$

The key point of Theorem 13.14 is that the height of the "settled water" is actually the height of the function f at some point c in the interval $[a, b]$.

❓ Question   In fact, in Figure 13.17 there are *two* values $c \in (a, b)$ with the property that $f(c)(a - b)$ is equal to the area under the graph of f on $[a, b]$. Can you find the second value of c in Figure 13.17? □

◈ Caution   Theorem 13.14 is another example of an "existence" theorem, like the Mean Value Theorem and the Intermediate Value Theorem; it tells you that at least one point $c \in (a, b)$ must exist that has the desired property, but it doesn't tell you how to *find* such a c. □

PROOF (THEOREM 13.14)   Define $F(x) := \int_a^x f(t)\,dt$ to be the function describing the area under the graph of f from $t = a$ to $t = x$. To prove Theorem 13.14 we will apply the Mean Value Theorem to the *area* function of f. By the Second Fundamental Theorem, we know that $F'(x) = f(x)$. Moreover, the Second Fundamental Theorem also tells us that $F(x)$ is continuous on $[a, b]$ and differentiable on (a, b), so the Mean Value Theorem applies to this function $F(x)$ and says that

(2)   *There exists a number $c \in (a, b)$ such that $F'(c) = \frac{F(b) - F(a)}{b - a}$.*

Since $F'(x) = f(x)$, this means that there is some $c \in (a, b)$ for which

$$f(c) = \frac{F(b) - F(a)}{b - a} \qquad \text{(by Mean Value Theorem, } F'(c) = f(c)\text{)}$$

$$= \frac{\int_a^b f(t)\,dt - \int_a^a f(t)\,dt}{b - a} \qquad \text{(definition of } F(x) \text{ as area function of } f\text{)}$$

$$= \frac{\int_a^b f(t)\,dt}{b - a}. \qquad \text{(since } \int_a^a f(t)\,dt = 0\text{)}$$

Multiplying both sides by $b - a$ tells us that there is some $c \in (a, b)$ for which

$$f(c)(b - a) = \int_a^b f(t)\,dt.$$

The variable t is a "dummy variable" here and can be replaced by x without confusion (since x is not used for anything else in the equation). Thus we are done. ■

You may have already noticed that the Mean Value Theorem for Integrals seems to have something to do with the average value of a function; we had the same "water wave" discussion when we defined average values in Section 12.4. In fact, the conclusion of the Mean Value Theorem for Integrals can be rewritten to say that there is some $c \in (a, b)$ for which $f(c)$ is equal to the average value of f on $[a, b]$:

$$f(c) = \frac{1}{b - a} \int_a^b f(x)\, dx.$$

By this discussion, the Mean Value Theorem for Integrals is equivalent to the following theorem.

THEOREM 13.15

Average Value Interpretation of the Mean Value Theorem for Integrals
If $f(x)$ is continuous on a closed interval $[a, b]$, then there exists some $c \in (a, b)$ such that $f(c)$ is the average value of $f(x)$ on $[a, b]$.

The key point here is that the average value of a function f on an interval $[a, b]$ is actually equal to the height of the graph of f at some point $c \in (a, b)$. Stated this way, it sounds really obvious; certainly the average value of f on $[a, b]$ should be between the highest and lowest values of f on that interval, and by the Intermediate Value Theorem, the function f must actually be equal to that average value at some point $c \in (a, b)$. This is in fact an alternative way to prove the Mean Value Theorem for Integrals, which you will do in the exercises.

EXAMPLE 13.17   **An illustration of the Mean Value Theorem for Integrals**

Consider the function $f(x) = x^2$ on the interval $[1, 3]$. Since this function is continuous, we can apply the Mean Value Theorem for Integrals, which says that there is some point $c \in (1, 3)$ where $f(c)$ is equal to the average value of f on $[1, 3]$. In other words, there is some $c \in (1, 3)$ for which $f(c)(3 - 1) = \int_1^3 x^2\, dx$, or, equivalently, for which

$$f(c) = \frac{1}{3 - 1} \int_1^3 x^2\, dx.$$

Of course the Mean Value Theorem for Integrals does not tell us how to find such a point $c \in (1, 3)$; it tells us only that one must exist. However, we can easily find such a point c, since we can integrate to find the average value of f on $[a, b]$:

$$\frac{1}{3 - 1} \int_1^3 x^2\, dx = \tfrac{1}{2} \left[\tfrac{1}{3} x^3\right]_1^3 = \tfrac{1}{2}\left(\tfrac{1}{3}(3^3) - \tfrac{1}{3}(1)^3\right) = \tfrac{13}{3}.$$

Now it is easy to find a $c \in (1, 3)$ for which $f(c) = c^2$ equals this average value:

$$c^2 = \tfrac{13}{3} \implies c = \pm\sqrt{\tfrac{13}{3}}.$$

Note that $-\sqrt{\tfrac{13}{3}}$ is *not* in the interval $(1, 3)$, so the only value $c \in (1, 3)$ for which $f(c)$ is the average value $\tfrac{13}{3}$ is $c = \sqrt{\tfrac{13}{3}}$. This value of c and the average value $13/3$ are shown with the graph of $f(x) = x^2$ in Figure 13.18.

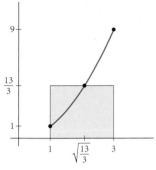

Figure 13.18

$f(\sqrt{\frac{13}{3}})$ is equal to the average value $\frac{13}{3}$.   □

What you should know after reading Section 13.3

► How to define an "area" function in terms of a definite integral, and how to interpret such functions graphically.

► The statement and proof of the Second Fundamental Theorem of Calculus, and why it is important.

► How to use the Second Fundamental Theorem of Calculus to differentiate functions that are defined in terms of definite integrals.

► The statement and proof of the Mean Value Theorem for Integrals, and how to interpret this theorem graphically. What does this theorem have to do with the average value of a function on an interval?

Exercises 13.3

Concepts

0. Read the section and make your own summary of the material. Include whatever you think will help you review the material later on.

1. Use a definite integral to express the function whose output at any real number x is the (signed) area between the graph of $f(t) = t^2$ and the t-axis on the interval $[2, x]$.

2. Consider the function $A(x) = \int_a^x f(t)\, dt$. What is the independent variable of this function? What does the dependent variable ("output") represent? Explain why we say that t is a "dummy" variable.

3. Suppose $A(x) = \int_0^x f(t)\, dt$, where $f(x)$ is positive and decreasing for $x > 0$. Is $A(x)$ increasing or decreasing, and why?

4. Suppose $A(x) = \int_0^x f(t)\, dt$, where $f(x)$ is positive and decreasing for $x > 0$. **(a)** Explain graphically why the graph of $A(x)$ should be concave down. **(b)** Then find

$A''(x)$ and argue that if $f(x)$ is decreasing, then $A(x)$ must be concave down.

5. State the Second Fundamental Theorem of Calculus. Why is this theorem important?

6. Explain how we get the inequality $f(m_h) \cdot h \le \int_x^{x+h} f(t)\, dt \le f(M_h) \cdot h$ in the proof of the Second Fundamental Theorem of Calculus. Make sure you clearly define m_h and M_h; it may help to draw some pictures.

7. The functions $A(x) = \int_0^x t^2\, dt$ and $B(x) = \int_3^x t^2\, dt$ differ by a constant. Explain why this is so by **(a)** comparing the derivatives of $A(x)$ and $B(x)$ and **(b)** showing that $A(x) - B(x)$ is a constant. Then **(c)** explain what this constant means graphically.

8. Let f be the function shown in the graph on the next page, and define $A(x) := \int_0^x f(t)\, dt$. List the following quantities in order from smallest to largest: **(a)** $A(1)$; **(b)** $A(3)$; **(c)** $A(6)$; **(d)** $A(7)$.

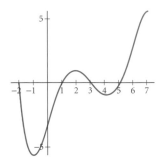

9. Let f be the function graphed in Problem 8, and define $A(x) := \int_0^x f(t)\, dt$. List the following quantities in order from smallest to largest: **(a)** $A(0)$; **(b)** $A(-1)$; **(c)** $A(-2)$; **(d)** $A(5)$.

10. Let f be the function shown below, and define $A(x) := \int_0^x f(t)\, dt$. On which interval(s) is $A(x)$ positive? Negative? Increasing? Decreasing?

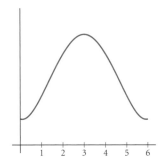

11. Let f be the function graphed in Problem 10, and define $A(x) := \int_0^x f(t)\, dt$. Sketch a rough graph of $A(x)$.

12. Let f be the function shown below, and define $A(x) := \int_0^x f(t)\, dt$. On which interval(s) is $A(x)$ positive? Negative? Increasing? Decreasing?

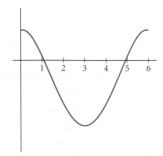

13. Let f be the function graphed in Problem 12, and define $A(x) := \int_0^x f(t)\, dt$. Sketch a rough graph of $A(x)$.

14. Is an indefinite integral the "opposite" of a derivative? Fill in the blanks below to answer this question.

(a) $\displaystyle \int f'(x)\, dx = $ _____

(b) $\displaystyle \frac{d}{dx} \int f(x)\, dx = $ _____

15. Is a definite integral the "opposite" of a derivative? Fill in the blanks below to answer this question.

(a) $\displaystyle \int_a^b f'(x)\, dx = $ _____

(b) $\displaystyle \frac{d}{dx} \int_a^b f(x)\, dx = $ _____

16. Is an area accumulation function the "opposite" of a derivative? Fill in the blanks below to answer this question.

(a) $\displaystyle \int_a^x f'(t)\, dt = $ _____

(b) $\displaystyle \frac{d}{dx} \int_a^x f(t)\, dt = $ _____

17. Write the function $F(x) = \int_3^{x^2} \sin t\, dt$ as a composition of two functions. What is the "inside" function? What is the "outside" function?

18. Explain why the function $f(x) = \int_1^x \frac{1}{t}\, dt$ has derivative $f'(x) = \frac{1}{x}$ and value $f(1) = 0$. A function that we use all the time has this same derivative and value. What is it?

19. Given that $\ln x = \int_1^x \frac{1}{t}\, dt$ (see Problem 18), express the area between the graph of $y = \frac{1}{x}$ and the x-axis from $x = e$ to $x = 10$ in terms of logarithms.

20. State the Mean Value Theorem for Integrals, and explain what this theorem means. Include a picture with your explanation. What does the Mean Value Theorem for Integrals have to do with average values?

21. Explain why the Mean Value Theorem for Integrals applies to the function $f(x) = x^3$ on the interval $[0, 2]$. Next state the conclusion of the Mean Value Theorem for Integrals in this particular case, and sketch a graph illustrating this conclusion. Then find all the values $c \in [0, 2]$ for which $f(c)$ is equal to the average value of $f(x)$ on $[0, 2]$, and indicate these values on your graph.

22. Explain why the Mean Value Theorem for Integrals applies to the function $f(x) = x(x - 6)$ on the interval $[1, 5]$. Next state the conclusion of the Mean Value Theorem for Integrals in this particular case, and sketch a graph illustrating this conclusion. Then find all the values $c \in [1, 5]$ for which $f(c)$ is equal to the average value of $f(x)$ on $[1, 5]$, and indicate these values on your graph.

■ **Determine whether each of the following statements is true or false. If a statement is true, explain why. If a statement is false, explain why or provide a counterexample.**

23. True or False: $\dfrac{d}{dv} \int_2^u f(v)\, du = f(u)$

24. True or False: $\dfrac{d}{dx} \int_a^{x^2} \sin t\, dt = \sin x^2$

25. True or False: If f is continuous on $[a, b]$, then there is exactly one $c \in [a, b]$ with $f(c) = \frac{1}{b-a} \int_a^b f(x)\, dx$.

26. True or False: The function $A(x) = \int_0^x (3 - t)\, dt$ is positive and increasing on $[0, 3]$.

27. True or False: The function $A(x) = \int_0^x (3 - t)\, dt$ is concave up.

28. True or False: Every continuous function has an antiderivative.

Skills

■ For each area accumulation function $A(x)$ below, (a) illustrate $A(2)$ graphically, (b) calculate $A(2)$ and $A(5)$, and (c) find an explicit "elementary" formula for $A(x)$.

29. $A(x) = \int_1^x (t^2 + 1)\, dt$ 30. $A(x) = \int_{-\pi}^x \sin t\, dt$

31. $A(x) = \int_3^x \dfrac{1}{t}\, dt$

■ Use the Second Fundamental Theorem of Calculus to write down three antiderivatives of each function $f(x)$ below.

32. $f(x) = \sqrt{1 + e^{4x}}$ 33. $f(x) = \sin^2(3^x)$

34. $f(x) = \dfrac{\ln(\sin x)}{x^2 - 1}$

■ Find the derivative, with respect to x, of each function below.

35. $\displaystyle\int_2^x \dfrac{\sin t}{t}\, dt$ 36. $\displaystyle\int_x^4 e^{t^2+1}\, dt$

37. $\displaystyle\int_0^{x^2} \cos t\, dt$ 38. $\displaystyle\int_1^{\sqrt{x}} x \ln t\, dt$

39. $\left(\displaystyle\int_{\sin x}^{\pi} \dfrac{t}{\cos t}\, dt \right)^3$ 40. $\displaystyle\int_x^{x+2} \sin(t^2)\, dt$

■ Express each of the following as an "elementary" function (i.e., without an integral). Pay close attention to the variables involved.

41. $\dfrac{d}{dx} \displaystyle\int_1^x \ln t\, dt$ 42. $\dfrac{d}{dx} \displaystyle\int_1^8 \ln t\, dt$

43. $\dfrac{d}{dx} \left(x^2 \displaystyle\int_0^{\sin x} \sqrt{t}\, dt \right)$ 44. $\dfrac{d^2}{dx^2} \displaystyle\int_2^{3x} (t^2 + 1)\, dt$

45. $\dfrac{d^2}{dx^2} \displaystyle\int_{x^2}^1 \ln |t|\, dt$ 46. $\dfrac{d^2}{dx^2} \displaystyle\int_1^{e^x} f(t)g(t)\, dt$

■ For each problem below, find a function $f(x)$ that has the given derivative $f'(x)$ and value $f(c)$. Find an antiderivative of $f'(x)$ by hand, if possible. If it is not possible to antidifferentiate by hand, use the Second Fundamental Theorem of Calculus to write down an antiderivative.

47. $f'(x) = \dfrac{1}{2x - 1}, \quad f(1) = 3$

48. $f'(x) = \dfrac{1}{x^2 + 1}, \quad f(-1) = 0$

49. $f'(x) = \dfrac{1}{x^3 + 1}, \quad f(2) = 0$

50. $f'(x) = e^{-x^2}, \quad f(1) = 0$

51. $f'(x) = 2 \sin(\pi x), \quad f(2) = 4$

52. $f'(x) = \sin(x^2), \quad f(0) = 0$

■ For each function $f(x)$ and interval $[a, b]$ given below, find a real number $c \in (a, b)$ such that $f(c)$ is the average value of $f(x)$ on $[a, b]$.

53. $f(x) = 2x^2 + 1, \quad [a, b] = [-1, 2]$

54. $f(x) = \sin x, \quad [a, b] = [0, \pi]$

55. $f(x) = \dfrac{1}{3x + 1}, \quad [a, b] = [2, 6]$

56. $f(x) = 4e^{3x}, \quad [a, b] = [0, \tfrac{1}{3} \ln 2]$

57. $f(x) = 100(1.12)^x, \quad [a, b] = [0, 10]$

58. $f(x) = \dfrac{1}{\sqrt{1 - 4x^2}}, \quad [a, b] = [-\tfrac{1}{2}, \tfrac{1}{2}]$

Applications

59. Suppose $f(t) = 0.36t^2$ describes the height, in centimeters, of a growing plant t days after it breaks through the soil.

   (a) What does the Mean Value Theorem (see Section 3.6.3) say about the height of the plant during the first four days of its growth? Be as specific as possible.

   (b) What does the Mean Value Theorem for Integrals (from this section) say about the height of the plant during the first four days of its growth? Be as specific as possible.

60. Just as you are driving past the clock tower on High Street, you notice a stop sign 50 feet ahead of you. You slam on your brakes and end up coming to a full stop exactly at the stop sign. Your distance from the stop sign (in feet) t seconds after stepping on the brakes is given by the function $s(t) = 50 - 9t^2$.

   (a) What does the Mean Value Theorem for Integrals say about your distance from the stop sign during the time that you are applying the brakes? Be as specific as possible.

   (b) What does the Mean Value Theorem for Integrals say about your velocity during the time that you are applying the brakes? Be as specific as possible.

Proofs

61. Prove that if $f(x)$ has an antiderivative (say $G(x)$), then $A(x) := \int_0^x f(t)\, dt$ must also be an antiderivative of $f(x)$. (*Hint:* Use the (first) Fundamental Theorem of Calculus.)

62. Use the definition of derivative to show that if f is a continuous function and $F(x) := \int_a^x f(t)\, dt$, then $F'(x) = f(x)$. (*Note:* The proof in this section assumed

$h \to 0^+$ in the definition of derivative. Write this proof in your own words, and then write a proof of what happens as $h \to 0^-$.)

63. Prove that if f is continuous on $[a, b]$ and we define $F(x) := \int_a^x f(t)\, dt$, then $F(x)$ is continuous on the closed interval $[a, b]$. (*Hint:* F is continuous on the interior of the interval because it is differentiable there; why? It then suffices to prove that $F(x)$ is left-continuous at a and right-continuous at b.)

64. Let f be any continuous function. Show that for all real numbers a and b, the functions $A(x) = \int_a^x f(t)\, dt$ and $B(x) := \int_b^x f(t)\, dt$ differ by a constant. Interpret this constant graphically.

65. Prove Theorem 13.13: If f is continuous on $[a, b]$ and $u(x)$ is a differentiable function, then for all $x \in [a, b]$ we have $\int_a^{u(x)} f(t)\, dt = f(u(x))u'(x)$.

66. Prove the Mean Value Theorem for Integrals by applying the Mean Value Theorem to the function $F(x) = \int_a^x f(t)\, dt$.

67. Prove the Mean Value Theorem for Integrals in its alternative form (Theorem 13.15), by following these steps:

   (a) Use the Extreme Value Theorem to argue that f has a maximum value M and a minimum value m on the interval $[a, b]$.

   (b) Use a Maximum Sum and a Minimum Sum with one rectangle to argue that $m(b - a) \le \int_a^b f(x)\, dx \le M(b - a)$, and thus that the average value of f on $[a, b]$ is between m and M.

   (c) Use the Intermediate Value Theorem to argue that there is some point $c \in (a, b)$ for which $f(c)$ is equal to the average value of $f(x)$ on $[a, b]$.

CHAPTER 14

Basic Integration Techniques

*I*n this chapter we will learn how to solve many different kinds of integrals, using four different integration techniques. Integration by substitution provides a method for "undoing" the chain rule. Integration by parts is related to "undoing" the product rule. Integrals involving powers and products of trigonometric functions can be solved by making use of trigonometric identities so that integration by substitution or integration by parts will apply. Finally, the method of trigonometric substitution makes use of the derivatives of inverse trigonometric functions and gives us a way to solve still more integrals.

Hedge maze at Longleat by Adrian Fisher Mazes Ltd, www.mazemaker.com

CONTENTS

14.1   Integration by Substitution

In this section we learn the most basic integration technique: integration "by substitution." We have already seen simple substitution problems (although we did not call them that). For example, we already know how to solve integrals like

$$\int \frac{1}{3x+1}\,dx \quad \text{and} \quad \int 2x\cos x^2\,dx,$$

with intelligent "guessing and checking" (and a little bit of luck, by recognizing the integrands as the "results" of some chain rule calculations). In this section we will formalize the process of "undoing" the chain rule.

14.1.1   Change of variables

The chain rule tells us that the derivative of a composition $f(u(x))$ of functions is

$$\frac{d}{dx}(f(u(x))) = f'(u(x))u'(x).$$

This means that $f(u(x))$ is an antiderivative of $f'(u(x))u'(x)$, that is, that

(1)
$$\int f'(u(x))u'(x)\,dx = f(u(x)) + C.$$

As we saw in Section 13.1.4, if we notice that an integrand is in the form $f'(u(x))u'(x)$, then we can easily find its antiderivative by doing the chain rule "backwards."

EXAMPLE 14.1

Recognizing the chain rule pattern in an integral

If $f(x) = \sin x$ and $u(x) = x^2$, then $f'(x) = \cos x$, $f'(u(x)) = \cos(u(x)) = \cos x^2$, and $u'(x) = 2x$. In this context, Equation (1) says that

$$\int \cos x^2(2x)\,dx = \sin x^2 + C.$$

You can check the integration step by differentiating: $\frac{d}{dx}(\sin x^2) = \cos x^2(2x)$.   □

The trouble is, the pattern $f'(u(x))u'(x)$ is not always so easy to recognize, and the formula in Equation (1) is difficult to work with. Luckily, there is an algorithm for handling integrals of this form, called **integration by substitution.** The key to integration by substitution is a **change of variables,** where we change an integral in terms of a variable x into an integral in terms of a different variable u. To do this, we must pay close attention to the **differential** "dx" that appears in integrals. In particular, as we will see in a moment, we *cannot* just replace dx with du when we change variables from x to u.

For any differentiable function $f(x)$, we know that $\int f'(x)\,dx = f(x) + C$. This is true even if we decide to use a different letter to represent the independent variable (for example, the letter u). Therefore, if $f(u)$ is a function of u, we have

(2)
$$\int f'(u)\,du = f(u) + C.$$

For example, if $f(u) = \sin u$ then $\int \cos u\,du = \sin u + C$, since $\frac{d}{du}(\sin u) = \cos u$. The "$du$" in the integral tells us that we are integrating with respect to the variable u, just as the "du" in $\frac{d}{du}(\sin u)$ tells us that we are differentiating with respect to u, that is, that we are considering u to be the independent variable.

Now compare Equations (1) and (2). Our goal is to turn an integral in the form of Equation (1) into an *equivalent* integral in the form of Equation (2). If u is a function of x, then $f(u) = f(u(x))$, so the integrals in Equations (1) and (2) are equal:

$$\int f'(u(x))u'(x)\,dx = \int f'(u)\,du.$$

Since $f'(u(x)) = f'(u)$, this means that we should have $du = u'(x)\,dx$. Visually, we can think of this equality as the result of "multiplying" both sides of $\frac{du}{dx} = u'(x)$ by dx. What this means is that the **differentials** dx and du must act exactly the way the Leibnitz notation $\frac{du}{dx}$ would suggest. If u is a function of x, then the differentials du and dx are related as follows:

DEFINITION 14.1

The Differential of a Function

If $u(x)$ is a function of x, then the differential du is equal to

$$du = u'(x)\,dx.$$

Now we can easily change an integral in terms of a variable x into an integral in terms of another variable u. The following example repeats the calculation in Example 14.1 with this new notation.

EXAMPLE 14.2

Changing variables to get a simpler integral

Consider again the integral $\int \cos x^2(2x)\,dx$. If we set $u(x) = x^2$, so that $du = u'(x)\,dx = 2x\,dx$, then

$$\int \cos x^2(2x)\,dx = \int \cos x^2(2x\,dx) = \int \cos u\,du.$$

The second integral is obviously simpler than the first, and we quickly see that it is equal to $\sin u + C$; therefore

$$\int \cos x^2(2x)\,dx = \int \cos u\,du = \sin u + C = \sin x^2 + C.$$

Note that in the last step, we wrote our answer in terms of the variable x. Also note that our answer is the same as the answer in Example 14.1. □

Although we did extra work in Example 14.2 to get the same answer we found in Example 14.1, it will be worth it. The process we went through is the key to the method of integration by substitution, which is a much more reliable and useful technique than the pattern recognition we used previously.

14.1.2 The method of integration by substitution

In general, to solve an integral in terms of a variable x by "changing variables," we must first write the integrand, *including* the differential dx, in terms of the new variable u and the corresponding differential du. Then we can solve the new, simpler integral and change variables back from u to x. The following algorithm provides a systematic way to do this.

ALGORITHM 14.1

Integration by Substitution

1. Choose some part of the integrand to be $u(x)$; usually a good choice for $u(x)$ is an "inside" function of a composition, whose derivative $u'(x)$ appears as a multiplicative factor of the integrand.

2. Differentiate the equation for $u(x)$ to find du; remember that $\frac{du}{dx} = u'(x)$, so $du = u'(x)\,dx$.

3. Change variables; that is, use the equations for $u(x)$ and du to rewrite the integrand, *including* the dx, *entirely* in terms of u and du.

4. Integrate the new, hopefully simpler integral.

5. Change variables back to x by substituting the formula for $u(x)$ for each u in the answer.

Algorithm 14.1 works for any integral in the form of Equation (1). The following example uses the algorithm of integration by substitution to find the integral from Example 14.1 and Example 14.2. (We keep repeating the same example so you can see how the substitution algorithm is derived from the chain rule.)

EXAMPLE 14.3

Using integration by substitution

Use integration by substitution to find $\int \cos x^2 (2x)\,dx$.

Solution   Since x^2 is the "inside" of a composition, and its derivative $2x$ appears elsewhere in the integrand, we'll try setting $u = x^2$. Now we must differentiate $u(x)$ to find du:

$$u = x^2 \quad \Longrightarrow \quad \frac{du}{dx} = 2x \quad \Longrightarrow \quad du = 2x\,dx \ .$$

By replacing x^2 by u (so $\cos(x^2) = \cos u$) and $2x\,dx$ by du (using the boxed equations), we have

$$\int \cos x^2 (2x)\,dx = \int \cos u\,du \qquad \text{(since } u = x^2 \text{ and } du = 2x\,dx\text{)}$$

$$= \sin u + C \qquad \text{(integrate the new integral)}$$

$$= \sin x^2 + C. \qquad \text{(change variables back to } x\text{)} \qquad \square$$

⊛ *Caution*   In Example 14.3, the "new" integral $\int \cos u\,du$ was written *entirely* in terms of u and du. If there had been any x's left in the integral after we changed variables, it would have meant that we had made a bad choice for $u(x)$. We *cannot* integrate if the integrand involves both x and u. $\qquad \square$

The method of integration by substitution works even if the integrand is not *exactly* in the form $f'(u(x))u'(x)$. The following example shows that we can use integration by substitution even when a constant multiple is "missing" from the integrand.

EXAMPLE 14.4

Using integration by substitution when a constant multiple is "missing"

Use integration by substitution to find $\int x^2\, e^{x^3+1}\,dx$.

Solution   Since $x^3 + 1$ is the "inside" of a composition, and its derivative $3x^2$ is *almost* a part of the integrand (that is, without the constant multiple of 3), we'll try setting $u = x^3 + 1$ and see what happens. With this choice of u we have

$$u = x^3 + 1 \implies \frac{du}{dx} = 3x^2 \implies du = 3x^2\,dx.$$

We don't have a $3x^2\,dx$ in the integrand (we have only $x^2\,dx$), but since $du = 3x^2\,dx$, we know that

$$\tfrac{1}{3}\,du = x^2\,dx\ .$$

(We have now solved the equation $du = 3x^2\,dx$ for the quantity we wish to replace in our integral, $x^2\,dx$.) Using the boxed equations, we can rewrite the integral in terms of u and du. The resulting integral is easy to solve.

$$\int x^2\, e^{x^3+1}\,dx = \int e^{x^3+1}(x^2\,dx)$$

$$= \int e^u(\tfrac{1}{3}\,du) \qquad \text{(change variables)}$$

$$= \tfrac{1}{3}\int e^u\,du \qquad \text{(rewrite)}$$

$$= \tfrac{1}{3}e^u + C \qquad \text{(integrate)}$$

$$= \tfrac{1}{3}e^{x^3+1} + C. \qquad \text{(write in terms of x)}$$

Although it was not necessary, we added parentheses in the first step of the calculation to make sure the substitution of $\tfrac{1}{3}\,du$ for $x^2\,dx$ was clear.   □

✓ *Checking the Answer*   We can differentiate to check the answer. Since $\frac{d}{dx}(\tfrac{1}{3}e^{x^3+1}) = \tfrac{1}{3}e^{x^3+1}(3x^2) = x^2 e^{x^3+1}$, we know we have integrated correctly.   □

❓ *Question*   Why didn't we choose $u = x^3$ (instead of $u = x^3 + 1$) in Example 14.4? What would the "new" integral have been if we had? In general, we usually take u to be as much of the integrand as possible, since that produces the simplest possible integral after changing variables.   □

◈ *Caution*   In Example 14.4, we factored the constant $\tfrac{1}{3}$ out of the integral. Don't forget that you can do this only with *constant multiples*. In particular, you cannot bring anything involving the variable x or u to the outside of the integral.   □

◈ *Caution*   Be sure that you don't just replace the differential dx with du when changing variables; remember that $du = u'(x)\,dx$.   □

14.1.3 More examples of integration by substitution

The first step in integration is always to choose a method of attack. Some integrals can be simplified with an appropriate "u" substitution; others are integrals we know from memory. Still other integrals might be backwards quotient or product rule problems, or it might be easy, after using algebra, to rewrite the integrand. Once you decide to use integration by substitution, you might have to try a few different choices of u before you find one that works. Don't expect to know how to solve an integral the minute you see it (as you do for derivatives); you may have to try a few different methods before you find one that works. If you do a lot of practice problems, you will begin to recognize which methods are most likely to work for particular types of integrals.

Example 14.5 illustrates that integration by substitution is not a good method for every integral and provides some practice for choosing u. Remember that if you try to change variables and end up with an integral that involves *both* u and x, then you have made a bad choice for u, or perhaps you need to try another integration method. The key to integration by substitution is to find a change of variables that allows you to rewrite the integral as a *simpler* integral (ideally, one that you can integrate).

EXAMPLE 14.5   **Deciding when to use integration by substitution**

For each integral, determine whether integration by substitution is appropriate, and if so, determine the substitution $u(x)$.

(a) $\displaystyle\int x^4 \sqrt{x^5 + 1}\, dx$ (b) $\displaystyle\int x^3 \sqrt{x^5 + 1}\, dx$

(c) $\displaystyle\int \cos x \sin^2 x\, dx$ (d) $\displaystyle\int \tan x\, dx$

(e) $\displaystyle\int \frac{e^x}{1 + e^x}\, dx$ (f) $\displaystyle\int \frac{1 + e^x}{e^x}\, dx$

Solution

(a) Integration by substitution will work for this integral. If we choose $u = x^5 + 1$, then $\frac{du}{dx} = 5x^4$, so $x^4\, dx = \frac{1}{5} du$. We can rewrite the integral as $\frac{1}{5} \int \sqrt{u}\, du$, which we know how to integrate.

(b) Integration by substitution won't work here. The most reasonable choice for u would be $u = x^5 + 1$, which makes $\frac{du}{dx} = 5x^4$. We cannot write the integral entirely in terms of u and du with this choice of u. For this integral, no choice of u will work.

(c) If we choose $u = \sin x$, then $\frac{du}{dx} = \cos x$, so $du = \cos x\, dx$. This changes the integral into $\int u^2\, du$, which is easy to integrate. The choice $u = \cos x$ would *not* work in this example; why not?

(d) This looks too simple to be an integration-by-substitution problem, until we rewrite $\tan x$ as $\frac{\sin x}{\cos x}$. We can then choose $u = \cos x$ so that $\frac{du}{dx} = -\sin x$, or $-du = \sin x\, dx$. This changes the integral into

$$\int \tan x\, dx = \int \frac{\sin x}{\cos x}\, dx = -\int \frac{1}{u}\, du,$$

which we can easily integrate. In this example the choice of $u = \sin x$ would *not* work; why not?

(e) This may not look like an integration-by-substitution problem at first glance, but remember that the derivative of e^x is itself—and in fact the derivative of $1 + e^x$ is also e^x. Thus we can let u be the denominator $u = 1 + e^x$ and get $\frac{du}{dx} = e^x$, that is, $du = e^x\, dx$. This changes the integral into $\int \frac{1}{u}\, du$, which is easy to integrate.

(f) The same method won't work here; we cannot use $u = 1 + e^x$ and $du = e^x\, dx$ to change this integral into a new integral involving only u and du. The problem for this substitution is that the derivative e^x is in the *denominator*; it is not a *multiplicative* factor of the integrand. This integral can be solved by using algebra to write the integrand $\frac{1+e^x}{e^x}$ as $\frac{1}{e^x} + \frac{e^x}{e^x} = e^{-x} + 1$. Try it!   □

Sometimes it is not so easy to recognize when an integrand is of the form $f'(u(x))u'(x)$ for some $u(x)$. This is one reason why the method of integration by substitution is so useful. We can try various substitutions $u(x)$ to see whether they work, instead of having to recognize the entire pattern $f'(u(x))u'(x)$ all at once.

EXAMPLE 14.6

Choosing a substitution

Use integration by substitution to find:

(a) $\displaystyle\int \frac{\ln x}{x}\,dx$
(b) $\displaystyle\int \frac{\sin\sqrt{x}}{\sqrt{x}}\,dx$

Solution

(a) Clearly $u = x$ will not be a good substitution for this integral (in fact, this is *never* a useful substitution; why?), but perhaps $u = \ln x$ will work. With this choice of u, we have

$$u = \ln x \quad\Longrightarrow\quad \frac{du}{dx} = \frac{1}{x} \quad\Longrightarrow\quad du = \frac{1}{x}\,dx \ .$$

At first you might think that we do not have a $\frac{1}{x}$ in the original integrand; we have only an $\ln x$ and an x. But notice that the quotient $\frac{\ln x}{x}$ is actually equal to the *product* $(\ln x)(\frac{1}{x})$; we *do* have a $\frac{1}{x}$ in the integrand. Changing variables gives us

$$\int \frac{\ln x}{x}\,dx = \int (\ln x)\left(\frac{1}{x}\,dx\right) \qquad \text{(rewrite integrand)}$$

$$= \int u\,du \qquad \text{(using the boxed equations)}$$

$$= \tfrac{1}{2}u^2 + C \qquad \text{(integrate the "new" integral)}$$

$$= \tfrac{1}{2}(\ln x)^2 + C. \qquad \text{(since } u = \ln x\text{)}$$

As always, it is a good idea to check by differentiating. Since $\frac{d}{dx}(\frac{1}{2}(\ln x)^2) = \frac{1}{2}(2\ln x)(\frac{1}{x}) = \frac{\ln x}{x}$, we know we did the integration correctly.

(b) A sensible first choice for u would be $u = \sqrt{x}$, since this is the "inside" of a composition in the integrand. With this choice of u, we have

$$u = \sqrt{x} \quad\Longrightarrow\quad \frac{du}{dx} = \tfrac{1}{2}x^{-\frac{1}{2}} = \frac{1}{2\sqrt{x}} \quad\Longrightarrow\quad 2\,du = \frac{1}{\sqrt{x}}\,dx \ .$$

We solved for the expression $\frac{1}{\sqrt{x}}\,dx$ because that is the part of the integrand that we need to rewrite in terms of du. (The $\sin\sqrt{x}$ part of the integrand will become $\sin u$.) With this change of variables, we have

$$\int \frac{\sin\sqrt{x}}{\sqrt{x}}\,dx = \int (\sin\sqrt{x})\left(\frac{1}{\sqrt{x}}\,dx\right) \qquad \text{(rewrite integrand)}$$

$$= \int (\sin u)(2\,du) \qquad \text{(using boxed equations)}$$

$$= 2\int \sin u\,du \qquad \text{(simplify new integral)}$$

$$= 2(-\cos u) + C \qquad \text{(integrate the new integral)}$$

$$= -2\cos\sqrt{x} + C. \qquad \text{(since } u = \sqrt{x}\text{)}$$

Again, we can easily check this answer: $\frac{d}{dx}(-2\cos\sqrt{x}) = -2(-\sin\sqrt{x})(\frac{1}{2}x^{-\frac{1}{2}})$
$= \frac{\sin\sqrt{x}}{\sqrt{x}}$. $\qquad\qquad\qquad\qquad\qquad\qquad\qquad\qquad\qquad\qquad\qquad\qquad\square$

Sometimes integration by substitution works even when the integrand is not at all in the form $f'(u(x))u'(x)$. The following example involves such an integral, where a change of variables produces an integrand that can be algebraically simplified (unlike the original integrand). Because the original integrand is not in anything like the form $f'(u(x))u'(x)$, we will also have to make a "back-substitution."

EXAMPLE 14.7

Integration by substitution, with back-substitution

Use integration by substitution to find $\int x\sqrt{x-1}\,dx$.

Solution   The integrand certainly does not appear to be in the form $f'(u(x))u'(x)$, but to see what happens, let's choose $u = x - 1$ (this is at least the "inside" function of a composition). With this choice of u, we have

$$u = x - 1 \quad \implies \quad \frac{du}{dx} = 1 \quad \implies \quad du = dx \;.$$

We can replace the $\sqrt{x-1}$ part of the integrand with $\sqrt{u}$, and we can replace the dx part of the integrand with du. However, if we do this, there will be an "x" left over that we haven't written in terms of u or du. Here's where "back-substitution" comes in; we can use the equation $u = x - 1$ to solve for x:

$$x = u + 1 \;.$$

Now, with these *three* boxed equations (the last of which is a "back-substitution"), we have

$$\int x\sqrt{x-1}\,dx = \int (u+1)\sqrt{u}\,du \qquad \text{(using the boxed equations)}$$

$$= \int \left(u^{\frac{3}{2}} + u^{\frac{1}{2}}\right)du \qquad \text{(algebra; note that } u\sqrt{u} = u^{\frac{3}{2}})$$

$$= \int u^{\frac{3}{2}}\,du + \int u^{\frac{1}{2}}\,du \qquad \text{(sum rule for integrals)}$$

$$= \tfrac{2}{5}u^{\frac{5}{2}} + \tfrac{2}{3}u^{\frac{3}{2}} + C \qquad \text{(integrate the "new" integrals)}$$

$$= \tfrac{2}{5}(x-1)^{\frac{5}{2}} + \tfrac{2}{3}(x-1)^{\frac{3}{2}} + C. \qquad \text{(since } u = x - 1)$$

Why was the "new" integral easier to solve than the original integral? The original integrand could not be algebraically expanded, but after the change of variables, we had an integral that we could multiply out, since the sum was then *outside*, rather than *inside*, the square root.   □

❓ Question   Check the answer found in Example 14.7. You'll need to do a bit of algebra after differentiating to show that $\frac{d}{dx}(\tfrac{2}{5}(x-1)^{\frac{5}{2}} + \tfrac{2}{3}(x-1)^{\frac{3}{2}}) = x\sqrt{x-1}$.   □

14.1.4   Finding definite integrals using substitution

There are two methods for finding a definite integral $\int_a^b f(x)\,dx$ using integration by substitution. Perhaps the most straightforward way is to use integration by substitution to find $\int f(x)\,dx$ in terms of x and then evaluate this antiderivative from $x = a$ to $x = b$. In other words, we can use the fact that, by the Fundamental Theorem of Calculus,

$$(3) \qquad \int_a^b f(x)\,dx = \left[\int f(x)\,dx\right]_a^b.$$

The following example illustrates how to use Equation (3) when applying integration by substitution; notice that we pay close attention to the limits of integration.

EXAMPLE 14.8

Using integration by substitution to solve a definite integral

Use integration by substitution to find $\int_1^4 \frac{x}{3x^2+1}\,dx$.

Solution   We can find the integral of $\frac{x}{3x^2+1}$ using integration by substitution with $u = 3x^2 + 1$. With this choice of u, we have

$$u = 3x^2 + 1 \implies \frac{du}{dx} = 6x \implies \tfrac{1}{6}\,du = x\,dx\ .$$

Now that we are about to change variables, we also have to be more precise about the limits of integration; the 1 and 4 on the integral sign really mean $x = 1$ and $x = 4$, and our new integral will be in terms of u, not x. Therefore, we write "$x = 1$" and "$x = 4$" as the limits of integration on the "new" integral:

$$\int_1^4 \frac{x}{3x^2+1}\,dx = \int_1^4 \frac{1}{3x^2+1}(x\,dx) = \int_{x=1}^{x=4} \frac{1}{u}\left(\tfrac{1}{6}\,du\right).$$

(Note that this is *not* the same as writing just "1" and "4" as the limits of integration on the new integral; since the new integral is in terms of u and du, that would implicitly mean $u = 1$ and $u = 4$, which is not what we want.) With this change of variables, we have

$$\int_1^4 \frac{x}{3x^2+1}\,dx = \frac{1}{6}\int_{x=1}^{x=4} \frac{1}{u}\,du \qquad \text{(change variables)}$$

$$= \tfrac{1}{6}\left[\ln|u|\right]_{x=1}^{x=4} \qquad \text{(integrate the "new" integral)}$$

$$= \tfrac{1}{6}\left[\ln|3x^2+1|\right]_1^4 \qquad \text{(since } u = 3x^2 + 1\text{)}$$

$$= \tfrac{1}{6}\left(\ln|3(4)^2+1| - \ln|3(1)^2+1|\right) \qquad \text{(evaluate from } x=1 \text{ to } x=4\text{)}$$

$$= \tfrac{1}{6}(\ln(49) - \ln(4)) \qquad \text{(exact answer)}$$

$$\approx 0.417588. \qquad \text{(approximate answer)}$$

□

? Question   Show that $\frac{1}{6}\int_1^4 \frac{1}{u}\,du = \frac{\ln 4}{6} \approx 0.23104$. Why is this different from the value of the definite integral $\frac{1}{6}\int_{x=1}^{x=4} \frac{1}{u}\,du$ that we computed in Example 14.8?   □

We could use a slightly different method of handling the limits of integration from that shown in Example 14.8. In Example 14.8, we kept the limits of integration ($x = 1$ and $x = 4$) in terms of x and then had to write the antiderivative in terms of x before we evaluated from $x = 1$ to $x = 4$. Instead, we could change the limits of integration to be in terms of the new variable u. In symbols, we can use integration by substitution to write

$$\int_a^b f(x)\,dx = \int_{u(a)}^{u(b)} g(u)\,du = \left[\int g(u)\,du\right]_{u=a}^{u=b}$$

for some functions $u(x)$ and $g(u)$. Example 14.9 repeats the calculation done in Example 14.8 but uses this alternative method.

EXAMPLE 14.9

Changing the limits of integration when using substitution

Use integration by substitution to find $\int_1^4 \frac{x}{3x^2+1}\,dx$.

Solution   The choice of *u* will still be the same as in Example 14.8. Again we have

$$u = 3x^2 + 1 \implies \frac{du}{dx} = 6x \implies \tfrac{1}{6}\,du = x\,dx \ .$$

We will now write the limits of integration ($x = 1$ and $x = 4$) in terms of the new variable *u*. When $x = 1$ we have $u = 3(1)^2 + 1 = 4$, and when $x = 4$ we have $u = 3(4)^2 + 1 = 49$; in other words, we have

$$u(1) = 4 \quad \text{and} \quad u(4) = 49 \ .$$

Using the information in all four boxed equations, we can change variables completely:

$$\int_1^4 \frac{x}{3x^2 + 1}\,dx = \frac{1}{6}\int_4^{49} \frac{1}{u}\,du \qquad \text{(change variables completely)}$$

$$= \frac{1}{6}\left[\ln|u|\right]_4^{49} \qquad \text{(integrate the "new" integral)}$$

$$= \frac{1}{6}(\ln|49| - \ln|4|) \qquad \text{(evaluate; exact answer)}$$

$$\approx 0.417588. \qquad \text{(approximate answer)}$$

Note that, of course, we got the same answer as we found in Example 14.8. Also notice that we did not have to write the antiderivative $\ln|u|$ in terms of *x* before the evaluation step.   □

❓ Question   Compare the calculations in Examples 14.8 and 14.9. How are they the same? How are they different? What work did you have to do in Example 14.8 that you did not have to do in Example 14.9? Where was the equivalent work done in Example 14.9?   □

✔ Checking the Answer   You can check the integration steps in Examples 14.8 and 14.9 by differentiating. In Example 14.8 you would check that $\frac{d}{dx}(\frac{1}{6}\ln|3x^2 + 1|) = \frac{x}{3x^2+1}$, whereas in Example 14.9 you would check that $\frac{d}{du}(\ln|u|) = \frac{1}{u}$. You can check the numerical answer with a graphing calculator by approximating the area between the graph of $f(x) = \frac{x}{3x^2+1}$ and the *x*-axis from $x = 1$ to $x = 4$.   □

What you should know after reading Section 14.1

▶ How to write *du* in terms of *dx* if $u(x)$ is a function of *x*. How to change variables in an integral, and the role played by *du* in this process.

▶ The method of integration by substitution: what it is for, how it works, how to use it, when to use it, and how to make a good choice of substitution $u(x)$. Be able to use integration by substitution even if the integrand is not in the form $f'(u(x))u'(x)$ (when it applies), and be able to recognize when integration by substitution would *not* be a good method to apply.

▶ How to solve definite integrals using the Fundamental Theorem of Calculus and the method of integration by substitution (there are two ways to do this).

Exercises 14.1

Concepts

0. Read the section and make your own summary of the material. Include whatever you think will help you review the material later on.

1. Write down five integrals that can be solved using the method of integration by substitution. Explain what to use as $u(x)$ in each case, and why.

2. Write down five integrals that *cannot* be solved using the method of integration by substitution. Explain why the various possible choices for $u(x)$ in each case will not work.

3. Write down three integrals that you are sure we cannot do using the methods we now know. Choose integrands that are as simple as possible.

4. Explain why the integrals $\int \frac{2x}{x^2+1}\,dx$ and $\int \frac{1}{x\ln x}\,dx$ are essentially the "same" integral after a change of variables.

■ For each integral below, write down *three* integrals that will have that form after a substitution of variables.

5. $\int u^2\,du$ **6.** $\int \sin u\,du$

7. $\int e^u\,du$ **8.** $\int \frac{1}{\sqrt{u}}\,du$

■ For each function $u(x)$ below, write the differential du in terms of the differential dx.

9. $u(x) = x^2 + 1$ **10.** $u(x) = x^2 + x + 1$

11. $u(x) = \sin x$ **12.** $u(x) = \dfrac{1}{x}$

13. For every integration problem in Problems 31–48, determine whether integration by substitution would be a good method, and if so, try to determine the substitution $u(x)$. (Don't actually solve the integrals; just try to come up with a strategy for solving each one.)

14. List some things that would suggest that a choice of substitution $u(x)$ is a "good" choice. (In other words, what do you look for when choosing $u(x)$?)

15. Suppose $u(x) = x^2$. Calculate and compare the values of $\int_{-1}^{5} u^2\,du$, $\int_{x=-1}^{x=5} u^2\,du$, and $\int_{u(-1)}^{u(5)} u^2\,du$.

16. Solve the integral $\int \sin x \cos x\,dx$ in two ways (using two different changes of variables), and show that your answers differ by a constant.

■ Determine whether each of the following statements is true or false. If a statement is true, explain why. If a statement is false, explain why or provide a counterexample.

17. True or False: $\int g'(h(x))h'(x)\,dx = g(h(x)) + C$.

18. True or False: If $v = u^2 + 1$, then
$$\int \sqrt{u^2 + 1}\,du = \int \sqrt{v}\,dv.$$

19. True or False: If $u = x^3$, then
$$\int x\sin(x^3)\,dx = \frac{1}{3x}\int \sin u\,du.$$

20. True or False: If $u(x)$ is a function of x, then
$$\int_0^3 u^2\,du = \int_{x=0}^{x=3} u^2\,du.$$

21. True or False: $\displaystyle\int_0^1 x^2\,dx = \int_0^1 u^2\,du.$

22. True or False: $\displaystyle\int_2^4 xe^{x^2-1}\,dx = \frac{1}{2}\int_2^4 e^u\,du.$

23. True or False: $\displaystyle\int_2^3 f(u(x))u'(x)\,dx = \int_{u(2)}^{u(3)} f(u)\,du.$

24. True or False: $\displaystyle\int_0^6 f(u(x))u'(x)\,dx = \left[\int f(u)\,du\right]_0^6.$

Skills

■ Solve the following integrals in two ways: **(a)** by using the method of integration by substitution, and **(b)** without using substitution (for example, by using algebra instead). Then **(c)** show that your two answers are the same (up to a constant).

25. $\int (3x + 1)^2\,dx$ **26.** $\int x^2(x^3 + 1)\,dx$

27. $\int x(x^2 - 1)^2\,dx$ **28.** $\int \dfrac{x^{-3} - 4}{x^2}\,dx$

29. $\int \dfrac{\sqrt{x} + 3}{2\sqrt{x}}\,dx$ **30.** $\int \dfrac{x^{\frac{2}{3}} + 1}{\sqrt[3]{x}}\,dx$

■ Solve the following integrals, using whatever method you like. Check each of your answers by differentiating. (*Hints:* Some of these integrals can be done with integration by substitution. Some integrals will benefit from a little bit of algebra. Also, some integrals involve the "back-substitution" method used in Example 14.7.)

31. $\int 3\cos(\pi x)\,dx$ **32.** $\int x^3 \cos x^4\,dx$

33. $\int x \csc^2 x^2\,dx$ **34.** $\int \dfrac{1}{3x+1}\,dx$

35. $\int \dfrac{x}{\sqrt{3x^2+1}}\,dx$ **36.** $\int \dfrac{\sqrt{x}+1}{x}\,dx$

37. $\int \sin \pi x \cos \pi x \, dx$

38. $\int \sec 2x \tan 2x \, dx$

39. $\int \cot^5 x \csc^2 x \, dx$

40. $\int x^4 (x^3 + 1)^2 \, dx$

41. $\int x^{\frac{1}{4}} \sin x^{\frac{5}{4}} \, dx$

42. $\int \sin x \, e^{\cos x} \, dx$

43. $\int 2x e^{3x^2} \, dx$

44. $\int x \left(2^{x^2+1}\right) dx$

45. $\int \frac{\sqrt{\ln x}}{x} \, dx$

46. $\int \frac{\cos(\ln x)}{x} \, dx$

47. $\int x \ln \left(e^{x^2+1}\right) dx$

48. $\int \frac{\ln \sqrt{x}}{x} \, dx$

49. $\int \frac{e^x}{2 - e^x} \, dx$

50. $\int \frac{2 - e^x}{e^x} \, dx$

51. $\int \frac{e^{\sqrt{x}}}{\sqrt{x}} \, dx$

52. $\int \frac{(\cos x + 1)^{\frac{3}{2}}}{\csc x} \, dx$

53. $\int \frac{x \cos x^2}{\sqrt{\sin x^2}} \, dx$

54. $\int \frac{\sin\left(\frac{1}{x}\right)}{x^2} \, dx$

55. $\int x \sqrt{x^2 + 1} \, dx$

56. $\int (x^2 + 1)\sqrt{x} \, dx$

57. $\int x\sqrt{x + 1} \, dx$

58. $\int x^2 \sqrt{x + 1} \, dx$

59. $\int \frac{x}{\sqrt{3x + 1}} \, dx$

60. $\int \frac{x^2}{\sqrt{x + 1}} \, dx$

61. $\int \frac{e^x}{\csc e^x} \, dx$

62. $\int \frac{1}{x \ln x} \, dx$

63. $\int \frac{\sec^2 x}{\tan x + 1} \, dx$

64. $\int x(x + 1)^{\frac{1}{3}} \, dx$

65. $\int x^2 (x - 1)^{100} \, dx$

66. $\int \frac{x}{1 + x^4} \, dx$

67. $\int \frac{\sin^{-1} x}{\sqrt{1 - x^2}} \, dx$

68. $\int \frac{1}{\sqrt{x} \sec \sqrt{x}} \, dx$

69. $\int \cot x \ln(\sin x) \, dx$

■ Solve each of the following definite integrals in two ways: (a) by using the method from Example 14.8 (i.e., without changing the limits of integration), and (b) by using the method from Example 14.9 (i.e., by changing the limits of integration). Check your numerical answers by estimating an area with your graphing calculator.

70. $\int_{-2}^{2} e^{5-3x} \, dx$

71. $\int_{2}^{4} \frac{x}{x^2 - 1} \, dx$

72. $\int_{0}^{3} x(x^2 + 1)^{\frac{1}{3}} \, dx$

73. $\int_{-3}^{-1} \frac{1}{\sqrt{1 - 3x}} \, dx$

74. $\int_{0}^{\sqrt{4\pi}} x \sin x^2 \, dx$

75. $\int_{-\pi}^{\pi} \sin x \cos x \, dx$

76. $\int_{2}^{4} \frac{e^{2x}}{1 + e^{2x}} \, dx$

77. $\int_{1}^{2} \frac{1}{x\sqrt{\ln x}} \, dx$

78. $\int_{0}^{1} x\sqrt{1 - x} \, dx$

79. Consider the function $f(x) = \sin x \cos x$.
   (a) Find the signed area of the region between the graph of $f(x)$ and the x-axis from $x = 0$ to $x = \pi$.
   (b) Find the "true" area of the region between the graph of $f(x)$ and the x-axis from $x = 0$ to $x = \pi$.
   (c) Find the area of the region between the graphs of $f(x)$ and $g(x) = \sin x$ on the interval $[0, \pi]$.
   (d) Find the average value of the function $f(x)$ on the interval $[0, \pi]$.

80. Consider the function $f(x) = \dfrac{x}{x^2 + 1}$.
   (a) Find the signed area of the region between the graph of $f(x)$ and the x-axis from $x = -1$ to $x = 3$.
   (b) Find the "true" area of the region between the graph of $f(x)$ and the x-axis from $x = -1$ to $x = 3$.
   (c) Find the area of the region between the graphs of $f(x)$ and $g(x) = \frac{1}{2}x$ on the interval $[-1, 3]$.
   (d) Find the average value of the function $f(x)$ on the interval $[-1, 3]$.

Proof

81. Use the chain rule and the Fundamental Theorem to prove that $\int_{a}^{b} f'(u(x))u'(x) \, dx = f(u(b)) - f(u(a))$.

14.2   Integration by Parts

Although we can solve many integrals using substitution, there are still many integrals that we don't know how to solve. For example, we still don't know how to integrate

$$\int \ln x \, dx \quad \text{or} \quad \int x \sin x \, dx.$$

Both of these integrals, and many more, can be solved using ***integration by parts.***

14.2.1 The integration-by-parts formula

Recall that the product rule says that the derivative of a product $u(x)v(x)$ of two functions $u(x)$ and $v(x)$ is

$$\frac{d}{dx}(u(x)v(x)) = u'(x)v(x) + u(x)v'(x).$$

Turning this around, this means that $u(x)v(x)$ is an antiderivative of $u'(x)v(x) + u(x)v'(x)$. In other words,

(1)
$$\int (u'(x)v(x) + u(x)v'(x))\,dx = u(x)v(x) + C.$$

In Section 13.1.4 we solved such integrals by recognizing the integrand as the "result" of a product rule calculation. For example, with $u(x) = x^2$ and $v(x) = \sin x$, we have

$$\int (2x\sin x + x^2 \cos x)\,dx = x^2 \sin x + C.$$

Unfortunately, it is not very common to run across integrands of the form $u'(x)v(x) + u(x)v'(x)$ shown in Equation (1). However, we can use Equation (1) to construct a very useful integration formula:

THEOREM 14.1

An Integration Formula Derived from the Product Rule

If $u(x)$ and $v(x)$ are differentiable functions, then

$$\int u(x)v'(x)\,dx = u(x)v(x) - \int v(x)u'(x)\,dx.$$

If the integral $\int v(x)u'(x)\,dx$ happens to be simpler than the integral $\int u(x)v'(x)\,dx$, then the formula in Theorem 14.1 can be very useful. Before proving Theorem 14.1 or discussing the traditional integration-by-parts notation, we'll do a simple example.

EXAMPLE 14.10

Using the formula in Theorem 14.1

Use Theorem 14.1 to find $\displaystyle\int x\sin x\,dx$.

Solution   There are a number of ways that we could write the integrand $x\sin x$ in the form $u(x)v'(x)$. For example, we could let $u(x) = x$ and $v'(x) = \sin x$ (this turns out to be the most useful choice). In order to use Theorem 14.1, we will need to know what $u'(x)$ and $v(x)$ are (since those functions appear in the right-hand side of the formula). Since $u(x) = x$, we must have $u'(x) = 1$. Since $v'(x) = \sin x$, $v(x)$ must be some function whose derivative is $\sin x$; the simplest choice is $v(x) = -\cos x$. Using Theorem 14.1, we have

$$\int x\sin x\,dx = (x)(-\cos x) - \int (-\cos x)(1)\,dx \qquad \text{(Theorem 14.1)}$$

$$= -x\cos x + \int \cos x\,dx. \qquad \text{(simplify)}$$

We have now managed to write the difficult integral $\int x\sin x\,dx$ in terms of the function $-x\cos x$ (which does not have to be integrated) and the *much* simpler integral

$\int \cos x \, dx$. In fact, we know that $\int \cos x \, dx = \sin x$ and therefore

$$\int x \sin x \, dx = -x \cos x + \int \cos x \, dx = -x \cos x + \sin x + C.$$

Of course, you can check the answer by differentiating:

$$\frac{d}{dx}(-x \cos x + \sin x) = (-1) \cos x - x(-\sin x) + \cos x = x \sin x. \qquad \square$$

PROOF (THEOREM 14.1)   The key to the proof will be the product rule. Suppose $u(x)$ and $v(x)$ are differentiable functions. Then by the product rule (written as an integral; see Equation (1)), we have

$$\int (u'(x)v(x) + u(x)v'(x)) \, dx = u(x)v(x) + C.$$

By the sum rule for integrals, this means that

$$\int u'(x)v(x) \, dx + \int u(x)v'(x) \, dx = u(x)v(x) + C.$$

Solving this equation for $\int u(x)v'(x) \, dx$, we have

$$\int u(x)v'(x) \, dx = u(x)v(x) - \int u'(x)v(x) \, dx.$$

(Of course, $u'(x)v(x)$ is equal to $v(x)u'(x)$, which is what is used in Theorem 14.1.) Note that in this last step we dropped the "$+ C$," since *both* sides of the equation involve indefinite integrals, which are by definition *families* of antiderivatives. ∎

We now develop a more compact notation for Theorem 14.1. Recall that since $u(x)$ and $v(x)$ are functions of x, the differentials of $u(x)$ and $v(x)$ are

$$du = u'(x) \, dx \quad \text{and} \quad dv = v'(x) \, dx.$$

Using this notation, and writing $u(x)$ and $v(x)$ as simply u and v, respectively, we get the traditional integration-by-parts formula.

THEOREM 14.2

The Integration-by-Parts Formula
If $u = u(x)$ and $v = v(x)$ are differentiable functions, then

$$\int u \, dv = uv - \int v \, du.$$

Using this formula is called integration "by parts" because we are splitting the integrand into two parts, u and dv. Then we differentiate u and integrate dv, thereby "switching" the parts and relating the original integral to $\int v \, du$, which we hope is a simpler integral.

◈ *Caution*   Be sure you understand that the integration-by-parts formula in Theorem 14.2 is just the formula in Theorem 14.1 written in a more compact notation. In particular, when using the integration-by-parts formula, you must keep in mind that dv represents $v'(x) \, dx$ and that du represents $u'(x) \, dx$. $\qquad \square$

14.2.2 Using integration by parts

Deciding when an integral can be done by parts, and then deciding which part of the integrand should be "u" and which part should be "dv," can sometimes be difficult. We will address these decisions in Section 14.2.3. First, we present the method of integration by parts as an algorithm.

ALGORITHM 14.2

Using Integration by Parts

1. Try to do the integral by substitution or algebraic simplification if possible; these methods are generally easier than integration by parts.
2. Once you decide to use integration by parts, split the integrand (*including* the "dx") into two parts: u and dv (which must include dx).
3. Differentiate u to find $du = u'dx$, and integrate $dv = v'dx$ to find v.
4. Apply the integration-by-parts formula from Theorem 14.2.
5. Determine how to solve the (hopefully simpler) integral $\int v\, du$.

We differentiate u to get du because $du = u'(x)\, dx$. The reason why we integrate dv to get v is that

$$\int dv = \int v'(x)\, dx = v(x).$$

Notice that we don't worry about including a "$+\,C$" here. After applying integration by parts, there is still another integral to do, and we can add the constant at the end. (Actually, there is a technical argument to make here concerning why you can drop the arbitrary constant from v when using integration by parts; you will do this in the exercises.)

It will be helpful if you always do integration-by-parts problems the same way, in the same notation and style. In particular, it is important to keep track of what you choose for u and dv and of the resulting v and du. Example 14.11 suggests one way that you can efficiently organize your work.

EXAMPLE 14.11

A basic integration-by-parts problem

Use integration by parts to find $\displaystyle\int xe^{2x}\, dx$.

Solution After a bit of trial and error (or experience; see Section 14.2.3) you will find that a good way to split up the integrand is by choosing $u = x$ and $dv = e^{2x}\, dx$. Notice that the "dx" is part of the dv; this must *always* be the case, since by definition $dv = v'\, dx$ (in other words, you're really choosing $v' = e^{2x}$ here, so $dv = e^{2x}\, dx$). We now have to differentiate u and integrate (i.e., antidifferentiate) dv. Since differentiation intuitively seems like a "forward" operation, whereas antidifferentiation goes "backwards," a sensible way to organize your work is by writing

$$
\begin{aligned}
u &= x &\longrightarrow\quad du &= \underline{\hspace{2cm}}\\
v &= \underline{\hspace{2cm}} &\longleftarrow\quad dv &= e^{2x}\, dx
\end{aligned}
$$

Then when you find the derivative of u and the integral of dv, you can put them in the blanks. In this example we have $\frac{du}{dx} = 1$, and thus $du = dx$. The integral of dv in this example is $\int dv = \int e^{2x}\, dx = \frac{1}{2}e^{2x}$. Therefore, we have

$$
\begin{aligned}
u &= x &\longrightarrow\quad du &= dx\\
v &= \tfrac{1}{2}e^{2x} &\longleftarrow\quad dv &= e^{2x}\, dx
\end{aligned}
$$

Now we have all the things we need (u, v, and du) on hand to use in the integration-by-parts formula:

$$\int xe^{2x}\,dx = uv - \int v\,du \qquad \text{(integration-by-parts formula)}$$

$$= x\left(\tfrac{1}{2}e^{2x}\right) - \int \tfrac{1}{2}e^{2x}\,dx \qquad (u=x, v=\tfrac{1}{2}e^{2x}, du=dx)$$

$$= \tfrac{1}{2}xe^{2x} - \tfrac{1}{2}\int e^{2x}\,dx \qquad \text{(simplify)}$$

$$= \tfrac{1}{2}xe^{2x} - \tfrac{1}{2}\left(\tfrac{1}{2}e^{2x}\right) + C. \qquad \text{(integrate the remaining integral)} \quad \square$$

✓ **Checking the Answer**   Of course, after all that complicated work you should always check your answer! Luckily, this is easy to do by differentiating: $\frac{d}{dx}\left(\tfrac{1}{2}xe^{2x} - \tfrac{1}{2}\left(\tfrac{1}{2}e^{2x}\right)\right) = \tfrac{1}{2}e^{2x} + xe^{2x} - \tfrac{1}{2}e^{2x} = xe^{2x}.$ $\quad\square$

14.2.3 Strategies for applying integration by parts

Integrals that look very similar can sometimes require very different methods of solution. In Example 14.11, we used integration by parts to find $\int xe^{2x}\,dx$. If the integral had instead been $\int xe^{(x^2)}\,dx$, we would have used integration by substitution. (Why?) If the integral had been simply $\int e^{(x^2)}\,dx$, we would not have been able to integrate it at all! Since integration by parts is more difficult than substitution or algebraic manipulation, you should consider algebra or integration by substitution before attempting integration by parts.

Once you decide to integrate by parts, how do you decide what to choose for u and dv? There are usually many possible choices, as illustrated in the following example.

EXAMPLE 14.12   **Possible choices for u and dv when using integration by parts**

Suppose you want to try integrating $\int x\cos x\,dx$ by parts. What are the possible choices for u and dv, and which choices are most likely to work?

Solution   Table 14.1 describes the possible choices for u and dv in this example and then lists the resulting du and v for each choice and the integral $\int v\,du$ that we hope will be simpler than $\int u\,dv = \int x\cos x\,dx$.

u	dv	du	v	$\int v\,du$
x	$\cos x\,dx$	dx	$\sin x$	$\int \sin x\,dx$
$\cos x$	$x\,dx$	$-\sin x\,dx$	$\tfrac{1}{2}x^2$	$-\tfrac{1}{2}\int x^2\sin x\,dx$
$x\cos x$	dx	$\cos x - x\sin x\,dx$	x	$\int(x\cos x - x^2\sin x)\,dx$
1	$x\cos x\,dx$	$0\,dx$	???	???

Table 14.1 Choices when $\int u\,dv = \int x\cos x\,dx$

There are four possible ways to split $x\cos x$ into u and dv. In particular, we can choose u to be everything but the dx (which will be dv); that's the third choice in Table 14.1. On the other hand, we could choose $u=1$, so that dv is the entire integrand (including the dx); that's the last choice in the table. However, that choice is *never* going to work, because we will have to integrate dv (which will be the same problem as solving the original integral). The second choice in the table produces a *more* complicated integral for $\int v\,du$, so certainly that choice would not be helpful.

The first choice produces the simplest integral $\int v\,du$, so that is the choice that is most likely to work. With this choice of u and dv, we can use integration by parts to show that

$$\int x\cos x\,dx = x\sin x - \int \sin x\,dx = x\sin x + \cos x + C.$$

☐

The reason why the choice $u = x$, $dv = \cos x\,dx$ produced the simplest integral $\int v\,du$ in Example 14.12 is that this choice produced the simplest du and v. One good strategy is to choose u so that it gets simpler (or at least not more complicated) when differentiated, and to choose dv so that it gets simpler (or at least not more complicated) when integrated. We'll summarize the strategies for choosing u and dv after we look at two more examples.

EXAMPLE 14.13 **Choosing u and dv when using integration by parts**

Use integration by parts to find:

(a) $\displaystyle\int \ln x\,dx$ 　　　　　　　　(b) $\displaystyle\int x^2 \ln x\,dx$

Solution

(a) At first glance this might not look like an integration-by-parts problem, since there is only one thing in the integrand. However, we cannot solve this integral with substitution or algebra, so trying integration by parts might be a good idea anyway. We have two choices: Let $u = \ln x$ and $dv = dx$, or let $u = 1$ and $dv = \ln x\,dx$. As discussed previously, the second choice *never* works; so we'll take $u = \ln x$ and $dv = dx$. Since the derivative of $\ln x$ is $\frac{1}{x}$, and the integral of 1 is x, we have

$$u = \ln x \quad \longrightarrow \quad du = \tfrac{1}{x}\,dx$$
$$v = x \quad \longleftarrow \quad dv = dx$$

Now the integration-by-parts formula gives us

$$\int \ln x\,dx = (\ln x)(x) - \int x\!\left(\tfrac{1}{x}\right)dx \qquad \text{(integration by parts)}$$

$$= x\ln x - \int 1\,dx \qquad\qquad\qquad \text{(simplify)}$$

$$= x\ln x - x + C. \qquad\qquad\qquad\quad \text{(integrate)}$$

(b) This time there are *six* ways to choose u and dv (can you list them?). Although it is true that the function x^2 gets simpler when differentiated (making it a good choice for u), with that choice we would have $dv = \ln x$, which, as we have just seen, has an antiderivative of $x\ln x - x$, which is pretty messy. After trying out various choices (do this), you should come to the conclusion that $u = \ln x$ and $dv = x^2\,dx$ are the best choices of u and dv.

$$u = \ln x \quad \longrightarrow \quad du = \tfrac{1}{x}\,dx$$
$$v = \tfrac{1}{3}x^3 \quad \longleftarrow \quad dv = x^2\,dx$$

Notice that $u = \ln x$ gets *much* simpler when we differentiate it. Even though $v = \frac{1}{3}x^3$ is a little more complicated than $dv = x^2\,dx$, it was worth it. There will be some nice canceling in the integral $\int v\,du$:

$$\int x^2 \ln x\,dx = (\ln x)\!\left(\tfrac{1}{3}x^3\right) - \int \left(\tfrac{1}{3}x^3\right)\!\left(\tfrac{1}{x}\right)dx \qquad \text{(integration by parts)}$$

$$= \tfrac{1}{3}x^3 \ln x - \tfrac{1}{3}\int x^2\,dx \qquad\qquad\qquad \text{(simplify)}$$

$$= \tfrac{1}{3}x^3 \ln x - \tfrac{1}{3}\!\left(\tfrac{1}{3}x^3\right) + C. \qquad\qquad\quad \text{(integrate)} \quad ☐$$

Part (a) of Example 14.13 proves the following integration formula. It is a good idea to memorize this formula, but be sure that you also know how to use integration by parts to derive it.

THEOREM 14.3

Integral of the Natural Logarithmic Function

$$\int \ln x \, dx = x \ln x - x + C$$

The following algorithm suggests a few strategies for choosing u and dv once you have decided to apply integration by parts.

ALGORITHM 14.3

Strategies for Choosing u and dv

▶ Sometimes it is useful to choose a u that gets simpler (or at least not worse) after differentiating. Similarly, it is sometimes useful to choose a dv that gets simpler (or at least not worse) after integrating. The point is that you want the components du and v of the "new" integral to be simpler than the components u and dv of the "old" integral.

▶ Sometimes it is useful to choose the entire integrand to be u (with $dv = dx$); this can often be a good "last resort" technique.

▶ Make sure that you can integrate whatever you choose for dv; in particular, you can *never* choose dv to be the entire integrand. Sometimes it is useful to choose dv to be the largest portion of the integrand that you know how to integrate.

▶ Don't forget that you can split an integrand into u and dv in many different ways, as long as you are writing the integrand as the *product* of u and dv.

❓ *Question*   List some functions that get "better" when differentiated and some functions that stay "about the same" in complexity. What functions don't get any "worse" when you integrate them?   ☐

✎ *Caution*   Whatever you choose for u and dv, the integrand of your original integral must be the *product* of u and dv. For example, if you wanted to use integration by parts to solve the integral $\int \frac{\ln x}{x^2} \, dx$, you could *not* choose $u = \ln x$ and $dv = x^2 \, dx$, since the product of these choices would be $(\ln x)(x^2) \, dx$, not $\frac{\ln x}{x^2} \, dx$. An allowable choice for parts in this example would be $u = \ln x$ and $dv = \frac{1}{x^2} \, dx$.   ☐

14.2.4 Using integration by parts more than once

Remember that, like integration by substitution, integration by parts does not actually solve the integral; it just enables us to rewrite it. After applying integration by parts, you may need to use algebra, substitution, or parts (again) to solve the *new* integral. In the following example, it is necessary to use integration by parts two times.

EXAMPLE 14.14   **Applying integration by parts twice**

Find $\int x^2 e^x \, dx$.

Solution First notice that we cannot use algebra or substitution to simplify or solve this integral, so we'll try integration by parts. A sensible choice of parts is $u = x^2$ and $dv = e^x \, dx$, since differentiation will "improve" x^2, and integration will not make e^x any worse:

$$u = x^2 \quad \longrightarrow \quad du = 2x \, dx$$
$$v = e^x \quad \longleftarrow \quad dv = e^x \, dx$$

Using integration by parts, we have

$$\int x^2 e^x \, dx = x^2 e^x - 2 \int x e^x \, dx.$$

To finish the problem, we now have to find $\int x e^x \, dx$; although it will take some work to solve this integral, it is at least a "simpler" integral than the one we started with. We choose

$$u = x \quad \longrightarrow \quad du = dx$$
$$v = e^x \quad \longleftarrow \quad dv = e^x \, dx$$

Continuing the calculation, we now have

$$\int x^2 e^x \, dx = x^2 e^x - 2 \left(x e^x - \int e^x \, dx \right)$$
$$= x^2 e^x - 2(x e^x - e^x) + C.$$

In some cases, doing integration by parts two times brings us back to the original integral; this produces an equation involving the original integral that we can solve. The following example illustrates this technique.

EXAMPLE 14.15

Applying integration by parts twice and then solving for the original integral

Find $\displaystyle\int e^x \sin x \, dx$.

Solution One good choice of parts is

$$u = e^x \quad \longrightarrow \quad du = e^x \, dx$$
$$v = -\cos x \quad \longleftarrow \quad dv = \sin x \, dx$$

With this choice, integration by parts gives us

$$\int e^x \sin x \, dx = -e^x \cos x + \int e^x \cos x \, dx.$$

This "new" integral is almost the same type of integral as what we started with, so integration by parts didn't seem to improve our situation. However, if we do integration by parts again, with the *same* kind of choice for u and dv (see the "Caution" on the next page) something good will happen. Choose u and dv as follows:

$$u = e^x \quad \longrightarrow \quad du = e^x \, dx$$
$$v = \sin x \quad \longleftarrow \quad dv = \cos x \, dx$$

This gives us

$$\int e^x \sin x \, dx = -e^x \cos x + e^x \sin x - \int e^x \sin x \, dx.$$

It might seem that we aren't getting anywhere at this point; we still have to find $\int e^x \sin x \, dx$ (the "new" integral), which is the same as the original integral! However, notice that the equation is of the form $I = -e^x \cos x + (e^x \sin x - I)$ (where I is the integral). We can solve this equation for I, thereby finding an expression for the original integral.

$$\int e^x \sin x \, dx = -e^x \cos x + e^x \sin x - \int e^x \sin x \, dx$$

$$\int e^x \sin x \, dx + \int e^x \sin x \, dx = -e^x \cos x + e^x \sin x + C$$

$$2 \int e^x \sin x \, dx = -e^x \cos x + e^x \sin x + C$$

$$\int e^x \sin x \, dx = -\tfrac{1}{2} e^x \cos x + \tfrac{1}{2} e^x \sin x + C$$

Since $\int e^x \sin x \, dx$ is the integral we were trying to find, we are done! Notice that we included a "$+ C$" in the first step, because at that point all of the integrals were on one side of the equation. (Notice also that the "C" in the last line is really one-half of the "C" in the previous line, but since C represents an arbitrary constant, we just rename $\tfrac{1}{2} C$ as C.) □

✅ *Checking the Answer*   After all that, we should check the answer by differentiating (with two applications of the product rule):

$$\frac{d}{dx}\left(-\tfrac{1}{2} e^x \cos x + \tfrac{1}{2} e^x \sin x\right)$$

$$= -\tfrac{1}{2} e^x \cos x + \left(-\tfrac{1}{2} e^x\right)(-\sin x) + \tfrac{1}{2} e^x \sin x + \tfrac{1}{2} e^x \cos x$$

$$= e^x \sin x. \qquad \square$$

💲 *Caution*   After choosing $u = e^x$ and $dv = \sin x \, dx$ in Example 14.15, we arrived at the "new" integral $\int e^x \cos x \, dx$. At this point we could have chosen $u = \cos x$ and $dv = e^x \, dx$ for the second application of integration by parts. However, if we do this, we will "undo" the first integration-by-parts step. When doing integration by parts more than once, always be careful not to undo what you have already done! □

❓ *Question*   We can also solve the integral in Example 14.15 by choosing $u = \sin x$ and $dv = e^x \, dx$ in the first integration by parts and then choosing a similar u and dv in the second integration by parts. Do this, and check that you get the same answer. □

14.2.5 Finding definite integrals using integration by parts

When using integration by parts to solve a definite integral $\int_a^b f(x) \, dx$, we must evaluate the *entire* right-hand side of the integration-by-parts formula from $x = a$ to $x = b$, as illustrated in the following theorem.

THEOREM 14.4

Integration by Parts for Definite Integrals
If $u(x)$ and $v(x)$ are differentiable functions, and $x = a$ and $x = b$ are any real numbers, then

$$\int_a^b u \, dv = [uv]_a^b - \int_a^b v \, du.$$

PROOF (**THEOREM 14.4**)   The proof of this theorem follows directly from the Fundamental Theorem of Calculus, the integration-by-parts formula, and the definition of "evaluation" from $x = a$ to $x = b$:

$$\int_a^b u\,dv = \int_a^b u(x)v'(x)\,dx = \left[\int u(x)v'(x)\,dx\right]_a^b = \left[u(x)v(x) - \int v(x)u'(x)\,dx\right]_a^b$$

$$= [u(x)v(x)]_a^b - \int_a^b v(x)u'(x)\,dx = [uv]_a^b - \int_a^b v\,du.$$

(In the exercises, you will fill in the "reasons" justifying each of these steps.)

The following example illustrates the use of Theorem 14.4. It also illustrates that sometimes you need to do a substitution after applying integration by parts.

EXAMPLE 14.16   **Solving a definite integral with integration by parts**

Calculate $\int_0^{\frac{\pi}{4}} x\sec^2 x\,dx$.

Solution   Notice that $\sec^2 x$ is easy to integrate (its antiderivative is $\tan x$), so we choose

$$u = x \quad\longrightarrow\quad du = dx$$
$$v = \tan x \quad\longleftarrow\quad dv = \sec^2 x\,dx$$

By integration by parts, we have

$$\int_0^{\frac{\pi}{4}} x\sec^2 x\,dx = [x\tan x]_0^{\frac{\pi}{4}} - \int_0^{\frac{\pi}{4}} \tan x\,dx.$$

We can do the "new" integral $\int_0^{\frac{\pi}{4}} \tan x\,dx = \int_0^{\frac{\pi}{4}} \frac{\sin x}{\cos x}\,dx$ using integration by substitution. Since we've already used the letter u, let's use the letter w here:

$$w = \cos x \quad\Longrightarrow\quad \frac{dw}{dx} = -\sin x \quad\Longrightarrow\quad -dw = \sin x\,dx\ .$$

Continuing the calculation, we have

$$\int_0^{\frac{\pi}{4}} x\sec^2 x\,dx = [x\tan x]_0^{\frac{\pi}{4}} - \int_0^{\frac{\pi}{4}} \tan x\,dx \qquad\text{(integration by parts)}$$

$$= [x\tan x]_0^{\frac{\pi}{4}} - \left(-\int_{x=0}^{x=\frac{\pi}{4}} \frac{1}{w}\,dw\right) \qquad\text{(substitution with } w = \cos x)$$

$$= [x\tan x]_0^{\frac{\pi}{4}} + [\ln|w|]_{x=0}^{x=\frac{\pi}{4}} \qquad\text{(integrate)}$$

$$= [x\tan x]_0^{\frac{\pi}{4}} + [\ln|\cos x|]_0^{\frac{\pi}{4}} \qquad\text{(since } w = \cos x)$$

$$= \left(\tfrac{\pi}{4}\tan\tfrac{\pi}{4} - 0\tan 0\right) + \left(\ln|\cos\tfrac{\pi}{4}| - \ln|\cos 0|\right) \qquad\text{(evaluate)}$$

$$= \tfrac{\pi}{4}(1) - 0 + \ln\left(\tfrac{\sqrt{2}}{2}\right) - \ln(1) \qquad\text{(simplify)}$$

$$= \tfrac{\pi}{4} + \ln\left(\tfrac{\sqrt{2}}{2}\right) \qquad\text{(exact answer)}$$

$$\approx 0.438825. \qquad\text{(approximate answer)}$$

Although that was quite a complicated calculation, there was only one antidifferentiation step, and it was very easy. (Can you find it?)

✓ **Checking the Answer** You can check the integration steps by showing that $\frac{d}{dx}(x \tan x + \ln |\cos x|)$ is equal to the original integrand $x \sec^2 x$. You can check the numerical answer by using a graphing calculator to approximate the area under the graph of $y = x \sec^2 x$ from $x = 0$ to $x = \frac{\pi}{4}$.   □

What you should know after reading Section 14.2

▶ The integration-by-parts formula, how to prove it, what it is for, and how to use it. In particular, be sure that you understand what "du" and "dv" represent.

▶ How to decide when to use integration by parts, and, if you do use it, how to determine what to choose for u and dv. You will get better at this after you do a lot of practice problems.

▶ How to use integration by parts or other methods multiple times in order to find an integral. In particular, what can you do when you integrate by parts twice and get back to the original integral?

▶ The formula for integration by parts with definite integrals, how to prove it, and how to use it.

Exercises 14.2

Concepts

0. Read the section and make your own summary of the material. Include whatever you think will help you review the material later on.

1. List five integrals that can be solved using integration by parts and five integrals that can be solved using substitution. In each case, identify u and dv (if parts) or u (if substitution).

2. List three integrals that we still cannot solve with the integration techniques we have learned so far. Choose integrands that are as simple as possible.

3. State the integration-by-parts formula for indefinite integrals (**a**) using the notation du and dv and (**b**) without using the notation du and dv.

4. State the integration-by-parts formula for definite integrals (**a**) using the notation du and dv and (**b**) without using the notation du and dv.

5. Write down an integral that can be solved using integration by parts with $u = x$ and another integral that can be solved using integration by parts with $dv = x\, dx$.

6. Write down an integral that can be solved using integration by parts with $u = \sin x$ and another integral that can be solved using integration by parts with $dv = \sin x\, dx$.

7. Write down an integral that can be solved using integration by parts by choosing u to be the entire integrand and $dv = dx$.

8. Explain why choosing $u = 1$ (and thus choosing dv to be the entire integrand, including dx) is *never* a good choice for integration by parts.

9. For each of the integrals in the odd-numbered problems among Problems 28–51, determine whether algebra, integration by substitution, or integration by parts would be the most successful strategy. In each case try to describe the algebra, choice of substitution u, or choice of parts u and dv. (You may want to do this problem after you have solved some of these integrals.)

10. Suppose $v(x)$ is a function of x. Explain why the integral of dv is equal to v (up to a constant).

11. Provide justification for each equality in the proof of Theorem 14.4.

12. If $u(x) = \sin 3x$ and $v(x) = x$, what are du and dv? Write down the integrals $\int u\, dv$ and $\int v\, du$. Which of these integrals would be easier to find? What does this have to do with integration by parts?

■ For each function $u(x)$ and $v(x)$, write down (**a**) the derivative of $u(x)v(x)$, (**b**) the same derivative written as an integral, and (**c**) the integration-by-parts formula for $\int u\, dv$. How are parts (**a**), (**b**), and (**c**) related?

13. $u(x) = x, \quad v(x) = \cos 2x$

14. $u(x) = x^3, \quad v(x) = e^{3x}$

15. $u(x) = \ln x, \quad v(x) = x$

■ Each expression below is the result of an integration-by-parts calculation; that is, each is an expression of the form $uv - \int v\,du$ for some functions u and v. Identify u, v, du, and dv, and determine the original integral $\int u\,dv$.

16. $-x^2 \cos x + 2 \int x \cos x\,dx$

17. $\dfrac{x2^x}{\ln 2} - \dfrac{1}{\ln 2} \int 2^x\,dx$ **18.** $-\dfrac{\ln x}{x^2} + \int \dfrac{1}{x^3}\,dx$

19. This problem concerns Example 14.15 in this section.

(a) Explain how we solved the integral in Example 14.15, even though, after using integration by parts twice, we got back to the original integral.

(b) Explain why choosing $u = \cos x$ and $dv = e^x\,dx$ in the second integration-by-parts step in Example 14.15 would *not* have worked. Then actually try to solve the problem with this choice of u and dv in the second application of integration by parts. What goes wrong?

(c) Redo Example 14.15 using a different choice of u and dv in the first application of integration by parts. Show that the answer you get differs from the answer in Example 14.15 by a constant.

■ Determine whether each of the following statements is true or false. If a statement is true, explain why. If a statement is false, explain why or provide a counterexample.

20. True or False: If $u = x^2 + 1$, then $du = 2x$.

21. True or False: We can apply integration by parts with $u = \ln x$ and $dv = x\,dx$ to the integral $\int \dfrac{\ln x}{x}\,dx$.

22. True or False: We can apply integration by parts with $u = x$ and $dv = \ln x\,dx$ to the integral $\int \dfrac{\ln x}{x}\,dx$.

23. True or False: $\int u(x)v(x)\,dx = u(x)v(x) - \int v(x)u(x)\,dx$.

24. True or False: $\int_0^3 xe^x\,dx = xe^x - \int_0^3 e^x\,dx$.

Skills

■ Solve each of the following integrals (a) with integration by parts and (b) without integration by parts. Show that your answers differ by a constant.

25. $\int x^2 \ln(x^3)\,dx$ **26.** $\int \dfrac{\ln x}{x}\,dx$

27. $\int x(x+1)^{100}\,dx$

■ Solve each of the following integrals. Check your answers by differentiating. (*Hints:* Some problems require integration by parts and others do not. Some problems may need integration by parts (or other methods) more than once, and a few require the technique used in Example 14.15.)

28. $\int x \sin x\,dx$ **29.** $\int xe^x\,dx$

30. $\int x \ln x\,dx$ **31.** $\int x \cos x\,dx$

32. $\int x \sin x^2\,dx$ **33.** $\int x^2 \cos x\,dx$

34. $\int x^2 e^{3x}\,dx$ **35.** $\int 3xe^{x^2}\,dx$

36. $\int (3-x)e^{2x}\,dx$ **37.** $\int \ln(3x)\,dx$

38. $\int \ln(x^3)\,dx$ **39.** $\int x \ln(x^2)\,dx$

40. $\int \dfrac{x^2+1}{e^x}\,dx$ **41.** $\int \dfrac{x}{e^{x^2+1}}\,dx$

42. $\int (x - e^x)^2\,dx$ **43.** $\int \ln\sqrt{x}\,dx$

44. $\int x \ln\sqrt{x}\,dx$ **45.** $\int \sqrt{x}\ln x\,dx$

46. $\int x^3 e^{-x}\,dx$ **47.** $\int x^3 e^{x^2}\,dx$

48. $\int x \csc^2 x\,dx$ **49.** $\int x^3 \cos x\,dx$

50. $\int x^3 \sin x^2\,dx$ **51.** $\int x^5 \cos x^3\,dx$

52. $\int \dfrac{x^2}{\sqrt{x-1}}\,dx$ **53.** $\int \arcsin x\,dx$

54. $\int \tan^{-1} 3x\,dx$ **55.** $\int e^{2x} \sin x\,dx$

56. $\int \sin \pi x \cos \pi x\,dx$ **57.** $\int e^x \ln(e^{2x})\,dx$

58. $\int x^3 e^{x^2}\,dx$ **59.** $\int e^{-x} \cos x\,dx$

60. $\int x \tan^{-1} x\,dx$

■ Find the exact value of each of the following definite integrals. Check your answers by (a) differentiating and (b) using a graphing calculator to approximate the appropriate area.

61. $\int_0^\pi x \sin x\,dx$ **62.** $\int_1^2 \ln x\,dx$

63. $\int_{-1}^1 \dfrac{x}{e^x}\,dx$ **64.** $\int_{\frac{\pi}{4}}^{\frac{\pi}{2}} x \csc^2 x\,dx$

65. $\int_{-\frac{3\pi}{4}}^{-\frac{\pi}{4}} x \sec x \tan x\,dx$ **66.** $\int_{-\frac12}^{\frac12} \sin^{-1} x\,dx$

67. Consider the function $f(x) = \ln x$.

(a) Find the signed area of the region between the graph of $f(x)$ and the x-axis from $x = \frac12$ to $x = 4$.

(b) Find the "true" area of the region between the graph of $f(x)$ and the x-axis from $x = \frac{1}{2}$ to $x = 4$.

(c) Find the area of the region between the graphs of $f(x)$ and $g(x) = 1$ on the interval $[\frac{1}{2}, 4]$.

(d) Find the average value of the function $f(x)$ on the interval $[\frac{1}{2}, 4]$.

68. Consider the function $f(x) = x \cos 2x$.

(a) Find the signed area of the region between the graph of $f(x)$ and the x-axis from $x = 0$ to $x = \frac{3\pi}{4}$.

(b) Find the "true" area of the region between the graph of $f(x)$ and the x-axis from $x = 0$ to $x = \frac{3\pi}{4}$.

(c) Find the area of the region between the graphs of $f(x)$ and $g(x) = -x$ on the interval $[0, \frac{3\pi}{4}]$.

(d) Find the average value of the function $f(x)$ on the interval $[0, \frac{3\pi}{4}]$.

Proofs

69. Prove the integration-by-parts formula in Theorem 14.2.

70. If $v(x)$ is a function of x, then $\int dv = \int v'(x)\, dx = v(x) + C$. However, when we use integration by parts, we just write $v(x)$ for the integral of dv (i.e., we drop the "plus C"). In this problem you will show that it is okay to drop the constant from $v(x)$ when using integration by parts.

(a) Explain why we need to show that $uv - \int v\, du = u(v + C) - \int (v + C)\, du$.

(b) Write the equation above using the fact that $u = u(x)$, $v = v(x)$, and $du = u'(x)\, dx$.

(c) Prove that the equation you wrote down in part (b) is always true. (*Hint:* Start from the left-hand side and simplify with algebra and integration rules until you get the right-hand side.)

14.3 Trigonometric Integrals

In this section we examine integrals that involve trigonometric functions. Many of these integrals become easier to handle after using trigonometric identities.

14.3.1 Using Pythagorean identities to set up a substitution

Many trigonometric identities, especially those that involve products or powers of trigonometric functions, can be rewritten in a way that makes integration by substitution convenient. The following algorithm lists six of the most common forms of these integrals.

ALGORITHM 14.4

Trigonometric Integrals That Are Set Up for Substitution

The following six types of trigonometric integrals are easy to solve using integration by substitution:

$\displaystyle\int \text{(expression in } \sin x\text{)}\, \cos x\, dx$ (choose $u = \sin x$, so $du = \cos x\, dx$)

$\displaystyle\int \text{(expression in } \cos x\text{)}\, \sin x\, dx$ (choose $u = \cos x$, so $du = -\sin x\, dx$)

$\displaystyle\int \text{(expression in } \tan x\text{)}\, \sec^2 x\, dx$ (choose $u = \tan x$, so $du = \sec^2 x\, dx$)

$\displaystyle\int \text{(expression in } \sec x\text{)}\, \sec x \tan x\, dx$ (choose $u = \sec x$, so $du = \sec x \tan x\, dx$)

$\displaystyle\int \text{(expression in } \cot x\text{)}\, \csc^2 x\, dx$ (choose $u = \cot x$, so $du = -\csc^2 x\, dx$)

$\displaystyle\int \text{(expression in } \csc x\text{)}\, \csc x \cot x\, dx$ (choose $u = \csc x$, so $du = -\csc x \cot x\, dx$)

Some integrals are already in the form of Algorithm 14.4. For example, the integral $\int \sin^2 x \cos x\, dx$ can be turned into the simpler integral $\int u^2\, du$ using the substitution $u = \sin x$. Many other trigonometric integrals can be written in the form of Algorithm 14.4 using Pythagorean identities.

The Pythagorean identity $\sin^2 x + \cos^2 x = 1$ can be used to write any even power of sine in terms of cosine, and vice versa. To do this, we use one of the following forms of this Pythagorean identity:

(1) $\sin^2 x = 1 - \cos^2 x$ and thus $\cos^2 x = 1 - \sin^2 x$.

For example, $\cos^4 x$ can be rewritten in terms of sines by writing $\cos^4 x = (\cos^2 x)^2 = (1 - \sin^2 x)^2 = 1 - 2\sin^2 x + \sin^4 x$. Similarly, the Pythagorean identities $\tan^2 x + 1 = \sec^2 x$ and $1 + \cot^2 x = \csc^2 x$ can be used to rewrite even powers of tangent or cotangent in terms of powers of secant or cosecant, respectively (and vice versa). Specifically, we have

(2) $\tan^2 x = \sec^2 x - 1$ and thus $\sec^2 x = \tan^2 x + 1$

(3) $\cot^2 x = \csc^2 x - 1$ and thus $\csc^2 x = \cot^2 x + 1$.

The following four examples illustrate how to use the Pythagorean identities and Algorithm 14.4 to solve trigonometric integrals.

EXAMPLE 14.17   **Integrating an odd power of sine**

Find $\int \sin^5 x\, dx$.

Solution   The key is to split off *one* factor of $\sin x$ and write the rest of the integrand in terms of cosine; that way we will be able to do an integration by substitution with $u = \cos x$ (and $du = \sin x\, dx$).

$$\int \sin^5 x\, dx = \int (\sin^4 x)(\sin x)\, dx = \int (\sin^2 x)^2 (\sin x)\, dx \qquad \text{(split up integrand)}$$

$$= \int (1 - \cos^2 x)^2 \sin x\, dx \qquad \text{(Pythagorean identity)}$$

This integral is of the second type shown in Algorithm 14.4. The entire integrand except for the $\sin x\, dx$ is written in terms of cosine, and if we let $u = \cos x$, then we will have $du = \sin x\, dx$. This means we will be able to write this integral entirely in terms of u and du.

$$u = \cos x \quad \Longrightarrow \quad \frac{du}{dx} = -\sin x \quad \Longrightarrow \quad -du = \sin x\, dx$$

We now have

$$\int \sin^5 x\, dx = -\int (1 - u^2)^2\, du \qquad \text{(change of variables)}$$

$$= -\int (1 - 2u^2 + u^4)\, du \qquad \text{(multiply out)}$$

$$= -\left(-u - 2\tfrac{1}{3}u^3 + \tfrac{1}{5}u^5\right) + C \qquad \text{(integrate)}$$

$$= \cos x + \tfrac{2}{3}\cos^3 x - \tfrac{1}{5}\cos^5 x + C. \qquad \text{(since } u = \cos x) \qquad \square$$

EXAMPLE 14.18

Integrating $\sin^n x \cos^m x$ when n or m is odd

Find $\displaystyle\int \sin^3 x \cos^4 x \, dx$.

Solution   Note that $\sin^3 x = (\sin^2 x)(\sin x)$. Using a Pythagorean identity, we can write the entire integrand as a product of $\sin x$ with a sum of powers of cosine:

$$\int \sin^3 x \cos^4 x \, dx = \int (\sin^2 x)(\sin x)\cos^4 x \, dx \qquad \text{(split off one } \sin x)$$

$$= \int (1 - \cos^2 x)\cos^4 x \sin x \, dx. \qquad \text{(Pythagorean identity)}$$

Now we can use integration by substitution with $u = \cos x$:

$$u = \cos x \quad \Longrightarrow \quad \frac{du}{dx} = -\sin x \quad \Longrightarrow \quad -du = \sin x \, dx \ .$$

With this substitution, we have

$$\int \sin^3 x \cos^4 x \, dx = -\int (1 - u^2)u^4 \, du \qquad \text{(change of variables)}$$

$$= -\int (u^4 - u^6) \, du \qquad \text{(multiply out)}$$

$$= -\left(\tfrac{1}{5}u^5 - \tfrac{1}{7}u^7\right) + C \qquad \text{(integrate)}$$

$$= -\tfrac{1}{5}\cos^5 x + \tfrac{1}{7}\cos^7 x + C. \qquad \text{(since } u = \cos x) \qquad \square$$

In order for us to use the method in Example 14.18, one of the powers of sine or cosine must be odd. Then, when we split off one copy of $\sin x$ (or $\cos x$) for the du in the substitution, there are an *even* number of copies of $\sin x$ (or $\cos x$), to which we can apply a Pythagorean identity (to get one of the forms in Algorithm 14.4).

❓ Question   Why can't we write the integral $\int \sin^4 x \cos^4 x \, dx$ in one of the forms from Algorithm 14.4? (*Hint:* In this integral, both powers are even. Why is this a problem?) $\square$

The method for integrating functions of the form $\sec^m x \tan^n x$ depends on whether m and n are odd or even. If m is even, then we can rewrite the integral so that we can use substitution with $u = \tan x$. The key is to "save" the derivative $\sec^2 x$ so that we can make this substitution.

EXAMPLE 14.19

Integrating $\sec^n x \tan^m x$ when n is even

Find $\displaystyle\int \sec^4 x \tan^3 x \, dx$.

Solution   Since the power of secant is even, we can split off $\sec^2 x$ and still be left with an even power of secant that can be converted into tangents with a Pythagorean identity. Then we will be able to use an integration by substitution with $u = \tan x$ and $du = \sec^2 x \, dx$.

$$\int \sec^4 x \tan^3 x \, dx = \int \sec^2 x \tan^3 x \sec^2 x \, dx \qquad \text{(save } \sec^2 x \, dx \text{ for } du)$$

$$= \int (\tan^2 x + 1)\tan^3 x \sec^2 x \, dx \qquad \text{(Pythagorean identity)}$$

Now, with $u = \tan x$, and thus $du = \sec^2 x \, dx$, we have

$$= \int (u^2 + 1) u^3 \, du \qquad \text{(change of variables)}$$

$$= \int (u^5 + u^3) \, du \qquad \text{(multiply out)}$$

$$= \tfrac{1}{6} u^6 + \tfrac{1}{4} u^4 + C \qquad \text{(integrate)}$$

$$= \tfrac{1}{6} \tan^6 x + \tfrac{1}{4} \tan^4 x + C. \qquad \text{(since } u = \tan x) \qquad \square$$

The method in Example 14.19 works for integrating $\int \sec^m x \tan^n x \, dx$ whenever the power m of secant is even (regardless of whether the power n of tangent is even or odd). If the power of secant is odd, we must use another method. The following example illustrates that we can make the substitution $u = \sec x$ if the power n of tangent is odd (regardless of whether the power of secant is even or odd). This time, the key is to "save" the derivative $\sec x \tan x$ so that we can make this substitution.

EXAMPLE 14.20

Integrating $\sec^n x \tan^m x$ when m is odd

Find $\int \sec^3 x \tan^3 x \, dx$.

Solution We cannot use the method of Example 14.19, because the power of secant is odd. Instead,

$$\int \sec^3 x \tan^3 x \, dx = \int \sec^2 x \tan^2 x \, (\sec x \tan x) \, dx \qquad \text{(save } \sec x \tan x \text{ for } du)$$

$$= \int \sec^2 x \, (\sec^2 x - 1) \, (\sec x \tan x) \, dx. \qquad \text{(Pythagorean identity)}$$

Now, with $u = \sec x$, and thus $du = \sec x \tan x \, dx$, we have

$$= \int u^2 (u^2 - 1) \, du \qquad \text{(change of variables)}$$

$$= \int (u^4 - u^2) \, du \qquad \text{(multiply out)}$$

$$= \tfrac{1}{5} u^5 - \tfrac{1}{3} u^3 + C \qquad \text{(integrate)}$$

$$= \tfrac{1}{5} \sec^5 x - \tfrac{1}{3} \sec^3 x + C. \qquad \text{(since } u = \sec x)$$

$$\square$$

? *Question* Example 14.19 can also be done using the method of Example 14.20 (since the power of tangent is odd). Do this, and show that the answer you get is the same (up to a constant); you will need to use some trigonometric identities. $\square$

If m is odd and n is even, then we won't be able to use either the method in Example 14.19 or the method in Example 14.20 to integrate $\int \sec^m x \tan^n x \, dx$. We will see how to solve integrals of this type (for example, $\int \sec^3 x \tan^2 x \, dx$) in Section 14.3.3.

14.3.2 Using double-angle identities to reduce powers

Not all trigonometric integrals can be solved using Pythagorean identities and substitution. For example, consider the integral $\int \sin^2 x \, dx$. Using a Pythagorean identity, we could change this integral into $\int (1 - \cos^2 x) \, dx$, but this new integral would be no easier to solve. However, a different kind of trigonometric identity can help simplify this integral. Recall the double-angle identities from Theorem 10.9 in Section 10.3.4:

(4)
$$\sin^2 x = \frac{1 - \cos 2x}{2} \quad \text{and} \quad \cos^2 x = \frac{1 + \cos 2x}{2}.$$

These double-angle identities can be used to reduce the powers in a trigonometric expression involving sine or cosine. In particular, these identities can help us solve integrals of the form $\int \sin^k x \, dx$ or $\int \cos^k x \, dx$ when k is a positive even integer. Sometimes the double-angle identities will need to be applied more than once. The following example illustrates such a calculation; we apply a double-angle identity, multiply out the result, and then apply a double-angle identity again to reduce any remaining even powers.

EXAMPLE 14.21   **Using double-angle identities to integrate an even power of cosine**

Find $\displaystyle\int \cos^4 x \, dx$.

Solution   This time we will have to apply the double-angle identity for cosine *twice:*

$$\int \cos^4 x \, dx = \int (\cos^2 x)^2 \, dx \qquad \text{(algebra)}$$

$$= \int \left(\frac{1 + \cos 2x}{2}\right)^2 dx \qquad \text{(double-angle identity)}$$

$$= \tfrac{1}{4} \int (1 + 2\cos 2x + \cos^2 2x) \, dx \qquad \text{(multiply out integrand)}$$

$$= \tfrac{1}{4} \left(\int 1 \, dx + 2 \int \cos 2x \, dx + \int \cos^2 2x \, dx \right). \qquad \text{(sum rule for integrals)}$$

The first two integrals are easy to solve. To solve the third integral, we will have to use the double-angle identity for cosine *again*. Continuing with the calculation, we have

$$\int \cos^4 x \, dx = \tfrac{1}{4} \left(x + 2(\tfrac{1}{2}\sin 2x) + \int \frac{1 + \cos 4x}{2} \, dx \right) \qquad \text{(integrate, use identity)}$$

$$= \tfrac{1}{4} \left(x + \sin 2x + \frac{1}{2} \int (1 + \cos 4x) \, dx \right) \qquad \text{(simplify)}$$

$$= \tfrac{1}{4} \left(x + \sin 2x + \tfrac{1}{2}(x + \tfrac{1}{4}\sin 4x) \right) + C. \qquad \text{(integrate)}$$

The second time we applied the double-angle identity, we had $\cos^2 2x = \frac{1+\cos 4x}{2}$; this is the double-angle identity $\cos^2 \theta = \frac{1+\cos 2\theta}{2}$ with $\theta = 2x$ (so $2\theta = 4x$).   □

✅ *Checking the Answer*   You can differentiate to check the answer. If you do so, you will find that

$$\frac{d}{dx}\left(\tfrac{1}{4}\left(x + \sin 2x + \tfrac{1}{2}\left(x + \tfrac{1}{4}\sin 4x\right)\right) \right) = \tfrac{1}{4}\left(1 + 2\cos 2x + \tfrac{1}{2}(1 + \cos 4x)\right).$$

This expression *is* equal to the original integrand $\cos^4 x$, but it takes a lot of algebra to get there. In particular, you will have to apply the double-angle identity $\cos 2x = 2\cos^2 x - 1$ a few times. Try it!   □

The technique of repeatedly using double-angle identities will work for integrating $\sin^k x$ or $\cos^k x$ whenever k is a positive even integer (although, as you can imagine, the process gets pretty tedious for large values of k). This technique will *not* work if k is a positive *odd* integer. (Why not?) However, if k is a positive odd integer, we can use Pythagorean identities and integration by substitution (Algorithm 14.4).

Double-angle identities can also be used to solve integrals of the form $\int \sin^m x \cos^n x \, dx$ where both m and n are even (this was the case that couldn't be handled by the methods in Section 14.3.1). One way to do this is to use a Pythagorean identity to write $\sin^m x \cos^n x$ in terms of a power of sine or cosine. This power will be even if m and n are even, so we will be able to use double-angle identities to solve the resulting integral. This technique is illustrated in the following example.

EXAMPLE 14.22

Integrating $\sin^n x \cos^m x$ when n and m are both even

Find $\displaystyle\int \sin^2 x \cos^2 x \, dx$.

Solution We can use the fact that $\cos^2 x = 1 - \sin^2 x$ to write this integral in terms of two integrals that we already know how to solve:

$$\int \sin^2 x \cos^2 x \, dx = \int \sin^2 x (1 - \sin^2 x) \, dx \qquad \text{(Pythagorean identity)}$$

$$= \int \sin^2 x \, dx - \int \sin^4 x \, dx. \qquad \text{(write as two integrals)}$$

Using the first trigonometric identity in Equation (4), we have $\int \sin^2 x \, dx = \frac{1}{2}(x - \frac{1}{2}\sin 2x)$. We can show that $\int \sin^4 x \, dx = \frac{1}{4}(x - \sin 2x + \frac{1}{2}(x - \frac{1}{4}\sin 4x))$ using the method of Example 14.21. The integral we are looking for is the difference of these two integrals:

$$\int \sin^2 x \cos^2 x \, dx = \frac{1}{2}\left(x - \frac{1}{2}\sin 2x\right) - \frac{1}{4}\left(x - \sin 2x + \frac{1}{2}\left(x - \frac{1}{4}\sin 4x\right)\right)$$

$$= \frac{1}{8}x + \frac{1}{32}\sin 4x + C.$$

Of course we could check this answer by differentiating, but we would have to use a number of trigonometric identities to show that the derivative of the answer is equal to the original integrand. □

? Question You can also solve the integral in Example 14.22 using the double-angle identities in Equation (4) to write both $\sin^2 x$ and $\cos^2 x$ in terms of $\cos 2x$. (Then multiply out and use a double-angle identity again to simplify $\cos^2 2x$.) Try it! □

? Question You can also solve the integral in Example 14.22 using the identity $\sin x \cos x = \frac{1}{2}\sin 2x$, which follows from the double-angle identity for sine in Section 10.3.4. (You'll also need a double-angle identity to simplify $\sin^2 2x$.) Try it! □

14.3.3 Integrating powers of tangent and secant

The following example illustrates the method for integrating powers of tangent. A similar method can be used to integrate powers of cotangent. The key is to "split off" two copies of tangent at a time and convert them into secants with a Pythagorean identity. This will produce two integrals, one of which can be solved using Algorithm 14.4, and another that will have a power of tangent smaller than that of the original integral. We can then repeat the process for this smaller power of tangent.

EXAMPLE 14.23

Integrating a power of tangent

Find $\displaystyle\int \tan^5 x \, dx$.

Solution   Since $\tan^2 x + 1 = \sec^2 x$, we can write $\tan^2 x = \sec^2 x - 1$. Using this identity on part of the integrand, we have

$$\int \tan^5 x \, dx = \int \tan^3 x \tan^2 x \, dx$$

$$= \int \tan^3 x (\sec^2 x - 1) \, dx \qquad \text{(Pythagorean identity)}$$

$$= \int \tan^3 x \sec^2 x \, dx - \int \tan^3 x \, dx.$$

The first integral can be evaluated using substitution (with $u = \tan x$). The second integral can be evaluated by repeating the process of "splitting off" two copies of tangent. Since $\tan^3 x = \tan x (\tan^2 x) = \tan x (\sec^2 x - 1)$, we have

$$\int \tan^5 x \, dx = \int \tan^3 x \sec^2 x \, dx - \int \tan x \sec^2 x \, dx + \int \tan x \, dx.$$

The first two integrals can be solved using substitution with $u = \tan x$, by Algorithm 14.4; the last integral can be solved using the different substitution $w = \cos x$ (since $\tan x = \frac{\sin x}{\cos x}$). We leave the details of these three integrals for the exercises. □

In general, powers of tangent (or cotangent) can *always* be integrated using the method in Example 14.23—that is, by using a Pythagorean identity for two copies of tangent (or cotangent) at a time. The method works whether the power involved is even or odd.

? Question   The technique used in Example 14.23 also works if the power of tangent is even. Use this technique to find $\int \tan^6 x \, dx$. What is different about the last steps of the computation when k is even? □

Powers of secant are perhaps the most difficult of all the trigonometric integrals. In fact, as the following example shows, even integrating $\sec x$ is a challenge.

EXAMPLE 14.24   **Integrating the secant function**

Find $\displaystyle\int \sec x \, dx$.

Solution   After thinking about this integral for a few minutes, you should be convinced that substitution and parts will not help (at least not at the start). However, if we multiply the integrand by a very special form of 1, we will have an integral that can be evaluated using substitution:

$$\int \sec x \, dx = \int \sec x \left(\frac{\sec x + \tan x}{\sec x + \tan x} \right) dx \qquad \left(\text{note that } \tfrac{\sec x + \tan x}{\sec x + \tan x} = 1\right)$$

$$= \int \frac{\sec^2 x + \sec x \tan x}{\sec x + \tan x} \, dx. \qquad \text{(algebra)}$$

Now, if we choose $u = \sec x + \tan x$, we can use integration by substitution, because

$$u = \sec x + \tan x \implies \frac{du}{dx} = \sec x \tan x + \sec^2 x$$

$$\implies du = (\sec x \tan x + \sec^2 x) \, dx \ .$$

With this change of variables, the integral becomes

$$\int \sec x \, dx = \int \frac{1}{u} \, du = \ln |u| + C = \ln |\sec x + \tan x| + C.$$

  □

❓ *Question* Find $\int \csc x \, dx$ using a similar method, by multiplying the integrand by $\frac{\csc x + \cot x}{\csc x + \cot x}$. What would you multiply by to solve the integral $\int \sec(3x) \, dx$? ☐

Of course, since $\frac{d}{dx}(\tan x) = \sec^2 x$, it is very easy to integrate $\sec^2 x$. Integrating other powers of secant is more complicated and involves integration by parts with $dv = \sec^2 x$. We briefly outline two examples here; you will fill in the details in the exercises.

EXAMPLE 14.25

Integrating a power of secant

Find $\displaystyle\int \sec^3 x \, dx$.

Solution Using integration by parts with $u = \sec x$ and $dv = \sec^2 x \, dx$, we have

$$\int \sec^3 x \, dx = \sec x \tan x - \int \sec x \tan^2 x \, dx.$$

Since $\tan^2 x = \sec^2 x - 1$, we have

$$\int \sec^3 x \, dx = \sec x \tan x - \int \sec^3 x \, dx + \int \sec x \, dx.$$

The last integral is equal to $\ln|\sec x + \tan x|$ (see Example 14.24). This means we have an equation of the form $I = \sec x \tan x - I + \ln|\sec x + \tan x|$, where $I = \int \sec^3 x \, dx$ is the original integral. Solving this equation for I, we have

$$2\int \sec^3 x = \sec x \tan x + \ln|\sec x + \tan x| + C$$

$$\int \sec^3 x \, dx = \tfrac{1}{2}\sec x \tan x + \tfrac{1}{2}\ln|\sec x + \tan x| + C.$$ ☐

EXAMPLE 14.26

Another example of integrating a power of secant

Find $\displaystyle\int \sec^4 x \, dx$.

Solution Again choose $dv = \sec^2 x \, dx$. Then $u = \sec^2 x$, and by integration by parts, we have

$$\int \sec^4 x \, dx = \sec^2 x \tan x - 2\int \sec^2 x \tan^2 x \, dx.$$

The "new" integral can be solved using substitution with $u = \tan x$ (since then $du = \sec^2 x \, dx$). We have

$$\int \sec^4 x \, dx = \sec^2 x \tan x - \tfrac{2}{3}\tan^3 x + C.$$ ☐

Finally, the techniques for integrating powers of secant can be used to solve integrals of the form $\int \sec^m x \tan^n x \, dx$, where m is odd and n is even (recall that this was a type of integral we could not solve with Algorithm 14.4). These types of integrals can be solved by using a Pythagorean identity to convert the even power of tangent into secants; this will produce a sum of integrals that are powers of secant, which we can handle by the methods outlined in Examples 14.25 and 14.26.

EXAMPLE 14.27

Integrating $\sec^n x \tan^m x$ when n is odd and m is even

Find $\displaystyle\int \sec x \tan^2 x \, dx$.

Solution   Using a Pythagorean identity, we can write

$$\int \sec x \tan^2 x \, dx = \int \sec x \, (\sec^2 x - 1) \, dx = \int \sec^3 x \, dx - \int \sec x \, dx.$$

We have already solved these integrals (see Examples 14.24 and 14.25); with those results, we have

$$\int \sec x \tan^2 x \, dx = \tfrac{1}{2} \sec x \tan x + \tfrac{3}{2} \ln |\sec x + \tan x| + C.$$ □

The integrals of the tangent, cotangent, secant, and cosecant functions appear often, so you may find it worthwhile to memorize the following formulas.

THEOREM 14.5

The Integrals of Tangent, Cotangent, Secant, and Cosecant

(a) $\displaystyle\int \tan x \, dx = -\ln |\cos x| + C$

(b) $\displaystyle\int \cot x \, dx = \ln |\sin x| + C$

(c) $\displaystyle\int \sec x \, dx = \ln |\sec x + \tan x| + C$

(d) $\displaystyle\int \csc x \, dx = -\ln |\csc x + \cot x| + C$

The proofs of these formulas are left to the exercises; the first two involve simple integration by substitution, and the last two were discussed earlier in this section. Although you can memorize these formulas, be sure that you also know how to use integration techniques to derive them.

☙ *Caution*   Don't try to memorize all the different cases and strategies in this section! Instead, try to understand how and why each strategy works, and do a lot of practice problems. When you encounter an integral involving products and/or powers of trigonometric functions, think about the different ways you could rewrite the integrand, and which would be most useful.   □

What you should know after reading Section 14.3

▶ How to solve trigonometric integrals by using Pythagorean identities to rewrite the integrand so that integration by substitution is possible (see Algorithm 14.4). When will such a technique apply?

▶ How to use double-angle identities to solve integrals of even powers of sine, even powers of cosine, or products of even powers of sine and cosine.

▶ How to integrate powers of tangent, secant, cotangent, and cosecant, and how to use powers of secant to integrate functions of the form $\sec^m x \tan^n x$ when m is odd and n is even (this is the case where Algorithm 14.4 does not apply).

▶ The formulas for integrating the tangent, cotangent, secant, and cosecant functions, and how to prove these integration formulas.

Exercises 14.3

Concepts

0. Read the section and make your own summary of the material. Include whatever you think will help you review the material later on.

1. List three integrals that can be improved by using one or both of the double-angle identities $\sin^2 x = \frac{1-\cos 2x}{2}$ and $\cos^2 x = \frac{1+\cos 2x}{2}$.

2. List three integrals that can be rewritten by using the Pythagorean identity $\sin^2 x + \cos^2 x = 1$ so that integration by substitution applies.

3. List three integrals that can be rewritten by using the Pythagorean identity $\tan^2 x + 1 = \sec^2 x$ so that integration by substitution applies.

4. For each of the integrals in the odd-numbered problems among Problems 16–39, determine a strategy for solving the integral. In each case, describe the trigonometric identities you would use and what kind of substitution or parts would be used after applying the identities. (You may want to do this problem after you have done some of these integrals.)

■ Describe in your own words a strategy for solving integrals of the forms listed below, where k, m, and n are positive integers. (*Hint:* Don't forget to consider all the different cases! Consider what happens when k, m, and n are even or odd.)

5. $\int \cos^k x\, dx$

6. $\int \csc^k x\, dx$

7. $\int \cot^k x\, dx$

8. $\int \sin^m x \cos^n x\, dx$

9. $\int \sec^m x \tan^n x\, dx$

10. $\int \csc^m x \cot^n x\, dx$

11. What goes wrong if you try to apply Algorithm 14.4 to the integral $\int \sin^4 x \cos^4 x\, dx$?

12. Explain how to integrate (**a**) $\sec x$ and (**b**) $\csc x$.

13. Finish the calculation in Example 14.23 in this section.

14. Fill in the missing details in Example 14.25 in this section.

15. Fill in the missing details in Example 14.26 in this section.

Skills

■ Calculate the following integrals.

16. $\int \cos^2 x\, dx$

17. $\int \sin^2 3x\, dx$

18. $\int \sin^3 x\, dx$

19. $\int \cos^5 x\, dx$

20. $\int \tan 2x\, dx$

21. $\int \sec 4x\, dx$

22. $\int \csc x\, dx$

23. $\int \tan^6 x\, dx$

24. $\int \cot^5 x\, dx$

25. $\int \csc^2 x\, dx$

26. $\int \sec^3 2x\, dx$

27. $\int \csc^4 x\, dx$

28. $\int \sec^6 x\, dx$

29. $\int \sin x \cos^4 x\, dx$

30. $\int \cos^3 x \sin^4 x\, dx$

31. $\int \sin^2 3x \cos^2 3x\, dx$

32. $\int \sin^5 x \cos^2 x\, dx$

33. $\int \sec^2 x \tan^5 x\, dx$

34. $\int \sec^8 x \tan x\, dx$

35. $\int \sec^3 x \tan^5 x\, dx$

36. $\int \tan^2 4x \sec 4x\, dx$

37. $\int \csc^4 x \cot^2 x\, dx$

38. $\int (\cot x \csc x)^3\, dx$

39. $\int \csc x \cot^2 x\, dx$

40. $\int \cos^3 x \sec^2 x\, dx$

41. $\int \tan^2 x \csc x\, dx$

42. $\int \sin x \tan x\, dx$

43. $\int \tan x \cos^5 x\, dx$

44. $\int \frac{3}{\sin^2(\pi x)}\, dx$

45. $\int \frac{\sin^2 x}{\cos^4 x}\, dx$

46. $\int \frac{\sin^3 x}{\cos x}\, dx$

47. $\int \sin x \cos x \ln(\cos x)\, dx$

48. $\int (\sin x \sqrt{\cos x})^3\, dx$

■ Calculate the integrals below, and then check your answer by differentiating (and using algebra and trigonometric identities, if necessary).

49. $\int \cos^2 3x\, dx$

50. $\int \sin^2 x \cos^3 x\, dx$

51. $\int \sec x \tan^3 x\, dx$

■ Calculate the integrals below in two different ways (as indicated); then use algebra and trigonometric identities to show that your answers are the same (up to a constant).

52. $\int \sin^2 x \cos^2 x\, dx$, (**a**) by using double-angle identities to write both $\sin^2 x$ and $\cos^2 x$ in terms of $\cos 2x$, and (**b**) by using the identity $\sin x \cos x = \frac{1}{2} \sin 2x$.

53. $\int \sec^2 x \tan^3 x \, dx$, (a) by integration by substitution with $u = \tan x$, and (b) by integration by substitution with $u = \sec x$.

■ **Calculate the exact values of the following definite integrals. Check each answer by approximating an appropriate area on your calculator.**

54. $\int_{-\pi}^{\pi} \sin^2 x \, dx$

55. $\int_{0}^{\pi} \sin^3 x \cos^2 x \, dx$

56. $\int_{0}^{\frac{\pi}{4}} \sec^4 x \tan^2 x \, dx$

Proofs

57. Prove that $\int \sec x \, dx = \ln|\sec x + \tan x| + C$ in two ways:

    (a) By calculating $\int \sec x \, dx$. (*Hint:* You'll need to do some algebra first.)

    (b) By showing that $\frac{d}{dx}(\ln|\sec x + \tan x|) = \sec x$.

58. Prove that $\int \cot x \, dx = \ln|\sin x| + C$.

59. Prove that $\int \csc x \, dx = -\ln|\csc x + \cot x| + C$ in two ways:

    (a) By calculating $\int \csc x \, dx$. (*Hint:* You'll need to do some algebra first.)

    (b) By showing that $\frac{d}{dx}(-\ln|\csc x + \cot x|) = \csc x$.

60. Prove that $\int \csc x \, dx = -\ln|\csc x + \cot x| + C$.

61. The techniques in this section can help us solve many integrals, but they don't tell us how to solve the very simple-seeming integral $\int \sin 2x \cos 3x \, dx$. The following trigonometric identity can help:

$$\sin \alpha \cos \beta = \tfrac{1}{2}(\sin(\alpha - \beta) + \sin(\alpha + \beta)).$$

    (a) Explain why the techniques covered in the reading do not apply to the integral $\int \sin 2x \cos 3x \, dx$.

    (b) Use the trigonometric identity above to solve $\int \sin 2x \cos 3x \, dx$.

    (c) Use the sum and difference identities for sine to prove the trigonometric identity above.

62. As you may have noticed, it is rather difficult to calculate integrals of powers of secant (especially if the power is large). The following "reduction formula" can help; for $k > 1$, we have

$$\int \sec^k x \, dx = \tfrac{1}{k-1} \sec^{k-2} x \tan x + \tfrac{k-2}{k-1} \int \sec^{k-2} x \, dx.$$

    (a) Use the reduction formula above to find $\int \sec^3 x \, dx$ and $\int \sec^7 x \, dx$. You may have to apply the formula more than once.

    (b) Why is this formula called a "reduction" formula?

    (c) Prove the reduction formula above using integration by parts. (*Hint:* Choose $dv = \sec^2 x \, dx$.)

14.4   Trigonometric Substitution

Our final integration technique involves using inverse trigonometric functions to change variables. The technique of ***trigonometric substitution*** employs a clever method of "backwards" substitution so that the change of variables seems to involve trigonometric (rather than inverse trigonometric) functions.

14.4.1   A substitution with an inverse trigonometric function

Consider the integral

$$\int \frac{1}{x^2\sqrt{1 - x^2}} \, dx.$$

How can we solve this integral? It does not seem to be an integral that we can solve using parts or substitution. However, as we show in the following example, a very nonintuitive substitution is possible.

EXAMPLE 14.28   **Using an inverse trigonometric function as a substitution**

Apply the unlikely substitution $u = \sin^{-1} x$ to the integral $\int \frac{1}{x^2\sqrt{1 - x^2}} \, dx.$

Solution   Although there are no inverse trigonometric functions in this integral, the derivative $\frac{d}{dx}(\sin^{-1} x) = \frac{1}{\sqrt{1-x^2}}$ is a part of the integrand, so we can at least hope that this substitution will work. We have

$$u = \sin^{-1} x \quad \Longrightarrow \quad \frac{du}{dx} = \frac{1}{\sqrt{1-x^2}} \quad \Longrightarrow \quad du = \frac{1}{\sqrt{1-x^2}}\, dx \ .$$

Moreover, since $u = \sin^{-1} x$, we also have

$$x = \sin u \ .$$

Using the last boxed equation, we can use a "back-substitution" to change variables:

$$\int \frac{1}{x^2\sqrt{1-x^2}}\, dx = \int \frac{1}{x^2}\frac{1}{\sqrt{1-x^2}}\, dx \qquad \text{(rewrite integrand)}$$

$$= \int \frac{1}{(\sin u)^2}\, du \qquad \text{(change variables)}$$

$$= \int \csc^2 u\, du \qquad \text{(simplify)}$$

$$= -\cot u\, du \qquad \text{(since } \frac{d}{dx}(\cot u) = -\csc^2 u)$$

$$= -\cot(\sin^{-1} x) + C. \qquad \text{(since } u = \sin^{-1} x) \qquad \square$$

Substitutions with inverse trigonometric functions can help solve a variety of integrals. However, it is easier to solve these integrals with a slightly different method of substitution called ***trigonometric substitution.*** Trigonometric substitution is a kind of "backwards" substitution, since we write x as a function of another variable u (rather than choosing a substitution u that is a function of x). In the following example, we will solve the integral from Example 14.28 by choosing $x = \sin u$ as our initial substitution (instead of choosing $u = \sin^{-1} x$).

EXAMPLE 14.29   **Substituting an expression for x**

Apply the "backwards" substitution $x = \sin u$ to the integral $\displaystyle \int \frac{1}{x^2\sqrt{1-x^2}}\, dx.$

Solution   Although the choice of substitution is essentially the same in this example as it was in Example 14.28, the mathematics is a little different. We can use $x = \sin u$ to find du in terms of dx by using implicit differentiation:

$$x = \sin u \quad \Longrightarrow \quad \frac{d}{dx}(x) = \frac{d}{dx}(\sin u) \qquad \textbf{(differentiate both sides)}$$

$$\Longrightarrow \quad 1 = \cos u\, \frac{du}{dx} \qquad \textbf{(chain rule; } u \textbf{ is a function of } x\textbf{)}$$

$$\Longrightarrow \quad dx = \cos u\, du \ . \qquad \textbf{(solve for } dx\textbf{)}$$

(We could also differentiate the equation $x = \sin u$ with respect to u and obtain $\frac{du}{dx} = \cos u$. This leads to the same result of $dx = \cos u\, du$.) We can now change variables, as follows:

$$\int \frac{1}{x^2\sqrt{1-x^2}}\, dx = \int \frac{1}{\sin^2 u\sqrt{1-\sin^2 u}}\cos u\, du \qquad \text{(change variables)}$$

$$= \int \frac{1}{\sin^2 u\sqrt{\cos^2 u}}\cos u\, du \qquad \text{(Pythagorean identity)}$$

$$= \int \frac{1}{\sin^2 u \cos u}\cos u\, du \qquad \text{(algebra)}$$

$$= \int \frac{1}{\sin^2 u}\, du \qquad \text{(algebra)}$$

$$= \int \csc^2 u\, du \qquad \text{(algebra)}$$

$$= -\cot u + C \qquad \left(\text{since } \tfrac{d}{dx}(\cot u) = -\csc^2 u\right)$$

$$= -\cot(\sin^{-1} x) + C. \qquad (\text{since } u = \sin^{-1} x)$$

The last step used the fact that since $x = \sin u$, we also have

$$u = \sin^{-1} x \ .$$

Notice that the answer here is exactly the same as the answer we found in Example 14.28. $\square$

The method in Example 14.29 involved more algebra (after changing variables) than the method used in Example 14.28, but at least it did not explicitly involve any inverse trigonometric functions. The key thing that made the substitution $x = \sin u$ work here was the expression $\sqrt{1-x^2}$ in the denominator of the integrand. Using a Pythagorean identity, we know that $\sqrt{1-\sin^2 u} = \cos u$ (which greatly simplifies the integral). As we will see in Section 14.4.3, the substitution $x = \sin u$ will often be useful for integrals involving the expression $\sqrt{1-x^2}$.

⊛ *Caution*   There are some important domain considerations to take into account when using trigonometric substitution. In Example 14.29 we decided to write x as the function $x = \sin u$. Since the range of the sine function is $[-1, 1]$, any work we do with $x = \sin u$ will assume that x is between -1 and 1. This is not a problem for the integrand $\frac{1}{x^2\sqrt{1-x^2}}$, since that function is defined only on $(-1, 1)$. However, if the integrand had been $\frac{1}{x^2(1-x^2)}$ instead, trigonometric substitution would not have been such a good idea, since that function is defined for values of x that are *not* between -1 and 1. $\square$

14.4.2 Rewriting trigonometric compositions

In the previous examples, we saw that $\int \frac{1}{x^2\sqrt{1-x^2}}\, dx = -\cot(\sin^{-1} x) + C$. However, as we saw in Section 11.2.4, we can always write the composition of a trigonometric function with an inverse trigonometric function as an *algebraic* function.

⊛ *Caution*   After solving integrals using trigonometric substitution, you are *always* expected to rewrite compositions like $-\cot(\sin^{-1} x)$ as algebraic functions (see Example 14.30). $\square$

EXAMPLE 14.30

Writing a trigonometric composition as an algebraic function

Simplify the answer to Example 14.29 by writing $\cot(\sin^{-1} x)$ as an algebraic function.

Solution In Example 14.29 we made the trigonometric substitution $x = \sin u$. That means that $u \in [-\frac{\pi}{2}, \frac{\pi}{2}]$. Figures 14.1 and 14.2 show reference triangles for the angle u in the first and fourth quadrants, respectively.

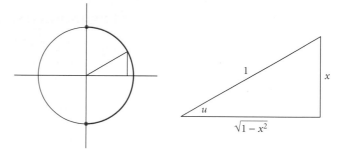

Figure 14.1 $u \in [0, \frac{\pi}{2}]$ and $\sin u = x$

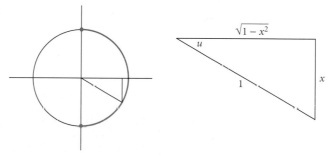

Figure 14.2 $u \in [-\frac{\pi}{2}, 0]$ and $\sin u = x$

Since in each case we wish to have $u = \sin x$, we label the sides of each triangle in such a way that the ratio $\frac{\text{opp}}{\text{hyp}}$ is equal to $\frac{x}{1} = x$. Using the Pythagorean Theorem, we can then find the length $\sqrt{1 - x^2}$ of the remaining side of the reference triangle for u (that quantity should look familiar; look at the original integrand!). Note that if $u \in [0, \frac{\pi}{2}]$, then x is a positive number (a positive height on the unit circle), whereas if $u \in [-\frac{\pi}{2}]$, then x is a negative number (a negative height on the unit circle). In each case the cotangent of the angle u is the ratio of the adjacent and opposite sides of the reference triangle for u; therefore,

$$\cot(\sin^{-1} x) = \cot u = \frac{\text{adj}}{\text{opp}} = \frac{\sqrt{1 - x^2}}{x}.$$

(Note that we do *not* write $\cot u = \frac{\sqrt{1-x^2}}{-x}$ in the second case, because x is already negative.) Now we can write the answer to Example 14.29 as

$$\int \frac{1}{x^2 \sqrt{1 - x^2}}\, dx = -\cot(\sin^{-1} x) + C = -\frac{\sqrt{1 - x^2}}{x} + C. \qquad \square$$

✓ *Checking the Answer* We can check the integral from Example 14.30 by differentiating (and doing quite a bit of algebra, which isn't included here):

$$\frac{d}{dx}\left(-\frac{\sqrt{1 - x^2}}{x}\right) = \frac{-\frac{1}{2}(1 - x^2)^{-\frac{1}{2}}(-2x)(x) - \left(-(1 - x^2)^{\frac{1}{2}}\right)(1)}{x^2} = \frac{1}{x^2 \sqrt{1 - x^2}}.$$

We could also check the answer by showing that $\frac{d}{dx}(-\cot(\sin^{-1} x)) = \frac{1}{x^2\sqrt{1-x^2}}$, but that would be much more difficult. $\qquad \square$

⊗ *Caution*   The method of using a triangle (and the unit circle) to simplify the composition of a trigonometric function with an inverse trigonometric function will *not* work for expressions like $\sin(2\cos^{-1} x)$. In this case, the constant multiple of 2 will cause a problem if we do not get rid of it first. We must use the double-angle identity first and then use triangles to write $\sin(\cos^{-1} x)$ as an algebraic function.

□

14.4.3 The method of trigonometric substitution

The technique developed in Section 14.4.2 is called ***trigonometric substitution*** and can be useful for solving many otherwise unsolvable integrals. The information in the box titled "What You Need for Trigonometric Substitution" below describes when to try trigonometric substitution with $x = a\sin u$, $x = a\tan u$, and $x = a\sec u$.

You don't need to *memorize* the three cases, since everything in each of the three columns makes sense in terms of the choice of substitution. For example, the choice of $x = a\sin u$ makes sense for integrals involving $\sqrt{a^2 - x^2}$ because we have a Pythagorean identity that makes $a^2 - a^2\sin^2 u$ a perfect square (namely $a^2\cos^2 u$). In this case, if $x = a\sin u$ we have $\sin u = \frac{x}{a}$, and so u can be thought of as an angle in a triangle with "opposite" side of length x and hypotenuse of length a. Note also that the domain and range of $x = a\sin u$ listed at the bottom of the first column come from the domain and range of the restricted sine function.

What You Need for Trigonometric Substitution			
When you see:	$\sqrt{a^2 - x^2}$	$x^2 + a^2$	$\sqrt{x^2 - a^2}$
Try the substitution:	$x = a\sin u$	$x = a\tan u$	$x = a\sec u$
Then use the identity:	$a^2 - a^2\sin^2 u$ $= a^2\cos^2 u$	$a^2\tan^2 u + a^2$ $= a^2\sec^2 u$	$a^2\sec^2 u - a^2$ $= a^2\tan^2 u$
May need the triangle:			
Domains and ranges:	$u \in \left[-\frac{\pi}{2}, \frac{\pi}{2}\right]$ $x \in [-a, a]$	$u \in \left(-\frac{\pi}{2}, \frac{\pi}{2}\right)$ $x \in (-\infty, \infty)$	$u \in \left[0, \frac{\pi}{2}\right) \cup \left(\frac{\pi}{2}, \pi\right]$ $x \in (-\infty, -a] \cup [a, \infty)$

Technical Note: Notice that we look for square roots in the first and last types of substitutions, but not in the second type. This is because we want the substitution to be compatible with the domain of the integrand. Note that $x = \sin u$ is defined only for $-1 \le x \le 1$, which is exactly the domain of $\sqrt{1 - x^2}$. On the other hand, $x = \sec u$ is defined only for $x \ge 1$ and $x \le -1$, which is exactly the domain of $\sqrt{x^2 - 1}$. The range of $x = \tan u$ is *all* real numbers, which is the domain of $x^2 + 1$; we don't need to see a square root in this case (although $x = \tan u$ works for integrals involving $\sqrt{x^2 + 1}$ as well).

Technical Note: The whole point of using trigonometric substitution is to be able to use a Pythagorean identity to write an expression like $a^2 - x^2$ as a perfect square (so that its square root will be a simple expression). For example,

$$\sqrt{a^2 - a^2\sin^2 u} = \sqrt{a^2\cos^2 u} = |a\cos u| = a\cos u.$$

The absolute value appears because for any number A, we have $\sqrt{A^2} = |A|$. Luckily, in this case we can drop the absolute value sign because we are assuming $a > 0$, and $\cos u$ is always positive for $u \in [-\frac{\pi}{2}, \frac{\pi}{2}]$ (the restricted domain of sine). Similarly, we can write

$$\sqrt{a^2 \tan^2 u + a^2} = \sqrt{a^2 \sec^2 u} = |a \sec u| = a \sec u,$$

because $\sec u$ is always positive for $u \in (-\frac{\pi}{2}, \frac{\pi}{2})$. We aren't so lucky in the third case, however:

$$\sqrt{a^2 \sec^2 u - a^2} = \sqrt{a^2 \tan u} = |a \tan u| = \begin{cases} a \tan u, & \text{if } x < -a \\ -a \tan u, & \text{if } x > a. \end{cases}$$

We need two cases this time because $\tan u$ is positive for $u \in [0, \frac{\pi}{2})$ (this corresponds to $x \in (-\infty, -a]$) and negative for $u \in (\frac{\pi}{2}, \pi]$ (this corresponds to $x \in [a, \infty)$). Therefore, when we use the substitution $x = a \sec u$, we will have to consider two cases: the case where $x < -a$ and the case where $x > a$; see Example 14.31 below.

Technical Note: At the end of many trigonometric substitution problems, you will have to rewrite compositions of trigonometric and inverse trigonometric functions as algebraic functions. When we did this in Example 14.30, we were careful to consider the appropriate quadrants of the unit circle. Luckily, it turns out that when using the substitutions $x = a \sin u$ and $x = a \tan u$, we do *not* have to consider the unit circle. For example, when using the substitution $x = a \sin u$ (as in Example 14.30), we must consider the first and fourth quadrants. Although x may be positive or negative, we don't have to "add" any signs when computing the trigonometric functions in either quadrant. A similar thing happens when we choose $x = a \tan u$. Unfortunately, we *do* have to take quadrants into account when using the substitution $x = a \sec u$ (see Example 14.31).

Trigonometric substitution is one of the more difficult integration techniques, so you should usually try a simpler substitution, algebra, or even integration by parts before resorting to trigonometric substitution. Once you have decided to apply trigonometric substitution, apply the following algorithm.

ALGORITHM 14.5

How to Apply Trigonometric Substitution

1. Choose $x = a \sin u$, $x = a \tan u$, or $x = a \sec u$, as appropriate.
2. Use the choice of x to write dx in terms of du.
3. Change variables using what you found above, and simplify as much as possible.
4. Solve the "new" integral in whatever way seems most appropriate, and write the answer in terms of x.
5. Rewrite any compositions of trigonometric and inverse trigonometric functions.

After making a trigonometric substitution, we often end up with an integral that involves products and/or powers of trigonometric functions (like the ones in Section 14.3). The following example involves such an integral and also illustrates how to handle the cases required when secant is used for trigonometric substitution.

EXAMPLE 14.31

A trigonometric substitution using secant

Use trigonometric substitution to find $\displaystyle\int \frac{1}{\sqrt{x^2 - 4}}\, dx$.

Solution   This integral involves the expression $x^2 - 4 = x^2 - 2^2$, so we will use the trigonometric substitution $x = 2 \sec u$:

$$x = 2 \sec u \quad \Longrightarrow \quad dx = 2 \sec u \tan u \, du \ .$$

With this substitution, we have

$$\int \frac{1}{\sqrt{x^2 - 4}} \, dx = \int \frac{1}{\sqrt{4 \sec^2 u - 4}} (2 \sec u \tan u) \, du \qquad \text{(trigonometric substitution)}$$

$$= \int \frac{1}{\sqrt{4 \tan^2 u}} (2 \sec u \tan u) \, du. \qquad \text{(Pythagorean identity)}$$

Remember that since we are making a trigonometric substitution with secant, we must consider the cases $x > 2$ and $x < -2$ separately; we will start with the case $x > 2$. If $x > 2$, then $2 \sec u > 2$ and thus $\sec u > 1$. This means that we must have $u \in [0, \frac{\pi}{2})$; think about the unit circle or the graph of secant to see why. In particular, this means that $\tan u$ is positive and thus that $\sqrt{4 \tan^2 u} = |2 \tan u| = 2 \tan u$. Continuing the calculation, we have

$$= \int \frac{1}{2 \tan u} (2 \sec u \tan u) \, du \qquad \text{(since } \tan u > 0)$$

$$= \int \sec u \, du \qquad \text{(algebra)}$$

$$= \ln | \sec u + \tan u| + C. \qquad \text{(integral of secant)}$$

Since $x = 2 \sec u$, we must have $\frac{x}{2} = \sec u$, and therefore $u = \sec^{-1} \frac{x}{2}$:

$$= \ln \left| \sec \left(\sec^{-1} \tfrac{x}{2} \right) + \tan \left(\sec^{-1} \tfrac{x}{2} \right) \right| + C \qquad \text{(since } u = \sec^{-1} \tfrac{x}{2})$$

$$= \ln \left| \tfrac{x}{2} + \tan \left(\sec^{-1} \tfrac{x}{2} \right) \right| + C. \qquad \text{(algebra)}$$

It now remains only to rewrite $\tan(\sec^{-1} \frac{x}{2})$ as an algebraic function. At the moment, we are considering the case where $x > 2$ and thus $u \in [0, \frac{\pi}{2})$. Therefore, u is an angle in the first quadrant, with $\sec u = \frac{x}{2}$. Figure 14.3 shows the angle u in the unit circle and the reference triangle for u such that $\sec u = \frac{x}{2}$. From the triangle, we see that

$$\tan \left(\sec^{-1} \frac{x}{2} \right) = \tan u = \frac{\text{opp}}{\text{adj}} = \frac{\sqrt{x^2 - 4}}{2}.$$

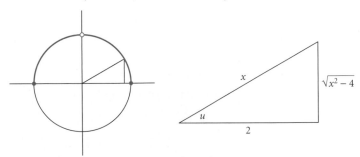

Figure 14.3 $u \in [0, \frac{\pi}{2})$ and $\sec u = \frac{x}{2}$

Therefore, we have (in the case where $x > 2$)

$$\int \frac{1}{\sqrt{x^2 - 4}} \, dx = \ln \left| \tfrac{x}{2} + \tan(\sec^{-1} \tfrac{x}{2}) \right| + C$$

$$= \ln \left| \frac{x}{2} + \frac{\sqrt{x^2 - 4}}{2} \right| + C.$$

The case where $x < -2$, and thus $u \in (\frac{\pi}{2}, \pi]$, is similar. In fact, there are only two differences: The first is that in this second case we have $\sqrt{4 \tan^2 u} = |2 \tan u| = -2 \tan u$, since $\tan u$ is negative for $u \in (\frac{\pi}{2}, \pi]$. This means that before integrating, we have to insert a negative sign. By the end of the integration step, we still have a negative sign, so for $x < -2$ we have

$$\int \frac{1}{\sqrt{x^2 - 4}}\, dx = -\ln|\sec u + \tan u| + C = -\ln\left|\frac{x}{2} + \tan\left(\sec^{-1}\frac{x}{2}\right)\right| + C.$$

The second difference is that since $u \in (\frac{\pi}{2}, \pi]$ we must work in the second quadrant. Figure 14.4 shows the angle u in the unit circle and the reference triangle for this angle (note that the reference angle v is not the same as the angle u). Note in particular that the "adjacent" side now represents a *negative* length. Therefore, in this case, we have $\tan(\sec^{-1}\frac{x}{2}) = \tan u = \frac{\sqrt{x^2-4}}{-2}$. Therefore, in the case where $x < -2$,

$$\int \frac{1}{\sqrt{x^2 - 4}}\, dx = -\ln\left|\frac{x}{2} + \tan\left(\sec^{-1}\frac{x}{2}\right)\right| + C$$

$$= -\ln\left|\frac{x}{2} - \frac{\sqrt{x^2 - 4}}{2}\right| + C.$$

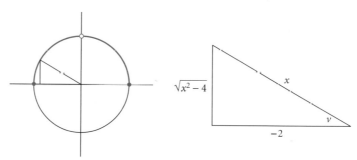

Figure 14.4 $u \in (\frac{\pi}{2}, \pi]$ and $\sec u = \frac{x}{2}$

Figures 14.5 and 14.6 show the graphs of the function $f(x) = \frac{1}{\sqrt{x^2-4}}$ and one of its antiderivatives $F(x)$ (defined by the answers to the two cases above).

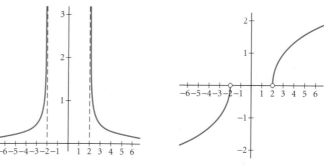

Figure 14.5
$$f(x) = \frac{1}{\sqrt{x^2 - 4}}$$

Figure 14.6
$$F(x) = \begin{cases} \ln\left|\frac{x}{2} + \frac{\sqrt{x^2-4}}{2}\right|, & \text{if } x > 2 \\ -\ln\left|\frac{x}{2} - \frac{\sqrt{x^2-4}}{2}\right|, & \text{if } x < -2 \end{cases}$$

You should be able to convince yourself from looking at these graphs that it is at least reasonable that the derivative of $F(x)$ is equal to $f(x)$. Notice also that it makes sense that we had to look at two separate cases to integrate $f(x)$, since $f(x)$ is defined on two different intervals (and not between them). □

◈ *Caution*   You do not have to consider separate cases when using trigonometric substitution with $x = a \sin u$ or with $x = a \tan u$. However, don't forget to consider the cases $x > a$ and $x < -a$ separately when using the trigonometric substitution $x = a \sec u$.   □

Since trigonometric substitution is such a complicated process, it is almost always better to use some other method of integration if you can. In the following example, each integral looks like it might be a good candidate for trigonometric substitution, but trigonometric substitution is necessary for only *one* of these integrals.

EXAMPLE 14.32   **Deciding when to use trigonometric substitution**

Which of the following integrals would be good candidates for trigonometric substitution?

(a) $\displaystyle\int \frac{x}{3 + x^2}\, dx$ (b) $\displaystyle\int \frac{x^2}{3 + x^2}\, dx$ (c) $\displaystyle\int \frac{3 + x^2}{x^2}\, dx$

Solution

(a) Although we *could* try a trigonometric substitution here, with $x = \sqrt{3}\tan u$, we could also use the more conventional substitution $u = 3 + x^2$ (since then $x\, dx = \frac{1}{2}\, du$). Conventional substitution is much easier; try both and see!

(b) There does not seem to be a conventional substitution that will work for this problem, and at first glance it does not seem that integration by parts would help. This integral is a good candidate for trigonometric substitution, with $x = \sqrt{3}\tan u$.

(c) Although you could certainly try trigonometric substitution (with $x = \sqrt{3}\tan u$), it's *much* easier to use algebra instead, using the fact that $\frac{3 + x^2}{x^2} = \frac{3}{x^2} + \frac{x^2}{x^2} = 3x^{-2} + 1$ (which is easy to integrate).   □

14.4.4 Trigonometric substitution after completing a square

Sometimes an integral can be converted into one that involves one of the forms $a^2 - x^2$, $x^2 + a^2$, or $x^2 - a^2$ by "completing the square" of a quadratic (this simply means writing the quadratic as a sum or difference of perfect squares). Example 14.33 involves such an integral. This example also illustrates that we can choose even more general trigonometric substitutions than $x = a \sin u$, $x = a \tan u$, and $x = a \sec u$.

EXAMPLE 14.33   **Completing the square so that we can make a trigonometric substitution**

Find $\displaystyle\int \frac{1}{\sqrt{x^2 - 6x + 13}}\, dx$.

Solution   After looking at this integral for a while, you should be convinced that parts and conventional substitution won't help. But to try trigonometric substitution, we would like to have $a^2 - x^2$, $x^2 + a^2$, or $x^2 - a^2$ in the integrand, which we don't. However, notice what happens when we complete the square for the quadratic $x^2 - 6x + 13$:

$$x^2 - 6x + 13 = (x^2 - 6x + 9) - 9 + 13 = (x - 3)^2 + 4.$$

We added and subtracted 9 from the quadratic (effectively adding 0), so that the quadratic would involve the perfect square $x^2 - 6x + 9 = (x - 3)^2$. In general, to complete the square of a quadratic $x^2 + bx + c$, you should add and subtract $(\frac{b}{2})^2$. This will enable you to write the quadratic in terms of the perfect square $(x - \frac{b}{2})^2$ (and some other constant). Now that we have completed the square, we have

$$\int \frac{1}{\sqrt{x^2 - 6x + 13}}\, dx = \int \frac{1}{\sqrt{(x-3)^2 + 4}}\, dx.$$

The part of the integrand inside the radical is *almost* of the form $x^2 + a^2$. Instead, it is of the form $(x - 3)^2 + a^2$ (with $a = 2$). Therefore, we will choose $x - 3$ to be equal to $2 \tan u$:

$$x - 3 = 2 \tan u \quad \Longrightarrow \quad dx = 2 \sec^2 u\, du \;.$$

With this trigonometric substitution, we have

$$\int \frac{1}{\sqrt{x^2 - 6x + 13}}\, dx = \int \frac{1}{\sqrt{(x-3)^2 + 4}}\, dx \qquad \text{(complete the square)}$$

$$= \int \frac{1}{\sqrt{(2 \tan u)^2 + 4}} (2 \sec^2 u)\, du \qquad \text{(with } x - 3 = 2 \tan u\text{)}$$

$$= \int \frac{1}{\sqrt{4 \tan^2 u + 4}} (2 \sec^2 u)\, du \qquad \text{(algebra)}$$

$$= \int \frac{1}{2 \sec u} (2 \sec^2 u)\, du \qquad \text{(Pythagorean identity)}$$

$$= \int \sec u\, du \qquad \text{(algebra)}$$

$$= \ln |\sec u + \tan u| + C. \qquad \text{(integral of secant)}$$

Since we chose $x - 3 = 2 \tan u$, we have

$$x - 3 = 2 \tan u \quad \Longrightarrow \quad \frac{x - 3}{2} = \tan u \quad \Longrightarrow \quad u = \tan^{-1}\left(\frac{x-3}{2}\right).$$

Continuing the calculation, this means that

$$\int \frac{1}{\sqrt{x^2 - 6x + 13}}\, dx = \ln \left| \sec\left(\tan^{-1}\left(\frac{x-3}{2}\right)\right) + \tan\left(\tan^{-1}\left(\frac{x-3}{2}\right)\right) \right| + C.$$

All that remains now is to write the expressions $\sec(\tan^{-1}(\frac{x-3}{2}))$ and $\tan(\tan^{-1}(\frac{x-3}{2}))$ as algebraic functions. Clearly the second expression is equal to $\frac{x-3}{2}$. The first expression can be found using a triangle, as follows: Since $\tan u = \frac{x-3}{2}$, we will draw a triangle with angle u that has "opposite" side of length $x - 3$ and "adjacent" side of length 2, as in Figure 14.7. (Remember that we don't have to think about the unit circle or quadrants when using trigonometric substitution with tangent.)

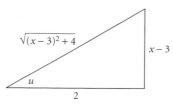

Figure 14.7
A triangle with $\tan u = \frac{x-3}{2}$

By the Pythagorean Theorem, the length of the hypotenuse of this triangle is $\sqrt{(x-3)^2 + 4}$. We can now see that

$$\sec\left(\tan^{-1}\left(\frac{x-3}{2}\right)\right) = \sec(u) = \frac{\text{hyp}}{\text{adj}} = \frac{\sqrt{(x-3)^2 + 4}}{2}.$$

Therefore, our final answer is

$$\int \frac{1}{\sqrt{x^2 - 6x + 13}}\,dx = \ln\left|\frac{\sqrt{(x-3)^2 + 4}}{2} + \frac{x-3}{2}\right| + C.$$

□

What you should know after reading Section 14.4

▶ What trigonometric substitution is, how it works, and why it works. What does trigonometric substitution have to do with inverse trigonometric functions? With Pythagorean identities? With implicit differentiation? With triangles?

▶ How to solve integrals using trigonometric substitutions ($x = a\sin u$, $x = a\tan u$, or $x = a\sec u$). This includes being able to write the answers to such integrals in as simple a form as possible. Which type of trigonometric substitution often involves an analysis of two cases, and why? For which type of trigonometric substitution must you consider the quadrants of the unit circle, and why?

▶ When to apply trigonometric substitution, and when *not* to. How to use trigonometric substitution to solve integrals that involve quadratics (after completing the square).

Exercises 14.4

Concepts

0. Read the section and make your own summary of the material. Include whatever you think will help you review the material later on.

1. Explain why the substitution $u = \sin^{-1} x$ in Example 14.28 was sensible (and worked), even though the integrand did not involve the inverse sine function.

2. Explain why the substitution $x = \sin u$ in Example 14.29 is equivalent to the substitution $u = \sin^{-1} x$ in Example 14.28.

3. Show that if $x = \tan u$, then $dx = \sec^2 u \, du$ in two different ways: **(a)** by using implicit differentiation, thinking of u as a function of x, and **(b)** by thinking of x as a function of u.

4. For each part below, write down an integral that satisfies the given criteria.

   (a) An integral to which you can apply trigonometric substitution with $x = \tan u$.

   (b) An integral to which you can apply trigonometric substitution with $x = 4\sec u$.

   (c) An integral to which you can apply trigonometric substitution with $x - 2 = 3\sin u$.

   (d) An integral for which trigonometric substitution is possible, but an easier method is available.

   (e) An integral that we don't know how to solve with the techniques we know at this point.

5. Show by differentiating (and then using algebra) that $-\cot(\sin^{-1} x)$ and $-\frac{\sqrt{1-x^2}}{x}$ are both antiderivatives of $\frac{1}{x^2\sqrt{1-x^2}}$. How can these two very different-looking functions be antiderivatives of the same function?

6. Explain how to tell when to use the trigonometric substitutions $x = \sin u$, $x = \tan u$, and $x = \sec u$. Describe the trigonometric identity and the triangle that will be needed in each case. What are the possible values for x and u in each case?

7. Explain how to tell when to use the trigonometric substitutions $x = a\sin u$, $x = a\tan u$, and $x = a\sec u$. Describe the trigonometric identity and the triangle that will be needed in each case. What are the possible values for x and u in each case?

8. Why do you think we don't ever bother to use the trigonometric substitution $x = \cos u$?

9. Explain why it makes sense to try the trigonometric substitution $x = \sec u$ if an integrand involves the expression $\sqrt{x^2 - 1}$.

10. Why don't we need to have a square root involved in order to apply trigonometric substitution with $x = a \tan u$? In other words, why can we use the substitution $x = a \tan u$ when we see $x^2 + a^2$, even though we can't use the substitution $x = a \sin u$ unless the integrand involves the *square root* of $a^2 - x^2$?

11. Explain why using trigonometric substitution with $x = a \tan u$ often involves a triangle with side lengths a and x and hypotenuse of length $\sqrt{x^2 + a^2}$.

12. Why can't we use trigonometric substitution to solve the integral $\int \dfrac{1}{x^2 - 4}\, dx$?

13. Why can we use trigonometric substitution to solve the integral $\int \dfrac{1}{x^2 + 4}\, dx$, even though a square root is not involved?

14. Could trigonometric substitution be applied to the integral $\int \dfrac{1}{(5 - x^2)^{\frac{3}{2}}}\, dx$, and why? (*Hint:* Does this integral involve a square root?) If so, what substitution would you use?

15. Explain why if $x = a \sin u$, then $\sqrt{a^2 - x^2} = a \cos u$. Your explanation should include a discussion of domains and absolute values.

16. Explain why if $x = a \sec u$, then
$$\sqrt{a^2 \sec^2 u - a^2} = \begin{cases} a \tan u, & \text{if } x < -a \\ -a \tan u, & \text{if } x > a. \end{cases}$$
Your explanation should include a discussion of domains and absolute values.

17. Why do we need to consider the cases $x > a$ and $x < -a$ separately when using trigonometric substitution with $x = a \sec u$?

18. Why is it okay to use a triangle *without* thinking about the unit circle when simplifying expressions that result from a trigonometric substitution of $x = a \sin u$ or $x = a \tan u$? Why do we need to think about the unit circle after trigonometric substitution with $x = a \sec u$?

19. Why do we need to consider the cases $x > a$ and $x < -a$ separately when using trigonometric substitution with $x = a \sec u$?

20. Why doesn't the definite integral $\int_2^3 \sqrt{1 - x^2}\, dx$ make sense? (*Hint:* Think about domains.)

21. Which of the following would be good candidates for trigonometric substitution? If trigonometric substitution is a good strategy, name the substitution. If another method is a better strategy, explain that method.

(a) $\int \dfrac{4 + x^2}{x}\, dx$ (b) $\int \dfrac{x}{4 + x^2}\, dx$

(c) $\int \dfrac{x^2}{4 + x^2}\, dx$

22. Determine strategies for solving each of the integrals in the odd-numbered problems among Problems 45–62. Describe the trigonometric substitution, conventional substitution, parts, or algebra that you would use for each problem. (You may want to do this problem after you have solved some of these integrals.)

Skills

■ Use triangles and the unit circle to solve each of the following problems.

23. If $x = \sin u$, write $\cos u$ as an algebraic function of x.

24. If $x = \sec u$, write $\cot u$ as an algebraic function of x.

25. If $x = 2 \tan u$, write $\sin u$ as an algebraic function of x.

26. If $x + 3 = 4 \tan u$, write $\sec u$ as an algebraic function of x.

27. If $x - 5 = 3 \sin u$, write $\tan^2 u$ as an algebraic function of x.

28. Write $\sin(2 \cos^{-1} x)$ as an algebraic function.

29. Write $\cos(2 \tan^{-1} x)$ as an algebraic function.

■ Complete the square for each quadratic below. Then describe the trigonometric substitution that would be appropriate if you were solving an integral that involved that quadratic.

30. $x^2 + 6x - 2$ **31.** $x^2 - 4x - 8$

32. $x^2 - 5x + 1$ **33.** $2x^2 - 4x + 1$

34. $x - 3x^2$ **35.** $4 - 8x - 2x^2$

■ Calculate the integrals below in two or more different ways (as indicated); then use algebra to show that your answers are the same (up to a constant).

36. $\int \dfrac{x^2}{1 + x^2}\, dx$, by (a) the substitution $u = \tan^{-1} x$, and (b) the trigonometric substitution $x = \tan u$.

37. $\int \dfrac{1}{\sqrt{1 - x^2}}\, dx$, by (a) recognizing the derivative of $\sin^{-1} x$, and (b) trigonometric substitution.

38. $\int \dfrac{1}{x^2 + 9}\, dx$, by (a) trigonometric substitution, and (b) algebra and the derivative of arctangent.

39. $\int \dfrac{x}{4 + x^2}\, dx$, by (a) conventional substitution, and (b) trigonometric substitution.

40. $\int \dfrac{x + 2}{(x^2 + 4x)^{\frac{3}{2}}}\, dx$, with (a) conventional substitution, and (b) trigonometric substitution.

41. $\int x^3 \sqrt{x^2 - 1}\, dx$, by **(a)** integration by parts, **(b)** conventional substitution, and **(c)** trigonometric substitution.

■ Calculate the integrals below, and then check your answers by differentiating (and using algebra, if necessary).

42. $\int \dfrac{1}{(x^2 + 4)^{\frac{3}{2}}}\, dx$ **43.** $\int \dfrac{1}{\sqrt{3 - x^2}}\, dx$

44. $\int \dfrac{\sqrt{x^2 - 4}}{2x}\, dx$

■ Solve the following integrals using whatever method is most appropriate. (*Hints:* Not all of these integrals require trigonometric substitution, and some integrals require algebraic manipulation.)

45. $\int x^3 \sqrt{x^2 + 1}\, dx$ **46.** $\int x\sqrt{x^2 + 1}\, dx$

47. $\int \dfrac{\sqrt{4 - x^2}}{x^2}\, dx$ **48.** $\int \dfrac{\sqrt{x^2 - 1}}{x}\, dx$

49. $\int \dfrac{x}{\sqrt{x^2 - 1}}\, dx$ **50.** $\int \dfrac{1}{x^2 \sqrt{4 - x^2}}\, dx$

51. $\int \dfrac{1}{x^2 \sqrt{x^2 - 9}}\, dx$ **52.** $\int \dfrac{1}{x^2 \sqrt{x^2 + 3}}\, dx$

53. $\int \dfrac{x}{\sqrt{1 - 4x^2}}\, dx$ **54.** $\int \dfrac{1}{\sqrt{x^2 + 9}}\, dx$

55. $\int (1 - x^2)^{-\frac{3}{2}}\, dx$ **56.** $\int \dfrac{1}{x^2 - 4x + 13}\, dx$

57. $\int \dfrac{3 + x}{\sqrt{9 - 4x^2}}\, dx$ **58.** $\int x^5 \sqrt{x^2 - 1}\, dx$

59. $\int \sqrt{x^2 - 8x + 25}\, dx$ **60.** $\int \dfrac{x^3}{3x^2 + 5}\, dx$

61. $\int (3 - x^2)^{\frac{3}{2}}\, dx$ **62.** $\int \dfrac{1}{(x^2 + 1)^{\frac{5}{2}}}\, dx$

63. $\int \sqrt{2 + x^2}\, dx$ **64.** $\int \sqrt{x^2 - 2}\, dx$

65. $\int \sqrt{2 - x^2}\, dx$ **66.** $\int \sqrt{x^2 + 6x + 18}\, dx$

67. $\int \ln(x^2 + 1)\, dx$ **68.** $\int x \sin^{-1} x\, dx$

69. $\int e^{2x} \sqrt{e^{2x} + 1}\, dx$ **70.** $\int e^{2x}(1 - e^{4x})^{\frac{3}{2}}\, dx$

71. $\int \dfrac{1}{e^x \sqrt{e^{2x} + 3}}\, dx$

■ Solve each of the following integrals by using polynomial long division to rewrite the integrand. (This is one way that you can sometimes avoid using trigonometric substitution; moreover, sometimes it works when trigonometric substitution does not apply.)

72. $\int \dfrac{x^3}{x^2 + 4}\, dx$ **73.** $\int \dfrac{x^2 - 1}{x^2 + 1}\, dx$

74. $\int \dfrac{x^4 - 3}{2 + 3x^2}\, dx$

■ Calculate the exact values of the following definite integrals. Check each answer by approximating an appropriate area on your calculator.

75. $\int_0^4 x\sqrt{x^2 + 4}\, dx$ **76.** $\int_{\frac{1}{4}}^{\frac{1}{2}} \dfrac{1}{x^2 \sqrt{1 - x^2}}\, dx$

77. $\int_3^5 \sqrt{x^2 - 9}\, dx$ **78.** $\int_{-1}^1 x^3 \sqrt{9x^2 - 1}\, dx$

79. $\int_0^4 x^3 \sqrt{x^2 + 4}\, dx$ **80.** $\int_1^2 \dfrac{3}{(9 - 2x^2)^{\frac{3}{2}}}\, dx$

Proof

81. Prove that the area of a circle of radius r is πr^2, as follows:

   (a) Write down a definite integral that represents the area of the circle of radius r centered at the origin. (*Hints:* The equation of such a circle is $x^2 + y^2 = r^2$; you can use this to find an equation for the top curve of the circle. The area of the circle is twice the area of the top half of the circle.)

   (b) Use trigonometric substitution to solve the definite integral above. (*Hint:* It will be easier if you change the limits of integration to match your substitution.)

CHAPTER 15

Applications of Integration

*I*n this chapter we will investigate various
geometrical and practical applications of the definite
integral. We will use the "subdivide, approximate,
and add" process to construct definite integrals that
represent the length of a curve, the volume of a solid
of revolution, and practical applications such as
work, mass, and hydrostatic force.

"Endless Ribbon,"
sculpture by Max Bill.
© Photo RMN/Art
Resource, NY

CONTENTS

<table>
<tr><td>**15.1**</td><td># Arc Length</td></tr>
</table>

In Chapter 12, we used definite integrals to define the area under the graph of a function on an interval. We will now use definite integrals to define the "length" of the graph of a function on an interval.

15.1.1 Approximating arc length

What is the circumference of a circle of radius 2? You probably know that the circumference C of a circle of radius r is given by $C = 2\pi r$; therefore, the circumference of a circle of radius 2 must be equal to $C = 2\pi(2) = 4\pi$. But where does that magical formula $C = 2\pi r$ come from, anyway? If we didn't know this formula, how could we approximate the circumference of a circle?

To make this problem simpler, let's consider only the top half of a circle of radius 2 and try to find the length of the curve that defines that half. Since the equation of a circle of radius 2 centered at the origin is $x^2 + y^2 = 2^2$, the function that defines the top half of the circle is $f(x) = \sqrt{4 - x^2}$. Figure 15.1 shows the graph of this function on $[-2, 2]$.

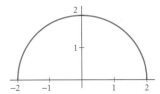

Figure 15.1
$f(x) = \sqrt{4 - x^2}$ on $[-2, 2]$

One way to approximate the length of this curve would be to follow the "subdivide, approximate, and add" strategy that we used for approximating areas; however, this time we will approximate lengths instead of areas. For example, we could divide the interval $[-2, 2]$ into four subintervals and approximate the length of the curve on each subinterval with a line segment. We do this in the following example.

<table>
<tr><td>EXAMPLE 15.1</td><td>**Using line segments to approximate the length of a curve**</td></tr>
</table>

Approximate the length of the curve $f(x) = \sqrt{4 - x^2}$ from $x = -2$ to $x = 2$ with four line segments.

Solution   Let's divide the interval $[-2, 2]$ into four subintervals of equal size: $[-2, -1]$, $[-1, 0]$, $[0, 1]$, and $[1, 2]$. To approximate the length of the curve over each subinterval, consider the line segments shown in Figure 15.2.

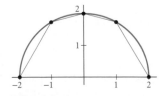

Figure 15.2
Length of $f(x) = \sqrt{4 - x^2}$ on $[-2, 2]$ approximated by four line segments

It is easy to determine the length of each of these line segments using the distance formula. For example, the first line segment begins at the point $(-2, f(-2)) = (-2, 0)$

and ends at the point $(-1, f(-1)) = (-1, \sqrt{3})$. Thus it has length

$$\sqrt{(-1 - (-2))^2 + (f(-1) - f(-2))^2} = \sqrt{(-1 - (-2))^2 + (\sqrt{3} - 0)^2}$$
$$= \sqrt{1 + 3} = \sqrt{4} = 2.$$

If we use each line segment to approximate the length of the curve over each subinterval, then the sum of the lengths of the four line segments will approximate the length of the curve over the entire interval $[-2, 2]$. Therefore, we have

$$\text{Length} \approx \sqrt{(-1 - (-2))^2 + (f(-1) - f(-2))^2} + \sqrt{(0 - (-1))^2 + (f(0) - f(-1))^2}$$
$$+ \sqrt{(1 - (0))^2 + (f(1) - f(0))^2} + \sqrt{(2 - (1))^2 + (f(2) - f(1))^2}$$
$$\approx 2 + 1.03528 + 1.03528 + 2 = 6.07056.$$

Clearly this approximation should be an underapproximation of the actual length, since the lengths of the lines are shorter than the lengths of the curves they approximate. We "know" that the length of the curve should be $\frac{1}{2}(2\pi(2)) = 2\pi \approx 6.28319$ (half the circumference of a circle of radius 2), so our approximation is an underapproximation by about 0.21263 unit. □

To get a better approximation of the length of the curve, we could use more line segments; that is, we could subdivide the interval $[-2, 2]$ into more pieces (see Figure 15.3).

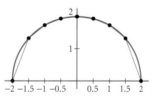

Figure 15.3
Length of $f(x) = \sqrt{4 - x^2}$
on $[-2, 2]$ approximated
by eight line segments

? *Question*   Subdivide the interval $[-2, 2]$ into eight pieces, each of length $\frac{1}{2}$, to approximate the length of $f(x) = \sqrt{4 - x^2}$ from $x = -2$ to $x = 2$. You should get an approximation of about 6.2338 units (note that this is closer to the exact answer than the approximation we found in Example 15.1). □

? *Question*   Why are the first and last line segments so much longer than the rest of the line segments in Figure 15.3 even though we are dividing the interval $[-2, 2]$ into equal pieces? As we move from each of the nine marked points on the circle to the next, the change in x-values is the same. Is the change in the corresponding y-values the same for each line segment? □

15.1.2 Defining arc length

Although you probably have a pretty good idea of what we mean when we say the "length" of a curve, at this point we do not have a mathematical description of "length" for anything but line segments. Since the approximation of a curve by line segments seems to approach what we would like to consider the "length" of that curve, this is how we will *define* that length. Before we state the formal definition of *arc length,* we need some notation.

Most of the notation involved in defining arc length is the same as the notation we used when we defined the area under a curve. If we divide an interval $[a, b]$ into n subintervals of equal width, then each subinterval will have width $\Delta x := \frac{b-a}{n}$. If we define $x_k := a + k \Delta x$ for $k = 0, 1, \ldots, n$, then each subinterval is of the form $[x_{k-1}, x_k]$.

Suppose we wish to approximate the length of a curve $f(x)$ on each subinterval using line segments. The kth line segment will connect the points $(x_{k-1}, f(x_{k-1}))$ and $(x_k, f(x_k))$. By the distance formula, the length of this line segment is

$$\sqrt{(x_k - x_{k-1})^2 + (f(x_k) - f(x_{k-1}))^2} = \sqrt{(\Delta x)^2 + (f(x_k) - f(x_{k-1}))^2}.$$

For *every* value k, the difference $x_k - x_{k-1}$ is equal to Δx. We will define Δy_k to be the difference $f(x_k) - f(x_{k-1})$ of the y-coordinates for each k. Note that, unlike Δx, the quantity Δy_k is *different* for each k (why?). The notation is represented visually in Figure 15.4.

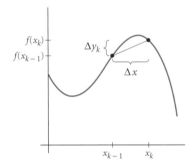

Figure 15.4

In this notation, the sum of the lengths of the k line segments can be written in sigma notation as follows:

$$\sum_{k=1}^{n} \sqrt{(\Delta x)^2 + (f(x_k) - f(x_{k-1}))^2}$$

$$= \sum_{k=1}^{n} \sqrt{(\Delta x)^2 + (\Delta y_k)^2} \qquad \text{(notation)}$$

$$= \sum_{k=1}^{n} \sqrt{(\Delta x)^2 \left(1 + \frac{(\Delta y_k)^2}{(\Delta x)^2}\right)} \qquad \text{(factor out } (\Delta x)^2)$$

$$= \sum_{k=1}^{n} \sqrt{1 + \left(\frac{\Delta y_k}{\Delta x}\right)^2} \; \Delta x. \qquad \text{(algebra)}$$

As n gets larger (and thus Δx gets smaller), this approximation gets closer to what we would like to consider the "arc length." Therefore, we *define* the arc length as the limit of this sum as $n \to \infty$.

DEFINITION 15.1

The Arc Length of a Function on an Interval

Suppose $f(x)$ is a continuous, differentiable function with a continuous derivative. The arc length of $f(x)$ from $x = a$ to $x = b$ is defined as a limit of sums, as follows:

$$\begin{array}{l} \text{Arc length of } f(x) \\ \text{from } x = a \text{ to } x = b \end{array} := \lim_{n\to\infty} \sum_{k=1}^{n} \sqrt{1 + \left(\frac{\Delta y_k}{\Delta x}\right)^2} \; \Delta x,$$

where $\Delta x := \frac{b-a}{n}$, $x_k := a + k\,\Delta x$, and $\Delta y_k := f(x_k) - f(x_{k-1})$ for all $k = 0, 1, 2, \dots, n$.

15.1.3 Arc length as an integral

The limit of sums in the definition of arc length should look familiar to you. It is in fact a kind of Riemann sum, and therefore it can be written as a definite integral.

THEOREM 15.1

Writing Arc Length as a Definite Integral

Suppose $f(x)$ is a continuous, differentiable function with a continuous derivative. The arc length of $f(x)$ from $x = a$ to $x = b$ can be written as a definite integral, as follows:

$$\text{Arc length of } f(x) \atop \text{from } x = a \text{ to } x = b = \int_a^b \sqrt{1 + (f'(x))^2}\, dx.$$

If you think about it for a moment, Theorem 15.1 says something very interesting: The length of a curve $f(x)$ on an interval $[a, b]$ is the same as the *area* under the graph of $\sqrt{1 + (f'(x))^2}$ on that same interval.

If you remember the definition of the definite integral, you should be able to see how the limit of sums in Definition 15.1 becomes the definite integral in Theorem 15.1, at least if you believe that $\frac{\Delta y_k}{\Delta x}$ becomes $f'(x)$ as $n \to \infty$. The key to showing this, and thus the key to the proof of Theorem 15.1, is the Mean Value Theorem.

PROOF (THEOREM 15.1) Suppose $f(x)$ is a continuous, differentiable function. Then the Mean Value Theorem applies to the function $f(x)$ on each subinterval $[x_{k-1}, x_k]$ and says that

(1)   *There exists some $x_k^* \in (x_{k-1}, x_k)$ such that $f'(x_k^*) = \dfrac{f(x_k) - f(x_{k-1})}{x_k - x_{k-1}}.$*

In the notation of the definition of arc length, this means that for each k between 1 and n, there is some x_k^* for which

$$f'(x_k^*) = \frac{f(x_k) - f(x_{k-1})}{x_k - x_{k-1}} = \frac{\Delta y_k}{\Delta x}.$$

Therefore, there is an x_k^* in each subinterval such that the definition of arc length can be written as

$$\lim_{n \to \infty} \sum_{k=1}^{n} \sqrt{1 + \left(\frac{\Delta y_k}{\Delta x}\right)^2}\, \Delta x = \lim_{n \to \infty} \sum_{k=1}^{n} \sqrt{1 + (f'(x_k^*))^2}\, \Delta x.$$

Since the function $f'(x)$ is assumed to be continuous, so is the function $\sqrt{1 + (f'(x))^2}$. Therefore, this limit of sums is exactly the definition of the definite integral of $\sqrt{1 + (f'(x))^2}$ on the interval $[a, b]$ (see Section 12.3.1). Therefore, we have

$$\lim_{n \to \infty} \sum_{k=1}^{n} \sqrt{1 + (f'(x_k^*))^2}\, \Delta x = \int_a^b \sqrt{1 + (f'(x))^2}\, dx.$$

Notice that this last step required that $\Delta x = \frac{b-a}{n}$ and $x_k = a + k\,\Delta x$. ∎

15.1.4 Calculating arc length

In the following example, we calculate the *exact* length of the graph of $f(x) = \sqrt{4 - x^2}$ on the interval $[-2, 2]$ and use it to find the exact circumference of a circle of radius 2.

EXAMPLE 15.2

Using a definite integral to calculate an arc length

Find the exact arc length of $f(x) = \sqrt{4 - x^2}$ from $x = -2$ to $x = 2$, and use it to find the circumference of a circle of radius 2.

Solution   Since we will need to calculate a definite integral involving $\sqrt{1 + (f'(x))^2}$, we first need to find the derivative of $f(x)$:

$$f'(x) = \tfrac{1}{2}(4 - x^2)^{-\frac{1}{2}}(-2x) = -\frac{x}{\sqrt{4 - x^2}}.$$

Now we can use Theorem 15.1 to write the arc length as the definite integral:

$$\int_{-2}^{2} \sqrt{1 + (f'(x))^2} \, dx$$

$$= \int_{-2}^{2} \sqrt{1 + \left(-\frac{x}{\sqrt{4 - x^2}}\right)^2} \, dx \qquad \text{(see above)}$$

$$= \int_{-2}^{2} \sqrt{1 + \frac{x^2}{4 - x^2}} \, dx = \int_{-2}^{2} \sqrt{\frac{(4 - x^2) + x^2}{4 - x^2}} \, dx \qquad \text{(algebra)}$$

$$= 2 \int_{-2}^{2} \frac{1}{\sqrt{4 - x^2}} \, dx. \qquad \text{(algebra)}$$

This is an integral that we know how to solve using trigonometric substitution.

Technical Note: The integrand $\frac{1}{\sqrt{4-x^2}}$ is not defined at the limits of integration, $x = -2$ and $x = 2$. (We call integrals such as this *improper integrals;* you may study such integrals in your next calculus course.) The Fundamental Theorem of Calculus applies only if the integrand is defined on the *entire* interval $[a, b]$ and thus does not apply to our "improper" integral. However, after trigonometric substitution and simplification, the integral will be "proper," and the Fundamental Theorem will apply.

We will choose the substitution

$$x = 2 \sin u \quad \Longrightarrow \quad dx = 2 \cos u \, du \quad \left(\text{and} \quad u = \sin^{-1}\left(\tfrac{x}{2}\right)\right).$$

With this trigonometric substitution, the limits of integration become

$$x = -2 \quad \Longrightarrow \quad u = \sin^{-1}\left(\tfrac{-2}{2}\right) = \sin^{-1}(-1) = -\frac{\pi}{2},$$

$$x = 2 \quad \Longrightarrow \quad u = \sin^{-1}\left(\tfrac{2}{2}\right) = \sin^{-1}(1) = \frac{\pi}{2}.$$

Now we solve the definite integral

$$2 \int_{-2}^{2} \frac{1}{\sqrt{4 - x^2}} \, dx = 2 \int_{-\frac{\pi}{2}}^{\frac{\pi}{2}} \frac{1}{\sqrt{4 - 4 \sin^2 u}} \, 2 \cos u \, du \qquad \textbf{(trigonometric substitution)}$$

$$= 2 \int_{-\frac{\pi}{2}}^{\frac{\pi}{2}} \frac{1}{2 \cos u} \, 2 \cos u \, du \qquad \textbf{(Pythagorean identity)}$$

$$= 2 \int_{-\frac{\pi}{2}}^{\frac{\pi}{2}} 1 \, du \qquad \textbf{(algebra)}$$

$$= 2 [u]_{-\frac{\pi}{2}}^{\frac{\pi}{2}} \qquad \textbf{(integrate)}$$

$$= 2 \left(\frac{\pi}{2} - \left(-\frac{\pi}{2}\right)\right) \qquad \textbf{(evaluate)}$$

$$= 2\pi. \qquad \textbf{(algebra)}$$

Therefore, the length of the graph of $f(x) = \sqrt{4 - x^2}$ from $x = -2$ to $x = 2$ is 2π units. This arc length is *half* the arc length of the circumference of a circle with radius 2, so the circumference of a circle of radius 2 is $2(2\pi) = 4\pi$ units. This is exactly the answer we expected, since by the formula for the circumference of a circle, we should have $C = 2\pi r = 2\pi(2) = 4\pi$. ☐

Unfortunately, calculating arc length exactly can be quite difficult, even for a function as simple as $f(x) = x^3$. This is because we need to integrate the often complicated function $\sqrt{1 + (f'(x))^2}$ (not the original function $f(x)$). When we cannot find the *exact* arc length, we can approximate it either by using a sum of lengths of line segments or by using a calculator to approximate a definite integral.

EXAMPLE 15.3   **Approximating arc length in two ways**

Approximate the arc length of one period of the function $f(x) = \sin x$ in two ways:

(a) By calculating the sum in Definition 15.1, with $n = 4$.

(b) By using Theorem 15.1 to write the arc length as a definite integral and then using a calculator to approximate the value of this definite integral.

Solution

(a) Since $\sin x$ has a period of 2π (and will have the same arc length on every interval of 2π), we will approximate the arc length of $f(x) = \sin x$ from $x = 0$ to $x = 2\pi$. Since we are using $n = 4$ line segments (see Figure 15.5), we have $\Delta x = \frac{2\pi - 0}{4} = \frac{\pi}{2}$. We could use the distance formula to find the length of each of these line segments (and thus not bother with sigma notation), but we were asked to use the form of the sum in Definition 15.1, so we will do so. This means we have to know the value of each of the four differences $\Delta y_k = f(x_k) - f(x_{k-1})$, where $x_0 = 0$, $x_1 = \frac{\pi}{2}$, $x_2 = \pi$, $x_3 = \frac{3\pi}{2}$, and $x_4 = 2\pi$. We have

$$\Delta y_1 = f(x_1) - f(x_0) = \sin\left(\tfrac{\pi}{2}\right) - \sin 0 = 1 - 0 = 1,$$

$$\Delta y_2 = f(x_2) - f(x_1) = \sin(\pi) - \sin\left(\tfrac{\pi}{2}\right) = 0 - 1 = -1,$$

$$\Delta y_3 = f(x_3) - f(x_2) = \sin\left(\tfrac{3\pi}{2}\right) - \sin(\pi) = -1 - 0 = -1,$$

$$\Delta y_4 = f(x_4) - f(x_3) = \sin(2\pi) - \sin\left(\tfrac{3\pi}{2}\right) = 0 - (-1) = 1.$$

Now we can approximate the arc length of $f(x) = \sin x$ on $[0, 2\pi]$ with the sum

$$\sum_{k=1}^{4} \sqrt{1 + \left(\tfrac{\Delta y_k}{\Delta x}\right)^2}\, \Delta x$$

$$= \sqrt{1 + \left(\tfrac{\Delta y_1}{\Delta x}\right)^2}\, \Delta x + \sqrt{1 + \left(\tfrac{\Delta y_2}{\Delta x}\right)^2}\, \Delta x + \sqrt{1 + \left(\tfrac{\Delta y_3}{\Delta x}\right)^2}\, \Delta x + \sqrt{1 + \left(\tfrac{\Delta y_4}{\Delta x}\right)^2}\, \Delta x$$

$$= \sqrt{1 + \left(\tfrac{1}{\pi/2}\right)^2}\left(\tfrac{\pi}{2}\right) + \sqrt{1 + \left(\tfrac{-1}{\pi/2}\right)^2}\left(\tfrac{\pi}{2}\right) + \sqrt{1 + \left(\tfrac{-1}{\pi/2}\right)^2}\left(\tfrac{\pi}{2}\right) + \sqrt{1 + \left(\tfrac{1}{\pi/2}\right)^2}\left(\tfrac{\pi}{2}\right)$$

$$\approx 1.8621 + 1.8621 + 1.8621 + 1.8621$$

$$= 7.4484.$$

(In this example, all four of the line segments happen to be the same length; of course that is not usually the case!) Figure 15.5 shows the four line segments used in the calculation.

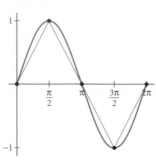

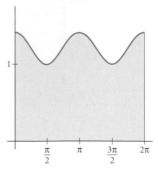

Figure 15.5
Arc length of $f(x) = \sin x$
on $[0, 2\pi]$ approximated by
four line segments

Figure 15.6
The area under the graph of
$\sqrt{1 + (f'(x))^2} = \sqrt{1 + \cos^2 x}$
on $[0, 2\pi]$

(**b**) To find the exact value of the arc length of one period of $f(x) = \sin x$, we would have to calculate the definite integral

$$\int_0^{2\pi} \sqrt{1 + (f'(x))^2}\, dx = \int_0^{2\pi} \sqrt{1 + \left(\tfrac{d}{dx}(\sin x)\right)^2}\, dx$$

$$= \int_0^{2\pi} \sqrt{1 + \cos^2 x}\, dx.$$

We do not know how to calculate this integral. However, the definite integral represents an area, and we can use a graphing calculator to approximate this area. The area in question is shown in Figure 15.6. This area is *not* the area under the graph of $f(x) = \sin x$. Rather, it is the area under the graph of $\sqrt{1 + (f'(x))^2} = \sqrt{1 + \cos^2 x}$ from $x = 0$ to $x = 2\pi$. Using a graphing calculator, we can approximate this area as 7.6404; therefore, the arc length of $f(x) = \sin x$ from $x = 0$ to $x = 2\pi$ is approximately 7.6404. □

? ***Question***   Which of the approximations in Example 15.3 do you think is more accurate, and why? □

? ***Question***   Do you think the approximations in Example 15.3 are overapproximations or underapproximations, and why? Is there any function $f(x)$ whose arc length approximation by line segments would be an overapproximation? □

What you should know after reading Section 15.1

▶ How to define the arc length of a curve by using a limit of sums of lengths of line segments. What does all the notation mean in the definition of arc length? How is it similar to the definition of the definite integral? How is it different?

▶ The formula for finding the *exact* arc length of a function on an interval, how to prove it from the definition of arc length, and how to use it to calculate arc lengths.

▶ How to approximate arc lengths by using line segments (with the distance formula or with the sum in the definition of arc length), as well as by using a graphing calculator to approximate a definite integral.

Exercises 15.1

Concepts

0. Read the section and make your own summary of the material. Include whatever you think will help you review the material later on.

1. What is the *definition* of the arc length of a continuous function $f(x)$ on an interval $[a, b]$? Explain this definition in words, and explain the meaning of all the notation in this definition.

2. Suppose $f(x)$ is a differentiable function with a continuous derivative. Describe the arc length of $f(x)$ on the interval $[a, b]$ in terms of a definite integral. Why is this definite integral equal to the definition of arc length for $f(x)$ on $[a, b]$? How is the Mean Value Theorem involved?

3. Figure 15.3 in this section shows an approximation of the arc length of $f(x) = \sqrt{4 - x^2}$ on $[-2, 2]$ (the top half of a circle of radius 2) using eight line segments. Explain why the changes Δy_k are different for each k, even though Δx is the same on each subinterval.

4. Which approximation in Example 15.3 do you think is more accurate, and why?

5. Fill in the blank: The length of the graph of $y = x^2$ on $[2, 4]$ is equal to the area under the graph of the function _____ on the same interval.

6. Fill in the blank: The area under the graph of $y = \sqrt{1 + e^{2x}}$ on $[0, 2]$ is equal to the length of the graph of the function _____ on the same interval.

7. In this problem you will investigate whether approximating the length of a curve with line segments results in an underapproximation or an overapproximation.

(a) Use a calculator to graph each of the functions $f(x) = x^3 - 4x + 1$, $g(x) = e^{-2x}$, and $h(x) = \cos 3x$ on the interval $[-2, 2]$.

(b) Suppose you wanted to approximate the arc length of each of these functions on $[-2, 2]$ using six line segments. Draw these line segments on each graph.

(c) Determine graphically whether each approximation would be greater than or less than the actual arc length for each function above.

(d) Explain why approximating the length of a curve with line segments will *always* yield an underapproximation.

8. Suppose you want to approximate the arc length of the function $f(x) = \tan x$ on the interval $[0, \pi]$. Would you get a good approximation if you subdivided the interval $[0, \pi]$ into five subintervals and added up the lengths of the line segments over each subinterval? Why or why not? (*Hint:* Look at the graph of $f(x) = \tan x$ on $[0, \pi]$ and draw the line segments in question.)

9. Use definite integrals to show that $f(x) = \frac{1}{x}$ and $g(x) = -\frac{1}{x}$ have the same arc length on $[1, 3]$. (*Hint:* You do *not* have to solve the integrals!) Why does this make sense in terms of transformations?

10. Use definite integrals to show that $f(x) = x^2 - 3$ and $g(x) = 5 - x^2$ have the same arc length on $[0, 2]$. (*Hint:* You do *not* have to solve the integrals!) Why does this make sense in terms of transformations?

11. Do you think that the function $f(x) = 2x^2$ has twice the arc length of $g(x) = x^2$ on the interval $[0, 3]$? Why or why not?

■ For each function $f(x)$ and interval $[a, b]$, write down the definite integral we would have to solve in order to find the exact arc length of $f(x)$ on $[a, b]$. Then determine whether this integral is one we know how to solve (but don't actually solve it).

12. $f(x) = x^3$, $[a, b] = [-3, 3]$

13. $f(x) = xe^x$, $[a, b] = [-2, 0]$

14. $f(x) = \sin(x^2)$, $[a, b] = [0, \sqrt{\pi}]$

15. $f(x) = \ln(\csc x)$, $[a, b] = [\frac{\pi}{4}, \frac{\pi}{2}]$

16. Write down a definite integral that describes the circumference of a circle of radius 5. Don't try to solve the integral; just write it down.

17. Write down a definite integral that describes the arc length around an ellipse with horizontal radius of 3 units and vertical radius of 2 units. (*Note:* The ellipse in this problem extends 6 units across and 4 units vertically and has equation $\frac{x^2}{9} + \frac{y^2}{4} = 1$.) Don't try to solve the integral; just write it down.

■ Each definite integral below represents the arc length of a function $f(x)$ on an interval $[a, b]$. Determine the function and interval.

18. $\displaystyle\int_0^\pi \sqrt{1 + \sin^2 x}\, dx$

19. $\displaystyle\int_{-2}^5 \sqrt{1 + 9e^{6x}}\, dx$

20. $\displaystyle\int_0^2 \sqrt{1 + 36x}\, dx$

21. $\displaystyle\int_2^3 \sqrt{\frac{x^4 + 1}{x^4}}\, dx$

22. $\displaystyle\int_1^4 \sqrt{1 + \frac{4x^2}{x^4 + 8x^2 + 16}}\, dx$

23. $\displaystyle\int_0^{\frac{\pi}{4}} \sec x\, dx$

24. In this problem you will approximate the arc length of $f(x) = x^2$ on $[0, 2]$ in two ways:

    (a) By using four line segments and the distance formula.

    (b) By using a Left-Hand Sum with four rectangles to approximate the area under the graph of the function $y = \sqrt{1 + 4x^2}$ on $[0, 2]$.

    (c) Why do the two approximations above both approximate the arc length of $f(x) = x^2$ on $[0, 2]$? Are the approximations the same? Which, if either, do you think is a better approximation?

Skills

■ Approximate the arc length of $f(x)$ on $[a, b]$ using n line segments and the distance formula. You do *not* have to use the x_k notation, or sigma notation, to do these problems. Your work should include a sketch of the function $f(x)$, together with the line segments that you are using.

25. $f(x) = x^3$, $[a, b] = [-2, 2]$, $n = 4$

26. $f(x) = \cos x$, $[a, b] = [0, \pi]$, $n = 3$

27. $f(x) = e^x$, $[a, b] = [-1, 3]$, $n = 4$

28. $f(x) = \ln x$, $[a, b] = [1, 3]$, $n = 4$

29. $f(x) = \sqrt{9 - x^2}$, $[a, b] = [-3, 3]$, $n = 3$

30. $f(x) = \sqrt{9 - x^2}$, $[a, b] = [-3, 3]$, $n = 6$

■ Approximate the arc length of $f(x)$ on $[a, b]$ using the approximation $\sum_{k=1}^{n} \sqrt{1 + \left(\frac{\Delta y_k}{\Delta x}\right)^2} \Delta x$ with the given value of n. In each problem, list the values of Δy_k for $k = 1, 2, \ldots, n$.

31. $f(x) = x^3$, $[a, b] = [-2, 2]$, $n = 4$

32. $f(x) = \cos x$, $[a, b] = [0, \pi]$, $n = 3$

33. $f(x) = e^x$, $[a, b] = [-1, 3]$, $n = 4$

34. $f(x) = \ln x$, $[a, b] = [1, 3]$, $n = 4$

35. $f(x) = \sqrt{9 - x^2}$, $[a, b] = [-3, 3]$, $n = 3$

36. $f(x) = \sqrt{9 - x^2}$, $[a, b] = [-3, 3]$, $n = 6$

■ Find the *exact* value of the arc length of $f(x)$ on $[a, b]$ by writing the arc length as a definite integral and solving that integral.

37. $f(x) = 3x + 1$, $[a, b] = [-1, 4]$

38. $f(x) = 4 - x$, $[a, b] = [2, 5]$

39. $f(x) = x^{\frac{3}{2}}$, $[a, b] = [0, 2]$

40. $f(x) = 2x^{\frac{3}{2}} + 1$, $[a, b] = [1, 3]$

41. $f(x) = (2x + 3)^{\frac{3}{2}}$, $[a, b] = [-1, 1]$

42. $f(x) = 2(1 - x)^{\frac{3}{2}} + 3$, $[a, b] = [-2, 0]$

43. $f(x) = \sqrt{1 - x^2}$, $[a, b] = [-1, 1]$

44. $f(x) = \sqrt{9 - x^2}$, $[a, b] = [-3, 3]$

45. $f(x) = \frac{1}{3}x^{\frac{3}{2}} - x^{\frac{1}{2}}$, $[a, b] = [0, 1]$

46. $f(x) = x^2 - \frac{1}{8}\ln x$, $[a, b] = [1, 2]$

47. $f(x) = \ln(\sin x)$, $[a, b] = \left[\frac{\pi}{4}, \frac{3\pi}{4}\right]$

48. $f(x) = \ln(\cos x)$, $[a, b] = \left[0, \frac{\pi}{4}\right]$

49. $f(x) = \dfrac{x^4 + 3}{6x}$, $[a, b] = [1, 3]$

50. $f(x) = x^2$, $[a, b] = [-1, 1]$

Proofs

51. Suppose $f(x)$ is a continuous function on an interval $[a, b]$, and let n be a positive integer. With the notation Δx, x_k, and Δy_k given in Definition 15.1, prove that

$$\sum_{k=1}^{n} \sqrt{(x_k - x_{k-1})^2 + (f(x_k) - f(x_{k-1}))^2}$$

$$= \sum_{k=1}^{n} \sqrt{1 + \left(\frac{\Delta y_k}{\Delta x}\right)^2} \Delta x.$$

52. Prove Theorem 15.1 using Definition 15.1 and the definition of the definite integral.

53. Prove that the arc length of a linear function $f(x) = mx + c$ on an interval $[a, b]$ is equal to $(b - a)\sqrt{1 + m^2}$ in two ways:

    (a) By using the distance formula.

    (b) By using Theorem 15.1.

54. Use Theorem 15.1 to prove that a circle of radius 5 has circumference 10π.

55. Use Theorem 15.1 to prove that a circle of radius r has circumference $2\pi r$.

56. Prove that if f is continuous on $[a, b]$ and C is any real number, then $f(x)$ and $f(x) + C$ have the same arc length on $[a, b]$. Also explain why this makes sense graphically.

15.2   Volumes by Slicing

In this section we investigate the volumes of *solids of revolution,* that is, solids obtained by rotating a region in the plane around an axis. As with area and arc length, the key to calculating these volumes will involve definite integrals.

15.2.1 Approximating volumes with discs

A ***solid of revolution*** is a three-dimensional object obtained by rotating a planar region around an axis, or line. For example, consider the planar region shown in Figure 15.7.

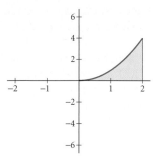

Figure 15.7
A planar region

We could revolve this region around the x-axis to obtain the solid in Figure 15.8. Alternatively, we could revolve this region around the y-axis to obtain the solid in Figure 15.9. Many of the solids of revolution we deal with in this section will have the x-axis or the y-axis as the ***axis of revolution.*** However, we could revolve the region around any line we like; for example, Figure 15.10 shows the solid that results when we revolve the region in Figure 15.7 around the line $y = -1$.

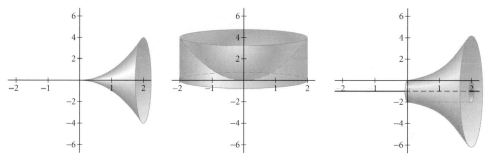

Figure 15.8
Revolved around x-axis

Figure 15.9
Revolved around y-axis

Figure 15.10
Revolved around $y = -1$

Perhaps the simplest solid of revolution is a cylinder, which can be obtained by revolving a rectangle around an axis. For example, the solid shown in Figure 15.12 is the result of rotating the rectangle in Figure 15.11 around the x-axis.

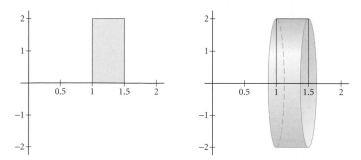

Figure 15.11 A rectangle

Figure 15.12 A cylinder

The volume of a cylinder with radius r and height h is given by the formula $V = \pi r^2 h$. This follows from the definition of volume for a solid with a homogeneous cross section: The volume of such a solid is the product of the area of the cross section with the "height" of the solid. A cylinder has a homogeneous cross section that is a circle of radius r, and each of these cross sections has area πr^2 (the area formula for a circle was proved in the exercises of Section 14.4), so the volume of the cylinder is $\pi r^2 h$. The cylinder in Figure 15.12 has a radius of 2 units and a "height" of 0.5 unit (since the cylinder is on its side in the figure, the "height" is really a "width"). Therefore, the cylinder has a volume of $V = 2\pi r^2 h = 2\pi(2)^2(0.5) = 4\pi$.

In Section 12.2 we used rectangles to approximate the area of a planar region, and in Section 15.1 we used line segments to approximate the length of a curve. We will now use cylinders, or "discs," to approximate the volumes of solids of revolution. Let's begin by examining a sphere of radius 2. Such a sphere can be constructed by rotating (around the x-axis) the region between the graph of $f(x) = \sqrt{4 - x^2}$ and the x-axis on $[-2, 2]$ (see Figures 15.13 and 15.14).

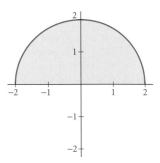

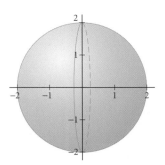

Figure 15.13
Area under $f(x) = \sqrt{4 - x^2}$
on $[-2, 2]$

Figure 15.14
Area in Figure 15.13 revolved
around the x-axis

In Chapter 7 we mentioned that the volume of a sphere of radius r is given by $V = \frac{4}{3}\pi r^3$; therefore, the volume of a sphere of radius 2 should be $V = \frac{4}{3}\pi(2)^3 = \frac{32\pi}{3} \approx 33.5103$. But so far we have been taking this volume formula on faith. We have not *proved* that a sphere has volume given by that formula. In this section we will finally be able to prove this formula, but first we will use cylinders, or discs, to approximate this volume.

Suppose we divide the interval $[-2, 2]$ into six subintervals. If we wanted to approximate the *area* of the region in Figure 15.13, we would consider a rectangle on each subinterval, like the one in Figure 15.15. Since we want to approximate a *volume*, we need to consider a three-dimensional object on each subinterval. If we revolve the rectangle in Figure 15.15 around the x-axis, we obtain the disc (i.e., the cylinder) in Figure 15.16.

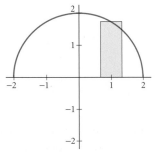

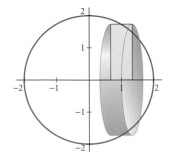

Figure 15.15
A rectangle can be used
to approximate area

Figure 15.16
A disc can be used to
approximate volume

Figure 15.17 shows the six rectangles you would use to approximate the area under the graph of $f(x) = \sqrt{4 - x^2}$ on $[-2, 2]$ using the Midpoint Sum. If we revolve each of these six rectangles around the *x*-axis, we obtain the row of six discs in Figure 15.18. The sum of the volumes of all these discs is a relatively good approximation for the volume of a sphere of radius 2.

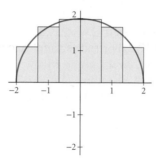

Figure 15.17
Six rectangles approximate
the area of the region

Figure 15.18
Six discs approximate the
volume of the sphere

EXAMPLE 15.4

Approximating a volume with discs

Approximate the volume of a sphere of radius 2 by finding the volumes of the six discs shown in Figure 15.18.

Solution   Each of the six discs in Figure 15.18 has a width of $\frac{2}{3}$, since $\frac{2-(-2)}{6} = \frac{4}{6} = \frac{2}{3}$. Thus the first disc begins at $x = -2 = -\frac{6}{3}$ and ends at $x = -\frac{4}{3}$. The radius of the first disc is equal to $f(-\frac{5}{3})$, which is the length of the line segment from the "middle" point $(-\frac{5}{3}, 0)$ of the disc to the "top" point $(-\frac{5}{3}, f(-\frac{5}{3}))$. (Examine the first rectangle in Figure 15.17 and the first disc in Figure 15.18 until you are convinced of this!) Since $f(x) = \sqrt{4 - x^2}$, we have $f(-\frac{5}{3}) = \sqrt{4 - (-\frac{5}{3})^2} = \frac{\sqrt{11}}{3}$. Therefore, the volume of the first disc is

$$\pi r^2 h = \pi \left(f(-\tfrac{5}{3}) \right)^2 \left(\tfrac{2}{3} \right) = \pi \left(\tfrac{\sqrt{11}}{3} \right)^2 \left(\tfrac{2}{3} \right) = \tfrac{22\pi}{27} \approx 2.55982.$$

We can find the volumes of all six discs in Figure 15.18 in a similar manner and then add up all these volumes to get an approximation for the volume *V* of the sphere:

$$V \approx \begin{array}{c} \text{volume of} \\ \text{first disc} \end{array} + \begin{array}{c} \text{volume of} \\ \text{second disc} \end{array} + \begin{array}{c} \text{volume of} \\ \text{third disc} \end{array} + \cdots + \begin{array}{c} \text{volume of} \\ \text{sixth disc} \end{array}$$

$$= \pi \left(f(-\tfrac{5}{3}) \right)^2 \left(\tfrac{2}{3} \right) + \pi \left(f(-1) \right)^2 \left(\tfrac{2}{3} \right) + \pi \left(f(-\tfrac{1}{3}) \right)^2 \left(\tfrac{2}{3} \right) + \pi \left(f(\tfrac{1}{3}) \right)^2 \left(\tfrac{2}{3} \right)$$

$$+ \pi \left(f(1) \right)^2 \left(\tfrac{2}{3} \right) + \pi \left(f(\tfrac{5}{3}) \right)^2 \left(\tfrac{2}{3} \right)$$

$$= \tfrac{22\pi}{27} + 2\pi + \tfrac{70\pi}{27} + \tfrac{70\pi}{27} + 2\pi + \tfrac{22\pi}{27}$$

$$\approx 2.55982 + 6.28319 + 8.14487 + 8.14487 + 6.28319 + 2.55982$$

$$\approx 33.9758.$$

Notice that this approximation is very close to what we anticipate will be the actual volume of a sphere of radius $r = 2$, namely, $\frac{4}{3}\pi r^3 = \frac{4}{3}\pi(2)^3 \approx 33.5103$.   □

15.2.2 Defining volume as the limit of a sum

We now give general notation for the process of approximating the volume of a solid of revolution obtained by revolution around the *x*-axis. Given a solid of revolution defined by rotating a curve $f(x)$ on an interval $[a, b]$ around the x-axis, we will subdivide the interval (thereby "slicing" the solid into pieces vertically), use discs to approximate the volume of each slice, and then add up all of these approximations.

Most of the notation will be the same as what we used in Sections 12.3 and 15.1 for area and arc length. If we divide the interval $[a, b]$ into n subintervals of equal width $\Delta x := \frac{b-a}{n}$, then each subinterval is of the form $[x_{k-1}, x_k]$ where $x_k := a + k\,\Delta x$ for $k = 0, 1, \ldots, n$. For each subinterval, choose a point $x_k^* \in [x_{k-1}, x_k]$ to use as the "center" of a disc. (In Example 15.4 we used the midpoint of each subinterval for x_k^*.) The kth disc will have radius $f(x_k^*)$ and "height" (here seen as a "width") of Δx; this is represented graphically in Figure 15.19.

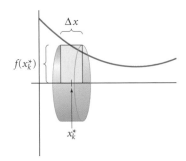

Figure 15.19

Therefore, the volume of the kth disc is

$$\begin{matrix} \text{Volume of} \\ \text{the } k\text{th disc} \end{matrix} = \pi r^2 h = \pi (f(x_k^*))^2\, \Delta x.$$

Now we use sigma notation to add the volumes of the discs and obtain an approximation for the volume V of the solid of revolution:

(1)
$$V \approx \sum_{k=1}^{n} \pi (f(x_k^*))^2\, \Delta x.$$

EXAMPLE 15.5

Using sigma notation to express an approximation of volume by discs

In Example 15.4 we had $f(x) = \sqrt{4 - x^2}$, $[a, b] = [-2, 2]$, and $n = 6$. Therefore, $\Delta x = \frac{2-(-2)}{6} = \frac{2}{3}$, and since $x_k = -2 + k(\frac{2}{3})$, we had

$$x_0 = -2, \quad x_1 = -\tfrac{4}{3}, \quad x_2 = -\tfrac{2}{3}, \quad x_3 = 0, \quad x_4 = \tfrac{2}{3}, \quad x_5 = \tfrac{4}{3}, \quad \text{and} \quad x_6 = 2.$$

In each subinterval $[x_{k-1}, x_k]$ we chose the midpoint $\frac{x_{k-1}+x_k}{2}$ to be the point x_k^* that determined the radius of the kth disc. Therefore, we had

$$x_1^* = -\tfrac{5}{3}, \quad x_2^* = -1, \quad x_3^* = -\tfrac{1}{3}, \quad x_4^* = \tfrac{1}{3}, \quad x_5^* = 1, \quad \text{and} \quad x_6^* = \tfrac{5}{3}.$$

In sigma notation, the sum we calculated to approximate the volume of the sphere of radius 2 was

$$V \approx \sum_{k=1}^{6} \pi \left(\sqrt{4 - (x_k^*)^2} \right)^2 \left(\tfrac{2}{3}\right) = \sum_{k=1}^{6} \pi (4 - (x_k^*)^2)\left(\tfrac{2}{3}\right).$$

$\square$

As n gets larger (and thus Δx gets smaller), the approximation in Equation (1) gets closer to what we would like to consider the "volume" of the solid of revolution. Therefore, we *define* the volume as the limit of the sum as $n \to \infty$.

DEFINITION 15.2

The Volume of a Solid of Revolution

Consider the region between the graph of a continuous function $f(x)$ and the x-axis on an interval $[a, b]$. The *volume* of the solid of revolution obtained by rotating this region around the x-axis is defined to be the limit of sums

$$\lim_{n \to \infty} \sum_{k=1}^{n} \pi (f(x_k^*))^2 \Delta x,$$

where $\Delta x := \frac{b-a}{n}$, $x_k := a + k \Delta x$, and $x_k^* \in [x_{k-1}, x_k]$ for all $k = 0, 1, 2, \ldots, n$.

? *Question*   It is important to point out that Definition 15.2 applies whether or not f is always positive. For example, it applies to the function $f(x) = x - 2$ on the interval $[0, 3]$. Draw the solid of revolution obtained by revolving this curve around the x-axis, and show a few representative discs on this graph.   □

15.2.3 Volume as a definite integral

The limit of sums in Definition 15.2 is a Riemann sum involving the function $\pi (f(x))^2$. In addition, since $f(x)$ is assumed to be continuous, so is the function $\pi (f(x))^2$. Therefore, the limit of the sum in Definition 15.2 as $n \to \infty$ will exist, so as $n \to \infty$, the Riemann sum in Definition 15.2 becomes a definite integral.

THEOREM 15.2

The Volume of a Solid Obtained by Revolving Around the x-axis

Consider the region between the graph of a continuous function $f(x)$ and the x-axis on an interval $[a, b]$. The volume of the solid of revolution obtained by rotating this region around the x-axis is given by the definite integral

$$\pi \int_a^b (f(x))^2 \, dx.$$

The proof of Theorem 15.2 follows directly from Definition 15.2, the definition of the definite integral, and the previous discussion. It is left for the exercises.

◈ *Caution*   Theorem 15.2 does not apply to *every* solid of revolution. It applies only when a region is rotated around the x-axis, and that region is the area between the function and the x-axis. If a different axis of rotation is used, or if one side of the region fails to be the x-axis, then Theorem 15.2 does not apply.   □

Instead of blindly applying the formula in Theorem 15.2, we will draw a representative rectangle in the region that is to be rotated, draw the corresponding representative disc, and compute its volume. Then we can write the volume of the solid as a limit of sums and write this quantity as a definite integral. We go through this process in the following example.

EXAMPLE 15.6   **Using a definite integral to find an exact volume**

Find the exact volume of a sphere of radius 2 by calculating the volume of the solid of revolution obtained by rotating the curve $f(x) = \sqrt{4 - x^2}$ around the x-axis on $[-2, 2]$.

Solution Instead of simply applying the formula in Theorem 15.2, we will take the advice in the "Caution" on page 861 and examine a representative rectangle and disc. Figure 15.20 shows what the kth rectangle and corresponding disc might look like.

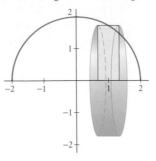

Figure 15.20
A representative rectangle and disc

If x_k^* is the point at which we determine the height of this rectangle, then the disc in question has radius $f(x_k^*) = \sqrt{4 - x_k^*}$ and width Δx. The volume of this representative disc is

$$\pi \left(\sqrt{4 - (x_k^*)^2} \right)^2 \Delta x.$$

Therefore, the volume of the solid resulting when we revolve $f(x) = \sqrt{4 - x^2}$ around the x-axis is approximated by

$$\sum_{k=1}^{n} \pi \left(\sqrt{4 - (x_k^*)^2} \right)^2 \Delta x.$$

As $n \to \infty$, this sum approaches the actual volume of the solid of revolution, given by the definite integral

$$\pi \int_{-2}^{2} (\sqrt{4 - x^2})^2 \, dx = \pi \int_{-2}^{2} (4 - x^2) \, dx \qquad \text{(algebra; see below)}$$

$$= \pi \left[4x - \tfrac{1}{3}x^3 \right]_{-2}^{2} \qquad \text{(integrate)}$$

$$= \pi \left(\left(4(2) - \tfrac{1}{3}(2)^3 \right) - \left(4(-2) - \tfrac{1}{3}(-2)^3 \right) \right) \qquad \text{(evaluate)}$$

$$= \pi \left(8 - \tfrac{8}{3} + 8 - \tfrac{8}{3} \right) = \frac{32\pi}{3}.$$

Notice that in the first step, $(\sqrt{4 - x^2})^2$ is equal to $4 - x^2$ (rather than $|4 - x^2|$) since $4 - x^2$ is positive on $[-2, 2]$. Notice also that $\frac{32\pi}{3}$ is exactly the volume predicted by the formula for the volume of a sphere with radius $r = 2$: $V = \frac{4}{3}\pi r^3 = \frac{4}{3}\pi(2)^3 = \frac{32\pi}{3}$. ☐

15.2.4 Finding volumes with washers

So far, we have considered only solids of revolution defined using planar regions that are between the graph of a function and the x-axis. We could also consider solids of revolution that are defined using a region *between* two functions. The following example illustrates how to deal with this situation.

EXAMPLE 15.7

A solid of revolution whose representative slice is a washer

Find the volume of the solid of revolution obtained by rotating the region between the graphs of $f(x) = x + 1$ and $g(x) = x^2 - 4x + 5$ on $[1, 4]$ around the x-axis.

Solution   Figure 15.21 shows the region in question. The graphs of $f(x)$ and $g(x)$ happen to meet at the endpoints $x = 1$ and $x = 4$ (although in general the graphs may not meet up at the endpoints of the interval). Figure 15.22 shows the solid of revolution whose volume we wish to find (the solid is *hollow*, like a megaphone with a straight outside and curvy inside). What does a representative "slice" look like in this example? It can't be a disc, because the solid we are interested in is "hollow." Figure 15.23 shows a representative rectangle in the region and the shape that results when we revolve this rectangle around the x-axis.

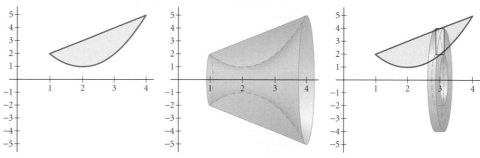

Figure 15.21
The region between $f(x) = x+1$ and $g(x) = x^2 - 4x + 5$ on $[1, 4]$

Figure 15.22
The solid of revolution obtained by rotating the region around the x-axis

Figure 15.23
A representative rectangle and washer

The representative slice in this case is a **washer,** that is, a disc with a smaller disc taken out of the center. A washer with a large radius of R, a small radius of r, and a "width" of Δx (or a "height" if you prefer, but the washer is on its side) will have a volume of

$$\pi R^2 \, \Delta x - \pi r^2 \, \Delta x = \pi (R^2 - r^2) \, \Delta x.$$

This volume formula follows from the fact that the washer is a large disc (of volume $\pi R^2 \, \Delta x$) with a smaller disc (of volume $\pi r^2 \, \Delta x$) removed.

If the representative washer in Figure 15.23 is at the x-value x_k^*, then the large radius of the washer is given by $f(x_k^*) = x_k^* + 1$ (the height of the upper curve at x_k^*), and the small radius of the washer is given by $g(x_k^*) = (x_k^*)^2 - 4(x_k^*) + 5$ (the height of the lower curve at x_k^*). The volume of the representative washer is

$$\pi((x_k^* + 1)^2 - ((x_k^*)^2 - 4(x_k^*) + 5)^2) \, \Delta x.$$

If we add up the volumes of all these representative washers, we get the sum

$$\sum_{k=1}^{n} \pi((x_k^* + 1)^2 - ((x_k^*)^2 - 4(x_k^*) + 5)^2) \, \Delta x.$$

As $n \to \infty$, this approximation approaches the true volume V of the solid of revolution in Figure 15.22 and becomes the definite integral

$$V = \pi \int_1^4 ((x + 1)^2 - (x^2 - 4x + 5)^2) \, dx \qquad \text{(limit as } n \to \infty \text{ of sum above)}$$

$$= \pi \int_1^4 (-x^4 + 8x^3 - 25x^2 + 42x - 24) \, dx \qquad \text{(multiply out and collect terms)}$$

$$= \pi \left[-\tfrac{1}{5}x^5 + \tfrac{8}{4}x^4 - \tfrac{25}{3}x^3 + \tfrac{42}{2}x^2 - 24x \right]_1^4 \qquad \text{(integrate)}$$

$$\approx 73.513. \qquad \text{(evaluate)}$$

Therefore, the volume of the solid of revolution in Figure 15.22 is approximately 73.513 cubic units.   □

? *Question*   In Example 15.7, we could have found the volume of the solid of revolution defined by rotating $f(x) = x + 1$ around the x-axis on $[1, 4]$ and found the volume of the solid of revolution defined by rotating $g(x) = x^2 - 4x + 5$ around the x-axis on $[1, 4]$, and then found the difference between those two volumes. Would this produce the same volume that we found in Example 15.7? Try it and see!   □

In general, the method used in Example 15.7 can be used to find the volume of a solid of revolution that arises from rotating the region between two curves around the x-axis.

THEOREM 15.3

The Volume of a Solid of Revolution Defined by Two Functions

Suppose $f(x)$ and $g(x)$ are continuous functions such that $f(x) \geq g(x)$ on $[a, b]$. The volume of the solid of revolution obtained by rotating the region between the graphs of $f(x)$ and $g(x)$ on $[a, b]$ around the x-axis is equal to the definite integral

$$\pi \int_a^b \left((f(x))^2 - (g(x))^2 \right) dx.$$

The proof of this theorem is a general version of the argument in Example 15.7 and is left for the exercises.

◈ *Caution*   When using Theorem 15.3, be careful that you do not accidentally use $(f(x) - g(x))^2$ as the integrand. It may help to remember that the formula in Theorem 15.3 comes from the formula $\pi(R^2 - r^2)$ of a washer.   □

◈ *Caution*   Theorem 15.3 requires that the function $f(x)$ be greater than or equal to the function $g(x)$ on the *entire* interval $[a, b]$. In general, you may have to split the integral into cases depending on whether $f(x)$ or $g(x)$ is greater on various parts of the interval $[a, b]$, much like we did when calculating the areas of such regions in Section 12.4.   □

15.2.5 Finding volumes by integrating along the y-axis

If a solid of revolution is obtained by rotating the graph of a function on an interval around the y-axis, then slicing vertically will not result in discs. For example, consider the region between the graphs of $f(x) = \frac{3}{2}x$ and $g(x) = 3$ on $[0, 2]$ shown in Figure 15.24. If we revolve this region around the y-axis, we obtain the cone in Figure 15.25.

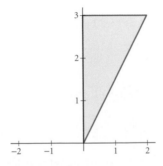

Figure 15.24
The region between $f(x) = \frac{3}{2}x$ and $g(x) = 3$ on $[0, 2]$

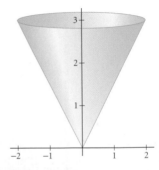

Figure 15.25
The cone obtained by rotating the region in Figure 15.24 around the y-axis

In this example it does not make sense to slice the region or the solid vertically, at least if we want to approximate the volume with discs. Vertical slices will become "shells," not discs, when revolved around the *y*-axis (we'll see how to approximate volumes with shells in the next section). However, if we slice the region (and thus the solid) *horizontally*, our horizontal slices will become discs when we revolve them around the *y*-axis. Figure 15.26 shows such a representative rectangle (horizontal slice) and disc.

Since we are slicing the solid horizontally in this example, we must subdivide the *y*-interval [0, 3] instead of the *x*-interval [0, 2]. Except for the fact that the variables *x* and *y* will be reversed, the notation is the same: Now $\Delta y := \frac{B-A}{n}$, where [*A*, *B*] is the *y*-interval [0, 3]. We will write $y_k := A + k\,\Delta y$ for each of the subdivision points and choose a point y_k^* in each subinterval $[y_{k-1}, y_k]$ that will determine the radius of the disc. This time the discs are stacked one on top of another instead of side-by-side; a picture for $n = 6$ is shown in Figure 15.27. In the following example, we find the *exact* volume of the cone.

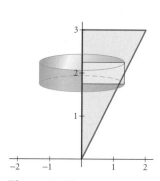

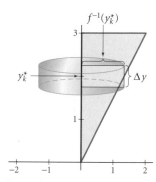

Figure 15.26
A representative rectangle and disc

Figure 15.27
Approximating the volume with $n = 6$ horizontal discs

Figure 15.28
Geometrical meanings of the notation Δy, y_k^*, and $f^{-1}(y_k^*)$

EXAMPLE 15.8

Finding a volume by integrating along the *y*-axis

Find the volume of the cone in Figure 15.25 obtained by rotating the region between $f(x) = \frac{3}{2}x$ and $g(x) = 3$ on [0, 2] around the *y*-axis.

Solution As you can see in Figure 15.28, the height of the *k*th disc is Δy. The radius of that disc is $f^{-1}(y_k^*)$, since this is the *x*-coordinate on the graph corresponding to the *y*-coordinate y_k^*. In this particular example we have $f(x) = \frac{3}{2}x$, and thus for every (x, y) on the graph of *f* we have $y = \frac{3}{2}x$, and therefore $x = \frac{2}{3}y$. In other words, $f^{-1}(y) = \frac{2}{3}y$, and thus the radius of the *k*th disc is $f^{-1}(y_k^*) = \frac{2}{3}y_k^*$. The volume of the *k*th disc is

$$\pi(f^{-1}(y_k^*))^2\,\Delta y = \pi\left(\tfrac{2}{3}y_k^*\right)^2\Delta y.$$

The sum of the volumes of all the discs is

$$\sum_{k=1}^{n} \pi\left(\tfrac{2}{3}y_k^*\right)^2\Delta y.$$

As $n \to \infty$, this sum approaches the actual area of the cone and becomes the definite integral

$$\pi \int_0^3 \left(\tfrac{2}{3}y\right)^2 dy.$$

Notice that this integral is in terms of the variable *y* rather than the variable *x*; this includes the limits of integration, which run from $y = 0$ to $y = 3$ as we "integrate along the *y*-axis." Other than the fact that we are using *y* instead of *x*, this construction is exactly the same as the one we did for vertical discs.

Using the definite integral we have constructed, the volume of the cone in Figure 15.25 is

$$\pi \int_0^3 \left(\tfrac{2}{3}y\right)^2 dy = \pi \int_0^3 \tfrac{4}{9}y^2 \, dy \qquad \text{(algebra)}$$

$$= \pi \left(\tfrac{4}{9}\right)\left[\tfrac{1}{3}y^3\right]_0^3 \qquad \text{(integrate)}$$

$$= \tfrac{4\pi}{9}\left(\tfrac{1}{3}(3)^3 - \tfrac{1}{3}(0)^3\right) \qquad \text{(evaluate)}$$

$$= 4\pi. \qquad \qquad \square$$

Notice that 4π is exactly the same as the volume of a cone with radius 2 and height 3 as predicted by the formula $V = \tfrac{1}{3}\pi r^2 h$. In the exercises you will prove this volume formula using a general version of the calculation in Example 15.8. The construction we did in that example is completely general and leads to the following theorem (which you will prove in the exercises):

THEOREM 15.4

The Volume of a Solid Obtained by Revolving Around the *y*-axis

Suppose $x = g(y)$ is a continuous function on an interval $[A, B]$. The volume of the solid of revolution obtained by rotating the graph of $g(y)$ around the y-axis is equal to the definite integral

$$\pi \int_A^B (g(y))^2 \, dy.$$

Theorem 15.4 is just another version of Theorem 15.2, with the roles of the independent and dependent variables reversed. If $y = f(x)$ is a one-to-one function of x that is revolved around the y-axis, then $g(y) = f^{-1}(y)$. We could also make a y-axis analogue of Theorem 15.3.

❖ *Caution*   Be very careful to use the formula in Theorem 15.4 only when it applies: when the axis of rotation is the y-axis, and the region being rotated is the area between a function and the y-axis. In more exotic examples, you will have to consider a representative slice of the solid and construct a definite integral from scratch. $\qquad \square$

15.2.6 Other axes of revolution

The methods we have developed can also be used to find the volumes of solids of revolution whose axes of revolution are something other than the x-axis or the y-axis. Of course, in this more general case, the formulas developed previously will not apply. However, we can still use the process of finding a representative disc or washer, finding its volume, and then using this volume to write down a definite integral representing the exact volume of the solid.

EXAMPLE 15.9   **Finding volume when the axis of revolution is not the *x*-axis or the *y*-axis**

Find the volume of the solid of revolution obtained by revolving the region between the graph of $f(x) = x^2 + 2$ and the line $y = 2$ on the interval $[0, 1]$ around the line $y = 1$.

Solution   The region in question is shown in Figure 15.29, and the resulting solid of revolution is shown in Figure 15.30. This solid has a horizontal cylinder of open space along its center axis, so we will need washers to slice this solid.

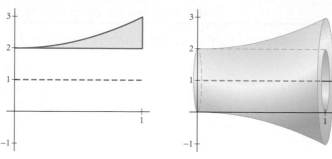

Figure 15.29   **Figure 15.30**

Drawing the representative washer (and the solid itself) can often be difficult. One way to make this easier is to start by *reflecting* (not rotating) the region in Figure 15.29 across the axis of revolution; see Figure 15.31. Then you can sketch a representative rectangle (horizontal or vertical, as appropriate) and revolve it around the axis of rotation to get a disc or washer (in this case a washer) to match up with the reflected region; see the rectangle in Figure 15.31.

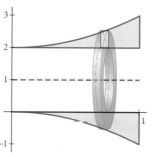   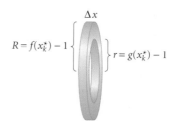

Figure 15.31   **Figure 15.32**

Now we must find the volume of the representative washer. This washer is shown on its own in Figure 15.32. Suppose the washer is located at the x-value x_k^*. Then the large radius of the washer is the vertical distance from the axis of rotation $y = 1$ to the height of the function $f(x)$ at x_k^*; in other words, we have

$$R = f(x_k^*) - 1 = (x_k^*)^2 + 2 - 1 = (x_k^*)^2 + 1.$$

The small radius is the vertical distance from the axis of rotation $y = 1$ to the function $g(x) = 2$; therefore, we have

$$r = g(x_k^*) - 1 = 2 - 1 = 1.$$

(Notice that r is the same for all choices of x_k^*; why?) Now the volume of the representative washer is

$$\pi(R^2 - r^2)\,\Delta x = \pi(((x_k^*)^2 + 1)^2 - 1^2)\,\Delta x.$$

Therefore, the actual volume of the solid of revolution in Figure 15.30 is given by the definite integral

$$\pi \int_0^1 ((x^2 + 1)^2 - 1)\,dx = \pi \int_0^1 (x^4 + 2x^2)\,dx \qquad \text{(algebra)}$$

$$= \pi \left[\tfrac{1}{5}x^5 + \tfrac{2}{3}x^3\right]_0^1 \qquad \text{(integrate)}$$

$$= \pi\left(\left(\tfrac{1}{5}(1)^5 + \tfrac{2}{3}(1)^3\right) - \left(\tfrac{1}{5}(0)^5 + \tfrac{2}{3}(0)^3\right)\right) \qquad \text{(evaluate)}$$

$$= \frac{13\pi}{15}.$$

The integration step was easy; the hard part was setting up the correct definite integral in the first place!   □

> **What you should know after reading Section 15.2**
>
> ▶ Why we can approximate the volume of a solid of revolution with discs or washers, and how to approximate volumes in this way. What does this have to do with rectangles?
>
> ▶ How to write a sum, in sigma notation, that approximates the volume of a solid of revolution, and then how to write that sum as a definite integral as $n \to \infty$. This includes choosing a representative rectangle and finding the volume of the corresponding disc or washer.
>
> ▶ How to handle solids of revolution that require *horizontal* rather than vertical slicing and solids of revolution whose axes of rotation are lines other than the x-axis or the y-axis.

Exercises 15.2

Concepts

0. Read the section and make your own summary of the material. Include whatever you think will help you review the material later on.

1. Consider the rectangle bounded by $y = 3$ and $y = 0$ on the x-interval $[2, 2.25]$.

   (a) What is the volume of the disc obtained by rotating this rectangle around the x-axis?

   (b) What is the volume of the washer obtained by rotating this rectangle around the line $y = 5$?

2. Consider the rectangle bounded by $x = 1$ and $x = 4$ on the y-interval $[3, 3.5]$.

   (a) What is the volume of the disc obtained by rotating this rectangle around the line $x = 4$?

   (b) What is the volume of the washer obtained by rotating this rectangle around the y-axis?

3. Consider the region between the function $f(x) = 5 - x^2$ and the x-axis on the interval $[0, 4]$. Draw a Riemann sum approximation of the area of this region using a Midpoint Sum with four rectangles. Explain how this Riemann sum is related to an approximation with four discs of the solid of revolution obtained by rotating the region around the x-axis.

4. Suppose you want to use four discs to approximate the volume of the solid of revolution obtained by rotating the region between the function $f(x) = 5 - x^2$ and the x-axis on the interval $[0, 4]$ around the x-axis.

   (a) Find the values of Δx, x_0, x_1, x_2, x_3, and x_4 and represent these values graphically.

   (b) Choose values for x_1^*, x_2^*, x_3^*, and x_4^* and represent these values graphically.

   (c) Determine $f(x_1^*)$, $f(x_2^*)$, $f(x_3^*)$, and $f(x_4^*)$ and represent these values graphically.

   (d) What is the volume of the second disc in your approximation?

   (e) What is the volume of the kth disc in your approximation? (Your answer should involve x_k^*.)

5. Suppose you want to use four discs to approximate the volume of the solid of revolution obtained by rotating the region between the function $f(x) = x^2$ and the y-axis between $y = 0$ and $y = 4$ around the y-axis.

   (a) Find the values of Δy, y_0, y_1, y_2, y_3, and y_4 and represent these values graphically.

   (b) Choose values for y_1^*, y_2^*, y_3^*, and y_4^* and represent these values graphically.

   (c) Determine $f^{-1}(y_1^*)$, $f^{-1}(y_2^*)$, $f^{-1}(y_3^*)$, and $f^{-1}(y_4^*)$ and represent these values graphically.

   (d) What is the volume of the second disc in your approximation?

   (e) What is the volume of the kth disc in your approximation? (Your answer should involve y_k^*.)

6. Suppose you want to use four washers to approximate the volume of the solid of revolution obtained by rotating the region between the function $f(x) = x^2$ and the y-axis between $y = 0$ and $y = 4$ around the x-axis.

   (a) Find the values of Δx, x_0, x_1, x_2, x_3, and x_4 and represent these values graphically.

   (b) Choose values for x_1^*, x_2^*, x_3^*, and x_4^* and represent these values graphically.

   (c) Determine $f(x_1^*)$, $f(x_2^*)$, $f(x_3^*)$, and $f(x_4^*)$ and represent these values graphically.

   (d) What is the volume of the second washer in your approximation?

   (e) What is the volume of the kth washer in your approximation? (Your answer should involve x_k^*.)

7. Suppose you want to use four washers to approximate the volume of the solid of revolution obtained by rotating the region between the function $f(x) = 5 - x^2$ and the x-axis between $x = 1$ and $x = 4$ around the y-axis.

(a) Find the values of Δy, y_0, y_1, y_2, y_3, and y_4 and represent these values graphically.

(b) Choose values for y_1^*, y_2^*, y_3^*, and y_4^* and represent these values graphically.

(c) Determine $f^{-1}(y_1^*)$, $f^{-1}(y_2^*)$, $f^{-1}(y_3^*)$, and $f^{-1}(y_4^*)$ and represent these values graphically.

(d) What is the volume of the second washer in your approximation?

(e) What is the volume of the kth washer in your approximation? (Your answer should involve y_k^*.)

■ Write each of the following limits of sums as a definite integral, and identify a solid of revolution whose volume is represented by that definite integral. (*Hints:* You will have to use the information about Δx and x_k^* to find the limits of integration. Finding the solid of revolution involves determining the region and the axis of rotation.)

8. $\lim\limits_{n \to \infty} \sum\limits_{k=1}^{n} \pi (1 + x_k^*)^2 \, \Delta x$, where $\Delta x = \frac{1}{n}$ and $x_k^* = x_k = 2 + k \, \Delta x$

9. $\lim\limits_{n \to \infty} \sum\limits_{k=1}^{n} \pi (1 + y_k^*)^2 \, \Delta y$, where $\Delta y = \frac{3}{n}$ and $y_k^* = y_k = 1 + k \, \Delta x$

10. $\lim\limits_{n \to \infty} \sum\limits_{k=1}^{n} \pi (4 - (x_k^*)^2) \, \Delta x$, where $\Delta x = \frac{2}{n}$ and $x_k^* = x_k = k \, \Delta x$

■ Each of the following definite integrals represents the volume of a solid of revolution. Determine the solid of revolution by finding a region in the plane and an axis of revolution that will produce that solid. (*Hint:* The axis of revolution is the x-axis or the y-axis in each of these problems.)

11. $\pi \displaystyle\int_0^\pi \sin^2 x \, dx$

12. $\pi \displaystyle\int_1^3 (16 - (x+1)^2) \, dx$

13. $\pi \displaystyle\int_1^3 (x^2 - 2x + 1) \, dx$

14. $\pi \displaystyle\int_1^5 \left(\frac{y-1}{2} \right)^2 dy$

15. $\pi \displaystyle\int_0^2 y \, dy$

16. $\pi \displaystyle\int_0^4 (2^2 - (\sqrt{y})^2) \, dy$

■ For each problem below, write down a definite integral (or sum of definite integrals) describing the volume of the solid of revolution obtained by rotating the region described around the given axis of revolution.

17. The solid of revolution obtained by revolving the region between the graphs of $f(x) = (x-2)^2$ and $g(x) = x$ on the x-interval $[0, 4]$ around the x-axis.

18. The solid of revolution obtained by revolving the region between the graphs of $f(x) = 3 + 4x - x^2$ and $g(x) = 4x$ on the interval $[0, 4]$ around the x-axis.

19. The solid of revolution obtained by revolving the region between the graphs of $f(x) = (x-2)^2$ and $g(x) = x$ on the x-interval $[1, 4]$ around the y-axis.

20. The solid of revolution obtained by revolving the region between the graphs of $f(x) = 3 + 4x - x^2$ and $g(x) = 4x$ on the interval $[0, \sqrt{3}]$ around the y-axis.

■ For each pair of definite integrals below, decide whether they are equal or one is larger than the other (and if so, which one). Do this *without* solving the integrals; instead, sketch a graph of the solids of revolution represented by the definite integrals, and make a geometric argument.

21. (a) $\pi \displaystyle\int_0^{\frac{\pi}{2}} \cos^2 x \, dx$,   (b) $\pi \displaystyle\int_{\frac{\pi}{2}}^{\pi} \cos^2 x \, dx$

22. (a) $\pi \displaystyle\int_0^1 e^{2x} \, dx$,   (b) $\pi \displaystyle\int_0^1 (e^{2x} - 1) \, dx$

23. (a) $\pi \displaystyle\int_0^3 x^2 \, dx$,   (b) $\pi \displaystyle\int_0^3 (9 - x^2) \, dx$

24. (a) $\pi \displaystyle\int_0^2 x^4 \, dx$,   (b) $\pi \displaystyle\int_0^2 (16 - x^4) \, dx$

25. (a) $\pi \displaystyle\int_0^2 x^4 \, dx$,   (b) $\pi \displaystyle\int_0^4 y \, dy$

26. (a) $\pi \displaystyle\int_0^4 y \, dy$,   (b) $\pi \displaystyle\int_4^8 (8 - y) \, dy$

27. Let V be the volume of the solid of revolution obtained by rotating around the x-axis the region between the graph of $y = \sqrt{x}$ and the x-axis on $[0, b]$ for some positive number b.

(a) Suppose you know that $V = 8\pi$. Use this information (and a graphical argument) to find the volume of the solid of revolution obtained by rotating around the x-axis the region bounded by the graphs of $y = \sqrt{x}$, $y = \sqrt{b}$, and $x = 0$. (You should be able to do this without writing down or solving *any* definite integrals and without finding the value of b.)

(b) If $V = 8\pi$, what is the value of b?

28. Describe a region that, when revolved around the x-axis, produces a solid of revolution for which *both* discs and washers are needed in order to find its volume.

29. Describe a region that, when revolved around the y-axis, produces a solid of revolution for which *both* discs and washers are needed in order to find its volume.

30. Describe a solid of revolution for which it is not possible to use the disc/washer method.

Skills

■ For each problem below, use four discs or washers to approximate the volume of the solid of revolution that results from rotating the described region around the given axis of revolution. Sketch the solid of revolution and the discs or washers that you are using.

31. The region between the graph of $f(x) = \sqrt{x}$ and the x-axis from $x = 0$ to $x = 4$, revolved around:

    (a) The x-axis (b) The y-axis
    (c) The line $y = 5$ (d) The line $x = -3$

32. The region between the graph of $f(x) = \sin x$ and the x-axis from $x = 0$ to $x = \frac{\pi}{2}$, revolved around:

    (a) The x-axis (b) The y-axis
    (c) The line $y = -1$ (d) The line $x = \pi$

33. The region between the graphs of $f(x) = x^2$ and $g(x) = x$ from $x = 0$ to $x = 1$, revolved around:

    (a) The x-axis (b) The y-axis
    (c) The line $y = 1$ (d) The line $x = 2$

34. The region between the graphs of $f(x) = \sqrt{x}$ and $g(x) = \frac{1}{2}x$ from $x = 0$ to $x = 4$, revolved around:

    (a) The x-axis (b) The y-axis
    (c) The line $y = -2$ (d) The line $x = 7$

■ For each region described below, sketch the solid of revolution that results from rotating the region around the given axis of revolution. Then choose a representative disc or washer, and determine its dimensions and volume (in terms of x_k^* and Δx or in terms of y_k^* and Δy).

35. The region and axes given in Problem 31

36. The region and axes given in Problem 32

37. The region and axes given in Problem 33

38. The region and axes given in Problem 34

■ For Problems 39–48, find the volume of the solid of revolution obtained by rotating the described region around the given axis (or axes) of revolution. Your work should include a sketch of the solid and a representative disc or washer.

39. The region and axis given in Problem 17

40. The region and axis given in Problem 18

41. The region and axis given in Problem 19

42. The region and axis given in Problem 20

43. The region and axes given in Problem 31

44. The region and axes given in Problem 32

45. The region and axes given in Problem 33

46. The region and axes given in Problem 34

47. The region between the graph of $f(x) = x^2 - 4$ and the x-axis from $x = -4$ to $x = 4$, revolved around:

    (a) The x-axis
    (b) The y-axis
    (c) The line $y = 12$

48. The region between the graph of $f(x) = x - 2$ and the x-axis from $x = 2$ to $x = 5$, revolved around:

    (a) The x-axis (b) The y-axis
    (c) The line $y = 3$ (d) The line $x = 2$
    (e) The line $y = -2$ (f) The line $x = 6$

49. Use the techniques of this section to show that a sphere of radius 3 has a volume of 36π (the volume predicted by the formula $V = \frac{4}{3}\pi r^3$).

50. Use the techniques of this section to show that a cone of radius 3 and height 5 has a volume of $V = 15\pi$ (the volume predicted by the formula $V = \frac{1}{3}\pi r^2 h$).

51. A "torus" is a donut-shaped object that is obtained by rotating a circle around an axis. For example, we could revolve the circle with equation $y^2 + (x - 3)^2 = 1$ (i.e., the circle with radius 1 and center at $(0, 3)$) around the y-axis to obtain a torus. Find the volume of this torus. (*Hints:* Find the volume of the top half of the torus (like the top half of a sliced bagel), and multiply this volume by 2. You will have to integrate along the y-axis with washers. The functions $\sqrt{1 - y^2} + 3$ and $-\sqrt{1 - y^2} + 3$ will be involved; why?)

Proofs

52. Prove that the limit of sums in Definition 15.2 is equal to the definite integral in Theorem 15.2.

53. Prove Theorem 15.3 by constructing a limit of sums (in the manner of Example 15.7) and writing this limit of sums as a definite integral.

54. Prove Theorem 15.4 by constructing a limit of sums (in the manner of Example 15.8) and writing this limit of sums as a definite integral.

55. State and prove a version of Theorem 15.3 in the case where f and g are revolved around the y-axis.

56. Prove that a sphere of radius r has a volume of $V = \frac{4}{3}\pi r^3$.

57. Prove that a cone of radius r and height h has a volume of $V = \frac{1}{3}\pi r^2 h$.

15.3 Volumes by Shells

In this section we develop an alternative method for calculating the volume of a solid of revolution. Although this alternative method is more complicated than the "slicing" method of discs or washers, it will enable us to find some volumes we cannot find using our previous methods.

15.3.1 Approximating volume with shells

Not all solids of revolution can be integrated using the "slicing" method from Section 15.2. For example, consider the region bounded by the graph of $f(x) = -x^2 - x + 8$ and the x-axis from $x = 0$ to $x = 2$ shown in Figure 15.33. If this region is revolved around the y-axis, it becomes the solid of revolution shown in Figure 15.34.

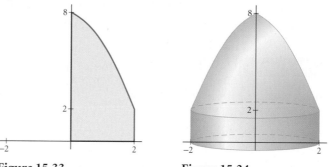

Figure 15.33   **Figure 15.34**

To use discs to find the volume of this solid, we would have to slice the solid horizontally and integrate along the y-axis. Figure 15.35 shows four rectangles that, when revolved about the y-axis, produce the four discs shown in Figure 15.36.

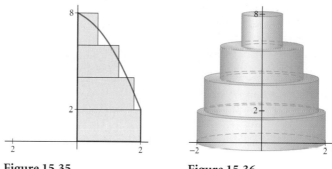

Figure 15.35   **Figure 15.36**

However, it is difficult to find the radii of these discs, since it is not easy to solve the equation $y = -x^2 - x + 8$ for x. (If the x^2 term were x^3 instead, it would be *impossible* to solve for x.) We cannot use the method of slicing to find the volume of the solid in Figure 15.34.

In Figure 15.35 we used horizontal rectangles. What happens if we try to use vertical rectangles? In Figure 15.37 the region is approximated with vertical rectangles (using a midpoint rule). If we revolve these rectangles around the y-axis, each rectangle makes a **shell** rather than a disc or washer. The four shells are shown in Figure 15.38.

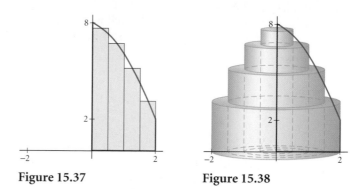

Figure 15.37   **Figure 15.38**

It is important to note the differences between the approximations in Figures 15.36 and 15.38. The four discs in Figure 15.36 are stacked atop one another like the layers of a wedding cake. The four shells in Figure 15.38 are nested, one *inside* the other (like the rings of a tree or the layers of an onion). For example, the innermost "shell" is really a tall cylinder sitting on the x-axis, and the outermost shell is a thick ring that surrounds all of the other shells.

Although we could not approximate the volume of the solid in Figure 15.34 using discs, we *can* approximate the volume using the shells in Figure 15.38; we do so in the following example.

EXAMPLE 15.10   **Approximating a volume with shells**

The solid in Figure 15.34 is the result of rotating the region between the graph of $f(x) = -x^2 - x + 8$ and the x-axis on $[0, 2]$ around the y-axis. Approximate the volume V of this solid by finding the volumes of the four shells in Figure 15.38.

Solution   The outermost shell in Figure 15.38 is the result of rotating the rightmost rectangle in Figure 15.37 around the y-axis. This rectangle extends from $x = 1.5$ to $x = 2$ on the x-axis, and thus the outermost shell is a tall washer with inner radius of $r_1 = 1.5$ and outer radius of $r_2 = 2$. We are using the midpoints of each subinterval to determine the rectangles, so the height of the outermost shell is $h = f(1.75) = -(1.75)^2 - (1.75) + 8 = 3.1875$. The volume of the outermost shell is therefore

$$\pi(r_2)^2 h - \pi(r_1)^2 h = \pi\left(r_2^2 - r_1^2\right)h = \pi((2)^2 - (1.5)^2)(3.1875) \approx 17.52.$$

(This formula comes from the fact that the shell whose volume we are computing is a cylinder with an inner cylinder removed.)

The volumes of the remaining shells can be computed in a similar manner (even the volume of the innermost shell, which can be thought of as a cylinder with a cylinder of radius zero removed). The sum of these four volumes gives us the approximation

$$V \approx \begin{array}{c}\text{volume of}\\\text{first shell}\end{array} + \begin{array}{c}\text{volume of}\\\text{second shell}\end{array} + \begin{array}{c}\text{volume of}\\\text{third shell}\end{array} + \begin{array}{c}\text{volume of}\\\text{fourth shell}\end{array}$$

$$= \pi((2)^2 - (1.5)^2)\, f(1.75) + \pi((1.5)^2 - (1)^2)\, f(1.25)$$

$$\quad + \pi((1)^2 - (0.5)^2)\, f(0.75) + \pi((0.5)^2 - (0)^2)\, f(0.25)$$

$$\approx 17.52 + 20.37 + 15.76 + 6.04$$

$$= 59.69.$$

Therefore, the volume of the solid in Figure 15.34 is approximately 59.69 cubic units.
□

15.3.2 Constructing a Riemann sum for the shell method

Before we can construct a definite integral to represent the *exact* volume of a solid of revolution using the shell method, we must find an easier way to calculate the volume of a shell. Figure 15.39 shows a shell with inner radius r_1, outer radius r_2, and height h.

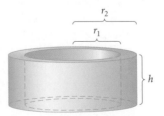

Figure 15.39

Since the shell in Figure 15.39 is just a cylinder (of radius r_2) with a smaller cylinder (of radius r_1) removed, its volume is given by the formula

(1)
$$V = \pi r_2^2 h - \pi r_1^2 h = \pi \left(r_2^2 - r_1^2 \right) h.$$

This volume formula would be difficult to use for constructing a definite integral, because it involves two different radii and thus two different x-values. The following theorem proposes a different method for calculating the volume of a shell.

THEOREM 15.5

The Volume of a Shell
A shell with average radius r, height h, and thickness Δx has volume
$$V = 2\pi r h \Delta x.$$

Two of the terms in Theorem 15.5 need explanation. The **average radius** of a shell is the average $r := \frac{r_1 + r_2}{2}$ of the inner radius r_1 and outer radius r_2 of the shell. The **thickness** of the shell is the difference $\Delta x = r_2 - r_1$ between the outer and inner radii.

To help you remember the formula in Theorem 15.5, notice the following: A shell is just a "fat" cylinder, and the formula $V = 2\pi r h \Delta x$ is just the product of the surface area of a cylinder (with radius r and height h) and the "thickness" of the cylinder (Δx). The proof of Theorem 15.5 follows easily from Equation (1).

PROOF **(THEOREM 15.5)** Given a shell with inner radius r_1, outer radius r_2, and height h, define $r := \frac{r_1 + r_2}{2}$ and $\Delta x = r_2 - r_1$. With this notation, we have

$$2\pi r h \Delta x = 2\pi \left(\frac{r_1 + r_2}{2} \right) h (r_2 - r_1) \qquad \text{(definitions of } r \text{ and } \Delta x\text{)}$$

$$= \pi (r_1 + r_2)(r_1 - r_2) h \qquad \text{(algebra)}$$

$$= \pi \left(r_2^2 - r_1^2 \right) h, \qquad \text{(algebra)}$$

which we know from Equation (1) to be the volume of the shell. ∎

To construct a definite integral for the shell method, we have to employ our "subdivision" notation. Suppose $f(x)$ is continuous on an interval $[a, b]$, and we are interested in the volume of the solid obtained by revolving the area between $f(x)$ and the x-axis on $[a, b]$ around the y-axis. As usual, we will divide the interval $[a, b]$ into n subintervals of the form $[x_{k-1}, x_k]$, where $\Delta x := \frac{b-a}{n}$ and $x_k := a + k \Delta x$. Then the kth shell will be the result of rotating the rectangle whose base is the subinterval $[x_{k-1}, x_k]$ and whose height is defined by the function $f(x)$ at the midpoint of the subinterval. In notation, this means we will choose x_k^* to be the midpoint $x_k^* = \frac{x_{k-1} + x_k}{2}$ of the subinterval $[x_{k-1}, x_k]$, and the height of the kth shell will be $f(x_k^*)$. A representative shell with this notation is shown in Figure 15.40.

The kth shell has inner radius x_{k-1} and outer radius x_k and thus average radius $x_k^* = \frac{x_{k-1} + x_k}{2}$. Since the height of the shell is $f(x_k^*)$ and the thickness of the shell is $\Delta x = x_k - x_{k-1}$, Theorem 15.5 tells us that the volume of the kth shell is

$$\text{Volume of the } k\text{th shell} = 2\pi x_k^* f(x_k^*) \Delta x.$$

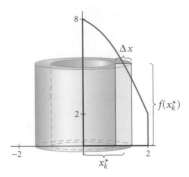

Figure 15.40

The volume V of the solid of revolution is approximated by the sum of the volumes of the shells:

(2)
$$V \approx \sum_{k=1}^{n} 2\pi x_k^* f(x_k^*) \, \Delta x.$$

EXAMPLE 15.11

Using a Riemann sum to express an approximation of a volume by shells

In Example 15.10 we had $f(x) = -x^2 - x + 8$, $[a, b] = [0, 2]$, and $n = 4$. Therefore, $\Delta x = \frac{2-0}{4} = \frac{1}{2}$, and the subdivision points $x_k = 0 + k(\frac{1}{2})$ were

$$x_0 = 0, \quad x_1 = 0.5, \quad x_2 = 1, \quad x_3 = 1.5, \quad \text{and} \quad x_4 = 2.$$

In Example 15.10 we used the formula $\pi(r_2^2 - r_1^2)h$ to calculate the volume of each shell, which led us to a sum of the form

$$\sum_{k=1}^{4} \pi((x_k)^2 - (x_{k-1})^2) f(x_k^*) = \sum_{k=1}^{4} \pi((x_k)^2 - (x_{k-1})^2)(-(x_k^*)^2 - x_k^* + 8).$$

This sum represents the approximation of the solid with four shells. However, if we were to write this sum in terms of an arbitrary number n of shells, we would not have a Riemann sum. (Why?) This means that we cannot interpret a sum of this form as a definite integral as $n \to \infty$.

Now we will use a sum like the one in Equation (2) to express the volume approximation. On each subinterval $[x_{k-1}, x_k]$ we will choose x_k^* to be the midpoint $\frac{x_{k-1}+x_k}{2}$. Therefore, we have

$$x_1^* = 0.25, \quad x_2^* = 0.75, \quad x_3^* = 1.25, \quad \text{and} \quad x_4^* = 1.75.$$

Using Theorem 15.5, the sum of the volumes of the shells can be written as

$$\sum_{k=1}^{4} 2\pi x_k^* f(x_k^*) \, \Delta x = \sum_{k=1}^{4} 2\pi x_k^* (-(x_k^*)^2 - x_k^* + 8)\left(\frac{1}{2}\right).$$

This *is* a Riemann sum, and if we wrote it in terms of an arbitrary number n of shells and let $n \to \infty$, we could construct a definite integral. (See Section 15.3.3.)   □

15.3.3 A definite integral for the shell method

The sum in Equation (2) is a Riemann sum involving the function $2\pi x f(x)$, which is continuous if the function $f(x)$ is continuous. Therefore, if we take the limit as $n \to \infty$, we get the definite integral in Theorem 15.6.

THEOREM 15.6

A Definite Integral for a Simple Case of Volume by Shells

Suppose $f(x)$ is a continuous function on an interval $[a, b]$, and R is the region between the graph of $f(x)$ and the x-axis. The volume of the solid of revolution obtained by rotating the region R around the y-axis is equal to the definite integral

$$2\pi \int_a^b x\, f(x)\, dx.$$

⑤ *Caution* The formula in Theorem 15.6 applies only in simple cases, so don't try to solve every "shell" problem by applying this formula. When finding a volume using the shell method, you should always sketch, and find the volume of, a representative shell. Then the expression for the volume of the representative shell can be used to construct a definite integral. This will enable you to do more general problems than the ones covered by Theorem 15.6.   □

Now we can use a definite integral to calculate the *exact* volume of the solid of revolution from Figure 15.34 and Example 15.10.

EXAMPLE 15.12   **Calculating a volume exactly using the shell method**

The solid in Figure 15.34 is the result of rotating the region between the graph of $f(x) = -x^2 - x + 8$ and the x-axis on $[0, 2]$ around the y-axis. Find the exact volume of this solid using the shell method.

Solution   The solid described is shown in Figure 15.34 on page 871, and Figure 15.40 on page 874 shows a representative shell for this region. The volume of the representative shell at x_k^* is

$$2\pi x_k^*\, f(x_k^*)\, \Delta x = 2\pi x_k^*(-(x_k^*)^2 - x_k^* + 8)\, \Delta x.$$

The volume of the solid is given by the definite integral

$$2\pi \int_0^2 x(-x^2 - x + 8)\, dx = 2\pi \int_0^2 (-x^3 - x^2 + 8x)\, dx$$

$$= 2\pi \left[-\tfrac{1}{4}x^4 - \tfrac{1}{3}x^3 + \tfrac{8}{2}x^2 \right]_0^2 \qquad \text{(integrate)}$$

$$= 2\pi \left(\left(-\tfrac{1}{4}(2)^4 - \tfrac{1}{3}(2)^3 + 4(2)^2 \right) \right.$$

$$\left. - \left(-\tfrac{1}{4}(0)^4 - \tfrac{1}{3}(0)^3 + 4(0)^2 \right) \right) \qquad \text{(evaluate)}$$

$$= \frac{56\pi}{3}.$$

This answer is approximately equal to 58.64, which means that our approximation in Example 15.10 of 59.69 cubic units was actually quite close.   □

⑤ *Caution* Notice that we integrate from the *center* of the solid to the outside when using the shell method, not from one side of the solid to the other. Why?   □

15.3.4 Integrating along the y-axis with shells

The shell method can also be useful for calculating the volumes of solids of revolution obtained by rotating a region around the x-axis. Although it is often easier to use the disc method than the shell method, there are occasions where the shell method is easier. The following example illustrates one such occasion.

EXAMPLE 15.13   **Using shells by integrating along the *y*-axis**

Consider the region bounded by the graph of $f(x) = 3 \ln x$, the line $y = 2$, and the *x*- and *y*-axes. Find the volume of the solid of revolution obtained by rotating this region around the *x*-axis.

Solution   The region in question is shown in Figure 15.41, and the resulting solid of revolution (a cylinder with a "cone" removed on one side) is shown in Figure 15.42. To find the volume of this solid by slicing, we would have to set up *two* integrals: one integral from $x = 0$ to $x = 1$ that uses discs, and one integral from $x = 1$ to $x = e^{\frac{2}{3}}$ that uses washers. Moreover, the second integral would be one that is difficult to solve (set it up to see why). In this case it is much easier to use the shell method along the *y*-axis.

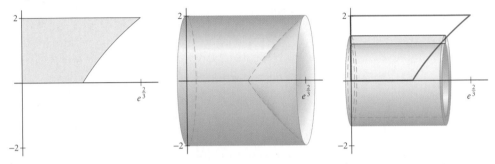

Figure 15.41
The region

Figure 15.42
Revolved around *x*-axis

Figure 15.43
Representative shell

A representative shell can be obtained by rotating a representative (horizontal) rectangle around the *x*-axis; such a shell is shown in Figure 15.43. The thickness of this shell is Δy, and its radius is y_k^* (the chosen point along the *y*-axis). The height of the shell is the horizontal distance from the point $(0, y_k^*)$ to the point where the rectangle meets the curve $y = 3 \ln x$. Therefore, we need to solve for *x* in terms of *y*:

$$y = 3 \ln x \implies \frac{y}{3} = \ln x \implies x = e^{\frac{y}{3}}.$$

This means that the (horizontal) height of the shell is $e^{\frac{y_k^*}{3}}$. Using the formula for the volume of a shell in Theorem 15.5, we have

$$\text{Volume of shell at } y_k^* = 2\pi \, y_k^* \, e^{\frac{y_k^*}{3}} \, \Delta y.$$

Summing the volumes of all the shells and taking the limit as $n \to \infty$, we get a definite integral that we can solve with integration by parts:

$$2\pi \int_0^2 y e^{\frac{y}{3}} \, dy = 2\pi \left(\left[3ye^{\frac{y}{3}} \right]_0^2 - 3 \int_0^2 e^{\frac{y}{3}} \, dy \right) \qquad \text{(parts with } u = y, \, dv = e^{\frac{y}{3}} \, dy\text{)}$$

$$= 2\pi \left(\left[3ye^{\frac{y}{3}} \right]_0^2 - \left[9e^{\frac{y}{3}} \right]_0^2 \right) \qquad\qquad \text{(integrate)}$$

$$= 2\pi \left(\left(6e^{\frac{2}{3}} - 0 \right) - \left(9e^{\frac{2}{3}} - 9e^0 \right) \right) \qquad\qquad \text{(evaluate)}$$

$$= 2\pi \left(-3e^{\frac{2}{3}} + 9 \right). \qquad\qquad\qquad\qquad \square$$

? Question   Write an analogue of the formula in Theorem 15.6 for using the shell method along the *y*-axis.   □

15.3.5 Choosing among the shell, disc, and washer methods

Most of the time it is easiest to use discs or washers to find volumes. Occasionally that is not possible (as in Example 15.12), or the shell method is easier (as in Example 15.13).

If you are good at both techniques, then you will have more tools at your disposal when you wish to find the volume of a solid of revolution.

In general, regions revolved around the x-axis, or any horizontal axis, should be integrated along the x-axis with discs or washers, unless it is necessary or easier to integrate along the y-axis with shells. On the other hand, regions revolved around the y-axis, or any vertical axis, should be integrated along the y-axis with discs or washers, unless it is necessary or easier to integrate along the x-axis with shells. Algorithms 15.1 and 15.2 summarize these possibilities.

ALGORITHM 15.1

Discs, Washers, and Shells for Regions Revolved Around the x-axis

Vertical slices of a region revolved around the x-axis will produce discs or washers, and an integral along the x-axis. Horizontal slices of a region revolved around the x-axis will produce shells, and an integral along the y-axis.

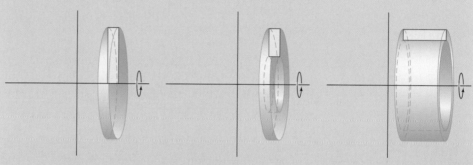

Discs along the x-axis   Washers along the x-axis   Shells along the y-axis

ALGORITHM 15.2

Discs, Washers, and Shells for Regions Revolved Around the y-axis

Horizontal slices of a region revolved around the y-axis will produce discs or washers, and an integral along the y-axis. Vertical slices of a region revolved around the y-axis will produce shells, and an integral along the x-axis.

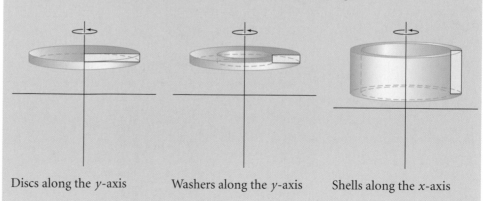

Discs along the y-axis   Washers along the y-axis   Shells along the x-axis

The definite integral that represents the volume of a solid of revolution will depend on the method you use (discs, washers, or shells), the axis you integrate along (the x-axis or the y-axis), and the functions involved. In general, definite integrals that represent volumes of surfaces of revolution always have one of the following general forms.

ALGORITHM 15.3

General Forms for Definite Integrals That Represent Volumes

The volume of a solid of revolution can be obtained using one or more of the following general types of definite integrals. In each integral, a and b represent values along the x-axis, and A and B represent values along the y-axis. The letters r and R represent radii (of discs, washers, or shells), and h represents a height (of a shell). The quantities r, R, and h may be constants, but much of the time they are *functions* of x or y. In particular, r, R, and h often depend on the functions that define the region to be revolved.

$$\pi \int_a^b r^2 \, dx \qquad \pi \int_a^b (R^2 - r^2) \, dx \qquad 2\pi \int_a^b r h \, dx$$

$$\pi \int_A^B r^2 \, dy \qquad \pi \int_A^B (R^2 - r^2) \, dy \qquad 2\pi \int_A^B r h \, dy$$

The key to setting up these types of integrals is to use the region and axis of rotation to determine the quantities r, R, and h that are needed.

Example 15.14 illustrates how to construct definite integrals to represent the volume of a solid of revolution. Sometimes the volume of a solid of revolution must be calculated differently for different parts of the solid (which means we must construct *two* integrals). This is the case in Example 15.14.

EXAMPLE 15.14

Setting up definite integrals for volumes

Consider the region bounded by the graph of $f(x) = e^x$ and the lines $y = 0$, $y = 3$, $x = 0$, and $x = 2$. **(a)** Consider the solid obtained by rotating this region around the x-axis. Write the volume V_1 of this solid in terms of definite integrals in two ways: first by integrating along the x-axis and then by integrating along the y-axis. **(b)** Do the same thing for the volume V_2 of the solid obtained by rotating the region around the y-axis.

Solution   The region in question is shown in Figure 15.44. This region changes character at $x = \ln 3$, since that is the solution to the equation $e^x = 3$. Figure 15.45 shows the solid that results from rotating this region around the x-axis, and Figure 15.46 shows the result of rotating the region around the y-axis.

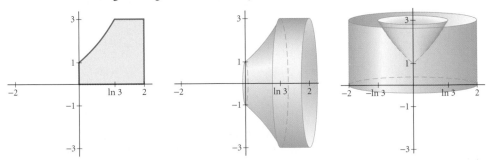

Figure 15.44
The region

Figure 15.45
Revolved around x-axis

Figure 15.46
Revolved around y-axis

(a)   We can write the volume of the solid in Figure 15.45 as the sum of two definite integrals along the x-axis using the disc method. From $x = 0$ to $x = \ln 3$, the radius of the disc at x_k^* will be given by $e^{x_k^*}$, and from $x = \ln 3$ to $x = 2$ the radius of the disc at x_k^* is 3. The volume can be written as

$$V_1 = \pi \int_0^{\ln 3} (e^x)^2 \, dx + \pi \int_{\ln 3}^2 (3)^2 \, dx.$$

Alternatively, we could use the shell method along the y-axis (from $y = 0$ to $y = 3$). Solving $y = e^x$ for x, we get $x = \ln y$. Therefore, from $y = 1$ to $y = 3$, the (horizontal) height of the shell at y_k^* will be the difference $2 - \ln y_k^*$. From $y = 0$ to $y = 1$, each shell will have a (horizontal) height of $2 - 0 = 2$. Therefore, we can write the volume of the solid in Figure 15.45 as

$$V_1 = 2\pi \int_0^1 y\,(2)\, dy + 2\pi \int_1^3 y\,(2 - \ln y)\, dy.$$

(b) Integrating along the x-axis produces shells in this case. From $x = 0$ to $x = \ln 3$, the shell at x_k^* will have a height of $e^{x_k^*}$, and from $x = \ln 3$ to $x = 2$, the shell will have a height of 3. Therefore, the volume of the solid in Figure 15.46 can be written as the following sum of definite integrals:

$$V_2 = 2\pi \int_0^{\ln 3} xe^x\, dx + 2\pi \int_{\ln 3}^2 x(3)\, dx.$$

On the other hand, to find the volume of the solid in Figure 15.46 by integrating along the y-axis, we have to use discs and washers. From $y = 0$ to $y = 1$, the disc at y_k^* will have a radius of 2, and from $y = 1$ to $y = 3$, a rectangle at y_k^* will produce a washer, with inner radius given by $\ln(y_k^*)$ (since $y = e^x$, we have $x = \ln y$) and an outer radius of 2. Therefore, we can write the volume of the solid in Figure 15.46 as

$$V_2 = \pi \int_0^1 (2)^2\, dy + \pi \int_1^3 (2 - \ln y)^2\, dy. \qquad \square$$

? *Question* Draw the discs, washers, and shells used in the various parts of Example 15.14, and label the locations, radii, heights, and/or widths of these objects in terms of x_k^* and Δx or y_k^* and Δy. $\qquad \square$

In Example 15.14 we had to write the volumes of the solids in Figures 15.45 and 15.46 in terms of *two* definite integrals, regardless of whether we chose to integrate along the x-axis or the y-axis. In this particular example, it happened to be easier to integrate with discs (along the x-axis) for the solid in Figure 15.45, and easier to integrate with shells (again along the x-axis) for the solid in Figure 15.46. If both disc and shell methods apply, you may decide which method is easiest.

What you should know after reading Section 15.3

▶ How to approximate the volume of a solid of revolution using shells, and how to write this approximation in notation involving x_k^* and Δx or y_k^* and Δy.

▶ The alternative method for finding the volume of a "shell," and how to prove it. Why do we need to use this alternative method?

▶ How to set up definite integrals that represent the volume of a solid of revolution using the shell method. What types of solids of revolution require the shell method when you integrate along the x-axis? What types require using discs or washers? What about when you integrate along the y-axis? What about other axes?

▶ When would you have to use *two* definite integrals to express the volume of a solid of revolution? What are the general forms of the definite integrals for using discs, washers, and shells to represent volumes?

Exercises 15.3

Concepts

0. Read the section and make your own summary of the material. Include whatever you think will help you review the material later on.

1. Consider the rectangle bounded by $y = 4$ and $y = 1$ on the x-interval $[1, 1.5]$.

   (a) Calculate the volume of the shell obtained by rotating this rectangle around the y-axis using the formula $\pi r_2^2 h - \pi r_1^2 h$.

   (b) Calculate the volume of the shell obtained by rotating this rectangle around the y-axis using the formula $2\pi r h \, \Delta x$.

   (c) Calculate the volume of the shell obtained by rotating this rectangle around the line $x = -1$.

2. Suppose a "shell" obtained by rotating a rectangle around an axis has an inner radius of r_1 and an outer radius of r_2. Define the "average radius" and the "thickness" of the shell in terms of r_1 and r_2.

3. Why is the shell volume formula $2\pi r h \, \Delta x$ better than the formula $\pi r_2^2 h - \pi r_1^2 h$ when we try to construct a definite integral that represents the volume of a solid of revolution?

4. Why do we integrate from the "center" to the "end" of a solid when using the shell method, rather than integrating along the entire span of the solid?

5. Consider the region between the graph of the function $f(x) = x^2$ and the x-axis on the interval $[0, 2]$. Draw a Riemann sum approximation of the area of this region using a Midpoint Sum with four rectangles. Explain how this Riemann sum is related to the approximation with four shells of the solid of revolution obtained by rotating the region around the y-axis.

6. Suppose you want to use four shells to approximate the volume of the solid of revolution obtained by rotating the region between the graph of $f(x) = x^2$ and the x-axis on the interval $[0, 2]$ around the y-axis.

   (a) Find the values of Δx, x_0, x_1, x_2, x_3, and x_4 and represent these values graphically.

   (b) Choose x_k^* to be the midpoint of each interval $[x_{k-1}, x_k]$. Find the values of x_1^*, x_2^*, x_3^*, and x_4^* and represent these values graphically.

   (c) Determine $f(x_1^*)$, $f(x_2^*)$, $f(x_3^*)$, and $f(x_4^*)$ and represent these values graphically.

   (d) What is the volume of the third shell (the third shell from the inside) in your approximation?

   (e) What is the volume of the kth shell in your approximation? (Your answer should include x_k^*.)

7. Suppose you want to use four shells to approximate the volume of the solid of revolution obtained by rotating the region between the graph of $f(x) = $

x^2 and the x-axis on the interval $[0, 2]$ around the x-axis.

   (a) Find the values of Δy, y_0, y_1, y_2, y_3, and y_4 and represent these values graphically.

   (b) Choose y_k^* to be the midpoint of each interval $[y_{k-1}, y_k]$. Find the values of y_1^*, y_2^*, y_3^*, and y_4^* and represent these values graphically.

   (c) Determine $f^{-1}(y_1^*)$, $f^{-1}(y_2^*)$, $f^{-1}(y_3^*)$, and $f^{-1}(y_4^*)$ and represent these values graphically.

   (d) What is the volume of the third shell (the third shell from the inside) in your approximation?

   (e) What is the volume of the kth shell in your approximation? (Your answer should include y_k^*.)

8. Describe a region that, when revolved around the x-axis, results in a solid of revolution for which both the shell method and the disc/washer method are possible, but the shell method is preferable.

9. Describe a region that, when revolved around the y-axis, results in a solid of revolution for which both the shell method and the disc/washer method are possible, but the shell method is preferable.

10. Describe a region that, when revolved around the x-axis, results in a solid of revolution for which the disc/washer method is impossible, whereas the shell method works.

11. Describe a region that, when revolved around the y-axis, results in a solid of revolution for which the disc/washer method is impossible, whereas the shell method works.

12. Describe a solid of revolution for which neither the disc/washer method nor the shell method is possible.

■ Write each of the following limits of sums as a definite integral, and identify a solid of revolution whose volume is represented by that definite integral. (*Hints:* You will have to use the information about Δy and y_k^* to find the limits of integration. Finding the solid of revolution involves determining the region and the axis of revolution it is revolved around.)

13. $\displaystyle\lim_{n\to\infty} \sum_{k=1}^{n} 2\pi (x_k^*)^{\frac{3}{2}} \Delta x$, where $\Delta x = \frac{4}{n}$, $x_k = k \, \Delta x$, and $x_k^* = \frac{x_{k-1} + x_k}{2}$

14. $\displaystyle\lim_{n\to\infty} \sum_{k=1}^{n} 2\pi (y_k^*)^3 \Delta y$, where $\Delta y = \frac{2}{n}$, $y_k = k \, \Delta y$, and $y_k^* = \frac{y_{k-1} + y_k}{2}$

■ Each of the following definite integrals represents the volume of a solid of revolution (using the shell method). Determine the solid of revolution by finding a region in the plane and an axis of revolution that will produce that solid. (*Hint:* The axis of revolution is the x-axis or the y-axis in each of these problems.)

15. $2\pi \displaystyle\int_0^1 x(x^2)\, dx$

16. $2\pi \displaystyle\int_0^{\frac{\pi}{4}} x \cos x\, dx$

17. $2\pi \displaystyle\int_0^1 (x^2 - x^3)\, dx$

18. $2\pi \displaystyle\int_4^5 y\sqrt{y-4}\, dy$

19. $2\pi \displaystyle\int_0^1 (3y - 3y^2)\, dy$

20. $2\pi \displaystyle\int_1^e y \ln y\, dy$

■ For each problem below, write down definite integrals (or sums of definite integrals) for (a) the shell method (if possible) and (b) the disc/washer method (if possible) that represent the volume of the solid of revolution obtained by rotating the region described around the given axis of revolution. Then (c) determine which method would be easier in each case, and why.

21. The region between the graph of $f(x) = 2x + 1$ and the x-axis from $x = 0$ to $x = 3$, revolved around the x-axis

22. The region between the graph of $f(x) = \frac{1}{x}$ and the lines $y = 0$, $y = 1$, $x = 0$, and $x = 2$, revolved around the y-axis

23. The region between the graph of $f(x) = |x|$ and the x-axis on $[-2, 2]$, revolved around the x-axis

24. The region between the graph of $f(x) = x^2 + 1$ and the x-axis on $[0, 1]$, revolved around the line $x = 3$

25. The region between the graph of $f(x) = x^2 + 1$ and the x-axis on $[0, 1]$, revolved around the line $y = -2$

Skills

■ For each problem below, use four shells to approximate the volume of the solid of revolution that results from rotating the described region around the given axis of revolution. Sketch the solid of revolution and the shells that you are using.

26. The region between the graph of $f(x) = \cos x$ and the x-axis on $[0, \frac{\pi}{2}]$, revolved around the y-axis

27. The region between the graph of $f(x) = \sqrt{x}$ and the x-axis on $[0, 4]$, revolved around the x-axis

28. The region between the graph of $f(x) = \sqrt{x}$ and the x-axis on $[0, 4]$, revolved around the y-axis

29. The region between the graphs of $f(x) = \sqrt{x}$ and $g(x) = 2$ on $[0, 4]$, revolved around the y-axis

■ For each region described below, sketch the solid of revolution that results from rotating the region around the

given axis of rotation. Then choose a representative shell and determine its dimensions and volume (in terms of x_k^* and Δx or in terms of y_k^* and Δy).

30. The region and axis given in Problem 26

31. The region and axis given in Problem 27

32. The region and axis given in Problem 28

33. The region and axis given in Problem 29

■ Find the volume of each solid of revolution described below in two ways (if possible): (a) using shells, and (b) using discs and/or washers. Your work should include a sketch of representative discs, washers, or shells and the volumes of these discs, washers, or shells.

34. The region and axis given in Problem 26

35. The region and axis given in Problem 27

36. The region and axis given in Problem 28

37. The region and axis given in Problem 29

■ Find the volume of each solid of revolution described below using the shell method. Your work for each problem should include a labeled sketch of a representative shell.

38. The region between the graph of $f(x) = e^{2x}$ and the x-axis on $[0, 4]$, revolved around the x-axis

39. The region between the graph of $f(x) = e^x$ and the line $y = e$ on $[0, 1]$, revolved around the x-axis

40. The region between the graph of $f(x) = 4 - x^2$ and the line $y = 4$ on $[0, 2]$, revolved around the y-axis

41. The region between the graph of $f(x) = (x - 3)^2 + 2$ and the x-axis on $[0, 6]$, revolved around the y-axis

42. The region between the graphs of $f(x) = \sqrt{x}$ and $g(x) = x^2$ on $[0, 1]$, revolved around the x-axis

43. The region between the graph of $f(x) = (x - 2)^2$ and the x-axis on $[0, 4]$, revolved around the x-axis

44. The region between the graph of $f(x) = x^2 + 2$ and the x-axis on $[0, 3]$, revolved around the line $x = -1$

45. The region between the graph of $f(x) = x^2$ and the x-axis on $[0, 3]$, revolved around the line $y = 10$

46. The region bounded by the graphs of $f(x) = 2 \ln x$, $y = 0$, $y = 3$, and $x = 0$, revolved around the x-axis

■ Find the volume of each solid of revolution described below in any way that you prefer.

47. The region between the graph of $f(x) = 9 - x^2$ and the x-axis on $[0, 3]$, revolved around the x-axis

48. The region between the graph of $f(x) = x^2 - 4x + 4$ and the x-axis on $[0, 2]$, revolved around the y-axis

49. The region bounded by the graphs of $f(x) = \frac{2}{x}$, $y = 0$, $y = 3$, $x = 0$, and $x = 2$, revolved around the x-axis

50. Use the shell method to show that a sphere of radius 3 has a volume of 36π (the volume predicted by the formula $V = \frac{4}{3}\pi r^3$).

51. Use the shell method to show that a cone of radius 3 and height 5 has a volume of $V = 15\pi$ (the volume predicted by the formula $V = \frac{1}{3}\pi r^2 h$).

Proofs

52. Prove that a shell with average radius r, height h, and thickness Δx has volume $V = 2\pi r h \, \Delta x$.

53. Prove Theorem 15.6 by constructing a limit of Riemann sums that represents the volume of a solid of revolution using the shell method.

54. Use the shell method to prove that a sphere of radius r has a volume of $V = \frac{4}{3}\pi r^3$.

55. Use the shell method to prove that a cone of radius r and height h has a volume of $V = \frac{1}{3}\pi r^2 h$.

15.4   Practical Applications

In this section we examine three types of practical applications that lend themselves to the "subdivide, approximate, and add" strategy that we have used to find areas, arc lengths, and volumes—and that therefore can be measured with definite integrals.

15.4.1   Work given a variable force

How much "work" does it take to lift an object? Clearly it takes more work to lift a heavy object (say, a car) then it takes to lift a lighter object (say, a book). It also takes more work to lift an object a long distance (such as lifting a car 30 feet) than a short distance (such as lifting a car 1 inch off the ground). In fact, the **work** involved in lifting an object is the *product* of the weight of the object and the distance it is to be lifted. Work is defined to be "force times distance," where "force" is the weight of the object (the force due to gravity).

DEFINITION 15.3

Work Given a Constant Force and Distance

The **work** involved in lifting an object weighing F pounds through a distance of d feet is defined to be

$$W := F \cdot d \text{ foot-pounds.}$$

Note that work involves units called "foot-pounds"; this is because the work is the product of the force (measured in pounds) and the distance (measured in feet). It is important to note that the concepts of "work" and "force" are much more general than what we have defined here. We are restricting our attention to the work involved in lifting an object. In physics, one might see force measured in "newtons" and distance measured in meters, in which case work would be measured in "newton-meters," which are also called "joules." We will keep the physics simple: All of our examples concerning work will involve the lifting of objects, and our units will always be measured in feet, pounds, and foot-pounds.

EXAMPLE 15.15   **Calculating work when force and distance are constant**

Suppose you wish to lift a large bucket full of water. The bucket is in the shape of a cylinder with a radius of 1 foot and a height of 2 feet. Find the work required to lift the bucket 3 feet off the ground (assume the weight of the bucket itself is negligible).

Solution   To find the work involved in lifting the water bucket, we need to know how much the water weighs. Many of the problems we do in this section require this information, and we state it now:

(1)   Water weighs 62.4 pounds per cubic foot.

This means that V cubic pounds of water will weigh $62.4V$ pounds. The water in our bucket has a volume of $V = \pi r^2 h = \pi(1)^2(2) = 2\pi$ cubic feet, and thus it weighs $62.4(2\pi)$ pounds. The work required to lift the bucket 3 feet off the ground is

$$W = Fd = (62.4(2\pi) \text{ pounds})(3 \text{ feet})$$

$$= 374.4\pi \text{ foot-pounds} \approx 1176.21 \text{ foot-pounds.} \qquad \square$$

If either the weight of the object or the distance that it is to be moved is variable, then finding the work involved in lifting that object is more complicated. For example, consider the task of pumping all the water out of a swimming pool. The water at the bottom of the pool has to be moved farther than the water at the top of the pool, so the distance involved is variable. On the other hand, consider the task of lifting a leaky water bucket to the top of a building by a rope. For every foot the bucket is lifted, some of the water leaks out. Therefore, the weight involved is variable. In some situations it could even be the case that *both* the weight and the distance are variable!

The general strategy will be to "subdivide, approximate, and add," much as we did for areas, arc lengths, and volumes. We will subdivide in such a way that we can assume without too much error that the weight and/or distance are constant, and then use these constants to approximate the work involved on each "piece" of the problem. Then we will add all these approximations to approximate the total work. As you will see in Example 15.16, we will then be able to recognize the approximating sum as a Riemann sum and write the *exact* amount of work as a definite integral.

EXAMPLE 15.16

Using a definite integral to calculate work when distance varies

Find the work involved in pumping all of the water out of the top of a cylindrical hot tub that is 4 feet deep and has a radius of 3.5 feet.

Solution   The water near the bottom of the hot tub requires more work to pump out than does the water near the surface, since the water near the bottom of the hot tub has to be moved a greater distance. If we consider a thin horizontal slice of the hot tub, we can assume without too much error that all the water in that slice has to be moved the same distance (we are considering only the *upward* distance that the water has to move, and if the slice is thin, then all the water in the slice will be at approximately the same depth).

We now need some notation. Suppose the hot tub is sitting on the x-axis, so that the bottom of the tub is at $y = 0$ and the top of the tub is at $y = 4$. We will divide the interval from $y = 0$ to $y = 4$ into n subintervals $[y_{k-1}, y_k]$ of thickness Δy. A representative slice of the hot tub at some height $y_k^* \in [y_{k-1}, y_k]$ is shown in Figure 15.47.

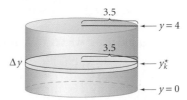

Figure 15.47

The slice at y_k^* needs to be moved upwards $4 - y_k^*$ units to be pumped out of the tub. This slice is a disc with volume $\pi r^2 h$, where r is its radius and h is its height (thickness). Here we have $r = 3.5$ and $h = \Delta y$; therefore, the volume of the water in the slice is $V = \pi(3.5)^2 \Delta y$ cubic feet. By the fact in Statement (1), this means that the weight of the water in the slice is $(62.4)(\pi)(3.5)^2 \Delta y$ pounds. The work involved in pumping out this slice of water is approximately

$$\text{Work to move } k\text{th slice} = F \cdot d = (62.4)(\pi)(3.5)^2 \Delta y \cdot (4 - y_k^*).$$

By repeating this approximation for each horizontal slice and then adding all these approximations, we can get an approximation for the total work W required to pump all of the water out of the hot tub:

$$(2) \qquad W \approx \sum_{k=1}^{n} (62.4)(\pi)(3.5)^2 (4 - y_k^*) \Delta y.$$

You should recognize this sum as a Riemann sum. As $n \to \infty$ (i.e., as we make thinner and thinner slices), the approximation approaches the *actual* work given by the definite integral

$$W = \int_0^4 (62.4)(\pi)(3.5)^2 (4 - y) \, dy$$
$$= (62.4)(\pi)(3.5)^2 \left[4y - \tfrac{1}{2}y^2 \right]_0^4 \qquad \text{(integrate)}$$
$$= (62.4)(\pi)(3.5)^2 \left((16 - \tfrac{1}{2}(16)) - (0 - 0) \right) \qquad \text{(evaluate)}$$
$$\approx 19,211.5 \text{ foot-pounds.} \qquad \square$$

? Question   Why is the task of pumping all of the water out of a hot tub different from the task of lifting *all* the water in a hot tub 4 feet in the air? How much work would the latter task take? $\square$

There are a few things that are worth mentioning about the process used in Example 15.16. First of all, the sum in Equation (2) could be used to approximate the work required to pump the water out of the hot tub. For a particular value of n, say $n = 6$, you could determine the values of y_k^* and Δy and use them to find the value of the sum. A larger value of n will produce a better approximation, because the slices of the hot tub will then be thinner (so our approximation that the entire slice is at the same depth will involve less error).

? Question   Approximate the work required to pump all of the water out of the hot tub in Example 15.16 using $n = 6$ slices. $\square$

The sum in Equation (2) is a Riemann sum because it is of the form $\sum_{k=1}^{n} f(y_k^*) \Delta y$ for some function $f(y)$ (can you name the function?). Moreover, since the function $f(y)$ in the Riemann sum is continuous, we know that the Riemann sum becomes a definite integral as $n \to \infty$. Finally, notice that the limits of integration are $y = 0$ and $y = 4$. You may want to review the definition of the definite integral in Section 12.3.1 to make sure that you understand this construction.

Example 15.16 was relatively simple because each "slice" of the hot tub had the same volume (and thus the same weight). In general this is not necessarily the case. For example, if the hot tub had been in the shape of a cone, then the volume of the slices at the bottom of the tub would be smaller than the volume of the slices at the top of the tub. Since a conical hot tub would be fairly inconvenient to use, let's think instead about a conical tank:

EXAMPLE 15.17   **Using a definite integral to calculate work when distance and volume vary**

Find the work involved in pumping all of the water out of the top of a cone-shaped tank that is 4 feet tall and has a radius of 3.5 feet at the top.

Solution   Again we will say that the bottom of the tank is at height $y = 0$ and the top of the tank is at height $y = 4$. A representative horizontal slice at some point y_k^* is shown in Figure 15.48.

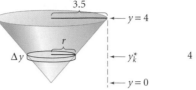

Figure 15.48   **Figure 15.49**

To find the volume of the water in this slice, we must know its radius. By the law of similar triangles (see Figure 15.49), we have

$$\frac{r}{y_k^*} = \frac{3.5}{4} \implies r = \tfrac{3.5}{4} y_k^*.$$

The radius depends on the height y_k^* of the slice (as you would expect). The volume of the slice is $\pi r^2 \, \Delta y$, and the weight of the slice is $(62.4)\pi (\tfrac{3.5}{4} y_k^*)^2 \, \Delta y$. The vertical distance that the water in the slice must move is approximately $4 - y_k^*$ (since we are assuming the slice is very thin). Therefore, the work involved in pumping out the slice of water in Figure 15.48 is

$$\begin{matrix} \text{Work to move} \\ k\text{th slice} \end{matrix} = F \cdot d = (62.4)(\pi)\left(\tfrac{3.5}{4} y_k^*\right)^2 \Delta y \cdot (4 - y_k^*).$$

This means that the work W involved in pumping out all of the water in the tank is approximately

$$W \approx \sum_{k=1}^{n} (62.4)(\pi)\left(\tfrac{3.5}{4}\right)^2 (y_k^*)^2 (4 - y_k^*) \, \Delta y.$$

This is a Riemann sum, so as $n \to \infty$, we obtain a definite integral that represents the *exact* amount of work required:

$$W = \int_0^4 (62.4)(\pi)\left(\tfrac{3.5}{4}\right)^2 y^2 (4 - y) \, dy$$

$$= (62.4)(\pi)\left(\tfrac{3.5}{4}\right)^2 \int_0^4 (4y^2 - y^3) \, dy \qquad\qquad \text{(algebra)}$$

$$= (62.4)(\pi)\left(\tfrac{3.5}{4}\right)^2 \left[\tfrac{4}{3} y^3 - \tfrac{1}{4} y^4\right]_0^4 \qquad\qquad \text{(integrate)}$$

$$= (62.4)(\pi)\left(\tfrac{3.5}{4}\right)^2 \left(\left(\tfrac{4}{3}(64) - \tfrac{1}{4}(256)\right) - (0 - 0)\right) \quad \text{(evaluate)}$$

$$\approx 3201.91 \text{ foot-pounds.}$$

Notice that it takes less work to pump the water out of the conical tank than it did to pump the water out of the hot tub in Example 15.16. Why?   □

15.4.2 Mass given a variable density

The ***mass*** of an object is defined to be the product of its density and its volume. Therefore, a small object that is very dense (like a small rock) could have the same mass as a larger object that is less dense (for example, a very large serving of cotton candy). If the object is "homogeneous" in the sense that it has the same density everywhere, then we simply multiply its volume by this density to obtain its mass.

DEFINITION 15.4

Mass Given a Constant Density and Volume

An object with volume V cubic centimeters and constant density of ρ grams per cubic centimeter has a *mass* of

$$m := \rho \cdot V \text{ grams.}$$

The symbol "ρ" is the lowercase Greek letter "rho." We could use units other than those in Definition 15.4. For example, an object that has volume V cubic inches and a density of ρ ounces per cubic centimeter would have a mass of ρV ounces.

EXAMPLE 15.18   **Calculating mass when density and volume are constant**

Gold has a density of approximately 19.3 grams per cubic centimeter, and silver has a density of approximately 10.5 grams per cubic centimeter. The mass of a gold coin of radius 1.5 centimeters and thickness 0.2 centimeter is

$$m = \rho V = \left(19.3 \tfrac{\text{grams}}{\text{cm}^3}\right)(\pi(1.5)^2(0.2) \text{ cm}^3) \approx 27.28 \text{ grams.}$$

A silver coin of the same dimensions has a smaller mass:

$$m = \rho V = \left(10.5 \tfrac{\text{grams}}{\text{cm}^3}\right)(\pi(1.5)^2(0.2) \text{ cm}^3) \approx 14.84 \text{ grams.} \qquad \square$$

If the density of an object varies, then finding the mass of the object is more complicated. The key is to "slice" the object in such a way that for each piece, we can assume that the density is approximately constant.

EXAMPLE 15.19   **Using a definite integral to calculate mass when density varies**

Suppose a cylindrical rod with a radius of 2 centimeters and a length of 24 centimeters is made of a combination of silver and gold in such a way that the density of the rod x centimeters from the left end is given by the function $\rho(x) = 10.5 + 0.01527x^2$ grams per cubic centimeter. Find the total mass of the rod.

Solution   If we look at a thin slice of the rod at some point x_k^* centimeters from the left end, we can assume without too much error that the density throughout the entire thin slice is $\rho(x_k^*) = 10.5 + 0.01527(x_k^*)^2$ grams per cubic centimeter. Figure 15.50 shows such a representative slice.

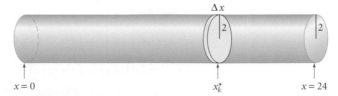

Figure 15.50

The slice in Figure 15.50 is a disc of radius 2 and "thickness" Δx, and therefore it has volume $\pi(2)^2 \Delta x$. The mass of the slice is approximately:

$$\text{Mass of } k\text{th slice} = (10.5 + 0.01527(x_k^*)^2)(\pi(2)^2 \Delta x).$$

The approximate mass of the entire rod is given by the sum

$$\sum_{k=1}^{n} 4\pi(10.5 + 0.01527(x_k^*)^2) \Delta x.$$

As $n \to \infty$, this Riemann sum approaches the actual mass of the rod, given by the definite integral

$$\int_0^{24} 4\pi(10.5 + 0.01527x^2)\, dx$$

$$= 4\pi\left[10.5x + \frac{0.01527}{3}x^3\right]_0^{24} \qquad \text{(integrate)}$$

$$= 4\pi\left((10.5(24) + \frac{0.01527}{3}(24)^3) - (0 + 0)\right) \qquad \text{(evaluate)}$$

$$\approx 4050.95 \text{ grams.} \qquad \qquad \Box$$

In Example 15.19, the density of each slice varied according to the distance from the left end of the rod, but the volume of each slice was the same. In the following example, the density *and* the volume are variable depending on the height of the "slice" of the object.

EXAMPLE 15.20

Using a definite integral to calculate mass when density and volume vary

Alina has made a jello mold that is 4.5 inches tall, 6 inches across its base, and in the shape of the top of a downward-pointing parabola that has been revolved around the y-axis. She put strawberry pieces into the jello mold, but they have tended to settle down to the bottom. This causes the density of the jello mold to vary linearly with height in such a way that the density at the very top of the mold is 0.25 ounce per cubic inch, whereas the density at the very bottom of the mold is 1.3 ounces per cubic inch. Find the mass of the entire jello mold.

Solution   We first need to find functions that describe the shape and density of the jello mold. This is not difficult to do using the information given in the problem; you will do so in the exercises. It turns out that the shape of the jello mold is the solid of revolution obtained by rotating the graph of $y = 4.5 - 0.5x^2$ on $[0, 3]$ around the x-axis and that the density of the jello mold at height y is approximately given by the function $\rho(y) = 1.3 - 0.233y$. (*Note:* Once again, $\rho(y)$ here means "ρ of y," not "ρ times y"; the density ρ of the jello mold is a *function* of the height y.)

Since the density of the jello depends only on height, if we make a thin horizontal slice at some height y_k^*, we will be able to assume that the entire slice has the same density without introducing too much error. Figure 15.51 shows such a representative slice.

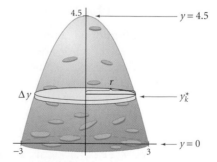

Figure 15.51
Strawberries settle to the bottom.

The volume of this slice is $\pi r^2 \Delta y$, where r is the radius corresponding to the height y_k^*. Since the radius is determined by the graph of $y = 4.5 - 0.5x^2$, we have $y_k^* = 4.5 - 0.5r^2$ and thus $r = \sqrt{9 - 2y_k^*}$. Since the entire slice is approximately at height y_k^*, we will assume that the density of the jello mold is $\rho(y_k^*) = 1.3 - 0.233y_k^*$ ounces per cubic inch everywhere in the slice. The approximate mass of the representative slice is the product of the density and the volume of the slice:

$$\text{Mass of the } k\text{th slice} = (1.3 - 0.233 y_k^*)\left(\pi \left(\sqrt{9 - 2y_k^*} \right)^2 \Delta y \right)$$

$$= \pi(1.3 - 0.233 y_k^*)(9 - 2y_k^*)\, \Delta y.$$

The total mass m of the jello mold can be approximated by the sum of the approximate masses of all of the slices:

$$m \approx \sum_{k=1}^{n} \pi(1.3 - 0.233 y_k^*)(9 - 2y_k^*)\, \Delta y.$$

This is a Riemann sum, and as $n \to \infty$, this sum approaches a definite integral that represents the actual mass of the jello mold:

$$\int_0^{4.5} \pi(1.3 - 0.233 y)(9 - 2y)\, dy$$

$$= \pi \int_0^{4.5} (11.7 - 4.697 y + 0.466 y^2)\, dy \qquad \text{(algebra)}$$

$$= \pi \left[11.7y - \tfrac{4.697}{2} y^2 + \tfrac{0.466}{3} y^3 \right]_0^{4.5} \qquad \text{(integrate)}$$

$$= \pi \left(\left(11.7(4.5) - \tfrac{4.697}{2}(4.5)^2 + \tfrac{0.466}{3}(4.5)^3 \right) - 0 \right) \qquad \text{(evaluate)}$$

$$\approx 60.4682 \text{ ounces.} \qquad \Box$$

? *Question*   Use the information in the description of the problem in Example 15.20 to show that the height of the jello mold is given by the function $y = 4.5 - 0.5x^2$ and that the density of the jello mold at height y is given by the linear function $\rho(y) = 1.3 - 0.233 y$. $\Box$

? *Question*   If the approximating sum in Example 15.20 runs from $k = 1$ to $k = n$, why does the definite integral have 0 and 4.5 as its limits of integration? (Think about the definitions of Δy and y_k.) $\Box$

15.4.3 Hydrostatic force given a variable depth

The bottom of any tank or swimming pool is under pressure from the water above it. The force exerted on the bottom surface of such a tank or pool is equal to the weight of the water above it. Since water weighs 62.4 pounds per cubic foot, the weight of the water will be 62.4 times its volume. Moreover, if the horizontal cross section of the water is the same everywhere, with area A, then the volume of the water will be A times the depth d, and the force on the bottom of the tank will be $62.4Ad$. This leads us to the following definition.

DEFINITION 15.5

Hydrostatic Force Given a Constant Depth and Area

The *hydrostatic force* on a horizontal wall of area A under a depth of d feet of water is defined to be

$$F = 62.4Ad \text{ pounds.}$$

EXAMPLE 15.21

Calculating hydrostatic force when depth and area are constant

How much hydrostatic force is exerted on the bottom of a rectangular swimming pool measuring 10 feet by 30 feet that has a constant depth of 5 feet?

Solution   The bottom of the swimming pool is a rectangular surface with area of $A = (10)(30) = 300$ square feet. Since the depth of the water is constantly $d = 5$ feet, the hydrostatic force on the bottom of the pool is

$$F = \left(62.4 \tfrac{\text{lbs}}{\text{ft}^3}\right)(300 \text{ ft}^2)(5 \text{ ft}) = 93{,}600 \text{ pounds}.$$

Notice that the hydrostatic force on the bottom of the pool is the same as the weight of the water in the pool.   □

In fact, the same principle applies to the force exerted against *any* wall (horizontal, vertical, or otherwise) by a mass of water. (This is due to the fact that the pressure, or force per square foot of area, exerted by the water is the same in all directions.) However, a nonhorizontal wall will be at different depths at various points on the wall, and the force from the water will be different at these various points. For example, consider one of the side walls of the swimming pool in Example 15.21. Near the bottom of the side wall the water will be exerting a great deal of force, but near the top of the side wall the water will exert less force. We will have to use a "slicing" technique to find the hydrostatic force exerted on nonhorizontal walls.

EXAMPLE 15.22

Using a definite integral to calculate hydrostatic force when depth varies

Find the hydrostatic force exerted on one of the shorter side walls of the swimming pool in Example 15.21.

Solution   The shorter side walls of the pool in Example 15.21 are 5 feet high and 10 feet wide. The force exerted by the water on one of these walls is different at different points on the wall, depending on the depth at each of those points. However, if we consider a thin horizontal slice of the wall, then we can assume without too much approximation error that the entire slice is at the same depth. Figure 15.52 shows a representative slice of the side wall at some height y_k^* from the bottom of the pool.

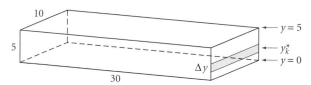

Figure 15.52

The slice shown in Figure 15.52 has an area of $10\Delta y$ square feet. Assuming that the entire thin slice of wall is at a depth of $d = 5 - y_k^*$ feet, the hydrostatic force on that slice of wall is given by:

$$\begin{matrix}\text{Hydrostatic force}\\ \text{on } k\text{th slice}\end{matrix} = 62.4(10\Delta y)(5 - y_k^*).$$

Therefore, the hydrostatic force on the entire side wall is approximately

$$\sum_{k=1}^{n} 62.4(10)(5 - y_k^*)\,\Delta y.$$

As $n \to \infty$, this Riemann sum approaches a definite integral that represents the actual amount of hydrostatic force on the side wall of the pool:

$$\int_0^5 62.4(10)(5 - y)\,dy = (62.4)(10)\big[5y - \tfrac{1}{2}y^2\big]_0^5 \qquad \text{(integrate)}$$

$$= (62.4)(10)\big((25 - \tfrac{25}{2}) - (0 - 0)\big) \qquad \text{(evaluate)}$$

$$= 7800 \text{ pounds.} \qquad \qquad \square$$

⟡ Caution   It may sound hard to believe, but the hydrostatic force exerted on the side wall of a swimming pool does not depend on how much water is behind that wall. This means we would get the same answer if there were 300, rather than 30, feet of water extending behind the short wall of the pool. Mathematically, this is due to the fact that hydrostatic force is calculated using the *area* of a slice of the pool wall, not the volume of a slice of the entire pool. $\qquad \square$

Example 15.21 was made relatively simple by the fact that each slice of the side wall of the pool had the same area. If the horizontal cross section of the side of the pool is different at different depths, then the area of each slice of that side will depend on the depth. This is the case in Example 15.23.

EXAMPLE 15.23   **Using a definite integral to calculate hydrostatic force when depth and area vary**

Consider a rectangular swimming pool that is 30 feet long and 10 feet wide, with a depth of 4 feet at the shallow end and 8 feet at the deep end. Suppose the shallow end of the pool comprises the first 10 feet of the pool, followed by a 10-foot linear ramp where the depth of the pool increases from 4 feet to 8 feet, and then a final length of 10 feet for the deep end. Find the hydrostatic force exerted on one of the long sides of the pool.

Solution   A side view of the pool, along with two thin horizontal slices, is shown in Figure 15.53. If y_k^* is between $y = 0$ and $y = 4$, then the length of the slice varies according to the depth. If y_k^* is between $y = 4$ and $y = 8$, then the length of the slice is always 30 feet.

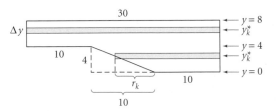

Figure 15.53

For y_k^* between $y = 0$ and $y = 4$, the width of the kth slice is Δy, and the length of the kth slice is $10 + r_k$, where r_k is the distance that the slice extends into the "ramp" part of the pool (we call it "r_k" because it depends on the choice of k; that is, it depends on what horizontal slice you are looking at.) Consider the right triangle with legs 4 and 10 feet shown with a dotted line in Figure 15.53. It contains a smaller triangle whose legs measure y_k^* and r_k feet. By the law of similar triangles, we have

$$\frac{r_k}{y_k^*} = \frac{10}{4} \quad \Longrightarrow \quad r_k = \tfrac{5}{2}y_k^*.$$

The area of the kth slice, assuming that the slice is between $y = 0$ and $y = 4$, is $\left(10 + \frac{5}{2} y_k^*\right) \Delta y$. Since we are assuming that this entire slice is at depth $8 - y_k^*$, the hydrostatic force on the slice is approximately

$$\begin{array}{c} \text{Hydrostatic force} \\ \text{on } k\text{th slice} \end{array} = 62.4\left(\left(10 + \frac{5}{2} y_k^*\right) \Delta y\right)(8 - y_k^*).$$

The total hydrostatic force F_1 on the part of the side wall from $y = 0$ to $y = 4$ (the bottom half of the wall) can be approximated by the sum

$$F_1 \approx \sum_{k=1}^{n} 62.4\left(10 + \frac{5}{2} y_k^*\right)(8 - y_k^*)\, \Delta y.$$

As $n \to \infty$, this sum approaches the definite integral

$$F_1 = \int_0^4 62.4\left(10 + \frac{5}{2} y\right)(8 - y)\, dy.$$

By an argument similar to the one used in Example 15.22, we can show that the hydrostatic force F_2 on the upper half of the side wall (from $y = 4$ to $y = 8$) is given by the definite integral

$$F_2 = \int_4^8 62.4(30)(8 - y)\, dy.$$

The total hydrostatic force F exerted on the side wall is

$$F = F_1 + F_2$$

$$= \int_0^4 62.4\left(10 + \frac{5}{2} y\right)(8 - y)\, dy + \int_4^8 62.4(30)(8 - y)\, dy$$

$$\approx 21{,}632 \text{ pounds} + 59{,}904 \text{ pounds}$$

$$= 81{,}536 \text{ pounds}.$$

(The solving of the integrals is left to the reader.) $\quad\Box$

❓ **Question** In each example in this section, we followed the same general procedure. Try to explain this procedure in your own words. Can you think of any other quantities that could be measured using this procedure? $\quad\Box$

What you should know after reading Section 15.4

▶ The formulas for work in terms of force and distance, mass in terms of density and volume, and hydrostatic force in terms of area and depth, and how to calculate work, mass, and hydrostatic force when the other quantities are constant. What are the units of these quantities?

▶ How to calculate work, mass, and hydrostatic force when the other quantities involved vary. This includes being able to "slice" in such a way that the quantities are approximately constant on each slice, obtaining a Riemann sum, and then finding and solving an appropriate definite integral. It may also include some mathematical modeling to describe information given in the problem in terms of functions or equations.

Exercises 15.4

Concepts

0. Read the section and make your own summary of the material. Include whatever you think will help you review the material later on.

1. Describe the "subdivide, approximate, and add" strategy that is used in each of the examples in this section. Why is it necessary? How does it work? What do Riemann sums and definite integrals have to do with it?

2. Why do we need to "subdivide, approximate, and add" in Example 15.16 in order to calculate the work, when in Example 15.15 we can just use the formula that says that work is force times distance? What is different in Example 15.16?

3. What do we mean when we say that a sum is a "Riemann sum"? What has to be true about a Riemann sum for us to be able to say that it approaches a definite integral when we take the limit as $n \to \infty$?

4. Explain how the values of Δx and x_k (or Δy and y_k) in a Riemann sum determine the limits of integration on the definite integral obtained by taking the limit of the Riemann sum as $n \to \infty$.

5. Why is the task of pumping all of the water out of the hot tub in Example 15.16 different from the task of lifting the entire hot tub 4 feet in the air (even assuming the hot tub itself doesn't weigh anything)? How much work will it require to lift the entire tub? Will that be more or less than the work required to pump out all the water?

6. Use the information given in Example 15.20 to show **(a)** that the jello mold in that example has the shape of the solid of revolution obtained by rotating the graph of $y = 4.5 - 0.5x^2$ on $[0, 3]$ around the y-axis, and **(b)** that the density of the mold at height y is given by the function $\rho(y) = 1.3 - 0.233y$.

Applications

7. You wish to lift a bathtub full of water 30 feet into the air. The bathtub itself weighs 150 pounds, and it holds a rectangular volume of water measuring 5 feet by 2 feet by 2 feet. Find the work required to lift the bathtub.

8. Copper has a density of 8.93 grams per cubic centimeter. Calculate the mass of a block of copper that has a square base with sides measuring 20 centimeters and a height of 13 centimeters.

9. Calculate the hydrostatic force on the bottom of a full cylindrical water glass with a radius of 2 inches and a height of 7 inches. (*Caution:* Watch your units!)

■ For each of the following problems, approximate the quantity described by using four appropriate "slices." In each case list the value of Δx (or Δy), list each value

of x_k and x_k^* (or y_k and y_k^*), and represent these values graphically.

10. The work required to pump all of the water out of a cylindrical hot tub with radius 3.5 feet and depth 4 feet.

11. The work required to pump all of the water out of a conical tank with radius 3.5 feet (at the top) and depth 4 feet.

12. The mass of a cylindrical rod with a radius of 2 centimeters and a length of 24 centimeters whose density at a point x centimeters from the left end of the rod is given by the function $\rho(x) = 10.5 + 0.01527x^2$ grams per cubic centimeter.

13. The mass of a jello mold in the shape of the solid of revolution obtained by rotating the graph of $y = 4.5 - 0.5x^2$ on $[0, 3]$ around the y-axis, whose density at height y is given by the function $\rho(y) = 1.3 - 0.233y$ ounces per cubic inch.

14. The hydrostatic force exerted on one of the shorter side walls of a swimming pool that is 30 feet long, 10 feet wide, and 5 feet deep.

15. The hydrostatic force exerted on one of the longer side walls of a swimming pool that is 30 feet long, 10 feet wide, 4 feet deep at the shallow end (along the first 10 feet of the pool), and 8 feet deep at the deep end (along the last 10 feet of the pool) if the middle section of the bottom of the pool is a straight ramp leading down from the shallow end to the deep end.

■ For each problem below, use a "slicing" technique to approximate the quantity described with a Riemann sum, and then find and solve a definite integral that represents the exact value of the quantity. Your work should include a picture of a representative slice, as well as the dimensions and work, mass, or hydrostatic force associated with this slice.

16. Alina owns a giant rectangular hot tub that is 5 feet wide, 8 feet long, and 3 feet deep.
    (a) Find the work involved in pumping all of the water out of the top of Alina's hot tub.
    (b) Find the hydrostatic force exerted on one of the long sides of Alina's hot tub.

17. Find the work required to pump all of the water out of the top of an upright conical tank that is 10 feet high and has a radius of 4 feet at the top.

18. Find the work required to pump all of the water out of the top of a spherical tank with a radius of 27 feet.

19. Linda has a large upright cylindrical tank buried in her back yard that has a radius of 4 feet and a height of 13 feet. The tank is buried so that its top is 3 feet below the surface.

(a) If Linda's tank is full of water, find the work required to pump all the water out of the top of the tank and up to ground level.

(b) If Linda's tank is full of a strange liquid that weighs 71.8 pounds per cubic foot and is only two-thirds full, find the work required to pump all the strange liquid out of the top of the tank and to a height 4 feet above ground level.

20. A cylindrical tank on a tanker truck (so that the cylinder is on its side) has a length of 100 feet and a radius of 8 feet and is full of a liquid that weighs 42.3 pounds per cubic foot. Find the work required to pump all of the liquid out of the tank.

21. Alina has tied a rope to a bucket containing 50 pounds of water, and she plans to hoist the bucket up to her upstairs bedroom window (which is 23 feet off the ground) by pulling up the rope.

(a) How much work will it take for Alina to lift the bucket up to her window, assuming that the weight of the rope is negligible?

(b) How much work will it take for Alina to lift the bucket up to her window, assuming that the rope weighs 0.25 pound per foot?

(c) How much work will it take for Alina to lift the bucket up to her window, assuming that the weight of the rope is negligible and the bucket leaks 2 pounds of water for each foot it is lifted?

22. Suppose a cylindrical rod with a radius of 3 centimeters and a length of 20 centimeters is made of a combination of copper and aluminum in such a way that the density of the rod x centimeters from the left end is given by the function $\rho(x) = 8.93 - 0.015x$ grams per cubic centimeter. Find the total mass of the rod.

23. Suppose a 12-inch rod with a square cross section of side length 1.5 inches is constructed in such a way that its density x inches from the left end is given by the function $\rho(x) = 4.2 + 0.4x - 0.03x^2$ grams per cubic inch. Find the total mass of the rod.

24. A rectangular block of metal is constructed in such a fashion that its density y centimeters from the base of the block is given by the function $\rho(y) = 6.8 + 0.14y$ grams per cubic centimeter. If the block is 30 centimeters wide, 21 centimeters long, and 14 centimeters tall, find the total mass of the block.

25. A solid cone of metal with a radius of 10 centimeters and a height of 15 centimeters is made of a combination of metals in such a way that, if the cone is pictured with its point up, then the density of the metal in the cone increases by 0.35 gram per cubic centimeter for each centimeter of height.

(a) Given that the density at the bottom of the cone is 7.25 grams per cubic centimeter, find a function $\rho(y)$ that describes the density of the cone y centimeters above its base.

(b) Using the density function you found above, find the total mass of the cone.

26. Last week Linda made an upside-down pudding cake that was 8 inches tall and 20 inches across its base, in the shape of the top of a downward-pointing parabola that has been revolved around the y-axis. The raisins she put in the cake have tended to drift toward the bottom of the cake over the past week (nobody seems to want to eat this cake!), in such a way that the density of the cake at the very top of the mold is 0.15 ounce per cubic inch, whereas the density at the very bottom of the cake is 1.12 ounces per cubic inch.

(a) Describe Linda's upside-down pudding cake as a solid of revolution (identify the region to be revolved and the axis of revolution).

(b) Describe the density of Linda's upside-down pudding cake as a linear function of the height y of a point in the cake.

(c) Find the mass of Linda's upside-down pudding cake.

27. Find the hydrostatic force exerted on one of the long sides of a rectangular swimming pool that is 8 feet long, 6 feet wide, and 4.5 feet deep.

28. Alina has a small birdbath in the shape of a cylinder with a radius of 0.4 foot that is 1 foot deep. Find the hydrostatic force exerted by the water in the birdbath on the cylindrical "side" of the birdbath.

29. A dam in the shape of an isosceles triangle with top of length 200 feet and remaining sides of length 100 feet is holding back a body of water that reaches to the top of the dam. Find the hydrostatic force exerted on the dam.

30. A dam in the shape of a trapezoid whose top is 400 feet long and whose base is 250 feet long is holding back a body of water that reaches all 200 feet from the bottom to the top of the dam. Find the hydrostatic force exerted on the dam.

31. Linda has a swimming pool that is 20 feet long and 8 feet wide, with a shallow end that has a depth of 3.5 feet (for the first 6 feet of the pool) and a deep end that has a depth of 12 feet (for the last 6 feet of the pool). The middle portion of the bottom of the pool is a straight ramp leading down from the shallow depth to the depth at the deep end.

(a) Find the hydrostatic force exerted on one of the long sides of the pool.

(b) Find the hydrostatic force exerted on the bottom of the pool (including the ramp).

■ **The following problems illustrate other applications of definite integrals. Some problems will require you to construct the definite integral, and in other problems the definite integral is given.**

32. Linda wants to hang some Christmas lights along the edge of the front side of her garage roof. The edge of the front side of the roof of the garage is a curve in the

shape of a downward-pointing parabola extending 3 feet above the ceiling of the garage and 12 feet across. How long a string of Christmas lights does Linda need?

33. A laboratory equipment manufacturer makes a specialty glass flask whose bottom half is part of a spherical "bulb" and whose top half is a cylindrical tube. The bulb consists of a sphere with a radius of 10 centimeters whose top and bottom are truncated by 2 centimeters (thus the bottom part of the flask makes up the first 16 centimeters of the height of the flask). The cylindrical tube has a radius just large enough to connect to the bulb and is 7 centimeters tall. Find the volume of the flask.

34. If $f(x)$ is a differentiable function with a continuous derivative, then the surface area of the solid of revolution obtained by rotating the region under the graph of $f(x)$ on the interval $[a, b]$ around the x-axis is given by the definite integral

$$SA = 2\pi \int_a^b f(x)\sqrt{1 + (f'(x))^2}\, dx.$$

(a) Use the formula above to calculate the surface area of the solid of revolution obtained by revolving the region between the graph of $f(x) = 2x^3$ and the x-axis on $[0, 3]$ around the x-axis.

(b) Use the formula above to prove that the surface area of a sphere of radius r is given by the formula $SA = 4\pi r^2$.

35. Suppose you cut a shape out of a piece of cardboard, and attempt to balance the shape on the end of a pencil. The point in the interior of the shape at which it will balance perfectly is called the ***centroid*** of the shape. If f is a positive, continuous function on an interval $[a, b]$, then the centroid of the region between the graph of $f(x)$ and the x-axis on $[a, b]$ is the point $(\bar{x}, \bar{y})$, where

$$\bar{x} = \frac{\int_a^b x f(x)\, dx}{\int_a^b f(x)\, dx} \quad \text{and} \quad \bar{y} = \frac{\frac{1}{2}\int_a^b (f(x))^2\, dx}{\int_a^b f(x)\, dx}.$$

(a) Use the formulas above to find the coordinates of the centroid of the region between the graph of $f(x) = x^2$ and the x-axis on $[0, 3]$.

(b) Use the formulas above to find the coordinates of the centroid of the region between the graph of $f(x) = \cos x$ and the x-axis on $[0, \pi]$.

ANSWERS TO SELECTED PROBLEMS

CHAPTER 0

SECTION 0.1

2. **(a)** 1.5; **(b)** $\sqrt{2}$; **(c)** 2; **(d)** $3 + 4i$

4. $n = \frac{n}{1}$ **7.** No **8.** Yes; $\frac{1.7}{2.3} = \frac{17}{23}$ **12.** Yes **13.** 1

18. **(a)** $\{x \mid a \le x < b\}$;
(b) The set of all real numbers greater than or equal to a and less than b.

21. **(a)** $\{x \mid x \ge a\}$;
(b) The set of all real numbers greater than or equal to a.

24. **(a)** $\{x \mid x < b\}$;
(b) The set of all real numbers less than b.

25. **(a)** $\{x \mid x \in \mathbb{R}\}$; **(b)** The set of all real numbers.

26. **(a)** $\{x \mid x = a\}$;
(b) The set consisting of one real number, the number a.

29. $A \cap B$ is always a subset of A, as well as a subset of $A \cup B$.

30. **(a)** The set of all red baseball hats;
(b) The set of all hats that are either red hats or baseball hats (or both).

32. **(a)** The set of all Elvis record albums;
(b) The set of all record albums.

37. The interval $(3, 5)$ and the point $(3, 5)$ in the coordinate plane.

39. True **40.** False **41.** False **42.** False **43.** True

44. True **45.** False **47.** $\frac{13}{100}$ **50.** $\frac{-117}{10}$

51. $\frac{-2}{15}$ (or $\frac{2}{-15}$) **53.** $\frac{0}{1}$ (or $\frac{0}{8}$, etc.) **54.** $\frac{-2}{3}$

57. **(a)** $\{x \mid -2 < x < 3\}$; **(b)** $(-2, 3)$

59. **(a)** $\{x \mid x \in \mathbb{R}\}$; **(b)** $(-\infty, \infty)$

60. **(a)** $\{x \mid x > 3 \text{ or } x < 2\}$; **(b)** $(-\infty, 2) \cup (3, \infty)$

62. $\{\ldots, -7, -5, -3, -1\}$

64. **(a)** $\{x \mid x = -3, 4, 7, \text{ or } 102\} = \{-3, 4, 7, 102\}$;
(b) $\{-3\} \cup \{4\} \cup \{7\} \cup \{102\}$

67. $(-1, 3]$ **68.** $[-1, \infty)$ **71.** $\{0\}$

73. $(-\infty, -2) \cup (-1, 3]$ **74.** $(-\infty, -2) \cup (-1, 0]$

76. $(-\infty, -2) \cup [-1, 2)$ **79.** 2 **81.** $-3b$

83. $-ab$ **85.** $-(ab - 1)$ **87.** 13 **89.** 5.53

92. Length is $\sqrt{13}$, midpoint is $(-\frac{1}{2}, -2)$

93. Length is $\sqrt{58}$, midpoint is $(-\frac{3}{2}, \frac{7}{2})$

97. Length is $\sqrt{30.33} \approx 5.50727$, midpoint is $(0.35, -0.6)$

SECTION 0.2

11. **(a)** Yes; **(b)** No; **(c)** $(0, 2), (2, 96), (-1, 0)$

14. $(4, 1)$ is not a solution; $(2, -1)$ is a solution.

16. False **17.** False **18.** False **19.** True

20. False **21.** True **22.** True

23. $(x + 5)(x - 3)$ **25.** $(x - 6)(x + 8)$

28. $2x(3x - 5)(x + 2)$ **29.** $(x + 3)(x - 3)$

33. $(9x^2 + 1)(3x + 1)(3x - 1)$

35. $(x + 1)(x + 1)(x + 1)$

37. $(3 - 4x)(16x^2 + 12x + 9)$

40. $(t - x)(t^6 + t^5 x + t^4 x^2 + t^3 x^3 + t^2 x^4 + t x^5 + x^6)$

41. $x = -1, x = \frac{3}{2}$ **44.** $x = \frac{3}{2}$

45. $x = -1 - \sqrt{6}, x = -1 + \sqrt{6}$

48. $x = 1, x = 6$ **51.** $x = 0, x = -\frac{1}{2}, x = 2$

52. $x = 0, x = \frac{1}{3}, x = 5$ **54.** $x = \frac{1 - \sqrt{7}}{2}, x = \frac{1 + \sqrt{7}}{2}$

56. $x = 2 - \sqrt{2}, x = 2 + \sqrt{2}$ **59.** $x = \frac{3}{2}, x = -\frac{3}{2}$

61. $x = 0, x = \frac{3}{2}, x = -\frac{3}{2}$ **63.** $-x(x - 1)(x + 1)^2$

64. $\frac{(x-2)(x-1)}{x+1}$ **66.** $\frac{x(5-4x)}{x^2-1}$ **68.** $x = -2$

70. $x = 0$ and $x = -2$ **71.** $x = -2$ **75.** $x = 0, x = \frac{7}{6}$

76. $\varnothing$ **77.** $(x, y) = (8, -12)$ **78.** $(a, b) = (-\frac{4}{7}, \frac{5}{7})$

81. $(t, u) = (-1, 0)$ and $(t, u) = (2, 3)$

83. $(a, b, c) = (-3, 1, -7)$ **85.** $\varnothing$

SECTION 0.3

5. The first set is $(-\infty, \infty)$ and the second set is $(0, 2)$.

6. $x = 2$ is a solution, but $x = 3$ is not.

8. $|x - 5| < 6$ **11.** $|r - 4| < 0.7$ **12.** $|x - 2| > 0.1$

14. $x \in (-\infty, c - \delta] \cup [c + \delta, \infty)$ **15.** $x \in (4 - c, 4 + c)$

18. $x \in (-\infty, -c - r] \cup [-c + r, \infty)$

19. $[c - \delta, c + \delta]$ **22.** $(-\infty, -\delta] \cup [\delta, \infty)$ **23.** $(1.9, 2.1)$

27. $A < D$ **29.** Cannot say how A is related to D.

30. **(a)** $|3 + 1| = 4$, $|3| + |1| = 4$, and $4 \le 4$
(b) $|3 - 1| = 2$, $|3| + |1| = 4$, and $2 \le 4$
(c) $|3 - 1| = 2$, $|3| - |1| = 2$, and $2 \ge 2$
(d) $|3 - 1| = 2$, $||3| - |1|| = 2$, and $2 \ge 2$

32. **(a)** $|3 + (-1)| = 2$, $|3| + |-1| = 4$, and $2 \le 4$
(b) $|3 - (-1)| = 4$, $|3| + |-1| = 4$, and $4 \le 4$
(c) $|3 - (-1)| = 4$, $|3| - |-1| = 2$, and $4 \ge 2$
(d) $|3 - (-1)| = 4$, $||3| - |-1|| = 2$, and $4 \ge 2$

35. **(a)** $|-1 + 3| = 2$, $|-1| + |3| = 4$, and $2 \le 4$
(b) $|-1 - 3| = 4$, $|-1| + |3| = 4$, and $4 \le 4$
(c) $|-1 - 3| = 4$, $|-1| - |3| = -2$, and $4 \ge -2$
(d) $|-1 - 3| = 4$, $||-1| - |3|| = 2$, and $4 \ge 2$

38. False **39.** False **40.** False **41.** False **42.** True

43. False **44.** False **48.** $(\frac{3}{2}, \infty)$ **50.** $(\frac{1}{3}, \infty)$

51. $(-\infty, -1) \cup (2, \infty)$ **52.** $[\frac{1}{2}, 3]$

58. $(-\infty, -3] \cup (-2, -1]$ **59.** $(-2, 1) \cup (4, \infty)$

63. $(-5, 11)$ **65.** $[-9, -7]$ **67.** $(-\infty, -1) \cup (5, \infty)$

69. $(-\infty, -2] \cup [2, \infty)$ **71.** $[0, 2]$ **72.** $(-\infty, 1.7247]$

74. $(-\infty, -0.72005] \cup [0.8233, 2.987]$

76. $(-\infty, -1.31607] \cup (1, 1.316]$

78. $|3x + 5| \leq |3x| + |5| = |3||x| + |5| = 3|x| + 5$

79. $|3x + 5| = |3x - (-5)| \geq |3x| - |-5| = 3|x| - 5$

83. $|3x - 5| \geq ||3x| - |5|| = |3|x| - 5|$

84. $|x^2 + x - 6| = |(x - 2)(x + 3)| = |x - 2||x + 3|$

SECTION 0.4

2. "A or B" is true in either case.

4. If C is true, then D must be true. If C is false, then D may or may not be true.

6. R is true but S is false.   **9.** (b) and (c)

11. ((soup or salad) and Not(soup and salad)) and ((potato or vegetable) and Not(potato and vegetable)).

12. (a) "For all $x > 0$, we have $x > -2$."
     (b) "If $x > 0$, then $x > -2$."

15. The contrapositive is "Not(Q) $\Rightarrow$ Not(P)," which is logically equivalent to $P \Rightarrow Q$.

16. The original statement is true. The converse is "Every rectangle is a square," which is false. The contrapositive is "Everything that is not a rectangle is not a square," which is true.

20. "There exists $y \in S$ such that P is false."

21. "P is false or Q is true."

24. "A is true and B is false and C is false."

26. "For all x, either A is false or B is false."

28. "There exists x such that, for all y, A is true and B is false.

29. True. The negation is "For all real numbers x, $x \leq 2$ and $x \geq 3$."

31. False. The negation is "There exists a real number that is neither greater than nor less than zero."

32. True. The negation is "There exists a real number that is both rational and irrational."

35. True. The negation is "There exists x such that, for all y, $y \neq x^2$." (In other words, "There exists x for which there is no y with $y = x^2$.")

38. True. The negation is "There is some integer x greater than 1 for which $x < 2$."

39. True. The negation is "There exist real numbers a and b such that $a < b$ but $3a + 1 \geq 3b + 1$."

40. "There exists $M > 0$ such that, for all $N > 0$, there is some x with $x > N$ but $x^2 \leq M$."

41. "There exists $\varepsilon > 0$ such that, for all $\delta > 0$, there is some x with $0 < |x - 2| < \delta$ but $|x^2 - 4| \geq \varepsilon$."

44. False. One counterexample is $x = -1$.

47. False. One counterexample is $x = 1.35$.

48. True. The negation is "There is some number x such that either $x^2 < 0$ or $|x| < 0$."

49. True. The negation is "There are some numbers x and y such that $x < y$ but $2x - 1 \geq 2y - 1$."

51. True. The negation is "There is some x such that, for all y, $x \geq y$."

54. False. The only counterexample is $x = 0$, $y = 0$.

56. True. One example is $x = 3$.

57. False. The negation is "For all x, either $x > 1$ or $x < 2$."

60. False. The negation is "For all $x < 0$ and $y < 0$, we have $xy \geq 0$.

63. True. One example is $x = 0$, since for all y we have $|y| > 0$.

65. (a) $B \Rightarrow$ (Not A);   (b) (Not B) $\Rightarrow A$

68. (a) $C \Rightarrow (A$ and $B)$
     (b) Not(C) $\Rightarrow$ (Not A) or (Not B)

70. (a) $(B$ and $C) \Rightarrow A$
     (b) ((Not B) or (Not C)) $\Rightarrow$ (Not A)

74. (a) The converse is "If $x \geq 1$, then $x \geq 2$," which is false; one counterexample is $x = 1.5$.
     (b) The contrapositive is "If $x < 1$, then $x < 2$," which is true.

76. (a) The converse is "If $x \geq 3$, then $x > 2$," which is true.
     (b) The contrapositive is "If $x < 3$, then $x \leq 2$," which is false. One counterexample is $x = 2.5$.

80. (a) The converse is "If $|x| = -x$, then $x \leq 0$, which is true.
     (b) The contrapositive is "If $|x| \neq -x$, then $x > 0$," which is true.

85. (a) The converse is "If there is some integer n such that $x = 2n + 1$, then x is odd," which is true.
     (b) The contrapositive is "If there is no integer n such that $x = 2n + 1$, then x is not odd," which is true.

87. Linda wears a red hat, Alina wears a blue hat, Phil wears a green hat, Stuart wears a yellow hat.

89. For a truth-teller, the statement "I am a truth-teller" is true, so a truth-teller can utter this statement. For a liar, the statement "I am a truth-teller" is a lie, so a liar can also utter this statement.

92. Liz and Rein are liars, and Zubin tells the truth.

SECTION 0.5

4. $\sqrt{2} + \pi$ is irrational, but $\sqrt{2} + (-\sqrt{2}) = 0$ is rational.

7. All integers greater than or equal to 4.

9. All odd integers greater than or equal to 3.

12. $2 = 1(1 + 1); 2 + 4 + 6 + 8 = 4(4 + 1); 2 + 4 + 6 + 8 + 10 + 12 + 14 + 16 + 18 = 9(9 + 1)$

13. $1 = 1^2; 1 + 3 + 5 + 7 = 4^2; 1 + 3 + 5 + 7 + 9 + 11 + 13 + 15 + 17 = 9^2$

15. $1 + 2 + 3 + \cdots + (n - 1) = \frac{(n-1)(n)}{2}$

16. $2 + 4 + 6 + \cdots + 2(n - 1) = (n - 1)((n - 1) + 1)$

19. **PROOF** A is true by hypothesis. Thus (Not B) is true (since $A \Rightarrow$ (Not B)). Therefore, C must be true (since (Not B) $\Rightarrow C$). ∎

22. **PROOF**   (by contradiction): Seeking a contradiction, suppose A is false. Since we are given that (Not A) $\Rightarrow B$, this means that B is true. Since $C \Rightarrow$ (Not B), we know that $B \Rightarrow$ (Not C), and thus C is false. However, we are given that C is true. Since C cannot be both true and false, we have a contradiction. Therefore, our original assumption that A is false must itself be false. Therefore, A must be true. ∎

23. **PROOF**   If x and y are rational numbers, then we can write $x = \frac{r}{s}$ and $y = \frac{t}{u}$ for some integers r, s, t, and u. Then we have $x + y = \frac{r}{s} + \frac{t}{u} = \frac{ru+st}{su}$; since $ru + st$ and su must be integers, $x + y$ is a rational number. ◼

24. **PROOF**   (by contradiction): Suppose x is irrational and r is rational. Seeking a contradiction, suppose $x - r$ is rational. By the previous problem, the sum of two rational numbers is rational, and thus $(x - r) + r = x$ must be rational. But this contradicts the hypothesis that x is irrational; therefore, $x - r$ must be an irrational number. ◼

28. If $M = (\frac{x_1+x_2}{2}, \frac{y_1+y_2}{2})$, the distance formula easily shows that dist$(M, P) =$ dist(M, Q).

31. **PROOF**   (a): If a, b, and c are any real numbers with $b \neq 0$ and $c \neq 0$, then $\frac{a/b}{c} = \frac{a/b}{c/1} = \frac{a(1)}{bc} = \frac{a}{bc}$. ◼

47. **PROOF**   (by induction): The statement is true for $n = 1$, since $2 = 1(1 + 1)$. Assume true for $n - 1$; i.e., assume that: $2 + 4 + 6 + \cdots + 2(n - 1) = (n - 1)((n - 1) + 1) = (n - 1)(n)$. Prove true for n given inductive hypothesis: $2 + 4 + 6 + \cdots + 2n = 2 + 4 + 6 + \cdots + 2(n - 1) + 2n = (2 + 4 + 6 + \cdots + 2(n - 1)) + 2n = (n - 1)n + 2n = n^2 - n + 2n = n^2 + n = n(n + 1)$. Thus, by induction, we have $2 + 4 + 6 + \cdots + 2n = n(n + 1)$. ◼

CHAPTER 1

SECTION 1.1

2. v is the independent variable, u is the dependent variable, and w is the "name" of the function.

4. $f(x) = x^2$ is a function, but $f(x) = \pm\sqrt{x}$ is not.

7. **(a)** $f(\text{person}) =$ the state that person is in right now. Range$(f) = S$.
    (b) $f(\text{person}) =$ the states that person has visited this year.
    (c) The function $f(\text{person}) =$ Kansas.
    (d) No identity function is possible.

10. **(a)** $f(\text{state}) =$ the state that comes next in alphabetical order (cycling around so that $f(\text{Wyoming}) =$ Alabama).
    (b) $f(\text{state}) =$ all the bordering states.
    (c) $f(\text{state}) =$ Alaska (so $f(\text{Kansas}) =$ Alaska, $f(\text{Alaska}) =$ Alaska, etc.).
    (d) $f(\text{state}) =$ itself (*e.g.*, $f(\text{Kansas}) =$ Kansas, $f(\text{Virginia}) =$ Virginia, etc.).

12. **(a)**

x	1	2	3	4
$f(x)$	6	2	2	10

Range is {2, 6, 10}.

(b)

x	1	2	3
$f(x)$	6	2	8

(c)

x	1	2	3	2
$f(x)$	6	10	8	4

13. **(a)**

x	2	4	6	8	10
$f(x)$	4	8	8	2	6

Range is {2, 4, 6, 8}.

(b)

x	2	6	8	10
$f(x)$	4	8	2	6

(c)

x	10	6	2	4	10
$f(x)$	2	4	6	8	10

15. **(a)** $f(2) = 5$ **(b)** There is no x for which $f(x) = 0$.
    (c) You can solve $y = x^2 + 1$ for x if and only if $y \in [1, \infty)$.

18. **(a)** $f(\text{person}) = 42$ (range is {42}).
    (b) No identity function is possible.

21. $[-\frac{1}{3}, \infty)$

23. $(-\infty, -\frac{3}{2}] \cup [1, \infty)$

25. $[0, \frac{5}{3}) \cup (\frac{5}{3}, \infty)$

26. $(-\infty, -\frac{3}{2}) \cup (1, \infty)$

27. $(-\infty, -3) \cup (-3, -1] \cup [1, 3) \cup (3, \infty)$

31. $(1, 2) \cup (2, \infty)$

33. **(a)** $-\frac{2}{3}$ **(b)** $\frac{q}{1-q}$ **(c)** $-\frac{a+1}{a}$ **(d)** $\frac{x^2+x}{1-x^2-x}$

35. **(a)** $\frac{(1+h)^2+1-2}{h}$ **(b)** $\frac{t^2+1-2}{t-1}$ **(c)** 37   **(d)** $(x^2 + 1)^2 + 1$

36. The area A is the dependent variable, the radius r is the independent variable; $A(r) = \pi r^2$.

39. The number g of groundhogs is the dependent variable, the time t (in years after 1990) is the independent variable; $g(t) = 3t + 25$.

SECTION 1.2

2. $\{(x, y) \mid x \geq 0 \text{ and } y = \sqrt{x}\}$

4. **(a)** Yes: $\sqrt{3+1} = 2$ **(b)** No: $\sqrt{1+1} \neq 1$
    (c) No: $\sqrt{-5+1} \neq 2$

7. **(a)** y-intercept at $y = 2$, no roots
    (b) Positive on $(-\infty, \infty)$
    (c) Increasing on $(-\infty, 0)$, decreasing on $(0, \infty)$
    (d) Concave up on $(-\infty, -1) \cup (1, \infty)$, concave down on $(-1, 1)$
    (e) Local maximum at $x = 0$ (with a value of $f(0) = 2$). This is also the global maximum. There is no global minimum.
    (f) Inflection points at $x = -1$ and $x = 1$
    (g) Horizontal asymptote at $y = 1$

10. These functions, when graphed, have the characteristics described: **(a)** 2^x; **(b)** -2^{-x}; **(c)** 2^{-x}; **(d)** -2^x; **(e)** not possible if defined on $\mathbb{R}$; **(f)** -2^{-x}.

12.

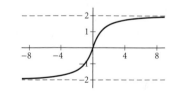

15.

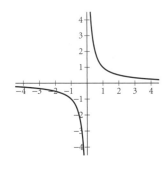

17. The graph of f is below or on the x-axis on all of I. Mathematically: for all $x \in I$, $f(x) \le 0$.

20. The graph of f is "moving up" as we go past the point $x = c$. Mathematically, there is a $\delta > 0$, so f is increasing on $(c - \delta, c + \delta)$.

23. There are three local extrema; we will discuss only the local maximum at $x = 2$. For this local extremum, $\delta = 1$ works. The *largest* possible δ for that extremum is approximately $\delta = 1.5$.

25. One good window for this graph is $x \in [-1, 1]$, $y \in [-0.3, 1]$.

28. $x = -1$ and $x = 3$

31. f is positive on $(-\infty, -1) \cup (3, \infty)$, and negative on $(-1, 3)$.

32. f is positive on $(-\infty, \infty)$.

33. $a < b \Rightarrow -3a > -3b \Rightarrow 1 - 3a > 1 - 3b \Rightarrow f(a) > f(b)$

SECTION 1.3

2. 3; 18

3. $\Delta y = -12$

6. Yes, since $y = kx$ is linear with $m = k$ and $b = 0$. No; for example, $y = 3x + 1$ is linear but not proportional.

8. Not proportional, could be linear ($y = x + 1$).

10. The missing entries in the first row are 4 and 9. The missing entries in the second row are -17 and -26.

11. Linear with slope $-\frac{1}{8}$, y-intercept $\frac{3}{8}$

12. Not linear

15. Linear with slope 2, y-intercept 13

16. Linear with slope $\frac{1}{6}$, y-intercept $\frac{2}{3}$

19. 4.2

21. 2

23. $\frac{\sqrt{10} - \sqrt{2}}{8}$

24. $y - 1 = 3(x - 0)$; another way: $y - 4 = 3(x - 1)$

28. $y + 2 = -(x - 3)$

29. $y = -x - 1$

31. $y = 2x - 2$

34. Let h be the number of hours each week spent watching television, and let d represent the amount, in dollars, of credit card debt. $h(d) = kd$ for some constant k.

37. Let t be the number of years after 1985, and let P represent the number of Whimsians on the island. $P(t) = \frac{56}{15}t + 35$.

39. 356

41. Let N be the number of CDs, and let t be time in months, with $t = 0$ representing January. Then $N(t) = 3t + 51$. $N = 200$ when $t \approx 49.667$; since December is $t = 11$, it will be 39 months after December before Alina has 200 CDs.

44. **PROOF**   Given any real number y, we can find an x in the domain of f such that $f(x) = y$: specifically, $f(\frac{y+1}{3}) = y$. ∎

46. **PROOF**   For all real numbers a and b, $\frac{f(b) - f(a)}{b - a} =$
$\frac{(-2b+4) - (-2a+4)}{b-a} = \frac{-2(b-a)}{b-a} = -2$. ∎

SECTION 1.4

2. (a) $3x + 1$ is linear, $x^2 + 1$ is not.
  (b) $3x^5$ is a power function, $3x^5 + 1$ is not.
  (c) $2x^3 - 1$ is a polynomial, $2x^3 - \sqrt{x}$ is not.
  (d) $\frac{1}{1+x}$ is rational, $\frac{1}{\sqrt{1+x}}$ is not.

7. No; yes   **8.** Rational

12. Algebraic (but *not* rational)

13. Linear (also can be thought of as power, polynomial, or rational)

14. If $x \ge 0$ then $\sqrt{x^2} = x$, and if $x < 0$ then $\sqrt{x^2} = -x$.

16.

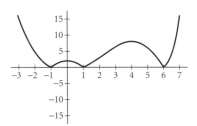

22. $(1, 1), (2, \frac{1}{2}), (-3, -\frac{1}{3}), (\frac{1}{2}, 2)$, and $(-0.1, -10)$ are points on the graph.

24.

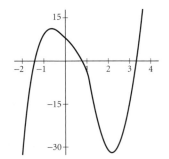

27. $f(-2) = 4$, $f(-1) = -1$, $f(0) = 1$, $f(1) = 3$, $f(2) = 5$

28. $f(-2) = -5$, $f(-1) = -2$, $f(0) = 1$, $f(1) = 4$, $f(2) = 8$

30. $|5 - 3x| = \begin{cases} 5 - 3x, & x \le \frac{5}{3} \\ 3x - 5, & x > \frac{5}{3} \end{cases}$

33. $|x^2 + 1| = x^2 + 1$ for all real numbers x.

34. $|x^2 - 3x - 4| = \begin{cases} x^2 - 3x - 4, & x \le -1 \\ -(x^2 - 3x - 4), & -1 < x < 4 \\ x^2 - 3x - 4, & x \ge 4 \end{cases}$

36.

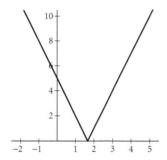

39.

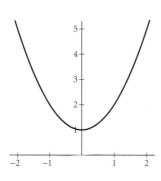

40.

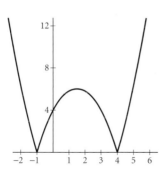

43.

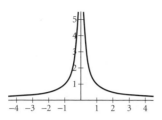

46.

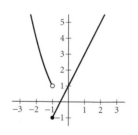

47.

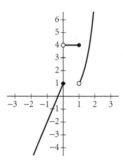

49. $I(t) = \begin{cases} 36{,}000, & 0 \le t < 4 \\ 38{,}500, & 4 \le t < 6 \\ 49{,}000, & t \ge 6 \end{cases}$

50. $M(t) = \begin{cases} 36{,}000t, & 0 \le t < 4 \\ 144{,}000 + 38{,}500(t-4), & 4 < t \le 6 \\ 221{,}000 + 49{,}000(t-6), & t \ge 6 \end{cases}$

You will have earned a total of one million dollars after approximately 21.9 years.

51. (a) \$2700.00; \$52,851.00

   (b) 15%; approximately 29.36%

   (c) $T(m)$ is equal to $0.15m$, if $0 \le m \le 26{,}250$;
$3937.5 + 0.28(m - 26{,}250)$, if $26{,}250 < m \le 63{,}550$;
$14{,}381.5 + 0.31(m - 63{,}550)$, if $63{,}550 < m \le 132{,}600$;
$35{,}787 + 0.36(m - 132{,}600)$, if $132{,}600 < m \le 288{,}350$;
$91{,}857 + 0.396(m - 288{,}350)$, if $m \ge 288{,}350$

SECTION 1.5

2. $[1, \infty) \cap [-4, 4] = [1, 4]$ **7.** $(1, 2) \cup (2, 4]$

8. Not enough information

10. $\text{Domain}(f) \cap \text{Domain}(g)$

12. $\{x \mid x \in \text{Domain}(f) \cap \text{Domain}(g), \ f(x) \ne 0, \text{ and } g(x) \ne 0\}$

15. $\{x \mid x \in \text{Domain}(f) \text{ and } f(x) \in \text{Domain}(f)\}$

17. $\{x \mid x \in \text{Domain}(h), \ h(x) \in \text{Domain}(f), \text{ and } f(h(x)) \in \text{Domain}(h)\}$

20. $\{x \mid x \in \text{Domain}(g) \cap \text{Domain}(h), \ g(x) \in \text{Domain}(f), \ g(x) \ne 0, \text{ and } h(x) \ne 0\}$

21. $(-\infty, -1] \cup [1, \infty)$ **23.** $(-\infty, -\frac{1}{5}) \cup (\frac{1}{5}, \infty)$

26. $f(x) + g(x)h(x); \ f(2) + g(2)h(2) = -8$

28. $2(h(x) + 3); \ 2(h(2) + 3) = 4$

29. 2   **32.** 1

34. $\{(x, \frac{f(x)}{g(x)}) \mid x \in \text{Domain}(f) \cap \text{Domain}(g), \text{ and } g(x) \ne 0\}$; the height of $\frac{f}{g}$ at $x = 2$ is the quotient of the height of f at $x = 2$ and the height of g at $x = 2$.

36. $g(-1) \approx -2, g(0) \approx -1, g(1) \approx 0$

38. The missing entries in the first row are $-3, -2, 0, 0$. The missing entries in the second row are $-2, 3, 3, -6$. The missing entries in the third row are $-1, 2, 2, -7$.

39. The missing entries in the first row are 1, 1, 1. The missing entries in the second row are 1, 4, 3, 1. The missing entries in the third row are 4, 1, 3, 1. The missing entries in the fourth row are 3, 3, 1, 3.

41. (a) $f(x) = \frac{1}{x+1}$ (b) $l(x) = \frac{1}{x^2}$ (c) $s(t) = t$
(d) $y(x) = x$

43. $(fgh)(x) = (x^2+1)(\frac{1}{x})(1-2x+x^2)$; domain is $x \neq 0$;
$(fgh)(1) = 0$

44. $(\frac{f}{g})(x) = \frac{x^2+1}{\frac{1}{x}}$; domain is $x \neq 0$; $(\frac{f}{g})(1) = 2$

46. $(\frac{f-1}{h})(x) = \frac{x^2}{1-2x+x^2}$; domain is $x \neq 1$; $(\frac{f-1}{h})(1)$ is
undefined

52. $(f \circ h)(x) = (1-2x+x^2)^2 + 1$; domain is $\mathbb{R}$;
$(f \circ h)(1) = 1$

55. $(g \circ g)(x) = \frac{1}{1/x}$; domain is $x \neq 0$; $(g \circ g)(1) = 1$

58. $((f+h) \circ g)(x) = \frac{2}{x^2} - \frac{2}{x} + 2$; domain is $x \neq 0$;
$((f+h) \circ g)(1) = 2$

61. $(\frac{f \circ g}{h \circ g})(x) = \frac{\frac{1}{x^2}+1}{1-\frac{2}{x}+\frac{1}{x^2}}$; domain is $x \neq 0$, $x \neq 1$; $(\frac{f \circ g}{h \circ g})(1)$ is
undefined

63. $-6, -5, -4, -2, 2, 13, 19$

65. $0, -3, -4, -1, 0, -5, -12$

66. $0, -\frac{1}{4}, -\frac{1}{3}, -\frac{1}{3}, -1, -\frac{5}{4}, -2$

68. $\frac{1}{2}, \frac{5}{3}, 4, 4,$ undefined$, 9, \frac{9}{2}$

70. $1, 5, 6, 6, 2, 4, 3$ **72.** $3, 6, 5, 4, 4, 0, 1$

75. $2, 4, 3, 0, 1, 0, 3$ **77.** $0, 5, 2, 4, 1, 3, 6$

78. $3, 3, 1, 3, 3, 3, 1$ **81.** $2, 3, 0, 2, 3, 2, 0$

83.

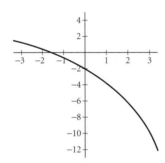

85.

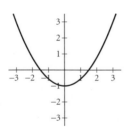

87.

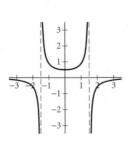

90.

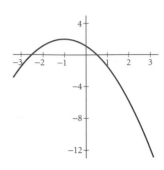

92.

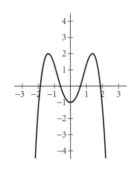

95.

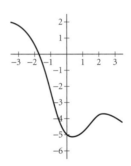

99. One way: $g(x) = 3x+5$, $h(x) = x^2$; another way: $g(x) = x+5$, $h(x) = 3x^2$

101. One way: $g(x) = \frac{6}{x}$, $h(x) = x+1$; another way: $g(x) = 6x$, $h(x) = \frac{1}{x+1}$

103. One way: $g(x) = \frac{x^2}{1+x^2}$, $h(x) = x-4$; another way: $g(x) = \frac{(x-2)^2}{1+(x-2)^2}$, $h(x) = x-2$

105. $g(x) = f(x)$, $h(x) = x+1$

106. $g(x) = 2x$, $h(x) = f(x)$

109. $(f+g)(-1) = 0$, $(f+g)(0) = 1$, $(f+g)(1) = 0$, $(f+g)(2) = 9$, $(f+g)(3) = 14$;

$$(f+g)(x) = \begin{cases} x+1, & x \leq 0 \\ x^2-x, & 0 < x < 2 \\ x^2+5, & x \geq 2. \end{cases}$$

111. $(g \circ f)(-1) = 1$, $(g \circ f)(0) = -1$, $(g \circ f)(1) = -1$, $(g \circ f)(2) = 5$, $(g \circ f)(3) = 5$;

$$(g \circ f)(x) = \begin{cases} -(2x+1), & x \leq 0 \\ -x^2, & 0 < x < \sqrt{2} \\ 5, & x \geq \sqrt{2}. \end{cases}$$

112. (a) $S(x) = x^2$; (b) $C(S) = 4.25S + 200$; (c) $C(x) = 4.25x^2 + 200$; $C(x)$ is the composition $C(S(x))$.

SECTION 1.6

2. $(4, 6)$

4. $(1, 2)$

6. $(2, 1)$

7. $(-2, -5)$

9. **(a)** The graph of g is C units to the left (or right) of the graph of f;

    (b) If (x, y) is on the graph of f, then $(x - C, y)$ is on the graph of g.

12. **(a)** The graph of g is squished vertically by a factor of A from the graph of f;

    (b) If (x, y) is on the graph of f, then $(x, \frac{1}{A} y)$ is on the graph of g.

14. $g(x) = x + C$, $h(x) = f(x)$; domain is Domain(f).

17. $g(x) = f(x)$, $h(x) = Ax$; domain is $\{x \mid Ax \in \text{Domain}(f)\}$.

18. $f(k(g(x))) = (-2x)^2$; from the graph of f, reflect over the y-axis and shrink horizontally by a factor of 2.

19. $k(f(g(x))) = -(2x)^2$; from the graph of f, reflect over the x-axis and shrink horizontally by a factor of 2.

23. $k(f(g(h(x)))) = -(2(x-3))^2$; from the graph of f, shrink horizontally by a factor of 2, *then* shift right three units, and reflect the graph over the x-axis.

24. $g(x) = x - 3$, $h(x) = 2x$, $k(x) = f(x)$; vertical stretch by a factor of 2 followed by vertical shift down three units.

25. $g(x) = 2x$, $h(x) = x - 3$, $k(x) = f(x)$; vertical shift down three units, followed by vertical stretch by a factor of 2.

29. $g(x) = f(x)$, $h(x) = 4x$, $k(x) = x + 2$; horizontal shrink by a factor of 4, followed by a horizontal shift two units to the left.

31. $y = -f(x) - 2$

32. No; if f is odd, then $f(0)$ must equal 0.

34. **(a)** $f(-2) = 1$, $f(-1) = -2$, $f(3) = 4$ ($f(0)$ could be anything).

    (b) $f(-2) = -1$, $f(-1) = 2$, $f(0) = 0$, and $f(3) = -4$.

35. The *only* such function is the constant function $f(x) = 0$.

36.

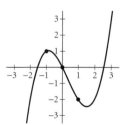

41.

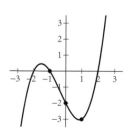

42.

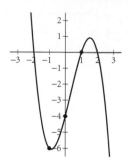

43.

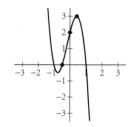

46.

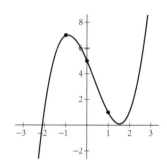

47.

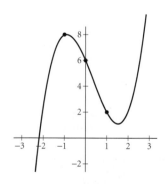

51. Shift the graph of $y = x^2$ down three units.

53. Shrink the graph of $y = x^2$ vertically by a factor of 3, and reflect the graph over the x-axis.

55. Stretch the graph of $y = \sqrt{x}$ vertically by a factor of 2, and then shift up one unit.

57. Shift the graph of $y = \frac{1}{x}$ to the left three units, and stretch the graph vertically by a factor of 2.

58. Shift the graph of $y = \frac{1}{x}$ to the right one unit and up one unit.

60.

x	-3	-2	-1	0	1	2	3
$\frac{1}{2} f(x)$	4	2.5	2	-1	0	1.5	2

62.

x	0	1	2	3	4	5	6
$f(x-3)$	8	5	4	-2	0	3	4

64.

x	-2	-1.5	-1	-0.5	0	0.5	1
$f(2x+1)$	8	5	4	-2	0	3	4

66. Neither

68. Odd

71. Even, since $f(-x) = f(x)$.

72. Neither, since $f(-x) \neq f(x)$ and $f(-x) \neq -f(x)$.

74. Odd, since $f(-x) = -f(x)$.

75. $-2f(x) = \begin{cases} -4x - 2, & x \le 1 \\ -2x^2, & x > 1 \end{cases}$

77. $f(x+3) = \begin{cases} 2(x+3) + 1, & x \le -2 \\ (x+3)^2, & x > -2 \end{cases}$

79. **PROOF**   If f is odd, then for all x in the domain of f we have $f(-x) = -f(x)$. In particular, $f(-0) = -f(0)$, so $f(0) = -f(0)$. This means we must have $f(0) = 0$. ∎

80. **PROOF**   If (x, y) is on the graph of f, then $y = f(x)$, so $Af(x) = Ay$; thus (x, Ay) is a point on the graph of $Af(x)$. ∎

81. **PROOF**   If (x, y) is on the graph of f, then $y = f(x)$, so $f(A(\frac{1}{A}x)) = f(x) = y$, so $(\frac{1}{A}x, y)$ is on the graph of $f(Ax)$. ∎

SECTION 1.7

2.

x	1	2	3	4
$f(x)$	2	3	5	8

This function is one-to-one.

x	1	2	3	4
$f(x)$	2	3	2	8

This function is not one-to-one.

4. $f(2) = f(-2)$ **7.** -1 **8.** $(3, 2)$

9. $(4, -2)$ **12.** $(f^{-1})^{-1} = f$

14. Yes; $f(x) = \frac{1}{x}$ is one example.

16. $f(a) = f(b) \Rightarrow 2 - 4a = 2 - 4b \Rightarrow -4a = -4b \Rightarrow a = b$

17. $f(a) = f(b) \Rightarrow a^3 + 1 = b^3 + 1 \Rightarrow a^3 = b^3 \Rightarrow a = b$

18. $f(a) = f(b) \Rightarrow a^2 = b^2$, and if $a, b \ge 0$ this means $a = b$.

22. If $a < b$, then $f(a) = 1 + 15a$ is less than $f(b) = 1 + 15b$.

24. $f(g(x)) = 2 - 3(-\frac{1}{3}x + \frac{2}{3}) = x$ and $g(f(x)) = -\frac{1}{3}(2 - 3x) + \frac{2}{3} = x$

27. $f(g(x)) = f(\frac{x}{1+x}) = \frac{\frac{x}{1+x}}{1 - \frac{x}{1+x}} = x$ (for $x \neq -1$); similarly, $g(f(x)) = x$ for $x \neq 1$.

29. $f^{-1}(x) = \frac{1-2x}{5}$

31. $f^{-1}(x) = \frac{1}{x-1}$

33. $f^{-1}(x) = \frac{-x-1}{x-1}$

34. $(1, 0)$, $(2, -1)$, and $(8, -3)$ are points on this graph.

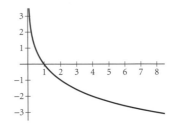

35. Not invertible; fails Horizontal Line Test at $y = 2$.

39. $(-\infty, 1.2]$ is one such restricted domain.

42. (a) $C(n) = 20 + 12n$

(b) Suppose there were a and b in the domain $(-\infty, \infty)$ of $C(n)$ with $C(a) = C(b)$. In this case $20 + 12a = 20 + 12b$, so $12a = 12b$, and therefore we must have $a = b$.

(c) $n(C) = \frac{C-20}{12}$. This function describes the number n of shirts that you can have made for C dollars.

(d) $n(150) \approx 10.83$, so you can make 10 full shirts.

43. (a) $g(t) = 92 - 6t$

(b) $g(t)$ is always decreasing.

(c) $t(g) = -\frac{1}{6}(g - 92)$ predicts the number of hours a student watched TV given that student's grade on the exam.

(d) $t(40) = \frac{52}{6} \approx 8.667$ hours; $t(85) = \frac{7}{6} \approx 1.1667$ hours

45. **PROOF**   Suppose f is increasing on its entire domain. To show that f is one-to-one, we will show that for all a and b in the domain of f, we have $a \neq b \Rightarrow f(a) \neq f(b)$, as follows: If $a \neq b$ then either $a < b$ or $a > b$. If $a < b$ then $f(a) < f(b)$ (since f is increasing), and thus $f(a) \neq f(b)$. If $a > b$ then $f(a) > f(b)$ (since f is increasing), and thus $f(a) \neq f(b)$. In any case, if $a \neq b$ then $f(a) \neq f(b)$, and thus f is by definition a one-to-one function. The proof that a decreasing function is one-to-one is similar. ∎

46. **PROOF**   If $f^{-1}(b) = a$ then $b = f(f^{-1}(b)) = f(a)$. ∎

CHAPTER 2

SECTION 2.1

3. $\lim_{x \to 2}(1 - x^2) = -3$; i.e., as $x \to 2$, $1 - x^2 \to -3$.

5. As we travel along the graph of f toward the left, the graph eventually gets higher and higher the farther left we go.

9.

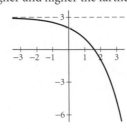

10. One such graph is

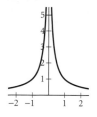

12.

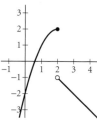

14. $\lim_{x \to 1} f(x) = 5$; cannot say anything about $f(1)$

16. $\lim_{x \to 2^-} f(x)$ is not equal to 8.

18. f has a horizontal asymptote at $y = 3$ and a vertical asymptote at $x = 1$.

20. Draw a function whose graph goes through each of the datapoints in the table and then decreases to $-\infty$ sometime after $x = 10,000$.

21. Draw a function whose graph goes through each of the datapoints in the table and then has a vertical asymptote at $x = 2$.

23. (a) -2; (b) 1; (c) -2; (d) 2; (e) 2; (f) -2; (g) 2; (h) does not exist

25. -93 **27.** Does not exist.   **29.** $\frac{1}{2}$ **31.** 7

32. Does not exist.   **34.** $-\infty$

SECTION 2.2

2. (a) $(1, 1.5) \cup (1.5, 2)$
(b) $(0, 0.5) \cup (0.5, 1)$
(c) $(0, 1) \cup (1, 2)$
(d) $(0, 0.25) \cup (0.25, 0.5)$

3. (a) $\sqrt{x + 7}$ approaches 2 as x approaches -3.
(b) We can make $\sqrt{x + 7}$ as close as we like to 2 by making x sufficiently close (but not equal) to -3.
(c) For any small distance ε around 2, there is a small distance δ around -3 such that when x is within δ of -3, $\sqrt{x + 7}$ is within ε of 2.
(d) For all $\varepsilon > 0$, there exists a $\delta > 0$ such that if $0 < |x + 3| < \delta$, then $|\sqrt{x + 7} - 2| < \varepsilon$.
(e) For all $\varepsilon > 0$, there exists a $\delta > 0$ such that if $x \in (-3 - \delta, -3) \cup (-3, -3 + \delta)$, then $\sqrt{x + 7} \in (2 - \varepsilon, 2 + \varepsilon)$.

5. (a) $\frac{(2+h)^2 - 4}{h}$ approaches 4 as h approaches 0.
(b) We can make $\frac{(2+h)^2 - 4}{h}$ as close as we like to 4 by making h sufficiently close (but not equal) to 0.
(c) For any small distance ε around 4, there is a small distance δ around 0 such that when h is within δ of 0, $\frac{(2+h)^2 - 4}{h}$ is within ε of 4.

(d) For all $\varepsilon > 0$, there exists a $\delta > 0$ such that if $0 < |h| < \delta$, then $|\frac{(2+h)^2 - 4}{h} - 4| < \varepsilon$.
(e) For all $\varepsilon > 0$, there exists a $\delta > 0$ such that if $h \in (-\delta, 0) \cup (0, \delta)$, then $\frac{(2+h)^2 - 4}{h} \in (4 - \varepsilon, 4 + \varepsilon)$.

7. $0 < |x - 5| < 0.01$; $|f(x) + 2| < 0.5$

9. $x \in (1.9, 2) \cup (2, 2.1)$

12. $x^2 - 1 \in (-3.5, -2.5)$

14. $f(x) \in (L - \varepsilon, L + \varepsilon)$

15. If $x \in (2 - \delta, 2) \cup (2, 2 + \delta)$, then $f(x) \in (5 - \varepsilon, 5 + \varepsilon)$.

17. If $x \in (N, \infty)$, then $f(x) \in (2 - \varepsilon, 2 + \varepsilon)$.

18. If $x \in (1, 1 + \delta)$, then $f(x) \in (M, \infty)$.

20. Sketch the graph of $y = x^2$ and show an ε interval around $y = 4$ with $\varepsilon = 0.4$ (i.e., show the horizontal "bar" between $y = 3.6$ and $y = 4.4$). Then draw the δ interval around $x = -2$ with $\delta = 0.075$, excluding $x = -2$ (i.e., show the vertical "bar" between $x = -2.075$ and $x = -1.925$, excluding $x = -2$). Indicate on your graph that every x-value in $(-2.075, -2) \cup (-2, -1.925)$ has an $f(x)$-value in $(3.6, 4.4)$.

23. $(2.25, 5.75)$

24. For all $\varepsilon > 0$, there exists a $\delta > 0$ such that if $0 < |x - 1| < \delta$, then $|(x^2 - 3) + 2| < \varepsilon$.

27. For all $\varepsilon > 0$, there exists a $\delta > 0$ such that if $x \in (3 - \delta, 3)$, then $|(4 - x^2) + 5| < \varepsilon$.

30. For all $M > 0$, there exists a $\delta > 0$ such that if $x \in (1, 1 + \delta)$, then $\frac{1}{1 - x} < -M$.

31. For all $\varepsilon > 0$, there exists an $N > 0$ such that if $x > N$, then $|\frac{x}{1 - 2x} + 0.5| < \varepsilon$.

32. For all $M > 0$, there exists an $N < 0$ such that if $x < N$, then $3x - 1 > M$.

33. There is some $\varepsilon > 0$ for which there is no $\delta > 0$ that would guarantee that if $0 < |x - 2| < \delta$, then $|f(x) - 1| < \varepsilon$.

36. There is some $\varepsilon > 0$ for which there is no $N > 0$ that would guarantee that if $x > N$, then $|f(x) - 3| < \varepsilon$.

38. There is some $M > 0$ for which there is no $N > 0$ that would guarantee that if $x > N$, then $\frac{1000}{x} > M$.

40. Given ε, we can find a δ.

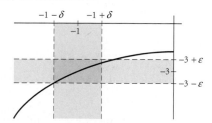

43. Given M, we can find a δ.

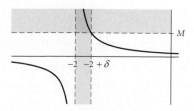

45. Given M, we can find an N.

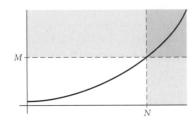

47. For all $M < 0$, there exists a $\delta > 0$ such that if $x \in (c, c+\delta)$, then $f(x) < M$.

49. For all $\varepsilon > 0$, there exists an $N < 0$ such that if $x < N$, then $|f(x) - L| < \varepsilon$.

51. For all $M > 0$, there exists an $N < 0$ such that if $x < N$, then $f(x) > M$.

52. $\delta = (8.5)^{\frac{1}{3}} - 2 \approx 0.0408$

55. $\delta = 1$

56. $\delta \approx 0.0005$

57. $N \approx 5.053$

SECTION 2.3

2. For all $\varepsilon > 0$, there exists a $\delta > 0$ such that if $0 < |x+2| < \delta$, then $|\frac{3}{x+1} + 3| < \varepsilon$.

3. $(2 - \frac{0.25}{3}, 2) \cup (2, 2 + \frac{0.25}{3})$

4. $(2, 3) \cup (3, 4)$ **5.** $(1, 1.25)$ **6.** $(\frac{1}{0.15}, \infty)$

7. $(3, 3.001)$

10. (a) $\delta = \frac{5}{325}$

    (b) $\delta = \frac{.01}{325}$

    (c) $\delta = 1$

11. $\delta = \varepsilon/2$ **13.** $\delta = \varepsilon/4$

15. $\delta = \sqrt{\varepsilon/3}$

18. $\delta = \min(1, \varepsilon/5)$

20. $\delta = \varepsilon^2$; $\lim\limits_{x \to -2^+} (1 + \sqrt{x+2}) = 1$

21. $\delta = \frac{1}{M}$; $\lim\limits_{x \to 1^-} \frac{1}{1-x} = \infty$

23. $N = \sqrt{M-2}$; $\lim\limits_{x \to \infty} (x^2 + 2) = \infty$

24. **PROOF**   Given $\varepsilon > 0$, choose $\delta = \varepsilon/2$. Then if $0 < |x - 1| < \delta$, we have $|(2x + 4) - 6| = 2|x - 1| < 2\delta = \varepsilon$. ∎

25. **PROOF**   Given $\varepsilon > 0$, choose $\delta = \varepsilon$. Then if $0 < |x + 6| < \delta$, we have $|(x + 2) - (-4)| = |x + 6| < \delta = \varepsilon$. ∎

28. **PROOF**   Given $\varepsilon > 0$, choose $\delta = \sqrt{\varepsilon/3}$. Then if $0 < |x| < \delta$, we have $|(3x^2 + 1) - 1| = |3x^2| = 3|x|^2 < 3\delta^2 = 3(\sqrt{\varepsilon/3})^2 = \varepsilon$. ∎

31. **PROOF**   Given $\varepsilon > 0$, choose $\delta = \varepsilon$. Then if $0 < |x - 1| < \delta$, we have $|\frac{x^2-1}{x-1} - 2| = |\frac{x^2-2x+1}{x-1}| = |x-1| < \delta = \varepsilon$. ∎

35. **PROOF**   Given $\varepsilon > 0$, choose $\delta = \min(1, \varepsilon/5)$. Then if $0 < |x-3| < \delta$, we have $|(x^2 - 2x - 3) - 0| = |x^2 - 2x - 3| = |x - 3||x + 1| < \delta|x + 1| < \delta(5) < \varepsilon$. ∎

38. **PROOF**   Given $\varepsilon > 0$, choose $\delta = \varepsilon^2$. Then if $x \in (-2, -2+\delta)$, we have $0 < x + 2 < \delta$, and thus $|(1 + \sqrt{x+2}) - 1| = \sqrt{x+2} < \sqrt{\delta} = \varepsilon$. ∎

40. **PROOF**   Given $M > 0$, choose $\delta = 1/M$. Then if $x \in (-2, -2 + \delta)$, we have $0 < x + 2 < \delta$, and thus $\frac{1}{x+2} > \frac{1}{\delta} = \frac{1}{1/M} = M$. ∎

41. **PROOF**   Given $\varepsilon > 0$, choose $N = \frac{1}{\varepsilon}$. Then if $x > N$, we have $|\frac{2x-1}{x} - 2| = |\frac{2x-1-2x}{x}| = \frac{1}{|x|} < \frac{1}{N} = \varepsilon$. ∎

SECTION 2.4

1. $\lim\limits_{x \to 0} \sqrt{x}$; $\lim\limits_{x \to 1} \frac{1}{x-1}$; $\lim\limits_{x \to 2} (3x + 1)^{\frac{1}{3}}$

3. For any real numbers c and k, $\lim\limits_{x \to c} k = k$; for example, $\lim\limits_{x \to 3} 5 = 5$.

6. For any real numbers c and k, if $\lim\limits_{x \to c} f(x)$ exists, then $\lim\limits_{x \to c} k f(x) = k \lim\limits_{x \to c} f(x)$; for example, $\lim\limits_{x \to 2} 3(x + 1) = 3 \lim\limits_{x \to 2} (x + 1)$.

9. If $\lim\limits_{x \to c} f(x)$ and $\lim\limits_{x \to c} g(x)$ exist, then $\lim\limits_{x \to c} (f(x)g(x)) = (\lim\limits_{x \to c} f(x))(\lim\limits_{x \to c} g(x))$; for example, $\lim\limits_{x \to c} x^2(x + 1) = (\lim\limits_{x \to c} x^2)(\lim\limits_{x \to c} (x + 1))$.

12. If k is a positive integer, then $\lim\limits_{x \to c} x^k = c^k$; for example, $\lim\limits_{x \to 4} x^3 = 64$.

16. Consider the limit $\lim\limits_{x \to 1} \frac{x-1}{x^2-1}$.

17. Constant functions and functions of the form x^k for a positive integer k.

18. $f(x) = \frac{1}{x}$, $g(x) = -\frac{1}{x}$, $c = 0$

20. For all $\varepsilon > 0$, there exists $\delta > 0$ such that if $0 < |x-c| < \delta$, then $|x - c| < \varepsilon$.

21. For all $\varepsilon > 0$, there exists $\delta > 0$ such that if $0 < |x-c| < \delta$, then $|(mx + b) - (mc + b)| < \varepsilon$.

24. For all $\varepsilon > 0$, there exists $\delta > 0$ such that if $0 < |x-c| < \delta$, then $|(f(x) - g(x)) - (L - M)| < \varepsilon$, where $L = \lim\limits_{x \to c} f(x)$ and $M = \lim\limits_{x \to c} g(x)$.

28. If k is a positive integer, then for all $\varepsilon > 0$, there exists $\delta > 0$ such that if $0 < |x - c| < \delta$, then $|x^k - c^k| < \varepsilon$.

29. The graph of $f(x) = 3$ is a horizontal line at $y = 3$. Given any ε bar you draw around $y = 3$, *any* δ bar around $x = 2$ will work (since the value of $f(x)$ is equal to 3 and thus within the ε bar, no matter how far away you are from $x = 2$).

31. Sketch $f(x) = 1 - 3x$ and draw an ε bar around the height $y = -11$. Look at the right triangle whose diagonal is the piece of the graph of $y = 1 - 3x$ that is trapped in this ε bar, and think about "rise over run."

33. -2

34. Not enough information. (We don't yet know what to do with limits of compositions.)

35. Not enough information.

37. $\frac{1}{2}$

38. (a) $\lim\limits_{x\to 1} 15(3-2x) = 15\lim\limits_{x\to 1}(3-2x) = 15(\lim\limits_{x\to 1}(3) + \lim\limits_{x\to 1}(-2x)) = 15(\lim\limits_{x\to 1}(3) - 2\lim\limits_{x\to 1}(x)) = 15(3-2(1)) = 15$

(b) $\lim\limits_{x\to 1} 15(3-2x) = \lim\limits_{x\to 1}(45-30x) = 45 - 30(1) = 15$

43. $\lim\limits_{x\to 0}(x^2 - 1) = \lim\limits_{x\to 0} x^2 - \lim\limits_{x\to 0} 1 = 0 - 1 = -1$

44. $\lim\limits_{x\to -3} \dfrac{2}{3x+1} = \dfrac{\lim\limits_{x\to -3} 2}{\lim\limits_{x\to -3}(3x+1)} = \dfrac{2}{\lim\limits_{x\to -3}(3x+1)}$

$= \dfrac{2}{\lim\limits_{x\to -3} 3x + \lim\limits_{x\to -3} 1} = \dfrac{2}{3\lim\limits_{x\to -3} x + \lim\limits_{x\to -3} 1}$

$= \dfrac{2}{3(-3)+1} = -\dfrac{1}{4}$

45. 6 **48.** 4 **49.** $-\frac{1}{3}$

50. **PROOF** For all real numbers m, b, and c, $\lim\limits_{x\to c}(mx+b) = \lim\limits_{x\to c} mx + \lim\limits_{x\to c} b = m\lim\limits_{x\to c} x + \lim\limits_{x\to c} b = m(c) + b.$ ■

51. **PROOF** $\lim\limits_{x\to c}(f(x) - g(x)) = \lim\limits_{x\to c}(f(x) + (-g(x)))$ $= \lim\limits_{x\to c} f(x) + \lim\limits_{x\to c}(-g(x)) = \lim\limits_{x\to c} f(x) - \lim\limits_{x\to c} g(x)$ ■

SECTION 2.5

1. False; counterexample: $\lim\limits_{x\to 1}\dfrac{x^2-1}{x-1}$ **2.** D

3. $f(x) = \sqrt{x+1}$ fails condition (a); $g(x) = \frac{1}{x-1}$ satisfies condition (a) but fails condition (b) at $x = 1$.

6. $(3x+1)(3x+1)(3x+1)$

8. (a) $f(x) = (x-1)^2$, $g(x) = x - 1$, $c = 1$
(b) $f(x) = x - 1$, $g(x) = (x-1)^2$, $c = 1$
(c) $f(x) = (x-2)(x-1)$, $g(x) = x - 2$, $c = 2$
(d) $f(x) = (x-2)(x+1)$, $g(x) = x - 2$, $c = 2$

11. Graph $h(x) = \frac{-2}{3-x}$. **12.** $x \neq 2$

14. (a) $\lim\limits_{x\to 3}(3x + x^2(2x+1)) = \lim\limits_{x\to 3}(3x) + \lim\limits_{x\to 3}(x^2(2x+1)) = \lim\limits_{x\to 3}(3x) + \lim\limits_{x\to 3}(x^2) \cdot \lim\limits_{x\to 3}(2x+1) = 3(3) + (3^2)(2(3)+1) = 72$

(b) $f(x) = 3x + x^2(2x+1)$ consists of linear and positive integer power functions that are added and multiplied together, and $f(3) = 72$ is a real number, so $\lim\limits_{x\to 3}(3x + x^2(2x+1)) = 3(3) + 3^2(2(3)+1) = 72$.

18. $\lim\limits_{x\to 0}\frac{x^2-1}{x-1} = \frac{0^2-1}{0-1} = 1$ **20.** 0 **22.** -2

24. DNE; $\lim\limits_{x\to 1^-}\dfrac{1}{1-x} = \infty$, $\lim\limits_{x\to 1^+}\dfrac{1}{1-x} = -\infty$

28. $-\frac{1}{3}$

29. DNE; $\lim\limits_{x\to 0^-}\frac{x^2+1}{x(x-1)} = \infty$, $\lim\limits_{x\to 0^+}\frac{x^2+1}{x(x-1)} = -\infty$

30. ∞ **32.** -1 **36.** $-1, 1, \text{DNE}$

37. $9, 9, 9$ **40.** $4, 0, \text{DNE}$ **41.** $a = 1$ **43.** $a = 1$

44. Not possible.

47. **PROOF** We know $\lim\limits_{x\to c}(mx+b) = mc + b$; therefore, if $f(x) = mx + b$, then $\lim\limits_{x\to c} f(x) = \lim\limits_{x\to c}(mx + b) = mc + b = f(c)$. ■

48. **PROOF** Suppose $f(x) = (mx+b)(kx+a)$ is a product of linear functions. Then for any real number c, we have $\lim\limits_{x\to c} f(x) = \lim\limits_{x\to c}(mx+b)(kx+a) = \lim\limits_{x\to c}(mx+b) \cdot \lim\limits_{x\to c}(kx+a) = (mc+b)(kc+a) = f(c)$. ■

SECTION 2.6

3. $f(x) = \frac{x-3}{(x-3)(x-2)}$; $f(x) = \begin{cases} x+1, & x < 3 \\ 2x-1, & x \geq 3 \end{cases}$; $f(x) = \frac{1}{x-3}$

4. f has a jump discontinuity at $x = 2$ and is right (but not left) continuous at $x = 2$.

6. f has a removable discontinuity at $x = 2$ and is neither right nor left continuous at $x = 2$.

9.

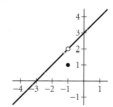

10.

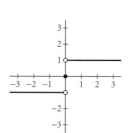

13. Not possible.

15.

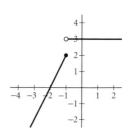

18. Look at the graph of $f(x) = \frac{1}{x-3}$.

20. Define $f(1)$ to be 0.

21. Define $f(1)$ to be 2.

23. For all $\varepsilon > 0$, there exists a $\delta > 0$ such that if $x \in (c, c+\delta)$, then $|f(x) - f(c)| < \varepsilon$.

24. $x = \sqrt{2}$ **25.** $x = 2$ **26.** $x = 3$

29. No. Try drawing a function whose graph has a jump discontinuity at $x = 2$ with a "closed dot" on the left piece of the graph and an "open dot" on the right piece of the graph.

30. False   **31.** False   **32.** $\lim_{x\to 1} f(x) = f(1)$

36. Discontinuous at $x = -1$ (removable) and $x = 1$ (infinite). Not left or right continuous at $x = -1$. Left but not right continuous at $x = 1$. Continuous on $(-\infty, -1)$, $(-1, 1]$, and $(1, \infty)$.

37. Discontinuous at $x = -1$ (removable) and at $x = 2$ (jump). Neither right nor left continuous at $x = -1$. Right but not left continuous at $x = 2$. Continuous on $(-\infty, -1)$, $(1, 2)$, and $[2, \infty)$.

38. Not continuous at $x = 0$.

39. Continuous at $x = 0$.

42. 4, 4, 4, DNE; removable discontinuity

43. $-\infty$, ∞, DNE, DNE; infinite discontinuity

46. 2, 4, DNE, 2; jump discontinuity

47. Right but not left continuous at $x = -1$.

49. Left and right continuous at $x = 2$.

52. Since the quotient of continuous functions is continuous, $f(x)$ is continuous on its domain $(-\infty, 2) \cup (2, \infty)$.

55. $\lim_{x\to 0} |3x^2 - 5| = |\lim_{x\to 0}(3x^2 - 5)| = |3(0)^2 - 5| = |-5| = 5$

57. Not continuous at $x = 2$.

58. Continuous at $x = 1$.

65. PROOF   Suppose $f(x) = mx + b$ is a linear function. Then for any value c in the domain $(-\infty, \infty)$ of $f(x)$, we have $\lim_{x\to c} f(x) = \lim_{x\to c} mx + b = mc + b = f(c)$. Since $\lim_{x\to c} f(x) = f(c)$ for all c, the function f is continuous. ∎

68. PROOF   The following argument proves that if $f(x) = 3x - 5$, then $\lim_{x\to 2} f(x) = f(2)$. Given $\varepsilon > 0$, let $\delta = \varepsilon/3$. If $0 < |x - 2| < \delta$, then $|(3x - 5) - (3(2) - 5)| = |3x - 6| = 3|x - 2| < 3\delta = \varepsilon$. ∎

69. We will do a delta–epsilon argument to show that $\lim_{x\to c} |x| = |c|$ for all real numbers c.

PROOF   Given $\varepsilon > 0$, choose $\delta = \varepsilon$. If $0 < |x - c| < \delta$, then (using the inequality given in the problem): $||x| - |c|| < |x - c| < \delta = \varepsilon$. ∎

SECTION 2.7

3. (a) The hypothesis is that f is continuous on a closed interval $[a, b]$. The conclusion is that for any number K between $f(a)$ and $f(b)$, there is at least one $c \in (a, b)$ such that $f(c) = K$.

(b) The theorem says that a continuous function on a closed interval $[a, b]$ takes on all *intermediate values* between $f(a)$ and $f(b)$.

(c) Converse: If for any number K between $f(a)$ and $f(b)$, there is at least one $c \in (a, b)$ such that $f(c) = K$, then f must be continuous on the closed interval $[a, b]$; False.

(d) Contrapositive: If there is a number K between $f(a)$ and $f(b)$ for which there is no $c \in (a, b)$ with $f(c) = K$, then f is not continuous on the closed interval $[a, b]$; True.

4. (a), (b), and **(c):** See the reading.

(d) One example, on the interval $[-1, 3]$, is

The existence of this example tells us that the converse of the Intermediate Value Theorem (IVT) must be false (this function takes on all intermediate values between $f(-1) = 2$ and $f(3) = -2$, but it is not continuous).

9. False   **10.** False   **11.** True   **12.** False   **13.** False

14. False   **15.** True   **16.** False

17.

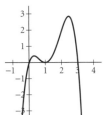

19. (a) This function is continuous on $[-2, 2]$ (since it is a polynomial), so the Extreme Value Theorem (EVT) guarantees that f will have global maximum and minimum values on $[-2, 2]$.

(b) The maximum value occurs twice, at $x = -2$ and at $x = 2$, and the minimum value occurs both at $x \approx -1.22$ and at $x \approx 1.22$.

23. (a) f is continuous on $[0, 2]$, so the Extreme Value Theorem (EVT) applies.

(b) The maximum value occurs at $x = \frac{4}{3}$, and the minimum value occurs at $x \approx 0.8165$.

27. (a) This function is continuous on $[-2, 4]$ (since it is a polynomial), and $K = -4$ is between $f(-2) = -22$ and $f(4) = 14$, so the Intermediate Value Theorem (IVT) guarantees that there is some $c \in (-2, 4)$ such that $f(c) = -4$.

(b) Using a graphing calculator we can approximate that $f(c) = -4$ at $c = 1$, $c \approx -0.732$, and $c \approx 2.732$.

28. (a) f is continuous on $[-2, 4]$, and $f(2) < -4 < f(4)$, so the IVT applies.

(b) $c \approx 2.732$.

33. (a) This function is continuous (since it is a polynomial), so the IVT applies. Since $f(-3) = (-3)^3 + 2 = -25 < -15$ and $f(0) = 2 > -15$, the IVT guarantees that there is some $c \in (-3, 0)$ for which $f(c) = -15$.

(b) With a graphing calculator we can approximate that $c \approx -2.57$ is such a value.

34. (a) f is continuous, and $f(0) > 2 > f(2)$, so the IVT applies.

(b) $c = 1$.

37. (a) The special case of the Intermediate Value Theorem applies because f is continuous on $[1, 2]$ (since it is a polynomial), and $f(1) = -2 < 0$ whereas $f(2) = 4 > 0$.

(b) Thus f has a root somewhere in $(1, 2)$.

(c) With a graphing calculator we can see that one such root is approximately $x \approx 1.56$.

42. (a) f is continuous on $[-1, 0]$, but $f(-1)$ and $f(0)$ are both positive; thus we can't use the special case of the IVT on the interval $[-1, 0]$.

44. $[-1, \frac{2}{3}] \cup [1, \infty)$ **45.** $(-\infty, -1) \cup (0, 3)$

48. Positive on $(-\infty, -2) \cup (-\frac{1}{2}, \infty)$, negative on $(-2, -\frac{1}{2})$

49. Positive on $(-1, 0) \cup (1, \infty)$, negative on $(-\infty, -1) \cup (0, 1)$

50. Positive on $(-4, -\frac{3}{2}) \cup (1, \infty)$, negative on $(-\infty, -4) \cup (-\frac{3}{2}, 1)$

53. Positive on $(-\infty, -2) \cup (5, \infty)$, negative on $(-2, 0) \cup (0, 5)$

54. Positive on $(-\infty, -2) \cup (-1, 1) \cup (2, \infty)$, negative on $(-2, -1) \cup (1, 2)$

56. Positive on $(2, \infty)$, negative on $(-\infty, 2)$

58. Positive on $(0, 2]$, negative on $(-\infty, 0) \cup (2, \infty)$

59. The length $H(t)$ of Alina's hair is a continuous function (her hair can't suddenly get longer or shorter without going through a continuous change), so the EVT and the IVT apply to this function. We are given that $H(0) = 2$ and $H(6) = 42$. The Extreme Value Theorem says that Alina's hair had a maximum and a minimum length sometime in the last 6 years. The Intermediate Value Theorem says that Alina's hair has been every length between 2 and 42 inches at some point during the last 6 years.

CHAPTER 3

SECTION 3.1

2. The line through the point $(c, f(c))$ whose slope is $f'(c)$ (provided that $f'(c)$ exists).

6. The tangent line and the secant line for $h = 1$ are shown below.

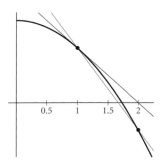

$h = 0.5$ and $h = -0.5$ produce better approximations.

8. (a) The x-coordinate of the first dot

(b) The horizontal distance between the first and second dots

(c) The x-coordinate of the second dot

(d) The y-coordinate of the first dot

(e) The y-coordinate of the second dot

(f) The vertical distance between the first and second dots

(g) The slope of the line connecting the two dots

(h) The slope of the line tangent to the graph at the first dot

11. (a) The x-coordinate of the first dot

(b) The horizontal distance between the first and second dots

(c) The x-coordinate of the second dot

(d) The y-coordinate of the first dot

(e) The y-coordinate of the second dot

(f) The vertical distance between the first and second dots

(g) The slope of the line connecting the two dots

(h) The slope of the line tangent to the graph at the first dot

16. Define $x := c + h$; then $h = x - c$, and as $h \to 0$, we have $x \to c$.

17. $f'_-(c) := \lim\limits_{x \to c^-} \dfrac{f(x) - f(c)}{x - c}$,

$f'_+(c) := \lim\limits_{x \to c^+} \dfrac{f(x) - f(c)}{x - c}$

18. Exist, equal

19. For all $\varepsilon > 0$, there exists $\delta > 0$ such that $0 < |h - 0| < \delta \Rightarrow |\frac{f(c+h)-f(c)}{h} - f'(c)| < \varepsilon$.

20. For all $\varepsilon > 0$, there exists $\delta > 0$ such that $0 < |x - c| < \delta \Rightarrow |\frac{f(x)-f(c)}{x-c} - f'(c)| < \varepsilon$.

22. -11.41 **23.** 2.01 **26.** 1.999 **28.** -11.41

30. 2.001 **33.** 2 **34.** 0.248756

37. When $h = 0.1$, the slope is ≈ -0.13158.
When $h = 0.001$, the slope is ≈ -0.12506.
When $h = -0.001$, the slope is ≈ -0.12494.
When $h = -0.1$, the slope is ≈ -0.11905.
$f'(-2)$ seems to be ≈ -0.125.

39. 0 **41.** $\lim\limits_{h \to 0} \dfrac{(1 + h)^2 - 1^2}{h} = 2$

43. $\lim\limits_{h \to 0} \dfrac{\frac{1}{2(-2+h)} - \frac{1}{2(-2)}}{h} = -\dfrac{1}{8}$

44. 0 **45.** 0 **49.** 1

51. $\lim\limits_{h \to 0} \dfrac{(-3 + h)^2 - (-3)^2}{h} = -6$

53. $\lim\limits_{h \to 0} \dfrac{(4(-2 + h) + 3) - (4(-2) + 3)}{h} = 4$

55. -3 **57.** -1 **60.** $\lim\limits_{x \to -3} \dfrac{x^2 - (-3)^2}{x - (-3)} = -6$

61. $\lim\limits_{x \to 1} \dfrac{x^2 - 1^2}{x - 1} = 2$ **63.** 0 **65.** -3

69. $y - 9 = -6(x + 3)$

71. $y + 1 = -3(x - 1)$ **72.** $y - 2 = -(x - 3)$

74. $f(x) = 3x^2 + 1, c = -1; f'(-1) = -6$

75. $f(x) = \frac{1}{x}, c = 2; f'(2) = -\frac{1}{4}$

78. $f(x) = x^4, c = 3; f'(3) = 108$

79. $f(x) = \frac{1}{x+1}, c = 0; f'(0) = -1$

80. $f'_-(2) = \lim\limits_{h \to 0^-} \dfrac{((2+h)+4)-3(2)}{h} = 1;$

$f'_+(2) = \lim\limits_{h \to 0^+} \dfrac{3(2+h)-3(2)}{h} = 3;$

$f'(2)$ does not exist

83. $f'_-(-1) = \lim\limits_{h \to 0^-} \dfrac{(-1+h)^2-(-1)^2}{h} = -2;$

$f'_+(-1) = \lim\limits_{h \to 0^+} \dfrac{(-1-2(-1+h))-(-1)^2}{h} =$

$-2; \; f'(-1) = -2$

84. $S(h)$ represents $\dfrac{((3+h)^5-26)-217}{h}$ in the table below.

h	-0.1	-0.01	-0.001	-	0.001	0.01	0.1
$S(h)$	432.9	407.7	405.3	-	404.7	402.3	378.9

$f'(3)$ seems to be $\approx 405.$

88. $S(h)$ represents $\dfrac{\frac{1-(1+h)}{1+(1+h)}-0}{h}$ in the table below.

h	-0.1	-0.01	-0.001	-	0.001	0.01	0.1
$S(h)$	-0.526	-0.503	-0.5002	-	-0.4998	-0.498	-0.476

$f'(1)$ seems to be $\approx -0.5.$

91. Left-sided estimate: 0.153; Right-sided estimate: 0.548; Two-sided estimate: 0.064

92. Left-sided estimate: -2; Right-sided estimate: -4; Two-sided estimate: -3

94. **PROOF**   The tangent line (if it exists) to f at $x = c$ is the line with slope $f'(c)$ that passes through the point $(c, f(c))$. The point-slope form of this line is $y - f(c) = f'(c)(x - c)$, that is, $y = f'(c)(x - c) + f(c)$. ∎

Section 3.2

2. Velocity is *directional* speed; speed is always positive, but velocity can be positive or negative.

5. (b) < (c) < (a) < (d)

6. (c) < (d) < (a) < (b)

7. $x = 0.5$; $x = -1$ (or $x = 2$); $x = -0.4$ and $x = 1.3$

8. $(0, \infty)$; $(-\infty, 0)$; $x = 0$

10. Your graph should have a horizontal tangent line at $x = 2$. The line connecting the points $(1, f(1))$ and $(2, f(2))$ on the graph should have slope 2. The slope from $(2, f(2))$ to $(4, f(4))$ should be 1.

11. -7 **13.** -1.0101 **15.** $-\dfrac{10}{9} \approx -1.111$

18. From the right: The average rate of change on $[1, 1.1]$ is -2.1; on $[1, 1.01]$ is -2.01; on $[1, 1.001]$ is -2.001; these values seem to approach -2. From the left: The average rate of change on $[0.9, 1]$ is -1.9; on $[0.99, 1]$ is -1.99; on $[0.999, 1]$ is -1.999; these values also seem to approach -2. The instantaneous rate of change seems to be -2.

20. $\lim\limits_{h \to 0} \dfrac{(3(-2+h)+1)-(-5)}{h} = 3$

24. 2   **25.** -1

28. 5 miles per hour

29. Miles; $\frac{mi}{hr}$; $\frac{mi}{hr^2}$

33. (a) Her average rate of change is 0 on $[0, 30]$; this means that after 30 minutes she is exactly as far away from the oak tree as she was at time $t = 0$.

(b) The second 10 minutes

(c) One approximation might be $\frac{140-280}{10-5} = -28$ feet per minute. The fact that the sign is negative indicates that she is moving *toward* the oak tree.

(d) $t \approx 4$ minutes, $t \approx 12$ minutes, $t \approx 25$ minutes. These are the times she changes direction.

(e) $(4, 12)$ and $(25, 30)$. These are the times she is moving toward the oak tree.

34. (a) $s(t) = 0$ at approximately $t = 1.62$ seconds, so the average velocity of the orange is $\frac{s(1.62)-s(0)}{1.62-0} \approx -25.93$ feet per second.

(b) Average rate of change (AROC) of $s(t)$ on $[1.6, 1.62]$ is ≈ -51.52; AROC of $s(t)$ on $[1.61, 1.62]$ is ≈ -51.68; AROC of $s(t)$ on $[1.619, 1.62]$ is ≈ -51.824. Thus instantaneous rate of change (IROC) of $s(t)$ at $x = 1.62$ (the time the orange hits the ground) might be approximately -51.825 feet per second.

(c) $-51.84 \frac{feet}{sec}$;   (d) $-32 \frac{feet}{sec^2}$

35. (a) $h(12)$ is the average height of a 12-year-old, in feet. $h'(12)$ is measured in feet per year and represents the instantaneous rate of change of the height of a 12-year-old person; that is, $h'(12)$ is the rate at which an average 12-year-old is growing (in feet per year).

(b) $h(12)$ is positive; $h'(12)$ is positive.

(c) $h(t)$ might have a maximum at $t \approx 60$ years (people may tend to slouch or get shorter as they get older); $h'(t)$ might have a maximum at $t \approx 14$ years (growth spurt).

Section 3.3

1. f is differentiable, and thus continuous, at $x = c$.

3. (a) No;   (b) No

4. (a) No;   (b) Yes

7. $\lim\limits_{h \to 0} \dfrac{(4(2+h)^3-5(2+h)+1)-23}{h}$ exists.

8. False   **9.** False   **10.** False   **11.** False   **12.** True

13. Not differentiable, but left and right differentiable, at $x = -1$. Not differentiable, and not left or right differentiable, at $x = 2$.

14. Not differentiable at $x = -1$. Left but not right differentiable at $x = -1$.

16. $\lim\limits_{h \to 0} \dfrac{(2(-2+h)-5)-(-9)}{h} = 2$, so differentiable

18. $f'_-(0) = -\infty$, $f'_+(0) = \infty$, so not differentiable

21. $\lim\limits_{h \to 0^-} \dfrac{|(-1+h)^2-(-1+h)-2|-0}{h} = -3,$

$\lim\limits_{h \to 0^+} \dfrac{|(-1+h)^2-(-1+h)-2|-0}{h} = 3$, so f is left *and* right differentiable at $x = -1$, but not differentiable at $x = -1$.

23. Not differentiable at $x = 0$

24. Differentiable at $x = 0$

25. (a) $\lim\limits_{x \to -2^-} (-x - 1) = -1$, $\lim\limits_{x \to -2^+} (1 - x^2) = -3$, and $f(-2) = -1$, so f is not continuous at $x = -2$.
(b) Since $f(x)$ is not continuous at $x = -2$, it cannot be differentiable at $x = -2$.

27. (a) Continuous; **(b)** differentiable

28. (a) Not continuous; **(b)** not differentiable

31. Not continuous and not differentiable

34. Continuous and differentiable

35. (a) $S(3) = 200 + 8(10)(3) = \$440$, $S(6) = 200 + 8(10)(6) = \$680$, and $S(8) = 200 + 8(10)(6) + 11.5(10)(2) = \$910$.
(b) $S'(3) = 80$ dollars per week, $S'(8) = 11.5(10) = \$115$ dollars per week. Cannot compute $S'(6)$ because the rate at which you are paid changes at that point.
(c) $S(t) = \begin{cases} 200 + 8(10)t, & t \le 6 \\ 680 + 11.5(10)(t - 6), & t > 6 \end{cases}$

36. **PROOF** f is differentiable at $x = c$ if and only if $\lim\limits_{h \to 0} \frac{f(c+h) - f(c)}{h}$ exists, which is true if and only if $\lim\limits_{h \to 0^-} \frac{f(c+h) - f(c)}{h}$ and $\lim\limits_{h \to 0^+} \frac{f(c+h) - f(c)}{h}$ both exist and are equal. This is equivalent to saying that $f'_-(c)$ and $f'_+(c)$ exist and are equal. ∎

SECTION 3.4

1. The height of the graph $y = f'(x)$ at $x = c$ is the same as the slope of the tangent line to the graph of $y = f(x)$ at $x = c$.

4. You could find $f'(c) = \lim\limits_{h \to 0} \frac{f(c+h) - f(c)}{h}$, or you could find $f'(x) = \lim\limits_{h \to 0} \frac{f(x+h) - f(x)}{h}$ and then evaluate the answer at $x = c$; usually the first way is easier if all you want is the derivative at one point.

6.

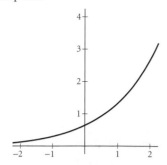

7. Your graph of $y = f'(x)$ should have zeros at $x = -1.5$, $x = 0$, and $x = 1.5$, should be positive on $(-\infty, -1.5)$ and $(0, 1.5)$, and should be negative on $(-1.5, 0)$ and $(1.5, \infty)$.

8. At $x = 1$ your graph should have a height of 2 and a horizontal tangent line. At $x = 3$ your graph should be drawn so that its tangent line has a slope of 2.

11. Your graph should go through the points $(-1, 2)$ and $(1, -2)$, and the tangent line at each of those points should have slope 3.

12. Graph I, f''; Graph II, f; Graph III, f'

13. Your graph should have horizontal tangent lines at $x = -2$ and at $x = 0$. Moreover, the tangent lines to the graph of your function should have a negative slope everywhere on $(-\infty, -2)$, a positive slope everywhere on $(-2, 0)$, and a negative slope everywhere on $(0, \infty)$.

15. (a) $f''''(-4)$ (or $f^{[4]}(-4)$); **(b)** $\frac{d^4 f}{dx^4}\Big|_{-4}$;
(c) $\frac{d^4}{dx^4}\Big|_{-4}(3x^2 - 1)$

18. 3   **20.** $2x + h$

23. $-\frac{1}{x(x+h)}$ **24.** 3

25. $t + x$ **28.** $t^3 + t^2 x + tx^2 + x^3$

30. $2x$ **32.** $\frac{-1}{x^2}$ **34.** $3x^2$

36. (a) $\lim\limits_{h \to 0} \frac{(5(x + h) - 2) - (5x - 2)}{h} = 5$
(b) $\lim\limits_{t \to x} \frac{(5t - 2) - (5x - 2)}{t - x} = 5$

38. (a) $2x$;   **(b)** $2x$

40. (a) $\lim\limits_{h \to 0} \frac{(-2(x + h)^2) - (-2x^2)}{h} = -4x$
(b) $\lim\limits_{t \to x} \frac{-2t^2 - (-2x^2)}{t - x} = -4x$
(c) $8, 0, -12$

41. (a) $1 + 2x$;   **(b)** $1 + 2x$;   **(c)** $-3, 1, 7$

42. (a) 0;   **(b)** 0;   **(c)** $0, 0, 0$

45. (a) $3x^2$;   **(b)** $3x^2$;   **(c)** $12, 0, 27$

46. (a) $4x^3$;   **(b)** $4x^3$;   **(c)** $-32, 0, 108$

48. (a) $3x^2 + 1$;   **(b)** $3x^2 + 1$;   **(c)** $13, 1, 28$

49. (a) $\frac{-1}{x^2}$;   **(b)** $\frac{-1}{x^2}$;   **(c)** $-\frac{1}{4}$, DNE, $-\frac{1}{9}$

50. (a) $-2\frac{1}{(x+1)^2}$;   **(b)** $-2\frac{1}{(x+1)^2}$;   **(c)** $-2, -2, -\frac{1}{8}$

52. $y + 8 = 8(x + 2)$; $y = 0$; $y + 18 = -12(x - 3)$

53. $y - 3 = -3(x + 2)$; $y - 1 = 1(x - 0)$; $y - 13 = 7(x - 3)$

54. The graph of f' is the graph of the constant function $y = -2$.

56. Clearly the graph of f' should have a root at $x = 0$ and should be positive on $(-\infty, 0)$ and negative on $(0, \infty)$. There is one more thing to notice here, however; the graph of f gets very "flat" at its ends, so its tangent lines have very shallow slopes as $x \to \infty$ and as $x \to -\infty$. Therefore, the graph of the derivative should be approaching zero as $x \to \infty$ and as $x \to -\infty$. Thus your graph of f' should also have a two-sided horizontal asymptote at $y = 0$.

58. $f(x) = 2x^3$; $f'(x) = 6x^2$

59. $f(x) = \frac{1}{3x}$; $f'(x) = -\frac{1}{3x^2}$

60. $f(x) = x^3 + 1$; $f'(x) = 3x^2$

63. (a) $6x^2$;   **(b)** $12x$;   **(c)** 12

64. (a) 54;   **(b)** 12;   **(c)** 12

65. (a) 13;   **(b)** $6x$;   **(c)** 12

67. (a) 2 hours;
(b) $v(t) = -20t - 40$;
(c) $a(t) = -20$, accelerating (speeding up) at a rate of 20 miles per hour per hour.

68. (a) The graph is an upside-down parabola with roots at $x = 0$ and at $x = 2$. The physical interpretation is that your velocity is always positive and that you start out with a zero velocity, speed up, and then eventually slow back down to a zero velocity.

  (b) The graph has a root at about $x = 1$ and is positive on $[0, 1)$ and negative on $(1, 2)$ (for example, a line with negative slope passing through $(1, 0)$). This represents the fact that you were speeding up, and then slowing down, during your trip.

70. **PROOF** $f'(x) = \lim_{h \to 0} \dfrac{(m(x + h) + b) - (mx + b)}{h} =$
$\lim_{h \to 0} \dfrac{mx + mh + b - mx - b}{h} = \lim_{h \to 0} \dfrac{mh}{h} = m.$ ∎

Section 3.5

3. In Leibnitz notation: If m and b are any real numbers, then $\frac{d}{dx}(mx + b) = m$. In "prime" notation: If $f(x) = mx + b$ is a linear function, then $f'(x) = m$.

6. $\frac{d}{dx}(f(x) + g(x)) = \frac{df}{dx} + \frac{dg}{dx}; (f(x) + g(x))' = f'(x) + g'(x)$

10. $\frac{1}{x}, \sqrt{x}, \sin x, (1 + 3x)^{\frac{1}{3}}$

12. At some point in your calculation, you should have $\frac{d}{dx}(3) - 2\frac{d}{dx}(x) + 5\frac{d}{dx}(x^2)$.

15. $\lim_{h \to 0} \dfrac{(3 - 2(x + h) + 5(x + h)^2) - (3 - 2x + 5x^2)}{h}$
$= -2 + 10x$

17. 5

18. Not possible (yet)

19. 4

21. -11

22. (a) and (c) only

23. $f(x) = x, g(x) = x + 1$; then $h'(x) = 2x + 1$ but $f'(x)g'(x) = 1$

24. $f(x) = 2x; f(x) = 2x + 5$

25. $f(x) = x - x^2; f(x) = x - x^2 - 1$

28. $f(x) = \frac{1}{18}x^{18}; f(x) = \frac{1}{18}x^{18} - 4$

29. $f(x) = \frac{1}{5}x^5 + \frac{1}{2}x^4 - \frac{1}{2}x^2 - 2x; f(x) = \frac{1}{5}x^5 + \frac{1}{2}x^4 - \frac{1}{2}x^2 - 2x - 3$

32. $f'(x) = 2(\frac{d}{dx}(1) + 3\frac{d}{dx}(x^2)) = 2(0 + 3(2x)) = 12x$

34. $f'(x) = -3(7x^6)$

35. $f'(x) = 2(3) - 4(5x^4)$

37. $f'(x) = 2x + 1$

39. $f'(x) = 81x^2 + 108x + 36$

40. $f'(x) = -6x^2$

42. $f'(x) = \frac{d}{dx}(x - 1) = 1$, for $x \neq -1$

43. $f'(x) = 2x - 3$

45. Can't do this one by rules (yet);
$f'(x) = \lim_{t \to x} \dfrac{\frac{1}{t} - \frac{1}{x}}{t - x} = -\frac{1}{x^2}$

46. Can't do this one by rules (yet);
$f'(x) = \lim_{h \to 0} \dfrac{\frac{x+h}{x+h+1} - \frac{x}{x+1}}{h} = \dfrac{1}{(x + 1)^2}$

47. $f'(x) = 24x^2 - 7$

50. $f''(x) = -24x - 4$

51. $f'''(x) = -24$ **53.** $f''(-1) = 20$

56. $3.4x - 0.4$ **58.** -240

60. $-80x^3 + 18x$ **62.** -12

65. (a) Continuous;
  (b) Not differentiable, but left and right differentiable

68. $f'(x) = \begin{cases} 3, & x < 2 \\ \text{DNE}, & x = 2 \\ 3x^2, & x > 2 \end{cases}$

71. $a = -\frac{1}{2}, b = \frac{3}{2}$

72. $a = 0, b = 0$

82. **PROOF** $\frac{d}{dx}(mx + b) = \frac{d}{dx}(mx) + \frac{d}{dx}(b) = m\frac{d}{dx}(x) + \frac{d}{dx}(b) = m(1) + 0 = m$ ∎

84. **PROOF** **(a):** $\dfrac{d}{dx}(ax^2 + bx + c) =$
$\lim_{h \to 0} \dfrac{(a(x + h)^2 + b(x + h) + c) - (ax^2 + bx + c)}{h} = 2ax + b$

  PROOF **(b):** $\dfrac{d}{dx}(ax^2 + bx + c) = a\dfrac{d}{dx}(x^2) + b\dfrac{d}{dx}(x) + c = a(2x) + b(1) + 0 = 2ax + b$ ∎

88. **PROOF** If f is differentiable at $x = c$, and $g(x) = kf(x)$, then by the constant multiple rule $g'(x) = kf'(x)$. Therefore $g'(c) = kf'(c)$, which exists because f is differentiable at $x = c$. Therefore $g(x) = kf(x)$ is differentiable at $x = c$. ∎

Section 3.6

1. Zero or does not exist

3. $x = -2, x = 0, x = 4, x = 5$

6. Your graph should have roots at $x = -2$ and $x = 2$ and horizontal tangent lines at three places between these roots.

11. If f is continuous on $[a, b]$ and differentiable on (a, b), then there exists some $c \in (a, b)$ such that the slope of the tangent line to the graph of f at c is equal to the slope of the line from $(a, f(a))$ to $(b, f(b))$.

12. True **13.** False **14.** False

15. True **16.** False **17.** False

19. One example is the graph of the function
$$f(x) = \begin{cases} -(x + 2)^2, & x \neq -2 \\ 3, & x = -2. \end{cases}$$

21. One example is an "upside-down V" with roots at $x = -2$ and $x = 2$ where the top point of the V occurs at $x = -1$.

25. *Hint:* Draw a graph that happens to have a slight "cusp" just at the place where its tangent line would have been equal to the average rate of change.

27. $f'(x)$ has at least two zeros in the interval $[-4, 2]$.

28. There is some $c \in (-2, 4)$ with $f'(c) = -\frac{1}{3}$.

30. When we applied the Extreme Value Theorem; when we stated that if $x = c$ is an extremum of f, then $f'(c) = 0$.

32. $c \approx -1.7$, $c \approx -0.5$, and $c \approx 1.2$; only the second and third values are used in the proof (since they are the global extreme values).

34. $f'(x) = 0$ at $x \approx 0.5$, $x \approx 2$, and $x \approx 3.5$. The first point is a local minimum, the last point is a local maximum, and the middle point is neither a maximum nor a minimum.

35. $f'(2)$ does not exist; $x = 2$ is not an extremum.

39. $x = -0.65$

40. $x = -1$, $x = \frac{2}{3}$

41. $x = -3$, $x = 0$, $x = 1$

44. From the graph, f appears to be continuous on $[0, 4]$ and differentiable on $(0, 4)$, and moreover $f(0) = f(4) = 0$, so Rolle's Theorem applies. Therefore, there is some $c \in (0, 4)$ such that $f'(c) = 0$. In this example there are three such values of c, namely $c \approx 0.5$, $c \approx 2$, and $c \approx 3.5$.

46. f is continuous and differentiable on $[-2, 2]$, and $f(-2) = f(2) = 0$, so Rolle's Theorem applies; $c \approx -1.27279$, $c \approx 0$, and $c \approx 1.27279$.

47. f is neither continuous nor differentiable on $[0, 4]$, so Rolle's Theorem does not apply.

50. Show that f satisfies the hypotheses of Rolle's Theorem. Then you know that there is some $c \in (-1, 4)$ such that $f'(c) = 0$. By solving $f'(x) = 0$, we can see that there is exactly one such value, $c = \frac{3}{2}$.

51. f is continuous and differentiable on $[0, 1.7]$ and $f(0) = f(1.7) = 0$; $c - 0.85$.

54. f appears to be continuous and differentiable on $[-3, 0]$; there are two values $x = c$ that satisfy the conclusion of the Mean Value Theorem (MVT), at $c \approx -2.8$ and $c \approx -0.9$.

56. f is not continuous or differentiable on $[-3, 2]$, so the Mean Value Theorem does not apply.

57. f is continuous and differentiable on $[-2, 3]$; $c \approx -0.5275$ and $c \approx 2.5275$.

60. f is continuous and differentiable on $[-3, 2]$; thus by the MVT there is some $c \in (-3, 2)$ such that $f'(c) = 1$ (the average rate of change of f on $[-3, 2]$). Solving $f'(x) = 1$, we see that there is exactly one such value, namely $x = -\frac{1}{2}$.

61. $C(h)$ is a differentiable function, and $C'(4) = 0.6 \neq 0$, so $h = 4$ cannot be a minimum of $C(h)$.

63. **(a)** Let $s(t)$ be Alina's distance from the grocery store (in miles) at time t. Then $s(0) = 20$ and $s(0.5) = 0$; since $s(t)$ is continuous and differentiable, the MVT applies and tells you that there is some time $c \in (0, 0.5)$ where $s'(c) = \frac{0-20}{0.5-0} = -40$ miles per hour (the negative sign means she is traveling toward the store).

67. **PROOF**   Suppose r_1, r_2, and r_3 are the roots of f. Since f is continuous and differentiable everywhere, Rolle's Theorem guarantees that f' will have at least one root on $[r_1, r_2]$ and at least one root on $[r_2, r_3]$; therefore, f' has at least two roots. ∎

1. True   **2.** True

3. True   **4.** False

5. False   **6.** False

7. False   **8.** False

11. Possible graph of f:

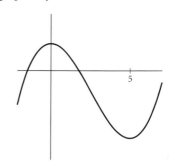

14. Your graph of f should be increasing on $(-\infty, 1)$, $(3, 8)$, and $(8, \infty)$ and decreasing on $(1, 3)$. It should also have horizontal tangent lines at $x = 1$ (a local maximum) and $x = 8$ (not an extremum), and be nondifferentiable at $x = 3$ (a local minimum). Your graph of f' should be positive on $(-\infty, 1)$, $(3, 8)$, and $(8, \infty)$ and negative on $(1, 3)$. It should also have roots at $x = 1$ and $x = 8$ and be undefined at $x = 3$.

17. Yes; Yes; No

18. Yes; Yes; Yes

20.

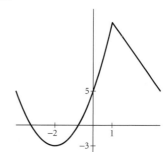

22. Possible graph of f':

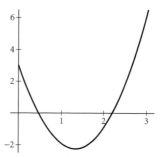

24. Your graph of f' should have roots at $x = 0$ and $x = 2$ and should be positive on $(0, \infty)$ and negative on $(-\infty, 0)$. In particular, this means that your graph of f' should "bounce" off the x-axis at $x = 2$.

26.

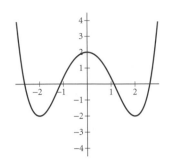

27. Your graph of f should have horizontal tangent lines at $x = 1$ at $x = 3$ and should be decreasing on $(-\infty, 1)$ and $(1, 3)$ and increasing on $(3, \infty)$. The point $x = 1$ is an inflection point (but not a local extremum) of f, while the point $x = 3$ is a local minimum of f.

28. $x = \pm\sqrt{15}$; f is increasing on $(-\infty, -\sqrt{15}) \cup (\sqrt{15}, \infty)$ and decreasing on $(-\sqrt{15}, \sqrt{15})$; $x = -\sqrt{15}$ is a local maximum, and $x = \sqrt{15}$ is a local minimum.

30. $f'(x) = 6x(x - 3) > 0$ when $(x > 0$ and $x - 3 > 0)$ or $(x < 0$ and $x - 3 < 0)$, so f is increasing on $(-\infty, 0) \cup (3, \infty)$.

33. Increasing on $(-\infty, 0)$ and $(3, \infty)$; decreasing on $(0, 3)$

35. Increasing on $(-\infty, -\frac{1}{3})$ and $(1, \infty)$; decreasing on $(-\frac{1}{3}, 1)$

36. Increasing on $(-\infty, \frac{-2-\sqrt{13}}{3})$ and $(\frac{-2+\sqrt{13}}{3}, \infty)$; decreasing on $(\frac{-2-\sqrt{13}}{3}, \frac{-2+\sqrt{13}}{3})$

38. $x = \frac{5}{2}$ is a local minimum.

40. $x = -5$ is a local maximum.

42. f has a critical point at $x = 2$ but does not have a local extremum there; thus f has no local extrema.

45. (a) $(0, 2.5)$; (b) $t = 2.5$; (c) $(0, 1)$ and $(3.3, 4)$;
(d) $(3.5, 3.8)$ is such an interval; Bubbles is moving toward the left side of the tunnel and is slowing down while doing so.

52. Given that $f(x) = mx + b$ where $m \neq 0$ (so $f(x)$ is nonconstant), show that $f'(x)$ is either always positive or always negative. Since $f'(x) = m$ is a (nonzero) constant, this is clearly the case.

SECTION 3.8

3. "If f is concave up on I, then f'' is positive on I"; this is *false* (f'' might be zero at some point).

5. Linear   **6.** False

7. True   **8.** True

9. False   **10.** True

11. True   **12.** False

13. True

14. Your graph should be positive, decreasing, and concave up everywhere.

16. Not possible

17. Your graph should be increasing everywhere, concave up on $(-\infty, -2)$, and concave down on $(-2, \infty)$ (with an inflection point at $x = -2$).

18. The graph of $f(x) = (x - 3)^3$ has these properties.

21. f' has a local maximum in the second and third graphs; f' has a local minimum in the first and fourth graphs.

23. 5, 4

25. (a) $f''(x) = 12x^2$, so $f''(0) = 0$.
(b) $f''(-1) > 0$ and $f''(1) > 0$, so $f''(x)$ does not change sign at $x = 0$.

26. Concave up on $(-\infty, \infty)$; never concave down

28. Concave up on $(1, \infty)$; concave down on $(-\infty, 1)$

30. Concave up on $(-\infty, 0) \cup (1, \infty)$; concave down on $(0, 1)$

32. No inflection points

34. $x = \frac{2}{3}$

35. No inflection points

37. $x = \frac{6+\sqrt{12}}{6}$, $x = \frac{6-\sqrt{12}}{6}$

38. Local minimum at $x = -\frac{3}{2}$ ($f'(-\frac{3}{2}) = 0$ and $f''(-\frac{3}{2}) = 2$)

42. Local minimum at $x = 0$

43. Local minimum at $x = -\frac{3}{2}$ (note that the first derivative test is needed to test the critical point $x = 0$).

44. Local maximum at $x = -2$, local minimum at $x = 1$; inflection point at $x = -\frac{1}{2}$

46. Local minimum at $x = 0$, no local maxima; no inflection points

49. The graph of $f(x) = 2 - (x - 2)^2$ has those number lines.

52. The graph of $f(x) = \frac{x^2(x-2)^2}{x-1} + 2$ has those number lines.

54. Among other things, your graph should have a local minimum at $x = -2$ and inflection points at $x = -1$ and $x = 0$.

59. No local extrema, inflection point at $x = -2$:

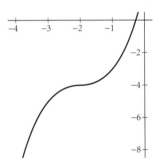

61. Local maximum at $x = 0$, local minima at $x = \pm\frac{1}{\sqrt{3}}$, inflection points at $x = \pm\frac{1}{3}$:

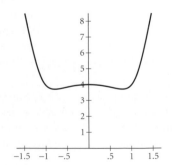

63. Graph of $f'(x)$:

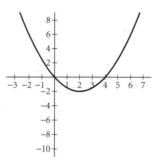

Graph of $f''(x)$:

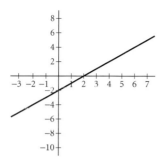

65. Your graph of $f'(x)$ should have roots at $x = -3$ and at $x = 0$ and should be positive on $(-\infty, -3)$ and negative on $(-3, 0)$ and $(0, \infty)$. Your graph of $f''(x)$ should have roots at $x \approx -2$ and $x = 0$ and should be negative on $(-\infty, -2) \cup (0, \infty)$ and positive on $(-2, 0)$.

66. Possible graph of $f(x)$:

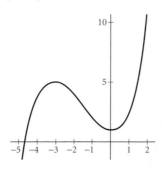

Graph of $f''(x)$:

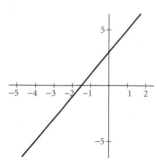

68. Your graph of $f(x)$ should have a horizontal tangent line (in fact, a local minimum) at $x = 1$ and should be decreasing on $(-\infty, 1)$ and increasing on $(1, \infty)$. Your graph of f should have a slope approaching -1 as $x \to -\infty$.

Your graph of $f''(x)$ should have no roots, should be always positive, and should have a horizontal asymptote at $y = 0$ as $x \to -\infty$ (since the slope of $f'(x)$ approaches zero as $x \to -\infty$).

69. Your graph of $f'(x)$ should be decreasing on $(-\infty, 2)$ and increasing on $(2, \infty)$, with a local minimum at $x = 2$. Your graph of $f(x)$ should be concave down on $(-\infty, 2)$ and concave up on $(2, \infty)$, with an inflection point at $x = 2$.

72. (a) $[0, 15]$; she is north of the corner, walking north, and slowing down.

(b) $[60, 80]$; she is south of the corner, walking south, and speeding up.

(c) $[15, 40]$; she is walking south and speeding up (first north and then south of the corner).

(d) At $t = 40$; she is south of the corner, walking south; she just finished speeding up (in the southward direction) and is now just starting to slow down (but still walking in the southward direction).

76. **PROOF** If $f(x) = ax^2 + bx + c$, then $f''(x) = 2a$. If $a > 0$ then $f(x)$ is always concave up, and if $a < 0$ then $f(x)$ is always concave down. ∎

CHAPTER 4

SECTION 4.1

3. Not a power function

4. $3x^0$ **5.** $\frac{3}{2}x^{-5}$ **7.** $x^{-\frac{3}{2}}$

12. (a) It is not automatically clear what would it mean to multiply a number by itself "zero times," so we choose to define $x^0 = 1$ if $x \neq 0$.

(b) If x^0 were zero, then $\frac{x^2}{x^2} = x^{2-2} = x^0$ would equal zero (instead of what we want it to equal: 1).

(c) If x^0 were zero, then we'd have $x^2 = x^{2+0} = x^2 x^0 = x^2(0) = 0$, and clearly it isn't true that $x^2 = 0$ for all x!

(d) Because the rules generalize what happens when we work with positive integer exponents.

13. (a) $\lim_{x \to 0} x^0 = \lim_{x \to 0} 1 = 1$ **(b)** $\lim_{k \to 0} 0^k = \lim_{k \to 0} 0 = 0$

19. $\frac{27}{8}$ **21.** Undefined   **22.** $\frac{4}{9}$ **24.** 1

26. Undefined   **27.** Undefined

28. 0   **32.** $x \neq -1$ **34.** $[-1, \infty)$ **36.** $(-\infty, \infty)$

39. $(-\infty, -1) \cup (1, \infty)$ **41.** $x \neq -1$

42. $(-\infty, -\sqrt{5}) \cup (\sqrt{5}, \infty)$ **44.** $f(x) = x^{-\frac{3}{10}}$

48. $f(x) = x^{-2}$ **49.** Not a power function

51. $f(x) = \frac{1}{16}x^2$ (for $x \neq 0$)

54. One counterexample is $x = -1$.

56. One counterexample is $x = -2$.

58. One counterexample is $x = 1$, $y = 1$.

63. $[-1, 0] \cup [1, \infty)$

65. $[2, \infty)$ **67.** $[0, \infty)$ **69.** $2x + 6 + h$

71. $\frac{1}{(x+1)(x+h+1)}$ **72.** $\frac{-1}{\sqrt{x}\sqrt{x+h}(\sqrt{x}+\sqrt{x+h})}$ **75.** $\frac{1}{\sqrt{t}+\sqrt{x}}$

78. $a^2 = b^5$; $(a, b) = (0, 0)$ is an extraneous solution.

79. $a = b^2$; $(a, b) = (4, -2)$ is an extraneous solution.

81. $a^3 = b$; no extraneous solutions

83. $1 = a^9 b$; no extraneous solutions

85. The graph of $f(x)$ is the graph of $y = x^3$ shifted right 2 units and up 1 unit.

89. The graph of $f(x)$ is the graph of $y = x^2$ shrunk vertically by a factor of 3 and shifted down 5 units.

90. The graph of $f(x)$ is the graph of $y = x^{\frac{2}{3}}$ shifted left 3 units and down 2 units.

92. The graph of $f(x)$ is the graph of $y = \sqrt{x}$ reflected over the y-axis.

94. The difference of two power functions is not necessarily a power function. One counterexample is $f(x) = x^2$, $g(x) = x^3$.

95. The quotient of two power functions is a power function (at least for $x \neq 0$).

> **PROOF**   If $f(x) = Ax^k$ and $g(x) = Bx^l$, then $\frac{f(x)}{g(x)} = \frac{Ax^k}{Bx^l} = \frac{A}{B}x^{k-l}$ (when $x \neq 0$). ■

103. **PROOF** $(\frac{x}{y})^a = (x \cdot \frac{1}{y})^a = x^a(\frac{1}{y})^a = x^a(y^{-1})^a = x^a y^{-a} = x^a \frac{1}{y^a} = \frac{x^a}{y^a}$ ■

SECTION 4.2

2. $f(x) = x^{\frac{1}{2}}$ is continuous on the interval $[0, \infty)$; in other words, it is continuous on $(0, \infty)$ and right but not left continuous at $x = 0$.

3. (a) $\lim_{x \to 0} 3x^{\frac{1}{3}} = 0$

(b) For all $\varepsilon > 0$, there exists $\delta > 0$ such that if $0 < |x| < \delta$, then $|3x^{\frac{1}{3}}| < \varepsilon$.

5. (a) $\lim_{x \to 2} x^{\frac{1}{4}} = 2^{\frac{1}{4}}$

(b) For all $\varepsilon > 0$, there exists $\delta > 0$ such that if $0 < |x - 2| < \delta$, then $|x^{\frac{1}{4}} - 2^{\frac{1}{4}}| < \varepsilon$.

8. (a) $\lim_{x \to -1} \sqrt[6]{x} \neq \sqrt[6]{-1}$

(b) There exists $\varepsilon > 0$ for which there is *no* $\delta > 0$ such that $0 < |x + 1| < \delta \Rightarrow |\sqrt[6]{x} - \sqrt[6]{-1}| < \varepsilon$.

9. $(-\infty, \infty)$ **11.** $(0, \infty)$ **12.** $x \neq 0$

15. Any x greater than $\sqrt[3]{10,000}$ will work.

18. Any x between -0.01 and 0 will work.

20. $\delta \approx 1.70834$ **21.** $\delta \approx 0.0388386$ **24.** $2\sqrt{3}$

25. Can't use continuity; $\lim_{x \to 0^+} x^{-\frac{1}{6}} = \infty$, $\lim_{x \to 0^-} x^{-\frac{1}{6}}$ does not exist.

26. Can use continuity only from the right; $\lim_{x \to 0^+} x^{\frac{3}{8}} = 0$, $\lim_{x \to 0^-} x^{\frac{3}{8}}$ does not exist.

29. ∞ **31.** $\lim_{x \to 0^+} -4x^{-3} = -\infty$, $\lim_{x \to 0^-} -4x^{-3} = \infty$

33. $\lim_{x \to 0^+} 2x^{\frac{3}{4}} = 0$, $\lim_{x \to 0^-} 2x^{\frac{3}{4}}$ does not exist.

35. 0   **36.** $-\infty$ **39.** 0   **41.** $-\frac{1}{6}$

42. $\lim_{x \to 1^+} \frac{x^2}{x-1} = \infty$, $\lim_{x \to 0^-} \frac{x^2}{x-1} = -\infty$

45. -184 **47.** $\sqrt{6} - 40$ **51.** ∞ **52.** $-\infty$ **54.** 0

57. Continuous everywhere except at $x = 1$

59. $\frac{1}{2\sqrt{3}}$ **60.** $-\frac{1}{4}$ **62.** $-\frac{1}{4}$

65. Give a delta–epsilon proof showing that for all $\varepsilon > 0$, there exists $\delta > 0$ such that if $0 < x < \delta$, then $|\sqrt{x} - \sqrt{0}| < \varepsilon$. You will end up choosing $\delta = \varepsilon^2$.

SECTION 4.3

3.

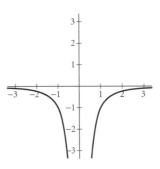

4.

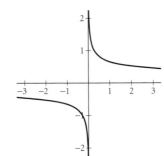

5. The power rule does not apply.

7. The power rule does not apply.

10. $-\frac{1}{2}x^{-2} - 2x^{-5}$ **13.** Yes; Yes

15. Any function of the form $f(x) = \frac{8}{3}x^{\frac{3}{2}} + C$, where C is a constant, will work.

16. $\lim_{h \to 0} \frac{3(2+h)^{-1} - 3(2)^{-1}}{h} = -\frac{3}{4}$

17. $\lim_{h \to 0} \frac{3(h)^{-1} - 3(3)^{-1}}{h} = -\frac{1}{3}$

21. $\lim_{h \to 0} \frac{(0+h)^{\frac{2}{3}} - 0^{\frac{2}{3}}}{h}$ approaches ∞ as $x \to 0^+$ and $-\infty$ as $x \to 0^-$ (so $f'(0)$ does not exist).

22. $\lim_{t \to x} \frac{\frac{2}{t} - \frac{2}{x}}{t - x} = -\frac{2}{x^2}$

24. $\lim_{h \to 0} \frac{3\sqrt{x+h} - 3\sqrt{x}}{h} = \frac{3}{2\sqrt{x}}$

26. $\lim\limits_{h\to 0} \dfrac{\frac{1}{\sqrt{x+h}} - \frac{1}{\sqrt{x}}}{h} = \dfrac{-1}{2x^{\frac{3}{2}}}$

29. $3^{-\frac{1}{3}}(-\frac{5}{3})x^{-\frac{8}{3}}$ **30.** 1

32. $\frac{3}{2}x^{\frac{1}{2}} + 18x^5$ **34.** $2\sqrt{5}x$

37. Can't do (yet).   **38.** Can't do (yet).

39. $\frac{18}{5}x^{-\frac{3}{5}} - \frac{24}{5}x^{-\frac{9}{5}} - 2x^{-3}$

41. $27(\frac{9}{2})x^{\frac{7}{2}} + 27(\frac{7}{2})x^{\frac{5}{2}} + 9(\frac{5}{2})x^{\frac{3}{2}} + \frac{3}{2}x^{\frac{1}{2}}$

43. $\frac{64}{15}x^{\frac{5}{3}}$ **44.** Can't do (yet).

48. $-2.6x^{-3.6} + 3x^{-4}$

49. Not continuous (so not differentiable) at $x = -1$; left continuous, left differentiable

50. Continuous but not differentiable at $x = 1$; left and right differentiable

55. $f'(x) = \begin{cases} 3x^2, & x < 1 \\ \text{DNE}, & x = 1 \\ 1, & x > 1 \end{cases}$

56. $f'(x) = \begin{cases} -2x, & x \le 0 \\ 2x, & x > 0 \end{cases}$

58. $a = 3, b = 0$

60. $a = -640, b = -72$

62. $f(x) = \frac{4}{x^2}, c = 2; f'(2) = -1$

63. $f(x) = x^{-\frac{1}{3}}, c = 8; f'(8) = -\frac{1}{48}$

64. $f(x) = x^{-\frac{1}{3}}, c = 8; f'(8) = -\frac{1}{48}$

68. $f(x) = \frac{10}{3}x^{\frac{3}{2}} + \frac{5}{3}$

70. $f(x) = -2x^{-\frac{1}{2}} + \frac{5}{3}$

71. $f(x) = \frac{15}{2}x^{\frac{1}{5}} - \frac{11}{2}$

74. **(a)** $s(t) = 6t^{\frac{1}{2}} + 100$;

(b) $a(t) = -\frac{3}{2}t^{-\frac{3}{2}}$

SECTION 4.4

3. Domain is $(-\infty, 0) \cup (0, \infty)$, range is $(-\infty, 0)$. f is negative on $(-\infty, 0) \cup (0, \infty)$ and does not exist at $x = 0$. f is decreasing on $(-\infty, 0)$ and increasing on $(0, \infty)$. f is concave down on $(-\infty, 0) \cup (0, \infty)$. f has a vertical asymptote at $x = 0$ and a horizontal asymptote at $y = 0$. f has no roots, no local or global extrema, and no inflection points.

4. Every power function $f(x) = Ax^k$ has $f(0) = 0$ or $f(0)$ undefined.

5.

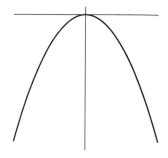

7.

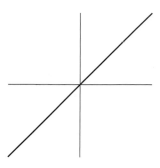

10.

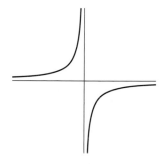

11. $(0, \infty)$ if k is even, $(-\infty, 0) \cup (0, \infty)$ if k is odd

13. $y = x^4$ is the graph with the "flattest" bottom part; $y = 2x^2$ is the graph that is closest to the y-axis; $y = x^2$ is the remaining graph.

16. **(a)** $\sqrt[k]{A}$;   **(b)** $\frac{1}{\sqrt[k]{A}}$

18.

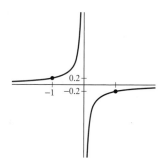

21.

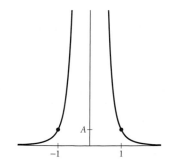

23.

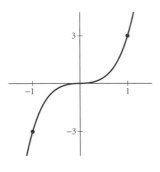

25.

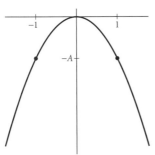

29. Neither; $f(-x) \neq f(x)$ and $f(-x) \neq -f(x)$

30. Neither; $f(-x) \neq f(x)$ and $f(-x) \neq -f(x)$

32. Odd; $f(-x) = -f(x)$

33. $f(x) = 3x^3$; $f(x) = 3x^5$

36. $f(x) = \frac{-2}{x}$; $f(x) = \frac{-2}{x^3}$

37. $f(x) = 2x^2$; $f(x) = \frac{1}{2}x^4$

39. $f(x) = 2x^4$ **42.** $-\frac{1}{2}x^{-3}$ **43.** $\frac{1}{5}x^2$

45. The function $g(x) = \frac{1}{f(x)}$ has values $g(-2) = -\frac{1}{4}, g(-1) = -2, g(0)$ DNE, $g(1) = 2,$ and $g(2) = \frac{1}{4}$. The graph of g has a horizontal asymptote at $y = 0$ and a vertical asymptote at $x = 0$.

46. The graph passes through the points $(-2, -\frac{1}{12}), (-1, -\frac{1}{3}), (1, -\frac{1}{3}),$ and $(2, -\frac{1}{12})$ and has a vertical asymptote at $x = 0$:

48. The graph of $f(x)$ is the graph of $y = x^2$ shifted 2 units to the right and 3 units up.

51. The graph of $f(x)$ is the graph of $y = \frac{1}{x}$ shifted 2 units to the right.

52. The graph of $f(x)$ is the graph of $y = x^{-4}$ stretched vertically by a factor of 3 and shifted to the right 2 units.

53. The graph of $f(x)$ is the graph of $y = x^2$ stretched vertically by a factor of 35 and reflected over the x-axis and *then* shifted 100 units up and 13 units to the right.

55. The graph of $f(x)$ is the graph of $y = x^2$ shifted up 2 units and left 1 unit and *then* shrunk horizontally by a factor of 3.

57. Domain is $(-\infty, \infty)$, range is $[0, \infty)$. $f(x) = 0$ at $x = 0$, $f(x) > 0$ on $(-\infty, 0) \cup (0, \infty)$. $f'(x) = 0$ at $x = 0$, $f'(x) > 0$ on $(0, \infty)$, $f'(x) < 0$ on $(-\infty, 0)$. $f''(x)$ is never zero and is always positive. $\lim_{x \to \infty} f(x) = \infty$, $\lim_{x \to -\infty} f(x) = \infty$.

59. Domain is $(-\infty, \infty)$, range is $(-\infty, \infty)$. $f(x) = 0$ at $x = 0$, $f(x) < 0$ on $(-\infty, 0)$, $f(x) > 0$ on $(0, \infty)$. $f'(x) = 0$ at $x = 0$, $f'(x) > 0$ on $(-\infty, 0) \cup (0, \infty)$. $f''(x) = 0$ at $x = 0$, $f''(x) < 0$ on $(-\infty, 0)$, $f''(x) > 0$ on $(0, \infty)$. $\lim_{x \to \infty} f(x) = \infty$, $\lim_{x \to -\infty} f(x) = -\infty$.

61. Domain is $(-\infty, 0) \cup (0, \infty)$, range is $(-\infty, 0) \cup (0, \infty)$. $f(x)$ does not exist at $x = 0$, $f(x) > 0$ on $(0, \infty)$, $f(x) < 0$ on $(-\infty, 0)$. $f'(x)$ does not exist at $x = 0$, $f'(x) < 0$ on $(-\infty, 0) \cup (0, \infty)$. $f''(x)$ does not exist at $x = 0$, $f''(x) > 0$ on $(0, \infty)$, $f''(x) < 0$ on $(-\infty, 0)$. $\lim_{x \to \infty} f(x) = 0$, $\lim_{x \to -\infty} f(x) = 0$, $\lim_{x \to 0^+} f(x) = \infty$, $\lim_{x \to 0^-} f(x) = -\infty$.

67. **PROOF** If k is an odd integer and $f(x) = x^k$, then $f(-x) = (-x)^k = (-1 \cdot x)^k = (-1)^k x^k = (-1)x^k = -x^k = -f(x)$, so f is an odd function. (The fourth equality uses the fact that k is odd, since in that case $(-1)^k = -1$.) ∎

68. **PROOF** $x^6 \leq x^4 \Longleftrightarrow x^6 - x^4 \leq 0 \Longleftrightarrow x^4(x^2 - 1) \leq 0 \Longleftrightarrow x^2 - 1 \leq 0 \Longleftrightarrow |x| < 1$. (The second to last "if and only if" follows from the fact that x^4 is always positive.) The proof of the second part is similar. ∎

SECTION 4.5

3. Domain is $(-\infty, \infty)$, range is $(-\infty, \infty)$. f is positive, decreasing, and concave down on $(-\infty, 0)$ and is negative, decreasing, and concave up on $(0, \infty)$. f has a vertical tangent line at $x = 0$. $\lim_{x \to \infty} f(x) = -\infty$ and $\lim_{x \to -\infty} f(x) = \infty$.

5.

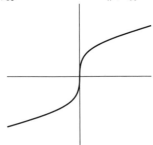

8.

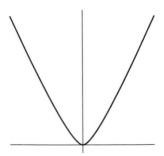

9.

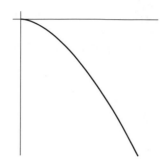

12. Every integer k can be written as a rational number $\frac{k}{1}$.

14. $y = x^{\frac{2}{3}}$ is the graph that is closest to the x-axis; $y = 2x^{\frac{2}{5}}$ is the graph that bends in closest to the y-axis; $y = 2x^{\frac{2}{3}}$ is the remaining graph.

15. If $k \leq 1$ or if $k = \frac{p}{q}$ and q is even

19.

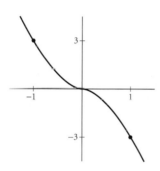

20.

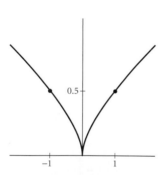

21.

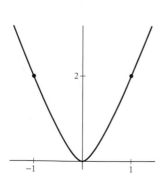

25.

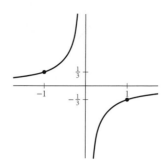

27. Neither; $f(x)$ is not defined for $x < 0$.

29. Even; $f(-x) = f(x)$

31. Odd; $f(-x) = -f(x)$

34. $\frac{1}{f(x)} = \frac{1}{3}x^{-\frac{2}{3}}$ has domain $(-\infty, 0) \cup (0, \infty)$ and range $(0, \infty)$.

35. $\frac{1}{f(x)} = 2x^{-1.75}$ has domain $(0, \infty)$ and range $(0, \infty)$.

36. Not one-to-one; is one-to-one on restricted domain $[0, \infty)$, with inverse $f^{-1}(x) = x^{\frac{1}{4}}$.

38. One-to-one; inverse is $f^{-1}(x) = x^{-7}$ (restricted to $(0, \infty)$).

40. Not one-to-one; is one-to-one on restricted domain $[0, \infty)$, with inverse $f^{-1}(x) = 3^{-\frac{5}{2}}x^{\frac{5}{2}}$.

42. $f(x) = 3x^{\frac{2}{3}}$; $f(x) = 3x^{\frac{4}{7}}$

44. $f(x) = -2x^{\frac{3}{2}}$, $f(x) = -2x^{\frac{5}{4}}$

47. $f(x) = -\frac{1}{4}x^{\frac{2}{3}}$; $f(x) = (-8^{-\frac{4}{5}})x^{\frac{4}{5}}$

48. $f(x) = 2x^{\frac{1}{4}}$

50. $f(x) = 3x^{\frac{1}{3}}$

52. The graph of $f(x)$ is the graph of $y = x^{\frac{1}{4}}$ reflected over the x-axis and *then* shifted 3 units up and 1 unit to the left.

54. The graph of $f(x)$ is the graph of $y = x^{-\frac{3}{2}}$ (i.e., the reciprocal of the graph of $y = x^{\frac{3}{2}}$) shifted 1 unit to the right.

56. The graph of $f(x)$ is the graph of $y = x^{-\frac{2}{5}}$ stretched vertically by a factor of 2 and *then* shifted up 3 units.

57. **(a)** It looks like there is a horizontal tangent line at $x = 0$.

   (b) $f'(0) = \lim\limits_{h \to 0} \dfrac{(0+h)^{\frac{6}{5}} - 0^{\frac{6}{5}}}{h} = 0$

59. **(a)** The tangent line seems to be vertical.

   (b) $\lim\limits_{h \to 0} \dfrac{(0+h)^{\frac{1}{3}} - 0^{\frac{1}{3}}}{h} = \infty$

60. **(a)** The graph of $f^{-1}(x)$ has values $f^{-1}(-6) = -8$, $f^{-1}(-3) = -1$, $f^{-1}(0) = 0$, $f^{-1}(3) = 1$, and $f^{-1}(6) = 8$. Its graph has a horizontal tangent line at $x = 0$ and can be obtained by reflecting the graph of $f(x)$ over the line $y = x$.

   (b) The graph of $g(x) = \frac{1}{f(x)}$ has values $g(-8) = -\frac{1}{6}$, $g(-1) = -\frac{1}{3}$, $g(1) = \frac{1}{3}$, and $g(8) = \frac{1}{6}$. This reciprocal function g is not defined at $x = 0$ and, in fact, has a vertical asymptote there (as well as a horizontal asymptote at $x = 0$).

61. (a)

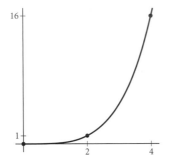

(b)

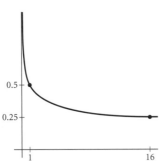

67. Domain is $(-\infty, \infty)$, range is $[0, \infty)$. $f(x) = 0$ at $x = 0$, $f(x) > 0$ on $(-\infty, 0) \cup (0, \infty)$. $f'(x)$ does not exist at $x = 0$, $f'(x) < 0$ on $(-\infty, 0)$, $f'(x) > 0$ on $(0, \infty)$. $f''(x)$ does not exist at $x = 0$, $f''(x) < 0$ on $(-\infty, 0) \cup (0, \infty)$. $\lim_{x \to -\infty} f(x) = \infty$, $\lim_{x \to \infty} f(x) = \infty$ (so no horizontal asymptotes). $\lim_{x \to 0^-} f'(x) = -\infty$, $\lim_{x \to 0^+} f'(x) = \infty$ (so f has a vertical cusp at $x = 0$).

71. **PROOF**   If p is even and q is odd, then $f(-x) = (\sqrt[q]{-x})^p = (-\sqrt[q]{x})^p = (\sqrt[q]{x})^p = f(x)$. If p and q are both odd, then $f(-x) = (\sqrt[q]{-x})^p = (\sqrt[q]{-x})^p = -(\sqrt[q]{x})^p = -f(-x)$. If q is even, then f is not defined for $x < 0$. ∎

CHAPTER 5

SECTION 5.1

2. f is a quartic polynomial with leading coefficient -1, leading term $-x^4$, degree 4, and constant term 1, where $a_1 = 0$ and $a_3 = 0$.

3. Not a polynomial

5. f is a polynomial with leading coefficient π, leading term πx^{17}, degree 17, and constant term 0, where $a_1 = 3$ and $a_3 = 0$.

6. f is a polynomial with leading coefficient 4, leading term 4, degree 0, and constant term 4, where $a_1 = 0$ and $a_3 = 0$.

7. Not a polynomial

11. (a) $f(x) = x(2x^2 - 3x + 8)$. The quadratic term is irreducible because $b^2 - 4ac = (-3)^2 - 4(2)(8) < 0$.
   (b) $f(x) = 2x(x^2 - \frac{3}{2}x + 4)$

13. If $f(x) = a_n x^n + a_{n-1} x^{n-1} + \cdots + a_1 x + a_0$, then the leading term of $f(x)$ is $a_n x^n$.

18. A polynomial of degree 4

20. A quadratic $ax^2 + bx + c$ that has no real roots (so $b^2 - 4ac < 0$)

23. $f(-1) = 0$, $f(1) = 0$, and $f(\frac{1}{2}) = 0$, but $f(2) = 9 \neq 0$ and $f(-3) = 224 \neq 0$.

24. $\pm 1, \pm 2, \pm 3, \pm 4, \pm 6,$ and ± 12

27. $f(x) = -2x^7 + 3x^5 + 2x - 1$

28. $f(x) = x(x-2)(x+5)$; $f(x) = 5x(x-2)(x+5)$; $f(x) = x(x-2)^2(x+5)^3$

31. $f(x) = x^4 + 1$

33. $f(x) = (x-1)^2(x+3)$

35. $f(x) = (x-2)(x+1)(x-\frac{1}{2})(x+\frac{1}{3})$

38. $f(x) = (x-\pi)(x+\sqrt{2})$

39. $f(x) = 2(x+1)(x-2)^2$

42. $(x-c)g(x)$

45. $(x^2+1)(x^2+x+4) = x^4 + x^3 + 5x^2 + x + 4$. Can't factor this because there are no real roots (and can't do grouping or use any factoring formulas); however, it is *not* irreducible (since it is of degree 4).

47. (a) One example is $f(x) + 2$;   **(b)** $f(x) + 1.5$;   **(c)** $f(x)$;   **(d)** $f(x) - 0.4$ or $f(x) - 1.8$;   **(e)** One example is $f(x) - 1$.

48. False   **49.** False   **50.** True   **51.** False

52. True   **53.** False   **54.** False

55. Possible integer roots: $\pm 1, \pm 3$; none of these is actually a root.

58. Possible integer roots: $\pm 1, \pm 2, \pm 3, \pm 4, \pm 6, \pm 8, \pm 12, \pm 24$; only $x = -2$ is actually a root.

59. Possible integer roots: ± 1 and ± 2; none of these is actually a root.

61. $(x+1)(x-1)(x-3)$

63. $(x-1)(x+1)(x+2)(x+5)$

65. $(x+1)(x-3)^2$ **67.** $(1+x^2)(x-2)$

68. $-(4x-1)(2x-3)$ **71.** $(x+4)(x-4)(1+2x)$

73. $2(x-0)(x-1)(x+3)$ **75.** $2(x-0)(x^2-x+\frac{1}{2})$

78. $3(x-\frac{2}{3})(x^2+4)$ **79.** $(x-3)(x-1)(x+2)$

81. $(x-1)^2(x+6)$ **84.** $2(x-1)(x+1)(4+x^2)$

86. $2x(x-2)^2(x+1)^3$

87. (a) $(2x^2 + 8x + 23) + \frac{64}{x-3}$
   (b) $(2x^2 + 8x + 23)(x-3) + 64$

89. (a) $(-x^3 + 3x^2 - 9x + 27) - \frac{80}{x+3}$
   (b) $(-x^3 + 3x^2 - 9x + 27)(x+3) - 80$

90. (a) $(5x^3 - 5x^2 + 5x - 8) + \frac{10}{x+1}$
   (b) $(5x^3 - 5x^2 + 5x - 8)(x+1) + 10$

93. Possible rational roots: $\pm 4, \pm 2, \pm 1, \pm \frac{1}{2}$; $(x^2 + 4)(2x + 1)$

95. Possible rational roots: $\pm 1, \pm \frac{1}{2}, \pm \frac{1}{4}, \pm \frac{1}{8},$ and $\pm \frac{1}{16}$; $(2x-1)^2(4x+1)$

97. **PROOF**   If $f(x)$ is a polynomial with $a_0 = 0$, then $f(x) = a_n x^n + a_{n-1} x^{n-1} + \cdots + a_2 x^2 + a_1 x = x(a_n x^{n-1} + a_{n-1} x^{n-2} + \cdots + a_2 x + a_1)$. Since $x = (x - 0)$ is a factor of $f(x)$, $x = 0$ is a root. ∎

SECTION 5.2

2. Polynomial functions never have asymptotes.

3. Polynomial functions are always differentiable everywhere.

6. The degree of f is odd. Since there are three roots and two turning points, the degree is at least 3 (in fact, it is at least 5 because of the inflection point at $x = 1$). The leading coefficient is negative.

7. The degree of f is even and at least 6. The leading coefficient is positive.

8.

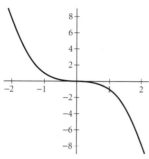

10.

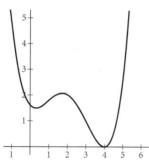

13. Such a polynomial cannot exist.

16. $\lim\limits_{x \to \infty} f(x) = -\infty$; $\lim\limits_{x \to \infty} g(x) = \infty$; $\lim\limits_{x \to \infty} h(x) = 0$

18. An odd-degree polynomial must have an even number of turning points.

19. True   **20.** False   **21.** True   **22.** True   **23.** False

24. True   **25.** False   **26.** False   **27.** 6

31. $-\infty$ **32.** $-\infty$ **33.** ∞ **34.** 0   **38.** 1

41. $60x^3 - 24x^2 + 6$ **42.** $-24x^7 - 7x^6 + 12$

43. $64x^3 - 96x^2 + 48x - 8$ **45.** $360x$ **46.** 0

48. $f(x) = \frac{1}{2}x^6 - \frac{2}{3}x^3 + 4x + 1$

50. $f(x) = -\frac{4}{7}x^7 + x + \frac{18}{7}$

53. $f(x) = -\frac{3}{10}x^{10} + 4x^6 + \frac{1}{5}x^5 - 8x + \frac{71}{10}$

55. Local minimum at $x = -2$; inflection points at $x = \pm 1$

57. Local minimum at $x = 3$; no inflection points

58. Local minimum at $x = -2$; no inflection points

62. Local minima at $x = -1$ and $x = 3$, local maximum at $x = 1$; inflection points at $x = \frac{3 \pm 2\sqrt{3}}{3}$

64. $s(t) = 10t^4 - \frac{100}{3}t^3 + 25t^2 + 45t + 550$
$a(t) = 120t^2 - 200t + 50$

66. **(a)** $a(t) = 0.024t$, $s(t) = 0.004t^3 + 400t$ (assuming initial position is $s_0 = 0$ at Venus). $a_0 = 0$ thousands of miles per hour per hour, $v_0 = 400$ miles per hour, $s_0 = 0$ miles.

(b) Yes, since the velocity is always positive.

(c) No, the acceleration is increasing (since $a(t) = 0.024t$ grows larger as t increases).

(d) Approximately 140 hours, at which time it will be going about 635.2 thousand miles per hour.

68. If $a(t) = -32$ then (by antidifferentiating) $v(t) = -32t + C$, and $v(0) = v_0$ implies that $C = v_0$. Now (by antidifferentiating) $s(t) = -16t^2 + v_0 t + K$, and $s(0) = s_0$ implies that $K = s_0$.

70. **PROOF**   Every polynomial function is a sum of power functions with positive integer powers. Every power function with a positive integer power is continuous, and the sum of continuous functions is always a continuous function. ∎

SECTION 5.3

1. $(x - c)^2 g(x)$

3. $x = -3$ is a double root, $x = -1$ is a (single) root, and $x = 2$ is a triple root.

5. $x = -2$ is a single root, and $x = 1$ appears to be a root of multiplicity 4.

6. $f(x) = x^2(x - 1)^2(x + 8)$

8. $f(x) = (x^2 + 1)(x - 2)(x - 3)(x - 4)(x - 5)^2$

9. $f(x) = (x - 2)(x - 3)(x^2 + 1)^2$

12. $f(x) = (x + 1)(x - 1)(x - 3) + 10$

14. False   **15.** True   **16.** True   **17.** True

18. True   **19.** False   **20.** False   **23.** Not possible

24. Your graph should have a quadruple root at $x = -2$, a root at $x = 1$, and a double root at $x = 3$, and it should point "up" at the left end and "down" at the right end. There is a y-intercept at $(0, 288)$.

26.

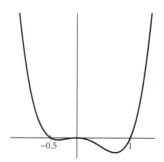

27. Cannot quickly sketch the graph.

28.

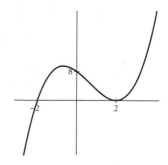

32. Root at $x = 2$, local maximum at $x = \frac{1}{3}$, local minimum at $x = 1$, inflection point at $x = \frac{2}{3}$:

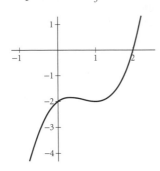

33. No roots, local minimum at $x = 0$, no inflection points

34. Roots at $x = 2$ (double root) and $x = -1$, local maximum at $x = 0$, local minimum at $x = 2$, inflection point at $x = 1$

38. Root at $x = 0$, no local extrema, inflection points at $x = 0$, $x = \frac{1}{2}$, and $x = 1$:

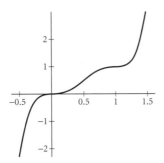

40. $f(x) = 2(x + 1)(x - 2)(x - 3)$

42. $f(x) = -(x + 2)(x - 1)^3$

44. $f(x) = -\frac{1}{4}(x - 3)^4(x + 1)$

46. $f(x) = -2(x + 2)^2 x(x - \frac{5}{2})$

48. $f(x) = 2x^3 - 3x^2 - 4$

49. $f(x) = \frac{5}{4}(x - 1)^2(x - 2)^2$

50. $f(x) = (x - 2)^3$

53. $f(x) = x^3 - x^2 - 3x - 5$

54. $f(x) = x^2 + 3x - 5$

56. (a) $x = -2$, $x = 0$, $x = 2$, and $x = 4$; 4; negative

(b) $f'(x) = -x(x + 2)(x - 2)(x - 4)$

(c) $f(x) = -\frac{1}{5}x^5 + x^4 + \frac{4}{3}x^3 - 8x^2 + C$

(d) $C = 25$

60. (a) $s(t) = -\frac{2}{5}t^3 - 17t^2 + 575$

(b) $s(5) = 100$, $s'(5) = -200$, $s''(5) = -46$, $s'(0) = 0$

(c) 575 feet

(d) $a(t) = -\frac{12}{5}t - 34$ is *not* constant.

63. **PROOF**   If $f(x) = ax^3 + bx^2 + cx + d$, then $f(0) = d$. Now $f'(x) = 3ax^2 + 2bx + c$, so $f'(0) = c$. Since $f''(x) = 6ax + 2b$, we have $f''(0) = 2b$, and thus $b = \frac{f''(0)}{2}$. Finally, $f'''(x) = 6a$, so $f'''(0) = 6a$, and thus $a = \frac{f'''(0)}{6}$. ∎

SECTION 5.4

2. (a) There is some $\delta > 0$ such that $f(2) \geq f(x)$ for all $x \in (2 - \delta, 2 + \delta)$.

(b) $f(-1) \geq f(x)$ for all $x \in [-3, 5]$.

(c) $f(c) \leq f(x)$ for all $x \in I$.

4. There will not be any endpoint extrema; there may or may not be a global maximum (or minimum) on the interval.

6. The derivative may or may not be zero (or fail to exist) at the endpoints of the interval.

9. (a) Global maximum at $x = -1$ with value $f(-1) = 7$, global minimum at $x = -3$ with value $f(-3) = -45$

(b) Global maximum at $x = 0$ with value $f(0) = 0$, global minimum at $x = 2$ with value $f(2) = -20$

(c) No global maximum, global minimum at $x = 2$ with value $f(2) = -20$

(d) Global maximum at $x = -1$ with value $f(-1) = 7$, no global minimum

(e) No global maximum, global minimum at $x = 2$ with value $f(2) = -20$

11. (a) Global maximum at $x = -1$ with value $f(-1) = 11$, global minimum at $x = 1$ with value $f(1) = -5$

(b) No global maximum or global minimum

(c) Global maximum at $x = -1$ with value $f(-1) = 11$, no global minimum

(d) Global maximum at $x = 0$ with value $f(0) = 0$, global minimum at $x = 3$ with value $f(3) = -117$

(e) Global maximum at $x = -1$ with value $f(-1) = 11$, no global minimum

14. $a = 50$, $b = 50$

16. The horizontal top and bottom edges should have length $\approx$ 19.6903 units, and the vertical edges should have length 13.1268 units.

18. $(-\frac{1}{5}, \frac{2}{5})$ **20.** $(\frac{2-\sqrt{6}}{2}, 1.75)$ (or $(\frac{2+\sqrt{6}}{2}, 1.75)$)

22. At $t = 4$ seconds; at $t = 0$ seconds

24. A square pen with sides of length $\frac{175}{2}$ feet will produce the maximum area of 7656.25 square feet.

27. The east and west fences (and two parallel interior fences) should be 300 feet long, and the north and south fences (and their one parallel interior fence) should be 400 feet long; this will produce the maximum area of 120,000 square feet.

28. (a) You would sell the most (975 paintings) at $5.00 apiece. You would sell the least (15 paintings) at $45.00 apiece.

(b) $R(c) = c(0.6c^2 - 54c + 1230)$

(c) You would earn the most money ($8327.10) by selling the paintings for $15.28 apiece. You would earn the least money ($672.90) by selling the paintings for $44.72 apiece.

(d) Although you sell the most paintings when charging $5.00 apiece, you earn only $5.00 for each of those paintings.

29. The volume is maximized when you cut a square of side length 0.78475 inch from each corner.

31. The largest possible volume is 417.482 cubic inches.

33. The largest possible volume is 11,664 cubic inches; the largest possible surface area is 3332.6 square inches.

34. The largest possible volume is 14,851 cubic inches; the largest possible surface area is 462.6 square inches.

35. The minimum area occurs when you cut the wire so that 4.399 inches of wire is used to make the circle. The maximum area occurs when you don't cut the wire and use all 10 inches to make the circle.

38. **PROOF** If the sides of the rectangle have length x and y, then $P = 2x + 2y$, so $y = \frac{1}{2}(P - 2x)$. Therefore, the area is $A = xy = x(\frac{1}{2}(P - 2x))$, which has derivative $A'(x) = \frac{1}{2}P - 2x$. Thus the only critical point of $A(x)$ is $x = \frac{P}{4}$. Since $A'(\frac{P}{4} - 1) > 0$ and $A'(\frac{P}{4} + 1) < 0$, the first derivative test tells us that $x = \frac{P}{4}$ is a local maximum of $A(x)$; in fact it is the global maximum of $A(x)$ on $[0, \frac{P}{2}]$. Therefore, A is maximized when $x = \frac{P}{4}$ (and thus $y = \frac{P}{4}$), i.e., when the rectangle is a square.

CHAPTER 6

SECTION 6.1

2. A rational function is improper if the degree of its numerator is greater than or equal to the degree of its denominator.

4. $f(x) = \frac{-x^3 + 2x + 3}{x^2 + 3x}$

6. Not a rational function

8. $f(x) = \frac{3x^2 - x + 2}{x^3}$

11. Domain$(f(x)) = \{x \mid q(x) \neq 0\}$

14. The graph should be just like $y = x + 1$, but with holes at $x = -2$ and $x = 3$.

15. f has roots at $x = 1$ and $x = -3$, and the domain of f is $(-\infty, 2) \cup (2, 3) \cup (3, 4) \cup (4, \infty)$.

18. $f(x) = \frac{(x-1)(x-3)}{(x+2)(x-4)}$ **21.** $f(x) = \frac{(x-1)^2(x-3)}{(x-3)(x+2)}$

23. $f(x) = \frac{(x-3)^2}{(x-3)}$

25. A common irreducible quadratic factor in the numerator and the denominator will not affect a rational function (since an irreducible quadratic factor is never zero).

28. $p(x) = q(x)(x^2 - x + 3) + (3x + 1)$; $\frac{p(x)}{q(x)} = (x^2 - x + 3) + \frac{3x - 1}{q(x)}$

30. True   **31.** False   **32.** True

33. False   **34.** True   **35.** False

37. Domain is $(-\infty, 0) \cup (0, 1) \cup (1, 2) \cup (2, \infty)$; roots at $x = -5$ and $x = 3$; a "hole" at $x = 1$

39. $f(x) = \frac{(x+2)(x-1)}{(x-2)(2x+3)}$. Domain is $(-\infty, -\frac{3}{2}) \cup (-\frac{3}{2}, 2) \cup (2, \infty)$; roots at $x = -2$ and $x = 1$; no "holes."

42. $f(x) = \frac{(x-2)^3}{-(x-3)(x-2)(x+2)}$. Domain is $(-\infty, -2) \cup (-2, 2) \cup (2, 3) \cup (3, \infty)$; no roots; a "hole" at $x = 2$.

43. $f(x) = \frac{(x-1)(x+2)(x+5)}{(x-1)(x+1)(x+3)(x^2+1)}$. Domain is $(-\infty, -3) \cup (-3, -1) \cup (-1, 1) \cup (1, \infty)$; roots at $x = -2$ and $x = -5$; a "hole" at $x = 1$.

44. $f(x) = \frac{(x+4)(x-1)(x-4)}{(x^2+4)(x+2)(x-2)}$. Domain is $(-\infty, -2) \cup (-2, 2) \cup (2, \infty)$; roots at $x = 1$ and $x = \pm 4$; no "holes."

46. $f(x) = x^2 + \frac{5}{x-2}$

48. $f(x) = x^3 - 2x - 4$ (when $x \neq \pm\sqrt{3}$)

49. $f(x) = (-x^2 + 2x - 5) + \frac{12x - 4}{x^2 + 2x - 1}$

53. $f(x) = (\frac{1}{2}x^2 - \frac{7}{4}) + \frac{\frac{15}{4}}{2x^2 + 1}$

55. $f(x) = (x^5 + 2x^4 + 3x^3 + 4x^2 + 5x + 6) + \frac{7x - 7}{x^2 - 2x + 1}$

57. **PROOF** The domain of a quotient $f(x) = \frac{p(x)}{q(x)}$ of functions is $\{x \mid x \in \text{Domain}(p(x)) \cap \text{Domain}(q(x))$ and $q(x) \neq 0\}$. Since $p(x)$ and $q(x)$ are polynomials, they are defined on $(-\infty, \infty)$; thus the domain of $f(x)$ is $\{x \mid q(x) \neq 0\}$.

SECTION 6.2

3. No; Yes

4. $f(x) = \frac{x}{(x+1)(x-1)}$

6. (a) $n \leq m$; (b) $n < m$; (c) $n = m + 1$; (d) $n \geq m + 2$; (e) $n = m + 4$

8. $f(x) = \frac{(x-3)}{(x-3)^2}$ **12.** $f(x) = \frac{3x^2 + 1}{(x+2)(x-2)}$

13. $f(x) = \frac{-5(x+2)^3}{(x+2)(x-1)(x+3)}$ **14.** $f(x) = \frac{x(x^2+1)}{x(x-2)}$

15. $f(x) = (x^3 + 2) + \frac{1}{x-1} = \frac{x^4 - x^3 + 2x + 1}{x-1}$

16. True   **17.** False   **18.** False   **19.** False

20. False   **21.** $-\frac{1}{3}$ **24.** $\frac{3}{4}$ **25.** 0

26. The limit is infinite; specifically, $-\infty$ as $x \to 2^-$ and ∞ as $x \to 2^+$.

28. 0 **35.** ∞ **36.** 1 **37.** ∞ **38.** 0

39. Roots at $x = 2$ and $x = 5$; hole at $x = 1$; vertical asymptote at $x = 3$; no horizontal asymptotes

42. Root at $x = 2$; no holes; vertical asymptotes at $x = 1$ and $x = -\frac{2}{3}$; horizontal asymptote at $y = 0$

43. Roots at $x = \frac{1}{2}$ and $x = -\frac{5}{3}$; no holes; vertical asymptotes at $x = 0$, $x = 1$, and $x = -2$; horizontal asymptote at $y = 0$

44. No roots; holes at $x = 2$ and $x = -2$; vertical asymptote at $x = 0$; horizontal asymptote at $y = -1$

46. Roots at $x = \pm 2$; no holes; vertical asymptote at $x = 1$; horizontal asymptote at $y = \frac{1}{2}$

49. Roots at $x = 1$ and $x = -\frac{3}{2}$; no holes; vertical asymptote at $x = -1$; slant asymptote with equation $y = 2x - 1$

52. Roots at $x = \pm 1$ and $x = -\frac{2}{3}$; no holes; slant asymptote with equation $y = 3x - 1$

53. Root at $x = -\frac{3}{2}$; hole at $x = \frac{3}{2}$; curve asymptote with equation $y = -8x^3 - 12x^2 - 18x - 27$

56. Root at $x = 0$; hole at $x = 1$; curve asymptote with equation $y = x^3 - 2x^2 + 2x - 2$

58. **PROOF** Suppose $f(x)$ is the rational function $\frac{a_n x^n + a_{n-1}x^{n-1} + \cdots + a_1 x + a_0}{b_m x^m + b_{m-1}x^{m-1} + \cdots + b_1 x + b_0}$. Then

$$\lim_{x \to \infty} f(x) = \lim_{x \to \infty} \frac{a_n x^n + a_{n-1}x^{n-1} + \cdots + a_1 x + a_0}{b_m x^m + b_{m-1}x^{m-1} + \cdots + b_1 x + b_0} \left(\frac{1/x^m}{1/x^m}\right) =$$

$$\lim_{x \to \infty} \frac{a_n x^{n-m} + a_{n-1}x^{n-1-m} + \cdots + a_1 x^{1-m} + a_0 x^{-m}}{b_m + b_{m-1}x^{-1} + \cdots + b_1 x^{1-m} + b_0 x^{-m}}.$$

(a) If $n < m$, then each of the terms in the numerator has a negative exponent (since $n < m$) and thus goes to zero as $x \to \infty$. Each term after the first one in the denominator also goes to zero as $x \to \infty$. Thus the limit above approaches $\frac{0}{b_m} = 0$ as $x \to \infty$. A similar argument shows that $\lim\limits_{x \to -\infty} f(x) = 0$. Therefore f has a horizontal asymptote at $y = 0$.
(b) and **(c)** are similar, using the same limit calculation. ∎

59. **PROOF** **(a)** We have $\frac{p(x)}{q(x)} = \frac{l(x)q(x)+R(x)}{q(x)}$, so $l(x)q(x)$ must have the same degree as $p(x)$ (say, degree n). Since we know that the degree of $q(x)$ is $n-1$, the degree of $l(x)$ must be 1 (i.e., $l(x)$ is a linear function).
(b) The polynomial long-division algorithm terminates when the degree of the remainder is less than the degree of $q(x)$; therefore, the degree of $R(x)$ is less than the degree of $q(x)$, so $\frac{R(x)}{q(x)}$ is a proper rational function.
(c) $\lim\limits_{x \to \infty}\left(\frac{p(x)}{q(x)} - l(x)\right) = \lim_{x \to \infty} \frac{R(x)}{q(x)} = 0$ because $\frac{R(x)}{q(x)}$ is a proper rational function. The calculation for the limit as $x \to -\infty$ is similar.
(d) We have just shown that as $x \to \infty$ (and as $x \to -\infty$), the difference between the rational function $f(x) = \frac{p(x)}{q(x)}$ and the linear function $l(x)$ goes to zero. This means that $f(x) = \frac{p(x)}{q(x)}$ gets closer and closer to the line $l(x)$ as $x \to \infty$ (and as $x \to -\infty$), that is, that $l(x)$ is a slant asymptote of $f(x) = \frac{p(x)}{q(x)}$. ∎

SECTION 6.3

2. $f'(x) = \frac{(\frac{1}{2}x^{-\frac{1}{2}}-6x^{-3})(x^{\frac{3}{2}}+1)-(x^{\frac{1}{2}}+3x^{-2})(\frac{3}{2}x^{\frac{1}{2}})}{(x^{\frac{3}{2}}+1)^2}$.

3. **(a)** If there is no cancellation in the expression $p'(x)q(x) - p(x)q'(x)$, then the numerator of f' will have degree $n+m-1$ (if there is cancellation, then the numerator of f' may have a lower degree). In any case, the denominator of f' will have degree $2m$.
(b) Yes, since if $n < m$, then $n+m-1 < m+m-1 = 2m-1 < 2m$.
(c) Yes, at $y = 0$.

4. $f(x) = \frac{x^3-x+5}{x^2+3x-1}$

6. **(a)** Global minimum at $x = -1$, global maximum at $x \approx \frac{1}{2}$
(b) No global minimum, global maximum at $x \approx \frac{1}{2}$
(c) No global minimum, global maximum at $x \approx \frac{1}{2}$
(d) No global maximum or minimum
(e) Global minimum at $x = 5$, no global maximum

7. $f'(2) = \lim\limits_{h \to 0} \frac{\frac{(2+h)-1}{(2+h)+3} - \frac{2-1}{2+3}}{h} = \frac{4}{25}$

10. $f'(1) = \lim\limits_{x \to 1} \frac{\frac{x-1}{(x+1)(x+2)} - \frac{1-1}{(1+1)(1+2)}}{x-1} = \frac{1}{6}$

11. $f'(x) = \lim\limits_{h \to 0} \frac{\frac{(x+h)-1}{(x+h)+3} - \frac{x-1}{x+3}}{h} = \frac{4}{(x+3)^2}$

13. $f'(x) = \lim\limits_{t \to x} \frac{\frac{t^2-3t}{t^2-2t+1} - \frac{x^2-3x}{x^2-2x+1}}{t-x} = \frac{(x-1)(x+3)}{(x-1)^4}$

15. $f'(x) = \lim\limits_{h \to 0} \frac{\frac{(x+h)^3}{(x+h)+1} - \frac{x^3}{x+1}}{h} = \frac{2x^3+3x^2}{(x+1)^2}$

17. $f'(x) = \frac{23}{(4+5x)^2}$

18. $f'(x) = \frac{3x^2+2x^3}{(1+x)^2}$

21. $f'(x) = \frac{(7x^6-15x^4)(1-3x^4)-(x^7-3x^5+4)(-12x^3)}{(1-3x^4)^2}$

22. $f'(x) = \frac{2x(x^3+5x^2-3x)-x^2(3x^2+10x-3)}{(x^3+5x^2-3x)^2}$

24. $f'(x) = \frac{(2+3x+x^2)-(x-1)(3+2x)}{(x+1)^2(x+2)^2}$

26. $x = -\frac{1}{2}, x = 1, x = -2$

30. $x = 0, x = 3 - \sqrt{5}, x = 3 + \sqrt{5}, x = 2$

31. $x = -\frac{1}{\sqrt{3}}, x = \frac{1}{\sqrt{3}}, x = 0, x = 1, x = -1$

32. Local maximum at $x = -\frac{1}{2}$

36. Local maximum at $x = 0$, local minima at $x = 3 - \sqrt{5}$ and $x = 3 + \sqrt{5}$

37. Local maximum at $x = \frac{1}{\sqrt{3}}$, local minimum at $x = -\frac{1}{\sqrt{3}}$

39. No local extrema

41. Global maximum at $x = -\frac{1}{2}$, no global minimum

42. Global minimum at $x = 5$, no global maximum

46. Global minimum at $x = 4$, no global maximum

50. **PROOF** If $f(x)$ is a rational function with numerator of degree n and denominator of degree m, then its derivative $f'(x)$ is a rational function with numerator of degree at most $n+m-1$ and denominator of degree $2m$. If $f(x)$ has a horizontal asymptote, then $n \le m$, which implies that the degree of the numerator of $f'(x)$ is less than the degree of the denominator of $f'(x)$, since $n+m-1 \le m+m-1 = 2m-1 < 2m$. Therefore, $f'(x)$ must have a horizontal asymptote. ∎

SECTION 6.4

3. Roots, holes, and vertical and horizontal asymptotes

4. The intervals on which the function is increasing or decreasing and concave up or concave down, as well as any local extrema or inflection points

5. The locations of any vertical asymptotes (and the behavior of the function near those asymptotes), the locations of any horizontal asymptotes (and in general the behavior of the function at the "ends"), and any holes in the graph of the function

6. **(a)** Graph the functions given in part **(b)** below.
(b) $f(x) = \frac{-(x-1)(x-3)(x+1)}{(x-2)^2(x+1)}$; $f(x) = \frac{-(x-1)(x-3)(x+1)^2}{(x-2)^2(x+1)^2}$;
$f(x) = \frac{-(x-1)^2(x-3)(x+1)}{(x-2)(x+1)(x^2+1)}$

10. $f(x) = \frac{(x+3)^2(x-2)(x-5)}{(x+1)(x-3)(x-5)}$

12. If you can factor the numerator of f, then you may or may not be able to factor the numerator of f'. If you can factor the denominator of f, then you can factor the denominator of f'.

14. The graph of $f(x)$ looks like the graph of $y = x + 3$ with a hole at $x = 2$.

17. The graph of $f(x)$ has roots at $x = 0$ and $x = -1$, vertical asymptotes at $x = 5$ and $x = -2$, and a horizontal asymptote at $y = 0$.

18. The graph of $f(x)$ has roots at $x = 0$, $x = -3$, and $x = 1$, vertical asymptotes at $x = \pm 2$, and a slant asymptote at $y = 2x + 4$.

19. The graph of $f(x)$ has a hole at $x = -1$, roots at $x = -\frac{3}{2}$ and $x = 1$, a vertical asymptote at $x = 3$, and a slant asymptote with equation $y = 2x + 7$.

23.

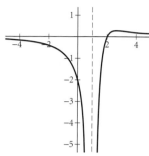

25. The graph of $f(x)$ has no roots or holes, a vertical asymptote at $x = 2$, and a horizontal asymptote at $y = 3$; moreover, the graph of $f(x)$ is always positive. $f(x)$ has a local minimum at $x = -\frac{2}{5}$ and an inflection point at $x = -\frac{8}{5}$.

26. The graph of $f(x)$ has a hole at $x = -1$, roots at $x = 0$ and $x = 1$, a vertical asymptote at $x = 2$, and a slant asymptote with equation $y = x$. $f(x)$ has a local maximum at $x = 2 - \sqrt{2}$, a local minimum at $x = 2 + \sqrt{2}$, and no inflection points. $\lim\limits_{x \to 2^-} f(x) = -\infty$, and $\lim\limits_{x \to 2^+} f(x) = \infty$.

30.

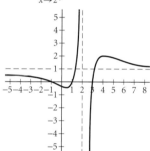

33. $f(x) = \frac{-4(x^2+1)(x-4)}{(x+2)(x-2)(x-4)}$

34. $f(x) = \frac{3(x-2)(x+2)(x-6)}{(x-3)(x-6)}$

35. $f(x) = \frac{-4(x-1)(x+1)}{(x^2+1)^2}$

39. $f(x) = (x+1) + \frac{-1}{x-1} = \frac{x^2-2}{x-1}$

CHAPTER 7

SECTION 7.1

3. Graph $f(x)$ and your candidate for $f'(x)$, and check that the slope of the graph of $f(x)$ at every point is given by the height of the graph of $f'(x)$.

5. $|x| = \sqrt{x^2}$.   **6.** True   **7.** True   **8.** True

9. False   **10.** True   **11.** False

12. $x \neq 0$, $x \neq 1$ **14.** $(-\infty, -1) \cup (-1, 3]$

18. $[\frac{2}{3}, 3) \cup (3, \infty)$ **19.** $(-3, -1] \cup (2, \infty)$

21. $-\frac{3}{\sqrt{2}}$ **23.** $\lim\limits_{x \to 1^-} \frac{1}{\sqrt{x-1}} = \infty$; $\lim\limits_{x \to 1^+} \frac{1}{\sqrt{x-1}} = \infty$

25. 0   **27.** $\frac{1}{8}$ **28.** $-\infty$ **31.** $-\infty$

35. $-\frac{1}{8}$ **37.** $f'(x) = \frac{1}{2}x^{-\frac{1}{2}} + \frac{1}{2}x^{-\frac{3}{2}}$

40. $f'(x) = \frac{-(1-5x^{-\frac{1}{2}})}{(x-10x^{\frac{1}{2}}+25)^2}$ **41.** Not possible (yet)

42. $f'(x) = \frac{7}{3}x^{\frac{4}{3}} + \frac{1}{6}x^{-\frac{7}{6}}$

45. $f'(x) = \frac{5}{3}x^{\frac{2}{3}} - 14x^{\frac{11}{3}} + \frac{1}{6}x^{-\frac{5}{6}}$

47. $f'(x) = \frac{3(-x^{-\frac{1}{2}})(1+x^{-\frac{1}{2}}) - 3(1-2\sqrt{x})(-\frac{1}{2}x^{-\frac{3}{2}})}{(1+x^{-\frac{1}{2}})^2}$

49. $\frac{1}{4}x^{-\frac{3}{2}}(1+\sqrt{x})^{-2} + \frac{1}{2}x^{-1}(1+\sqrt{x})^{-3}$

50. $\frac{7}{864}$ **52.** -1.22579 **54.** Does not exist   **55.** -1

56. $\frac{\sqrt{3}}{3}$ **58.** $-\frac{1}{2}$ **60.** -0.08588 **62.** $-\frac{1}{12\sqrt{6}}$

64. $f'(x) = \lim\limits_{h \to 0} \frac{((x+h)^2-4)^{-\frac{1}{2}} - (x^2-4)^{-\frac{1}{2}}}{h} = \frac{-x}{(x^2-4)^{\frac{3}{2}}}$

68. $x = \frac{1}{6^4}$, $x = \frac{1}{3^4}$, and $x = 0$ **69.** $x = 0$

70. $x = 1$, $x = 0$, and $x = \frac{16}{9}$ **74.** $x = 0$, $x = \frac{1}{4}(\frac{7}{2})^{\frac{2}{3}}$

SECTION 7.2

2. $(kf(x))' = kf'(x)$; $\frac{d}{dx}(kf(x)) = k\frac{df}{dx}$; for example, $\frac{d}{dx}(3x^2) = 3\frac{d}{dx}(x^2) = 3(2x)$.

6. $(f(u(x)))' = f'(u(x))u'(x)$; $\frac{df}{dx} = \frac{df}{du}\frac{du}{dx}$; for example, if $f(u) = u^2$ and $u(x) = 3x+1$, then $\frac{df}{dx} = (2u)(3) = 2(3x+1)(3)$.

7. $f(x) = x^2$, $g(x) = x + 1$

9. $\frac{d}{dx}((3x + \sqrt{x})^2) = 2(3x + \sqrt{x})(3 + \frac{1}{2}x^{-\frac{1}{2}})$;
$\frac{d}{dx}((3x + \sqrt{x})(3x + \sqrt{x})) = (3 + \frac{1}{2}x^{-\frac{1}{2}})(3x + \sqrt{x}) + (3x + \sqrt{x})(3 + \frac{1}{2}x^{-\frac{1}{2}})$; $\frac{d}{dx}(9x^2 + 6x^{\frac{3}{2}} + x) = 18x + 9x^{\frac{1}{2}} + 1$

12. $f'(x)g(x)h(x) + f(x)g'(x)h(x) + f(x)g(x)h'(x)$

14. $(f(u(v(x))))' = f'(u(v(x)))u'(v(x))v'(x)$

15. The quotient rule is first. You also need the chain, sum, constant multiple, and power rules.

17. The chain rule is first. You also need the power, sum, and constant multiple rules.

20. The sum rule is first. You also need the chain, sum, constant multiple, and power rules.

22. $f'(x^{-2})(-2x^{-3})$ **23.** $-2(f(x))^{-3}f'(x)$ **25.** $\frac{-f'(x)}{(f(x))^2}$

28. $f'(g(x)h(x))(g'(x)h(x) + g(x)h'(x))$

30. $2xf(g(x)) + x^2 f'(g(x))g'(x)$

31. $f'(g(x))g'(x)(h(x))^2 + f(g(x))(2h(x))h'(x)$

33. 12   **35.** 21   **37.** 1   **38.** -2 **40.** 6

43. 4   **45.** -12 **46.** True   **47.** True

48. True   **49.** False   **50.** True

52. $f'(x) = 87(1 - x^4)^{86}(-4x^3)$ (chain, power, and difference rules)

54. $f'(x) = (2x + 3)(x - 2)^{\frac{3}{2}} + (x^2 + 3x - 1)(\frac{3}{2})(x - 2)^{\frac{1}{2}}(1)$ (product, sum, constant multiple, chain, and power rules)

56. $f'(x) = \frac{\frac{1}{2}x^{-\frac{1}{2}}(x\sqrt{x-1})-\sqrt{x}(\sqrt{x-1}+x(\frac{1}{2})(x-1)^{-\frac{1}{2}})}{x^2(x-1)}$

(quotient, power, product, chain, and difference rules)

57. $f'(x) = (3x^2+1)^9 + 54x^2(3x^2+1)^8$

59. $f'(x) = -2(x\sqrt{x+1})^{-3}(\sqrt{x+1}+x(\frac{1}{2}(x+1)^{-\frac{1}{2}}))$

61. $f'(x) = 3(-\frac{2}{3})((x^2+1)^8 - 7x)^{-\frac{5}{3}}(8(x^2+1)^7(2x) - 7)$

63. $f'(x) = \frac{1}{2}(3x - 4(2x+1)^6)^{-\frac{1}{2}}(3 - 24(2x+1)^5(2))$

64. $f'(x) = \frac{(-x^{-2}-6x)(x^5-x^{-\frac{1}{2}})-(x^{-1}-3x^2)(5x^4+\frac{1}{2}x^{-\frac{3}{2}})}{(x^5-x^{-\frac{1}{2}})^2}$

65. $f'(x) = \frac{5}{2}x^{\frac{3}{2}} - \frac{5}{6}x^{-\frac{1}{6}} + 5(\frac{19}{2})x^{\frac{17}{2}}$

67. $f'(x) = 7(5x^4 - 3x^2)^6(20x^3 - 6x)(2x^3 + 1) + (5x^4 - 3x^2)^7(6x^2)$

71. $f'(x) = 8(x^4 - \sqrt{3-4x})^7(4x^3 - \frac{1}{2}(3-4x)^{-\frac{1}{2}}(-4)) + 5$

72. $f'(x) = \frac{25}{6}x^{\frac{19}{6}}$

75. $\frac{(-3)(x^5+2x^4+x^3)-(-2-3x)(5x^4+8x^3+3x^2)}{(x^5+2x^4+x^3)^2}$

79. 0   **80.** 162   **81.** $x = -\frac{1}{3}, x = -\frac{2}{7}, x = 0$

82. $x = -1, x = 0, x = 1$ **85.** $x = \frac{1}{2}, x = 0, x = 1$

88. (a) $\frac{dA}{dr} = 2\pi r$ (b) No; Yes

(c) $\frac{dA}{dt} = \frac{d}{dt}(\pi(r(t))^2) = 2\pi r(t)r'(t) = 2\pi r\frac{dr}{dt}$

(d) Yes; Yes   (e) $\frac{dA}{dt}\big|_{r=24} = 2\pi(24)(2) = 96\pi$

90. **PROOF** $\frac{d}{dx}\left(\frac{f(x)}{g(x)}\right) = \frac{d}{dx}(f(x)(g(x))^{-1}) = \frac{d}{dx}(f(x)) \cdot$
$(g(x))^{-1} + f(x) \cdot \frac{d}{dx}((g(x))^{-1}) = f'(x) \cdot (g(x))^{-1} + f(x) \cdot$
$(-g(x))^{-2}g'(x) = \frac{f'(x)}{g(x)} - \frac{f(x)g'(x)}{(g(x))^2} = \frac{f'(x)g(x)-f(x)g'(x)}{(g(x))^2}$ ∎

92. **PROOF** $\frac{d}{dx}(f(u(v(x)))) = f'(u(v(x))) \cdot \frac{d}{dx}(u(v(x)))$
$= f'(u(v(x))) \cdot u'(v(x)) \cdot v'(x)$. Similarly, $\frac{d}{dx}(f(u(v(w(x))))) =$
$f'(u(v(w(x)))) \cdot u'(v(w(x))) \cdot v'(w(x)) \cdot w'(x)$. ∎

SECTION 7.3

3. Combine the graphs of $y = \sqrt{\frac{1}{3}x^2 - \frac{16}{3}}$ and $y = -\sqrt{\frac{1}{3}x^2 - \frac{16}{3}}$.

4. Graph III

5. Graph I

8. $3s^2s'$ **9.** 0   **11.** $s^2 + 2rss'$

15. y is a function of x: $y = \sqrt[5]{2x-3}$.

16. y is *not* a function of x, and it is not possible to solve for y.

18. y is a function of x: $y = \frac{x+1}{1-x}$.

21. $\frac{dy}{dx} = \frac{4x}{y}$ **23.** $\frac{dy}{dx} = \frac{-6x-y^2}{2xy}$ **25.** $\frac{dy}{dx} = \frac{-3(y^2-y+6)}{(3x+1)(2y-1)}$

26. $\frac{dy}{dx} = \frac{5y}{\frac{3}{2}(3y-1)^{-\frac{1}{2}}-5x}$ **27.** $\frac{dy}{dx} = \frac{(3y-1)^2}{2y(3y-1)-3(y^2+1)}$

30. $\frac{dy}{dx} = \frac{2x^3y^2(y+1)-y^2(y+1)^2}{x^4y^2-x^2(y+1)^2}$ **32.** $(\frac{1}{2}, \frac{\sqrt{3}}{2}), (\frac{1}{2}, -\frac{\sqrt{3}}{2})$

33. Slope is $-\frac{1}{\sqrt{3}}$ at the point $(\frac{1}{2}, \frac{\sqrt{3}}{2})$ and is $\frac{1}{\sqrt{3}}$ at the point $(\frac{1}{2}, -\frac{\sqrt{3}}{2})$.

34. $(\frac{\sqrt{2}}{2}, \frac{\sqrt{2}}{2})$ and $(-\frac{\sqrt{2}}{2}, \frac{\sqrt{2}}{2})$.

35. Slope is -1 at the point $(\frac{\sqrt{2}}{2}, \frac{\sqrt{2}}{2})$ and is 1 at the point $(-\frac{\sqrt{2}}{2}, \frac{\sqrt{2}}{2})$.

36. $(1, 0)$ and $(-1, 0)$ **37.** $(\frac{\sqrt{2}}{2}, \frac{\sqrt{2}}{2})$ and $(-\frac{\sqrt{2}}{2}, -\frac{\sqrt{2}}{2})$

44. $(1, -1)$ **45.** Slope is $\frac{1}{4}$ at $(1, -1)$.   **46.** $(-3, 1)$

47. Tangent line is vertical at $(-3, 1)$ (undefined slope).

48. No such points exist.

49. Only at the point $(-3, 1)$.
The graph of $y^3 + xy + 2 = 0$ looks like

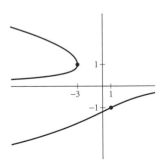

57. **PROOF**   If $y = x^{\frac{3}{5}}$, then $y^5 = x^3$. Differentiating both sides (with respect to x), we get $5y^4y' = 3x^2$, so $y' = \frac{3x^2}{5y^4} = \frac{3x^2}{5(x^{\frac{3}{5}})^4} = \frac{3x^2}{5x^{\frac{12}{5}}} = \frac{3}{5}x^{2-\frac{12}{5}} = \frac{3}{5}x^{-\frac{2}{5}}$. ∎

SECTION 7.4

3. (a) $V = \pi y^2 s$, $SA = 2\pi ys + 2\pi y^2$

(b) $V = \frac{1}{3}\pi u^2 w$, $SA = \pi u\sqrt{u^2+w^2} + \pi u^2$

(c) $V = \frac{4}{3}\pi(\frac{x}{2})^3$, $SA = 4\pi(\frac{x}{2})^2$

(d) $V = \frac{\pi}{4}h^3$, $SA = \frac{3}{2}\pi h^2$

(e) $V = \pi r^3$, $SA = (\sqrt{10}+1)\pi r^2$

5. $\frac{dV}{dt} = 4\pi r^2\frac{dr}{dt}$

7. $\frac{dV}{dt} = \frac{c^2}{2\pi^2}\frac{dc}{dt}$, or, equivalently, $\frac{dV}{dt} = 2r^2\frac{dc}{dt}$

9. (a) $\frac{dV}{dt} = 4\pi r^2\frac{dr}{dt}$

(b) $\frac{dS}{dt} = 8\pi r\frac{dr}{dt}$

(c) $\frac{dV}{dt} = \frac{1}{2}r\frac{dS}{dt}$

10. (a) $\frac{dV}{dt} = 2\pi rh\frac{dr}{dt} + \pi r^2\frac{dh}{dt}$

(b) $\frac{dV}{dt} = 2\pi rh\frac{dr}{dt}$

(c) $\frac{dV}{dt} = \pi r^2\frac{dh}{dt}$

(d) $\frac{dV}{dt} = \frac{3\pi}{4}h^2\frac{dh}{dt}$

12. $f' = 2uu' + kv'$

13. $f' = v + tv' + kv'$

14. $f' = 2v'\sqrt{u+w} + \frac{v(u'+w')}{\sqrt{u+w}}$

18. $f' = w'(u+t)^2 + 2w(u+t)(u'+1)$

19. $f' = \frac{1}{k}(u't + u + w')$

20. (a) $\frac{dA}{dt}\big|_{r=12} = 96\pi \frac{\text{in.}^2}{\text{sec}}$; $\frac{dA}{dt}\big|_{r=24} = 192\pi \frac{\text{in.}^2}{\text{sec}}$; $\frac{dA}{dt}\big|_{r=100} = 800\pi \frac{\text{in.}^2}{\text{sec}}$

(b) $\frac{dA}{dt}\big|_{A=100} = 80\sqrt{\pi} \frac{\text{in.}^2}{\text{sec}}$; $\frac{dA}{dt}\big|_{A=200} = 8\sqrt{200\pi} \frac{\text{in.}^2}{\text{sec}}$;
$\frac{dA}{dt}\big|_{A=1000} = 8\sqrt{1000\pi} \frac{\text{in.}^2}{\text{sec}}$

21. (a) $\frac{dV}{dt}\big|_{s=8} = 384$ cubic inches per minute

(b) $\frac{dV}{dt}\big|_{V=55} = 6\sqrt[3]{55}$ cubic inches per minute

23. (a) $\frac{dr}{dt}\big|_{r=12} = \frac{5}{24\pi}\,\frac{\text{in.}}{\text{sec}}$

(b) $\frac{dr}{dt}\big|_{V=300} = \frac{120}{(4\pi)^{\frac{1}{3}}\,900^{\frac{2}{3}}}\,\frac{\text{in.}}{\text{sec}}$

(c) $16\,\frac{\text{in.}^2}{\text{sec}}$

24. Suppose x is Alison's distance from the streetlight, l is the length of her shadow, and $y = x + l$ is the distance from the tip of her shadow to the streetlight.

(a) $\frac{dl}{dt}\big|_{x=10} = -\frac{12}{7}\,\frac{\text{ft}}{\text{sec}}$; this does not depend on x.

(b) $\frac{dy}{dt}\big|_{x=10} = -\frac{40}{7}\,\frac{\text{ft}}{\text{sec}}$; this does not depend on x.

26. (a) $\frac{dr}{dt}\big|_{r=2} = \frac{9}{8\pi}\,\frac{\text{in.}}{\text{sec}}$

(b) $\frac{dr}{dt}\big|_{h=4} = \frac{1}{8\pi}\,\frac{\text{in.}}{\text{sec}}$

(c) $\frac{dh}{dt}\big|_{r=2} = \frac{3}{4\pi}\,\frac{\text{in.}}{\text{sec}}$

(d) $\frac{dh}{dt}\big|_{h=4} = \frac{1}{12\pi}\,\frac{\text{in.}}{\text{sec}}$

28. $\frac{dh}{dt}\big|_{h=3} = -\frac{25}{36\pi}\,\frac{\text{in.}}{\text{sec}}$

SECTION 7.5

2. Solve $f'(x) = 0$, and find the values of x for which $f'(x)$ does not exist.

5. Solve $f(x) < 0$ by marking a number line with the solutions to $f(x) = 0$ and the numbers at which $f(x)$ does not exist or is discontinuous, and then checking the sign of $f(x)$ on each subinterval.

8. If $f'(c)$ does not exist because $\lim\limits_{x\to c^+} f'(x)$ and $\lim\limits_{x\to c^-} f'(x)$ both exist but are not equal to each other, then you have a non-vertical cusp or corner (a cusp if there is curvature involved and a corner if $f''(x) = 0$ in a punctured interval around $x = c$).

11. $\lim\limits_{x\to c^-} f(x)$, or $\lim\limits_{x\to c^+} f(x)$, or both are infinite.

14. $\lim\limits_{x\to c^-} f(x)$ and $\lim\limits_{x\to c^+} f(x)$ both exist but are not equal.

15. $f(x)$ might have a vertical asymptote on I.

18. $x \neq -\frac{1}{3}$ **20.** $(-\infty, 1) \cup (3, \infty)$

23. Infinite discontinuity at $x = 1$ (limit of $f(x)$ from the left and the right is ∞)

24. Not differentiable at $x = 0$ (vertical cusp)

26. Not differentiable at $x = 0$ (vertical tangent line)

27. Horizontal asymptote at $y = 3$, no vertical asymptotes

28. Horizontal asymptote at $y = 0$ as $x \to \infty$, horizontal asymptote at $y = 2$ as $x \to -\infty$; vertical asymptote at $x = -\frac{1}{3}$

31. $x = -3$, $x = 1$

33. Positive on $(0, \infty)$, negative on $(-\infty, -3) \cup (-3, 0)$

35. Positive on $(-\infty, -\sqrt{5}) \cup (1, \sqrt{5})$, negative on $(-\sqrt{5}, 1) \cup (\sqrt{5}, \infty)$

37. Always decreasing

39. Concave down on $(-\infty, 0) \cup (2, \infty)$, concave up on $(0, 2)$

41. Concave up on $(-\infty, 1) \cup (3, \infty)$, concave down on $(1, 3)$

44. $x = 1$ is the only local extremum of f (a local minimum).

45. $x = 1$, $x = 3$ **46.** $x = \frac{1}{\sqrt{2}}$, $x = -\frac{1}{\sqrt{2}}$

49. Global maximum at $x = 0$, no global minimum

51. Global minimum at $x = 10$, no global maximum

52. Local minimum at $x = 1$, vertical asymptotes at $x = 0$ and $x = 3$:

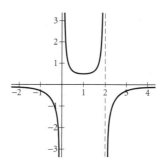

56. Domain $[0, \infty)$, global minimum at $x = 1$:

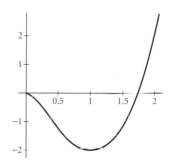

57. Domain $[-1, \infty)$, global minimum at $x = 1$:

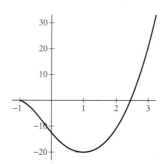

59. Two-sided horizontal asymptote at $x = 3$, global maximum at $x = 0$:

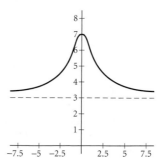

61. Vertical tangent line at $x = 1$, inflection points at $x = 1$ and $x = 3$ (the second inflection point is difficult to see in the graph):

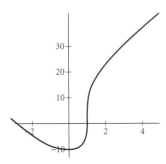

63. Vertical asymptotes at $x = -2$ and $x = 2$, horizontal asymptotes at $y = -1$ and $y = 1$, inflection point at $x = 0$:

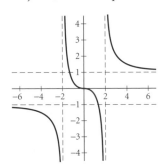

64. To get the minimum surface area, the oil drums should be constructed so they are $\frac{40}{20^{\frac{2}{3}}\pi^{\frac{1}{3}}} \approx 5.4288$ feet high with a radius of $(\frac{20}{\pi})^{\frac{1}{3}} \approx 2.7144$ feet. There is no global maximum.

65. To minimize the cost, the cans should be made with a radius of $\frac{20}{(10\pi)^{\frac{1}{3}}} \approx 2.335$ inches and a height of $2(\frac{10}{\pi})^{\frac{1}{3}} \approx 11.6754$ inches. It is not possible to maximize the cost.

67. $(1, \sqrt{2})$ **68.** 200

70. The steam pipe should be buried along the 800-foot side of the parking lot for 425 feet and then diagonally under the parking lot to the opposite corner.

74. **PROOF** $\lim\limits_{x \to -2} \frac{(x^2-4)^{\frac{2}{5}}}{(x+2)^{\frac{2}{5}}} = \lim\limits_{x \to -2}(x-2)^{\frac{2}{5}} = (-4)^{\frac{2}{5}} \approx$ 1.7411, but $f(-2)$ does not exist. ∎

76. **PROOF** If $f(x) = x(x-2)^{\frac{1}{3}}$, then $f'(x) = \frac{4x-6}{3(x-2)^{\frac{2}{3}}}$. Therefore, $\lim\limits_{x \to 2^+} f'(x) = \infty$ and $\lim\limits_{x \to 2^-} f'(x) = \infty$, and thus $f(x)$ has a vertical tangent line at $x = 2$. ∎

CHAPTER 8

SECTION 8.1

2. $f(x) = 3^x$ is an exponential function (the base is constant and the exponent is the independent variable), whereas $f(x) = x^3$ is a power function (the base is the independent variable and the exponent is a constant).

3. Neither

6. $\sqrt{3} \approx 1.73205$. We have $2^{1.7} \approx 3.2490$, $2^{1.73} \approx 3.3173$, $2^{1.732} \approx 3.3219$, $2^{1.7320} \approx 3.3219$, $2^{1.73205} \approx 3.3220$, and so on. Each of these approximations gets closer to the value of $2^{\sqrt{3}}$.

7. $f(3) = 2^9 = 512$ and $g(3) = 8^2 = 64$. The solutions to $2^{(x^2)} = (2^x)^2$ are $x = 0$ and $x = 2$.

8. Since $f(x) = b^x$ is a function, $x = y \Rightarrow f(x) = f(y)$. Since $f(x) = b^x$ is one-to-one, $f(x) = f(y) \Rightarrow x = y$.

10. False   **11.** True   **12.** False   **13.** True

14. This is an exponential function: $\frac{3}{2^x} = 3(\frac{1}{2})^x$.

16. This is not an exponential function.

18. This is an exponential function: $5^x 2^{3-x} = 8(\frac{5}{2})^x$.

21. This is not an exponential function; it cannot be written in the form Ab^x.

23. $(-\infty, \infty)$ **25.** $(-\infty, 0) \cup (0, \infty)$

28. $[-1, 0) \cup (0, \infty)$ **29.** $x = -3$

30. $x = 6$ **32.** $x = 0$ and $x = 3$ **33.** $\emptyset$

36. $x = 0$ and $x = 2$

37. $x = 0$ and $x = -2$

SECTION 8.2

1. $e := \lim\limits_{h \to 0}(1 + h)^{\frac{1}{h}}$; $e \approx (1 + 0.000001)^{\frac{1}{0.000001}} \approx 2.71828$

4. $f(x) = e^x$; domain $(-\infty, \infty)$, range $(0, \infty)$; it is invertible.

7. (a) $\ln y = x$, $(-\infty, \infty)$, $(0, \infty)$; (b) x, $(0, \infty)$; (c) x, $(-\infty, \infty)$

9. $k = \ln b$, $b = e^k$

11. True   **12.** False   **13.** False   **14.** False

15. (a) 1;   (b) e;   (c) DNE;   (d) 0;   (e) 1

17. 9   **18.** 2   **21.** 6   **23.** ex^2 **24.** $(-\infty, \infty)$

27. $(-\infty, -1) \cup (-1, 2) \cup (2, \infty)$

28. $[-\frac{1}{2}, \frac{\ln 5}{3}) \cup (\frac{\ln 5}{3}, \infty)$ **31.** $x = \frac{1}{3}\ln 5 \approx 0.536479$

33. $x = \frac{\ln 2}{\ln 3} \approx 0.63093$ **36.** $x = 2$, $x = -3$

37. $x = \pm\sqrt{\frac{e^{\frac{7}{4}}+3}{2}} \approx \pm 2.0922$

38. $x = \frac{\sqrt{6}}{e^2} \approx 0.331502$

40. $f(x) = 2e^{(\ln 3)x} \approx 2e^{1.0986x}$

41. $f(x) = 1.2e^{(\ln 0.72)x} \approx 1.2e^{-0.328504}$

44. $f(x) = 1.2(e^{-7.2})^x \approx 1.2(0.000747)^x$

46. $f(x) = \frac{2}{e^4}(e^3)^x \approx 0.03663(20.0855)^x$

47. $f(x) = 9e^{\frac{\ln(\frac{2}{3})}{2}x} \approx 9e^{-0.2027x}$

49. $f(x) = -3e^{\frac{\ln(\frac{7}{3})}{5}x} \approx -3e^{0.16946x}$

51. $f(x) = 2e^{-(\frac{\ln 32}{5})x} \approx 2e^{-0.693147x}$

54. 1999 (after approximately $t \approx 9.006$ years)

60. **PROOF** Given any exponential function $f(x) = Ae^{kx}$ (this assumes that $k \neq 0$), define $b := e^k$ (note that since $k \neq 0$, we have $b \neq 1$; moreover, $b > 0$). Now $f(x) = Ae^{kx} = A(e^k)^x$ (by algebra) $= Ab^x$ (by our definition of b). ∎

SECTION 8.3

2. For all $c > 0$, $\lim\limits_{x \to c} \ln x = \ln c$.

5. $\lim\limits_{x \to \infty} e^x = \infty$

8. For all $\varepsilon > 0$, there exists $\delta > 0$ such that $0 < |x - c| < \delta \Rightarrow |e^x - e^c| < \varepsilon$.

10. For all $\varepsilon > 0$, there exists $\delta > 0$ such that $0 < |x - 3| < \delta \Rightarrow |5(2^x) - 40| < \varepsilon$.

12. For all $\varepsilon > 0$, there exists $N < 0$ such that $x < N \Rightarrow |e^{3x}| < \varepsilon$.

13. For all $M > 0$, there exists $N > 0$ such that $x > N \Rightarrow 2^x > M$.

18. 0   **20.** $(0, \infty)$ **21.** $(0, 1)$ **24.** ∞

26. (a) $\lim\limits_{x \to -\infty} \dfrac{\left(\frac{1}{2}\right)^x - 1}{\left(\frac{3}{4}\right)^x} \to \dfrac{\infty}{\infty}$, which is indeterminate

(b) $\lim\limits_{x \to -\infty} \dfrac{3^x \left(\left(\frac{2}{3}\right)^x - \left(\frac{4}{3}\right)^x\right)}{3^x} = \infty$

28. 48   **29.** $-\frac{1}{4}$ **31.** $\frac{7}{27}$ **33.** ∞ **35.** $\frac{1}{2}$ **36.** $\frac{2}{5}$

39. 4   **41.** ∞ **43.** 0   **45.** ∞ **49.** 0   **50.** 2

51. Horizontal asymptote at $y = 25$ on the right

53. Horizontal asymptotes at $y = \frac{1}{2}$ on the right and $y = 0$ on the left

54. Horizontal asymptotes at $y = 0$ on the right and $y = \frac{1}{2}$ on the left

62. **PROOF**   If $k > 0$, then $\lim\limits_{x \to \infty} e^{kx} = \lim\limits_{x \to \infty} (e^x)^k$

$= (\lim\limits_{x \to \infty} e^x)^k \to \infty^k \to \infty$ ■

SECTION 8.4

2. $f'(x) = 5x^4$ (power rule); $g'(x) = (\ln 5)5^x$ (exponential rule)

3. No; No   **6.** $e^{u(x)} u'(x)$; $(\ln b) b^{u(x)} u'(x)$

8. Without algebra first: $\dfrac{0(\sqrt{3^x}) - 1\left(\frac{1}{2}(3^x)^{-\frac{1}{2}} (\ln 3)3^x\right)}{(\sqrt{3^x})^2}$

With algebra first: $\dfrac{d}{dx}\left(3^{-\frac{1}{2}x}\right) = -\dfrac{\ln 3}{2\sqrt{3^x}}$

11. $\dfrac{d}{dx}\left(2^{(x^2)}\right) = (\ln 2)2^{(x^2)}(2x)$

12. $f(x) = \frac{2}{3}e^{3x} + C$ for any constant C

13. $f(x) = \frac{1}{\ln 2} 2^x + C$ for any constant C

16. $f'(x) = -2e^{-2x}$

18. $f'(x) = 5(3 - e^x)^4(-e^x)$

22. $f'(x) = -(2 - e^{5x})^{-2}(-5e^{5x})$

23. $f'(x) = \dfrac{(\ln 2)2^{3x}(3)(x^2 - 1) - 2^{3x}(2x)}{(x^2 - 1)^2}$

24. $f'(x) = 2e^{\frac{3}{2}x + \frac{1}{2}}\left(\frac{3}{2}\right)$

26. $f'(x) = 3\left(\frac{2}{3}\right)^x + 3(\ln \frac{2}{3})x\left(\frac{2}{3}\right)^x$

28. $f'(x) = (\ln 2)2^{1 - 3^x}(-(\ln 3)3^x)$

31. $f'(x) = e^{ex + 1}$ **34.** $x = 0$, $x = \frac{-2}{\ln 3}$

36. $x = \ln(\frac{3}{2})$ **38.** $x = \frac{\ln 2 - 1}{3}$

39. The number of fruit flies in the population

40. (a) $r(t) = 0.2667t + 45$; $r(99) \approx 71.4$ rats

(b) $r(t) = 45e^{0.00545t}$; $r(99) \approx 77.2$ rats

41. (a) $A(t) = 1000e^{0.077t}$;

(b) $A(30) \approx \$10{,}074.42$;

(c) $t \approx 18$ years

44. **PROOF** $\dfrac{d}{dx}(b^x) = \dfrac{d}{dx}(e^{(\ln b)x}) = (\ln b)e^{(\ln b)x} = (\ln b)b^x$ ■

SECTION 8.5

1. A

3. If $f(x) = 2(3^x)$, then $f(-x) = 2(\frac{1}{3})^x$; therefore, the graph of $y = 2(\frac{1}{3})^x$ can be obtained from the graph of $y = 2(3^x)$ by reflection over the y-axis.

6. $(-\infty, 0)$ **7.** $(1, \infty)$ **10.** $f(x) = 2e^{-x} - 3$

11. $f(x) = -5e^{-x} + 10$

12. True   **13.** True   **14.** False   **15.** False

16. True   **17.** True   **18.** False   **19.** False

21. The graph passing through $(1, 2)$ is 2^x; the graph passing through $(-1, 2)$ is $(\frac{1}{2})^x$; the graph passing through $(-1, 5)$ is $(\frac{1}{5})^x$.

23. The graph passing through $(0, 2)$ is $2(2^x)$; the graph passing through $(0, 1)$ and $(1, 4)$ is 4^x; the graph passing through $(2, 4)$ is 2^x.

25.

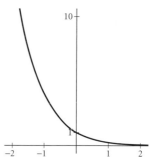

27.

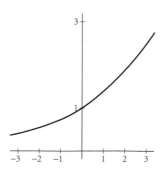

28.

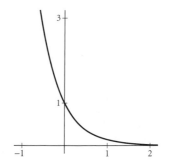

30.

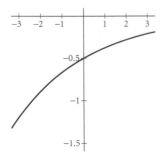

33.

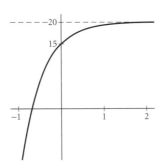

35. Start with the graph of $y = e^{3x}$; then shift left two units and stretch vertically by a factor of 4.

37. Start with the graph of $y = (\frac{1}{2})^x$; then reflect over the x-axis and shift up ten units.

38. Domain is $(-\infty, \infty)$, always increasing and concave down, $\lim\limits_{x \to \infty} f(x) = 400$, $\lim\limits_{x \to -\infty} f(x) = -\infty$

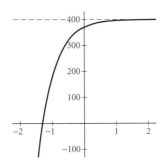

40. Domain is $(-\infty, \infty)$, always increasing, concave up on $(-\infty, -0.604)$, concave down on $(-0.604, \infty)$, $\lim\limits_{x \to \infty} f(x) = 2$, $\lim\limits_{x \to -\infty} f(x) = 0$

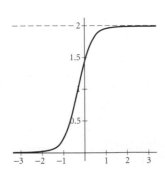

42. Domain is $(-\infty, 0]$, always decreasing and concave down, $\lim\limits_{x \to -\infty} f(x) = \infty$

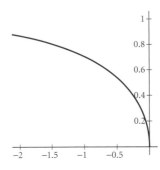

Section 8.6

2. $Q(t)$ **3.** Constant; exponential   **6.** 1.16583%

7. 15% compounded daily is better.

10. $Q(17) = \frac{1}{2}Q_0$, $Q(51) = \frac{1}{8}Q_0$, $Q(-34) = 4Q_0$

11. Yes, since $\frac{dW}{dt}$ is proportional to $W(t)$, and thus $W(t)$ is an exponential function.

13. e^6 **14.** $\frac{1}{e}$ **18.** $e^{-\frac{3}{2}}$

20. $\ln(1.24) \approx 0.215$

21. 31.71%   **23.** $\frac{1}{32}\ln(\frac{1}{2}) \approx -0.02166$

25. $\frac{\ln 2}{\ln 1.13} \approx 5.67$ years

26. (a) 16.98 years; (b) 9.93 years; (c) $\ln(0.96) \approx -0.0408$

30. (a) If yearly, \$98,711.43; if monthly, \$131,571.79; if daily, \$135,428.50; if four times a day, \$135,529.50

(b) If continuously, \$135,563.19

(c) If yearly, 22% a year; if monthly, 24.3597% a year; if daily, 24.5994% a year; if continuously, 24.6077% a year

(d) If yearly, 26.64 years; if monthly, 24.3 years; if daily, 24.09 years; if continuously, 24.08 years

31. (a) 11.869%;   (b) $\ln(\frac{840}{600})/3 \approx 0.11216$;
(c) 12.36 days;   (d) 66.14 days

32. (a) 21.328 hours;   (b) 2.5237 hours;
(c) 31.607%;   (d) 72,800%

33. (a) 2.36%;
(b) $\ln(\frac{1}{2})/29 \approx -0.0239$;
(c) 156 years;
(d) 53.6 years ago;
(e) 0.19233 gram will be left, which is 0.0769% of 250 grams.
(f) 125.34 years

35. **PROOF**   If $Q(12) = 2Q_0$ then $Q_0 e^{k(12)} = 2Q_0$, and thus $e^{12k} = 2$. Then $Q(t + 12) = Q_0 e^{k(t+12)} = Q_0 e^{kt+12k} = Q_0 e^{kt} e^{12k} = Q_0 e^{kt}(2) = 2(Q_0 e^{kt}) = 2Q(t)$.   ■

38. $\alpha = \frac{\ln 2}{k}$;

PROOF   If α is the doubling time, and $Q(t) = Q_0 e^{kt}$, then $2Q_0 = Q_0 e^{k\alpha}$. Solve for α: $2 = e^{k\alpha} \Rightarrow \ln 2 = k\alpha \Rightarrow \alpha = \frac{\ln 2}{k}$.   ■

SECTION 8.7

2. (a) $\pm\infty$; **(b)** 0; **(c)** Indeterminate; **(d)** ∞; **(e)** 0;
(f) Indeterminate; **(g)** Indeterminate; **(h)** ∞;
(i) 0; **(j)** ∞; **(k)** ∞; **(l)** Indeterminate

4. (a) $f(x) = x - 2$, $g(x) = (x-2)^3$
(b) $f(x) = (x-2)^2$, $g(x) = x - 2$
(c) $f(x) = x - 2$, $g(x) = x - 2$
(d) $f(x) = 3(x-2)$, $g(x) = x - 2$

6. $\lim\limits_{x\to\infty} \frac{x}{2^x} \to \frac{\infty}{\infty}$, and $\lim\limits_{x\to\infty} \frac{1/2^x}{1/x} \to \frac{0}{0}$; the first case makes L'Hôpital's Rule simpler to apply.

9. $\lim\limits_{x\to\infty} \frac{2^x}{3^x}$; using algebra, this is equal to $\lim\limits_{x\to\infty} (\frac{2}{3})^x = 0$.

11. $\lim\limits_{x\to\infty} \frac{f(x)}{g(x)} = \infty$

12. The only error is in the second application of L'Hôpital's Rule (it does not apply).

13. False   **14.** False   **15.** False   **16.** False   **17.** False
18. False   **19.** True   **20.** True   **21.** False

22. (a) $\lim\limits_{x\to 1} \frac{2x+1}{1} = 3$; **(b)** $\lim\limits_{x\to 1}(x+2) = 3$

23. (a) $\lim\limits_{x\to\infty} \frac{1}{-6x} = 0$; **(b)** $\lim\limits_{x\to\infty} \frac{\frac{1}{x}-\frac{1}{x^2}}{\frac{2}{x^2}-3} = \frac{0-0}{0-3} = 0$

26. (a) $\lim\limits_{x\to\infty} \frac{3e^{3x}}{-2e^{2x}} = \lim\limits_{x\to\infty} \frac{3e^x}{-2} = -\infty$
(b) $\lim\limits_{x\to\infty} \frac{1}{\frac{1}{e^{3x}}-\frac{1}{e^x}} \to \frac{1}{0^-} \to -\infty$

28. $-8\ln 2$ **30.** $\frac{1}{2}$ **32.** 0   **34.** 0   **36.** ∞ **38.** 0
39. 0   **40.** 0   **44.** g dominates f.   **45.** g dominates f.
48. Neither function dominates the other.
49. 0   **51.** 0

CHAPTER 9

SECTION 9.1

1. The inverse of the function b^x

3. (a) Domain $(0, \infty)$, range $(-\infty, \infty)$
(b) Domain $(3, \infty)$, range $(-\infty, \infty)$

4. $(0, \infty)$; 2^y **7.** 2; 3

9. $\log_b(\frac{x}{y})$ **11.** $a\log_b x$

13. The graph that passes through $(2, 1)$ is $\log_2 x$; the graph that passes through $(2, \frac{1}{2})$ is $\log_4 x$.

15. $f(x) = 2\ln x$ is the graph of $\ln x$ stretched vertically by a factor of 2; $g(x) = -\ln x$ is the graph of $\ln x$ reflected over the x-axis.

16. False   **17.** False   **18.** False
19. False   **20.** False   **21.** False

24. $\log_2 x = \log_3 x \Longleftrightarrow \frac{\ln x}{\ln 2} = \frac{\ln x}{\ln 3} \Longleftrightarrow (\ln 3)(\ln x) = (\ln 2)(\ln x) \Longleftrightarrow (\ln x)(\ln 3 - \ln 2) = 0 \Longleftrightarrow \ln x = 0 \Longleftrightarrow x = 1$

25. 3   **27.** $-\frac{1}{2}$ **30.** 1   **32.** -2 **33.** 0
35. $-\frac{1}{6}$ **38.** 1   **40.** 4   **42.** $\frac{1}{2}$ **43.** -2
45. 6.69897   **47.** -3.16993

49. $(-2, -1) \cup (-1, \infty)$
51. $(-\frac{2}{3}, -\frac{1}{3}) \cup (-\frac{1}{3}, 0) \cup (0, \infty)$
53. $(-\infty, -1) \cup (2, \infty)$
54. $(2, 3) \cup (3, \infty)$
55. $[3, \infty)$
59. $x = 2$
60. $x = -\frac{17}{15}$
64. $x = \frac{1}{2}(\log_7(\frac{8}{5}) + 1) \approx 0.752$
65. $x = \frac{-\ln 3}{\ln 2 - \ln 3} \approx 2.70951$
68. $x = \log_3 2 \approx 0.63093$
69. $x = \log_2(24) \approx 4.585$
70. Stretch $y = \log_4 x$ vertically by a factor of 3; then shift down eight units.

73. **PROOF** $y = \log_b x$ if and only if $b^y = x$. Since the only solution to $b^y = 1$ is $y = 0$ (if $b \neq 0$), we know that $\log_b 1 = 0$. ▪

79. **PROOF** $\log_b(\frac{x}{y}) = \log_b(x \cdot y^{-1}) = \log_b x + \log_b(y^{-1}) = \log_b x - \log_b y$ ▪

SECTION 9.2

1. The limit does not exist; specifically, $\lim\limits_{x\to 1^-} \ln(x - 1)$ does not exist, and $\lim\limits_{x\to 1^+} \ln(x - 1) = -\infty$.

3. (a) $\lim\limits_{x\to\infty} \frac{(\frac{2x^4}{x^5})}{(\frac{4x^3}{x^4})} = \lim\limits_{x\to\infty} \frac{5}{4} = \frac{5}{4}$
(b) $\lim\limits_{x\to\infty} \frac{5\ln x}{4\ln x} = \lim\limits_{x\to\infty} \frac{5}{4} = \frac{5}{4}$

4. (a) $\frac{\frac{5x^4}{x^5}\ln(x^4)-\ln(x^5)\frac{4x^3}{x^4}}{(\ln(x^4))^2}$ (which equals 0 after simplification)
(b) $\frac{d}{dx}(\frac{5\ln x}{4\ln x}) = \frac{d}{dx}(\frac{5}{4}) = 0$

7. $\lim\limits_{h\to 0} \frac{\log_b(x+h)-\log_b x}{h} = \frac{1}{(\ln b)x}$

8. (a) $-\infty$; **(b)** ∞; **(c)** $-\infty$; **(d)** ∞
12. 0
14. $f(x) = \frac{3}{2}\ln(2x+1) + (6 - \frac{3}{2}\ln 5)$
16. $\log_3 21$
18. $-\infty$ **20.** 0   **21.** 0   **22.** ∞
25. 1   **26.** 1   **30.** 0
32. $f'(x) = \frac{1}{2x}$; $x = 0$ is the only critical point.
33. $f'(x) = \frac{1}{2x\sqrt{\ln x}}$; critical points $x = 0$, $x = 1$
34. $f'(x) = \frac{\ln 3}{\ln 2}$; no critical points
36. $f'(x) = (\ln 2)2^x$; no critical points
38. $f'(x) = \frac{x^2+2x+3}{x(x^2+x+1)}$; only critical point is $x = 0$
40. $f'(x) = 2x\log_2 x + \frac{x}{\ln 2} + 3x^2$
42. $f'(x) = \frac{(e^x\ln x+\frac{e^x}{x})(x^2-1)-2xe^x\ln x}{(x^2-1)^2}$
43. $f'(x) = 2x\ln(\ln x) + x^2\frac{1}{\ln x}\cdot\frac{1}{x}$
48. g dominates f.
49. Neither function dominates.
51. Neither function dominates.

52. Domain $(0, \infty)$, local minimum at $x = \frac{1}{e}$, always concave up, $\lim\limits_{x\to 0^+} f(x) = 0$, $\lim\limits_{x\to\infty} f(x) = \infty$

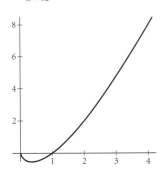

54. Domain $(1, \infty)$, always concave up and increasing, root at $x = e$

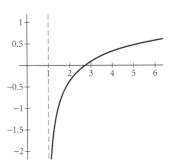

56. Domain $(-\infty, -1) \cup (1, \infty)$, roots at $x = \pm\sqrt{2}$, $\lim\limits_{x\to -1^-} f(x) = \infty$, $\lim\limits_{x\to 1^+} f(x) = \infty$, $\lim\limits_{x\to -\infty} f(x) = -\infty$, $\lim\limits_{x\to\infty} f(x) = -\infty$

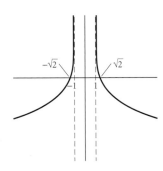

60. PROOF   Differentiate both sides of $e^{\ln x} = x$: $e^{\ln x}\frac{d}{dx}(\ln x) = 1$, so $\frac{d}{dx}(\ln x) = \frac{1}{e^{\ln x}} = \frac{1}{x}$. ∎

Section 9.3

1. **(a)** $3^{x-1} = 2 \Rightarrow \ln(3^{x-1}) = \ln 2 \Rightarrow (x-1)(\ln 3) = (\ln 2) \Rightarrow x = \frac{\ln 2}{\ln 3} + 1$
(b) $3^{x-1} = 2 \Rightarrow \log_3(3^{x-1}) = \log_3 2 \Rightarrow x - 1 = \log_3 2 \Rightarrow x = (\log_3 2) + 1$
The answers are the same by the base conversion formula.

4. **(a)** 0; **(b)** Indeterminate; **(c)** ∞; **(d)** ∞;
(e) Indeterminate; **(f)** 1; **(g)** 0; **(h)** Indeterminate;
(i) ∞; **(j)** ∞; **(k)** 0; **(l)** Indeterminate

6. **(a)** $\lim\limits_{x\to 3} f(x) = 1$;   **(b)** $\lim\limits_{x\to 1} f(x) = \infty$;
(c) $\lim\limits_{x\to 0} f(x) = 0$

8. $f'(2) = -12$ **12.** $x = \frac{\ln 3}{\ln 2 - \ln 3}$ **16.** $x = \frac{\ln(\frac{5}{3})}{\ln 3}$

17. $x = -\frac{\ln 4}{\ln 3} - 2$ **19.** ∞ **23.** 1   **24.** e **25.** 1

27. $\lim\limits_{x\to\infty} \left(\frac{1}{x+1}\right)^x \to 0^\infty \to 0$ **28.** e **29.** 1

30. **(a)** $(\lim\limits_{x\to\infty}(1+\frac{3}{x})^x)^2 = (e^3)^2 = e^6$
(b) $\lim\limits_{x\to\infty}(2x\ln(1+\frac{3}{x})) = 6$, so $\lim\limits_{x\to\infty}(1+\frac{3}{x})^{2x} = e^6$

32. $f'(x) = \frac{6x}{3x^2+4}$

34. $f'(x) = \frac{1}{2}(x\ln|2^x+1|)^{-\frac{1}{2}}(\ln|2^x+1| + \frac{(\ln 2)x2^x}{2^x+1})$

37. $f'(x) = (\frac{e^{2x}(x^3-2)^4}{x(3e^{5x}+1)})(2 + \frac{4(3x^2)}{x^3-2} - \frac{1}{x} - \frac{15e^{5x}}{3e^{5x}+1})$

38. $f'(x) = \frac{2^x\sqrt{x^3-1}}{\sqrt{x}(2x-1)}(\ln 2 + \frac{3x^2}{2(x^3-1)} - \frac{1}{2x} - \frac{2}{2x-1})$

39. $f'(x) = \sqrt{\frac{x^2(3x+1)^{99}}{4x^2-3x+2}}(\frac{1}{2})(\frac{2}{x} + \frac{99(3)}{3x+1} - \frac{8x-3}{4x^2-3x+2})$

42. $f'(x) = x^{\ln x}(2)(\ln x)(\frac{1}{x})$

45. $f'(x) = (\frac{x}{x-1})^x(\ln x - \ln(x-1) + 1 - \frac{x}{x-1})$

46. $f'(x) = (\ln x)^{\ln x}(\frac{\ln(\ln x)}{x} + \frac{1}{x})$

49. PROOF $\lim\limits_{x\to\infty}\ln((1+\frac{r}{x})^x) = \lim\limits_{x\to\infty} x\ln(1+\frac{r}{x}) =$
$\lim\limits_{x\to\infty}\frac{\ln(1+\frac{r}{x})}{\frac{1}{x}} \overset{L'H}{=} \lim\limits_{x\to\infty}\frac{\frac{1}{1+\frac{r}{x}}(\frac{-r}{x^2})}{\frac{-1}{x^2}} = \lim\limits_{x\to\infty}\frac{r}{1+\frac{r}{x}} = r$, and thus $\lim\limits_{x\to\infty}(1+\frac{r}{x})^x = e^r$ ∎

50. PROOF   Let $y = Ax^k$. We want to show that $y' = kAx^{k-1}$. Taking logs of both sides gives us $\ln|y| = \ln|Ax^k| = \ln|A| + k\ln|x|$. Differentiating both sides yields $\frac{1}{y}y' = \frac{k}{x}$, so $y' = y(\frac{k}{x}) = (Ax^k)(\frac{k}{x}) = kAx^{k-1}$. ∎

54. PROOF   Let $y' = \frac{f(x)}{g(x)}$. Taking logs of both sides gives us $\ln|y| = \ln|f(x)| - \ln|g(x)|$; differentiating yields $\frac{1}{y}y' = \frac{f'(x)}{f(x)} - \frac{g'(x)}{g(x)}$, so $y' = \frac{f(x)}{g(x)}(\frac{f'(x)}{f(x)} - \frac{g'(x)}{g(x)}) = \frac{f(x)}{g(x)}(\frac{f'(x)g(x)-g'(x)f(x)}{f(x)g(x)}) = \frac{f'(x)g(x)-g'(x)f(x)}{(g(x))^2}$. ∎

CHAPTER 10

Section 10.1

2. They are *all* equal to $\cot\theta$.

3. All six trigonometric functions are ratios of side lengths, and side lengths are always positive.

4. The length of the hypotenuse of a triangle is always greater than the length of either leg of the triangle; thus $\sin\theta = \frac{\text{opp}}{\text{hyp}} < 1$ for all triangles. $\tan\theta$ is *not* always less than 1, since the length of the side of a triangle opposite to an angle θ is *not* always less than the length of the adjacent side.

7. **(a)** False; **(b)** True; **(c)** False

8. $\frac{5}{13}; \frac{12}{13}; \frac{5}{12}; \frac{13}{12}; \frac{13}{5}; \frac{12}{5}$

10. 0.981627   **13.** 3.23607   **15.** 0.017455

16. 1.439557   **20.** $\frac{1}{\sqrt{3}}$ **22.** 2   **23.** 1

25. $\sqrt{2}$ **27.** $\frac{5}{\sqrt{3}}$ (leg); $\frac{10}{\sqrt{3}}$ (hypotenuse)

29. $x \approx 3.38095$

31. $x \approx 19.6261$

33. $A = \frac{9\sqrt{3}}{4}$ square units

35. 211.97 feet

36. 2.0946 light-years apart

37. **PROOF**   **(a)** Suppose θ is an acute angle. Then $\csc\theta = \frac{\text{hyp}}{\text{opp}} = \frac{1}{\text{opp/hyp}} = \frac{1}{\sin\theta}$. The remaining three parts are similar. ▪

40. **PROOF**   Suppose a and b are the lengths of the legs of the triangle (with a as the side "opposite" θ). Then $(\sin\theta)^2 + (\cos\theta)^2 = (\frac{a}{1})^2 + (\frac{b}{1})^2 = a^2 + b^2$, which by the Pythagorean Theorem is equal to $a^2 + b^2 = 1^2 = 1$. ▪

Section 10.2

1. The circumference of a circle with radius $r = 1$ is $C = 2\pi r = 2\pi(1) = 2\pi$.

3. $\frac{10\pi}{3}, \frac{16\pi}{3}, -\frac{2\pi}{3}, -\frac{8\pi}{3}$

6. Because the distance around the unit circle is 2π; No

7. If θ is an angle in standard position, then $\sin\theta$ is the vertical coordinate y of the point (x, y) where the terminal edge of θ intersects the unit circle.

9. $\frac{\sqrt{15}}{4}; -\frac{1}{4}; -\sqrt{15}$

10. $\cos\theta = -\frac{\sqrt{8}}{3}$ if θ is in the third quadrant; $\cos\theta = \frac{\sqrt{8}}{3}$ if θ is in the fourth quadrant; θ cannot be in the first or second quadrant.

12. The terminal edges of the angles $\frac{\pi}{4}, \frac{9\pi}{4}$, and $-\frac{7\pi}{4}$ all meet the unit circle at the same point (and, in particular, at the same y-coordinate).

14. $\frac{3\sqrt{5}}{7}; -\frac{2}{7}; -\frac{3\sqrt{5}}{2}$ **16.** 30   **18.** Second or fourth

20. True   **21.** False   **22.** True   **23.** False

24. True   **25.** False   **26.** False   **27.** False

28. True   **29.** True   **30.** True   **31.** True

32. False   **34.** $-\frac{\pi}{6}$ **36.** 4π

38. $150°$ **40.** $-135°$

42. $\frac{720}{\pi} \approx 229.18°$

44. Note that this angle goes around the unit circle more than once and in the negative directon:

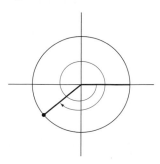

46. Note that $5(\frac{180}{\pi}) \approx 286.5°$:

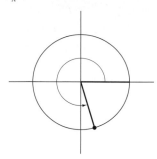

50. The reference angle is $30°$:

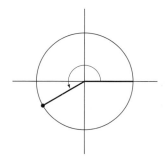

52. The reference angle is $60°$:

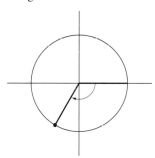

53. The reference angle is $45°$:

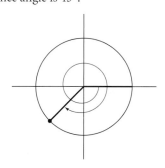

55. $\frac{201\pi}{2}$ terminates at the same place as $\frac{\pi}{2}$ radians; no reference triangle exists.

56. 0   **58.** 0   **59.** $-\frac{\sqrt{3}}{2}$ **61.** -1 **63.** $\sqrt{2}$ **64.** $-\frac{1}{\sqrt{3}}$

66. -1 **70.** Does not exist   **71.** $(\frac{1}{2}, \frac{\sqrt{3}}{2})$ **73.** $(0, -1)$

75. $(-\frac{3\sqrt{2}}{2}, -\frac{3\sqrt{2}}{2})$ **77.** 0.91162   **79.** -1.0257

80. 0.84147   **84.** $-\frac{\sqrt{24}}{5}$ **86.** $-\sqrt{8}$

SECTION 10.3

1. No; Yes   **3.** $\theta = \frac{\pi}{3}$

5. $\{\theta \mid \theta \neq \frac{\pi}{2} + \pi k\}; (-\infty, -1] \cup [1, \infty)$

7. $\sin \theta$ is the y-coordinate of the point where the terminal edge of θ meets the unit circle; every angle θ has such a well-defined y-coordinate, and each of those y-coordinates must be between -1 and 1 (inclusive).

10. $\cos \theta = 0$ if the terminal edge of θ intersects the unit circle at $(0, 1)$ or $(0, -1)$, that is, if $\theta = \frac{\pi}{2} + \pi k$ for any integer k.

12. The equation fails for $\theta = 0$.

15. Cosine and secant are even functions; the remaining four are odd.

17. $\theta = \frac{\pi}{2}$

20. $\sin(\frac{4\pi}{3}) = -\frac{\sqrt{3}}{2}$ and $\cos(\frac{5\pi}{6}) = -\frac{\sqrt{3}}{2}$; $\cos(\frac{4\pi}{3}) = -\frac{1}{2}$ and $\sin(\frac{11\pi}{6}) = -\frac{1}{2}$

21. $\sec^2 \theta$ **22.** $-\sin \theta$ **25.** $\cos 2\theta$ **28.** $\cos \theta$

30. The way the proof is written, it says that *if* $\sin 2\theta = 2 \sin \theta \cos \theta$ is true (which is what we meant to *prove*, not *assume*), then something obvious is true (namely that $2 \sin \theta \cos \theta = 2 \sin \theta \cos \theta$). This is bad logic; rewrite the proof so that it is a chain of equalities starting with $\sin 2\theta$ and ending with $2 \sin \theta \cos \theta$.

31. (a) $\theta = \frac{\pi}{2}$
    (b) $\theta = \frac{\pi}{2} + 2\pi k$, where k is any integer

33. (a) $\theta = \frac{\pi}{3}, \theta = \frac{5\pi}{3}$
    (b) $\theta = \frac{\pi}{3} + 2\pi k$ and $\theta = \frac{5\pi}{3} + 2\pi k$

35. (a) $\theta = \frac{\pi}{3}, \theta = \frac{2\pi}{3}$
    (b) $\theta = \frac{\pi}{3} + 2\pi k, \theta = \frac{2\pi}{3} + 2\pi k$

36. (a) $\theta = \frac{5\pi}{6}, \theta = \frac{11\pi}{6}$
    (b) $\theta = \frac{5\pi}{6} + \pi k$

37. (a) $\theta = \frac{\pi}{4}, \theta = \frac{3\pi}{4}, \theta = \frac{5\pi}{4}, \theta = \frac{7\pi}{4}$
    (b) $\theta = \frac{\pi}{4} + \frac{\pi}{2} k$

40. $(-\infty, \infty)$ **42.** $\theta \neq \frac{\pi}{2} k$, for any integer k

45. $\cdots \cup (-\frac{5\pi}{2}, -\frac{3\pi}{2}) \cup (-\frac{\pi}{2}, \frac{\pi}{2}) \cup (\frac{3\pi}{2}, \frac{5\pi}{2}) \cup \cdots$

46. Neither: $f(-\theta) \neq f(\theta)$, $f(-\theta) \neq -f(\theta)$

47. Odd: $f(-\theta) = -f(\theta)$

51. Even: $f(-\theta) = f(\theta)$

52. $\frac{\sqrt{2}-\sqrt{6}}{4}$ (use $-\frac{\pi}{12} = \frac{\pi}{4} - \frac{\pi}{3}$)

54. $\sqrt{3} - 2$ **56.** $\frac{2\sqrt{2}}{1-\sqrt{3}}$

63. **PROOF** $\cot(-\theta) = \frac{\cos(-\theta)}{\sin(-\theta)} = \frac{\cos \theta}{-\sin \theta} = -\cot(\theta)$ ∎

67. **PROOF** $\frac{1-\cos 2\theta}{2} = \frac{1-(1-2\sin^2 \theta)}{2} = \frac{2\sin^2 \theta}{2} = \sin^2 \theta$.
The proof of the second identity is similar. ∎

70. **PROOF**   First of all, $\frac{2\tan \theta}{1-\tan^2 \theta} = \frac{2\frac{\sin \theta}{\cos \theta}}{1 - \frac{\sin^2 \theta}{\cos^2 \theta}} = \frac{2\sin \theta \cos \theta}{\cos^2 \theta - \sin^2 \theta}$.

On the other hand, $\tan 2\theta = \frac{\sin 2\theta}{\cos 2\theta} = \frac{2\sin \theta \cos \theta}{\cos^2 \theta - \sin^2 \theta}$. Thus $\tan 2\theta = \frac{2\tan \theta}{1-\tan^2 \theta}$. ∎

72. **PROOF** $(\sin \theta + \cos \theta)^2 = \sin^2 \theta + 2 \sin \theta \cos \theta + \cos^2 \theta = 2 \sin \theta \cos \theta + 1 = \sin 2\theta + 1$ ∎

SECTION 10.4

2. 2   **4.** Removable discontinuity

7. (a) Shift up two units;   (b) shift left two units;   (c) vertical stretch by a factor of 2;   (d) horizontal shrink by a factor of 2

8. Tangent and cotangent have period π; the remaining four have period 2π.

10. $\lim_{x\to 0}(\cos x - 1) = 0$ implies that $\lim_{x\to 0} \cos x = 1$; therefore $\lim_{x\to 0} \cos x = \cos 0$, so $\cos x$ is continuous at $x = 0$.

14. 0   **15.** Does not exist (oscillates)

16. False   **17.** True   **18.** False   **19.** True

20. True   **21.** True   **22.** $-\frac{1}{2}$ **24.** ∞

26. $-\infty$ from the left, ∞ from the right

29. 0   **30.** 4   **31.** 0   **34.** 0   **35.** ∞ **36.** -1

38. 1   **40.** 4   **42.** 0   **48.** 0   **49.** Does not exist   **50.** 0

52. Discontinuous at $x = 0$: $\lim_{x\to 0} f(x)$ does not exist.

53. Continuous at $x = 0$: $\lim_{x\to 0} f(x) = f(0) = 0$

55. (a) $\lim_{t\to\infty} s(t)$ does not exist, since both sine and cosine oscillate between -1 and 1 as $x \to \infty$.
    (b) There is no friction to damp the oscillations of the spring.
    (c) $s(t) = \sqrt{2} \sin(\sqrt{\frac{9}{2}}\, t) + 2 \cos(\sqrt{\frac{9}{2}}\, t)$:

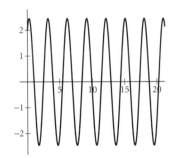

56. (a) $\lim_{t\to\infty} s(t) = 0$
    (b) Friction causes each oscillation to be smaller than the last.
    (c) $s(t) = e^{-\frac{3}{2}t}(4 \sin(\frac{3}{2}t) + 2 \cos(\frac{3}{2}t))$:

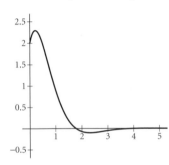

61. **PROOF** $\sec x = \frac{1}{\cos x}$ is the reciprocal of the continuous function $\cos x$, and thus $\sec x$ is continuous for all x where $\cos x \neq 0$ (i.e., for all x in the domain of secant). ∎

SECTION 10.5

2. The height of $y = \cos x$ is the slope of $y = \sin x$ at $x = \frac{\pi}{2}$; the height of $y = \sin x$ is the negative of the slope of $y = \cos x$ at $x = \frac{\pi}{2}$.

4. Bad differentiation in the denominator when applying L'H; also $\frac{-\sin\frac{\pi}{2}}{-\cos\pi} = -1$. The correct calculation shows that the limit is equal to 0.

6. If x is in degrees, then the slope of the graph of $\sin x$ at $x = 0$ is very small and, in particular, not equal to $\cos 0 = 1$. To convince yourself of this, graph $\sin x$ (in degrees) together with the line $y = x$ (which has slope 1 at $x = 0$).

7. $f'(x) = \frac{2x\cos x + (x^2+1)\sin x}{\cos^2 x}$

9. $f'(x) = 3 + 2\csc x \cot x$

11. $f'(x) = \sec(x^3) + x\sec(x^3)\tan(x^3)(3x^2)$

13. $f'(x) = \cos x \ln x + \frac{\sin x}{x}$

14. $f'(x) = \cos(\ln x)(\frac{1}{x})$

15. $f'(x) = \frac{\cos x}{\sin x}$

21. $f'(x) = \cos(\cos(\sec x))(-\sin(\sec x))(\sec x \tan x)$

22. $f'(x) = 2\csc(e^x)(-\csc(e^x)\cot(e^x))(e^x)$

23. $f'(x) = e^{\csc^2 x}(2\csc x)(-\csc x \cot x)$

25. $f'(x) = \sqrt{\sin x \cos x} + \frac{1}{2}x(\sin x \cos x)^{-\frac{1}{2}}(\cos^2 x - \sin^2 x)$

26. $f'(x) = \frac{\cos^3 x + 2\sin^2 x \cos x}{\cos^4 x}$

30. $f'(x) = (\sin x)^{\cos x}(-\sin x \ln(\sin x) + \frac{\cos^2 x}{\sin x})$

33. (a) $\lim_{x \to 0}\left(\frac{\sin 3x}{3x} \cdot \frac{\sin 3x}{3x} \cdot \frac{9}{2}\right) = \frac{9}{2}$

    (b) $\lim_{x \to 0}\frac{2\sin 3x(\cos 3x)(3)}{4x} = \frac{9}{2}$

34. 1   **35.** 0   **37.** 1   **39.** 3   **40.** 0   **41.** 1
46. 0   **47.** 1   **48.** 1   **50.** 2   **52.** 1   **54.** 1

55. $f(x) = -\cos x + 1$ **57.** $f(x) = -\csc x + 2$

59. $f(x) = \sin(x^2) + 1$

62. Not differentiable at $x = 0$: $f(x)$ is continuous, but $\lim_{x \to 0} f'(x) = \lim_{x \to 0}(\sin(\frac{1}{x}) - \frac{1}{x}\cos(\frac{1}{x}))$ does not exist.

63. Differentiable at $x = 0$: $f(x)$ is continuous, and $\lim_{x \to 0} f'(x) = \lim_{x \to 0}(3x^2 \sin(\frac{1}{x}) - x\cos(\frac{1}{x})) = 0$.

64. (a) $s''(t) = -A(\frac{k}{m})\sin\left(\sqrt{\frac{k}{m}}t\right) - B(\frac{k}{m})\cos\left(\sqrt{\frac{k}{m}}t\right)$, and $s''(t) + \frac{k}{m}s(t) = 0$

    (b) $s_0 = s(0) = B$, $v_0 = s'(0) = A\sqrt{\frac{k}{m}}$

67. Mimic the proof of Theorem 10.15(a).

69. Mimic the proof for the derivative of tangent.

72. **PROOF** $\frac{d}{dx}(\cos x) = \frac{d}{dx}(\sin(x + \frac{\pi}{2})) = \cos(x + \frac{\pi}{2})$
$(1) = \cos x \cos\frac{\pi}{2} - \sin x \sin\frac{\pi}{2} = (\cos x)(0) - (\sin x)(1) = -\sin x$

SECTION 10.6

3. If $\sin c = 0$, then $\lim_{x \to c}\csc x = \lim_{x \to c}\frac{1}{\sin x} \to \frac{1}{0}$.

4. For every x, $\sin x$ is the vertical coordinate of the point where the terminal edge of the angle x meets the unit circle; this vertical coordinate oscillates between -1 and 1.

9. $f(x) = 4\sin(\frac{2}{3}x)$, $f(x) = 4\sin(\frac{2}{3}(x - \pi))$, $f(x) = -4\sin(\frac{2}{3}x) + 5$

10. $f(x) = \frac{1}{2}\cos(\frac{\pi}{2}x)$, $f(x) = \frac{1}{2}\cos(\frac{\pi}{2}(x + 2))$, $f(x) = -\frac{1}{2}\cos(\frac{\pi}{2}x) - 4$

12. The graphs should look like the graphs of $y = \sin x$ and $y = \cos x$, stretched or shrunk horizontally so that they have **(a)** period 8π;   **(b)** period 1;   **(c)** period 6 (also, for part (c), the sine graph has been reflected over the x-axis, but the cosine graph has not).

14. $(2, 8)$, $(10, 8)$, and $(-6, 8)$

15. $(\frac{\pi}{2}, -2)$, $(\frac{3\pi}{2}, -2)$, and $(-\frac{\pi}{2}, -2)$

17. The coordinates $(\frac{1}{2} + k, -3)$ are inflection points for every integer k.

19. False   **20.** True   **21.** True   **22.** True   **23.** False
24. False   **25.** False   **26.** False   **28.** 4   **29.** $\frac{\pi}{2}$ **30.** 1

33. $\frac{\pi}{2} + 2\pi k$ (where k is any integer)

36. $\frac{\pi}{4} + \frac{\pi}{2}k$ **38.** $\frac{\pi}{4} + \pi k$

41. One way is $\frac{\pi}{2}k$, where k is any integer.

42. $\frac{\pi}{2} + \pi k$

44. Period $\frac{2\pi}{3}$, amplitude 1, center point $(-\frac{\pi}{2}, 0)$:

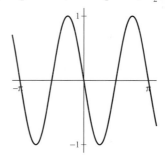

45. Period $\frac{2\pi}{3}$, amplitude 1, center point $(-\frac{\pi}{4}, 0)$:

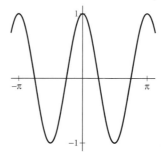

46. Period 4, amplitude 2, center point $(0, 0)$:

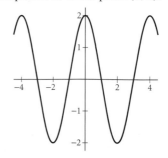

50. Period 2, amplitude 1, center point $(1, -3)$:

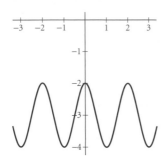

51. Period 6π, amplitude 2, center point $(\pi, -1)$:

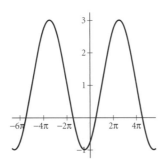

52. Period 4, amplitude π, center point $(2, -\pi)$:

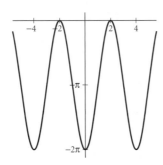

56. Expanded to period 2, then shifted right one unit:

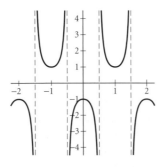

58. $f(x) = 3\sin(\frac{1}{2}x) + 2$; $f(x) = 3\cos(\frac{1}{2}(x - \pi)) + 2$

59. $f(x) = -0.5\sin(\pi x) + 1$; $f(x) = 0.5\cos(\pi(x - 0.5)) + 1$

61. $f(x) = 2\sin(\frac{\pi}{4}(x + 1)) - 1$; $f(x) = 2\cos(\frac{\pi}{4}(x - 1)) - 1$

63. $f(x) = 0.5\sin(\frac{1}{4}(x + \pi)) - 0.5$; $f(x) =$
$0.5\cos(\frac{1}{4}(x - \pi)) - 0.5$

65. All three of $f(x)$, $f'(x)$, and $f''(x)$ do not exist for $x = \frac{\pi}{2} + \pi k$ and are zero for $x = 2\pi k$. $f(x)$ is positive, decreasing, and concave up on $(-\frac{\pi}{2}, 0)$ and negative, decreasing, and concave down on $(0, \frac{\pi}{2})$, and then this pattern repeats.

74. ▶ The domain of $f(x)$ is $(-\infty, \infty)$.

    ▶ $f(x) = e^x \sin x$ is zero if $x = \pi k$ for any integer k.

    ▶ $f'(x) = e^x(\sin x + \cos x)$ always exists and is zero only if x is an angle that terminates at $\frac{3\pi}{4}$ or $-\frac{\pi}{4}$, that is, if $x = \frac{3\pi}{4} + \pi k$ for any integer k.

    ▶ $f''(x) = 2e^x \cos x$, which always exists and is zero at $x = \frac{\pi}{2} + \pi k$ for any integer k.

    ▶ From the information above, make number lines for f, f', and f''.

    ▶ $\lim\limits_{x \to \infty} e^x \sin x \to (\infty)(\text{bounded}) = \text{DNE}$

    ▶ $\lim\limits_{x \to -\infty} e^x \sin x \to (0)(\text{bounded}) = 0$

    ▶ Now calculate some of the values of $f(x)$ at the "important" points, and sketch a graph.

76. (a) $P(t) = 150\sin(\frac{2\pi}{365}(x - \frac{365}{4})) + 350$

    (b) $P(t) = 150\cos(\frac{2\pi}{365}(x - \frac{365}{2})) + 350$

77. (b) $M = 7.175$ feet; $N = 0.4$ foot; $P = 739$ minutes; $Q = 738$ minutes

    (c) Amplitude is 3.3875 feet, period is approximately 738.5 minutes. One possible "center" point is $(563.5, 3.7875)$. Then $H(t) = 3.3875\sin(\frac{2\pi}{738.5}(x - 563.5)) + 3.7875$ should be a good model for the data.

CHAPTER 11

SECTION 11.1

1. $[0, \pi]$; $[-\pi, 0]$; $[\frac{\pi}{4}, \frac{\pi}{2}]$

4. The fourth and first quadrants of the unit circle (the interval $[-\frac{\pi}{2}, \frac{\pi}{2}]$); the first and second quadrants (the interval $[0, \pi]$).

6. $\sin^{-1} x$ is defined to be the inverse of the restricted sine function; $\arctan x$ is defined to be the inverse of the restricted tangent function.

9. $[-\frac{\pi}{2}, 0) \cup (0, \frac{\pi}{2}]$

11. $y = \cos^{-1} x$ is an angle. $x = \cos y$ is the horizontal coordinate of the point on the unit circle corresponding to the angle y.

14. $\frac{5\pi}{6}$ is not in the range of $\sin^{-1} x$.

16. Only (a) and (c) are defined.

17. (a) Negative; (b) Positive; (c) Positive; (d) Positive

18. $[0, \pi]$; cosine   **21.** $\cos^{-1} x$; $[-1, 1]$; $[0, \pi]$

24. Fourth **26.** $\frac{1}{x}$ **27.** $\frac{1}{3x}$ **29.** $\frac{1}{3}(\sin\theta - 1)$

31. $\frac{2\pi}{3} + 2\pi k$ and $\frac{4\pi}{3} + 2\pi k$; $\frac{2\pi}{3}$ **33.** $\frac{\pi}{4} + \pi k$; $\frac{\pi}{4}$

34. False **35.** False **36.** False **37.** True **38.** False

39. False **40.** True **41.** 0 **43.** $\frac{\pi}{2}$ **46.** π **48.** $\frac{2\pi}{3}$

51. $-\frac{\pi}{3}$ **53.** $\frac{2\pi}{3}$ **56.** $\frac{3\pi}{4}$ **58.** $-\frac{\pi}{4}$ **59.** $\frac{\pi}{3}$

61. 0.201358 **63.** Not defined **66.** 1.77215

67. Not defined **70.** 0.312751 **72.** $\frac{\pi}{5} \approx 0.628319$

74. **PROOF** If $\theta = \csc^{-1} x$, then $\csc \theta = x$, and thus $\frac{1}{\sin \theta} = x$. Therefore $\sin \theta = \frac{1}{x}$, so $\theta = \sin^{-1}(\frac{1}{x})$. ■

SECTION 11.2

3. Domain$(\sin^{-1} x) = [-1, 1]$, Domain$(\frac{1}{\sqrt{1-x^2}}) = (-1, 1)$

6. $y = \sin x$ has horizontal tangent lines at $(-\frac{\pi}{2}, -1)$ and $(\frac{\pi}{2}, 1)$, so $y = \sin^{-1} x$ has vertical tangent lines at $(-1, -\frac{\pi}{2})$ and $(1, \frac{\pi}{2})$.

8. The range of $\sin^{-1} x$ is $[-\frac{\pi}{2}, \frac{\pi}{2}]$.

10. By similar triangles, the ratios of the side lengths will be the same if we scale down the triangle to have a hypotenuse of length 1.

13. $\sqrt{1 - x^2}$ **15.** $x^2 + 1$

16. $\sqrt{x^2 - 1}$ if $x > 1$, $-\sqrt{x^2 - 1}$ if $x < 1$

17. $\frac{1}{x}$ (restricted to $x \in [-1, 1]$)

21. $f'(x) = \frac{-2}{\sqrt{1-4x^2}}$ **22.** $f'(x) = \frac{2 \sin 2x}{\cos^2 2x}$

25. $f'(x) = \frac{6x}{\sqrt{1-9x^4}}$ **27.** $f'(x) = \frac{e^x}{|e^x|\sqrt{(e^x)^2-1}} = \frac{1}{\sqrt{e^{2x}-1}}$

28. $f'(x) = 2x \arctan x^2 + x^2 (\frac{2x}{1+x^4})$

32. $f'(x) = \frac{1}{\mathrm{arcsec}(\sin^2 x)} \frac{1}{\sin^2 x \sqrt{\sin^4 x - 1}} (2 \sin x \cos x)$

34. $f'(x) = \sec(1 + \tan^{-1} x) \tan(1 + \tan^{-1} x) \times (\frac{1}{1+x^2})$

36. $f(x) = \frac{2}{3} \tan^{-1}(3x) + C$ (for any constant C)

37. $f(x) = \frac{1}{\sqrt{3}} \sin^{-1}(\sqrt{3}x) + C$

39. $f(x) = \frac{1}{4}(2) \tan^{-1} \frac{x}{2} = \frac{1}{2} \tan^{-1} \frac{x}{2} + C$

40. $f(x) = \ln(1 + x^2) + C$

44. $f(x) = \tan^{-1}(e^x) + C$

CHAPTER 12

SECTION 12.1

5. The terms in the sum do not appear to have a consistent pattern.

7. The sum of the squares of the integers greater than or equal to 3 and less than or equal to 87.

9. (a) $b_p + b_{p+1} + \cdots + b_{q-1} + b_q$
  (b) Index is i; starting value is p; ending value is q; b_i describes each term in the sum.
  (c) p and q must be nonnegative integers; b_i can be any real number; also, we must have $p \le q$.

10. One possible answer is $a_k = k^2$, $m = 3$, $n = 7$.

11. $a_2 = -2$, $a_3 = -\frac{3}{2}$, $a_4 = -\frac{4}{3}$, $a_5 = -\frac{5}{4}$

13. The first sum is $1 + \frac{1}{2} + \frac{1}{5} + \frac{1}{10} + \cdots + \frac{1}{65}$; the second sum is $2(\frac{1}{2} + \frac{1}{4} + \frac{1}{10} + \frac{1}{20} + \cdots + \frac{1}{130})$.

15. (a) $\displaystyle\sum_{k=1}^{5} \frac{2}{k^2}$; **(b)** $\displaystyle\sum_{k=2}^{6} \frac{2}{(k-1)^2}$; **(c)** $\displaystyle\sum_{k=0}^{4} \frac{2}{(k+1)^2}$

16. True   **17.** False   **18.** False   **19.** False

20. False   **21.** False

25. $0.5 + 0.625 + 1 + 1.625 = 3.75$ square units

27. 4.625 square units

28. $16 + 25 + 36 + 49 + 64 + 81 = 271$

31. $0 + \frac{1}{8} + \frac{1}{2} + \frac{9}{8} + 2 + \frac{25}{8} = \frac{55}{8}$

33. Approximately $0.17321 + 0.17607 + 0.18166 + 0.18439 + 0.18708 + 0.18973 + 0.19235 + 0.19494 + 0.19748 \approx 1.67691$

35. $\displaystyle\sum_{k=1}^{8} 3$; $m = 1$, $n = 8$, $a_k = 3$

37. $\displaystyle\sum_{k=1}^{5} \frac{k+2}{k^3}$ and $\displaystyle\sum_{k=3}^{7} \frac{k}{(k-2)^3}$ are the most obvious ways.

38. $\displaystyle\sum_{k=2}^{10} (k^2 + 1)$

40. $2a_0 + 2a_1 + 2a_2 - a_{101} + \displaystyle\sum_{k=3}^{100} a_k$

42. $\displaystyle\sum_{k=2}^{25} (3k^2 + 2k - 1) - 52$ **43.** 11   **45.** 48

46. $3n - \frac{n(n+1)}{2}$; -4750; $-123,750$; $-497,500$

48. $\frac{n(n+1)(2n+1)}{6} + 2\frac{n(n+1)}{2} + n - 13$; 348,537; 42,042,737; 334,835,487

50. $\frac{1}{n^4}\left(\frac{n^2(n+1)^2}{4} - n\right)$; 0.255024; 0.251001; 0.2505

53. $\displaystyle\lim_{n\to\infty} \frac{2n^3 + 6n^2 + 10n}{6n^3} = \frac{1}{3}$

55. $\displaystyle\lim_{n\to\infty} \left(n + n(n+1) + \frac{n(n+1)(2n+1)}{6}\right) = \infty$

57. $\displaystyle\lim_{n\to\infty} \left(1 + \frac{n^2 + n}{n^2} + \frac{2n^3 + 3n^2 + n}{6n^3}\right) = 1 + 1 + \frac{2}{6} = \frac{7}{3}$

58. **PROOF** $\displaystyle\sum_{k=0}^{n} 3a_k = 3a_0 + 3a_1 + 3a_2 + \cdots + 3a_n = 3(a_0 + a_1 + a_2 + \cdots + a_n) = 3\displaystyle\sum_{k=0}^{n} a_k$ ■

SECTION 12.2

1. True   **2.** False   **3.** False   **4.** False   **5.** True

7. Any linear function will have this property.

9. (a) 7.85; **(b)** No;
  (c) Upper Sum is ≥ 8.2, Lower Sum is ≤ 7.5; **(d)** No;
(e) No.

10. The heights of the rectangles used with the Upper Sum are by definition always greater than or equal to the heights of the rectangles used for any other Riemann sum.

13. Right-Hand Sum with $f(x) = x^2$, $N = 4$, $\Delta x = \frac{1}{2}$, and $x_k = 1 + \frac{k}{2}$; thus $a = 1$ and $b = 3$.

15. Trapezoid Sum with $f(x) = \sin x$, $N = 100$, $\Delta x = 0.05$, and $x_k = 0 + 0.05k$ (so $x_{k-1} = 0 + 0.05(k - 1)$); thus $a = 0$ and $b = 0 + 0.05(100) = 5$.

17. Part of the sum would indicate that $\Delta x = \frac{1}{2}$, whereas another part would give us $\Delta x = 0.25$.

18. If $N = 100$, $\Delta x = 0.1$, and $a = 2$, then $b = 2 + 100(0.1) = 12$, so we can't have $[a, b] = [2, 5]$.

19. (a) $M_1 = 0$, $M_2 = 2$
   (b) $M_1 = 0$, $M_2 = \frac{2}{3}$ (or $\frac{4}{3}$), $M_3 = 2$
   (c) $M_1 = 0$, $M_2 = \frac{1}{2}$, $M_3 = \frac{3}{2}$, $M_4 = 2$

23. 1.219   **24.** 51.88   **25.** $\frac{8}{3}$

28. (a) 5;   (b) 6.875

29. (a) 1.16641;   (b) 1.26997

30. (a) 51.3434;   (b) 52.9562

31. (a) 2;   (b) 2.148;   (c) 2.25

35. $\displaystyle\sum_{k=0}^{2} k^2 \left(\text{or } \sum_{k=1}^{3} (k-1)^2 \right)$

39. $\displaystyle\sum_{k=0}^{99} \ln\left(2 + \tfrac{3}{100}k\right)\left(\tfrac{3}{100}\right) \left(\text{or } \sum_{k=1}^{100} \ln\left(2 + \tfrac{3}{100}(k-1)\right)\left(\tfrac{3}{100}\right) \right)$

40. $\displaystyle\sum_{k=1}^{20} \sqrt{\left(-1 + \tfrac{2k-1}{20}\right)^2 - 1}\left(\tfrac{1}{10}\right)$

42. Exact $= 26$ (triangle and rectangle); LHS $= 24.5$; RHS $= 27.5$; Midpoint $= 26$; Upper $= 27.5$; Lower $= 24.5$; Trapezoid $= 26$

43. Exact $= \frac{\pi}{2}$ (half of circle); LHS ≈ 1.366; RHS ≈ 1.366; Midpoint ≈ 1.630; Upper ≈ 1.866; Lower ≈ 0.866; Trapezoid ≈ 1.366

SECTION 12.3

2. $\int_a^b f(x)\,dx$; the definite integral of $f(x)$ on $[a, b]$

3. $\int_a^b f(x)\,dx = \lim\limits_{N \to \infty} \sum\limits_{k=1}^{N} f(x_k^*)\,\Delta x$, where $\Delta x := \frac{b-a}{N}$, $x_k = a + k\,\Delta x$, and $x_k^* \in [x_{k-1}, x_k]$

6. The region between the graph of $f(x)$ and the x-axis from $x = a$ to $x = a$ has a width of zero.

8. False   **9.** True   **10.** False   **11.** False

12. False   **13.** False   **14.** False

15. (a) $\int_a^b (f(x) + g(x))\,dx = \int_a^b f(x)\,dx + \int_a^b g(x)\,dx$, but $\int_a^b f(x) \cdot g(x)\,dx \neq \int_a^b f(x)\,dx \cdot \int_a^b g(x)\,dx$.
   (b) With $f(x) = x$ and $g(x) = 1$, $\int_0^1 f(x)\,dx = \frac{1}{2}$, $\int_0^1 g(x)\,dx = 1$, and $\int_0^1 (f(x) + g(x))\,dx = \frac{3}{2}$.
   (c) With $f(x) = x$ and $g(x) = x$, $\int_0^1 f(x)\,dx = \frac{1}{2}$ and $\int_0^1 g(x)\,dx = \frac{1}{2}$, but $\int_0^1 f(x)g(x) = \frac{1}{3}$.

16. (a) $\displaystyle\sum_{k=1}^{N} \left(\tfrac{3k}{N}\right)^2 \left(\tfrac{3}{N}\right) = \frac{27}{N^3}\left(\frac{N(N+1)(2N+1)}{6}\right)$
   (b) $\displaystyle\sum_{k=1}^{N} \left(\tfrac{3}{N}(k-1)\right)^2 \left(\tfrac{3}{N}\right) =$
$\frac{27}{N^3}\left(\frac{N(N+1)(2N+1)}{6} - 2\frac{N(N+1)}{2} + N\right)$
   (c) For $N = 100$, RHS ≈ 9.13545 and LHS ≈ 8.86545; for $N = 1000$, RHS ≈ 9.0135 and LHS ≈ 8.9865.
   (d) The expressions in parts (a) and (b) are both equal to 9 if we let $N \to \infty$.

18. 2   **20.** 264   **21.** $\frac{\pi}{2}$

25. $\displaystyle\sum_{k=1}^{N} \left(5 - \left(2 + \tfrac{3k}{N}\right)\right)\left(\tfrac{3}{N}\right) = \frac{3}{N}(3N) - \frac{9}{N^2}\left(\frac{N(N+1)}{2}\right)$. When $N = 100$, the sum is ≈ 4.455; when $N = 1000$, the sum is ≈ 4.4955; the limit as $N \to \infty$ is $\frac{9}{2}$.

26. $\displaystyle\sum_{k=1}^{N} 2\left(\tfrac{k}{N}\right)^2 \left(\tfrac{1}{N}\right) = \frac{2}{N^3}\left(\frac{N(N+1)(2N+1)}{6}\right)$. When $N = 100$, the sum is ≈ 0.6767; when $N = 1000$, the sum is ≈ 0.667667; the limit as $N \to \infty$ is $\frac{2}{3}$.

29. $\displaystyle\sum_{k=1}^{N} \left(\left(2 + \tfrac{k}{N}\right) + 1\right)^2 \left(\tfrac{1}{N}\right) = \frac{1}{N}(9N) + \frac{6}{N^2}\left(\frac{N(N+1)}{2}\right) +$
$\frac{1}{N^3}\left(\frac{N(N+1)(2N+1)}{6}\right)$. When $N = 100$, the sum is ≈ 12.3684; when $N = 1000$, the sum is ≈ 12.3368; the limit as $N \to \infty$ is $9 + \frac{6}{2} + \frac{2}{6} = \frac{37}{3}$.

30. $\frac{62}{3}$ **32.** -93 **34.** $\frac{136}{3}$ **37.** 5   **38.** 14

39. Not enough information   **43.** Not enough information

45. -8 **47.** $-\frac{52}{3}$

48. (a) **PROOF** $\displaystyle\lim_{N \to \infty} \sum_{k=1}^{N} \left(3\left(1 + \tfrac{2k}{N}\right) + 4\right)\left(\tfrac{2}{N}\right) =$
$\displaystyle\lim_{N \to \infty} \left(\tfrac{2}{N}(7N) + \tfrac{12}{N^2} \cdot \frac{N(N+1)}{2}\right) = 20$ ▪
   (b) **PROOF**   The area is a rectangle with base of length 2 and height of length 7, with a triangle on top with base 2 and height 6; therefore, the area is $2(7) + \frac{1}{2}(2 \cdot 6) = 20$. ▪
   (c) **PROOF** $\int_1^3 (3x + 4)\,dx = 3\int_1^3 x\,dx + 4\int_1^3 1\,dx$
$= 3 \cdot \frac{1}{2}(3^2 - 1^2) + 4 \cdot 1(3 - 1) = 20$ ▪

49. **PROOF**   If c is any real number, $\Delta x = \frac{b-a}{N}$, $x_k = a + k\Delta x$, and $x_k^* \in [x_{k-1}, x_k]$, then $\int_a^b cf(x)\,dx =$
$\displaystyle\lim_{N \to \infty} \sum_{k=1}^{N} cf(x_k^*)\,\Delta x = \lim_{N \to \infty} c\sum_{k=1}^{N} f(x_k^*)\,\Delta x =$
$c \displaystyle\lim_{N \to \infty} \sum_{k=1}^{N} f(x_k^*)\,\Delta x = c\int_a^b f(x)\,dx$. ▪

SECTION 12.4

2. (a) $\int_{-3}^{4} f(x)\,dx$
   (b) $\int_{-3}^{-1} f(x)\,dx - \int_{-1}^{2} f(x)\,dx + \int_{2}^{4} f(x)\,dx$

4. Three (although two of them will be equal); positive; negative

7. Sketch the graphs of   (a) $f(x) = x + 3$;
   (b) $f(x) = 3 - x$;   (c) $f(x) = -3$;   (d) $f(x) = x + 1.5$.

10. $\displaystyle\lim_{N \to \infty} \sum_{k=1}^{N} (f(x_k^*) - g(x_k^*))\,dx$, where $\Delta x = \frac{b-a}{N}$, $x_k = a + k\,\Delta x$, and $x_k^* \in [x_{k-1}, x_k]$

11. The average value is the average *height* of the graph of $f(x)$ on $[a, b]$, whereas the average rate of change is the average *slope* of the graph on the same interval.

13. The average height of the function should be 10, and the slope from $(-2, f(-2))$ to $(5, f(5))$ should be -3.

15. (a) Negative; **(b)** negative

17. One approximation is an average value of ≈ -3.

18.

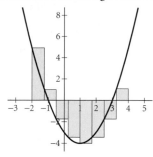

21. (a) $\int_{-2}^{4} f(x)\,dx$

   (b) $\int_{-2}^{-1} f(x)\,dx - \int_{-1}^{3} f(x)\,dx + \int_{3}^{4} f(x)\,dx$

22. (a) $\int_{-3}^{7} f(x)\,dx$

   (b) $-\int_{-3}^{-2} f(x)\,dx + \int_{-2}^{1} f(x)\,dx - \int_{1}^{6} f(x)\,dx + \int_{6}^{7} f(x)\,dx$

25.

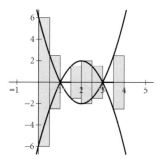

27. $\int_{-2}^{1}(f(x) - g(x))\,dx + \int_{1}^{3}(g(x) - f(x))\,dx$

29. $\int_{-2.5}^{-1.5}(f(x) - g(x))\,dx + \int_{-1.5}^{0.5}(g(x) - f(x))\,dx + \int_{0.5}^{2}(f(x) - g(x))\,dx$

33. 0   **34.** 4π **35.** 0   **36.** False   **37.** False

38. True   **39.** False   **40.** False   **41.** False

42. Using a Left-Hand Sum with $N = 8$, we have
(a) -11.1973;   **(b)** 12.2229.

44. Using a Left-Hand Sum with $N = 8$, we have 8.84399.

47. $(f(-1) + f(-0.75) + f(-0.5) + f(-0.25) + f(0) + f(0.25) + f(0.5) + f(0.75))(\frac{1}{8}) \approx 0.990978$

49. (a) $\frac{5}{2}$;   **(b)** $\frac{13}{2}$

53. (a) $-\frac{4}{3}$;   **(b)** $\frac{109}{12}$

54. (a) $\frac{25}{2}$;   **(b)** 25.2 **56.** $\frac{13}{2}$ **58.** $\frac{38}{6}$

59. 15   **61.** 7   **62.** 4   **65.** -1

67. (a) $\int_{1.51}^{5.29}(55 - f(x))\,dx + \int_{5.29}^{18}(r(x) - f(x))\,dx + \int_{18}^{30.71}(s(x) - g(x))\,dx + \int_{30.71}^{34.49}(55 - g(x))\,dx$

   (b) 561.9 square feet (too big!)

68. (a) $t = \frac{5\sqrt{2}}{3} \approx 2.357$ seconds

   (b) Approximately 33.3337 feet

   (c) Average velocity is $\frac{s(2.357) - s(0)}{2.357 - 0} = \frac{0 - 50}{2.357} \approx -21.213$ feet per second.

   (d) $v(t) = -18t$, so average velocity is $\frac{1}{2.357 - 0} \cdot \int_{0}^{2.357}(-18t)\,dt \approx -21.213$ feet per second.

   (e) The area under the velocity graph is the total change in distance! (In this case, 50 feet.)

69. (a) 3.75 feet;   **(b)** 3.4 feet;   **(c)** $\frac{1}{5-0}\int_{0}^{5}(0.25t^2 + 1)\,dt$;
   (d) 3.08 feet

CHAPTER 13

SECTION 13.1

3. (a) $\int x^6\,dx = \frac{1}{7}x^7 + C$;

   (b) $\frac{1}{7}x^7$ is an antiderivative of x^6;

   (c) The derivative of $\frac{1}{7}x^7$ is x^6, that is, $\frac{d}{dx}(\frac{1}{7}x^7) = x^6$.

4. (a) $\int 6x^5\,dx = x^6 + C$;

   (b) x^6 is an antiderivative of $6x^5$;

   (c) The derivative of x^6 is $6x^5$, that is, $\frac{d}{dx}(x^6) = 6x^5$.

7. At this point we don't know of a function whose derivative is $\sec x$, but we do know of a function whose derivative is $\sin x$ (it is $-\cos x$).

9. True   **10.** False   **11.** False   **12.** True

13. False   **14.** False   **15.** False   **16.** False

18. $\frac{d}{dx}(\frac{1}{2}e^{(x^2)}(x^2 - 1)) = x^3 e^{(x^2)}$

19. $\frac{d}{dx}(x(\ln x - 1)) = 1(\ln x - 1) + x(\frac{1}{x} - 0) = \ln x$

22. One easy example is $f(x) = x$, $g(x) = x^2$.

25. $f(x) = \frac{1}{9}x^9 - \frac{1}{3}x^3 + x - 1$ **26.** $f(x) = 3x^{\frac{2}{3}} - 1$

29. $f(x) = \frac{5}{3}\ln|3x - 2| + 8$

32. $\frac{3}{4}x^4 + \frac{5}{3}x^3 - \frac{3}{2}x^2 - 5x + C$

36. $\frac{2}{3}x^{\frac{3}{2}} + 2x^{\frac{1}{2}} + C$ **37.** $\frac{3.2}{\ln(1.43)}(1.43)^x - 50x + C$

38. $\frac{4}{3}e^{2x-3} + C$ **39.** $2e^{2x-6} + C$

40. $-\frac{3}{\pi}\csc(\pi x) + C$ **41.** $3(\tan x - x) + C$

44. $\frac{7}{2}\sin^{-1}(2x) + C$ **45.** $\frac{1}{3}\sec^{-1}(3x) + C$

47. $\frac{a}{b}e^{bx+c} + C$ **50.** $\frac{1}{ab}\sec(ax) + C$

51. $\frac{a}{b}\tan^{-1}(bx) + C$ **52.** $x^2 e^x + C$

55. $\ln|3x^2 + 1| + C$ **57.** $\ln x \cos x + C$

58. $\frac{1}{18}(x^3 + 1)^6 + C$ **59.** $\frac{x^2}{\ln x} + C$

61. PROOF $\frac{d}{dx}(\frac{1}{k+1}x^{k+1}) = \frac{1}{k+1}(k+1)x^{(k+1)-1} = x^k$ ▨

64. PROOF $\frac{d}{dx}(\frac{1}{k}e^{kx}) = \frac{1}{k}(ke^{kx}) = e^{kx}$ ▨

68. PROOF $\frac{d}{dx}(-\cot x) = -(-\csc^2 x) = \csc^2 x$ ▨

71. PROOF $\frac{d}{dx}(-\frac{a}{b}\cos(bx)) = -\frac{a}{b}(-\sin(bx))(b) = a\sin(bx)$ ▨

72. PROOF $\frac{d}{dx}(\sin^{-1} x) = \frac{1}{\sqrt{1-x^2}}$ ▨

77. PROOF **(a)** $\frac{d}{dx}(F(x)) = \frac{d}{dx}(G(x) + C) \Rightarrow \frac{dF}{dx} = \frac{dG}{dx} + 0 \Rightarrow \frac{dF}{dx} = \frac{dG}{dx}$

   (b) $F'(x) = G'(x) \Rightarrow F'(x) - G'(x) = 0 \Rightarrow \frac{d}{dx}(F(x) - G(x)) = 0 \Rightarrow F(x) - G(x)$ is a constant (say C).

   (c) *All* antiderivatives of a function differ from each other by a constant, so writing $\int f(x)\,dx = F(x) + C$ describes *all* the antiderivatives of $f(x)$ and makes sense for *any* antiderivative $F(x)$. ▨

SECTION 13.2

1. $\frac{1}{b-a}\int_a^b v(t)\,dt$ would be equal to $\frac{s(b)-s(a)}{b-a}$.

4. The Fundamental Theorem of Calculus (FTC) relates the concept of area under a curve to antiderivatives, which enables us to find the exact area under a curve quickly.

5. The Mean Value Theorem

9. Equivalent to FTC: (a), (c), (f), and (g)

10. (a) 7.455 (Right-Hand Sum)

(b) $\lim_{N\to\infty}\left(\frac{3}{N}(N)+\frac{6}{N^2}\frac{N(N+1)}{2}+\frac{3}{N^3}\frac{N(N+1)(2N+1)}{6}\right)=7$

(c) $3(\frac{1}{3})(2^3-1^3)=7$

(d) $\left[3(\frac{1}{3})x^3\right]_1^2=7$

12. (a) 343.36 (Right-Hand Sum)

(b) $\lim_{N\to\infty}\left(\frac{96N(N+1)(2N+1)}{N^3}+\frac{96N(N+1)}{N^2}+\frac{16N}{N}\right)=304$

(c) $9(\frac{1}{3})(4^3-0^3)+12(\frac{1}{2})(4^2-0^2)+4(4-0)=304$

(d) $\left[3x^3+6x^2+4x+\frac{8}{9}\right]_0^4=304$

14. $\frac{42}{\ln 2}$ **16.** $-\frac{2}{3}(5)^{-\frac{3}{2}}+\frac{2}{3}(2)^{-\frac{3}{2}}$ **17.** $-\frac{1}{2e}+\frac{1}{2}$

19. $\frac{\pi}{4}$ **23.** $\ln(32)-\ln(8)$

24. $\frac{\pi}{6}$ **27.** $\frac{1}{e}+e$ **28.** $\frac{25}{2}$

30. $\frac{343}{12}$ **31.** $f(4)\approx 4.125$ **34.** $f(-2)\approx 2.5$

35. (a) 0;   (b) 4

39. (a) $3-e$;   (b) $e+2\ln 4-5$

40. (a) $\frac{9}{4}-2\ln 4$;   (b) $-\frac{1}{4}-4\ln(\frac{3}{2})-2\ln(4)$

42. $\frac{19}{3}$ **43.** $\frac{145}{12}$ **46.** $-2+\sqrt{3}+\frac{\pi}{3}$ **47.** $\frac{16\sqrt{2}}{5}$

50. $\frac{1}{9}(\ln(16)-\ln(7))$ **52.** $-\frac{1}{9}\cos 9+\frac{1}{9}$

54. **PROOF**   If $F(x)-G(x)=C$ then $F(x)=G(x)+C$, so $\left[F(x)\right]_a^b=\left[G(x)+C\right]_a^b=(G(b)+C)-(G(a)+C)=G(b)-G(a)=\left[G(x)\right]_a^b$.   ∎

SECTION 13.3

1. $\int_2^x t^2\,dt$

2. x is the independent variable; $A(x)$ is the dependent variable, which represents the area under the graph of $f(t)$ from $t=0$ to $t=x$; t does not affect the dependent or independent variable.

4. (a) As x increases, the area $A(x)$ accumulated also increases, but at a rate that decreases (so the rate of change of the rate of change of $A(x)$ is negative, and thus $A(x)$ is concave down).

(b) $A''(x)=\frac{d}{dx}(\frac{d}{dx}\int_0^x f(t)\,dt)=\frac{d}{dx}(f(x))=f'(x)$; therefore, f decreasing $\Rightarrow$ f' negative $\Rightarrow$ A'' negative $\Rightarrow$ A concave down.

7. (a) $A'(x)$ and $B'(x)$ are both equal to x^2, so $A(x)$ and $B(x)$ differ by a constant.

(b) $A(x)-B(x)=\int_0^x t^2\,dt-\int_3^x t^2\,dt=\int_0^3 t^2\,dt$, which is a constant.

(c) $\int_0^3 t^2\,dt$ is the area under the graph of $y=x^2$ from $x=0$ to $x=3$.

9. $A(5)<A(0)<A(-1)<A(-2)$

11. The graph of $A(x)$ should start at $(0,0)$ and then increase on all of $[0,6]$. The graph should be concave up on $[0,3]$ and concave down on $[3,6]$ (so there is an inflection point at $x=3$).

12. $A(x)$ is positive on approximately $[0,2]$ and negative on $[2,6]$. $A(x)$ is increasing on $[0,1]$ and $[5,6]$ and decreasing on $[1,5]$.

14. (a) $f(x)+C$;   (b) $f(x)$ **15.** (a) $f(b)-f(a)$;   (b) 0

16. (a) $f(x)-f(a)$;   (b) $f(x)$

17. $F(x)=A(g(x))$, where $A(x)=\int_3^x \sin t\,dt$ ("outside") and $g(x)=x^2$ ("inside").

19. $\ln 10-1$

22. f is continuous on $[1,5]$, so the Mean Value Theorem for integrals applies and says that there is some $c\in(1,5)$ for which $f(c)=c(c-6)$ is equal to the average value $\frac{1}{5-1}\int_1^5 x(x-6)\,dx$ of the function f. You can compute this average value; it is equal to $-\frac{23}{3}$. Therefore, there is some $c\in(1,5)$ such that $f(c)=-\frac{23}{3}$. We can solve the equation $f(c)=-\frac{23}{3}$ to find such a value of c. Actually there are two such values: $c\approx 1.8453$ and $c\approx 4.1547$.

23. False   **24.** False   **25.** False

26. True   **27.** False   **28.** True

29. (a) $A(2)$ is the area under the graph of t^2+1 from $t=1$ to $t=2$.

(b) $A(2)=\frac{10}{3}$, $A(5)=\frac{136}{3}$ (c) $A(x)=\frac{1}{3}x^3+x-\frac{4}{3}$

31. (a) $A(x)$ is the area under the graph of $\frac{1}{t}$ from $t=3$ to $t=x$.

(b) $A(2)=\ln 2-\ln 3$, $A(5)=\ln 5-\ln 3$

(c) $A(x)=\ln x-\ln 3$

33. $\int_0^x \sin^2(3^t)\,dt$, $\int_2^x \sin^2(3^t)\,dt$, $\int_{-1}^x \sin^2(3^t)\,dt$

36. $-e^{x^2+1}$ **37.** $\cos(x^2)(2x)$

38. $\int_1^{\sqrt{x}}\ln t\,dt+x\ln(\sqrt{x})(\frac{1}{2}x^{-\frac{1}{2}})$

40. $\sin((x+2)^2)-\sin(x^2)$

42. 0   **44.** $54x$ **45.** $-2\ln|x^2|-4$

47. $f(x)=\frac{1}{2}\ln|2x-1|+3$

49. $f(x)=\int_2^x\frac{1}{t^3+1}\,dt$ **50.** $f(x)=\int_1^x e^{-t^2}\,dt$

53. $c=1$ **56.** $c=-\frac{1}{3}\ln(\ln 2)$ **58.** $c=\pm\frac{\sqrt{\pi^2-4}}{2\pi}$

59. (a) There is some time $c\in(0,4)$ when $f'(c)=\frac{f(4)-f(0)}{4-0}=1.44$; therefore there is some time c in the first four days when the rate of growth of the plant is 1.44 centimeters per day.

(b) There is some time c during the first four days when the height of the plant is equal to the average height of the plant over those four days. Since this average height is 1.92 feet, we know that there is some $c\in(0,4)$ such that $f(c)=1.92$. (We can solve for $c=2.3094$.)

61. **PROOF**   If $G(x)$ is an antiderivative of $f(x)$, then by the Fundamental Theorem of Calculus we have $\int_a^b f(x)\,dx=G(b)-G(a)$. Therefore, $A(x)=\int_0^x f(t)\,dt=G(x)-G(0)$. Since $A(x)$ and $G(x)$ differ by a constant, they have the same derivative (namely $f(x)$); thus $A(x)$ is an antiderivative of $f(x)$.   ∎

CHAPTER 14

SECTION 14.1

3. $\int \frac{1}{1-x^2}\,dx$, $\int x\sin x\,dx$, $\int \sin(x^2)\,dx$

4. Both integrals turn into $\int \frac{1}{u}\,du$ after a change of variables ($u = x^2 + 1$ in the first case, $u = \ln x$ in the second).

6. $\int 3x^2\sin(x^3)\,dx$, $\int \frac{\sin(\ln x)}{x}\,dx$, $\int \frac{\sin(\sqrt{x}+2)}{2\sqrt{x}}\,dx$

8. $\int \frac{3}{\sqrt{3x+1}}\,dx$, $\int \frac{\cos x}{\sqrt{\sin x}}\,dx$, $\int \frac{1}{x\sqrt{\ln x}}\,dx$

10. $du = (2x+1)\,dx$ **11.** $du = \cos x\,dx$

15. $\int_{-1}^{5} u^2\,du = \frac{126}{3} = 42$; $\int_{x=-1}^{x=5} u^2\,du = \frac{1}{3}(5)^6 - \frac{1}{3}(-1)^6 = 5208$; $\int_{u(-1)}^{u(5)} u^2\,du = \int_1^{25} u^2\,du = \frac{1}{3}(25)^3 - \frac{1}{3}(1)^3 = 5208$

17. True **18.** False **19.** False **20.** False

21. True **22.** False **23.** True **24.** False

25. (a) $\frac{1}{3}\int u^2\,du = \frac{1}{9}(3x+1)^3 + C$

(b) $\int(9x^2 + 6x + 1)\,dx = 3x^3 + 3x^2 + x + C$

(c) $\frac{1}{9}(3x+1)^3 = (3x^3 + 3x^2 + x) + 1$

26. (a) $\frac{1}{3}\int u\,du = \frac{1}{6}(x^3+1)^2 + C$

(b) $\int(x^5 + x^2)\,dx = \frac{1}{6}x^6 + \frac{1}{3}x^3 + C$

(c) $\frac{1}{6}(x^3+1)^2 = (\frac{1}{6}x^6 + \frac{1}{3}x^3) + \frac{1}{6}$

29. (a) $\int \frac{\sqrt{x}+3}{2\sqrt{x}}\,dx = \frac{1}{2}(\sqrt{x}+3)^2 + C$

(b) $\int(\frac{1}{2} + \frac{3}{2}x^{-\frac{1}{2}})\,dx = \frac{1}{2}x + 3x^{\frac{1}{2}} + C$

(c) $\frac{1}{2}(\sqrt{x}+3)^2 = (\frac{1}{2}x + 3x^{\frac{1}{2}}) + \frac{9}{2}$

33. $-\frac{1}{2}\cot x^2 + C$ **34.** $\frac{1}{3}\ln|3x+1| + C$

37. $\frac{1}{2\pi}\sin^2 \pi x + C$ **38.** $\frac{1}{2}\sec 2x + C$

39. $-\frac{1}{6}\cot^6 x + C$ **40.** $\frac{1}{11}x^{11} + \frac{1}{4}x^8 + \frac{1}{5}x^5 + C$

41. $-\frac{4}{5}\cos(x^{\frac{5}{4}}) + C$ **44.** $\frac{1}{2\ln 2}2^{x^2+1} + C$

45. $\frac{2}{3}(\ln x)^{\frac{3}{2}} + C$ **48.** $\frac{1}{4}(\ln x)^2 + C$

49. $-\ln|2 - e^x| + C$ **50.** $-2e^{-x} - x + C$

53. $(\sin x^2)^{\frac{1}{2}} + C$ **54.** $\cos(\frac{1}{x}) + C$

55. $\frac{1}{3}(x^2+1)^{\frac{3}{2}} + C$ **57.** $\frac{2}{5}(x+1)^{\frac{5}{2}} - \frac{2}{3}(x+1)^{\frac{3}{2}} + C$

59. $\frac{2}{27}(3x+1)^{\frac{3}{2}} - \frac{2}{9}(3x+1)^{\frac{1}{2}} + C$ **61.** $-\cos(e^x) + C$

65. $\frac{1}{103}(x-1)^{103} + \frac{2}{102}(x-1)^{102} + \frac{1}{101}(x-1)^{101} + C$

66. $\frac{1}{2}\tan^{-1}(x^2) + C$ **67.** $\frac{1}{2}(\sin^{-1} x)^2 + C$

70. (a) $-\frac{1}{3}\int_{x=-2}^{x=2} e^u\,du = -\frac{1}{3}[e^{5-3x}]_{-2}^{2} = -\frac{1}{3}(e^{-1} - e^{11})$

(b) $-\frac{1}{3}\int_{11}^{-1} e^u\,du = -\frac{1}{3}[e^u]_{11}^{-1} = -\frac{1}{3}(e^{-1} - e^{11})$

72. (a) $\frac{1}{2}\int_{x=0}^{x=3} u^{\frac{1}{3}}\,du = [\frac{3}{8}(x^2+1)^{\frac{4}{3}}]_0^3 = \frac{3}{8}(10^{\frac{4}{3}} - 1)$

(b) $\frac{1}{2}\int_1^{10} u^{\frac{1}{3}}\,du = [\frac{3}{8}u^{\frac{4}{3}}]_1^{10} = \frac{3}{8}(10^{\frac{4}{3}} - 1)$

75. (a) $\int_{x=-\pi}^{x=\pi} u\,du = [\frac{1}{2}\sin^2 x]_{-\pi}^{\pi} = 0$

(b) $\int_0^0 u\,du = 0$

77. (a) $\int_{x=1}^{x=2} u^{-\frac{1}{2}}\,du = 2(\ln 2)^{\frac{1}{2}}$

(b) $\int_0^{\ln 2} u^{-\frac{1}{2}}\,du = 2(\ln 2)^{\frac{1}{2}}$

80. (a) $\frac{1}{2}(\ln 10 - \ln 2)$ **(b)** $\frac{1}{2}(\ln 2 + \ln 10)$

(c) $\frac{3}{2} + \frac{3}{2}\ln 2 - \frac{1}{2}\ln 10$ **(d)** $\frac{1}{8}(\ln 10 - \ln 2)$

SECTION 14.2

3. (a) $\int u\,dv = uv - \int v\,du$

(b) $\int u(x)v'(x)\,dx = u(x)v(x) - \int v(x)u'(x)\,dx$

5. $\int x\sin x\,dx$; $\int x\ln x\,dx$ **7.** $\int \ln x\,dx$

8. If $u = 1$, then integrating dv is equivalent to solving the original integral.

12. $du = 3\cos 3x\,dx$ and $dv = dx$; $\int u\,dv = \int \sin 3x\,dx$, and $\int v\,du = \int 3x\cos 3x\,dx$. The first integral is clearly easier, and the second integral can be rewritten in terms of the first using integration by parts.

13. (a) $\frac{d}{dx}(x\cos 2x) = \cos 2x - 2x\sin 2x$

(b) $\int(\cos 2x - 2x\sin 2x)\,dx = x\cos 2x + C$

(c) $\int(-2x\sin 2x)\,dx = x\cos 2x - \int \cos 2x\,dx$

15. (a) $\frac{d}{dx}(x\ln x) = \ln x + 1$

(b) $\int(\ln x + 1)\,dx = x\ln x + C$

(c) $\int \ln x\,dx = x\ln x - \int 1\,dx$

17. $u = x$, $v = \frac{1}{\ln 2}2^x$ **18.** $u = \ln x$, $v = -x^{-2}$

20. False **21.** False **22.** False **23.** False **24.** False

25. (a) $x^3\ln x - \frac{1}{3}x^3 + C$

(b) $\frac{1}{3}(x^3\ln x^3 - x^3) + C$

27. (a) $\frac{1}{101}x(x+1)^{101} - \frac{1}{101}(\frac{1}{102})(x+1)^{102} + C$

(b) $\frac{1}{102}(x+1)^{102} - \frac{1}{101}(x+1)^{101} + C$

29. $xe^x - e^x + C$ **30.** $\frac{1}{2}x^2\ln x - \frac{1}{4}x^2 + C$

32. $-\frac{1}{2}\cos x^2 + C$ **34.** $\frac{1}{3}x^2e^{3x} - \frac{2}{9}xe^{3x} + \frac{2}{27}e^{3x} + C$

35. $\frac{3}{2}e^{x^2} + C$ **38.** $3(x\ln x - x)$

40. $-e^{-x}(x^2 + 1) - 2xe^{-x} - 2e^{-x} + C$

42. $\frac{1}{3}x^3 - 2(xe^x - e^x) + \frac{1}{2}e^{2x} + C$

45. $\frac{2}{3}x^{\frac{3}{2}}\ln x - \frac{4}{9}x^{\frac{3}{2}} + C$ **47.** $\frac{1}{2}x^2e^{x^2} - \frac{1}{2}e^{x^2} + C$

48. $-x\cot x + \ln|\sin x| + C$

49. $x^3\sin x + 3x^2\cos x - 6x\sin x - 6\cos x + C$

50. $-\frac{1}{2}x^2\cos x^2 + \frac{1}{2}\sin x^2 + C$

54. $x\tan^{-1} 3x - \frac{1}{6}\ln|1 + 9x^2| + C$

55. $-\frac{1}{5}e^{2x}\cos x + \frac{2}{5}e^{2x}\sin x + C$

59. $\frac{1}{2}e^{-x}\sin x - \frac{1}{2}e^{-x}\cos x + C$

60. $\frac{1}{2}x^2\tan^{-1} x + \frac{1}{2}\tan^{-1} x - \frac{1}{2}x + C$

62. $2\ln 2 - 1$ **63.** $-\frac{2}{e}$ **64.** $\frac{1}{4}(\pi + 2\ln 2)$

68. (a) $-\frac{1}{4} - \frac{3\pi}{8}$; **(b)** $\frac{5\pi}{8} - \frac{1}{4}$;

(c) $(-\frac{1}{4} - \frac{3\pi}{8}) + \frac{9\pi^2}{32}$; **(d)** $-\frac{1}{3\pi} - \frac{1}{2}$

SECTION 14.3

1. $\int \sin^2 x\,dx$, $\int \cos^4 x\,dx$, $\int \sin^2 x\cos^2 x\,dx$

2. $\int \sin^3 x\,dx$, $\int \sin^3 x\cos^2 x\,dx$, $\int \sin^2 x\cos^5 x\,dx$

3. $\int \sec^3 x\tan^3 x\,dx$, $\int \sec^4 x\tan x\,dx$, $\int \sin^2 x\cos^5 x\,dx$

6. If $k = 1$, multiply by $\frac{\csc x + \cot x}{\csc x + \cot x}$; if $k = 2$, just antidifferentiate; if $k > 2$, then use parts with $u = \csc^{k-2} x$ and $dv = \csc^2 x$.

8. If one of m or n is odd, use a Pythagorean identity to rewrite the integral so that substitution is possible (with $u = \sin x$ if m is odd and $u = \cos x$ if n is odd); if both m and n are even, use half-angle identities.

12. (a) Multiply the integrand by $\frac{\sec x + \tan x}{\sec x + \tan x}$.

   (b) Multiply the integrand by $\frac{\csc x + \cot x}{\csc x + \cot x}$.

17. $\frac{1}{2}x - \frac{1}{12}\sin(6x) + C$

19. $\sin x - \frac{2}{3}\sin^3 x + \frac{1}{5}\sin^5 x + C$

20. $-\frac{1}{2}\ln|\cos 2x| + C$

21. $\frac{1}{4}\ln|\sec 4x + \tan 4x| + C$

24. $-\frac{1}{4}\cot^4 x + \frac{1}{2}\cot^2 x + \ln|\sin x| + C$

25. $-\cot x + C$

26. $\frac{1}{4}\sec 2x \tan 2x + \frac{1}{4}\ln|\sec 2x + \tan 2x| + C$

27. Use parts with $u = \csc^2 x$ and $dv = \csc^2 x\, dx$ to get $-\csc^2 x \cot x - 2\int \csc^2 x \cot^2 x\, dx$. Then use substitution with $u = \cot x$ to get $-\csc^2 x \cot x + \frac{2}{3}\cot^3 x + C$.

30. $\frac{1}{5}\sin^5 x - \frac{1}{7}\sin^7 x + C$

31. $-\frac{1}{12}x - \frac{1}{96}\sin(12x) + C$ **34.** $\frac{1}{8}\sec^8 x + C$

35. $\frac{1}{7}\sec^7 x - \frac{2}{5}\sec^5 x + \frac{1}{3}\sec^3 x + C$

39. $-\frac{1}{2}\csc x \cot x + \frac{1}{2}\ln|\csc x + \cot x| + C$

40. $\sin x + C$ **41.** $\frac{1}{\cos x} + C$

43. $-\frac{1}{5}\cos^5 x + C$ **45.** $\frac{1}{3}\tan^3 x + C$

47. $-\frac{1}{2}\cos^2 x \ln(\cos x) + \frac{1}{4}\cos^2 x + C$

48. $-\frac{2}{5}(\cos x)^{\frac{5}{2}} + \frac{2}{7}(\cos x)^{\frac{7}{2}} + C$ **51.** $\frac{1}{3}\sec^3 x - \sec x + C$

52. (a) $\int (\frac{1 - \cos 2x}{2})(\frac{1 + \cos 2x}{2})\, dx = \frac{1}{4}(\frac{1}{2}x - \frac{1}{8}\sin 4x) + C$

   (b) $\int (\frac{1}{2}\sin 2x)^2\, dx = \frac{1}{8}(x - \frac{1}{4}\sin 4x) + C$

55. $\frac{4}{15}$ **56.** $\frac{8}{15}$

61. (a) The "inputs" of sine and cosine don't agree in the function $\sin 2x \cos 3x$.

   (b) $\int \frac{1}{2}(\sin(-x) + \sin(5x))\, dx = \frac{1}{2}\cos(-x) - \frac{1}{10}\cos(5x) + C$

   (c) PROOF $\frac{1}{2}(\sin(\alpha - \beta) + \sin(\alpha + \beta)) = \frac{1}{2}((\sin\alpha\cos\beta - \sin\beta\cos\alpha) + (\sin\alpha\cos\beta + \sin\beta\cos\alpha)) = \frac{1}{2}(2\sin\alpha\cos\beta) = \sin\alpha\cos\beta$ ∎

62. (a) $\frac{1}{2}\sec x \tan x + \frac{1}{2}\ln|\sec x + \tan x| + C$; $\frac{1}{6}\sec^5 x \tan x + \frac{5}{6}(\frac{1}{4}\sec^3 x \tan x + \frac{3}{4}(\frac{1}{2}\sec x \tan x + \frac{1}{2}\ln|\sec x + \tan x|)) + C$

   (b) Each time you use the formula, the power of secant *reduces* by 2.

   (c) PROOF By integration by parts with $u = \sec^{k-2} x$ and $dv = \sec^2 x\, dx$, we have $\int \sec^k x\, dx = \tan x \sec^{k-2} x - (k-2)\int \sec^{k-2} x \tan^2 x\, dx = \tan x \sec^{k-2} x - (k-2)\int \sec^{k-2} x(\sec^2 x - 1)\, dx$. Multiplying out the "new" integral and then solving an equation for $\int \sec^k x\, dx$, we get the desired reduction formula. ∎

Section 14.4

2. $u = \sin^{-1} x$ if and only if $x = \sin u$ (as long as $x \in [-1, 1]$ and $u \in [-\frac{\pi}{2}, \frac{\pi}{2}]$).

3. (a) $1 = \sec^2 u \frac{du}{dx} \Rightarrow dx = \sec^2 u\, du$

   (b) $\frac{dx}{du} = \sec^2 u \Rightarrow dx = \sec^2 u\, du$

4. (a) $\int \frac{1}{\sqrt{x^2+1}}\, dx$;   **(b)** $\int \frac{\sqrt{x^2-16}}{x}\, dx$;

   (c) $\int \sqrt{9 - (x-2)^2}\, dx$;   **(d)** $\int \frac{x}{x^2+1}\, dx$;

   (e) $\int \frac{1}{x^2-1}\, dx$

8. In terms of the Pythagorean identity, sine and cosine behave the same way; we can always use $x = \sin u$ instead of $x = \cos u$.

10. $x = a \sin u$ makes sense only for $x \in [-1, 1]$, but the domain of $a^2 - x^2$ is $(-\infty, \infty)$. On the other hand, $x = a \tan u$ makes sense for $x \in (-\infty, \infty)$, i.e., everywhere that $x^2 + a^2$ is defined.

14. Yes; $x = \sqrt{5}\sin u$.

20. $\sqrt{1 - x^2}$ is not defined on $[2, 3]$; its domain is $[-1, 1]$.

21. (a) Algebra;

   (b) Conventional substitution with $u = 4 + x^2$;

   (c) Trigonometric substitution with $x = 2 \tan u$

23. $\cos u = \sqrt{1 - x^2}$ **25.** $\sin u = \frac{x}{\sqrt{x^2+4}}$

27. $\tan^2 u = \frac{(x-5)^2}{9 - (x-5)^2}$ **28.** $\sin(2\cos^{-1} x) = 2x\sqrt{1 - x^2}$

30. $(x+3)^2 - 11$; $x + 3 = \sqrt{11}\sec u$

32. $(x - \frac{5}{2})^2 - \frac{21}{4}$; $x - \frac{5}{2} = \frac{\sqrt{21}}{2}\sec u$

35. $2(x+2)^2 + 12$; $x + 2 = \sqrt{6}\tan u$

36. (a) $\int \tan^2 u\, du = x - \tan^{-1} x + C$

   (b) $\int \frac{\tan^2 u}{1 + \tan^2 u}\sec^2 u\, du = x - \tan^{-1} x + C$

38. (a) $\int \frac{1}{9\tan^2 u + 9}(3\sec^2 u)\, du = \frac{1}{3}\tan^{-1}\frac{x}{3} + C$

   (b) $\frac{1}{9}\int \frac{1}{(\frac{x}{3})^2+1}\, dx = \frac{1}{3}\tan^{-1}\frac{x}{3} + C$

39. (a) $\frac{1}{2}\int \frac{1}{u}\, du = \frac{1}{2}\ln|4 + x^2| + C$

   (b) $\int \frac{2\tan u}{4 + 4\tan^2 u}(2\sec^2 u)\, du = -\ln|\frac{2}{\sqrt{x^2+4}}| + C$

43. $\sin^{-1}\frac{x}{\sqrt{3}} + C$ **47.** $-\frac{\sqrt{4-x^2}}{x} - \sin^{-1}\frac{x}{2} + C$

48. If $x > 1$, we have $\sqrt{x^2 - 1} - \sec^{-1} x + C$; if $x < -1$, we have $\sqrt{x^2 - 1} + \sec^{-1} x + C$.

49. $(x^2 - 1)^{\frac{1}{2}} + C$ **51.** $\frac{\sqrt{x^2-9}}{9x} + C$

55. $\frac{x}{\sqrt{1-x^2}} + C$ **56.** $\frac{1}{3}\tan^{-1}(\frac{x-2}{3}) + C$

57. $-\frac{1}{4}\sqrt{9 - 4x^2} + \frac{3}{2}\sin^{-1}\frac{2x}{3} + C$

59. $\frac{1}{2}(x - 4)\sqrt{(x-4)^2 + 9} + \frac{9}{2}\ln|\frac{1}{3}\sqrt{(x-4)^2 + 9} + \frac{1}{3}(x - 4)| + C$

61. $\sqrt{3 - x^2}(\frac{15}{8}x - \frac{1}{4}x^3) + \frac{27}{8}\sin^{-1}(\frac{x}{\sqrt{3}})$

65. $\frac{1}{2}x\sqrt{2 - x^2} + \sin^{-1}(\frac{x}{\sqrt{2}})$

67. $x\ln(1 + x^2) - 2x + 2\tan^{-1} x + C$

69. $\frac{1}{3}(e^{2x} + 1)^{\frac{3}{2}} + C$ **71.** $-\frac{1}{3}\frac{\sqrt{e^{2x}+3}}{e^x} + C$

72. $\frac{1}{2}x^2 - 2\ln(x^2 + 4) + C$

74. $-\frac{2}{9}x + \frac{1}{9}x^3 - \frac{23}{9\sqrt{6}}\tan^{-1}(\sqrt{\frac{3}{2}}x) + C$

75. $\frac{1}{3}(20)^{\frac{3}{2}} - \frac{1}{3}(4)^{\frac{3}{2}}$ **76.** $-\sqrt{3} + \sqrt{15}$ **80.** $\frac{2}{3} - \frac{1}{3\sqrt{7}}$

81. PROOF   (a) $2\int_{-r}^{r}\sqrt{r^2 - x^2}\, dx$

   (b) $2\int_{-\frac{\pi}{2}}^{\frac{\pi}{2}}\sqrt{r^2 - r^2\sin^2 u}\,(r\cos u)\, du = 2r^2\int_{-\frac{\pi}{2}}^{\frac{\pi}{2}}\cos^2 u\, du$
   $= 2r^2(\frac{1}{2})\int_{-\frac{\pi}{2}}^{\frac{\pi}{2}}(1 + \cos 2u)\, du = r^2[u + \frac{1}{2}\sin 2u]_{-\frac{\pi}{2}}^{\frac{\pi}{2}} = \pi r^2$ ∎

CHAPTER 15

SECTION 15.1

3. The rate of change of the function $f(x)$ is nonconstant.

5. $y = \sqrt{1 + 4x^2}$ **6.** $y = e^x$

8. No; $f(x)$ has a vertical asymptote at $x = \frac{\pi}{2}$.

10. Both arc lengths are given by the definite integral $\int_0^2 \sqrt{1 + 4x^2}\, dx$, so they are equal. This makes sense because $f(x) = x^2 - 3$ and $g(x) = 5 - x^2$ differ by a vertical flip and shift only; these transformations should not affect the arc length.

12. $\int_{-3}^3 \sqrt{1 + 9x^4}\, dx$; we don't yet know how to solve this integral.

15. $\int_{\frac{\pi}{4}}^{\frac{\pi}{2}} \sqrt{1 + \cot^2 x}\, dx$; we can solve this integral by using a trigonometric identity.

16. $2\int_{-5}^5 \sqrt{1 + (\frac{-x}{\sqrt{25 - x^2}})^2}\, dx$

19. $f(x) = e^{3x}$, $[a, b] = [-2, 5]$

21. $f(x) = \frac{1}{x}$, $[a, b] = [2, 3]$

23. $f(x) = \ln|\cos x|$, $[a, b] = [0, \frac{\pi}{4}]$

25. 16.97   **28.** 2.3   **30.** 9.25

31. $\Delta y_1 = 7$, $\Delta y_2 = 1$, $\Delta y_3 = 1$, $\Delta y_4 = 7$; 16.97

32. $\Delta y_1 = -\frac{1}{2}$, $\Delta y_2 = -1$, $\Delta y_3 = -\frac{1}{2}$; 3.77

35. $\Delta y_1 = \sqrt{8}$, $\Delta y_2 = 0$, $\Delta y_3 = -\sqrt{8}$; 8.928

37. $\int_{-1}^4 \sqrt{10}\, dx = 5\sqrt{10}$

40. $\int_1^3 \sqrt{1 + 9x}\, dx = \frac{2}{27}(28)^{\frac{3}{2}} - \frac{2}{27}(10)^{\frac{3}{2}}$

41. $\int_{-1}^1 \sqrt{28 + 18x}\, dx = \frac{1}{27}(46)^{\frac{3}{2}} - \frac{1}{27}(10)^{\frac{3}{2}}$

44. $\int_{-3}^3 \frac{3}{\sqrt{9 - x^2}}\, dx = 3\pi$ **45.** $\frac{1}{2}\int_0^1 \frac{1 + x}{\sqrt{x}}\, dx = \frac{4}{3}$

46. $\int_1^2 \frac{1 + 16x^2}{8x}\, dx = 3 + \frac{\ln 2}{8} \approx 3.087$

47. $\int_{\frac{\pi}{4}}^{\frac{3\pi}{4}} \csc x\, dx = \ln(2\sqrt{2} + 3)$

50. $\int_{-1}^1 \sqrt{1 + 4x^2}\, dx \approx 2.958$

53. (a) PROOF   The distance from $(a, f(a)) =$ $(a, ma + c)$ to $(b, f(b)) = (b, mb + c)$ is given by
$$\sqrt{(b - a)^2 + ((mb + c) - (ma + c))^2} =$$
$$\sqrt{(b - a)^2 + m^2(b - a)^2} = (b - a)\sqrt{1 + m^2}.$$
(Since the graph of $f(x) = mx + c$ is a line, this is the exact arc length.) ▪

(b) PROOF   If $f(x) = mx + c$, then $f'(x) = m$ (a constant), so the arc length is given by $\int_a^b \sqrt{1 + m^2}\, dx = [\sqrt{1 + m^2}]_a^b = \sqrt{1 + m^2}(b - a)$. ▪

55. PROOF   If $f(x) = \sqrt{r^2 - x^2}$ (the top half of the circle), then $f'(x) = -\frac{x}{\sqrt{r^2 - x^2}}$. Thus the circumference of the circle is twice the arc length of the top half: $2\int_{-r}^r \sqrt{1 + (-\frac{x}{\sqrt{r^2 - x^2}})^2}\, dx = 2\int_{-r}^r \frac{r}{\sqrt{r^2 - x^2}}\, dx$. Using trigonometric substitution with $x = r \sin u$ (and changing the limits of integration accordingly), this is equal to $2r\int_{-\frac{\pi}{2}}^{\frac{\pi}{2}} 1\, du = 2r[u]_{-\frac{\pi}{2}}^{\frac{\pi}{2}} = 2r(\frac{\pi}{2} - (-\frac{\pi}{2})) = 2\pi r$. ▪

SECTION 15.2

1. (a) 2.25π;   **(b)** 5.25π **2. (a)** 4.5π;   **(b)** 7.5π

5. (a) $\Delta y = 1$, $y_0 = 0$, $y_1 = 1$, $y_2 = 2$, $y_3 = 3$, $y_4 = 4$

(b) One choice is $y_1^* = 0.5$, $y_2^* = 1.5$, $y_3^* = 2.5$, and $y_4^* = 3.5$.

(c) $f^{-1}(y_1^*) = \sqrt{0.5}$, $f^{-1}(y_2^*) = \sqrt{1.5}$, $f^{-1}(y_3^*) = \sqrt{2.5}$, $f^{-1}(y_4^*) = \sqrt{3.5}$. (These values are locations on the x-axis.)

(d) $\pi(\sqrt{1.5})^2(1) = 1.5\pi$ **(e)** $y_k^*\,\pi$

6. (a) $\Delta x = \frac{1}{2}$, $x_0 = 0$, $x_1 = 0.5$, $x_2 = 1$, $x_3 = 1.5$, $x_4 = 2$

(b) One choice is $x_1^* = 0.25$, $x_2^* = 0.75$, $x_3^* = 1.25$, and $x_4^* = 1.75$.

(c) $f(x_1^*) = 0.0625$, $f(x_2^*) = 0.5625$, $f(x_3^*) = 1.5625$, $f(x_4^*) = 3.0625$

(d) $(16\pi - \pi((0.75)^2)^2)(\frac{1}{2}) \approx 24.6357$

(e) $\frac{\pi}{2}(16 - (x_k^*)^4)$

8. $\pi\int_2^3 (1 + x)^2\, dx$; the solid obtained by rotating the region between the graph of $f(x) = 1 + x$ and the x-axis on $[2, 3]$ around the x-axis

10. $\pi\int_0^2 (4 - x^2)\, dx$; the solid obtained by rotating the region between the graph of $f(x) = \sqrt{4 - x^2}$ and the x-axis on $[0, 2]$ around the x-axis

12. The region between $f(x) = x + 1$ and the line $y = 4$ from $x = 1$ to $x = 3$, rotated around the x-axis

13. The region between $f(x) = x - 1$ and the x-axis on $[1, 3]$, rotated around the x-axis

14. The region between the graph of $f(x) = 2x + 1$ and the line $y = 5$ on the x-interval $[0, 2]$, rotated about the y-axis

18. $\pi\int_0^{\sqrt{3}} ((3 + 4x - x^2)^2 - (4x)^2)\, dx + \pi\int_{\sqrt{3}}^4 ((4x)^2 - (3 + 4x - x^2)^2)\, dx$

19. $\pi\int_0^1 ((2 + \sqrt{y})^2 - (2 - \sqrt{y})^2)\, dy + \pi\int_1^4 ((2 + \sqrt{y})^2 - y^2)\, dy$

21. (a) and **(b)** are equal.   **24. (b)** is larger.

25. (b) is larger.   **27. (a)** $\pi(b^2 - 8)$;   **(b)** $b = 4$

28. The region bounded by the graph of $y = \sqrt{x}$, the graph of $y = x - 2$, and the x-axis, rotated around the x-axis

30. The region bounded by the graph of $y = x^3 + x + 1$ and the x-axis on $[0, 4]$, rotated around the y-axis

31. (a) Using the right side of each interval: 10π
(b) Using the "top" of each interval: 65.78
(c) Using the right side of each interval: 161.68
(d) Using the "top" of each interval: 145.89

33. (a) Using the right side of each interval: 0.387
(b) Using the "top" of each interval: 0.491
(c) Using the right side of each interval: 5.473
(d) Using the "top" of each interval: 5.2387

35. (a) $\pi x_k^*\,\Delta x$ **(b)** $\pi(16 - (y_k^*)^4)\,\Delta y$
(c) $\pi(25 - (5 - \sqrt{x_k^*})^2)\,\Delta x$
(d) $\pi(49 - (3 + (y_k^*)^2)^2)\,\Delta y$

38. (a) $\pi(x_k^* - \frac{1}{4}(x_k^*)^2)\,\Delta x$
(b) $\pi(4(y_k^*)^2 - (y_k^*)^4)\,\Delta y$
(c) $\pi((\sqrt{x_k^*} + 2)^2 - (\frac{1}{2}x_k^* + 2)^2)\,\Delta x$
(d) $\pi((7 - (y_k^*)^2)^2 - (7 - 2y_k^*)^2)\,\Delta y$

40. $\frac{1}{5}(1216 + 48\sqrt{3})\pi$ **41.** $\frac{45\pi}{2}$

43. (a) $\pi \int_0^4 x\,dx = 8\pi$

(b) $\pi \int_0^2 (16 - y^4)\,dy = \frac{128\pi}{5}$

(c) $\pi \int_0^4 (25 - (5 - \sqrt{x})^2)\,dx = \frac{136\pi}{3}$

(d) $\pi \int_0^2 (49 - (3 + y^2)^2)\,dy = \frac{288\pi}{5}$

45. (a) $\pi \int_0^1 (x^2 - x^4)\,dx = \frac{2\pi}{15}$

(b) $\pi \int_0^1 (y - y^2)\,dy = \frac{\pi}{6}$

(c) $\pi \int_0^1 ((1 + x)^2 - (1 + x^2)^2)\,dx = \frac{7\pi}{15}$

(d) $\pi \int_0^1 ((2 - y)^2 - (2 - \sqrt{y})^2)\,dy = \frac{\pi}{2}$

46. (a) $\pi \int_0^4 (x - \frac{1}{4}x^2)\,dx = \frac{8\pi}{3}$

(b) $\pi \int_0^2 (4y^2 - y^4)\,dy = \frac{64\pi}{15}$

(c) $\pi \int_0^4 ((\sqrt{x} + 2)^2 - (\frac{1}{2}x + 2)^2)\,dx = \frac{100\pi}{3}$

(d) $\pi \int_0^2 ((7 - y^2)^2 - (7 - 2y)^2)\,dy = \frac{1352\pi}{15}$

48. (a) $\pi \int_2^5 (x - 2)^2\,dx = 9\pi$

(b) $\pi \int_0^3 (25 - (y + 2)^2)\,dy = 36\pi$

(c) $\pi \int_2^5 (9 - (3 - (x - 2))^2)\,dx = 18\pi$

(d) $\pi \int_0^3 (9 - ((y + 2) - 2)^2)\,dy = 18\pi$

(e) $\pi \int_2^5 ((2 + (x - 2))^2 - 4)\,dx = 27\pi$

(f) $\pi \int_0^3 ((6 - (y + 2))^2 - 1)\,dy = 18\pi$

50. The cone can be obtained by rotating the region between $y = \frac{3}{5}x$ and the x-axis around the x-axis; this solid has volume $\frac{9\pi}{25} \int_0^5 x^2\,dx = 15\pi$.

56. **PROOF**   A sphere of radius r can be obtained by rotating the region under the graph of $y = \sqrt{r^2 - x^2}$ on $[-r, r]$ around the x-axis. The volume of this solid is given by the definite integral $\pi \int_{-r}^r (\sqrt{r^2 - x^2})^2\,dx = \frac{4}{3}\pi r^3$.

SECTION 15.3

1. (a) $\pi(1.5)^2(3) - \pi(1)^2(3) = 3.75\pi$

(b) $2\pi(1.25)(3)(0.5) = 3.75\pi$

(c) 6.75π

2. Average radius is $\frac{r_1 + r_2}{2}$, thickness is $r_2 - r_1$.

3. It involves Δx, which enables us to construct a Riemann sum (and thus a definite integral).

6. (a) $\Delta x = 0.5$, $x_0 = 0$, $x_1 = 0.5$, $x_2 = 1$, $x_3 = 1.5$, $x_4 = 2$

(b) $x_1^* = 0.25$, $x_2^* = 0.75$, $x_3^* = 1.25$, $x_4^* = 1.75$

(c) $f(x_1^*) = (0.25)^2 = 0.0625$, $f(x_2^*) = (0.75)^2 = 0.5625$, $f(x_3^*) = (1.25)^2 = 1.5625$, $f(x_4^*) = 3.0625$

(d) $2\pi(1.25)f(1.25)(0.5) \approx 6.136$

(e) $2\pi x_k^* f(x_k^*)(0.5) = \pi(x_k^*)^3$

8. The region bounded by the graphs of $y = \ln x$, $y = 0$, $y = 1$, and $x = 0$

11. The region between the graph of $y = x^2 + x + 1$ and the x-axis on $[0, 2]$

14. $\int_0^2 2\pi y^3\,dy = 2\pi \int_0^2 y(y^2)\,dy$; the solid obtained by rotating the region between the graph of $f(x) = \sqrt{x}$, the y-axis, and the line $y = 2$ around the x-axis

16. The region between the graph of $f(x) = \cos x$ and the x-axis on the interval $[0, \frac{\pi}{4}]$, rotated around the y-axis

18. The region between $y = x^2 + 4$ and $y = 5$ from $x = 0$ to $x = 1$, rotated around the x-axis

20. The region between $y = e^x$ and $y = e$ from $x = 0$ to $x = 1$, rotated around the x-axis

21. (a) $2\pi \int_0^1 3y\,dy + 2\pi \int_1^7 y(3 - \frac{y-1}{2})\,dy$

(b) $\pi \int_0^3 (2x + 1)^2\,dx$

(c) Discs are easier.

23. (a) $2(2\pi \int_0^2 y(2 - y)\,dy)$

(b) $2(\pi \int_0^2 x^2\,dx)$

(c) Discs are easier.

24. (a) $2\pi \int_0^1 (3 - x)(x^2 + 1)\,dx$

(b) $\pi \int_0^1 (3^2 - 2^2)\,dy + \pi \int_1^2 ((3 - \sqrt{y - 1})^2 - 2^2)\,dy$

(c) Shells are easier.

27. With midpoints for y_k^*: 25.9181

28. With midpoints for y_k^*: 79.7425

31. $2\pi y_k^*(4 - (y_k^*)^2)\,\Delta y$

33. $2\pi x_k^*(2 - \sqrt{x_k^*})\,\Delta x$

35. (a) $2\pi \int_0^2 y(4 - y^2)\,dy = 8\pi$

(b) $\pi \int_0^4 (\sqrt{x})^2\,dx = 8\pi$

37. (a) $2\pi \int_0^4 x(2 - \sqrt{x})\,dx = \frac{32\pi}{5}$

(b) $\pi \int_0^2 y^4\,dy = \frac{32\pi}{5}$

40. $2\pi \int_0^2 x(4 - (4 - x^2))\,dx = 2\pi \int_0^2 x^3\,dx = 8\pi$

41. $2\pi \int_0^6 x((x - 3)^2 + 2)\,dx = 180\pi$

42. $2\pi \int_0^1 y(\sqrt{y} - y^2)\,dy = \frac{3\pi}{10}$

45. $2\pi \int_0^9 (10 - y)(3 - \sqrt{y})\,dy = \frac{657\pi}{5}$

47. With discs: $\pi \int_0^3 (9 - x^2)^2\,dx = \frac{648\pi}{5}$

48. With shells: $2\pi \int_0^2 x(x^2 - 4x + 4)\,dx = \frac{8\pi}{3}$

49. With discs: $\pi \int_0^{2/3} 9\,dx + \pi \int_{2/3}^2 (\frac{2}{x})^2\,dx = 10\pi$.

With shells: $2\pi \int_0^1 2y\,dy + 2\pi \int_1^3 (\frac{2}{y})y\,dy = 10\pi$.

51. The region between the graph of $y = \frac{5}{3}x$ and the lines $y = 5$ and $x = 0$, rotated around the y-axis, is a cone of radius 3 and height 5. The volume of this solid is $2\pi \int_0^3 x(5 - \frac{5}{3}x)\,dx = 15\pi$.

54. **PROOF**   A sphere of radius r can be obtained by rotating the region between the graph of $y = \sqrt{r^2 - x^2}$ and the x-axis on $[-r, r]$ around the x-axis. The volume of this solid is $2\pi \int_0^r x(\sqrt{r^2 - x^2} - (-\sqrt{r^2 - x^2}))\,dx = 2\pi \int_0^r x(2\sqrt{r^2 - x^2}) = \frac{4}{3}\pi r^3$ (using integration by substitution with $u = r^2 - x^2$). Technical note: Usually we can't do the shell method along the x-axis when we've rotated a region around the x-axis; it works here only because the solid is a perfect sphere, so it is equivalent to rotating the region between the graph of $x = \sqrt{r^2 - y^2}$ and the y-axis from $y = -r$ to $y = r$ around the y-axis.

SECTION 15.4

2. The distance varies depending on the depth of the water.

4. If you know x_k, Δx, and N, then you can solve for a and b, since $x_k = a + k\,\Delta x$ and $\Delta x = \frac{b-a}{N}$.

5. It would require 38,422.9 foot-pounds of work to lift the entire full hot tub; it would take significantly less work to pump the water out of the tub, since only the water at the bottom of the tub has to move 4 feet (the rest of the water has a shorter distance to travel).

8. 46,436 grams   **9.** 3.1765 pounds

11. $\Delta y = 1$, $y_0 = 0$, $y_1 = 1$, $y_2 = 2$, $y_3 = 3$, $y_4 = 4$. If we choose $y_1^* = 0.5$, $y_2^* = 1.5$, $y_3^* = 2.5$, and $y_4^* = 3.5$, then the work is approximately 3301.97 foot-pounds.

12. $\Delta x = 6$, $x_0 = 0$, $x_1 = 6$, $x_2 = 12$, $x_3 = 18$, $x_4 = 24$. If we choose $x_1^* = 0$, $x_2^* = 6$, $x_3^* = 12$, and $x_4^* = 18$, then the mass is approximately 3746.95 grams.

15. $\Delta y = 2$, $y_0 = 0$, $y_1 = 2$, $y_2 = 4$, $y_3 = 6$, $y_4 = 8$. If we choose $y_1^* = 1$, $y_2^* = 3$, $y_3^* = 5$, and $y_4^* = 7$, then the force is approximately 78,624 pounds.

16. **(a)** $W = (62.4)(40) \int_0^3 (3 - y)\, dy = 11{,}232$ foot-pounds

   (b) $F = (62.4)(8) \int_0^3 (3 - y)\, dy = 2246.4$ pounds

17. $W = (62.4)\left(\frac{4\pi}{25}\right) \int_0^{10} y^2 (10 - y)\, dy = 26{,}138.1$ foot-pounds

19. **(a)** $W = (62.4)(16\pi) \int_0^{13} (16 - y)\, dy = 387{,}366$ foot-pounds

   (b) $W = (71.8)(16\pi) \int_0^{\frac{26}{3}} (20 - y)\, dy = 490{,}030$ foot-pounds

23. $m = (1.5)^2 \int_0^{12} (4.2 + 0.4x - 0.03x^2)\, dx = 139.32$ grams

26. **(a)** The solid of revolution obtained by revolving the region between the graph of $y = 8 - \frac{2}{25} x^2$ and the x-axis on $[0, 10]$ around the y-axis

   (b) $\rho(y) = -\frac{0.97}{8} y + 1.12$

   (c) $m = \pi \int_0^8 \left(-\frac{0.97}{8} y + 1.12\right)\left(100 - \frac{25}{2} y\right) dy = 1001.12$ ounces

28. $F = (62.4)(0.8\pi) \int_0^1 (1 - y)\, dy = 78.4142$ pounds

30. $F = 62.4 \int_0^{200} \left(\frac{150}{200} y + 250\right)(200 - y)\, dy = 3.744 \times 10^8$ pounds

33. $V = \pi \int_{-8}^8 \left(\sqrt{10^2 - y^2}\right)^2 dy + \pi \int_8^{15} 6^2\, dy = 4745.9$ cubic centimeters

34. **(a)** $SA = 2\pi \int_0^3 (2x^3)\sqrt{1 + 36x^4}\, dx = \frac{4\pi}{216}\left((2917)^{\frac{3}{2}} - 1\right) \approx 9165.54$

   (b) $SA = 2\pi \int_{-r}^r \sqrt{r^2 - x^2}\sqrt{1 + \frac{x^2}{r^2 - x^2}}\, dx = 2\pi \int_{-r}^r r\, dx = 4\pi r^2$

INDEX

A

a_k, *see* sigma notation, 718
absolute value, 9, 29, 93, 94, 250
acceleration, 242, 266, 405
accumulation function, 787, 789
acute angle, 630
adjacent side, 631
algebraic function, 91, 483
algebraic rules
  for exponents, 328, 533
  for fractions, 19
  for inequalities, 25
  for logarithms, 545, 601, 603
amplitude, 682, 685
angle, 630, 639
annulus, 863
antiderivative, 356, 767, 768, 779,
    790, 802
  constructing, 792
  elementary, 775
  of exponential function, 565, 770
  by guess-and-check, 773
  involving logarithms, 619
  of logarithmic function, 818
  of polynomial function, 405
  of power function, 356, 769
  of trigonometric functions, 770,
    830, 832
approximating area, 715
  with rectangles, 728, 732–734
  with trapezoids, 735
approximating length, 848
approximating volume
  with discs, 859, 860
  with shells, 874
  with washers, 863
arbitrary constant, 768
arc length, 850, 851
arccos x, *see* inverse cosine
    function, 695
arcsec x, *see* inverse secant
    function, 695
arcsin x, *see* inverse sine
    function, 695
arctan x, *see* inverse tangent
    function, 695
area of a circle, 715

area under a curve, 742
  signed, 753
  true, 753
area between curves, 755
area function, 787, 789
arithmetic function operations, 99
asymptote, 78
  curve, 457
  horizontal, 78, 532, 554
  slant, 455, 456
  vertical, 78, 541, 664
average radius, 873
average rate of change, 82, 236, 778
average value, 759, 760, 778, 796
average velocity, 235, 778
axis of revolution, 866

B

b^x, *see* exponential function, 531
back-substitution, 808, 835
bad proof, 48
bad window, 69
base, 531, 600
base conversion formula, 605

C

cancellation, 19, 185, 441
Cartesian plane, 10
center point, 684, 685
chain rule, 492, 494, 673, 793,
    794, 802
change of variables, 224, 802, 803
checking answers
  antiderivatives, 356
  area under a curve, 750
  asymptotes, 455
  continuity, 353
  critical points, 486
  curve sketching, 317
  definite integral, 750, 810
  δ given ε, 159
  derivative, 225
  differentiability, 353
  domain, 483
  equations, 18, 334

extrema, 407
factorization, 397
global extrema, 468
for hole in graph, 444
increasing and decreasing, 298
indefinite integral, 805
inequalities, 28
inflection points, 407
inverse functions, 130
limits, 181
local extrema, 467
polynomial long division, 448
synthetic division, 396
systems of equations, 23
circular reference, *see* reference,
    circular, 894
circumference of a circle, 848, 856
closed endpoint, 427, 520
closed interval, 6
coefficient, 332, 387, 425
common factor, 19, 441
complex number, 3
composition, 102, 103
concavity, 75, 308, 309, 314, 374, 518
conclusion, 40
cone, 511
conjugate, 330
constant, 15
constant function, 64, 388
constant multiple rule
  for definite integrals, 745
  for derivatives, 274
  for indefinite integrals, 771
  for limits, 170
  for sums, 719
constant term, 387
constraint, 431
constructing antiderivatives, 792
continuity, 253
  of algebraic functions, 484
  of area functions, 790
  of a composition, 201
  of exponential functions, 551
  of a function, 198
  on an interval, 193
  of logarithmic functions, 609
  of natural exponential function,
    549, 550

REFERENCE PAGES
(continued from front inside cover)

Logic

- The *negation* of a statement is a statement that means the exact logical opposite of the original statement. (Section 0.4.2)
- Suppose the statement "For all x, we have P" is false. A *counterexample* to this statement is a value of x for which P is false. (Section 0.4.2)
- The *contrapositive* of "$A \Rightarrow B$" is the statement "(Not B) $\Rightarrow$ (Not A)." The contrapositive is always logically equivalent to the original statement. (Section 0.4.5)
- The *converse* of the statement "$A \Rightarrow B$" is the statement "$B \Rightarrow A$." In general, the converse can have a different meaning than the original statement. (Section 0.4.5)
- To prove a statement P by *contradiction*, we suppose that P is false, and then show that this supposition leads to a contradiction. Thus the statement P cannot be false, and must be true. (Section 0.5.4)
- *The Induction Axiom:* Suppose P is a statement concerning positive integers n such that P is true for $n = 1$, and such that for all positive integers $n > 1$, if P is true for $n - 1$, then P is true for n. Then P is true for all positive integers. (Section 0.5.5)

Functions

BASIC DEFINITIONS

- A *function* $f\colon A \to B$ is a rule that assigns to each element of A a unique element of B. (Section 1.1.1)
- The *domain* of a function f is the set of values for which f is defined. The *range* of a function f is the set of possible outputs:
  $\{y \in \mathrm{Target}(f) \mid y = f(x) \text{ for some } x \in \mathrm{Domain}(f)\}$. (Section 1.1.1)
- f is an *even* function if $f(-x) = f(x)$ for all $x \in \mathrm{Domain}(f)$. f is an *odd* function if $f(-x) = -f(x)$ for all $x \in \mathrm{Domain}(f)$. (Section 1.6.5)
- f is a *one-to-one* function if, for all a, $b \in \mathrm{Domain}(f)$, we have $a \neq b \implies f(a) \neq f(b)$. (Section 1.7.1)
- f^{-1} is the *inverse* of a one-to-one function f if $f(f^{-1}(x)) = x$ for all $x \in \mathrm{Domain}(f^{-1})$, and $f^{-1}(f(x)) = x$ for all $x \in \mathrm{Domain}(f)$. (Section 1.7.2)

TYPES OF FUNCTIONS

- *Constant:* $f(x) = c$, for $c \in \mathbb{R}$. (Section 1.1.5)
- *Identity:* $f(x) = x$. (Section 1.1.5)
- *Linear:* $f(x) = mx + b$, for m, $b \in \mathbb{R}$. (Section 1.3.1)
- *Proportional:* $f(x) = kx$, for $k \in \mathbb{R}$. (Section 1.3.3)

- *Power:* $f(x) = Ax^k$, for $A \neq 0$ in $\mathbb{R}$ and k rational. (Section 4.1.6)
- *Quadratic:* $f(x) = ax^2 + bx + c$, for a, b, $c \in \mathbb{R}$, $a \neq 0$. (Sections 0.2.2 and 5.1.1)
- *Cubic:* $f(x) = ax^3 + bx^2 + cx + d$, for a, b, c, $d \in \mathbb{R}$, $a \neq 0$. (Section 5.1.1)
- *Polynomial:* $f(x) = a_n x^n + a_{n-1} x^{n-1} + \cdots + a_2 x^2 + a_1 x + a_0$, for n a positive integer and $a_i \in \mathbb{R}$. (Section 5.1.1)
- *Rational:* $f(x) = \frac{p(x)}{q(x)}$, for $p(x)$ and $q(x)$ polynomial functions. (Section 6.1.1)
- *Algebraic:* Any function that can be expressed using only arithmetic operations and rational powers. (Sections 1.4.1 and 7.1.1)
- *Transcendental:* Any function that is not algebraic. (Sections 1.4.2 and 8.1.1)
- *Exponential:* $f(x) = Ae^{kx}$, for $A \neq 0$ and $k \neq 0$ in $\mathbb{R}$; or $f(x) = Ab^x$, for $A \neq 0$ and $b > 0$, $b \neq 1$ in $\mathbb{R}$. (Sections 8.1.1 and 8.2.3)
- *Logarithmic:* $f(x) = \ln x$ and $f(x) = \log_b x$ for $b > 0$, $b \neq 1$ in $\mathbb{R}$. (Sections 8.2.2 and 9.1.1)
- *Trigonometric:* $\sin x$, $\cos x$, $\tan x$, $\sec x$, $\csc x$, and $\cot x$. (Sections 10.1.1 and 10.2.2)
- *General Sine:* $f(x) = A\sin(B(x + C)) + D$, for A, B, C, $D \in \mathbb{R}$. (Section 10.6.2)
- *General Cosine:* $f(x) = A\cos(B(x + C)) + D$, for A, B, C, $D \in \mathbb{R}$. (Section 10.6.2)
- *Inverse Trigonometric:* $\sin^{-1} x$, $\cos^{-1} x$, $\tan^{-1} x$, $\sec^{-1} x$, $\csc^{-1} x$, and $\cot^{-1} x$. (Section 11.1.2)
- *Functions Defined By Integrals:* $F(x) = \int_a^x f(t)\, dt$, for $a \in \mathbb{R}$, f a continuous function. (Section 13.3.1)

DOMAINS

- Functions with domain $\mathbb{R}$: constant, identity, linear, proportional, quadratic, cubic, polynomial, exponential, sine and cosine, general sine and cosine, inverse tangent. (Sections 1.1.5, 1.3.1, 1.3.3, 5.2.1, 8.1.3, 10.3.1, and 11.1.2)
- If k is an integer, then Ax^k has domain $\mathbb{R}$ if k is positive, and $(-\infty, 0) \cup (0, \infty)$ if k is negative. (Section 4.4.1)
- If $\frac{p}{q}$ is rational and positive, then $Ax^{\frac{p}{q}}$ has domain $\mathbb{R}$ if q is odd, and $[0, \infty)$ if q is even. (Section 4.5.2)
- If $\frac{p}{q}$ is rational and negative, then $Ax^{\frac{p}{q}}$ has domain $(-\infty, 0) \cup (0, \infty)$ if q is odd, and $(0, \infty)$ if q is even. (Section 4.5.2)
- The domain of a rational function $\frac{p(x)}{q(x)}$ is the set $\{x \mid q(x) \neq 0\}$. (Section 6.1.3)
- $\ln x$ and $\log_b x$ have domain $(0, \infty)$.
- $\tan x$ and $\sec x$ have domain $\{x \mid x \neq \frac{\pi}{2} + \pi k\}$.
- $\csc x$ and $\cot x$ have domain $\{x \mid x \neq \pi k\}$.
- $\sin^{-1} x$ and $\cos^{-1} x$ have domain $[-1, 1]$.
- $\sec^{-1} x$ has domain $(-\infty, -1] \cup [1, \infty)$.

Graphical Behavior

BEHAVIOR AT A POINT

- f has a *root* (*i.e.* a *zero*, or an *x-intercept* at $x = c$ if $f(c) = 0$. (Section 1.2.3)
- f has a *y-intercept* at $y = k$ if $f(0) = k$. (Section 1.2.3)
- f has a *global maximum* at $x = c$ if $f(c) \geq f(x)$ for all $x \in$ Domain(f). (Sections 1.2.7, 5.4.2 and 7.5.2)
- f has a *global minimum* at $x = c$ if $f(c) \leq f(x)$ for all $x \in$ Domain(f). (Sections 1.2.7, 5.4.2 and 7.5.2)
- f has a *local maximum* at $x = c$ if, for some $\delta > 0$, $f(c) \geq f(x)$ for all $x \in (c - \delta, c + \delta)$. (Sections 1.2.7, 3.7.3, and 3.8.4)
- f has a *local minimum* at $x = c$ if, for some $\delta > 0$, $f(c) \leq f(x)$ for all $x \in (c - \delta, c + \delta)$. (Sections 1.2.7, 3.7.3, and 3.8.4)
- f has an *inflection point* at $x = c$ if f changes concavity at $x = c$, that is, if f'' changes sign at $x = c$. (Sections 1.2.7 and 3.8.3)
- f has a *critical point* at $x = c$ if $f'(c) = 0$ or if $f'(c)$ does not exist. (Section 3.6.1)
- f has a *turning point* at $x = c$ if $x = c$ is a local minimum or maximum of f. (Section 3.7.3)

BEHAVIOR ON AN INTERVAL

- f is *positive* on an interval I if $f(x) > 0$ for all $x \in I$. (Section 1.2.4)
- f is *negative* on an interval I if $f(x) < 0$ for all $x \in I$. (Section 1.2.4)
- f is *increasing* on an interval I if, for all $b > a$ in I, $f(b) > f(a)$. If f' is positive on the interior of I, then f is increasing on all of I. (Sections 1.2.5 and 3.7.1)

- f is *decreasing* on an interval I if, for all $b > a$ in I, $f(b) < f(a)$. If f' is negative on the interior of I, then f is decreasing on all of I. (Sections 1.2.5 and 3.7.1)
- f is *concave up* on an interval I if f' is increasing on I. If f'' is positive on I, then f is concave up on I. (Sections 3.8.1 and 3.8.2)
- f is *concave down* on an interval I if f' is decreasing on I. If f'' is negative on I, then f is concave down on I. (Sections 3.8.1 and 3.8.2)
- The *average rate of change* of f on an interval $[a, b]$ is $\frac{f(b) - f(a)}{b - a}$. (Sections 1.3.2 and 3.2.2)

TRANSFORMATIONS

Suppose the graph of $f(x)$ contains the point (x, y). The table at the bottom of this page describes the effect of various transformations of $f(x)$.

ASYMPTOTES

- f has a *vertical asymptote* at $x = c$ if at least one of $\lim\limits_{x \to c^-} f(x)$ or $\lim\limits_{x \to c^+} f(x)$ is infinite. (Sections 1.2.8 and 2.1.4)
- f has a *horizontal asymptote* at $y = k$ if at least one of $\lim\limits_{x \to \infty} f(x)$ or $\lim\limits_{x \to -\infty} f(x)$ is equal to k. (Sections 1.2.8 and 2.1.5)
- f has a *slant asymptote* of a line $l(x)$ if at least one of $\lim\limits_{x \to \infty} (f(x) - l(x))$ or $\lim\limits_{x \to -\infty} (f(x) - l(x))$ is 0. (Section 6.2.4)
- f has a *curve asymptote* of a curve $c(x)$ if at least one of $\lim\limits_{x \to \infty} (f(x) - c(x))$ or $\lim\limits_{x \to -\infty} (f(x) - c(x))$ is 0. (Section 6.2.5)

Transformation	Graphical Effect	New Point	
$f(x) + C$	Graph shifts vertically by C units (up if $C > 0$, down if $C < 0$).	$(x, y + C)$	(Section 1.6.1)
$f(x + C)$	Graph shifts horizontally by C units (left if $C > 0$, right if $C < 0$).	$(x - C, y)$	(Section 1.6.1)
$Af(x)$	Graph stretches vertically if $A > 1$, compresses vertically if $0 < A < 1$.	(x, Ay)	(Section 1.6.2)
$f(Ax)$	Graph compresses horizontally if $A > 1$, stretches horizontally if $0 < A < 1$.	$\left(\frac{x}{A}, y\right)$	(Section 1.6.2)
$-f(x)$	Graph is reflected over the x-axis.	$(x, -y)$	(Section 1.6.4)
$f(-x)$	Graph is reflected over the y-axis.	$(-x, y)$	(Section 1.6.4)
$f^{-1}(x)$	Graph is reflected over the line $y = x$.	(y, x)	(Section 1.7.3)
$\frac{1}{f(x)}$	Small values become large, and vice-versa. Values of zero become asymptotes, and vice-versa.	$\left(x, \frac{1}{y}\right)$	(Section 4.4.5)

CONTINUITY AND DISCONTINUITY

- f is *continuous* at $x = c$ if $\lim\limits_{x \to c} f(x) = f(c)$. (Section 2.6.1)
- f is *left-continuous* at $x = c$ if $\lim\limits_{x \to c^-} f(x) = f(x)$. (Section 2.6.2)
- f is *right-continuous* at $x = c$ if $\lim\limits_{x \to c^+} f(x) = f(x)$. (Section 2.6.2)
- f is *continuous on an open interval* (a, b) if f is continuous for all $c \in (a, b)$. (Section 2.6.2)
- f is *continuous on a closed interval* $[a, b]$ if f is continuous for all $c \in (a, b)$, right continuous at $x = a$, and left continuous at $x = b$. (Section 2.6.2)
- f has a *removable discontinuity* at $x = c$ if $\lim\limits_{x \to c} f(x)$ exists, but is not equal to $f(c)$. (Section 2.6.3)
- f has a *jump discontinuity* at $x = c$ if $\lim\limits_{x \to c^-} f(x)$ and $\lim\limits_{x \to c^+} f(x)$ both exist, but are not equal to each other. (Section 2.6.3)
- f has an *infinite discontinuity* at $x = c$ if at least one of $\lim\limits_{x \to c^-} f(x)$ or $\lim\limits_{x \to c^+} f(x)$ is infinite. (Section 2.6.3)

DIFFERENTIABILITY AND NONDIFFERENTIABILITY

- f is *differentiable* at $x = c$ if $\lim\limits_{h \to 0} \frac{f(c+h) - f(c)}{h}$ exists. (Section 3.3.1)
- f is *left differentiable* at $x = c$ if $\lim\limits_{h \to 0^-} \frac{f(c+h) - f(c)}{h}$ exists. (Section 3.3.1)
- f is *right differentiable* at $x = c$ if $\lim\limits_{h \to 0^+} \frac{f(c+h) - f(c)}{h}$ exists. (Section 3.3.1)
- f is *differentiable on an open interval* (a, b) it f is differentiable for all $c \in (a, b)$. (Section 3.3.1)
- f is *differentiable on a closed interval* $[a, b]$ if f is differentiable for all $c \in (a, b)$, right differentiable at $x = a$, and left differentiable at $x = b$. (Section 3.3.1)
- A continuous function f has a *corner* or *cusp* at $x = c$ if $\lim\limits_{x \to c^-} f'(x)$ and $\lim\limits_{x \to c^+} f'(x)$ both exist, but are not equal to each other. (Section 7.5.1)
- A continuous function f has a *vertical cusp* at $x = c$ if $\lim\limits_{x \to c^-} f'(x)$ and $\lim\limits_{x \to c^+} f'(x)$ are infinite, with opposite signs. (Section 7.5.1)
- A continuous function f has a *vertical tangent line* at $x = c$ if $\lim\limits_{x \to c^-} f'(x)$ and $\lim\limits_{x \to c^+} f'(x)$ are infinite, with the same sign. (Section 7.5.1)

Major Theorems

THEOREMS ABOUT CONTINUOUS FUNCTIONS

- *The Extreme Value Theorem:* If f is continuous on a closed interval $[a, b]$, then f has both a maximum and a minimum value on $[a, b]$. (Section 2.7.1)
- *The Intermediate Value Theorem:* If f is continuous on a closed interval $[a, b]$, then for any K between $f(a)$ and $f(b)$, there is at least one $c \in (a, b)$ such that $f(c) = K$. (Section 2.7.2)
- *Special Case of the Intermediate Value Theorem:* If f is continuous on a closed interval $[a, b]$, then if $f(a)$ and $f(b)$ have opposite signs, there exists at least one $c \in (a, b)$ such that $f(c) = 0$. (Section 2.7.3)
- *L'Hôpital's Rule:* Suppose $f(x)$ and $g(x)$ are differentiable, and $g(x)$ is nonzero, near (but not necessarily at) $x = c$. If $\lim\limits_{x \to c} f(x)$ and $\lim\limits_{x \to c} g(x)$ are both zero, or both infinite, then $\lim\limits_{x \to c} \frac{f(x)}{g(x)} = \lim\limits_{x \to c} \frac{f'(x)}{g'(x)}$.

THEOREMS ABOUT DIFFERENTIABLE FUNCTIONS

- *Differentiability Implies Continuity:* If f is differentiable at $x = c$, then f is continuous at $x = c$. (Section 3.3.5)
- *The Derivative at a Local Extremum:* If f has a local minimum or maximum at $x = c$, then either $f'(c)$ does not exist or $f'(c) = 0$. (Section 3.6.1)
- *Rolle's Theorem:* Let f be a function that is continuous on $[a, b]$ and differentiable on (a, b). If $f(a) = f(b) = 0$, then there is at least one value $c \in (a, b)$ for which $f'(c) = 0$. (Section 3.6.2)
- *The Mean Value Theorem:* If f is continuous on $[a, b]$ and differentiable on (a, b), then there exists at least one value $c \in (a, b)$ such that $f'(c) = \frac{f(b) - f(a)}{b - a}$. (Section 3.6.3)

THEOREMS ABOUT INTEGRALS

- *The Fundamental Theorem of Calculus:* If $f(x)$ is continuous on $[a, b]$, and $F(x)$ is any antiderivative of $f(x)$, then $\int_a^b f(x)\,dx = F(b) - F(a)$. (Section 13.2.1)
- *Alternative Form of the Fundamental Theorem of Calculus:* If $f(x)$ is differentiable on $[a, b]$, and its derivative $f'(x)$ is continuous on $[a, b]$, then $\int_a^b f'(x)\,dx = f(b) - f(a)$. (Section 13.2.1)
- *The Second Fundamental Theorem of Calculus:* If f is continuous on $[a, b]$ and we define $F(x) := \int_a^x f(t)\,dt$ for all $x \in [a, b]$, then $F(x)$ is continuous on $[a, b]$ and differentiable on (a, b), with a derivative of $F'(x) = f(x)$. (Section 13.3.2)
- *Alternative Form of the Second Fundamental Theorem of Calculus:* If f is continuous on $[a, b]$, then for all $x \in [a, b]$ we have $\frac{d}{dx} \int_a^x f(t)\,dt = f(x)$. (Section 13.3.3)
- *The Mean Value Theorem for Integrals:* If $f(x)$ is continuous on a closed interval $[a, b]$, then there exists some $c \in (a, b)$ such that $\int_a^b f(x)\,dx = f(c)(b - a)$. (Section 13.3.4)

Applications

POSITION, VELOCITY, AND ACCELERATION

- Velocity is the derivative of position ($v(t) = s'(t)$). Acceleration is the derivative of velocity ($a(t) = v'(t)$), that is, the second derivative of position ($a(t) = s''(t)$). (Sections 3.2.3–3.2.4)
- Suppose an object falls from an initial height of s_0 feet with an initial velocity of v_0 feet per second, and that acceleration due to gravity is $-g$ feet per second per second (on Earth, $g = 32$ feet per second per second). The object has position, velocity, and acceleration given by:
$$s(t) = -\frac{g}{2}t^2 + v_0 t + s_0 \qquad v(t) = -gt + v_0$$
$$a(t) = -g \qquad \text{(Exercises to Section 5.2)}$$

EXPONENTIAL GROWTH AND DECAY

- $Q(t) = e^{kt}$ represents exponential growth if $k > 0$, and exponential decay if $k < 0$. $Q(t) = b^t$ represents exponential growth if $b > 1$, and exponential decay if $0 < b < 1$. (Section 8.5.1)
- $Q(t)$ is exponential if and only if $Q'(t) = kQ(t)$ for some constant k. (Section 8.4.4)
- $Q(t)$ is exponential if and only if $Q(t)$ grows at a fixed percentage each time period. (Section 8.6.1)
- $Q(t)$ is exponential if and only if it has a constant doubling time or half-life. (Section 8.6.4)
- *Percentage growth:* If $Q(t)$ grows at a fixed percentage r every time period t, then $Q(t) = Q_0(1 + r)^t$. (Section 8.6.1)
- *Percentage growth with multiple compounding:* If $Q(t)$ grows at a r percent, compounded n times per time period t, then $Q(t) = Q_0(1 + \frac{r}{n})^{nt}$. (Section 8.6.2)
- *Continuous growth:* If $Q(t)$ has a continuous growth rate of k, then $Q(t) = Q_0 e^{kt}$. (Section 8.6.3)

PHYSICS

- The *work* involved in lifting an object weighing F pounds through a distance of d feet is $W = F \cdot d$ foot-pounds. (Section 15.4.1)
- The *mass* of an object with volume V cubic centimeters and constant density of ρ grams per cubic centimeter is $m = \rho \cdot V$ grams. (Section 15.4.2)
- The *hydrostatic force* exerted on a horizontal wall of area A under a depth of d feet of water (which weighs 62.4 pounds per cubic foot) is $F = 62.4A \cdot d$ pounds. (Sections 15.4.1 and 15.4.3)

Derivatives

DEFINITIONS

- The derivative at a point: $f'(c) = \lim\limits_{h \to 0} \frac{f(c+h)-f(c)}{h}$ (Section 3.1.2)
- Alternative definition of the derivative at a point: $f'(c) = \lim\limits_{x \to c} \frac{f(x)-f(c)}{x-c}$ (Section 3.1.2)
- Left derivative: $f'_-(c) = \lim\limits_{h \to 0^-} \frac{f(c+h)-f(c)}{h}$ or $f'_-(c) = \lim\limits_{x \to c^-} \frac{f(x)-f(c)}{x-c}$ (Section 3.1.4)

- Right derivative: $f'_+(c) = \lim\limits_{h \to 0^+} \frac{f(c+h)-f(c)}{h}$ or $f'_+(c) = \lim\limits_{x \to c^+} \frac{f(x)-f(c)}{x-c}$ (Section 3.1.4)
- The derivative of a function: $f'(x) = \lim\limits_{h \to 0} \frac{f(x+h)-f(x)}{h}$ (Section 3.4.1)
- Alternative definition of the derivative of a function: $f'(x) = \lim\limits_{t \to x} \frac{f(t)-f(x)}{t-x}$ (Section 3.4.1)

DIFFERENTIATION RULES

Constant functions	$\frac{d}{dx}(c) = 0$	(Section 3.5.1)
Identity function	$\frac{d}{dx}(x) = 1$	(Section 3.5.1)
Linear functions	$\frac{d}{dx}(mx + b) = m$	(Section 3.5.1)
Power rule	$\frac{d}{dx}(x^n) = nx^{n-1}$	(Sections 3.5.2, 4.3.2, 7.3.4)
Constant multiple rule	$(kf)' = kf'$	(Section 3.5.3)
Sum rule	$(f + g)' = f' + g'$	(Section 3.5.4)
Difference rule	$(f - g)' = f' - g'$	(Section 3.5.4)
Quotient rule	$\left(\frac{f}{g}\right)' = \frac{f'g - fg'}{g^2}$	(Section 6.3.2)
Product rule	$(fg)' = f'g + fg'$	(Section 7.2.1)
Chain rule	$(f(g(x)))' = f'(g(x))g'(x)$	(Section 7.2.2)
Chain rule in Leibniz notation	$\frac{df}{dx} = \frac{df}{du}\frac{du}{dx}$	(Section 7.2.2)
Chain rule with power rule	$\frac{d}{dx}((u(x))^k) = k(u(x))^{k-1}u'(x)$	(Section 7.2.2)
Exponential functions	$\frac{d}{dx}(e^x) - e^x$	(Section 8.4.1)
	$\frac{d}{dx}(e^{kx}) = ke^{kx}$	(Section 8.4.2)
	$\frac{d}{dx}(b^x) = (\ln b)b^x$	(Section 8.4.2)
Chain rule with exponential rule	$\frac{d}{dx}\left(e^{u(x)}\right) = e^{u(x)}u'(x)$	(Section 8.4.2)
Logarithmic functions	$\frac{d}{dx}(\ln x) = \frac{1}{x}$	(Section 9.2.3)
	$\frac{d}{dx}(\log_b x) = \frac{1}{(\ln b)x}$	(Section 9.2.3)
	$\frac{d}{dx}(\ln |x|) = \frac{1}{x}$	(Section 9.3.3)
Trigonometric functions	$\frac{d}{dx}(\sin x) = \cos x$	(Section 10.5.1)
	$\frac{d}{dx}(\cos x) = -\sin x$	(Section 10.5.1)
	$\frac{d}{dx}(\tan x) = \sec^2 x$	(Section 10.5.2)
	$\frac{d}{dx}(\sec x) = \sec x \tan x$	(Section 10.5.2)
	$\frac{d}{dx}(\cot x) = -\csc^2 x$	(Section 10.5.2)
Inverse trigonometric functions	$\frac{d}{dx}(\csc x) = -\csc x \cot x$	(Section 10.5.2)
	$\frac{d}{dx}(\sin^{-1} x) = \frac{1}{\sqrt{1-x^2}}$	(Section 11.2.1)
	$\frac{d}{dx}(\cos^{-1} x) = -\frac{1}{\sqrt{1-x^2}}$	(Section 11.2.1)
	$\frac{d}{dx}(\tan^{-1} x) = \frac{1}{1+x^2}$	(Section 11.2.1)
	$\frac{d}{dx}(\sec^{-1} x) = \frac{1}{|x|\sqrt{x^2-1}}$	(Section 11.2.1)

Sums

Sigma notation: $\displaystyle\sum_{k=1}^{n} a_k := a_1 + a_2 + a_3 + \cdots + a_{n-1} + a_n$

(Section 12.1.2)

THE ALGEBRA OF SUMS (Section 12.1.3)

Constant multiple rule for sums $\displaystyle\sum_{k=m}^{n} c a_k = c \sum_{k=m}^{n} a_k$

Sum rule for sums $\displaystyle\sum_{k=m}^{n} (a_k + b_k) = \sum_{k=m}^{n} a_k + \sum_{k=m}^{n} b_k$

Splitting a sum $\displaystyle\sum_{k=m}^{n} a_k = \sum_{k=m}^{p-1} a_k + \sum_{k=p}^{n} a_k$

SUM FORMULAS (Section 12.1.4)

$\displaystyle\sum_{k=1}^{n} 1 = n \qquad \sum_{k=1}^{n} k = \frac{n(n+1)}{2}$

$\displaystyle\sum_{k=1}^{n} k^2 = \frac{n(n+1)(2n+1)}{6} \qquad \sum_{k=1}^{n} k^3 = \frac{n^2(n+1)^2}{4}$

RIEMANN SUMS (Sections 12.2.2–12.2.3)

General Riemann sum $\displaystyle\sum_{k=0}^{n} f(x_k^*)\,\Delta x$, where

$\Delta x := \frac{b-a}{n}$, $x_k := a + k\Delta x$,

and $x_k^* \in [x_{k-1}, x_k]$

Right Hand Sum $\displaystyle\sum_{k=1}^{n} f(x_k)\,\Delta x$

Left Hand Sum $\displaystyle\sum_{k=0}^{n-1} f(x_k)\,\Delta x = \sum_{k=1}^{n} f(x_{k-1})\,\Delta x$

Midpoint Sum $\displaystyle\sum_{k=0}^{n} f\left(\frac{x_{k-1}+x_k}{2}\right)\Delta x$

Upper Sum $\displaystyle\sum_{k=0}^{n} f(M_k)\,\Delta x$, where M_k is a global
maximum of $f(x)$ on $[x_{k-1}, x_k]$

Lower Sum $\displaystyle\sum_{k=0}^{n} f(m_k)\,\Delta x$, where m_k is a global
minimum of $f(x)$ on $[x_{k-1}, x_k]$

Random Sum $\displaystyle\sum_{k=0}^{n} f(r_k)\,\Delta x$, where r_k is chosen
randomly in $[x_{k-1}, x_k]$

Trapezoid Sum $\displaystyle\sum_{k=0}^{n} \frac{f(x_{k-1})+f(x_k)}{2}\,\Delta x$

Definite Integrals

The *definite integral* of a continuous function $f(x)$ from $x = a$ to $x = b$ is:

$$\int_a^b f(x)\,dx := \lim_{n\to\infty} \sum_{k=1}^{n} f(x_k^*)\,\Delta x,$$

where $\Delta x := \frac{b-a}{n}$, $x_k := a + k\,\Delta x$, and $x_k^* \in [x_{k-1}, x_k]$.

PROPERTIES OF DEFINITE INTEGRALS
(Section 12.3.3)

Definite integral over $[a, a]$ $\quad \int_a^a f(x)\,dx = 0$

Definite integral over $[b, a]$ $\quad \int_b^a f(x)\,dx = -\int_a^b f(x)\,dx$

Constant multiple rule for definite integrals $\quad \int_a^b kf(x)\,dx = k\int_a^b f(x)\,dx$

Sum rule for definite integrals

$\int_a^b (f(x) + g(x))\,dx = \int_a^b f(x)\,dx + \int_a^b g(x)\,dx$

Splitting a definite integral

$\int_a^b f(x)\,dx = \int_a^c f(x)\,dx + \int_c^b f(x)\,dx$

DEFINITE INTEGRAL FORMULAS (Section 12.3.3)

$\int_a^b c\,dx = c(b-a) \qquad \int_a^b x\,dx = \frac{1}{2}(b^2 - a^2)$

$\int_a^b x^2\,dx = \frac{1}{3}(b^3 - a^3)$

DEFINITE INTEGRALS THAT REPRESENT GEOMETRIC QUANTITIES

- Signed area under the graph of $f(x)$ on $[a, b]$: $\int_a^b f(x)\,dx$ (Section 12.4.1)
- True area under the graph of $f(x)$ on $[a, b]$: $\int_a^b |f(x)|\,dx$ (Section 12.4.1)
- Area between the graphs of $f(x)$ and $g(x)$ on $[a, b]$: $\int_a^b |f(x) - g(x)|\,dx$ (Section 12.4.2)
- Average value of $f(x)$ on $[a, b]$: $\frac{1}{b-a} \int_a^b f(x)\,dx$ (Section 12.4.3)
- Arc length of $f(x)$ on $[a, b]$: $\int_a^b \sqrt{1 + (f'(x))^2}\,dx$ (Section 15.1.3)
- General forms for finding volumes of solids of revolution by integrating along the x-axis from $x = a$ to $x = b$ with discs of radius $r = r(x)$, washers of radii $R = R(x)$ and $r = r(x)$, or shells with radius $r = r(x)$ and height $h = h(x)$:

  $\pi \int_a^b r^2\,dx \qquad \pi \int_a^b (R^2 - r^2)\,dx \qquad 2\pi \int_a^b rh\,dx$

  (Sections 15.2.3–15.2.4, 15.3.3, and 15.3.5)
- General forms for finding volumes of solids of revolution by integrating along the y-axis from $x = A$ to $x = B$ with discs of radius $r = r(y)$, washers of radii $R = R(y)$ and $r = r(y)$, or shells with radius $r = r(y)$ and height $h = h(y)$:

  $\pi \int_A^B r^2\,dy \qquad \pi \int_A^B (R^2 - r^2)\,dy \qquad 2\pi \int_A^B rh\,dy$

  (Sections 15.2.5, 15.3.4, and 15.3.5)